Annual Review of
Astronomy and Astrophysics

Production Editor: Roselyn Lowe-Webb
Managing Editors: Linley E. Hall, Nina G. Perry
Bibliographic Quality Control: Mary A. Glass
Electronic Content Coordinators: Suzanne K. Moses, Erin H. Lee
Illustration Editor: Alan Stonebraker

Annual Review of Astronomy and Astrophysics

Volume 52, 2014

S.M. Faber, *Co-Editor*
Lick Observatory, University of California, Santa Cruz

Ewine van Dishoeck, *Co-Editor*
Leiden Observatory, Leiden University and Max-Planck Institute for Extraterrestrial Physics, Garching

John Kormendy, *Associate Editor*
University of Texas, Austin

www.annualreviews.org • science@annualreviews.org • 650-493-4400

Annual Reviews
4139 El Camino Way • P.O. Box 10139 • Palo Alto, California 94303-0139

Annual Reviews
Palo Alto, California, USA

International Standard Serial Number: 0066-4146
International Standard Book Number: 978-0-8243-0952-7
Library of Congress Control Number: 63008846

∞ The paper used in this publication meets the minimum requirements of American National Standards for Information Sciences—Permanence of Paper for Printed Library Materials, ANSI Z39.48-1992.

TYPESET BY APTARA
PRINTED AND BOUND BY SHERIDAN BOOKS, INC., CHELSEA, MICHIGAN

Annual Review of Astronomy and Astrophysics

Volume 52, 2014

Contents

Indexes

Errata

An online log of corrections to *Annual Review of Astronomy and Astrophysics* articles may be found at http://www.annualreviews.org/errata/astro

Related Articles

From the ***Annual Review of Earth and Planetary Sciences***, Volume 42 (2014)

From the ***Annual Review of Nuclear and Particle Science***, Volume 63 (2013)

Annual Reviews is a nonprofit scientific publisher established to promote the advancement of the sciences. Beginning in 1932 with the *Annual Review of Biochemistry*, the Company has pursued as its principal function the publication of high-quality, reasonably priced *Annual Review* volumes. The volumes are organized by Editors and Editorial Committees who invite qualified authors to contribute critical articles reviewing significant developments within each major discipline. The Editor-in-Chief invites those interested in serving as future Editorial Committee members to communicate directly with him. Annual Reviews is administered by a Board of Directors, whose members serve without compensation.

Annual Reviews of

Analytical Chemistry
Animal Biosciences
Anthropology
Astronomy and Astrophysics
Biochemistry
Biomedical Engineering
Biophysics
Cell and Developmental Biology
Chemical and Biomolecular Engineering
Clinical Psychology
Condensed Matter Physics
Earth and Planetary Sciences
Ecology, Evolution, and Systematics
Economics
Entomology
Environment and Resources
Financial Economics
Fluid Mechanics
Food Science and Technology
Genetics
Genomics and Human Genetics
Immunology
Law and Social Science
Marine Science
Materials Research
Medicine
Microbiology
Neuroscience
Nuclear and Particle Science
Nutrition
Organizational Psychology and Organizational Behavior
Pathology: Mechanisms of Disease
Pharmacology and Toxicology
Physical Chemistry
Physiology
Phytopathology
Plant Biology
Political Science
Psychology
Public Health
Resource Economics
Sociology
Statistics and Its Application
Virology

George Field

Wondering About Things

George B. Field

Astronomy Department, Harvard University, Cambridge, Massachusetts 02138;
email: gfield@cfa.harvard.edu

Annu. Rev. Astron. Astrophys. 2014. 52:1–42

The *Annual Review of Astronomy and Astrophysics* is online at astro.annualreviews.org

This article's doi:
10.1146/annurev-astro-081913-040050

Keywords

biography, molecular clouds, magnetohydrodynamics, interstellar medium, cosmic microwave background, intergalactic medium, Chern-Simons electrodynamics

Abstract

Here you will find facts about and the opinions of an American astrophysicist who practiced in the second half of the twentieth century. The title explains why I did it. I invented some new ideas, I applied them to some astro objects, I computed things with pen and paper; I ended up thinking that I had succeeded in pushing the field ahead a bit.

Attracted by Newtonian theory, I did some experiments too. I love hydrodynamics and magnetic fields in space. The math is beautiful, and the objects are stupendous in their brilliant displays. For some reason I meditated on gases between the stars, their pressures and motions. I left the stars to others, believing that their physics was under control.

As I grew older, I had to decide whether to direct others rather than just myself and ended up at the Harvard–Smithsonian Center for Astrophysics doing both. It was thrilling because I had never had management experience and was flying by the seat of my pants, as I guess other astrodirectors do. In the process, I advised the US government on future directions in astronomy, chairing a number of committees. It is astonishing that the government is interested in astronomy, and it is exciting to interact with the people in the National Academy of Sciences (NAS), the Congress, and the Executive branch who have dedicated their lives to enable the expansion of our knowledge of astronomy.

Along the way I studied more abstract concepts in physics, including magnetic helicity and its relation to the winding numbers of nonabelian particle physics. These are topological concepts that I should have learned in grad school but did not.

This review has two parts. The first part is for scientists, and covers my life in chronological order. The second part is for laymen who are interested in science. It gives a flavor of my scientific work with no math and a minimum of jargon.

1. PART ONE

1.1. Early Days

My family were Yankees, my father from Lowell, Massachusetts, and my mother from a nearby town, Chelmsford. My great grandfather, whose name was the same as mine, was a country doctor whose forebears were farmers in the Connecticut River valley, and his son, my grandfather James Field, graduated from Harvard in 1880. He became an doctor and opened his practice in Lowell, where my father Winthrop met my mother Pauline Woodworth; they married in 1916. Pauline was unusual at the time in that she graduated in Classics from Radcliffe College, then the woman's college associated with Harvard. My dad graduated from Harvard in 1915 with a degree in math. Very unusually, instead of following his classmates to Wall Street or the like, my dad—and mom—decided to rent a farm in northern New Hampshire and go into sheep and dairy farming. That was a tough way to earn a living in the days before tractors. At one point, their draft horse broke through the floor of the barn and ended up straddling a beam. I can just imagine Dad in the barn at 3 AM, wondering how to solve that problem. His solution was to put a load of hay on the floor below and then saw through the beam. When the beam fell, so did the horse, who then walked away unharmed. They endured many such incidents over 11 years, but in 1927, with the birth of my sister Sarah, they threw in the towel and moved to the city, where they knew their children would get a better education.

I was born on October 25, 1929, which happened to be the week of the 1929 stock market crash. Our home was in Edgewood, Rhode Island, a subdivision of Cranston, itself a suburb of Providence, where Dad had his office. Why Providence? Dad had found a job with a New York company specializing in advising gentlemen farmers, as businessmen were known who had bought farms as tax write-offs in the new age of the income tax. The company wanted to open a new office with easy access to New England farms, and Providence provided such access.

Three things were in his favor in landing the job: In New Hampshire he had served the US Department of Agriculture as a County Agent, he had a Harvard degree in math, and the owner of the company was a Harvard alumnus. In any case, his job was to visit the farms and analyze how to cut their losses. Dad loved visiting the farms and chatting with the farmers all around New England, having just left that situation himself. I loved it too on those few occasions he invited me along. On one of those occasions Dad again demonstrated his ability to solve practical problems. We were tootling along a highway when he slammed on the brakes and threw the car into reverse. I saw why; he had noticed a farmer quite far from the road, prostrate on the ground with a riderless tractor circling nearby. We ran down, hauled the fellow up, and called the police. Apparently something had knocked him off his seat. He survived.

During this period I was attending elementary and middle schools in Cranston. I loved math, of course, but also English, taught by the sister of the children's librarian at the William H. Hall Free Library in Edgewood. It is an endowed library in a marble building set on spacious grounds. Mom and dad were voracious readers, so Mom would pick up a new load of books every week, and she ended up as a member of the library board. Both Mom's sister Rachel and my sister Sarah were librarians, so it ran in the family. One time, I went to the library with her and looked through the science section, where I found copies of *The Universe Around Us* and *The Stars in Their Courses* by Sir James Jeans (1929, 1931), a prominent theoretical astrophysicist. It was during that period that I decided to become an astrophysicist myself, so I owe a lot to William H. Hall.

At that time I was a nerd, but something happened that added another string to my bow. As I came and went, I would pass by boys playing touch football in the street. One day, Frank Roque, who became a lifelong friend, asked if I would like to play too. I did, and I became a member of the local group of guys playing football; when Spring came around, we played baseball and tennis too. Those were halcyon days. I found I had some talent in all three sports. For example, I could throw a pass further than the other guys and could hit a baseball well. The respect I got from this enabled me to climb out of the pure nerd category and feel comfortable with other fellows in the neighborhood. I think this made a difference when I was later asked to step into leadership positions in science, where group action is necessary. However, I now see that instead of sandlot ball, joining a team would have been even better. Learning to play a sport properly is a lifelong gift, and even more, learning the give and take of team sports is good training for all the activities of life to follow.

In my early period, I was very shy around girls. I never attended dances at school, and I learned to dance only later when I was a student at MIT. I enrolled at MIT in 1947 when I graduated from high school. As everyone at MIT did, I took the introductory courses in physics and calculus. In those days calculus was not taught in high school, so we started from the beginning. I was hooked by the beauty of those two subjects. In essence, we were studying the accomplishments of Galileo and Newton in the seventeenth century, but as I recall, there was little mention of those two great scientists by name. Only later, after reading about Einstein, did I read their histories. Heilbron's (2010) biography of Galileo shows how wonderfully he entered into the lives of the writers, philosophers, and musicians of his time in Florence and Venice. Maybe that should be assigned reading for budding physicists who want to be humanists on the side, as Galileo was.

MIT at that time didn't stress the humanities; there wasn't time with all the science courses. The Institute was just beginning to change this situation. They appointed an architect as Dean of Humanities and started a course required of all undergraduates taught by a historian named Georgio de Santillana. I can't recall the reading assignments now, but he made a valiant effort to convert all of us into regular people. Now, as a retired professor from Harvard, I see what a fine reputation is enjoyed by Alan Lightman, who from a perch in the MIT Physics and Humanities Departments, has become a widely read creative writer. Alan was on the staff of the Harvard-Smithsonian Center for Astrophysics, where his work on theoretical high-energy astrophysics was outstanding. He moved to MIT to share his knowledge of science while also doing creative writing.

Fortunately, my parents instilled a love of classical music in me, and my brother John, who was studying Social Relations at Harvard at the same time I was at MIT, debated politics with me. He himself was a socialist and applied his studies of society to try to convince me that socialism was the way to go. This had the effect of at least making me aware of the injustices in American society, racial discrimination in particular. The Civil Rights Movement was taking shape in those years, and I tried to support it when I could. Later, when I was a faculty member at Princeton, I was shocked by the television images of sheriff deputies beating black children in Selma, Alabama. I got on a plane to Selma and arrived in time to accompany a student on a march the next day led by Martin Luther King, Jr. Later, stimulated by a sermon given by the Reverend Benjamin Anderson, my wife Sylvia and I joined a historically black church in Princeton, New Jersey, where quixotically I became an elder and a committee chairman even though I did not intellectually accept the doctrines of the church.

In the induction ceremony as an elder, I had to say that I accepted the doctrines, like Galileo, who, though formally abjuring belief in the Copernican system during the Inquisition, according to legend said under his breath, "Nevertheless it [Earth] moves." Later I learned that my good friend and astronomer at the University of California, Berkeley (UCB), Dave Cudaback, had joined a church there as "the resident atheist," in a move somewhat like mine. Somewhere I acquired a love of the visual arts, with Vincent van Gogh as my favorite. So in spite of the limited MIT curriculum, I emerged as a part-time humanist.

The MIT Physics Department required a senior thesis. One of my classmates produced a novel for this purpose and later became a very successful businessman. Note the "man" in that word. In my day there was an extremely small number of girls at MIT. That had the effect that most of my classmates spent the weekends with girls from Wellesley and Radcliffe—but not I; I was too shy. But a breakthrough occurred when my friend Howie Bernstein suggested that I go on a double date with him and his girlfriend, a student at Boston University across the Charles River from MIT. My date Shirley was a down-to-earth nursing student who ended up teaching me to dance, sort of. It was the era of Glenn Miller, and I was carried away by the romance of it all. I fell in love with her, and I gave her my first kiss. I recall that following that event, I ran all the way back to MIT.

One effect of this experience was that my grades plummeted in my junior year. But I recovered enough to complete an experimental thesis on the alignment of dust particles in an electric field. I chose that topic because I had read that interstellar dust must be aligned somehow, because the starlight passing through it is polarized. I developed writer's cramp the night before the thesis was due. My dad told me to "just start writing," and it worked.

At the senior bash in 1951, I got drunk for the first time. My companion at that event was Irwin Manning, another physics student I had met as a freshman. Irv had graduated from Brooklyn Tech high school and was a ham radio operator. We shared a deep love of physics. One day, when we were walking along the Charles River and discussing our class in atomic physics, we realized that a quantum state is a vector in an infinite dimensional space, constrained to have a unit length. All of this seemed like such a natural extension of what we had learned about three-dimensional vector space, in particular, about the orthogonal matrices that rotate coordinates in such systems, which represent the operators of quantum mechanics. This stuff was fantastic! The chutzpah of gliding from finite to infinite in one simple step!

I learned the formalism of nonrelativistic quantum mechanics, but when I encountered quantum field theory in graduate school, I was slowed down considerably. I know that it is an even more beautiful subject, but I never got around to studying it in depth. As I will explain in the next section, I ended up with magnetohydrodynamics (MHD), which is based on Newtonian fluid dynamics and Maxwell's electromagnetic theory. A remnant of graduate school sits on my bookshelf in the form of Steve Weinberg's magisterial book, *The Quantum Theory of Fields*. I have not studied it, but every once in a while, when I learn of a new discovery in particle physics, I take down his book and read the relevant pages. Nowadays, MIT physicist Bert Zweibach teaches an undergraduate course on string theory, which at present seems to be at the limits of physicists' imagination. Bravo!

1.2. A Graduate Student at Princeton

Entering Princeton as a graduate student in the fall of 1952, I was assigned to a desk under the eaves of the Old Observatory on the corner of Prospect Avenue and Washington Street. The observatory building was referred to as the "scope and dome" by the undergraduates who walked past every day on the way to the dining clubs on Prospect Avenue, which had similar names. The observatory was part of the family home of the Director, Professor Lyman Spitzer, so my desk was virtually in the home of the well-known astrophysicist who was to become my thesis advisor. (Lyman was always known to the community by his first name, and as his student who talked to him frequently over 40 years, I will do the same here, respectfully and fondly.)

The observatory was a Victorian building with charming nooks and crannies. Walking in the front door, one found oneself in a short hall with the department office on the left, the library straight ahead, and a tearoom on the right. In the tearoom was a large table, some chairs, and the most used journals, like the *Astrophysical Journal*, reminding us of the fact that one of the first scientists to publish in "ApJ" in 1895 was Henry Norris Russell, the renowned astronomer who had preceded Lyman as director of the observatory. Russell, who is known today as the originator of the Hertzsprung-Russell diagram and Russell-Saunders coupling, was still living in Princeton at the time and sometimes showed up for lawn parties at the New Observatory, which housed a 24-inch refractor telescope, only about a quarter of a mile away.

There was a classroom down a hallway to the left, where Lyman taught his course on the interstellar medium and Martin Schwarzschild taught his course on stellar interiors. Louis Green, a Princeton PhD in 1937, once regaled me with an amusing tale about Professor Russell, who had lectured in that room when Louis was a student. Louis noticed that Russell appeared to be sleepy, and shortly into the lecture, he laid his head on the table and went to sleep. The students discussed how they should react and decided to file out quietly to avoid waking Professor Russell. Naturally, the table became a famous relic of the past for future students. I myself lectured from it when I became a member of the faculty.

I am also not immune to drowsiness after lunch. At Harvard many years later, I was discussing his thesis with a student when a fit of drowsiness hit. Desperate to avoid a "Russell Event" with an esteemed student, I excused myself to use the bathroom, where I doused my face with cold water. Returning to my office, I was able to carry on as before.

It was exciting to be at Princeton in those years. All over the world, science was awaking from the desolation of World War II, and Princeton was no exception; prominent theorists were digesting new data from powerful optical telescopes like those at Mt. Wilson and Mt. Palomar. Among the colloquium speakers, I recall Subrahmanyan Chandrasekhar ("Chandra" to all), a theorist at the University of Chicago whose reputation rivaled those of Lyman and Martin, Walter Baade from Mt. Wilson, who brought the discovery of the two stellar populations, Rudolph Schildt from Yale, who had just announced his finding that the chief opacity in the solar photosphere is the negative hydrogen ion, Alan Sandage and his age dating of globular clusters, and Fred Hoyle with his steady-state Universe, to name just a few.

Lyman was a physicist who was influenced by Chandra and Sir Arthur Eddington during a visit to Cambridge University after graduating from Yale. He went on to obtain a PhD in astronomy at Princeton in 1938 under Henry Norris Russell, an example of the continuing tradition of physicists entering astronomy. Although an able mathematician, his strongest suit was his physical intuition, which enabled him to craft theoretical solutions to observed phenomena, such as the optical interstellar lines on which he gathered data on annual visits to Mt. Wilson. Martin was admired for his humane concern for his students and the enthusiasm he brought to his work on the well-established field of the theory of stellar interiors. The latter was necessarily largely numerical, aided by calculations by his assistant, Richard Harm.

Here is a place to state my attitude about numerical work. Nowadays such work has become the creation and study of numerical simulations. They are essential parts of theoretical work, but I don't personally enjoy working with them. I like working with pen and paper and seeing key ideas emerging from the equations.

The story of how Lyman and Martin came together at Princeton, which is perhaps apocryphal, is this: After World War II, in which Lyman served as an analyst in antisubmarine warfare and Martin as an observer in an artillery unit on the European front, each of them returned to science as postdoctoral fellows at the Harvard College Observatory (HCO), then directed by Harlow Shapley. When Russell retired, both of them were being considered as his successor as both Director and Professor at Princeton. No doubt Shapley, as a leading astronomer at the time and a former student of Russell's, supported one or both of the candidates. The tale continues that Lyman was offered the directorship at Princeton, and he proposed his colleague at Harvard, Martin, to join him as Professor. Whatever the truth, the presence of both eminent astronomers at Princeton more than fulfilled my dream of a first-rate education in astrophysics.

Martin was engaged in writing an authoritative monograph on the theory of stellar structure. Lyman was more interested in the new field of interstellar matter, which was opening up to the new techniques of radio and UV astronomy, in which I was developing interest.

During my student days at Princeton, a new project, dubbed Matterhorn, was developing behind the scenes. The hydrogen bomb had recently been developed and tested. It consisted of deuterium ignited into explosive fusion by a fission bomb. Lyman and Martin realized that if deuterium gas could somehow be heated to the extraordinary temperature of about 100 million degrees while being confined in an appropriate vessel for a long enough time, the fusion process could be controlled and electric power could be produced and used peacefully. The challenge of controlled fusion is to confine the hot gas, or plasma, in spite of the pressure it exerts.

A plasma is a gas heated to the point that the electrons of the constituent atoms are torn from their nuclei, and so the hot gas contemplated by Lyman and Martin is a plasma. Both electrons and ions are electrically charged, so they can be trapped in a magnetic field, hence the appropriate vessel for Lyman and Martin's power station is a vacuum tube of steel in which the plasma is trapped by a strong magnetic field. They dubbed this concept the Stellarator in reference to the stars, which are the only known sites of continuing controlled fusion at this time. As I recall, their plan was to carry out the work in stages: In Stage 1, they would construct a model of the Stellarator and the necessary electromagnets in the basement. Stage 2 would be ignition; that is, fusion of the deuterium ions to release energy. Stage 3 would be breakeven; that is, the amount of power being generated by fusion would be equal to the power used to heat the plasma. And stage 4 would be a giant power station large enough to supply the entire East Coast. The project, started by Lyman in 1950, is still underway at the Princeton Plasma Physics Laboratory. There the Tokamak Fusion Test Reactor has achieved a temperature of 600 million degrees and the target plasma pressure. It produced 6 Megawatts of power from nuclear fusion, still a small amount of power by current power station standards of 1,000 MW (**Figure 1**), so the project seems to still be in stage 3.

Project Matterhorn, as it was called after an Alpine ascent that Lyman had made, was initially classified, no doubt because fusion is the same process used in the highly classified hydrogen bomb. When it was learned in 1958 that the Soviets were also pursuing controlled fusion, the program was declassified, three years after I completed my PhD thesis. It was only then that I understood Lyman's frequent absences from the observatory.

In the meantime, I undertook research on my thesis topic, which I had chosen in complete ignorance of the plasma physics being done a mile or so away as part of Project Matterhorn. I had read about a puzzle in solar physics, revealed by radio studies of the Sun. Solar flares up to that time had been studied only by their optical emission; but along with studying exotic objects like supernova remnants and radio galaxies, radio astronomers were discovering remarkable radio bursts in the solar corona initiated by solar flares. The subject was advanced enough at that time that radio bursts had been classified into Types 1, 2, and 3. Type 2 bursts caught my attention. They consist of intense radiation in the meter band, at any moment tightly confined to two different frequencies. Two characteristics demanded explanation: The higher frequency is twice the lower one, and both frequencies drift from high to low as the minutes pass. Observers noted that the frequencies were close to the plasma frequencies expected in the plasma of the solar corona, 8,100 $\sqrt{n_e}$ $\sec^{-1}$, where n_e is the electron density where the radiation originates. The frequency drift could be understood if there were a disturbance propagating outward through the corona, exciting plasma oscillations at the local plasma frequency, because the electron density, and therefore the frequency, falls as the disturbance moves outward through the corona. And the frequency doubling suggests that the oscillations are so violent that not only the fundamental frequency is excited, but also the first overtone.

Figure 1

The Tokomak Fusion Test Reactor at the Princeton Plasma Physics Laboratory. The magnetic field circling in the device traps a plasma that is at an extremely high temperature. Photo taken by Dietmar Krause/ Princeton Plasma Physics Laboratory.

So far so good. But there is a problem: Theory shows that the normal modes of excitation in a uniform unmagnetized plasma are either plasma oscillations, which are longitudinal oscillations confined to the plasma, or modified electromagnetic radiation, which typically propagates at a fraction of the speed of light. How could the energy in a plasma oscillation emit radiation? I set forth to solve this problem. Soviet physicists Ginzburg & Zheleznyakov (1959) applied advanced theory to this problem, with encouraging results. Not knowing of their work, I proceeded in a different direction. I reasoned that a nonuniformity in the electron density might serve to couple the two. It turned out that a sharp discontinuity does so, but I was unable to calculate how sharp a discontinuity would be necessary. Lyman suggested that I calculate the result if the plasma is embedded in a magnetic field, and again that proved to be interesting. Because Matterhorn's attempt to magnetically confine very hot plasma was still classified when I submitted my thesis in 1955, I did not know why Lyman would be interested in that case, but of course I realized why when Matterhorn was declassified. In any case, the Soviet work was more interesting than mine, because it posited electrostatic turbulence, which would not require an unrealistically sharp edge. Such turbulence would be a natural consequence of the propagation of a collisionless shock wave in the corona, which would also explain the frequency drift as well as the second harmonic produced by the nonlinear interaction of strong plasma waves.

I passed the oral exam in 1955, and as the American Astronomical Society (AAS) was scheduled to meet in Princeton that spring, I prepared an oral presentation of my thesis for the AAS meeting. I was delighted and somewhat unnerved to see Chandra, a leading theorist and, at that time, editor of the *Astrophysical Journal* in the audience. Later Lyman told me that he approved of the work. Then I relaxed, and the thesis was published (Field 1956).

At the same AAS meeting, I met a recent PhD from Harvard, Dave Heeschen. Dave was in charge of the radio telescope at Agassiz Station in Harvard, Massachusetts, where he and Frank Drake were conducting research on 21-cm hyperfine emission by galactic atomic hydrogen, or HI. After the discovery of this emission in 1951 by Harold "Doc" Ewen and Edward Purcell at Harvard, Bart Bok, Professor of Astronomy at Harvard and a well-known investigator of dark clouds in the Milky Way, secured funding for a 60-ft dish for galactic research. Dave and Frank, electronic officers in World War II, were among the first of Bok's students to use the 60-ft dish for 21-cm radio astronomy.

I had an idea in this field that I discussed with Dave. I had computed that if there is atomic hydrogen between the galaxies as well as in them, even a tiny mean density, 10^{-5} atoms cm^{-3}, would make a significant contribution to the total mass density of the Universe. That was a hot topic at the time and still is, because in a Friedman model of the Universe there is a critical value of the density that separates geometrically closed from open models. Lilley & McLain (1956) had recently detected the 21-cm line in absorption in the redshifted spectrum of the bright radio galaxy, Cygnus A. Stimulated by that discovery, I calculated that a critical density of HI between the galaxies might also be detectable in absorption. I proposed observing Cyg A for an absorption "trough" due to intergalactic HI between us and Cyg A. I say trough rather than "line" because the atoms would be spread out in velocity by the expansion of the Universe.

There was a trick involved, though. As with any astronomical absorption line, absorptions from the ground state tend to be cancelled by stimulated emissions from the upper state. In optical astronomy this effect is usually negligible, because the downward radiative transitions tend to be faster (about 10^8 sec^{-1}) than the atomic collisions that populate the upper level, with the result that stimulated emissions from the upper state are rare compared to absorptions from the ground state.

This situation is reversed with the 21-cm line, because the downward radiative transitions are at a tiny rate, so stimulated emission normally nearly cancels absorption. This situation is described by the so-called spin temperature T_S, the temperature that could be assigned to the two-level system by describing the populations with a Boltzmann factor. High values correspond to large populations of the upper level, and hence nearly complete cancellation and weak absorption, and the opposite correspondence occurs for low values. Calculations of T_S appropriate for interstellar clouds with densities of order 1 cm^{-3} find that collisional excitations are fast enough that T_S is normally very close to the kinetic temperature T_K, which describes the random motions of the atoms. Completely independent theoretical calculations of T_K based on a balance of heating and cooling yield values near 100 K and observations of T_S agree with that. But intergalactic densities are vastly smaller, so new calculations are necessary.

While I was waiting in Princeton for the formal commencement ceremony, I got a delightful surprise in the form of a letter from the Society of Fellows at Harvard, inviting me to apply for a postdoctoral fellowship. The Society was founded by President Lowell of Harvard in the 1930s as a way of supporting promising scholars to start their careers without necessarily acquiring a PhD degree, although many applicants do have the PhD. In any case, I applied and was interviewed and accepted as a so-called Junior Fellow, starting my fellowship in the fall of 1955.

The story behind my acceptance is interesting for the light it casts on what is sometimes called the old boy network. It happened that at that time, Professor Edward Purcell was a Senior Fellow in the Society. Senior Fellows choose the Junior Fellows and have dinner with all of the Junior Fellows once a week during term time. In preparing for the interview, I dusted off an idea that

I had but had not worked on, namely, that grains of interstellar dust, which were calculated to be charged and rapidly rotating, would emit radio waves that, if detected, would yield important information about these enigmatic contributors to interstellar material. The interview committee, composed largely of nonscientists, were able to suppress any enthusiasm for my idea, but one, Ed Purcell, was interested. It turned out he had read about interstellar dust and thought about its properties, consulting with the leading expert on interstellar matter at that time, Lyman Spitzer. Lyman and Ed seemed to be good friends, so it appears to me now that the fix may have been in even before I stepped on the train to Boston.

Recall that independently I had talked to Dave Heeschen about using the Harvard radio telescope for my intergalactic project; so when I became a Junior Fellow, I dropped the rotating grains idea and went for intergalactic atomic hydrogen. As a footnote, the grains idea was conceived independently by Bill Erickson at the University of Maryland, who went on to get a PhD on that basis (Erickson 1957). In the past few years, the effect has been detected and is now studied, in part because it occurs in a wave band covered by studies of the cosmic microwave background (CMB), and it is a foreground that must be removed in CMB studies.

1.3. The Society of Fellows

The Society of Fellows experience was a precious one. In addition to the weekly dinners with the Senior Fellows, the Junior Fellows met for lunch, and as you can imagine, the conversation on those occasions was of an entirely different character, with gossip about the Senior Fellows allowed. During my two years as a Junior Fellow, the chair was Crane Brinton, a historian and the author of *The Anatomy of Revolution*. My favorites among the Junior Fellows included Donald Hall, who became Poet Laureate of the United States; Kurt Gottfried, a quantum physicist who joined the faculty at Cornell and like me fought against the crazy proposal for Reagan's Star Wars; Marvin Minsky, who started research on artificial intelligence at MIT; and Henry Rosovsky, who later became Dean of the Faculty at Harvard. There were also frequent visitors, one of whom became a famous legal scholar and whom I recall debating with after a meeting over cigars. How else would I ever have met Ronald Dworkin? Of course, my most productive relationship was with Ed Purcell, who it turned out had earlier calculated a collision cross section essential for evaluating the spin temperature in intergalactic space. We joined together to write a paper on that subject that is still cited today (Purcell & Field 1956).

One contribution to the excitation of the 21-cm line is the UV fluorescence owing to Lyman-alpha (Lyα) radiation. Wouthuysen had presented a brief paper to the AAS (Wouthuysen 1952) pointing out that the absorption of a Lyα photon by an atom in the ground hyperfine state, followed by a radiative decay to the excited hyperfine state, would on average increase the population of the upper state, and thereby increase the spin temperature T_S. As this would decrease the net absorption coefficient at 21-cm wavelength, it would make it harder to detect intergalactic hydrogen; it was crucial input when interpreting my observations.

What value of T_S can be expected if the Wouthuysen mechanism is dominant? Wouthuysen pointed out that it would depend on the number of photons that can excite the upper hyperfine state relative to the number that deexcite it, and therefore the shape of the Lyα profile. He argued that if the system were in thermodynamic equilibrium, then the Lyα profile, the radiation field, and the matter all have the same temperature, namely, the kinetic temperature T_K of the atomic hydrogen gas, so having the same temperature should be true in general.

The Wouthuysen mechanism deserved study, because it is likely that UV photons between the Lyman limit and Lyα line from distant galaxies would reach random points in the intergalactic medium, and because they would be redshifted by the expansion of the Universe as they traveled, all those photons would ultimately reach a point where they would have the wavelength of the Lyα line. And because both collisions and absorption of 21-cm photons from other sources would also affect the value of T_S, it was important to know the relative rates of these processes, a topic that Wouthuysen did not address.

At that time, little was known about the amounts of UV radiation because it was only later that space missions could detect radiation of such short wavelengths. But even if one had only a crude estimate of the amount of radiation, there would still be a problem evaluating its effects, because any photon in the vicinity of the Lyα line would be scattered by the local H atoms and trapped for a time before it could escape. Shifted back and forth within the Lyα profile by the Doppler effect of the moving atoms, a photon might reach so far to the red that the opacity is small enough for it to escape and continue to redshift indefinitely. The challenge was to calculate how the distribution of photons in frequency changes with time. Note that the distribution is determined by the Doppler effect, which depends upon the velocities of the atoms, which in turn depends on the kinetic temperature T_K. Recall that Wouthuysen had argued that in thermodynamic equilibrium, $T_S = T_K$; so here was a hint of what the final result of my work might be.

Wouthuysen had stated that, in any encounter between an atom and an energetic photon, there would be momentum transferred by the photon to the atom, thereby resulting in a gradual reddening of the radiation. After many such encounters, there would be more red photons than blue ones, and this would determine the value of T_S in the absence of other processes.

How long would it take for T_S to approach T_K? I was able to solve my simplified problem analytically for the specific case that the broadening of the line is due solely to the Doppler effect (Field 1959). A key realization was that the probability that a photon absorbed at frequency ν_1 is reemitted at frequency ν_2 (graphed in **Figure 2**) has a discontinuous derivative along the edge at $\nu_2 = \nu_1$. This enabled me to recognize that an integral operator that occurs in the radiative transfer problem is the inverse of a differential operator, which resulted in an analytically solvable differential equation (**Figure 3**). I have to admit, however, that this occurred to me while I was at the observatory and my wife Sylvia was at the hospital awaiting the birth of our son, Christopher. Unfortunately that was not the only time that I chose science over family concerns. Incidently, we gave Christopher the middle name Lyman in honor of Lyman Spitzer.

The solution for the line shape as a function of time is a gradual broadening on the timescale of the mean free time of a photon at the line center. The asymptotic line shape is mesa function as shown in **Figure 4**. Because the incoming photon has to share some of its momentum with the atom, the outgoing photon has a lower frequency than the incoming one. The result is that instead of the mesa being symmetric around the rest frequency, it is tilted slightly upward to the red. When I put in the numbers, I found that $T_S = T_K$ as Wouthuysen had predicted, after a number of photon free times that I calculated to equal the optical depth at the line center to ∞, rather than the square of that quantity, which applies to coherent scattering.

The application to the astronomical situation is that any photon emitted in the band of wavelengths between the Lyman limit and Lyα line would proceed freely through the intergalactic medium until it was redshifted to the wavelength of the Lyα line by the expansion of the Universe. At this point it would start to scatter with the local intergalactic atoms, bouncing from one side of the line to the other, and slowly drift toward the red, leaving a tilted mesa profile. After the number of photon mean free times that I calculated, it would happen to have moved so far to the red that the Lyα opacity is negligible and it no longer scatters. At that point, the expansion of the Universe would take over, shifting the photon ever further to the red and away from the Lyα resonance of the local atoms. A reasonable analogy for the history of a photon would be a bull rider in a rodeo, who lurches violently from side to side until he or she is thrown off and can escape from the ring.

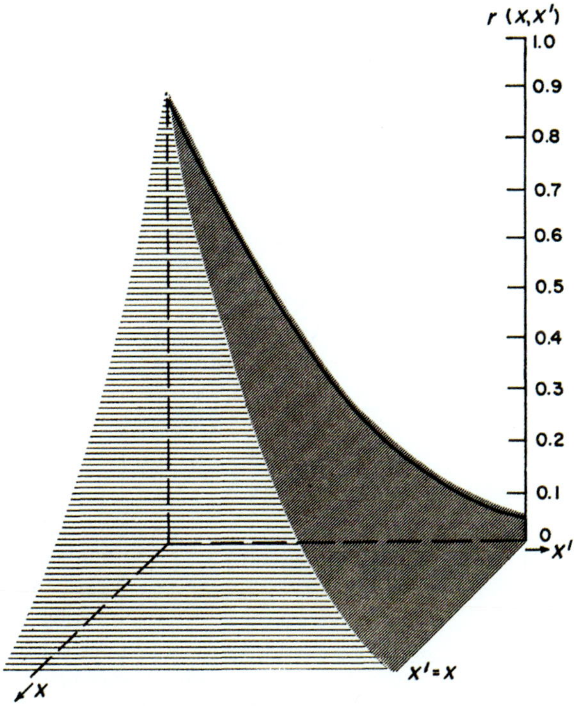

Figure 2

$r(x, x')$, the redistribution transition probability for isotropic scattering of Lyman α photons, with zero damping and recoil. Only one quadrant is shown. Note the edge along $|x| = |x'|$. Reprinted from Field (1959) with permission of the AAS.

While I was working on this theoretical problem, I was also observing Cyg A for signs of intergalactic absorption. Because Cyg A was visible only at night at that time, I drove out to the Agassiz Station every night at about seven and tuned the receiver to a frequency below the lab frequency of the 21-cm line, 1420.406 MHz.

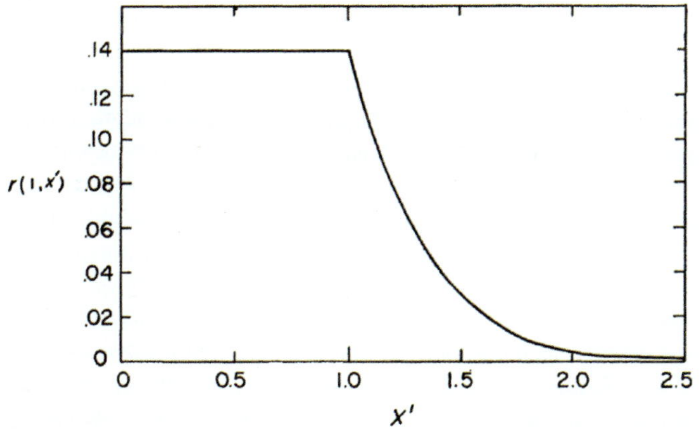

Figure 3

$r(x, x')$, the redistribution transition probability for $x = 1$. Reprinted from Field (1959) with permission of the AAS.

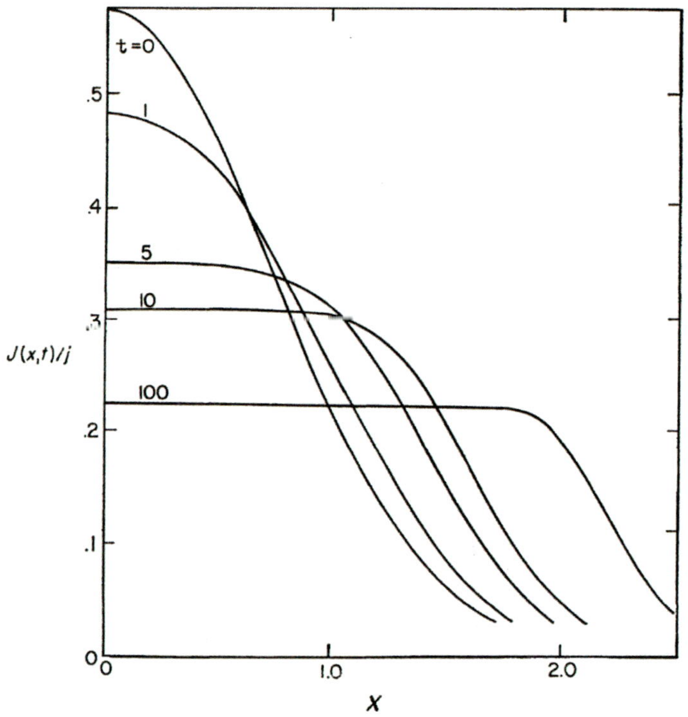

Figure 4

Solutions to the initial-value problem, for $t = 0, 1, 5, 10$, and 100 in units of mean free time at the line center. Reprinted from Field (1959) with permission of the AAS.

The procedure for measuring the absorption at a frequency chosen from the twenty or so frequencies in the expected absorption trough was to point the telescope at the radio galaxy Cyg A and record the intensity, and then, because the gain of the receiver varies slowly with time, point the telescope to the bright galactic radio source Cassiopeia A—which cannot experience intergalactic absorption—for a comparable time. Thus I had a ratio of Cyg A to Cas A that should be independent of gain fluctuations. This ratio covered a frequency interval extending from somewhat below the redshifted frequency of Cyg A to a frequency even higher than the lab frequency. Thus there were measurements both in the interval where the absorption trough was expected and beyond it, allowing a comparison at frequencies where absorption is expected to those where it is not.

Needless to say, no absorption was detected; otherwise we would read about it in books about cosmology. Therefore I was content to estimate the largest optical depth that would have escaped detection, of the order of 1%, and translate this into an upper limit on the intergalactic density of HI. This latter step required me to estimate the value of T_S using estimates of UV intensity for the Wouthuysen effect, which were very approximate at that time.

Another mechanism that would excite the hydrogen hyperfine line is absorption of 21-cm photons from whatever source. Such photons would reach Earth also and in principle be measurable. However, at that time radio astronomers tabulated observed antenna temperatures whose zero points were uncertain by at least several degrees, so lacking an absolute measurement at about 21 cm, I proposed to make one with the antenna at Agassiz Station. I approached Doc Ewen, the codiscoverer of the astronomical 21-cm line, with my proposal, and he said that it would be difficult to control the zero point of temperature with equipment available at that time, so I had to

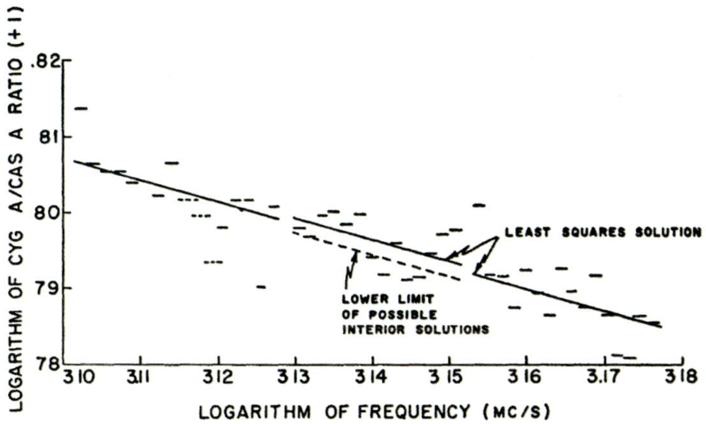

Figure 5

Final spectrum of the ratio of the flux from the extragalactic source Cygnus A to that of the galactic source Cassiopeia A on a log-log plot. Solid lines are least-squares solutions, whereas the dotted line represents what would have been observed if the optical depth of intergalactic HI were $\tau = 0.25\%$, the adopted upper limit. Reprinted from Field (1962) with permission of the AAS.

be satisfied with extrapolating many measurements of the higher intensities at lower frequencies of the galactic synchrotron radiation, of which the result was an uncertain 1 K. That was in 1955 to 1957, and the CMB was not discovered until 1965. We now know my estimate should have been at least equal to the CMB result of 2.73 K. But more on this in the next section.

Fast forwarding a moment, I was appointed to the faculty at Princeton in 1957 and decided to try my hand at intergalactic matter once more, this time with the 85-ft dish of the National Radio Astronomy Observatory (NRAO) in Green Bank, West Virginia. Once again there was no sign of the desired effect, but I was able to strengthen the upper limit a bit (Field 1962 and **Figure 5**). A side effect was that I got to know Dave Heeschen, who had become the Director of NRAO, much better and to watch in real time while Frank Drake discovered the microwave emission from Venus, showing for the first time that its surface, hidden from view since the time of Galileo by a cloud deck, had a temperature of 700 K. I met most of the staff and learned to appreciate the dedication of the telescope operators, young people from the neighborhood.

This turn at observational astronomy served me well as a theorist. However, I now confess publicly that while observing at Agassiz Station, I came frighteningly close to destroying the 60-ft dish. Here is what happened. Like many other users, to gain an hour of observing time, I used the weight of the Nautical Almanac to disable the limit switch on the console, which prevented the alt-az–mounted dish from swinging too close to the western horizon. One night I heard an ominous grinding noise. Running outside, I could see that the dish had run into the pier. The total disaster of the drive motor grinding the antenna to pieces was prevented only by the fact that the bar attached to the pier, which swings down to lash the dish to the pier in a storm, had been jolted by the crash to fall down on the microswitch that was attached to the pier to prevent such an accident and turn off the drive motor. The damage was minor, but I compounded my error by not telling anyone that I was responsible for it. I interpreted this event as a message from the god of experimental physics to return to theoretical work.

1.4. Princeton Faculty

In the fall of 1957, Lyman invited me to return to Princeton as an Assistant Professor. The Junior Fellowship would have continued for another year, but it seemed to me at the time that the offer

Video 1

The *Hubble Space Telescope*, a lifelong dream of my colleague, Lyman Spitzer. To view the video, access this article on the Annual Reviews website at **http://www.annualreviews.org**. Reprinted with permission; credit: NASA and M. Estacion (STScI) (see HubbleSite 2010).

of a faculty position was a better choice. I never regretted it, because in those days, and still today, Princeton was the place to be for theoretical astrophysics. I had been married in the meantime, and my wife Sylvia had our first child, Christopher Lyman Field, in March of 1957.

Sputnik was launched in October of 1957, opening the space age and setting the stage for space astronomy, an activity that engaged me a lot in the years to follow. I was very concerned about the US response, which was weak and confused, with the failure of Project Vanguard at the Naval Research Laboratory. An issue at the time was whether the US program should be classified, as the Soviet program was. I wrote a letter to the New York Times urging that our space efforts be unclassified. Fortunately, that was also the conclusion of the Congress, and the National Aeronautics and Space Administration (NASA) was established the next year on that basis.

Returning to Princeton, I was surprised to learn that Lyman had already proposed a large space telescope (LST), in 1946, and that he had already been working on a proposal to NASA to launch a smaller telescope dedicated to high-resolution spectroscopy that could lead to an LST in the future. His dream was realized 44 years later with the 1990 launch of what became the *Hubble Space Telescope* (HST) (**Video 1**). As I explain later, while all this was going on, Lyman was also engaged in a classified project to develop a controlled fusion machine. In retrospect, his scientific vision was extraordinary.

As a theorist, I did not seek a role in the space science area, but before I knew it, I was invited by NASA to serve on an advisory committee on lunar exploration. NASA had been told to put a man on the Moon before the Soviets could do it, and had initiated Project Apollo to do so. NASA's mission statement included space science, so the committee was tasked to develop a program to carry out science on the Moon. My contributions to the committee were modest at best, but it was a precious learning experience that led ultimately to service on many similar NASA committees. I particularly remember Harold Urey, who had a Nobel Prize in Chemistry for his discovery of deuterium. A tough-minded scientist, he had a passionate interest in lunar exploration, in particular its chemical composition, a key to understanding its origin. I was impressed by the fact that NASA would support fundamental science, not only the show of engineering skill that was the Apollo Program.

That motif emerged often in discussions of the manned program by scientists. Most of us felt that, from a purely scientific point of view, scientific goals could be achieved far more cheaply by unmanned missions. I recall a meeting in which Tommy Gold of Cornell vigorously advocated this position together with the advantage of using robots rather than humans, thereby supporting the budding industry of robotics. But most of us accepted the argument that manned activities were what excited the public and therefore enabled Congress to vote for the large sums required to fund them. The argument continued that science would profit most by using the more modest funds made available by NASA for science almost as an afterthought. This position is no doubt a tainted one, but one that I encountered throughout my career. For example, the Templeton Foundation is interested in finding scientific support for religion, and although most physicists are at least agnostics if not atheists, some choose to accept funding from Templeton anyway.

In the Astronomy Department I taught a course on the Introduction to Astrophysics for undergraduates, and I took over Lyman's course on the interstellar medium. I don't recall whether it was a result of the undergraduate course, but I ended up writing two papers with undergraduates, one with Bruce Partridge (Field & Partridge 1961) on detecting a radio frequency spectral line in the interstellar medium and one with Bill Saslaw (Field & Saslaw 1965) on a model of clouds being accelerated by HII regions and agglomerating to form massive clouds that become gravitationally unstable. The former paper was based on the assumption that the Lyα radiation in HII regions is unattenuated, but later research concluded that Lyα in HII regions is lost in absorption by interstellar dust. The latter paper is still cited today.

My undergraduate teaching encountered an unexpected problem. I wrote a final exam and averaged its grade with the homework to get a final course grade. Then I translated the numerical grade into a letter grade using a formula I got I don't know where. Some of them were Cs. After the grade reached the students, they sent a delegation to Lyman's office, complaining about the Cs. Lyman explained to me that giving a C in an undergraduate elective course was just not done. OK, so he upgraded the students, so to speak. Lyman forwarded the corrected grades to the appropriate committee, which had the original grades in hand. As required, Lyman gave the reason for the change in grades. The reason he gave was "incorrect weighting of course components," which of course was not true. Neither Lyman nor I was concerned about our white lie.

We should have been. The whole matter of the change in grades had to be approved by the entire faculty at its monthly meeting. Naturally I had to be present to answer questions. Believe it or not, there was such a question from a clever professor who had noticed a logical flaw in Lyman's explanation. "How," he asked, "is it possible to reweight course components so that all the new grades were better than the old ones?" Fair point, Professor X! My heart was in my mouth for reasons I'll explain in a moment. Very fortunately, the point was greeted by laughter as the faculty began to understand what had happened. I was free!

Why was I terrified? I had crept into the faculty meeting at the back of the room, hoping to escape observation. As I stood there, I thought I sensed cool air around my bottom, so I felt the rear of my trousers. Egad! The seam was open about 8 inches! Wouldn't you be apprehensive in that situation? A lowly Assistant Professor at his first faculty meeting, taking questions with a gaping hole in his pants!

Was it the god of junior faculty who had stepped in? Or was it the laughter that intervened in an otherwise disastrous situation? Or was it the fact that the grade change was signed by a highly respected professor, who was trying to cover up a mistake that they themselves had made at one time or other? I will never know.

Princeton was a perfect place for research. In the past, Martin and Lyman had met once a week with faculty from other departments for lunch, particularly Bob Dicke and John Wheeler from Physics and John Tukey from Mathematics. But after I arrived, the Institute for Advanced Study (IAS) in Princeton decided to appoint a senior member in astrophysics. Their choice was Bengt Stromgren from Denmark, a genial astronomer whom everyone liked and respected. He was well known for solving the problem of the effect of ionizing radiation by a hot star on the surrounding atomic interstellar medium. His solution was not intuitively obvious. It consisted of an almost completely ionized HII region bounded by a thin boundary layer where the transition to HI takes place. Lyman admired his invention of the "Stromgren Sphere," as it came to be known, and it may have been the reason he recommended Bengt to be the first astrophysicist at the IAS. In any event, the appointment proved to be felicitous. Bengt fostered communication among all of the astronomers in Princeton by establishing a weekly lunch at the IAS cafeteria. It was a special experience, because the tradition developed that, without warning, any one of the attendees could be called upon to explain their current research. Having the likes of Lyman, Martin, and Bob in attendance served to concentrate the minds of the younger people attending.

My first project at Princeton was to sharpen my Harvard result on intergalactic HI by using the 85-ft antenna at NRAO. My observations drove the upper limit down by a factor of three or so (Field 1962 and **Figure 5**), no big deal; but one issue still bothered me: the absolute temperature of radiation at the 21-cm wavelength, because such radiation contributes to the excitation of the line in intergalactic space, and it was necessary that I know it for interpreting my results. As I explained above, I had asked Doc Ewen if that could be measured. He believed not, because an absolute measurement was not feasible at that time. Actually, such a measurement had been done by Dicke et al. (1946) at a different wavelength, which resulted in an upper limit of 20 K. Unfortunately, in the preADS era I had not come across that paper, so the issue was unresolved even at the time of my 1962 paper, and as a result the density of intergalactic HI was not limited as precisely as it might have been. Since then there have been a number of papers on the problem. Penzias & Wilson (1969) looked for the emission step rather than the absorption trough and found an interesting limit that was independent of the complicated problem of the intergalactic spin temperature. Of course, as I discuss further below, the radiation temperature in space was finally accurately determined in 1965.

At that time, Lyman was making sense of the interstellar absorption lines of sodium and calcium, which showed sharp lines at various discrete velocities, by positing that each line of sight to a star at about 1 kpc has on average about 7 clouds of HI. These clouds would have densities of about 10 cm^{-3}, and because 21-cm observations indicated a global average of 1 cm^{-3}, the filling factor would be 10% and the radius of each cloud about 10 pc. This model also fitted the observed variations in the extinction by interstellar dust.

At that point, grad student Carl Heiles presented himself and asked me if I could suggest an observational problem in radio astronomy. Carl had a BS from Cornell and was more familiar with engineering than I. Together we decided that if the calcium clouds observed by Lyman also contained HI, we should be able to resolve their 21-cm emission with a radio telescope of high angular resolution, particularly at high galactic latitudes where the typical clouds would have larger angular diameters. Soon Carl was observing with the Green Bank 300-ft antenna and bringing me huge contour maps in l and b that exhibited what I at least interpreted as Lyman's clouds. Carl, who is now recognized as an expert in the atomic interstellar medium, did not agree with that interpretation (Heiles 1967). I wonder what he thinks now.

Lyman's line widths indicated internal cloud velocities of the order of 1 km s^{-1} or 1 pc per megayear, so he reasoned that without some confining mechanism, a cloud would expand into the background in 10 Myr. That seemed to be a problem, so he looked at possible confinement mechanisms. Self-gravitation is a possibility, but with the known densities and radii, Lyman showed that self-gravitation is too weak. Maybe there is an intercloud medium that would confine the clouds, but such a medium had not been observed.

At this point, I became interested in the problem. Because the internal pressure of the clouds is thermal motions, $p_i = kn_iT_i$, or, to save writing, we refer to pressure without the factor of k, as simply n_iT_i. His model required that n_1 equals 10, and 21-cm observations suggested that T_i equals 100 K, so $n_iT_i \simeq 1{,}000$ cm^{-3} K. Could there be an external medium with the same pressure, $n_eT_e = 1{,}000$, for example, with $n_e = 0.1$ and $T_e = 10{,}000$? I calculated that such a medium could have escaped detection and began to wonder what it could be made of. At that point, I began

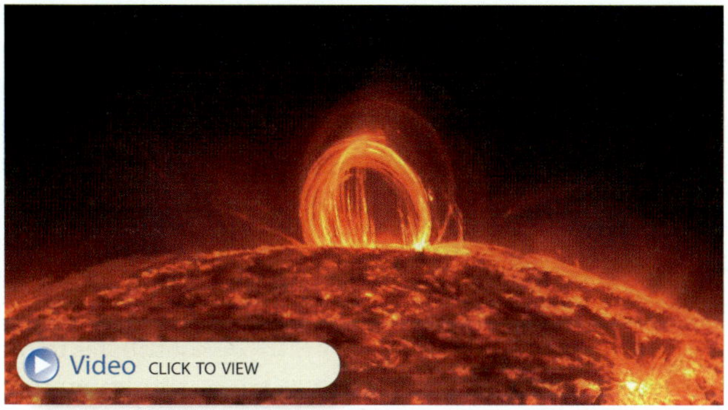

Video 2

The Sun filtered through the Balmer alpha line of hydrogen showing cool gas in coronal magnetic loops raining down on the surface of the Sun. To view the video, access this article on the Annual Reviews website at **http://www.annualreviews.org**. Reprinted with permission; created by the Solar Dynamics Observatory, Scientific Visualization Studio, Goddard Space Flight Center, NASA (see NASA Solar Dyn. Obs. 2013).

to think of a medium with two or more phases, like water and ice, but instead both mediums in phases of gas though at different densities. That seemed reasonable, because in the solar corona, which is largely at a million degrees one often observes condensations at a temperature of the order of 10,000 K, which solar observers seemed to accept as pressure equilibrium. In **Video 2** is an image of the Sun taken through a filter at the Balmer alpha line of atomic hydrogen. The corona itself does not show because at a million degrees all of the hydrogen is ionized. However, the gas in the magnetic loops shines brightly because at the temperature there, 10,000 K, the gas is mainly in atomic form, which emits the Balmer alpha line. The density of the gas in the loops is larger than that in the corona, so there may be approximate pressure equilibrium between the gas in the loops and the coronal gas, even though the temperatures differ by a factor of 100.

Lyman at that time was also thinking of pressure equilibrium in the interstellar medium because it is needed to confine clouds. I was also inspired by papers from Parker, Zanstra, and Pikelner, who had thought about an instability of a hot medium that could lead to transitions to a new condensed phase.

So I began to work on what I called thermal instability. I showed that, for certain types of the dependence of the heating and cooling rates on density and temperature, such an instability would occur on the timescale of the cooling rates. This work was largely done during meetings of the Summer Program in Geophysical Fluid Dynamics at the Woods Hole Oceanographic Institution in Woods Hole, MA, sponsored by the National Science Foundation (NSF). Woods Hole is a small seaport on Cape Cod that had pleasant diversions for my wife Sylvia and my kids Christopher and Natasha. For several summers, we rented an old fashioned cabin located on Crooked Pond in Falmouth, MA. Tasha played by the shore while Chris roamed the pine woods nearby and got into the rowboat for some fishing. Sylvia cooked the fish after I had cleaned them, and Sylvia and I attended performances of Gilbert and Sullivan operas in Falmouth as often as we could find baby sitters.

The participants in the program were mostly theoretical oceanographers, meteorologists, and applied mathematicians, with one theoretical astrophysicist, an old friend, Ed Spiegel from New York University. I profited a lot from the insights of all of these people, particularly Willem Malkus of MIT. The resulting paper, titled *Thermal Instability* (Field 1965) gained some notice from those who found new applications of the concepts in the paper. For example, while I was attending a conference in the USSR, the well-known physicist Yakov Zel'dovich came up, smiling, and said that he had recommended it to his students. This paper stimulated the community to think of multiple gas phases in various astronomical situations.

The year 1965 was a turning point in other ways too. In the previous year, I had given a class on radio astronomy at Caltech that was attended by my new friends John Bahcall and Jim Gunn, and I took the opportunity to visit UCB, at the invitation of Dave Cudaback, a research associate there doing radio astronomy. The scientists there greeted me warmly, and the time we spent there was especially pleasant.

1.5. Berkeley

When I returned to Princeton, I was surprised to receive a call from John Phillips, chair of astronomy at UCB, offering me a full professorship in the Astronomy Department. At that time I was Associate Professor at Princeton, and happy enough. However, I enjoyed the prospect of a change to a beautiful location on San Francisco Bay and the possibility of working at a location different from that of my thesis advisor, Lyman Spitzer. I admired Lyman, but it was hard to compete for good students because we were working in the same field. UCB had impressed me with the quality of its research programs, particularly in radio astronomy. I accepted the offer by UCB. I did not ask for a counteroffer from Princeton, but simply went to Lyman's office and told him what I had done. He was surprised and responded that as a courtesy he would seek to promote me to full professor at Princeton. I appreciated his response, and I explained my reasons for leaving, which I think he understood. My place was taken by Jerry Ostriker, a student of Chandra's, who as expected became an intellectual leader of the first rank.

There were other factors at play. One was the famous antiwar demonstration at UCB led by Mario Savio from the top of a police car, which gave us pause as we considered leaving the peaceful environs of Princeton. Although we thought the Vietnam War was immoral, we were also opposed to violent protests. In the end, in 1965 we decided to move to Berkeley and to buy a house in the hills with a view across the Bay to the Golden Gate Bridge. To find a house, I asked the advice of my good friend Jon Collier, a fellow graduate student at Princeton and an architect who at that time happened to be living with his family in Berkeley. Jon volunteered to check out the possibilities. We ended up buying a home on Senior Avenue that he recommended, although we had never seen it. Thanks to Jon's good taste and hard work, we stayed happily in that home until leaving for Harvard in 1972.

The very first year at UCB was eventful. At one of the first informal department lunches at the Golden Bear restaurant on the campus, Professor Harold Weaver, the founder of the radio observatory at Hat Creek, relayed the information that an interstellar source of OH emission had been observed to vary rapidly. If I recall correctly, I immediately said "MASER!," as it later proved to be. Perhaps that instant response was related to my paper with Bruce Partridge (Field & Partridge 1961), in which we had commented that our dependence on stimulated emission for our result was the first application of the MASER (microwave amplification by stimulated emission of radiation) principle in astronomy.

A little later, I got a call from Professor Edward Teller of the Physics Department asking me to visit him at his home in Berkeley. I found Teller's home easily and spent a pleasant hour or so with him and his wife. He was interested in what this new astrophysicist he had heard about was like. It turned out that he felt responsible for a physics graduate student, Christopher McKee, who had expressed interest in theoretical astrophysics. He wondered if I would take him on.

Chris and I hit it off and, perhaps because of Teller's interest in explosions, we talked about supernova explosions and their interactions with the surrounding interstellar medium. Chris jumped at the problem and proposed that he could simulate the interaction on one of the powerful computers at Livermore. The result was very interesting. As the ions in the supernova gas streamed through the interstellar medium, an instability grew furiously and turned all of the directed energy into turbulence. Chris got the PhD and later wrote a classic paper with Jerry Ostriker on all the types of explosions of interest in astronomy. In his spare time, he became the Chair of the UCB Physics Department, as well as Director of the Space Science Laboratory. Chris is now recognized as an authority in the theory of the interstellar medium.

One day I got a call from Bernie Burke at MIT saying that Penzias and Wilson at Bell Labs had discovered background radiation with a temperature of 3.5 K at a wavelength of 7.35 cm that might be a remnant of the Big Bang (Penzias & Wilson 1965). At that news, I had an "Aha!" moment.

As an interstellar guy while still at Princeton, I had read a footnote in Herzberg's (1950) book on diatomic molecules that stated that Walter Adams at Mt. Wilson had obtained high-resolution spectra of the interstellar molecule CN, called cyanogen, which exhibited an unusual feature. Unlike other interstellar molecules, its first excited rotational state, $J = 1$, is populated with an excitation temperature of about 3 K. According to that reference, the responsible mechanism was unknown. As I was familiar with the problem of excitation of the hyperfine levels of atomic hydrogen, it was simple to write down the appropriate equation of statistical equilibrium including radiation and collisions for the first rotational state of CN. Putting in the densities and temperatures of interstellar clouds, I concluded that collisions could not explain the observations; so I began to think about radiation at the wavelength of the rotational transition, 2.6 mm. In order to proceed, I needed to know the Einstein B for the transition, which depends on the permanent dipole moment of the molecule. I could come up only with a rough estimate of its value, so Lyman concluded that the paper would be too speculative, so I never published it.

When I heard about the radiation discovered by Penzias and Wilson, I immediately looked for the draft of the paper that I had written in Princeton. Maybe the radiation that I had postulated to explain the excitation of CN is related to that of Penzias and Wilson! I looked high and low for the draft, but it was nowhere to be found; perhaps it was lost in the switch from Princeton to Berkeley, so I would have to start all over. Once again I looked for the dipole moment in the literature but could not find it. A remarkable coincidence yielded its value. It happened that on that same day I had been working on the proofs of an unrelated article that I had submitted to the *Annual Review of Astronomy and Astrophysics*. They had provided a copy of the proof of another article as a guide to markup. I had just thrown it away, as I had finished with it. In a flash, I recalled that the proof was of an article on molecules in comets by C. Arpigny (1965). Retrieving the article from the wastebasket, I scanned it for mention of CN. Yes, it was there! The article concerned the excitation of CN by sunlight and subsequent deexcitation by rotational transitions, which depends upon its dipole moment. The value of the latter was not mentioned, but it could be inferred from the resulting calculations. Presto! The missing information I needed was in my wastebasket!

As if that were not enough, in conversation with John Hitchcock, a grad student whose office was down the hall, I learned that as part of an observational project he was doing with George Herbig, he had high-resolution spectra of interstellar CN in his office, from which he could deduce the population of the excited state. In a beer joint at the northern edge of the UCB campus, we drafted a joint letter to the *Physical Review Letters* based on my calculations and his measurement of the population of the CN excited state (Field et al. 1966). Our result was that the excitation of CN must be due to radiation at a wavelength of 2.6 mm, whose temperature is >2.7 K, in agreement with the data of Penzias and Wilson at 7.35 cm. Overnight the fit to the blackbody spectrum expected by Peebles and Dicke for the cosmic background radiation had been extended by a factor of 28, and relatively close to the maximum of a 3 K blackbody at a wavelength of 1 mm.

1.6. Harvard, the Smithsonian, and the Center for Astrophysics

Professor Edward Purcell, codiscoverer of the interstellar 21-cm line, had been a friend and collaborator ever since I came to Harvard as a Junior Fellow in 1955, so it was not surprising to get a phone call from him at our summer home in New Hampshire. He invited us to lunch at his summer home, which happened to be fairly nearby. His home had a wonderful view of the White Mountains, so while my wife chatted inside with Ed's wife Beth, Ed and I sat on the porch talking about the days long ago when my grandmother played the piano in a nearby house for a sing-along with some Harvard professors. Suddenly the conversation turned to the Harvard Observatory. Ed was on a search committee for a new director because Leo Goldberg, the then director, had accepted an appointment as director of the Kitt Peak National Observatory in Tucson, Arizona. The next sentence was a shock: "George, we want you to become the director."

What an honor! Harvard Observatory was the place that Edward Pickering had created the Henry Draper Catalogue of stellar spectra, and Harlow Shapley had located the center of the Milky Way. But why me? I never learned, but in preparing this review, I came up with a plausible explanation: I had been recommended to the Harvard Society of Fellows by Lyman Spitzer, who I later learned was a close friend of Ed's. Ed was a Senior Fellow in the Society, so my appointment there was no doubt due to Ed. So my career would be heavily influenced by Ed, and as a Nobel Prize winner in physics, his opinion was valued highly by the Dean of the Faculty of Arts and Sciences, to which the Director of the Observatory reports. Thus, it was no surprise that my name would come up in discussing the directorship. But there is more to the story. It happened that during that period, the NAS had initiated decade reviews in both physics, chaired by Ralph Bromley at Yale, and astronomy, chaired by Jesse Greenstein at Caltech. I had been asked to join both, and I accepted both invitations. Now Ed, as a member of the Bromley Committee, was in a position to observe my contributions to that report. Would they measure up?

I suggested that I chair a panel on Astrophysics and Relativity, which would report to both committees. I intended to try to convince the physicists that astronomy was ready to make new contributions to cosmology. My own contributions to the field were practically nonexistent, but having been at Princeton, I was keenly interested in the LST, which would certainly contribute to cosmology.

My suggestion was accepted by both Ralph and Jesse, and I went about recruiting members. I was able to get both Steve Weinberg, author of the reigning text on general relativity (Weinberg 1972), and Maarten Schmidt, the discoverer of the first quasar (Schmidt 1963), as well as experts on the instruments that would be critical to the new effort and of interest particularly to the Greenstein Committee.

At this point, I must explain my own attitude toward observational astronomy. It is the critical core of the field. Furthermore, the rate of discovery has always been determined by the development of innovative instruments. In the period in question, the innovative instruments were LST, the Very Large Array, high-energy (X rays and gamma rays), and infrared telescopes. The panel made specific recommendations as to the instrumental programs needing support. I submitted the report early in the deliberations of the Greenstein Committee (Field 1973). Imagine my pleasure when Jesse leaned back and said, "George has already written our report!" I think that the report of the Panel on Astrophysics and Relativity may have influenced Ed as the Harvard search committee deliberated about the directorship of HCO, because he had been a member of the Bromley Committee to which the report had been submitted.

As an aside, the final meeting of the Greenstein Committee was a cliffhanger. Don Morton and I felt that the draft report was not sufficiently supportive of LST and got Jesse's permission to revise the draft to be more supportive. Then at the truly last minute, Jesse made an astonishing statement to the Committee that went something like this: "I am sorry, but I cannot sign the report. It shifts the bulk of the funding to space programs, and it will be the death of ground-based astronomy." We listened in stunned silence as we contemplated the waste of two years of committee work. It fell to Bernie Burke to suggest that we take a break and consider what to do. Bernie used the break to call Lyman Spitzer, whose counsel Bernie thought Jesse would listen to. We were not privy to what happened next, but Jesse called the meeting to order, and it went on as before. The Greenstein Committee Report was published in 1973 (Greenstein 1973) and, in my opinion, had a very positive effect on astronomical research. It is interesting to think about whether Jesse's concerns were valid. Certainly not in detail; for example, private donors later funded the powerful Keck telescopes serving California astronomers, including those at Caltech where Jesse taught.

Back to the Harvard directorship. Naturally, despite the honor involved, I studied the situation to find out what funding would be available to pursue the new astronomy I had in mind, as indicated by the report to the Greenstein Committee. The answer was precious little beyond a few endowed professorships, remnants of the glory days at HCO. Leo Goldberg, a solar physicist, had used the Harvard Observatory as a base for solar research using the Orbiting Solar Observatories funded by NASA. That made available some very able experimenters, but of course they were interested in the Sun, not cosmology. Leo had raised money for a beautiful new building, the Perkin Laboratory for Astrophysics, which could accommodate new enterprises. But the elephant in the room was the Smithsonian Astrophysical Observatory (SAO), which had moved from Washington to Cambridge in 1955 at the invitation of Donald Menzel, a solar physicist who preceded Leo as the director of HCO.

A little history is useful here. Fred Whipple, a Harvard professor and preeminent expert on comets and interplanetary bodies, had volunteered to direct SAO when it arrived in Cambridge. When I was a postdoc at Harvard from 1955 to 1957, there was only a small group at SAO studying meteors and meteorites. Fred Whipple, who had been serving as a member of the so-called Rocket Panel with the likes of Werner von Braun, was asked by the infant NASA to set up a network of telescopes to track Russian and US satellites. Fred had experience in doing this because of his Prairie Network, which tracked meteors. This work, undertaken on a rushed basis, succeeded in being ready in time to track the first Russian satellite, *Sputnik*. *Sputnik*, in beating the US into space, induced an almost hysterical response in the American public, so Whipple ended up on the cover of *Life* magazine. As a result of this success, Fred was able to build up a staff of some 300 people, the majority on NASA funds, to provide the computer and engineering support for the tracking network. On the scientific side, Fred himself continued his personal work on comets, and important results on Earth's atmosphere, geodesy, and meteoritics emerged.

As time went on, Fred was able to diversify the SAO staff into other fields by adding scientists to the so-called federal roll. As an accident of its founding as a national institution in 1846 on the basis of a gift to the US government by a British subject, James Smithson, the Smithsonian Institution, in addition to its complement of federal civil service employees, also enjoyed status as a private entity. Because there were not enough civil servants to staff the tracking network, Fred was able to contract with NASA as a private institution to provide the funds to do so. When I began to study the situation in 1972, SAO had a federal staff of about 40 scientists working on a range of subjects from atmospheric science to early type stars, together with a much larger NASA-supported staff on the so-called private side. To support both the federal and private sides at SAO, there was an engineering group, a library, and other facilities.

As I thought about the situation, it seemed to me that if a way could be found to use federal positions at SAO to attract tenure scientists, perhaps I could fund initiatives more in keeping with contemporary astronomy. But Fred Whipple had already set the goals of SAO. What could be done? One very important opportunity occurred at about this time. Leo told me that at one point he had talked to Riccardo Giacconi, a scientist at a defense contractor in Cambridge called American Science and Engineering, about joining HCO and doing research in X-ray astronomy. I had met Riccardo and knew that he was a very active physicist interested in cosmic science, so this was of very great interest to me. I met with Riccardo, and he was interested in joining us.

There was a story behind the relationship between HCO and SAO that bore on the directorships of both. As I understood it from some faculty members, the scientific direction of HCO was the responsibility of the HCO director, but with the advice of the HCO Council, which was made up of tenured faculty in the Department of Astronomy, of which Fred was one. At the same time, Fred, as the director of SAO, could act independently using Smithsonian funding. Fred was interested in developing an optical observatory, now named for him, on Mt. Hopkins, south of Tucson, Arizona, while Leo wanted HCO to develop an optical observatory somewhere else. It is not hard to imagine that Leo and Fred might come to loggerheads over these plans, and that is what happened. When I asked Leo why he decided to leave HCO, he did not mention this as a reason. However, I could see that if I became director of HCO, rivalry between HCO and SAO could become a problem. Therefore I told Ed Purcell that I would accept the HCO position only if I could also be the director of SAO. That concept had not come up before, but my suggestion was not rejected out of hand. On the contrary, I heard rather quickly that it had been discussed with Fred and he was open to taking retirement as director of SAO. So it came about that I would direct both observatories.

Thus began the Center for Astrophysics, later to be called the Harvard-Smithsonian Center for Astrophysics (CfA) to distinguish it from other institutions with similar names. Of course, there were many decisions to be made before CfA became a reality. I decided to take leave for a year from UCB and to move to our summer home in New Hampshire from which I could commute every two weeks or so to Cambridge. I planned to speak with all the senior scientists at HCO and SAO about future arrangements. These conversations were enlightening, particularly regarding the aspirations of each individual and how they could be met in the new institution. One senior professor was unhappy with the prospects, but most individuals seemed reasonably content with them.

As I visited with people it became clear to me that it would be helpful to have individuals who are in daily contact with the 100 scientists to organize the research in each area and to arrange for needed support. People liked the idea, so we set up several research divisions, each headed by an Associate Director (AD): High-Energy Astrophysics (Riccardo Giacconi), Radio Astronomy (Ed Lilley), Theoretical Astrophysics (Alex Dalgarno), Solar Physics (Bob Noyes), Planetary Sciences (Al Cameron), Geoastronomy (George Weiffenbach), and Optical and Infrared Astronomy (Nat Carleton). I stipulated that any scientist was free to join any division and to associate informally with any other one. I met with the ADs on a regular basis to discuss budgets and personnel. When the AD positions became vacant, I solicited the opinion of division members as to whom would be a suitable replacement. This arrangement worked well; it is still in place 30 years later.

Complex administrative arrangements were necessary, because different people drew salaries from three different sources: the Harvard Faculty of Arts and Sciences, the Smithsonian Federal Civil Service, or from contracts and grants on each side, the terms and rules for which were all different. And the services available to each scientist were different for historical reasons. As time went by, we figured out how to keep the books in such a way that all the services were available to all. An interesting example is library services. Each observatory had been funding its own library, reporting to either the Harvard College Library or to the Smithsonian Library in Washington, DC. Over time, both libraries were physically merged, providing access to a wider range of materials to each user, in a single space and under the name of a long-time benefactor of HCO, John Wolbach.

I had no experience as a manager, so I delegated all of these problems to an Assistant Director, John Gregory, who later took over the HCO management when Bob Reed retired as head of HCO Administration. Without John's expert help, things would have been in a hopeless muddle. John had been recruited as a Program Manager from NASA by Fred Whipple, who suggested him for the position. Fred, by the way, was assigned to a corner office comparable to his previous one and continued to work on his research for many years after retiring as director of SAO.

Riccardo is a dynamic scientist and leader. Of course he wanted to bring many of his team with him from American Science and Engineering. I had to find civil service positions for about a half dozen of his senior staff. With the backing of the Assistant Secretary of Science at the Smithsonian, Dave Challinor, this was done. However, I had to find the funds to support those positions out of the SAO federal budget, which was under stress at that time. The only option was to phase out activities that I thought were less important. With the Associate Directors, I drew up a plan of what activities could be dropped without serious damage to our goals. Needless to say, this was a painful exercise for those whose positions would be affected. At SAO, job descriptions were very carefully worded, so that if funds were not available to support a certain position, the individual holding that position could be let go without his/her "bumping" another employee, because in general he/she would not be qualified for it. That protected the Director's ability to preserve the program that had been approved.

While all this was going on my wife Sylvia and I were having difficulty in our marriage. I moved out in 1976, and the divorce became final in 1978. It was painful all around. In retrospect, I don't know how I maintained a measure of good sense during that period. I sought the help of a therapist and saw her weekly for over five years. The therapist had been recommended by Babette Whipple, Fred's wife, and a therapist in her own right.

Babbie was an old friend, and one day when we were together she asked, "How is your social life?" I replied, "Nonexistent." Being the wife of an astronomer herself, she understood this rather technical way of putting it. A few weeks later, she called to say that she had a friend that I might like to meet, Susan DiMeo. I called her; we talked, met, and began to go out together. I found that Susan is a very special person, with several qualities that were perfect for me. She is always thinking how she can help or cheer up someone else. She is intuitive and bright, and ready to share her opinions with anyone at a moment's notice. Later I found that she is a talented writer, whose essays on the life around her are usually accepted by whatever newspaper is available. Unlike mine, her stories never undergo revision—they go straight from her laptop to the editor.

Soon we were living together, and I suggested that she join me at the IAU General Assembly in Montreal in 1979. She arrived in time for the banquet, so I ushered her into the hall with about 1,000 astronomers, probably a daunting experience for her. No problem; soon she was dancing with my old buddy Hugo van Woerden from Groningen. Thus the transition from a regular person to an astronomer's lady friend was accomplished in less than an hour.

Later, on a visit to the Whipple Observatory in Tucson, we decided to get married. While I was at the University of Arizona, Sue called to see what the requirements for getting married would be. The lady said that we would have to be tested for syphilis before we could get a marriage license and that it would take three days to complete the test. As is her habit, Sue entered into a friendly conversation with the lady, in which Sue explained that we could not wait that long because we were due in Phoenix the next day. After some more chatting, the lady said, "You don't sound like you have syphilis, so why don't you come down now." We did, and we were married by a judge, who happened to be a Native American woman, and that evening Susan was introduced as Mrs. Field. The previous evening she was introduced as "my traveling companion," a phrase that the emcee made up and used without consulting me, so Phoenix was a welcome change. There was a nice party for us when we returned to our apartment in Boston, and we began our 32 years of happy married life together. Sue has two delightful daughters, Cristina and Stefanie.

At Harvard I continued my previous research programs. In Part 2, I explain them in qualitative terms, but one of them, based on a suggestion by Roman Jackiw, a theoretical physicist at MIT, has enough mathematics that it deserves more explanation here in Part 1.

Stimulated by his work in other fields, Roman suggested a new term in the Lagrangian for electromagnetism. It would seem that the classical term describing the field, $B^2 - E^2 = F_{\alpha\beta}F^{\alpha\beta}$, should suffice, as it has been tested experimentally countless times. (I omit numerical factors hereafter.) There is another candidate for an electromagnetic Lagrangian, also quadratic in the fields, the so-called Chern-Simons (C-S) term, $\mathbf{E}\cdot\mathbf{B} = F_{\alpha\beta}\tilde{F}^{\alpha\beta}$. Although some texts don't mention this fact, it can be shown that the C-S term is the divergence of a four-vector, $H^{\alpha} = (\mathbf{A}\cdot\mathbf{B}, \mathbf{E}\times\mathbf{A} + \psi\mathbf{B})$, called the magnetic helicity (Carter 1979, Berger & Field 1984, Field 1986). $\mathbf{A}$ is the vector potential and ψ is the scalar potential. The density of magnetic helicity is the time component $\mathbf{A}\cdot\mathbf{B}$, whereas the space components represent its flux. The concept of magnetic helicity was introduced into astrophysics by Woltjer (1958) who used it to prove a theorem in MHD. A fluid dynamicist at Cambridge University, Keith Moffatt, had thought about a similar quantity that arises in fluid mechanics and wondered how to interpret it physically. His resulting papers (Moffatt 1969, 1981) showed, with the help of a paper by Moreau (1961), that it is a topological quantity that represents the degree to which two magnetic lines of force are linked. Woltjer (1958) had proved that it is conserved in ideal MHD (IMHD) in which the resistivity is zero. In relativistic notation, $\partial_{\alpha}H^{\alpha} = 0$. If we measure it in a flow of magnetized plasma at one moment, it is the same at all later times.

How can we conceive of this result? Alfvén (1950) had proved that in IMHD the magnetic lines of force are frozen into the plasma as it moves. By this I mean that if a particular line passes through an element of plasma at one moment, it will do so for all later times. Therefore, if one imagines two loops of magnetic field linked together, they can become unlinked only if at one or more points one element of plasma moves through another, which is not possible. (It takes some persistence to convince yourself that this is true.) Therefore any change in the magnetic helicity requires unlinking some lines of force, which is not possible in IMHD, so magnetic helicity is conserved.

It is well known that $\mathbf{E}\cdot\mathbf{B}$ is a pseudoscalar quantity, and therefore violates parity. Experimentally we know that is not true of electromagnetism, so textbooks state that this term is not acceptable as a Lagrangian. But as indicated above, it is also a divergence, and divergences in a Lagrangian have no effect on the classical equations of motion. For those more interested in the mathematics behind magnetic helicity conservation, there is a sophisticated version by Carter (1979), who used the exterior calculus to prove that it is conserved in arbitrary manifolds, even those without a metric. His result on p. 288 of that paper is that $\tilde{\eta} = (*\mathbf{F})\cdot\mathbf{A}$.

Of course Roman knew that $\partial_{\alpha}H^{\alpha}$ cannot be used as a Lagrangian because it is a divergence, so his proposal for a new term was different, $p_{\alpha}H^{\alpha}$, where p_{α} is a constant four-vector. He suggested that this term might lead to observable effects in astronomy. I worked out the effect on polarized radiation, such as the synchrotron radiation from radio galaxies. A Harvard graduate student, Sean Carroll, volunteered to assemble the observational data about this in the literature, and from that data he determined that the predicted effect is absent. To avoid conflict with the observed isotropy of the Universe, we assumed that the space components of p_{α} vanish identically, leaving only the time component p_0, so the new term became simply $p_0\mathbf{A}\cdot\mathbf{B}$. From the observations, Sean put an upper limit on p_0 equal to 10^{-42} GeV, which might as well be zero as far as current physics is concerned. As of this writing, this paper has been cited over 500 times. (Carroll et al. 1990).

Later on, Sean and I considered another term related to helicity, $\varphi\partial_{\alpha}H^{\alpha}$, where φ is a dimensionless field representing a new degree of freedom (Carroll & Field 1991). Because $\partial_{\alpha}H^{\alpha}$ is pseudoscalar, φ must also be pseudoscalar, to avoid parity violation of the Lagrangian. The term $\partial_\alpha H^\alpha$, analogous to $\partial_t \rho + \nabla \cdot (\rho \mathbf{v})$ in nonrelativistic physics, represents a source of magnetic helicity. When we analyzed the effects of this Lagrangian, we found two interesting things: (*a*) that φ corresponds to the angle by which the plane of polarization of an electromagnetic wave is rotated as it proceeds from the source to the observer, and (*b*) that our term had been analyzed earlier (Ni 1977) in the context of verifying the Schiff conjecture that any Lagrangian satisfying the strong equivalence principle must also obey the Einstein equivalence principle. The first finding allowed us to conclude from the observations that the rotation angle of a wave crossing most of the visible Universe is less than a few degrees, a tiny amount. If this angle is in fact zero, the violation of parity that would be the result if the polarization of the wave rotated either to the right or left is absent. Thus our proposed term must be absent, or nearly so in the Universe out to a redshift of a few, but the early Universe may be another story, as I explain below.

The second finding is of interest to experts in general relativity (GR), which include Sean but not me. Most of what I know about GR is from lectures Sean gave to the grad students at Harvard when he was himself a student. They are the basis of Sean's textbook *Spacetime and Geometry* (Carroll 2004).

Finally, there is an unexpected relationship between all this and IMHD. If there is no resistance, electric fields short out, and in particular, their component along the magnetic field must vanish, so $\mathbf{E} \cdot \mathbf{B} = 0$. Thus in IMHD there is no source of magnetic helicity. If one integrates the equation $\partial_\alpha H^\alpha = 0$ over a volume, one finds that $\partial_t \int_V \mathbf{A} \cdot \mathbf{B} d^3x = 0$, which is the equation of magnetic helicity conservation for IMHD (Woltjer 1958, Moffatt 1969). This result is not fundamental because it applies only to systems with negligible resistivity. But as we see in Part 2, it is key to understanding where the magnetic fields observed by astronomers come from.

1.7. Astronomy Survey

Starting in 1960 and continuing every ten years thereafter, the NAS has published reports produced by an Astronomy Survey Committee, which reviews and prioritizes projects planned for the next decade. The purpose of doing so is to advise the government on what budgets would be needed to support the field for the next decade.

Each survey is carried out by an Astronomy Survey Committee appointed by the Academy. Normally, the Academy appoints the Chairs, who in consultation with the Academy appoints the remaining members. The Committee must be sure that its recommendations are in line with the aspirations of the astronomy community, and as far as possible be within the limits set by prevailing budgets of the agencies involved, primarily the NSF and NASA, but also the Department of Energy (DOE). It is also common sense to appoint members who are broadly representative of the various subfields of astronomy and of the various institutions around the country.

Jerry Ostriker, a professor at Princeton and an active member of the Academy, called me and asked if I would consider being the chair of the committee that would recommend programs for the 1980s. Prior committees had been chaired by Albert Whitford of the University of Wisconsin (1960s), and Jesse Greenstein of Cal Tech (1970s); both reports had positive impacts on astronomy. I had been a member of the Greenstein Committee. I thought about it a bit and decided it would be interesting.

Curiously, I myself was not a member of the Academy at the time, having been elected in the 70s and declined. Moreover, I had reservations about honorary societies that take it upon themselves to decide whom to honor. In the immortal words of Groucho Marx, "Why would I join a club that would have me as a member?" Much later, when I was nominated again, Don Osterbrock, an old friend from my Princeton days and a member of the Academy, argued that I should accept, because otherwise those who had accepted membership would justifiably feel put out. That was a fair argument and I felt like joining, so I accepted. Fortunately the astronomy members forgave and forgot, and all was well. But all this was much later, so at the time of the survey committee I was not a member; anomalous, perhaps, but nobody cared.

Soon I found myself in Washington at the Academy building with a small group of senior astronomers who proposed members for the survey committee. Many interesting suggestions were made. I got on the phone and asked the opinion of other astronomers and, thus, narrowed the list to a group to submit to the Academy for their approval. There were no objections, so I called each of the candidates and requested their participation. Fortunately there were few if any refusals, perhaps because like me they recognized an unusual opportunity to contribute to the vitality of American astronomy, and thus to world astronomy.

Meeting in Washington, we discussed how to organize our work. Of course we were aided by the fact that previous surveys had already done much of that work for us, but innovations were introduced by Paul Blanchard, an astronomy PhD who had signed on as Executive Secretary. And compared with previous surveys, we were aided by the universal use of email, which allowed us to get more done with fewer meetings. (Even so, I crossed the country 34 times.)

We agreed that the report would be concise, the total budget would be limited to approximately the amounts recommended by previous Surveys (adjusted for inflation), programs would be presented in priority order, and each subfield would be studied by a separate panel, whose results would be presented to the committee. The areas to be covered were ultraviolet, optical, and infrared (UVOIR) astronomy; radio and high-energy (X-ray and gamma rays) astronomy; laboratory astrophysics; theoretical astrophysics; solar physics; planetary astronomy; education; and relations with the public. The chairs of all of the panels became ex officio members of the committee. Solar physics was highlighted because of the unique importance of our nearest star, and planetary astronomy as a field was included despite the fact that it is the responsibility of the Space Science Board. The latter emphasizes space missions to the planets, whereas we were concerned about the vitality of observing planets from Earth or Earth orbit.

There were programs recommended by previous surveys but not completed, notably the HST, previously referred to as the LST. After considerable debate, the Committee decided to give high priority to the HST Program, which was in Phase B at the time. This action assured that it would receive attention at the highest levels of the government, which turned out to be essential when opposition to this relatively expensive program developed in Congress. The program was delayed for budget reasons for two years, and it took major efforts by the AAS and by individual astronomers, particularly John Bahcall at the IAS, to get it through Congress.

Perhaps the HST Program has had more impact on astronomy than any other, not only because of its unique discoveries but because of the founding of the Space Telescope Science Institute with the support of NASA and operated by the scientific community. For years, a major part of the support for astronomy was through NASA grants to support the large number of observers at home institutions. Particularly innovative was the Hubble Postdoctoral Fellowship Program, which supports a cadre of promising young astronomers. These innovations squarely supported the committee's recommendations.

An effort was made for the committee to meet at institutions that carry out astronomical research and educate astronomers in order to hear and respond to their concerns. Astronomers throughout the country were invited to participate in the work of the panels and to communicate with the committee. Our meetings were not "all work and no play." One of our members, Bob Gehrz, amazed us at one meeting by hanging a coffee spoon from his nose. This occult art caught on, and ultimately a picture of the whole committee appeared in the magazine *Discover* with spoons on our noses.

Our results were published by the NAS (Field 1982). The greatest impact of the report was doubtless the few pages listing the programs with their estimated costs, including both construction and operations, in priority order. Of course program officers at the NSF, NASA, and DOE were not surprised by what they saw, as they were present at all our meetings, where their input was essential. But the printed version was hard to ignore in meetings of program officers with higher-ups. The Office of Management and Budget serving the US President, we noticed, took heed of our recommendations and, in general, followed our priorities in constructing the President's Budget. The Ways and Means Committee of the House of Representatives of course had final say over whether funds would be allocated to the programs we recommended, and it in general followed the recommendations of the subcommittees responsible for all of the different departments and agencies of the government.

It happened that astronomy fell into the purview of the Subcommittee on Veterans Affairs, then chaired by Congressman Boland of western Massachusetts. I made a presentation to the Boland Committee and met from time to time with his staff. I found them to be knowledgeable and supportive; in many ways, they were key to our success in Congress. As they are not elected officials, they are not directly responsible to the constituents of the elected members of the Boland Committee. However, the members of the Boland Committee are and they never forget it. Hence, when I made the presentation to the Boland Committee, I would address not only the substantive issues, like costs, of interest to the staff but also the concepts that would be useful to the Congessman in explaining the program to his or her constituents. For example, even busy Congressmen can feel the excitement of taking an image of the most distant objects in the Universe, or discovering signs of life on Mars.

This part of my responsibilities was an enjoyable challenge. I recall once being in the departure lounge for the DC–Boston flight when I spotted "Tip" O'Neill, then the Speaker of the House and famous for fashioning a compromise with President Reagan on Social Security. His motto was "All politics is local." I never understood exactly what that meant but it surely applied to him, an Irish politician in Cambridge, where his photo is hanging from every other light pole even now long after his death. I sensed an opportunity for astronomy, where the fight in Congress over the HST was in progress. As one of his Cambridge constituents, I introduced myself to him, and after a little small talk I mentioned the name of Congressman Boland, who, it happened, was an old friend and (at the time) his DC roommate. His eyes lit up (or at least I thought they did), and I spoke about the issue that was before the Boland Committee. I expressed the opinion of the Astronomy Survey Committee, which I chaired, that the HST deserved the highest priority. Did that message ever reach Boland, perhaps over a beer? Who knows, but it was fun trying.

I must add a word about John Bahcall, a scientist for whom I had the highest respect. He was a quintessential theorist, but for some reason he was passionate about the ST, as it was called around NASA. Early on he reached an understanding with the head of Astrophysics at NASA that if he ever needed support for the ST from the scientific community, John would organize it for him. (NASA frowned on such activity, but it was tolerated if it was kept quiet.) Several times at the height of the ST problem in Congress, John was called into action. I was a member of his team and would get a request to call so and so in Congress or arrange a letter writing campaign for such and such. I know that John talked to lobbyists and industrial people, but he had a special friend in Senator Barbara Mikulski, who was a rabid supporter of the Goddard Space Flight Center, which happened to be in her state of Maryland. He would visit her in Washington and follow her around the Capitol. On one such occasion, he related to me, he saw Congressional logrolling first hand. As I recall, John was on an elevator reserved for Members of Congress when a Senator stepped on. They made small talk until the door opened again. On stepped another Senator, who started to talk to the first one. The subject was a bill that would benefit the state of the second

Senator. The first one replied that he was sympathetic but he needed help on another bill. They shook hands and the deal was done. Of course John himself shook some hands too, always for ST.

2. PART TWO

Yes, I'm a theoretical astrophysicist. What's that? I know it is a mouthful. Let's look at the three words that make it up: astro means star, and a physicist is a person who works with the science of physics, and theoretical refers to studies of theories about things. So a theoretical astrophysicist is a person who uses physics to propose theories about stars. Does that help?

Why did I want to be one? Well, when I was a kid I wondered what the stars are. And I never stopped wondering. As I learned in college, to answer my question I had to learn the science of physics. Modern physics started when Galileo conducted his experiments in about 1604, and theoretical physics began when Isaac Newton put the experiments into mathematical form in 1667. Now it is a flourishing field that can predict the behavior of matter if you understand the forces on it.

When I reached this point in college, I was taking courses in physics and mathematics. The approach started by Galileo, which was to assemble apparatus like weights and lenses to do experiments, did not appeal to me as much as Newton' s approach, which was to assemble mathematical equations to describe what is happening when one does experiments. Unlike many people, I like math, perhaps because my father did.

When I applied to become a grad student at Princeton University in 1952, the professors were doing research in theoretical astrophysics. This was no accident, because I had checked out the programs at various universities and chosen Princeton. In addition to astronomy, I studied various fields of physics, like mechanics, relativity, gravitation, and quantum theory. In particular, I became very interested in plasma physics, a new branch of physics that one of my professors, Lyman Spitzer, a physicist in the Astronomy Department, was developing in connection with Project Matterhorn, which I discuss later. He wrote a brief book about it that became a classic (Spitzer 1956). He kept everything as simple as possible, with a minimum of complex formulas, so as to appeal to our intuition.

Plasma physics has nothing to do with giving blood, as you might have thought. Instead, it is the study of physics plasmas, gases that are so hot that the electrons that normally are trapped in atoms are ripped away and travel around at random. The electrons are negatively charged, and the atoms left behind—called ions—are positively charged. Even though the electrons have escaped from the atoms, they can't escape completely because they are pulled back by the ions in the plasma.

Plasmas are important to understanding the stars and galaxies, the job of theoretical astrophysics, because virtually all of the matter in them is in the form of plasma. Plasmas are very different from the gases in the air that surrounds us, because they are able to conduct electricity as efficiently as the copper wires in your home. Therefore you can expect to understand something about plasmas if you can understand how electricity works in your home. In fact, the laws that apply to home appliances are exactly the same as the ones that apply to plasmas in space. I'll use several laws to explain how plasmas work.

Ampère's law states that every electric current, wherever it may be, creates a magnetic field in its vicinity. That is how the magnets in the electric motors that power home appliances work. According to Ohm's law, to get a current you have to apply a voltage. In the United States, we do that by plugging the appliance into a household circuit that is maintained at 120 volts by a nearby power station.

Ever wondered how the power station produces a voltage? Michael Faraday discovered that if you move a coil of wire through the field of a magnet, the coil produces a voltage. In a power station there are electric generators that have big stationary magnets, and within the magnet assembly there are coils of wire that are mounted on a rotating shaft. The shaft is coupled to a steam turbine that is forced to turn by steam pressure from a boiler. When the operator turns a valve, the pressure on the turbine goes up, the shaft begins to rotate, and the coils of wire begin to move through the field of the magnets. According to Faraday's law, a voltage appears on the wires connecting the generator to your home. This voltage drives a current that enters your appliance and drives a motor. It does this by creating a local magnetic field in the motor in accordance with Ampère's law. The net effect of all of this is that the heat energy released by the boiler in the power station ends up driving the motor in your appliance.

This system replaces an earlier system in which if you needed to drive a machine in a factory, you had to have a steam engine there, which is in itself a motor. Surely the new system is better. Otherwise every factory would have to have a steam engine in the basement.

Believe it or not, the system described above is operating everywhere in the Universe with no human intervention. No astronomer has found a wire in space, let alone a motor or a generator, so how can that be? Here is how it works: Take the Sun, for example, where magnetic fields were discovered by George Ellery Hale in 1908. Astronomers have found that the gases in the solar atmosphere are in the form of plasma. Moreover, they appear to be moving randomly about at high speeds, approximately 1 km sec^{-1} or 2,000 mph. When they move in a region with a magnetic field, Faraday's law kicks in and creates a voltage; this constitutes an electric generator on the Sun. But as in our example above, Ohm's law says that the voltage results in a current, and Ampère's law says the the current will create a magnetic field. When this is added to the original magnetic field, the new magnetic field is different from the old one. That is what solar astronomers observe; the magnetic fields on the Sun are constantly changing.

I have been assuming that in fact the Sun has magnetic fields, simply because they are observed. As theoretical astrophysicists, we should try to figure out why the fields are there in the first place. After a lot of thinking and calculations, physicists have proved that in any body of plasma—like a star that is rotating, as most stars do—in which the plasma is locally moving about in a special way, even an extremely small value of a magnetic field automatically grows to the point that we can see its effects. When we recall that the Sun and other stars as well as our own Galaxy are rotating and that the plasma is moving about, we have the beginning of an answer. We'll come back to this in Section 7, but for now we'll accept the existence of magnetic fields in space as an observational fact and now look into their effects.

Most of the effects we observe are a consequence of yet another law, discovered by Lorentz. Lorentz's law is that a magnetic field causes a force on a wire carrying an electric current. The force is perpendicular to the wire and proportional to both the current and to the strength of the magnetic field. Since the magnetic field of the Sun is created by an electric current in the plasma, there must be a force there as the magnetic field interacts with the current that created it. Newton's law of motion is that if a force acts on an object, it accelerates it in the direction of the force.

Let's summarize. We observe magnetic fields on the Sun, and we observe that the plasma there is moving. Therefore there is a voltage and a current. The current interacts with the magnetic field to create a force, which accelerates the plasma. The result of this is that the motion of the plasma that caused all of this is changed in magnitude and direction. Such changes are observed. It is the task of theoretical astrophysicists to see if the observations agree with the theory that we have described.

Here I have tried to explain it in words, but following Newton, theoretical astrophysicists need to use the mathematical equations, such as Newton's law of motion, to express our words in formulas. These formulas can be found on the web under the title magnetohydrodynamics. When we consider a particular problem, like the phenomenon of coronal rain described below, we solve the equations that apply to that case. We then compare our theoretical results to the observations. If they correspond pretty well, we have a better understanding of the phenomenon.

Let's dissect this word, magnetohydrodynamics: Magneto is obvious, hydro refers to water or any fluid including plasmas, and dynamics refers to forces and motion. Whatever the words, I studied the applicable formulas and started to use them in my research. Because of this, sometimes I call myself a magnetohydrodynamicist, although to many people this sounds like gobbledygook.

In case you don't believe me, **Video 2** is a video of the region surrounding the Sun, called its corona. If you have ever done the experiment of bringing a toy magnet up to a paper with iron filings on it, you might recognize the pattern. Yes, those are the lines of force of a magnetic field, stretching into space. If you view the video, you can see gas flowing into the Sun, called coronal rain. Notice that the rain seems to be contained by the magnetic field. That's because of the Lorentz force that I explained earlier, which acts to resist motion across the magnetic field. The rain might fall on the Sun directly, like the rain on Earth, but it can't because the Lorentz force prevents that. The Lorentz force doesn't act along the magnetic field, so the rain falls along it.

The Sun was the first star to have its magnetic field discovered. We now realize that its magnetic field has great influence on Earth. When it changes shape for some reason, its energy is propelled outward at 200,000 mph. When it hits Earth's magnetic field—yes, we have one too—there are all sorts of effects, like cutting off radio communications. For that reason among others, NASA sponsors research into solar physics.

Not surprisingly, almost all stars also have magnetic fields. Not only that but the gases that lie between the stars, called interstellar matter, do also. Look at **Figure 6**, a map of the Milky Way with the interstellar magnetic field lines indicated. It took years of patient measurements to make that map.

Later I describe more about interstellar matter. It is made up mostly of hydrogen in atomic or molecular form. If it were the only constituent, interstellar matter would not be a plasma, which

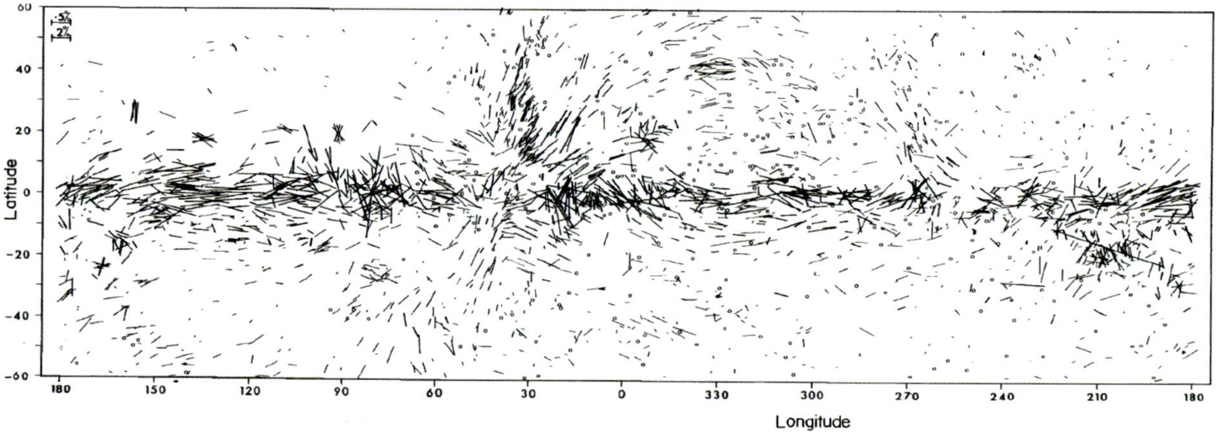

Figure 6

The E-vectors of the optical polarization measurements at Siding Spring Observatory and of Northern Hemisphere observers plotted on the sky. Each measurement is of the light from a star, which is polarized as it comes to us through dust in interstellar space. The dust grains, which are elongated, are aligned by the interstellar magnetic field. The pattern of the lines tells us the direction of the field. All galaxies like our own show patterns like this. Reprinted with permission from the Royal Astronomical Society; figure originally published in Mathewson & Ford (1970) (*Memoirs of the Royal Astronomical Society*), Volume 74.

is made up of ions and electrons. But there are also small amounts of other chemical elements present, like carbon, that can lose their electrons more easily than hydrogen, so interstellar matter is also a plasma and can support a magnetic field.

When I studied for the PhD at Princeton, I became interested in plasma physics and ended up writing a thesis on a phenomenon known as a solar Type 2 radio burst. When radio astronomy was developed by radar physicists after World War II, radio astronomers looked around for radio sources in the sky. When they directed their radio telescopes toward the Sun, they found that it emits radio waves all right, but superimposed on a steady signal were all sorts of bursts. The Type 2 burst that interested me had a fascinating structure. First of all, at any moment the frequency of the radiation is confined to two narrow bands, one twice the other. Secondly, both frequencies drift downward as time goes by. Astronomers quickly figured out that the underlying phenomenon is a shock wave traveling outward through the solar corona at speeds like the one I mentioned above and that the lower of the two frequencies is the plasma frequency appropriate for the location of the shock wave.

I have not yet explained the plasma frequency, so here goes. Plasmas exhibit complex behavior. One of the simplest is called a plasma oscillation. Irving Langmuir and Lewi Tonks discovered that if the plasma is disturbed, it begins to oscillate at a frequency that depends on the density of the electrons, called the plasma frequency. In Type 2 radio bursts the outgoing shock waves disturb the electrons in the plasma, and they oscillate at the plasma frequency appropriate for the current position of the shock wave. As the shock wave moves out through the corona, it encounters lower and lower electron densities, thus accounting for the observed downward drift in frequencies.

All this was understood when I started my research, but there was a nagging problem: The astronomers were observing radiation from these oscillations, but at the time it was not understood how the oscillations would emit radiation. I solved the problem for my doctoral thesis. If the plasma is not distributed very uniformly, the equations tell us that the plasma oscillations emit radio waves. The second harmonic of the plasma frequency is observed because the oscillation is very violent, a phenomenon that occurs often in physics. Evidently the shock wave excites very strong plasma oscillations.

After I got my PhD in 1955, I obtained a postdoctoral appointment at Harvard. Unlike many postdocs, there were very few constraints on what I decided to study next. At Princeton I had learned something about the field referred to as cosmology, the study of the entire Universe. At that time cosmology was in a primitive state because the only observational facts known about it then were that galaxies are distributed more or less uniformly in space and that they are moving apart in an surprising way called Hubble's law. Hubble's law is, first of all, that all galaxies are moving away from us, leading to the popular concept of the expanding Universe. Moreover, the speed at which the galaxies are moving is proportional to their distance from us, suggesting that they must have started moving all at the same time; this is sometimes referred to as the Big Bang. This extraordinary fact caught the attention of astrophysicists, and because it is an unprecedented discovery, many of them tried to understand it. Finally it was understood that only Einstein's theory of general relativity could possibly deal with the concept of the Big Bang, and a number of experts in that theory proposed models that were consistent with general relativity and also agreed with the data.

At the end of the day, most astrophysicists agreed with the models that the gravity of all the matter in the Universe would cause the velocity with which it is expanding to slow down over time. One intriguing thing about the models is that they mimick the behavior of a rocket launched straight up from the surface of Earth. Since the time of Newton, we have known what happens. If the launch speed is low, the rocket falls back to Earth. If it is very high, it escapes the gravitational field of Earth and keeps moving forever. There is a speed at which it does neither, called the escape velocity, which for Earth is 7 km s^{-1}. If the rocket is launched at precisely that speed, it just barely escapes, its velocity approaching zero as time goes on.

The equations tell us that the expanding Universe behaves in exactly the same way. Suppose I am in a galaxy, say our Galaxy, and you are in another one far away. I see you moving away in accordance with Hubble's law. I am wondering if you are moving away fast enough to continue moving away forever. I draw an imaginary sphere centered on me and passing through you. That sphere contains matter, so it attracts you and your galaxy, slowing you down exactly as Earth does the rocket. As in the case of a rocket, there must be an escape velocity that tells me whether you will escape or not. But there is a difference between the galaxy case and the rocket case. We know the mass of Earth, so we can calculate that the escape velocity is 7 km sec^{-1}, but we are not sure what the mass is within the sphere centered on us. So cosmologists turn the problem around and ask what is the value of the mass that makes the actual velocity equal to the escape velocity. The result is usually stated in terms of the density of matter within our imaginary sphere that is required to achieve this. This density is called the critical density and is estimated to be about 10^{-29} times the density of water.

By the way, you may wonder whether this estimate depends on my choice of the center of the sphere and the distance to your galaxy. Remarkably enough, it does not. I can choose any center and any location for you and I always get the same answer. The reason for this is that the model of the Universe that fits the distribution of the galaxies and Hubble's law is extraordinarily simple. When one considers the vast number of possible arrangements consistent with the laws of general relativity, the model we are discussing is by far the simplest. It is this fact that enables us to understand the Universe solely in terms of a few numbers. As we have seen, one of those numbers is the critical density, which has been determined solely from the constant ratio of the velocity of expansion of any galaxy to its distance, known as Hubble's constant.

Now the question becomes an observational one: How does the actual density compare with the critical density? In analogy to the case of the rocket, if the density is less than the critical value, the Universe will expand forever, whereas if it is greater, the Universe will return to its initial state at some time in the future. If it is exactly the same as the critical value, the Universe will expand forever, but its velocity will continually decrease, reaching zero only after an infinite time.

During the time I was working at Harvard, astronomers were able to conclude that the contribution of galaxies to the density of the Universe is much less than the critical density, so continued expansion seemed to be the answer. But I realized that there might be additional matter in space between the galaxies that would be invisible to astronomers using optical telescopes. Like my thesis advisor Lyman Spitzer, I had been studying the gases between the stars, called interstellar matter, which is composed mostly of the simplest chemical element, hydrogen. Before I started my postdoc in 1955, Ewen and Purcell at Harvard had discovered 21-cm radio emission by interstellar hydrogen gas in 1951. Bart Bok in the Astronomy Department had put their discovery to work in astronomy by having a 60-foot microwave dish constructed for 21-cm observations at Harvard's Agassiz Station, where a number of grad students, including Dave Heeschen and Frank Drake, did their thesis research. I reasoned that if there was also hydrogen gas between the galaxies, it could be detected by its effect in absorbing the radio waves coming from distant galaxies. I proposed searching for intergalactic hydrogen to Dave and started observing in 1955. To interpret the results, I needed to know something called the spin temperature of intergalactic hydrogen. This involved some calculations that I carried out in the daytime at the Observatory while conducting my observations at night. Doing so required me to analyze an effect predicted by Wouthuysen (1952), in which UV light from stars and galaxies is absorbed by hydrogen atoms in such a way as to affect the spin temperature. This turned out to be a math problem that I solved as part of my project. **Figure 4** is a page from the article in the *Astrophysical Journal* from 1959 reporting my efforts to compute the spin temperature.

Unfortunately, the absorption effect I was looking for in the radio telescope output was so weak that it could not be seen in the data, shown in **Figure 5**. Hence I could only state that the effect could not have been stronger than a certain amount, else it would have been observed. When I converted this limit to the density of intergalactic hydrogen using my best estimate of the spin temperature, my result was in the ballpark of the critical density mentioned above, but the uncertainties in my results were so large that I could not conclude whether the density of intergalactic atomic hydrogen is less than or greater than the critical density. Moreover, my office mate at Harvard, a theorist and friend named Dennis Sciama, had already pointed out that any intergalactic hydrogen atoms would probably have lost their electrons, so the atoms would be too rarified to detect anyway. I leave it to the reader to judge whether I should have given up my project then and there. One effect of my project was that it called attention to the possible existence of intergalactic hydrogen, which has since been investigated by many astronomers.

Incidently, since that time new techniques have found that there is atomic hydrogen in intergalactic space after all, but its amount is far too small to contribute to the critical density. By contrast, there seems to be much more ionized hydrogen, as Dennis had predicted, but again, the amount is much smaller than the critical density.

In the past few decades the subject has become much more interesting. Optical astronomers working on the structure of galaxies have found that to keep galaxies themselves from falling apart, there must be a component of matter in them that exerts a gravitational attraction like ordinary matter we find on Earth, but for some reason it is invisible at any wavelength we use, including X-rays, ordinary light, or radio waves. We still have no idea what this "dark matter" could be. Not only that but, in the very recent past, there has been evidence of a completely new physical effect called "dark energy." I won't even try to describe what this concept of dark energy is all about.

When I finished the Harvard postdoc in 1957, I was appointed to the faculty at Princeton. Lyman Spitzer, the director of the observatory there, was most interested in interstellar matter and had made observations at Mt. Wilson Observatory in Pasadena showing that the gas in interstellar space is very unevenly distributed. In analogy with the clouds of water droplets we see every day floating in our atmosphere, he suggested that most of the interstellar hydrogen is condensed into clouds. When he estimated the typical size and mass of these clouds, he calculated that the gravitational force that the clouds would exert on atoms in them was far too weak to keep the atoms from escaping from them. The random velocities of the atoms are so great that a typical cloud would expand away in a short time. He began to wonder if there is some medium, such as rarified hot gas between the clouds, that presses on them and keeps them together but might not be easy to observe.

I picked up this idea as my next project; I ended up writing a paper on a phenomenon called thermal instability, probably my most useful contribution to astronomy. My idea was that under certain conditions, interstellar gas that had been heated up some way might spontaneously produce clouds of cooler gas. If the whole region were kept under constant pressure (perhaps because of the gravitational attraction to the Galaxy), any cold gas would be forced to form clouds of higher density. It is not obvious why this should this be the case. Here is why: If for a moment we suppose that all the gas remains uniformly distributed while cooling off, the conservation of mass implies that the density of the cold gas is the same as that of the hot gas with which we started. The perfect gas law states that the pressure exerted by any gas is equal to its density times its temperature, so the pressure of the cold gas is less than we started with, which is contrary to our assumption that the pressure remains constant. The only other possibility is that the condensation process takes place in only a portion of the volume, forming cold clouds. This is just what Spitzer's observations required, so the idea seemed to be promising.

The next problem is whether the heating and cooling processes that determine the temperature of interstellar gas are able to produce both a hot phase and a cold phase at the same pressure. I found that if the heating mechanism is independent of the density of hydrogen atoms while the cooling mechanism is proportional to the density, any hot gas would produce clouds of cold gas at the same pressure, as required.

Heating and cooling mechanisms satisfying these requirements had been investigated earlier by Spitzer, so it was necessary only to choose the parameters to fit the observations. My choice of the heating mechanism was controversial, so let me say a little about it. There is a phenomenon in the Galaxy called cosmic rays—consisting of the ions of hydrogen and other elements that move through space at high speeds, and therefore high energies—ranging from a few million electron volts (eV) to 10^{19} eV. For comparison, the most powerful particle accelerator on Earth reaches 10^{12} eV, whereas the energy of the atoms in a flame are less than 1 eV. Cosmic rays are observed individually as they hit Earth, but there is also strong astronomical evidence that they fill the entire Galaxy. If a cosmic ray hits a hydrogen atom as it moves through the interstellar medium, it rips off the electron, leaving the proton behind. In the process, the electron speeds away with several electron volts of energy, ultimately dissipating its energy as heat. This is the mechanism that I chose to heat the interstellar medium and set the stage for forming cool clouds. There is a problem, though. The most effective heating comes from low-energy cosmic rays, which are the hardest to observe; so to make it work I had to assume that I could extrapolate the data a bit beyond the direct measurements to meet our requirements. Then other astronomers showed that, if my estimates were correct, the numbers of protons that initiate reactions that would produce observed complex interstellar molecules would be about 10 times too many. For that reason, advocates of another heating mechanism that does not produce protons began to dominate the debate. The new mechanism behaves the same way as mine, but does not produce protons. More on this below.

Finally, I wanted to show that a cloudy medium is the only solution. I therefore considered an alternative solution in which the gas simply stays in the hot phase, and nothing happens. I then imagined a small perturbation away from that state, such as would always occur in nature. My result was that such a perturbation would grow exponentially in time; so the configuration I just described is unstable and, therefore, unlikely to occur in nature. I did not follow it into the regime in which complicated terms enter, but let it go at that.

When I moved from Princeton to UCB in 1965, I took on a postdoc, Harm Habing, and a grad student, Don Goldsmith. Together we considered the special case in which the heating is due to low-energy cosmic rays. We found that indeed this heating mechanism can support an equilibrium with both a hot and a cold phase and computed what observers would see at various wavelengths. The comparison with actual data was quite successful (Field et al. 1969).

One particular comparison sticks in my mind because it constitutes a direct test of cosmic-ray heating. It was known at the time that long radio waves from radio galaxies on a line of sight through our interstellar medium are being absorbed by the interstellar medium. We had calculated the expected amount by which the radio waves would be absorbed if our model were correct. It agreed roughly with the observations, which indicated to us that even in the cold clouds of predominantly atomic hydrogen, there are free electrons as well as the corresponding protons created by collisions with cosmic rays. When electrons pass near protons, the combination absorbs long radio waves. This fact was important to us, because we thought that the clouds were heated by cosmic rays passing through them, and the heating process would necessarily produce electron-proton pairs, which would account for the observed absorption of radio waves.

Nowadays, a completely different process has been accepted as the main heat source in interstellar atomic hydrogen clouds. In this process, UV light hits particles of interstellar dust and releases energetic electrons from them. This process works well, but does not result in free protons and would therefore not account for the observed absorption of radio waves. Can the new process explain the observed absorption?

In the meantime, there has been a debate about cosmic-ray heating, with some data seeming to contradict our model and more recent data seeming to confirm it. So far I have not been able to get a student to help settle the problem definitively. If you are interested, I am too and can be reached at gravturb@yahoo.com.

Several authors pointed out that supernova explosions stir up the interstellar medium and affect our conclusions in a big way. Don Cox, Chris McKee, and Jerry Ostriker especially concluded that there must be a third phase of a much higher temperature present. Our model omitted the effects of supernovas and resulted in only two stable phases, a cold phase at about 100 K and a warm phase at about 10,000 K. The supernova-heated phase would be at about 1,000,000 K. The very hot phase is interesting in that it is able to conduct heat easily. McKee and Ostriker showed that the very hot phase cools effectively by conducting heat into any cold clouds present, where it is dissipated by radiation. This changes the scenario we proposed substantially. [Any student interested in this problem should read their papers. Look at the Astrophysical Data System (ADS), for those papers and the later ones that cite them.] Perhaps the most significant effect of my paper Field et al. (1969) on thermal instability is that astronomers often observe gases in various phases and that theorists consider that they may be the result of thermal instability.

All this occurred while I was a Professor at UCB; I greatly enjoyed my years in Berkeley, because of both the beauty of the campus and the Bay area and the excitement there, which is now referred to as "The 60s." I particularly valued a Sunday discussion group organized by Dave Cudaback and his wife Dot in which scientists discussed topics far from science. For example, I recall a moment in which one of our members maintained that a statement could be both true and false. I responded by quoting Aristotle and only later realized how out of date such a statement would be perceived there. Ever since the student protests in 1965, there were constant agitations about one injustice or another, most of which I supported. However, the protests sometimes interrupted the quiet thinking we treasure at a university. For example, I was teaching Astronomy 1 one day during a student strike, and a group burst into the hall shouting that I was a slave master. Taken aback, I said the first thing that came to mind: "I don't feel like a slave master." I had no thought that this was an adequate response, but it got a positive reception, and the strikers left. In retrospect, I think they may have left because in the 60s "feelings" were considered to be more real than "thoughts."

In 1972, I was invited to join the Astronomy Department of Harvard University. As I describe in Section 5, I moved from Berkeley because of an opportunity to establish a new institution there, the Harvard-Smithsonian Center for Astrophysics. I also began to teach a new undergraduate course on cosmic evolution, which described a conceptually unified history of the Universe from the Big Bang to the emergence of the human species. Since my youth, I had always assumed that this is what happened, in contrast to many accounts that begin with supernatural creation. The latter extraordinarily imaginative accounts satisfied the needs of the prescientific age, but are often at odds with what we now know to be true.

Of course, this brings up the role of religion in modern life. I had studied the Bible when I was a teenager and recognized that much of it, particularly the Ten Commandments, Jesus's Sermon on the Mount, and the Golden Rule, is the source of inspired truths about how humans can and should behave toward each other, but these are truths that still elude the method of science. It is not necessary to believe that a human rose from the dead or started life from a virgin birth to accept these insights by wise people. There is a very important difference between religion and science. In the former, it is acceptable to propose new ideas that not only have not been tested experimentally but actually are contradicted by previous experiments. I concluded that I cannot consistently pursue both methods, religious thinking and scientific thinking, so I stick with the latter.

Of course, my concept of cosmic evolution, based on scientific observations of the Universe by astronomers, the discoveries of geology and paleobiology, and modern molecular biology, is far from being widely accepted, but Harvard students were open to it—so open that I wrote a book with that title with Cyril Ponnamperuma and Gerritt Verschuur (Field et al. 1978). It was meant as a text for Astronomy 1, but it was a commercial failure, perhaps because it went into too much detail. However, a junior faculty member named Eric Chaisson took over the course and taught it for many years to large classes at Harvard.

My true love was plasma astrophysics, and I taught a graduate course on it at Harvard for many years. Oddly, many of the grad students I had fostered over those years never took the course. Nowadays, it is recognized that lecture courses like that are not as effective as other methods. I should have adopted the approach of my colleague, Prof. David Layzer, who taught probability and relativity to undergraduates in a seminar format in which each student prepared oral presentations to the others on the assigned topics of the course. The few comments I heard from the undergraduates who took astronomy with David put his classes among the best they had taken at Harvard.

Another research project I took up was the origin of the so-called X-ray background. I had encountered this problem at UCB, where my colleague Stu Bowyer built X-ray detectors to be flown on rockets by NASA. One of his instruments discovered a strong diffuse background at photon energies of about 0.25 keV, called soft X rays. In astronomy the term background refers to radiation that is smoothly distributed over the sky, either because it is the result of so many faint point sources they cannot be distinguished as such or because the radiation is truly diffuse, like the blue sky in the daytime. Because Stu's instrument had a wide field of view, it was hard to distinguish between, on the one hand, a relatively diffuse emission in the sky and, on the other hand, a cloud of point sources. I hoped it might be a diffuse emission like that detected at other X-ray wavelengths. By that time, I had realized that intergalactic hydrogen, if any, had to be in the form of a hot plasma. I had been impressed by previous experiments at higher energies by Herb Friedman at the Naval Reseach Laboratory, and I had calculated the X-ray flux to be expected if the intergalactic medium were a very hot plasma, with a temperature of about 500 million K. Friedman's results to some degree fitted that hypothesis, and Bowyer's results seemed to do so also. However, Bowyer's results, pertaining to a lower photon energy, also displayed a new phenomenon: absorption by the interstellar medium. That was not unexpected, and we showed (Bowyer et al. 1968) that the observations are consistent with a cloudy distribution of interstellar atomic hydrogen that was favored by optical and 21-cm observations. That finding neatly supported my earlier work on a 2-phase interstellar medium, so for a while Stu and I felt happy, but further experiments showed that our model of a diffuse X-ray background from the intergalactic medium being absorbed by foreground interstellar medium was too simple, because evidence was found that the interstellar medium emits as well as absorbs soft X-rays. At that point, disentangling the various components of low-energy X-rays became too difficult for simple models, and I moved on to other projects.

At Harvard, I interested several grad students in the problem of the X-ray background and hot intergalactic gas. Steve Perrenod studied the literature and compared the data at many X-ray energies with a simple model that I suggested: a certain density and temperature of hot gas in intergalactic space, which would be expanding along with the galaxies, whose history of heating and cooling over the billions of years was specified by the laws of physics, given that its temperature at the present time was about 500 million degrees. It turned out that the model fit the data within the errors (Field & Perrenod 1977). We were excited, particularly because the required density of gas was about equal to the critical density that I explained above.

In the meantime my colleague at Harvard, Prof. Riccardo Giacconi, was developing X-ray imaging telescopes in space of ever increasing resolving power with the support of NASA. One of his goals was to find out whether the X-ray background is due to unresolved X-ray sources at great distances or rather the intergalactic gas that I advocated. With each new mission, he was able to account for a larger fraction of the background using resolved sources that had been unresolved in previous missions. Finally, he was able to account for 90% of the background by resolved sources, and my model was cooked.

It was also cooked for other reasons. My student Bob Rogers applied the information gathered on these flights to show that any physical mechanism that could heat the gas to such high temperatures would have shown up in the data, but did not. And I found that to heat the gas with quasar outbursts would require unreasonable fractions of their energy budget. Of course, Steve and I were disappointed that our model did not work. But the whole point of science is to find out the truth, not to fulfill our dreams. Some dreaming helps to guide the work, but it is reality we are after. Riccardo and I are friends, and I rejoiced when he won the Nobel Prize in Physics for his discovery of stellar-mass black holes.

At Harvard I also pursued other research interests, including magnetically dominated accretion disks with my student Bob Rogers (Field & Rogers 1993), limits on a new version of electrodynamics with Roman Jackiw of MIT and my student Sean Carroll (Carroll et al. 1990), an explanation of high velocity HI clouds as a fountain of galactic gas with my student Paul Shapiro (Field & Shapiro 1976), and a theory of galaxy formation with my student Boqi Wang (Wang & Field 1989). I have been reformulating mean-field dynamo theory with my former student Eric Blackman at the University of Rochester (Field & Blackman 2002). The latter program has unfolded over many years and is still in progress. As I discuss above in this Part, the effects of magnetic fields are understood with MHD, but that leaves the issue of the origin of the magnetic fields in question.

This issue had been studied by many authors, and the result was converging to this: The magnetic fields of stars and galaxies are created in their observed forms by a process known as a large-scale dynamo. Dynamo theory predicts that certain types of plasma motion known as helical turbulence can take a very weak field, whose length-scale is about the same as the object in question, and rapidly increase its strength. Standard textbook dynamo theory is called linear because the output large-scale magnetic field is proportional to the input field. But in fact, as the output field grows, it reacts back on the dynamo action that created it so that it is no longer proportional to the input field. In attempting to address this shortcoming of the linear theory, I suggested to my student Hong Song Chou that he start with a small large-scale field and arrange to force the kind of plasma flow required and then let both the large-scale field and the plasma flow evolve in a computer simulation. The computer program that Hong Song wrote would then automatically take the back reaction into account. Hong Song's program worked and his results were very interesting, but I had difficulty in interpreting them in a useful way. So Eric and I went back to study the derivation of standard dynamo theory. We noticed that it omits a potentially important term. When we derived the new term and solved the resulting equations, we noticed that a large-scale magnetic field starts to grow as expected from the textbook equations, but as time goes on it saturates at a certain value determined by the new term we had derived. Returning to the literature, we found that our new term in fact had been discovered earlier by Annick Pouquet (Pouquet et al. 1976) in France, but we had missed it because we had not properly understood some complicated mathematical manipulations that she had done.

Then we realized that our results had a subtle explanation. Annick had taken into account a new quantity that is missing from the standard dynamo theory. This quantity is important because like energy it is conserved in any IMHD system. This quantity is called magnetic helicity, referring to magnetic fields that are helical or screw-like in nature. The amount of magnetic helicity can be calculated using mathematical equations, and the fact that it is conserved is a mathematical theorem that I discuss in Part 1. But its physical meaning remained obscure until 1969, when Moffatt showed that it represents the linkage between loops of magnetic field. It is a finite number if they are linked and zero if they are not. Moreover, the number is positive if the linkage is right-handed and negative if it is left-handed.

How does magnetic helicity affect dynamo theory? Suppose the large-scale field is growing and that the underlying plasma motion is helical in nature, as in fact it must be for even the standard theory to work. It is natural that the helical motion will endow the magnetic field with helicity also. But where is this magnetic helicity coming from? Because it is conserved, if it is zero to begin with, it must remain zero. This apparent contradiction is solved by Annick's new term, which has the effect of transferring magnetic helicity from small to large scales. So if the large-scale field has positive magnetic helicity, the small-scale field has negative magnetic helicity. The result for astronomy is that the magnetic fields that we observe on the scale of the object has helicity that has been borrowed from fields that are at such small scales they cannot be observed. This has all been confirmed by computer simulations and has spawned a new generation of solar observations that track the flow of magnetic helicity.

As I was studying magnetic helicity, I received a remarkable suggestion by Roman Jackiw, a theoretical physicist at MIT. The electromagnetic theory formulated by Maxwell had ruled all of physics for a century, and no one had suggested any major changes because this theory accounted for every one of jillions of experiments. But Roman, coming from the direction of particle physics, suggested a modification to electromagnetic theory that seemed at least superficially plausible. Coincidentlly, his suggestion included magnetic helicity, the very thing we had been studying, as a basic new ingredient. To do this in a way consistent with Einstein's theory of special relativity, Roman proposed a new constant of nature that had the character of an arrow in four-dimensional space time. If such an arrow in fact existed, it would violate key principles of modern cosmology, so his proposal presented an interesting challenge. Like all new ideas in theoretical physics, it would survive only if experiments are consistent with it.

Roman suggested that astronomers find a way to test his idea, and I took him up on it. My idea was based on the study of the characteristics of the electromagnetic waves from distant galaxies being studied by radio astronomers. A Harvard graduate student named Sean Carroll volunteered to study the results of many such observations to see if there were signs of Roman's effect. Perhaps it is not surprising, but there were no such effects observed. We could rule out such effects with amazing accuracy. As radio waves cross a substantial fraction of the Universe, the effect predicted by Roman's new term on the radio waves turned out to be less than a few percent.

This was so interesting that Sean and I went on to investigate a general case of Roman's arrow in which the arrow is not necessarily a constant but instead points in a special direction at each point in space time (Field & Carroll 2000). This direction is familiar to hikers. It is the direction in which the ground is rising most steeply. Of course there is no "ground" in space time, but there is something conceptually like it. The height of the ground where the hiker is standing is replaced by an arbitrary number at each point of space time. Then the arrow in space time that we are talking about points in the direction that this number is increasing fastest. In this way of looking at it, Sean and I found that the number assigned to each point has direct physical significance: It is the angle through which the electric field in the radio wave is turned as it propagates across the Universe from the point in question. There is another way of thinking of this effect: that the angle represents what is called a scalar field. In physics, a field is an effect that depends on space and time. There are different kinds of fields, depending on the complexity of the effect. If the effect can be represented by a single number at every point, for example, the local temperature, the field is called a scalar field. If it involves arrows at every point, like a magnetic field, it is called a vector field. If it involves a two-dimensional set of numbers, it is called a tensor field. The electromagnetic field is a tensor field, but the components of the set of numbers can be regarded as a mixture of two vector fields, electric and magnetic. By this perspective, the angle of rotation is a scalar field.

Thinking about this recently, I realized that our measurement is related to the concept of parity. To explain this concept, suppose we carry out an experiment in the lab and have a certain result. Normally we assume that the result does not depend on the coordinate system that we use to describe our results. In particular, if we use the ordinary Cartesian coordinates x, y, z in the order specified by the right-hand rule (hold up your right hand and extend your thumb, index finger, and the third finger, then you identify the thumb with x, the index finger with y, and the third finger with z), we get the same answer that we would get if we use the same coordinate system but based on the left hand. Note that the left hand is the mirror image of the right hand, so that the standard statement is that physical results are the same independent of the handedness of the coordinate system used to describe them. In theoretical physics this statement is known as the conservation of parity.

Almost all physical effects observed conform to the standard statement above, but a remarkable exception to this rule was discovered in 1957, called parity violation of the weak interactions. I recall the reaction of my colleague Ed Purcell when he heard this news and explained it to us. For him, a Nobel Prize winner as well as a person with a deep knowledge of nature, it was an astonishing discovery. I was overwhelmed to learn that there is something in nature that recognizes the difference between right and left. That is understood in biology as an accident of evolution, but here we are talking about the nature of space and time. I was impressed that Ed found it as amazing as I did.

All of this is background to my realization that the upper limit on the angle of rotation of the electric field is a limit on parity violation. If Roman's new term is finite, the electric field would have rotated one direction or the other. This would indicate that nature senses the difference in right and left in a new way, not only in the weak interactions, a truly sensational finding.

But that is not what happened. There was no rotation either to the right or left, so we were left with the somewhat uninteresting conclusion that Roman's term does not occur in nature. Uninteresting perhaps, but fundamental in some sense.

Now get this: Magnetic helicity, which I have explained above in connection with MHD, is related to our measurement that the rotation is zero. How can those two phenomena possibly be connected? Unfortunately, the answer is mathematical in nature, so if you don't like math, skip these two paragraphs. The term that Roman proposed to modify the equations of electromagnetism is the product of two parts; one is the constant; the second is an expression composed of products of the electromagnetic field tensors that is considered in physics textbooks and is rejected on the grounds that it would violate parity, which has been tested with great accuracy for electromagnetism. Let's ignore that for a moment and reflect on what the second part means. I realized that the second part represents a source of magnetic helicity, and that would profoundly affect our understanding of IMHD, that is, MHD with high electrical conductivity. If we accept the textbook reasons for setting it equal to zero, then that might actually be the case in the real world. But the textbook argument ignores an important possibility, that Roman's term might be multiplied by another one that also violates parity. It is a mathematical fact the product of two parity-violating terms does not violate parity. In retrospect now, I realize that is what Sean and I were doing when we introduced a new field, one that we discussed above that is the angle of rotation of the electric field in an electromagnetic wave. In short, Roman's second term alone would violate parity, but that of mine and Sean's would not, consistent with experiments on

Figure 7

"Blue Buckets"—my wife, Susan G. Field.

electromagnetism. This resolution of the situation is achieved only by assuming the existence of our new field. As explained above, our new field does not exist, so there is no fundamental source of magnetic helicity that, if it did exist, would violate the equations of IMHD.

All of this activity I have been discussing occurred while I was still a Harvard professor. In 1999 at the age of 70, I decided to retire, but being an enthusiastic scientist, I could not stop wondering about things. It happened that my wife Susan, having been introduced to the British Virgin Islands in the Caribbean Sea by Fred and Babbie Whipple in the 1970s, suggested that we spend the winters away from Cambridge in the British Virgin Islands. That proved to be a good choice, because as I've aged and had more physical problems, we can avoid the ice and snow that are difficult for older people. We bought a boat and started to go fishing along with going to the beach. It is not all fun and games. We joined the British Red Cross (**Figure 7**) and helped with their annual prostate clinics. Susan is a writer and contributes to the local newspapers' articles about life and times. A couple of years ago a close friend of hers and I were able to sneakily collect most of her articles in a book, which astonished her when we presented her with copies.

But of course, I continued to think about astronomy. Nowadays a theorist can do things even on a small island by using ADS. I scan it every day and read about one paper a week. So what happened? I stumbled on papers by Ron Allen, at the Space Telescope Science Institute, in which he presents data on atomic hydrogen (HI) in the spiral arms of other galaxies that suggest that as the interstellar medium rotates into a spiral arm it is molecular, and as it rotates out, it is largely HI. In later papers, he shows that the conversion from molecular to atomic hydrogen correlates well with far UV (FUV) radiation observed with satellites. FUV is known from IR studies to penetrate the outer layers of molecular clouds, where it dissociates the hydrogen molecules, so his picture of the interstellar medium as constantly changing from molecular to atomic and back is an appealing concept.

With my colleagues Eric Blackman and Eric Keto I am trying to understand the impact of Allen's discovery on a model of molecular clouds that we are working on. I have postponed my work on this problem to write this contribution to Annual Reviews, but now I can go back to it. Nice talking to you.

DISCLOSURE STATEMENT

The author is not aware of any affiliations, memberships, funding, or financial holdings that might be perceived as affecting the objectivity of this review.

LITERATURE CITED

Alfvén H. 1950. *Cosmical Electrodynamics*. Oxford, UK: Clarendon Press

Arpigny C. 1965. *Annu. Rev. Astron. Astrophys.* 3:351

Berger MA, Field GB. 1984. *J. Fluid Mech.* 147:153

Bowyer CS, Field GB, Mack JE. 1968. *Nature* 217:32

Brinton C. 1938. *The Anatomy of Revolution*. New York: Prentice-Hall

Carroll SM. 2004. *Spacetime and Geometry*. San Francisco: Addison-Wesley

Carroll SM, Field GB. 1991. *Phys. Rev. D* 43:3789

Carroll SM, Field GB, Jackiw R. 1990. *Phys. Rev. D* 41:1231

Carter B. 1979. Perfect fluid and magnetic field conservation laws in the theory of black hole accretion rings. In *Active Galactic Nuclei*, ed. C Hazard, S Mitton, pp. 273–300. Cambridge, UK: Cambridge Univ. Press

Dicke RH, Beringer R, Vane AB. 1946. *Phys. Rev.* 70:340

Erickson WC. 1957. *Ap. J.* 123:172

Field GB. 1956. *Ap. J.* 124:555

Field GB. 1959. *Ap. J.* 129:551

Field GB. 1962. *Ap. J.* 135:684

Field GB. 1965. *Ap. J.* 142:531

Field GB, Goldsmith DW, Habing HJ. 1969. *Ap. J.* 158:173

Field GB. 1973. Astrophysics and relativity, Chapter 8. In *Astronomy and Astrophysics for the 1970s, Reports of the Panels, Astronomy Survey Committee*. Washington, DC: Natl. Acad. Sci.

Field GB. 1982. *Astronomy and Astrophysics for the 1980s, Vol.1: Report of the Astronomy Survey Committee*. Washington, DC: Natl. Acad.

Field GB. 1986. In *Magnetospheric Processes in Astrophysics*, ed. RI Epstein, WC Feldman. *AIP Conf. Proc.* 144:324

Field GB, Blackman EG. 2002. *Ap. J.* 572:685

Field GB, Carroll SM. 2000. *Phys. Rev. D* 60:3008

Field GB, Goldsmith DW, Habing HJ. 1969. *Ap. J.* 155:149

Field GB, Herbig GB, Hitchcock J. 1966. *A. J.* 71:161

Field GB, Partridge RB. 1961. *Ap. J.* 134:959

Field GB, Perrenod SC. 1977. *Ap. J.* 215:717

Field GB, Rogers RD. 1993. *Ap. J.* 403:84

Field GB, Saslaw WC. 1965. *Ap. J.* 142:568

Field GB, Shapiro PR. 1976. *Ap. J.* 205:762

Field GB, Verschuur GL, Ponnamperuma C. 1978. *Cosmic Evolution: An Introduction to Astronomy*. Boston, MA: Houghton Mifflin

Ginzburg VL, Zheleznyakov VV. 1959. In *Paris Symposium on Radio Astronomy, IAU Symp. 9*, ed. RN Bracewell, p. 574. Stanford, CA: Stanford Univ. Press

Greenstein JL. 1972. *Astronomy and Astrophysics for the 1970s. Report of the Astronomy Survey Committee*. Washington, DC: Natl. Acad.

Heilbron JL. 2010. *Galileo*. Oxford, UK: Oxford Univ. Press

Heiles C. 1967. *Ap. J. Suppl.* 15:97

Herzberg G. 1950. *Molecular Spectra and Molecular Structure. Vol. 1: Spectra of Diatomic Molecules.* New York: Van Nostrand Reinhold, 2nd ed.

Jeans J. 1929. *The Universe Around Us*. Cambridge, UK: Cambridge Univ. Press

Jeans J. 1931. *The Stars in Their Courses*. Cambridge, UK: Cambridge Univ. Press

Lilley AE, McLain EF. 1956. *Ap. J.* 123:172

Mathewson DS, Ford VL. 1970. *MmRAS* 74:139

Moffatt HK. 1969. *J. Fluid Mech.* 35:117

Moffatt HK. 1981. *J. Fluid Mech.* 106:27

Moreau JJ. 1961. *C. R. Acad. Sci. Paris* 252:2810

HubbleSite. 2010. *Starry-eyed Hubble celebrates 20 years of awe and discovery*. News Release Number: STScI-2010-13, April 22. **http://hubblesite.org/newscenter/archive/releases/2010/13/video/a/**

NASA Solar Dyn. Obs. 2013. Coronal rain on the sun. *Astronomy Picture of the Day*, Feb. 26. **http://apod.nasa.gov/apod/ap130226.html**

Ni W-T. 1977. *Phys. Rev. Lett.* 38:301

Penzias AA, Wilson RW. 1965. *Ap. J.* 142:419

Penzias AA, Wilson RW. 1969. *Ap. J.* 156:799

Pouquet A, Frisch U, Le'orat J. 1976. *J. Fluid Mech.* 76:321

Purcell EM, Field GB. 1956. *Ap. J.* 124:542

Schmidt M. 1963. *Nature* 197:1040

Spitzer L Jr. 1956. *Physics of Fully Ionized Gases*. New York: Intersci. Publ.

Wang B, Field GB. 1989. *Ap. J.* 346:3

Weinberg S. 1972. *Gravitation and Cosmology: Principles and Applications of the General Theory of Relativity*. New York: Wiley

Weinberg S. 1995. *The Quantum Theory of Fields*, Vol. 1. Cambridge, UK: Cambridge Univ. Press

Woltjer L. 1958. *Proc. Natl. Acad. Sci. USA* 44:489

Wouthuysen SA. 1952. *Astron. J.* 57:31

Short-Duration Gamma-Ray Bursts

Edo Berger

Harvard-Smithsonian Center for Astrophysics, Cambridge, Massachusetts 02139;
email: eberger@cfa.harvard.edu

Annu. Rev. Astron. Astrophys. 2014. 52:43–105

First published online as a Review in Advance on June 4, 2014

The *Annual Review of Astronomy and Astrophysics* is online at astro.annualreviews.org

This article's doi:
10.1146/annurev-astro-081913-035926

Keywords

gamma rays: observations, theory; gravitational waves; radiation mechanisms: nonthermal; relativistic processes; stars: neutron

Abstract

Gamma-ray bursts (GRBs) display a bimodal duration distribution with a separation between the short- and long-duration bursts at about 2 s. The progenitors of long GRBs have been identified as massive stars based on their association with Type Ic core-collapse supernovae (SNe), their exclusive location in star-forming galaxies, and their strong correlation with bright UV regions within their host galaxies. Short GRBs have long been suspected on theoretical grounds to arise from compact object binary mergers (neutron star–neutron star or neutron star–black hole). The discovery of short GRB afterglows in 2005 provided the first insight into their energy scale and environments, as well as established a cosmological origin, a mix of host-galaxy types, and an absence of associated SNe. In this review, I summarize nearly a decade of short GRB afterglow and host-galaxy observations and use this information to shed light on the nature and properties of their progenitors, the energy scale and collimation of the relativistic outflow, and the properties of the circumburst environments. The preponderance of the evidence points to compact object binary progenitors, although some open questions remain. On the basis of this association, observations of short GRBs and their afterglows can shed light on the on- and off-axis electromagnetic counterparts of gravitational wave sources from the Advanced LIGO/Virgo experiments.

1. INTRODUCTION

Gamma-ray bursts (GRBs) are short, intense, and nonrepeating flashes of approximately megaelectronvolt gamma-rays with a wide range of spectral and temporal properties. GRBs were discovered serendipitously by the *Vela* satellites starting in 1967 (Klebesadel, Strong & Olson 1973), leading to an intense investigation of their origin, energy source, and progenitor systems that is still ongoing. Despite the broad diversity in their gamma-ray emission properties (peak energy, spectral shape, variability timescale, duration), it was recognized primarily on the basis of data from the Burst and Transient Source Experiment (BATSE) onboard the *Compton Gamma-Ray Observatory* that GRBs can be generally divided into two groups depending on their duration and spectral hardness: the short-hard and long-soft bursts with a separation at about 2 s (Norris et al. 1984, Dezalay et al. 1992, Kouveliotou et al. 1993).

For both GRB classes, the uniform projected distribution of the bursts on the sky and the fluence distribution (the so-called $\log N - \log S$ or $V/V_{\max}$ distributions) pointed to a cosmological origin (e.g., van den Bergh 1983; Goodman 1986; Paczynski 1986; Schmidt, Higdon & Hueter 1988; Meegan et al. 1992; Piran 1992; Schmidt 2001; Guetta & Piran 2005). Despite these statistical tests, the distance scale of GRBs continued to be debated at least into the mid-1990s (Lamb 1995, Paczynski 1995). However, the nonrepeating nature of GRBs, their harder nonthermal spectra, and non-Euclidean space distribution separated them from the soft gamma-ray repeaters (SGRs; e.g., Kouveliotou et al. 1987, Laros et al. 1987, Norris et al. 1991), which originate from magnetars in the Milky Way and nearby galaxies (Kouveliotou et al. 1998); it is important to note, however, that some events classified as short GRBs may represent extragalactic giant SGR flares with long recurrence timescales (Abbott et al. 2008, Ofek et al. 2008, Hurley et al. 2010, Abadie et al. 2012b).

In the framework of a cosmological origin, the large energy release (up to an isotropic-equivalent value of $\sim 10^{54}$ erg), short variability timescale (down to milliseconds), and observed nonthermal gamma-ray spectra led to the "compactness problem"; namely, the observed properties result in an enormous optical depth to the pair-production process and, hence, an expected thermal emission (Ruderman 1975). A key result in GRB research was the realization that this problem can be resolved by invoking relativistic expansion with a large bulk Lorentz factor of $\Gamma \gtrsim 10^2$ (e.g., Paczynski 1986; Krolik & Pier 1991; Fenimore, Epstein & Ho 1993; Baring & Harding 1997). The large Lorentz factor in turn requires exceedingly clean explosions with ejecta masses of $\lesssim 10^{-5}$ $M_{\odot}$, the so-called baryon loading problem (Shemi & Piran 1990). In addition, it was recognized that the interaction of the relativistic outflow with the surrounding medium will generate synchrotron emission ranging from radio waves to X-rays with a longer duration than the prompt gamma-ray emission—the afterglow (Rees & Meszaros 1992; Meszaros & Rees 1993; Paczynski & Rhoads 1993; Meszaros & Rees 1997; Sari, Piran & Narayan 1998). The detection of such afterglow emission became an intense pursuit because it held the key to precise positions and, hence, a distance scale to the bursts.

Motivated in part by the $\sim$0.1 Gyr merger timescale of the Hulse-Taylor binary (Hulse & Taylor 1975), seminal studies by Paczynski (1986) and Eichler et al. (1989) pointed out that neutron star binary mergers may lead to gamma-ray emission with an energy scale of $\gtrsim 10^{50}$ erg, which is typical of a cosmological GRB. Such mergers, owing to the absence of a dense surrounding environment, are also expected to result in an outflow with a low baryon load and can therefore drive relativistic outflows. A more detailed investigation of the neutron star binary merger model was carried out by Narayan, Paczynski & Piran (1992), who pointed out that testable predictions include significant offsets from the host galaxies, owing to natal kicks imparted to the binaries at birth, and gravitational wave (GW) emission in the frequency range of the Laser Interferometer Gravitational-Wave Observatory (LIGO). Lattimer et al. (1977) and Eichler et al. (1989) also proposed these mergers as potential sites for *r*-process nucleosynthesis. Subsequent to these initial studies, a wide range of investigations of neutron star binary (NS-NS) and neutron star–black hole binary (NS-BH) mergers were carried out with the aim of exploring their GW emission, the production of *r*-process-enriched ejecta, and the possible production of GRBs. In parallel, an alternative model of GRBs from core-collapse supernovae (SNe) (the collapsar model) was developed (MacFadyen & Woosley 1999) with clear predictions about the resulting population that included an exclusive origin in star-forming galaxies and association with core-collapse SNe.

The major breakthrough in the study of GRBs, their distance and energy scales, environments, and progenitors, came from the eventual discovery in 1997 of afterglows from long GRBs (Costa et al. 1997, Frail et al. 1997, van Paradijs et al. 1997). The localization of the bursts to arcminute scale (in X-rays) and to subarcsecond scale (in optical and radio waves) led to redshift measurements, from optical spectroscopy of the afterglows and host galaxies, that directly demonstrated a cosmological origin (Metzger et al. 1997, Kulkarni et al. 1999). The afterglow emission also provided observational evidence for relativistic expansion (Waxman, Kulkarni & Frail 1998; Taylor et al. 2004), jet collimation with typical opening angles of $\sim$3–10° (Harrison et al. 1999, Stanek et al. 1999, Frail et al. 2001), a beaming-corrected energy scale of $\sim 10^{51}$ erg (Bloom, Frail & Sari 2001; Frail et al. 2001; Berger, Kulkarni & Frail 2003), and a typical circumburst density of $\sim$1–10 cm^{-3} with evidence for mass loss from the massive progenitor stars in some cases (Chevalier & Li 2000, Panaitescu & Kumar 2002, Yost et al. 2003). In addition, follow-up studies of the hosts demonstrated an exclusive location in star-forming galaxies (Bloom et al. 1998; Djorgovski et al. 1998; Christensen, Hjorth & Gorosabel 2004; Wainwright, Berger & Penprase 2007) while high-resolution imaging with the *Hubble Space Telescope* (HST) showed that long GRBs follow the radial distribution expected for star formation in disk galaxies (Bloom, Kulkarni & Djorgovski 2002) and are spatially correlated with bright star-forming regions in their hosts (Fruchter et al. 2006). Moreover, long GRBs were shown to be associated with Type Ic SNe on the basis of both photometric and spectroscopic observations (Galama et al. 1998, Hjorth et al. 2003, Stanek et al. 2003, Woosley & Bloom 2006, Hjorth & Bloom 2012). Taken together, the environments and SN associations indicated that long GRBs arise from the death of massive stars and not from compact object binary mergers.

Despite the impressive pace of discovery for long GRBs, the study of short GRBs proved much more challenging; however, compact object binary mergers remained an attractive progenitor model. Kehoe et al. (2001) carried out rapid but shallow optical follow-up of three short GRBs, with nondetections to a limit of $\sim$13–15 mag. Hurley et al. (2002) presented deeper optical follow-up (to limits of 16–20 mag with a delay of $\sim$2 days) and the first radio follow-up (to limits of about 0.5–1.5 mJy with a similar delay) for four short GRBs, but none were detected. In light of the existing afterglow detections summarized in this review, these early searches were woefully inadequate. At the same time, Panaitescu, Kumar & Narayan (2001) argued that a lower energy scale, and potentially lower circumburst densities, will result in afterglows dimmer by at least an order of magnitude compared with long GRB afterglows (e.g., typical optical brightness of $\sim$23 mag at 10 hours post-burst). Similarly, Perna & Belczynski (2002) argued based on compact object binary population synthesis models that, due to kicks, such mergers will tend to occur in lower density environments than long GRBs ($\sim$0.1 cm^{-3}) and, hence, lead to fainter afterglows, if they indeed produce short GRBs.

As in the case of long GRBs, the watershed moment in the study of short GRBs came with the discovery of the first afterglows in May–July 2005, following bursts from the *Swift* (Gehrels et al. 2004) and *High Energy Transient Explorer* (HETE)-2 (Ricker et al. 2003) satellites. Rapid *Swift* X-ray Telescope follow-up of GRB 050509B led to the discovery of the first X-ray counterpart, localized to about 9 arcsec radius (Gehrels et al. 2005). Deep follow-up optical observations

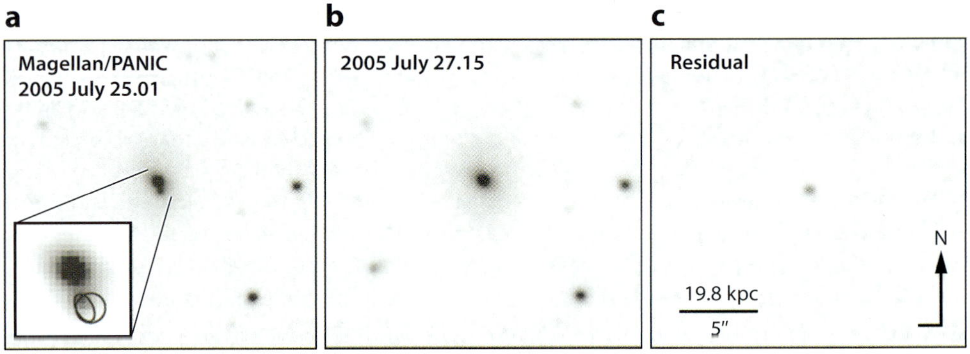

Figure 1

Near-infrared images of the afterglow and host galaxy of the short GRB 050724, obtained (*a*) 0.49 days and (*b*) 2.63 days postburst, reveal (*c*) a fading afterglow on the outskirts of an elliptical galaxy. The inset in panel *a* shows the locations of the X-ray afterglow detected with *Chandra* (*circle*) and the radio afterglow detected with the Very Large Array (*ellipse*). This was the first short GRB with a definitive association to an elliptical galaxy and a broadband afterglow spanning from radio waves to X-rays. From Berger et al. (2005b).

revealed no afterglow emission, but instead uncovered a massive elliptical galaxy at $z = 0.225$ near the X-ray error circle with a chance coincidence probability of $\sim 10^{-3}$ (Castro-Tirado et al. 2005, Gehrels et al. 2005, Bloom et al. 2006). Assuming that the association with the galaxy (and hence the redshift) was correct, it was also shown that the burst lacked an SN (Hjorth et al. 2005a). Shortly thereafter, HETE-2 discovered the short GRB 050709 (Villasenor et al. 2005), and follow-up observations with the *Chandra X-ray Observatory* precisely localized the X-ray afterglow (Fox et al. 2005); subsequent observations revealed the first short GRB optical afterglow (Hjorth et al. 2005b). The resulting subarcsecond position pinpointed the origin of the burst to the outer regions of a star-forming galaxy at $z = 0.160$ (Fox et al. 2005) while optical follow-up ruled out the presence of an associated SN (Hjorth et al. 2005b). Finally, the detection of the short GRB 050724 by *Swift* (Barthelmy et al. 2005) led to the discovery of X-ray, optical/near-IR, and the first radio afterglows, and a definitive localization in an elliptical galaxy at $z = 0.257$ (**Figure 1**; Berger et al. 2005b). The combination of radio to X-ray afterglow emission also demonstrated that both the energy and the density scale were lower than those of long GRBs (Berger et al. 2005b). Taken together, these three early events demonstrated that short GRBs are cosmological in origin, that they produce afterglow emission similar to that of long GRBs but with a lower energy and density scale, and that their progenitors are not massive stars.

In the decade following the initial discovery of short GRB afterglows, over 70 short GRBs were discovered by *Swift* and other gamma-ray satellites, including a substantial detection fraction of X-ray and optical afterglows and a few radio afterglow detections. The resulting localizations have also led to the identification of nearly 40 host galaxies and have enabled studies of the subgalactic locations of short GRBs. This is the first substantial sample of short GRBs with extensive afterglow and host-galaxy data. The primary goal of this review is to utilize and synthesize the information from short GRB afterglows and environments for the first time in order to (*a*) delineate the basic observational properties of short GRBs and their environments (from parsec to galactic scales); (*b*) use this information to confront theoretical progenitor models; (*c*) determine some of the basic properties of the progenitors and the resulting explosions; and (*d*) explore the expected electromagnetic (EM) counterparts of Advanced LIGO/Virgo GW sources in the context of compact object binary progenitors.

In this vein, the outline of the review is as follows. In Section 2, I summarize the basic properties of the prompt gamma-ray emission, compare these with the gamma-ray properties of long GRBs, and discuss the issue of short versus long GRB classification. In Section 3, I describe current short GRB progenitor models. In Section 4, I define the existing sample of short GRBs, including the completeness for afterglow and host-galaxy detections. In Section 5, I summarize basic observational arguments for a distinct origin from the massive star explosions that produce long GRBs. In Section 6, I provide a detailed analysis of short GRB host-galaxy properties and use this information to shed light on the age distribution of the progenitors. In Section 7, I zoom in on the subgalactic locations of short GRBs in terms of their radial offset distribution and their spatial relation to the underlying UV and optical light distribution of their hosts and then compare these results with the locations of long GRBs and SNe and with predictions from compact object binary population synthesis models. In Section 8, I turn to the afterglow properties in X-rays, optical bands, and radio waves and use a comparison to both long GRB afterglows and afterglow theory to provide insight on the energy scale of short GRBs, their collimation, and their circumburst environments. In Section 9, I discuss the expected production and optical/near-IR emission from *r*-process ejecta in compact object binary mergers and compare these predictions to observations of the short GRB 130603B, the first burst to show evidence for such emission. In Section 10, I use the inferred properties of short GRBs and their afterglows to explore on- and off-axis EM emission from GW sources detected by the upcoming Advanced LIGO/Virgo detectors. I conclude with the key results and a discussion of future directions in Section 11.

In addition to the comprehensive review of short GRBs presented here, I also refer the reader to several previous reviews that provide additional insight on GRB theory and observations. These include a review of GRBs in the BATSE era (Fishman & Meegan 1995); an early review of long GRB afterglow discoveries, the emerging GRB-SN connection, and afterglow theory (van Paradijs, Kouveliotou & Wijers 2000); comprehensive reviews of the prompt emission and afterglow theory (Mészáros 2002, Piran 2004); a detailed review of the long GRB-SN connection (Woosley & Bloom 2006); a preliminary review of short GRBs, focused primarily on theoretical issues, and including data from the first few well-localized events in 2005 (Nakar 2007); a comprehensive review of short GRB progenitor models, primarily compact object mergers and collisions (Lee & Ramirez-Ruiz 2007); a review of the primary early results from the *Swift* satellite, including in part the detections of short GRBs (Gehrels, Ramirez-Ruiz & Fox 2009); and finally, an initial review of short GRB galactic and subgalactic environments (Berger 2011).

2. THE PROMPT EMISSION AND THE SHORT-LONG DIVIDE

The defining feature of GRBs among the wide variety of known astrophysical transients is their dominant, nonrepeating, and nonperiodic prompt gamma-ray emission, which displays a rich phenomenology in terms of duration, variability, spectral parameters, fluence, peak flux, temporal and spectral evolution, and various correlations between these parameters. Within this broad range of properties, the clearest subclasses are the short-hard and long-soft bursts. It is important to note that generally the short end of the short GRB duration distribution is affected by detector trigger times (tens of milliseconds) while at the long end of the short GRB duration distribution there is some overlap with the duration distribution of long GRBs, such that some events belonging to the short burst progenitor population will have durations of $\gtrsim 2$ s, whereas some events with durations of $\lesssim 2$ s will have long GRB progenitors (e.g., Bromberg et al. 2013). In addition, because various instruments measure durations in different energy bands, it is possible that the value of T_{90} separating short and long GRBs varies for each sample; for example, Bromberg et al. (2013) suggest that for *Swift* the division is at $T_{90} \approx 0.8$ s. The effect of cosmological time dilation adds another potential complication, as indicated for example by the high-redshift GRBs 080913 ($z = 6.7$) and 090423 ($z = 8.3$), whose observed durations place them in the long GRB category, whereas their rest-frame durations of $\approx$1 s are nominally in the short GRB range (Zhang et al. 2009, Belczynski et al. 2010).

Such effects can be potentially mitigated by the use of auxiliary information, motivated in part by our knowledge of long GRBs (e.g., their association with Type Ic SNe, their exclusive location in star-forming galaxies) and partly by expectations for short GRB progenitors. In this vein, Zhang et al. (2007, 2009) advocate a new classification scheme using the designations Type I GRBs (compact object mergers) and Type II GRBs (massive star progenitors), motivated by the nomenclature for SNe. These authors argue for the use of information such as redshift, SN association, host-galaxy type and star-formation rate (SFR), burst offset relative to the host galaxy, the density profile of the circumburst medium, the gamma-ray and afterglow energy scale, and a combination of prompt emission properties (see Zhang et al. 2009, their figure 8) to determine the type of each burst. Lü et al. (2010) instead advocate the use of a new prompt emission empirical parameter, $E_{\gamma,\mathrm{iso}}/E_{\mathrm{p}}^{5/3}$, with a value of $\lesssim$0.03 coupled with a rest-frame $T_{90} \lesssim 5$ s separating Type I and II events; here $E_{\gamma,\mathrm{iso}}$ is the isotropic-equivalent gamma-ray energy and E_{p} is the spectral peak of the prompt emission. Bloom, Butler & Perley (2008) propose a classification scheme based on physical properties of the progenitor systems, for example destructive (Type I) versus nondestructive (Type II) events, with further separation into subclasses made on the basis of the presence or absence of a degenerate object (and whether it is a neutron star or black hole) and membership in a binary system.

Although such classification schemes are potentially more powerful than relying on duration and hardness alone, and in the long run may be essential for a complete mapping of GRB progenitors, they also carry the risk of biasing the results. For example, the requirement of elliptical hosts for Type I GRBs obscures the fact that their presumed compact object binary progenitors occur in both star-forming and elliptical galaxies and may, therefore, impede the ability to determine properties such as the delay time distribution from the host-galaxy demographics (e.g., Zheng & Ramirez-Ruiz 2007, Berger 2009, Leibler & Berger 2010, Fong et al. 2013). Similarly, the requirement of a lower energy scale or larger physical offsets for Type I GRBs restricts our understanding of the full range of properties spanned by the progenitors and explosions. In the case of a progenitor-based classification scheme, it is currently unclear from either observations or theory how different systems manifest in terms of unique observables, particularly in the context of the prompt gamma-ray emission for which even the radiation mechanism has not been conclusively identified. Finally, because complete knowledge of the prompt emission, distance scale, afterglow parameters, and host-galaxy properties is only available for a small subset of events, it is currently more profitable to begin the investigation of short GRBs with a broad sample that may guide future classification schemes and observational follow-up efforts. Ultimately, it is important to recognize that while the analogy with the SN classification scheme is tempting, the SN classification is based on spectroscopic signatures that already provide critical (though not always definitive) physical insight into the nature of the explosions (e.g., Filippenko 1997). For GRBs, however, the starting point for any classification scheme is the prompt emission, which exhibits many similarities between long and short GRBs (as I discuss below), and even after four decades of investigation is not fully understood.

Thus, in this review I pursue the simplest approach of studying a sample of short GRBs on the basis of a duration cut of $T_{90} \lesssim 2$ s. It is possible that some events in the sample are produced by long GRB progenitors and that some events with $T_{90} \gtrsim 2$ s that belong to the short GRB progenitor population are not included in the sample. However, as I discuss in detail in subsequent sections, the overall differences in the resulting short and long GRB populations, in terms of afterglow and host-galaxy properties, are sufficiently clear that they both shed light on the nature of short GRB progenitors (despite any marginal contamination) and can serve as the basis for future classification schemes by revealing the most relevant and critical differences.

2.1. Comparative Studies of the Prompt Emission

Within the context of a duration-based classification scheme, there have been multiple comparative studies of the prompt emission properties of short and long GRBs, on the basis of samples from various instruments, primarily BATSE, *Swift*, and most recently *Fermi*. These studies have uncovered both differences and commonalities in the prompt emission of short and long GRBs. Below I provide a short overview of these results, and I refer the reader to Nakar (2007) and Gehrels, Ramirez-Ruiz & Fox (2009) for a more detailed discussion.

In the framework of the phenomenological Band function, which was constructed to fit the spectrum of the prompt emission (Band et al. 1993), a key finding is that short GRBs generally have harder spectra than long GRBs owing to a combination of a shallower low-energy spectral slope and a higher spectral peak (e.g., Paciesas et al. 2003; Ghirlanda et al. 2009; Ghirlanda, Ghisellini & Nava 2011; Goldstein et al. 2012). In particular, the short GRB sample detected by the *Fermi*/Gamma-Ray Burst Monitor (GBM) has a median $E_{\rm p} \sim 580$ keV, compared with the catalog average of about 200 keV (Goldstein et al. 2012). An analysis of the three brightest short GRBs from early in the *Fermi* mission (Guiriec et al. 2010) indicates that short GRBs have spectra with $E_{\rm p} > 1$ MeV for at least part of the emission period, but also exhibit strong spectral evolution so that their spectra appear stretched from low to higher energies over their duration. The *Fermi*/GBM short GRBs also appear to have steeply falling spectra above the peak energy, often showing exponential cutoffs in preference to the Band function power law. However, it has also been claimed that the differences between short and long GRBs are less evident when the sample is restricted to short GRBs with the highest peak fluxes (Paciesas et al. 2003, Kaneko et al. 2006) or when considering only the first $\sim$2 s of long GRB light curves (Ghirlanda, Ghisellini & Celotti 2004). Similarly, the variability timescales of short GRBs are comparable to those in the first $\sim$2 s of long GRBs (Nakar & Piran 2002).

Another difference between short and long GRBs is evident in the distribution of spectral lags, which are generally $\gtrsim$0 s for long GRBs (i.e., they exhibit a hard-to-soft evolution), whereas they are $\sim$0 s ($\pm$20 ms) for short GRBs (Norris & Bonnell 2006; Norris, Gehrels & Scargle 2011). Some long GRBs also exhibit negligible lags, but those are events with the highest peak luminosities, which are distinct from the zero lag, low peak luminosity short GRBs (Norris & Bonnell 2006). The difference in spectral lags between short and long GRBs may ultimately provide another useful prompt emission discriminator in the context of classification schemes (e.g., Gehrels et al. 2006).

Finally, it has been noted that short GRBs occupy a distinct region in the $E_{\rm p} - E_{\gamma,\rm iso}$ phase-space as opposed to long GRBs; namely they have lower $E_{\gamma,\rm iso}$ values for a given $E_{\rm p}$ than long GRBs (Amati et al. 2002, Ghirlanda et al. 2009, Qin & Chen 2013, Tsutsui et al. 2013). It remains to be seen, however, whether both subclasses follow a similar relation if only the first $\sim$2 s of long GRBs are considered when determining their $E_{\rm p}$ and $E_{\gamma,\rm iso}$ values.

By contrast, an overall similarity between short and long GRBs has been found in terms of a common inverse correlation between the duration and intensity for individual pulses (Hakkila & Preece 2011; Norris, Gehrels & Scargle 2011). In addition, short and long GRBs appear to follow the same relation between $E_{\rm p}$ and $L_{\rm peak}$ (Yonetoku et al. 2004, Ghirlanda et al. 2009).

To date, only two short GRBs have been detected at gigaelectronvolt energies by the *Fermi* Large Area Telescope. GRB 081024B was detected at $\sim$0.1–3 GeV with a slight delay and a duration of about 3 s, compared with $T_{90} \approx 0.8$ s at megaelectronvolt energies (Abdo et al. 2010). Even more extreme gigaelectronvolt emission (up to ~30 GeV) was observed in GRB 090510 having a delay of about 0.6 s and extending to ~200 s (Abdo et al. 2009, Ackermann et al. 2010). The delayed onset and longer duration of the gigaelectronvolt emission in both long and short GRBs have been interpreted as evidence for a hadronic origin (Asano, Guiriec & Mészáros 2009), a synchrotron self-Compton origin (Zou, Fan & Piran 2009; Ackermann et al. 2010), or an afterglow origin (Gao et al. 2009; Ackermann et al. 2010; Ghirlanda, Ghisellini & Nava 2010; Kumar & Barniol Duran 2010).

2.2. Extended Emission

An interesting feature of some short GRBs is the presence of extended gamma-ray emission that is softer than the prompt spike, lasts ~10–100 s, and sometimes exhibits a delayed onset. The presence of such an emission component was first noted in a stacking analysis of BATSE short GRBs by Lazzati, Ramirez-Ruiz & Ghisellini (2001), who demonstrated that the emission was softer than both the prompt spike and long GRBs of similar durations. These authors interpreted the soft extended component as the onset of the X-ray afterglow. Subsequent studies confirmed the presence of extended emission in at least some BATSE and Konus short bursts (Connaughton 2002, Frederiks et al. 2004).

The presence of soft extended emission in individual bursts garnered attention with the discovery of the first short GRB afterglows. GRB 050709 exhibited an initial spike with $T_{90} \approx 0.07$ s in the 30–400 keV band, followed by a pronounced bump in the 2–25 keV range with a duration of about 130 s, and a larger fluence by about a factor of 2.5 compared to the initial spike (Villasenor et al. 2005). A similar component was observed in GRB 050724; emission in the 15–25 keV band lasted about 100 s, but accounted for only 10% of the fluence of the initial spike (Barthelmy et al. 2005). A particularly extreme example is GRB 080503, in which the extended emission was dominant over the fluence of the prompt spike by a factor of 32 (Perley et al. 2009).

A study by Norris & Bonnell (2006) identified eight BATSE short bursts with individual detections of extended emission, a counts ratio relative to the initial spike ranging up to ~40 and an estimated overall dynamic range of this ratio of $\sim 10^4$. These authors also noted that short GRBs with and without extended emission share a negligible spectral lag in the initial spike. A follow-up study of extended emission in *Swift* short bursts by Norris, Gehrels & Scargle (2010) indicated that about one-quarter of the events clearly exhibit extended emission, whereas the bulk of the remaining events lack such a component. However, I note that not all of the bursts in their sample have a genuine initial short spike, and the incidence rate of extended emission may be closer to ~15%. Building on this sample, Norris, Gehrels & Scargle (2011) argued that bursts with extended emission have durations and pulse profiles in the prompt spike that are ~2–3 times longer than events lacking extended emission, potentially indicative of differences in the central engine.

Two other noteworthy events in the context of extended emission and the T_{90} divide between long and short bursts are GRBs 060505 and 060614, with durations of 4 and 102 s and low redshifts of 0.089 and 0.125, respectively, and yet no SN associations to limits of about 6–7 mag fainter than typical long GRB-SNe (Della Valle et al. 2006, Fynbo et al. 2006, Gal-Yam et al. 2006, Gehrels et al. 2006). The prompt emission of GRB 060505 exhibited a significant spectral lag, typical of long GRBs, and the burst coincided with a low-metallicity and young star-forming region of its late-type host galaxy (Ofek et al. 2007, Thöne et al. 2008), matching the properties of long GRB environments. The gamma-ray emission of GRB 060614 exhibited an initial 5-s spike with negligible spectral lag, typical of short GRBs, followed by extended emission with significant variability, a somewhat softer spectrum, and a larger fluence than the prompt spike by about a factor of five (Gehrels et al. 2006). The host galaxy of GRB 060614 has a low luminosity typical of long GRB hosts, but a lower-than-average specific star-formation rate (SSFR; Gal-Yam et al. 2006). Moreover, the burst itself was located at a significant offset from the host in a region with little evidence for underlying UV emission (Gal-Yam et al. 2006). Thus, GRBs 060505 and 060614 present a challenge to any GRB classification scheme; their prompt emission shares properties typical of both long and short GRBs, their host environments similarly span locations typical of long and short GRBs, and the lack of an SN association suggests a nonmassive star progenitor.

There have been attempts to discern other differences between the populations of short GRBs with and without extended emission. Norris, Gehrels & Scargle (2011) claim that short bursts with extended emission have larger X-ray fluxes at $\sim$100 s postburst and longer-lasting X-ray afterglows. However, it is not obvious that the X-ray emission at such early times is not directly related to the soft extended emission itself. Troja et al. (2008) used a sample of short GRBs through mid-2007 to argue that events with extended emission have systematically smaller spatial offsets from their hosts and, hence, distinct progenitors. However, subsequent studies of the offset distribution indicate that short GRBs with extended emission span as wide an offset range as events with no extended emission (Fong, Berger & Fox 2010; Fong & Berger 2013); a Kolmogorov-Smirnov (K-S) test comparing the two offset distributions leads to a p-value of 0.9, indicating that there is no evidence for two distinct populations (Fong & Berger 2013). In addition, Fong et al. (2013) find that the distributions of host-galaxy types for short GRBs with and without extended emission are currently indistinguishable.

Motivated by the discovery of extended emission, several attempts have been made to provide a theoretical explanation. Metzger, Quataert & Thompson (2008) propose that short GRBs with extended emission are produced by the formation of a rapidly rotating protomagnetar through accretion-induced collapse of a white dwarf, the merger of a white dwarf binary, or the merger of an NS-NS. In this framework, the prompt spike is due to the standard picture of accretion onto the central object, whereas the extended emission is powered by a relativistic wind that extracts the protomagnetar's rotational energy. Bucciantini et al. (2012) discuss a slight variation of the magnetar model, in which the delay in the onset of extended emission results from the breakout of the relativistic outflow through a baryon-loaded wind from the protomagnetar. Metzger et al. (2010a) suggest instead that the gap between the prompt spike and extended emission may be due to heating from r-process nucleosynthesis, which momentarily halts fall-back accretion onto the central object. In this scenario, events lacking extended emission are due to a timescale for the r-process heating of $\gtrsim$1 s, which leads to a complete cutoff in fallback accretion, whereas those with extended emission resume fall-back accretion after a delay. A.I. MacFadyen, E. Ramirez-Ruiz, W. Zhang (unpublished manuscript, astro-ph/0510192) proposed a model in which the short GRBs with extended emission are due to the accretion-induced collapse of a neutron star (generating the prompt spike) followed by interaction of the relativistic outflow with the nondegenerate stellar companion (leading to a delay followed by soft extended emission).

2.3. X-Ray Flares

The discovery of early X-ray flares following the prompt gamma-ray emission was enabled by the rapid follow-up capabilities of *Swift* (Burrows et al. 2005, Nousek et al. 2006, Chincarini et al. 2007, Falcone et al. 2007). The X-ray flares generally follow a fast rise exponential decay profile with $\Delta t/t \ll 1$ and share spectral and temporal properties with the prompt emission, which points to a common origin in activity from the central engine (Chincarini et al. 2010, Margutti et al. 2010); here, Δt is the flare duration and t is the time since the onset of the burst. In the case of short

GRBs, Margutti et al. (2011) carried out a systematic search for flares and found eight *Swift* events that exhibit flaring activity. The sample includes bursts with and without extended emission and in both early- and late-type host galaxies. The resulting flares follow some of the characteristics of long GRB flares, for example, a positive correlation between the flare width and time since the burst. However, the short GRB flares are generally weaker in terms of flux contrast relative to the underlying X-ray emission (with $\Delta F/F \lesssim 2$), have about 1% of the peak luminosity and isotropic-equivalent energy of long GRB flares, and (like the prompt emission) do not follow the lag-luminosity relation of long GRB flares. The lower peak luminosity compared with long GRB flares remains true even when scaled by the peak luminosity of the prompt emission, but short and long GRB flares track the same negative correlation between the ratio of peak flare to prompt luminosity as a function of the ratio of flare peak time to burst duration. Potentially related to this point, the ratio of flare to prompt isotropic-equivalent energy is similar for both short and long GRBs. Given these results, Margutti et al. (2011) conclude that flares in both short and long GRBs are related to the prompt emission.

Flares and rebrightenings at late time (~hours) have also been detected in some short GRBs, with distinct properties from the early flares described above. GRB 050724 exhibited a significant X-ray flare centered at ≈14 h with $\Delta F/F \approx 40$ and $\Delta t/t \approx 1$, with apparent associated emission in the optical/near-IR bands (Campana et al. 2006, Grupe et al. 2006, Malesani et al. 2007). The fact that the flare is superposed on an underlying single power-law decline suggests a distinct, though currently not understood, origin from the afterglow. A delayed peak was found in the optical (and potentially X-ray) emission from GRB 080503 at ~1 day, although the subsequent evolution matched typical afterglow emission rather than a distinct flare (Perley et al. 2009). Finally, excess X-ray emission at $\gtrsim$1 day was found in GRB 130603B; it had no corresponding emission in the optical or radio bands, and it did have an apparent single power-law behavior compared with the early X-ray emission (Fong et al. 2013). Thus, the late flares and rebrightenings are likely to have a distinct origin from the early rapid X-ray flares.

Several ideas have been proposed to explain flares in short (and long) GRBs. Perna, Armitage & Zhang (2006) suggest that fragmentation in the outer accretion disk, caused by gravitational instabilities, could lead to large amplitude variations of the central engine. Rosswog (2007) proposes that the flares can be caused by fall-back accretion of material launched into eccentric orbits during a compact object merger (see also Lee, Ramirez-Ruiz & López-Cámara 2009). Dai et al. (2006) propose a model in which an NS-NS merger leads to the formation of a massive, differentially rotating neutron star whose interior field is wound up and amplified so that it can break through the neutron star surface and reconnect to produce flares. Gao & Fan (2006) propose the same basic model, but appeal to X-ray flares from dipole radiation. Clearly, these models predict that flares should only occur in NS-NS mergers, and not in NS-BH mergers. Proga & Zhang (2006) suggest that the accretion can be episodically stopped and restarted through variations in the accumulated magnetic flux around the accreting central engine. Finally, Giannios (2006) proposes that the flares can be produced by delayed magnetic reconnection as a strongly magnetized outflow decelerates through interaction with the circumburst medium. Unlike the first set of proposed models, the latter two models do not require a reactivation of the central engine.

2.4. Precursors

Precursor gamma-ray emission, preceding the main event by tens of seconds, has been found in ~20% of long GRBs (Lazzati 2005) and has been attributed either to the transition of the expanding fireball to the optically thin regime or to the breakout of the relativistic jet from the surface of the progenitor star (e.g., Lazzati & Begelman 2005). If the former scenario is correct, precursors should also be present in short GRBs. In addition, precursors in short GRBs may be produced by other mechanisms, for example magnetospheric interaction between the compact objects (Hansen & Lyutikov 2001) or resonant shattering of the neutron star crusts (Tsang et al. 2012) prior to the merger. Troja, Rosswog & Gehrels (2010) conducted a search for precursors in 49 *Swift* short GRBs with and without extended emission and uncovered five potential candidates (with 2–5.5 σ significance), of which three candidates (GRBs 081024A, 090510, and 091117) had likely independent detections in other satellite data. In those three cases, the precursor candidates occurred 0.5–2.7 s prior to the main spike, and it is therefore not clear whether these emission episodes are indeed distinct precursors or a part of the prompt emission complex.

3. SHORT GAMMA-RAY BURST PROGENITOR MODELS

The bimodality of GRB durations is highly suggestive of two dominant progenitor populations. The short durations of short GRBs, down to tens of milliseconds, point to compact progenitor systems with a dynamical timescale of milliseconds. In this context, the most popular progenitor model is the merger of compact object binaries comprised of an NS-NS or an NS-BH (Eichler et al. 1989; Narayan, Paczynski & Piran 1992). The merger occurs because of angular momentum and energy losses by GW radiation. In the NS-NS case, the expected remnant is a black hole surrounded by a hyperaccreting disk of debris, whereas an NS-BH merger can lead to the same configuration if the neutron star is tidally disrupted outside of the black hole's horizon. The resulting interplay of high accretion rate and rapid rotation can lead to energy extraction via neutrino-antineutrino annihilation or magnetohydrodynamic (MHD) processes (e.g., Blandford & Znajek 1977, Rosswog & Ramirez-Ruiz 2003, Lee & Ramirez-Ruiz 2007), which in turn drive a collimated relativistic outflow. As discussed in Section 1, NS-NS/NS-BH mergers were proposed as GRB progenitors even prior to the discovery of the duration bimodality because they provided a known source population with roughly the correct event rate, the requisite rapid release of a large energy reservoir, and a clean environment to avoid significant baryon loading.

The compact object merger model leads to several testable observational predictions. First, the delay time between the binary formation and eventual merger is expected to span a wide range that depends on the initial separation and constituent masses, $\tau_{GW} \propto a^4/(\mu M^2)$, where a is the initial binary separation, $M \equiv M_1 + M_2$ is the total binary mass and $\mu \equiv M_1 M_2/M$ is the reduced mass. As a result of the wide delay time distribution, the resulting short bursts will occur in both early- and late-type galaxies (e.g., Belczynski et al. 2006, Zheng & Ramirez-Ruiz 2007). This is indeed the case for the small observed population of Galactic NS-NS binaries, which have coalescence timescales of tens of millions of years to much longer than a Hubble time (e.g., Burgay et al. 2003, Champion et al. 2004). Second, natal kicks imparted to the binary system during the SN explosions that gave rise to the neutron stars and/or black hole, coupled with the broad range of merger timescales, should lead to some mergers at large offsets from their birth sites and host galaxies (tens to hundreds of kiloparsecs) (Narayan, Paczynski & Piran 1992; Bloom, Sigurdsson & Pols 1999; Fryer, Woosley & Hartmann 1999; Perna & Belczynski 2002; Belczynski et al. 2006). A broad spatial distribution is also expected if some short GRBs result from dynamically formed binaries in globular clusters (Grindlay, Portegies Zwart & McMillan 2006; Lee, Ramirez-Ruiz & van de Ven 2010). Third, the mergers will be accompanied by strong GW emission, detectable with the Advanced LIGO/Virgo detectors to about 200 Mpc for NS-NS mergers (Abramovici et al. 1992, Harry & LIGO Scientific Collaboration 2010, Accadia et al. 2011). Fourth, the mergers will produce neutron-rich ejecta, which will in turn lead to *r*-process nucleosynthesis; the decay of the resulting radioactive elements may be detectable at optical/near-IR wavelengths (Lattimer & Schramm 1976, Li & Paczyński 1998, Metzger et al. 2010b, Barnes & Kasen 2013).

Finally, the mergers will not be accompanied by SN explosions. Most of these predictions can now be tested with existing observations of short GRBs, as delineated in subsequent sections of this review.

In principle, some of these properties may differ between NS-NS and NS-BH mergers. For example, numerical simulations suggest that mergers with a larger mass ratio (i.e., NS-BH) may produce larger ejecta masses with more pronounced asymmetry that may lead to brighter optical/near-IR counterparts (Piran, Nakar & Rosswog 2013). Similarly, because neutron star disruption outside the horizon of a typical black hole ($\sim 10\,M_{\odot}$) requires significant black hole spin, the resulting accretion disk will undergo significant Lense-Thirring precession that may imprint detectable variability on the prompt emission light curve (Dichiara et al. 2013; Stone, Loeb & Berger 2013). Finally, it is possible that the larger masses of NS-BH systems may lead to systematically smaller effective kicks and, hence, smaller offsets than for NS-NS mergers (e.g., Belczynski et al. 2006, Dhawan et al. 2007). However, it is important to keep in mind that to date no NS-BH binaries have been identified in nature, and it is therefore not clear whether they contribute to the short GRB population at all, if any of the predicted differences match the actual properties of NS-BH binaries, and whether the differences relative to NS-NS binaries are significant enough to be discernible in the existing short GRB sample. These questions may be resolved with joint GW and short GRB detections (Section 10).

In the NS-NS merger model it is generally assumed that a black hole is rapidly formed, with the resulting accretion powering the short GRB and its afterglow. However, the recent discovery of neutron stars with masses of about 2 $M_{\odot}$ (Demorest et al. 2010, Antoniadis et al. 2013) suggests that some mergers may lead instead to a transitory or stable rapidly spinning and highly magnetized neutron star (a magnetar) (Duncan & Thompson 1992; Metzger, Quataert & Thompson 2008). The magnetar can in turn power a short GRB through its spin-down energy. In the transitory case, as the magnetar spins down it will eventually collapse to a black hole when differential rotation can no longer support its large mass. This transition could in principle lead to an observable signature; for example, it has been recently claimed that short-lived X-ray plateaus in some short GRBs are due to magnetar spin-down energy injection and subsequent collapse to a black hole (Rowlinson et al. 2013). However, fall-back accretion may also explain the plateaus (Rosswog 2007). Similarly, it is not clear if a stable magnetar can be distinguished from an accreting black hole engine in terms of the resulting prompt and afterglow emission.

Given the potential ability of magnetars to provide the required energy source, it has also been proposed that such objects can form and power short GRBs or SGRs following the accretion-induced collapse of a white dwarf or white dwarf binary merger (Levan et al. 2006b; Metzger, Quataert & Thompson 2008). Unlike formation through core collapse, such delayed magnetar formation will lead to events in both early- and late-type galaxies, similar to the prediction for NS-NS/NS-BH mergers. However, such systems are not expected to experience significant natal kicks, to be accompanied by GW emission in the Advanced LIGO/Virgo band, or to produce r-process radioactive elements.

Another proposed progenitor model is accretion-induced collapse of a neutron star to a black hole (Qin et al. 1998; A.I. MacFadyen, E. Ramirez-Ruiz, W. Zhang, unpublished manuscript, astro-ph/0510192), although this model has not been explored in detail. Finally, Lazzati, Morsony & Begelman (2010) suggested that short GRBs with extended emission may be produced from the same massive star progenitors as long GRBs, but with a wide off-axis viewing angle that leads to predominant emission from the cocoon surrounding the jet. In this model, short GRBs with extended emission will occur only in star-forming galaxies and will be accompanied by Type Ic SNe. These requirements are violated in several events (e.g., Berger et al. 2005b, Hjorth et al. 2005b).

4. DEFINING THE SAMPLE

In this review, I address the properties of short GRBs, their afterglows, and environments by focusing on a sample of 70 events discovered primarily by the *Swift* satellite (Gehrels et al. 2004) in the eight-year period spanning January 2005 to January 2013, supplemented by the recent well-studied benchmark event GRB 130603B (Berger, Fong & Chornock 2013; Cucchiara et al. 2013; Tanvir et al. 2013; de Ugarte Postigo et al. 2014). The sample includes three short bursts discovered by HETE-2 (GRBs 050709 and 060121; Fox et al. 2005, Hjorth et al. 2005b, Villasenor et al. 2005, de Ugarte Postigo et al. 2006, Levan et al. 2006a) and INTEGRAL (GRB 070707; McGlynn et al. 2008, Piranomonte et al. 2008) for which afterglows and host galaxies have been identified. This is the first sample of short GRBs with a substantial fraction of afterglow detections, enabling studies of both the burst properties and the host galaxies. A basic summary of the prompt and afterglow emission of these GRBs is provided in **Table 1**.

The most detailed information, in terms of both explosion properties and host-galaxy and subgalactic environments, comes from events localized to subarcsecond precision through optical afterglow detections or *Chandra* detections of the X-ray afterglow (e.g., Berger et al. 2005b, Fox et al. 2005, Hjorth et al. 2005b, Fong et al. 2012, Margutti et al. 2012); radio detections are also a route to subarcsecond positions (Berger et al. 2005b), but to date all short GRBs detected in the radio band have also been detected in the optical band. *Swift* X-ray Telescope (XRT) positions alone, generally with a precision of $\sim$1.5–5 arcsec, are also useful for host identifications (Berger et al. 2007a, Bloom et al. 2007, Fong et al. 2013). However, for these bursts the availability of only X-ray data limits the extraction of afterglow properties, and the positions are not precise enough to locate the bursts within their hosts.

In this vein, of the 67 *Swift* short GRBs considered here, 53 events were rapidly followed up using the on-board XRT on a timescale of $\lesssim$100 s, leading to 47 detections (89%). This is only slightly lower than the detection rate for long GRBs. It is important to note that the 6 undetected events had a lower-than-average gamma-ray fluence of $\langle F_\gamma \rangle \approx 2.5 \times 10^{-8}$ erg cm^{-2} and may therefore simply be fainter events. Two of the three non-*Swift* events (GRBs 060121 and 070707) were also followed up using the XRT and both were detected, enabling the subsequent identification of optical afterglows. The third non-*Swift* event (GRB 050709) was detected in the X-rays with *Chandra* (Fox et al. 2005). Of the 50 total X-ray detections, 28 events (56%) exhibit long-term X-ray emission beyond $\sim$10^3 s, whereas the remaining 22 events rapidly fade below the *Swift*/XRT detection threshold at $\lesssim$10^3 s. The *Swift*/XRT positions span a range of 1.4–5.5 arcsec radius and have a median value of 1.8 arcsec (90% confidence; UK *Swift* Sci. Data Cent. 2011). These positions are generally sufficient for a robust identification of host galaxies with a probability of chance coincidence of $\sim$1–10% (Bloom et al. 2007, Berger et al. 2007a, Fong et al. 2013). Finally, the 14 bursts lacking rapid XRT follow-up include 8 events with *Swift* observing constraints and 6 events with delayed follow-up ($\gtrsim$1 h postburst). Thus, the lack of X-ray detections for these events is not expected to bias the sample considered here.

Deep optical follow-up observations were obtained for 37 of the 47 *Swift* bursts with XRT positions, leading to 21 detections (57%) with a range of about 21–26 mag at $\delta t \approx$ 0.4–30 h. It is remarkable that the optical afterglow detection fraction is comparable to that for long GRBs despite the relative faintness of short GRB afterglows. This is a testament to the concerted follow-up effort undertaken by the GRB community over the past decade, utilizing the largest ground-based telescopes. The three non-*Swift* events were also detected in the optical. Of the 10 events lacking deep optical follow-up, 6 events were located along constrained Galactic sight lines with large extinction and/or contaminating bright stars, whereas the other 4 events lack deep afterglow and host-galaxy searches at the present time. Thus, as in the case of missing

Table 1 Prompt emission and afterglow properties of short gamma-ray bursts (GRBs)[a]

GRB	Satellite	T_{90} (s)	F_γ (erg cm^{-2})	z	$F_{X,11}$ (erg cm^{-2} s^{-1})	t_{opt} (h)	$F_{\nu,opt}$ (μJy)	t_{radio} (d)	$F_{\nu,radio}$ (μJy)
1	2	3	4	5	6	7	8	9	10
050202	*Swift*	0.27	3.0×10^{-8}	NA	NA[b]	NA	NA	NA	NA
050509B	*Swift*	0.04	2.3×10^{-8}	0.225	$<1.95 \times 10^{-14}$	2.1	<0.8	NA	NA
050709	HETE-2	0.07	4.0×10^{-7}	0.161	1.92×10^{-14}	34.0	2.3	1.57	<65
050724	*Swift*	3.0/EE	6.3×10^{-7}	0.257	9.55×10^{-14}	12.0	8.4	1.69	465
050813	*Swift*	0.6	1.2×10^{-7}	NA	$<2.12 \times 10^{-14}$	12.8	<1.9	1.64	<55
050906	*Swift*	0.26	6.0×10^{-9}	NA	NA[c]	NA	NA	NA	NA
050925	*Swift*	0.07	7.6×10^{-8}	NA	NA[c]	NA	NA	NA	NA
051210	*Swift*	1.3	8.3×10^{-8}	1.3	$<2.70 \times 10^{-14}$	19.2	<1.6	NA	NA
051105A	*Swift*	0.09	2.2×10^{-8}	NA	NA[c]	NA	NA	NA	NA
051221A	*Swift*	1.4	1.2×10^{-6}	0.546	1.08×10^{-12}	3.1	17.4	0.91	173
060121	HETE-2	2.0	4.7×10^{-6}	NA	1.09×10^{-12}	7.4	8.8	NA	NA
060313	*Swift*	0.7	1.1×10^{-6}	NA	3.85×10^{-13}	2.8	10.8	2.12	<110
060502B	*Swift*	0.09	4.0×10^{-8}	0.287	$<1.47 \times 10^{-14}$	16.8	<0.7	NA	NA
060801	*Swift*	0.5	8.1×10^{-8}	1.130	$<9.80 \times 10^{-15}$	12.4	<0.8	0.49	<105
061006	*Swift*	0.4/EE	1.4×10^{-6}	0.438	2.27×10^{-13}	14.9	2.9	NA	NA
061201	*Swift*	0.8	3.3×10^{-7}	0.111	1.92×10^{-13}	8.6	2.9	NA	NA
061210	*Swift*	0.2/EE	3.0×10^{-7}	0.409	1.36×10^{-13}	2.1	<1.4	1.90	<102
061217	*Swift*	0.2	4.6×10^{-8}	0.827	NA	2.8	2.0	NA	NA
070209	*Swift*	0.09	2.2×10^{-8}	NA	NA[c]	NA	NA	NA	NA
070406	*Swift*	1.20	3.6×10^{-8}	NA	NA[b]	NA	NA	NA	NA
070429B	*Swift*	0.5	6.3×10^{-8}	0.902	1.13×10^{-13}	4.8	<0.6	NA	NA
070707	INTEGRAL	1.1	1.4×10^{-6}	NA	5.04×10^{-13}	11.0	1.9	NA	NA
070714B	*Swift*	2.0	7.2×10^{-7}	0.923	6.30×10^{-14}	23.6	0.7	15.5	<139
070724A	*Swift*	0.4	3.0×10^{-8}	0.457	1.28×10^{-13}	3.3	3.8	1.06	<250
070729	*Swift*	0.9	1.0×10^{-7}	0.8	$<4.71 \times 10^{-14}$	0	<0	NA	NA
070809	*Swift*	1.3	1.0×10^{-7}	0.473	5.30×10^{-13}	11.2	0.7	NA	NA
070810B	*Swift*	0.08	1.2×10^{-8}	NA	NA[c]	NA	NA	NA	NA
070923	*Swift*	0.05	5.0×10^{-8}	NA	NA[b]	NA	NA	NA	NA
071112B	*Swift*	0.30	4.8×10^{-8}	NA	$<2.00 \times 10^{-13}$	6.3	<1.4	NA	NA
071227	*Swift*	1.8	2.2×10^{-7}	0.381	3.20×10^{-14}	7.0	1.6	NA	NA
080121	*Swift*	0.7	3.0×10^{-8}	NA	NA[b]	NA	NA	NA	NA
080426	*Swift*	1.7	3.7×10^{-7}	NA	2.24×10^{-13}	7.5	<2.6	NA	NA
080503	*Swift*	0.3/EE	6.1×10^{-8}	NA	$<3.05 \times 10^{-14}$	25.9	0.3	3.05	<55
080702A	*Swift*	0.5	3.6×10^{-8}	NA	$<1.06 \times 10^{-13}$	NA	NA	0.83	<155
080905A	*Swift*	1.0	1.4×10^{-7}	0.122	$<6.70 \times 10^{-14}$	8.5	0.8	NA	NA
080919	*Swift*	0.6	7.2×10^{-8}	NA	$<2.12 \times 10^{-14}$	NA[e]	NA	NA	NA
081024A	*Swift*	1.8	1.2×10^{-7}	NA	$<4.56 \times 10^{-14}$	NA[e]	NA	NA	NA
081226A	*Swift*	0.4	9.9×10^{-8}	NA	$<2.57 \times 10^{-14}$	0.37	1.3	NA	NA
090305	*Swift*	0.4	7.5×10^{-8}	NA	$<8.50 \times 10^{-14}$	0.56	2.0	NA	NA

(Continued)

Table 1 ***(Continued)***

GRB	Satellite	T_{90} (s)	F_γ (erg cm^{-2})	z	$F_{X,11}$ (erg cm^{-2} s^{-1})	t_{opt} (h)	$F_{\nu,opt}$ (μJy)	t_{radio} (d)	$F_{\nu,radio}$ (μJy)
090417A	*Swift*	0.07	1.9×10^{-8}	NA	NA[b]	NA	NA	NA	NA
090426	*Swift*	1.2	1.8×10^{-7}	2.609	2.63×10^{-13}	0.35	14.7	NA	NA
090510	*Swift*	0.3	3.4×10^{-7}	0.903	5.04×10^{-14}	9.0	2.3	1.46	<145
090515	*Swift*	0.04	2.1×10^{-8}	0.403	$<8.43 \times 10^{-14}$	1.9	0.1	0.87	<60
090621B	*Swift*	0.14	7.0×10^{-8}	NA	2.70×10^{-14}	0.9	<3.3	0.61	<54
090715	*Swift*	0.5/EE	9.7×10^{-7}	NA	NA[b]	NA	NA	NA	NA
090815C	*Swift*	0.6	4.4×10^{-8}	NA	$<7.00 \times 10^{-14}$	NA	NA	NA	NA
090916	*Swift*	0.3/EE	9.5×10^{-7}	NA	$<3.00 \times 10^{-14}$	NA	NA	NA	NA
091109B	*Swift*	0.3	1.9×10^{-7}	NA	1.26×10^{-13}	6.0	0.7	NA	NA
100117A	*Swift*	0.3	9.3×10^{-8}	0.915	$<2.50 \times 10^{-14}$	8.4	0.2	NA	NA
100206A	*Swift*	0.1	1.4×10^{-7}	0.408	$<1.07 \times 10^{-14}$	15.7	<0.5	NA	NA
100625A	*Swift*	0.3/EE	2.3×10^{-7}	0.453	3.95×10^{-15}	12.7	<4.4	NA	NA
100628A	*Swift*	0.04	2.5×10^{-8}	NA	NA[c]	NA	NA	NA	NA
100702A	*Swift*	0.16/EE	1.2×10^{-7}	NA	$<6.63 \times 10^{-14}$	NA[e]	NA	NA	NA
101129A	*Swift*	0.35	9.0×10^{-8}	NA	NA[b]	NA	NA	NA	NA
101219A	*Swift*	0.6	4.6×10^{-7}	0.718	$<2.00 \times 10^{-14}$	1.0	<3.2	NA	NA
101224A	*Swift*	0.2	5.8×10^{-7}	NA	NA[d]	NA	NA	NA	NA
110112A	*Swift*	0.5	3.0×10^{-8}	NA	8.22×10^{-14}	15.4	2.8	1.90	<75
110420B	*Swift*	0.08	5.3×10^{-8}	NA	$<1.40 \times 10^{-13}$	NA	NA	0.46	<45
111020A	*Swift*	0.4	6.5×10^{-8}	NA	4.87×10^{-13}	17.7	<0.6	0.7	<39
111117A	*Swift*	0.5	1.4×10^{-7}	1.3	3.21×10^{-14}	13.2	<0.2	0.50	<18
111121A	*Swift*	0.45/EE	2.2×10^{-6}	NA	4.80×10^{-13}	NA[e]	NA	0.80	<78
111126A	*Swift*	0.8	7.0×10^{-8}	NA	NA[b]	NA	NA	NA	NA
120229A	*Swift*	0.22	4.1×10^{-8}	NA	NA[b]	NA	NA	NA	NA
120305A	*Swift*	0.10	2.0×10^{-7}	NA	2.40×10^{-14}	NA	NA	2.25	<18
120403A	*Swift*	1.25	1.0×10^{-7}	NA	NA[b]	NA	NA	NA	NA
120521A	*Swift*	0.45	7.8×10^{-8}	NA	$<1.50 \times 10^{-13}$	17.7	<0.9	NA	NA
120630A	*Swift*	0.6	6.1×10^{-8}	NA	NA[c]	NA	NA	NA	NA
120804A	*Swift*	0.81	8.8×10^{-7}	1.3	5.86×10^{-13}	5.5	5.4	1.95	<18
121226A	*Swift*	1.0	1.4×10^{-7}	NA	2.24×10^{-13}	11.1	<1.9	NA	NA
130603B	*Swift*	0.18	6.3×10^{-7}	0.356	6.00×10^{-13}	8.2	8.6	0.37	119

[a]Prompt emission and afterglow properties of the short GRB sample spanning January 2005 to January 2013, with the addition of GRB 130603B. The columns are: (1) GRB name, (2) discovery satellite, (3) duration (EE, extended emission), (4) fluence in the 15–150 keV band for *Swift* events, (5) redshift, (6) observed X-ray flux at 11 h postburst, (7) time of first optical detection, (8) flux density in the optical *r*-band, (9) time of radio observation for nondetections or of peak radio brightness for detections, and (10) radio flux density (at 5 or 8.5 GHz).
[b]No rapid *Swift*/XRT follow-up due to observing constraints.
[c]No *Swift*/XRT detection despite rapid follow-up.
[d]No *Swift*/XRT observations at $\gtrsim 10^3$ s.
[e]Sight line with significant Milky Way extinction.
Abbreviation: NA, not available.

X-ray follow-up, these events are not expected to bias the sample considered here. The 16 bursts with deep follow-up and no detected optical afterglows typically have limits of ~23 mag at ~1–20 h, comparable to the median brightness of the detected afterglows (e.g., Berger 2010). As I demonstrate in Section 8, these events do not appear to be distinct from those with detected afterglows, suggesting that the searches were likely too shallow. In addition, 2 of the 16 events with optical nondetections (GRBs 111020A and 111117A) were subsequently detected using *Chandra*, leading to subarcsecond positions (Fong et al. 2012, Margutti et al. 2012).

Finally, radio observations have been obtained for 28 short GRBs (including some non-*Swift* bursts and *Swift* bursts with only gamma-ray positions), leading to only 3 detections to date (11%; GRBs 050724A, 051221A, and 130603B: Berger et al. 2005b, Soderberg et al. 2006a, Fong et al. 2013). This is a low detection fraction compared with long GRBs (with ~30%; Chandra & Frail 2012), but it is not surprising considering that radio follow-up is generally sensitivity-limited even for long GRBs. As I discuss in Section 8, despite the low detection fraction, the radio limits are generally useful for placing constraints on the circumburst densities of short GRBs.

In addition to afterglow follow-up observations, there has been an intensive effort to characterize the environments of short GRBs from subgalactic to galaxy cluster scales (e.g., Berger et al. 2007b; Berger 2009; D'Avanzo et al. 2009; Fong, Berger & Fox 2010; Berger 2011; Fong et al. 2013; Fong & Berger 2013). The sample presented in this review includes nearly 40 identified host galaxies along with about 30 redshift measurements. A subset of these galaxies have detailed measurements that include morphologies, stellar masses, stellar population ages, SFRs, and metallicities (see **Table 2**). In addition, for about 20 events (those with subarcsecond afterglow localizations), there are precise measurements of the subgalactic environments based on HST observations, including projected offsets and the brightness at the GRB location relative to the underlying light distribution of the host galaxy (Berger 2010; Fong, Berger & Fox 2010; Fong & Berger 2013).

5. NON-MASSIVE STAR ORIGIN

Two key observations helped to establish the connection between long GRBs and the deaths of massive stars (the collapsar model; MacFadyen & Woosley 1999): (*a*) the association of long GRBs with Type Ic core-collapse SNe (Galama et al. 1998, Hjorth et al. 2003, Stanek et al. 2003, Woosley & Bloom 2006) and (*b*) the exclusive locations of long GRBs in star-forming galaxies (Bloom et al. 1998; Djorgovski et al. 1998; Christensen, Hjorth & Gorosabel 2004; Wainwright, Berger & Penprase 2007), as well as their spatial offsets relative to their hosts' overall light distribution and their coincidence with bright star-forming regions within their hosts (Bloom, Kulkarni & Djorgovski 2002; Fruchter et al. 2006). The SN associations are based on spectroscopic observations, mainly at $z \lesssim 0.5$, where the spectroscopic signatures can be discerned with sufficient significance (e.g., Hjorth et al. 2003, Stanek et al. 2003, Pian et al. 2006, Berger et al. 2011), and on photometric rebrightenings on a timescale of ~15–20 days postburst that match Type Ic SN light curves in both brightness and color (to $z \sim 1$; e.g., Bloom et al. 1999, Bloom et al. 2002, Woosley & Bloom 2006). The observed distribution of long GRB-SN peak magnitudes is relatively narrow, spanning only about 1 mag for the bulk of the sample (Woosley & Bloom 2006, Hjorth & Bloom 2012). Moreover, long GRB-SNe are generally more luminous than normal Type Ib/c SNe (Drout et al. 2011), although there is some overlap in the distributions. These results are summarized in **Figure 2** in which I plot the peak absolute magnitudes of GRB-SNe and Type Ib/c SNe relative to the canonical SN 1998bw associated with GRB 980425. In terms of spatial locations within their hosts, long GRBs radially track an exponential light distribution, typical of star formation in disk galaxies, with a median offset of about one half-light radius (Bloom, Kulkarni & Djorgovski

Table 2 Properties of short gamma-ray burst (GRB) host galaxies[a]

GRB	z	Type	L_B (L^*)	SFR ($M_\odot$ year^{-1})	τ_* (Gyr)	log(M_*) ($M_\odot$)	12 + log(O/H)
1	2	3	4	5	6	7	8
Subarcsecond localizations							
050709	0.161	L	0.1	0.15	0.26	8.8	8.5
050724A	0.257	E	1.0	<0.1	0.94	10.8	NA
051221A	0.546	L	0.3	0.95	0.17	9.4	8.8
060121	<4.1	?	NA	NA	NA	NA	NA
060313	<1.7	?	NA	NA	NA	NA	NA
061006	0.438	L	0.1	0.24	0.24	9.0	8.6
061201	0.111	H/L	0.1	0.14	NA	NA	NA
070429B	0.902	L	0.6	1.1	0.46	10.4	NA
070707	<3.6	?	NA	NA	NA	NA	NA
070714B	0.922	L	0.1	0.44	0.22	9.4	NA
070724A	0.457	L	1.4	2.5	0.30	10.1	8.9
070809	0.473	H/E	1.9	<0.1	3.10	11.4	NA
071227	0.381	L	1.2	0.6	0.49	10.4	8.5
080503	<4.2	H/?	NA	NA	NA	NA	NA
080905A	0.122	L	NA	NA	NA	NA	8.7
081226A	<4.1	?	NA	NA	NA	NA	NA
090305	<4.1	H/?	NA	NA	NA	NA	NA
090426A	2.609	L	NA	NA	NA	NA	NA
090510	0.903	L	0.3	0.3	0.14	9.7	NA
090515	0.403	H/E	1.0	0.1	4.35	11.2	NA
091109B	<4.4	?	NA	NA	NA	NA	NA
100117A	0.915	E	0.5	<0.2	0.79	10.3	NA
110112A	<5.3	H/?	NA	NA	NA	NA	NA
111020A	NA	?	NA	NA	NA	NA	NA
111117A	1.2	L	0.6	6.0	0.09	9.6	NA
120804A	1.3	L	0.8	8.0	0.13	10.8	NA
130603B	0.356	L	1.0	1.7	NA	9.7	8.7
Swift/XRT Positions Only							
050509B	0.225	E	5.0	<0.15	3.18	11.6	NA
050813	NA	E	NA	NA	NA	NA	NA
051210	>1.4	?	NA	NA	NA	NA	NA
060502B	0.287	E	1.6	0.8	1.3	11.8	NA
060801	1.130	L	0.6	6.1	0.03	9.1	NA
061210	0.409	L	0.9	1.2	0.38	9.6	8.8
061217	0.827	L	0.4	2.5	0.03	9.1	NA
070729	0.8	E	1.0	<1.5	0.98	10.6	NA
080123	0.495	L	1.2	NA	0.31	10.1	NA

(*Continued*)

Table 2 (*Continued*)

GRB	z	Type	L_B (L^*)	SFR ($M_\odot$ year^{-1})	τ_* (Gyr)	log(M_*) ($M_\odot$)	12 + log(O/H)
100206A	0.407	L	1.0	30	0.10	10.8	9.2
100625A	0.452	E	0.2	0.3	0.79	10.3	NA
101219A	0.718	L	1.3	16	0.03	9.2	NA

[a]Properties of short GRB host galaxies. The columns are: (1) GRB name, (2) redshift, (3) galaxy type (E, early-type; L, late-type; ?, unknown; H, host-less), (4) optical B-band luminosity normalized by L_B^* at the appropriate redshift, (5) star-formation rate, (6) dominant stellar population age, (7) logarithm of the stellar mass, and (8) metallicity. Data are from Berger (2009), Leibler & Berger (2010), Berger (2010), Fong et al. (2013) and references therein.

2002). Moreover, long GRBs are spatially correlated with bright star-forming regions, even in comparison with normal core-collapse SNe (Fruchter et al. 2006, Svensson et al. 2010).

5.1. Lack of Supernova Associations

In the short GRB sample defined in Section 4 there are several events at sufficiently low redshifts to allow clear detections of associated SNe, yet none have been found to date (Fox et al. 2005; Hjorth et al. 2005a,b; Soderberg et al. 2006a; D'Avanzo et al. 2009; Kocevski et al. 2010; Rowlinson et al. 2010; Berger, Fong & Chornock 2013). In **Figure 2**, I plot all existing upper limits for associated SNe (for 7 short bursts), measured relative to the peak absolute magnitude of the canonical long GRB-SN 1998bw. I also plot the limits for the peculiar events GRBs 060505 and 060614. As can be seen from the figure, SNe associated with long GRBs span a narrow peak brightness range and have a median and standard deviation relative to SN 1998bw of $+0.18 \pm 0.45$ mag. By contrast, the upper limits on SN associations for short GRBs range from 0.6 to 7.4 mag fainter than SN 1998bw; GRBs 060505 and 060614 have limits of 6.9 and 6.4 mag fainter than SN 1998bw, respectively. Thus, in all cases associations with SNe that are drawn from the same distribution as long GRB-SNe can be ruled out. This demonstrates that short and long GRBs do not share a common progenitor system and that at least the short GRBs with deep SN limits are not produced by massive star explosions. It is also important to note that of the 7 short bursts with limits on associated SNe, 6 events are located in star-forming galaxies, indicating that although the hosts exhibit ongoing star-formation activity, the short GRB progenitors themselves do not belong to a young population of massive stars. I return to this point in Section 7 when discussing the locations of short GRBs within and around their host galaxies.

The dominance of star-forming hosts in the sample of short GRBs lacking SN associations indicates that the use of galaxy type as an indicator of progenitor type (as advocated, for example, by Zhang et al. 2007, 2009) can severely bias the resulting GRB classification. In particular, a star-forming host does not indicate a young massive star progenitor. In a similar vein, I note that two of the short bursts with SN nondetections (GRBs 051221A and 070724A) are claimed to be likely collapsars according to the duration-hardness analysis of Bromberg et al. (2013), and a third event (GRB 080905A) has a duration that is longer than their nominal divide between short and long GRBs (see *inset* of **Figure 2**). The fact that these events lack associated SNe casts doubt on the claimed statistical significance of a collapsar origin assigned to individual events in their analysis.

5.2. A Mix of Ellipticals and Spirals

The second clear distinction between the short and long GRB populations is the occurrence of some short GRBs in elliptical galaxies (**Figure 3**). The current short GRB sample includes two secure cases of elliptical host galaxies on the basis of subarcsecond afterglow positions and spatially

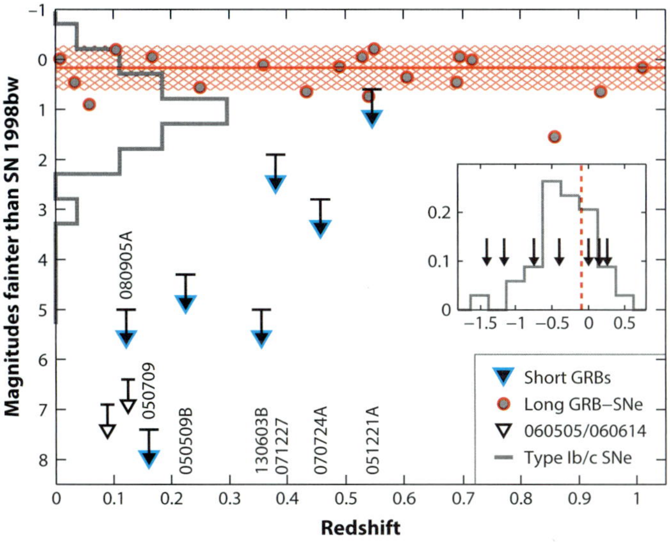

Figure 2

Limits on supernovae (SNe) associated with short GRBs (*filled triangles*) relative to the peak absolute magnitude of the canonical long GRB-SN 1998bw. Also shown are the distribution of SN peak magnitudes for long GRBs (*filled circles*; *hatched region* marks the median and standard deviation for the population; Hjorth & Bloom 2012), local Type Ib/c SNe (*histogram*; Drout et al. 2011), and two unusual long GRBs that lacked associated SNe (*open triangles*; 060505 and 060614; Della Valle et al. 2006, Fynbo et al. 2006, Gal-Yam et al. 2006, Gehrels et al. 2006). The latter may represent a long duration or extended emission tail of the short GRB population. With the exception of GRB 050509B, all short bursts with limits on associated SNe occurred in star-forming galaxies, indicating that despite the overall star-formation activity, the short GRB progenitors were not massive stars. The inset shows the overall duration distribution of the short GRBs considered in this review (*histogram*); the durations of the 7 short GRBs with SN limits are marked by arrows. The dotted vertical line marks the claimed duration separating *Swift* noncollapsar and collapsar progenitors according to the analysis by Bromberg et al. (2013), and yet three of the short GRBs lacking SN associations have longer durations.

coincident hosts (GRBs 050724A and 100117A; Berger et al. 2005b, Fong et al. 2011), two additional cases with subarcsecond afterglow positions and likely elliptical hosts with large projected offsets (GRBs 070809 and 090515; Berger 2010), and four additional likely cases (probabilities of about 1–5%) based on *Swift*/XRT positions alone (GRBs 050509B, 060502B, 070729, 100625A) (Gehrels et al. 2005; Bloom et al. 2006, 2007; Fong et al. 2013). Overall, about 20% of short GRBs are associated with early-type host galaxies (Fong et al. 2013). In nearly all cases, the identification of the hosts as early-type galaxies is based on spectroscopic observations that reveal no star-formation activity (to $\lesssim 0.1\,M_\odot$ year^{-1}), optical/near-IR spectral energy distributions that are matched by a single stellar population with an age of $\gtrsim 1$ Gyr, and/or morphological information based on HST observations. I explore the host-galaxy demographics distribution, and its implications for the progenitor population, in the next section, but it is clear from the occurrence of at least some short GRBs in elliptical galaxies that the progenitors belong to an old stellar population.

6. SHORT GAMMA-RAY BURST GALAXY-SCALE ENVIRONMENTS

Having established that the progenitors of short GRBs are generally distinct from those of long GRBs on the basis of the lack of SN associations and their occurrence in elliptical galaxies, I now turn to the question of what the progenitors are and what we can infer about their nature from the properties of the host galaxies. In general, the galaxy-scale and local environments of astrophysical transients provide critical insight into the nature of their progenitors, and this has been used to establish the progenitor properties of various SN types and long GRBs (e.g., van den Bergh & Tammann 1991; Bloom, Kulkarni & Djorgovski 2002; Fruchter et al. 2006; Li et al. 2011). In this section, I contrast the environments of short and long GRBs to further demonstrate their distinct origins and compare the short GRB environments with field galaxy populations to establish some of the basic properties of their progenitors. I also provide a comparison with the hosts of Type Ia SNe because these white dwarf thermonuclear explosions are known to arise from an evolved stellar population that exhibits a broad range of delay times (Mannucci, Della Valle & Panagia 2006; Sullivan et al. 2006; Maoz et al. 2011), which is potentially similar to compact object binary mergers.

A summary of the key properties of short GRB host galaxies is provided in **Table 2**. The data are primarily drawn from Berger (2009), Leibler & Berger (2010), Berger (2010), Fong et al. (2013) and references therein.

6.1. The Redshift Distribution

At the most basic level, host-galaxy associations are critical because essentially all short GRB redshifts to date (spectroscopic or photometric) have been obtained from the associated hosts (**Figure 3**; e.g., Fox et al. 2005; Berger et al. 2005b; Bloom et al. 2006; Berger 2009, 2011). The

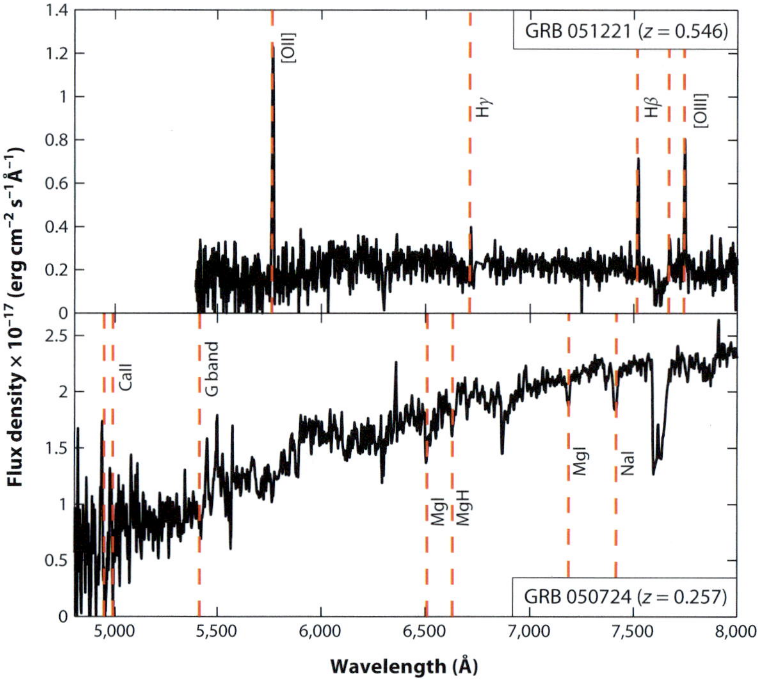

Figure 3

Representative optical spectra of a late-type short GRB host galaxy exhibiting emission lines typical of star-forming galaxies (*top*: GRB 051221 at $z = 0.546$) and an early-type short GRB host galaxy with no evidence for ongoing star-formation activity (*bottom*: GRB 050724 at $z = 0.257$). Primary absorption and emission lines seen in the spectra are marked. The spectra are from Soderberg et al. (2006a) and Berger et al. (2005b), respectively.

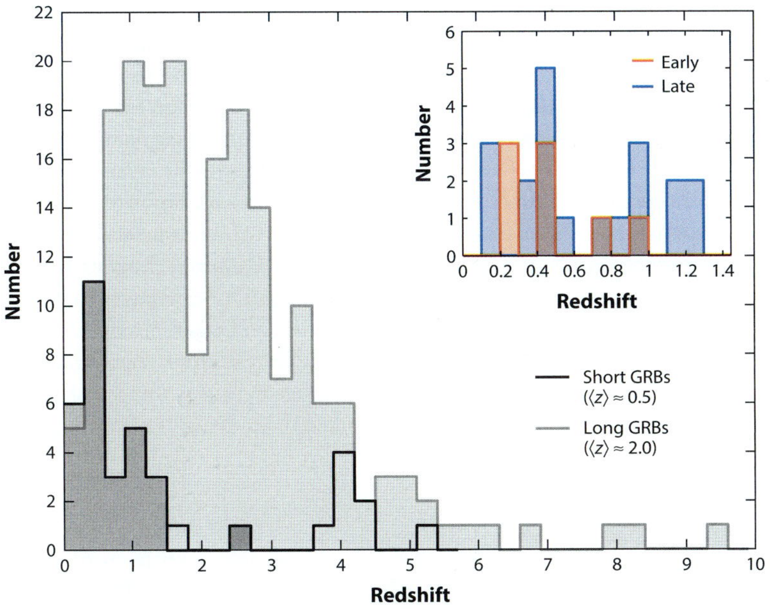

Figure 4

The redshift distribution of short GRBs (*black*) and long GRBs (*gray*). The open histogram marks redshift upper limits based on the lack of a Lyman break in afterglow and/or host-galaxy optical detections. The inset shows the redshift distribution of short GRBs separated by host-galaxy type, which exhibits no discernible difference between early-type (*red*) and late-type (*blue*) hosts.

sole exceptions are GRB 090426 at $z = 2.609$ (Antonelli et al. 2009, Levesque et al. 2010a) and GRB 130603B at $z = 0.356$ (Cucchiara et al. 2013, de Ugarte Postigo et al. 2014) for which redshifts have been determined from afterglow absorption spectra. The bulk of the measured redshifts span $z \approx 0.1$–1.3, but it is likely that at least some of the faintest host galaxies (with optical magnitudes of ≈24–27), which lack redshift measurements, are located at $z \gtrsim 1$ (Berger et al. 2007a). The redshift distributions of short and long GRBs are shown in **Figure 4**. The median redshift of the short GRB population with established redshifts is $\langle z \rangle \approx 0.48$, whereas the addition of faint hosts that are presumably located at $z \gtrsim 1$ increases the median to $\langle z \rangle \approx 0.63$; the use of redshift upper limits from afterglow and/or host-galaxy detections in optical bands (i.e., the lack of a Lyman break) leads to an upper bound on the median redshift of $\langle z \rangle \lesssim 0.83$. Because only 6 short bursts lack XRT positions despite rapid follow-up (Section 4), and hence the ability to identify hosts, the observed redshift distribution robustly represents the redshifts of *Swift* short GRBs. Assuming that the median redshift is $\langle z \rangle \sim 0.5$–0.8, and that it is not affected by the sensitivity threshold of *Swift*, the resulting median progenitor age relative to the peak of the cosmic star-formation history ($z \sim 3$) is $\tau \sim 5$ Gyr. Because the observed sample is almost certainly limited by the detector sensitivity, this inferred age can be used as an upper bound.

There is no clear trend between redshift and host-galaxy type; both early- and late-type hosts span the same redshift range with similar median values (**Figure 4**; Fong et al. 2013). Because the progenitors are expected to be systematically older in early-type galaxies and, hence, to occur at lower redshifts (e.g., Zheng & Ramirez-Ruiz 2007), the similar redshift distributions for early- and late-type galaxies suggest that the *Swift* sensitivity threshold indeed plays a role in the redshift distribution. The short GRB population is substantially more local than *Swift* long GRBs, which extend to $z \sim 9.4$ (Cucchiara et al. 2011) with a median value of $\langle z \rangle \approx 2$ (Berger et al. 2005a, Jakobsson et al. 2006). This is partly a reflection of the lower energy scale of short GRBs, but also a result of longer delay times between star-formation activity and the occurrence of short GRBs.

Several authors have attempted to constrain the progenitor age distribution and the short GRB luminosity function using the observed redshift and log$N$-log$S$ distributions (Guetta & Piran 2006, Nakar et al. 2006, Virgili et al. 2011). They generally find that a delay relative to the cosmic star-formation history is required to explain the observed distributions. However, Virgili et al. (2011) point out that to reproduce both the log$N$-log$S$ and the two-dimensional luminosity-redshift distribution with a simple luminosity function requires either a mix of progenitors with negligible and long delays or a narrow delay time distribution centered at ~2 Gyr. Such distributions may be at odds with the results of population synthesis models and the Galactic NS-NS population, although those are also uncertain. I also note that while the *Swift* detector sensitivity affects the redshift distribution, it is not the only effect that plays a role in the resulting observed distribution. This may in turn have an effect on parameters extracted from the redshift distribution.

6.2. Demographics

An additional constraint on the progenitors and their age distribution is provided by the demographics of the host-galaxy sample. Although long GRBs only occur in star-forming galaxies (Bloom, Kulkarni & Djorgovski 2002; Fruchter et al. 2006; Wainwright, Berger & Penprase 2007), as expected for young massive star progenitors, short GRBs occur in a mixed population of early-type and star-forming galaxies (Berger et al. 2005b, Fox et al. 2005, Bloom et al. 2006, Fong et al. 2011, Fong et al. 2013), indicating that their progenitors span a wide range of ages. More broadly, because a roughly equal fraction of the cosmic stellar mass density is in early-type and star-forming galaxies at $z \sim 0$–1 (Bell et al. 2003, Ilbert et al. 2010), which is the redshift range relevant for the existing short GRB population, an exclusively old progenitor population (i.e., tracking stellar mass alone) will also exhibit an equal fraction of early- and late-type hosts. However, a progenitor population skewed to relatively short delay times relative to star-formation activity (tens to hundreds of millions of years) will exhibit an overabundance of late-type hosts owing to their recent star formation (Zheng & Ramirez-Ruiz 2007; O'Shaughnessy, Belczynski & Kalogera 2008). Studies of the first few short GRB host galaxies from 2005 led several groups to conclude that early-type galaxies dominate the sample at a ratio of about 2:1 and that the progenitors are therefore exceedingly old, having a characteristic age of ~5–10 Gyr (Prochaska et al. 2006, Nakar 2007, Gal-Yam et al. 2008).

A different conclusion is reached by Fong et al. (2013) in a study of the short GRB host-galaxy demographics using a much larger sample of 36 events. These authors find that regardless of various cuts on the sample, only ~20% of the hosts are early-type galaxies (**Figure 5**; see also Leibler & Berger 2010). This result is robust when considering only events with subarcsecond positions, when including those with XRT positions, or when restricting the sample according to the duration-hardness analysis of Bromberg et al. (2013). The dominance of late-type galaxies indicates that the short GRB rate does not depend on stellar mass alone and is instead influenced by recent star-formation activity. Comparing the observed relative fraction of early- and late-type hosts to the theoretical predictions of Zheng & Ramirez-Ruiz (2007) leads to a power-law delay time distribution of $P(\tau) \propto \tau^{-1}$.

An additional clue to the role of star formation in the short GRB rate is the identification of two short GRB hosts as ultraluminous and luminous infrared galaxies (U/LIRGs). GRB 100206A is associated with a dusty LIRG at $z = 0.407$ (Perley et al. 2012), whereas GRB 120804A is associated with a dusty ULIRG at a photometric redshift of $z \approx 1.3$ (Berger et al. 2013). The

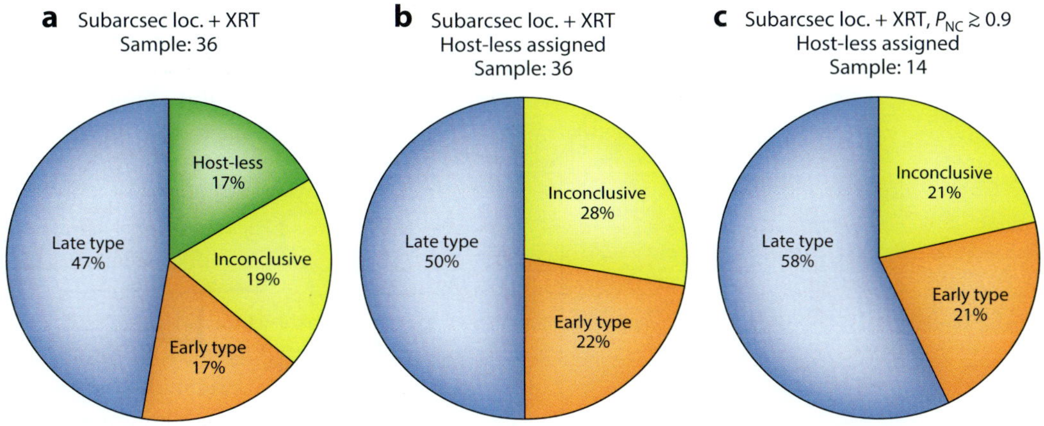

Figure 5

Demographics of the galaxies hosting short GRBs. (*a*) A breakdown into late type (*blue*), early type (*orange*), host-less (*green*), and inconclusive (*yellow*) for all identified hosts based on subarcsecond positions and *Swift*/XRT positions (**Table 2**). (*b*) Same as panel *a*, but with the host-less events assigned to the other categories based on the galaxies with the lowest probability of chance coincidence in each case (Berger 2010, Fong & Berger 2013). (*c*) Same as panel *b*, but for short GRBs with a probability of a noncollapsar origin of $P_{\rm NC} \gtrsim 0.9$ on the basis of analysis by Bromberg et al. (2013). Regardless of the sample selection, late-type galaxies dominate the host sample. This indicates that star-formation activity plays a role in the short GRB rate. Adapted from Fong et al. (2013).

expected U/LIRG fraction for a progenitor population that tracks stellar mass alone is $\sim$1% (Caputi et al. 2006), whereas for progenitors that track only star formation it is $\sim$25% (Le Floc'h et al. 2005, Caputi et al. 2007). The observed fraction in the short GRB sample is $\sim$5–10% percent, suggesting that the progenitor population is influenced by both stellar mass and star-formation activity and indicating a broad age distribution.

6.3. Stellar Masses and Stellar Population Ages

The distribution of host-galaxy stellar masses and stellar population ages can also shed light on the progenitor age distribution. The stellar masses inferred from modeling of the host optical/near-IR spectral energy distributions with single stellar population models span $M_* \approx 10^{8.5-11.8}\,{\rm M}_\odot$. The median for the full sample is $\langle M_* \rangle \approx 10^{10.0}\,{\rm M}_\odot$, whereas for the star-forming hosts alone it is $\langle M_* \rangle \approx 10^{9.7}\,{\rm M}_\odot$ (**Figure 6**; Leibler & Berger 2010). The stellar masses of long GRB hosts are substantially lower, having a median value of about $10^{9.2}\,{\rm M}_\odot$ (Savaglio, Glazebrook & Le Borgne 2009; Leibler & Berger 2010). This indicates that even the star-forming hosts of short GRBs are typically more massive than the hosts of long GRBs and points to stellar mass playing a more dominant role in determining the rate of short GRBs.

A comparison with the mass function of field galaxies is even more illuminating. In **Figure 6**, I compare the cumulative distributions of stellar masses for the early- and late-type hosts of short GRBs with the expected distributions for mass selection from the field galaxy mass function (Bell et al. 2003, Ilbert et al. 2010). For a progenitor population that tracks stellar mass alone, we expect that the observed stellar mass distribution of short GRB hosts will closely track the mass-weighted mass distribution of field galaxies. Although this is indeed the case for the early-type mass function, the late-type hosts of short GRBs have systematically lower stellar masses than expected from mass selection alone (Leibler & Berger 2010). This indicates that in late-type galaxies the short GRB rate per unit stellar mass is higher than in early-type galaxies, owing to the presence

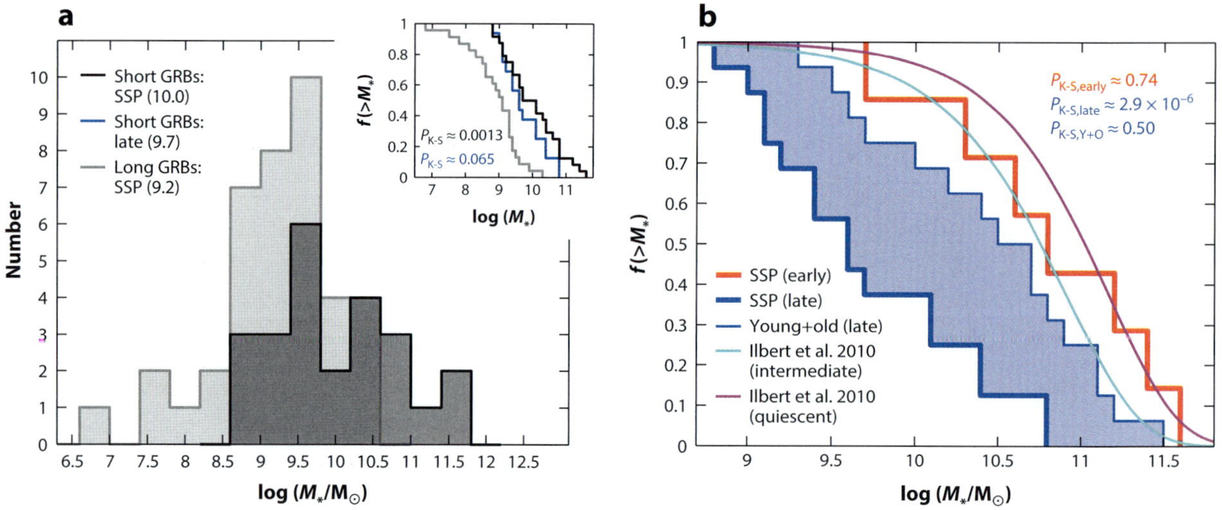

Figure 6

(*a*) Histogram of host-galaxy stellar masses for short GRBs (*black with dark gray shading*) and long GRBs (*gray with light gray shading*). Median values for each population (and separately for short GRB late-type hosts) are quoted in parentheses. The inset shows the cumulative distributions along with Kolmogorov-Smirnov (K-S) probabilities that the short and long GRB hosts are drawn from the same parent population. (*b*) Comparison of the cumulative distributions of stellar masses for late-type (*blue*) and early-type (*red*) short GRB hosts to the expected distributions for a mass-weighted selection from the field galaxy mass function (*cyan* and *magenta lines*, respectively). The resulting K-S probabilities indicate that the early-type hosts are consistent with pure mass selection, whereas the late-type hosts have lower than expected stellar masses. This indicates that star-formation activity plays a role in the short GRB rate. Adapted from Leibler & Berger (2010). Abbreviation: SSP, single stellar population.

of star-formation activity. This agrees with the observed overabundance of late-type galaxies in the short GRB host population. Quantitatively, for a range of host stellar mass determinations, the short GRB rate per unit stellar mass in late-type galaxies is about 2–9 times higher than in early-type galaxies (Leibler & Berger 2010).

The difference between short and long GRB hosts is also evident in the distribution of stellar population ages. The distribution for short GRB hosts spans a wide range, from a few tens of millions of years to about 4 Gyr (with a median value of $\langle\tau_*\rangle \approx 0.25$ Gyr), whereas the long GRB hosts have a median stellar population age of $\langle\tau_*\rangle \approx 60$ Myr (**Figure 7**; Leibler & Berger 2010). Coupled with the stellar mass distribution discussed above, this distribution indicates that the short GRB rate per $10^{10}\,M_\odot$ in late-type hosts (with $\langle\tau_*\rangle \approx 0.15$ Gyr) and early-type hosts (with $\langle\tau_*\rangle \approx 1.6$ Gyr) is 0.7 and 0.075, respectively, leading to a power-law delay time distribution of $P(\tau) \propto \tau^{-0.9}$, which is in good agreement with the distribution inferred from the early- to late-type fraction. Similarly, separating the sample into galaxies with stellar population ages younger and older than ~0.3 Gyr, I find that the typical ages are about 0.1 and 1 Gyr, respectively, with a resulting rate per $10^{10}\,M_\odot$ of 0.8 and 0.12, respectively, which leads to a delay time distribution of $P(\tau) \propto \tau^{-0.8}$. The inferred delay time distribution is in good agreement with that of Galactic neutron star binaries with a power-law index of $n \sim -1$ (e.g., Piran 1992).

6.4. Specific Star-Formation Rates

The interplay of stellar mass and star-formation activity in short GRB hosts is also evident in the distribution of SSFRs, relative to both long GRB hosts and field star-forming galaxies. In **Figure 8**,

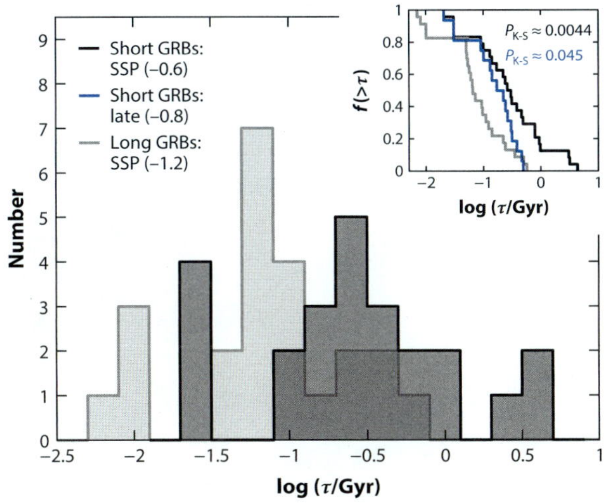

Figure 7

Histogram of host-galaxy stellar population ages for short GRBs (*black with dark gray shading*) and long GRBs (*gray with light gray shading*). Median values for each population (and, separately, for short GRB late-type hosts) are quoted in parentheses. The inset shows the cumulative distributions along with Kolmogorov-Smirnov (K-S) probabilities that the short and long GRB hosts are drawn from the same parent population. The results indicate that short GRB hosts, even the late-type galaxies, have a systematically older stellar population than long GRB hosts. Adapted from Leibler & Berger (2010). Abbreviation: SSP, single stellar population.

I plot the SFRs as a function of rest-frame optical B-band luminosity (L_B) for short and long GRB hosts. For the short GRB hosts the luminosities span $L_B \approx 0.1$–$5L^*_B$, and the SFRs span $\lesssim 0.1$ to $\approx 5\ M_\odot$ year^{-1}. The upper limits correspond to the early-type hosts. Here L^*_B is the characteristic luminosity in the Schechter function. However, though both short and long GRB hosts exhibit a clear trend between SFR and L_B (at least for the star-forming short GRB hosts), the overall normalization (i.e., the SSFR) is lower for the short GRB hosts by about a factor of 5 with $\langle \mathrm{SFR}/L_B \rangle \approx 2\ M_\odot$ year$^{-1} L^*_B$. The K-S test gives a p-value of 0.003 for the null hypothesis that the short and long GRB hosts are drawn from a galaxy population with the same underlying distribution of SSFR.

Because the SSFR is the inverse of the characteristic timescale to build up a galaxy's stellar mass, the lower values for short GRB hosts indicates that while the majority are star-forming galaxies, their star-formation activity is more moderate and has a longer characteristic timescale than in long GRB hosts. Thus, the massive star progenitors of long GRBs track recent star formation (tens of millions of years), whereas the progenitors of short GRBs track star formation with a delay of hundreds of millions of years to several gigayears. Moreover, the short GRB hosts clearly track the distribution of SSFRs in field galaxies of similar luminosities (Kobulnicky & Kewley 2004) with a K-S p-value of 0.8, indicating that short GRBs select galaxies from the general field sample.

6.5. Metallicities

Short GRB hosts span a wide range of metallicities, with $12 + \log(\mathrm{O/H}) \approx 8.5$–$9.2$ and a median value of $\langle 12+\log(\mathrm{O/H}) \rangle \approx 8.8 \approx 1\ Z_\odot$ (**Figure 9**; Berger 2009, D'Avanzo et al. 2009). In addition, the measured metallicities track a positive trend with host-galaxy luminosity, as generally observed for field galaxies (Tremonti et al. 2004). The metallicities of short GRB hosts are significantly higher than those of long GRB hosts, for which the median value is $\langle 12 + \log(\mathrm{O/H}) \rangle \approx 8.3$

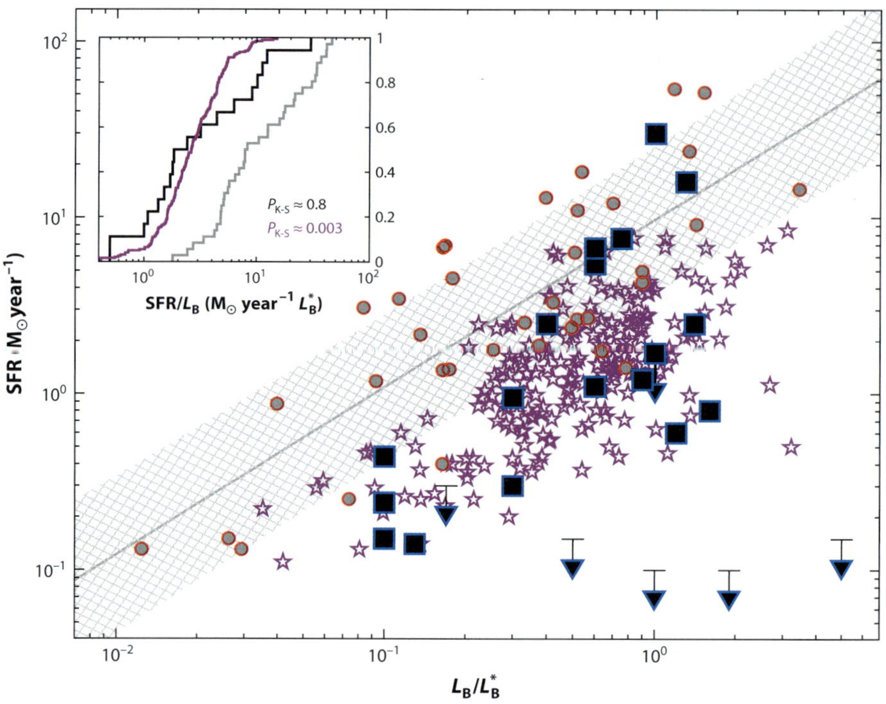

Figure 8

Star-formation rate (SFR) as a function of rest-frame B-band luminosity for the host galaxies of short GRBs (*squares*), long GRBs (*circles*), and field star-forming galaxies at similar redshifts to short GRB hosts (*stars*; Kobulnicky & Kewley 2004). The gray line (and *hatched region*) delineates the correlation (and standard deviation) for long GRB host galaxies. Short GRB host galaxies have substantially lower SFRs as a function of luminosity than long GRBs hosts (i.e., they have lower specific SFRs), but they closely track the field galaxy population (*inset*). Abbreviation: K-S, Kolmogorov-Smirnov.

(Stanek et al. 2006, Modjaz et al. 2008, Levesque et al. 2010b). By contrast, short GRB hosts track the metallicity distribution of field star-forming galaxies at similar redshifts and with similar luminosities (Kobulnicky & Kewley 2004). Combining the host stellar masses and SFRs according to the fundamental metallicity relation of Mannucci et al. (2010), $\mu = \log(M_*) - 0.32\ \log(\mathrm{SFR})$, I find that short GRB hosts track the relation for field galaxies (**Figure 9**). Thus, though the sample of short GRB hosts with metallicities is still small, it demonstrates that the hosts follow the relations for field galaxies, but are clearly distinct from long GRB hosts. Based on this result, it does not appear that short GRB progenitors are directly affected by metallicity.

6.6. Galaxy Clusters

The first short GRB with an afterglow detection, GRB 050509B, was most likely associated with a galaxy cluster at $z = 0.225$ (Gehrels et al. 2005, Pedersen et al. 2005, Bloom et al. 2006, Dahle et al. 2013). This association led to several investigations of galaxy clusters hosting short GRBs. Ghirlanda et al. (2006) cross-correlated BATSE short GRBs with catalogs of X-ray selected clusters to $z \sim 0.45$, claiming an association at a 2σ confidence level on an angular scale of $\lesssim 3°$. Berger et al. (2007b) searched for galaxy clusters in association with 4 *Swift* short GRBs using multiobject optical spectroscopy, as well as in association with 15 *Swift* short GRBs using stacked

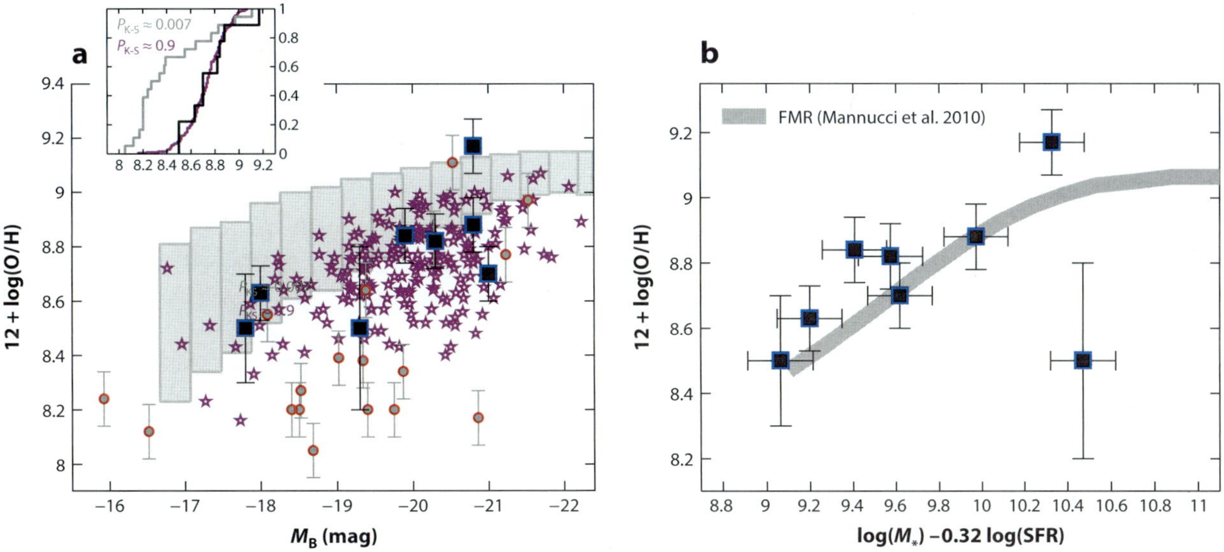

Figure 9

(*a*) Metallicity as a function of host-galaxy rest-frame B-band luminosity for short GRBs (*squares*), long GRBs (*circles*), field galaxies at similar redshifts to short GRB hosts (*stars*; Kobulnicky & Kewley 2004), and the Sloan Digital Sky Survey luminosity-metallicity relation (Tremonti et al. 2004). Short GRB host galaxies have higher metallicities than long GRB hosts, but they closely track the luminosity-metallicity relation for the field galaxy population (*inset*). Adapted from Berger (2009). (*b*) Comparison of short GRB hosts to the fundamental metallicity relation of star-forming galaxies (*gray line*; Mannucci et al. 2010), which uses a combination of stellar mass and star-formation rate (SFR), $\log(M_*) - 0.32 \log(\mathrm{SFR})$. The short GRB hosts follow the relation along a wide span of the combined stellar mass and star-formation parameter. Abbreviations: FMR, fundamental metallicity relation; K-S, Kolmogorov-Smirnov.

X-ray observations from *Swift*/XRT. The search yielded a galaxy cluster at $z = 0.165$, on the basis of redshift clustering and diffuse X-ray emission, in the field of the short GRB 050911 (BAT-only position) with a probability of chance coincidence of 0.1–1%. The limit on galaxy clusters in the fields of the remaining bursts is $M \lesssim 5 \times 10^{13}\,\mathrm{M}_\odot$ based on the lack of diffuse X-ray emission (Berger et al. 2007b).

Because the stellar populations in galaxy clusters are systematically older than in field galaxies, the fraction of short GRBs in clusters can be used as an indicator of the age distribution (Shin & Berger 2007). At present, the fraction of short GRBs in clusters appears to be ~10% (Berger et al. 2007b), which is at the low end of the fraction of cosmic stellar mass in clusters of about 10–20% (Fukugita, Hogan & Peebles 1998; Eke et al. 2005). There is therefore no indication from galaxy cluster associations for an exceedingly old progenitor population.

6.7. Comparison with the Host Galaxies and Delay Time Distribution of Type Ia Supernovae

Type Ia SNe result from the thermonuclear explosions of CO white dwarfs that approach the Chandrasekhar mass in single- or double-degenerate binary systems. Unlike core-collapse SNe, Type Ia events are found in both early- and late-type galaxies (e.g., van den Bergh 1990, van den Bergh & Tammann 1991, Li et al. 2011), reflecting the wide range of timescales for the accretion or merger process that eventually leads to an SN. As a result, the environments and delay time distribution of Type Ia SNe provide a useful comparison to short GRBs, particularly in the context of compact object merger progenitors. Mannucci et al. (2005) found that for nearby

Type Ia SNe about 30% occur in E/S0 galaxies, whereas the remaining events occur in spirals (63%) and irregular galaxies (7%). This distribution is similar to the short GRB demographics (**Figure 5**; Fong et al. 2013). The Type Ia SNe also exhibit a clear increase in the rate per unit stellar mass from early- to late-type galaxies, ranging from about 0.044 (E/S0) to 0.065 (Sa/b) to 0.17 (Sc/d) per $10^{10}\,M_\odot$ per century (Mannucci et al. 2005), similar to the inference for short GRBs (Section 6.3; Leibler & Berger 2010).

The Type Ia SN rate as a function of galaxy type and properties (stellar mass, SFR) led Mannucci, Della Valle & Panagia (2006) and Sullivan et al. (2006) to propose a bimodal or two-component delay time distribution with a prompt population (~0.1–0.3 Gyr) whose rate depends primarily on the host-galaxy SFR and a delayed population (~3 Gyr with a broad range) whose rate depends on the host-galaxy stellar mass (see also Maoz & Badenes 2010, Maoz et al. 2011). Indeed, Sullivan et al. (2006) find that the Type Ia SN rate per unit stellar mass increases by at least an order of magnitude with higher SSFR (over the range of 0.02–1 Gyr^{-1}). They conclude that the Type Ia SN rate per year is about $5 \times 10^{-4}\,(10^{10}\,M_\odot)^{-1} + 4 \times 10^{-4}\,(M_\odot/\text{year})^{-1}$. More recent work on the delay time distribution points instead to a continuous (rather than bimodal) distribution, following a power law of about $\tau^{-1.1}$ (Maoz, Mannucci & Brandt 2012). Thus, the results for short GRBs and Type Ia SNe exhibit close similarities in terms of both the distribution of host-galaxy types and the enhanced rate in late-type galaxies, with typical delays of ~0.1–0.3 Gyr. This indicates that short GRBs and Type Ia SN progenitors share a similar delay time distribution.

7. THE LOCATIONS OF SHORT GAMMA-RAY BURSTS IN AND AROUND THEIR HOST GALAXIES

As demonstrated in the previous section, the properties of short GRB hosts help to reveal the short GRB redshift distribution, shed light on the progenitor age distribution, and establish the differences between long and short GRB progenitors. Studies of short GRB subgalactic environments can further this insight by providing information on their spatial association (or lack thereof) with star formation or stellar mass. Similar studies of long GRBs have established that their projected spatial offsets relative to the host-galaxy centers follow the expected distribution for star formation in exponential disk galaxies, having a median value of $\approx 1\,r_e$ (Bloom, Kulkarni & Djorgovski 2002), where r_e is the host-galaxy half-light radius. Moreover, studies of the rest-frame UV brightness at the locations of long GRBs relative to the overall UV light distribution of the hosts have shown that long GRBs tend to occur in unusually bright star-forming regions, significantly more so than core-collapse SNe (Fruchter et al. 2006, Svensson et al. 2010).

What is the expectation for the locations of short GRBs? As I already demonstrated from the lack of SN associations, the occurrence of some short GRBs in elliptical galaxies, and the systematically older stellar population of short GRB hosts compared with long GRB hosts, short GRBs are not expected to coincide with young star-forming environments. By contrast, on the basis of the distribution of host-galaxy stellar masses and the host-galaxy demographics, I concluded that short GRB progenitors have a broad delay time distribution and that their rate is influenced in part by star-formation activity (i.e., some systems have delays of only ~0.1 Gyr). In addition, one of the key predictions of the compact object merger model is systemic natal kicks that may lead to substantial offsets between the birth and explosion sites of these systems. Indeed, offset distributions predicted from population synthesis models indicate a typical median projected offset of 5–7 kpc, with about 10–25% of events at offsets of $\gtrsim$20 kpc and up to ~10% beyond 100 kpc (Bloom, Sigurdsson & Pols 1999; Fryer, Woosley & Hartmann 1999; Perna & Belczynski 2002; Belczynski et al. 2006). Observationally, short GRBs at such large offsets will appear to be host-less because their projected locations will extend much beyond the visible extent of typical galaxies. A progenitor population of young magnetars, however, will exhibit a spatial coincidence with star-forming regions and offsets that track an exponential disk distribution. Delayed magnetars will not track young star-forming regions but will track the overall light distribution of their hosts owing to the lack of natal kicks. Finally, a dominant population of dynamically formed NS-NS binaries in globular clusters will track the overall globular cluster distribution around galaxies, extending to tens of kiloparsecs (Salvaterra et al. 2010).

7.1. The Offset Distribution

Determining the locations of short GRBs relative to their host centers (offsets) and relative to the underlying light distribution in the rest-frame optical (stellar mass) and UV (star formation) bands requires the high angular resolution and superior depth of the HST. A comprehensive study based on HST observations of 32 short GRB host galaxies was carried out by Fong, Berger & Fox (2010) and Fong & Berger (2013; see also Church et al. 2011). The HST data, combined with ground-based optical afterglow observations (and, in two cases, *Chandra* observations), provide accurate offsets at the subpixel level and reveal the broad distribution shown in **Figure 10**. The projected offsets span 0.5–75 kpc with a median of about 5 kpc. These are about four times greater than the median offset for long GRBs (Bloom, Kulkarni & Djorgovski 2002) and about 1.5 times greater than the median offsets of core-collapse and Type Ia SNe (Prieto, Stanek & Beacom 2008). In addition, though no long GRBs and only ~10% of SNe have offsets of $\gtrsim$10 kpc, the fraction of short GRBs with such offsets is about 25%. For offsets of $\gtrsim$20 kpc, the fraction of short GRBs is about 10%, but essentially no SNe exhibit such large offsets.

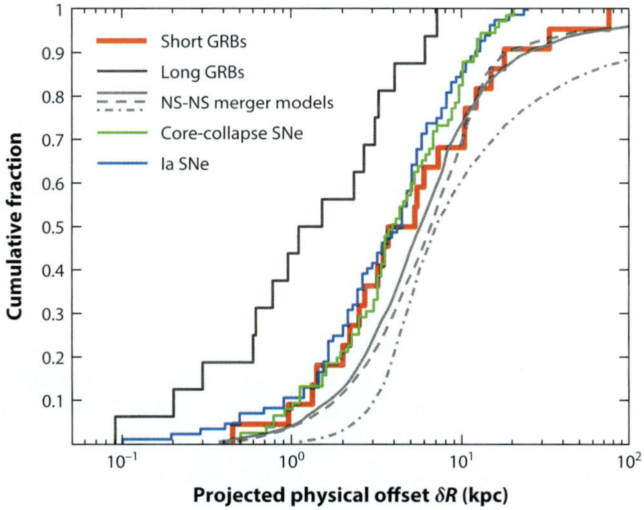

Figure 10

Cumulative distribution of projected physical offsets for short GRBs with subarcsecond positions (*red*) (Fong, Berger & Fox 2010; Fong & Berger 2013), compared with the distributions for long GRBs (*black*; Bloom, Kulkarni & Djorgovski 2002), core-collapse supernovae (SNe) (*green*; Prieto, Stanek & Beacom 2008), Type Ia SNe (*blue*; Prieto, Stanek & Beacom 2008), and predicted offsets for neutron star (NS) binaries from population synthesis models (*gray*) (Bloom, Sigurdsson & Pols 1999; Fryer, Woosley & Hartmann 1999; Belczynski et al. 2006). Short GRBs have substantially larger offsets than long GRBs and match the predictions for compact object binary mergers. Reprinted from Fong & Berger (2013) with permission.

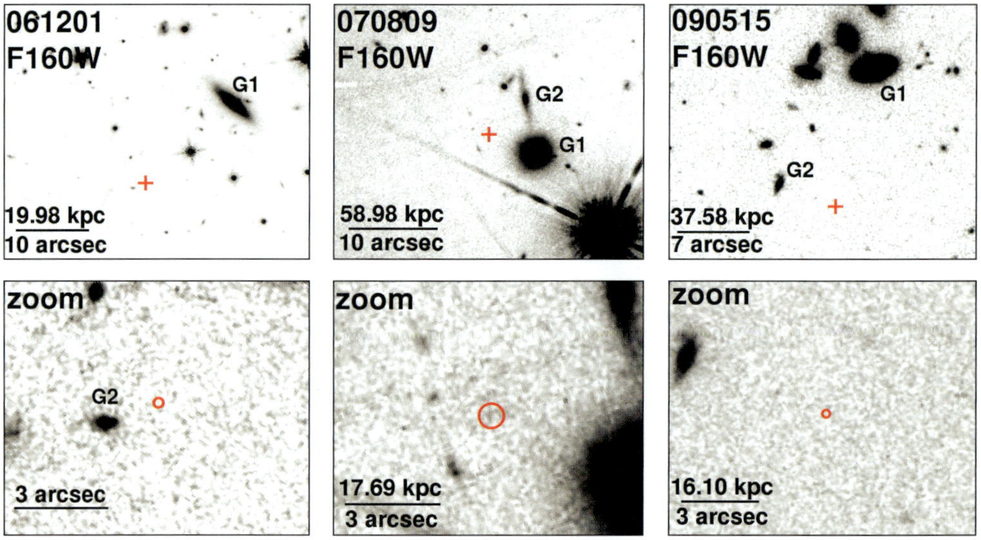

Figure 11

Hubble Space Telescope Wide-Field Camera 3 images of the locations of three short GRBs with subarcsecond positions and no coincident host galaxies (host-less bursts). In each case the top panel shows a wide field, whereas the bottom panel is an enlargement focusing on the GRB location (*red circle*). The galaxies marked "G1" and "G2" represent the objects with the lowest and second lowest probabilities of chance coincidence in each field, respectively. The magnitude limits at the GRB positions are $m_{\mathrm{F160W}} \gtrsim 26$ mag, ruling out the presence of galaxies typical of short GRB hosts at $z \lesssim 3$. Adapted from Fong & Berger (2013).

Equally importantly, the observed offset distribution is in remarkable agreement with the population synthesis predictions for compact object mergers, particularly the fraction of events with large offsets. I note that the observed distribution is mainly based on short GRBs with optical afterglows, which may be skewed on average to higher circumburst densities, and hence to smaller offsets. It is therefore conceivable that events with only XRT or BAT positions (14 and 6 *Swift* short GRBs, respectively; Section 4), whose precise offsets cannot be ascertained, in reality have systematically larger offsets than a few kiloparsecs.

The subset of short GRBs with offsets of $\gtrsim$20 kpc require specific mention because these offsets are larger than the typical visible extent of galaxies. These bursts lack coincident host galaxies at angular separations of $\lesssim$1 arcsec in deep optical and near-IR HST observations and have been termed host-less. Systematic studies of the host-less bursts, their afterglows, and environments were carried out by Berger (2010) and Fong & Berger (2013), although individual cases were noted previously (Bloom et al. 2007, Stratta et al. 2007, Perley et al. 2009). A few examples based on deep HST near-IR observations are shown in **Figure 11**. In all cases, there is a clear absence of coincident galaxies to optical limits of $\gtrsim$27 mag and near-IR limits of $\gtrsim$26 mag (Berger 2010, Fong & Berger 2013). Any undetected coincident host galaxies with properties similar to the known host-galaxy sample (**Figure 8**) will have to reside at $z \gtrsim 3$ to evade detection, leading to an unexpected bimodal redshift distribution (**Figure 4**; Berger 2010, Fong & Berger 2013). Similarly, if we impose a similar redshift range for any undetected hosts, then the resulting host luminosities will be $\lesssim 0.01\ L^*$, at least an order of magnitude below the known host population (**Figure 8**), making this explanation equally unlikely.

A more likely alternative is that these bursts are associated with nearby galaxies in their field; their resulting projected offsets of tens of kiloparsecs are as expected for a subset of compact object mergers. This possibility can be assessed by analyzing the probabilities of chance coincidence for field galaxies in the HST images. The probability for a given galaxy of brightness m to be located at a separation δR from a short GRB position is given by (Bloom, Kulkarni & Djorgovski 2002; Berger 2010)

$$P_{cc} = 1 - e^{-\pi(\delta R)^2 \Sigma(\leq m)}, \tag{1}$$

where the galaxy number counts are given by $\Sigma(\leq m) = 1.3 \times 10^{0.33(m-24)-2.44}$ arcsec^{-2} (Hogg et al. 1997, Beckwith et al. 2006). Applying this approach, Berger (2010) and Fong & Berger (2013) find that the host-less short GRBs exhibit nearby field galaxies with low chance-coincidence probabilities of $P_{cc} \approx$ few percent, indicating a likely association with resulting projected separations of $\sim$10 arcsec. A similar analysis for short GRBs with coincident host galaxies does not reveal a similar effect (Berger 2010), demonstrating that the offset galaxies with a low probability of chance coincidence in the fields of host-less short GRBs are indeed the hosts. Although a definitive association will benefit from independent redshift measurements for the afterglow and galaxy, the observed excess of galaxies at moderate separations is highly indicative of a progenitor population capable of occurring at large offsets. The redshifts and properties of these likely host galaxies reveal that they are similar to the known population of short GRB hosts, and the resulting projected physical offsets are tens of kiloparsecs (Berger 2010).

Although the projected physical offsets are already indicative of compact object mergers, it is important to account for the range of host-galaxy sizes and any systematic trends in these sizes between the various GRB and SN populations. This can be accomplished by normalizing the projected offsets in units of host-galaxy effective radii. Short GRB hosts tend to be larger than long GRB hosts, commensurate with their larger luminosities and stellar masses (Fong, Berger & Fox 2010; Fong & Berger 2013). The cumulative distributions of host-normalized offsets for short GRBs, long GRBs, core-collapse SNe, and Type Ia SNe are shown in **Figure 12**. The median host-normalized offsets for long GRBs and for both Type Ia and core-collapse SNe is $\delta R/r_e \approx 1$ (Fong & Berger 2013), as expected from the definition of the half-light radius and the fact that long GRB and SN progenitors do not migrate from their birth sites. Short GRBs, however, have a median offset of $\delta R/r_e \approx 1.5$; 20% of the population occur at $\delta R/r_e \gtrsim 5$ (compared with only a few percent of stellar light), and only 20% of the bursts are located at $\delta R/r_e \lesssim 1$. A K-S test relative to the Type Ia SN population yields a p-value of about 10^{-3} for the null hypothesis that both populations are drawn from the same underlying distribution of host-normalized offset (Fong & Berger 2013). Thus, the offset distribution of short GRBs indicates that they are located at larger distances than expected for a progenitor population that radially tracks stellar light, particularly when accounting for their host-galaxy sizes. The offsets are larger than even those of Type Ia SNe, but are in good agreement with compact object population synthesis predictions that include kicks.

7.2. Relation to the Underlying Ultraviolet and Optical Light Distribution

Independent of the offset distribution, an additional test of the progenitor population is the spatial correlation (or lack thereof) between short GRBs and the underlying light distribution of their hosts. In particular, for progenitors that track star-formation activity with a short delay (i.e., massive stars) we expect direct spatial correlation with the underlying rest-frame UV light (as found for long GRBs and core-collapse SNe; Fruchter et al. 2006; Kelly, Kirshner & Pahre 2008; Svensson et al. 2010; Kelly & Kirshner 2012). By contrast, an old progenitor population without a kick mechanism will spatially track rest-frame optical light (as is the case for Type Ia SNe; Wang

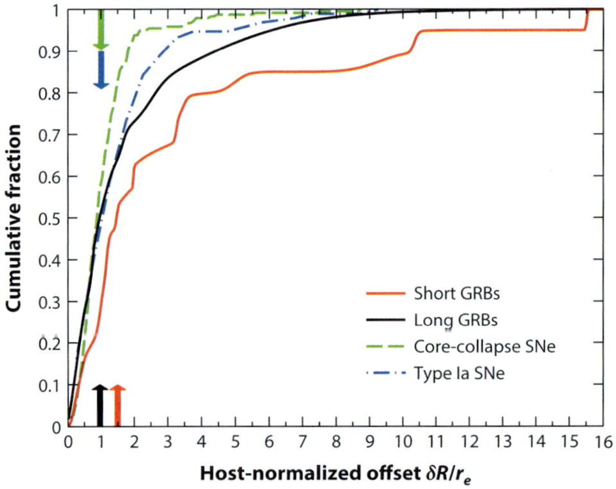

Figure 12

Cumulative host-normalized offset distributions for short GRBs, long GRBs, core-collapse supernovae (SNe), and Type Ia SNe (colors are as in **Figure 10**). The host-normalized offsets are determined relative to the effective radius of each host, and the cumulative distributions take into account the uncertainty in each offset measurement. The distributions for core-collapse SNe are from Kelly & Kirshner (2012), whereas those for Type Ia SNe are from Galbany et al. (2012). The arrows mark the median host-normalized offset for each type of transient; long GRBs and SNe match the expected value of about 1 r_e. Short GRBs, however, have a median offset of about 1.5 r_e, and only 20% of events occur within 1 r_e. Reprinted from Fong & Berger (2013) with permission.

et al. 2013). Finally, compact object mergers with significant kicks will result in a distribution that is only weakly correlated with the underlying rest-frame optical light.

Fong, Berger & Fox (2010) and Fong & Berger (2013) determined the light fraction at the subpixel locations of 20 short GRBs relative to the overall light distribution of each host galaxy [following the methodology of Fruchter et al. (2006)]. These measurements are available only for bursts with subarcsecond positions and HST follow-up, but as I discussed above, the requirement of an optical afterglow detection might only bias the resulting distribution to locations with higher ambient densities, and hence to brighter regions within the hosts. The cumulative distributions of the light fraction at short GRB locations relative to the underlying rest-frame UV and optical light are shown in **Figure 13**. The results are striking; specifically, about half of the short GRBs are located in regions of their hosts that are fainter in both the UV and optical bands than any other location within these galaxies (i.e., they have a fractional flux value of zero). This essentially means that even with the superior depth of HST, no stellar light is detected at the short GRB locations. This result remains true even if the short GRBs at projected offsets of tens of kiloparsecs (the host-less events) are ignored (Fong & Berger 2013).

A K-S test relative to the null hypothesis that the progenitors track the underlying UV or optical light results in p-values of 0.01 and 0.04, respectively (Fong & Berger 2013). Similarly, a comparison to long GRBs in terms of UV light leads to $p \approx 2 \times 10^{-3}$, whereas a comparison with Type Ia SNe in terms of optical light indicates $p \approx 0.02$. This indicates that short GRBs occur in distinct environments from other explosive transients whose progenitors are not expected to experience kicks (massive stars, single- or double-degenerate white dwarf systems) and are overall weakly correlated with the underlying distribution of star formation and even stellar mass in their hosts (Fong & Berger 2013). Given this clear discrepancy, the implication is that short GRB

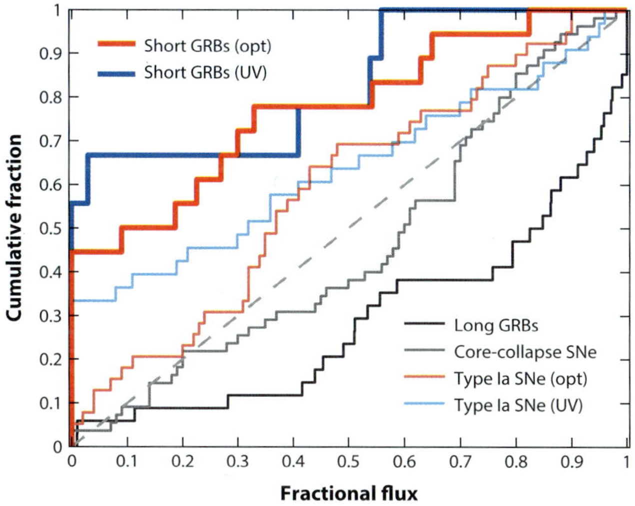

Figure 13

Cumulative distribution of the fractional flux at short GRB locations relative to the underlying light distributions of their hosts. Shown are the distributions for rest-frame optical (*red*) and rest-frame UV (*blue*) from *Hubble Space Telescope* observations of short GRBs with subpixel positions (Fong, Berger & Fox 2010; Fong & Berger 2013). Also shown are the distributions for long GRBs (UV, *black*) (Fruchter et al. 2006, Svensson et al. 2010), Type Ia supernovae (SNe) (UV, *light blue*; optical, *pink*) (Wang et al. 2013), and core-collapse SNe (UV, *gray*) (Svensson et al. 2010). The diagonal dashed line marks the expected distribution for a population that linearly tracks the underlying stellar light. Long GRBs and core-collapse SNe strongly track UV light (star formation), whereas Type Ia SNe track optical light (stellar mass). Short GRBs, by contrast, exhibit a poor correlation with UV or optical light, indicating that the progenitors migrate from their birth sites to the eventual explosion sites. Reprinted from Fong & Berger (2013) with permission.

explosion sites are not representative of the progenitor birth sites, and therefore require significant migration (kicks). This is a strong line of argument in favor of compact object mergers.

7.3. Kick Velocities

The offset distribution and the locations of short GRBs relative to the stellar light of their hosts are indicative of systemic kicks. The required kick velocities can be inferred from a comparison of the projected physical offsets and the host-galaxy stellar population ages, taking into account the host masses. Although projection effects and uncertainty about the age of any specific short GRB progenitor relative to the host mean stellar population age prevent the extraction of a detailed distribution, it is still possible to determine characteristic kick velocities. Making the assumption (Bloom et al. 2007) that the required kick velocities are roughly the geometric mean between the minimum required velocity ($v_{\rm kick,min} = \delta R/\tau_*$) and the host velocity dispersion ($v_{\rm disp} \approx 120$ km s^{-1} for late-type hosts and ≈ 250 km s^{-1} for the more massive early-type hosts; Battaglia et al. 2005, Xue et al. 2008), the resulting projected kick velocities span $v_{\rm kick} \sim 20$–140 km s^{-1} with a median value of about 60 km s^{-1} (Fong & Berger 2013). This range is consistent with the kick velocities derived for Galactic NS-NS binaries based on population synthesis models, $v_{\rm kick} \approx 5$–500 km s^{-1} (Fryer & Kalogera 1997; Fryer, Burrows & Benz 1998; Wang, Lai & Han 2006; Wong, Willems & Kalogera 2010). It is important to note that the inferred distribution of kicks for short GRBs is consistent with some progenitors having negligible kicks, and it does not strictly require velocities that will unbind the progenitors from their hosts.

7.4. Globular Clusters

An alternative explanation for the broad offset distribution and weak correlation with stellar light is that some short GRB progenitors originate in globular clusters, owing to dynamically formed compact object binaries, or to the tidal capture and collisions of compact objects (e.g., Bhattacharya & van den Heuvel 1991; Grindlay, Portegies Zwart & McMillan 2006; Lee, Ramirez-Ruiz & van de Ven 2010; Salvaterra et al. 2010). An estimate of the event rate for these processes compared to primordial binaries is $\lesssim$10% (Grindlay, Portegies Zwart & McMillan 2006; Lee, Ramirez-Ruiz & van de Ven 2010). This is roughly similar to the ~10% of short GRBs with inferred offsets of $\gtrsim$20 kpc, but it is not sufficient to explain the weak correlation with stellar light, i.e., ~50% of short GRBs with fractional flux values of zero.

A substantial fraction of short GRBs originating in globular clusters will also lead to the prediction of a dominant population of early-type hosts because the globular cluster specific frequency is several times higher in early-type than in late-type galaxies (Harris 1991, Brodie & Strader 2006). This is in contrast to the observed low fraction of early-type hosts in the short GRB sample, which is already deficient relative to the cosmic stellar mass budget (Section 6). In addition, formation of binaries in globular clusters generally requires core-collapsed clusters, thereby leading to a significant time delay relative to the cluster formation epoch (Grindlay, Portegies Zwart & McMillan 2006; Lee, Ramirez-Ruiz & van de Ven 2010; Salvaterra et al. 2010). As a result, a dominant globular cluster population will lead to systematically lower redshifts; for example, Lee, Ramirez-Ruiz & van de Ven (2010) predict a median redshift of $z \approx 0.5$ and essentially no short GRBs at $z \gtrsim 1.5$. This result is not in strong conflict with the current redshift distribution, although the median redshift of short GRBs is probably somewhat higher than $z \sim 0.5$ (Section 6). Thus, though it is unlikely that most short GRBs originate in globular clusters, it is possible that ~10% of the events originate in these environments.

8. THE AFTERGLOWS AND EXPLOSION PROPERTIES OF SHORT GAMMA-RAY BURSTS

Having explored the galactic and subgalactic environments of short GRBs, and the implications for the nature of their progenitors, I now turn to an investigation of the short GRB explosion properties and their parsec-scale circumburst environments. The key parameters of interest that can be extracted from the prompt and afterglow emission are the energy scale in gamma-rays (E_γ) and in the blastwave powering the afterglow ($E_{\rm K}$), the geometry of the outflow (characterized by a jet half opening angle, $\theta_{\rm j}$), and the density of the ambient medium (n). The standard formulation of the afterglow synchrotron emission includes additional free parameters related to the relativistic shock microphysics: the fraction of postshock energy in the magnetic fields ($\epsilon_{\rm B}$) and in the radiating relativistic electrons ($\epsilon_{\rm e}$), which are assumed to follow a power-law distribution, $N(\gamma) \propto \gamma^{-p}$, above a minimum Lorentz factor, $\gamma_{\rm m}$ (Sari, Piran & Narayan 1998). Determining the values of the various parameters requires multiwavelength, multiepoch afterglow observations spanning radio waves to X-rays.

The synchrotron emission spectrum resulting from the interaction of the relativistic outflow with the circumburst medium is characterized by three break frequencies, corresponding to self-absorption ($\nu_{\rm a}$), the minimum Lorentz factor ($\nu_{\rm m}$), and synchrotron cooling ($\nu_{\rm c}$), as well as an overall flux normalization (Sari, Piran & Narayan 1998). The instantaneous values of these four parameters and the slope of the spectrum at $\nu > \nu_{\rm m}$ can be converted to $E_{\rm K,iso}$, n, $\epsilon_{\rm e}$, $\epsilon_{\rm B}$, and p; here $E_{\rm K,iso}$ is the isotropic-equivalent blastwave kinetic energy. In addition, the time evolution of the spectrum can be used to extract $\theta_{\rm j}$ and p. This simple formulation has been confirmed with observations of multiple long GRBs (e.g., Panaitescu & Kumar 2002, Yost et al. 2003), although some deviations from the standard model are apparent at $\lesssim 10^3$ s, particularly in the X-ray band (e.g., initial steep decline, flares, plateaus; Nousek et al. 2006, Chincarini et al. 2007). Using the formulation of the afterglow model in Granot & Sari (2002), I employ the following relations to determine the locations of the break frequencies (relevant for the dominant slow-cooling regime, with $\nu_{\rm a} < \nu_{\rm m} < \nu_{\rm c}$):

$$\nu_{\rm a} = 1.24 \times 10^9 \text{ Hz} \frac{(p-1)^{3/5}}{(3p+2)^{3/5}} (1+z)^{-1} \bar{\epsilon}_{\rm e}^{-1} \epsilon_{\rm B}^{1/5} n_0^{3/5} E_{52}^{1/5}, \tag{2}$$

$$\nu_{\rm m} = 3.73 \times 10^{15} \text{ Hz } (p-0.67)(1+z)^{1/2} \bar{\epsilon}_{\rm e}^{2} \epsilon_{\rm B}^{1/2} E_{52}^{1/2} t_d^{-3/2}, \tag{3}$$

$$\nu_{\rm c} = 6.37 \times 10^{13} \text{ Hz } (p-0.46) e^{-1.16p} (1+z)^{-1/2} \epsilon_{\rm B}^{-3/2} n_0^{-1} E_{52}^{-1/2} t_d^{-1/2}, \tag{4}$$

and

$$F_{\nu_{\rm a}} = 0.65 \text{ mJy} \frac{(p-1)^{6/5}}{(3p-1)(3p+2)^{1/5}} (1+z)^{1/2} \bar{\epsilon}_{\rm e}^{-1} \epsilon_{\rm B}^{2/5} n_0^{7/10} E_{52}^{9/10} t_d^{1/2} d_{\rm L,28}^{-2}, \tag{5}$$

where $\bar{\epsilon}_{\rm e} \equiv \epsilon_{\rm e}(p-2)/(p-1)$, E_{52} is the isotropic-equivalent blastwave kinetic energy in units of 10^{52} erg, $t_{\rm d}$ is the observer-frame time in days, and $d_{\rm L,28}$ is the luminosity distance in units of 10^{28} cm. For fiducial parameters relevant to short GRBs ($z \sim 0.5$, $E_{52} \sim E_{\gamma,\rm iso,52} \sim 0.1$, $n_0 \sim 0.1$), and assuming the typical microphysical parameters from long GRB afterglow data ($\epsilon_{\rm e} \sim 0.1$, $\epsilon_{\rm B} \sim 0.01$, $p \sim 2.4$; Panaitescu & Kumar 2002, Yost et al. 2003), the typical values of the synchrotron parameters are $\nu_{\rm a} \sim 1$ GHz, $\nu_{\rm m} \sim 200\ t_d^{-1.5}$ GHz, $\nu_{\rm c} \sim 2 \times 10^{17}\ t_d^{-0.5}$ Hz, and $F_{\nu_{\rm a}} \sim 20$ μJy. These values confirm that observations from radio waves to X-rays are required to probe the full set of synchrotron parameters. They also indicate that the typical brightness level in the radio band (~1–10 GHz) is quite challenging for detection, even with the upgraded Karl G. Jansky Very Large Array (JVLA). Similarly, these parameters predict a faint optical afterglow brightness level of ~23 AB mag at ~7 h, the mean timescale for existing searches (see also Panaitescu, Kumar & Narayan 2001).

In reality, the afterglow data for short GRBs are much sparser than for long GRBs, and the complete set of physical parameters for most individual events cannot be determined uniquely. For example, only three short GRBs to date have been detected in the radio band (Berger et al. 2005b, Soderberg et al. 2006a, Fong et al. 2013), providing meaningful constraints on the location of the self-absorption frequency. Still, using basic constraints, including expectations for the values of the microphysical parameters, and a combination of detections and upper limits in the radio, optical, and X-ray bands, it is possible to determine the distributions of some key short GRB properties.

Of particular interest for our understanding of the central engine and energy extraction mechanism is the beaming-corrected energy, which requires the determination of jet opening angles (see Section 8.4). The jet collimation also influences the true short GRB event rate, which in turn provides an additional constraint on the progenitor model, and potentially useful information on the rate of GW detections in the specific context of compact object merger progenitors. Finally, the circumburst density provides insight on the progenitors, through constraints on their parsec-scale environments. Beyond the direct properties of short GRBs, the energy scale and circumburst densities can also be used to shed light on the expected brightness of EM counterparts to GW sources, both on- and off-axis; I discuss this point in detail in Section 10.

To date there have been only a few studies of short GRB afterglows. Berger et al. (2005b) determined the properties of GRB 050724 from radio, optical/near-IR, and X-ray observations, whereas Grupe et al. (2006) placed lower bounds on its collimation. Soderberg et al. (2006a) determined the explosion properties of GRB 051221A and along with Burrows et al. (2006) measured its opening angle. Fong et al. (2013) presented the radio, optical, and X-ray afterglows of GRB 130603B and determined its properties and jet collimation angle. Berger (2007) presented an analysis of the gamma-ray and X-ray observations of 16 short GRBs that revealed a clear correlation between $E_{\gamma,\mathrm{iso}}$ and $L_{\mathrm{X,iso}}$ (a rough proxy for $E_{\mathrm{K,iso}}$), as well as a broad range of isotropic-equivalent energies in the prompt emission and afterglows (see also Nakar 2007, Gehrels et al. 2008). Similarly, Nysewander, Fruchter & Pe'er (2009) investigated correlations between the gamma-ray, X-ray, and optical emission.

8.1. X-Ray Afterglow Emission

The most extensive afterglow data set for short GRBs is in the X-ray band from *Swift*/XRT (**Table 1**). As shown in Section 4, of the 50 short GRBs with X-ray detections, 28 events exhibit detectable long-term X-ray emission beyond $\sim 10^3$ s, as expected for afterglow emission, whereas the remaining events rapidly fade below the XRT detection threshold at $\lesssim 10^3$ s and therefore yield only upper limits on the timescale dominated by the afterglow. The initial rapid decline phase is generally attributed to high latitude emission associated with the prompt phase (e.g., Nousek et al. 2006) and therefore cannot be reliably used to constrain the afterglow properties. In **Figure 14**, I plot the isotropic-equivalent X-ray luminosity of short and long GRBs at a fiducial rest-frame time of 11 h in a rest-frame band of 0.3–10 keV ($L_{\mathrm{X},11}$) as a function of the isotropic-equivalent gamma-ray energy ($E_{\gamma,\mathrm{iso}}$). I use $E_{\gamma,\mathrm{iso}}$ as a proxy for the more directly relevant, but generally unknown $E_{\mathrm{K,iso}}$. This is a reasonable approach for each GRB sample individually because variations in the

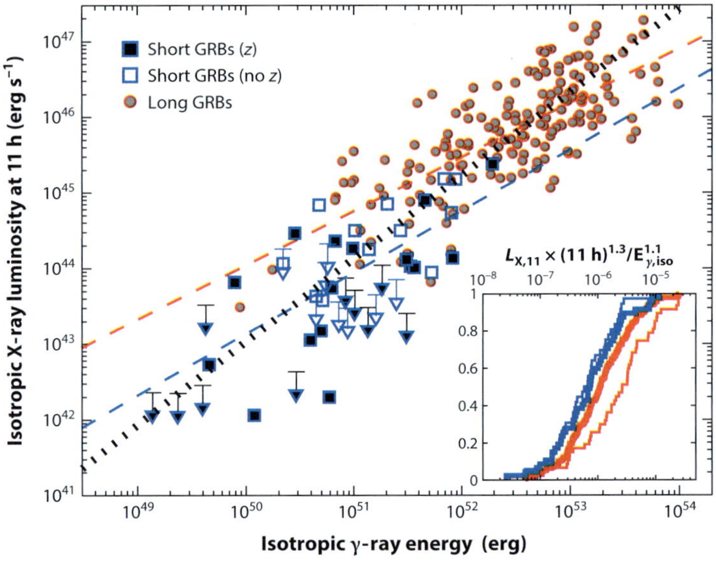

Figure 14

Isotropic-equivalent afterglow X-ray luminosity at a rest-frame time of 11 h ($L_{\mathrm{X},11}$) versus the isotropic-equivalent gamma-ray energy ($E_{\gamma,\mathrm{iso}}$) for short GRBs (*blue*) and long GRBs (*red circles*). Open symbols for short GRBs indicate events without a known redshift, for which a fiducial value of $z = 0.75$ is assumed. The dashed blue and red lines are the best-fit power-law relations to the trends for short and long GRBs, respectively, whereas the dotted black line is the expected correlation based on the afterglow synchrotron model with $\nu_{\mathrm{X}} > \nu_{\mathrm{c}}$ and $p = 2.4$ ($L_{\mathrm{X},11} \propto E_{\mathrm{K,iso}}^{1.1}$). The inset shows the distribution of the ratio $L_{\mathrm{X},11} \times (11\ \mathrm{h})^{1.3}/E_{\gamma,\mathrm{iso}}^{1.1}$ for the full samples (*thick lines*) and for bursts in the region of $E_{\gamma,\mathrm{iso}}$ overlap (*thin lines*). The lower level of $L_{\mathrm{X},11}$ relative to $E_{\gamma,\mathrm{iso}}$ for short GRBs is evident from these various comparisons.

radiative gamma-ray efficiency, $\epsilon_\gamma \equiv E_{\gamma,\rm iso}/(E_{\gamma,\rm iso} + E_{\rm K,iso})$, will increase the scatter but should not introduce systematic trends. By contrast, any systematic trends in ϵ_γ between short and long GRBs may lead to systematic effects when using $E_{\gamma,\rm iso}$ as a proxy for $E_{\rm K,iso}$. I return to this point in Section 8.5.

As expected, both the short and long GRB samples exhibit a clear correlation between $L_{\rm X,11}$ and $E_{\gamma,\rm iso}$ (**Figure 14**; Berger 2007; Gehrels et al. 2008; Nysewander, Fruchter & Pe'er 2009). For long GRBs the best-fit power-law relation is $L_{\rm X,11} \approx 5.6 \times 10^{44}\ E^{0.72}_{\gamma,\rm iso,51}$ erg s^{-1}, whereas for short GRBs it is systematically lower, with $L_{\rm X,11} \approx 8.5 \times 10^{43}\ E^{0.83}_{\gamma,\rm iso,51}$ erg s^{-1}. In the region of overlap between the two GRB populations ($E_{\gamma,\rm iso} \approx 3 \times 10^{50}$–$10^{52}$ erg), the short GRB isotropic X-ray afterglow luminosity is on average about seven times lower than for long GRBs.

Comparing the short GRBs with afterglow detections to those with only a rapid decline phase and no X-ray detections beyond $\sim 10^3$ s, I find no clear difference in the ratio $L_{\rm X,11}/E_{\gamma,\rm iso}$. In particular, short GRBs with X-ray afterglow detections span $L_{\rm X,11}/E_{\gamma,\rm iso} \approx 10^{-8}$–$10^{-6}$, with 40% of these events having $L_{\rm X,11}/E_{\gamma,\rm iso} \lesssim 10^{-7}$. The upper limits are generally $L_{\rm X,11}/E_{\gamma,\rm iso} \lesssim 10^{-7}$ (although some shallower limits exist as well). Thus, it appears that events without detected X-ray emission beyond $\sim 10^3$ s are generally due to inadequate searches rather than to intrinsically weaker X-ray afterglows.

In the context of the standard afterglow synchrotron model, the X-ray band is expected to be located near or above the synchrotron cooling frequency ($\nu_{\rm X} > \nu_{\rm c}$; see discussion following Equation 5). In this spectral regime the afterglow X-ray flux is given by $F_{\nu,\rm X} \propto E^{(p+2)/4}_{\rm K,iso}\,\epsilon_{\rm e}^{p-1}\epsilon_{\rm B}^{(p-2)/4} \approx \epsilon_{\rm e} E_{\rm K,iso}$ (Granot & Sari 2002). Assuming that $E_{\gamma,\rm iso}$ is a close proxy for $E_{\rm K,iso}$, it is evident that the observed correlations for short and long GRBs are flatter than the theoretical expectation. At the same time, the observed dispersion around this relation is at least a factor of a few, suggestive of dispersion in the values of p, $\epsilon_{\rm e}$, and $\epsilon_{\rm B}$, as well as cases with $\nu_{\rm X} < \nu_{\rm c}$ for which the density also influences the observed flux.

In **Figure 14**, I also plot the distribution of the ratio $L_{\rm X,11} \times (11\ \rm h)^{1.3}/E^{1.1}_{\gamma,\rm iso}$, which is the expected relation for $p = 2.4$ and $E_{\gamma,\rm iso} = E_{\rm K,iso}$ on the basis of the afterglow model. For the full samples of short and long GRBs I find that the distributions of this ratio are similar, although with a systematically lower value for short GRBs; a K-S test yields a p-value of about 0.23, indicating that the two samples are compatible with the null hypothesis of being drawn from the same underlying distribution. However, using only events in the overlap region of $E_{\gamma,\rm iso} \approx 3 \times 10^{50}$–$10^{52}$ erg yields a substantial difference (K-S test value of $p \approx 1.1 \times 10^{-4}$). I therefore conclude that short GRBs have systematically weaker X-ray afterglow emission compared with long GRBs, relative to both the observed correlation between $L_{\rm X,11}$ and $E_{\gamma,\rm iso}$, and the relation expected from afterglow theory.

8.2. Optical Afterglow Emission

I next turn to the optical afterglow emission, which has been detected in 24 short GRBs, with upper limits for an additional 16 bursts (Section 4). Most detections and limits have been obtained on a timescale of about 1 h to 1 day after the burst with a magnitude range of about 21–26 AB mag (**Table 1**). The median r-band brightness level at a fiducial time of 7 h is $\langle m_r(7\ \rm h)\rangle \approx 23.2$ AB mag, in excellent agreement with the expectation for the fiducial set of parameters discussed in relation to Equations 2–5 and with the prediction by Panaitescu, Kumar & Narayan (2001). For comparison, long GRB afterglows are generally an order of magnitude brighter, with $\langle m_r(7\ \rm h)\rangle \approx 20.8$ AB mag (e.g., Kann et al. 2011) despite their significantly larger redshifts (**Figure 4**).

Converting to rest-frame properties, in **Figure 15** I plot the isotropic-equivalent optical luminosity in the rest-frame r-band at a fiducial rest-frame time of 7 h ($L_{\rm opt,7}$) as a function of $E_{\gamma,\rm iso}$. As in the case of the isotropic X-ray luminosity, there is a clear correlation between the

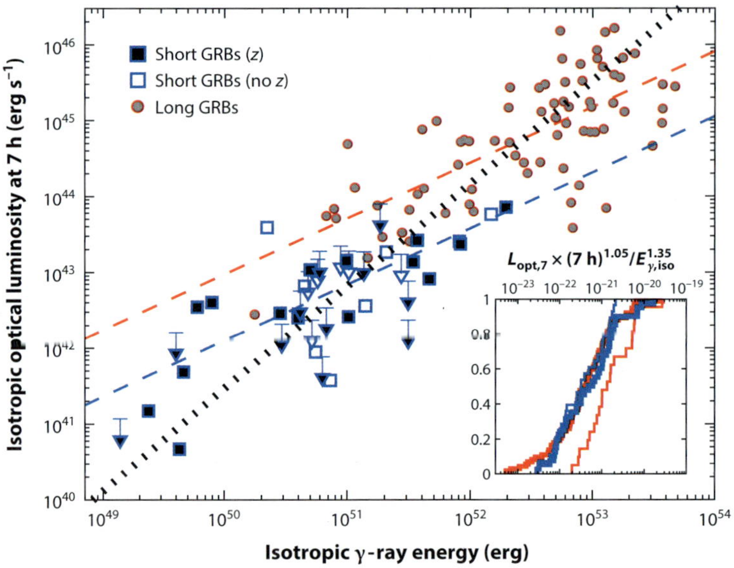

Figure 15

Same as **Figure 14** but for the isotropic-equivalent afterglow optical luminosity at a rest-frame time of 7 h ($L_{opt,7}$). The dotted black line is the expected correlation based on the afterglow model for $\nu_m < \nu_{opt} < \nu_c$ and $p = 2.4$ ($L_{opt,7} \propto E^{1.35}_{K,iso}$). The inset shows the distribution of the ratio $L_{opt,7} \times (7\ h)^{1.05}/E^{1.35}_{\gamma,iso}$ for the full samples (*thick lines*) and for bursts in the region of $E_{\gamma,iso}$ overlap (*thin lines*). The lower level of $L_{opt,7}$ relative to $E_{\gamma,iso}$ for short GRBs is evident from these various comparisons.

optical luminosity and gamma-ray energy. For short GRBs, the best-fit power-law relation is $L_{opt,7} \approx 6.9 \times 10^{42}\ E^{0.74}_{\gamma,iso,51}$ erg s^{-1}, whereas for long GRBs the relation is systematically higher, $L_{opt,7} \approx 5.0 \times 10^{43}\ E^{0.73}_{\gamma,iso,51}$ erg s^{-1}. Thus, in the range of $E_{\gamma,iso}$ in which the short and long GRB populations overlap, the optical afterglow luminosity of short GRBs is on average about seven times fainter than for long GRBs, just as in the case of the X-ray luminosity.

The expected trend, on the basis of the afterglow synchrotron model (for $\nu_m < \nu_{opt} < \nu_c$), is $F_{\nu,opt} \propto E^{(p+3)/4}_{K,iso} n_0^{1/2} \epsilon_e^{p-1} \epsilon_B^{(p+1)/4}$. For a typical value of $p = 2.4$, this indicates a steeper slope than inferred from the data, with $F_{\nu,opt} \propto E^{1.35}_{K,iso}$. This trend is shown in **Figure 15** with a dotted line and indeed seems to be a reasonable representation of the full range of $E_{\gamma,iso}$ spanned by short and long GRBs. This is also borne out in the nearly identical cumulative distributions of the ratio $L_{opt,7} \times (7\ h)^{1.05}/E^{1.35}_{\gamma,iso}$ for both short and long GRBs (**Figure 15**). However, despite the overall similarity, the short GRBs have systematically lower values of this ratio in the overlap range of $E_{\gamma,iso} \approx 3 \times 10^{50}$–$10^{52}$ erg. A K-S test yields a *p*-value of about 0.016, indicating that the two samples are not likely to be drawn from the same underlying distribution. Thus, despite the larger scatter in the optical band compared to the X-ray band, it appears that short GRBs have systematically weaker optical afterglow emission compared with long GRBs, relative to both the observed correlation between $L_{opt,7}$ and $E_{\gamma,iso}$, and the relation expected from afterglow theory.

8.2.1. X-ray/optical comparison. The X-ray and optical luminosities of short GRB afterglows appear to be fainter relative to $E_{\gamma,iso}$ than for long GRBs by a similar factor of about seven. This indicates that the ratio $L_{opt,7}/L_{X,11}$ should have a similar distribution for both GRB samples. In **Figure 16**, I plot $L_{opt,7}$ versus $L_{X,11}$ for short and long GRBs and indeed find that both populations

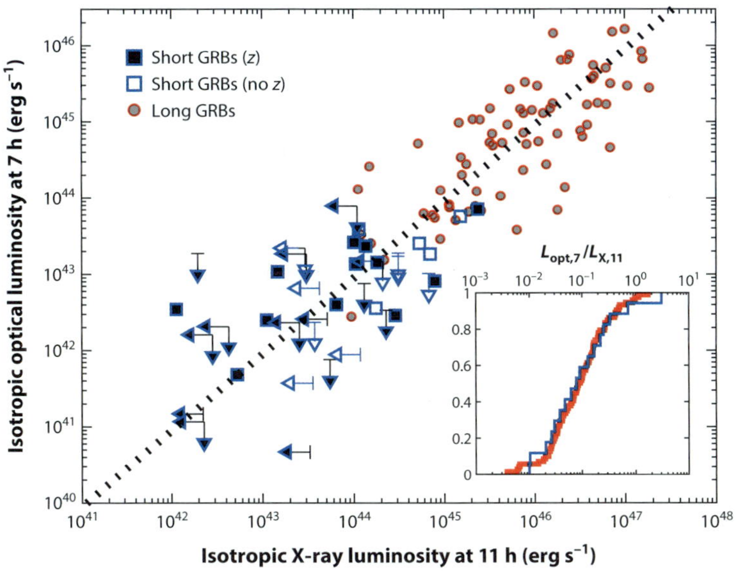

Figure 16

Isotropic-equivalent afterglow optical luminosity at a rest-frame time of 7 h ($L_{\rm opt,7}$) versus isotropic-equivalent afterglow X-ray luminosity at a rest-frame time of 11 h ($L_{\rm X,11}$). Symbols are as in **Figure 14**. The dotted black line marks a linear correlation, expected for $\nu_{\rm X} \lesssim \nu_{\rm c}$. The inset shows the distribution of the ratio $L_{\rm opt,7}/L_{\rm X,11}$, indicating that both short and long GRBs exhibit a similar ratio and that in general $L_{\rm opt,7}/L_{\rm X,11} \lesssim 1$, indicative of $\nu_{\rm X} \lesssim \nu_{\rm c}$ for short GRBs.

follow a similar trend, consistent with a linear correlation (see also Nysewander, Fruchter & Pe'er 2009). The median value for both the short and long GRB samples is $\langle L_{\rm opt,7}/L_{\rm X,11}\rangle \approx 0.08$, and nearly all events have $L_{\rm opt,7}/L_{\rm X,11} \lesssim 1$. In the context of the synchrotron model, the expected ratio for the case in which both the optical and X-ray bands are located below $\nu_{\rm c}$ is $L_{\rm opt,7}/L_{\rm X,11} \approx 0.2$ (for $p = 2.4$). By contrast, for the case when $\nu_{\rm c}$ is intermediate between the optical and X-ray bands (e.g., $\nu_{\rm c} \sim 10^{16}$ Hz) the expected ratio is $L_{\rm opt,7}/L_{\rm X,11} \approx 1$, which is higher than most observed values. Thus, the comparison of optical and X-ray luminosities suggests that generally ν_c is located near or above the X-ray band.

Because the ratio of optical to X-ray luminosity for $\nu_{\rm c} \sim \nu_{\rm X}$ and for $\nu_{\rm c} \gg \nu_{\rm X}$ is essentially the same, and because in the latter case both the X-ray and optical luminosity depend on density (with $F_\nu \propto n^{0.5}$), the lower values of $L_{\rm opt,7}/E_{\gamma,\rm iso}$ and $L_{\rm X,11}/E_{\gamma,\rm iso}$ for short GRBs can be interpreted as being due to a lower density scale; this will also naturally lead to $\nu_{\rm c} \gg \nu_{\rm X}$ because $\nu_{\rm c} \propto n^{-1}$. However, long GRBs generally have higher densities and hence $\nu_{\rm c} \sim \nu_{\rm X}$. Matching the observed weaker optical and X-ray afterglows of short GRBs requires a density scale that is about 50 times lower than for long GRBs or equivalently $n \sim 0.1$ cm^{-3}.

An alternative possibility is that the systematic offset in the trends of $L_{\rm opt,7}$ and $L_{\rm X,11}$ versus $E_{\gamma,\rm iso}$ for short and long GRBs is due to systematic differences between the gamma-ray and blastwave kinetic energies. In particular, if ϵ_γ for short GRBs is systematically higher than for long GRBs, this would lead to values of $E_{\gamma,\rm iso}$ that overpredict $E_{\rm K,iso}$ and hence to lower ratios of $L_{\rm opt,7}/E_{\gamma,\rm iso}$ and $L_{\rm X,11}/E_{\gamma,\rm iso}$ compared with those for long GRBs. Given the overall shift of the best-fit relations shown in **Figures 14** and **15**, this scenario would require an efficiency that is about an order of magnitude larger in short GRBs. With a fiducial value of $\epsilon_\gamma \sim 0.5$ for long GRBs (e.g., Panaitescu & Kumar 2002), this requires $\epsilon_\gamma \sim 0.95$ for short GRBs, which is an unlikely scenario.

A possible variation of this scenario is that the blastwave kinetic energies of long GRBs are systematically larger relative to $E_{\gamma,\rm iso}$ owing to significant energy injection into the forward shock following the prompt emission phase, which is absent in short GRBs. Indeed, the X-ray plateaus in long GRBs, spanning $\sim 10^3$–10^4 s, have been interpreted as evidence for significant energy injection (Nousek et al. 2006), but a smaller fraction of short GRBs exhibit such plateaus (Margutti et al. 2013). Effectively, larger energy injection in long GRBs will translate to a lower value of ϵ_γ than for short GRBs. An energy injection factor in long GRBs of about an order of magnitude will bring the relations for short and long GRBs into alignment.

8.2.2. The energy and density scale extracted from optical observations. Independent of the relative behavior of short and long GRB afterglows, each optical detection or upper limit can be converted to an allowed range in the phase-space of $E_{\rm K,iso}$ and n. For this purpose, I use fixed fiducial values of the microphysical parameters, $\epsilon_{\rm e} = 0.1$, $\epsilon_{\rm B} = 0.01$, and $p = 2.4$. In addition, I restrict the range of $E_{\rm K,iso}$ for each burst to $(0.3–3) \times E_{\gamma,\rm iso}$ for a reasonable expected range of ϵ_γ. The resulting distribution is shown in **Figure 17**. For the bulk of the short GRB population, the inferred density range is $n \sim 10^{-3}$–1 cm^{-3} with a median for the full distribution of $\langle n \rangle \lesssim 0.15$ cm^{-3}; this is an upper limit owing to the substantial fraction of optical nondetections. This result confirms our inference of a lower density scale compared to long GRBs based on the ratios of optical and X-ray luminosity to $E_{\gamma,\rm iso}$.

There are three bursts that may stand out from this distribution with inferred densities of ~ 50–10^3 cm^{-3} (if $E_{\rm K} \approx E_\gamma$ and if their microphysical parameters are as listed above). Two of these events (GRBs 070724A and 110112A) have low gamma-ray fluences of $F_\gamma \approx 3 \times 10^{-8}$ erg cm^{-2}, which may indicate that the *Swift*/BAT energy band severely underestimates the value of $E_{\gamma,\rm iso}$ and, hence, their likely $E_{\rm K,iso}$ values. With larger energies, the required densities will be lower than the values inferred above. By contrast, the third event (GRB 050709) does not fall in this category. Clearly, the inferred distribution of circumburst densities is affected by dispersions in the microphysical parameters and in the ratio of $E_{\gamma,\rm iso}$ to $E_{\rm K,iso}$, and may therefore be somewhat wider or narrower than inferred here. However, the median of the distribution is fairly robust. The relatively low density is not unexpected considering that most short GRB offsets are larger than 1 r_e and that nearly half of all short GRBs reside in regions with no detectable stellar light (Section 7).

8.3. Radio Afterglow Emission

Radio afterglow observations have been undertaken for 28 short GRBs and have, so far, led to only 3 detections (GRBs 050724, 051221A, and 130603B; Berger et al. 2005b, Soderberg et al. 2006b, Fong et al. 2013; see **Table 1**). With the exception of a few short GRBs observed with the improved sensitivity of the JVLA starting in 2011, most upper limits are in the range of $F_\nu \sim 60$–150 μJy (3σ). It is clear from a comparison to the fiducial synchrotron model listed above (with $F_\nu \sim 40$ μJy at ~ 10 GHz; Equation 5) that most of the existing limits are too shallow compared with the expected flux density scale for $n \sim 0.1$ cm^{-3}. By contrast, because in the relevant regime ($\nu_{\rm a} < \nu_{\rm radio} < \nu_{\rm m}$) the flux density scales as $F_\nu \propto n^{0.5}$, the nondetections generally indicate $n \lesssim$ a few particles per cubic centimeter. This result is supported by the three events with detected radio afterglows (Berger et al. 2005b, Soderberg et al. 2006a, Fong et al. 2013) and agrees with the more constraining results from the optical afterglow data. I also note that of the three bursts with potential high density on the basis of optical observations, I find that radio observations of GRB

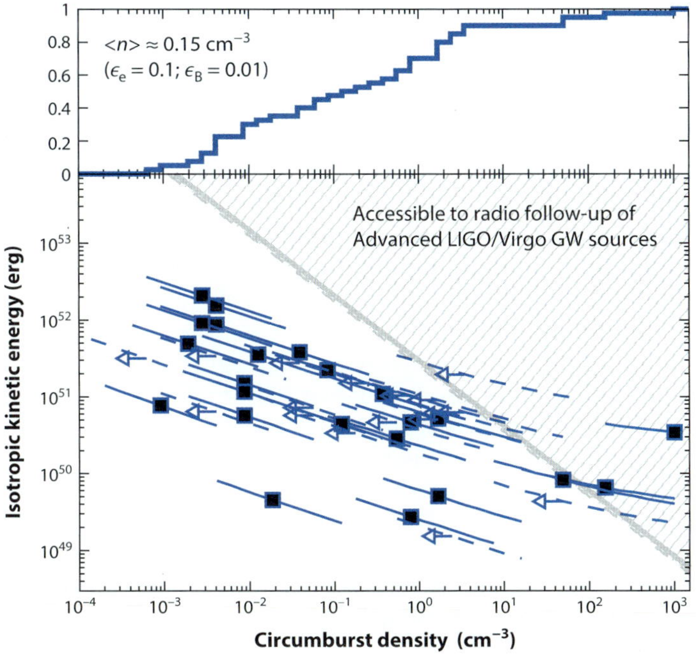

Figure 17

(*Bottom*) Blastwave isotropic-equivalent kinetic energy ($E_{K,iso}$) versus circumburst density (n) for individual short GRBs based on their optical afterglow emission (*arrows* indicate optical nondetections). The lines span $E_{K,iso} = (0.3\text{–}3) \times E_{\gamma,iso}$. The gray diagonal thick solid and thin dashed lines delineate the phase-space required for the detections of radio counterparts to Advanced LIGO/Virgo gravitational wave sources (*hatched gray region*; based on the models of Nakar & Piran 2011). The gray diagonal thick solid line is for an off-axis short GRB afterglow with fiducial parameters (Section 8), whereas the gray diagonal thin dashed line is for an isotropic mildly relativistic outflow. The short GRB results point to an energy-density scale that will generally prevent radio detections of Advanced LIGO/Virgo sources in searches that cover a full error region (Section 10). (*Top*) Cumulative distribution of the circumburst densities (for $E_{K,iso} = E_{\gamma,iso}$). The median circumburst density is $\lesssim 0.15$ cm^{-3}, which is at least an order of magnitude lower than for long GRBs and points to explosions in more tenuous environments.

050709 instead indicate $n \lesssim 0.05$ cm^{-3} (see also Fox et al. 2005); the radio limits for GRBs 070724A and 110112A are not constraining beyond the information available from the optical detections.

8.4. Jets and Implications for the Energy Scale and Event Rates

The collimation of GRB outflows has direct implications for the true energy scale and event rate, as well as for our understanding of the energy extraction mechanism. The signature of collimation in GRB afterglows is the so-called jet break that occurs at the time (t_j) when the bulk Lorentz factor of the outflow declines to $\Gamma(t_j) \approx 1/\theta_j$ (Rhoads 1999; Sari, Piran & Halpern 1999). The break is due to a combination of an edge effect, when the entire emitting surface of the jet becomes visible, and sideways expansion of the outflow. Together, these effects lead to a change in the evolution of the synchrotron spectrum. In particular, at frequencies above the spectral peak ($\nu \gtrsim \nu_m$) the jet break is manifested as a steepening in the decline rate from roughly $F_\nu \propto t^{-1}$ to $F_\nu \propto t^{-p}$. This is relevant for the optical and X-ray bands. In the radio band, which is located below ν_m for typical jet break timescales, the evolution transitions from a mild rise to a shallow decline, subsequently followed by $F_\nu \propto t^{-p}$ when ν_m declines below the radio band. Because the jet break is due to relativistic and hydrodynamic effects, it is expected to be achromatic from radio waves to X-rays. The relation between the jet break time and opening angle is given by (Sari, Piran & Halpern 1999)

$$\theta_j = 0.13 \left(\frac{t_{j,d}}{1+z}\right)^{3/8} \left(\frac{n_0}{E_{52}}\right)^{1/8}. \tag{6}$$

In the case of long GRBs, there is ample evidence for jet collimation on the basis of the detection of jet breaks in X-rays, optical, and/or radio waves on a wide range of timescales (Harrison et al. 1999; Bloom, Frail & Sari 2001; Frail et al. 2001; Panaitescu & Kumar 2002; Berger, Kulkarni & Frail 2003). The resulting jet opening angles are mainly in the range of $\theta_j \approx 3$–$10°$; some events extend to about 20° (Bloom, Frail & Sari 2001; Frail et al. 2001; Berger, Kulkarni & Frail 2003). The typical beaming correction factor is therefore $f_b = [1 - \cos(\theta_j)] \approx 6 \times 10^{-3}$. This leads to a two-orders-of-magnitude-downward correction in the energy scale with a resulting typical value of $E_\gamma \sim E_K \sim$ few $\times 10^{51}$ erg (Bloom, Frail & Sari 2001; Frail et al. 2001; Berger, Kulkarni & Frail 2003), albeit with a spread of about an order of magnitude (Soderberg et al. 2006b, Cenko et al. 2011). Similarly, the upward correction to the long GRB rate is $f_b^{-1} \sim 10^2$.

The information on jet breaks in short GRBs is more sparse, and with the exception of GRB 130603B (Fong et al. 2013), the claimed breaks have not been detected in multiple wave bands that can confirm the expected achromatic behavior. The key challenge for jet break detections is the faintness of short GRB afterglows, which typically fade below detectable levels on a day timescale, translating to a weak constraint of $\theta_j \gtrsim 3°$ if no break is observed (Fong et al. 2012).

Despite this observational challenge, there are a few credible detections of jet breaks so far. GRB 051221A was the first short GRB showing evidence for collimation on the basis of a break in its X-ray light curve at about 5 days and with the expected steepening to a power-law index of $\alpha_X \lesssim -2$ (Burrows et al. 2006, Soderberg et al. 2006a). The inferred opening angle is $\theta_j \approx 6$–$8°$ (Soderberg et al. 2006a), similar to the opening angles of long GRBs. GRB 111020A similarly exhibited a break in its X-ray light curve, at about 2 days, leading to an inferred opening angle of $\theta_j \approx 3$–$8°$ (Fong et al. 2012); the larger uncertainty compared with GRB 051221A is due to the unknown redshift of this burst. Most recently, GRB 130603B exhibited a significant break in its optical light curve at about 0.45 days that was accompanied by a decline in the radio band that matches the expectations of a post-jet break behavior (Fong et al. 2013). The inferred opening angle is $\theta_j \approx 4$–$8°$ (Fong et al. 2013). A potential break was also noted in the optical light curve of GRB 090426 at about 0.4 days, with a steep decline typical of post-jet break evolution (Nicuesa Guelbenzu et al. 2011) and an inferred opening angle of $\theta_j \approx 4°$.

In addition to these likely breaks, there are also several meaningful lower limits on jet opening angles. X-ray observations of GRB 050724 from *Swift*/XRT and *Chandra* revealed no break to about 22 days, leading to an inferred lower limit of $\theta_j \gtrsim 20°$ (Grupe et al. 2006). The optical afterglow of GRB 050709 exhibited potential steepening at about 10 days on the basis of a single data point (Fox et al. 2005), which if attributed to a jet break leads to $\theta_j \approx 15°$; however, this interpretation has been disputed (Watson et al. 2006), and here I conservatively use $\theta_j \gtrsim 15°$ as a lower limit. X-ray observations of GRB 111117A from *Swift*/XRT and *Chandra* revealed no break to about 3 days, leading to an inferred lower limit of $\theta_j \gtrsim 6°$ (Margutti et al. 2012). Finally, the lack of a break in the X-ray afterglow of GRB 120804A to about 46 days indicates an opening angle of $\theta_j \gtrsim 11°$ (Berger et al. 2013).

There are also several claimed breaks at much earlier times, ~0.5–2 h, with a postbreak steep decline rate that is reminiscent of jet break behavior (GRBs 061201, 090305, and 090510). If these are indeed jet breaks, then the resulting opening angles are even narrower than for most long GRBs, $\theta_j \sim 1$–$2°$. However, in the case of GRB 090305 the claimed break is based on a

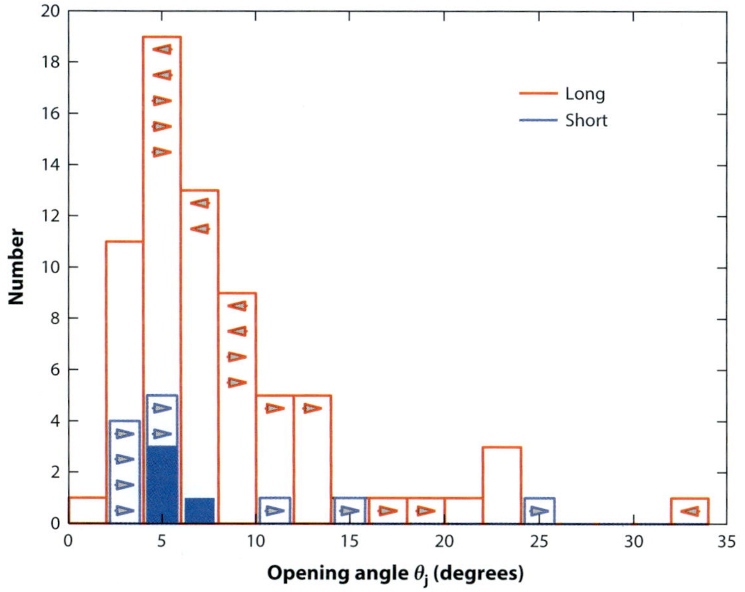

Figure 18

Distributions of jet opening angles for short (*blue*) and long (*red*) GRBs on the basis of breaks in their afterglow emission. Arrowheads mark lower or upper limits on the opening angles. The observations are summarized in Section 8.4. From Fong et al. (2013) and references therein.

single optical data point (Nicuesa Guelbenzu et al. 2012a), whereas for GRB 090510 there is no corresponding break in the optical band despite simultaneous coverage (Nicuesa Guelbenzu et al. 2012b). In addition to these putative breaks, several short GRBs exhibit no break in their X-ray light curves to ~1 day, leading to typical limits of $\gtrsim 3°$ (e.g., GRBs 070714B, 070724A, 071227, 081226, and 101219A; Fong et al. 2012). The distribution of jet opening angles for short GRBs along with a comparison to that of long GRB jets is shown in **Figure 18**.

Using the most robust detections and constraints described above (3 detections and 4 lower limits), I find that the mean opening angle is $\langle\theta_j\rangle \gtrsim 10°$. The resulting mean beaming factor is $f_b \gtrsim 0.015$, indicating that the correction to the observed short GRB rate is a factor of $\lesssim 70$ or a rate of up to $\sim 10^3$ Gpc^{-3} year^{-1} (Nakar, Gal-Yam & Fox 2006; Coward et al. 2012; Fong et al. 2012, 2013). In the context of compact object mergers, this value coincides with the middle of the range on the basis of estimates from the Galactic NS-NS binary population and from population synthesis models (Kalogera et al. 2004, Abadie et al. 2010). When converted to the rate within the expected sensitivity volume of Advanced LIGO (~200 Mpc), the resulting value is ~25 year^{-1}. I note that if claimed values of $\theta_j \sim 1$–$2°$ are indeed correct, then it is possible that the correction to the short GRB rate is up to $\sim 3 \times 10^3$, leading to an Advanced LIGO detection rate of $\sim 10^3$ year^{-1}. Thus, in the framework of compact object binary progenitors, the discovery rate of NS-NS binaries with Advanced LIGO will provide insight on short GRB beaming.

The inferred opening angles also determine the true energy scale of short GRBs. For the three events with likely jet breaks, the inferred beaming-corrected gamma-ray energies are $E_\gamma \approx (0.5$–$5) \times 10^{49}$ erg. The four events with lower limits on the opening angles lead to a similar range for their minimum beaming-corrected gamma-ray energies (with the upper bounds determined by the isotropic-equivalent values). Because the short GRB sample spans $E_{\gamma,\mathrm{iso}} \sim 10^{49}$–$10^{52}$ erg

(**Figures 14** and **15**), it is likely that the beaming-corrected energies are generally $E_\gamma \lesssim 10^{50}$ erg (using the typical beaming correction of $f_b \sim 0.015$). It remains to be seen, when a larger sample of opening angles becomes available, whether the true energies span a narrow range, centered on $E_\gamma \sim 10^{49}$ erg, about two orders of magnitude lower than the energy scale of long GRBs.

8.5. Implications for the Progenitors and Energy Source

The comparison of short and long GRB afterglows and the comparison of short GRB afterglow data with the standard synchrotron model reveal several important trends. First, the ratios $L_{X,11}/E_{\gamma,iso}$ and $L_{opt,7}/E_{\gamma,iso}$ are systematically lower for short GRBs, by about a factor of seven. By contrast, the ratio $L_{opt,7}/L_{X,11}$ is similar for both GRB samples. One implication of these results is that the afterglows of short and long GRBs are overall similar in their properties and are generated by the same mechanism, namely an external shock with the circumburst environment that amplifies magnetic fields and accelerates electrons to relativistic velocities. However, short GRBs appear to explode in lower-density environments. Using the optical afterglows, I find a typical density scale of $\lesssim 0.15$ cm^{-3}. This is supported by radio afterglow limits, which indicate $n \lesssim 1$ cm^{-3}, and by the few events with radio to X-ray band detections (Berger et al. 2005b, Soderberg et al. 2006a, Fong et al. 2013). The relatively low circumburst densities are consistent with the larger offsets of short GRBs compared with long GRBs and with their weak spatial correlation with the underlying distribution of stellar light in their hosts. Although it is less likely, the difference in the ratios of $L_{X,11}$ and $L_{opt,7}$ to $E_{\gamma,iso}$ for short and long GRBs may also reflect a systematic difference in the gamma-ray efficiency, owing either to a higher radiative efficiency in the prompt emission of short GRBs or to substantial energy injection in the early afterglow phase of long GRBs. In either case, a difference by about an order of magnitude is required to reconcile the two samples.

Beyond the relative trends, it is clear from **Figures 14** and **15** that the distribution of $E_{\gamma,iso}$ for short GRBs is just as broad as for long GRBs but with systematically lower values of $\sim 10^{49}$–10^{52} erg. The beaming-corrected energies in the several cases with likely jet opening angle measurements or meaningful lower limits are $E_\gamma \sim (0.5\text{–}5) \times 10^{49}$, about two orders of magnitude lower than for long GRBs. The opening angles are generally broader than those for long GRBs, with a range of $\theta_j \sim 5°$ to $\gtrsim 20°$ and a median of $\langle\theta_j\rangle \gtrsim 10°$.

It is instructive to compare these inferences to the results of hydrodynamic and MHD simulations of compact object mergers as they pertain to estimates of the geometry, energy extraction mechanism, and the energy scale of relativistic outflows. Several researchers carried out hydrodynamic simulations of jets launched by the neutrino-antineutrino annihilation mechanism and found jets with typical opening angles of $\theta_j \sim 5$–$20°$ and total isotropic-equivalent energies of $\sim 10^{51}$ erg (or typical beaming-corrected energies of up to $\sim 10^{49}$ erg; Janka et al. 1999; Rosswog, Ramirez-Ruiz & Davies 2003; Setiawan, Ruffert & Janka 2004; Aloy, Janka & Müller 2005; Birkl et al. 2007; Dessart et al. 2009). However, only $\sim 10\%$ of this energy is available to produce the prompt emission, leading to a typical true energy of $E_\gamma \sim 10^{48}$ erg. This value, and the estimated upper bound on E_{iso}, is about an order of magnitude lower than that observed in short GRBs, although the jet opening angle estimates agree with the observed values. This suggests that for at least some short GRBs, the neutrino-antineutrino annihilation mechanism is not likely to provide sufficient energy to power the observed emission.

Instead, recent general relativistic MHD simulations by Rezzolla et al. (2011) suggest the formation of a $\sim 10^{15}$ G magnetic field both in the torus (toroidal) and along the rotation axis (poloidal) following a compact object merger and the formation of a black hole. The typical opening angle of the poloidal component is ~ 10–$30°$. If an outflow is then powered by the

Blandford-Znajek mechanism (Blandford & Znajek 1977), the typical isotropic-equivalent energy scale is $\sim$few $\times 10^{51}(B/2\times10^{15}\ \mathrm{G})^2$ erg (e.g., Lee, Wijers & Brown 2000; Rosswog, Ramirez-Ruiz & Davies 2003; Shibata et al. 2011); this assumes that 10% of the available energy is channeled into the relativistic outflow and that the typical opening angle is $\sim$10° based on the existing observations. Thus, these simulations suggest that compact object mergers can lead to conditions that give rise to opening angles and an energy scale (via the Blandford-Znajek mechanism) that are well matched to the observations of short GRBs.

9. "KILONOVA" EMISSION AND SHORT GRB 130603B

One of the key predictions of the compact object merger model is the ejection of neutron-rich matter, either dynamically during the merger or by winds from an accretion disk (Lattimer & Schramm 1974, Eichler et al. 1989). The typical ejecta masses and velocities inferred from merger simulations are in the range of $M_{\mathrm{ej}} \sim 10^{-3}$ to a few $\times\ 10^{-2}\ \mathrm{M}_{\odot}$ and $v_{\mathrm{ej}} \sim 0.1$–$0.3c$ (Rosswog et al. 1999; Rosswog et al. 2000; Ruffert & Janka 2001; Rosswog 2005; Etienne et al. 2008; Bauswein, Goriely & Janka 2013; Piran, Nakar & Rosswog 2013; Rosswog, Piran & Nakar 2013); the ejecta masses generally increase with larger asymmetry in the mass ratio (i.e., they are larger for NS-BH mergers than for NS-NS mergers; Hotokezaka et al. 2013; Kyutoku, Ioka & Shibata 2013; Rosswog, Piran & Nakar 2013). The resulting kinetic energy of the outflow is therefore in the range $E_{\mathrm{K}} \sim 10^{49}$–$3\times10^{51}$ erg. The ejecta will generate radio emission through interaction with the ambient medium; the ejecta velocity and kinetic energy, as well as the ambient density, determine the peak brightness and timescale of the radio emission (see Section 10).

The rapid decompression of the neutron-rich matter from nuclear densities will lead to the formation of heavy radioactive elements via the r-process (Lattimer & Schramm 1974, Eichler et al. 1989). The radioactive decay of these elements spans a wide range of half-lives as the elements settle on the region of β stability and provide a source of heating (Li & Paczyński 1998). Given the low ejecta mass and rapid expansion velocity, the ejecta will become optically thin on a shorter timescale and with a lower peak luminosity than SN explosions. This has led to terminology of "mini-SN," "macronova," or "kilonova" to describe these counterparts (Li & Paczyński 1998; Metzger et al. 2010b; Goriely, Bauswein & Janka 2011; Roberts et al. 2011; Piran, Nakar & Rosswog 2013); I use the last term here. In addition to the dynamical properties of the ejecta, the timescale, luminosity, and characteristic spectral energy distribution of a kilonova also depend on the opacity of the r-process material, because the peak of the light curve is expected when the opacity becomes sufficiently low that photons can diffuse on the expansion timescale (e.g., Arnett 1982). The diffusion timescale is given by (e.g., Metzger et al. 2010b, Barnes & Kasen 2013)

$$t_{\mathrm{d}} = \frac{B\kappa M_{\mathrm{ej}}}{cR}, \tag{7}$$

where $B \approx 0.07$ is a geometric factor, κ is the opacity, and $R = vt$ is the radius of the expanding ejecta. In addition, the peak luminosity is given by the radioactive heating rate (Q) at the diffusion timescale

$$L_{\mathrm{p}} \approx \frac{Q_{\mathrm{p}}}{t_{\mathrm{d}}(R_{\mathrm{p}})}, \tag{8}$$

where R_{p} is the radius of the ejecta at the diffusion timescale. Finally, the characteristic blackbody temperature of the resulting emission is given by

$$T_{\mathrm{p}} \approx \left(\frac{L_{\mathrm{p}}}{4\pi R_{\mathrm{p}}^2\sigma}\right)^{1/4}. \tag{9}$$

For an opacity of $\kappa \sim 0.1$ cm^2 g^{-1} at $\sim$0.3–1 μm, typical of the Fe-peak elements synthesized in SNe, the resulting peak timescale is $t_p \sim 0.3$ days, with a luminosity of $L_p \sim$ a few $\times 10^{41}(f_{-6})$ erg s^{-1} (where $f \sim$ a few $\times 10^{-6}$ is the fraction of ejecta rest-mass energy that goes into radioactive heating; Li & Paczyński 1998, Metzger et al. 2010b), and the photospheric temperature is $T_p \approx 10^4$ K, leading to a peak in the UV/optical band.

However, recent investigations of the opacities caused by r-process elements lead to a different conclusion. In a pioneering effort, Kasen, Badnell & Barnes (2013) and Barnes & Kasen (2013) demonstrated that the opacities of r-process matter, particularly the lanthanides, are about two orders of magnitude higher than for Fe-peak elements, with $\kappa \sim$ 10–100 cm^2 g^{-1} at $\lambda \sim$ 0.3–3 μm. With these opacity values, the expected peak timescale becomes longer by about an order of magnitude, $t_p \sim$ a few days, the peak luminosity becomes correspondingly lower, $L_p \sim$ a few $\times 10^{40}$ erg s^{-1}, and the characteristic temperatures shift to $T_p \sim 3,000$ K or a peak in the near-IR. Initial calculations of the r-process opacities and the resulting light curves and spectral energy distributions are provided in several recent papers (Barnes & Kasen 2013; Bauswein, Goriely & Janka 2013; Grossman et al. 2013; Tanaka & Hotokezaka 2013).

From an observational point of view, the effect of the larger opacities is dramatic. Instead of a transient that peaks in the blue part of the optical spectrum on a timescale of about a day and with an absolute magnitude of about -15, the kilonova is expected to peak with a similar absolute magnitude in the near-IR ($\sim$1.5 μm) on a timescale of about a week. In the optical bands, however, the peak absolute magnitude and timescale are about -12 mag and $\sim$1 day, respectively. This has significant implications for the detectability of kilonovae in blind surveys or in optical follow-up of GW sources (Section 10).

Optical searches for kilonova emission have been carried out in association with the short GRBs 050509B (Bloom et al. 2006), 070724A (Berger et al. 2009, Kocevski et al. 2010), 080503A (Perley et al. 2009), 080905A (Rowlinson et al. 2010), and 100206A (Perley et al. 2012), but none has yielded detections. These searches were guided by the early predictions for bright UV/optical emission with $\sim$1 day timescale, but unfortunately they do not provide strong constraints in the context of the current opacity estimates.

The first credible search focusing on the near-IR band was carried out for GRB 130603B at $z = 0.356$ using a combination of ground-based observations at $\lesssim$2 days and HST observations at about 9 and 30 days (Berger, Fong & Chornock 2013; Tanvir et al. 2013; see **Figure 19**). The afterglow of this burst was relatively bright at early time, but faded rapidly with $F_\nu \propto t^{-2.6}$ starting at about 0.45 days (indicative of a jet break as discussed in Section 8.4). Based on the decline rate and spectral energy distribution, the afterglow was predicted to have an exceedingly low flux at the time of the first HST observation, $m_{F606W} \approx 31$ mag and $m_{F160W} \approx 29.3$ mag. Instead, the HST observations revealed a source with $m_{F160W} \approx 25.8$ mag, but with no corresponding counterpart in the optical data (to $m_{F606W} \approx 27.7$ mag). The excess emission in the near-IR by a factor of about 25, coupled with the unusually red color, matches the predictions for a kilonova in the context of r-process opacities. At the time of writing, this is one of the strongest lines of evidence in favor of compact object binary progenitors.

The near-IR luminosity of the likely kilonova in GRB 130603B provides a rough estimate of the mass of r-process ejecta produced in this event. Utilizing the r-process opacity models, Berger, Fong & Chornock (2013) estimate $M_{ej} \approx$ 0.03–0.08 $M_\odot$ for ejecta velocities of $v_{ej} \approx$ 0.1–0.3 c. Clearly, the theoretical models are preliminary and there are variations between the calculations carried out by different groups (Barnes & Kasen 2013; Bauswein, Goriely & Janka 2013; Grossman et al. 2013; Tanaka & Hotokezaka 2013), but the results for GRB 130603B suggest that at least some short GRB progenitors may eject as much as $\sim$a few $\times 10^{-2}$ $M_\odot$ of r-process material.

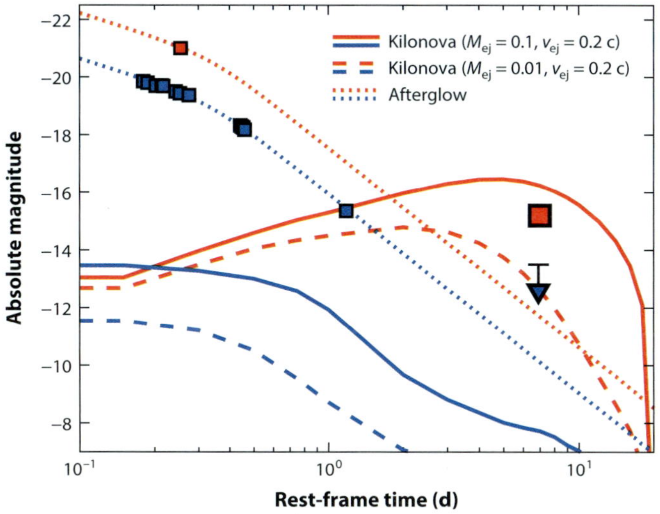

Figure 19

Optical (*blue*) and near-IR (*red*) observations of short GRB 130603B from ground-based telescopes ($\lesssim$2 days) and from the *Hubble Space Telescope* (HST) (Berger, Fong & Chornock 2013; Cucchiara et al. 2013; Fong et al. 2013; Tanvir et al. 2013; de Ugarte Postigo et al. 2014). The dotted lines are a fit to the early afterglow evolution, which underestimates the near-IR flux detected in the HST observations. The excess emission can be explained as emission from an *r*-process kilonova, resulting from a compact object binary merger. Theoretical kilonova models from Barnes & Kasen (2013) that take into account the opacities of *r*-process elements are shown as dashed and solid lines for ejecta masses of 0.01 and 0.1 $M_\odot$, respectively. From Berger, Fong & Chornock (2013).

Coupled with current estimates for the rate of NS-NS mergers, this indicates that compact object binary mergers may be the dominant site of *r*-process nucleosynthesis.

Detailed multiepoch and multiband follow-up observations of future nearby short GRBs will establish the ubiquity of kilonova associations and determine the range of ejecta masses. In particular, near-IR spectroscopy may reveal broad absorption features that will uniquely establish a nonafterglow origin and will also provide a rough measure of the typical ejecta velocities. The key challenges for such searches are the need for short GRBs at $z \lesssim 0.3$, which occur at a rate of $\lesssim$1 year^{-1}; a fast-fading afterglow to avoid significant contamination; and rapid follow-up with HST in the optical and near-IR. A key strategy to confirm a kilonova origin for any late-time excess emission is the use of multiple near-IR and optical filters to trace the expected sharp spectral cutoff at $\sim$1 μm.

10. SHORT GAMMA-RAY BURSTS AS ELECTROMAGNETIC COUNTERPARTS OF GRAVITATIONAL WAVE SOURCES

Several lines of evidence, presented in the previous sections, point to compact object mergers as the progenitors of short GRBs. As a result, existing observations of short GRBs and their afterglows can uniquely shed light on the on- and off-axis EM signatures that will accompany GW sources detected by Advanced LIGO/Virgo. The frequency range of these GW detectors is designed to uncover the final inspiral and merger of compact object binaries (NS-NS, NS-BH, BH-BH). The expected EM counterparts of Advanced LIGO/Virgo sources, and their detectability with existing and upcoming facilities, are a topic of growing interest, which has been explored in several recent papers (e.g., Metzger & Berger 2012; Piran, Nakar & Rosswog 2013). In assessing the importance and detectability of various EM counterparts, it is important to note that the Advanced LIGO/Virgo volume-averaged distance for NS-NS is about 200 Mpc, whereas ideally placed and oriented binaries (overhead relative to the detector network and with a face-on binary orientation) can be detected to about 450 Mpc (Abadie et al. 2010). The face-on orientation may also be ideal for the detection of an on-axis short GRB because the jet axis is expected to be oriented perpendicular to the plane of the binary's orbit. The typical localization region for the Advanced LIGO/Virgo network (three detectors) is $\sim$100 deg^2, although this may improve with future detectors in Japan and India (Fairhurst 2009). The poor localization capability presents a unique challenge for EM follow-up, and for counterpart discrimination and identification, compared for example with studies of short GRBs from *Swift*, which are localized to better than a few arcminutes radius.

Despite the obvious observational challenges, the information that can be gleaned from EM counterparts is invaluable (Kochanek & Piran 1993, Metzger & Berger 2012). In particular, the joint detection of GWs and a short GRB will definitively establish the compact object merger model and may also shed light on any differences in the short GRB population that are due to mergers of NS-NS versus NS-BH binaries. Even more broadly, EM counterparts are essential for pinpointing the exact locations of the mergers, thereby providing an association with host galaxies, measurements of a distance scale, and a relation to specific stellar populations that will shed light on the kick and merger timescale distributions. Similarly, the properties of the EM counterparts will shed light on the behavior of matter following the merger, namely accretion disks, jets, and dynamical or wind-driven outflows. A summary of potential counterparts is provided in **Figure 20**.

Metzger & Berger (2012) examined a wide range of putative on- and off-axis counterparts, and examined their potential impact for joint GW/EM studies in relation to four critical virtues: (*a*) detectability with present or upcoming telescope facilities using a reasonable allocation of resources; (*b*) the potential to accompany a high fraction of GW detections; (*c*) distinguishability from other astrophysical transients that may contaminate the large search areas; and (*d*) allowing for a localization to arcsecond precision. Other studies of EM counterparts have also explored the issue of detectability and background contamination, generally reaching similar conclusions, although some studies employ different criteria for "detectability" (e.g., Nakar & Piran 2011; Piran, Nakar & Rosswog 2013).

10.1. An On-Axis Short Gamma-Ray Burst Counterpart

At least some Advanced LIGO/Virgo sources are expected to be accompanied by on-axis short GRBs with detectable gamma-ray emission and a bright multiwavelength afterglow. Because existing gamma-ray satellites and large ground-based telescopes are capable of detecting short GRBs and their afterglows to $z \sim 1$, such an event at $\lesssim$450 Mpc ($z \lesssim 0.1$) will be exceedingly bright. In particular, using the measured X-ray and optical luminosities of even the least energetic short GRBs (with $E_{\gamma,\rm iso} \sim 10^{49}$ erg; **Figures 14** and **15**), I predict an X-ray flux at $\sim$0.5 days of $F_X \sim 5 \times 10^{-14}$ erg s^{-1} cm^{-2} and an optical brightness of $r \sim 23$ mag. These are detectable levels assuming that the gamma-ray emission can be pinpointed to an area typical of *Swift*/BAT localizations. For a short GRB with a median energy scale, the expected X-ray flux is $F_X \sim 5 \times 10^{-12}$ erg s^{-1} cm^{-2} and the optical brightness is $r \sim 20$ mag. Such a bright optical counterpart may be detectable even if the gamma-ray position is as poor as the GW localization, for example as is typical of *Fermi*/GBM positions. However, such a large area cannot be effectively searched with *Swift*/XRT, which has a small field of view of only 0.25 deg^2.

The temporally coincident detection of a short GRB and a GW source will serve as the clearest smoking gun for the compact object merger model. It may also help to improve the positional

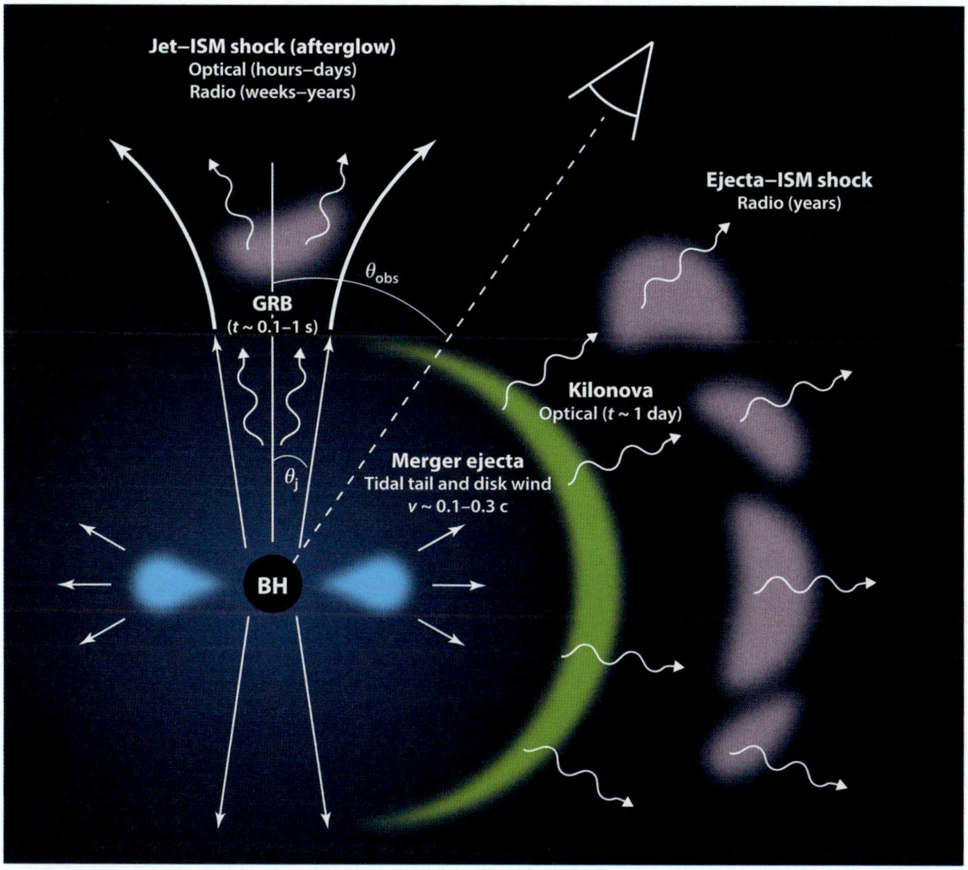

Figure 20

Potential electromagnetic counterparts of compact object binary mergers as a function of the observer viewing angle (θ_{obs}). Rapid accretion of a centrifugally supported disk (*blue*) powers a collimated relativistic jet, which produces a short GRB. Owing to relativistic beaming, the gamma-ray emission is restricted to observers with $\theta_{obs} \lesssim \theta_j$. Afterglow emission results from the interaction of the jet with the circumburst medium (*pink*). Optical afterglow emission is detectable for observers with $\theta_{obs} \lesssim 2\theta_j$. Radio afterglow emission is observable from all viewing angles once the jet decelerates to mildly relativistic velocities on a timescale of months-years and can also be produced on timescales of years from subrelativistic ejecta. Short-lived isotropic optical/near-IR emission lasting a few days (kilonova; *green*) can also accompany the merger, powered by the radioactive decay of r-process elements synthesized in the ejecta. Reprinted from Metzger & Berger (2012) with permission.

accuracy of the GW source (if detected by an instrument such as *Swift*/BAT), but even a poor gamma-ray localization will provide a convincing association based on the temporal coincidence. However, because the current estimate of the beaming fraction is $f_b \sim 70$ (Section 8.4), such joint detections will be rare. The occurrence rate can be estimated using the observed short GRB redshift distribution (Metzger & Berger 2012). In particular, there are currently no known short GRBs within the Advanced LIGO/Virgo maximum detection distance for NS-NS binaries of $z \approx 0.1$. Extrapolating the observed redshift distribution to $z \lesssim 0.1$, and correcting from the *Swift*/BAT field of view to roughly all-sky coverage (e.g., Interplanetary Network, *Fermi*/GBM), the expected coincidence rate is $\sim$0.3 year^{-1} (Metzger & Berger 2012). Thus, though joint GW and short GRB detections will play a critical role in establishing the identity of the progenitors, and will provide particularly bright counterparts, such detections are expected to be rare.

Another profitable approach in the context of joint gamma-ray and GW detections is to carry out a systematic search for GW emission in temporal coincidence with short GRBs (Abadie et al. 2012a; Metzger & Berger 2012; Kelley, Mandel & Ramirez-Ruiz 2013). The key advantage is that a search around a narrow time window, and with a position reconstruction that is consistent with the short GRB location, will allow for a reduced significance threshold. Such a search was carried out during LIGO Science Run 6 and Virgo Science Runs 2 and 3 (Abadie et al. 2012a), but unsurprisingly it did not yield any detections because the short GRB sample extends much beyond the reach of these initial detectors (**Figure 4**). However, the improved sensitivity of Advanced LIGO/Virgo may reach the nearest short GRBs.

10.2. An Off-Axis Optical Afterglow Counterpart

A coincident optical afterglow is more likely to be detected than a gamma-ray counterpart because even with an initial off-axis viewing angle the jet will eventually decelerate and spread into the observer's line of sight. However, larger off-axis viewing angles result in reduced peak brightness, owing to the longer timescale for the jet to spread into the line of sight. Using the existing sample of short GRB optical afterglows, it is possible to predict their appearance for off-axis observers. Metzger & Berger (2012) utilized the off-axis afterglow models of van Eerten, Zhang & MacFadyen (2010) with the parameters appropriate for on-axis short GRB afterglows (Section 8) and found that for an off-axis angle that is twice the jet opening angle ($\theta_{\rm obs} = 2\theta_{\rm j}$), the peak brightness of the afterglows at a typical distance of 200 Mpc is expected to be $\sim$23–25 mag with a peak time of $\sim$1–10 days, which is barely within the reach of the largest wide-field optical telescopes. At even larger off-axis angles the optical afterglow is essentially undetectable ($\sim$27–29 mag for $\theta_{\rm obs} = 4\theta_{\rm j}$).

The fraction of GW events that will be oriented within an angle of $2\theta_{\rm j}$ is $f_{\rm opt} \approx 6.8\langle\theta_{\rm j}\rangle^2 \sim 0.1$ (Metzger & Berger 2012), where the prefactor of 6.8 takes into account the nearly face-on orientation of the binary, which leads to a larger detection volume than for random orientations. Thus, a few GW sources per year may be accompanied by off-axis optical afterglow emission with a peak brightness of $\sim$23–25 mag and a peak timescale of $\sim$1–10 days. Finding such counterparts will require deep and rapid searches with wide-field optical telescopes, which is within the reach of existing and upcoming facilities.

10.3. An Off-Axis Radio Afterglow Counterpart

The underlying reason for the faintness of optical afterglows at large off-axis viewing angles is that by the time the jet spreads into the observer's line of sight (weeks to months to years after the event), the outflow becomes only mildly relativistic and the peak of the synchrotron spectrum is shifted to the radio gigahertz band. In principle, this means that radio afterglow emission can serve as an ideal counterpart for all GW sources owing to its detectability from any off-axis angle (Nakar & Piran 2011). In reality, the detectability of such radio emission depends on the kinetic energy of the blastwave and the ambient density. In addition to the initially collimated and relativistic outflow, radio emission will also be produced by any nonrelativistic isotropic outflow as it interacts with the ambient medium and decelerates (Nakar & Piran 2011). In the latter scenario, the initial velocity of the outflow is another critical parameter that determines the detectability of the radio emission. Such outflows are expected in compact object mergers from tidally unbound debris or from accretion disk winds (Rosswog et al. 1999, 2000; Ruffert & Janka 2001; Rosswog 2005; Etienne et al. 2008; Bauswein, Goriely & Janka 2013; Piran, Nakar & Rosswog 2013; Rosswog, Piran & Nakar 2013).

For the case of an initially collimated and relativistic outflow it is again useful to consider the observed properties of short GRB afterglows to investigate the expected brightness and detectability of an off-axis radio afterglow. The timescale for the blastwave to decelerate, isotropize, and reach peak brightness is $t_{\rm dec} \approx 30\,{\rm d}\; E_{\rm K,49}^{1/3} n_0^{-1/3} \beta_0^{-5/3}$, where $E_{\rm K,49}$ is the kinetic energy in units of 10^{49} erg and β_0 is the blastwave velocity in units of the speed of light; for an initially collimated and relativistic jet $\beta_0 \approx 1$ (Nakar & Piran 2011). Thus, for the typical energy and density scales inferred for short GRBs ($E_{\rm K,49} \sim$ a few and $n \sim 0.1$ cm^{-3}; Section 8), the resulting peak time is $t_{\rm dec} \sim 100$ days. The expected peak brightness (occurring at $t = t_{\rm dec}$) is $F_{\nu,\rm p} \approx 0.15$ mJy $E_{\rm K,49}\, n_0^{0.8} \epsilon_{\rm B,-2}^{0.8} \epsilon_{\rm e,-1}^{1.2} \beta_0^2 \nu_{\rm GHz}^{-0.6}$ (Nakar & Piran 2011), where I have scaled to a distance of 200 Mpc and to the fiducial values of $\epsilon_{\rm e} = 0.1$ and $\epsilon_{\rm B} = 0.01$ used in Section 8 to infer the short GRB afterglow properties. For the typical energy and density of short GRBs, this leads to $F_{\nu,\rm p} \sim 0.1$ mJy.

For dynamical or wind-driven ejecta the typical mass scale is $\sim 0.01\,{\rm M}_\odot$ with a velocity of $\beta \sim 0.2$, corresponding to $E_{\rm K,49} \approx 40$ (Rosswog, Piran & Nakar 2013). Because the density of $\sim$0.1 cm^{-3} inferred from short GRB observations is still relevant for this case, this indicates $t_{\rm dec} \approx 2$ year and $F_{\nu,\rm p} \approx 0.03$ mJy. For the most extreme outflow parameters indicated by numerical simulations ($\sim 0.1\,{\rm M}_\odot$ and $\beta \sim 0.3$), the peak brightness is $F_{\nu,\rm p} \approx 1.5$ mJy, but it will only occur on a timescale of $t_{\rm dec} \approx 10$ year. It is important to note that $t_{\rm dec}$ represents not only the timescale to reach peak brightness but also the typical timescale for the source to display significant brightness variability. From the point of view of a counterpart search, a source that does not vary significantly on a decade timescale, about a decade following a GW trigger is unlikely to be identified as an associated transient with any level of confidence.

The key question, therefore, is whether a radio transient with a typical peak brightness of $\sim$0.03–0.1 mJy and a peak/variability timescale of $\sim$1 year can be detected and identified in a $\sim$100 deg^2 region with existing and upcoming facilities. To assess this question, I note that for the JVLA, with its 0.45 deg^2 field of view at 1 GHz, full coverage of such an area requires at least 200 individual pointings. With a synthesized beam of about 5″ diameter, there are about 10^8 independent beams within this region, indicating that the minimum required significance level for a detection is 6σ (making the unrealistically optimistic assumption that the noise in a 100 deg^2 region is purely Gaussian with no systematic effects). Detecting such a source as a transient requires at least a factor of two change in brightness, which means that the observations need to reach a typical 1σ noise level of $\sim$3–10 μJy. With the JVLA this requires a few hours of integration per pointing or about a month of continuous observing per epoch. This unrealistic follow-up scenario is further exacerbated by the fact that at 1 GHz source confusion leads to a sensitivity floor of $\sim$0.03–0.1 mJy (depending on the JVLA configuration), which is comparable to the level of the expected signal. Thus, it is unlikely that radio emission will be detectable with JVLA follow-up of a typical Advanced LIGO/Virgo error region.

The Australian Square Kilometre Array Pathfinder (ASKAP) wide-field interferometer is expected to image $\sim 10^2$ deg^2 in only a few pointings, and is therefore better matched to typical GW error regions. However, ASKAP's limited angular resolution of about 10 arcsec will lead to source confusion at the level of the anticipated signal, $\sim$0.05–0.1 mJy (Johnston et al. 2008). Finally, lower-frequency instruments such as the Low-Frequency Array and the Murchison Widefield Array have sufficiently wide fields to instantaneously cover a full GW error region, but they are even more severely hampered by source confusion owing to their lower angular resolution and the increased background at lower radio frequencies ($\sim$0.1–0.3 GHz). Moreover, at these frequencies the typical timescale for the outflow to reach peak brightness will be years to decades, instead of months at gigahertz frequencies.

I therefore conclude that though radio counterparts are in principle detectable from all viewing angles, existing facilities are not well matched to the faintness of the anticipated signals. It is possible that in the long term, multidetector GW networks will localize some events to $\lesssim 10$ deg^2 (Fairhurst 2009), making radio searches more feasible. In addition, because the radio signal is delayed compared to counterparts at other wave bands, a more profitable approach may be to use radio observations to follow up candidate counterparts from gamma-ray, X-ray, or optical/near-IR searches, potentially as a way of distinguishing a true counterpart from unrelated sources (e.g., SNe, AGN).

10.4. A Kilonova Counterpart

As I discussed in Section 9, NS-NS and NS-BH mergers are also expected to be accompanied by isotropic optical/near-IR emission produced by the radioactive decay of r-process elements in a dynamical or accretion disk wind-driven outflow (Rosswog et al. 1999, 2000; Ruffert & Janka 2001; Rosswog 2005; Etienne et al. 2008; Bauswein, Goriely & Janka 2013; Piran, Nakar & Rosswog 2013; Rosswog, Piran & Nakar 2013). To date, the primary insight on the expected properties of such counterparts comes from numerical simulations coupled with r-process opacity calculations (Barnes & Kasen 2013; Bauswein, Goriely & Janka 2013; Grossman et al. 2013; Tanaka & Hotokezaka 2013) and from the near-IR counterpart associated with the short GRB 130603B (Section 9; Berger, Fong & Chornock 2013; Tanvir et al. 2013). In this context, the peak of the kilonova spectral energy distribution is expected at $\sim$1.5–3 μm (the near-IR *HK* bands) with an apparent brightness of $\sim$21.5–24 AB mag, and a timescale of $\sim$1 week. In the optical *izy* bands (0.7–1 μm), the peak brightness is $\sim$22–24 AB mag and the timescale is $\sim$1–3 days (e.g., Barnes & Kasen 2013).

Although the kilonova spectral energy distribution is expected to peak in the near-IR, deep wide-field searches in this band are not currently feasible. For example, the ESO's Visible and Infrared Survey Telescope for Astronomy (VISTA) has an instantaneous field of view of about 0.6 deg^2, requiring about 170 pointings to cover a typical GW error region. This limits the exposure time per pointing to $\sim$100 s for a single-night visit. The 5σ point-source sensitivity in such an exposure time is $H \approx 20.5$ AB mag, about a magnitude too shallow even for a bright kilonova. A better chance to detect the near-IR emission from a kilonova counterpart may be presented by a future space-based observatory, for example *Wide Field Infrared Survey Telescope* (WFIRST). The current design reference document (D. Spergel, N. Gehrels, J. Breckinridge, M. Donahue, A. Dressler, et al. unpublished manuscript, arXiv:1305.5425) indicates an instantaneous field of view of about 0.3 deg^2, requiring several hundred pointings to cover a typical GW error region. However, the point-source sensitivity in a 1-min exposure is expected to be $H \approx 25$ AB mag (5σ), allowing for the detection of a wide range of kilonovae.

A more profitable approach, especially prior to the launch of a facility such as WFIRST, is to use wide-field optical telescopes in the reddest bands accessible to CCD detectors (*izy*). In this context, there are several existing and upcoming facilities capable of mounting sensitive follow-up campaigns [e.g., the Panoramic Survey Telescope & Rapid Response System (Pan-STARRS), the Blanco Dark Energy Camera (DECam), the Subaru Hyper Suprime-Cam (HSC), the Large Synoptic Survey Telescope (LSST)]. Detecting the faint and fast kilonova signal with sufficient significance requires at least nightly coverage of the full GW error region. For example, Pan-STARRS has a field of view of 7 deg^2 and can therefore cover a typical error region with about 15 min per pointing, achieving a 5σ point-source sensitivity of $iz \approx 23.5$ AB mag. Similarly, with the DECam's 3 deg^2 field of view, a typical GW error region can be covered to a depth of $iz \approx 24.5$ AB mag (5σ) in a single night. A similar depth can be reached with Subaru/HSC, which has a larger aperture but a smaller field of view. Finally, LSST will be able to provide rapid and efficient coverage to $iz \approx 24.5$ AB mag (5σ) in $\sim$0.5 h or alternatively to $iz \approx 26$ AB mag in a single night. These sensitivities are sufficient to detect kilonovae over a broad range of expected ejecta masses provided that a systematic campaign lasting several days is undertaken. Equally importantly, nondetections with such searches will yield meaningful upper bounds on the ejecta properties.

Unlike in gamma-rays, and to some extent even in the radio band, optical searches with the required area and depth to robustly detect a kilonova will undoubtedly yield a wide range of unrelated variable and transient sources. To robustly identify or rule out a kilonova will therefore require the ability to reject background and foreground contaminants. One possibility is to restrict the search area to regions around galaxies within the expected Advanced LIGO/Virgo range of 200 Mpc (Nuttall & Sutton 2010; LIGO Sci. Collab. et al. 2012). Achieving this goal requires all-sky spectroscopic information for galaxies down to at least $\sim$0.1 L^*, which represent the hosts of short GRBs (**Figure 8**; Berger 2009). Such spectroscopic information can be obtained from a narrow-band Hα imaging survey or using an HI emission line radio survey.

The potential of such surveys was investigated by Metzger, Kaplan & Berger (2013) using information from existing galaxy surveys, as well as guidance from the distribution of short GRB host-galaxy properties (e.g., stellar masses, SFRs). The basic finding is that realistic Hα and HI surveys can uncover about 80–90% of the star formation within 200 Mpc, but only $\sim$30–50% of the stellar mass. The high incompleteness for stellar mass is due to the fact that about half of the stellar mass in the local Universe is contained in elliptical galaxies, which produce weak or no Hα emission and contain little neutral hydrogen gas. The level of completeness with respect to the population of short GRB hosts is about 50% for both methods (Metzger, Kaplan & Berger 2013). Thus, using galaxy catalogs to reduce the search area to locations around galaxies within 200 Mpc may result in a factor of 2–3 loss in the counterpart identification rate. By contrast, this approach may reduce the background and foreground contamination dramatically (Nuttall & Sutton 2010; LIGO Sci. Collab. et al. 2012). For NS-BH mergers, which can be detected to much larger distances, this method is likely to be inefficient.

The mitigation of contaminating sources may also benefit from color information. In the existing kilonova models, the spectral energy distribution is unusually red compared with most known transients. Therefore, the combined use of *iz* band photometry could substantially reduce the contamination from AGN flares, SNe, SN shock breakouts, or Galactic stellar flares, which generally peak in the optical/UV bands (Metzger & Berger 2012). Finally, the shorter duration of kilonova events, though still $\sim$1–3 days, can help to distinguish these sources from foreground stellar flares (generally about hour durations) and background SNe (generally about weeks durations). Clearly, the challenge of background/foreground rejection should be investigated using observations that match the anticipated signal in depth, timescale, and color. These parameters are not currently covered by ongoing time-domain surveys such as Pan-STARRS, Palomar Transient Factory (PTF), or Dark Energy Survey (DES).

10.5. Speculative Counterparts

In addition to potential EM counterparts that are motivated by existing short GRB observations, other speculative counterparts have been theorized in recent years. The validity of these theoretical models and the nature of any resulting EM radiation are difficult to assess without observational evidence, whereas the proposed energy scale for some counterparts is likely too low to be detectable with existing facilities. Zrake & MacFadyen (2013) argue that compact object mergers lead to the amplification of $\sim 10^{16}$ G magnetic fields on the merger timescale and speculate that if even $\sim 0.1\%$ of the available energy is channeled into gamma-rays, the resulting isotropic signal will be detectable with *Swift*/BAT. Palenzuela et al. (2013) consider the interaction of the NS magnetospheres and show that it can power EM emission with $\sim 10^{40}$–10^{43} erg s^{-1} prior to the merger. However, the nature of the resulting radiation (e.g., radio, gamma-ray) is unclear. Several researchers considered the formation of a massive millisecond magnetar as a merger remnant, which could power an isotropic counterpart with an optical/near-IR luminosity approaching those of SNe and a typical timescale of hours to days, as well as bright X-ray emission on a timescale of hours (Fan & Xu 2006; Gao et al. 2013; Zhang 2013; Yu, Zhang & Gao 2013). Lehner et al. (2012) consider the collapse of a massive, rapidly rotating magnetar remnant to a black hole and find that the detached magnetosphere can power a bright EM counterpart. Kyutoku, Ioka & Shibata (2014) suggest that the shock produced by the merger may accelerate the outer layer of the neutron star to relativistic velocities with an energy of $\sim 10^{47}$ erg, which may produce an isotropic radio to X-ray afterglow through interaction with the ambient medium. Another proposed energy source for EM radiation is the failure or cracking of the neutron star crust prior to the merger, with a potential energy release of up to $\sim 10^{47}$ erg, although the nature of any resulting radiation is unclear (Penner et al. 2012, Tsang et al. 2012).

11. THE PROGENITORS OF SHORT GAMMA-RAY BURSTS AND FUTURE DIRECTIONS

The discovery of short GRB afterglows in 2005 was a watershed event that led to the first identifications of host galaxies, to a determination of the distance scale, and to a delineation of the burst properties. In the span of less than a decade, intense follow-up efforts aimed at studying the afterglow emission and the host-galaxy population have led to several key findings that I summarized and explored in this review:

- Short GRBs span a wide range of redshifts, at least $z \sim 0.1$–1.5. The redshift distribution appears to be affected by the sensitivity threshold of *Swift*/BAT, as evidenced by the similar redshift distributions in early- and late-type hosts, despite the anticipated longer time delay in the former. The redshift distribution indicates typical progenitor delay times of less than a few gigayears.
- Short GRBs at $z \lesssim 0.5$, for which sensitive SN searches have been carried out, lack associations with Type Ic SNe similar to those that accompany long GRBs. Nearly all of these limits are for short GRBs in star-forming galaxies, indicating that such environments are not indicative of an association with massive stars.
- Short GRBs occur in both early- and late-type galaxies, with the former accounting for about 20% of the sample. The subdominant fraction of early-type hosts indicates that the short GRB rate is influenced by both stellar mass and star-formation activity. The inferred delay time distribution on the basis of the host demographics is roughly $P(\tau) \propto \tau^{-1}$.
- The distribution of host-galaxy stellar masses follows the expectation from the field galaxy mass function for early-type hosts, but is lower than expected for late-type hosts. This indicates that star-formation activity results in an elevated short GRB rate per unit stellar mass. Combined with the broad distribution of host-galaxy stellar population ages, the inferred delay time distribution roughly follows $P(\tau) \propto \tau^{-1}$.
- The host galaxies of short GRBs follow the general properties and trends of the field galaxy population (in terms of stellar masses, SFRs, SSFRs, and metallicities), but have systematically higher luminosities, larger stellar masses, older stellar population ages, higher metallicities, and lower SFRs and SSFRs than the hosts of long GRBs.

- Short GRBs have systematically larger radial offsets from their host galaxies than long GRBs, but match the predicted offset distribution of compact object binaries from population synthesis models that include natal kicks. About 10% of short GRBs have offsets of $\gtrsim$20 kpc, extending beyond the typical visible extent of their host galaxies. These bursts appear host-less in deep optical/near-IR imaging from HST, but exhibit nearby galaxies with a low probability of chance coincidence. The short GRB offsets normalized by host-galaxy size are similarly larger than those of long GRBs, core-collapse SNe, and Type Ia SNe, with only 20% located at $\lesssim 1\ r_e$ and about 20% located at $\gtrsim 5\ r_e$. These results are indicative of natal kicks or an origin in globular clusters, both of which point to compact object binary mergers. The inferred kick velocities are $\sim$20–140 km s^{-1}, which is in reasonable agreement with Galactic NS-NS binaries and population synthesis models.
- Short GRBs exhibit a weak spatial correlation with the underlying distribution of UV and optical light in their host galaxies; $\sim$50% of all bursts are located in the faintest regions of their hosts, often with no detectable underlying stellar light despite deep imaging with HST. This distribution is distinct from the strong correlation of long GRBs and core-collapse SNe with UV light and the correlation of Type Ia SNe with optical light. This is indicative of natal kicks that displace the progenitors from their birth sites to their eventual explosion sites.
- The afterglows of short GRBs are systematically less luminous than those of long GRBs. This is partly a result of the lower isotropic-equivalent energy scale of short GRBs, but it also reflects a lower density scale by about an order of magnitude. The distribution of optical afterglow luminosities indicates a typical density scale of $\sim$0.1 cm^{-3}, which is supported by radio limits that require a density scale of $\lesssim$1 cm^{-3}.
- Short GRBs exhibit evidence for collimation based on breaks in their X-ray, optical, and radio light curves that closely match the expectations for jet breaks. The inferred opening angles span $\sim5°$ to $\gtrsim 20°$, with a resulting mean beaming factor of about 70. This leads to a beaming-corrected energy scale of $\sim(0.5\text{–}5) \times 10^{49}$ erg and an event rate of $\sim 10^3$ Gpc^{-3} year^{-1}. The inferred rate within the Advanced LIGO sensitivity volume is $\sim$25 year^{-1}.
- There is evidence for a kilonova association from late near-IR excess emission in the short GRB 130603B, with an inferred *r*-process ejecta mass of $\sim 0.05\ M_\odot$. If confirmed with future kilonova detections, this result provides strong evidence for compact object binary progenitors, indicates that compact object mergers are the dominant site for *r*-process nucleosynthesis, and demonstrates that Advanced LIGO/Virgo GW sources will be accompanied by detectable optical/near-IR emission.
- In the context of compact object binary progenitors, short GRBs provide unique insight into the expected EM counterparts of Advanced LIGO/Virgo GW sources, both on- and off-axis. Coincident detections of GWs and on-axis short GRBs are expected to be rare, $\sim$0.3 year^{-1}. Off-axis optical afterglow emission is expected to be detectable for $\sim$10% of GW sources. Radio emission from an off-axis jet or from an isotropic mildly relativistic outflow is in principle detectable for all viewing angles, but the predicted fluxes are too low to be detected with existing or upcoming facilities. Finally, kilonova emission can be detected within the Advanced LIGO/Virgo volume using large wide-field optical telescopes in the reddest optical bands.

Within the range of proposed progenitor systems for short GRBs, the binary compact object merger scenario provides the best agreement with the broad range of observations. A dominant channel of young magnetars conflicts with the occurrence of short GRBs in early-type galaxies, with the large offsets and with the lack of spatial correlation with host UV light. A dominant channel of delayed magnetars, formed through white dwarf accretion-induced collapse or white dwarf binary mergers, can explain the host demographics, but does not naturally lead to the observed offset distribution or to the weak spatial correlation with optical light. Finally, models in which short GRBs are produced by massive stars in a similar fashion to long GRBs are ruled out by the lack of SN associations, the occurrence of short GRBs in early-type galaxies, the clear differences in the host-galaxy population, the lack of spatial association with UV light, and the large differences in energy and circumburst density scales. It is indeed quite remarkable that the underlying theoretical motivations that led to the compact object binary merger model prior to the identification of short GRBs as a distinct class (e.g., Eichler et al. 1989; Narayan, Paczynski & Piran 1992) have been borne out by the extensive observational effort over the past decade.

At the same time, there are clear challenges for the compact object binary merger model, which may require a more careful assessment of the physical conditions during the merger or perhaps a minor contribution from a distinct progenitor channel. For example, the ~100-s extended emission observed in about 15% of short GRBs does not naturally fit in the basic merger framework, although several theoretical explanations exist for this emission component (A.I. MacFadyen, E. Ramirez-Ruiz, W. Zhang, unpublished manuscript, astro-ph/0510192; Rosswog 2007; Metzger, Quataert & Thompson 2008; Metzger et al. 2010a; Bucciantini et al. 2012). Similarly, there are potential clues in the commonalities and differences between the prompt emission properties of long and short GRBs (with and without extended emission) that may shed light on the nature of the progenitors and/or the engine. In addition, it remains to be seen if NS-BH mergers contribute to the short GRB population, and if so, what observational signatures distinguish them from NS-NS binary mergers. The answer may have to await GW detections that can shed light on the relative fraction of occurrence of NS-NS and NS-BH binaries.

Continued studies of short GRBs from *Swift*, *Fermi*, and future gamma-ray missions remain a top priority. The existing sample already provides deep insight into the nature of the progenitors and the explosion properties, but larger samples, combined with individual high-quality events (e.g., GRB 130603B) are essential for addressing some of the open questions listed above. An operational gamma-ray satellite (preferably with all-sky coverage) in the Advanced LIGO/Virgo era is also critical for establishing a direct connection between compact object mergers and short GRBs.

DISCLOSURE STATEMENT

The author is not aware of any affiliations, memberships, funding, or financial holdings that might be perceived as affecting the objectivity of this review.

ACKNOWLEDGMENTS

I thank my primary collaborators in the study of short GRBs: Wen-fai Fong, Ryan Chornock, Raffaella Margutti, Brian Metzger, Ashley Zauderer, and Derek Fox. I also acknowledge helpful discussions and occasional collaborations with Alicia Soderberg, Ramesh Narayan, Josh Grindlay, Christopher Fryer, Enrico Ramirez-Ruiz, Eliot Quataert, Lars Bildsten, Daniel Kasen, Camille Leibler, S. Bradley Cenko, Ehud Nakar, Re'em Sari, Nial Tanvir, Andrew Levan, Krzysztof Belczynski, Duncan Brown, Luis Lehner, Daniel Holz, Alessandra Bounano, David Kaplan, Neil Gehrels, Jochen Greiner, Ryan Foley, Dale Frail, Avishay Gal-Yam, Antonino Cucchiara, Shrinivas Kulkarni, and Paul Price. I also thank the participants of the "Chirps, Mergers and Explosions: The Final Moments of Coalescing Compact Binaries" workshop at the Kavli Institute for Theoretical Physics, Santa Barbara, for stimulating discussions about short GRBs and GW astrophysics. Finally, I acknowledge support for some of this work from the National Science Foundation under grant AST-1107973, and from several NASA *Swift* and *Fermi* grants.

LITERATURE CITED

Abadie J, Abbott BP, Abbott R, Abbott TD, Abernathy M, et al. 2012a. *Ap. J.* 760:12
Abadie J, Abbott BP, Abbott R, Abernathy M, Accadia T, et al. 2010. *Class. Quantum Gravity* 27:173001
Abadie J, Abbott BP, Abbott TD, Abbott R, Abernathy M, et al. 2012b. *Ap. J.* 755:2
Abbott B, Abbott R, Adhikari R, Agresti J, Ajith P, et al. 2008. *Ap. J.* 681:1419–30
Abdo AA, Ackermann M, Ajello M, Asano K, Atwood WB, et al. 2009. *Nature* 462:331–34
Abdo AA, Ackermann M, Ajello M, Asano K, Atwood WB, et al. 2010. *Ap. J.* 712:558–64
Abramovici A, Althouse WE, Drever RWP, Gursel Y, Kawamura S, et al. 1992. *Science* 256:325–33
Accadia T, Acernese F, Antonucci F, Astone P, Ballardin G, et al. 2011. *Class. Quantum Gravity* 28:114002
Ackermann M, Asano K, Atwood WB, Axelsson M, Baldini L, et al. 2010. *Ap. J.* 716:1178–90
Aloy MA, Janka HT, Müller E. 2005. *Astron. Astrophys.* 436:273–311
Amati L, Frontera F, Tavani M, in't Zand JJM, Antonelli A, et al. 2002. *Astron. Astrophys.* 390:81–89
Antonelli LA, D'Avanzo P, Perna R, Amati L, Covino S, et al. 2009. *Astron. Astrophys.* 507:L45–48
Antoniadis J, Freire PCC, Wex N, Tauris TM, Lynch RS, et al. 2013. *Science* 340:448
Arnett WD. 1982. *Ap. J.* 253:785–97
Asano K, Guiriec S, Mészáros P. 2009. *Ap. J. Lett.* 705:L191–94
Band D, Matteson J, Ford L, Schaefer B, Palmer D, et al. 1993. *Ap. J.* 413:281–92
Baring MG, Harding AK. 1997. *Ap. J.* 491:663
Barnes J, Kasen D. 2013. *Ap. J.* 775:18–26
Barthelmy SD, Chincarini G, Burrows DN, Gehrels N, Covino S, et al. 2005. *Nature* 438:994–96
Battaglia G, Helmi A, Morrison H, Harding P, Olszewski EW, et al. 2005. *MNRAS* 364:433–42
Bauswein A, Goriely S, Janka HT. 2013. *Ap. J.* 773:78
Beckwith SVW, Stiavelli M, Koekemoer AM, Caldwell JAR, Ferguson HC, et al. 2006. *Astron. J.* 132:1729–55
Belczynski K, Holz DE, Fryer CL, Berger E, Hartmann DH, O'Shea B. 2010. *Ap. J.* 708:117–26
Belczynski K, Perna R, Bulik T, Kalogera V, Ivanova N, Lamb DQ. 2006. *Ap. J.* 648:1110–16
Bell EF, McIntosh DH, Katz N, Weinberg MD. 2003. *Ap. J. Suppl.* 149:289–312
Berger E. 2007. *Ap. J.* 670:1254–59
Berger E. 2009. *Ap. J.* 690:231–37
Berger E. 2010. *Ap. J.* 722:1946–61
Berger E. 2011. *New Astron. Rev.* 55:1–22
Berger E, Cenko SB, Fox DB, Cucchiara A. 2009. *Ap. J.* 704:877–82
Berger E, Chornock R, Holmes TR, Foley RJ, Cucchiara A, et al. 2011. *Ap. J.* 743:204
Berger E, Fong W, Chornock R. 2013. *Ap. J. Lett.* 774:L23–27
Berger E, Fox DB, Price PA, Nakar E, Gal-Yam A, et al. 2007a. *Ap. J.* 664:1000–10
Berger E, Kulkarni SR, Fox DB, Soderberg AM, Harrison FA, et al. 2005a. *Ap. J.* 634:501–8
Berger E, Kulkarni SR, Frail DA. 2003. *Ap. J.* 590:379–85
Berger E, Price PA, Cenko SB, Gal-Yam A, Soderberg AM, et al. 2005b. *Nature* 438:988–90
Berger E, Shin MS, Mulchaey JS, Jeltema TE. 2007b. *Ap. J.* 660:496–503
Berger E, Zauderer BA, Levan A, Margutti R, Laskar T, et al. 2013. *Ap. J.* 765:121
Bhattacharya D, van den Heuvel EPJ. 1991. *Phys. Rep.* 203:1–124
Birkl R, Aloy MA, Janka HT, Müller E. 2007. *Astron. Astrophys.* 463:51–67
Blandford RD, Znajek RL. 1977. *MNRAS* 179:433–56
Bloom JS, Butler NR, Perley DA. 2008. In *Gamma-Ray Bursts 2007: Proc. Santa Fe Conf.*, ed. M Galassi, D Palmer, E Fenimore. *AIP Conf. Proc.* 1000:11. Melville, NY: AIP
Bloom JS, Djorgovski SG, Kulkarni SR, Frail DA. 1998. *Ap. J. Lett.* 507:L25–28
Bloom JS, Frail DA, Sari R. 2001. *Astron. J.* 121:2879–88
Bloom JS, Kulkarni SR, Djorgovski SG. 2002. *Astron. J.* 123:1111–48
Bloom JS, Kulkarni SR, Djorgovski SG, Eichelberger AC, Côté P, et al. 1999. *Nature* 401:453–56
Bloom JS, Kulkarni SR, Price PA, Reichart D, Galama TJ, et al. 2002. *Ap. J. Lett.* 572:L45–49
Bloom JS, Perley DA, Chen HW, Butler N, Prochaska JX, et al. 2007. *Ap. J.* 654:878–84
Bloom JS, Prochaska JX, Pooley D, Blake CH, Foley RJ, et al. 2006. *Ap. J.* 638:354–68
Bloom JS, Sigurdsson S, Pols OR. 1999. *MNRAS* 305:763–69

Brodie JP, Strader J. 2006. *Annu. Rev. Astron. Astrophys.* 44:193–267
Bromberg O, Nakar E, Piran T, Sari R. 2013. *Ap. J.* 764:179
Bucciantini N, Metzger BD, Thompson TA, Quataert E. 2012. *MNRAS* 419:1537–45
Burgay M, D'Amico N, Possenti A, Manchester RN, Lyne AG, et al. 2003. *Nature* 426:531–33
Burrows DN, Grupe D, Capalbi M, Panaitescu A, Patel SK, et al. 2006. *Ap. J.* 653:468–73
Burrows DN, Romano P, Falcone A, Kobayashi S, Zhang B, et al. 2005. *Science* 309:1833–35
Campana S, Tagliaferri G, Lazzati D, Chincarini G, Covino S, et al. 2006. *Astron. Astrophys.* 454:113–17
Caputi KI, Lagache G, Yan L, Dole H, Bavouzet N, et al. 2007. *Ap. J.* 660:97–116
Caputi KI, McLure RJ, Dunlop JS, Cirasuolo M, Schael AM. 2006. *MNRAS* 366:609–23
Castro-Tirado AJ, de Ugarte Postigo A, Gorosabel J, Fathkullin T, Sokolov V, et al. 2005. *Astron. Astrophys.* 439:L15–18
Cenko SB, Frail DA, Harrison FA, Haislip JB, Reichart DE, et al. 2011. *Ap. J.* 732:29
Champion DJ, Lorimer DR, McLaughlin MA, Cordes JM, Arzoumanian Z, et al. 2004. *MNRAS* 350:L61–65
Chandra P, Frail DA. 2012. *Ap. J.* 746:156
Chevalier RA, Li ZY. 2000. *Ap. J.* 536:195–212
Chincarini G, Mao J, Margutti R, Bernardini MG, Guidorzi C, et al. 2010. *MNRAS* 406:2113–48
Chincarini G, Moretti A, Romano P, Falcone AD, Morris D, et al. 2007. *Ap. J.* 671:1903–20
Christensen L, Hjorth J, Gorosabel J. 2004. *Astron. Astrophys.* 425:913–26
Church RP, Levan AJ, Davies MB, Tanvir N. 2011. *MNRAS* 413:2004–14
Connaughton V. 2002. *Ap. J.* 567:1028–36
Costa E, Frontera F, Heise J, Feroci M, in't Zand J, et al. 1997. *Nature* 387:783–85
Coward DM, Howell EJ, Piran T, Stratta G, Branchesi M, et al. 2012. *MNRAS* 425:2668–73
Cucchiara A, Levan AJ, Fox DB, Tanvir NR, Ukwatta TN, et al. 2011. *Ap. J.* 736:7
Cucchiara A, Prochaska JX, Perley D, Cenko SB, Werk J, et al. 2013. *Ap. J.* 777:94
Dahle H, Sarazin CL, Lopez LA, Kouveliotou C, Patel SK, et al. 2013. *Ap. J.* 772:23
Dai ZG, Wang XY, Wu XF, Zhang B. 2006. *Science* 311:1127–29
D'Avanzo P, Malesani D, Covino S, Piranomonte S, Grazian A, et al. 2009. *Astron. Astrophys.* 498:711–21
de Ugarte Postigo A, Castro-Tirado AJ, Guziy S, Gorosabel J, Jóhannesson G, et al. 2006. *Ap. J. Lett.* 648:L83–87
de Ugarte Postigo A, Thoene CC, Rowlinson A, Garcia-Benito R, Levan AJ, et al. 2014. *Astron. Astrophys.* 563:A62
Della Valle M, Chincarini G, Panagia N, Tagliaferri G, Malesani D, et al. 2006. *Nature* 444:1050–52
Demorest PB, Pennucci T, Ransom SM, Roberts MSE, Hessels JWT. 2010. *Nature* 467:1081–83
Dessart L, Ott CD, Burrows A, Rosswog S, Livne E. 2009. *Ap. J.* 690:1681–705
Dezalay J-P, Barat C, Talon R, Syunyaev R, Terekhov O, Kuznetsov A. 1992. In *Gamma-Ray Bursts*, eds. W Paciesas, GJ Fishman. *AIP Conf. Proc.* 265:304. Melville, NY: AIP
Dhawan V, Mirabel IF, Ribó M, Rodrigues I. 2007. *Ap. J.* 668:430–34
Dichiara S, Guidorzi C, Frontera F, Amati L. 2013. *Ap. J.* 777:132–37
Djorgovski SG, Kulkarni SR, Bloom JS, Goodrich R, Frail DA, et al. 1998. *Ap. J. Lett.* 508:L17–20
Drout MR, Soderberg AM, Gal-Yam A, Cenko SB, Fox DB, et al. 2011. *Ap. J.* 741:97
Duncan RC, Thompson C. 1992. *Ap. J. Lett.* 392:L9–13
Eichler D, Livio M, Piran T, Schramm DN. 1989. *Nature* 340:126–28
Eke VR, Baugh CM, Cole S, Frenk CS, King HM, Peacock JA. 2005. *MNRAS* 362:1233–46
Etienne ZB, Faber JA, Liu YT, Shapiro SL, Taniguchi K, Baumgarte TW. 2008. *Phys. Rev. D* 77:084002
Fairhurst S. 2009. *New J. Phys.* 11:123006
Falcone AD, Morris D, Racusin J, Chincarini G, Moretti A, et al. 2007. *Ap. J.* 671:1921–38
Fan YZ, Xu D. 2006. *MNRAS* 372:L19–22
Fenimore EE, Epstein RI, Ho C. 1993. *Astron. Astrophys. Suppl. Ser.* 97:59–62
Filippenko AV. 1997. *Annu. Rev. Astron. Astrophys.* 35:309–55
Fishman GJ, Meegan CA. 1995. *Annu. Rev. Astron. Astrophys.* 33:415–58
Fong W, Berger E. 2013. *Ap. J.* 776:18
Fong W, Berger E, Chornock R, Margutti R, Levan AJ, et al. 2013. *Ap. J.* 769:56
Fong W, Berger E, Chornock R, Tanvir NR, Levan AJ, et al. 2011. *Ap. J.* 730:26

Fong W, Berger E, Fox DB. 2010. *Ap. J.* 708:9–25
Fong W, Berger E, Margutti R, Zauderer BA, Troja E, et al. 2012. *Ap. J.* 756:189
Fong W, Berger E, Metzger BD, Margutti R, Chornock R, et al. 2014. *Ap. J.* 780:118
Fox DB, Frail DA, Price PA, Kulkarni SR, Berger E, et al. 2005. *Nature* 437:845–50
Frail DA, Kulkarni SR, Nicastro L, Feroci M, Taylor GB. 1997. *Nature* 389:261–63
Frail DA, Kulkarni SR, Sari R, Djorgovski SG, Bloom JS, et al. 2001. *Ap. J. Lett.* 562:L55–58
Frederiks DD, Aptekar RL, Golenetskii SV, Il'inskii VN, Mazets EP, et al. 2004. In *3rd Rome Workshop on Gamma-Ray Bursts in the Afterglow Era*, ed. M Feroci, F Frontera, N Masetti, L Piro, *ASP Conf. Ser.* 312:197. San Francisco: ASP
Fruchter AS, Levan AJ, Strolger L, Vreeswijk PM, Thorsett SE, et al. 2006. *Nature* 441:463–68
Fryer C, Burrows A, Benz W. 1998. *Ap. J.* 496:333
Fryer C, Kalogera V. 1997. *Ap. J.* 489:244
Fryer CL, Woosley SE, Hartmann DH. 1999. *Ap. J.* 526:152–77
Fukugita M, Hogan CJ, Peebles PJE. 1998. *Ap. J.* 503:518–30
Fynbo JPU, Watson D, Thöne CC, Sollerman J, Bloom JS, et al. 2006. *Nature* 444:1047–49
Gal-Yam A, Fox DB, Price PA, Ofek EO, Davis MR, et al. 2006. *Nature* 444:1053–55
Gal-Yam A, Nakar E, Ofek EO, Cenko SB, Kulkarni SR, et al. 2008. *Ap. J.* 686:408–16
Galama TJ, Vreeswijk PM, van Paradijs J, Kouveliotou C, Augusteijn T, et al. 1998. *Nature* 395:670–72
Galbany L, Miquel R, Östman L, Brown PJ, Cinabro D, et al. 2012. *Ap. J.* 755:125
Gao H, Ding X, Wu XF, Zhang B, Dai ZG. 2013. *Ap. J.* 771:86
Gao WH, Fan YZ. 2006. *Chin. J. Astron. Astrophys.* 6:513–16
Gao WH, Mao J, Xu D, Fan YZ. 2009. *Ap. J. Lett.* 706:L33–36
Gehrels N, Barthelmy SD, Burrows DN, Cannizzo JK, Chincarini G, et al. 2008. *Ap. J.* 689:1161–72
Gehrels N, Chincarini G, Giommi P, Mason KO, Nousek JA, et al. 2004. *Ap. J.* 611:1005–20
Gehrels N, Norris JP, Barthelmy SD, Granot J, Kaneko Y, et al. 2006. *Nature* 444:1044–46
Gehrels N, Ramirez-Ruiz E, Fox DB. 2009. *Annu. Rev. Astron. Astrophys.* 47:567–617
Gehrels N, Sarazin CL, O'Brien PT, Zhang B, Barbier L, et al. 2005. *Nature* 437:851–54
Ghirlanda G, Ghisellini G, Celotti A. 2004. *Astron. Astrophys.* 422:L55–58
Ghirlanda G, Ghisellini G, Nava L. 2010. *Astron. Astrophys.* 510:L7
Ghirlanda G, Ghisellini G, Nava L. 2011. *MNRAS* 418:L109–13
Ghirlanda G, Magliocchetti M, Ghisellini G, Guzzo L. 2006. *MNRAS* 368:L20–24
Ghirlanda G, Nava L, Ghisellini G, Celotti A, Firmani C. 2009. *Astron. Astrophys.* 496:585–95
Giannios D. 2006. *Astron. Astrophys.* 455:L5–8
Goodman J. 1986. *Ap. J. Lett.* 308:L47–50
Goldstein A, Burgess JM, Preece RD, Briggs MS, Guireic S, et al. 2012. *Ap. J. Suppl.* 199:19–36
Goriely S, Bauswein A, Janka HT. 2011. *Ap. J. Lett.* 738:L32
Granot J, Sari R. 2002. *Ap. J.* 568:820–29
Grindlay J, Portegies Zwart S, McMillan S. 2006. *Nat. Phys.* 2:116–19
Grossman D, Korobkin O, Rosswog S, Piran T. 2013. *MNRAS*. In press. arXiv:1307.2943
Grupe D, Burrows DN, Patel SK, Kouveliotou C, Zhang B, et al. 2006. *Ap. J.* 653:462–67
Guetta D, Piran T. 2005. *Astron. Astrophys.* 435:421–26
Guetta D, Piran T. 2006. *Astron. Astrophys.* 453:823–28
Guiriec S, Briggs MS, Connaughton V, Kara E, Daigne F, et al. 2010. *Ap. J.* 725:225–41
Hakkila J, Preece RD. 2011. *Ap. J.* 740:104
Hansen BMS, Lyutikov M. 2001. *MNRAS* 322:695–701
Harris WE. 1991. *Annu. Rev. Astron. Astrophys.* 29:543–79
Harrison FA, Bloom JS, Frail DA, Sari R, Kulkarni SR, et al. 1999. *Ap. J. Lett.* 523:L121–24
Harry GM, LIGO Scientific Collaboration. 2010. *Class. Quantum Gravity* 27:084006
Hjorth J, Bloom JS. 2012. *The Gamma-Ray Burst—Supernova Connection*. In *Gamma-Ray Bursts*, ed. C Kouveliotou, RAMJ Wijers, SE Woosley. *Camb. Astrophys. Ser.* 51:169–90. Cambridge, UK: Cambridge Univ. Press
Hjorth J, Sollerman J, Gorosabel J, Granot J, Klose S, et al. 2005a. *Ap. J. Lett.* 630:L117–20
Hjorth J, Sollerman J, Møller P, Fynbo JPU, Woosley SE, et al. 2003. *Nature* 423:847–50

Hjorth J, Watson D, Fynbo JPU, Price PA, Jensen BL, et al. 2005b. *Nature* 437:859–61
Hogg DW, Pahre MA, McCarthy JK, Cohen JG, Blandford R, et al. 1997. *MNRAS* 288:404–10
Hotokezaka K, Kiuchi K, Kyutoku K, Okawa H, Sekiguchi Y, et al. 2013. *Phys. Rev. D* 87:024001
Hulse RA, Taylor JH. 1975. *Ap. J. Lett.* 195:L51–53
Hurley K, Berger E, Castro-Tirado A, Castro Cerón JM, Cline T, et al. 2002. *Ap. J.* 567:447–53
Hurley K, Rowlinson A, Bellm E, Perley D, Mitrofanov IG, et al. 2010. *MNRAS* 403:342–52
Ilbert O, Salvato M, Le Floc'h E, Aussel H, Capak P, et al. 2010. *Ap. J.* 709:644–63
Jakobsson P, Levan A, Fynbo JPU, Priddey R, Hjorth J, et al. 2006. *Astron. Astrophys.* 447:897–903
Janka HT, Eberl T, Ruffert M, Fryer CL. 1999. *Ap. J. Lett.* 527:L39–42
Johnston S, Taylor R, Bailes M, Bartel N, Baugh C, et al. 2008. *Exp. Astron.* 22:151–273
Kalogera V, Kim C, Lorimer DR, Burgay M, D'Amico N, et al. 2004. *Ap. J. Lett.* 601:L179–82
Kaneko Y, Preece RD, Briggs MS, Paciesas WS, Meegan CA, Band DL. 2006. *Ap. J. Suppl.* 166:298–340
Kann DA, Klose S, Zhang B, Covino S, Butler NR, et al. 2011. *Ap. J.* 734:96
Kasen D, Badnell NR, Barnes J. 2013. *Ap. J.* 774:25
Kehoe R, Akerlof C, Balsano R, Barthelmy S, Bloch J, et al. 2001. *Ap. J. Lett.* 554:L159–62
Kelley LZ, Mandel I, Ramirez-Ruiz E. 2013. *Phys. Rev. D* 87:123004
Kelly PL, Kirshner RP. 2012. *Ap. J.* 759:107
Kelly PL, Kirshner RP, Pahre M. 2008. *Ap. J.* 687:1201–7
Klebesadel RW, Strong IB, Olson RA. 1973. *Ap. J. Lett.* 182:L85
Kobulnicky HA, Kewley LJ. 2004. *Ap. J.* 617:240–61
Kocevski D, Thöne CC, Ramirez-Ruiz E, Bloom JS, Granot J, et al. 2010. *MNRAS* 404:963–74
Kochanek CS, Piran T. 1993. *Ap. J. Lett.* 417:L17
Kouveliotou C, Dieters S, Strohmayer T, van Paradijs J, Fishman GJ, et al. 1998. *Nature* 393:235–37
Kouveliotou C, Meegan CA, Fishman GJ, Bhat NP, Briggs MS, et al. 1993. *Ap. J. Lett.* 413:L101–4
Kouveliotou C, Norris JP, Cline TL, Dennis BR, Desai UD, et al. 1987. *Ap. J. Lett.* 322:L21–25
Krolik JH, Pier EA. 1991. *Ap. J.* 373:277–84
Kulkarni SR, Djorgovski SG, Odewahn SC, Bloom JS, Gal RR, et al. 1999. *Nature* 398:389–94
Kumar P, Barniol Duran R. 2010. *MNRAS* 409:226–36
Kyutoku K, Ioka K, Shibata M. 2014. *MNRAS* 437:L6
Kyutoku K, Ioka K, Shibata M. 2013. *Phys. Rev. D* 88:041503(R)
Lamb DQ. 1995. *Publ. Astron. Soc. Pac.* 107:1152
Laros JG, Fenimore EE, Klebesadel RW, Atteia JL, Boer M, et al. 1987. *Ap. J. Lett.* 320:L111–15
Lattimer JM, Mackie F, Ravenhall DG, Schramm DN. 1977. *Ap. J.* 213:225–33
Lattimer JM, Schramm DN. 1974. *Ap. J. Lett.* 192:L145–47
Lattimer JM, Schramm DN. 1976. *Ap. J.* 210:549–67
Lazzati D. 2005. *MNRAS* 357:722–31
Lazzati D, Begelman MC. 2005. *Ap. J.* 629:903–7
Lazzati D, Morsony BJ, Begelman MC. 2010. *Ap. J.* 717:239–44
Lazzati D, Ramirez-Ruiz E, Ghisellini G. 2001. *Astron. Astrophys.* 379:L39–43
Le Floc'h E, Papovich C, Dole H, Bell EF, Lagache G, et al. 2005. *Ap. J.* 632:169–90
Lee HK, Wijers RAMJ, Brown GE. 2000. *Phys. Rep.* 325:83–114
Lee WH, Ramirez-Ruiz E. 2007. *N. J. Phys.* 9:17
Lee WH, Ramirez-Ruiz E, López-Cámara D. 2009. *Ap. J. Lett.* 699:L93–96
Lee WH, Ramirez-Ruiz E, van de Ven G. 2010. *Ap. J.* 720:953–75
Lehner L, Palenzuela C, Liebling SL, Thompson C, Hanna C. 2012. *Phys. Rev. D* 86:104035
Leibler CN, Berger E. 2010. *Ap. J.* 725:1202–14
Levan AJ, Tanvir NR, Fruchter AS, Rol E, Fynbo JPU, et al. 2006a. *Ap. J. Lett.* 648:L9–12
Levan AJ, Wynn GA, Chapman R, Davies MB, King AR, et al. 2006b. *MNRAS* 368:L1–5
Levesque EM, Bloom JS, Butler NR, Perley DA, Cenko SB, et al. 2010a. *MNRAS* 401:963–72
Levesque EM, Kewley LJ, Berger E, Zahid HJ. 2010b. *Astron. J.* 140:1557–66
Li L, Paczyński B. 1998. *Ap. J. Lett.* 507:L59–62
Li W, Chornock R, Leaman J, Filippenko AV, Poznanski D, et al. 2011. *MNRAS* 412:1473–507

LIGO Sci. Collab., Virgo Collab., Abadie J, Abbott BP, Abbott R, Abbott TD, Abernathy M, et al. 2012. *Astron. Astrophys.* 539:A124
Lü HJ, Liang EW, Zhang BB, Zhang B. 2010. *Ap. J.* 725:1965–70
MacFadyen AI, Woosley SE. 1999. *Ap. J.* 524:262–89
Malesani D, Covino S, D'Avanzo P, D'Elia V, Fugazza D, et al. 2007. *Astron. Astrophys.* 473:77–84
Mannucci F, Cresci G, Maiolino R, Marconi A, Gnerucci A. 2010. *MNRAS* 408:2115–27
Mannucci F, Della Valle M, Panagia N. 2006. *MNRAS* 370:773–83
Mannucci F, Della Valle M, Panagia N, Cappellaro E, Cresci G, et al. 2005. *Astron. Astrophys.* 433:807–14
Maoz D, Badenes C. 2010. *MNRAS* 407:1314–27
Maoz D, Mannucci F, Brandt TD. 2012. *MNRAS* 426:3282–94
Maoz D, Mannucci F, Li W, Filippenko AV, Della Valle M, Panagia N. 2011. *MNRAS* 412:1508–21
Margutti R, Berger E, Fong W, Zauderer BA, Cenko SB, et al. 2012. *Ap. J.* 756:63
Margutti R, Chincarini G, Granot J, Guidorzi C, Berger E, et al. 2011. *MNRAS* 417:2144–60
Margutti R, Guidorzi C, Chincarini G, Bernardini MG, Genet F, et al. 2010. *MNRAS* 406:2149–67
Margutti R, Zaninoni E, Bernardini MG, Chincarini G, Pasotti F, et al. 2013. *MNRAS* 428:729–42
McGlynn S, Foley S, McBreen S, Hanlon L, O'Connor R, et al. 2008. *Astron. Astrophys.* 486:405–10
Meegan CA, Fishman GJ, Wilson RB, Horack JM, Brock MN, et al. 1992. *Nature* 355:143–45
Mészáros P. 2002. *Annu. Rev. Astron. Astrophys.* 40:137–69
Mészáros P, Rees MJ. 1993. *Ap. J. Lett.* 418:L59
Mészáros P, Rees MJ. 1997. *Ap. J.* 476:232
Metzger BD, Arcones A, Quataert E, Martínez-Pinedo G. 2010a. *MNRAS* 402:2771–77
Metzger BD, Berger E. 2012. *Ap. J.* 746:48
Metzger BD, Kaplan DL, Berger E. 2013. *Ap. J.* 764:149
Metzger BD, Martínez-Pinedo G, Darbha S, Quataert E, Arcones A, et al. 2010b. *MNRAS* 406:2650–62
Metzger BD, Quataert E, Thompson TA. 2008. *MNRAS* 385:1455–60
Metzger MR, Djorgovski SG, Kulkarni SR, Steidel CC, Adelberger KL, et al. 1997. *Nature* 387:878–80
Modjaz M, Kewley L, Kirshner RP, Stanek KZ, Challis P, et al. 2008. *Astron. J.* 135:1136–50
Nakar E. 2007. *Phys. Rep.* 442:166–236
Nakar E, Gal-Yam A, Fox DB. 2006. *Ap. J.* 650:281–90
Nakar E, Piran T. 2002. *MNRAS* 330:920–26
Nakar E, Piran T. 2011. *Nature* 478:82–84
Narayan R, Paczynski B, Piran T. 1992. *Ap. J. Lett.* 395:L83–86
Nicuesa Guelbenzu A, Klose S, Greiner J, Kann DA, Krühler T, et al. 2012a. *Astron. Astrophys.* 548:A101
Nicuesa Guelbenzu A, Klose S, Krühler T, Greiner J, Rossi A, et al. 2012b. *Astron. Astrophys.* 538:L7
Nicuesa Guelbenzu A, Klose S, Rossi A, Kann DA, Krühler T, et al. 2011. *Astron. Astrophys.* 531:L6
Norris JP, Bonnell JT. 2006. *Ap. J.* 643:266–75
Norris JP, Cline TL, Desai UD, Teegarden BJ. 1984. *Nature* 308:434
Norris JP, Gehrels N, Scargle JD. 2010. *Ap. J.* 717:411–19
Norris JP, Gehrels N, Scargle JD. 2011. *Ap. J.* 735:23
Norris JP, Hertz P, Wood KS, Kouveliotou C. 1991. *Ap. J.* 366:240–52
Nousek JA, Kouveliotou C, Grupe D, Page KL, Granot J, et al. 2006. *Ap. J.* 642:389–400
Nuttall LK, Sutton PJ. 2010. *Phys. Rev. D* 82:102002
Nysewander M, Fruchter AS, Pe'er A. 2009. *Ap. J.* 701:824–36
Ofek EO, Cenko SB, Gal-Yam A, Fox DB, Nakar E, et al. 2007. *Ap. J.* 662:1129–35
Ofek EO, Muno M, Quimby R, Kulkarni SR, Stiele H, et al. 2008. *Ap. J.* 681:1464–69
O'Shaughnessy R, Belczynski K, Kalogera V. 2008. *Ap. J.* 675:566–85
Paciesas WS, Briggs MS, Preece RD, Mallozzi RS. 2003. In *Gamma-Ray Burst and Afterglow Astronomy 2001: A Workshop Celebrating the First Year of the HETE Mission*, eds. GR Ricker, RK Vanderspek. *AIP Conf. Proc.* 662:248. Melville, NY: AIP
Paczynski B. 1986. *Ap. J. Lett.* 308:L43–46
Paczynski B. 1995. *Publ. Astron. Soc. Pac.* 107:1167
Paczynski B, Rhoads JE. 1993. *Ap. J. Lett.* 418:L5
Palenzuela C, Lehner L, Ponce M, Liebling SL, Anderson M, et al. 2013. *Phys. Rev. Lett.* 111:061105

Panaitescu A, Kumar P. 2002. *Ap. J.* 571:779–89
Panaitescu A, Kumar P, Narayan R. 2001. *Ap. J. Lett.* 561:L171–74
Pedersen K, Elíasdóttir Á, Hjorth J, Starling R, Cerón JMC, et al. 2005. *Ap. J. Lett.* 634:L17–20
Penner AJ, Andersson N, Jones DI, Samuelsson L, Hawke I. 2012. *Ap. J. Lett.* 749:L36
Perley DA, Metzger BD, Granot J, Butler NR, Sakamoto T, et al. 2009. *Ap. J.* 696:1871–85
Perley DA, Modjaz M, Morgan AN, Cenko SB, Bloom JS, et al. 2012. *Ap. J.* 758:122
Perna R, Armitage PJ, Zhang B. 2006. *Ap. J. Lett.* 636:L29–32
Perna R, Belczynski K. 2002. *Ap. J.* 570:252–63
Pian E, Mazzali PA, Masetti N, Ferrero P, Klose S, et al. 2006. *Nature* 442:1011–13
Piran T. 1992. *Ap. J. Lett.* 389:L45–48
Piran T. 2004. *Rev. Mod. Phys.* 76:1143–210
Piran T, Nakar E, Rosswog S. 2013. *MNRAS* 430:2121–36
Piranomonte S, D'Avanzo P, Covino S, Antonelli LA, Beardmore AP, et al. 2008. *Astron. Astrophys.* 491:183–88
Prieto JL, Stanek KZ, Beacom JF. 2008. *Ap. J.* 673:999–1008
Prochaska JX, Bloom JS, Chen HW, Foley RJ, Perley DA, et al. 2006. *Ap. J.* 642:989–94
Proga D, Zhang B. 2006. *MNRAS* 370:L61–65
Qin B, Wu XP, Chu MC, Fang LZ, Hu JY. 1998. *Ap. J. Lett.* 494:L57
Qin YP, Chen ZF. 2013. *MNRAS* 430:163–73
Rees MJ, Mészáros P. 1992. *MNRAS* 258:41P–43P
Rezzolla L, Giacomazzo B, Baiotti L, Granot J, Kouveliotou C, Aloy MA. 2011. *Ap. J. Lett.* 732:L6
Rhoads JE. 1999. *Ap. J.* 525:737–49
Ricker GR, Atteia JL, Crew GB, Doty JP, Fenimore EE, et al. 2003. In *Gamma-Ray Burst and Afterglow Astronomy 2001: A Workshop Celebrating the First Year of the HETE Mission*, eds. GR Ricker, RK Vanderspek. *AIP Conf. Proc.* 662:3. Melville, NY: AIP
Roberts LF, Kasen D, Lee WH, Ramirez-Ruiz E. 2011. *Ap. J. Lett.* 736:L21
Rosswog S. 2005. *Ap. J.* 634:1202–13
Rosswog S. 2007. *MNRAS* 376:L48–51
Rosswog S, Davies MB, Thielemann FK, Piran T. 2000. *Astron. Astrophys.* 360:171–84
Rosswog S, Liebendörfer M, Thielemann FK, Davies MB, Benz W, Piran T. 1999. *Astron. Astrophys.* 341:499–526
Rosswog S, Piran T, Nakar E. 2013. *MNRAS* 430:2585–604
Rosswog S, Ramirez-Ruiz E. 2003. *MNRAS* 336:L7–11
Rosswog S, Ramirez-Ruiz E, Davies MB. 2003. *MNRAS* 345:1077–90
Rowlinson A, O'Brien PT, Metzger BD, Tanvir NR, Levan AJ. 2013. *MNRAS* 430:1061–87
Rowlinson A, Wiersema K, Levan AJ, Tanvir NR, O'Brien PT, et al. 2010. *MNRAS* 408:383–91
Ruderman M. 1975. *Ann. NY Acad. Sci.* 262:164
Ruffert M, Janka HT. 2001. *Astron. Astrophys.* 380:544–77
Salvaterra R, Devecchi B, Colpi M, D'Avanzo P. 2010. *MNRAS* 406:1248–52
Sari R, Piran T, Halpern JP. 1999. *Ap. J. Lett.* 519:L17–20
Sari R, Piran T, Narayan R. 1998. *Ap. J. Lett.* 497:L17
Savaglio S, Glazebrook K, Le Borgne D. 2009. *Ap. J.* 691:182–211
Schmidt M. 2001. *Ap. J. Lett.* 559:L79–82
Schmidt M, Higdon JC, Hueter G. 1988. *Ap. J. Lett.* 329:L85–87
Setiawan S, Ruffert M, Janka HT. 2004. *MNRAS* 352:753–58
Shemi A, Piran T. 1990. *Ap. J. Lett.* 365:L55–58
Shibata M, Suwa Y, Kiuchi K, Ioka K. 2011. *Ap. J. Lett.* 734:L36
Shin MS, Berger E. 2007. *Ap. J.* 660:1146–50
Soderberg AM, Berger E, Kasliwal M, Frail DA, Price PA, et al. 2006a. *Ap. J.* 650:261–71
Soderberg AM, Kulkarni SR, Nakar E, Berger E, Cameron PB, et al. 2006b. *Nature* 442:1014–17
Stanek KZ, Garnavich PM, Kaluzny J, Pych W, Thompson I. 1999. *Ap. J. Lett.* 522:L39–42
Stanek KZ, Gnedin OY, Beacom JF, Gould AP, Johnson JA, et al. 2006. *Acta Astron.* 56:333–45
Stanek KZ, Matheson T, Garnavich PM, Martini P, Berlind P, et al. 2003. *Ap. J. Lett.* 591:L17–20
Stone N, Loeb A, Berger E. 2013. *Phys. Rev. D* 87:084053

Stratta G, D'Avanzo P, Piranomonte S, Cutini S, Preger B, et al. 2007. *Astron. Astrophys.* 474:827–35
Sullivan M, Le Borgne D, Pritchet CJ, Hodsman A, Neill JD, et al. 2006. *Ap. J.* 648:868–83
Svensson KM, Levan AJ, Tanvir NR, Fruchter AS, Strolger LG. 2010. *MNRAS* 405:57–76
Tanaka M, Hotokezaka K. 2013. *Ap. J.* 775:113
Tanvir NR, Levan AJ, Fruchter AS, Hjorth J, Hounsell RA, et al. 2013. *Nature* 500:547–49
Taylor GB, Frail DA, Berger E, Kulkarni SR. 2004. *Ap. J. Lett.* 609:L1–4
Thöne CC, Fynbo JPU, Östlin G, Milvang-Jensen B, Wiersema K, et al. 2008. *Ap. J.* 676:1151–61
Tremonti CA, Heckman TM, Kauffmann G, Brinchmann J, Charlot S, et al. 2004. *Ap. J.* 613:898–913
Troja E, King AR, O'Brien PT, Lyons N, Cusumano G. 2008. *MNRAS* 385:L10–14
Troja E, Rosswog S, Gehrels N. 2010. *Ap. J.* 723:1711–17
Tsang D, Read JS, Hinderer T, Piro AL, Bondarescu R. 2012. *Phys. Rev. Lett.* 108:011102
Tsutsui R, Yonetoku D, Nakamura T, Takahashi K, Morihara Y. 2013. *MNRAS* 431:1398–404
UK *Swift* Sci. Data Cent. 2011. Table of *Swift* XRT GRB positions. Last updated May 24. **http://www.swift.ac.uk/xrt_positions/**
van den Bergh S. 1983. *Ap. Space Sci.* 97:385–88
van den Bergh S. 1990. *Publ. Astron. Soc. Pac.* 102:1318–20
van den Bergh S, Tammann GA. 1991. *Annu. Rev. Astron. Astrophys.* 29:363–407
van Eerten H, Zhang W, MacFadyen A. 2010. *Ap. J.* 722:235–47
van Paradijs J, Groot PJ, Galama T, Kouveliotou C, Strom RG, et al. 1997. *Nature* 386:686–89
van Paradijs J, Kouveliotou C, Wijers RAMJ. 2000. *Annu. Rev. Astron. Astrophys.* 38:379–425
Villasenor JS, Lamb DQ, Ricker GR, Atteia JL, Kawai N, et al. 2005. *Nature* 437:855–58
Virgili FJ, Zhang B, O'Brien P, Troja E. 2011. *Ap. J.* 727:109–22
Wainwright C, Berger E, Penprase BE. 2007. *Ap. J.* 657:367–77
Wang C, Lai D, Han JL. 2006. *Ap. J.* 639:1007–17
Wang X, Wang L, Filippenko AV, Zhang T, Zhao X. 2013. *Science* 340:170–73
Watson D, Hjorth J, Jakobsson P, Xu D, Fynbo JPU, et al. 2006. *Astron. Astrophys.* 454:L123–26
Waxman E, Kulkarni SR, Frail DA. 1998. *Ap. J.* 497:288
Wong TW, Willems B, Kalogera V. 2010. *Ap. J.* 721:1689–701
Woosley SE, Bloom JS. 2006. *Annu. Rev. Astron. Astrophys.* 44:507–56
Xue XX, Rix HW, Zhao G, Re Fiorentin P, Naab T, et al. 2008. *Ap. J.* 684:1143–58
Yonetoku D, Murakami T, Nakamura T, Yamazaki R, Inoue AK, Ioka K. 2004. *Ap. J.* 609:935–51
Yost SA, Harrison FA, Sari R, Frail DA. 2003. *Ap. J.* 597:459–73
Yu YW, Zhang B, Gao H. 2014. *Ap. J. Lett.* 776:L40
Zhang B. 2013. *Ap. J. Lett.* 763:L22
Zhang B, Zhang BB, Liang EW, Gehrels N, Burrows DN, Mészáros P. 2007. *Ap. J. Lett.* 655:L25–28
Zhang B, Zhang BB, Virgili FJ, Liang EW, Kann DA, et al. 2009. *Ap. J.* 703:1696–724
Zheng Z, Ramirez-Ruiz E. 2007. *Ap. J.* 665:1220–26
Zou YC, Fan YZ, Piran T. 2009. *MNRAS* 396:1163–70
Zrake J, MacFadyen AI. 2013. *Ap. J. Lett.* 769:L29

Observational Clues to the Progenitors of Type Ia Supernovae

Dan Maoz,[1] Filippo Mannucci,[2] and Gijs Nelemans[3,4]

[1]School of Physics and Astronomy, Tel-Aviv University, Tel Aviv 69978, Israel; email: maoz@astro.tau.ac.il

[2]INAF, Arcetri Astrophysical Observatory, 50125 Firenze, Italy; email: filippo@arcetri.astro.it

[3]Department of Astrophysics/IMAPP, Radboud University Nijmegen, Nijmegen 6500 GL, The Netherlands; email: nelemans@astro.ru.nl

[4]Institute for Astronomy, KU Leuven, 3000 Leuven, Belgium

Annu. Rev. Astron. Astrophys. 2014. 52:107–70

First published online as a Review in Advance on June 16, 2014

The *Annual Review of Astronomy and Astrophysics* is online at astro.annualreviews.org

This article's doi:
10.1146/annurev-astro-082812-141031

Keywords

intermediate-mass stars, white dwarfs, interacting binaries, novae, supersoft X-ray sources, explosive nucleosynthesis, supernova rates, supernova remnants, stellar populations, stellar mergers, stellar rotation, stellar winds, chemical evolution, cosmological distance indicators, gravitational wave sources

Abstract

Type Ia supernovae (SNe Ia) are important distance indicators, element factories, cosmic-ray accelerators, kinetic-energy sources in galaxy evolution, and end points of stellar binary evolution. It has long been clear that a SN Ia must be the runaway thermonuclear explosion of a degenerate carbon-oxygen stellar core, most likely a white dwarf (WD). However, the specific progenitor systems of SNe Ia, and the processes that lead to their ignition, have not been identified. Two broad classes of progenitor binary systems have long been considered: single-degenerate (SD), in which a WD gains mass from a nondegenerate star; and double-degenerate (DD), involving the merger of two WDs. New theoretical work has enriched these possibilities with some interesting updates and variants. We review the significant recent observational progress in addressing the progenitor problem. We consider clues that have emerged from the observed properties of the various proposed progenitor populations, from studies of SN Ia sites—pre- and postexplosion—from analysis of the explosions themselves and from the measurement of event rates. The recent nearby and well-studied event, SN 2011fe, has been particularly revealing. The observational results are not yet conclusive and sometimes prone to competing theoretical interpretations. Nevertheless, it appears that DD progenitors, long considered the underdog option, could be behind some, if not all, SNe Ia. We point to some directions that may lead to future progress.

Because I have contemplated explaining what I think, not only about the location and movement of this light, but also about its substance and origin, and believing that I have found an explanation that, for lack of evident contradictions, may well be true, I have finally arrived at the belief of being capable of knowing something about this wonder, beyond the point where pure conjecture ends.

—Letter by Galileo to O. Castelli regarding the *Stella Nova* of 1604

1. INTRODUCTION

1.1. Type Ia Supernovae: A Brief Overview

The observations of the supernovae (SNe) of 1572 and 1604 and the attempts by Tycho, Kepler, Galileo, and others to understand their natures and locations were transformational events in the history of science. These attempts signal the beginnings of modern astrophysics. Among the handful of historical SN events in the Galaxy recorded over the past two millennia, we now know that at least some were core-collapse (CC) SNe, involving the explosion of massive ($M \gtrsim 8\,M_\odot$) stars, whereas others were the thermonuclear explosions of lower-mass stars, now known as Type Ia supernovae (SNe Ia). In modern-day surveys that search for stellar explosions, SNe Ia are often the ones most frequently discovered, owing mainly to their large optical luminosities, $\sim 10^{43}$ erg s^{-1} near the time of maximum light.

SNe Ia are classified (see review by Filippenko 1997) by a lack of hydrogen and helium signatures in their spectra and by kinematically broadened and blueshifted ($\sim 10^3$–10^4 km s^{-1}) features, mainly of silicon, iron, and calcium, around the time of maximum light. Contrary to CC SNe (e.g., Smartt 2009), SNe Ia are observed to explode both in young and in old stellar populations. The spectra and light curves of most SNe Ia are remarkably uniform. The light curves rise to maximum light within ~15–20 days, decline by about 3 mag in one month, and then decline by a further mag each month, with the more specific behavior depending on the observed photometric band. On the basis of the light-curve shape, optical spectroscopy, and X-ray analysis of SN Ia remnants, the observed luminosity of SNe Ia is driven by inverse-β and electron-capture radioactive decays, first of ^{56}Ni to ^{56}Co, and then of ^{56}Co to stable ^{56}Fe, with exponential timescales of 9 days and 114 days, respectively. Gamma rays and positrons from these decays are reprocessed by the optically thick ejecta into the optical photons that dominate the luminosity (Colgate & McKee 1969). SNe Ia are a decisively optical-range phenomenon, with ~85% of the luminosity emitted between the U and I bands (e.g., Howell et al. 2009).

The tight correlation between the luminosity of SNe Ia at maximum light and various measures of light-curve development speed (Phillips 1993, and see Howell 2011 for a review) has made them excellent cosmological distance indicators. As such, they provided the first evidence for acceleration of the cosmic expansion (Riess et al. 1998, Perlmutter et al. 1999), and hence for the current, dark-energy-dominated cosmic mass-energy inventory. SNe Ia continue being a major cosmographic tool, with future efforts focusing on determining the evolution, on cosmological timescales, of the dark-energy equation of state (Howell et al. 2011). The yet-unknown identity of the progenitor systems of SNe Ia, the subject of this review, is a concern for SN Ia cosmography, given that changes with cosmic time or environment, in the progenitor populations or in some details of the explosions, could, in principle, lead to systematic errors in distances deduced based on calibrations of nearby SNe Ia. For example, systematic differences have been reported in the residuals in the Hubble diagram, depending on host galaxy mass and metallicity (e.g., D'Andrea et al. 2011, Hayden et al. 2013, Pan et al. 2013). These differences could impact the measurements of cosmological parameters, as low-mass and low-metallicity galaxies dominate the production of SNe Ia at progressively higher redshifts.

Every normal SN Ia enriches the interstellar medium (ISM) with roughly $0.7\,M_\odot$ of iron (to within a factor of a few) and a similar total amount in other elements, including mainly carbon, oxygen, neon, magnesium, silicon, sulfur, argon, and calcium. Combined with their large numbers, this means that SNe Ia play an important role in chemical evolution, and understanding such evolution requires knowing the dependences, on cosmic time and environment, of SN Ia rates and metal yields (e.g., Wiersma, Schaye & Theuns 2011).

The kinetic energy of SN Ia ejecta also affects gas dynamics and star formation in galaxies, contributing to the ejection of matter from galaxies to the intergalactic medium, and thus SNe Ia are a factor in galaxy evolution (e.g., Davé, Oppenheimer & Finlator 2011). As is the case for other SN types, the remnants of SNe Ia are the likely sites for the acceleration of cosmic rays up to energies of $\sim 10^{15}$ eV (e.g., Drury 2012). And, as detailed further in this review, SNe Ia may or may not signal two important end points in binary stellar evolution—accretion onto, and/or the mergers of, white dwarfs (WDs). The final stages of the latter events are expected to be the main source of foreground, though also interesting signals, for space-based gravitational-wave observatories (e.g., Amaro-Seoane et al. 2013). For all of these reasons, it is important to understand the physics of all stages of the development of SNe Ia and, particularly, to identify their progenitor systems.

1.2. Basic Physics of SNe Ia

Physical models of SN Ia explosions have been reviewed by Hillebrandt & Niemeyer (2000), with updates by Hillebrandt et al. (2013). Here we only skim the surface. The energetics and chemical composition of SN Ia explosions, as measured in the course of the explosions and in their remnants, suggest (Hoyle & Fowler 1960) that they involve the thermonuclear combustion of a degenerate stellar core. The absence of hydrogen and helium in the spectrum, and the occurrence of some SNe Ia in old stellar populations, indicates that this core is most likely a WD of carbon-oxygen (C/O) composition. The burning is partly into iron-group elements and partly into intermediate-mass elements, with some residual unburnt carbon and oxygen, as indicated by the stratification and composition deduced from observations. The spectral evolution of SNe Ia indicates the presence of $\sim 0.6\,M_\odot$ of radioactive ^{56}Ni in a typical explosion, along with a similar amount of total mass distributed among stable iron-group elements and among the lighter elements mentioned above, for a total ejecta mass similar to the Chandrasekhar WD mass limit, $M_{\mathrm{Ch}} = 1.44\,M_\odot$ (e.g., Mazzali et al. 2007). This chemical makeup and stratification is also deduced from X-ray analysis of the ejecta of historical SN Ia remnants (e.g., Badenes et al. 2006) such as Tycho's SN of 1572.

Looking at energetics, the nuclear binding energy released by burning $0.6\,M_\odot$ of C and O into ^{56}Ni is 1.1×10^{51} erg. A comparable mass of C and O burning to intermediate-mass elements releases additional energy. The gravitational binding energy of a WD of mass close to M_{Ch} is $\sim 0.5 \times 10^{51}$ erg. The thermonuclear energy release is thus sufficient to unbind the WD and to give the ejecta the kinetic energy indicated by the deduced ejecta mass and its observed expansion velocities, of order 10^4 km s^{-1}, i.e., $\sim 10^{51}$ erg. Furthermore, the equation of state of the degenerate electron gas in a WD is just what is needed for an unstable thermonuclear runaway once carbon is somehow ignited. Nuclear reaction rates have steep rising dependences on temperature. A highly degenerate gas, when heated, does not expand and cool as a classical gas would. As a result, an explosive burning front can, in principle, quickly consume a WD. The carbon ignition threshold can be crossed via an increase in some combination of pressure and temperature. That increase has generally been attributed to the accumulation of accreted mass on the WD, up to the neighborhood of M_{Ch}, with a corresponding decrease in radius. Specifically, the carbon in the core of a nonrotating C/O WD ignites at a WD mass of $\sim 1.38\,M_\odot$ (Arnett 1969, Nomoto 1982). The fact that the "bomb" is always a C/O WD with $M \sim M_{\mathrm{Ch}}$ would naturally explain the limited range of SN Ia luminosities, which has been so instrumental for cosmology.

Beyond this simplified picture, however, much about SNe Ia is still poorly understood. At the most fundamental level, it is unknown what are the initial conditions and evolutionary paths that, in practice, take a WD away from stable equilibrium and lead it to mass growth, ignition, and explosion. In other words, we do not know the identity and nature of the systems that explode as SNe Ia. This is the "SN Ia progenitor problem."

2. PROGENITOR MODELS AND OPEN QUESTIONS

Progenitor scenarios have traditionally focused on getting a C/O WD to ignite by having it approach or exceed M_{Ch}. This can be accomplished either through accretion from a nondegenerate star in the single-degenerate (SD) model (Whelan & Iben 1973) or by merging two WDs in a tight binary (through loss of energy and angular momentum to gravitational waves) in the double-degenerate (DD) model (Tutukov & Yungelson 1981, Iben & Tutukov 1984, Webbink 1984; see Wang & Han 2012 for a recent review).

2.1. Single-Degenerate Models

Accretion from a nondegenerate secondary can assume many configurations. Mass transfer can proceed through Roche-lobe overflow or through a strong wind from the companion (Li & van den Heuvel 1997). The secondary can be a main-sequence star (e.g., van den Heuvel et al. 1992), a subgiant (e.g., Han & Podsiadlowski 2004), a He star (Tutukov & Yungelson 1996) (often called a "hot subdwarf", i.e., a star that has been stripped of its hydrogen envelope by binary interaction, e.g., Geier et al. 2013), or a red giant (sometimes called a "symbiotic" companion, e.g., Patat et al. 2011). For each of these cases, different configurations are possible, depending on previous common envelope and mass transfer phases (e.g., Wang & Han 2012).

In the SD scenario, the challenge is to get an accreting WD to actually grow in mass. In a narrow range of about a factor of three in mass accretion rate, centered around $\dot{M} = 10^{-7}\,\mathrm{M}_\odot\,\mathrm{year}^{-1}$ for a 0.8-$\mathrm{M}_\odot$ WD, and around $\dot{M} = 5 \times 10^{-7}\,\mathrm{M}_\odot\,\mathrm{year}^{-1}$ for a 1.4-$\mathrm{M}_\odot$ WD, stable nuclear burning of hydrogen to helium on the WD surface can take place (Nomoto 1982). WDs in such an accretion mode should resemble an observed class of objects known as supersoft X-ray sources (van den Heuvel et al. 1992; see Section 3.1.2 below). Accretion above this range likely leads to expansion of the accretor to a red-giant-like configuration, engulfment of the donor in a common envelope, and thus cessation of the growth process toward M_{Ch} and toward explosion as a SN Ia (Iben & Tutukov 1984). In this high-accretion-rate regime, however, Hachisu, Kato & Nomoto (1996, 1999) have proposed that self-regulation of the accretion flow occurs, keeping the mass flow that actually accumulates on the WD at the steady-burning rate. This is achieved by means of an optically thick wind emerging from the accretor, driven by the luminosity of the nuclear burning on the WD surface. However, questions have been raised as to whether this requires too much fine tuning (e.g., Piersanti et al. 2000, Shen & Bildsten 2007, Woosley & Kasen 2011) or requires contradictory assumptions about the spherical symmetry of the accretion flow—on the one hand, an asymmetric Roche-lobe overflow, on the other hand a spherical inflow that can be regulated by the spherical outflowing wind (Hillebrandt & Niemeyer 2000). In any case, the efficiency of the accretion mode is limited to the ratio of the steady burning rate to the mass transfer rate.

Accretion at rates lower than the stable-burning range lead to accumulation of the accreted hydrogen in a cold degenerate layer on the accretor, up to its ignition and burning in a thermonuclear-runaway "nova" eruption (Starrfield et al. 1972). Because hydrogen burning produces more energy per unit mass than required to eject that mass from the surface of a WD, most of the accreted material is expected to leave the accretor. Over the accumulation timescale, hydrogen can diffuse inward and mix with the WD's C and O layer. As a result, during the burning, some of the original WD material is blown away, along with the ashes of the accreted material (e.g., Yaron et al. 2005). However, with increasing accretion rates and WD masses, ignition conditions are reached at lower accumulated mass and less degenerate conditions, so there may be a parameter range where mass gain is possible after all.

Recent studies of the long-term evolution of WDs accreting hydrogen at different rates have had some conflicting conclusions. Idan, Shaviv & Shaviv (2013), studying accretion at a high rate ($\dot{M} = 10^{-6}\,M_\odot\,\mathrm{year}^{-1}$, i.e., above the steady burning regime), find recurrent thermonuclear runaway flashes on ten-year timescales (rather than steady hydrogen burning), during which most of the accreted mass is retained. However, the accumulated He ash is ignited after several thousand hydrogen flashes and ejects essentially all of the accreted mass. At a lower accretion rate of $\dot{M} = 10^{-7}\,M_\odot\,\mathrm{year}^{-1}$, some mass is retained, and the WD mass grows, albeit inefficiently. Newsham, Starrfield & Timmes (2013) and Wolf et al. (2013), by contrast, find steady burning and WD mass growth, but again that a hot He layer ignites at some point, possibly ejecting much of the accumulated mass.

Alternatively, the donor may be a He-rich star, avoiding the challenges involved in mass gain through hydrogen-rich accretion (see Wang et al. 2009a). The accumulated He layer ignites at a larger accumulated mass, but may suffer from the same problem as described above, that of a He nova in which much of the accreted material is ejected. The efficiency of hydrogen and helium accretion is thus theoretically uncertain, leading to a large range of possibilities for WD growth (see Yungelson 2005; Bours, Toonen & Nelemans 2013).

2.2. Double-Degenerate Models

In the DD scenario, the more-massive WD is generally thought to tidally disrupt and accrete the lower-mass WD, initially through an accretion disk configuration, that then likely takes on a more spherical geometry (e.g., Lorén-Aguilar, Isern & García-Berro 2009; Pakmor et al. 2012; Raskin et al. 2012; Schwab et al. 2012; Shen et al. 2012; Moll et al. 2014). Accretion of carbon and oxygen at a high rate should prevent the problems of inefficient mass growth, encountered in novae, and eventually lead to carbon ignition in the core. Against this concept, however, it has long been argued that the large accretion rate leads to off-center ignition and burning of carbon in the accretor to oxygen and neon. Further accretion and approach to M_{Ch} drives electron capture on Mg and Ne, leading to gravitational instability and to an "accretion-induced collapse" to a neutron star (Nomoto & Iben 1985, Saio & Nomoto 1998, Shen et al. 2012). The collapse could manifest itself as some kind of non-SN Ia, low-luminosity, hydrogen-free transient [Darbha et al. (2010) and Piro & Kulkarni (2013) consider the signatures of such an event]. By contrast, rotation of the merger remnant has been considered as a way to slow down the accretion and to avoid off-center ignition (Piersanti et al. 2003, Tornambé & Piersanti 2013; see also Section 2.5, below).

Recent 3D hydrodynamical merger models have cast doubt on the concept of a merger remnant as an unperturbed primary-mass WD that accretes the remains of the secondary WD. Instead, complex structures with regions of high temperature and density are found, in which explosive carbon ignition may occur. Such "violent mergers" may lead to a SN Ia, even if the total mass of the merger is below M_{Ch} (Guillochon et al. 2010; van Kerkwijk, Chang & Justham 2010; Pakmor et al. 2013; Raskin et al. 2014; Zhu et al. 2013; Dan et al. 2014). Details depend on the WD masses (a primary WD mass $\gtrsim 0.9\,M_\odot$ is likely required) and on the presence of He, which can facilitate ignition.

2.3. Collisional Double-Degenerate Models

A special case of a WD merger configuration is the "collisional" one, whereby two WDs collide head-on rather than spiraling in owing to gravitational-wave losses. Benz, Thielemann & Hills (1989), Lorén-Aguilar, Isern & García-Berro (2010), Raskin et al. (2009, 2010), Rosswog et al. (2009), and Hawley, Athanassiadou & Timmes (2012) have invoked this scenario, generally in the context of some SNe Ia possibly arising in dense stellar environments such as globular clusters and galactic nuclei. A different implementation of the collisional DD scenario was raised by Thompson (2011), who proposed that many binary WDs may actually be in "hierarchical triple" systems with a third, low-mass star in orbit about the inner WD binary. The tertiary could induce Kozai (1962)–Lidov (1962) oscillations in the eccentricity of the inner binary. A high eccentricity increases gravitational wave losses, and thus decreases the time until merger, and so enlarges the population of tight WD binaries that merge (leading to an enhanced rate of SNe Ia). Rare head-on collisions could sometimes also occur among the most eccentric systems. Prodan, Murray & Thompson (2013) considered also tidal dissipation of orbital energy of the inner binary (Mazeh & Shaham 1979), further enhancing the merger rate. However, it is not clear that triple systems, with the postcommon-envelope orbit ratios needed for this mechanism to operate, can be realized.

Katz & Dong (2012) discovered that one of the approximations used in previous Kozai-Lidov calculations is invalid for "mild" hierarchical triples, i.e., those in which the tertiary orbit is $\sim$3–10 times the inner orbit. Among such cases, they found, through direct numerical integration for a range of initial tertiary orbit inclinations, a rather high probability, $\sim$5%, for a Kozai-Lidov-induced head-on collision between the inner WD pair. Katz & Dong (2012) argue that, if a large fraction of all intermediate-mass stars are in the appropriate triple systems, most or all SNe Ia could come from such collisions. Kushnir et al. (2013) and García-Senz et al. (2013) performed the latest hydrodynamic thermonuclear simulations of such collisions and obtained promising agreement with some of the phenomenology of observed SN Ia explosions. Kushnir et al. (2013), using high spatial resolution ($\sim$1 km) in their simulations and avoiding some numerical artifacts that affected previous work, find that a collision between, e.g., two 0.7-$M_\odot$ WDs produces an explosion with a ^{56}Ni yield of 0.56 $M_\odot$, similar to a typical SN Ia.

However, it seems unlikely that a large fraction of all WDs are in such triples—only $\sim$10–20% of stars are triples (Duchêne & Kraus 2013, Leigh & Geller 2013), and probably only a fraction of these have the required special orbit ratios. A further problem with this model is how to avoid a collision of the inner binary early on, when the components are still on the main sequence and their geometrical cross sections for collision are 100 times greater than after they have become WDs. Hamers et al. (2013) used "binary population synthesis" calculations (see Section 2.6 below) to confirm that, for this reason alone, the enhancement of SN Ia rates is already minimal for initially wide inner binaries that would not interact without a tertiary component. Katz & Dong (2012) suggest that the relative orientations of the inner and outer orbits could be "reset" by small kicks during the mass loss phases of the stars of the inner binary or by a passing perturbing star. Colliding WD cases could then arise from among those systems that avoided collisions while on the main sequence.

2.4. Double Detonations and Rotating Super-Chandrasekhar-Mass Models

WDs accreting He at low rates (from either nondegenerate or He-WD donors) can accumulate a He layer that is sufficiently massive and degenerate such that the resulting He ignition becomes explosive (Taam 1980). This detonation drives a shock into the underlying C/O WD, which then ignites carbon at or near the center, even in sub-$M_{\rm Ch}$ WDs (Livne 1990, Shen & Bildsten 2009, Fink et al. 2010). Originally, these "double-detonation" or "edge-lit detonation" models were considered for nondegenerate He donors in which the ignition masses of the He layers were found to be $\sim 0.2\,M_\odot$. Models for such explosions typically predicted strong signatures of carbon and oxygen (from the burned He) in the SN spectra, which were inconsistent with observed spectra. However, Bildsten et al. (2007) found that, for He WD donors, the higher mass transfer rates lead to smaller He layers at explosion time and better agreement with observations (e.g., Fink, Hillebrandt & Röpke 2007). Recently, Guillochon et al. (2010), Pakmor et al. (2013), and Shen & Bildsten (2014) find that also in "violent mergers" (see Section 2.2 above) there may first be a surface He detonation, triggering the C detonation, if the merger is between a He and a C/O WD or in mergers between two C/O WDs where there is a thin surface layer of He present.

At the other extreme, the nickel mass deduced for some very bright SN Ia explosions is suggestive of a super-Chandrasekhar-mass progenitor, with mass $\gtrsim 2\,M_\odot$ (e.g., Howell et al. 2006; Scalzo et al. 2010, 2012; Silverman et al. 2011; Taubenberger et al. 2011, 2013a; Kamiya et al. 2012). WD rotation has been proposed as a means of supporting such massive progenitors. However, Hillebrandt, Sim & Röpke (2007) and Hachinger et al. (2013) argue that the explosion of a rotating super-$M_{\rm Ch}$ WD does not necessarily produce the inferred nickel mass, or other characteristics, of such bright events, and they suggest asymmetric explosions instead (see also Moll et al. 2014). Chamel, Fantina & Davis (2013) show that another proposed means of supporting super-$M_{\rm Ch}$ WDs, via ultrastrong quantizing magnetic fields, is impractical owing to electron-capture reactions that would make the WD unstable.

More generally, WD rotation has been invoked in several so-called spin-up/spin-down scenarios. Di Stefano, Voss & Claeys (2011), Justham (2011), Yoon & Langer (2004, 2005), and Hachisu, Kato & Nomoto (2012) have argued, in the context of the SD model, that a WD that has grown in mass, even beyond $M_{\rm Ch}$, could be rotation-supported against collapse and ignition, perhaps for a long time, during which the accretion could run its course and end and the traces of the process (or even of the donor itself) could disappear. DD mergers whose explosions are delayed by rotational support have also been proposed (Piersanti et al. 2003, Tornambé & Piersanti 2013). A WD undergoing maximal solid-body rotation is stable against carbon ignition up to a mass of $1.47\,M_\odot$, i.e., only $\approx 0.1\,M_\odot$ more than in the case of a nonrotating WD. In the context of SD models, where a typical WD needs to accrete $\sim 1\,M_\odot$ before exploding, it would appear that this extra $0.1\,M_\odot$ provides not much of an opportunity for the accretion process to conclude. Yoon & Langer (2005) therefore constructed models of differentially rotating WDs in which, for some radially increasing profiles of angular rotation speed, the maximal stable WD mass reaches $\sim 2\,M_\odot$. Saio & Nomoto (2004) and Piro (2008), however, argue that "baroclinic" instabilities, and/or the shear growth of small magnetic fields, provide torques that quickly bring a differentially rotating WD to solid-body rotation, with its limited rotational support. Hachisu et al. (2012) countered that these instabilities might not occur, as only the necessary conditions for them, but not the sufficient ones, are satisfied. The theoretical viability of massive, rotation-supported, WDs is thus still unclear. Observationally, a spin-up/spin-down scenario could potentially "erase" many of the clues that we discuss in this review.

2.5. Alternative Models

Kashi & Soker (2011) have introduced a "core-degenerate" model, in which a WD and the core of an asymptotic-giant-branch (AGB) star merge already in the common-envelope phase. After ejection of the envelope, the merged core is supported by rotation, potentially for long times, until it spins down, e.g., via magnetic dipole radiation, and only then explodes (Ilkov & Soker 2011). Wheeler (2012) sketches a SD progenitor model consisting of a WD and an M-dwarf donor, in which accretion by the WD and its growth toward $M_{\rm Ch}$ are enhanced by magnetic channeling, self-excited mass loss from the M dwarf, and magnetic inhibition of mixing of the WD surface layer, thus avoiding excessive mass loss in nova events. Single-star SN Ia progenitor models have also been occasionally attempted (Iben & Renzini 1983, Tout 2005) in which the degenerate carbon-oxygen core of an AGB star is somehow ignited after it has lost its hydrogen envelope (as it must, if the SN is to appear as a Type Ia).

2.6. Linking Theory and Observations

As already noted briefly above, apart from the issue of the identity of the progenitors and the intrinsic problems of each of the progenitor scenarios, many gaps remain in our understanding of the phases that precede each "progenitor setup" that will eventually lead to an observed SN Ia. These gaps of knowledge include the details of binary evolution and particularly the enigmatic common-envelope phases that a pre-SN-Ia binary system undergoes in almost all models (see Ivanova et al. 2013 for a review).

An important theoretical tool for obtaining observational predictions from the various scenarios, despite these obstacles, has been the calculation of binary population synthesis (BPS) models. In BPS, one begins with a large population of binaries having a chosen distribution of initial parameters (component masses, separations), and one models the various stages of their stellar and binary evolution, including transfer and loss of both mass and angular momentum, and possibly multiple common-envelope phases, with the physics of each stage parameterized in some way (e.g., Jorgensen et al. 1997; Yungelson & Livio 2000; Wang, Li & Han 2010; Bogomazov & Tutukov 2011; Meng & Yang 2011; Toonen, Nelemans & Portegies Zwart 2012; Mennekens et al. 2013; Ruiter et al. 2013; Claeys et al. 2014). At the end of such a simulation, one can see what fraction of the initial stellar population, and from which specific progenitors, has ended up at the conditions thought to lead to a SN Ia explosion through a particular channel (e.g., DD, Chandrasekhar-mass SD, etc.). Furthermore, as discussed in Section 3.5 below, for each progenitor channel one can obtain the distribution of "delay times" between star formation and SN Ia explosion, a distribution that can be compared with observations.

Even after the conditions for a SN Ia explosion have been met, many additional questions remain as to the ensuing phases: the trigger and the locations of WD ignition; the mode in which the burning front consumes the WD (see Hillebrandt & Niemeyer 2000; and see Section 3.3.4, below); uncertainties in nuclear cross sections (Parikh et al. 2013); and the transfer of radiation through the expanding ejecta (e.g., Pinto & Eastman 2000; Piro & Nakar 2013, 2014; Dessart et al. 2014b; Mazzali et al. 2014). There is thus a multitude of theoretical paths to a SN Ia explosion, some of this multiplicity arising from real physical possibilities and some of it due to uncertainties in the often-complex physics. Among these many theoretical paths, the ones that are actually realized and seen in nature likely encrypt observational clues to the solution of the progenitor question. With this in mind, we attempt an overview of the state of the art of observations that may provide such clues. We arrange our review according to the various possible observational approaches to the problem.

3. EVIDENCE FROM THE OBSERVATIONS

A variety of observational approaches have been brought to bear on the SN Ia progenitor problem: The existing populations of potential progenitors can be studied (Section 3.1); pre-explosion data at the sites of nearby events may reveal the progenitors (Section 3.2); the observed properties of the events themselves may carry clues to the progenitors (Section 3.3); the remnants of presumed SNe Ia can be searched for remaining traces of the progenitor systems (Section 3.4); and the rates at which SNe Ia explode as a function of time and environment provide another avenue to address the problem (Section 3.5). We review each of these approaches in turn.

3.1. Clues from Potential Progenitor Populations

Because a typical galaxy hosts on the order of 10^7 SNe Ia over a Hubble time (see Section 3.5.7), a viable type of progenitor system should be present in significant numbers and therefore observable, whether as individual objects or through their collective emission. Thus, the first observational approach to the progenitor problem that we consider is to look for specific potential progenitor systems, to measure their properties and numbers, and to see if those properties and numbers conform to expectations, given known SN Ia properties.

3.1.1. Recurrent novae. The Milky Way and similar large galaxies have populations of $\sim 10^6$ "cataclysmic variables"—WDs accreting from nondegenerate donor stars through Roche-lobe overflow or through a wind, with orbital periods down to about 80 min (e.g., Warner 2003). Among the ways in which these systems reveal themselves are "nova eruptions" (see Section 2.1, above), which occur in the Galaxy and in M31 at rates of ~ 35 year^{-1} and ~ 65 year^{-1}, respectively (Della Valle et al. 1994, Darnley et al. 2006). As already noted, it is thought that few, if any, novae gain more mass during accretion than mass they lose in the eruptions, in which case such systems are not SN Ia progenitors. [See, however, Zorotovic, Schreiber & Gänsicke (2011), who find that accreting WDs, on average, have higher masses than WDs with low-mass companions in postcommon-envelope binaries that are destined to become semidetached systems with stable mass transfer. This suggests either that WD masses do grow or that the majority of cataclysmic variables form from systems with higher mass companions, that by some selection effect, harbor more massive WDs.]

However, members of a subclass of novae called recurrent novae have long been suspected as possible SN Ia progenitors (e.g., Starrfield, Sparks & Truran 1985; Della Valle & Livio 1996; Schaefer 2010; Kato & Hachisu 2012). Recurrent novae are defined as novae with more than a single outburst over the past century or so. There are only 10 recurrent novae known in the Galaxy. Some of the donor stars are main-sequence, some subgiants, and some red giants. Recurrent novae have eruptions every few decades on an irregular basis. Because the mass needed for ignition scales inversely with both WD mass and accretion rate (Fujimoto 1982, Truran & Livio 1986, Shen & Bildsten 2009), these short recurrence times have been interpreted to indicate that the WD mass is close to $M_{\rm Ch}$, specifically $M \gtrsim 1.2\,{\rm M}_\odot$, and that the accretion rate is relatively high, i.e., the same parameters for which mass gain may be possible. A number of observational mass estimates are indeed suggestive of large WD masses, but uncertainties are large and there are only two double-lined eclipsing systems that have reliable mass estimates: U Sco, with a WD mass of $1.55 \pm 0.24\,{\rm M}_\odot$ (Thoroughgood et al. 2001), and CI Aql, with $1.00 \pm 0.14\,M_\odot$ (Sahman et al. 2013). Observationally, little is known about the formation of massive WDs in binaries, so it is unclear if these high masses indicate that the WDs have accumulated mass. Recurrent nova eruptions may in fact be caused by instabilities in the accretion discs, leading to periodic accretion at the steady-burning rate, during which the WDs do grow in mass (Alexander et al. 2011). Then again, Hachisu & Kato (2012) point out that it is unclear if steady burning can ignite for such short accretion episodes.

Yet another view is that recurrent novae are systems in which the WD mass is decreasing with time and which will thus never become SNe Ia (Patterson et al. 2013, Schaefer 2013; and see below). As noted, there are few known recurrent novae and even fewer that have had well-studied multiple outbursts. The debate regarding their being SN Ia progenitors has therefore focused on individual objects and on individual outbursts.

For example, after the latest outburst of U Sco in 2010 (e.g., Schaefer et al. 2010), Schaefer (2013) deduces, based on the period change, that much more material was ejected than had been previously accreted, challenging the idea of WD mass growth (e.g., Diaz et al. 2010). Whether or not the massive WD has a C/O composition (as required of a SN Ia) or an O-Ne one, is also debated (e.g., Kato & Hachisu 2012, Mason 2013). RS Oph is a symbiotic binary consisting of a massive WD and a red giant in a 454-day orbit (see Kato & Hachisu 2012 and references therein). Its latest outburst in 2006, from which Hachisu et al. (2006) infer a high WD mass, also revealed a shock from the interaction of the ejecta with the red-giant wind (Sokoloski et al. 2006) and radio emission suggested the launch of a jet (Rupen, Mioduszewski & Sokoloski 2008). Patat et al. (2011) studied the evolution of the absorption features and find similarities with those found in the spectra of some SNe Ia (see Section 3.3.8). T Pyx is a short-period system ($P = 1.8$ h; Uthas, Knigge & Steeghs 2010), which implies a low donor mass (if it is to fit in the Roche lobe at this period), and is more typical of systems harboring classical novae; its most recent outburst was in 2011. Schaefer, Pagnotta & Shara (2010) suggest a high mass-transfer rate that would imply a high WD mass. However, Uthas, Knigge & Steeghs (2010) derive a mass ratio of 0.2, which for a realistic donor mass implies a WD mass of only 0.7 $M_\odot$. In any case, Selvelli et al. (2008) and Patterson et al. (2013) conclude that this system, again, is ejecting more mass than it accretes. T CrB is another wide symbiotic recurrent nova with two known outbursts (see Anupama 2013), for which Luna, Sokoloski & Mukai (2008) argue a high WD mass based on the detection of hard X-ray emission. V407 Cyg is formally not a recurrent nova, as only one nova outburst (in 2010) has been observed, but the similarity of the eruption to that of RS Oph (e.g., Shore et al. 2011) implies a massive WD (Hachisu & Kato 2012, Nelson et al. 2012). By contrast, Chomiuk et al. (2012a) use extensive radio observations to argue that the environment of V407 Cyg is not one typical of SNe Ia.

Could recurrent novae be the phase during which accreting WDs achieve a significant fraction of their growth toward $M_{\rm Ch}$ and explosion as SNe Ia? The answer is no, unless the number of recurrent novae is orders of magnitude larger than estimated. Beyond the 10 known systems, Schaefer (2010) estimates that as many as 60–100 Galactic sources that have been classified as classical novae are, in reality, recurrent novae whose repeated outbursts have been missed. Furthermore, his analysis of the Galactic spatial distribution of both types of novae suggests that their true numbers are several times larger than the numbers currently known. Thus, the Galactic population of recurrent novae could conceivably number $\sim$300. However, to get a Galactic SN Ia rate of at least once per 200 years (see Section 3.5.7) and have the systems accrete at the very least $\sim$0.2 $M_\odot$ at the limiting accretion rate onto a WD for a recurrent nova, of $\dot{M} < 3 \times 10^{-7}\, M_\odot\, {\rm year}^{-1}$, over 3,300 systems are needed. The recurrent nova phase can constitute $\lesssim$9% of the time of the final 0.2 $M_\odot$ of WD mass growth. If we consider a more realistic $0.4-0.9$-$M_\odot$ mass gain required of C/O WDs (which have masses of $\sim$0.5–1 $M_\odot$) in order to reach $M_{\rm Ch}$, then only 2–5% of that mass gain can take place during recurrent nova phases.

3.1.2. Supersoft X-ray sources. In the evolutionary scenario proposed by Hachisu, Kato & Nomoto (1999), the recurrent nova phase, with accretion rates estimated at $\dot{M} \sim 1$–$3 \times 10^{-7}\, M_\odot\, {\rm year}^{-1}$, follows a phase when a SD progenitor system is a supersoft X-ray source, growing via a larger accretion rate, $\dot{M} \sim 3-6 \times 10^{-7}\, M_\odot\, {\rm year}^{-1}$. The high accretion rate (and hence temperature), compared with those of novae, lifts the degeneracy of the accreted hydrogen layer, and it burns stably to helium on the WD surface.

Observationally, "persistent" or "permanent" supersoft X-ray sources are identified, as their name implies, by their soft X-ray spectra, peaking at 30–100 eV, with typical luminosities of 10^{37-38} erg s^{-1}. Optical and X-ray follow-up of the best-studied of these objects (e.g., Lanz et al.

2005, Rajoelimanana et al. 2013) has shown that they indeed consist of a hot WD in a close orbit with a nondegenerate donor, where hydrogen accreted from the donor, at roughly the above rates, burns more or less stably into helium on the WD surface (van den Heuvel et al. 1992). However, not all supersoft source spectra are easily interpreted in this way, as some show P-Cygni profiles indicative of a wind rather than a WD atmosphere (Bearda et al. 2002). Because of interstellar absorption of their soft X-ray spectrum, supersoft sources have been discovered mainly in external nearby galaxies. Only two are known in the Galaxy (MR Vel and Q And; e.g., Simon 2003), but there are 15 in the Magellanic Clouds and of order 10 in M31 (Orio et al. 2010). Classical and recurrent novae, discussed in Section 3.1.1 above, also have supersoft phases, but these are transient, occurring after their outbursts, and lasting on the order of a month. As a result, only a minority of the supersoft sources that turn up in X-ray surveys of nearby galaxies are of the persistent kind, rather than being the transient supersoft phases of postoutburst novae (Orio et al. 2010).

Because the persistent supersoft X-ray sources in an $\sim L_*$ galaxy such as M31 again number only in the few tens of objects, the same argument, used above in the case of recurrent novae, holds as well: If supersoft sources are accreting WDs growing toward $M_{\rm Ch}$ and such SD systems are to explain the bulk of the SN Ia rate, then <1% of the WD's growth time is spent in this phase. This is an order of magnitude less than the mean ~5% fraction of the time in the supersoft phase that is found by model calculations (Meng & Yang 2011). This paucity of supersoft sources has been pointed out by Di Stefano (2010), based on the numbers of such sources observed in six nearby galaxies, and by Gilfanov & Bogdán (2010), based on the integrated X-ray flux observed from nearby elliptical galaxies. Hachisu, Kato & Nomoto (2010) and Meng & Yang (2011) (see also Lipunov, Panchenko & Pruzhinskaya 2011) countered that the nuclear-burning accreting WDs spend the majority of the time in a third possible phase, hidden within optically thick outflows (e.g., Nielsen et al. 2013, Wheeler & Pooley 2013, Woods & Gilfanov 2013). We discuss this possibility in Section 3.1.3. Although proto-DD systems, at the time that only the first WD has formed, may also undergo mass transfer and a SD-like supersoft phase, its duration is much shorter than in SD progenitor systems (Nielsen et al. 2014). As a result, fewer supersoft sources are expected in the DD scenario by at least an order of magnitude, and this is consistent with their observed rarity. Considering also the delay, between the supersoft phase of proto-DD systems and the eventual WD merger and SN Ia explosion, further lowers the number of expected supersoft sources from DD progenitors.

3.1.3. Rapidly accreting white dwarfs. As noted above, Hachisu, Kato & Nomoto (1996, 1999) proposed that the rapid accretion phase is self-regulated by an optically thick wind. The wind drives away the excess mass, effectively keeping the WD growth and the stable nuclear burning on the WD surface at the same rates as during the supersoft phase (but see Idan, Shaviv & Shaviv 2013; and see Section 2.1, above). The excess mass that is blown off the WD could create an optically thick photosphere that reprocesses the X-rays to UV emission. Nielsen et al. (2013), Nomoto et al. (2007), Wheeler & Pooley (2013), and Lepo & van Kerkwijk (2013) consider the possible appearances of such "rapidly accreting WDs" and argue that they might appear as undersized OB stars, i.e., with effective temperatures of $10^{4.5-5}$K, but with sizes of only a few $R_\odot$, dictated by the WD's Roche-lobe radius. Another possibility is an appearance similar to Wolf-Rayet (WR) stars or WR planetary nebulae.

Among the objects possibly considered to be such rapidly accreting WDs are V-Sagittae-type nova-like cataclysmic variables, a class consisting of a handful of known objects (Steiner & Diaz 1998). The prototype, V Sge, is a double-lined eclipsing binary with a rich high-ionization emission-line spectrum, a period of 0.5 day, quasi-periodic high and low states lasting ~180 days, and supersoft X-ray emission in its low state. Hachisu & Kato (2003a) have modeled V Sge as a 1.25-$M_\odot$ WD accreting at a high rate from a Roche-lobe-filling ~3-$M_\odot$ companion. Other moderately well-studied objects in the V Sge class are RX 0513.9-6950 (Hachisu & Kato 2003b) and QU Car (Kafka, Honeycutt & Williams 2012). A different type of object that could be a rapidly accreting WD is the peculiar planetary nebula LMC N66 in the Large Magellanic Cloud (LMC), whose WR-like core has undergone two optical outbursts over the past 60 years, each lasting several years, during which its luminosity was comparable to WN-type WR stars. There is no evidence yet for binarity in LMC N66, and no X-ray emission has been detected. Hamann et al. (2003) have modeled it, again, as a WD with mass inflow from a companion at rates of 10^{-6} to 10^{-5} $M_\odot$ year^{-1} during its low and high states.

Lepo & van Kerkwijk (2013) performed a search in the Small Magellanic Cloud for the ~100 rapidly accreting WD systems that are expected there, if such progenitors are to produce the bulk of the SN Ia rate. They obtained optical spectra for about 750 sources, selected to be UV-bright at 1,600 Å and to have unusual optical colors. They failed to find any object with unusual spectral signatures such as strong HeII lines, or with optical variability, based on the long-term OGLE monitoring database (Udalski, Kubiak & Szymanski 1997). From this, they deduce a $\lesssim$10 % contribution to the SN Ia progenitor population of systems that are similar to LMC N66 or to WR stars. The color selection of targets employed by Lepo & van Kerkwijk (2013) included only a limited number of objects with colors similar to V Sge, and therefore such systems cannot yet be ruled out as SN Ia progenitors by this experiment.

Taking a more integral approach, Woods & Gilfanov (2013) argue that, in any scenario where the X-ray emission from nuclear-burning WDs is shielded by $T = 10^{5-6}$K photospheres, the emerging UV radiation will still be quite hard and capable of ionizing interstellar He atoms. In elliptical galaxies, a substantial mass of extended neutral hydrogen is often detected as well as diffuse regions of low-ionization line emission, which is thought to arise through ionization of some of the gas by evolved stars. The integrated radiation from a population of rapidly accreting WDs could lead to ionization of some of the He in this gas to He^{2+} and to a detectable signal of diffuse HeII λ4686 recombination line emission. Woods & Gilfanov (2013) combine stellar spectral synthesis models with photoionization models to calculate the line emission as a function of galaxy age, gas covering fraction, and WD photospheric temperature. They predict, for an elliptical galaxy with a SD-dominated progenitor population with $T = 2 \times 10^5$K photospheres, a typical line equivalent width of a few tenths of an angstrom. In the absence of nuclear-burning SD systems, with ionization only by evolved stellar populations, a line flux one order of magnitude lower is expected. Woods & Gilfanov (2014) extend these predictions to excitation of forbidden emission lines of C, N, and O by nuclear-burning WDs with photospheres in the upper part of the $T = 10^{5-6}$K range. Johansson, Woods & Gilfanov (2014) searched for the HeII λ4686 signal in high-signal-to-noise stacks of 11,600 Sloan Digital Sky Survey (SDSS; Abazajian et al. 2009) spectra of early-type galaxies. The galaxies are selected to have weak but measurable line emission in order to include galaxies with gas, but to exclude, on the basis of line diagnostics, galaxies with ongoing star formation or with active nuclei. In all four galaxy age-group stacks that they produce, they detect HeII λ4686 emission only at the level expected from ionization by the known evolved stellar populations, but not from ionization by rapidly accreting WDs. This sets a 10% upper limit on the contribution to the SN Ia rate by rapidly accreting WDs.

3.1.4. Helium-rich donors. No clear-cut cases are known of systems that could be SN Ia progenitors through the He-rich SD donor channel. However, the only known He nova, V445 Pup, shows a preoutburst magnitude (Ashok & Banerjee 2003) that suggests it is a WD accreting from an evolved He star (Kato et al. 2008, Woudt et al. 2009). Modeling the nova outburst, Kato et al. (2008) find a high WD mass, but the predicted distance disagrees somewhat with the one determined from the expansion parallax of the bipolar outflow from Woudt et al. (2009). At least one case is known of a close detached binary consisting of a C/O pre-WD and a lower-mass He star, CD-30° 11223 (Geier et al. 2013). Its future helium accretion rate is expected to be in the right range to set off an edge-lit detonation, possibly leading to a double-detonation SN Ia (Yungelson 2008; see Section 2.4). The same could be true for some fraction of the so-called AM CVn systems, a class of objects in which a WD accretes helium from a very low-mass companion [see Solheim (2010) for a review], although the majority of the observed systems have evolved to too-low mass transfer rates for a double detonation.

3.1.5. Binary white dwarfs. As a test of the DD scenario, one can search our Galactic Neighborhood within a few kiloparsecs of the Sun for short-period WD binaries whose orbits will gravitationally decay and merge within, say, a Hubble time, surpassing (or not) the Chandrasekhar mass and perhaps thus producing SNe Ia. For the classic DD scenario to work, the Galactic SN Ia rate should match the WD merger rate (or some fraction of it involving mergers of the suitable masses). For the range of masses of WDs with a C/O composition, $0.5-1.0\,M_\odot$, pairs merging within a Hubble time will have separations of $a \lesssim 0.015$ AU, orbital periods of $P \lesssim 12$ h, and circular orbital velocities of $v \gtrsim 100$ km s^{-1}.

Marsh (2011) and Kaplan, Bildsten & Steinfadt (2012) present recent compilations from searches for binary WDs. In the current century, the first large survey to search systematically for such WD pairs was SPY (ESO SN Ia Progenitor surveY; Napiwotzki et al. 2004, Nelemans et al. 2005, Geier et al. 2007). SPY obtained high-resolution [$\sim$2 km s^{-1} radial velocity (RV) resolution] VLT spectra for a sample of $\sim$1,000 WDs, with two epochs per target, separated by at least one day. Given the $\lesssim$12-h periods of close WD binaries, this essentially gives two samplings at random phases in the orbit. They searched for RV variations between epochs that could result from the motion of an observed WD about a binary center of mass. [Note that a WD's luminosity can have a large range, depending on the WD effective temperature, which decreases with WD age, and on its surface area, which decreases with increasing WD mass. Therefore, in a WD binary, one of the WDs, not necessarily the least or more massive one, will often be much fainter than the other WD and thus will remain undetected.] Candidate systems discovered by the survey were then followed up to obtain RV curves and constraints on the binary parameters. SPY discovered $\sim$100 candidate binary WDs, and among the $\sim$10 systems for which parameters are measured, a handful will merge within a Hubble time. None of these future mergers have total masses $M_{\rm tot}$ that are unambiguously above $M_{\rm Ch}$, although for one system (WD 2020-425) the masses are difficult to measure and the best estimate is $M_{\rm tot} = 1.35\,M_\odot$ (Napiwotzki et al. 2007). Then again, $M_{\rm tot} \geq M_{\rm Ch}$ may not necessarily be a precondition for a DD-scenario SN Ia (see Sections 2.2 and 2.4, above). Furthermore, a statistical interpretation of these results that accounts for the selection effects and efficiencies of the survey is still lacking. It is therefore yet unclear what these results imply for the merger rate of WDs in general and for $M_{\rm tot} \geq M_{\rm Ch}$ mergers in particular.

A number of tight binaries, found by other searches, some that are not quite yet bona fide double WD systems, may possibly become $M_{\rm tot} \geq M_{\rm Ch}$, Hubble-time-merging, WD systems (Maxted, Marsh & North 2000; Geier et al. 2007, 2010; Rodríguez-Gil et al. 2010; Tovmassian et al. 2010). Brown et al. (2011) have even discovered a binary WD system with a 13-min orbit that will merge within <1 Myr, and Hermes et al. (2012) have detected, over an $\sim$1-year baseline, the gravitational decay of its orbit. However, with component WD masses of only 0.26 and $0.50\,M_\odot$, this merger is unlikely to result in a SN Ia.

In terms of analysis of the observed binary WD population, Nelemans, Yungelson & Portegies Zwart (2001), and Toonen, Nelemans & Portegies Zwart (2012) used BPS to simulate, under the various parameterizations for the physics of the common-envelope phase, the properties of WD binaries that would be observed, such as masses and periods. They compare them qualitatively to the masses and periods of a compilation of known binary WD systems, as well as quantitatively with the WD space density and WD production rate, finding reasonable correspondence. The Galactic WD merger rates indicated by these BPS models, $1–2 \times 10^{-2}$ year^{-1} (or $\sim 2–3 \times 10^{-13}$ year^{-1} $M_\odot^{-1}$ for a Milky Way stellar mass of 6×10^{10} $M_\odot$), agree with other BPS studies (e.g., Iben, Tutukov & Yungelson 1997; Han 1998). This merger rate is similar to recent estimates of the SN Ia rate per unit mass in Sbc-type galaxies of Milky-Way mass, $\sim 1 \times 10^{-13}$ year^{-1} $M_\odot^{-1}$ (Mannucci et al. 2005, Li et al. 2011b; see Section 3.5.7 below). However, only about one-third of the BPS mergers involve two C/O WDs, and an even smaller fraction have $M_{tot} \geq M_{Ch}$. Among known binary WDs, indeed only a handful are C/O + C/O, but this is partly due to the fact that some surveys are designed specifically to discover low-mass WD binaries (e.g., Kilic et al. 2012). The merger rates predicted by BPS thus may or may not be at the level required to explain SN Ia rates, depending on the ranges of M_{tot} assumed to lead to a SN Ia. More on this issue in Section 3.5.10 below.

Badenes et al. (2009) have been searching for close binary WDs among the spectra in the SDSS. The large number of WDs in SDSS, and the fact that SDSS spectra are always split into at least two subexposures separated by at least 30 min, make this possible. The $\sim$70-km-s^{-1} RV resolution possible with SDSS spectra of WDs (the resolution varies with WD type and brightness) permits discovering candidate short-period WD binaries. The observed distribution of maximum RV differences between any two epochs, $\Delta_{RV\,max}$, can be compared to model distributions predicted for a range of binary population parameters (Maoz, Badenes & Bickerton 2012). Badenes & Maoz (2012) have applied the method to a sample of 4,000 DA-type WDs in SDSS with the best signal-to-noise ratio. The model combinations of binary fraction and separation-distribution parameters that are consistent with the observed $\Delta_{RV\,max}$ distribution have a WD merger rate, per unit stellar mass (the analysis naturally gives the specific merger rate per WD, i.e., the reciprocal of the mean time until merger; this can be converted to a rate per unit stellar mass, using estimates for the stellar mass density and the WD number density in the Solar Neighborhood) of $\sim 1 \times 10^{-13}$ year^{-1} $M_\odot^{-1}$, which is a bit below the BPS estimates and again quite similar to the expected SN Ia rate per unit mass in the Milky Way. The fraction of $M_{tot} \geq M_{Ch}$ mergers is unconstrained by these observations—$\Delta_{RV\,max}$ is sensitive foremost to the WD separation rather than to M_{tot}. As noted above, on the basis of known double WD systems and from BPS calculations, the fraction may be small, but $M_{tot} \geq M_{Ch}$ may not be required for a SN Ia.

3.1.6. Rapidly rotating massive white dwarfs. In the various spin-up/spin-down scenarios (see Section 2, above), one could expect the existence of a progenitor population of rapidly rotating, super-M_{Ch} WDs, with rotation periods on the order of 10 seconds, either single or in postinteraction binaries. To achieve stability at significantly super-M_{Ch} masses, the WDs would need to have differential rotation profiles.

WD rotation is difficult to measure but, with few or no exceptions, WDs with measured rotations are observed to spin extremely slowly (see, e.g., Kawaler 2004). The typical period is $\sim$1 day, and most periods range from a few hours to a few days. For example, Berger et al. (2005), using the core of the Ca II K line to measure rotational broadening in 38 DAZ-type WDs, find a projected rotation velocity $v_{rot} \sin i < 30$ km s^{-1} and typically just a fraction of this velocity. This shows that there is efficient radial angular momentum transport in WDs, at least between the core and the pre-WD envelope, or that the WDs were formed slowly spinning (Spruit 1998, Suijs et al. 2008). Charpinet, Fontaine & Brassard (2009) and Fontaine, Brassard & Charpinet (2013) use asteroseismology to measure the inner rotation profiles of four newly formed pulsating WDs of the GW Vir class. They find slow, solid-body rotation throughout $\sim$99% of the WD mass. Although still few, these cases suggest strong coupling and efficient angular-momentum transfer between the layers of the WD, which again argues against the differential WD rotation that is essential for the spin-up/spin-down scenario. Córsico et al. (2011), applying the same methods to another GW-Vir-type WD, find that the WD core may be spinning up to four times faster than the surface, but within the uncertainties, solid-body rotation cannot be excluded in this case.

Magnetic fields in WDs are also difficult to measure, unless the fields are strong. Some 10–20% of WDs have fields of 10^4–10^9 G (Kawka et al. 2007), and ~10% have 10^3–10^4 G (Jordan et al. 2007, Landstreet et al. 2012). Nordhaus et al. (2011), Tout et al. (2008), and García-Berro et al. (2012) argue that strong magnetic fields in WDs are always the result of binary interactions, be it through accretion, common-envelope evolution, or mergers. In magnetic WDs, photometric variability due to spots and nonuniform dichroic polarization open two additional avenues to measure rotation. Although magnetic WDs tend to be significantly more massive than normal WDs (e.g., Kepler et al. 2013), they rotate just as slowly (Kawaler 2004), suggesting that WDs are unable to spin persistently, even after binary interactions. Several cases of nonvariable magnetic WDs, suspected as possibly unrecognized fast rotators, have been shown actually to be very slow rotators (Friedrich & Jordan 2001, Beuermann & Reinsch 2002). However, Boshkayev et al. (2013) propose that soft-gamma-ray repeaters and anomalous X-ray pulsars could actually be massive, rapidly rotating, magnetic WDs.

If spin-up/spin-down is a dominant SN Ia progenitor channel, we can estimate the number of WDs in Solar-Neighborhood samples that could reveal evidence of being such progenitors, again using a SN Ia rate per unit mass in Sbc spirals of Milky Way mass, 1×10^{-13} year^{-1} $M_\odot^{-1}$. The local stellar density is 0.085 $M_\odot$ pc^{-3} (McMillan 2011). If spin-down requires a Hubble time to occur, there should be, within a distance of 100 pc, about 400 WDs with super-$M_{\rm Ch}$ masses and rotations on the order of 1,000 km s^{-1}. If spin-down and explosion occur on timescales of $\lesssim 10^8$ year, only about four such nearby fast rotators are expected. Nonetheless, in the latter case, being young and hence hot and luminous, they should still be quite conspicuous (Di Stefano, Voss & Claeys 2011). Upcoming large and complete WD samples, such as those that will be discovered by *Gaia*, will further test this scenario.

3.2. Pre-Explosion Evidence

The most direct way to resolve the progenitor problem would be to see the progenitor system of an actual SN Ia before it exploded. Unfortunately, and contrary to the situation for CC SNe, where a good number of progenitor stars have been identified in pre-explosion images (see review by Smartt 2009), no such progenitor has ever been convincingly detected for a SN Ia. Pre-explosion optical images for several Virgo-distance SNe Ia, at the pre-explosion sites, have set upper limits on progenitor luminosities, corresponding to supergiants evolved from stars of initial mass $\gtrsim 8\ M_\odot$ (Maoz & Mannucci 2008; Nelemans et al. 2008; Voss & Nelemans 2008; Graur & Maoz 2012a,b; see also summary by Li et al. 2011a). Nielsen, Voss & Nelemans (2012, 2013) place upper limits on the pre-explosion progenitor X-ray luminosities at the sites of 13 nearby SNe Ia, in several cases limits that are comparable to the luminosities of supersoft X-ray sources. Voss & Nelemans (2008) did find a possible pre-explosion X-ray source at the site of SN Ia 2007on. However, the best-fit position of the X-ray source (based on 14 photons) is offset by 1.1 arcsec from the optical SN position, making this detection ambiguous. However, the source was not detected after the event (albeit in shallower images) and chance alignment of such a soft source, that almost certainly originates in the same galaxy as the SN, is unlikely as well (Roelofs et al. 2008).

The strongest pre-explosion limit on the presence of a SD donor star was set in the case of SN 2011fe in the nearby (6.4 Mpc; Shappee et al. 2011) galaxy M101. Being the nearest SN Ia event in 25 years, and having been discovered very early, SN 2011fe was extremely well studied and has permitted many tests relevant to the progenitor question (see Chomiuk 2013 and Kasen & Nugent 2013 for reviews). SN 2011fe was about as typical as a SN Ia can be, in all of its observed characteristics (Mazzali et al. 2014), making it particularly relevant for addressing the question of the progenitors of SNe Ia, as a population. As a final bonus, it had negligible Milky Way and intrinsic line-of-sight extinction (Nugent et al. 2011).

Li et al. (2011a) analyze deep (2σ-limiting Vega magnitudes $\sim$27–28) pre-explosion *Hubble Space Telescope* (HST) images of the site of the event in four optical bands, corresponding roughly to B, V, I (all from 2002), and R (from 1998). They used adaptive-optics imaging of the SN, obtained with the Keck II telescope, to determine the SN location on the HST images to 21-mas precision. *Spitzer Space Telescope* images at 3.6 μm to 8 μm, and 142 ks of *Chandra X-ray Observatory* images in the 0.3–8 keV band, both from 2004, were also examined. No source was detected at the SN position. In addition, they searched about 3,000 epochs of shallower, ground-based, monitoring data of M101 from 12 years preceding the explosion. These data set optical flux limits, reaching $\sim$20–23.5 mag, on any pre-SN variable or transient events at the SN location.

From the HST data, Li et al. (2011a) strongly rule out the presence of a red giant at the location of the SN in the decade prior to the explosion. Also excluded are any stars more massive than $3.5\,M_\odot$ that have evolved off the main sequence. Two Galactic symbiotic recurrent novae, RS Oph and T CrB, would have been detected in the data if placed, at their quiescent luminosities, at the distance of M101. The "helium nova" V445 Pup in quiescence, where the donor is a He star, would have likewise been detected in the HST data. Conversely, main-sequence and subgiant donors with masses below $3.5\,M_\odot$ are allowed, and a system like the recurrent nova U Sco, in which the donor is a main-sequence star, would be undetected in quiescence. **Figure 1** shows these results.

Li et al. (2011a) found that a pre-explosion eruption at the site of SN 2011fe, with a luminosity typical of nova outbursts, could have been detected by the decade of optical monitoring data, but there is a 37% probability that such an eruption would have been missed, given the actual cadence of those observations. Analysis of the X-ray data sets an upper limit on the pre-explosion bolometric X-ray luminosity of $(4\text{–}25) \times 10^{36}$ erg s^{-1} (depending on the assumed spectrum). Nielsen, Voss & Nelemans (2012) use a deeper stack of 800-ks *Chandra* data to set a somewhat stronger X-ray upper limit in the case of SN 2012fe. A weaker limit was set by Liu et al. (2012a) on the basis of these *Chandra* data, but Nielsen, Voss & Nelemans (2012) point out possible errors in the analysis. In any event, the X-ray data rule out the pre-explosion presence, at the location of SN 2011fe, of typical persistent supersoft X-ray sources, but allow the presence of somewhat fainter ones.

Shara et al. (2013) imaged M101 with HST, including the location of SN 2011fe, in 2010, about 1 year prior to the event, in a narrow band centered on the HeII λ4686 line. The hard photon flux from a nuclear-burning accreting WD (see Section 3.1.3) should, in principle, produce a 1–30-parsec-sized (i.e., unresolved or resolved, depending on the outflow rate and the surrounding gas density) HeIII Stromgren sphere or shell in the ISM of the galaxy (Rappaport et al. 1994). Graur, Maoz & Shara (2014) use the HST data to set a 2σ upper limit of $L_{\mathrm{HeII}} < 3.4 \times 10^{34}$ erg s^{-1} on the pre-explosion HeII line luminosity at the SN position for an unresolved HeIII region. From photoionization models for a supersoft X-ray source ionizing a surrounding ISM (Rappaport et al. 1994), the results imply a limit on the X-ray luminosity of $L_x < 3 \times 10^{37}$ erg s^{-1}. For rapidly accreting WDs, the data exclude wind-blown shells with inner radii 2–6 pc. As opposed to the limits based on direct X-ray observations, detailed above, on the presence of supersoft progenitors during the decade before the explosion, the HeII line emission limit is an indicator of the mean ionizing luminosity over $\sim 10^5$ years (the HeIII region's recombination time) prior to the explosion.

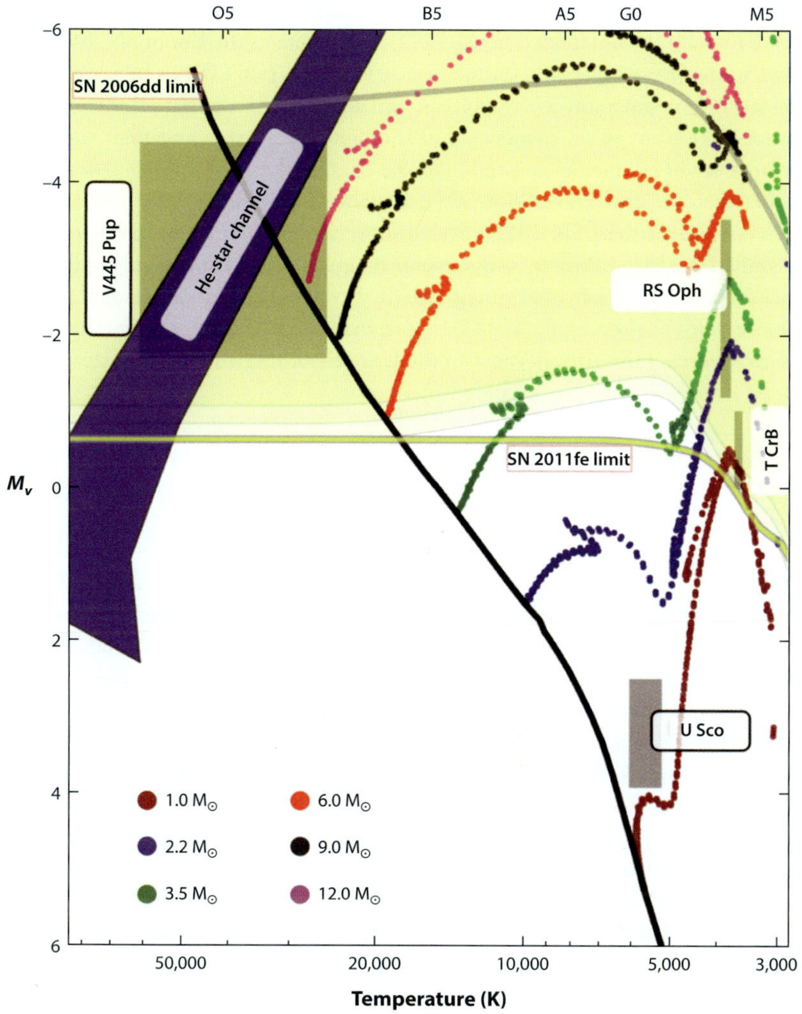

Figure 1

Hertzsprung-Russell diagram (absolute V magnitude versus effective temperature) showing the 2σ upper limits (*thick yellow line*) on the presence of progenitors in pre-explosion *Hubble Space Telescope* images of SN 2011fe in M101, from Li et al. (2011a). Also shown are theoretical evolution tracks of isolated stars with a range of masses, theoretical location of a SD He-star donor, and location on the diagram of several known recurrent novae. The data rule out red giants, and any evolved star more massive than 3.5 $M_\odot$, as well as the recurrent nova systems above the limit. Gray curve is the corresponding limit by Maoz & Mannucci (2008) for the more distant SN 2006dd. Reproduced by permission of Nature publishing group.

However, there is only one known case, CAL 83, of a supersoft X-ray source that has a detected ionization nebula, whereas nine others that have been searched for such extended line emission have yielded only upper limits, at luminosity levels an order of magnitude lower than that of CAL 83 (Remillard, Rappaport & Macri 1995). Furthermore, the X-ray luminosity of CAL 83 is $L_x = 3 \times 10^{37}$ erg s^{-1}, but its HeII line luminosity is only $L_{\rm HeII} \approx 2 \times 10^{33}$ erg s^{-1}, an order of magnitude below model expectations (Gruyters et al. 2012). Contrary to the Hα and [OIII] emission, which is roughly symmetrical around the source, the HeII emission is concentrated on one side within ~1 pc. The reasons for the discrepancy between the observed supersoft ionization nebula luminosities and model expectations are unclear, but, in any case, a HeIII region as faint as seen in CAL 83 would be undetected in the HST M101 data at the site of SN 2011fe.

In summary, the pre-explosion data for SN 2011fe rule out SD systems with red giant and He-star donors, bright supersoft X-ray sources, and accreting WDs that produce significant HeII ISM ionization, but allow main-sequence and subgiant SD donors and faint supersoft sources. DD progenitors are of course not limited by these data, nor are spin-up/spin-down models with long delays between the end of accretion and explosion.

A SN Ia even nearer than SN 2011fe was discovered early in 2014: SN 2014J in M82, at a distance of only 3.5 Mpc, although with several magnitudes of extinction by dust. Kelly et al. (2014) again use pre-explosion HST images in B through H bands to limit the presence of a red giant donor star. Nielsen et al. (2014) use pre-explosion *Chandra* observations to rule out a supersoft X-ray source. However, because of the large absorbing column density, a cool supersoft source with $kT < 80$ eV cannot be excluded.

3.3. Clues from During the Supernova Events

The observed characteristics of the emission of a SN Ia explosion itself, at various wavelengths and times, contain information on the exploding system, including its progenitor aspect. We now review this approach to the progenitor problem.

3.3.1. Early light curve and spectral evolution. Among the clues to the progenitor that, in principle, can be obtained from observables obtained during a SN event itself, the early development of the light curve and the spectra can be particularly revealing. The early light-curve evolution of a SN Ia, starting from the time of explosion, has been recently studied in a number of theory papers (Höflich & Schaefer 2009; Kasen 2010; Nakar & Sari 2010, 2012; Piro, Chang & Weinberg 2010; Rabinak & Waxman 2011; Piro & Nakar 2012, 2013; Rabinak, Livne & Waxman 2012). Briefly, the expected behavior is as follows.

Assuming that the explosion, which begins near the center of the WD, at some point expands outward as a supersonic detonation (see Section 3.3.4, below), the shock wave propagates toward the surface, heating and igniting the WD material and giving it an outward bulk velocity. The shock propagates most of the way at $\sim 10^4$ km s^{-1}, approaching the surface after ~ 1 s. The shock accelerates to relativistic velocities in the dropping density profile near the WD surface. When it emerges, it produces a millisecond-timescale gamma-ray flash of energy $E_\gamma \sim 10^{40\text{–}41}$ erg, constituting "shock breakout" (Nakar & Sari 2012). Following breakout, the radiation emerging from the explosion comes from the matter that is outside of a "diffusion front," advancing from the surface into the expanding ejecta below it. The radius of the diffusion front is defined by the diffusion time of photons through the outer envelope, at the time of observation. The expanding shock-heated ejecta quickly cools adiabatically. In the outer regions, where pressure is radiation dominated, the temperature decreases with growing radius R as $T \propto R^{-1}$, whereas in the inner matter-dominated regions, the cooling is faster, $T \propto R^{-2}$. The luminosity drops steeply for about 1 s after breakout until the diffusion front reaches material with subrelativistic bulk velocity. From this point on, the escaping bolometric luminosity initially falls with time as $L_{\rm bol} \sim t^{-0.4}$ while the effective temperature observed in the ejecta falls as $T_{\rm eff} \sim t^{-0.6}$.

The peak of the spectral energy distribution thus transits within minutes from gamma rays and X-rays to the UV band. Optical-UV-band observations will, at first, be on the Rayleigh-Jeans side of this thermal spectrum and will therefore see an optical luminosity that rises as $L_{\rm opt} \sim t^{1.5}$, peaks after $\sim$1,000 s at $L_{\rm opt} \sim 10^{39}$ erg s^{-1}, and then falls as the Wien peak enters the bandpass. After a time that is proportional to the initial stellar radius ($\sim$1 hr in the case of a WD), the diffusion front has traversed the outer $\sim 10^{-4}\,M_\odot$ of the WD mass and enters the cooler, matter-dominated region of the adiabatically expanding shock-heated ejecta, and therefore there is a sharp drop-off in luminosity (Rabinak, Livne & Waxman 2012). The luminosity may drop below detection limits, initiating a "dark phase" in the light curve, lasting until photons powered by radioactive decay can diffuse to the surface. This dark phase can last up to a few days, depending on the depth of the radioactive material.

As the diffusion front enters the ejecta, the material contributing to the observed luminosity includes progressively larger amounts of ^{56}Ni from the explosion. The energy from the gamma rays formed by the radioactive decay of ^{56}Ni diffuses through the line-blanketed ejecta and emerges in the optical. This causes the classic observed rise in the optical light curve, all the way until the peak of the bolometric light curve at ~18 days, which occurs roughly when the diffusion front has traversed all of the ejecta (and therefore the luminosity from all of the ^{56}Ni is observable). Beyond maximum light, the fall of the light curve is determined by the exponentially decreasing number of ^{56}Ni nuclei and their radioactive ^{56}Co daughter nuclei, the changing UV opacity of the ejecta, and the increasing fraction of gamma rays and positrons that can escape the remnant owing to the falling density.

Observations of the early light curve of any type of SN can provide constraints on the progenitor in several ways. First, the time between explosion and shock breakout, presuming the epoch of the explosion can be estimated, provides a measurement of the pre-explosion radius of the exploding star. (Shock breakout has been seen, to date, in several CC SNe; see, e.g., Tominaga et al. 2011). Nondetection of the postbreakout thermal emission from shock-heated material can also set upper limits on the pre-explosion radius, as further detailed below. Second, the observed time dependence of the ^{56}Ni-powered rise indicates the amount of ^{56}Ni as a function of depth into the ejecta and can thus constrain the ignition and combustion scenarios, and the amount of mixing in the ejecta. Third, in any configuration of accretion onto a WD via Roche-lobe overflow from a donor star, the L1 Lagrange point, and hence the donor, will be within a few donor-star radii of the explosion. Impact of the ejecta on the donor is therefore expected within minutes to hours (for main-sequence and giant donors, respectively). The consequences of the impact should be visible for up to a few days. This third possibility is discussed in Section 3.3.2 below.

The actual gamma-ray flash of the shock breakout in a SN Ia is expected to be too dim, too brief, and at too-high photon energies to be detected anywhere but in our Galaxy and perhaps in the Magellanic Clouds. However, the UV/optical emission from the adiabatically cooling, shock-heated, ejecta, on timescales $\lesssim$1 hr, as well as the earliest parts of the ^{56}Ni-powered light curve, is detectable in more distant events. Among the best observations to date, in this context, have again been those of SN 2011fe, which was detected 18 days before maximum light and was quickly followed up in many wavebands. Nugent et al. (2011) found that the g-band luminosity of the event is well fit by a $L_g \propto t^2$ dependence. Such a dependence has often been used to characterize the ^{56}Ni-powered phase of the light curve, both theoretically, under some simplifying assumptions (Arnett 1982), and observationally (e.g., Hayden et al. 2010b, who found $t^{1.8\pm0.2}$ by fitting the average light curves of SDSS-II SNe). Assuming also that this dependence can be extrapolated back in time, the actual time of explosion was determined by Nugent et al. (2011) to $\pm$20 min accuracy. They argued that their earliest detection of the SN, corresponding to 0.5 day after the derived explosion time, was already dominated by the ^{56}Ni-powered rise and was therefore past the shock-heated phase of the light curve. By comparing their photometry to models by Piro, Chang & Weinberg (2010), Kasen (2010), and Rabinak & Waxman (2011), they translate the limit on the duration of the shock-heated phase to a limit on the radius of the exploding star of $R_* < 0.1\,R_\odot$. Bloom et al. (2012) use upper limits on the observed luminosity from a nondetection at the SN location obtained 8 h earlier, i.e., just 4 h after the presumed explosion time (see **Figure 2**), to argue that the light curve was past the point of the drop-off in the shock-heated phase, occurring

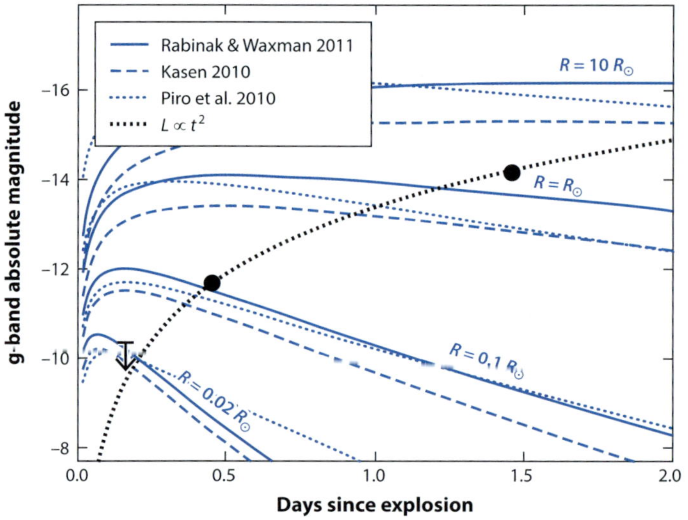

Figure 2

Early optical light curve for SN 2011fe compared with theoretical models. The two detections are from Nugent et al. (2011), and the 5σ upper limit is from Bloom et al. (2012). Black dotted curve shows the t^2 dependence due to radioactive heating expected in the simplified "fireball" approximation. The Rabinak & Waxman (2011) and Piro, Chang & Weinberg (2010) models are for the shock-heated luminosity from an object of a given radius, as marked. Kasen (2010) models are for a shock impacting a companion (see Section 3.3.2 below), with R being the separation distance, for the case of an observer aligned with the collision axis. The data indicate both a small exploding star *and* the absence of a shocked companion, but details depend on the assumed time of explosion. Reprinted from Bloom et al. (2012) with permission of the AAS.

when the diffusion front reaches the shells with gas-dominated pressure. This strengthened the limit on the initial stellar radius to $R_* < 0.02\,R_\odot$. Setting aside, for a moment, the progenitor scenarios being discussed in this review (SD, DD, etc.), these limits are particularly interesting, as they constitute the most direct evidence that the exploding object in a SN Ia is, in fact, a WD.

However, Piro & Nakar (2013, 2014) and Mazzali et al. (2014) question the accuracy of these conclusions. Piro & Nakar (2013) show that, in general for a SN Ia, the $L_{\rm opt} \propto t^2$ "fireball model" dependence is not expected under realistic conditions for the photospheric velocity (assumed to be constant in time in the fireball model) and for the radial distribution of ^{56}Ni in the ejecta [assumed to be uniform by Nugent et al. (2011)]. Piro & Nakar (2013) show that, for a range of possible conditions, the light curve is poorly approximated by a single power law, and the back-extrapolated $L_{\rm opt} \propto t^2$ dependence of the optical luminosity does not necessarily give the correct explosion time. Piro & Nakar (2014) fit the observed photospheric velocity evolution in SN 2011fe (as measured by Parrent et al. 2012) with a $v \propto t^{-0.22}$ dependence (shown to be expected theoretically by Piro & Nakar 2013) in order to constrain the explosion time, and they fit models with varying ^{56}Ni radial distributions to the observed optical light curve. They conclude that the true time of explosion is uncertain by $\sim\pm0.5$ day and could be $\sim$1 day earlier than concluded by Nugent et al. (2011). This, in turn, weakens the upper limit on the stellar radius, obtained by Bloom et al. (2012) to $R_* < 0.1\,R_\odot$. Mazzali et al. (2014) reach a similar conclusion by estimating the explosion time from their detailed fitting of photospheric temperatures, abundances, and velocities to early-time spectra (see Section 3.3.3 below). They also favor an explosion time of 0.9 day earlier than that deduced by Nugent et al. (2011) and find a limit of $R_* < 0.06\,R_\odot$. These somewhat weaker limits on the stellar radius are still quite useful in ruling out most options for the exploding object in SN

2012fe. Bloom et al. (2012) show that, apart from a WD, only pure carbon-oxygen stars, burning carbon stably, could have radii consistent even with these weaker limits.

Modeling of SN 2011fe by both Mazzali et al. (2014) and Piro & Nakar (2014) lead them to conclude that significant amounts of ^{56}Ni, $\sim$0.1 of the total amount, existed in the outer 0.1 $M_\odot$ of the ejecta. Piro & Nakar (2014) find this not only for SN 2011fe but also for SN 2012cg, another SN Ia with early time data. The presence or absence of ^{56}Ni in the outer layers is relevant to understanding the explosion details, especially the possible transition from a deflagration to a detonation (see Section 3.3.4 below).

Zheng et al. (2013) discovered a SN Ia, 2013dy, and caught it even earlier in its light evolution. With good light-curve coverage in the first few days, they find that, indeed, a t^2 behavior fits the data poorly. Instead, they obtain a best fitting broken power law that begins with an index of 0.9 and transits to an index of 1.8 after about 3 days. Extrapolating this fit back in time to zero flux implies that the SN's "first light" occurred just 2.4 $\pm$ 1.2 h before the first detection. The nondetections at earlier times again argue for upper limits of $\lesssim$0.25 $R_\odot$ on the progenitor radius.

The recent and very-nearby SN 2014J in M82 also provides input on early light curve and spectral evolution. First results are presented by Zheng et al. (2014) and Goobar et al. (2014).

3.3.2. Shocks from ejecta impact on companion stars and debris. The recent and very nearby SN 2014J in M82 also provides input on early light curve and spectral evolution. First results are presented by Zheng et al. (2014) and Goobar et al. (2014).

As already noted, in the classic SD picture, shocks from the SN ejecta hitting the companion star are unavoidable and may be observable. Kasen (2010) calculated the observational signatures of such events as a function of the companion size (giants naturally produce the largest signals) and as a function of the angle between the observer, the WD, and the donor. When a giant star is between the observer and the WD, the signatures of the impact are expected in X-rays within minutes to hours of the explosion. Emission from the shocked region in UV/blue bands develops on a timescale of several days. Bianco et al. (2011), Hayden et al. (2010a), Tucker (2011), and Ganeshalingam, Li & Filippenko (2011) have set observational upper limits in blue optical bands in the light curves of SNe Ia with good early-time optical coverage and have used the Kasen (2010) models to limit the presence of red-giant donors to <20% of SN Ia progenitor systems. Foley et al. (2012a) and Brown et al. (2012b) have extended this approach using space-UV measurements to obtain such limits in several cases.

This type of limit is strongest, here again, in the case of the nearby and promptly observed SN 2011fe in M101. Nugent et al. (2011) and Bloom et al. (2012) use the same early optical observations (discussed in Section 3.3.1 above, in the context of emission from shock-heated ejecta) to exclude also shocks from ejecta hitting a companion. Brown et al. (2012a) use prompt UV observations of SN 2011fe, obtained with *Swift* starting 18 h after the first Nugent et al. (2011) epoch, to do the same. All three studies rule out red giant donors; Brown et al. (2012a) also exclude main-sequence companions of radius $R_c \gtrsim R_\odot$, and Bloom et al. (2012) sharpen this limit to $R_c \lesssim 0.1\, R_\odot$. **Figure 2** shows also these limits. The conclusions, like those in Section 3.3.1 above, are sensitive to the exact time of the explosion.

This limitation has been circumvented by R.P. Olling, R. Mushotzky & E.J. Shaya (submitted), who used the *Kepler* mission to monitor continuously over several years, with 30-min exposures, 400 galaxies in the Kepler field. Among these galaxies, they discovered four SNe, two or three of them likely SNe Ia based on their light curves. The continuous coverage reveals no signatures of companion impacts within the few days before first light, when the actual explosion must have occurred. Olling and colleagues again use the Kasen (2010) models to limit the presence of red giant and main-sequence companions for a range of viewing angles. The extension of the *Kepler* mission, particularly in sky directions that will allow ground-based follow-up and classification, will permit obtaining such data for many additional events, which will set strong constraints on the fraction of SNe Ia in which a SD donor star is present.

Interestingly, Nugent et al. (2011) used the absence of impact signatures in SN 2011fe to argue also against a DD progenitor system. This argument is based on calculations by Fryer et al. (2010), who simulated DD mergers and found that a potentially large mass of debris, 0.1–0.7 $M_\odot$, forms an inhomogeneous envelope around the merged object, with a steeply falling density profile extending out to $\sim 1\ R_\odot$. This is based, in turn, on BPS calculations (Ruiter, Belczynski & Fryer 2009) that predict the majority of DD mergers to have masses around 2 $M_\odot$, whereas most other BPS studies predict lower masses (see also Section 3.3.4). Fryer et al. (2010) calculate the shock breakout and early-time spectral evolution in this model and find behavior very distinct from that expected of a "bare" exploding WD, with UV emission lasting for days, similar to the early phases of Type II SNe. However, Fryer et al. (2010) also find that, in general, the spectra and light curves produced in the context of this model bear little resemblance to those of normal SNe Ia. The results of the calculations, if taken at face value, can therefore be considered evidence against the DD scenario, irrespective of the lack of shock signatures in the Nugent et al. (2011) data for a very typical SN Ia. However, DD merger simulations differ regarding the amount of debris at small distances at the time of explosion (Dan et al. 2012; Raskin & Kasen 2013; Shen, Guillochon & Foley 2013). As explained by Pakmor et al. (2012), one critical ingredient is the time elapsed from merger until explosion. In the violent-merger scenarios, where ignition occurs within minutes of the merger, any merger debris moving at the escape velocity of a few thousand km s^{-1} will be overtaken by the ejecta and will cease to interact with it minutes after the explosion (see also, in this context Soker, Garcia-Berro & Althaus 2014). Also important for the amount and location of debris are the WD mass ratio and the assumed WD spin initial conditions (Dan et al. 2014).

3.3.3. Spectral and light-curve modeling. As the ejecta of a SN expands, the photosphere recedes into progressively deeper layers, revealing the chemical stratification and velocity structure of the ejecta. These in turn, may help distinguish among progenitor models. A way to reconstruct this physical structure is through calculation of synthetic spectra and light curves that match the observations.

In one such modeling approach (e.g., Stehle et al. 2005, Mazzali et al. 2007, Tanaka et al. 2011, Hachinger et al. 2013), an ejecta density profile is assumed on the basis of a hydrodynamical explosion model. A Monte Carlo radiative transfer calculation is used to propagate photon packets from a blackbody spectrum through the model photosphere. The element abundances and the velocity at each radius are varied in order to obtain the best fit to the set of observed spectra and photospheric velocities. Mazzali et al. (2014) recently applied the method to the combined UV spectra from HST and ground-based optical spectra for SN 2011fe. A good match to the data is found using a hybrid between the density profiles from the SD one-dimensional (1D) fast-deflagration W7 model by Nomoto, Thielemann & Yokoi (1984) and from a 1D delayed-detonation model by Iwamoto et al. (1999) (more on deflagrations and detonations in Section 3.3.4 below).

Blondin et al. (2011) carry out a more "first-principles" approach by taking a set of 44 hydrodynamical 2D M_{Ch} SD delayed-detonation models from Kasen, Röpke & Woosley (2009) and comparing them with observed light curves and multiepoch spectra of 251 SNe Ia. The details of the nucleosynthesis and the resulting element distributions were calculated by means of tracer particles distributed throughout the volume; nuclear reaction networks were calculated for the conditions at each location. Finally, photon packets were again Monte Carlo–propagated through the ejecta to obtain synthetic light curves and spectra at the epochs corresponding to the observations. Overall good agreement is found between the data and some of the models, particularly those that satisfy the Phillips (1993) relation, but some details do not match, such as too-high model velocities in the maximum-light spectra.

Röpke et al. (2012) extend this approach to test example models from the two main progenitor scenarios, SD and DD, against optical spectra of SN 2011fe. One model is a $M_{\rm Ch}$ SD delayed-detonation model by Seitenzahl et al. (2013b) and the other is a violent-merger DD model by Pakmor et al. (2012), merging 1.1 $M_\odot$ and 0.9 $M_\odot$ WDs. For each model, Röpke et al. (2012) used a 3D hydrodynamics code to calculate the evolution of the density and the temperature. Nucleosynthetic abundances are again calculated via tracer particles and Monte Carlo radiative transfer is performed. Both progenitor models produce reasonable matches to the observations, but again both have shortcomings in matching the details. Overall, the DD model fares somewhat better in this comparison. The differences in the predictions of the two models and the discriminating power of the data suggest that future applications of this approach, with a larger range of 3D models, should be very useful.

The amount and velocity distribution of unburnt carbon observed in SN Ia ejecta can be a diagnostic that distinguishes scenarios from one another. Folatelli et al. (2012), Parrent et al. (2011), Thomas et al. (2011), and Silverman & Filippenko (2012) analyze several samples with pre- or near-maximum spectra and estimate that the C II λ6580 absorption feature is present in about 30% of SNe Ia. In all studies, the feature tends to appear in events with "low velocity gradients" (see Section 3.3.4 below). The implied carbon mass is 10^{-3}–10^{-2} $M_\odot$. In SN 2011fe, Nugent et al. (2011) clearly detect this line in the earliest high-resolution spectrum along with high-velocity O I absorption. Both features almost disappeared in a spectrum obtained only 8 h later. From their modeling of these data, Mazzali et al. (2014) deduce the existence of an almost-pure carbon outer layer of $\sim$0.01 $M_\odot$. They speculate that this layer reflects the properties of the accreted material, whether it is hydrogen or helium that has burned to carbon. Soker, Garcia-Berro & Althaus (2014) propose, instead, that the outer-layer carbon enrichment is a by-product of crystallization in a WD that has been rotation-supported against collapse and has cooled, for $\sim$1.4 Gyr, in a spin-up/spin-down scenario such as a core-degenerate scenario. Zheng et al. (2013) similarly see strong C II absorption in SN 2013dy, within the first few days after explosion, weakening and becoming undetectable within a week.

3.3.4. Diversity and correlations among spectral and environmental observables. Observers have long sought, and sometimes found, relations among the observed properties of SN Ia events. These include relations among the various spectral and photometric properties and between those properties and the properties of the host galaxies as a whole or of the specific locations of SNe Ia within a galaxy. An observed correlation, assuming it is not a result of selection effects, may be merely a consequence of some basic physics of SN Ia explosions in general rather than of a specific progenitor channel leading up to the explosion. Nevertheless, in such relations there is a potential for clues to the progenitor problem.

3.3.4.1. The width-luminosity relation. The strongest and clearest correlation seen among SN Ia properties is the so-called Phillips (1993) relation, connecting the maximum-light luminosity of an event to its light-curve evolution (as parameterized in a variety of ways) and to the observed color at maximum light (e.g., Kattner et al. 2012, Hillebrandt et al. 2013). Less-luminous SNe Ia, which have synthesized less ^{56}Ni, evolve more quickly. Less-luminous SNe Ia are also redder owing to some combination of intrinsically redder color and reddening by dust, a combination that has proved hard to disentangle. The problem is compounded by evidence, often contradictory, that the dust involved in the effect might have properties distinct from Milky Way dust, with a low ratio of extinction to reddening, $R_V \equiv A_V/E(B-V) \approx 1.5-2$, as opposed to the typical Galactic $R_V \approx 3.1$ (see Howell 2011, but see Scolnic et al. 2014 for a different view).

The Phillips relation itself may be a consequence of SN Ia physics, rather than of a particular progenitor scenario. The mass of synthesized ^{56}Ni largely determines the total energy of the explosion. The larger the ^{56}Ni mass, the higher is the peak luminosity owing to the radioactive decay of ^{56}Ni. In parallel, iron-group elements also cause the bulk of the opacity, slowing the evolution of the light curve, i.e., leading to a higher light-curve "stretch" factor [e.g., Hoeflich & Khokhlov 1996, Kasen & Woosley 2007; see Kasen, Röpke & Woosley (2009) for additional factors, such as viewing angle, that can affect the relation]. Nevertheless, progenitor models invoking explosions always near $M_{\rm Ch}$ must explain the diversity, by a factor of $\sim$4, in explosion energy among "normal" SNe Ia (Branch & van den Bergh 1993; i.e., excluding underluminous, SN-1991bg-like, and overluminous, SN-1991T-like, events).

Timmes, Brown & Truran (2003) proposed progenitor metallicity as the main driver for explosion energy. Main-sequence stars with higher abundances of CNO produce, during He burning, higher abundances of neutron-rich nuclei, particularly ^{22}Ne, that end up in the WD. In a WD undergoing combustion as a SN Ia, neutron richness leads to more synthesis of stable elements such as ^{54}Fe and ^{58}Ni, at the expense of ^{56}Ni, and hence to a lower luminosity. Although Timmes, Brown & Truran (2003) show that a large-enough range in progenitor metallicity could reproduce the observed range in normal SN Ia explosion energies, Piro & Bildsten (2008) and Howell et al. (2009) show that the observed range of SN Ia host-galaxy metallicities does not reach the extreme values required for this mechanism to have a significant effect on SN Ia explosion energy. The observed metallicity range of SN Ia host galaxies, even accounting for local extremes within a galaxy, would lead to $<$10% variation in explosion energy. Piro & Bildsten (2008) further show that pre-explosion neutron richness is not controlled solely by metallicity but rather can be dominated by neutronization through carbon burning during the $\sim$1,000 years "simmering" phase that $M_{\rm Ch}$-explosion models generally undergo before explosion. Mazzali & Podsiadlowski (2006) showed that, even if metallicity variations produce the range in peak luminosity, they do not affect light-curve shape, i.e., the Phillips relation is not reproduced. Foley & Kirshner (2013) analyze two "twin" SNe Ia (2011fe and 2011by) that have identical light-curve shapes and optical spectra, but differ in their UV spectra and in their peak luminosity. Modeling the difference in UV spectral opacity by means of progenitor metallicity differences, the Timmes, Brown & Truran (2003) model correctly predicts the difference in peak luminosity. Metallicity effects may thus be at work, even if they do not explain the explosion energy diversity or the Phillips relation.

Another way to explain the observed diversity in explosions of $M_{\rm Ch}$ models is through the so-called deflagration-to-detonation transition. It has long been known that the energetics and spectra of $M_{\rm Ch}$ models do not match observations unless they are finely (and artificially) tuned as an initial subsonic deflagration that allows the WD to expand and, at the right time, spontaneously to evolve into a supersonic detonation (Khokhlov 1991, Hillebrandt & Niemeyer 2000, Blondin et al. 2013, Ma et al. 2013). Different progenitor metallicities and masses, and different accretion histories, could affect the number and location of ignition "kernels" and, thus, the transition to detonation (e.g., Umeda et al. 1999; Kasen, Röpke & Woosley 2009; Blondin et al. 2011; Seitenzahl et al. 2013b). Different transition times lead, in the final ejecta composition, to different proportions of unburnt carbon and oxygen, intermediate-mass elements, stable iron-peak elements, and light-curve-driving radioactive ^{56}Ni. This, in turn, could perhaps produce the observed diversity in SN Ia explosion energy (see, in this context, also "pulsational delayed detonation" models, e.g., Dessart et al. 2014a).

DD models can also readily produce a range of explosion energies. Ruiter et al. (2013) find, for "violent" mergers that include detonation of small He layers on top of the C/O WDs (Pakmor et al. 2013), that the explosion energy varies with the mass of the primary WD in the binary and can reproduce the full observed range of SN Ia luminosities by means of primary masses between 0.9 and 1.3 $M_{\odot}$. Combining this with a BPS calculation, they find that the relative frequency of SN Ia luminosities can also be reproduced, if the primary WD accretes a substantial amount of helium from the companion star when it is a He giant, prior to becoming the secondary WD. In their own violent-merger simulations, Moll et al. (2014) also reproduce the range of SN Ia luminosities and the Phillips relation through a combination of primary WD mass and viewing angle of the highly asymmetric explosion. Taubenberger et al. (2013a) analyze the late-time "nebular"-epoch spectra of several "super-Chandrasekhar" SNe Ia to estimate $\sim 2\,M_{\odot}$ of ejecta, about half of it is in ^{56}Ni, which they interpret as possibly arising from the merger of two massive WDs. In nebular-phase spectra of the subluminous SN Ia 2010lp, Taubenberger et al. (2013b) find oxygen emission lines that, they argue, support a violent merger scenario. In their collisional DD model, Kushnir et al. (2013) also reproduce the observed range of ^{56}Ni masses via different combinations of WD masses. The typical observed ^{56}Ni mass of $0.6\,M_{\odot}$ requires collisions of WDs of at least $0.7\,M_{\odot}$ each (see also Piro, Thompson & Kochanek 2014). S. Dong, B. Katz, D. Kushnir, & J.L. Prieto (unpublished, arXiv:1401.3347) present double-peaked line profiles in three SNe Ia among a sample of 20 events with high-quality nebular-phase spectra. This result compares favorably with a 3D collisional DD simulation they perform.

3.3.4.2. Explosion energy, colors, spectral features, and velocities. A second, potentially strong correlation between SN Ia properties that has been claimed is between luminous energy (as measured by the light-curve "stretch") and the kinematic width of iron emission features, particularly [FeIII] λ4701, in late-time, nebular-phase spectra (Mazzali et al. 1998). Blondin et al. (2012) and Silverman, Ganeshalingam & Filippenko (2013) argue that the relation is driven only by very underluminous events and that no correlation remains once they are excluded. Kushnir et al. (2013), however, have remeasured the velocities in the late-time spectra by fitting broadened narrow-line templates rather than measuring the width of individual lines. With this procedure, they recover a correlation between ^{56}Ni mass and late-time velocity width. Physically, such a correlation may simply mean that more energetic explosions produce larger ejecta velocity distributions without having particular relevance for the progenitor issue.

Other correlations that have been reported between observed SN Ia parameters tend to be rather weak trends, whose details and strengths sometimes vary among different samples. The blueshifted velocity of the SiII λ6150 and λ6355 absorptions at maximum light, v_{Si}, tracing the photospheric expansion velocity of the ejecta, has been crossed with various other observables. Foley & Kasen (2011) found that v_{Si} correlates with stretch, but that this is driven solely by the least-luminous events with $\Delta_{m15,B} > 1.5$ (where $\Delta_{m15,B}$ is the decrease in B magnitudes at 15 days after maximum light and is inversely related to stretch). Without those events, there is no significant trend between stretch and ejecta velocity. Wang et al. (2009b), Foley & Kasen (2011), and Foley, Sanders & Kirshner (2011) do find a weak trend for events with faster ejecta to have redder intrinsic colors, and Foley (2012) finds a weak anticorrelation between v_{Si} and host-galaxy mass. Maguire et al. (2012) show a weak trend of increasing near-UV CaII H&K absorption velocity and stretch [but Foley (2013) argues that this is an artifact]. They also find a rather significant dependence on stretch of the blueshift of the "λ_2" pseudoemission feature at 3,180 Å. The λ_2 feature is thought to arise from a spectral interval of slightly lower relative opacity in the heavily line-blanketed near-UV region (Walker et al. 2012). Wang et al. (2013) show that v_{Si} correlates with a number of environmental parameters. High-v_{Si} events tend to occur at relatively small normalized galactic radii, whereas the lower-v_{Si} events ($<$12,000 km s^{-1}) occur at all radii. Furthermore, the stellar surface brightness distribution at the locations of high-v_{Si} events is similar to that of CC SNe, suggestive of a young progenitor population, whereas lower-v_{Si} events tend to come from regions of relatively lower surface brightness. Finally, high-v_{Si} events tend to be in larger-diameter and more luminous galaxies [in apparent contradiction to the inverse relation with galaxy mass observed by Foley (2012)]. Wang et al. (2013) interpret these results in terms of two different SN Ia progenitor populations, distinguished by age and metallicity. The observed correlations are intriguing, but the conclusion may be premature given the limited direct evidence for a dependence on age and metallicity and given the different phenomenology that is observed regarding metallicity (Howell et al. 2009; see Section 3.3.4.1 above) and regarding progenitor age (see Section 3.3.4.3 below). We note here, in passing, that Foley et al. (2012b) find a correlation between v_{Si} and the probability for the presence of a blueshifted absorbing NaI D system in the SN spectrum, a result to which we return in Section 3.3.8 below.

Benetti et al. (2005) have separated SNe Ia into two classes—those with "low velocity gradients" and those with "high velocity gradients"—based on the decrease with time in the velocity of the SiII λ6355 absorptions, $\dot{v}_{Si}$, with the border at 70 km s^{-1} day^{-1}. Events with high velocity gradients tend to also have high velocities at maximum light (Foley, Sanders & Kirshner 2011). Maeda et al. (2010) show, for a sample of 20 SNe Ia, that high-velocity-gradient objects, in their late nebular phase, have nebular lines of [FeII] λ7155 and [NiII] λ7378 that are redshifted relative to the higher-excitation line of [FeIII] λ4701, whereas low-velocity-gradient objects tend to have velocities that are blueshifted. Maeda et al. (2010) interpret this result, in the context of the deflagration-to-detonation transition, by means of a combination of an off-center-ignited explosion and a viewing angle effect. In their picture, the nebular [FeII] and [NiII] lines come from the ashes of the initial deflagration phase, whereas the [FeIII] line, which requires low density and heating by ^{56}Ni decay, traces the detonation ashes. If the explosion is ignited off-center, e.g., within the WD hemisphere on the observer's side, the deflagration-ash lines will be blueshifted with respect to the detonation-ash lines. The observer will see a low velocity gradient in the photospheric lines because of the density and the velocity structure encountered by the photosphere as it moves inward. Maeda et al. (2011) and Cartier et al. (2011) both find that the nebular line shifts also correlate with the SN colors near maximum light, strengthening the same idea of a dependence of observed properties on viewing angle.

Another spectral signature in SNe Ia, which tends to appear in early-time spectra of some events, is the "high-velocity features" (e.g., Gerardy et al. 2004, Mazzali et al. 2005), absorptions in SiII λ6355 or the CaII NIR triplet, whose velocities are higher by several thousands of kilometers per second than those of the normal photospheric absorptions in these and other lines. The origin of these features is unclear. Recently, Childress et al. (2014) studied them in a sample of 58 SNe. They find that the existence and strength of high-velocity features correlates with SN Ia stretch and anticorrelates with v_{Si}. In a subset of SNe with weak high-velocity features, stretch and v_{Si} are correlated.

3.3.4.3. Supernova stretch/luminosity and host-galaxy age/star-formation rate. A correlation between peak SN Ia luminosity, or light-curve stretch, and the age of the host galaxy has been noted for some time: The oldest hosts, with little star formation, tend to host faint, low-stretch SNe Ia, whereas star-forming galaxies are more likely to host bright-and-slow SNe Ia (e.g., Hamuy et al. 2000, Hicken et al. 2009, Howell et al. 2009, Lampeitl et al. 2010, Smith et al. 2012, Pan et al. 2013, Xavier et al. 2013). Rigault et al. (2013) show this trend at the local level in SN Ia host galaxies using local Hα surface brightness as a star-formation tracer. Although all of these studies observe the same general trend, the scatter in the relation is large and the various samples give different pictures of how the trend arises. For example, Hamuy et al. (2000) observe the trend to be driven both by a lack of bright SNe in early-type galaxies and a lack of faint SNe in late-type galaxies. Howell et al. (2009) find that galaxies with luminosity-weighted ages $\lesssim$4 Gyr host SNe Ia of all explosion energies, but galaxies with ages $\gtrsim$4 Gyr host almost only faint SNe Ia. Conversely, Hicken et al. (2009) and Smith et al. (2012) find that early-type galaxies host the full range of light-curve widths, but late-type galaxies host only SNe Ia with broad light-curve shapes. This situation likely arises from a combination of effects. First, the trend, as noted, has a large scatter to start with. Second, different samples may have different selection effects that could bias for or against the inclusion of more or less luminous SNe Ia hosted by different types of galaxies. This can become an issue particularly when SN samples are small such that regions of the age-luminosity plane that have low SN rates are sparsely populated. The appearance of these scatter plots is likely affected by the different ways of defining light-curve parameters [this is evident from comparing the plot by Hicken et al. (2009), which uses the x_1 light-curve stretch parameter, with their plot that uses the Δ light-curve-width parameter] or by the surrogates used for the explosion energies (as in Howell et al. 2009). Similarly, there are different ways of defining galaxy morphological type or of considering a galaxy "age," which is a poorly defined concept (galaxies have composite populations, and their full star-formation histories are the more relevant observable). Finally, galaxies are characterized by multiple correlations between morphology, age, size, metallicity, star-formation rate, extinction, and more (e.g., Mannucci et al. 2010), and it is not clear what the main driver of the correlation with SN Ia luminosity is. We revisit this issue when discussing SN Ia rates and the delay-time distributions for SNe Ia of different stretches in Section 3.5 below. In the meantime, we note that this trend between explosion energy and host-population youth could well be an important clue to the progenitor question.

3.3.4.4. Synopsis of correlations. From all of the above, there are relations among the observed properties of SNe Ia, and one can try to sketch a rough trend of properties that go together (see, e.g., Maguire et al. 2013). At one extreme are events that are luminous, have broad light curves, red intrinsic colors, high-velocity ejecta, large velocity gradients, redshifted and broad nebular lines, no carbon signatures, and star-forming hosts. At the other extreme are events with the opposite properties. We further discuss the possible source of such a trend in Section 4 below.

3.3.4.5. Peculiar SNe Ia. Apart from the diversity among normal SNe Ia, there is a growing "zoo" of abnormal events that may nonetheless belong to the SN Ia class, though with varying degrees of certainty. The numbers of known peculiar SNe Ia have grown as a result of recent untargeted field transient surveys with large SN number yields, fainter flux limits, and faster cadences than previous surveys. Their peculiarities could shed light on the progenitor question if, e.g., the peculiarities can be identified with some particular progenitor channels or with specific deviations from the progenitor parameters of normal events.

Of particular recent interest is the class of explosions, sometimes called "SNe Iax" (e.g., Foley et al. 2013), that are similar to the prototype event SN 2002cx. Although spectroscopically similar to normal SNe Ia, SNe Iax are characterized by low photospheric velocities at maximum light (2,000–8,000 km s^{-1}), hot photospheres (based on the presence of high-ionization lines), and peak luminosities that are typically several magnitudes faint for the observed light-curve width. Their light curves lack the "second IR bump" of normal events. At late times, instead of the nebular-phase broad forbidden lines seen in normal SNe Ia, the spectrum shows narrow permitted lines, indicating a high gas density, $n_e > 10^9$ cm^{-3}, even over a year after explosion (e.g., McCully et al. 2014). Their spectra always show carbon features, and sometimes helium. Only a few tens of SNe Iax are known, but accounting for their detectability, Foley et al. (2013) estimate that there are ~20–50 such events for every 100 normal SNe Ia, which would make SNe Iax the most common type of peculiar SN Ia.

Jordan et al. (2012), Kromer et al. (2013), and Fink et al. (2014) model SNe Iax as "failed deflagrations"—SN Ia explosions in which a transition from deflagration to detonation fails to occur, and furthermore the explosion fails to completely unbind the WD, leaving behind an $\sim 1\,M_\odot$ bound remnant. This could explain the low ^{56}Ni yields, the low velocities, the unburnt C and He in the ejecta, the high degree of mixing, and the clumps of high-density material. However, SNe Iax occur predominantly in star-forming galaxies (but there is one case, SN 2008ge, occurring in an S0 galaxy, with no signs of star formation or pre-explosion massive stars at the explosion site; Foley et al. 2010b), and their locations within these galaxies track the SFR similarly to the common Type IIP CC SNe (Lyman et al. 2013). Indeed, Valenti et al. (2009) have argued that SNe Iax are actually CC SNe with low ejecta velocities derived from $7-9\,M_\odot$ or $25-30\,M_\odot$ progenitors, with cores collapsing into black holes. Conversely, Foley et al. (2010a) point to the presence of sulfur in the spectra of some SNe Iax as evidence for thermonuclear burning in a C/O WD and against a CC SN. It thus remains to be seen whether or not SNe Iax are the manifestations of incomplete pure deflagrations of $M_{\rm Ch}$ WDs.

3.3.5. Polarization and symmetry. Polarization can reveal deviations from circular symmetry of a source, as projected on the sky. In the context of SNe, linear polarization arises from electron and line scattering of photons emerging from an asymmetric source. The percentage of total flux that is polarized is proportional to the projected geometrical axis ratio, and the angle of polarization shows the asymmetry's orientation on the sky. The bandpass dependence of continuum polarization, and the polarization across individual line profiles, can provide further information on asymmetry and orientation for specific emitting elements and velocities and can show how they change with time as a SN evolves. Wang & Wheeler (2008) review spectropolarimetry of SNe, including SNe Ia. Spectropolarimetry can guide the SN Ia progenitor problem, as most SN Ia models have some built-in asymmetry, such as the presence of a donor or an accretion disk, a recently merged or collided secondary WD, rotational flattening, or an off-center ignition (e.g., Röpke et al. 2012).

As summarized by Wang & Wheeler (2008), SNe Ia tend to have low or zero continuum polarizations, typically fractions of a percent, and polarization is generally detected only pre- or near-maximum light. This indicates that the expanding photospheres are slightly asymmetric on the outside, at levels of up to of order 10%, but quite round in the inner layers of the explosion. Nonetheless, line polarizations of up to a few percent are sometimes seen in some line transitions, sometimes with different orientations than those of the continuum. Although, again, line polarization tends to be seen at early times, there are exceptions (e.g., Zelaya et al. 2013). The line polarizations have been interpreted in terms of clumpy distributions of the ejecta of particular elements (Kasen et al. 2003; Hole, Kasen & Nordsieck 2010; Kasen, Röpke & Woosley 2009). Smith et al. (2011) present spectropolarimetry of the nearby SN 2011fe. Like its other properties, its polarization behavior is typical of normal SNe Ia.

Wang, Baade & Patat (2007) show a correlation between Si II λ6355 polarization and Δm_{15}, suggesting that more luminous SNe Ia tend to be more symmetric, but this relation is driven by one or two objects, after excluding some outliers. Tanaka et al. (2010) extend this result to the case of the proposed super-$M_{\rm Ch}$ event SN 2009dc. Although this SN's line polarization is somewhat above that expected from the Wang, Baade & Patat (2007) trend, the low continuum polarization argues for little asphericity. Tanaka et al. (2010) use this to argue against the explanation by Hillebrandt, Sim & Röpke (2007) that this was an off-center $M_{\rm Ch}$ explosion and in favor of the view that it was a super-$M_{\rm Ch}$ progenitor. However, asymmetry would be expected also from a super-$M_{\rm Ch}$ rotation-supported configuration. In the normal SNe 2012fr and 2011fe, Maund et al. (2013) and Soker, Garcia-Berro & Althaus (2014), respectively, have used the very low continuum polarization level ($<0.1\%$) and the implied circular symmetry to argue against the violent DD-merger scenario

(Pakmor et al. 2012), in which an asymmetric ^{56}Ni distribution is predicted. At the other extreme of the same trend, Howell et al. (2001) and Patat et al. (2012) have used spectropolarimetry to deduce significant, coaligned asphericity in both lines and continuum in the subluminous SNe Ia 1999by and 2005ke, respectively, supporting fast rotation or a merger origin in these cases.

Leonard et al. (2005), Wang et al. (2006), and Patat et al. (2009) show that high SiII λ6355 line polarizations and line velocities are correlated. Maund et al. (2010) find, for a sample of normal SNe Ia, that the polarization of the SiII λ6355 line at 5 days premaximum is correlated with the line velocity's time gradient $\dot{v}_{Si}$ for a given SN. They suggest that normal SNe Ia have a single, asymmetric distribution of intermediate mass elements, with the diversity of observed properties arising partly from orientations effects.

At the two extremes in trends in SN Ia properties, discussed in Section 3.3.4 above, it is not clear into which "family" SNe Ia with relatively high polarization fall. According to Maund et al. (2010), it would be in the high-velocity, high-stretch family. According to Wang, Baade & Patat (2007), it would be in the opposite family.

3.3.6. Searches for emission from hydrogen. In the SD scenario, almost by definition, some signs of hydrogen or helium from the nondegenerate companion should be visible at some point. Apart from the shock from the ejecta impacting a SD companion (Section 3.3.2 above), the ejecta are expected to strip and entrain material from the companion. This stripped material of mass $\sim 0.1 - 0.2\, M_{\odot}$, which is excited by energy from the radioactive decay of ^{56}Co, is then expected to become visible in the form of emission lines of width $\sim$1,000 km s^{-1} during the nebular phase of the explosion when the ejecta are optically thin (Marietta, Burrows & Fryxell 2000; Mattila et al. 2005; Liu et al. 2012b). Mattila et al. (2005), Leonard (2007), and Lundqvist et al. (2013) use high signal-to-noise nebular spectra of five SNe Ia to set upper limits on the presence of any Hα emission. These limits translate to upper limits on hydrogen mass of 0.01 $M_{\odot}$ to 0.03 $M_{\odot}$. The most stringent application of this test has been, yet again, in the case of the nearby SN 2011fe, where Shappee et al. (2013) set a limit of 0.001 $M_{\odot}$ on the stripped H mass (see **Figure 3**). These results strongly suggest that, at least in the six cases examined, there was no main-sequence or red-giant SD companion in the system at the time of explosion.

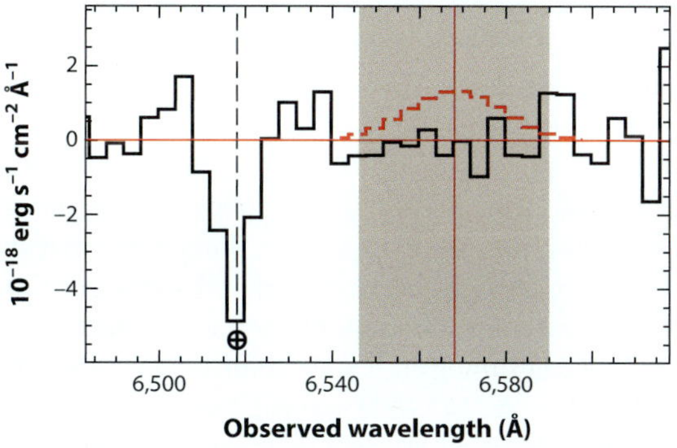

Figure 3

Nebular-phase (day 274 after maximum) Hα region of the spectrum of SN 2011fe, showing upper limits on Hα emission from any hydrogen stripped from a companion and entrained in the ejecta. The vertical red line and gray band show, respectively, the expected wavelength of Hα emission and $\pm$1,000 km s^{-1} velocity width. Black solid histogram is the observed continuum-subtracted spectrum and red dashed curve is the 3σ upper limit on line emission, which implies a hydrogen mass $<0.001\, M_{\odot}$. Reprinted from Shappee et al. (2013) with permission of the AAS.

Although the absence of H and He is a defining characteristic of SNe Ia, there are exceptions. Until recently, H and He in emission was reported in only two SNe Ia: SN 2002ic (Hamuy et al. 2003) and SN 2005gj (Aldering et al. 2006, Prieto et al. 2007). These events displayed a SN-Ia-like spectrum, but were topped by strong variable Hα line emission that was visible during most epochs and reminiscent of a CC Type IIn spectrum. The spectra were interpreted as showing the interaction of the SN ejecta with the circumstellar medium (CSM) of an evolved SD companion (Han & Podsiadlowski 2006, Wood-Vasey & Sokoloski 2006). Another possibility raised was that they were prompt DD mergers that encountered a recently ejected common envelope (Livio & Riess 2003). Chugai & Yungelson (2004) speculated that such events are examples of "SNe 1.5" (Iben & Renzini 1983), i.e., thermonuclear runaways in the C/O cores of single AGB stars that have not yet lost their envelopes. Alternatively, Chugai & Yungelson proposed that such "strongly CSM-interacting" SNe Ia are SD systems in which the donor is a massive red supergiant. In either case, the CSM envelope has a mass of at least several solar masses. Some doubts were initially raised as to whether these rare events were, in fact, true SNe Ia. Benetti et al. (2006) argued that these explosions are actually a subclass of Type Ic CC SNe, in which the ejecta are encountering the collapsed star's previously ejected envelope in a variant of what is thought to be the case for Type IIn SNe. Trundle et al. (2008) pointed out the resemblance of the absorption features in such events to those of the class of "luminous blue variables," which are thought to be the progenitors of Type Ib/c SNe.

The most recent and well-studied object of this type is PTF11kx (Dilday et al. 2012, Silverman et al. 2013a), further discussed below (Section 3.3.8) in the context of SNe Ia that have intervening NaI D absorption. In PTF11kx, the Hα emission is somewhat weaker and develops somewhat later than in the above two cases of CSM-interacting SNe Ia, and the light curve is more similar to those of normal SNe Ia. From analysis of the evolving spectra, Dilday et al. (2012) deduce the progenitor was a SD system with a red-giant donor and multiple shells of equatorially distributed circumstellar material (swept up by ejecta from pre-explosion nova eruptions) similar to those in the symbiotic recurrent nova RS Oph, in which a WD accretes through a wind from a red giant (see Section 3.1.1 above). Soker et al. (2013), however, estimate that the circumstellar mass in PTF11kx is at least $\sim 0.1 - 0.6\,M_{\odot}$, which is higher than the CSM expected in SD models. Instead, they advocate a variant of the Livio & Riess (2003) picture (ejecta from a DD merger encountering the previously ejected common envelope). In the context of the Soker (2013) core-degenerate scenario, Soker et al. (2013) propose that during the common envelope phase, the WD disrupts and accretes the less-dense AGB core (rather than vice versa), resulting in a prompt (rather than long-delayed) explosion.

Silverman et al. (2013b) compile a sample of 16 such strongly CSM-interacting events, some by reclassifying old events (one of them, SN 2008J, was concurrently studied by Taddia et al. 2012) and some from new discoveries from the PTF survey. They show that such events tend to have higher peak luminosities and slower light-curve evolution than normal SNe Ia and that their hosts are always galaxies with ongoing star formation. Dilday et al. (2012) estimate that the fraction of strongly CSM-interacting SNe Ia among all SNe Ia could be 0.1% (based on the discovery rate in the PTF survey) to 1% (based on the SDSS-II survey), but it could be potentially higher if cases with weaker H signatures have been overlooked or if some cases with stronger signatures have been misclassified as Type IIn SNe.

For the two prototype CSM-interacting cases, SN 2002ic and SN 2005gj, late-time (2.2–3.8 years and 0.4–1.4 years postexplosion, respectively) mid-IR *Spitzer* photometry analyzed by Fox & Filippenko (2013) shows variable emission by warm (500–800 K) dust at these late times. For SN 2005gj, the dust luminosity increases monotonically by a factor ~ 2 over the observed period. The inferred dust masses are of order $M_{\rm d} \sim 10^{-2}\,M_{\odot}$ at distances of $r \sim 10^{17}$ cm. It is unclear if this is newly formed dust condensing in the SN ejecta, pre-existing CSM dust that is being heated by UV and X-ray radiation from CSM interaction (see Section 3.3.7), or light echoes by dust that is far from and unrelated to a CSM (Patat 2005; see Section 3.3.9). In any event, similar late-time IR emission, implying similar dust parameters, is seen in many Type IIn SNe studied by Fox et al. (2013).

Thus, in several respects, CSM-interacting SNe Ia bear a strong resemblance to Type IIn SNe, in which ejecta impact a large mass of CSM on the order of 1 $M_\odot$, thus converting kinetic energy to optical-band energy that accrues significantly to the radioactively powered part of the light curve. This suggests that CSM-interacting SNe Ia are a rare subclass in which the ejecta interact with the CSM of a relatively massive star, rather than the CSM of a donor star in a traditional SD scenario. The presence of a massive star would explain why all such events identified to date have occurred in star-forming galaxies (Silverman et al. 2013b).

Returning to SN-2002cx-like events ("SNe Iax", see Section 3.3.4 above), Liu et al. (2013a) predict that if such events are indeed failed pure deflagrations in $M_{\rm Ch}$ SD systems, the low kinetic energy of the explosion means that only small amounts of H I, $\lesssim 0.01\, M_\odot$, are expected to be seen in the nebular-phase spectra of such objects.

3.3.7. Radio and X-ray emission from circumstellar medium interactions. Interaction of SN ejecta with a pre-explosion CSM or ISM is expected to produce radio synchrotron emission from shock-accelerated electrons in an amplified magnetic field and X-ray emission through inverse Compton upscattering, by those electrons, of the optical emission of the SN. The CSM could be from previous mass loss from a donor star or from losses from the accretion flow onto the WD. The physics and signatures of such interaction have been computed by Chevalier (1982, 1998) and Chevalier & Fransson (2006); see Chomiuk et al. (2012b) and Margutti et al. (2012), for recent generalizations. Briefly, mass conservation dictates that a CSM produced by a constant pre-explosion mass-loss rate, $\dot{M}$, driving a wind of constant velocity, v_w, will have a radial density profile $\rho_{\rm csm}(r) = (\dot{M}/4\pi v_w) r^{-2}$. As the SN shock advances through the CSM with speed v_s, it accelerates particles to relativistic energies and amplifies the ambient magnetic field. From the postshock energy density, $\rho_{\rm csm} v_s^2$, the fractions in relativistic electrons and in magnetic fields are usually parameterized as ϵ_e and ϵ_B, respectively. Postshock, the electrons assume a power-law energy distribution, $N(E) \propto E^{-p}$, above a minimum energy $E_{\rm min}$ that, in turn, depends on p, v_s, and ϵ_e.

The electrons gyrate along magnetic field lines, emitting a synchrotron spectrum with specific luminosity $L_\nu \propto \nu^{5/2}$, shaped by synchrotron self-absorption, up to a frequency, ν_S (generally in centimeter-wave radio bands), and an optically thin synchrotron spectrum, $L_\nu \propto \nu^{-(p-1)/2}$, at higher frequencies. As the shock advances through progressively less-dense gas, the peak emission shifts to lower frequencies. Measurements at a fixed frequency will therefore first see an intensity rising with time, t, when that frequency is in the self-absorbed part of the spectrum, up to a maximum, followed by a declining signal when observing at optically thin frequencies. In those Type Ib/c SNe in which synchrotron emission has been detected, the electron energy distribution has an index $p \approx 3$, producing an optically thin synchrotron power-law index of ≈ -1 (Chevalier & Fransson 2006, Soderberg et al. 2012). For these SNe, $\epsilon_e \approx 0.1$, and ϵ_B values in the range of 0.01–0.1 have been inferred. The shock velocity, v_s, which is set by the density profiles of the CSM and of the ejecta, has a weak time dependence. In broad terms, the optically thin synchrotron luminosity then behaves as

$$L_\nu \propto v_s^3\ E_{\rm min}\ \nu^{-1}\ t^{-1}\ \epsilon_e\ \epsilon_B\ \dot{M}^2\ v_w^{-2}. \qquad (1)$$

Given an estimate of v_s, a radio measurement of L_ν (or an upper limit on it) can therefore constrain the product $\epsilon_e \epsilon_B \dot{M}^2/v_w^2$. A value of $\epsilon_e = 0.1$ is generally assumed. Values of ϵ_B, as noted, span one order of magnitude, and stellar wind velocities, v_w, are generally in the range of tens to a few hundred kilometers per second. Thus some useful information can be derived regarding the interesting parameter $\dot{M}$. In the case of a uniform ISM [$\rho_{\rm csm}(r) = $ const.], L_ν grows slowly with time roughly as $t^{0.35}$ (Chomiuk et al. 2012b), and the observations constrain the product $\rho_{\rm csm}\,\epsilon_B^{0.9}$.

In X-rays, the inverse-Compton emission for an electron population with $p = 3$ is

$$L_\nu \propto v_s\ E_{\rm min}\ \nu^{-1}\ t^{-1}\ \epsilon_e\ \dot{M}\ v_w^{-1}\ L_{\rm SN}(t), \tag{2}$$

where $L_{\rm SN}(t)$ is the bolometric optical-UV luminosity from the SN photosphere (Chevalier & Fransson 2006, Horesh et al. 2012, Margutti et al. 2012). The inverse-Compton emission, in contrast to the synchrotron emission, does not depend on the uncertain ϵ_B parameter. In the case of a shock expanding into a constant-density ISM, rather than a CSM, the ratio of the inverse-Compton luminosity to the optical-UV luminosity is roughly constant with time.

No radio or X-ray emission from a SN Ia has ever been detected. In X-rays, Hughes et al. (2007) set upper limits on flux from four SNe Ia, including two strongly CSM-interacting cases, SN 2002ic, and SN 2005gj (see Section 3.3.6 above). Russell & Immler (2012) set X-ray flux upper limits on 53 individual SNe Ia, and on their stacked images, and concluded that winds from evolved donor stars can be excluded. In the radio, Panagia et al. (2006) obtained VLA observations at wavelengths of 0.7 to 20 cm of 27 nearby SNe Ia at 46 epochs, roughly ~10–100 days after explosion, and used them to set limits on CSM interaction in each case. Hancock, Gaensler & Murphy (2011) performed a stacking analysis of these data to obtain deeper limits. Assuming typical values for the other parameters involved, they set a limit of $\dot{M} \lesssim 10^{-7}\,{\rm M_\odot\ year^{-1}}$. In radio, for SN 2006X, which showed variable NaI D lines (see Section 3.3.8 below), VLA nondetection constrains any CSM at $r \sim 10^{17}$ cm to arise from a wind with $\dot{M} \lesssim 3 \times 10^{-7}\,{\rm M_\odot\ year^{-1}}$, assuming $v_w = 100\ {\rm km\ s^{-1}}$ (Patat et al. 2007). This suggests a limit on the mass in the region of $M \lesssim \dot{M}\ r/v_w = 10^{-4}\,{\rm M_\odot}$.

In the case of the nearby and well-studied SN 2011fe, Horesh et al. (2012) obtained radio data from 0.3 to 20 cm, using the CARMA, EVLA, and WSRT telescopes, as early as 1.3 days after the explosion time deduced by Nugent et al. (2011) (see Section 3.3.1 above). *Swift* and *Chandra* observations were secured at 1.2 days and 4 days, respectively. Comparing the upper limits on radio and X-ray flux to the expected light curves, Horesh et al. (2012) set a limit of $\dot{M} \lesssim 10^{-8}\,{\rm M_\odot\ year^{-1}}$, assuming $v_s = 4 \times 10^4\ {\rm km\ s^{-1}}$, $v_w = 100\ {\rm km\ s^{-1}}$, and $\epsilon_e = 0.1$, and also assuming $\epsilon_B = 0.1$ in the radio case. This excludes the presence of a circumstellar wind from a giant donor. Margutti et al. (2012) used additional *Swift* and *Chandra* epochs, including stacks of epochs that are not far from maximum optical light (which corresponds also to inverse-Compton X-ray maximum light) to obtain deeper and later-time X-ray limits. They use them to refine the mass-loss limit to $\dot{M} \lesssim 2 \times 10^{-9}\,{\rm M_\odot\ year^{-1}}$. These stronger limits are used to argue, beyond giant donors, against any stable hydrogen-burning Roche-lobe overflow scenario (i.e., with $\dot{M} > 3 \times 10^{-7}\,{\rm M_\odot\ year^{-1}}$; see Section 3.1.2) in which more than 1% of the accreted material is lost, e.g., through the outer Lagrange points (i.e., other than L1). A uniform-density CSM is constrained by Margutti et al. (2012) to have a particle density $<150\ {\rm cm^{-3}}$. These constraints are shown in **Figure 4**. Finally, deeper limits have been set also in the radio by Chomiuk et al. (2012b), using additional EVLA epochs between 2 and 19 days past explosion, mostly about 5 times deeper than those of Horesh et al. (2012). Using also a higher assumed $v_s = 0.35c$, Chomiuk et al. interpret their radio flux limits as a stringent limit on CSM mass loss, $\dot{M} \lesssim 6 \times 10^{-10}\,{\rm M_\odot\ year^{-1}}$, during the last 100–1,000 years leading up to the explosion. With this limit, a 1% mass loss through outer Lagrange points can be excluded even for recurrent novae, unless the wind speed v_w is particularly high, with a correspondingly rarified CSM. A uniform ISM density $n_{\rm ism} \gtrsim 6\ {\rm cm^{-1}}$ can be excluded (for $\epsilon_B = 0.1$) or $n_{\rm ism} \gtrsim 44\ {\rm cm^{-1}}$ can be excluded (for $\epsilon_B = 0.01$) at radii $r \sim 10^{15} - 10^{16}$ cm from the

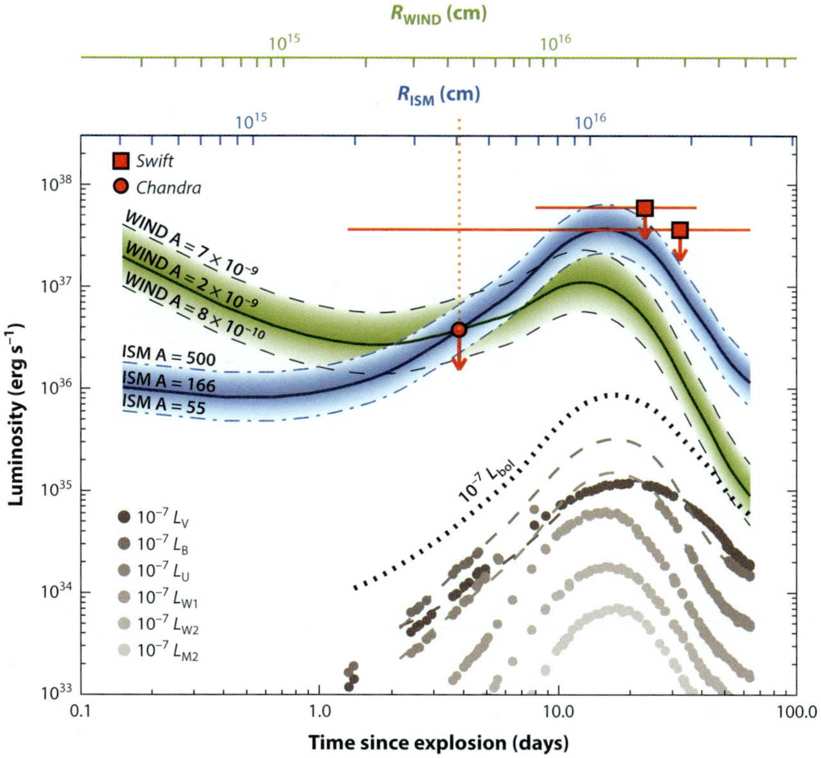

Figure 4

X-ray 3σ upper limits (*points*) from *Swift* and *Chandra* on the luminosity of SN 2011fe compared with the expected X-ray evolution from inverse Comptonization of the UV-optical light curve (shown by *gray filled circles* for the various UV-optical bands, and with a *black dotted line* for the estimated bolometric light curve.) Green band is for Comptonization by a circumstellar medium (CSM) wind, with $\dot{M}$ values as marked in units of solar mass per year, for a wind speed of 100 km s^{-1}. Blue band is for Comptonization by a uniform-density ISM, with densities in units per cubic centimeter, as marked. The data constrain the progenitor system's mass loss through a wind to $\dot{M} \lesssim 2 \times 10^{-9}$ $M_{\odot}$ $year^{-1}$ and a uniform-density CSM to have $\lesssim$150 cm^{-3}. Reprinted from Margutti et al. (2012) with permission of the AAS.

explosion. **Figure 5** shows these limits. We note that all of the above limits are slightly weakened if the explosion was in fact earlier by ~0.5–1.5 days than deduced by Nugent et al. (2011; Mazzali et al. 2014, Piro & Nakar 2014; see Section 3.3.1, above).

In summary, sensitive X-ray and radio observations of a good number of SNe Ia have so far failed to detect a CSM, including in the cases of several SNe Ia that show ejecta-CSM interactions and variable NaI D absorptions. The limits make giant SD donors unlikely. For SN 2011fe, these data show a particularly "clean" environment around a SN Ia and argue against most of the traditional SD scenarios.

3.3.8. Intervening absorption. In the SD scenario, SN Ia spectra may be expected to show variable narrow blueshifted absorption lines from a circumstellar wind from the companion, from accretion overflows, or from pre-explosion nova-like outbursts (e.g., Patat et al. 2007). As detailed below, absorptions interpreted as such signatures have been detected in a few SNe Ia. For the DD scenario, in turn, as already discussed in Section 3.3.2 above, DD merger simulations differ on the amount and radius of merger-related debris at the time of explosion. Furthermore, Shen,

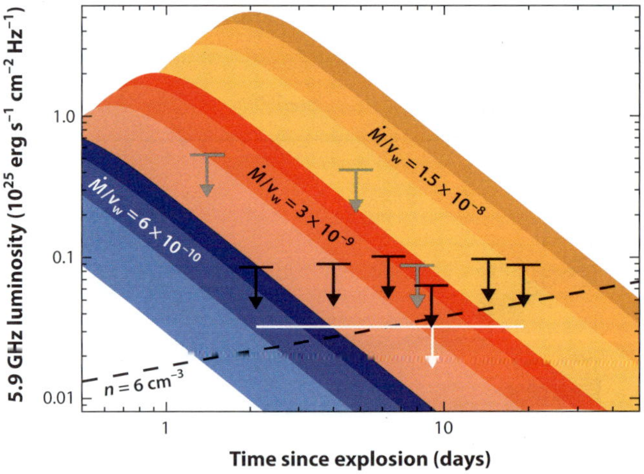

Figure 5

Expanded Very Large Array 3σ upper limits (*black arrows*) on the 5.9-GHz radio luminosity of SN 2011fe, compared with expectations from synchrotron emission due to interaction of its ejecta with a circumstellar medium. Also shown (*gray arrows*) are upper limits by Horesh et al. (2012), scaled to 5.9 GHz, and a time-stacked limit (*white arrow*). Colored swaths are the light curves expected from interaction with a wind, for the marked values of solar mass per year per unit velocity, in units of $M_\odot$ year^{-1} (100 km s^{-1})$^{-1}$; shades correspond to ϵ_B in the range of 0.01–0.1, and $\epsilon_e = 0.1$. The dashed black line is the expected evolution for a uniform-density ISM of density $n = 6$ cm^{-3}. The mass loss limit of $\dot{M} \lesssim 6 \times 10^{-10}$ $M_\odot$ year^{-1} argues against any Roche-lobe overflow single-degenerate scenario with $\gtrsim$1 % mass loss in the system. Reprinted from Chomiuk et al. (2012b) with permission of the AAS.

Guillochon & Foley (2013) have simulated a CSM that is shaped prior to DD mergers that involve a C/O primary WD and a He WD secondary (a combination raised in the context of sub-Chandrasekhar, or double-detonation, DD models, e.g., Pakmor et al. 2013, Shen & Bildsten 2014). A thin hydrogen layer, expected on the He WD's surface, accretes onto the C/O WD and leads to nova-like eruptions on timescales of $\sim 10^2$–10^3 years prior to the SN Ia explosion. In the ISM densities present in spiral galaxies, these pre-SN-Ia eruptions can shape the CSM so as to produce circumstellar absorptions similar to SD model expectations and to those sometimes observed in SNe Ia. Raskin & Kasen (2013) likewise use simulations to show that, under some circumstances, tidal debris from DD mergers could produce observable absorptions in SN Ia spectra.

In terms of the observations, variable, blueshifted ($\sim$10–100 km s^{-1}) absorption in the NaI D doublet ($\lambda\lambda$5890, 5896) has been detected in the high-resolution, multiepoch spectra of three SNe Ia (SN 2006X, Patat et al. 2007; SN 2007le, Simon et al. 2009; and the CSM-interacting case PTF11kx, Dilday et al. 2012; see Section 3.3.6 above). In the case of PTF11kx, the absorption is also seen in narrow lines of FeII, TiII, and HeI. A fourth variable NaI D case was found in low-resolution, but high-signal-to-noise, spectra of SN 1999cl by Blondin et al. (2009). Because of the low resolution, in this case one cannot determine whether the absorption is blueshifted or otherwise. However, Blondin et al. (2009) noted that absorption variations of the observed magnitude exist for only two out of 31 SNe Ia with data that could have revealed them (SN 1999cl, and SN 2006X, whose variations Blondin et al. could "rediscover" at low resolution). Sternberg et al. (2013) give statistics for the detection of variable NaI D at high spectral resolution of 3/17 (or 2/16, if excluding the unusual PTF11kx). In all four known cases, the NaI D absorption generally grows in strength on a timescale of about 10 days. In the three high-resolution cases, it is seen that the growth is in individual blueshifted velocity components and that the absorption

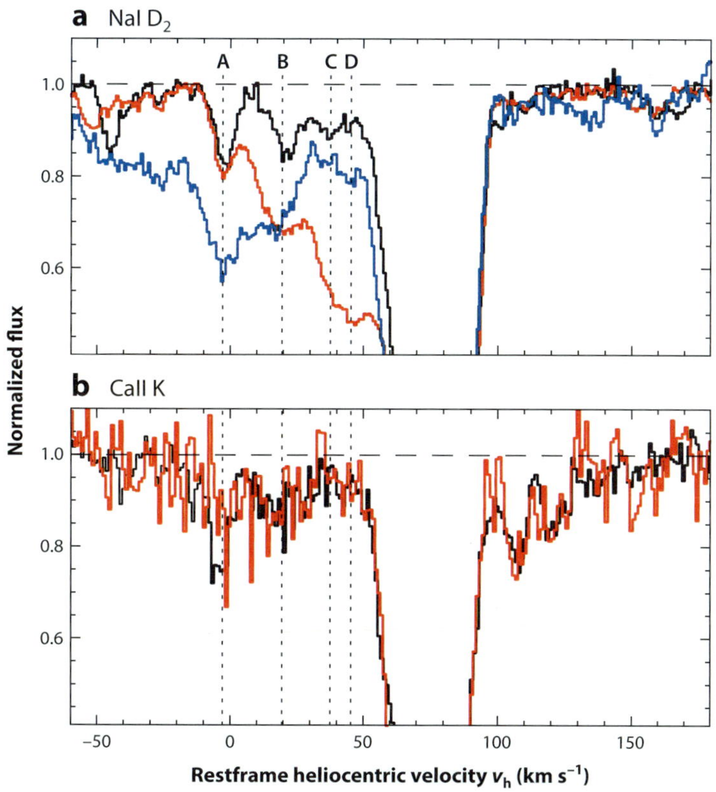

Figure 6

Spectra of SN 2006X showing time evolution of the NaI D_2 absorption feature (*a*) between days −2 (*black*), +14 (*red*), and +61 (*blue*). Individual variable absorption components are labeled A–D. For comparison, the nonvariable absorptions in the CaII K line (*b*; only first two epochs) are also shown. The growing absorption strength is thought to be due to recombining NaII ions in a circumstellar medium ionized earlier by UV photons from the explosion. The strongly saturated NaI and CaII absorptions come from unrelated interstellar gas clouds in the disk of the host galaxy. Reprinted from Patat et al. (2007) with permission of *Science* magazine.

in the CaII H&K lines (λλ3968, 3933) does not change in strength during the same period. **Figure 6** shows SN 2006X, as an example. At least three of the cases of SNe Ia with time-variable NaI D absorption belong to the subclass of SNe Ia with "high velocity gradients" (Benetti et al. 2005) (see Section 3.3.4 above). In all four cases the hosts are late-type spirals.

Using order-of-magnitude arguments (Patat et al. 2007) and detailed photoionization modeling (Simon et al. 2009), the growing NaI D absorption is interpreted as being due to recombining NaII ions in circumstellar material of density $n_{\rm csm} \sim 10^7$ cm^{-3} at distances of $r \sim 10^{16} - 10^{17}$ cm from the explosion, with a total mass of 10^{-5} to 10^{-2} $M_\odot$. Estimates of the UV spectrum and light curve of a SN Ia, based on model templates and extrapolations of the few SNe Ia with UV data, suggest that the SN flux can ionize NaI atoms (ionization potential 5.1 eV) out to such distances, but not CaII ions (ionization potential 11.9 eV) for which the required ionizing photons are orders of magnitude rarer in the SN spectrum. The short recombination timescale of 10 days further requires a high electron density, which implies a significant hydrogen ionization fraction and hence limits the distance to the absorbing clouds. In SN 2006X, part of the NaI D absorption actually weakens again 2 months after maximum light, which Patat et al.

(2007) interpret as the SN ejecta reaching and reionizing some components of the CSM (the corresponding weakening expected also in the Ca II absorption, due to such a process, could not be tested at that epoch because of low signal-to-noise ratio).

The absorbing gas may be a clumpy wind from a red-giant donor or an equatorial outflow from pre-explosion recurrent nova eruptions during the decades preceding the SN Ia explosion. The asymmetric nature of such outflows could introduce a viewing angle dependence that would explain the apparent rarity of SNe Ia with variable Na I D. An alternative explanation, invoking variable line-of-sight absorption by ISM material far from the SN (Chugai 2008), explains the nonvariability of the Ca II lines through abundance effects by means of differing fractions of Ca and Na locked in dust grains in intervening clouds. Patat et al. (2011) have noted the similarities of these variable absorptions with postoutburst features in the recurrent nova RS Oph (see Section 3.1.1 above). However, U Sco did not show any variable absorption during its outburst in 2010 (Kafka & Williams 2011). In the well-studied normal SN Ia 2011fe, despite the high signal-to-noise ratio of the available spectra, Patat et al. (2013) find no evidence for variable intervening absorption that could be associated with the CSM of the progenitor system.

Sternberg et al. (2011) compare the incidence of narrow blueshifted and redshifted Na I D absorptions in the single-epoch high-resolution spectra of 35 SNe Ia. They assigned a zero velocity to the strongest Na I D absorption, associating it with the host galaxy's ISM, and then identified additional absorption lines. They found 12 cases with blueshifted absorption, 5 with redshifted absorption, 5 with either single or symmetric blue/red Na I D absorption, and 13 cases with no detected Na I D absorption (most of them in early-type galaxies). Sternberg et al. (2011) interpret the 12-over-5 excess of blueshifted versus redshifted absorptions as signatures of the CSM and conclude that $>$20–25% of SNe Ia in spirals derive from SD progenitors (although, as noted, such absorptions may also arise in a DD scenario; Shen, Guillochon & Foley 2013).

Foley et al. (2012b) analyze a somewhat modified sample of 23 SNe Ia with high-resolution spectra, finding 10 cases with blueshifts and 5 with redshifts. They further find an association between ejecta velocity, as measured by Si II λ6355 near maximum, and the presence of blueshifted Na I D; SNe Ia with ejecta velocities $v_{\rm Si} \gtrsim 12{,}000$ km^{-1} s^{-1} are more likely to display blueshifted Na I D absorption. Foley et al. (2012b) point out that the run of ejecta velocities and their possible association with the presence of a CSM outflow suggest asymmetric progenitors and explosions, such that higher velocity ejecta are aligned with higher density CSM. Alternatively, the result could imply a variety of progenitor systems, such that progenitors with denser or closer CSMs tend to produce more energetic explosions. However, as already discussed, explosion energy and host-galaxy type are correlated, which could provide an indirect link between explosion energy and ISM (rather than CSM) absorption.

Maguire et al. (2013) reanalyze spectra for 16 of the Sternberg et al. (2011) events together with new data for 16 additional SNe Ia while defining differently the zero absorption velocity. In this combined sample, the excess of blueshifted over redshifted Na I D absorptions is 11 versus 4, again suggesting ~20% CSM-related absorption. Maguire et al. (2013) confirm the tendency for $v_{\rm Si} \gtrsim 12{,}000$ km^{-1} s^{-1} to have blueshifted Na I D and also find a correlation between $B-V$ color at maximum and the equivalent width of the Na I D absorption (see also Förster et al. 2013).

Phillips et al. (2013) study a sample of 32 SNe Ia at high spectral resolution and find that a quarter of the objects show anomalously large Na I D absorption relative to the dust extinction implied by the SN colors. The same extinction is indicated by the SN colors and by the diffuse interstellar band at 5,780 Å, suggesting that the dust, despite its usual low values of R_V, is in the host's ISM rather than in the CSM of the progenitor. However, all of the objects with excess Na I absorption have Na I D blueshifts, hinting that the Na is indeed in the CSM, which was perhaps pre-enriched with this element [Phillips et al. (2013) outline several scenarios for this]. Then again, three SNe Ia in their sample having variable NaI D, which are therefore strong candidates for having a CSM, do not show unusually strong NaI D lines.

3.3.9. Circumstellar dust and light echoes. Spectra of light echoing off interstellar dust clouds have permitted spectacular classification of several historical Galactic SNe centuries after their explosions, including two SNe Ia, SNR 0509–67.5 (Rest et al. 2008a) and Tycho's SN (Krause et al. 2008, Rest et al. 2008b). In principle, however, monitoring of SN light echoes at much earlier times, whether by spatially resolving the echo or by inferring its integrated effect on spectra and light curves, can provide a 3D mapping of the CSM and the ISM around the SN (Patat 2005, Patat et al. 2006; see Rest, Sinnott & Welch 2012 for a review). To date there have been only four clear detections of SN Ia light echoes: SN 1991T (Schmidt et al. 1994, Sparks et al. 1999); SN 1998bu (Cappellaro et al. 2001, Garnavich et al. 2001); SN 1995E (Quinn et al. 2006); and SN 2006X (Crotts & Yourdon 2008, Wang et al. 2008). In each of these cases, the optical light curve's decline flattened abruptly after a few hundred days, and the late-time spectra showed features from earlier phases. The late-time spectra could be modeled well as luminosity-weighted averages of the past spectral evolution of the SNe. Finally, HST imaging showed the actual resolved light-echo rings expanding around the SN sites. In several of these cases, polarimetry provided further constraints on the scattering geometry. Modeling of these data showed, in all cases, that the scattering dust is at distances of a few tens to a few hundred parsecs, generally in front of the SNe. Wang et al. (2008) raised the possibility that there exists an additional inner circumstellar echo component, but analysis of the same data by Crotts & Yourdon (2008) argues against this option. The presently known cases of light echoes most likely probe dusty regions far from the event and have no direct bearing on the progenitor problem.

As noted above (Section 3.3.6), signatures of possibly circumstellar dust have been seen in several CSM-interacting SNe Ia (Fox & Filippenko 2013). In three normal SNe Ia (including SN 2011fe), Johansson, Amanullah & Goobar (2013) use *Herschel* far-IR nondetections to set upper limits on the circumstellar dust mass of $\lesssim 10^{-2} - 10^{-1}\, M_\odot$.

3.4. Post-Explosion Evidence in Supernova Remnants

Long after the explosions, the spatially resolved remnants of SNe can provide detailed dissected views of the exploded systems, with additional possible clues to the progenitors. We now discuss this approach to the problem.

3.4.1. Searches for surviving companions. The donor star in a SD scenario will survive the explosion. It is likely to be identifiable by virtue of its anomalous velocity, rotation, composition, temperature, or luminosity (e.g., Ruiz-Lapuente 1997; Marietta, Burrows & Fryxell 2000; Canal, Méndez & Ruiz-Lapuente 2001; Wang, Li & Han 2010; Liu et al. 2013b; Pan, Ricker & Taam 2013; Shappee, Kochanek & Stanek 2013).

Searches for a surviving donor star in Tycho's SN of 1572 (as noted, a SN Ia confirmed with a light echo spectrum; Krause et al. 2008, Rest et al. 2008b), based on chemical abundances, RVs, proper motions, and rotation velocities, have not been able to reach a consensus (Ruiz-Lapuente et al. 2004, Fuhrmann 2005, Ihara et al. 2007, González Hernández et al. 2009, Kerzendorf et al. 2009). The individual studies have pointed out distinct preferred candidates and argue against those of the other studies. Most recently, Kerzendorf et al. (2012), on the basis of Keck spectra and proper motions from HST imaging, have concluded that there are no good candidates for SD survivors of any type (giant, subgiant, or main-sequence) in this remnant. Conversely, based largely on the same data, Bedin et al. (2014) continue to advocate a star labeled Tycho-G as the likely surviving SD-scenario donor star. Although the two groups broadly agree regarding the measured abundances, proper motion, and RV of Tycho-G, they differ regarding the degree to which these parameters are unusual or indicative of a donor-star origin for the star.

SN 1006 is widely considered a likely SN Ia on the basis of its 500-pc height above the Galactic plane, its symmetry, and the absence of a neutron star (Stephenson & Green 2002). However, the total iron mass in its remnant is apparently very low for a SN Ia, $\sim$0.06 $M_\odot$, based on X-ray emission lines of Fe-L (Vink et al. 2003) and Fe-K (Yamaguchi et al. 2008) and on UV and optical absorption lines of cold FeII along multiple lines of sight to background sources (Wu et al. 1993, Winkler et al. 2005). Conversely, the strong X-ray emission seen in Si, S, Ar, C, and O seems consistent with expectations from the layered outer parts of a dynamically young SN Ia remnant in which the reverse shock has yet to reach and heat the inner, iron-rich, regions (Badenes et al. 2007). Those regions might nonetheless be photoionized by the X-ray emission from the reverse shock (Hamilton & Sarazin 1984) and, hence, invisible to the UV-optical low ionization-Fe absorption studies. Be that as it may, Kerzendorf et al. (2013) and González Hernández et al. (2012), for this SN remnant, agree on the absence of any apparent surviving companions with luminosities greater than $L_V \sim 0.5\, L_{\odot,V}$. Thus, if SN1006 was a SN Ia, all evolved SD donors are ruled out. Normal main-sequence donors are still allowed, unless their luminosities are significantly increased by the interaction with the ejecta, as predicted by the models for the postexplosion appearance of donor stars, cited above.

SNR 0509–67.5 in the LMC is the remnant of an overluminous SN Ia from circa 1600 (Badenes et al. 2008), confirmed as such with a light-echo spectrum by Rest et al. (2008a). Fortuitously, the remnant is far from the center of the LMC in a region with low stellar density, and it has been imaged to great depth with HST. Schaefer & Pagnotta (2012) show (**Figure 7**) that there are no stars, down to V-band luminosities as low as $L_V = 0.04\, L_{\odot,V}$, in the area around the remnant's geometrical center that could be populated by a runaway donor star. This luminosity corresponds to late-K-type main-sequence stars of mass $\sim$0.5 $M_\odot$ and essentially rules out all traditional SD companions. Some diffuse emission seen in the center of the remnant is an unrelated background galaxy (Pagnotta, Walker & Schaefer 2014).

Another LMC remnant, SNR 0519–67.5, is projected on a denser stellar region and therefore provides weaker limits on a surviving companion. A light echo found by Rest et al. (2005) indicates an explosion that is 600 $\pm$ 200 years old, and Rest, Sinnott & Welch (2012) indicate that the echo's spectrum confirms a SN Ia classification (but the spectrum is still unpublished). In this case, Edwards, Pagnotta & Schaefer (2012), again using HST imaging, are able to rule out the presence of surviving post-main-sequence donor stars, but 27 main-sequence stars that are close enough to the explosion center cannot be excluded as surviving companions. However, as already noted, the interaction with the ejecta would likely brighten such stars significantly, and hence the main-sequence donor option is also in doubt.

Kepler's SN of 1604 is thought to have been a SN Ia, but this has not been established beyond doubt (see further discussion in Section 3.4.2 below). Kerzendorf et al. (2014) obtained spectroscopy for the 24 stars with $L > 10\, L_\odot$ in the central 38 arcsec of the remnant, none of which show any signatures expected of a former donor star. There are also no potential red giant, AGB (as has been proposed from hydrodynamical models, see below), or post-AGB surviving donor stars in this region.

The SD scenario can perhaps circumvent the above constraints via the spin-up/spin-down ideas (Justham 2011; Di Stefano, Voss & Claeys 2011; Hachisu, Kato & Nomoto 2012; see Section 2.4), if the delay between the end of accretion and the explosion is long enough for the donor star to evolve to a dim, undetectable, WD itself. However, as discussed above in Sections 2 and 3.1.6, the viability of fast WD rotation in general, and differential rotation in particular, faces theoretical and observational challenges.

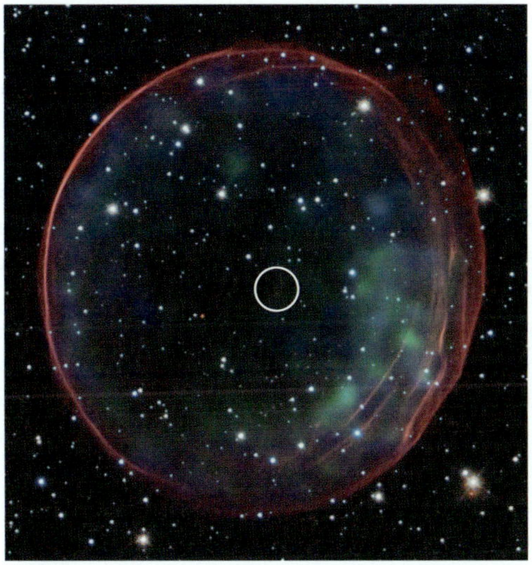

Figure 7

Composite *Hubble Space Telescope/Chandra* image, in *B*, *V*, *I*, Hα, and X-rays, of the ~400-year-old SN Ia remnant 0509–67.5 in the Large Magellanic Cloud. The central 1.4-arcsec-radius circle is the 3σ-confidence region that could host a surviving donor star, considering uncertainties in the geometrical center and in the remnant age and assuming the maximum plausible proper motion. As argued by Schaefer & Pagnotta (2012), the nondetection of any potential surviving stars within this region, down to $L_V = 0.04\,L_{\odot,V}$, essentially rules out the possibility of a surviving single-degenerate companion. Credit: NASA, ESA, CXC, SAO, the Hubble Heritage Team (STScI/AURA), and J. Hughes (Rutgers University).

3.4.2. Searches for circumstellar medium interaction in remnants. Another approach to search for clues of the progenitor systems in SN remnants is through hydrodynamical models for specific remnants that are thought to have been SN Ia events. The modeling elements in such analyses have included combinations of a pre-explosion environment (possibly shaped by outflows from the system); evolution of the temperature, density, and ionization of the ejecta and the CSM; and attempts to reproduce the observed geometry, dynamics, and X-ray spectrum of a remnant (e.g., Badenes et al. 2006, 2008; Kosenko et al. 2008; Kosenko, Blinnikov & Vink 2011; Patnaude et al. 2012; Sorokina et al. 2004; and see Vink 2012 for a recent review).

Badenes et al. (2007) modeled the X-ray emission in seven young SN Ia remnants to test for the presence of large (3–30 pc) wind-blown cavities in the ISM, with densities $n_{\rm csm} \lesssim 10^{-3}$ cm^{-3}, centered on the explosion sites. They show that such swept-out cavities are expected in the context of the optically thick, ~1,000 km s^{-1}, outflows from rapidly accreting WDs in the wind-regulated accretion picture (Hachisu, Kato & Nomoto 1996; see Section 3.1.3). In all of the seven cases examined, the observations do not support this scenario. In fact, Badenes et al. (2007) find that line fluxes and energy centroids in the X-ray spectra and the observed radii of the forward shocks, at the known ages of the remnants, are always as expected for ejecta advancing through a uniform-density ISM of $n_{\rm csm} \approx 1$ cm^{-3}. The fast pre-explosion outflows, leading to rarified CSMs, would have resulted in larger forward-shock radii and weaker lines, by factors of a few. This result shows that, if these SNe had SD progenitors, the growth of the WDs proceeded in some way other than wind-regulated accretion or that such growth ceased long before the explosion, as in spin-up/spin-down models. A CSM with a $\rho \propto r^{-2}$ radial density profile is expected from any constant-velocity outflow, fast or slow, whether it is from an evolved donor star or from the accreting WD itself.

Badenes et al. (2007) find that such density profiles also give poor correspondence with the data, but only a limited range of these $\rho \propto r^{-2}$ models was studied. Support for a uniform-density CSM comes also from modeling by Raymond et al. (2007) of the Hα filaments in the remnant of SN 1006, as seen in HST imaging. They conclude that $n_{\rm csm} \approx 0.25-0.4\ {\rm cm}^{-3}$, with variations of only $\sim$20% on parsec scales.

An exception to the absence of cavities in SN Ia remnants could be RCW 86, a proposed remnant for the SN that exploded in 185 C.E. (Vink et al. 2006), which is possibly a SN Ia based on the large observed mass of Fe and the absence of a neutron star. Vink, Kaastra & Bleeker (1997), Badenes et al. (2007), and most recently Williams et al. (2011) do find evidence for a 12-pc-radius central cavity in this case. Conversely, this remnant, which is in an OB association, has traditionally been considered a CC SN (e.g., Ghavamian et al. 2001).

Among potential SN Ia remnants, the remnant of Kepler's SN of 1604 has been studied perhaps in the most detail, but has also led to some confusing results. It was once thought to have been a CC SN on the basis of an asymmetrically bright, dense, and nitrogen-rich shell of material in the remnant's northern side. More recently, however, it is generally assumed to have been a SN Ia based on its high altitude above the Galactic plane (at least 350 pc, but possibly >800 pc, given persistent incompatible estimates of its distance; see Patnaude et al. 2012), its high Fe and low O content, and the lack of a detected neutron star (Reynolds et al. 2007). Chiotellis, Schure & Vink (2012) model the remnant assuming a SN Ia explosion in a SD configuration with an AGB-star donor of initial mass 4–5 $M_\odot$. The binary system likely had a high velocity of 250 km s^{-1} away from the Galactic disk (Borkowski, Sarazin & Blondin 1994), a velocity that may be difficult to explain for a runaway binary (Vink 2012). In this picture, the northern shell arises from the forward shock encountering a pre-explosion CSM of mass $\sim$1 $M_\odot$; the CSM is previously shaped by its movement through the ISM at the said high velocity (Borkowski, Blondin & Sarazin 1992; Vink 2008). Burkey et al. (2013) find further evidence for an AGB progenitor by using *Spitzer* 24-μm data to distinguish shocked CSM from shocked ejecta emission. They identify an equatorial CSM component that they associate with an edge-on disk-like outflow from an AGB donor.

However, as noted in Section 3.4.1 above, there is no surviving AGB star in the remnant within the central 38 arcsec (Kerzendorf et al. 2014). A post-AGB star or young WD that is only a few hundred years old would have a luminosity of $\sim 10^4\ L_\odot$ (Bloecker 1995), but the brightest star, even within a conservative 60-arcsec radius of the geometrical center, has a V-band luminosity of only 330 $L_\odot$ (Kerzendorf et al. 2014). Furthermore, Patnaude et al. (2012), modeling the south side of the remnant, report that a simple $\rho \propto r^{-2}$ wind-shaped CSM, as favored by Chiotellis, Schure & Vink (2012), cannot simultaneously fit the observed dynamics and the X-ray spectrum. To fit all the data, Patnaude et al. (2012) find that a 0.1-pc low-density central cavity needs to be included in the CSM model. Alternatively, a model that also works is one with no wind-shaped CSM at all, but rather with a remnant that expands into a more-or-less uniform-density ISM. A north-south gradient in this ISM [a gradient that has some support from direct 160-μm emission maps of the region by Blair et al. (2007)] could explain the northern bow shock. Each of these solutions (small central cavity or uniform ISM) corresponds to a different and incompatible distance to the object, which in turn implies a subenergetic or superenergetic SN Ia event (if, indeed, Kepler's SN was a SN Ia). With its multiple puzzles and contradictions, Kepler's SN remnant probably does not illuminate the SN Ia progenitor issue much at this point in time, but it could if its type and distance became known.

Borkowski, Hendrick & Reynolds (2006) modeled two remnants in the LMC with strong Fe-L line emission in their interiors, which they conclude requires a large interior density, as would be expected from a slow pre-explosion wind. Although distinct from the fast accretion-driven winds of the Hachisu, Kato & Nomoto (1996) scenario, this result may be considered as evidence for the presence of pre-explosion SD companions. However, without good X-ray spectra and well-determined ages, the typing of these two remnants as SNe Ia is not secure.

Another interesting constraint comes from the existence of "Balmer-dominated shocks" primarily in remnants thought to be of SN Ia origin (see Vink 2012). Optical spectra of such regions show pure Balmer-line emission with a narrow core and a broad base. The narrow emission is understood to arise by collisional excitation, by electrons and protons, of cold neutral hydrogen just being overrun by the forward shock. The broad component is due to hot postshock protons that undergo charge transfer with atoms, to become hydrogen atoms at excited levels, or subsequent collisional excitation of those hydrogen atoms (Chevalier, Kirshner & Raymond 1980; Heng 2010). CC remnants generally do not show Balmer-dominated shocks, which is reasonable given that their SN explosions are bright in the UV and therefore can photoionize a large surrounding region. Conversely, Ghavamian et al. (2003) have pointed out that the existence of Balmer-dominated shocks, and hence neutral H, at radii of <30 pc in SN Ia remnants (and often below 5 or 10 pc; see Vink 2012) limits the pre-explosion photoionizing flux of the progenitors during the $t \sim 10^5/n$ years prior to the explosion (corresponding to the recombination time at ambient densities of $n/(1\ \mathrm{cm}^{-3})$. Supersoft X-ray sources in particular, with their UV/X-ray luminosities of up to $\sim 10^{38}$ erg s^{-1}, would have carved out ionized Stromgren spheres of radius ~ 30 pc around the remnants (Rappaport et al. 1994), in conflict with the observed Balmer-dominated emission at smaller radii.

3.5. SN Ia Rates and the Delay-Time Distribution

SN Ia rates and their dependence on environment and cosmic time can provide further clues to the progenitor problem. In essence, finding the dependence of the SN rate on the age distribution of the host stellar population can reveal the age distribution of the actual SN Ia progenitors. Different progenitor scenarios involve different timescales that control the production rate of SN Ia events and will thus predict different SN Ia age distributions. For a detailed review on surveying for SNe for rate purposes, the derivation of SN Ia rates, and their application to the progenitor question, see Maoz & Mannucci (2012).

A fundamental function for progenitor-question purposes is the distribution of times between star formation and SN Ia explosion, usually called the delay-time distribution (DTD). The DTD is the hypothetical SN rate versus time that would follow a brief burst of star formation, with the burst having one unit of total mass in formed stars. It is the "impulse response" that embodies the physical information of the system, free of nuisances—in the present context–the diverse star-formation histories (SFHs) of the galaxies hosting the SNe. Determining the DTD observationally has increasingly become the objective of SN Ia rate measurements.

3.5.1. Theoretical expectations. As noted in Section 2, theoretical forms for the DTD can be derived from BPS calculations, obtained by numerically evolving simulated populations of binaries with chosen distributions of initial parameters or, alternatively, by following analytic approximations (e.g., Greggio 2005). The DTDs for each progenitor channel can be compared with observationally derived DTDs.

The DD model naturally gives rise to a broad range of delay times (Yungelson & Livio 2000). With some simplifying assumptions, this DTD form can be seen to result from the strong dependence of the merger time on the postcommon-envelope separation of the WD pair (Greggio 2005, Totani et al. 2008). Suppose the postcommon-envelope separation, a, is distributed as a power law, $dN/da \propto a^{\alpha}$, over the range of separations that will merge over a Hubble time. Suppose further that the time until merger depends on the separation to some power, $t \propto a^{\gamma}$. Then the form of the DTD will be given by $dN/dt \propto t^\beta$ with $\beta = -1 + (\alpha + 1)/\gamma$. For the angular momentum loss due to gravitational waves, $\gamma = 4$. The postcommon-envelope distribution of WD separations is generally found by BPS simulations to be similar in form to the input distribution of main-sequence binaries, which is about flat in log separation, i.e., a power law with $\alpha \approx -1$ (Toonen, Nelemans & Portegies Zwart 2012; Yungelson 2013). A $\sim t^{-1}$ DTD is thus expected even if the separation distribution is not exactly flat in log a, as long as the value of γ is sufficiently large. The "collisional DD" model by Katz & Dong (2012) (see Section 2.3) with $\gamma = 5/2$ would give a similar DTD.

The $\sim t^{-1}$ DTD behavior breaks down at short delays. A "bottleneck" in the process is the production rate of WDs that can serve as SN Ia progenitors. The production rate is zero for at least 30–40 Myr until main sequence stars with masses below $\sim 8\,M_\odot$ evolve into the first WDs to emerge from the stellar population. The WD production rate then rises to a maximum, and falls as $\sim t^{-1/2}$ (e.g., Pritchet, Howell & Sullivan 2008), up to a cutoff time, t_c, corresponding to the main-sequence lifetime of the least massive stars whose descendant WDs contribute to a DD-channel SN Ia. Because this main-sequence mass is usually considered to be 2–3 $M_\odot$, t_c is generally $\sim$1 Gyr. The DTD will be the convolution of the WD production rate and the merger rate dependences, resulting in a broken power law, t^β, with indices $\beta \sim -1/2$ at $t < t_c$ and $\beta \sim -1$ at $t > t_c$. More realistically, the supply of WDs that will eventually merge is more complex, given that two WDs are required and that the time until they are "ready" to evolve solely via gravitational decay depends on the initial binary parameters of their progenitors, which determine the ensuing interactions among the pair. BPS simulations indeed show complex behavior (that varies among models, according to the different physical recipes assumed) for the pre t_c DTD shape. However, as expected, the DD models generally do show the generic t^{-1} behavior after $\sim$1 Gyr. This can be seen in **Figure 8**, which includes a compilation by Nelemans, Toonen & Bours (2013) of some recent BPS theoretical DTDs for the DD channel.

For the standard SD model, the defining property is that the accretion rate onto the WD needs to be in a rather narrow range, near $\dot{M} \sim 3 \times 10^{-7}\,M_\odot\,\text{year}^{-1}$ (see Section 2 above). Such a rate is realized only by main-sequence, or slightly evolved subgiant, donor stars that can transfer of the order of 1 $M_\odot$ to the WD on a thermal (i.e., Kelvin-Helmholtz) timescale, t_{th}, of the donor star (van den Heuvel et al. 1992). A donor star, upon losing mass, will contract adiabatically on a dynamical timescale. Thermal equilibrium is regained on a thermal timescale, causing re-expansion and a renewed mass supply on that timescale (e.g., Pols & Marinus 1994). The accretion rate is thus

$$\dot{M} \sim \frac{1\,M_\odot}{t_{\text{th}}} \sim \frac{1\,M_\odot}{GM^2/RL} \sim 3 \times 10^{-7}\,M_\odot\,\text{year}^{-1} \left(\frac{M}{2\,M_\odot}\right)^3, \qquad (3)$$

assuming that stellar main-sequence radius scales with mass as $R \propto M$ and luminosity as $L \propto M^4$. There is therefore a limited range of donor masses, 2–3 $M_\odot$, that can supply material at the rate required for stable hydrogen burning on the WD surface. This is confirmed by detailed calculations (e.g., Langer et al. 2000, Han & Podsiadlowski 2004). Because these systems explode during, or shortly after, the main sequence, in many models the DTD is concentrated between a few-hundred million years and 1–2 Gyr, dropping off sharply before and after (Yungelson 2005). **Figure 8** illustrates this for a set of BPS theoretical DTDs for the SD channel.

Similar exponential cutoffs are seen in analytic approximations of SD model DTDs (e.g., Greggio 2005). A limiting factor of such predictions is that the binary evolution calculations involve many uncertain assumptions, and therefore the theoretical DTD predictions vary among different groups. This is apparent in particular for the SD channel in **Figure 8**.

Hachisu, Kato & Nomoto (1996) have suggested that an optically thick wind, driven from the accreting WD, can stabilize the mass transfer in binaries where a low-mass giant fills its

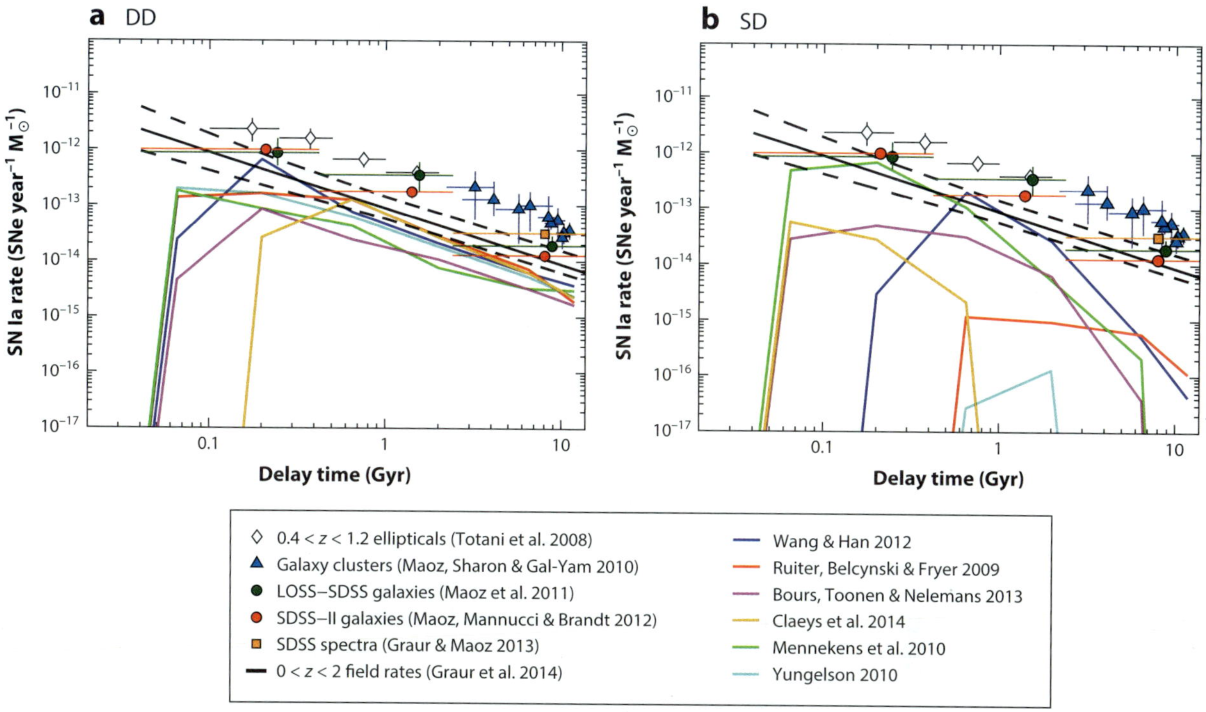

Figure 8

Observed versus theoretical delay-time distributions (DTDs). In both panels, points and the straight black lines are some of the observationally derived DTDs discussed here. Colored curves are theoretical DTDs adapted from the compilation of binary population synthesis predictions for (*a*) double-degenerate (DD) models and (*b*) single-degenerate (SD) models by Nelemans, Toonen & Bours (2013), who adjusted all models to have the same physical input parameters. All DTDs, observed and theoretical, are shown with a consistently assumed definition (SNe per year per formed stellar mass) and a consistent initial mass function. The Toonen, Nelemans & Portegies Zwart (2012) curve shown here corrects a misprint in the plot shown in Nelemans, Toonen & Bours (2013).

Roche lobe. The long evolutionary time of the low-mass donor can produce a SD model with long delay times. Hachisu, Kato & Nomoto (2008b) proposed that this same optically thick wind can strip mass from donors in the traditional main-sequence model, and thereby raise the mass range for donor stars up to 8 $M_\odot$, thus extending the SD model also to short delay times. Studies using these extended parameter ranges to calculate the DTD on the basis of BPS (Meng & Yang 2010; Mennekens et al. 2010; Wang, Li & Han 2010; Bours, Toonen & Nelemans 2013) have typically found DTDs that are indeed broader and (depending on how massive the donors are that are allowed) extend to short delay times, but find very few systems with giant donors. The reason is that the required period range for such systems, of a few hundreds of days, is sparsely populated in BPS models by postcommon-envelope binaries (see Yungelson & Livio 2000). There are known examples of symbiotic binaries (e.g., T CrB, RS Oph) and post-AGB-star binaries (see van Winckel et al. 2009) observed in that period range, but their statistical fraction is unclear. Hachisu, Kato & Nomoto (1999) proposed that many wide systems can evolve into this period range by interaction of the binary with the slow wind of the giant, although the efficiency of this process is debated. Combining all of these effects, Hachisu, Kato & Nomoto (2008a) have suggested that a combination of SD main-sequence and red-giant channels could together give a $\sim t^{-1}$-shaped DTD. The He SD channel is expected to contribute at short delay times ($\lesssim 10^8$ year), owing to the fact that the He donors evolve from rather massive main-sequence stars with little delay (and the WDs need to have formed even earlier, e.g., Wang et al. 2009a).

3.5.2. Observed delay-time distribution from supernova rates versus color and Hubble type. The first observational indications for the existence of a range of SN Ia delay times (i.e., a DTD) was found by Mannucci et al. (2005) and Mannucci, Della Valle & Panagia (2006), who, analyzing a local SN sample (Cappellaro, Evans & Turatto 1999), demonstrated that the SN Ia rate per unit stellar mass changes with host-galaxy color and Hubble type—parameters that trace the SFR. Nonetheless, early-type galaxies with no current star formation were seen to have a nonzero SN Ia rate (as was already well known). A similar SN Ia rate dependence on host color and specific SFR was found by Sullivan et al. (2006) for the Supernova Legacy Survey (SNLS) sample of SNe Ia, and more recently by Li et al. (2011b) for the Lick Observatory SN survey (LOSS) sample (Leaman et al. 2011, Li et al. 2011c), and by Smith et al. (2012) for the SDSS-II SN sample (Frieman et al. 2008, Sako et al. 2008).

Early interpretations of these results argued for the coexistence of two SN Ia populations: a "prompt" population, with rates proportional to the CC SN rates (Mannucci et al. 2005) or to the SFR (Sullivan et al. 2006), that explodes within $\sim$100–500 Myr; and a delayed channel, proportional to the stellar mass, that produces SNe Ia on timescales of order 5 Gyr. In the similar "$A + B$" formulation (Scannapieco & Bildsten 2005), the SN Ia rate in a galaxy is proportional to the stellar mass (through the A parameter) and to the galaxy's SFR (through B). In retrospect, however, the "prompt-plus-delayed" interpretation is, in essence, a DTD with two coarse time bins, and these two "channels" appear to represent the two sides of a broad DTD. More quantitative information on the DTD can be obtained by comparing the rates with the stellar age distribution of each galaxy type and indicates that, indeed, to reproduce the SN rate versus host-color relations, a broad DTD is required, with delays below 0.5 Gyr to trace the SFR, and an extended tail to $\gtrsim$5 Gyr delay times, to produce SNe Ia in galaxies with low current SFR. Mannucci, Della Valle & Panagia (2006) and Mannucci (2008) show this using several SD models, and Maoz & Mannucci (2012) show that a $\sim t^{-1}$ DTD can do so as well. Although the "$A + B$" parameterization is still sometimes used to analyze SN Ia rates, it is a poor physical metric. For example, the A component depends on the mean stellar age of the monitored population; younger stellar systems produce more SNe for any declining DTD.

3.5.3. Delay-time distribution from rates in galaxy clusters and field ellipticals. Measuring the SN rate in galaxy clusters as a function of redshift is probably the most direct way to recover the DTD. Optical spectroscopy and multiwavelength photometry of cluster galaxies has shown that most of their stars were formed within short episodes ($\sim$100 Myr) at high redshift ($z \sim 3$; Jimenez et al. 2007, Rettura et al. 2011), a good approximation to the δ-function burst defining the DTD. Thus, an almost-direct measurement of the DTD is provided by the observed SN Ia rate per unit stellar mass in galaxy clusters as a function of the time since the cluster's star-formation epoch (i.e., the cosmic epoch at the observed cluster redshift minus the stellar formation epoch). Cluster SN Ia rates have been measured over the past decade in the redshift range $0 < z < 1.5$, corresponding to DTD delays of 2–10 Gyr (Gal-Yam, Maoz & Sharon 2002; Germany et al. 2004; Maoz & Gal-Yam 2004; Sharon et al. 2007; Gal-Yam et al. 2008; Graham et al. 2008; Mannucci et al. 2008; Dawson et al. 2009; Dilday et al. 2010a; Maoz, Sharon & Gal-Yam 2010; Barbary et al. 2012; Sand et al. 2008, 2012). These measurements are still limited by significant Poisson errors due to small SN numbers.

Maoz, Sharon & Gal-Yam (2010) derive a DTD based on most of these galaxy-cluster SN Ia rate measurements, which is shown in **Figure 8**. They find consistency with a $\sim t^{-1}$ form at

$t > 2$ Gyr and show that the $\sim t^{-1}$ conclusion is not critically dependent on uncertainties in the input parameters involved, such as the precise redshift of star formation in clusters, whether it was a brief or extended burst or the contribution of ongoing low-level star formation in clusters, as long as these are at the levels, redshifts, and cluster locations allowed by direct measurements of star-formation tracers in clusters.

A related approach to DTD recovery was taken by Totani et al. (2008) using field elliptical galaxies instead of galaxy clusters. Comparing SN Ia rates in early-type galaxies of different luminosity-weighted ages, seen at $z = 0.4$–1.2 as part of the Subaru/XMM-Newton Deep Survey, they were the first to show observationally that the DTD is consistent with a t^{-1} form. This observed DTD is also shown in **Figure 8**. Like the elliptical galaxies in clusters, field ellipticals are dominated by old stellar populations created in a short burst at early times, allowing for a direct derivation of the DTD. However, in general, the concept of a "typical age" for a host galaxy, interpreted as a SN Ia progenitor age, is a risky zeroeth-order approximation to the full SFH of a galaxy (see Mannucci 2009, Maoz & Mannucci 2012). The SN rate of a galaxy with a wide distribution of ages is likely to be dominated by the youngest stellar population, even when those young stars are present only at low levels that may escape detection (Mannucci 2008). As opposed to ellipticals in the central regions of clusters, which appear to be very quiescent, noncluster ellipticals often host non-negligible amounts of late-time star formation (e.g., Zhu, Blanton & Moustakas 2010).

In a possible instance of this, Della Valle et al. (2005) found that radio-loud early-type galaxies have higher SN Ia rates than otherwise-similar radio-quiet galaxies. A possible confirmation of the effect was found by Graham et al. (2010). Mannucci, Della Valle & Panagia (2006) proposed that quiescent early-type galaxies sometimes acquire new gas in a minor merger and that this material leads to both radio activity (through accretion onto the central black hole) and star formation which, in turn, produces prompt SNe Ia. Greggio, Renzini & Daddi (2008) have pointed to the absence of the expected CC SNe in the same radio galaxies or in early-type galaxies in general (e.g., Hakobyan et al. 2008). Nonetheless, CC SNe have been found in some early-type galaxies showing clear signs of ongoing star formation based on UV data (Suh et al. 2011). The observed rate trend in radio galaxies still lacks confirmation, and furthermore the constraints on the DTD would be quite model-dependent, whereas other explanations have also been proposed (Capetti 2002; Livio, Riess & Sparks 2002).

Other applications of the "rate versus age" approach have been made by Aubourg et al. (2008), Cooper, Newman & Yan (2009), Raskin et al. (2009), Schawinski (2009), and Thomson & Chary (2011). They confirm the wide distribution of delays and, in particular, the existence of "prompt" SNe Ia, though with a wide range defining the age of that population. Uncertainties are again dominated by assumptions regarding the parent stellar populations.

3.5.4. Delay-time distribution from iron abundances in galaxy clusters. Constraints on the DTD can also come from considering element abundances in galaxy clusters. As reviewed by Maoz, Sharon & Gal-Yam (2010), X-ray observations of the intracluster medium (ICM) and optical observations of the stellar populations determine the ratio of iron-to-stellar masses. As iron is a main product of SN Ia explosions (with a mean of $\sim 0.7\,M_\odot$ per event), and nothing escapes the deep cluster potential, this ratio sets the time-integrated number of SNe Ia in clusters per formed stellar mass (after estimating and subtracting the iron contributions by CC SNe). This integral constraint, minus an integral over the DTD at delays $t > 2$ Gyr from direct cluster rate measurements, can provide an indirect estimate of the DTD at delays $0 < t < 2$ Gyr. Although the DTD value thus obtained in this early bin tends to be somewhat high, it is consistent with the $\sim t^{-1}$ form found directly from the SN Ia rates at lower redshifts. Uncertainties on the mass of the iron in the stellar and ICM components, and on the contribution of CC SNe to iron production, propagate into this estimate (Loewenstein 2013). The combined constraint on the DTD power-law index is $\beta = -1.2 \pm 0.3$.

The time evolution of iron abundance in clusters can, in principle, also be used to measure the shape of the DTD. Calura, Matteucci & Tozzi (2007) show that a broad DTD that is peaked at short delays, when combined with assumptions about the winds that transfer iron from the galaxies to the ICM, can reproduce the observed evolution of iron abundance in the ICM between $z = 0.2$ and $z = 1.2$.

3.5.5. Delay-time distribution from volumetric rate evolution. Another observational approach to recovering the DTD has been to compare the volumetric SN rate from field surveys, as a function of redshift, with the cosmic SFH. On the one hand, an advantage of this approach is the averaging-out, over large cosmic volumes, of the uncertainties in SN rates and SFHs that are based on few galaxies. Furthermore, field surveys are not biased in favor of SNe in particular types of galaxies, as opposed to galaxy-targeted surveys. On the other hand, uncertainties in the cosmic SFH have remained a limiting factor in deriving a DTD (e.g., Graur et al. 2014). As the DTD is the SN "response" to a short burst of star formation, the volumetric SN rate versus cosmic time, $R_{Ia}(t)$, will be the convolution of the DTD with the SFH [i.e., the SFR per unit comoving volume versus cosmic time, $\dot{\rho}(t)$],

$$R_{Ia}(t) \propto \int_0^t \dot{\rho}(t-\tau)\dot{N}(\tau)\mathrm{d}\tau. \tag{4}$$

Starting with Ruiz-Lapuente & Canal (1998) and Sadat et al. (1998), numerous volumetric rate measurements, progressively more accurate and at higher-redshift, have accumulated and been compared with the cosmic SFH [see Graur et al. (2014) for a recent compilation]. Early on, Dahlen, Strolger & Riess (2008), Dahlen et al. (2004), Strolger et al. (2004), and Strolger, Dahlen & Riess (2010) argued for a DTD that is peaked at a delay of $\sim$3 Gyr, with little power at either shorter or longer delays. This conclusion was based on their measurements of the SN Ia rate at $z > 1$ with HST. However, this conclusion was contested on the basis of both reanalysis of the same data (Kuznetsova & Connolly 2007; Mannucci, Della Valle & Panagia 2007; Greggio, Renzini & Daddi 2008; Blanc & Greggio 2008; Horiuchi & Beacom 2010) and additional $z > 1$ SN Ia rates (Poznanski et al. 2007). The main problems highlighted were the small number of SNe, and hence the large statistical errors in the rates, compounded by systematic uncertainties due to the necessary assumptions regarding SFHs and dust extinction. Mannucci, Della Valle & Panagia (2006) further pointed out that such a narrow DTD does not reproduce the dependence of local SN Ia rates on galaxy colors (see Section 3.5.2, above).

Graur et al. (2011) discovered a larger sample of 37 likely SNe Ia at $1 < z < 2$ by monitoring the Subaru Deep Field. Additional new high-z SN Ia rate measurements also have emerged from the recent CLASH (Graur et al. 2014) and CANDELS (Rodney et al. 2014) surveys with HST, with the highest-redshift SN Ia in the latter survey at $z = 2.15 \pm 0.1$. Several authors (Graur et al. 2011, 2014; Perrett et al. 2012; Graur & Maoz 2013; Kistler et al. 2013; Rodney et al. 2014) analyze some or all of these new high-z SN rate measurements, together with precise new rates at lower redshifts from SDSS-II (Dilday et al. 2010b) and SNLS (Perrett et al. 2012). They all find that a DTD with a power-law form, $\sim t^{-1}$, when convolved with a wide range of plausible SFHs, gives an excellent fit to the observed SN rates. For example, the best-fit power-law index found by Graur et al. (2014) is $\beta = -1.00 \pm 0.06$ (random error, due to the uncertainties in the SN rates), ± 0.10 (systematic error, due to the range of possible SFHs). This power law is shown in **Figure 8**. The precise $z < 1$ rate measurements are particularly constraining for the DTD. Graur et al. (2014)

show that the rates are incompatible with the functional forms of DTDs from SD-model BPS calculations, which have cutoffs at delays beyond several gigayears. Rodney et al. (2014) find that, assuming the DTD is a t^{-1} power law beyond 500 Myr, 30–65% of SNe Ia must have delays shorter than 500 Myr, consistent with a t^{-1} power law that extends all the way down to 40 Myr delays.

3.5.6. Delay-time distribution from star-formation histories of individual galaxies. The convolution in Equation 4 between SFH and the DTD, giving the SN Ia rate versus time, applies not only to a cosmological volume but also to individual galaxies. Brandt et al. (2010) and Maoz et al. (2011) introduced similar methods for recovering the DTD from SN Ia samples for which the monitored galaxies have individual estimated SFHs based on spectral synthesis. In essence, the SFH of every individual galaxy is convolved with a trial universal DTD, and the resulting current SN Ia rate is compared to the number of SNe the galaxy hosted in the survey (generally none, sometimes one, rarely more). Maoz et al. (2011) applied the method to a subsample of the galaxies in LOSS, and the SNe that they hosted with spectral-synthesis-based SFH reconstructions by Tojeiro et al. (2009), based on spectra from the SDSS. The limited time-resolution of these SFH reconstructions dictates that the recovered DTD has only three coarse time bins, with a prompt SN Ia component, that explode within $\tau < 420$ Myr of star formation, an intermediate component, at $0.42 < \tau < 2.4$ Gyr, and a delayed component that explodes after $\tau > 2.4$ Gyr. Brandt et al. (2010) apply their method to a sample of SNe Ia from the SDSS-II survey. Maoz, Mannucci & Brandt (2012) apply the Maoz et al. (2011) algorithm to an SDSS-II sample that is larger and more complete than the Brandt et al. (2010) sample, and correct several oversights in the previous analysis. Finally, Graur & Maoz (2013) apply the method to a sample of SNe Ia discovered in archival SDSS-I galaxy spectra (the galaxies happened to host a SN within the spectroscopic aperture during the observations), again using the spectral-synthesis-based SFHs of the galaxies. All of these three-time-bin DTD reconstructions give similar monotonically decreasing DTDs, consistent with $\sim t^{-1}$. For example, Maoz, Mannucci & Brandt (2012) find $\beta = -1.07 \pm 0.07$. These DTD reconstructions are included in **Figure 8**.

Brandt et al. (2010) used the "stretch parameter" of the SN light curves to divide their SN Ia sample into a "high-stretch" subsample and a "low-stretch" one, and derived the DTD for each subsample. They found that luminous, high-stretch, SNe Ia have most of their DTD signal at short delays, whereas low-stretch SNe Ia have a DTD that peaks in the longest-delay bin. However, this trend largely disappears in the analysis of the more-complete SDSS-II SN Ia sample by Maoz, Mannucci & Brandt (2012). Perrett et al. (2012) similarly fail to see any strong stretch dependence in their DTD analysis of volumetric SN Ia rates. Although such a dependence is expected on the basis of the observed trends between light-curve stretch and mean host galaxy stellar age (see Section 3.3.4), apparently the DTD analyses still lack the precision needed to reveal it.

3.5.7. Delay-time distribution, downsizing, and the rate-mass relation. Indirect evidence for a broad DTD comes also from an observed dependence of SN Ia rate per unit mass on galaxy mass. Li et al. (2011c) first noted that the SN Ia rate per unit mass in the LOSS sample decreases for more massive galaxies of a given type. The rate-size relation has been confirmed by Smith et al. (2012) for SDSS-II SNe Ia and by Graur & Maoz (2013) for the SNe Ia discovered in archival SDSS spectra. Kistler et al. (2013) show that this "rate-size relation" can be explained by a combined effect of "downsizing" in galaxy formation—more massive galaxies are formed, on average, earlier and on shorter timescales (e.g., Gallazzi et al. 2005)—combined with a $\sim t^{-1}$ DTD. Downsizing produces different mean ages, and therefore SNe in more massive galaxies come from the low-rate, long-delay tail of the DTD. Graur & Maoz (2013) independently confirm that downsizing plus a $\sim t^{-1}$ DTD explain the effect for their sample as well. Kistler et al. (2013) investigate also the possible effect of metallicity on the rate-mass relation. Less massive galaxies have lower metallicities (Tremonti et al. 2004, Mannucci et al. 2010). Stellar evolution at low metallicity can produce more massive WDs, and the SN Ia rates could be affected if such WDs are easier to bring to M_{Ch} via accretion. However, for their model's assumed influence of metallicity on WD mass and SN Ia production, Kistler et al. (2013) find only a weak effect of metallicity on the rate-mass relation.

We note in passing that the rate-mass relation permits a reliable estimate of the Milky Way's SN Ia rate. For Sbc spirals of Milky Way mass [$(6.4 \pm 0.6) \times 10^{10}\,\mathrm{M}_\odot$; McMillan 2011], the specific SN Ia rate is $(1.12 \pm 0.35) \times 10^{-13}\,\mathrm{year}^{-1}\,\mathrm{M}_\odot^{-1}$ (Li et al. 2011c), so if the Milky Way is not atypical, the Galactic SN Ia rate is $(7.2 \pm 2.3) \times 10^{-3}\,\mathrm{year}^{-1}$. The mean time between Galactic SNe Ia is thus 100–200 years.

3.5.8. Element abundances in stars. The different timescales of production of CC SNe and SNe Ia, coupled with the different chemical yields of the two populations of SNe, produce diverse abundance ratios in galaxies having different SFHs [see Nomoto, Kobayashi & Tominaga (2013) for a review]. If a galaxy forms most of its stars in a single, brief episode of star formation, the Fe-peak-element yields from most SNe Ia will not be incorporated into subsequent generations of stars, whereas α elements, such as O, Mg, Si, and Ca from CC SNe, with their much smaller delay time, will do so. As a consequence, the distribution of [α/Fe] (typically [O/Fe] versus [Fe/H]) is sensitive to the DTD (Tinsley 1979, Matteucci & Greggio 1986, Matteucci 2001, De Donder & Vanbeveren 2004). Matteucci et al. (2006) find that DTDs with ~10–30% of the explosions occurring at <100 Myr are consistent with the abundance ratios of stars in the Solar Neighborhood, but it is not clear if such a fraction of short delays is required by the data.

Mennekens et al. (2013) use BPS calculations combined with a Galactic chemical evolution model to investigate which combinations of progenitor scenarios, binary evolution parameters, SFHs, and initial mass functions (IMFs), can simultaneously match the observed DTD and the distribution of [Fe/H] in Solar-Neighborhood G-dwarfs. Their main conclusion is that reproducing the observed [Fe/H] distribution requires a high DTD normalization (i.e., models with efficient SN Ia production), as indicated independently by SN-Ia-rate-based DTD estimates (see Section 3.5.10 below), and a significant early contribution to the DTD. This can be achieved with a combination of SD and DD progenitors, with binary parameters tuned to boost their SN Ia production efficiency. Recent applications of chemical evolution modeling by Kobayashi & Nomoto (2009), Matteucci et al. (2009), and Tsujimoto & Shigeyama (2012) take the inverse approach and use the DTD from SN Ia rate measurements to constrain chemical evolution. With improved stellar abundance data, chemical mapping could provide stronger constraints on the early-time DTD.

Seitenzahl et al. (2013a) recently used the [Mn/Fe] abundance ratios in Solar-Neighborhood stars to constrain the explosion scenario directly (i.e., not via the DTD). They find that manganese is produced efficiently only in the high-density conditions of near-M_{Ch} SN Ia explosions—sub-M_{Ch} SNe Ia and CC SNe have yields that make [Mn/Fe] ratios that are about 3 times below Solar. Consequently, the observed distribution of stellar [Mn/Fe] requires that roughly 50% of SNe Ia come from near-M_{Ch} explosions, be they from SD, DD, or other progenitors.

3.5.9. Delay-time distributions from supernova remnants. An unconventional observational DTD reconstruction has been carried out by Maoz & Badenes (2010) by applying the Maoz et al. (2011) algorithm to a sample of SN remnants in the Magellanic Clouds (Badenes, Maoz & Draine 2010) and treating this as an effective SN sample, one from a survey in which the SNe are detectable for ~10 kyr (the visibility time of a SN remnant). The advantage of this approach is that the nearness of these galaxies permits derivation of detailed and accurate stellar age distributions of the individual galactic regions that host the SNe. The stellar age distributions are found by comparing the densities of the resolved stars in color-magnitude diagrams with model stellar isochrones (Harris & Zaritsky 2004, 2009). The small number of remnants, with the attendant large statistical errors, allows only the coarsest DTD resolution, with two SN Ia time bins. Nevertheless, Maoz & Badenes (2010) find a significant detection of a prompt (this time $35 < \tau < 330$ Myr) SN Ia component and an upper limit on the DTD level at longer delays that is consistent with the long-delay DTD levels measured with other methods. Larger samples can be produced in the future via ongoing and proposed deep radio surveys for the SN remnant populations in additional nearby galaxies, such as M33 and M31, and by using their spatially specific stellar age distributions, again based on the resolved stellar populations.

3.5.10. Normalization of the delay-time distribution. **Figure 8** compiles a number of the observational DTDs described above. Despite the consistent $\sim t^{-1}$ DTD shape, obtained by numerous different methods and samples, there is a clear tension among the normalizations of the DTD, i.e., among the Hubble-time-integrated ("total") numbers of SNe Ia formed by a stellar generation. The DTDs based on volumetric SN Ia rates (which are not biased toward bright galaxies and contain a significant number of SNe in dwarf galaxies; Childress et al. 2013, Leaman et al. 2011) typically indicate a time-integrated SN Ia production efficiency of about one SN Ia per 1,000 $M_\odot$ formed [for an assumed "diet-Salpeter" IMF, Bell et al. (2003)]. Galaxy-targeted SN surveys seem to find values about two times higher than this. Finally, DTDs based on galaxy clusters are higher yet, perhaps by another factor of 2–3. In fact, there have been previous indications for higher rates in cluster ellipticals than in field ellipticals (Mannucci et al. 2008, Dilday et al. 2010a). The highest observed DTDs are a factor of 3–10 higher than the lowest ones. Although the range in normalizations appears real, more work is required to confirm that some or all of it is not due to systematic errors, such as incorrect global or individual galaxy SFHs for particular galaxy types, surveys, or redshifts.

The lower-normalization DTDs are within reach of some of the DD-scenario DTDs from some recent BPS models (Mennekens, Vanbeveren & De Greve 2012; Toonen, Nelemans & Portegies Zwart 2012; Ruiter et al. 2013). For example, Toonen, Nelemans & Portegies Zwart (2012) find a time-integrated number of SNe Ia in their models of $2\text{–}3.3 \times 10^{-4} M_\odot^{-1}$, i.e., an order of magnitude lower than observed. However, this comes from considering as SN Ia progenitors only C/O-C/O WD mergers with total masses above $M_{\rm Ch}$. Considering all C/O-C/O WD mergers, they find a time-integrated rate of $\sim 0.8 \times 10^{-3} M_\odot^{-1}$, and this assumes 50% binarity of the initial stellar population. A higher binary fraction would give a correspondingly higher result, one quite similar to some of the observed DTD normalizations above, as apparent in **Figure 8**.

As also seen in **Figure 8**, the BPS predictions for SD models tend to be much lower than DD predictions, both at short delays (although there is a large diversity among models), and certainly beyond the cutoffs that SD models generically have at delays of a few gigayears. Their normalizations thus miss the observed ones by at least an order of magnitude, and often by several.

A higher SN Ia number per formed stellar mass observed in the most massive galaxies could perhaps be related to the recently noted dependence of IMF on galaxy mass (Cappellari et al. 2012, Conroy & van Dokkum 2012, Geha et al. 2013), or to the well-known excess of metals in clusters, compared with expectations from SN yields and the present-day stellar population (e.g., Maoz, Sharon & Gal-Yam 2010; Loewenstein 2013). However, it is unclear how the IMF variations, which are seen in the subsolar-mass region of the IMF, could affect the production of SNe Ia whose progenitors are WDs descended from intermediate-mass stars, on the part of the IMF where variations have not been claimed.

As already noted in Section 2 above, head-on collisions of WDs in dense environments, such as globular clusters and galactic centers, have been proposed as a way of boosting SN Ia production, which would raise the DTD normalization. Apart from collisions between unbound cluster WDs, dynamical encounters between binaries and other cluster stars will harden the binaries (Shara & Hurley 2002); such an effect has been used to explain the large enhancement in the number of low-mass X-ray binaries observed in globular clusters (Sarazin et al. 2003). Rosswog et al. (2009) show that, even if such an enhancement of WD collisions occurs, the small fraction of all WDs that are in globular clusters means that this could explain only of order 1% of all SN Ia events. Observationally, the lack of globular clusters at the locations of SNe Ia observed with HST (Voss & Nelemans 2012, Washabaugh & Bregman 2013) rules out globulars as a significant overall rate-enhancement mechanism for SNe Ia.

3.5.11. Summary of the observed delay-time distribution. In summary, SN Ia rate measurements quite consistently indicate a $\sim t^{-1}$ DTD. The coarse time resolution of many studies, the systematic uncertainties discussed above, and the sparse information below ~1 Gyr presently allow more structured shapes, but the overall behavior of the DTD appears to be similar to that generic to the DD model. Long delay times are possible in SD models with red-giant donors, but BPS models generally produce very few of them. Nonetheless, there is observational room for a SD contribution at delays $\lesssim 1$ Gyr, where the DTDs are not well constrained. As a result, the DTD data presently cannot rule out comparable SD and DD contributions. However, in terms of total numbers, BPS predictions for SD models are generally lower than DD predictions, which themselves barely reach some of the observed DTDs.

4. SUMMARY

4.1. The Emerging Picture

The SN Ia progenitor question is far from resolved, despite the major efforts to address it (e.g., there are over 800 papers in NASA-ADS over the past six years with the words "Type-Ia supernova progenitor" in their abstracts). Nevertheless, this large body of work has provided a wealth of relevant information. Progress has been made in the observational domain, but also in theoretical work where, e.g., recent 3D hydrodynamical simulations of mergers have revived the possibility that double-detonation merger scenarios could produce normal SNe Ia. We summarize some of the main observational results that have emerged.

- Censii in the Milky Way and nearby galaxies of the potential progenitor populations of SNe Ia in the standard SD model—accreting WDs growing in mass toward $M_{\rm Ch}$—show that they cannot be recurrent novae or supersoft X-ray sources for more than a small fraction of their growth time. A hidden population of rapidly accreting WDs is also limited to being no more than a minority of the progenitors based on the apparent absence of their expected ionizing radiation. A population of differentially spun-up super-$M_{\rm Ch}$ WDs, waiting to spin down and explode, may be out there, but this has theoretical problems and seems at odds with the very slow spins observed generically in WDs. As for close double WD binaries that could be DD-model progenitors, both BPS calculations and surveys for such systems indicate that their total merger rate is of the same order of magnitude as the SN Ia rate. The traditional $M_{\rm tot} > M_{\rm Ch}$, C/O + C/O mergers are probably too rare, by a factor of at least a few, but if lower-mass WD mergers, or even C/O + He WD mergers, lead to a normal SN Ia, then their numbers are likely sufficient. Which ones, if any, actually lead to SNe Ia, and how, is still unclear.

- SN 2011fe, the best-studied SN Ia and a very typical event to boot, has provided a wealth of progenitor constraints, essentially all of them based on null results. This event's immediate environment was remarkably "clean." There are strong limits on a pre-explosion companion star, on any signatures of ejecta interaction with a donor star or with a CSM, on any X-ray upscattering or optical absorption signatures by a CSM, and on any hydrogen that was stripped from a companion and entrained by the ejecta. All of these are consistent with a DD origin and challenge the SD picture for this event, unless spin-up is invoked.
- Variable NaI absorption, which could come from a CSM, has been seen, but only in three normal SNe Ia and one CSM-interacting SN Ia. The 2/31 or 2/16 discovery statistics in well-defined samples suggest an $\sim$5–15% occurrence fraction. Statistical studies using single-epoch high-resolution spectra deduce, on the basis of an excess of systems with blueshifted versus redshifted NaI absorption, an $\sim$20% fraction of progenitors with a CSM, but these estimates still suffer from small numbers and systematics in defining samples and zero reference velocities. Although a CSM producing the variable Na absorption is expected in the SD scenario, some DD merger models also predict such CSM absorptions.
- No normal SN Ia has shown signs of any entrained hydrogen from a companion, sometimes up to stringent limits. However, a rare subclass constituting $\sim$0.1–1% (maybe somewhat more) of SNe Ia does show prominent variable hydrogen emission and other signs of CSM interaction, such as kinetic energy input to the light curves. However, the implied CSM masses may be of order a solar mass, and it is presently unclear how such CSM masses fit into any of the traditional progenitor pictures.
- Among SN remnants known or suspected to have been SNe Ia, none have thus far revealed any obvious surviving SD-scenario donors, in several cases with strict upper limits. Hydrodynamic models of remnants having detailed X-ray observations almost always indicate ejecta expanding through a uniform-density medium, rather than through a low-density cavity from a pre-explosion SD-scenario wind.
- Derivations of the SN Ia DTD from a variety of SN Ia rate measurements and techniques, in different environments and redshifts, are all consistent with a t^{-1} dependence at delays $\gtrsim$1 Gyr (although DTDs from different samples may be showing somewhat different normalizations of this distribution). Such a dependence is typically found for DD models, whereas most BPS SD models predict few or no SNe Ia at long delays. A SD contribution to the DTD at short delays is not yet constrained.
- From the diversity and correlations among observed spectral properties of SNe Ia and of their host galaxies, a rough general trend has emerged. At one extreme are events with high luminosities and slow light-curve evolution occurring in star-forming galaxies, with high-velocity spectral features, no signs of unburnt carbon, higher symmetry indicators, and a relatively high chance of showing intervening absorption. At the other extreme are events with the opposite properties. As some of the former properties can be plausibly associated with SD-model explosions, it is tempting to suggest the existence of two "families" of progenitors, SD progenitors that produce the first type of events and DD progenitors that make the second type of SNe Ia. However, there are some problems with this picture. First, as reviewed in Section 3.3.4, the observed trends are weak, with large scatter, and sometimes have conflicting claims on the sense of the relations and on selection effects. Some combination of progenitor properties, e.g., age, metallicity, and others, may explain the trend rather than two separate families. Second, if two very diverse progenitor channels, e.g., a $M_{\rm Ch}$ SD channel and a double-detonation sub-$M_{\rm Ch}$ DD channel, were operating in parallel with comparable numbers, it would be remarkable that observed SN Ia properties

are so uniform and continuous—the very features of SNe Ia that have made them so useful for cosmology.

From a recent-historical point of view, it is noteworthy that there has been something of a paradigm shift. The DD scenario has traditionally been the underdog, with review articles (e.g., Livio 2000) often concluding that the SD model will eventually be confirmed. The main argument against the DD scenario has been theoretical—that a DD merger would never lead to anything akin to a SN Ia explosion. This theoretical objection is no longer very strong, as some merger simulations now seem to successfully produce SNe Ia. And for several SN Ia events and remnants, the observations suggest that, among the various models, only DD progenitors are not ruled out. Thus, not only can the DD model no longer be rejected offhand, but rather it could turn out to be the main, or even the only, scenario behind SNe Ia.

4.2. Future Outlook

Future observational data, and progress on the theoretical issues, will undoubtedly provide new clues to the progenitor problem. Apart from obvious breakthroughs, such as the direct detection of a progenitor or the explosion of a previously known candidate progenitor, there are several developments that, in the near or far future, could significantly advance the field.

On the theoretical side, open issues still are how exactly the accretion onto a WD proceeds in a SD model and what the parameter ranges are (mass transfer rate, composition, rotation, accretion history) that lead a C/O WD to grow in mass? Ignition in DD mergers is another unsolved problem. Detailed 3D hydrodynamical models, including nucleosynthesis, exploring the run of merger-model parameter space, are now being published. However, the finding that rather subtle effects may be important, such as residual hydrogen or helium on the WD or residual carbon in the centers of O-Ne WDs, suggests that robust results will require higher-resolution calculations that take the previous stellar evolution into account in detail.

Progress is within reach also in the area of progenitor populations and the theoretical DTD. A recent detailed comparison of four different BPS codes (Toonen et al. 2014) shows that their outputs differ not because of numerical accuracy or treatment but because of different physical assumptions. One clear conclusion, in particular for the DD and He-accreting SD models, is that the phase in which one of the two stars is a He star needs more detailed investigation. In parallel, comparison of the BPS predictions with observed local populations of binaries could calibrate out many of the uncertainties in BPS codes.

Turning to observations, the understanding of potential progenitor populations has advanced thanks to wide-field spectroscopic surveys, such as SDSS, that led to large and homogeneous samples of observed binaries. However, proper understanding of selection effects is still a challenge. The successful launch of Gaia has signaled the next big improvement with the discovery of large samples of WD binaries, for which distances and proper motions are also forthcoming.

The DTD has proved to be a powerful tool for testing models. A better knowledge of the parent stellar populations, and site-specific measurements of SN Ia rate versus the stellar age distributions at the SN locations, can improve the accuracy of observational DTDs. The multivariate distribution of delay times, explosion energies, host metallicities, and more (as opposed to the single-variable DTD), can provide physical links between these parameters, which will provide sharper discrimination among models.

SN 2011fe was a watershed event, not only because of its nearness but also because of the wealth of pre-explosion data and the very early and multiwavelength coverage. With the rising prominence of time-domain, multimessenger surveys, it is likely that similar or even better data will soon be available for additional SNe Ia.

Gamma-ray observations of SNe Ia by future MeV-range missions should provide excellent diagnostics not available at other energies. Horiuchi & Beacom (2010) simulate the gamma-ray spectra and light curves of SNe Ia, out to Virgo distances, that could be obtained by the proposed ACT mission, showing that various SD models would have distinct signatures in such gamma-ray data. Summa et al. (2013) conduct this exercise for the proposed GRIPS and ASTRO-H missions. They show that the same two SD and DD models, compared by Röpke et al. (2012) to the optical data for SN 2011fe (see Section 3.3, above), could be distinguished by gamma-ray data from these missions. The & Burrows (2014) have made predictions, for an array of explosion models, for the gamma-ray spectrum of the recent nearby SN 2014J in M82. Churazov et al. (2014) report, for this event, the first gamma-ray detection of a SN Ia in the 847-keV line of ^{56}Co, thus opening up this new portal to explosion physics.

In the more distant future, a space-based gravitational wave interferometer will be able to detect signals individually from several thousand Galactic double WD systems with orbital periods below 10 min, i.e., that are within ~1 Myr of merging (Amaro-Seoane et al. 2013). This will directly probe the WD merger rate. Together with the millions of longer-period binaries that are detectable as a foreground, this will provide quantitative information about the binary WD population and on its relevance as SN Ia progenitors. The gravitational signal of an actual merger will be detectable only for a Galactic (or perhaps Magellanic Cloud) event.

Indeed, we are arguably overdue for the privilege that Tycho, Kepler, and Galileo enjoyed, that of a Galactic SN Ia. Of course, we may well have to wait a few more centuries for this occasion, but when it comes, it will surely be another transformational event, this time for the understanding of the workings of SNe Ia.

DISCLOSURE STATEMENT

The authors are not aware of any affiliations, memberships, funding, or financial holdings that might be perceived as affecting the objectivity of this review.

ACKNOWLEDGMENTS

We thank R. Aloisio, C. Badenes, R. Foley, O. Graur, E. Nakar, P. Tozzi, and J. Vink for valuable discussions and input. D.M. acknowledges support by a grant from the Isreal Science Foundation (ISF) and by I-Core Program grant 1829/12 of the Planning and Budgeting Committee and the ISF.

LITERATURE CITED

Abazajian KN, Adelman-McCarthy JK, Agüeros MA, Allam SS, Allende Prieto C, et al. 2009. *Ap. J. Suppl.* 182:543–58

Aldering G, Antilogus P, Bailey S, Baltay C, Bauer A, et al. 2006. *Ap. J.* 650:510–27

Alexander RD, Wynn GA, King AR, Pringle JE. 2011. *MNRAS* 418:2576–83

Amaro-Seoane P, Aoudia S, Babak S, Binétruy P, Berti E, et al. 2013. *GW Notes* 6:4–110

Anupama GC. 2013. See Di Stefano, Orio & Moe. 2013, pp. 154–61

Arnett WD. 1969. *Ap. Space Sci.* 5:180–212

Arnett WD. 1982. *Ap. J.* 253:785–97

Ashok NM, Banerjee DPK. 2003. *Astron. Astrophys.* 409:1007–15

Aubourg É, Tojeiro R, Jimenez R, Heavens A, Strauss MA, Spergel DN. 2008. *Astron. Astrophys.* 492:631–36

Badenes C, Borkowski KJ, Hughes JP, Hwang U, Bravo E. 2006. *Ap. J.* 645:1373–91

Badenes C, Hughes JP, Bravo E, Langer N. 2007. *Ap. J.* 662:472–86

Badenes C, Hughes JP, Cassam-Chena G, Bravo E. 2008. *Ap. J.* 680:1149–57
Badenes C, Maoz D. 2012. *Ap. J. Lett.* 749:L11
Badenes C, Maoz D, Draine BT. 2010. *MNRAS* 407:1301–13
Badenes C, Mullally F, Thompson SE, Lupton RH. 2009. *Ap. J.* 707:971–78
Barbary K, Aldering G, Amanullah R, Brodwin M, Connolly N, et al. 2012. *Ap. J.* 745:32
Bearda H, Hartmann W, Ebisawa K, Heise J, Kaastra J, et al. 2002. *Astron. Astrophys.* 385:511–16
Bedin LR, Ruiz-Lapuente P, Gonzalez Hernandez JI, Canal R, Filippenko AV, Mendez J. 2014. *MNRAS* 439:354–71
Bell EF, McIntosh DH, Katz N, Weinberg MD. 2003. *Ap. J. Suppl.* 149:289
Benetti S, Cappellaro E, Mazzali PA, Turatto M, Altavilla G, et al. 2005. *Ap. J.* 623:1011–16
Benetti S, Cappellaro E, Turatto M, Taubenberger S, Harutyunyan A, Valenti S. 2006. *Ap. J. Lett.* 653:L129–32
Benz W, Thielemann FK, Hills JG. 1989. *Ap. J.* 342:986–98
Berger L, Koester D, Napiwotzki R, Reid IN, Zuckerman B. 2005. *Astron. Astrophys.* 444:565–71
Beuermann K, Reinsch K. 2002. *Astron. Astrophys.* 381:487–90
Bianco FB, Howell DA, Sullivan M, Conley A, Kasen D, et al. 2011. *Ap. J.* 741:20
Bildsten L, Shen KJ, Weinberg NN, Nelemans G. 2007. *Ap. J. Lett.* 662:L95–98
Blair WP, Ghavamian P, Long KS, Williams BJ, Borkowski KJ, et al. 2007. *Ap. J.* 662:998–1013
Blanc G, Greggio L. 2008. *New Astron.* 13:606–18
Bloecker T. 1995. *Astron. Astrophys.* 299:755
Blondin S, Dessart L, Hillier DJ, Khokhlov AM. 2013. *MNRAS* 429:2127–42
Blondin S, Kasen D, Röpke FK, Kirshner RP, Mandel KS. 2011. *MNRAS* 417:1280–302
Blondin S, Matheson T, Kirshner RP, Mandel KS, Berlind P, et al. 2012. *Astron. J.* 143:126
Blondin S, Prieto JL, Patat F, Challis P, Hicken M, et al. 2009. *Ap. J.* 693:207–15
Bloom JS, Kasen D, Shen KJ, Nugent PE, Butler NR, et al. 2012. *Ap. J. Lett.* 744:L17
Bogomazov AI, Tutukov AV. 2011. *Astron. Rep.* 55:497–504
Borkowski KJ, Blondin JM, Sarazin CL. 1992. *Ap. J.* 400:222–37
Borkowski KJ, Hendrick SP, Reynolds SP. 2006. *Ap. J.* 652:1259–67
Borkowski KJ, Sarazin CL, Blondin JM. 1994. *Ap. J.* 429:710–25
Boshkayev K, Izzo L, Rueda Hernandez JA, Ruffini R. 2013. *Astron. Astrophys.* 555:A151
Bours MCP, Toonen S, Nelemans G. 2013. *Astron. Astrophys.* 552:A24
Branch D, van den Bergh S. 1993. *Astron. J.* 105:2231–35
Brandt TD, Tojeiro R, Aubourg É, Heavens A, Jimenez R, Strauss MA. 2010. *Astron. J.* 140:804–16
Brown PJ, Dawson KS, de Pasquale M, Gronwall C, Holland S, et al. 2012a. *Ap. J.* 753:22
Brown PJ, Dawson KS, Harris DW, Olmstead M, Milne P, Roming PWA. 2012b. *Ap. J.* 749:18
Brown WR, Kilic M, Hermes JJ, Allende Prieto C, Kenyon SJ, Winget DE. 2011. *Ap. J. Lett.* 737:L23
Burkey MT, Reynolds SP, Borkowski KJ, Blondin JM. 2013. *Ap. J.* 764:63
Calura F, Matteucci F, Tozzi P. 2007. *MNRAS* 378:L11–15
Canal R, Méndez J, Ruiz-Lapuente P. 2001. *Ap. J. Lett.* 550:L53–56
Capetti A. 2002. *Ap. J. Lett.* 574:L25–27
Cappellari M, McDermid RM, Alatalo K, Blitz L, Bois M, et al. 2012. *Nature* 484:485–88
Cappellaro E, Evans R, Turatto M. 1999. *Astron. Astrophys.* 351:459–66
Cappellaro E, Patat F, Mazzali PA, Benetti S, Danziger JI, et al. 2001. *Ap. J. Lett.* 549:L215–18
Cartier R, Förster F, Coppi P, Hamuy M, Maeda K, et al. 2011. *Astron. Astrophys.* 534:L15
Chamel N, Fantina AF, Davis PJ. 2013. *Phys. Rev. D* 88:081301
Charpinet S, Fontaine G, Brassard P. 2009. *Nature* 461:501–3
Chevalier RA. 1982. *Ap. J.* 258:790–97
Chevalier RA. 1998. *Ap. J.* 499:810
Chevalier RA, Fransson C. 2006. *Ap. J.* 651:381–91
Chevalier RA, Kirshner RP, Raymond JC. 1980. *Ap. J.* 235:186–95
Childress M, Aldering G, Antilogus P, Aragon C, Bailey S, et al. 2013. *Ap. J.* 770:107
Childress MJ, Filippenko AV, Ganeshalingam M, Schmidt BP. 2014. *MNRAS* 437:338–50
Chiotellis A, Schure KM, Vink J. 2012. *Astron. Astrophys.* 537:A139

Chomiuk L. 2013. *Publ. Astron. Soc. Aust.* 30:46
Chomiuk L, Krauss MI, Rupen MP, Nelson T, Roy N, et al. 2012a. *Ap. J.* 761:173
Chomiuk L, Soderberg AM, Moe M, Chevalier RA, Rupen MP, et al. 2012b. *Ap. J.* 750:164
Chugai NN. 2008. *Astron. Lett.* 34:389–96
Chugai NN, Yungelson LR. 2004. *Astron. Lett.* 30:65–72
Churazov E, Sunyaev R, Grebenev S, Isern J, Jean P, et al. 2014. *Astron. Telegr.* 5992
Claeys JSW, Pols OR, Izzard RG, Vink J, Verbunt FWM. 2014. *Astron. Astrophys.* Accepted. arXiv:1401.2895
Colgate SA, McKee C. 1969. *Ap. J.* 157:623
Conroy C, van Dokkum P. 2012. *Ap. J.* 747:69
Cooper MC, Newman JA, Yan R. 2009. *Ap. J.* 704:687–704
Córsico AH, Althaus LG, Kawaler SD, Miller Bertolami MM, García-Berro E, Kepler SO. 2011. *MNRAS* 418:2519–26
Crotts APS, Yourdon D. 2008. *Ap. J.* 689:1186–90
Dahlen T, Strolger LG, Riess AG. 2008. *Ap. J.* 681:462–69
Dahlen T, Strolger LG, Riess AG, Mobasher B, Chary RR, et al. 2004. *Ap. J.* 613:189–99
Dan M, Rosswog S, Brueggen M, Podsiadlowski P. 2014. *MNRAS* 438:14–34
Dan M, Rosswog S, Guillochon J, Ramirez-Ruiz E. 2012. *MNRAS* 422:2417–28
D'Andrea CB, Gupta RR, Sako M, Morris M, Nichol RC, et al. 2011. *Ap. J.* 743:172
Darbha S, Metzger BD, Quataert E, Kasen D, Nugent P, Thomas R. 2010. *MNRAS* 409:846–54
Darnley MJ, Bode MF, Kerins E, Newsam AM, An J, et al. 2006. *MNRAS* 369:257–71
Davé R, Oppenheimer BD, Finlator K. 2011. *MNRAS* 415:11–31
Dawson KS, Aldering G, Amanullah R, Barbary K, Barrientos LF, et al. 2009. *Astron. J.* 138:1271–83
De Donder E, Vanbeveren D. 2004. *New Astron. Rev.* 48:861–975
Della Valle M, Livio M. 1996. *Ap. J.* 473:240
Della Valle M, Panagia N, Padovani P, Cappellaro E, Mannucci F, Turatto M. 2005. *Ap. J.* 629:750–56
Della Valle M, Rosino L, Bianchini A, Livio M. 1994. *Astron. Astrophys.* 287:403–9
Dessart L, Blondin S, Hillier DJ, Khokhlov A. 2014a. *MNRAS.* 441(1):532–50
Dessart L, Hillier DJ, Blondin S, Khokhlov A. 2014b. *MNRAS.* Accepted. arXiv:1308.6352
Di Stefano R. 2010. *Ap. J.* 712:728–33
Di Stefano R, Orio M, Moe M. 2013. *Binary Paths to Type Ia Supernovae Explosions, Proc. IAU Symp. 281.* Cambridge, UK: Cambridge Univ. Press
Di Stefano R, Voss R, Claeys JSW. 2011. *Ap. J. Lett.* 738:L1
Diaz MP, Williams RE, Luna GJ, Moraes M, Takeda L. 2010. *Astron. J.* 140:1860–67
Dilday B, Bassett B, Becker A, Bender R, Castander F, et al. 2010a. *Ap. J.* 715:1021–35
Dilday B, Howell DA, Cenko SB, Silverman JM, Nugent PE, et al. 2012. *Science* 337:942–45
Dilday B, Smith M, Bassett B, Becker A, Bender R, et al. 2010b. *Ap. J.* 713:1026–36
Drury LO. 2012. *Astropart. Phys.* 39:52–60
Duchêne G, Kraus A. 2013. *Annu. Rev. Astron. Astrophys.* 51:269–310
Edwards ZI, Pagnotta A, Schaefer BE. 2012. *Ap. J. Lett.* 747:L19
Filippenko AV. 1997. *Annu. Rev. Astron. Astrophys.* 35:309–55
Fink M, Hillebrandt W, Röpke FK. 2007. *Astron. Astrophys.* 476:1133–43
Fink M, Kromer M, Seitenzahl IR, Ciaraldi-Schoolmann F, Röpke FK, et al. 2014. *MNRAS* 438:1762–83
Fink M, Röpke FK, Hillebrandt W, Seitenzahl IR, Sim SA, Kromer M. 2010. *Astron. Astrophys.* 514:A53
Folatelli G, Phillips MM, Morrell N, Tanaka M, Maeda K, et al. 2012. *Ap. J.* 745:74
Foley RJ. 2012. *Ap. J.* 748:127
Foley RJ. 2013. *MNRAS* 435:273–88
Foley RJ, Brown PJ, Rest A, Challis PJ, Kirshner RP, Wood-Vasey WM. 2010a. *Ap. J. Lett.* 708:L61–65
Foley RJ, Challis PJ, Chornock R, Ganeshalingam M, Li W, et al. 2013. *Ap. J.* 767:57
Foley RJ, Challis PJ, Filippenko AV, Ganeshalingam M, Landsman W, et al. 2012a. *Ap. J.* 744:38
Foley RJ, Kasen D. 2011. *Ap. J.* 729:55
Foley RJ, Kirshner RP. 2013. *Ap. J. Lett.* 769:L1
Foley RJ, Rest A, Stritzinger M, Pignata G, Anderson JP, et al. 2010b. *Astron. J.* 140:1321–28
Foley RJ, Sanders NE, Kirshner RP. 2011. *Ap. J.* 742:89

Foley RJ, Simon JD, Burns CR, Gal-Yam A, Hamuy M, et al. 2012b. *Ap. J.* 752:101
Fontaine G, Brassard P, Charpinet S. 2013. *Eur. Phys. J. Web Conf.*, 43:Article No. 05011
Förster F, González-Gaitán S, Folatelli G, Morrell N. 2013. *Ap. J.* 772:19
Fox OD, Filippenko AV. 2013. *Ap. J. Lett.* 772:L6
Fox OD, Filippenko AV, Skrutskie MF, Silverman JM, Ganeshalingam M, et al. 2013. *Astron. J.* 146:2
Friedrich S, Jordan S. 2001. *Astron. Astrophys.* 367:577–81
Frieman JA, Bassett B, Becker A, Choi C, Cinabro D, et al. 2008. *Astron. J.* 135:338–47
Fryer CL, Ruiter AJ, Belczynski K, Brown PJ, Bufano F, et al. 2010. *Ap. J.* 725:296–308
Fuhrmann K. 2005. *MNRAS* 359:L35–36
Fujimoto MY. 1982. *Ap. J.* 257:767
Gallazzi A, Charlot S, Brinchmann J, White SDM, Tremonti CA. 2005. *MNRAS* 362:41–58
Gal-Yam A, Maoz D, Guhathakurta P, Filippenko AV. 2008. *Ap. J.* 680:550–67
Gal-Yam A, Maoz D, Sharon K. 2002. *MNRAS* 332:37–48
Ganeshalingam M, Li W, Filippenko AV. 2011. *MNRAS* 416:2607–22
García-Berro E, Lorén-Aguilar P, Aznar-Siguán G, Torres S, Camacho J, et al. 2012. *Ap. J.* 749:25
García-Senz D, Cabezón RM, Arcones A, Relaño A, Thielemann FK. 2013. *MNRAS* 436:3413–29
Garnavich PM, Kirshner RP, Challis P, Jha S, Branch D, et al. 2001. *Bull. Am. Astron. Soc.* 33:1370 (Abstr.)
Geha M, Brown TM, Tumlinson J, Kalirai JS, Simon JD, et al. 2013. *Ap. J.* 771:29
Geier S, Heber U, Kupfer T, Napiwotzki R. 2010. *Astron. Astrophys.* 515:A37
Geier S, Marsh TR, Wang B, Dunlap B, Barlow BN, et al. 2013. *Astron. Astrophys.* 554:A54
Geier S, Nesslinger S, Heber U, Przybilla N, Napiwotzki R, Kudritzki RP. 2007. *Astron. Astrophys.* 464:299–307
Gerardy CL, Höflich P, Fesen RA, Marion GH, Nomoto K, et al. 2004. *Ap. J.* 607:391–405
Germany LM, Reiss DJ, Schmidt BP, Stubbs CW, Suntzeff NB. 2004. *Astron. Astrophys.* 415:863–78
Ghavamian P, Rakowski CE, Hughes JP, Williams TB. 2003. *Ap. J.* 590:833–45
Ghavamian P, Raymond J, Smith RC, Hartigan P. 2001. *Ap. J.* 547:995–1009
Gilfanov M, Bogdán Á. 2010. *Nature* 463:924–25
González Hernández JI, Ruiz-Lapuente P, Filippenko AV, Foley RJ, Gal-Yam A, Simon JD. 2009. *Ap. J.* 691:1–15
González Hernández JI, Ruiz-Lapuente P, Tabernero HM, Montes D, Canal R, et al. 2012. *Nature* 489:533–36
Goobar A, Johansson J, Amanullah R, Cao Y, Perley DA, et al. 2014. *Ap. J.* Accepted. arXiv:1402.0849
Graham ML, Pritchet CJ, Sullivan M, Gwyn SDJ, Neill JD, et al. 2008. *Astron. J.* 135:1343–49
Graham ML, Pritchet CJ, Sullivan M, Howell DA, Gwyn SDJ, et al. 2010. *Astron. J.* 139:594–605
Graur O, Maoz D. 2012a. *Astron. Telegr.* 4226:1
Graur O, Maoz D. 2012b. *Astron. Telegr.* 4535:1
Graur O, Maoz D. 2013. *MNRAS* 430:1746–63
Graur O, Maoz D, Shara MM. 2014. *MNRAS.* Accepted. arXiv:1403.1878
Graur O, Poznanski D, Maoz D, Yasuda N, Totani T, et al. 2011. *MNRAS* 417:916–40
Graur O, Rodney SA, Maoz D, Riess AG, Jha SW, et al. 2014. *Ap. J.* 783:28
Greggio L. 2005. *Astron. Astrophys.* 441:1055–78
Greggio L, Renzini A, Daddi E. 2008. *MNRAS* 388:829–37
Gruyters P, Exter K, Roberts TP, Rappaport S. 2012. *Astron. Astrophys.* 544:A86
Guillochon J, Dan M, Ramirez-Ruiz E, Rosswog S. 2010. *Ap. J. Lett.* 709:L64–69
Hachinger S, Mazzali PA, Sullivan M, Ellis RS, Maguire K, et al. 2013. *MNRAS* 429:2228–48
Hachisu I, Kato M. 2003a. *Ap. J.* 598:527–44
Hachisu I, Kato M. 2003b. *Ap. J.* 590:445–59
Hachisu I, Kato M. 2012. *Baltic Astron.* 21:68–75
Hachisu I, Kato M, Kiyota S, Kubotera K, Maehara H, et al. 2006. *Ap. J. Lett.* 651:L141–44
Hachisu I, Kato M, Nomoto K. 1996. *Ap. J. Lett.* 470:L97
Hachisu I, Kato M, Nomoto K. 1999. *Ap. J.* 522:487–503
Hachisu I, Kato M, Nomoto K. 2008a. *Ap. J. Lett.* 683:L127–30
Hachisu I, Kato M, Nomoto K. 2008b. *Ap. J.* 679:1390–404
Hachisu I, Kato M, Nomoto K. 2010. *Ap. J. Lett.* 724:L212–16

Hachisu I, Kato M, Nomoto K. 2012. *Ap. J. Lett.* 756:L4
Hachisu I, Kato M, Saio H, Nomoto K. 2012. *Ap. J.* 744:69
Hakobyan AA, Petrosian AR, McLean B, Kunth D, Allen RJ, et al. 2008. *Astron. Astrophys.* 488:523–31
Hamann WR, Peña M, Gräfener G, Ruiz MT. 2003. *Astron. Astrophys.* 409:969–82
Hamers AS, Pols OR, Claeys JSW, Nelemans G. 2013. *MNRAS* 430:2262–80
Hamilton AJS, Sarazin CL. 1984. *Ap. J.* 287:282–94
Hamuy M, Phillips MM, Suntzeff NB, Maza J, González LE, et al. 2003. *Nature* 424:651–54
Hamuy M, Trager SC, Pinto PA, Phillips MM, Schommer RA, et al. 2000. *Astron. J.* 120:1479–86
Han Z. 1998. *MNRAS* 296:1019–40
Han Z, Podsiadlowski P. 2004. *MNRAS* 350:1301–9
Han Z, Podsiadlowski P. 2006. *MNRAS* 368:1095–100
Hancock PJ, Gaensler BM, Murphy T. 2011. *Ap. J. Lett.* 735:L35
Harris J, Zaritsky D. 2004. *Astron. J.* 127:1531–44
Harris J, Zaritsky D. 2009. *Astron. J.* 138:1243–60
Hawley WP, Athanassiadou T, Timmes FX. 2012. *Ap. J.* 759:39
Hayden BT, Garnavich PM, Kasen D, Dilday B, Frieman JA, et al. 2010a. *Ap. J.* 722:1691–98
Hayden BT, Garnavich PM, Kessler R, Frieman JA, Jha SW, et al. 2010b. *Ap. J.* 712:350–66
Hayden BT, Gupta RR, Garnavich PM, Mannucci F, Nichol RC, Sako M. 2013. *Ap. J.* 764:191
Heng K. 2010. *Publ. Astron. Soc. Aust.* 27:23–44
Hermes JJ, Kilic M, Brown WR, Winget DE, Allende Prieto C, et al. 2012. *Ap. J. Lett.* 757:L21
Hicken M, Challis P, Jha S, Kirshner RP, Matheson T, et al. 2009. *Ap. J.* 700:331–57
Hillebrandt W, Kromer M, Röpke FK, Ruiter AJ. 2013. *Front. Phys.* 8:116–43
Hillebrandt W, Niemeyer JC. 2000. *Annu. Rev. Astron. Astrophys.* 38:191–230
Hillebrandt W, Sim SA, Röpke FK. 2007. *Astron. Astrophys.* 465:L17–20
Hoeflich P, Khokhlov A. 1996. *Ap. J.* 457:500
Höflich P, Schaefer BE. 2009. *Ap. J.* 705:483–95
Hole KT, Kasen D, Nordsieck KH. 2010. *Ap. J.* 720:1500–12
Horesh A, Kulkarni SR, Fox DB, Carpenter J, Kasliwal MM, et al. 2012. *Ap. J.* 746:21
Horiuchi S, Beacom JF. 2010. *Ap. J.* 723:329–41
Howell DA. 2011. *Nat. Commun.* 2:350
Howell DA, Höflich P, Wang L, Wheeler JC. 2001. *Ap. J.* 556:302–21
Howell DA, Sullivan M, Brown EF, Conley A, LeBorgne D, et al. 2009. *Ap. J.* 691:661–71
Howell DA, Sullivan M, Nugent PE, Ellis RS, Conley AJ, et al. 2006. *Nature* 443:308–11
Hoyle F, Fowler WA. 1960. *Ap. J.* 132:565
Hughes JP, Chugai N, Chevalier R, Lundqvist P, Schlegel E. 2007. *Ap. J.* 670:1260–74
Iben I Jr, Renzini A. 1983. *Annu. Rev. Astron. Astrophys.* 21:271–342
Iben I Jr, Tutukov AV. 1984. *Ap. J. Suppl.* 54:335–72
Iben I Jr, Tutukov AV, Yungelson LR. 1997. *Ap. J.* 475:291
Idan I, Shaviv N, Shaviv G. 2013. *MNRAS* 433:2884–92
Ihara Y, Ozaki J, Doi M, Shigeyama T, Kashikawa N, et al. 2007. *Publ. Astron. Soc. Jpn.* 59:811–26
Ilkov M, Soker N. 2011. *MNRAS*:1798
Ivanova N, Justham S, Chen X, De Marco O, Fryer CL, et al. 2013. *Astron. Astrophys. Rev.* 21:59
Iwamoto K, Brachwitz F, Nomoto K, Kishimoto N, Umeda H, et al. 1999. *Ap. J. Suppl.* 125:439–62
Jimenez R, Bernardi M, Haiman Z, Panter B, Heavens AF. 2007. *Ap. J.* 669:947–51
Johansson J, Amanullah R, Goobar A. 2013. *MNRAS* 431:L43–47
Johansson J, Woods TE, Gilfanov M, Sarzi M, Chem Y-M, Oh K. 2014. *MNRAS* 442:1079–89
Jordan GC IV, Perets HB, Fisher RT, van Rossum DR. 2012. *Ap. J. Lett.* 761:L23
Jordan S, Aznar Cuadrado R, Napiwotzki R, Schmid HM, Solanki SK. 2007. *Astron. Astrophys.* 462:1097–101
Jorgensen HE, Lipunov VM, Panchenko IE, Postnov KA, Prokhorov ME. 1997. *Ap. J.* 486:110
Justham S. 2011. *Ap. J. Lett.* 730:L34
Kafka S, Honeycutt RK, Williams R. 2012. *MNRAS* 425:1585–90
Kafka S, Williams R. 2011. *Astron. Astrophys.* 526:A83
Kamiya Y, Tanaka M, Nomoto K, Blinnikov SI, Sorokina EI, Suzuki T. 2012. *Ap. J.* 756:191

Kaplan DL, Bildsten L, Steinfadt JDR. 2012. *Ap. J.* 758:64

Kasen D. 2010. *Ap. J.* 708:1025–31

Kasen D, Nugent P. 2013. *Annu. Rev. Nucl. Part. Sci.* 63:153–74

Kasen D, Nugent P, Wang L, Howell DA, Wheeler JC, et al. 2003. *Ap. J.* 593:788–808

Kasen D, Röpke FK, Woosley SE. 2009. *Nature* 460:869–72

Kasen D, Woosley SE. 2007. *Ap. J.* 656:661–65

Kashi A, Soker N. 2011. *MNRAS* 417:1466–79

Kato M, Hachisu I. 2012. *Bull. Astron. Soc. India* 40:393

Kato M, Hachisu I, Kiyota S, Saio H. 2008. *Ap. J.* 684:1366–73

Kattner S, Leonard DC, Burns CR, Phillips MM, Folatelli G, et al. 2012. *Publ. Astron. Soc. Pac.* 124:114–27

Katz B, Dong S. 2012. arXiv:1211.4584

Kawaler SD. 2004. In *Stellar Rotation, Proc. IAU Symp. 215*, ed. A Maeder, P Eenens, p. 561. San Francisco: ASP

Kawka A, Vennes S, Schmidt GD, Wickramasinghe DT, Koch R. 2007. *Ap. J.* 654:499–520

Kelly PL, Fox OD, Filippenko AV, Cenko SB, Prato L, et al. 2014. *Ap. J.* Submitted. arXiv:1403.4250

Kepler SO, Pelisoli I, Jordan S, Kleinman SJ, Koester D, et al. 2013. *MNRAS* 429:2934–44

Kerzendorf WE, Childress M, Scharwachter TJ, Schmidt BPK. 2014. *Ap. J.* 782:27

Kerzendorf WE, Schmidt BP, Asplund M, Nomoto K, Podsiadlowski P, et al. 2009. *Ap. J.* 701:1665–72

Kerzendorf WE, Schmidt BP, Laird JB, Podsiadlowski P, Bessell MS. 2012. *Ap. J.* 759:7

Kerzendorf WE, Yong D, Schmidt BP, Simon JD, Jeffery CS, et al. 2013. *Ap. J.* 774:99

Khokhlov AM. 1991. *Astron. Astrophys.* 245:114–28

Kilic M, Brown WR, Allende Prieto C, Kenyon SJ, Heinke CO, et al. 2012. *Ap. J.* 751:141

Kistler MD, Stanek KZ, Kochanek CS, Prieto JL, Thompson TA. 2013. *Ap. J.* 770:88

Kobayashi C, Nomoto K. 2009. *Ap. J.* 707:1466–84

Kosenko D, Blinnikov SI, Vink J. 2011. *Astron. Astrophys.* 532:A114

Kosenko D, Vink J, Blinnikov S, Rasmussen A. 2008. *Astron. Astrophys.* 490:223–30

Kozai Y. 1962. *Astron. J.* 67:591

Krause O, Tanaka M, Usuda T, Hattori T, Goto M, et al. 2008. *Nature* 456:617–19

Kromer M, Pakmor R, Taubenberger S, Pignata G, Fink M, et al. 2013. *Ap. J. Lett.* 778:L18

Kushnir D, Katz B, Dong S, Livne E, Fernández R. 2013. *Ap. J. Lett.* 778:L37

Kuznetsova NV, Connolly BM. 2007. *Ap. J.* 659:530–40

Lampeitl H, Smith M, Nichol RC, Bassett B, Cinabro D, et al. 2010. *Ap. J.* 722:566–76

Landstreet JD, Bagnulo S, Valyavin GG, Fossati L, Jordan S, et al. 2012. *Astron. Astrophys.* 545:A30

Langer N, Deutschmann A, Wellstein S, Höflich P. 2000. *Astron. Astrophys.* 362:1046–64

Lanz T, Telis GA, Audard M, Paerels F, Rasmussen AP, Hubeny I. 2005. *Ap. J.* 619:517–26

Leaman J, Li W, Chornock R, Filippenko AV. 2011. *MNRAS* 412:1419–40

Leigh NWC, Geller AM. 2013. *MNRAS* 432:2474–79

Leonard DC. 2007. *Ap. J.* 670:1275–82

Leonard DC, Li W, Filippenko AV, Foley RJ, Chornock R. 2005. *Ap. J.* 632:450–75

Lepo K, van Kerkwijk M. 2013. *Ap. J.* 771:13

Li W, Bloom JS, Podsiadlowski P, Miller AA, Cenko SB, et al. 2011a. *Nature* 480:348–50

Li W, Chornock R, Leaman J, Filippenko AV, Poznanski D, et al. 2011b. *MNRAS* 412:1473–507

Li W, Leaman J, Chornock R, Filippenko AV, Poznanski D, et al. 2011c. *MNRAS* 412:1441–72

Li XD, van den Heuvel EPJ. 1997. *Astron. Astrophys.* 322:L9–12

Lidov ML. 1962. *Planet. Space Sci.* 9:719–59

Lipunov VM, Panchenko IE, Pruzhinskaya MV. 2011. *New Astron.* 16:250–52

Liu J, Di Stefano R, Wang T, Moe M. 2012a. *Ap. J.* 749:141

Liu ZW, Kromer M, Fink M, Pakmor R, Röpke FK, et al. 2013a. *Ap. J.* 778:121

Liu ZW, Pakmor R, Röpke FK, Edelmann P, Hillebrandt W, et al. 2013b. *Astron. Astrophys.* 554:A109

Liu ZW, Pakmor R, Röpke FK, Edelmann P, Wang B, et al. 2012b. *Astron. Astrophys.* 548:A2

Livio M. 2000. In *Type Ia Supernovae: Theory and Cosmology*, ed. JC Niemeyer, JW Truran, pp.33–48. Cambridge, UK: Cambridge Univ. Press

Livio M, Riess A, Sparks W. 2002. *Ap. J. Lett.* 571:L99–102

Livio M, Riess AG. 2003. *Ap. J. Lett.* 594:L93–94
Livne E. 1990. *Ap. J. Lett.* 354:L53–55
Loewenstein M. 2013. *Ap. J.* 773:52
Lorén-Aguilar P, Isern J, García-Berro E. 2009. *Astron. Astrophys.* 500:1193–205
Lorén-Aguilar P, Isern J, García-Berro E. 2010. *MNRAS* 406:2749–63
Luna GJM, Sokoloski JL, Mukai K. 2008. In *RS Ophiuchi (2006) and the Recurrent Nova Phenomenon*, ed. A Evans, MF Bode, TJ O'Brien, MJ Darnley, *ASP Conf. Ser.* 401:342–46. San Francisco: ASP
Lundqvist P, Mattila S, Sollerman J, Kozma C, Baron E, et al. 2013. *MNRAS* 435:329–45
Lyman JD, James PA, Perets HB, Anderson JP, Gal-Yam A, et al. 2013. *MNRAS* 434:527–41
Ma H, Woosley SE, Malone CM, Almgren A, Bell J. 2013. *Ap. J.* 771:58
Maeda K, Benetti S, Stritzinger M, Röpke FK, Folatelli G, et al. 2010. *Nature* 466:82–85
Maeda K, Leloudas G, Taubenberger S, Stritzinger M, Sollerman J, et al. 2011. *MNRAS* 413:3075–94
Maguire K, Sullivan M, Ellis RS, Nugent PE, Howell DA, et al. 2012. *MNRAS* 426:2359–79
Maguire K, Sullivan M, Patat F, Gal-Yam A, Hook IM, et al. 2013. *MNRAS* 436:222–40
Mannucci F. 2008. *Chin. J. Astron. Astrophys. Suppl.* 8:143–54
Mannucci F. 2009. In *Probing Stellar Populations Out to the Distant Universe: Cefalu 2008*, ed. G Giobbi, A Tornambe, G Raimondo, M Limongi, LA Antonelli, N Menci, E Brocato. *AIP Conf. Ser.* 1111:467–76. Melville, NY: AIP
Mannucci F, Cresci G, Maiolino R, Marconi A, Gnerucci A. 2010. *MNRAS* 408:2115–27
Mannucci F, Della Valle M, Panagia N. 2006. *MNRAS* 370:773–83
Mannucci F, Della Valle M, Panagia N. 2007. *MNRAS* 377:1229–35
Mannucci F, Della Valle M, Panagia N, Cappellaro E, Cresci G, et al. 2005. *Astron. Astrophys.* 433:807–14
Mannucci F, Maoz D, Sharon K, Botticella MT, Della Valle M, et al. 2008. *MNRAS* 383:1121–30
Maoz D, Badenes C. 2010. *MNRAS* 407:1314–27
Maoz D, Badenes C, Bickerton SJ. 2012. *Ap. J.* 751:143
Maoz D, Gal-Yam A. 2004. *MNRAS* 347:951–56
Maoz D, Mannucci F. 2008. *MNRAS* 388:421–28
Maoz D, Mannucci F. 2012. *Publ. Astron. Soc. Aust.* 29:447–65
Maoz D, Mannucci F, Brandt TD. 2012. *MNRAS* 426:3282–94
Maoz D, Mannucci F, Li W, Filippenko AV, Della Valle M, Panagia N. 2011. *MNRAS* 412:1508–21
Maoz D, Sharon K, Gal-Yam A. 2010. *Ap. J.* 722:1879–94
Margutti R, Soderberg AM, Chomiuk L, Chevalier R, Hurley K, et al. 2012. *Ap. J.* 751:134
Marietta E, Burrows A, Fryxell B. 2000. *Ap. J. Suppl.* 128:615–50
Marsh TR. 2011. *Class. Quantum Gravity* 28:094019
Mason E. 2013. *Astron. Astrophys.* 556:C2
Matteucci F. 2001. *The Chemical Evolution of the Galaxy*, Vol. 253, *Ap. Space Sci. Libr.* Dordrecht, Neth.: Kluwer
Matteucci F, Greggio L. 1986. *Astron. Astrophys.* 154:279–87
Matteucci F, Panagia N, Pipino A, Mannucci F, Recchi S, Della Valle M. 2006. *MNRAS* 372:265–75
Matteucci F, Spitoni E, Recchi S, Valiante R. 2009. *Astron. Astrophys.* 501:531–38
Mattila S, Lundqvist P, Sollerman J, Kozma C, Baron E, et al. 2005. *Astron. Astrophys.* 443:649–62
Maund JR, Höflich P, Patat F, Wheeler JC, Zelaya P, et al. 2010. *Ap. J. Lett.* 725:L167–71
Maund JR, Spyromilio J, Höflich PA, Wheeler JC, Baade D, et al. 2013. *MNRAS* 433:L20–24
Maxted PFL, Marsh TR, North RC. 2000. *MNRAS* 317:L41–44
Mazeh T, Shaham J. 1979. *Astron. Astrophys.* 77:145–51
Mazzali P, Sullivan M, Hachinger S, Ellis R, Nugent PE, et al. 2014. *MNRAS* 439:1959–79
Mazzali PA, Benetti S, Altavilla G, Blanc G, Cappellaro E, et al. 2005. *Ap. J. Lett.* 623:L37–40
Mazzali PA, Cappellaro E, Danziger IJ, Turatto M, Benetti S. 1998. *Ap. J. Lett.* 499:L49
Mazzali PA, Podsiadlowski P. 2006. *MNRAS* 369:L19–22
Mazzali PA, Röpke FK, Benetti S, Hillebrandt W. 2007. *Science* 315:825–28
McCully C, Jha SW, Foley RJ, Chornock R, Holtzman JA, et al. 2014. *Ap. J.* 786:134
McMillan PJ. 2011. *MNRAS* 414:2446–57
Meng X, Yang W. 2010. *Ap. J.* 710:1310–23
Meng XC, Yang WM. 2011. *Res. Astron. Astrophys.* 11:965–73

Mennekens N, Vanbeveren D, De Greve JP. 2012. arXiv:1212.0313
Mennekens N, Vanbeveren D, De Greve JP, De Donder E. 2010. *Astron. Astrophys.* 515:A89
Mennekens N, Vanbeveren D, De Greve JP, De Donder E. 2013. See Di Stefano, Orio & Moe. 2013, pp. 232–35
Moll R, Raskin C, Kasen D, Woosley S. 2014. *Ap. J.* 785:105
Nakar E, Sari R. 2010. *Ap. J.* 725:904–21
Nakar E, Sari R. 2012. *Ap. J.* 747:88
Napiwotzki R, Karl CA, Lisker T, Heber U, Christlieb N, et al. 2004. *Ap. Space Sci.* 291:321–28
Napiwotzki R, Karl CA, Nelemans G, Yungelson L, Christlieb N, et al. 2007. In *15th Eur. Workshop White Dwarfs*, ed. R Napiwotzki, MR Burleigh, *ASP Conf. Ser.* 372:387–92. San Francisco: ASP
Nelemans G, Napiwotzki R, Karl C, Marsh TR, Voss B, et al. 2005. *Astron. Astrophys.* 440:1087–95
Nelemans G, Toonen S, Bours M. 2013. See Di Stefano, Orio & Moe. 2013, pp. 232–35
Nelemans G, Voss R, Roelofs G, Bassa C. 2008. *MNRAS* 388:487–94
Nelemans G, Yungelson LR, Portegies Zwart SF. 2001. *Astron. Astrophys.* 375:890–98
Nelson T, Donato D, Mukai K, Sokoloski J, Chomiuk L. 2012. *Ap. J.* 748:43
Newsham G, Starrfield S, Timmes F. 2013. In *Stella Novae: Past and Future Decades*, ed. P Woudt, VARM Ribeiro, *ASP Conf. Ser.* In press. arXiv:1303.3642
Nielsen MTB, Dominik C, Nelemans G, Voss R. 2013. *Astron. Astrophys.* 549:A32
Nielsen MTB, Gilfanov M, Bogdan A, Woods TE, Nelemans G. 2014. *MNRAS* 442:3400–6
Nielsen MTB, Nelemans G, Voss R, Toonen S. 2014. *Astron. Astrophys.* 563:A16
Nielsen MTB, Voss R, Nelemans G. 2012. *MNRAS* 426:2668–76
Nielsen MTB, Voss R, Nelemans G. 2013. *MNRAS* 435:187–93
Nomoto K. 1982. *Ap. J.* 253:798–810
Nomoto K, Iben I Jr. 1985. *Ap. J.* 297:531–37
Nomoto K, Kobayashi C, Tominaga N. 2013. *Annu. Rev. Astron. Astrophys.* 51:457–509
Nomoto K, Saio H, Kato M, Hachisu I. 2007. *Ap. J.* 663:1269–76
Nomoto K, Thielemann FK, Yokoi K. 1984. *Ap. J.* 286:644–58
Nordhaus J, Wellons S, Spiegel DS, Metzger BD, Blackman EG. 2011. *Proc. Natl. Acad. Sci. USA* 108:3135–40
Nugent PE, Sullivan M, Cenko SB, Thomas RC, Kasen D, et al. 2011. *Nature* 480:344–47
Orio M, Nelson T, Bianchini A, Di Mille F, Harbeck D. 2010. *Ap. J.* 717:739–65
Pagnotta A, Walker ES, Schaefer BE. 2014. *Ap. J.* 788:173
Pakmor R, Kromer M, Taubenberger S, Sim SA, Röpke FK, Hillebrandt W. 2012. *Ap. J. Lett.* 747:L10
Pakmor R, Kromer M, Taubenberger S, Springel V. 2013. *Ap. J. Lett.* 770:L8
Pan KC, Ricker PM, Taam RE. 2013. *Ap. J.* 773:49
Pan YC, Sullivan M, Maguire K, Hook IM, Nugent PE, et al. 2014. *MNRAS* 438:1391–416
Panagia N, Van Dyk SD, Weiler KW, Sramek RA, Stockdale CJ, Murata KP. 2006. *Ap. J.* 646:369–77
Parikh A, José J, Seitenzahl IR, Röpke FK. 2013. *Astron. Astrophys.* 557:A3
Parrent JT, Howell DA, Friesen B, Thomas RC, Fesen RA, et al. 2012. *Ap. J. Lett.* 752:L26
Parrent JT, Thomas RC, Fesen RA, Marion GH, Challis P, et al. 2011. *Ap. J.* 732:30
Patat F. 2005. *MNRAS* 357:1161–77
Patat F, Baade D, Höflich P, Maund JR, Wang L, Wheeler JC. 2009. *Astron. Astrophys.* 508:229–46
Patat F, Benetti S, Cappellaro E, Turatto M. 2006. *MNRAS* 369:1949–60
Patat F, Chandra P, Chevalier R, Justham S, Podsiadlowski P, et al. 2007. *Science* 317:924–26
Patat F, Chugai NN, Podsiadlowski P, Mason E, Melo C, Pasquini L. 2011. *Astron. Astrophys.* 530:A63
Patat F, Cordiner MA, Cox NLJ, Anderson RI, Harutyunyan A, et al. 2013. *Astron. Astrophys.* 549:A62
Patat F, Höflich P, Baade D, Maund JR, Wang L, Wheeler JC. 2012. *Astron. Astrophys.* 545:A7
Patnaude DJ, Badenes C, Park S, Laming JM. 2012. *Ap. J.* 756:6
Patterson J, Oksanen A, Monard B, Rea R, Hambsch FJ, et al. 2013. In *Stella Novae: Past and Future Decades*, ed. P Woudt, VARM Ribeiro, *ASP Conf. Ser.* In press. arXiv:1303.0736
Perlmutter S, Aldering G, Goldhaber G, Knop RA, Nugent P, et al. 1999. *Ap. J.* 517:565–86
Perrett K, Sullivan M, Conley A, González-Gaitán S, Carlberg R, et al. 2012. *Astron. J.* 144:59
Phillips MM. 1993. *Ap. J. Lett.* 413:L105–8
Phillips MM, Simon JD, Morrell N, Burns CR, Cox NLJ, et al. 2013. *Ap. J.* 779:38

Piersanti L, Cassisi S, Iben I Jr, Tornambé A. 2000. *Ap. J.* 535:932–42
Piersanti L, Gagliardi S, Iben I Jr, Tornambé A. 2003. *Ap. J.* 583:885–901
Pinto PA, Eastman RG. 2000. *Ap. J.* 530:757–76
Piro AL. 2008. *Ap. J.* 679:616–25
Piro AL, Bildsten L. 2008. *Ap. J.* 673:1009–13
Piro AL, Chang P, Weinberg NN. 2010. *Ap. J.* 708:598–604
Piro AL, Kulkarni SR. 2013. *Ap. J. Lett.* 762:L17
Piro AL, Nakar E. 2013. *Ap. J.* 769:67
Piro AL, Nakar E. 2014. *Ap. J.* 784:85
Piro AL, Thompson TA, Kochanek CS. 2014. *MNRAS* 438:3456
Pols OR, Marinus M. 1994. *Astron. Astrophys.* 288:475–501
Poznanski D, Maoz D, Yasuda N, Foley RJ, Doi M, et al. 2007. *MNRAS* 382:1169–86
Prieto JL, Garnavich PM, Phillips MM, DePoy DL, Parrent J, et al. 2007. arXiv:0706.4088
Pritchet CJ, Howell DA, Sullivan M. 2008. *Ap. J. Lett.* 683:L25–28
Prodan S, Murray N, Thompson TA. 2013. *Ap. J.* Submitted. arXiv:1305.2191
Quinn JL, Garnavich PM, Li W, Panagia N, Riess A, et al. 2006. *Ap. J.* 652:512–17
Rabinak I, Livne E, Waxman E. 2012. *Ap. J.* 757:35
Rabinak I, Waxman E. 2011. *Ap. J.* 728:63
Rajoelimanana AF, Charles PA, Meintjes PJ, Odendaal A, Udalski A. 2013. *MNRAS* 432:2886–94
Rappaport S, Chiang E, Kallman T, Malina R. 1994. *Ap. J.* 431:237–46
Raskin C, Kasen D. 2013. *Ap. J.* 772:1
Raskin C, Kasen D, Moll R, Schwab J, Woosley S. 2014. *Ap. J.* 788(1):75
Raskin C, Scannapieco E, Fryer C, Rockefeller G, Timmes FX. 2012. *Ap. J.* 746:62
Raskin C, Scannapieco E, Rhoads J, Della Valle M. 2009. *Ap. J.* 707:74–78
Raskin C, Scannapieco E, Rockefeller G, Fryer C, Diehl S, Timmes FX. 2010. *Ap. J.* 724:111–25
Raymond JC, Korreck KE, Sedlacek QC, Blair WP, Ghavamian P, Sankrit R. 2007. *Ap. J.* 659:1257–64
Remillard RA, Rappaport S, Macri LM. 1995. *Ap. J.* 439:646–51
Rest A, Matheson T, Blondin S, Bergmann M, Welch DL, et al. 2008a. *Ap. J.* 680:1137–48
Rest A, Sinnott B, Welch DL. 2012. *Publ. Astron. Soc. Aust.* 29:466–81
Rest A, Suntzeff NB, Olsen K, Prieto JL, Smith RC, et al. 2005. *Nature* 438:1132–34
Rest A, Welch DL, Suntzeff NB, Oaster L, Lanning H, et al. 2008b. *Ap. J. Lett.* 681:L81–84
Rettura A, Mei S, Stanford SA, Raichoor A, Moran S, et al. 2011. *Ap. J.* 732:94
Reynolds SP, Borkowski KJ, Hwang U, Hughes JP, Badenes C, et al. 2007. *Ap. J. Lett.* 668:L135–38
Riess AG, Filippenko AV, Challis P, Clocchiatti A, Diercks A, et al. 1998. *Astron. J.* 116:1009–38
Rigault M, Copin Y, Aldering G, Antilogus P, Aragon C, et al. 2013. *Astron. Astrophys.* 560:A66
Rodney S, Riess AG, Strolger L-G, Dahlen T, Graur O, et al. 2014. *Astron. J.* 148(1):13
Rodríguez-Gil P, Santander-García M, Knigge C, Corradi RLM, Gänsicke BT, et al. 2010. *MNRAS* 407:L21–25
Roelofs G, Bassa C, Voss R, Nelemans G. 2008. *MNRAS* 391:290–96
Röpke FK, Kromer M, Seitenzahl IR, Pakmor R, Sim SA, et al. 2012. *Ap. J. Lett.* 750:L19
Rosswog S, Kasen D, Guillochon J, Ramirez-Ruiz E. 2009. *Ap. J. Lett.* 705:L128–32
Ruiter AJ, Belczynski K, Fryer C. 2009. *Ap. J.* 699:2026–36
Ruiter AJ, Sim SA, Pakmor R, Kromer M, Seitenzahl IR, et al. 2013. *MNRAS* 429:1425–36
Ruiz-Lapuente P. 1997. *Science* 276:1813–14
Ruiz-Lapuente P, Canal R. 1998. *Ap. J. Lett.* 497:L57
Ruiz-Lapuente P, Comeron F, Méndez J, Canal R, Smartt SJ, et al. 2004. *Nature* 431:1069–72
Rupen MP, Mioduszewski AJ, Sokoloski JL. 2008. *Ap. J.* 688:559–67
Russell BR, Immler S. 2012. *Ap. J. Lett.* 748:L29
Sadat R, Blanchard A, Guiderdoni B, Silk J. 1998. *Astron. Astrophys.* 331:L69–72
Sahman DI, Dhillon VS, Marsh TR, Moll S, Thoroughgood TD, et al. 2013. *MNRAS* 433:1588–98
Saio H, Nomoto K. 1998. *Ap. J.* 500:388
Saio H, Nomoto K. 2004. *Ap. J.* 615:444–49
Sako M, Bassett B, Becker A, Cinabro D, DeJongh F, et al. 2008. *Astron. J.* 135:348–73

Sand DJ, Graham ML, Bildfell C, Zaritsky D, Pritchet C, et al. 2012. *Ap. J.* 746:163
Sand DJ, Zaritsky D, Herbert-Fort S, Sivanandam S, Clowe D. 2008. *Astron. J.* 135:1917–33
Sarazin CL, Kundu A, Irwin JA, Sivakoff GR, Blanton EL, Randall SW. 2003. *Ap. J.* 595:743–59
Scalzo R, Aldering G, Antilogus P, Aragon C, Bailey S, et al. 2012. *Ap. J.* 757:12
Scalzo RA, Aldering G, Antilogus P, Aragon C, Bailey S, et al. 2010. *Ap. J.* 713:1073–94
Scannapieco E, Bildsten L. 2005. *Ap. J. Lett.* 629:L85–88
Schaefer BE. 2010. *Ap. J. Suppl.* 187:275–373
Schaefer BE. 2013. *Am. Astron. Soc. Meet.* 221:233.06 (Abstr.)
Schaefer BE, Pagnotta A. 2012. *Nature* 481:164–66
Schaefer BE, Pagnotta A, Shara MM. 2010. *Ap. J.* 708:381–402
Schaefer BE, Pagnotta A, Xiao L, Darnley MJ, Bode MF, et al. 2010. *Astron. J.* 140:925–32
Schawinski K. 2009. *MNRAS* 397:717–25
Schmidt BP, Kirshner RP, Leibundgut B, Wells LA, Porter AC, et al. 1994. *Ap. J. Lett.* 434:L19–23
Schwab J, Shen KJ, Quataert E, Dan M, Rosswog S. 2012. *MNRAS* 427:190–203
Scolnic DM, Riess AG, Foley RJ, Rest A, Rodney SA, et al. 2014. *Ap. J.* 780:37
Seitenzahl IR, Cescutti G, Röpke FK, Ruiter AJ, Pakmor R. 2013a. *Astron. Astrophys.* 559:L5
Seitenzahl IR, Ciaraldi-Schoolmann F, Röpke FK, Fink M, Hillebrandt W, et al. 2013b. *MNRAS* 429:1156–72
Selvelli P, Cassatella A, Gilmozzi R, González-Riestra R. 2008. *Astron. Astrophys.* 492:787–803
Shappee BJ, Kochanek CS, Stanek KZ. 2013. *Ap. J.* 765:150
Shappee BJ, Stanek KZ, Pogge RW, Garnavich PM. 2013. *Ap. J. Lett.* 762:L5
Shara MM, Bibby JL, Zurek D, Crowther PA, Moffat AFJ, Drissen L. 2013. *Astron. J.* 146:162
Shara MM, Hurley JR. 2002. *Ap. J.* 571:830–42
Sharon K, Gal-Yam A, Maoz D, Filippenko AV, Guhathakurta P. 2007. *Ap. J.* 660:1165–75
Shen KJ, Bildsten L. 2007. *Ap. J.* 660:1444–50
Shen KJ, Bildsten L. 2009. *Ap. J.* 699:1365–73
Shen KJ, Bildsten L. 2014. *Ap. J.* 785:61
Shen KJ, Bildsten L, Kasen D, Quataert E. 2012. *Ap. J.* 748:35
Shen KJ, Guillochon J, Foley RJ. 2013. *Ap. J. Lett.* 770:L35
Shore SN, Augusteijn T, Ederoclite A, Uthas H. 2011. *Astron. Astrophys.* 533:L8
Silverman JM, Filippenko AV. 2012. *MNRAS* 425:1917–33
Silverman JM, Ganeshalingam M, Filippenko AV. 2013. *MNRAS* 430:1030–41
Silverman JM, Ganeshalingam M, Li W, Filippenko AV, Miller AA, Poznanski D. 2011. *MNRAS* 410:585–611
Silverman JM, Nugent PE, Gal-Yam A, Sullivan M, Howell DA, et al. 2013a. *Ap. J.* 772:125
Silverman JM, Nugent PE, Gal-Yam A, Sullivan M, Howell DA, et al. 2013b. *Ap. J. Suppl.* 207:3
Simon JD, Gal-Yam A, Gnat O, Quimby RM, Ganeshalingam M, et al. 2009. *Ap. J.* 702:1157–70
Simon V. 2003. *Astron. Astrophys.* 406:613–21
Smartt SJ. 2009. *Annu. Rev. Astron. Astrophys.* 47:63–106
Smith M, Nichol RC, Dilday B, Marriner J, Kessler R, et al. 2012. *Ap. J.* 755:61
Smith PS, Williams GG, Smith N, Milne PA, Jannuzi BT, Green EM. 2011. arXiv:1111.6626
Soderberg AM, Margutti R, Zauderer BA, Krauss M, Katz B, et al. 2012. *Ap. J.* 752:78
Soker N. 2013. See Di Stefano, Orio & Moe. 2013, pp. 72–75
Soker N, García-Berro E, Althaus LG. 2014. *MNRAS* 437:L66–70
Soker N, Kashi A, García-Berro E, Torres S, Camacho J. 2013. *MNRAS* 431:1541–46
Sokoloski JL, Luna GJM, Mukai K, Kenyon SJ. 2006. *Nature* 442:276–78
Solheim JE. 2010. *Publ. Astron. Soc. Pac.* 122:1133–63
Sorokina EI, Blinnikov SI, Kosenko DI, Lundqvist P. 2004. *Astron. Lett.* 30:737–50
Sparks WB, Macchetto F, Panagia N, Boffi FR, Branch D, et al. 1999. *Ap. J.* 523:585–92
Spruit HC. 1998. *Astron. Astrophys.* 333:603–12
Starrfield S, Sparks WM, Truran JW. 1985. *Ap. J.* 291:136–46
Starrfield S, Truran JW, Sparks WM, Kutter GS. 1972. *Ap. J.* 176:169
Stehle M, Mazzali PA, Benetti S, Hillebrandt W. 2005. *MNRAS* 360:1231–43
Steiner JE, Diaz MP. 1998. *Publ. Astron. Soc. Pac.* 110:276–82

Stephenson FR, Green DA. 2002. *Historical Supernovae and Their Remnants. Int. Ser. Astron. Astrophys.*, Vol. 5. Oxford: Clarendon Press
Sternberg A, Gal-Yam A, Simon JD, Leonard DC, Quimby RM, et al. 2011. *Science* 333:856–59
Sternberg A, Gal-Yam A, Simon JD, Patat F, Hillebrandt W, et al. 2013. *MNRAS*. Accepted. arXiv:1311.3645
Strolger LG, Dahlen T, Riess AG. 2010. *Ap. J.* 713:32–40
Strolger LG, Riess AG, Dahlen T, Livio M, Panagia N, et al. 2004. *Ap. J.* 613:200–23
Suh H, Yoon S-C, Jeong H, Yi SK. 2011. *Ap. J.* 730:110
Suijs MPL, Langer N, Poelarends AJ, Yoon S-C, Heger A, Herwig F. 2008. *Astron. Astrophys.* 481:L87–90
Sullivan M, Le Borgne D, Pritchet CJ, Hodsman A, Neill JD, et al. 2006. *Ap. J.* 648:868–83
Summa A, Ulyanov A, Kromer M, Boyer S, Röpke FK, et al. 2013. *Astron. Astrophys.* 554:A67
Taam RE. 1980. *Ap. J.* 237:142–47
Taddia F, Stritzinger MD, Phillips MM, Burns CR, Heinrich-Josties E, et al. 2012. *Astron. Astrophys.* 545:L7
Tanaka M, Kawabata KS, Yamanaka M, Maeda K, Hattori T, et al. 2010. *Ap. J.* 714:1209–16
Tanaka M, Mazzali PA, Stanishev V, Maurer I, Kerzendorf WE, Nomoto K. 2011. *MNRAS* 410:1725–38
Taubenberger S, Benetti S, Childress M, Pakmor R, Hachinger S, et al. 2011. *MNRAS* 412:2735–62
Taubenberger S, Kromer M, Hachinger S, Mazzali PA, Benetti S, et al. 2013a. *MNRAS* 432:3117–30
Taubenberger S, Kromer M, Pakmor R, Pignata G, Maeda K, et al. 2013b. *Ap. J. Lett.* 775:L43
The L-S, Burrows A. 2014. *Ap. J.* 786:141
Thomas RC, Aldering G, Antilogus P, Aragon C, Bailey S, et al. 2011. *Ap. J.* 743:27
Thompson TA. 2011. *Ap. J.* 741:82
Thomson MG, Chary RR. 2011. *Ap. J.* 731:72
Thoroughgood TD, Dhillon VS, Littlefair SP, Marsh TR, Smith DA. 2001. *MNRAS* 327:1323–33
Timmes FX, Brown EF, Truran JW. 2003. *Ap. J. Lett.* 590:L83–86
Tinsley BM. 1979. *Ap. J.* 229:1046–56
Tojeiro R, Wilkins S, Heavens AF, Panter B, Jimenez R. 2009. *Ap. J. Suppl.* 185:1–19
Tominaga N, Morokuma T, Blinnikov SI, Baklanov P, Sorokina EI, Nomoto K. 2011. *Ap. J. Suppl.* 193:20
Toonen S, Claeys JSW, Mennekens N, Ruiter AJ. 2014. *Astron. Astrophys.* 562:14
Toonen S, Nelemans G, Portegies Zwart S. 2012. *Astron. Astrophys.* 546:A70
Tornambé A, Piersanti L. 2013. *MNRAS* 431:1812–22
Totani T, Morokuma T, Oda T, Doi M, Yasuda N. 2008. *Publ. Astron. Soc. Jpn.* 60:1327–46
Tout CA. 2005. In *The Astrophysics of Cataclysmic Variables and Related Objects*, ed. J-M Hameury, JP Lasota, *ASP Conf. Ser.* 330:279. San Francisco: ASP
Tout CA, Wickramasinghe DT, Liebert J, Ferrario L, Pringle JE. 2008. *MNRAS* 387:897–901
Tovmassian G, Yungelson L, Rauch T, Suleimanov V, Napiwotzki R, et al. 2010. *Ap. J.* 714:178–93
Tremonti CA, Heckman TM, Kauffmann G, Brinchmann J, Charlot S, et al. 2004. *Ap. J.* 613:898–913
Trundle C, Kotak R, Vink JS, Meikle WPS. 2008. *Astron. Astrophys.* 483:L47–50
Truran JW, Livio M. 1986. *Ap. J.* 308:721–27
Tsujimoto T, Shigeyama T. 2012. *Ap. J. Lett.* 760:L38
Tucker BE. 2011. *Ap. Space Sci.* 335:223–30
Tutukov A, Yungelson L. 1996. *MNRAS* 280:1035–45
Tutukov AV, Yungelson LR. 1981. *Nauchn. Informatsii* 49:3
Udalski A, Kubiak M, Szymanski M. 1997. *Acta Astron.* 47:319–44
Umeda H, Nomoto K, Kobayashi C, Hachisu I, Kato M. 1999. *Ap. J. Lett.* 522:L43–47
Uthas H, Knigge C, Steeghs D. 2010. *MNRAS* 409:237–46
Valenti S, Pastorello A, Cappellaro E, Benetti S, Mazzali PA, et al. 2009. *Nature* 459:674–77
van den Heuvel EPJ, Bhattacharya D, Nomoto K, Rappaport SA. 1992. *Astron. Astrophys.* 262:97–105
van Kerkwijk MH, Chang P, Justham S. 2010. *Ap. J. Lett.* 722:L157–61
van Winckel H, Lloyd Evans T, Briquet M, De Cat P, Degroote P, et al. 2009. *Astron. Astrophys.* 505:1221–32
Vink J. 2008. *Ap. J.* 689:231–41
Vink J. 2012. *Astron. Astrophys. Rev.* 20:49
Vink J, Bleeker J, van der Heyden K, Bykov A, Bamba A, Yamazaki R. 2006. *Ap. J. Lett.* 648:L33–37
Vink J, Kaastra JS, Bleeker JAM. 1997. *Astron. Astrophys.* 328:628–33
Vink J, Laming JM, Gu MF, Rasmussen A, Kaastra JS. 2003. *Ap. J. Lett.* 587:L31–34

Voss R, Nelemans G. 2008. *Nature* 451:802–4
Voss R, Nelemans G. 2012. *Astron. Astrophys.* 539:A77
Walker ES, Hachinger S, Mazzali PA, Ellis RS, Sullivan M, et al. 2012. *MNRAS* 427:103–13
Wang B, Han Z. 2012. *New Astron. Rev.* 56:122–41
Wang B, Li XD, Han ZW. 2010. *MNRAS* 401:2729–38
Wang B, Meng X, Chen X, Han Z. 2009a. *MNRAS* 395:847–54
Wang L, Baade D, Höflich P, Wheeler JC, Kawabata K, et al. 2006. *Ap. J.* 653:490–502
Wang L, Baade D, Patat F. 2007. *Science* 315:212–14
Wang L, Wheeler JC. 2008. *Annu. Rev. Astron. Astrophys.* 46:433–74
Wang X, Filippenko AV, Ganeshalingam M, Li W, Silverman JM, et al. 2009b. *Ap. J. Lett.* 699:L139–43
Wang X, Li W, Filippenko AV, Foley RJ, Smith N, Wang L. 2008. *Ap. J.* 677:1060–68
Wang X, Wang L, Filippenko AV, Zhang T, Zhao X. 2013. *Science* 340:170–73
Warner B. 2003. *Cataclysmic Variable Stars*. Cambridge, UK: Cambridge Univ. Press
Washabaugh PC, Bregman JN. 2013. *Ap. J.* 762:1
Webbink RF. 1984. *Ap. J.* 277:355–60
Wheeler JC. 2012. *Ap. J.* 758:123
Wheeler JC, Pooley D. 2013. *Ap. J.* 762:75
Whelan J, Iben I Jr. 1973. *Ap. J.* 186:1007–14
Wiersma RPC, Schaye J, Theuns T. 2011. *MNRAS* 415:353–71
Williams BJ, Blair WP, Blondin JM, Borkowski KJ, Ghavamian P, et al. 2011. *Ap. J.* 741:96
Winkler PF, Long KS, Hamilton AJS, Fesen RA. 2005. *Ap. J.* 624:189–97
Wolf WM, Bildsten L, Brooks J, Paxton B. 2013. *Ap. J.* 777:136
Woods TE, Gilfanov M. 2013. *MNRAS* 432:1640–50
Woods TE, Gilfanov M. 2014. *MNRAS* 439:2351–63
Wood-Vasey WM, Sokoloski JL. 2006. *Ap. J. Lett.* 645:L53–56
Woosley SE, Kasen D. 2011. *Ap. J.* 734:38
Woudt PA, Steeghs D, Karovska M, Warner B, Groot PJ, et al. 2009. *Ap. J.* 706:738–46
Wu CC, Crenshaw DM, Fesen RA, Hamilton AJS, Sarazin CL. 1993. *Ap. J.* 416:247
Xavier HS, Gupta RR, Sako M, D'Andrea CB, Frieman JA, et al. 2013. *MNRAS* 434:1443–59
Yamaguchi H, Koyama K, Katsuda S, Nakajima H, Hughes JP, et al. 2008. *Publ. Astron. Soc. Jpn.* 60:141
Yaron O, Prialnik D, Shara MM, Kovetz A. 2005. *Ap. J.* 623:398–410
Yoon SC, Langer N. 2004. *Astron. Astrophys.* 419:623–44
Yoon SC, Langer N. 2005. *Astron. Astrophys.* 435:967–85
Yungelson LR. 2005. In *White Dwarfs: Cosmological and Galactic Probes*, ed. EM Sion, S Vennes, HL Shipman, 332:163–73. Dordrecht, Neth.: Springer
Yungelson LR. 2008. *Astron. Lett.* 34:620–34
Yungelson LR. 2013. See Di Stefano, Orio & Moe. 2013, pp. 244–47
Yungelson LR, Livio M. 2000. *Ap. J.* 528:108–17
Zelaya P, Quinn JR, Baade D, Clocchiatti A, Höflich P, et al. 2013. *Astron. J.* 145:27
Zheng W, Shivvers I, Filippenko AV, Itagaki K, Clubb KI, et al. 2014. *Ap. J. Lett.* 783:L24
Zheng W, Silverman JM, Filippenko AV, Kasen D, Nugent PE, et al. 2013. *Ap. J. Lett.* 778:L15
Zhu C, Chang P, van Kerkwijk MH, Wadsley J. 2013. *Ap. J.* 767:164
Zhu G, Blanton MR, Moustakas J. 2010. *Ap. J.* 722:491–519
Zorotovic M, Schreiber MR, Gänsicke BT. 2011. *Astron. Astrophys.* 536:A42

Tidal Dissipation in Stars and Giant Planets

Gordon I. Ogilvie

Department of Applied Mathematics and Theoretical Physics, University of Cambridge, Cambridge CB3 0WA, United Kingdom; email: gio10@cam.ac.uk

Annu. Rev. Astron. Astrophys. 2014. 52:171–210

The *Annual Review of Astronomy and Astrophysics* is online at astro.annualreviews.org

This article's doi:
10.1146/annurev-astro-081913-035941

Keywords

celestial mechanics, fluid dynamics, internal waves, binary stars, extrasolar planets, giant planets: satellites

Abstract

Astrophysical fluid bodies that orbit close to one another induce tidal distortions and flows that are subject to dissipative processes. The spin and orbital motions undergo a coupled evolution over astronomical timescales, which is relevant for many types of binary star, short-period extrasolar planetary systems, and the satellites of the giant planets in the Solar System. I review the principal mechanisms that have been discussed for tidal dissipation in stars and giant planets in both linear and nonlinear regimes. I also compare the expectations based on theoretical models with recent observational findings.

1. INTRODUCTION

Body 1: the body in which the tide is raised

Body 2: the body that raises the tide

The orbital motion of a gravitating system of extended fluid bodies, such as stars and giant planets, differs from that of a set of point masses. They are nonspherical as a result of both their rotation and their tidal deformation due to the nonuniform gravitational attraction of their companions. In a Newtonian system of two such bodies, the Keplerian orbital elements evolve in time. The largest effects are nondissipative in character and include orbital precession. Of greater interest and subtlety is the irreversible evolution of the size and shape of the orbit driven by dissipative processes.

The study of tides has a long and fascinating history (Cartwright 1999, Deparis et al. 2013). It involves an interesting combination of celestial mechanics, which is nontrivial but can be explored relatively easily, and fluid dynamics, which involves some deeper physical issues that are touched on in this article. Indeed, this review is devoted to stars and giant planets, which are wholly or predominantly fluid, rather than terrestrial bodies, which are predominantly solid, although deformable. In other words, our interest is in fluid dynamics rather than solid mechanics, in which the dissipative properties have a different nature. However, there is not a clean separation between these areas. Neutron stars have solid crusts, giant planets may contain solid cores, and terrestrial bodies may contain oceans and atmospheres with important tidal phenomena. In the case of Earth, a fluid layer that is only 0.023% of the planet's mass dominates the tidal dissipation.

Our focus is also on purely gravitational tides, which can be regarded as a subset of interactions between neighboring astrophysical bodies. Thermal and magnetic tides are also possible, in which one body affects the other in a way that is not spherically symmetric and may depend periodically on time owing to the spin and orbital motion. As for gravitational tides, both wave-like and nonwave-like disturbances may be generated.

A useful dimensionless parameter that can be taken as a simple estimate of the tidal deformation of one body by another is the tidal amplitude parameter $\epsilon = (M_2/M_1)(R_1/d)^3$, where M_1 and M_2 are the masses of the two bodies, R_1 is the (mean) radius of body 1 and d is the orbital separation (**Figure 1**). This is a measure of the ratio, at the surface of body 1, of the tidal gravity due to body 2, GM_2R_1/d^3, to the gravity of body 1 itself, GM_1/R_1^2; it is also an estimate of the ratio of the height of the tide compared with the radius of body 1.

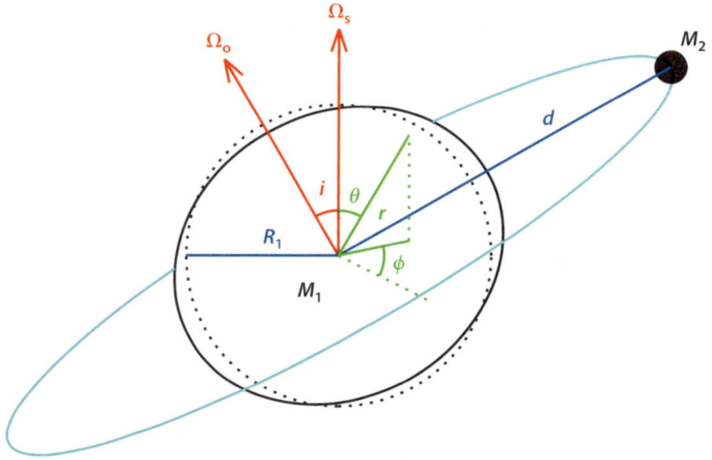

Figure 1

Geometry of tidal interaction. When considering the tide raised in body 1 by body 2, it is convenient to use spherical polar coordinates (r, θ, ϕ) centered on body 1 and aligned with its rotation axis. Ω_s and Ω_o are the spin and orbital angular velocity vectors, respectively. The figure represents the tidal bulge but not the rotational bulge.

Tidal interactions are therefore strongly dependent on the separation of the bodies in comparison with their sizes. The most familiar example (although beyond the scope of this review) is of course the Earth-Moon system, in which the Sun also plays a significant role. Tidal dissipation on Earth leads to a lengthening of both the day and the month as angular momentum is transferred from the planetary spin to the orbit. Tidal evolution, including the effects of tidal dissipation in giant planets, has been similarly influential for many of the regular satellites of the other planets in the Solar System (Peale 1999). Tidal interactions have many applications to binary stars, including pre-main-sequence, main-sequence, giant, and degenerate stars, whenever the ratio of orbital separation to stellar radius is sufficiently small. Much recent work on tides has been stimulated by the discovery since 1995 of numerous extrasolar planets that orbit very close to their host stars, and also by the prospect of detecting gravitational radiation from merging double-degenerate binary stars, in which tidal effects may alter the predicted wave signals (e.g., Kochanek 1992, Bildsten & Cutler 1992).

Tidal torque: the rate of transfer of (axial) angular momentum from the orbit to the spin of body 1

In this review we concentrate on situations in which the tidal amplitude parameter $\epsilon \ll 1$, although important internal nonlinearities may still be present, as discussed in Section 4. (There are of course more extreme situations in which the body is tidally disrupted, resulting in a significant loss of mass; this includes Roche-lobe overflow in binary stars with circular orbits and the tidal disruption of stars on eccentric orbits around the black hole at the center of a galaxy, or of planets scattered close to their host stars, e.g., Guillochon et al. 2011). For the giant planets of the Solar System, the largest values of ϵ are approximately 2×10^{-7} (Jupiter, due to Io), 3×10^{-8} (Saturn, due to Titan), 4×10^{-8} (Uranus, due to Ariel), and 8×10^{-8} (Neptune, due to Triton). Among exoplanetary systems confirmed so far, the largest values are approximately 2×10^{-4} for the tide in a star (WASP-18; Hellier et al. 2009) and 6×10^{-2} for the tide in a planet (WASP-19 b; Hebb et al. 2010). The tides in giant planets forced by small satellites and by nearby host stars may be in very different regimes, which motivates a study of linear and nonlinear tides.

The dominant tidal interaction is usually between the gravitational quadrupole moment of one body and the monopole moment of the other. The second body can then be thought of as a point mass, whereas the first is endowed with an ellipsoidal bulge (**Figure 1**). The monopole field of body 2 varies over the volume occupied by body 1, creating a tidal field that is aligned with the axis joining the centers of the bodies. This tidal field deforms body 1, generating a quadrupole moment. In the simplest case a spheroidal bulge arises, aligned with the axis, which gives rise to an attractive net force in that radial direction.

The process of deformation involves fluid dynamics, as discussed in detail later in this review, and is generally accompanied by dissipation. As a result, the quadrupole moment of body 1 is not instantaneously related to the tidal field and is not generally aligned with the radial direction. This effect is often modeled crudely by allowing for a certain time lag in the response. The dissipative part of the net force between the bodies has two aspects. One is that the radial force is retarded and therefore resists the oscillatory radial motion associated with any eccentricity of the orbit. The other is that an angular force arises that resists the relative angular motion associated with any inequality of the spin and orbital angular velocity vectors. This force also gives rise to a tidal torque that couples the spin and orbit, allowing an exchange of angular momentum. Both forces are resistive in nature and are accompanied by dissipation of energy.

The responses of spin and orbit to changes in angular momentum are different. Although spin angular momentum increases with angular velocity, orbital angular momentum decreases, which can lead to paradoxical behavior. The tidal torque seeks to equalize the spin and orbital angular velocity vectors through an exchange of angular momentum and the dissipation of energy. In many cases the spin moment of inertia is much smaller than the orbital moment of inertia, and the first effect of tidal dissipation is a tendency toward equalization of the spin and orbital angular velocity vectors, involving only a small change in the orbit. In the case of a circular orbit this means both synchronization of the spin with the orbit and alignment of the spin axis with the orbit normal, with the final angular velocity vector being determined by the total angular momentum. If the orbit is eccentric, however, then the orbital angular velocity is not constant and there is a tendency toward pseudosynchronization, i.e., spin at a rate close to, but slightly less than, the instantaneous orbital angular velocity at the pericenter, where the tidal interaction is strongest (Hut 1981).

Obliquity: the angle between the rotation axis of body 1 and the orbit normal; also known as the spin-orbit misalignment

Tidal equilibrium: a double synchronous state in which two bodies have a circular orbit with synchronized, aligned spins and no further tidal evolution occurs

Both the radial force and the tidal torque contribute to the evolution of eccentricity, which can either increase or decrease as a result. Similarly, the obliquity, which is the angle between the spin and orbital angular momentum vectors, can either increase or decrease. Circularization and spin-orbit alignment may be regarded as the normal outcomes of tidal dissipation; in order to increase e or i, the energy source associated with an asynchronous spin is needed. This possibility is discussed in more detail in Section 2.3 below.

In most binary stars the theoretical endpoint of tidal evolution, if other processes such as magnetic braking, gravitational radiation, and stellar evolution are ignored, is a tidal equilibrium or double synchronous state, in which both stars are in aligned, synchronous rotation with a circular orbit, as occurs with Pluto and Charon. There is then only a static tidal deformation and no dissipation. (However, internal processes such as convection and meridional circulation may induce a differential rotation in the stars, which would imply ongoing tidal dissipation.) The double synchronous state has not yet been verified observationally for a binary star. Two candidates can be found in Meibom et al. (2006, their table 1), although in each binary only one spin period and neither obliquity is measured. Albrecht et al. (2007) give the first observations of both stellar obliquities (consistent with zero) in a main-sequence binary, but the orbit is eccentric.

In systems of extreme mass ratio, however, such as the satellite systems of the giant planets and many short-period extrasolar planetary systems, this endpoint cannot be reached. Tidal equilibria may be characterized by their angular velocity Ω. The total angular momentum $L(\Omega)$ of a tidal equilibrium is the sum of the spin and orbital contributions, $L_s(\Omega)$ and $L_o(\Omega)$, which are increasing and decreasing functions, respectively, such that $L(\Omega)$ has a minimum, critical value L_c at $\Omega = \Omega_c$. If $L < L_c$, then no tidal equilibrium is accessible. If $L > L_c$, then two such states exist. The less compact one with $\Omega < \Omega_c$ has $L_o > 3L_s$ and is stable, being a minimum of the energy subject to the angular momentum being conserved (Counselman 1973, Hut 1980). The more compact one with $\Omega > \Omega_c$ has $L_o < 3L_s$ and is unstable; it may in any case be inaccessible if the orbit would be too small relative to the sizes of the bodies. The expressions for L_c and Ω_c are

$$L_c = 4I\Omega_c, \quad \Omega_c = (GM)^{1/2}\left(\frac{\mu}{3I}\right)^{3/4}, \tag{1}$$

where $M = M_1 + M_2$ is the total mass, $\mu = M_1M_2/M$ is the reduced mass, and $I = I_1 + I_2$ is the sum of the spin moments of inertia (assumed constant).

Many of the short-period extrasolar planetary systems for which tidal interactions are most important (having orbital periods less than about five days) have $L < L_c$ and therefore no tidal equilibrium; the orbit shrinks until the planet is destroyed. A few short-period systems with very massive planets are estimated to have $L > L_c$ (Levrard et al. 2009, Matsumura et al. 2010). In two of these, CoRoT-3 b and τ Boo b, the stellar spin appears to be approximately synchronized with the orbit (Husnoo et al. 2012), and these systems may already be in, or close to, a stable tidal equilibrium. In the case of τ Boo b (not mentioned by Levrard et al. or Matsumura et al.) it is likely that $L > L_c$ by only a narrow margin; the same is true of KELT-1 (Siverd et al. 2012) and CoRoT-15 (Bouchy et al. 2011), whose companions would usually be classified as brown dwarfs rather than planets. Some other short-period systems with $L > L_c$ could be evolving toward a stable equilibrium, although magnetic braking may eventually reduce L below L_c and destroy the equilibrium. Many longer-period systems with massive planets also have $L > L_c$ but are not expected to undergo significant tidal evolution.

All satellite systems of solar-system planets have $L > L_c$. For the giant planets, $L \gg L_c$ and $L_o \ll 3L_s$; if a synchronously spinning satellite were placed in a circular orbit near the corotation radius (synchronous orbit) of a giant planet, it would migrate away from that radius and the planetary spin would not be greatly affected.

Tidal potential: the part of the gravitational potential due to body 2 that deforms body 1

Tidal dissipation generates heat in astrophysical bodies, which in some cases may be important for their structure and evolution. Apart from the well-known applications to the volcanic activity of Jupiter's moon Io and Saturn's moon Enceladus, this effect has been investigated mainly for short-period extrasolar planets undergoing tidal circularization, in an attempt to explain the unexpectedly large radii of many transiting planets (e.g., Bodenheimer et al. 2001, Ibgui & Burrows 2009). In fact, the orbital energy that must be dissipated to circularize a short-period planet can easily exceed the planet's internal binding energy, which suggests that rapid circularization, if possible, either makes the planet very bright or threatens to destroy it. Gu et al. (2003) identified a tidal inflation instability that may lead to the disruption of gas giants.

2. TIDAL DYNAMICS

2.1. Tidal Potential

Consider two bodies orbiting about their mutual center of mass. Let their masses be M_1 and M_2 and their centers of mass $\boldsymbol{R}_1(t)$ and $\boldsymbol{R}_2(t)$. Suppose that body 2 is a point mass or can be treated as such for the purposes of determining the motion and tidal deformation of body 2. When its gravitational potential $-GM_2/|\boldsymbol{r} - \boldsymbol{R}_2|$ is expanded in a Taylor series about $\boldsymbol{r} = \boldsymbol{R}_1$, we obtain

$$-\frac{GM_2}{|\boldsymbol{d}|}\left[1 + \frac{\boldsymbol{d}\cdot\boldsymbol{x}}{|\boldsymbol{d}|^2} + \frac{3(\boldsymbol{d}\cdot\boldsymbol{x})^2 - |\boldsymbol{d}|^2|\boldsymbol{x}|^2}{2|\boldsymbol{d}|^4} + \cdots\right] = -\frac{GM_2}{|\boldsymbol{d}|}\sum_{l=0}^{\infty}\frac{|\boldsymbol{x}|^l}{|\boldsymbol{d}|^l}P_l\left(\frac{\boldsymbol{d}\cdot\boldsymbol{x}}{|\boldsymbol{d}||\boldsymbol{x}|}\right), \qquad (2)$$

where $\boldsymbol{d} = \boldsymbol{R}_2 - \boldsymbol{R}_1$ is the orbital separation, $\boldsymbol{x} = \boldsymbol{r} - \boldsymbol{R}_1$ is the position vector with respect to the center of body 1, and P_l is the Legendre polynomial of degree l. The first term in this series is a uniform potential and has no effect, whereas the second term gives rise to a uniform acceleration, which causes the basic Keplerian orbital motion of body 1. The remaining part of the expansion defines the tidal potential Ψ, which gives rise to a nonuniform acceleration that deforms body 1. It can also be represented using solid spherical harmonic functions of the second degree and higher, i.e., $r^l Y_l^m(\theta, \phi)$ with $l \geq 2$ and $|m| \leq l$, where (r, θ, ϕ) are spherical polar coordinates with their origin at the center of body 1.

If body 1 is rotating, then for the purposes of computing the tidal deformation it is helpful to choose the axis of the coordinate system to coincide with the rotation axis. Let the orbit have semimajor axis a, eccentricity e, and inclination i with respect to the equatorial plane of body 1. In celestial mechanics i is known as the obliquity of body 1 (with respect to the orbit of bodies 1 and 2). The periodic variation of the separation vector due to the basic Keplerian orbital motion can be expressed through Fourier series, leading to an expansion of the complete tidal potential in a nonrotating frame in the form (cf. Kaula 1961, Polfliet & Smeyers 1990)

$$\Psi = \mathrm{Re}\sum_{l=2}^{\infty}\sum_{m=0}^{l}\sum_{n=-\infty}^{\infty}\frac{GM_2}{a}A_{l,m,n}(e, i)\left(\frac{r}{a}\right)^l Y_l^m(\theta, \phi)\mathrm{e}^{-\mathrm{i}n\Omega_o t}. \qquad (3)$$

Here $\Omega_o = (GM/a^3)^{1/2}$ is the mean orbital angular velocity, which we refer to as the orbital frequency (and is known as the mean motion in celestial mechanics). The dimensionless complex coefficients $A_{l,m,n}$ depend in a complicated way on e and i. The integers l and m are the degree

Table 1 Quadrupolar components of the tidal potential, correct to first order in eccentricity and obliquity

l	m	n	$\lvert A \rvert$	Description
2	0	0	$\sqrt{\frac{\pi}{5}}$	Static tide
2	2	2	$\sqrt{\frac{6\pi}{5}}$	Asynchronous tide
2	0	1	$3e\sqrt{\frac{\pi}{5}}$	Eccentricity tides
2	2	1	$\frac{1}{2}e\sqrt{\frac{6\pi}{5}}$	
2	2	3	$\frac{7}{2}e\sqrt{\frac{6\pi}{5}}$	
2	1	0	$i\sqrt{\frac{6\pi}{5}}$	Obliquity tides
2	1	2	$i\sqrt{\frac{6\pi}{5}}$	

and order of the spherical harmonic, respectively, and m is also referred to as the azimuthal wave number. The integer n labels temporal harmonics of the orbital motion.

In most applications the bodies are sufficiently well separated that the quadrupolar components ($l = 2$) are strongly dominant. In the special case of a circular, coplanar orbit ($e = i = 0$), the only terms present have $n = m$, and $l - m$ must be even. In the case of a circular, inclined orbit ($e = 0$), n is restricted to the range $[-l, l]$, and $l - n$ must be even (e.g., Ogilvie 2013). For the complete representation of an eccentric orbit, all values of n are required. However, if terms smaller than $O(e^p)$ can be neglected, then the largest value of $\lvert n \rvert$ that needs to be considered is $l + p$. **Table 1** gives the amplitudes, but not the phases, of the quadrupolar components of the tidal potential, correct to first order in e and i.

Because $Y_l^m(\theta, \phi) \propto \mathrm{e}^{\mathrm{i}m\phi}$, the phase of each tidal component is $\arg A_{l,m,n} + m\phi - n\Omega_o t$. When $m \neq 0$, the phase rotates with angular velocity $n\Omega_o/m$. The angular frequency of each component measured in a nonrotating frame is $\omega = n\Omega_o$, which may be called the tidal frequency in the inertial frame. Of greater importance for the physical response of the fluid is the angular frequency measured in a frame that rotates with the spin angular velocity Ω_s (spin frequency) of body 1, $\hat{\omega} = n\Omega_o - m\Omega_s$, which may be called the tidal frequency in the fluid frame. When $m \neq 0$, the difference between ω and $\hat{\omega}$ is due to an angular Doppler shift. (In general, body 1 may rotate differentially, in which case $\hat{\omega}$ depends on position.)

Tidal component: a single component of the tidal potential when it is decomposed into spherical harmonic functions and Fourier-analyzed in time

Inertial frame: a nonrotating frame of reference

Fluid frame: a frame of reference rotating with the spin angular velocity of body 1

The tidal frequencies in the fluid frame are therefore integer linear combinations of the spin and orbital frequencies; for the seven components listed in **Table 1**, these are plotted in **Figure 2**. As is discussed in Section 3, the Coriolis force plays a dominant role in the wave-like part of the tidal response when $\lvert\hat{\omega}/\Omega_s\rvert < 2$ (and may still play an important role outside this interval), and it can be seen that all seven components typically have frequencies in this range, unless the body is far from the synchronous state.

As the eccentricity is increased from 0 toward 1, the time-dependence of the tidal forcing changes from a sinusoidal variation to one that is strongly peaked at the pericenter of the orbit. This is a result of the sensitivity of the tidal force to the orbital separation. The above expansion is still valid for large eccentricities, but a broad spectrum of frequencies is obtained (**Figure 3**). In a highly eccentric orbit, the tidal interaction has an impulsive character, consisting of a series of tidal encounters, each of which might be approximated as a parabolic orbit.

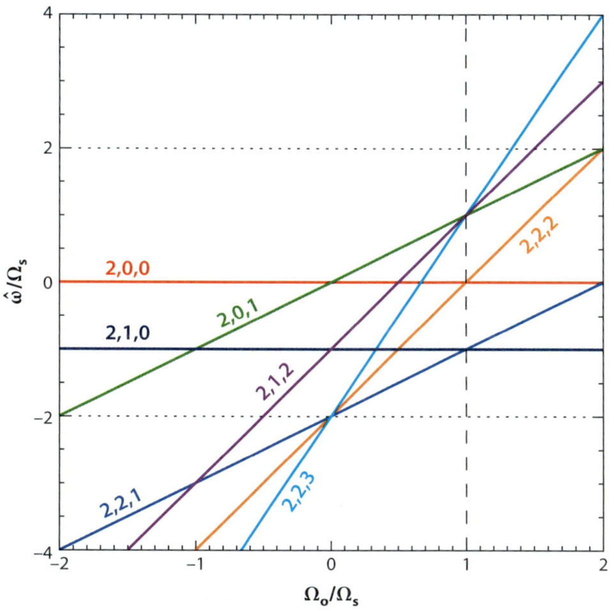

Figure 2

Tidal forcing frequencies for the seven lowest-order components listed in **Table 1**, labeled by l, m, n. The vertical axis is the ratio of the tidal frequency in the fluid frame to the spin frequency, whereas the horizontal axis is the ratio of the orbital frequency to the spin frequency. Each line has slope n and y-intercept $-m$. The dotted horizontal lines indicate the frequency range of inertial waves in a uniformly rotating body (see Section 3.5 below), whereas the dashed vertical line indicates the synchronous state.

2.2. Tidal Torque, Power, and Dissipation

If the tidal amplitude parameter ϵ is sufficiently small, the tidal response of a body can be determined using linear theory, as discussed in Section 3 below. In this approach, the tidal disturbance is treated as a small perturbation of a basic state, which is usually assumed to be steady and axisymmetric. To each component of the tidal potential there is an independent and linearly related response, which can be quantified using a response function. If a tidal potential component $\mathrm{Re}[\mathcal{A}(r/R)^l Y_l^m(\theta,\phi)\mathrm{e}^{-\mathrm{i}\omega t}]$ is applied to a body of nominal (e.g., mean or equatorial) radius R, and the resulting deformation of the body generates an external gravitational potential perturbation $\mathrm{Re}[\mathcal{B}(R/r)^{l+1} Y_l^m(\theta,\phi)\mathrm{e}^{-\mathrm{i}\omega t}]$ (possibly in addition to other components that are orthogonal to the applied one), then the complex dimensionless ratio $k_l^m(\omega) = \mathcal{B}/\mathcal{A}$ defines the potential Love number (named after A.E.H. Love). This is the most useful response function, because it is only through gravity that the tidal communication between the two bodies occurs. Of particular interest is the imaginary part $\mathrm{Im}[k_l^m(\omega)]$, which quantifies the part of the response that is out of phase with the tidal forcing and is associated with transfers of energy and angular momentum.

Potential Love number: a complex-valued, dimensionless, frequency-dependent response function relating the self-gravitational perturbation of a body to the applied tidal potential

Tidal power: the rate of transfer of energy from the orbit to body 1

Let the tidal power P and the tidal torque T be defined as the rates of transfer of energy and of the axial component of angular momentum from the orbit to the body (measured in the inertial frame, and averaged in time in the case of $m = 0$). It can be shown (e.g., Ogilvie 2013) that $P = \omega\mathcal{T}$ and $T = m\mathcal{T}$, where

$$\mathcal{T} = \frac{(2l+1)R|\mathcal{A}|^2}{8\pi G}\mathrm{Im}[k_l^m(\omega)]. \tag{4}$$

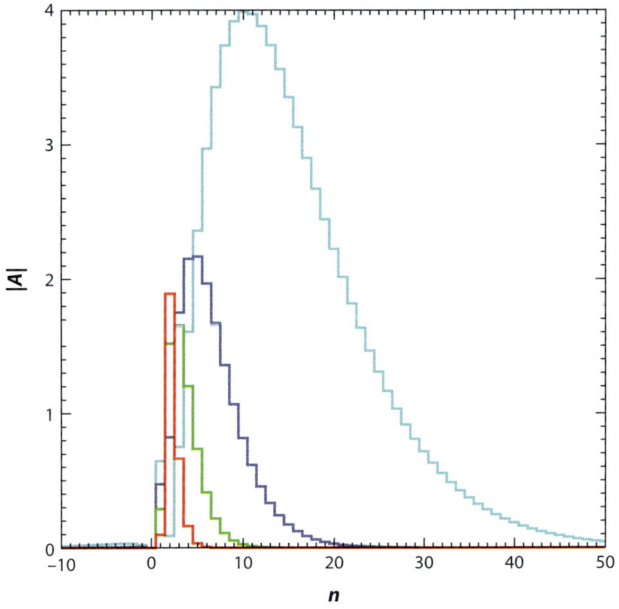

Figure 3

Amplitude $|A_{2,2,n}|$ of the $l = m = 2$ component of the tidal potential for obliquity $i = 0$ and eccentricities $e = 0.1, 0.3, 0.5$, and 0.7 (*red*, *green*, *blue*, and *cyan lines*, respectively).

In a frame that rotates with angular velocity Ω, the tidal torque is again T but the tidal power is $P - T\Omega$. When $m \neq 0$, the fact that $P = (\omega/m)T$ means that the tidal power vanishes in a rotating frame in which the tidal potential is stationary.

If the body rotates uniformly with angular velocity Ω_s, then the (frame-independent) tidal dissipation rate is $D = P - T\Omega_s = \hat{\omega}\mathcal{T}$. Although the torque, power, and dissipation rate are all quantities of second order in ϵ, they can be calculated within a linear analysis and are linearly proportional to the imaginary part of the response function. The difference between the dissipation rate and the power is the rate of change of the spin energy of the body, which vanishes in the fluid frame. Because D is a positive-definite quantity, we have the important result that $\mathrm{Im}[k_l^m(\omega)]$ has the same sign as $\hat{\omega}$. [A possible exception is that convection could in principle give energy to the tidal disturbance rather than dissipating it (Ogilvie & Lesur 2012).] $\mathrm{Im}[k_l^m(\omega)]$ should also vanish when $\hat{\omega} = 0$, because then the deformation is static in the fluid frame and should be in phase with the tidal forcing.

If the body rotates nonuniformly, then there is no simple global relation between D, P, and T. The dissipation rate cannot then be deduced from $\mathrm{Im}[k_l^m(\omega)]$ and ceases to be of such direct importance, except of course for tidal heating.

Several other parameterizations of the tidal response are in common use. Typically $\mathrm{Re}[k_l^m(\omega)]$ is a quantity of order unity, only weakly dependent on m and ω, and can be well approximated by its hydrostatic value. The evaluation of the hydrostatic Love number k_l (which is real and does not depend on m) for a spherical fluid body is a classical problem involving the equations of Clairaut and Radau (e.g., Kopal 1978); for a homogeneous body, $k_l = k_l^{\mathrm{hom}} = 3/[2(l-1)]$. [The apsidal motion constants long used in the theory of binary stars (e.g., Kopal 1978) are denoted by the same symbol k_l and are equivalent except that they are smaller by a factor of 2.] It is then common to write $|\mathrm{Im}[k_l^m(\omega)]|$ as k_l/Q, where Q is the tidal quality factor. It is also common to write k_l/Q as k_l^{hom}/Q', where Q' is a modified tidal quality factor; this has the advantage of combining Q with k_l, which may not be known accurately as it depends on the interior structure, whereas $k_l^{\rm hom}$ has a simple analytical expression. [For giant planets, it is estimated that $0.1 < k_2 < 0.4$ (Gavrilov & Zharkov 1977, Kramm et al. 2012), although smaller values are possible for very centrally condensed models. For the Sun, $k_2 \approx 0.0351$, making it even more important to distinguish clearly between Q and Q'.] The phase lag in the tidal response is $1/Q$ and the time lag is $\tau = 1/(Q|\hat{\omega}|)$. To summarize, in the case of quadrupolar tides ($l = 2$) the various parameterizations of tidal dissipation are related by

Tidal quality factor (modified); phase lag; time lag: alternative parameterizations of the tidal response functions, not to be taken literally

$$\mathrm{Im}[k_2^m(\omega)] = \sigma\frac{3}{2Q'} = \sigma\frac{k_2}{Q} = k_2\tau\hat{\omega}, \tag{5}$$

where $\sigma = \mathrm{sgn}\,\hat{\omega} = \pm 1$.

The expressions quality factor, phase lag, and time lag should not be taken literally. The above relations are based on a simplified conceptual model in which the tidal response resembles the hydrostatic one, but is slightly retarded in phase and unaffected in amplitude. However, if the tidal forcing resonates with a weakly damped normal mode of oscillation, especially if it is the fundamental mode as discussed in Section 3.1, it is possible in principle for $|\mathrm{Im}[k_l^m(\omega)]|$ to attain values much larger than unity because the amplitude of the tidal response greatly exceeds the hydrostatic one. No amount of phase lag could describe such a response. However, we can continue to use the above formal relations if we allow Q and Q' to be less than unity in such a situation (which is, however, unlikely to arise in practice).

Potential Love numbers can still be used to describe the tidal deformation of a body in a nonlinear regime, but then there can be interference between different components of the tide, and the Love numbers are no longer linear response functions of the body but depend on the amplitudes and relative phases of all applied tidal components.

It is particularly important to note that the tidal quality factor Q is not a fundamental material property of a body but is a dimensionless parameterization of a response function, which is partly why we favor the notation $\mathrm{Im}[k_l^m(\omega)]$. As well as depending on l, m, and ω, $\mathrm{Im}(k)$ depends on the internal structure and angular velocity of the body. In his first published paper, Goldreich (1963, pp. 258, 264) issues words of caution about this parameterization: "Q will in general vary with the frequency and amplitude of the tide and the size of the sphere in addition to its composition . . . In our discussions we shall use the language of linear tidal theory, but we must keep in mind that our numbers are really only parametric fits to a nonlinear problem."

2.3. Rate and Direction of Tidal Evolution

To lowest order in e and i, the rates of change of the orbital semimajor axis a, the spin angular velocity Ω_s of body 1, the orbital eccentricity e, and the obliquity (spin-orbit misalignment) i of body 1 due to tidal dissipation in body 1 are given by

$$\frac{1}{a}\frac{\mathrm{d}a}{\mathrm{d}t} = -3\kappa_{2,2,2}\frac{M_2}{M_1}\left(\frac{R_1}{a}\right)^5\Omega_\mathrm{o}, \tag{6}$$

$$\frac{1}{\Omega_\mathrm{s}}\frac{\mathrm{d}\Omega_\mathrm{s}}{\mathrm{d}t} = \frac{3}{2}\kappa_{2,2,2}\frac{L_\mathrm{o}}{L_\mathrm{s}}\frac{M_2}{M_1}\left(\frac{R_1}{a}\right)^5\Omega_\mathrm{o}, \tag{7}$$

$$\frac{1}{e}\frac{\mathrm{d}e}{\mathrm{d}t} = \frac{3}{16}(4\kappa_{2,2,2} - 6\kappa_{2,0,1} + \kappa_{2,2,1} - 49\kappa_{2,2,3})\frac{M_2}{M_1}\left(\frac{R_1}{a}\right)^5\Omega_\mathrm{o}, \tag{8}$$

$$\frac{1}{i}\frac{\mathrm{d}i}{\mathrm{d}t} = \frac{3}{4}\left[\kappa_{2,2,2}\left(1 - \frac{L_\mathrm{o}}{L_\mathrm{s}}\right) + (\kappa_{2,1,0} - \kappa_{2,1,2})\left(1 + \frac{L_\mathrm{o}}{L_\mathrm{s}}\right)\right]\frac{M_2}{M_1}\left(\frac{R_1}{a}\right)^5 \Omega_\mathrm{o}, \tag{9}$$

where we introduce the notation

$$\kappa_{l,m,n} = \mathrm{Im}[k_l^m(n\Omega_\mathrm{o})] \tag{10}$$

for the imaginary part of the Love number corresponding to the tidal component labeled by l, m, and n, and

$$\frac{L_\mathrm{o}}{L_\mathrm{s}} = \frac{\mu[GMa(1-e^2)]^{1/2}}{I_1\Omega_\mathrm{s}} = \frac{GM_1M_2(1-e^2)^{1/2}}{I_1\Omega_\mathrm{s}\Omega_\mathrm{o}a} \tag{11}$$

is the ratio of the orbital angular momentum to the spin angular momentum of body 1 (if it has a fixed moment of inertia I_1). Note that these formulae are valid only for sufficiently small values of e and i. Our notation assumes that Ω_o and L_o are positive by definition, as are a, e, and i, whereas $L_\mathrm{s} = I_1\Omega_\mathrm{s}$ can be either positive or negative. The case $\Omega_\mathrm{s} < 0$ allows us to consider situations in which spin and orbit are (nearly) counter aligned. Results of this type can be found in the classic work by Darwin (1880), although expressed in a different notation; they are derived by considering the effects of the tidal torque and power on the spin and orbit. The effects of tidal dissipation in body 2 on the orbital semimajor axis and eccentricity, and on its spin angular velocity and obliquity, are simply obtained by interchanging the roles of the two bodies in this calculation.

Without considering yet the details of the tidal responses of astrophysical bodies, but assuming that $|\kappa| \lesssim 1$ (usually $\ll 1$) for each component, a number of observations can be made. First, tidal evolution occurs on a timescale that is much longer than the orbital timescale, if $\epsilon \ll 1$ as we have assumed. Second, tidal dissipation in the two bodies may be of comparable importance; the ratio of the factor $(M_2/M_1)(R_1/a)^5$ to the corresponding factor $(M_1/M_2)(R_2/a)^5$ when the roles are reversed can be expressed in terms of the mean densities of the bodies as $(\bar{\rho}_2/\bar{\rho}_1)^2(R_2/R_1)$, which suggests that dissipation in the smaller body can dominate the evolution of e (Goldreich 1963) and also of a before body 2 becomes synchronized. (Among exoplanetary systems, this ratio has values both less than and greater than unity.) Third, in systems for which $L_\mathrm{s} \ll L_\mathrm{o}$, spin evolution proceeds much more rapidly than orbital evolution.

If body 1 rotates uniformly with angular velocity Ω_s and the tidal dissipation rate is positive, then, as noted above, $\kappa_{l,m,n}$ must have the same sign as the sign of the tidal frequency in the fluid frame, $\hat{\omega} = n\Omega_\mathrm{o} - m\Omega_\mathrm{s}$, and must vanish when $\hat{\omega} = 0$. Note that $\kappa_{2,2,2}$ vanishes in the synchronous case $\Omega_\mathrm{s} = \Omega_\mathrm{o}$. Even in the lowest-order analysis presented here, numerous possibilities arise for the signs of the above rates of change. The signs of the contributions of the lowest-order tidal components to the rates of change of a, Ω_s, e, and i are summarized in **Figure 4**. In most cases there are competing contributions to de/dt and di/dt, and the net result is not obvious (Jeffreys 1961, Goldreich 1963). The fourth contribution to de/dt comes with the largest weighting factor, but this might be offset by a weaker tidal response at the corresponding frequency.

In the unlikely event that all components have the same time lag τ, which is the case investigated in detail by Darwin (1880), Alexander (1973), Hut (1981), and many others, we have $\kappa_{l,m,n} = k_l\tau(n\Omega_\mathrm{o} - m\Omega_\mathrm{s})$ and obtain the simpler results,

$$\frac{1}{a}\frac{\mathrm{d}a}{\mathrm{d}t} = -6k_2\tau(\Omega_\mathrm{o} - \Omega_\mathrm{s})\frac{M_2}{M_1}\left(\frac{R_1}{a}\right)^5 \Omega_\mathrm{o}, \tag{12}$$

$$\frac{1}{\Omega_\mathrm{s}}\frac{\mathrm{d}\Omega_\mathrm{s}}{\mathrm{d}t} = 3k_2\tau(\Omega_\mathrm{o} - \Omega_\mathrm{s})\frac{L_\mathrm{o}}{L_\mathrm{s}}\frac{M_2}{M_1}\left(\frac{R_1}{a}\right)^5 \Omega_\mathrm{o}, \tag{13}$$

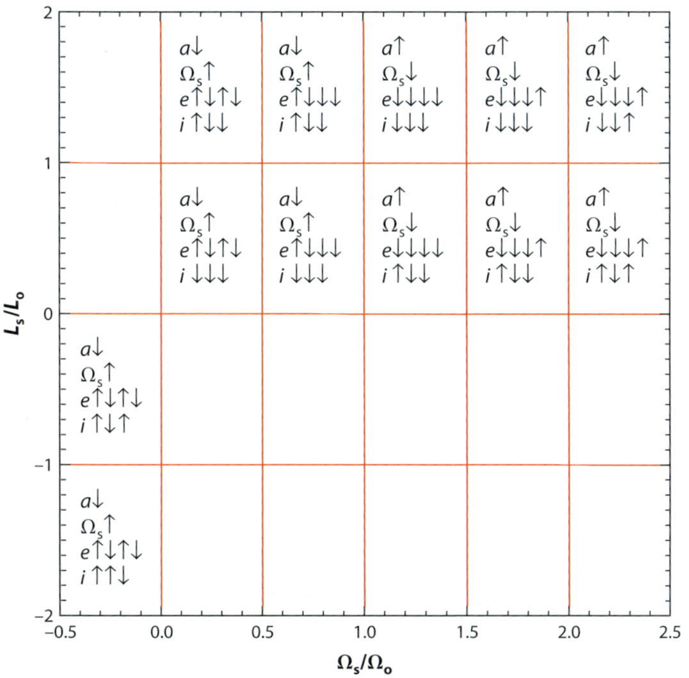

Figure 4

Signs of the contributions of the lowest-order tidal components to the rates of change of a, Ω_s, e, and i. The horizontal axis is the ratio of the spin frequency to the orbital frequency, whereas the vertical axis is the ratio of the spin angular momentum to the orbital angular momentum. Only the first and third quadrants are physically accessible. Upward and downward arrows indicate positive and negative rates of change, respectively. In the cases of e and i, the contributions are shown in the same order as that given in Equations 8 and 9. No further sign changes occur beyond the region plotted here.

$$\frac{1}{e}\frac{\mathrm{d}e}{\mathrm{d}t} = -\frac{3}{2}k_2\tau(18\Omega_o - 11\Omega_s)\frac{M_2}{M_1}\left(\frac{R_1}{a}\right)^5\Omega_o, \tag{14}$$

$$\frac{1}{i}\frac{\mathrm{d}i}{\mathrm{d}t} = -\frac{3}{2}k_2\tau\left[\Omega_s + (2\Omega_o - \Omega_s)\frac{L_o}{L_s}\right]\frac{M_2}{M_1}\left(\frac{R_1}{a}\right)^5\Omega_o, \tag{15}$$

which would imply that the eccentricity is excited for $\Omega_s/\Omega_o > 18/11$ (Darwin 1880), whereas the obliquity is excited for $\Omega_s/\Omega_o > 2L_o/(L_o - L_s)$, if $0 < L_s < L_o$ (Hut 1981), or always, if $L_s < 0$ (i.e., $\Omega_s < 0$). The extension of these equations to arbitrary e and i under the assumption of a common time lag can be found in Leconte et al. (2010) and Matsumura et al. (2010) and was first derived by Alexander (1973).

3. LINEAR TIDES

3.1. Linear Response of a Star or Giant Planet to Harmonic Forcing

Let us consider in more detail the linear response of an astrophysical fluid body to periodic forcing by a single tidal component. A simple explicit example is provided by the highly idealized model of a homogeneous, incompressible, nonrotating, spherical fluid body with a viscosity proportional to the pressure (cf. Sridhar & Tremaine 1992). Let the body have mass M, radius R, and dynamical frequency $\omega_{\mathrm{d}} = (GM/R^3)^{1/2}$, and let the dynamic viscosity be $\alpha p/\omega_{\mathrm{d}}$, where p is the pressure and α is a dimensionless viscosity coefficient. Then the linearized equations are solved by an irrotational displacement proportional to (but out of phase with) the tidal acceleration $-\nabla\Psi$, and the potential Love number can be shown to be

f mode: a fundamental oscillation mode relying on the surface gravity of a body

$$k_l^m(\omega) = \frac{3}{2(l-1)}\left[1 - \frac{\omega^2}{\omega_l^2} - \left(\frac{2l+1}{l}\right)\mathrm{i}\alpha\left(\frac{\omega}{\omega_d}\right)\right]^{-1}, \tag{16}$$

which is independent of m because of the spherical symmetry of the model. Here $\omega_l = [2l(l-1)/(2l+1)]^{1/2}\omega_{\mathrm{d}}$ is the natural frequency of the f mode of degree $l > 1$, which is the only oscillation mode that can be excited in this simple model. [This is the fundamental mode or surface gravity mode, first derived by Kelvin (Thomson 1863).] Note that the hydrostatic response $k_l = k_l^{\mathrm{hom}} = 3/[2(l-1)]$ is obtained in the low-frequency limit. The dynamics is exactly equivalent to that of a forced damped harmonic oscillator with natural frequency ω_l and damping coefficient $2(l-1)\alpha\omega_{\mathrm{d}}$. Now

$$\mathrm{Im}[k_l^m(\omega)] = \frac{3(2l+1)}{2l(l-1)}\alpha\left(\frac{\omega}{\omega_{\mathrm{d}}}\right)\left[\left(1 - \frac{\omega^2}{\omega_l^2}\right)^2 + \left(\frac{2l+1}{l}\right)^2\alpha^2\left(\frac{\omega}{\omega_{\mathrm{d}}}\right)^2\right]^{-1}. \tag{17}$$

For a weakly damped oscillator ($\alpha \ll 1$) this response function has a strong peak close to the natural frequency (and attains $|\mathrm{Im}[k_l^m(\omega)]| \gg 1$, as mentioned in Section 2.2). Well below this frequency, the factor in square brackets is close to unity, corresponding to a frequency-dependent tidal quality factor given by $Q^{-1} \approx [(2l+1)/l]\alpha(\omega/\omega_{\mathrm{d}})$ or a frequency-independent tidal time lag $\tau \approx [(2l+1)/l](\alpha/\omega_{\mathrm{d}})$. (For this model, Q' and Q are equivalent because the body is homogeneous, and $\hat{\omega}$ and ω are equivalent because the body is nonrotating.) Note that even a damped simple harmonic oscillator has a frequency-dependent quality factor for forced oscillations. In the case of $l = 2$, the low-frequency response of the body is simply a hydrostatic spheroidal bulge with a small lag due to viscosity. If instead the body has a uniform kinematic viscosity ν, as in the model of Darwin (1880), then for $\omega \ll \omega_{\mathrm{d}}$ we have a similar result with α replaced by $(2l+1)(\nu/R^2\omega_{\mathrm{d}})$.

More realistic models of stars and planets support larger families of oscillation modes, allowing more opportunities for resonances with tidal forcing. (See also the sidebar, Tides in Black Holes.) Oscillation modes of the Sun and other stars have been studied in great detail for the purposes of helioseismology and asteroseismology (e.g., Aerts et al. 2010). Generally, this work either neglects rotation or treats it as a perturbation: The stellar or planetary model is spherically symmetric. Following Cowling (1941), the modes are classified as f modes (fundamental; i.e., surface gravity

TIDES IN BLACK HOLES

Tides can also be raised on black holes. Since the original work by Hartle (1973), a number of calculations of tidal torques on black holes have been carried out, making a variety of approximations that ensure that the tidal interaction is weak and of low frequency (e.g., Comeau & Poisson 2009). These are all consistent with the viewpoint that the horizon of the black hole behaves equivalently to a massive viscous shell with a viscous damping timescale comparable with its light-crossing time, as in the membrane paradigm (Thorne et al. 1986). The tidal torques that have been calculated then resemble equilibrium tides on such a shell, with an appropriate time lag (Poisson 2009). In reality, of course, energy and angular momentum are exchanged through gravitational waves crossing the event horizon. It may be expected that a richer response occurs when the tidal frequency is comparable with that of the quasi-normal modes of the black hole.

waves), p modes (pressure; i.e., acoustic waves), and g modes (gravity; i.e., internal gravity waves). The natural frequencies of f modes and p modes ($\gtrsim 10^{-4}$ Hz for the Sun) are generally too high ($\gtrsim \omega_\mathrm{d}$) to be excited directly by tidal forcing, because the spin and orbital frequencies are usually small compared to the dynamical frequency of the body. The g modes, however, which propagate only in stably stratified (radiative) regions, form a dense spectrum at low frequencies and are good candidates for resonant excitation by tidal forcing. This is the basis of Zahn's theory of the dynamical tide (see Section 3.4 below).

As discussed in Section 2, tidal frequencies are often comparable to spin frequencies. This low-frequency part of the spectrum has received little attention in asteroseismology. The Coriolis force cannot be regarded as a small effect, and the wave equations are not separable in spherical harmonics. New types of oscillation mode (inertial waves) appear, which owe their existence to the Coriolis force. More interestingly, these low-frequency waves may not form classical normal modes and the tidal response may not be describable as a superposition of normal-mode resonances (see Section 3.5 below).

Internal waves supported by buoyancy and Coriolis forces play an important role in Earth's ocean and atmosphere, and have been widely studied through theory and experiment, but are less well known in astrophysics. They have properties quite different from acoustic or electromagnetic waves, being highly anisotropic and dispersive. In the fluid frame, the wave frequency depends only on the direction, and not on the magnitude, of the wave vector. Special directions are defined by gravity or rotation. Energy propagates at the group velocity, which is perpendicular to the wave vector and proportional to the wavelength. Internal waves are highly dispersive and do not steepen like acoustic waves, but when they exceed a critical amplitude they break, undergoing a local instability that causes their energy to be transferred to smaller scales and dissipated. In a closed domain, internal waves generally do not form normal modes with regular eigenfunctions, although the simplest cases are exceptions to this rule. The frequencies of internal waves are bounded and form a dense or continuous spectrum. The question then arises of how a rotating or stably stratified fluid responds to a tidal force with a frequency that lies within this spectrum.

p mode: an acoustic mode relying on compressibility

g mode: a global internal gravity mode relying on stable stratification

Internal gravity wave: an internal wave relying on buoyancy forces due to stable stratification

Dynamical tide: a tidal response that satisfies the time-dependent equations of fluid dynamics, and usually includes internal waves

Inertial wave: an internal wave relying on the Coriolis force due to rotation

Internal wave: a low-frequency, quasi-incompressible wave restored by buoyancy and/or Coriolis forces

Equilibrium tide: a hydrostatic approximation to the perturbations of density and pressure, accompanied by a displacement and velocity that may not be uniquely defined

3.2. Equilibrium and Dynamical Tides

The equilibrium tide is an approximation to the tidal response of a fluid body. It consists of a large-scale deformation of the body, which in the case of a quadrupolar tide ($l = 2$) is a spheroidal bulge. The density and pressure are distorted in order to achieve hydrostatic equilibrium in the instantaneous potential, which includes the distorted self-gravitational potential of the body in addition to the applied tidal potential. This hydrostatic balance is usually a good first approximation if the tidal frequency is small compared to the frequency of the f mode. In the model considered in Section 3.1, the equilibrium tide corresponds to the low-frequency limit $\omega \to 0$ of the tidal response.

A displacement of the fluid is required in order to bring about the tidal distortion. However, this is not uniquely defined in general and is a potential source of confusion or error, as discussed below. In any case, if the tidal frequency in the fluid frame is nonzero, the equilibrium tide does not satisfy the equation of motion exactly because the acceleration of the fluid is neglected in computing it, as is the effect of any damping forces. Corrections to the equilibrium tide generally include internal waves as well as nonwave-like components. Some or all of these corrections, or the complete tide calculated without a hydrostatic approximation, have been called the dynamical tide.

An unperturbed body in hydrostatic equilibrium satisfies

$$0 = -\frac{1}{\rho}\nabla p - \nabla\Phi, \tag{18}$$

where Φ is the self-gravitational potential of the body, but may also include the centrifugal potential corresponding to a uniform rotation (in which case the body is oblate). The curl of this equation implies that $\nabla\rho$ must be parallel to ∇p, which is also parallel to $\nabla\Phi$. Therefore ρ and p are functions of Φ only; we can also consider ρ to be a function of p only.

Let the body be perturbed by a tidal potential Ψ (satisfying $\nabla^2\Psi = 0$ inside the body) such that it maintains hydrostatic equilibrium. The linearized hydrostatic equation

$$0 = \frac{\rho'}{\rho^2}\nabla p - \frac{1}{\rho}\nabla p' - \nabla\Phi' - \nabla\Psi \tag{19}$$

is equivalent to

$$0 = -\nabla W + \left(\rho' - \frac{\mathrm{d}\rho}{\mathrm{d}p}p'\right)\frac{1}{\rho^2}\nabla p, \tag{20}$$

where $W = p'/\rho + \Phi' + \Psi$, and primes denote Eulerian perturbations. The curl of this equation implies that $\nabla(\rho' - \frac{\mathrm{d}\rho}{\mathrm{d}p}p')$ must be parallel to ∇p, as must ∇W. Therefore $\rho' - \frac{\mathrm{d}\rho}{\mathrm{d}p}p'$ must be a function of Φ only, and so must W. For an oscillatory tide (or tidal component) in which all perturbations have zero mean, they must in fact vanish. We deduce that

$$p' = -\rho(\Phi' + \Psi), \quad \rho' = -\frac{\mathrm{d}\rho}{\mathrm{d}p}\rho(\Phi' + \Psi). \tag{21}$$

The linearized Poisson equation then implies

$$\nabla^2\Phi' = -4\pi G\frac{\mathrm{d}\rho}{\mathrm{d}p}\rho(\Phi' + \Psi) \tag{22}$$

inside the body, whereas $\nabla^2\Phi' = 0$ outside. Given Ψ, this inhomogeneous Helmholtz-type equation (which can be related to the Clairaut and Radau equations if the unperturbed body is spherical) can be solved, subject to suitable conditions at the surface of the body, to determine Φ' and hence p' and ρ'. This is the equilibrium tide.

More problematic, though, is the tidal displacement $\boldsymbol{\xi}$ required to bring about these perturbations. Assuming adiabatic perturbations, we have

$$p' = -\gamma p\nabla\cdot\boldsymbol{\xi} - \boldsymbol{\xi}\cdot\nabla p, \quad \rho' = -\rho\nabla\cdot\boldsymbol{\xi} - \boldsymbol{\xi}\cdot\nabla\rho, \tag{23}$$

and so

$$0 = \rho' - \frac{\mathrm{d}\rho}{\mathrm{d}p}p' = -\left(\rho - \gamma p\frac{\mathrm{d}\rho}{\mathrm{d}p}\right)\nabla\cdot\boldsymbol{\xi}, \tag{24}$$

where $\gamma = (\partial\ln p/\partial\ln\rho)_s$ is the adiabatic exponent. If the body is stably stratified, having a nonzero (Brunt-Väisälä) buoyancy frequency N, then the quantity in brackets is nonzero, and we deduce that $\nabla\cdot\boldsymbol{\xi} = 0$: The displacement is incompressible. It then follows that

$$\boldsymbol{\xi}\cdot\boldsymbol{g} = \Phi' + \Psi, \tag{25}$$

where $\boldsymbol{g} = -\nabla\Phi = (\nabla p)/\rho$ is the (effective) gravitational acceleration. This relation determines the vertical displacement, and the horizontal displacement follows (nonlocally) from the condition $\nabla\cdot\boldsymbol{\xi} = 0$.

This solution is the standard equilibrium tidal displacement (Zahn 1966a, Goldreich & Nicholson 1989). It is the simplest form of displacement that achieves hydrostatic equilibrium in the distorted potential; under this volume-preserving motion, fluid elements retain their density and pressure and simply move them to a different location. However, if the body (or part of it) is adiabatically stratified (as in a convective zone, to a good approximation), then it is possible to add an arbitrary internal displacement satisfying the anelastic constraint $\nabla\cdot(\rho\boldsymbol{\xi}) = 0$ without changing ρ' or p'. Moreover, as discussed below, such an additional displacement is required in order to satisfy the equation of motion in the low-frequency limit. The total displacement then satisfies neither $\nabla \cdot \boldsymbol{\xi} = 0$ nor $\nabla \cdot (\rho\boldsymbol{\xi}) = 0$, except in the case of a homogeneous body.

In an adiabatically stratified region, the linearized inviscid equation of motion for an arbitrary tidal frequency $\hat{\omega}$ in the fluid frame is exactly

$$-\mathrm{i}\hat{\omega}\boldsymbol{u} + 2\boldsymbol{\Omega}_\mathrm{s} \times \boldsymbol{u} = -\nabla W, \tag{26}$$

where $\boldsymbol{u} = -\mathrm{i}\hat{\omega}\boldsymbol{\xi}$ is the velocity perturbation. In the case of a nonrotating body, or in the limit $\hat{\omega}^2 \gg 4\Omega_\mathrm{s}^2$, this equation implies that $\nabla \times \boldsymbol{u} = 0$ and $\nabla \times \boldsymbol{\xi} = 0$: The tidal flow and displacement are irrotational (Goodman & Dickson 1998, Terquem et al. 1998). If we write $\boldsymbol{\xi} = \nabla U$, then the expression for ρ' implies that

$$\nabla \cdot (\rho \nabla U) = \frac{\mathrm{d}\rho}{\mathrm{d}p}\rho(\Phi' + \Psi). \tag{27}$$

This Poisson-type equation can be solved for U, subject to suitable boundary conditions, to determine $\boldsymbol{\xi}$.

However, in the limit $\hat{\omega}^2 \ll 4\Omega_\mathrm{s}^2$, Equation 26 implies that the tidal flow is in geostrophic balance and has a different form again. If $\hat{\omega}^2$ is comparable to $4\Omega_\mathrm{s}^2$, then the solution generally includes inertial waves (see Section 3.5 below).

To summarize, although the Eulerian perturbations of density and pressure are well defined in the hydrostatic limit $\hat{\omega}^2 \ll \omega_\mathrm{d}^2$, the form of the tidal displacement and velocity in the low-frequency limit depends significantly on the ordering of $\hat{\omega}^2$, N^2, and $4\Omega_\mathrm{s}^2$. (At sufficiently low tidal frequencies the assumption of adiabatic perturbations may also break down.) The standard equilibrium tidal displacement is obtained only in the limit $\hat{\omega}^2 \ll N^2$ and applies only in radiative zones.

3.3. Equilibrium Tides

As described above, a time-dependent equilibrium tide is associated with a large-scale velocity field. Any process that resists this flow leads to dissipation. Because the ordinary molecular viscosity is much too small to be of interest, the main candidate for providing linear damping of this flow is turbulent convection; nonlinear possibilities are discussed in Section 4.1 below. In the simplest approach, the effect of convection on the equilibrium tide is modeled as an effective viscosity, which can be estimated using an extension of the same mixing-length theory that is used to describe the convective heat flux in the theory of stellar structure.

It has long been recognized that the effective viscosity should be reduced by some factor when the tidal period is short compared with the typical convective timescale, as is often the case. (In the Sun, this timescale ranges from a month or longer near the base of the convective zone to a few minutes near the photosphere. In giant planets the convective timescale is generally longer because of the much smaller heat flux, although the applicability of mixing-length theory is uncertain because of the rapid rotation.) Different reduction factors were proposed by Zahn (1966b) and by Goldreich & Nicholson (1977) and Goldreich & Keeley (1977). Zahn's reduction factor is weaker, with $\nu \propto |\hat{\omega}|^{-1}$ and so $\mathrm{Im}(k) \propto Q'^{-1} \propto |\hat{\omega}|^0$ for large $|\hat{\omega}|$. Goldreich's reduction factor is stronger, with $\nu \propto |\hat{\omega}|^{-2}$ and so $\mathrm{Im}(k) \propto Q'^{-1} \propto |\hat{\omega}|^{-1}$ for large $|\hat{\omega}|$. Subsequent theoretical arguments (Goodman & Oh 1997, Ogilvie & Lesur 2012) favor this stronger reduction, which gives a frequency dependence similar to a Maxwellian viscoelastic model with a relaxation time related to the convective timescale.

Figure 5 illustrates the differences between these models in the case of the solar convective zone. The values of $\mathrm{Im}[k_2^m(\omega)]$ are found by calculating the total dissipation rate in the convective zone using the formulae quoted by Zahn (1989) and assuming that the velocity field is either

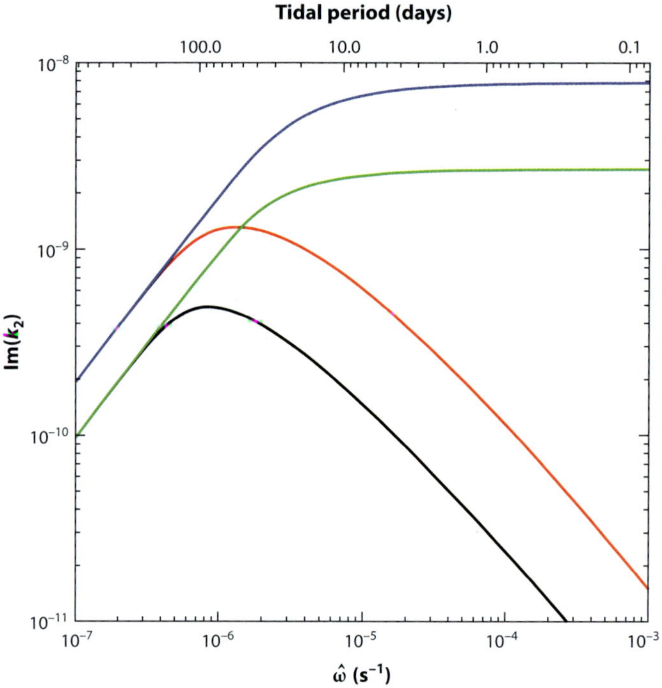

Figure 5

Frequency dependence of the dissipation of the nonwave-like tide in the convective region of a standard solar model. The turbulent viscosity is estimated using mixing-length theory, but is reduced using the prescriptions of Zahn (1966b) (*blue* and *green curves*) and Goldreich & Nicholson (1977) (*red* and *black curves*) in the forms quoted by Zahn (1989). The blue and red curves are computed assuming a standard equilibrium tide, whereas the green and black curves use the irrotational nonwave-like tide appropriate for a nonrotating convective region.

the standard equilibrium tide or the irrotational nonwave-like tide appropriate for a nonrotating convective region (discussed near Equation 27). For tidal periods of a few days or less, relevant for tidally interacting binary stars and extrasolar planets, the reduction factors can differ by more than one order of magnitude and the irrotational nonwave-like tide also leads to less dissipation. It must be emphasized that the results for other stars, or for the early Sun, could be very different.

In stars with deep convective envelopes, the scaling laws for the convective timescale t_c and for the effective viscosity ν and resulting tidal time lag τ, before any reduction factor is applied and omitting factors of order unity, are

$$t_c \sim \left(\frac{M_{\mathrm{env}} R^2}{L}\right)^{1/3}, \quad \nu \sim \left(\frac{L R^4}{M_{\mathrm{env}}}\right)^{1/3}, \quad \tau \sim \left(\frac{L M_{\mathrm{env}}^2 R^7}{G^3 M^6}\right)^{1/3}, \tag{28}$$

where M_{env} is the mass of the envelope, and M, R, and L are the stellar mass, radius, and luminosity, respectively. In a star that evolves to become a giant, τ can increase enormously, although this phase is relatively short-lived.

More recently, numerical simulations of turbulent convection have been used to address this issue. The results of Penev et al. (2007, 2009a,b) suggest that something closer to the prescription by Zahn (1966b) may be appropriate for intermediate frequencies, not for the reasons originally suggested but probably because the power spectrum of the convection in their simulations is less steep than the Kolmogorov spectrum assumed by Goldreich & Nicholson (1977). However, in a more idealized model, Ogilvie & Lesur (2012) obtained results that favor the steeper reduction factor, and they even found that negative effective viscosities are possible, meaning that the convection can do work on the tidal disturbance. Further investigation is clearly required.

In giant planets the complex nature of the fluid may allow other forms of dissipation. Stevenson (1983) has argued that a large anomalous bulk viscosity can arise in a two-phase medium, such as that thought to occur in Saturn and possibly Jupiter, when the planet has cooled sufficiently that helium becomes immiscible with hydrogen and forms droplets. The action of this bulk viscosity on the tidal flow could provide an important amount of dissipation. (There is some uncertainty in Stevenson's analysis concerning the amount of compression involved in the tidal flow. Contrary to the argument presented above, he states that the tidal displacement is not nearly incompressible because the region of helium separation should be stably stratified.) The properties of the droplets, however, remain very uncertain. In the same paper, Stevenson mentions another possibility, which is that the equilibrium tide could induce dissipation as fluid cycles across a first-order phase transition at the interface between molecular and metallic hydrogen. However, current models indicate that a first-order phase transition, although possible in principle (McMahon et al. 2012), is not expected in Jupiter (French et al. 2012).

A controversial suggestion by Tassoul (1987), that tidal synchronization in close binary stars could proceed efficiently by means of Ekman pumping involving a viscous boundary layer, was refuted by Rieutord (1992) and Rieutord & Zahn (1997), but resurfaces occasionally in the literature. In fact, Tassoul's proposed mechanism cannot be described within the framework we have presented, or classified as linear or nonlinear, because it produces a torque that is of first order in the tidal amplitude parameter ϵ. This is unphysical and apparently relies on an artificial transfer of angular momentum across the surface of the star by a viscous torque, rather than by gravity. Internal viscous boundary layers could contribute to the dissipation of the equilibrium tide, however, in giant planets containing a solid core or other internal interfaces.

3.4. Dynamical Tides in Radiative Regions

Cowling (1941), who first analyzed the oscillation modes of a stably stratified (polytropic) stellar model, found that the $l = 2$ g modes form an infinite sequence with frequencies tending to zero. He noted that the tidal forcing in a binary star could resonate with the higher members of this sequence, but also suggested that nonlinear effects would limit the importance of such resonances.

Zahn (1970, 1975, 1977) considered the tidal forcing of low-frequency internal gravity waves in early-type stars, which have convective cores and radiative envelopes. Tidal dissipation occurs because of the radiative damping of the waves near the stellar surface, which is more effective for lower tidal frequencies (shorter radial wavelengths). For sufficiently low frequencies, the waves are strongly attenuated and do not reflect to form global g modes. In this regime the tidal response scales as $|\mathrm{Im}(k)| \propto |\hat{\omega}|^{8/3}$ and increases strongly with the size of the convective core. The dynamical tide is expected to be more important than the equilibrium tide for higher frequencies. The combined description of equilibrium and dynamical tides by Zahn (1977) has been very influential (Langer 2009).

A physical interpretation (Goldreich & Nicholson 1989) is that the waves are launched in a narrow region near the base of the radiative envelope, where the buoyancy frequency matches the tidal frequency and the radial wavelength is long enough to connect to the tidal forcing. They propagate to the surface where they are damped and deposit their angular momentum flux. For higher frequencies, partial reflection occurs near the surface, and the response function contains discrete resonant peaks. Savonije & Papaloizou (1983, 1984) made explicit numerical calculations, including nonadiabatic effects in full, for particular stellar models.

A similar treatment was applied to solar-type stars, which have a radiative core and a convective envelope, by Terquem et al. (1998) and Goodman & Dickson (1998). Internal gravity waves are launched near the outer boundary of the radiative core and propagate inward. Again, for sufficiently low frequencies, the waves are strongly attenuated by radiative damping and $|\mathrm{Im}(k)| \propto |\hat{\omega}|^{8/3}$; for higher frequencies, the response function contains discrete resonant peaks.

Recently, related calculations have been made for carbon-oxygen white dwarfs by Fuller & Lai (2012b). These stars are mostly radiative, but may contain narrow convective zones and/or sharp features in the profile of buoyancy frequency associated with abrupt changes in composition. Fuller & Lai find that the tidal response is approximately $|\mathrm{Im}(k)| \propto |\hat{\omega}|^5$, but modulated by peaks and troughs owing to interference effects involving a wave cavity. It appears that the waves are mainly excited in regions where the buoyancy frequency changes sharply, even if it remains above the tidal frequency.

Lubow et al. (1997) applied Zahn's dynamical tide to hot Jupiters, short-period extrasolar giant planets in which stellar irradiation makes the atmosphere stably stratified to a depth on the order of 100 bars. This is a promising mechanism for tidal dissipation in such planets that requires further investigation, as the analysis has many uncertainties related to rotation, winds, radiative damping, wave reflection, nonlinearity, etc. This problem would best be treated through numerical simulations that can treat simultaneously the nonlinear atmospheric dynamics and the tidally forced waves. [Earlier, Ioannou & Lindzen (1993) proposed that rotationally modified g modes could be excited by tidal forcing in Jupiter itself, but their model relies on the existence of a deep, weakly stable stratification, which is not generally accepted.]

The effects of uniform rotation on the dynamical tide in radiative regions have been examined for early-type stars by Savonije et al. (1995), Savonije & Papaloizou (1997), Papaloizou & Savonije (1997), and Witte & Savonije (1999a), and for solar-type stars by Savonije & Witte (2002) and Ogilvie & Lin (2007). In this work the centrifugal distortion of the star is neglected, being of second order in the ratio of the spin frequency to the dynamical frequency, but the Coriolis force is retained, either in full or in the so-called traditional approximation. This approximation [originally introduced by Laplace, and given its name by Eckart (1960)] retains only the part of the Coriolis force that acts on the horizontal motion, which may be assumed to be large compared to the vertical motion in a radiative zone if the tidal frequency is small compared to the buoyancy frequency.

Although each oscillation mode of a nonrotating star or planet is described by a pure spherical harmonic and can be driven linearly only by a tidal component of the same form, the Coriolis force in a rotating body breaks the spherical symmetry and couples spherical harmonics of different degrees l. Retention of the full Coriolis force, even in spherical geometry, requires two-dimensional computations, whereas the traditional approximation allows a convenient separation of variables that is analogous to, but more complicated than, the nonrotating case. The horizontal structure of the separated solutions is described by Hough modes, which are eigenfunctions of the Laplace tidal equations. Laplace (1775; see also Cartwright 1999, Deparis et al. 2013) originally derived these equations for a shallow, incompressible ocean of uniform depth on a rotating planet, and Hough (1897, 1898) studied their solutions in detail. Because each Hough mode contains multiple values of l, coupled by the Coriolis force, each tidal component can excite a richer variety of modes than in the nonrotating case.

Examples of frequency-dependent linear tidal responses of uniformly rotating stellar models are shown in **Figure 6** for 10 $M_\odot$ and in **Figure 7** for 1 $M_\odot$. In both cases the response is dominated by resonances with rotationally modified g modes at higher frequencies and with quasi-toroidal modes at lower frequencies. These latter modes owe their existence to the Coriolis force and are related to the r modes of Papaloizou & Pringle (1978) as well as the Rossby waves well known in geophysical fluid dynamics.

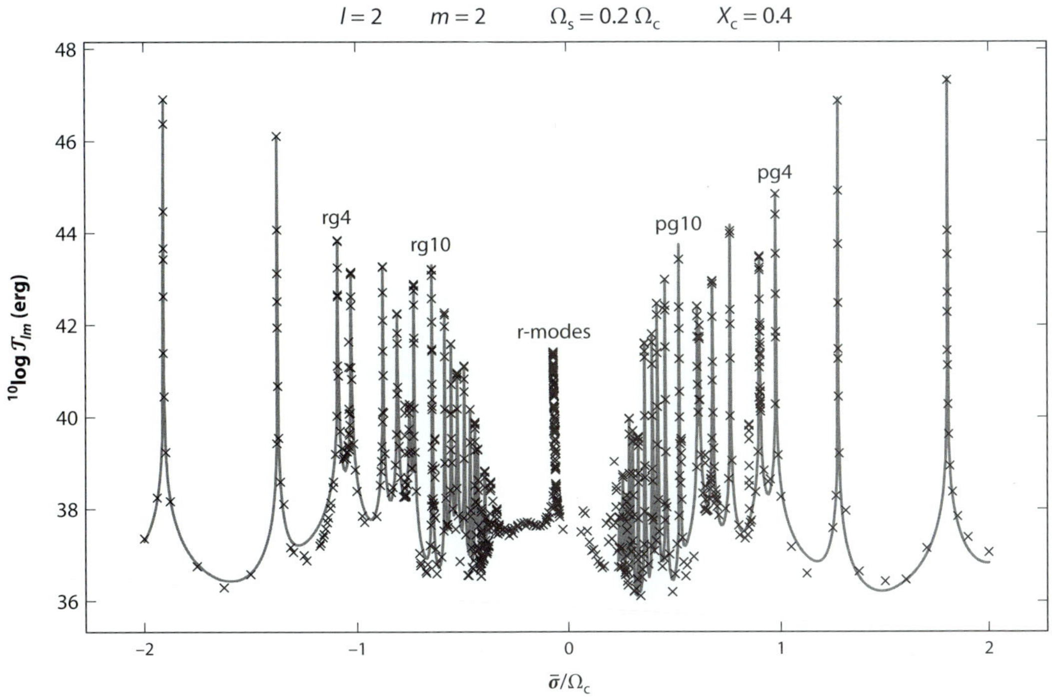

Figure 6

Linear response of a 10-$M_\odot$ star, uniformly rotating with $\Omega_s = 0.2\omega_d$, to $l = m = 2$ tidal forcing (reproduced, with permission, from Witte & Savonije 1999a). In our notation, the horizontal axis is $\hat{\omega}/\omega_d$. The quantity $|\mathcal{T}_{lm}| \approx 1.3 \times 10^{44}|\mathrm{Im}(k_2^2)|$ for this figure. Resonances with prograde and retrograde rotationally modified g modes of different radial orders are labeled. The low-frequency response involves quasi-toroidal modes.

Although the calculations of Savonije & Witte (2002) include the convective zone, the traditional approximation breaks down there and gives an inaccurate description of the inertial waves (Ogilvie & Lin 2004). Dynamical tides in convective zones are discussed in Section 3.5 below. The baseline value between resonances in **Figure 7** can perhaps be interpreted as being due to turbulent viscosity acting on the tide in the convective zone, although the role of the Coriolis force in this region under the traditional approximation is unclear. It corresponds to $|\mathrm{Im}(k_2^2)| \gtrsim 1 \times 10^{-7}$, which appears inconsistent with **Figure 5**, even under optimistic assumptions. It turns out that Savonije & Witte's implementation of mixing-length theory (and of the reduction factor) means that the turbulent viscosity is much larger than that estimated by Ogilvie & Lin (2007) and may also be much larger than the turbulent thermal diffusivity, contrary to expectations.

3.5. Dynamical Tides in Convective Regions

Models of stellar structure treat convective zones as being very slightly unstably stratified (i.e., superadiabatic), and interior models of giant planets are usually neutrally stratified (i.e., adiabatic). This is because, according to mixing-length theory, only a very small superadiabatic gradient is usually sufficient to provide the required heat flux through the body.

A simple approach to studying tides in convective regions is to neglect the convection per se and consider linear disturbances to a neutrally stratified basic state. If uniform rotation is also

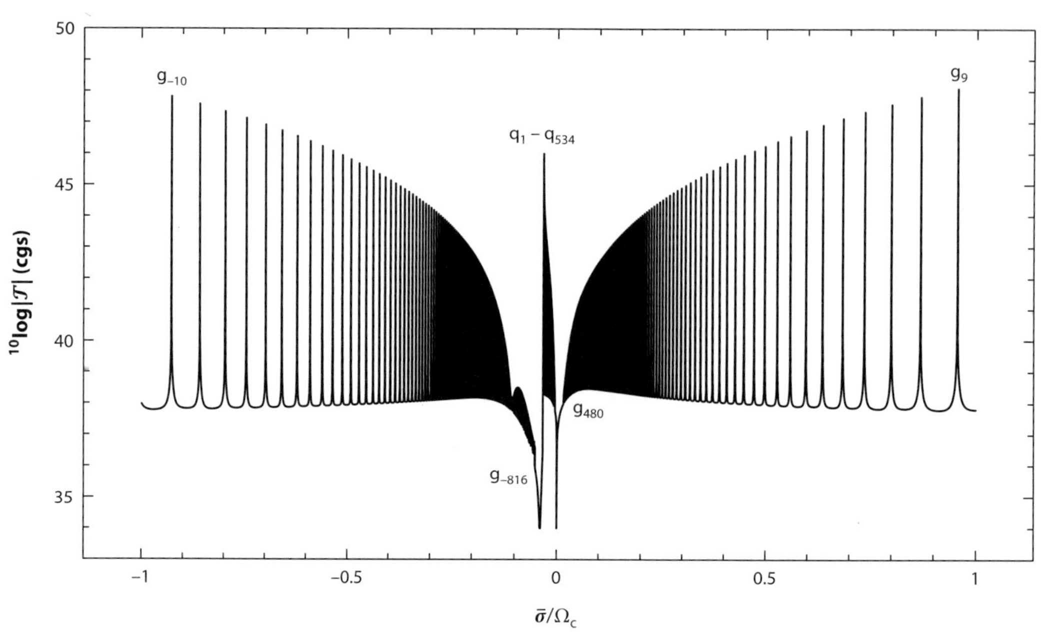

Figure 7

Linear response of a 1-$M_\odot$ star, uniformly rotating with $\Omega_s = 0.1\,\omega_d$, to $l = m = 2$ tidal forcing (reproduced, with permission, from Savonije & Witte 2002). In our notation, the horizontal axis is $\hat{\omega}/\omega_d$. The quantity $|\mathcal{T}| \approx 6.9 \times 10^{44}(R_\odot/R_1)|\mathrm{Im}(k_2^2)|$ for this figure. Resonances with prograde and retrograde rotationally modified g modes of different radial orders are labeled. The low-frequency response involves quasi-toroidal modes.

assumed for simplicity, then the low-frequency waves that comprise the dynamical tide are pure inertial waves restored by the Coriolis force (e.g., Greenspan 1968). If the body is fully convective, then the inertial waves propagate in a full sphere (or spheroid, more accurately). In other cases the waves may propagate only in a shell if they are excluded from either a solid planetary core or a stably stratified fluid core that the inertial waves cannot enter directly.

Linear inertial waves in a uniformly rotating, inviscid, incompressible fluid have been studied by Kelvin (Thomson 1880), Poincaré (1885), Bryan (1889), Cartan (1922), and many others (e.g., Greenspan 1968). They have remarkable mathematical properties. Their frequencies in the fluid frame satisfy $-2\Omega_s < \hat{\omega} < 2\Omega_s$, and the spatial structure of an inertial wave with a frequency in this range is governed by a second-order partial differential equation (named after Poincaré) that is hyperbolic; information propagates along characteristic surfaces that are inclined at an angle $\arcsin|\hat{\omega}/2\Omega_s|$ to the rotation axis. Waves with a given azimuthal wave number m satisfy an equation with similar properties in the meridional plane, with characteristic curves that are inclined at the same angle, which is also the inclination of the group velocity of the waves. Hyperbolic equations with conditions on the boundary of a closed domain generally give rise to ill-posed problems and the formation of singularities.

The global behavior of inertial waves in a reflective container is determined in part by the properties of the ray circuits, which depend sensitively on the angle of propagation and, therefore, on the wave frequency. In a full sphere or spheroid, the rays are space filling for most frequencies. It turns out that the waves can form smooth global normal modes at certain frequencies that are dense in the interval $(-2\Omega_s, 2\Omega_s)$. In a spherical shell, however, for most frequencies the rays are focused toward a small number of limit cycles known as wave attractors (Maas & Lam 1995,

Rieutord et al. 2001), which exist in certain bands of frequency, depending on the radius ratio of the shell. Generally speaking, inertial waves do not form smooth eigenmodes in a shell, although modes can be found localized around attractors when viscosity is introduced. In generalizations of this work, Dintrans et al. (1999) have found curved wave attractors for gravito-inertial waves in spherical geometry, as have Baruteau & Rieutord (2013) for inertial waves in a differentially rotating fluid.

In the past decade several studies have appeared examining linear dynamical tides in such configurations, showing that the Coriolis force gives rise to an enhanced tidal response in the frequency range $-2\Omega_s < \hat{\omega} < 2\Omega_s$, in which inertial waves can propagate and which is commonly found in applications (see **Figure 2**). Ogilvie & Lin (2004) considered a polytrope with a solid core and found that the tidal dissipation rate has a strong dependence on frequency in this range. This effect has been explored further by Ogilvie (2009) and Rieutord & Valdettaro (2010). The frequency dependence arises because of the sensitivity of the global propagation of inertial waves in a spherical shell to the inclination of the rays. It is more pronounced when the Ekman number $\mathrm{Ek} = \nu/2\Omega_s R^2$ is smaller and the rays can propagate further before dissipating. The behavior in the astrophysical limit of a very low Ekman number is complicated, although there seem to be intervals of frequency in which the dissipation rate becomes independent of or weakly dependent on the Ekman number. In a two-dimensional version of the same problem, it was shown that the dissipation rate is asymptotically independent of the magnitude and form of the small-scale dissipative mechanism (Ogilvie 2005); the wave attractor simply absorbs whatever energy flux is focused toward it. Although this result would be very appealing for astrophysical applications, its direct applicability is limited because of the competition between wave attractors and a singularity at the critical latitude, where the rays are tangent to the core (Goodman & Lackner 2009, Ogilvie 2009, Rieutord & Valdettaro 2010).

In contrast, studies of inertial waves in a full sphere, where the ray dynamics is much simpler, do not show this behavior. Instead, global normal modes, similar to those found by Bryan (1889), are excited, albeit rather weakly (Wu 2005a,b; Ivanov & Papaloizou 2010).

Examples of the frequency-dependent tidal responses are shown in **Figure 8** for an $n = 1$ polytrope with a perfectly rigid solid core, based on Ogilvie (2013). Responses are calculated for different values of the Ekman number in an effort to understand the astrophysical limit $\mathrm{Ek} \ll 1$; realistic values of Ek for giant planets are beyond current computational capabilities. For $|\hat{\omega}/\Omega_s| > 2$ the response is uninteresting and similar to an equilibrium tide, with $\mathrm{Im}(k) \propto \mathrm{Ek}$. A greatly enhanced, but strongly frequency-dependent, tidal response occurs for $|\hat{\omega}/\Omega_s| < 2$ because of the excitation of inertial waves. For a polytrope with a small core the response is dominated by near-resonances with two large-scale, global inertial modes of the full sphere, which are approximate solutions in the presence of a small core. When a larger core is introduced, normal modes are not excited and the response is dominated by wave singularities associated with the critical latitude on the inner boundary and with wave attractors (**Figure 9**; Ogilvie 2009).

A synthesis of these findings was attempted by Ogilvie (2013), who devised a method to obtain a frequency-averaged tidal response by means of an impulse calculation. The quantity computed is $\int \mathrm{Im}(k_l^m)\mathrm{d}\hat{\omega}/\hat{\omega}$, integrated over the low-frequency part of the spectrum where inertial waves are found. It can be taken as a gross measure of the tidal response in this range of frequencies, when the details of the response curve are washed out. In particular, for the $l = m = 2$ tidal component most often considered, the analytical result for a uniformly rotating homogeneous fluid body with a perfectly rigid core of fractional size f is

$$\int \mathrm{Im}(k_2^2)\frac{\mathrm{d}\hat{\omega}}{\hat{\omega}} = \frac{100\pi}{63}\left(\frac{\Omega_s}{\omega_d}\right)^2 \frac{f^5}{1 - f^5}. \tag{29}$$

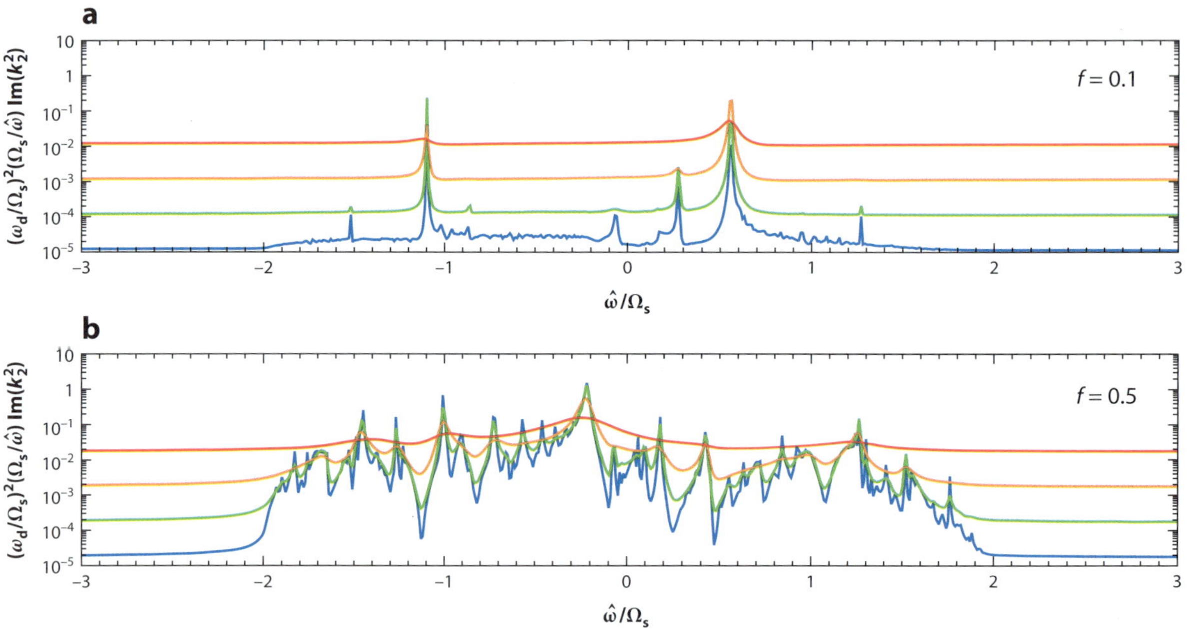

Figure 8

Linear response of a uniformly (and slowly) rotating $n = 1$ polytrope, with a perfectly rigid solid core of fractional size (*a*) $f = 0.1$ or (*b*) $f = 0.5$, to $l = m = 2$ tidal forcing, as in Ogilvie (2013). The fluid is viscous and the Ekman number is $\mathrm{Ek} = 10^{-3}$ (*red line*), 10^{-4} (*orange line*), 10^{-5} (*green line*), or 10^{-6} (*blue line*).

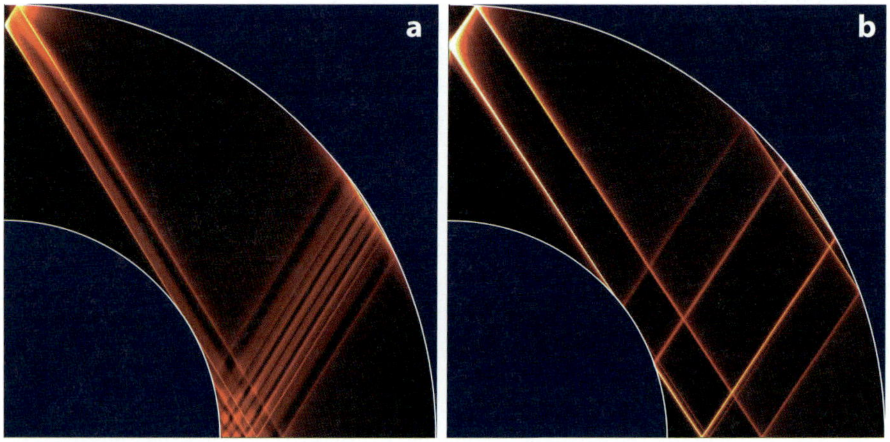

Figure 9

Structure of the velocity perturbation $|\boldsymbol{u}|$ in a meridional quarter-plane for the $l = m = 2$ tide forced in a uniformly (and slowly) rotating homogeneous fluid body with a perfectly rigid solid core of fractional size $f = 0.5$, as in Ogilvie (2009). The fluid experiences a frictional damping force with a damping coefficient $\gamma = 10^{-3}\ \Omega_s$. In the case of (*a*) $\hat{\omega}/\Omega_s = 1.05$ the response is dominated by a beam of inertial waves emitted from the critical latitude on the inner boundary. In the case of (*b*) $\hat{\omega}/\Omega_s = 1.10$ it is dominated by an inertial wave attractor.

Several points are noteworthy. First, the result is independent of the viscosity, which is consistent with the fact that the response curve involves more features, with higher peaks and lower troughs, as the viscosity is reduced. Second, the tidal response of inertial waves increases with the square of the spin frequency, which is an important general property (valid for $\Omega_s \ll \omega_d$; tidal dissipation in a rapidly rotating body has not yet been calculated). Third, the response vanishes for a full sphere and increases strongly with the core size, even diverging in the limit of a thin shell, although it should be noted that the core is assumed to be perfectly rigid and the planet to be of uniform density. (Tides in shallow rotating fluid layers, first considered by Laplace, are worthy of further investigation.)

It turns out that the dependence f^{2l+1} for $f \ll 1$ occurs for all sectoral harmonics ($m = l$), whereas for tesseral harmonics ($m \neq l$) the frequency-integrated response remains nonzero as $f \to 0$. This difference in behavior occurs because tesseral harmonics are able to resonate with the global inertial modes of a homogeneous full sphere, whereas the sectoral harmonics are not. Care is required, though, because the $l = 2$ tesseral harmonics have only formal resonances with the spin-up and spin-over modes, which involve rigid rotations and have no associated dissipation, and these dominate the integrated responses.

When the body is inhomogeneous, global modes can be excited even by sectoral harmonics in a full sphere. The strong dependence of the frequency-averaged response on the core size is not seen in polytropes of higher index that resemble centrally condensed stars. If the solid core is replaced by a fluid core such that the density is discontinuous at the interface between fluid layers, then behavior is found that is qualitatively similar but generally weaker, especially if the discontinuity is small. This occurs because of the different responses of the layers and because the inertial waves undergo partial internal reflection at the interface.

To summarize, as the size of the core is increased, a transition occurs from a response involving classical resonances with global normal modes to a richer one in which wave singularities are dominant. Generally a greater tidal response in inertial waves is obtained by having a larger core or a greater density contrast. It is expected that differential rotation will also increase the richness of the response (Ogilvie & Lin 2004, Baruteau & Rieutord 2013).

3.6. Tidal Encounters

Press & Teukolsky (1977) performed a classic calculation of the energy deposited in the oscillation modes of a nonrotating star during a parabolic tidal encounter. Although their study was motivated by the binary capture mechanism of Fabian et al. (1975), it is also relevant to the process by which extrasolar planets can reach short-period orbits through tidal interaction with the star if they start from a highly elliptical orbit. [Errors in the numerical results of Press & Teukolsky (1977) were corrected by Giersz (1986), Lee & Ostriker (1986), and McMillan et al. (1987).] An important dimensionless parameter in the problem is, in our notation,

$$\eta = \frac{\omega_d}{\Omega_{max}} = \left(\frac{M_1}{M}\right)^{1/2} \left(\frac{d_{min}}{R_1}\right)^{3/2}, \qquad (30)$$

where $\Omega_{max} = (GM/d_{min}^3)^{1/2}$ is the circular orbital angular velocity at the minimum orbital separation, d_{min}. The parameter η is a measure of the duration of periastron passage, relative to the dynamical timescale of the star, and is not directly related to the tidal amplitude parameter ϵ. It is expected that $\eta \gtrsim 1$ in order to avoid tidal disruption; in the case of an incompressible fluid body with $M_1 \ll M$, tidal disruption occurs for $\eta \lesssim 2.2$ (Kosovichev & Novikov 1992, Sridhar & Tremaine 1992). In a neutrally stratified star, the energy transferred, ΔE, is found to decline exponentially with η for large η; this is because the orbital motion at pericenter is too slow to excite the f mode and p modes of the star efficiently. In a stably stratified star, however, ΔE declines as an inverse power of η, because of the excitation of low-frequency g modes.

Lai (1997) has extended these calculations to include uniform rotation within the traditional approximation, whereas Papaloizou & Ivanov (2005) and Ivanov & Papaloizou (2007) have included the full Coriolis force in the case of a coreless model with neutral stratification. Here, global inertial modes dominate the response for large η, and the transfers of energy and angular momentum in the tidal encounter, ΔE and ΔJ, depend on the rotation rate through the parameter Ω_s/Ω_{max}, as well as on η. Papaloizou & Ivanov (2010) were able to include a solid core, even though global inertial modes do not occur, by solving an initial-value problem for linear tidal disturbances in a parabolic encounter. They found that ΔE increases with the size of the core, which is broadly in line with the findings described in Section 3.5 for periodic tides.

Indeed, a synthesis of the results on periodic and aperiodic tides can be achieved through the frequency-dependent response function $k_l^m(\omega)$. The energy and angular momentum transferred in a parabolic encounter are equal to integrals of $\mathrm{Im}[k_l^m(\omega)]$ over all frequencies, with two different weightings involving the temporal Fourier transform of the tidal potential for a parabolic orbit. This produces a viscosity-independent result that integrates out all the complicated structure in the response curve.

This calculation of ΔE and ΔJ, which implicitly assumes that the body is unperturbed before the encounter, is strictly relevant for a highly eccentric orbit only if the tidal disturbance is efficiently damped within an orbital period. If there is interference between waves excited by successive encounters, then the coupled dynamics of the orbit and oscillation modes can be initially chaotic (Mardling 1995a), although it becomes regular under the action of dissipation (Mardling 1995b).

3.7. Discussion of Linear Theory

We have described several situations in which linear theory predicts a tidal response whereby $\mathrm{Im}(k)$ varies strongly as a function of $\hat{\omega}$, often with peak and trough values that differ by several orders of magnitude. It should be borne in mind that these response functions are based on linear theory with various idealizations concerning the basic state. (Even so, they are computationally demanding because of the need for very high resolution in the spatial and frequency domains.) They also assume a steady-state response; the peak values, or tidal resonances, require multiple wave reflections and tidal periods to become established.

Nevertheless, it is important to ask how an astrophysical system evolves in such cases. The tidal frequency in the fluid frame for a particular tidal component, $\hat{\omega} = n\Omega_o - m\Omega_s$, is an integer linear combination of the spin and orbital frequencies, both of which evolve as a consequence of tidal dissipation (except in the case of $m = 0$). The evolution of $\hat{\omega}$ is most rapid at the peaks of the response curve, and the system therefore spends most of its time evolving slowly through the troughs. This suggests that strong tidal resonances are short-lived and of little importance.

However, Witte & Savonije (1999b, 2001) have found that systems can sometimes lock into tidal resonances and undergo relatively rapid evolution. In general, evolution through the troughs of the response curve of one tidal component may be accelerated or reversed by the effects of other tidal components, or by magnetic braking or stellar evolution, which can change the shape of the response curve and the locations of peaks and troughs. If these other effects cause an evolution toward a tidal resonance that itself causes an evolution in the opposite direction, then a balance can be attained in which the resonance is entered into to a certain degree. Strong resonant locking is most likely to happen when there is a near cancellation between the rates of change of $n\Omega_o$ and of $m\Omega_s$; for binary stars, in which the spin moment of inertia is small compared with the orbital moment of inertia, this condition can be met for a component with a large value of n, which may have appreciable amplitude if the eccentricity is not small.

It should be borne in mind that models of stellar and planetary interiors have been developed, for the most part, in order to understand their gross properties and long-term evolution, not for a detailed study of their dynamical perturbations. In fact, stellar interior models are increasingly being used for seismology, but this mostly involves acoustic waves. There are numerous uncertainties concerning the behavior of low-frequency waves in stellar and planetary interiors. In particular, it is not clear how to model correctly the behavior of inertial waves in a convective medium. In models of the Sun, for example, $-N^2$ is positive in the convective zone and rises upward, exceeding $4\Omega_s^2$ above about 0.9 $R_\odot$. Should the linearized equations include this unstable stratification, which would significantly alter the properties of the waves? Or is the unstable stratification effectively neutralized by the existence of the convection? Does convection damp the waves like a viscosity or scatter or affect them in some other way? What are the effects of magnetic fields, differential rotation, and imperfect reflections of the waves from the boundaries of the convective region?

4. NONLINEAR TIDES

4.1. Nonlinear Equilibrium Tides

As described in Section 3.3, the equilibrium tide consists of a spheroidal bulge that is usually time dependent in the fluid frame and is then associated with a large-scale velocity field. In Section 3.3, we considered linear damping mechanisms, such as the interaction of the tidal flow with turbulent convection, which lead to a dissipation rate proportional to ϵ^2, where ϵ is the tidal amplitude parameter. It is also possible that the flow could be damped as a result of its intrinsic instability. It would be natural for such a mechanism to set in, or to dominate, above a critical tidal amplitude and to lead to a dissipation rate that is no longer proportional to ϵ^2. This would mean that $|\mathrm{Im}(k)|$ depends on ϵ, although it is then not a linear response function but a parameterization of a nonlinear response (Goldreich 1963).

Press et al. (1975) noted that the Reynolds numbers of tidal flows in close binary stars are very large. They proposed that this leads to turbulence and to a turbulent viscosity ν_t, such that the effective Reynolds number of the tidal flow based on ν_t is a modest value of order unity, Re_t. The resulting scaling relations are $\nu_t \sim \mathrm{Re}_t^{-1}\epsilon R^2|\hat{\omega}|$ and $|\mathrm{Im}(k)| \sim \mathrm{Re}_t^{-1}\epsilon(\hat{\omega}/\omega_d)^2$. However, it remains to be demonstrated whether this mechanism works in the presence of either pre-existing turbulence (in convective zones) or stable stratification (in radiative zones).

A more sophisticated viewpoint is that the instability would take the form of a resonant coupling of the internal wave modes of a rotating and/or stably stratified body. This takes into account the fact that the tidal flow has a well-defined oscillation frequency and spatial structure. If two internal modes can be found with frequencies and spatial structures that are linked by those of the tidal flow, then that pair of modes may be destabilized. The density of the spectrum of internal waves makes such a coincidence more probable. A perfectly tuned pair of undamped modes grows at a rate proportional to ϵ, but this parametric instability may be reduced or suppressed by damping or detuning of the modes. It is much more difficult to determine the nonlinear outcome of such instabilities and the resulting dissipation rate, because a multitude of interacting internal waves may be involved.

An important example is the elliptical instability, which is a parametric resonance of internal (usually inertial) waves due to an elliptical deformation of the streamlines of a rotating flow

(Kerswell 2002). This is precisely the situation when a spheroidal tidal bulge is generated in a rotating body. Elliptical instability has been investigated by means of laboratory experiments, linear analysis, and numerical simulations, mainly for flows without stable stratification, in which pairs of inertial waves are destabilized through their coupling to the deformation. Its application to astrophysical systems, where it has sometimes been called tidal instability, has been discussed by Rieutord (2004), Le Bars et al. (2010), and Cébron et al. (2010, 2013). A necessary condition for elliptical instability involving a pair of inertial waves in a uniformly rotating body is that the primary tidal flow has $|\hat{\omega}| < 4\Omega_s$ (so that it can couple to two secondary inertial waves, each having $|\hat{\omega}| < 2\Omega_s$). For the $l = m = n = 2$ tide with $\hat{\omega} = 2(\Omega_o - \Omega_s)$ this condition is $-1 < \Omega_o/\Omega_s < 3$, which is a wider range than that for tidal forcing to excite inertial waves directly. Even so, the instability would not be allowed in the majority of stars hosting hot Jupiters, where $\Omega_o/\Omega_s > 3$.

In order to observe the elliptical instability in experiments or simulations, where the Ekman number $\mathrm{Ek} = \nu/2\Omega R^2$ is much larger than in astrophysical bodies, the tidal amplitude must also be made much larger than is usually realistic. This situation favors the dominance of large-scale modes in the nonlinear outcome of the instability. Although attempts have been made to deduce scaling relations, and it is claimed that an order-unity change in the basic flow is commonly obtained, it is not yet clear what the nonlinear outcome would be for realistic values of Ek and ϵ.

A limitation of much of this work is that it focuses on the case in which the deformation is stationary in the inertial frame. This means that the elliptical instability takes the special form of the spin-over mode, which has the same property. A typical nonlinear outcome is a steady or recurrent tilting of the spin axis of the body, and this has been proposed as an explanation of the spin-orbit misalignments observed in exoplanetary systems (see Section 5.3 below). However, a simple tilting of the stellar spin axis is not a permissible outcome; in order to conserve the total angular momentum of the system, the orbital angular momentum would have to increase in magnitude, requiring an increase in the total energy of the system, whereas in fact the total energy can only decrease as a result of dissipation.

In an astrophysical system, where the tidal potential and deformation rotate at the orbital frequency and the values of Ek and ϵ are much smaller, it appears more likely that the elliptical instability, if present, generates turbulence through the excitation of smaller-scale waves. In a local computational model, the instability either saturates through the formation of long-lived columnar vortices (Barker & Lithwick 2013a) or generates three-dimensional turbulence, in the presence of a magnetic field (Barker & Lithwick 2013b).

Another interesting line of research has investigated nonlinear couplings of tides and global oscillation modes (f, p, and g) of nonrotating stellar models (Kumar & Goodman 1996; Weinberg et al. 2012, 2013). It provides a weakly nonlinear theory that includes quadratic terms that couple different modes. A conventional application of this work is that, in radiative zones, the equilibrium tide can excite pairs of g modes by parametric resonance and may thence find a route to dissipation. Kumar & Goodman (1996) showed that this mechanism could be effective in damping the f mode excited in a tidal encounter. For a solar-type binary with a moderately eccentric orbit, Weinberg et al. (2012) find that parametric instability of the equilibrium tide occurs only for orbital periods less than about one day, because the unstable modes are concentrated near the core of the star and connect very weakly with the equilibrium tide. Using the same formalism, Weinberg et al. (2013) find a quite different, nonresonant instability of a p–g mode pair, even for a static (time-independent) equilibrium tide. This result has been challenged (Venumadhav et al. 2013) and appears to violate the energy principle; it may indicate the inconsistencies of using truncated expansions. As yet, this approach has not included the important effects of rotation. The assumption of a mode expansion may be problematic if the waves do not form global modes in the linear regime or if the waves are of sufficient amplitude to break.

4.2. Nonlinear Dynamical Tides

In Sections 3.4 and 3.5 we discussed the tidal excitation of internal waves and their dissipation through linear damping mechanisms. Both gravity waves and inertial waves may be dissipated more effectively through nonlinear processes. Internal waves break if they exceed a critical amplitude, which is approximately when the displacement in the direction parallel to gravity (or perpendicular to the rotation axis, in the case of inertial waves) is equal to the wavelength in that direction divided by 2π. Under these conditions, the wave so distorts the background stratification or vorticity that a localized convective or shear instability occurs, and turbulence may be generated. Below the breaking amplitude, internal waves may still be unstable through mode couplings and may transfer their energy to other waves, usually of smaller scale. In an unbounded medium or periodic box, in the absence of dissipative effects, a plane internal wave of any amplitude is unstable in this way (Mied 1976).

Goodman & Dickson (1998) noted that the dynamical tide in a star similar to the present Sun, forced by a binary companion in an eccentric orbit, can easily exceed the breaking amplitude in the innermost wavelengths near the stellar center and is then bound to be significantly damped. In fact, the level of nonlinearity in the wave increases strongly toward the center because of the concentration of the wave energy into a small volume. Ogilvie & Lin (2007) applied a similar idea to the host stars of hot Jupiters, where the stellar rotation is slow compared to the orbit and the eccentricity is negligible; they found that, for a model based on the present Sun, a companion of more than about three Jupiter masses would generate waves that could break near the center, but also noted that the threshold depends on the stellar model. More detailed calculations were made by Barker & Ogilvie (2010), who also performed numerical simulations of the breaking process. Among solar-type stars, the threshold companion mass decreases with increasing stellar mass and with increasing age. Older or more massive stars have a steeper stratification near the center because of their chemical evolution.

Numerical simulations of internal gravity waves approaching the stellar center (Barker & Ogilvie 2010, Barker 2011) show that, if the breaking amplitude is exceeded, the waves are almost completely dissipated. Angular momentum is deposited, and an expanding core is formed in which material is spun up to the orbital frequency. Waves approaching the boundary of this core encounter a critical layer (at which the angular velocity of the fluid matches that of the wave) that causes them to be dissipated there. The planet donates its orbital angular momentum to the stellar core. The resulting dissipation is equivalent to $Q' \approx 1.5 \times 10^5\ (P/\mathrm{day})^{8/3}$ for the Sun, where P is the orbital period; this estimate varies by no more than a factor of five for solar-type stars between 0.5 and 1.1 $M_\odot$. This mechanism should lead to the destruction of sufficiently massive hot Jupiters with orbital periods of about one day within a few million years of the commencement of wave breaking. (However, if the companion is sufficiently massive, then a tidal equilibrium may be approached instead, because the star can be spun up to match the increasing orbital frequency; see Section 5.3 below.)

In fact, this process is similar to one described by Goldreich & Nicholson (1989) in which early-type stars with close binary companions become synchronized from the outside in, as internal gravity waves generated by tidal forcing near the base of the radiative envelope are absorbed by a critical layer that advances inward from the stellar surface. This can be seen as a nonlinear modification of Zahn's dynamical tide, which is more effective in damping the waves and producing robust dissipation.

In principle, the process of an advancing critical layer could be initiated below the breaking amplitude as waves deposit angular momentum gradually by linear radiative damping near the stellar center (or surface). However, various processes may prevent the formation of the required differential rotation, such as hydrodynamic instability or magnetic coupling. It is not clear how this mechanism applies to eccentricity tides, where there are multiple tidal components with different angular pattern speeds (**Table 1**).

Linear theories of dynamical tides often exhibit strong resonant peaks in the tidal response at the frequencies of global internal modes, where tidally forced waves reflect and interfere constructively (Section 3.4). If such a resonance occurs, the waves are much more likely to break, in which case the torque is reduced to a nonresonant value. The passage through such a tidal resonance could therefore initiate an advancing critical layer. Models that rely on locking into a linear resonance may be compromised by the breaking of the waves. This was acknowledged by Witte & Savonije (1999b), but appears not to have been estimated accurately; Witte & Savonije (2002) capped their resonant torques by limiting the velocity perturbation to the sound speed at the stellar center, but in fact the relevant comparison is to the much smaller wave speed of internal (gravity) waves.

Nonlinear processes have also been discussed for inertial waves in convective regions. Goodman & Lackner (2009) argued that a fraction of the inertial wave flux generated at the critical latitude on the inner boundary of a spherical shell would be subject to dissipation through wave breaking. By preventing multiple reflections within the shell, this process would strongly reduce the frequency dependence of the tidal response. Jouve & Ogilvie (2014) have simulated an inertial wave attractor in linear and nonlinear regimes; as waves are focused toward the attractor, they may exceed the breaking amplitude and produce waves of smaller scale before they themselves can reach the scale of viscous dissipation, but the total dissipation rate is not greatly reduced compared with the linear theory. Favier et al. (2014) have performed numerical simulations of forced inertial waves in spherical shells. The results for low amplitudes agree with the linear theory, and for higher amplitudes the dominant nonlinear effect they find is the development of zonal flows, in which the angular velocity depends on cylindrical radius in a complicated way that is related to the locations at which the inertial waves are preferentially dissipated. As these zonal flows build up, they increasingly alter the propagation and dissipation of the inertial waves. This interaction of waves and a mean flow may be self-regulating, although in some cases hydrodynamic instability is seen to occur.

An important nonlinear aspect of tidal interactions, therefore, is the development of differential rotation through the nonuniform deposition of angular momentum by tidal disturbances, and the interaction of this differential rotation with the tides. This adds another layer of complexity to the problem, which is only just beginning to be explored in detail.

5. APPLICATIONS TO OBSERVED SYSTEMS

5.1. Direct Observations of Tides in Stars

Direct observations of tides in stars and giant planets are limited, even within the Solar System. For decades, though, ellipsoidal variations in the light curves of close binary stars have been understood as being due to the changing aspect of tidally deformed stars (e.g., Kopal 1978). Recently, Welsh et al. (2010) used *Kepler* to detect the much smaller (37 ppm) ellipsoidal variations in the star HAT P-7 due to its companion, a hot Jupiter in a circular orbit. *Kepler* has also been used to observe dynamical tides in several eccentric binary stars (Thompson et al. 2012). The light curves of these "heartbeat stars" resemble electrocardiograms, with a brightening near periastron and persistent oscillations throughout the orbit. A particularly interesting example is KOI-54 (HD 187091) (Welsh et al. 2011, who also mention three pre-*Kepler* cases). This system contains two very similar A stars in a nearly face-on orbit of eccentricity 0.83 and period 42 days and with a minimum separation of about 6 stellar radii. Fourier analysis of the light curve shows that several harmonics of the orbital frequency (notably the 90th and 91st) are preferentially excited, as well as some nonharmonic frequencies. Detailed dynamical analyses of this system identify these oscillations as tidally excited g modes in the two stars and reveal evidence for three-mode coupling (Fuller & Lai 2012a, Burkart et al. 2012, O'Leary & Burkart 2014).

5.2. Tidal Evolution in Binary Stars

Indirect evidence of tidal dissipation in stars and giant planets comes from a comparison of the observed distributions of orbital and spin parameters with the expected outcomes of tidal evolution, given reasonable assumptions about the initial conditions. For spectroscopic binary stars, the measured joint distribution of orbital eccentricity e and orbital period P provides good evidence for tidal circularization, showing a clear preference for smaller e at small P. The most detailed study is that of Meibom & Mathieu (2005; see also Mathieu 2005), who plot e versus P for eight populations of solar-type binaries of different ages from pre-main-sequence to late-main-sequence and define a characteristic circularization period (below which orbits are more circular) for each based on a fit to the distribution. The circularization period generally increases with age, suggestive of ongoing tidal dissipation, in a way that is very roughly consistent with $Q' \approx 10^6$ (Ogilvie & Lin 2007; see also Hansen 2010, discussed in Section 5.3 below). It is clear from **Figure 5** that convective damping of the equilibrium tide in a star similar to the present Sun is much too inefficient to explain this result, even with optimistic assumptions ($Q' \approx 10^{10}$ being a more reasonable estimate). (It should be noted that the primary stars span a range of masses around 1 $M_{\odot}$, and that the secondary stars, which might make an important or even dominant contribution to circularization, are mostly not observed.) If the stars are assumed to be synchronized, then the frequencies of the first-order eccentricity tides (**Table 1**) in the fluid frame are $\hat{\omega} = \pm\Omega_s$ (**Figure 2**) and can excite inertial waves in the convective zone. Studies of the dynamical tide in a solar model can then explain $Q' \approx 10^7$–10^8 (Ogilvie & Lin 2007), but this is still too inefficient to explain the observations. Witte & Savonije (2002) found that the dynamical tide in the radiative zone, together with the effects of rotation and resonant locking, could also provide much more dissipation than the equilibrium tide, but still not enough to explain the observations. A gap still exists, therefore, between theory and observations in this area. [Zahn & Bouchet (1989) previously argued on theoretical grounds that most of the circularization occurs during the pre-main-sequence phase, when the stars are larger and fully convective, but the observations do imply an ongoing process.]

Observations of stellar rotation are more difficult than those of orbital eccentricity, and most diagnose only the rotation of the surface. Meibom et al. (2006) have measured (from periodic photometric variability) the spin periods of the primary stars in several solar-type binaries in two open clusters. Although they do detect two short-period systems with circular orbits and synchronized primaries, their other findings do not appear consistent with the simple expectation that synchronization proceeds much more rapidly than circularization. A partial explanation may be that, as synchronism is approached, the tidal frequency (in the fluid frame) of the asynchronous tide tends to zero, whereas that of the eccentricity tides tends to $\pm\Omega_s$, as noted above. Therefore $\kappa_{2,2,2}$ in Equation 7 tends to zero, and the timescale for synchronization continues to increase, whereas the timescale for circularization tends to a nonzero constant.

For early-type main-sequence stars with radiative envelopes, Khaliullin & Khaliullina (2010) have tested theories of tidal synchronization and circularization in a set of 101 eclipsing binaries, going beyond previous work by Claret & Cunha (1997). Because of concerns that the rotation of the stellar surface might be different from that of the deep interior, they determine the rotation periods using the observed apsidal motion together with theoretical models of stellar structure. After correcting an important numerical error in Khaliullin & Khaliullina (2007), based on a misreading of Zahn's papers, they find that Zahn's theory of the dynamical tide is compatible with these observations.

For giant stars, success was also found by Verbunt & Phinney (1995), who showed that Zahn's theory of the equilibrium tide, including the effects of stellar evolution, could account for the circularization of spectroscopic binaries containing a giant. The circularization period (more than 200 days) is sufficiently long that the controversial reduction factor for high tidal frequencies is not relevant.

5.3. Tidal Evolution in Extrasolar Planetary Systems

Observations of extrasolar planets by the radial velocity method allow measurements of the orbital period, the orbital eccentricity, and the "projected" mass of the planet, and observations of transiting planets also provide information on the planetary radius. In some cases the spin period of the star can be measured and its (sky-projected) obliquity (spin-orbit misalignment) determined from the Rossiter-McLaughlin effect; there are no observations yet of the spin periods or obliquities of exoplanets. This is a rapidly developing field that is stimulating interest in tidal interactions. Early applications of tidal theory to exoplanets were made by Rasio et al. (1996), Marcy et al. (1997), and Lubow et al. (1997).

As for binary stars, the clearest evidence for tidal interactions between extrasolar planets and their host stars comes from the joint distribution of orbital eccentricity and orbital period. (Interactive plots of up-to-date exoplanetary data can be generated at **http://exoplanet.eu** or **http://exoplanets.org**.) Orbits of shorter period are more likely to be of low eccentricity (e.g., Kipping 2013). This can be explained by tidal circularization due to dissipation in the planet and is very roughly consistent with $Q'_\mathrm{p} \approx 10^{6.5}$ (Jackson et al. 2008, Husnoo et al. 2012; but see also Hansen 2010, discussed below). In some systems, especially those with more massive planets such as CoRoT-3, dissipation in the star may also be important for circularization.

As noted in Section 1, in many short-period exoplanetary systems, tidal dissipation in the star leads to inward orbital migration and ultimately to the tidal disruption of the planet rather than to the synchronization of the stellar spin with the orbit as in binary stars. **Figure 10** shows the evolutionary paths (but not the evolutionary timescale) for a planet interacting tidally with a star in the circular, aligned case ($e = i = 0$) in the absence of magnetic braking and other influences. If the total angular momentum L is less than the critical value L_c (Equation 1), then the orbit decays and the star is spun up; if $L > L_\mathrm{c}$, then a stable tidal equilibrium may be approached.

To interpret the observations, we need theories of both planet formation and planet destruction and should perhaps be concerned with the planets that are not observed as much as with those that are. Evidence for inward migration and/or destruction is much less clear than that for circularization. Indeed, the existence of numerous planets in orbits with periods of about 1 d or less implies that the relevant values of Q'_* must greatly exceed the empirical estimate of 10^6 deduced from the circularization of solar-type binary stars; otherwise the remaining lifetimes of the very short-period planets would be implausibly short. An extreme example is WASP-18 b (Hellier et al. 2009), whose remaining lifetime would be about 1×10^6 year if $Q'_* = 10^6$; in this case the orbital decay should be detectable within a few years through transit-time variations (Hellier et al. 2009), although Watson & Marsh (2010) suggest that the effects could be masked by orbital period changes resulting from a variation of the quadrupole moment of the star during its magnetic activity cycle. [Perhaps surprisingly, planets of shorter orbital period such as KOI-1843.03 (Rappaport et al. 2013) and Kepler-78 b (Sanchis-Ojeda et al. 2013) provide a weaker constraint on Q'_* because of the lower planetary masses and higher stellar densities in those systems.] By reexamining the system OGLE-TR-56 b, Adams et al. (2011) have constrained its period derivative

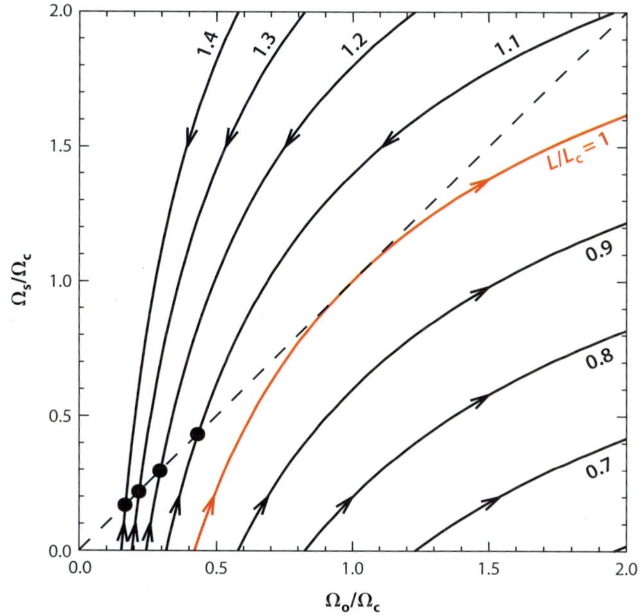

Figure 10

Evolutionary paths for a planet interacting tidally with a star in the circular, aligned case ($e = i = 0$) in the absence of magnetic braking and other influences. The axes represent the (stellar) spin and orbital angular velocities in units of the critical value (Equation 1). Contours of the total angular momentum L are plotted, and the direction of evolution is such that energy is dissipated. The planetary spin is assumed to be unimportant. On the dashed line, the stellar spin is synchronized with the orbit. Black points represent stable tidal equilibria when L exceeds the critical value L_c (Equation 1). Systems with $L < L_c$ evolve toward planetary destruction. The equation of each contour is $(\Omega_s/\Omega_c) + 3(\Omega_o/\Omega_c)^{-1/3} = 4(L/L_c)$.

to be $\dot{P} = -2.9 \pm 17$ ms year^{-1}, i.e., consistent with zero. Their analysis of this result is confusing; in fact, at the 3σ confidence level ($\dot{P} > -54$ ms year^{-1}), it implies that $Q'_* > 1 \times 10^5$.

Despite various attempts to estimate or constrain Q'_* by statistical modeling of the observed orbital semimajor axes and eccentricities, either of populations or of single systems (Jackson et al. 2008, 2009; Hansen 2010; Schlaufman et al. 2010; Brown et al. 2011), there is as yet no convincing evidence of any orbital migration due to tidal dissipation, because the initial conditions are poorly constrained. However, the host stars of several hot Jupiters are said to be spinning faster than expected (Pont 2009, Husnoo et al. 2012), suggesting a tidal transfer of angular momentum from the orbit to the stellar rotation in these systems, and two stars (τ Boo and CoRoT-3) show rotation compatible with synchronization with the orbits of their hot Jupiters (as do KELT-1 and CoRoT-15 with their brown dwarf companions). As mentioned in Section 1, these apparently synchronized systems probably have sufficient total angular momentum to attain a tidal equilibrium. It is not clear what dissipation mechanism can bring this about within the age of the system, though, because these are F stars with shallow convective zones.

Let us recall that Q' is not a fundamental property of a body; even in linear tidal theory, it is expected to depend on the tidal frequency as well as on the integers l and m that label the different spherical harmonics. Even though theory does not yet explain why $Q'_* \approx 10^6$ for the circularization of solar-type binary stars, it can plausibly explain why Q'_* should be much larger for the orbital decay of hot Jupiters. The host stars involved are mostly spinning more slowly than those in close binary stars, and the tidal frequency of the asynchronous tide in the fluid frame is usually such that $|\hat{\omega}| \gg \Omega_s$. But the eccentricity tides in a synchronized binary star have $|\hat{\omega}| = \Omega_s$ and can excite inertial waves in the convective zone, producing a much stronger tidal response (Ogilvie & Lin 2007). In the radiative zone of a solar-type star, the asynchronous tide can break and produce rapid orbital decay only if the planet and star are sufficiently massive and the star sufficiently old (Barker & Ogilvie 2010; see Section 4.2). The predictions of the wave-breaking theory appear to be in fair agreement with observations (T. Guillot, D.N.C. Lin, P. Morel, in preparation).

The most detailed comparison to date between (highly simplified) tidal theories and exoplanetary observations is that by Hansen (2010, 2012). He evolves interior models of the star and planet, including the effects of tidal heating, together with the equations of tidal evolution of the spin and orbit under the assumption of a common time lag τ, which allows arbitrary eccentricities to be considered. In the first paper (Hansen 2010), he makes the (arbitrary) assumption that all the stars, and all the planets, have fixed values of the parameter (Eggleton et al. 1998)

$$\sigma = \frac{2}{3} k_2 \tau \frac{G}{R^5} = \frac{1}{Q'|\hat{\omega}|} \frac{G}{R^5}. \tag{31}$$

The value of σ_* that he fits from the circularization of solar-type binary stars (see Section 5.2), which corresponds to $Q'_* = 10^6$ in the case of the present Sun for a tidal period of 6 days, would cause much too rapid an evolution of exoplanetary systems, and he finds it necessary to scale it down by more than two orders of magnitude. His preferred value of σ_p corresponds to $Q'_p = 10^8$ in the case of Jupiter for a tidal period of 4 days or to $10^7 \lesssim Q'_p \lesssim 10^8$ for the circularization of hot Jupiters.

In the second paper (Hansen 2012), he instead calculates the value of σ for each star or planet using Zahn's theory of the equilibrium tide, admitting also an empirical correction factor ν_0 for the turbulent viscosity. He finds that a mild correction $\nu_0 \approx 1.5$ is sufficient to explain the circularization of binaries containing a giant star (Verbunt & Phinney 1995) while also being compatible with the nonzero orbital eccentricities of several massive, short-period exoplanets. In fact, as illustrated in **Figure 5**, Zahn's theory is not expected to be applicable to the high tidal frequencies encountered in the latter systems.

Schlaufman & Winn (2013) find evidence for planet destruction by inward migration due to tidal dissipation in subgiant stars. The paucity of planets with $P \lesssim 200$ days around subgiants, in comparison with their presumed main-sequence progenitor stars, is taken as evidence that the shorter-period planets were consumed as the star swelled and developed a deep convective envelope after leaving the main sequence. However, it is difficult to explain the absence of planets out to 200 days by this means unless Q'_* is very small, i.e., of order 10^2 (see also Hansen 2010, 2012).

Measurements of (sky-projected) stellar obliquities provide interesting constraints on scenarios for planet formation, and tidal evolution may play a role in their interpretation. Surprisingly, many systems with hot Jupiters have large spin-orbit misalignments, exceeding 90° in some cases. Winn et al. (2010) argued that most of the cooler stars ($T_{\mathrm{eff}} < 6{,}250$ K) are aligned, whereas the hotter stars have widely spread obliquities. They suggested that hot Jupiters are formed in such a way that the stellar obliquities are broadly distributed, but those of the cooler stars are subsequently reduced by tidal dissipation. Albrecht et al. (2012) compiled a set of 53 observed systems and argued that the aligned systems are those in which the timescale for spin-orbit alignment is shorter than the age. They used the estimates by Zahn (1977) for tidal synchronization timescales due to equilibrium and dynamical tides for cooler and hotter stars, respectively. (The alignment timescale might be assumed to be similar to the synchronization timescale in systems for which $L_s \lesssim L_o$; see Equations 7 and 9.) In the case of a planet of 1 Jupiter mass in a 3-day orbit around a star of 1 $M_\odot$

that is slowly rotating with a period of 30 days, the alignment timescale can be estimated roughly as 2,000 Q'_* years, although several different tidal components are potentially involved. Albrecht et al.'s figure 24 (in which they confusingly reduced the timescales by a factor of 5×10^9) suggests, in fact, that tidal alignment is possible within the age of the system for only one or two of the stars, using Zahn's theory of the equilibrium tide.

If stellar tidal dissipation is effective in reducing the obliquity, then there is a danger that the planet will also undergo orbital decay and be destroyed. From Equations 6 and 9 and their generalizations to arbitrary obliquity (e.g., Lai 2012), we can see two situations in which alignment can proceed more rapidly than orbital migration. First, if $L_s \ll L_o$, then alignment can be faster because it involves mainly a change in the stellar spin axis, whereas the orbit is hardly affected. Second, if the tidal response to one or more of the $n = 0$ components (which are associated with a torque but no power, and so do not affect a) is much stronger, then alignment can be accelerated. Indeed, Lai (2012) has noted that the 2, 1, 0 obliquity tide has frequency $\hat{\omega} = -\Omega$ in the fluid frame, within the spectral range of inertial waves, and may thereby elicit a much stronger tidal response. However, Lai (2012) and Rogers & Lin (2013) point out that boosting this component could produce retrograde or even polar orbits, as well as aligned prograde orbits, because the associated $\mathrm{d}i/\mathrm{d}t$ is proportional to $-\sin i \cos^2 i[1 + (L_o/L_s)\cos i]$. Another possible objection is that there is an $m = 1$ mode at exactly $\hat{\omega} = -\Omega$, but it is the spin-over mode rather than a genuine inertial mode and has zero dissipation because it involves a uniform rotation. However, there may be nontrivial dynamics involving a core or involving precessional motion of the spin and orbit due to this secular component of the tidal potential. Further investigation is required.

Transit observations reveal a wide spread of planetary radii, many of which are too large to be explained by conventional models. In principle, tidal dissipation in the planet, due to the synchronization of its spin and the circularization of its orbit, can deposit a very large amount of heat, and the planet can be significantly inflated if the deposition is sufficiently deep. However, this effect does not persist long after the orbit is circularized. Leconte et al. (2010) and Hansen (2010) have concluded that tidal heating cannot provide a general explanation of the radius anomalies.

Nevertheless, Arras & Socrates (2010), following earlier studies of Venus by Gold & Soter (1969) and Correia et al. (2003), have argued that a thermal tide, caused by asymmetric irradiation of the planet by the star, causes a torque that drives the planet away from synchronous spin. The competition between this torque and the gravitational tidal torque could result in a sustained asynchronous state with ongoing tidal dissipation. Although the theory of thermal tides in giant planets is in an embryonic (and controversial) state, A. Socrates (unpublished, arXiv:1304.4121) finds that the expected scaling laws for the resulting tidal heating appear compatible with the observed radius anomalies.

As well as understanding the statistical distributions of properties of extrasolar planets, it is important to consider systems on an individual basis, because of the varying interior structures of stars and planets. Future work ought to take into account the dependence of the tidal response on the extent of a stellar convective zone, for example, or on the size of a planetary core, where this is constrained by observations of the planetary mass and radius.

5.4. Tidal Evolution in the Solar System

Tidal dissipation in the giant planets of the Solar System leads to orbital migration of their regular satellites, which have nearly circular and equatorial orbits. Except for the small, innermost moons that orbit within the planet's corotation radius (synchronous orbit), the migration is outward and its rate is proportional to the mass of the satellite and to the imaginary part of the relevant Love number, $\kappa_{2,2,2}$, at the appropriate tidal frequency. Even without considering the frequency dependence of the Love number, there is a strong dependence of the migration rate on the orbital separation or period (see Equation 6). Goldreich (1965) proposed that convergent differential migration could cause pairs of satellites to enter the stable commensurabilities (mean-motion resonances) that are observed in numerous cases, including the famous Laplace resonance between Io, Europa, and Ganymede. Two or more satellites that are locked into resonance migrate at a rate determined by their combined angular momentum and the combined tidal torque that the planet exerts on them. Details of the various resonances reveal many interesting puzzles and constraints on tidal dissipation in both the planets and their satellites (Peale 1999).

Goldreich & Soter (1966) deduced lower bounds on the planets' modified tidal quality factors, $Q' = 3/(-2\kappa_{2,2,2})$, from the existence of close satellites (outside corotation), assuming they are as old as the Solar System. It is assumed that Q' does not depend significantly on frequency. The solution for a, starting from $a = a_0$ at $t = t_0$, is then

$$a^{13/2} = a_0^{13/2} + \frac{117}{4}\frac{M_2}{M_1}\frac{R_1^5}{Q'}(GM)^{1/2}(t - t_0). \tag{32}$$

In the case of Jupiter, to move Io alone from the corotation radius to its present location in 4.5×10^9 years requires $Q'_J = 1.2 \times 10^6$. Once Io, Europa, and Ganymede are locked into resonance, however, their combined angular momentum is 4.3 times that of Io alone, and the combined tidal torque on them is scarcely larger than that on Io alone. This means that Q'_J could be as small as 2.7×10^5 if the resonance is formed early. Combining a similar constraint with a detailed study of the Laplace resonance and the heating of Io, Yoder & Peale (1981) concluded that $2.4 \times 10^5 \lesssim Q'_J \lesssim 8 \times 10^6$.

In the case of Saturn, to move Mimas alone from the corotation radius to its present location in 4.5×10^9 years requires $Q'_S = 8.0 \times 10^4$. Once Mimas and Tethys are locked into resonance, however, their combined angular momentum is 22 times that of Mimas alone, and the combined tidal torque on them is 18 times that on Mimas alone (assuming Q'_S is independent of frequency). This means that Q'_S could be as small as 6.6×10^4 if the resonance is formed early.

In the cases of Uranus and Neptune, the existence of Ariel and Proteus provides lower bounds of $Q'_U \gtrsim 9.1 \times 10^4$ and $Q'_N \gtrsim 6.7 \times 10^4$, respectively. Based on detailed evolutionary scenarios for Miranda, Ariel, and Umbriel, Tittemore & Wisdom (1990) concluded that $1.6 \times 10^5 \lesssim Q'_U \lesssim 5.6 \times 10^5$, and most probably toward the lower end of this range.

Recently, however, these constraints have been challenged. Lainey et al. (2009) have fitted a sophisticated dynamical model, including parameterized tidal dissipation, to astrometric observations of the Galilean satellites since 1891, and they deduce $(k_2/Q)_J = (1.1 \pm 0.2) \times 10^{-5}$, i.e., $Q'_J = (1.4 \pm 0.3) \times 10^5$, for the asynchronous tide due to Io. This value, which represents only a snapshot of the tidal evolution, is below the lower bounds quoted above for the average over astronomical time and might suggest that Q'_J is in fact frequency dependent.

Applying a similar analysis to Saturn, Lainey et al. (2012) deduce $(k_2/Q)_S = (2.3 \pm 0.7) \times 10^{-4}$, i.e., $Q'_S = (7.2 \pm 2.2) \times 10^3$, for the asynchronous tides due to Enceladus, Tethys, Dione, and Rhea. This value is far below the lower bound quoted above, based on the migration of Mimas (and Tethys), and could help to explain the heating rate of Enceladus (cf. Meyer & Wisdom 2007). However, this result is controversial, and Mimas is especially problematic in the model of Lainey et al. because it is apparently migrating inward and would require a large negative torque corresponding to $Q'_S = -570$. This suggests that the dynamical solution of Lainey et al. may not be correct. The most plausible origin of a negative torque on Mimas's orbit is Tethys, with which it is resonantly locked. It is possible that a nonuniform tidal evolution, due to a strong frequency dependence of Q'_S, could cause the resonant torque between Mimas and Tethys to fluctuate in such a way that Mimas is currently being repelled from Tethys.

The unexpectedly rapid orbital migration deduced by Lainey et al. (2012) has been taken as support for a model in which planetary satellites are formed at the outer edge of a planetary ring system (close to the Roche limit) and migrate outward through a combination of ring torques and planetary tidal torques. Initially, Charnoz et al. (2010) proposed this for the small inner moons (Atlas–Janus) of Saturn. Charnoz et al. (2011) extended this idea to Mimas–Rhea, requiring a small value of Q'_S as suggested by Lainey et al. (2012). (Beyond the 2:1 Lindblad resonance with the outer edge of the rings, only the planetary tidal torque is effective.) However, it has not been demonstrated that the current configuration can be reached while avoiding other stable commensurabilities such as 3:2. More recently, Crida & Charnoz (2012) have applied the same idea to the other planets except Jupiter. For Uranus and Neptune, however, the Roche limit (however reasonably defined) is well inside the corotation radius, and the mechanism may not work; indeed, the inner moons migrate inward and may be disrupted to form rings (Leinhardt et al. 2012).

It is natural to ask what range of values of Q'_S can be explained theoretically and whether Q'_S should be smaller than Q'_J by an order of magnitude or more, as required to cause a significant migration of Saturn's inner moons. The frequency of the asynchronous tide satisfies $|\hat{\omega}/\Omega_s| < 2$ for $0 < \Omega_o/\Omega_s < 2$ (**Figure 2**), which is true for almost all regular satellites. Inertial waves can therefore be excited in the planet. The tidal response can be expected to depend strongly on the tidal frequency and on the size of the core or on other features of the internal structure. Despite considerable uncertainty in the interior models, values of Q' of the right order of magnitude to provide significant tidal evolution of Io and Mimas can certainly be obtained (G.I. Ogilvie, in preparation), and Jupiter's smaller core could explain the difference between the planets. (Models of Uranus and Neptune are even more uncertain at the present time.) An alternative model involving viscoelastic dissipation in the planetary core (Dermott 1979, Remus et al. 2012) could potentially also produce values of Q' of the right order (and strongly increasing with core size, but only weakly dependent on frequency), although the rheological parameters are very uncertain.

We should not be surprised that the values of Q' for giant planets inferred from the orbital migration of their satellites are smaller (perhaps much smaller) than those inferred from the orbital circularization of hot Jupiters, because the tidal frequencies are very different in the two situations. Furthermore, hot Jupiters are expected to be more slowly rotating because of tidal synchronization, and their interiors may differ in an important way from Jupiter and Saturn; the tides in extrasolar planets are also of much larger amplitude and may be in a more nonlinear regime. In both cases the tidal frequency is expected to lie in the spectral range of inertial waves. We have seen that the frequency-averaged response in inertial waves scales such that $Q'^{-1} \propto (\Omega_s/\omega_d)^2$, which naturally provides a difference of about two orders of magnitude between the two situations. The alternative Maxwellian viscoelastic model could potentially explain a difference of about one order of magnitude between the two situations (Storch & Lai 2014), but it also requires that the Maxwell frequency of the core happens to be comparable to the tidal forcing frequencies due to satellites.

SUMMARY POINTS

1. The irreversible evolution of the spin and orbital parameters of an astrophysical fluid body with a close and massive companion is related (at least in the case of uniform rotation) to the rate of dissipation of energy associated with the response of the body to various components of the tidal potential.

2. The response of a fluid body to tidal forcing generally consists of a quasi-hydrostatic bulge and an associated large-scale flow, together with internal (gravity and inertial) waves, often of much smaller scale. Tidal forcing frequencies very often coincide with those of inertial waves.
3. The nonwave-like part of the tide can be dissipated by its interaction with convection or, in some cases, as a result of its own instability or possibly through the action of irreversible effects in multiphase fluids.
4. The wave-like part of the tide has a very different character in radiative and convective zones. Dissipation is effective when the waves achieve a sufficiently small scale to be damped by radiative diffusion or (turbulent) viscosity, when the wave amplitude is enhanced by resonances with global modes, or when nonlinearity causes the waves to break.
5. Idealized calculations of linear tidal responses show a greatly enhanced, but highly frequency-dependent, dissipation rate owing to the excitation of internal waves. Nonideal linear effects and nonlinearity may tend to smooth out the features in the response curves. The evolution of a system with a highly frequency-dependent tidal response is complicated and may involve resonance locking.
6. Tidal dissipation induces differential rotation in fluid bodies, which in turn affects the properties of the internal waves in a way that remains to be fully understood.
7. Observations provide indirect evidence of tidal evolution in binary stars, extrasolar planetary systems, and the satellite systems of the giant planets in the Solar System. In each case useful constraints on theories of tidal dissipation are obtained.

FUTURE ISSUES

1. Further studies of the interaction between oscillatory tidal disturbances and turbulent convection are needed to clarify the role of stellar and planetary convection in tidal dissipation.
2. Linear calculations of the tidal responses of stars and planets should be carried out for a wide range of realistic interior models, including the effects of rotation.
3. Nonlinear aspects of both wave-like and nonwave-like tides should be studied in more detail and under more realistic conditions.
4. The role of differential rotation, and its limitation by various processes inside stars and giant planets, needs to be assessed.
5. Atmospheric tides, both gravitational and thermal, in exoplanets should be studied using specialized numerical simulations.
6. Future observations, especially of transiting exoplanets, are likely to provide improving constraints on theories of tidal dissipation and will require systems to be modeled on an individual basis, taking into account their probable internal structure and evolutionary history.

DISCLOSURE STATEMENT

The author is not aware of any affiliations, memberships, funding, or financial holdings that might be perceived as affecting the objectivity of this review.

ACKNOWLEDGMENTS

I am grateful to Adrian Barker, Harry Braviner, Dong Lai, Doug Lin, John Papaloizou, and Stan Peale for comments and discussions that led to improvements in this review.

LITERATURE CITED

Adams ER, López-Morales M, Elliot JL, et al. 2011. *Ap. J.* 741:102

Aerts C, Christensen-Dalsgaard J, Kurtz DW. 2010. *Asteroseismology*. Dordrecht: Springer

Albrecht S, Reffert S, Snellen I, Quirrenbach A, Mitchell DS. 2007. *Astron. Astrophys.* 474:565–73

Albrecht S, Winn JN Johnson JA, et al. 2012. *Ap. J.* 757:18

Alexander ME. 1973. *Ap. Space Sci.* 23:459–510

Arras P, Socrates A. 2010. *Ap. J.* 714:1–12

Barker AJ. 2011. *MNRAS* 414:1365–78

Barker AJ, Lithwick Y. 2013a. *MNRAS* 435:3614–26

Barker AJ, Lithwick Y. 2013b. *MNRAS* 437:305–15

Barker AJ, Ogilvie GI. 2010. *MNRAS* 404:1849–68

Baruteau C, Rieutord M. 2013. *J. Fluid Mech.* 719:47–81

Bildsten L, Cutler C. 1992. *Ap. J.* 400:175–80

Bodenheimer P, Lin DNC, Mardling RA. 2001. *Ap. J.* 548:466–72

Bouchy F, Deleuil M, Guillot T. 2011. *Astron. Astrophys.* 525:A68

Brown DJA, Collier Cameron A, Hall C, Hebb L, Smalley B. 2011. *MNRAS* 415:605–18

Bryan GH. 1889. *Philos. Trans. R. Soc. Lond. A* 180:187–219

Burkart J, Quataert E, Arras P, Weinberg NN. 2012. *MNRAS* 421:983–1006

Cartan E. 1922. *Bull. Sci. Math. France* 46:317–52

Cartwright DE. 1999. *Tides: A Scientific History*. Cambridge: Cambridge Univ. Press

Cébron D, Le Bars M, Le Gal P, et al. 2013. *Icarus* 226:1642–53

Cébron D, Le Bars M, Leontini J, Maubert P, Le Gal P. 2010. *Phys. Earth Planet. Inter.* 182:119–28

Charnoz S, Crida A, Castillo-Rogez JC, et al. 2011. *Icarus* 216:535–50

Charnoz S, Salmon J, Crida A. 2010. *Nature* 465:752–54

Claret A, Cunha NCS. 1997. *Astron. Astrophys.* 318:187–97

Comeau S, Poisson E. 2009. *Phys. Rev. D* 80:087501

Correia ACM, Laskar J, Néron de Surgy O. 2003. *Icarus* 163:1–23

Counselman CC. 1973. *Ap. J.* 180:307–16

Cowling TG. 1941. *MNRAS* 101:367–75

Crida A, Charnoz S. 2012. *Science* 338:1196–99

Darwin GH. 1880. *Philos. Trans. R. Soc. Lond.* 171:713–891

Deparis V, Legros H, Souchay J. 2013. In *Tides in Astronomy and Astrophysics*, ed. J Souchay, S Mathis, T Tokieda, pp. 31–82. Heidelberg: Springer

Dermott SF. 1979. *Icarus* 37:310–21

Dintrans B, Rieutord M, Valdettaro L. 1999. *J. Fluid Mech.* 398:271–97

Eckart C. 1960. *Hydrodynamics of Oceans and Atmospheres*. Oxford: Pergamon

Eggleton PP, Kiseleva LG, Hut P. 1998. *Ap. J.* 499:853–70

Fabian AC, Pringle JE, Rees MJ. 1975. *MNRAS* 172:15p–18p

Favier B, Barker AJ, Baruteau C, Ogilvie GI. 2014. *MNRAS* 493:845–60

French M, Becker A, Lorenzen W, et al. 2012. *Ap. J. Suppl.* 202:5

Fuller J, Lai D. 2012a. *MNRAS* 420:3126–38

Fuller J, Lai D. 2012b. *MNRAS* 421:426–45
Gavrilov SV, Zharkov VN. 1977. *Icarus* 32:443–49
Giersz M. 1986. *Acta Astron.* 36:181–209
Gold T, Soter S. 1969. *Icarus* 11:356–66
Goldreich P. 1963. *MNRAS* 126:257–68
Goldreich P. 1965. *MNRAS* 130:159–81
Goldreich P, Keeley DA. 1977. *Ap. J.* 211:934–42
Goldreich P, Nicholson PD. 1977. *Icarus* 30:301–4
Goldreich P, Nicholson PD. 1989. *Ap. J.* 342:1079–84
Goldreich P, Soter S. 1966. *Icarus* 5:375–89
Goodman J, Dickson ES. 1998. *Ap. J.* 507:938–44
Goodman J, Lackner C. 2009. *Ap. J.* 696:2054–67
Goodman J, Oh SP. 1997. *Ap. J.* 486:403–12
Greenspan H. 1968. *The Theory of Rotating Fluids*. Cambridge: Cambridge Univ. Press
Gu P-G, Lin DNC, Bodenheimer PH. 2003. *Ap. J.* 588:509–34
Guillochon J, Ramirez-Ruiz E, Lin D. 2011. *Ap. J.* 732:74
Hansen BMS. 2010. *Ap. J.* 723:285–99
Hansen BMS. 2012. *Ap. J.* 757:6
Hartle JB. 1973. *Phys. Rev. D* 8:1010–24
Hebb L, Collier Cameron A, Triaud AHMJ, et al. 2010. *Ap. J.* 708:224–31
Hellier C, Anderson DR, Collier Cameron A, et al. 2009. *Nature* 460:1098–100
Hough SS. 1897. *Philos. Trans. R. Soc. Lond. A* 189:201–57
Hough SS. 1898. *Philos. Trans. R. Soc. Lond. A* 191:139–85
Husnoo N, Pont F, Mazeh T, et al. 2012. *MNRAS* 422:3151–77
Hut P. 1980. *Astron. Astrophys.* 92:167–70
Hut P. 1981. *Astron. Astrophys.* 99:126–40
Ibgui L, Burrows A. 2009. *Ap. J.* 700:1921–32
Ioannou PJ, Lindzen RS. 1993. *Ap. J.* 406:266–78
Ivanov PB, Papaloizou JCB. 2007. *MNRAS* 376:682–704
Ivanov PB, Papaloizou JCB. 2010. *MNRAS* 407:1609–30
Jackson B, Barnes R, Greenberg R. 2009. *Ap. J.* 698:1357–66
Jackson B, Greenberg R, Barnes R. 2008. *Ap. J.* 678:1396–406
Jeffreys H. 1961. *MNRAS* 122:339–43
Jouve L, Ogilvie GI. 2014. *J. Fluid Mech.* 745:223–50
Kaula WM. 1961. *Geophys. J. Int.* 5:104–33
Kerswell RR. 2002. *Annu. Rev. Fluid Mech.* 34:83–113
Khaliullin KhF, Khaliullina AI. 2007. *MNRAS* 382:356–66
Khaliullin KhF, Khaliullina AI. 2010. *MNRAS* 401:257–74
Kipping D. 2013. *MNRAS* 434:L51–55
Kochanek CS. 1992. *Ap. J.* 398:234–47
Kopal Z. 1978. *Dynamics of Close Binary Systems*. Dordrecht: Reidel
Kosovichev AG, Novikov ID. 1992. *MNRAS* 258:715–24
Kramm U, Nettelmann N, Fortney JJ, Neuhaüser R, Redmer R. 2012. *Astron. Astrophys.* 538:A146
Kumar P, Goodman J. 1996. *Ap. J.* 466:946–56
Lai D. 1997. *Ap. J.* 490:847–62
Lai D. 2012. *MNRAS* 423:486–92
Lainey V, Arlot J-E, Karatekin Ö, van Hoolst T. 2009. *Nature* 459:957–59
Lainey V, Karatekin Ö, Desmars J, et al. 2012. *Ap. J.* 752:14
Langer N. 2009. *Astron. Astrophys.* 500:133–34
Laplace PS. 1775. *Mém. Acad. R. Sci.* 88:75–182
Le Bars M, Lacaze L, Le Dizès S, Le Gal P, Rieutord R. 2010. *Phys. Earth Planet. Inter.* 178:48–55
Leconte J, Chabrier G, Baraffe I, Levrard B. 2010. *Astron. Astrophys.* 516:A64
Lee HM, Ostriker JP. 1986. *Ap. J.* 310:176–88

Leinhardt ZM, Ogilvie GI, Latter HN, Kokubo E. 2012. *MNRAS* 424:1419–31
Levrard B, Winisdoerffer C, Chabrier G. 2009. *Ap. J. Lett.* 692:L9–13
Lubow SH, Tout CA, Livio M. 1997. *Ap. J.* 484:866–70
Maas LRM, Lam F-PA. 1995. *J. Fluid Mech.* 300:1–41
McMahon JM, Morales MA, Pierleoni C, Ceperley DM. 2012. *Rev. Mod. Phys.* 84:1607–53
McMillan SLW, McDermott PN, Taam RE. 1987. *Ap. J.* 318:261–77
Marcy GW, Butler RP, Williams E, et al. 1997. *Ap. J.* 481:926–35
Mardling RA. 1995a. *Ap. J.* 450:722–31
Mardling RA. 1995b. *Ap. J.* 450:732–47
Mathieu RD. 2005. In *Tidal Evolution and Oscillations in Binary Stars: Third Granada Workshop on Stellar Structure*, ed. A Claret, A Giménez, J-P Zahn. *ASP Conf. Ser.* 333:26–38. San Francisco: ASP
Matsumura S, Peale SJ, Rasio FA. 2010. *Ap. J.* 725:1995–2016
Meibom S, Mathieu RD. 2005. *Ap. J.* 620:970–83
Meibom S, Mathieu RD, Stassun KG. 2006. *Ap. J.* 653:621–35
Meyer J, Wisdom J. 2007. *Icarus* 188:535–39
Mied RP. 1976. *J. Fluid Mech.* 78:763–84
Ogilvie GI. 2005. *J. Fluid Mech.* 543:19–44
Ogilvie GI. 2009. *MNRAS* 396:794–806
Ogilvie GI. 2013. *MNRAS* 429:613–32
Ogilvie GI, Lesur G. 2012. *MNRAS* 422:1975–87
Ogilvie GI, Lin DNC. 2004. *Ap. J.* 610:477–509
Ogilvie GI, Lin DNC. 2007. *Ap. J.* 661:1180–91
O'Leary RM, Burkart J. 2014. *MNRAS* 440:3036–50
Papaloizou JCB, Ivanov PB. 2005. *MNRAS* 364:L66–70
Papaloizou JCB, Ivanov PB. 2010. *MNRAS* 407:1631–56
Papaloizou JCB, Pringle JE. 1978. *MNRAS* 182:423–42
Papaloizou JCB, Savonije GJ. 1997. *MNRAS* 291:651–57
Peale SJ. 1999. *Annu. Rev. Astron. Astrophys.* 37:533–602
Penev K, Barranco J, Sasselov D. 2009a. *Ap. J.* 705:285–97
Penev K, Sasselov D, Robinson F, Demarque P. 2007. *Ap. J.* 655:1166–71
Penev K, Sasselov D, Robinson F, Demarque P. 2009b. *Ap. J.* 704:930–36
Poincaré H. 1885. *Acta Math.* 7:259–380
Poisson E. 2009. *Phys. Rev. D* 80:064029
Polfliet R, Smeyers P. 1990. *Astron. Astrophys.* 237:110–24
Pont F. 2009. *MNRAS* 396:1789–96
Press WH, Teukolsky SA. 1977. *Ap. J.* 213:183–92
Press WH, Wiita PJ, Smarr LL. 1975. *Ap. J. Lett.* 202:L135–37
Rappaport S, Sanchis-Ojeda R, Rogers LA, Levine A, Winn JN. 2013. *Ap. J. Lett.* 773:L15
Rasio FA, Tout CA, Lubow SH, Livio M. 1996. *Ap. J.* 470:1187–91
Remus F, Mathis S, Zahn J-P, Lainey V. 2012. *Astron. Astrophys.* 541:A165
Rieutord M. 1992. *Astron. Astrophys.* 259:581–84
Rieutord M. 2004. In *Stellar Rotation, Proc. IAU Symp. 215*, ed. A Maeder, PRJ Eenens, pp. 394–403. San Francisco: ASP
Rieutord M, Georgeot B, Valdettaro L. 2001. *J. Fluid Mech.* 435:103–44
Rieutord M, Valdettaro L. 2010. *J. Fluid Mech.* 643:363–94
Rieutord M, Zahn J-P. 1997. *Ap. J.* 474:760–67
Rogers TM, Lin DNC. 2013. *Ap. J. Lett.* 769:L10
Sanchis-Ojeda R, Rappaport S, Winn JN, et al. 2013. *Ap. J.* 774:54
Savonije GJ, Papaloizou JCB. 1983. *MNRAS* 203:581–93
Savonije GJ, Papaloizou JCB. 1984. *MNRAS* 207:685–704
Savonije GJ, Papaloizou JCB. 1997. *MNRAS* 291:633–50
Savonije GJ, Papaloizou JCB, Alberts F. 1995. *MNRAS* 277:471–96
Savonije GJ, Witte MG. 2002. *Astron. Astrophys.* 386:211–21

Schlaufman KC, Lin DNC, Ida S. 2010. *Ap. J. Lett.* 724:L53–58
Schlaufman KC, Winn JN. 2013. *Ap. J.* 772:143
Siverd RJ, Beatty TG, Pepper J, et al. 2012. *Ap. J.* 761:123
Sridhar S, Tremaine S. 1992. *Icarus* 95:86–99
Stevenson DJ. 1983. *J. Geophys. Res.* 88:2445–55
Storch NI, Lai D. 2014. *MNRAS* 438:1526–34
Tassoul J-L. 1987. *Ap. J.* 322:856–61
Terquem C, Papaloizou JCB, Nelson RP, Lin DNC. 1998. *Ap. J.* 502:788–801
Thompson SE, Everett M, Mullally F, et al. 2012. *Ap. J.* 753:86
Thomson W. 1863. *Philos. Trans. R. Soc. Lond.* 153:583–616
Thomson W. 1880. *Philos. Mag.* 10:155–68
Thorne KS, Price RH, MacDonald DA. 1986. *Black Holes: The Membrane Paradigm*. New Haven: Yale Univ. Press
Tittemore WC, Wisdom J. 1990. *Icarus* 85:394–443
Venumadhav T, Zimmerman A, Hirata CM. 2014. *Ap. J.* 781:23
Verbunt F, Phinney ES. 1995. *Astron. Astrophys.* 296:709–21
Watson CA, Marsh TR. 2010. *MNRAS* 405:2037–43
Weinberg NN, Arras P, Burkart J. 2013. *Ap. J.* 769:121
Weinberg NN, Arras P, Quataert E, Burkart J. 2012. *Ap. J.* 751:136
Welsh WF, Orosz JA, Aerts C, et al. 2011. *Ap. J. Suppl.* 197:4
Welsh WF, Orosz JA, Seager S, et al. 2010. *Ap. J. Lett.* 713:L145–49
Winn JN, Fabrycky D, Albrecht S, Johnson JA. 2010. *Ap. J. Lett.* 718:L145–49
Witte MG, Savonije GJ. 1999a. *Astron. Astrophys.* 341:842–52
Witte MG, Savonije GJ. 1999b. *Astron. Astrophys.* 350:129–47
Witte MG, Savonije GJ. 2001. *Astron. Astrophys.* 366:840–57
Witte MG, Savonije GJ. 2002. *Astron. Astrophys.* 386:222–36
Wu Y. 2005a. *Ap. J.* 635:674–87
Wu Y. 2005b. *Ap. J.* 635:688–710
Yoder CF, Peale SJ. 1981. *Icarus* 47:1–35
Zahn J-P. 1966a. *Ann. Astrophys.* 29:313–30
Zahn J-P. 1966b. *Ann. Astrophys.* 29:489–506
Zahn J-P. 1970. *Astron. Astrophys.* 4:452–61
Zahn J-P. 1975. *Astron. Astrophys.* 41:329–44
Zahn J-P. 1977. *Astron. Astrophys.* 57:383–94
Zahn J-P. 1989. *Astron. Astrophys.* 220:112–16
Zahn J-P, Bouchet L. 1989. *Astron. Astrophys.* 223:112–18

Gamma-Ray Pulsar Revolution

Patrizia A. Caraveo

Istituto di Astrofisica Spaziale e Fisica Cosmica (IASF) - Istituto Nazionale di Astrofisica (INAF), 20133 Milano, Italy; email: pat@iasf-milano.inaf.it

Annu. Rev. Astron. Astrophys. 2014. 52:211–50

First published online as a Review in Advance on June 4, 2014

The *Annual Review of Astronomy and Astrophysics* is online at astro.annualreviews.org

This article's doi:
10.1146/annurev-astro-081913-035948

Keywords

neutron stars, gamma-ray detectors, gamma-ray sources, radio-quiet pulsars, millisecond pulsars

Abstract

Isolated neutron stars (INSs) were the first sources identified in the field of high-energy gamma-ray astronomy. In the 1970s, only two sources had been identified, the Crab and Vela pulsars. However, although few in number, these objects were crucial in establishing the very concept of a gamma-ray source. Moreover, they opened up significant discovery space in both the theoretical and phenomenological fronts. The need to explain the copious gamma-ray emission of these pulsars led to breakthrough developments in understanding the structure and physics of neutron star (NS) magnetospheres. In parallel, the 20-year-long chase to understand the nature of Geminga unveiled the existence of a radio-quiet, gamma-ray-emitting INS, adding a new dimension to the INS family.

We are living through an extraordinary time of discovery. The current generation of gamma-ray detectors has vastly increased the population of known gamma-ray-emitting NSs. The 100 mark was crossed in 2011, and we are now over 150. The gamma-ray-emitting NS population exhibits roughly equal numbers of radio-loud and radio-quiet young INSs, plus an astonishing, and unexpected, group of isolated and binary millisecond pulsars (MSPs). The number of MSPs is growing so rapidly that they are on their way to becoming the most numerous members of the family of gamma-ray-emitting NSs. Even as these findings have set the stage for a revolution in our understanding of gamma-ray-emitting NSs, long-term monitoring of the gamma-ray sky has revealed evidence of flux variability in the Crab Nebula as well as in the pulsed emission from PSR J2021+4026, challenging a four-decades-old, constant-emission paradigm. Now we know that both pulsars and their nebulae can, indeed, display variable emission.

1. INTRODUCTION

Neutron stars (NSs) are very efficient natural particle accelerators owing to their rapidly rotating, hugely intense magnetic fields. Moreover, the accelerated particles (mostly electrons and positrons) are delivered into a highly magnetized surrounding, ideal for making them radiate high-energy gamma rays that bear the timing signature of their parent isolated neutron star (INS). In a field such as high-energy gamma-ray astronomy, which is hampered by poor angular resolution compounded by relatively low numbers of detected photons, the presence of an unambiguous timing signature has been crucial for allowing identification of gamma-ray sources with crude positions. The combination of the ideal physical conditions with the telltale time signature makes INSs the most prominent class of high-energy gamma-ray emitters in our Galaxy.

However, progress in this field has been hampered by all sorts of experimental difficulties, stemming from the paucity of gamma-ray photons. Gamma-ray pulsars cannot be detected in real time, like radio pulsars. To see pulsations, i.e., to build a statistically significant light curve, photons collected over weeks to years have to be properly phased, folding their arrival times according to precise timing parameters. Prior to doing so, however, photon arrival times must be converted to the Solar System barycenter, an ancillary yet inescapable operation upon which rests the success of the folding technique. Because such a correction is very position sensitive, precise source position is an essential bit of information for performing a search for pulsations in gamma rays. Thus, methods had to be devised to properly correct and fold the gamma-ray photon arrival times while testing the statistical significance of the results.

In the meantime, the performance of gamma-ray telescopes has improved generation after generation, going from NASA's pioneering SAS-2 (Fichtel et al. 1975) and ESA's COS-B (Bignami et al. 1975), to NASA's EGRET (Kanbach et al. 1988), to the current generation, encompassing Agenzia Spaziale Italiana's (ASI's) AGILE (Tavani et al. 2009) and NASA's *Fermi* (née GLAST, Atwood et al. 2009). Four decades of unrelenting efforts in hardware and software were needed to go from the first firm detection of a gamma-ray pulsar to a family portrait encompassing far more than 100 objects. The growth, stemming primarily from the dramatic acceleration of recent years, is impressive, as shown in **Figure 1**. In parallel, the overall quality of the data, namely the angular, spectral, and time resolution achieved for each photon as well as the overall sensitivity of the instruments, also significantly improved. This can be also seen in **Figure 1**, which shows a compilation of five sets of Vela pulsar light curves measured over a span of four decades by the five missions mentioned above.

2. THE PAST: WHEN SOURCES WERE FEW AND FAR BETWEEN, AND EACH ONE COUNTED

2.1. Pulsars as Gamma-Ray Sources

All the sky maps produced since the beginning of gamma-ray astronomy are dominated by three sources near the plane of the Galaxy. Folding the gamma photon arrival times using the radio ephemerides, SAS-2 identified first the Crab (Kniffen et al. 1974) and, later, the Vela pulsars (Thompson et al. 1975), i.e., the two brightest sources in the gamma-ray sky. The third source, shining next to the Crab in the Galactic anticenter, was named $\gamma 195+5$ from its Galactic coordinates and could not be identified with any notable celestial object.

For the Crab, the SAS-2 detection was the confirmation of early, contradictory balloon claims (e.g., Vasseur et al. 1970, 1971), whereas for the older and less energetic Vela it was a genuine novelty. In the 1970s pulsars were astronomical newcomers, and the gamma-ray detections of Crab and Vela added an important new piece of information in the struggle to understand those extreme

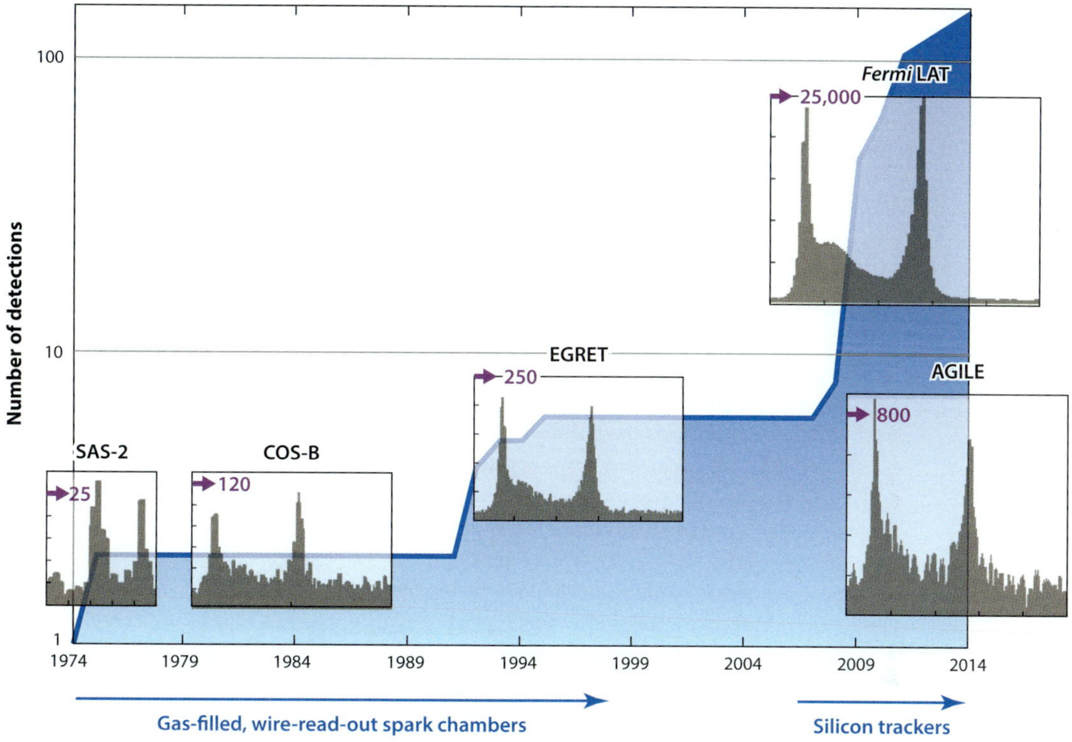

Figure 1

Evidence of the beneficial effects of sensitivity increase. Over four decades of gamma-ray astronomy, going from gas-and-wire to silicon detectors, the total number of gamma-ray pulsars is seen to increase by a factor of hundreds (*pale blue background area*). The phase histograms (*light curves*) of gamma-ray photons collected from the Vela pulsar by SAS-2 (>35 MeV), COS-B (>50 MeV), EGRET (>100 MeV), AGILE (>100 MeV) and *Fermi* LAT (>100 MeV) are also shown. The quality and detail level obtained (i.e., number of bins that can be afforded per graph and number of photons per bin) correlate directly with the increased photon statistics, i.e., mission sensitivity. The photon counting of the highest bin in the best light curve published by each mission is shown in magenta.

stars. The seminal work by Goldreich & Julian (1969) showed that a pulsar magnetosphere, far from being a vacuum, should be filled with plasma because the induced electric force wins over the gravitational pull on surface charges. Building on this result, Sturrock (1971) laid the foundations of pulsar electrodynamics and proposed the polar cap (PC) acceleration zone. Theoretical interpretations of the SAS-2 results elaborated on this idea, with contributions from Ruderman & Sutherland (1975) and Harding et al. (1978), and were further developed by Daugherty & Harding (1982). In PC models, particles, accelerated by rotation-induced electric fields above the PC, move along the dipole magnetic (B) field lines and produce curvature radiation. In this environment, photons above 1 GeV are absorbed by the B field and produce $e^+ e^-$ pairs, which radiate synchrotron photons and produce a second generation of pairs. Such a cascade will continue until the synchrotron photons fail to meet the energetic requirements to pair produce and can escape to contribute to the high-energy pulsar emission. The remaining pairs may supply particles to a coherent process that is responsible for the radio emission. This process takes place at low altitudes ($<1\ R_{NS}$, neutron star radius) above the stellar magnetic poles, where the beam of radio emission is also originating.

The premature termination of the SAS-2 mission interrupted its stream of discoveries, which was then taken up by COS-B, launched by the ESA in 1975. Similar to SAS-2 in dimensions (and, thus, in sensitivity), COS-B lasted much longer and had ample time to confirm and significantly improve the SAS-2 findings while also discovering a population of unidentified Galactic gamma-ray sources (Swanenburg et al. 1981).

The COS-B's detailed study of Crab and Vela behavior made it clear that the measured gamma-ray emission accounted for at least 10^{-3} of their rotation energy loss $\dot{E}$ ($\dot{E}_{\rm rot} = -4\pi^2 IP\dot{P}^{-3}$, where P and $\dot{P}$ are the pulsar period and period derivative, respectively, whereas I is the moment of inertia assumed to be 10^{45} g cm^{-2}). With a gamma-ray yield clearly dominating the pulsar emission, the gamma channel stands out as the most energetically demanding (Buccheri et al. 1978, Kanbach et al. 1980). To meet such requirements, Arons (1983) developed the slot gap (SG) model where pair creation takes place along favorably curved B-field lines above the PCs and far from the NS surface. This model works well for short period pulsars, such as Crab and Vela. Although their double-peaked gamma-ray light curves appear similar, it was then immediately clear that their ratios between the pulsar rotational energy loss $\dot{E}$ and their gamma-ray luminosity $L_\gamma = 4\pi d^2 F_\gamma f_\Omega$ were different (where d is the pulsar distance, F_γ the measured flux, and f_Ω is the beaming factor, which depends on pulsar geometry and is assumed to be 1 steradian, i.e., $f_\Omega = 1/4\pi$). The older and less energetic Vela was more efficient than the younger and more energetic Crab. The spectral shapes of the sources were different: Although the Crab could be fitted with a single power law[$\frac{dN}{dE} = K(\frac{E}{E_0})^{-2.1+/-0.3}$], for Vela a flattening at low energies combined with a steepening at high energies made a single power law fit much more challenging, if not impossible (Bennett et al. 1977). Moreover, the multiwavelength behavior of the two objects appeared vastly different. Although the Crab exhibits similar light curves at all wavelengths, Vela's light curves are radically different at different wavelengths (e.g., Bignami & Hermsen 1983). Such macroscopic effects, which could not be accounted for by the PC model, prompted Cheng et al. (1986) to propose an outer magnetosphere model, wherein particles are accelerated within the vacuum outer gap (OG) extending from the null surface (i.e., where $\Omega{\cdot}B = 0$, which spatially separates the opposing charges) to the light cylinder (a virtual cylindrical surface or radius $R_{\rm LC}$ where corotation stops, because $\Omega\, R_{\rm LC} = c$) and produce gamma rays far from the NS surface, mainly by curvature and synchrotron radiation. The OG model applies only to gamma-ray emission and disentangles the gamma-emitting region from the radio one. The PC and OG models produce gamma rays in totally different regions of the pulsar magnetosphere, relatively near the surface for the PC scenario and far away for the OG, which approaches the light cylinder. The two different locations imply different ambient B fields, and thus different emission-absorption combinations, as well as different geometry and thus different beaming. Moreover, geographically different emitting regions could come into view during the pulsar rotation, thus contributing at different pulsar phases to yield spectral shapes varying as a function of the pulsar rotation phase. The composite geometry interpretation was supported by the varying spectral shapes found in different phase intervals for the Vela pulsar by Grenier et al. (1988) in their attempt to perform phase-resolved spectroscopy.

2.2. More Gamma-Ray Sources: Geminga and the Rest of the Crowd

In addition to Crab and Vela, COS-B detected two dozen gamma-ray sources (Swanenburg et al. 1981) for which there were no obvious identifications. Owing to COS-B's uneven coverage of the sky, such sources were mainly located near the Galactic plane, with the notable exception of an excess positionally coincident with 3C273, the first extragalactic gamma-ray source (Bignami et al. 1981a). Particular attention was devoted to γ195+5, readily dubbed Geminga (a name inspired by the gamma-ray source position within the Gemini constellation, but also a pun in Milanese dialect meaning "is not there" or "there is nothing").

Both SAS-2 (Ogelman et al. 1976) and COS-B (Buccheri et al. 1978) data were searched for pulsed signals from known radio pulsars and both claimed low significance detections. Although such results could not be confirmed, the statistical evaluation of the search outcome led to the development of the Z_n^2 method of gauging the statistical significance of a light curve (Buccheri et al. 1983). de Jager et al. (1989; further elaborated by de Jager & Büsching 2010) proposed a different statistical test known as the H-test. Both Z_n^2 and the H-test are now widely used.

However, no statistical test can overcome the irregularities usually present in pulsar timing behavior that hamper, and sometimes prevent, the use of archival, years-old timing information to phase-fold the meager gamma-ray photon harvest. It was quickly realized that, if the radio timing parameters are not contemporary to the gamma-ray observations, the folding must be performed over a range of P and $\dot{P}$ values sampling the extrapolated parameter space. By multiplying the trials performed, this procedure weakens the significance of any tentative detection. Such a drawback was an important lesson learned for future instruments and pointed to the need for contemporaneous radio monitoring of promising pulsars. In the meantime, the SAS-2 and COS-B claims, though unconfirmed, spurred the discussion on the evolution of gamma-ray emission efficiency as a function of pulsar age. If true, they would have implied a steady growth of the gamma-ray yield as pulsars age. Although all those tentative detections have long been forgotten, the evolution of the gamma-ray emissivity as a function of pulsar age is still debated.

Meanwhile, by exploiting the latitude distribution of the unidentified sources (Swanenburg et al. 1981) as well as the shape of their log$N$-log$S$ distribution (Bignami & Caraveo 1980), it was possible to compute the average source luminosity and distance, showing that young, energetic pulsars at a few-kiloparsecs distance and with efficiencies between those of the Crab and Vela could account for at least a fraction of the newly found sources. This finding spurred the search for radio pulsars within the COS-B error boxes. However, the effort took some time and its results came too late to be useful to identify COS-B sources, but it paved the way for successful identifications with the following gamma-ray mission.

In parallel, by exploiting the imaging capability of the newly launched *Einstein Observatory*, a program to cover the error boxes of several COS-B unidentified sources was successfully carried out as an alternative way to search for gamma-ray source counterparts (Caraveo 1982). Much interest was focused on Geminga (Bignami et al. 1983), by far the brightest among the unidentified sources that had already defeated radio searches, but seminal results were obtained for 2CG135+01 (Bignami et al. 1981b), for which the peculiar binary system LSI 61° 303 was proposed as a counterpart.

The chase for Geminga went on after the demise of COS-B, exploiting all the space and ground instruments available at all possible wavelengths and bridging the hiatus between COS-B and the launch of EGRET. Bignami & Caraveo (1996) have summarized the long and checkered story that led to the discovery of the first bona fide INS pulsating in X- and gamma rays but not at radio wavelengths. Indeed, the gamma-ray pulsation was found only when *Rosat* secured the X-ray periodicity detection (Halpern & Holt 1992), making it possible to fold the gamma-ray data collected almost simultaneously by EGRET (Bertsch et al. 1992) on board the *Compton Gamma Ray Observatory* (CGRO). Geminga had been pulsating all the time, of course, but the paucity of photons detected over COS-B's seven-year lifetime, together with the poor source localization, hampered the search for pulsations that could be found only a posteriori (Bignami & Caraveo 1992). In retrospect, the discovery of a radio-quiet INS should not have been totally unexpected. In view of the geometry-driven emission mechanisms believed to be at work in radio pulsars, with radio-emitting regions probably detached from the gamma-ray regions, radio-quiet objects could have been expected. However, finding the first radio-quiet INS made it clear that gamma-ray astronomy had significant discovery space of its own.

2.3. EGRET: More Pulsars, at Last

Apart from studying the Crab (Nolan et al. 1993), Vela (Kanbach et al. 1994), and Geminga (Mattox et al. 1992, Mayer-Hasselwander et al. 1994), EGRET, which could count on a significant contemporary pulsar radio monitoring campaign, detected three more pulsars, namely PSR B1706-44 (Thompson et al. 1992, 1996), PSR B1055-52 (Fierro et al. 1993) and PSR B1951+32 (Ramanamurthy et al. 1995), whereas PSR B1509-58 was only detected at low energy by the Comptel instrument (Kuiper et al. 1999), also on board CGRO (Gehrels et al. 1994). Exploiting Comptel data, Kuiper et al. (2001) assessed the evolution of the CRAB light curve from soft X-rays up to high-energy gamma-rays. We note that PSR B1706-44 coincides with the COS-B source 2CG342-02, but the radio pulsar was discovered after the end of the COS-B mission.

Although Nel et al. (1996), using 3.5 years of EGRET observations, computed upper limits for 350 pulsars for which radio monitoring was available, it is worth mentioning that ad hoc searches did show that three more pulsars, PSR B0656+14 (Ramanamurthy et al. 1996), PSR B1046-58 (Kaspi et al. 2000), and PSR J0218+4232 (Kuiper et al. 2000), were worthy of further investigation; however, their detections were not yet compelling. While PSR B0656+14 and PSR B1046-58 were normal radio pulsars, similar to those already detected, PSR J0218+4232 was an old recycled millisecond pulsar (MSP) characterized by extremely fast rotation coupled with a magnetic field significantly lower than that of normal pulsars. Harding et al. (2002, 2005) proposed a model for acceleration in the open field line region above the PC that seemed particularly well suited for MSPs.

Thompson (2004) provides a comprehensive review of the EGRET pulsar results. Ordering the known pulsars on the basis of their overall energy output $\dot{E}$ divided by their distance factor ($4\pi d^2$), it became immediately clear that the pulsars so far detected in gamma rays were those ranking at the top of the list, i.e., those with the most favorable combination of energetics ($\dot{E} > 10^{34}$ erg sec^{-1}) and distance (Thompson et al. 1999). Although this ordering does not account for important variables, such as different inclinations and viewing angles as well as different efficiencies, and its value is only as good as the (usually uncertain) distance estimate, the $\frac{\dot{E}}{4\pi d^2}$ ranking (and its variant, $\frac{\sqrt{\dot{E}}}{4\pi d^2}$) proved to be an extremely useful tool. By plotting the pulsars' gamma-ray luminosities L_γ as a function of the open field line voltage à la Goldreich & Julian (1969), a trend can be seen pointing to a proportionality between L_γ and the open field line voltage, itself proportional to $\sqrt{\dot{E}}$ (Thompson 2004).

All pulsars exhibit a power law spectral shape with a high-energy cutoff. Light curves are usually double peaked; and the peak ratio varies with energy, with the second peak usually harder than the first (by definition, the "first" gamma-ray peak is the one that comes in phase immediately after the main radio peak). Reproducing such a double-peaked structure turns out to be an important test for pulsar models. Chiang & Romani (1992, 1994) argued that the EGRET pulsars' light curves, as well as their spectra, arise naturally from a modified version of the OG model. Working in 3D and accounting for the angle ζ between the observer's line of sight and the NS rotation axis as well as the dipole field inclination angle α (see the scheme of **Figure 2**), Romani & Yadigaroglu (1995) succeeded in reproducing the Vela light curve as measured at radio, optical, X-ray, and gamma-ray wavelengths for a very inclined geometric combination of $\alpha = 65°$ and $\zeta = 90°$. Indeed, to produce double-peaked light curves, OG models do prefer highly inclined rotators (see also Romani 1996), whereas PC models require nearly aligned geometry, allowing magnetic inclination angles comparable to the angular extent of the PC. To overcome the PC requirements on pulsar alignment, Daugherty & Harding (1996), elaborating on the seminal work by Arons (1983) on the SG idea, extrapolated the PC acceleration region to higher altitudes, which led to a full SG model by Muslimov & Harding (2003, 2004). Here particle acceleration takes place in thin SGs along the last open field line connecting the NS surface to the light cylinder. By extending the

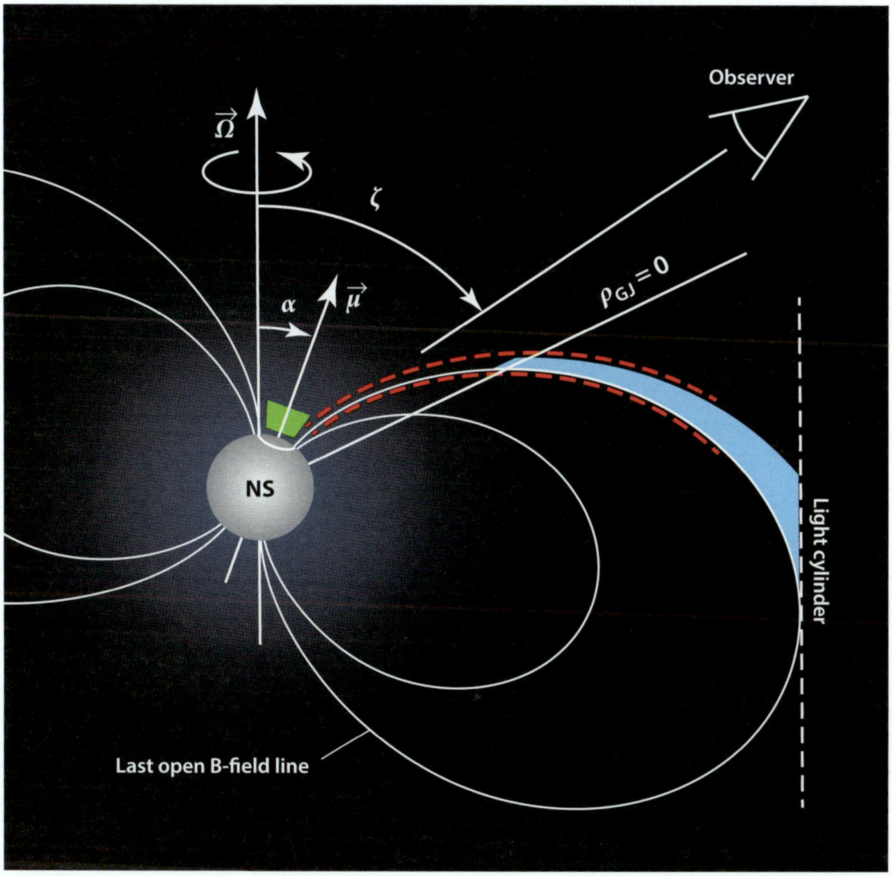

Figure 2

Scheme of a neutron star (NS) magnetosphere with the internal emission regions highlighted: polar cap (PC) model is in green, outer gap (OG) model in dark blue, and slot gap (SG) model in red (courtesy of J. Dyks). The striped wind scenario (outside the light cylinder) is not shown.

acceleration, and thus the gamma-ray production region, the SG can be adjusted to any magnetic inclination angle. Including special relativity effects, such as aberration and time of flight delay, in the SG framework, Dyks & Rudak (2003) developed a two-pole caustic (TPC) model that, in their opinion, can overcome some of the shortcomings of both PC and OG. To add freedom to the OG, Hirotami et al. (2003) proposed an OG that extends beyond the null surface. Production inside the light cylinder, however popular, is not the only option for pulsar modeling: Coroniti (1990) proposed a totally different approach with the striped wind model, where gamma rays are produced outside the light cylinder.

Apart from fitting light curves and spectra, models must account for the measured efficiency in converting rotational energy into gamma rays because a pulsar's gamma-ray yield is by far the dominant component of its multiwavelength emission. Romani (1996) as well as Arons (1996) discuss how to get high yields from OG and SG models. However, no model could account for a gamma-ray luminosity exceeding $\dot{E}$, as seemed to be the case for PSR B1055-52 (Thompson et al. 1999). That pulsar's gamma-ray luminosity, computed assuming a beaming factor of 1 steradian and using the distance derived from its radio dispersion measure (DM), required an efficiency of more than 100%, which points to a wrong distance estimate and, possibly, too large of a beaming factor. Clearly, coupling a few percent "true" pulsar efficiency with an overestimated distance and an uncertain beaming could produce an unreasonably high gamma-ray yield.

Because the light curves of all the newly discovered EGRET pulsars appeared to be different at different wavelengths and generally not aligned in phase (see e.g., Thompson 2004), seemingly following the Vela as opposed to the Crab paradigm, the need to disentangle the radio emission region (almost certainly related to the PC) from the gamma-ray region certainly supported OG or SG emission models rather than the classical PC model.

However, though both near and far emission models could be adjusted to produce the double-peaked light curves seen by EGRET, they differed in the gamma-ray spectral shape; the PC model was unable to produce very high energy photons. Because of the absorption in high-B fields, the PC model predicts sharp, superexponential cutoffs in the observed spectra at energies of a few gigaelectronvolts. However, high-altitude models, such as OG and SG, predict a simple exponential cutoff owing to the radiation reaction limit of the accelerated particles. The EGRET sensitivity above a few gigaelectronvolts, unfortunately, did not allow discrimination between the two classes of models.

2.4. EGRET: Hundreds of Gamma-Ray Sources and Many Interesting Candidates

Although EGRET had increased the number of gamma-ray-emitting INSs, pulsars were no longer the dominant celestial population in the gamma-ray sky. Extragalactic sources, many of them strongly variable, were now counted by the dozens. They greatly outnumbered the INSs in the final EGRET catalog, which listed 278 sources (Hartman et al. 1999), half of which remained unidentified. However, with two-thirds of the unidentified sources clustered around the Galactic plane, pulsars, both radio-loud and radio-quiet, continued to be natural candidates to account for nonvariable sources with no obvious counterpart. Repeating the geometric exercise with the EGRET low-latitude sources, Mukherjee et al. (1995) confirmed the findings of Swanenburg et al. (1981). The unidentified low-latitude EGRET sources lay at distances between 1.2 and 6 kpc, and their luminosities range from 0.7×10^{35} to 16.7×10^{35} erg s^{-1}, values indicating rather young and energetic INSs. Gehrels et al. (2000) elaborated on the log$N$-log$S$ distribution of the unidentified EGRET sources to claim the existence of a population of faint gamma-ray sources at mid-latitude.

Although Geminga-like INSs were ideal, but elusive, potential counterparts (e.g., Yadigaroglu & Romani 1995), deep radio searches started to detect promising young and energetic pulsars within EGRET error boxes. However, finding a radio pulsar in a gamma-ray error box does not guarantee the detection of pulsations in gamma rays. As mentioned above, if the gamma-ray observations are separated in time from the radio ones, the pulsar's period and period derivative must be extrapolated while also accounting for their uncertainties, such that the periodicity search in gamma rays must cover a vast parameter space, reducing the significance of any tentative detection. Moreover, young pulsars often exhibit timing noise and glitches that render the extrapolation much more uncertain. Thus, it should come as no surprise that the promising pulsars discovered at the end of the EGRET mission could not qualify for a solid detection of gamma-ray pulsations. As already mentioned, Kaspi et al. (2000) reported evidence of an association between the 20,000-year-old PSR B1046-58 and 3EG J1048-5840, adding this pulsar to the interesting candidates list. D'Amico et al. (2001), using Parkes data, found two young, promising radio pulsars inside the error boxes of 3EG J1420-6038 and 3EG J1837-0606. More young and energetic pulsars were found while exploring EGRET error boxes: Roberts et al. (2002) uncovered PSR J2021+3615 within the source 3EG J2021+3716 in the rich and complex Cygnus region. Yet, the pulsar timing noise prevented a meaningful back-extrapolation search for pulsations in the limited number of gamma-ray photons EGRET had collected. Similarly, PSR J2229+6114 is a young radio and X-ray pulsar whose energetics make it a plausible counterpart for 3EG 2227+6122, within which it was found, though its DM pointed to a distance value in excess of 10 kpc. If true, such a distance would have made PSR J2229+6114 too faint to be the gamma-ray source counterpart (Halpern et al. 2001b). Not surprisingly, a search for pulsations in the gamma-ray data was inconclusive. In more general terms, by cross-correlating 1,300 known radio pulsars with the EGRET catalog, Kramer et al. (2003) estimated that 19 ± 6 radio pulsar associations could have been genuine.

Although the new young and energetic pulsars went onto the exploration wish list of high-energy astrophysicists waiting for the next generation of gamma-ray instruments, these examples illustrate the complex interplay between gamma-ray and radio astronomy. Indeed, the discovery of radio pulsars within gamma-ray error boxes is a story that shall recur in our narrative and grow as time goes by.

Meanwhile, a multiwavelength approach, exploiting the sequence of X-ray and optical observations that had been successfully applied to Geminga, was pursued for a number of bright, unidentified sources at medium to low Galactic latitudes. Notable examples are 3EG J1835+5918, 3EG J2020+4017, and 3EG J0010+7309. 3EG J1835+5918, a steady EGRET source with a hard spectrum and high-energy cutoff, was dubbed "Next Geminga" owing to its similarities with the prototype radio-quiet INS. *Chandra* coverage unveiled an X-ray counterpart, RX J1836.2+5925, with both thermal and nonthermal emission but without optical and radio detection, pointing, yet again, to a radio-quiet NS (Mirabal & Halpern 2001, Reimer et al. 2001, Halpern et al. 2002).

Similarly, 3EG J2020+4017 (Brazier et al. 1996) and 3EG J0010+7309 (Brazier et al. 1998) appeared to be positionally associated with the supernova remnants (SNRs) γ Cygni and CTA-1, respectively. The two gamma-ray sources are nonvariable and have flat spectra similar to other gamma-ray pulsars. 3EG J0010+7309 has an X-ray counterpart that, again, looks like the young INS responsible for CTA-1 embedded in its plerion (Halpern et al. 2004).

Although identified, Geminga continued to attract attention, becoming one of the most scrutinized INSs in the soft X-ray domain. Both ESA's *XMM-Newton* and NASA's *Chandra* X-ray observatories devoted significant observing time to this source. A long *XMM-Newton* observation unveiled a nebula trailing the NS as it moves in the interstellar medium (Caraveo et al. 2003). The same data set also allowed for phase-resolved spectroscopy to be performed, making it possible to disentangle the nonthermal, power-law component from the surface thermal emission, which was divided into hot and cool components that were seen to vary as a function of the pulsar rotational phase (Caraveo et al. 2004, de Luca et al. 2005). Using the precise knowledge of the distance to convert X-ray fluxes into luminosities, the emitting areas were computed to show that the hot component (probably linked to the PC heating by return current) comes from a surface much smaller than that of any dipole-like PC. This pointed to a quasi-aligned rotator almost perpendicular to the line of sight. The nebula detected by *XMM-Newton* was confirmed by *Chandra* (Pavlov et al. 2010), which also resolved a comet-like structure trailing the pulsar (de Luca et al. 2006).

2.5. The EGRET Legacy: Open Points for a New Millennium

When CGRO was deorbited in June 2000, the EGRET mission legacy in pulsar astronomy amounted to as many as ten INSs (seven firmly established detections and three probable ones). Although the objects had somewhat different phenomenology, they undoubtedly channeled the major share of their rotational energy loss into gamma rays. As shown by the energy-per-decade plot (Thompson 2004) all the EGRET pulsar spectral energy distributions peak in the gamma-ray band. Their high efficiencies, coupled with extremely diverse multifrequency behaviors, clearly point to composite emission models, wherein different emitting regions at different locations in the pulsar magnetosphere contribute all at once. However, though the various models could account for gamma-ray emission below 1 GeV, the Rosetta stone of pulsar modeling lay in the few to 10-GeV region wherein different models predicted very different spectral shapes. Because the PC model predicts a sharp turnover at few to several gigaelectronvolts (as a result of the attenuation of the gamma-ray flux in the magnetic field above the star surface), the lack of such a turnover would rule out the already troubled PC model in favor of the OG or SG models, which produce gamma rays far from the pulsar's surface. In particular, the expected sensitivity of the Large Area Telescope (LAT), on board NASA's GLAST mission, to gigaelectronvolt photons brought hope for solving this spectral conundrum, making it possible for scientists to pin down the best emission model [as shown by Razzano et al. (2009) using simulated LAT data].

EGRET had also left a number of promising unidentified sources for which a direct search for periodicity was beyond reach, but that could be investigated at other wavelengths following the lessons learned in the chase for Geminga. Indeed, radio and X-ray searches were being actively pursued, yielding a growing number of potential gamma-ray source counterparts that were just waiting for the next gamma-ray observatories.

Expectations of pulsar detections for the coming AGILE and GLAST missions were rather uncertain, mainly due to considerable uncertainty as to which pulsar model to use. Back-extrapolating the log*N*-log*S* number-flux relation of the EGRET pulsars, Thompson (2004) predicted a grand total ranging from 30 to 100 GLAST detections, considering both radio-quiet and radio-loud INSs. On the basis of Monte Carlo simulations for PC-emitting pulsars, Harding et al. (2002) predicted the gamma-ray detection of 90 radio-loud and 101 radio-quiet pulsars. However, less than 10% of the radio-quiet pulsars were expected to yield a pulsed signal with the techniques available at the time. Simulations including OG emission models (Harding et al. 2007) yielded a large number of radio-quiet gamma-ray pulsars coupled to a comparatively small number of radio-loud ones, making it clear that the ratio of radio-quiet to radio-loud pulsars detected by GLAST would provide a useful clue in discriminating between the two classes of models. Moreover, Harding et al. (2005), on the basis of their PC model, predicted that several MSPs could be detectable by AGILE and GLAST, both yet to be launched.

Building on the lessons learned during past missions, careful campaigns involving all the major radio observatories were orchestrated to maximize the chances of success for detection of radio pulsars. Smith et al. (2008) describe the observing campaign organized by the GLAST collaboration together with all the major radio observatories (now known as the Pulsar Timing Consortium) to monitor about 200 pulsars selected for their large spin-down power ($\dot{E} > 10^{34}$ erg sec^{-1}) and ranked high according to their $\frac{\sqrt{\dot{E}}}{4\pi d^2}$. This effort also included careful calibration of the spacecraft clock (Smith & Thompson 2009, Abdo et al. 2009a) to allow for precise phase alignment of the radio and gamma-ray light curves. This rendered possible phase-folding of MSPs, a class of NSs often in binary systems, certainly interesting but not considered a prime target for the mission, in view of their rather low surface *B* field. For the radio-quiet Geminga, Jackson & Halpern (2005) undertook the task of maintaining phase-coherent timing parameters through biannual *XMM-Newton* measurements.

Preparation efforts were also carried out in the software domain. Extensive "data challenges," relying on massive Monte Carlo simulations (Baldini et al. 2006) of the gamma-ray sky, were conducted within the GLAST collaboration to test and debug the analysis software, with the goal of having a fully functional analysis pipeline ready at the time of launch.

Meanwhile, to meet the challenge of detecting pulsations using only the gamma-ray photons, Atwood et al. (2006) devised a new strategy to optimize the computing power needed to perform comprehensive blind searches covering a wide range of P and $\dot{P}$. Because very long exposure times, which are mandatory in gamma-ray astronomy, make complete Fourier analysis computationally prohibitive, the newly proposed method analyzes the differences of photon arrival times (up to a window of the order of weeks) rather than the time series itself and succeeds in maintaining good sensitivity while greatly reducing the effects of frequency derivatives and glitches. The time-difference method was successfully tried on EGRET data (Ziegler et al. 2008) and was ready to be tested on new gamma-ray data.

3. THE PRESENT: THE SILICON ERA IN GAMMA-RAY ASTRONOMY

The current new era in gamma-ray astronomy is characterized by significant hardware improvement owing to the introduction of the silicon-strip detector as the core constituent of a "solid-state" spark chamber (as opposed to the gas-filled, wire-read-out spark-chambers on board SAS-2, COS-B and EGRET) used to detect and track photons and to discriminate between them and the much more numerous charged particles. Silicon trackers are self-triggering devices that reduce the dead time to almost zero, enhancing the instrument's timing and count-rate capabilities while also improving its spectral and spatial resolution. Moreover, their very compact structure allows for the design of gamma-ray detectors with steradian-sized fields of view (FoVs), thus increasing the effective observing time for any given direction in the sky. In recent years, we have been living in an ideal time for high-energy astrophysics with two such instruments operating in orbit allowing, for the first time since the beginning of gamma-ray astronomy, almost continuous coverage of the sky, with the added possibility for each to independently prove (or disprove) the other's claims.

The two satellites are AGILE (Tavani et al. 2009), a small Italian mission of the Italian Space Agency (ASI) in close collaboration with the Istituto Nazionale di Astrofisica (INAF) and the Istituto Nazionale di Fisica Nucleare (INFN), and *Fermi* (the name given to GLAST in orbit; Atwood et al. 2009), a much bigger NASA mission with important international participation from Japan, France, Italy (again with ASI, INAF, and INFN) and Sweden.

3.1. April 2007: Enter AGILE

AGILE features a silicon tracker made of 12 40×40 cm trays and a thin calorimeter. The mass of the calorimeter is a limitation for the detection of photons with energies greater than tens of giga-electronvolts, but AGILE excels at low energies ($E < 100$ MeV). Above the tracker, another layer of silicon acts as a hard X-ray detector and is known as SuperAGILE. Although the dimensions of AGILE are about a quarter of those of EGRET, the effective areas of the two instruments are comparable thanks to the superior performance of the silicon tracker and analog read-out, with AGILE enjoying significantly better angular and time resolution, as well as a much smaller dead time.

Launched in April 2007 from the Indian base of Sriharikota, AGILE was put into a nearly equatorial orbit and started its observing program as a pointing mission characterized by a very large FoV, covering one-sixth of the sky. As usual, AGILE carried out its "on-orbit" calibration using the Vela and Crab pulsars as targets and tested its capabilities on Geminga and PSR B1706-44 (Pellizzoni et al. 2009a). After subtracting the pulsed photons from the Vela source, a faint diffuse emission appeared, yielding the first clear detection of a resolved pulsar wind nebula in high-energy gamma rays (Pellizzoni et al. 2010).

The first new gamma-ray pulsar was discovered as part of the AGILE Guest Observer Program by Halpern et al. (2008), who detected gamma-ray emission from PSR J2021+3651 showing that its radio distance was indeed overestimated. Thus, PSR J2021+3651 entered the stage as one of the contributors of the notoriously complex Cygnus region. Next Geminga was also closely scrutinized to ascertain its spectral shape as well as search for possible variability (Bulgarelli et al. 2008).

Folding the gamma-ray photon arrival times using contemporary ephemerides for the 35 top-ranking radio pulsars (for which radio monitoring had been organized), AGILE detected three new pulsars (Pellizzoni et al. 2009b), namely PSR J2229+6114, discovered by Halpern et al. (2001a) within the 3EG 2227+6122 error box; PSR B1509-58, a very energetic pulsar that had been seen by COMPTEL but not by EGRET owing to its very soft spectrum; and PSR J1824-2452, an MSP detected clearly for only a fraction of the observing time. PSR B1509-58 was further investigated by Pilia et al. (2010). Moreover, Pellizzoni et al. (2009b) found tantalizing signals for four more objects, namely PSRs J1016-5857, J1357-6429, J2043+2740, and J1524-5625. Of these, only the latter, which was also the least significant, was not independently detected by *Fermi*. By doubling the EGRET pulsar harvest, including the detections of the second youngest as well as the million-year-old PSR J2043+2740, AGILE was showing that gamma-ray emission is a common feature of high-ranking radio-loud pulsars, be they young or old.

3.2. June 2008: Enter *Fermi* and the Discovery of a New Geminga

In June 2008 GLAST was launched, and thus the *Fermi* observatory was born featuring two major instruments: the LAT (Atwood et al. 2009) and the Gamma Burst Monitor (GBM; Meegan et al. 2009). The LAT's modular structure, containing 16 towers, each featuring an 18-tray tracker sitting on a heavy calorimeter, makes it the most powerful gamma-ray telescope ever. Very well suited to detecting photons in the gigaelectronvolt range, where angular resolution is at its best, *Fermi* aimed at obtaining the sharpest (and deepest) vision of the gamma-ray sky. With a FoV of 2.4 steradian, *Fermi* was designed to primarily operate in scanning mode, covering the whole sky every three hours, i.e., once every two 90 min. orbits. Such an operating mode provides reasonably homogeneous sky coverage and guarantees that any given point in the sky is within the instrument FoV for approximately one-sixth of the time. Once in orbit, *Fermi* executed a planned calibration sequence of pointed observations during which the satellite pointed first at the Vela pulsar and, when the target was occulted by the Earth, at the EGRET unidentified source 3EG J0010+7309, coincident with the CTA-1 SNR.

Thanks to the instrument performance, the software readiness, the differencing technique, and the availability of a precise X-ray position, *Fermi's* on-orbit verification phase yielded the long-sought detection of periodicity from 3EG J0010+7309, making it the first high-visibility result (Abdo et al. 2008) from the new *Fermi* observatory. While 3EG J0010+7309 was quite a robust INS candidate, with a faint X-ray source surrounded by diffuse emission pointing to an NS embedded in a pulsar wind nebula (Halpern et al. 2004), the speed of the discovery was amazing, boding well for the future capability of *Fermi* as a hunter of radio-quiet pulsars (Bignami 2008).

The pulsar timing parameters yielded a characteristic age ($\tau_c = \frac{P}{2\dot{P}}$) of 14,000 years (comparable to the CTA-1 SNR age) and a rotational energy loss of 4.5×10^{35} erg s^{-1}, i.e., a radio-quiet INS 50 times younger and 10 times more energetic than Geminga. Its dipole B field is rather high (1.1×10^{13} G), which makes PRS J0007+7303 the second-highest magnetic field pulsar after PSR B1509-58. However, unlike PSR B1509-58, the newly discovered INS does emit high-energy photons. The pulsar light curve displays a double-peaked structure similar to that of PSR B1706-44, one of the historical EGRET pulsars that has a similar age. For radio-quiet pulsars, distance estimates rely on indirect methods such as X-ray line-of-sight absorption or their association with an SNR. Combining the radio and X-ray pieces of information, the best distance to PSR J0007+7303 is 1.4 ± 0.3 kpc (Pineault et al. 1993). This makes it possible to compute the pulsar luminosity—and thus its efficiency—that turns out to be about 1%. Interestingly, the efficiency of PSR J0007+7303 in converting its rotational energy loss into gamma rays is similar to that of PSR B1706-44, but smaller than that of the much older Geminga. Although PSR J0007+7303 is a Geminga-like gamma-ray pulsar with an X-ray counterpart, we note that, unlike Geminga, the gamma rays were the driver behind the periodicity discovery. The source faintness in X-rays prevented the detection of pulsations in that band. Only a long *XMM-Newton* observation, coupled with the precise knowledge of the gamma-ray timing parameters, yielded the X-ray pulsation detection (Caraveo et al. 2010). The X-ray counterpart of PSR J0007+7303 is a faint source, excellent for getting an accurate position, but of hardly any use for unveiling a hidden periodicity.

Pulsations from the radio pulsar PSR J1028-5819, freshly discovered by Keith et al. (2008) within 3EG J1027-5817, followed quickly (Abdo et al. 2009h), identifying the pulsar as being responsible for at least part of the flux of 3EG J1027-5817. Next came the detection of gamma-ray pulsations from PSR J0205+6449 in 3C58, a Crab-like 800-year-old SNR (Abdo et al. 2009b). The pulsar light curve is also Crab-like, and its two peaks are aligned with the X-ray ones (but not with the single radio pulse).

An impressive light curve of the Vela pulsar, including 32,400 pulsed photons collected during the verification phase, was quickly published (Abdo et al. 2009i) with a revealing spectral study of the phase-averaged gamma-ray emission. The shape of the light curve as a function of energy confirms the dramatic energy evolution hinted at by previous experiments with the appearance of a third peak in the bridge region above 1 GeV and the disappearance of the first peak at energies above 10 GeV. The *Fermi* spectrum can be described as a power law with exponential cutoff $\frac{dN}{dE} = K(\frac{E}{E_0})^{-\Gamma} exp(-\frac{E}{E_{cut}})^b$, spectral index $\Gamma = 1.5$, an energy cutoff $E_{cut} = 2.9$ GeV, and a b of 0.88 ± 0.4. This rules out any superexponential absorption that would have been the signature of attenuation owing to the interaction of high-energy gamma-ray photons with the strong magnetic field at low altitude in the pulsar magnetosphere. Thus, the lack of hyperexponential absorption points to high-latitude emission. This finding is strengthened by the detection of pulsed photons up to 17 GeV, an emission that must arise at $R > 3.8\ R_{NS}$ (from Baring 2004).

The first high-quality *Fermi* pulsar spectrum set the stage for gamma-ray emission far from the NS surface in the outer magnetosphere, near the light cylinder, ruling out the PC model in favor of the OG or SG models.

3.3. Surprises: Expected and Unexpected

Whereas gamma-ray emission from young, energetic pulsars (both radio-loud and radio-quiet) was widely expected, the detection of the MSP PSR J0030+0451 (Abdo et al. 2009j) came as a real surprise, mainly because such old, recycled pulsars were not supposed to be ideal gamma-ray emitters given that their surface B fields are 10^4 times weaker than those of young NSs. Conversely, the pulsar gamma-ray phenomenology was pretty normal looking: two narrow peaks, separated by 0.44 in phase, a spectrum well described by a power law with an exponential cutoff. However, its rather low rotational energy loss of 3.5×10^{33} erg s^{-1}, coupled with the parallactic distance of 300 pc and the measured flux ($E > 100$ MeV) of approximately 7×10^{-8} cm^2 sec^{-1}, implied a huge 15% gamma-ray production efficiency.

The detection of PSR J0030+0451 opened the way for a systematic search for gamma-ray emission from MSPs. Owing to the radio monitoring campaign (Smith et al. 2008), MSPs with $\dot{E}$ above 10^{34} erg sec^{-1} had contemporary ephemerides to be used for phase folding. The search immediately yielded interesting results; 7 more MSPs were found in a matter of months (Abdo et al. 2009c). Although MSPs are often in binary systems, we continue to refer to gamma-ray pulsars as INSs, because their emission is powered only by their rotational energy loss. With gamma-ray light curves and spectra similar to those of young pulsars, the emission regions for MSPs should also have been far from the NS surface. Although young pulsars and MSPs have vastly different B-fields at the stellar surface, the value of the B field at the light cylinder is similar, indicating a region where similar conditions naturally occur.

All the eight gamma-ray emitting MSPs were field objects, both isolated and in binary systems. The detection of the globular cluster 47 Tuc (Abdo et al. 2009d) was therefore attributed to the integrated emission of multiple MSPs within it. The case of 47 Tuc is not unique; *Fermi* has detected a dozen more globular clusters as point sources (Abdo et al. 2010a, Kong et al. 2010). In parallel, blind searches over a five-month observing interval yielded 15 more radio-quiet INSs, making it clear that radio-loud and radio-quiet sources were evenly contributing to the *Fermi* pulsar harvest. Abdo et al. (2009e) discuss the general properties of the population of 16 radio-quiet INSs discovered by *Fermi*. Because 13 of them were found within formerly unidentified EGRET sources that were among the set of suspected pulsars, deep X-ray investigations were already available. Indeed, 6 of the 16 pulsars were discovered by assuming a counterpart position derived from previous X-ray coverage or from newly obtained observations, using mainly the Swift X-ray Telescope. Apart from CTA-1, the newly discovered gamma-ray pulsars featured Next Geminga (Halpern et al. 2004), Gamma Cyg (Brazier et al. 1996), the Rabbit (Ng et al. 2005), Taz (Roberts & Brogan 2008), and the Eel (Roberts et al. 2001). However, the faintness of the X-ray counterparts made it clear that, for gamma-ray pulsars, the LAT's potential as a pulsation discoverer was vastly superior to that of X-ray telescopes.

The timing parameters of the 16 gamma-ray-selected pulsars point to rather young INSs, with energetics similar to that of radio-loud ones. Also, their averaged fluxes, light-curves, and spectra were no different than that of radio-loud pulsars. Subsequent deep radio observations yielded detections for three of the new pulsars—two rather normal looking and one exceedingly faint (Camilo et al. 2009, Abdo et al. 2010h).

A closer look at bright sources such as Vela (Abdo et al. 2010e), Crab (Abdo et al. 2010c), and Geminga (Abdo et al. 2010f), as well as Next Geminga (Abdo et al. 2010d), for which phase-resolved analysis was possible, clearly revealed that both spectral index and cutoff energy vary as a function of the pulsar phase, confirming early findings and adding a wealth of fine structure details. The detection of the Crab pulsar at $E > 100$ GeV by Magic and Veritas (Aliu et al. 2008, 2011; Aleksic et al. 2012) is well above any reasonable extrapolation of the *Fermi* LAT best spectral fit, pointing to a different emission mechanism for that high-energy component.

The first *Fermi* catalog of gamma-ray pulsars (1PC) lists 46 high-confidence pulsars detected within the first 6 months of the mission (Abdo et al. 2010b). Out of 46 NSs, 29 were detected in radio (further divided between 8 MSPs and 21 young pulsars) and 17 seen only in gamma rays (i.e., 16 discovered by LAT + Geminga). As expected, a significant fraction of the newly discovered LAT pulsars were found within the error boxes of high-interest EGRET unidentified sources. 1PC contains 15 such cases, 13 of which are radio-quiet and 2 radio-loud, namely PSR J1028-5918 (Abdo et al. 2009h) and PSR J2021+3651 (Abdo et al. 2009f).

A power law with an exponential cutoff can fit the spectra of all 46 gamma-ray pulsars. Their light curves are usually double peaked (with peak separation of 0.4–0.6), but a nonnegligible minority of single-peaked pulsars is also present. With very few exceptions, the gamma-ray peaks are not aligned with the radio ones, confirming early EGRET findings and pointing to an emission region far from the pulsar's surface.

To assess the luminosity of gamma-ray pulsars ($L_\gamma = 4\pi d^2 F_\gamma f_\Omega$), $f_\Omega = 1$ was used for all pulsars. It is a big change compared with the past, when this parameter was assumed to be $1/4\pi$, and is a direct consequence of the new preference for the OG model (see, e.g., Watters et al. 2009). The gamma-ray light curves and spectral shapes point to high-altitude emission regions producing fan beams that cover a large fraction of the celestial sphere. However, when computing the luminosity of LAT pulsars, the major source of uncertainty remains the assumed distance, because few INSs have a measured parallax. Although the majority of radio pulsars can rely on DM, distances for radio-quiet pulsars can be obtained only from X-ray absorption data, when it is available. This limits the number of radio-quiet pulsars in the *Fermi* luminosity plot.

Abdo et al. (2010b) show that the evolution of the gamma-ray luminosity as a function of the pulsar rotational energy loss cannot be fitted by a single function. Even considering the distance uncertainties, a substantial scatter is present, possibly arguing against the assumption of a common beaming factor for all objects. MSPs seem to climb more steeply in luminosity than the young pulsars, which evolve more gently.

3.4. *Fermi*'s Treasure Hunt

Numerous multiwavelength studies were triggered by the first wave of *Fermi* results, directly linked to both pulsars and, more generally, newly detected sources with no identification, first in the Fermi Bright Source Catalog (often referred to as 0FGL; Abdo et al. 2009g), and later in the First List of *Fermi* Sources, known as 1FGL (Abdo et al. 2010g). On the high-energy side, both exploratory and in-depth X-ray observations were carried out targeting the newly discovered radio-quiet INSs as well as the promising candidates that were emerging from the *Fermi* data, but that needed better position information to secure a statistically significant pulsation detection. On the low-energy side, the time-honored exercise of searching for radio pulsars within unidentified source error boxes was started anew. By exploiting the smaller *Fermi* error boxes, as well as more powerful analysis systems, it was possible to significantly reduce the time needed to cover each source. This allowed for multiple visits, a strategy that proved crucial to detecting radio emission from MSPs in binary systems. To maximize the chances of success, sources to be studied were selected on the basis of their "pulsarness," a parameter quantifying the lack of variability coupled to a suitably curved spectral shape (Ackermann et al. 2012a).

Having searched 25 such pulsar-like unidentified sources, Ransom et al. (2011) reported the detection of 3 new MSPs. Such a high success rate triggered the chase for field MSPs in unidentified pulsar-like *Fermi* sources, preferably at medium to high Galactic latitude. Keith et al. (2011) found 2 MSPs and a young pulsar in their coverage of 11 *Fermi* sources. Cognard et al. (2011) unveiled two MSPs, whereas Kerr et al. (2012) found 5 MSPs after having searched 14 unidentified *Fermi* sources. Once the timing parameters of the often-binary new MSPs were accurate enough, they were also detected in gamma rays, increasing the share of this class of rather faint gamma-ray emitters (both isolated and in binary systems). More searches focusing on the unidentified sources with high pulsarness listed in the second *Fermi* source catalog (hereafter referred to as 2FGL; Nolan et al. 2012) are being conducted, and the rate of discovery is impressive, as shown in **Figure 3**.

Although this is a major result on its own (because it increases significantly the number of known field MSPs; e.g. Caraveo 2010), the sudden jump in the MSP number revived the interest of the radio-astronomical community in using these super stable clocks to detect nanoHertz gravitational waves (GWs). The slight variations GWs may induce in the time of arrival of signals coming from widely separated pulsars will be correlated, thus allowing for the direct detection of GWs (e.g., Jenet et al. 2005). Recent re-evaluation (Cordes & Shannon 2012) of the method sensitivity shows that 5 years of monitoring of (at least) 20 carefully selected MPSs stands a fair chance of detecting the GW background, allowing us to study its spectrum. Of course, a higher number of MSPs, ranging from 50 to 100 objects, would provide firmer results. Indeed, by finding so many new MSPs, *Fermi* increases the number of targets to be monitored, searching for GW induced variations. The high success rate of *Fermi*'s treasure hunt also implies that MSPs could

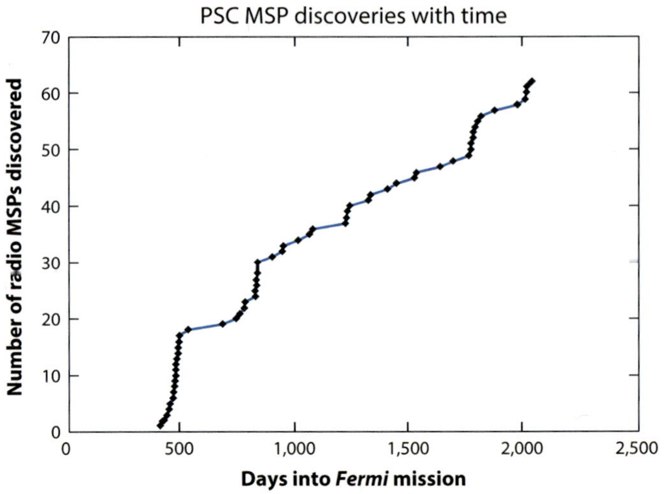

Figure 3

Cumulative view of the millisecond pulsars (MSPs) discovered in radio by the Pulsar Search Consortium (PSC), observing promising unidentified *Fermi* sources. Since the potential for discovery of such a technique was realized, the MSP discoveries have been continuous and no flattening in the growth curve is in sight. (Courtesy of Elizabeth Ferrara.)

play a major role in accounting for unidentified Galactic gamma-ray sources, especially the faint, high-Galactic latitude sources.

Normal young pulsars, however, should not be totally forgotten. Camilo et al. (2012), for instance, during the radio coverage of 1FGL J2030.0+3641, found a middle-aged radio pulsar that was also immediately detected in gamma rays. Indeed, as stated by Camilo et al. (2012), the fact that so few young isolated pulsars have been found in LAT sources (which are mostly along the plane) is a testament to how good the radio surveys were.

Meanwhile, blind searches were conducted in newly discovered, low-latitude pulsar-like LAT sources, selected on the basis of their pulsarness. Eight radio-quiet INSs were rapidly found (Saz Parkinson et al. 2010), and two more were added later (Saz Parkinson et al. 2011).

Although optimized through the time-differencing techniques, blind searches on ever fainter LAT sources had to cover progressively longer time spans, becoming very computer intensive and thus hampering the *Fermi* LAT's potential for discovery. To overcome the limitation in computing power, the Albert Einstein Institute in Hannover brought their supercomputer (mainly devoted to the search for GWs) into play, coupled with a new hierarchical search method originally aimed at detecting continuous GWs from rapidly rotating NSs. The reward was immediate. Nine INSs were quickly found (Pletsch et al. 2012a). PSR J1838-0537 was added later (Pletsch et al. 2012b), and its timing analysis shows that in September 2009 the pulsar suffered the largest glitch seen so far in any gamma-ray-only pulsar. Also, standard analysis using radio pulsar timing parameters continued, relentlessly following each potentially interesting pulsar because, if a pulsed signal is present, its significance grows with time and eventually reaches the 5-sigma level needed to announce a detection.

In view of the results obtained, the selection criteria used to build the list of pulsars under radio monitoring were revised, lowering the threshold $\dot{E}$ with a proportional increase in the number of pulsars under continuous investigation. A lot of work was also devoted to improving the gamma-ray analysis technique. A weighting algorithm was introduced to assign to each photon a weight according to its (energy-dependent) probability to come from a given pulsar (Kerr 2011). Such a weighting algorithm reduces the trials previously needed to optimize the extraction region together with the energy range and results in an enhanced sensitivity to the pulsation detection.

In parallel, Ray et al. (2011) improved the analysis techniques by applying a maximum likelihood method to extract pulse times of arrival from unbinned photon data. Using new phase-connected pulse timing solutions for the first group of 16 gamma-ray-selected pulsars, they were able to improve the source positioning, rivaling the X-ray in localization accuracy. Moreover, continuous folding unveiled the presence of glitches for PSRs J0007+7303, J1124-5916, and J1813-1246 (Ray et al. 2011), joining several other radio pulsars for which gamma-ray folding had already highlighted glitches and starting with the very first detection of a glitch from PSR B1706-44 in just 10 weeks of data (Saz Parkinson 2009). Indeed, it turns out that continuous gamma-ray coverage, coupled with the sensitivity of folding to pulsar parameters, is a powerful way to unveil pulsar glitches.

4. TOWARD THE *FERMI* GAMMA-RAY PULSAR REVOLUTION

After so many advancements, it was time to publish a second pulsar catalog on the basis of three years of *Fermi* data. The number of pulsars detected was already beyond the most optimistic guesses published prior to launch. Although Thompson (2001) bracketed the expectations between 30 and 100 objects, when the *Fermi* three-year observation database was frozen, 117 pulsars met the 5-sigma pulsation significance threshold (hereafter 2PC; Abdo et al. 2013). **Figure 4** shows the

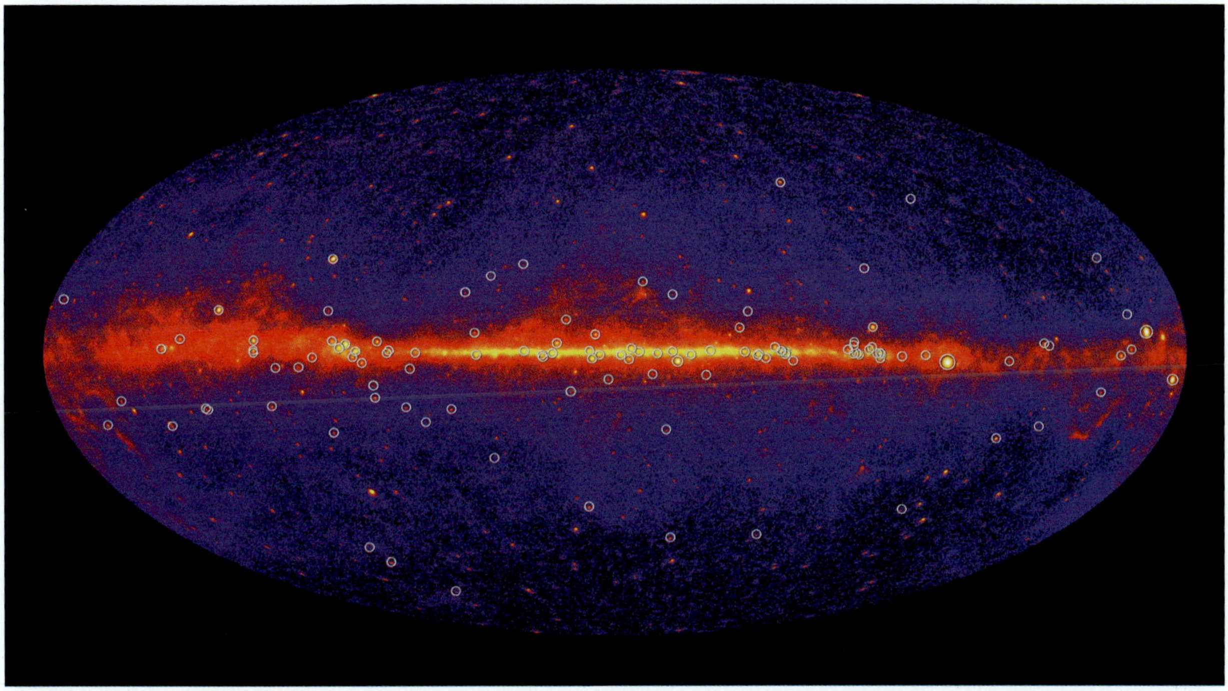

Figure 4

Fermi five-year sky map showing the positions of the 117 pulsars listed in the 2PC (Abdo et al. 2013). Image from NASA/DOE/Fermi LAT Collaboration (**http://svs.gsfc.nasa.gov/vis/a010000/a011300/a011342/**).

pulsar positions against the 5-year *Fermi* image of the gamma-ray sky. Of these, 42 are radio-loud pulsars and 35 are radio-quiet; 40 are MSPs, 20 of which were found through radio searches within unassociated *Fermi* LAT sources. Indeed, by the time the catalog was ready for publication, the number of MSPs discovered within unidentified sources had grown to 46 (Ray et al. 2012), 34 of which have been seen to pulsate in gamma rays.

Regarding the radio-quiet pulsars, a little clarification is in order. Out of the 36 INSs discovered by *Fermi* through blind searches, two (PSR J1741-2054 and PSR J2032+4127) were subsequently detected in radio (Camilo et al. 2009) and are thus counted as radio pulsars. By contrast, PSR J1907+0602 (Abdo et al. 2010h) and PSR J0106+4855 (Pletsch et al. 2012a), both of which have been detected with exceedingly low radio fluxes, remain in the radio-quiet (or radio-faint) class. This class contains 34 *Fermi* pulsars to which Geminga (as a prototype example of a truly radio-quiet pulsar) should be added.

The 2PC (Abdo et al. 2013) represents a milestone in pulsar astronomy. First of all, it establishes *Fermi* LAT as a powerful pulsar discoverer. Indeed, half of the pulsars listed in the 2PC (Abdo et al. 2013) were not known prior to the launch of *Fermi*, and they have been either discovered by *Fermi* through blind searches or, thanks to *Fermi*, by radio targeted searches of *Fermi* LAT unassociated sources. Moreover, the 117 entries are divided almost exactly *in partes tres* between young radio-loud pulsars, young radio-quiet pulsars, and MSPs. Indeed, the most dramatic advancements have been achieved in the field of MSPs. Before *Fermi*, 70 MSPs were known outside globular clusters; now 120 have been found, 39 of which are part of the 2PC (Abdo et al. 2013). The 40th *Fermi* MSP is J1823-3121A, located within the globular cluster NGC6624, which represents the first detection of gamma-ray pulsations from an MSP in a globular cluster (Freire et al. 2011).

MSPs found in *Fermi* error boxes are a shorter-period, more energetic population than radio-selected ones (Ray et al. 2012). Indeed gamma-ray MSPs dominate for $P < 0.003$ s. In addition, Roberts (2013) remarks that MSPs found in *Fermi* sources have dramatically increased the number of black widow–type systems where the MSP is driving the evaporation of its tightly-bound very light binary companion.

Thus, the unexpected split into three pulsar classes makes 2PC (Abdo et al. 2013) noteworthy both in quantity and diversity. This is highlighted in **Tables 1–3**, which summarize the parameters of the 117 2PC (Abdo et al. 2013) pulsars divided into radio-loud pulsars, radio quiet pulsars, and MSPs.

When looking at the 2PC (Abdo et al. 2013) numbers, one must also consider the sensitivity issue because different search techniques, as well as different sky locations, do result in different sensitivities. As shown by Dormody et al. (2011), blind searches are about 2.5 times less sensitive than folding gamma rays with a known ephemeris. Thus, the similar number of detections among radio-loud and radio-quiet pulsars tells us that the parent population must contain more radio-quiet pulsars, as expected based on the OG model (e.g., Romani & Yadigaroglu 1995 foresaw that the radio-quiet gamma-ray-emitting INSs should account for half of the young NSs, but that only 19% should be visible both in gamma-ray and radio wavelengths).

4.1. The Galactic Distribution of Pulsars Revealed by *Fermi*

Although hard to distinguish from their gamma-ray light curves and spectra, the three pulsar families have markedly different Galactic distributions as well as different average fluxes. The Galactic distribution of the different classes of 2PC (Abdo et al. 2013) pulsars, given in **Figure 5**, clearly shows that young pulsars (both radio-loud and radio-quiet) are clustered around the Galactic plane, where they are much harder to detect owing to higher Galactic background radiation, whereas MSPs (be they gamma-ray pulsars or new radio pulsars discovered in *Fermi*

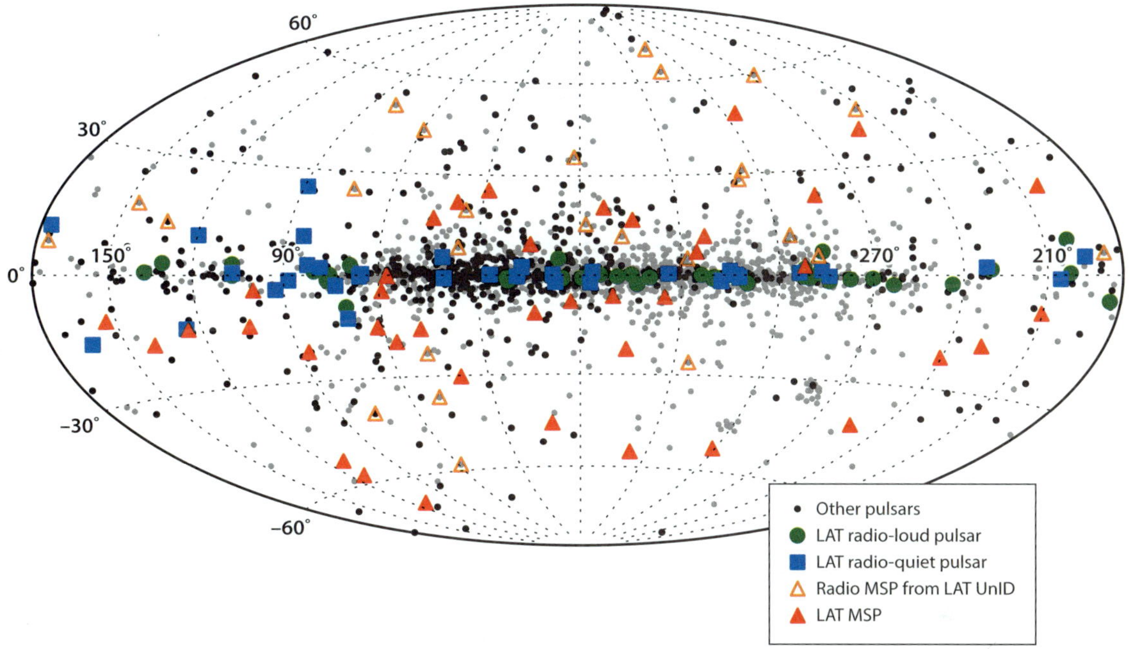

Figure 5

The Galactic distribution of 117 *Fermi* pulsars: radio-loud pulsars in green circles, radio-quiet in blue squares, and millisecond pulsars (MSPs) in red triangles. Empty triangles represent radio MSPs discovered within *Fermi* unidentified sources, whereas filled triangles are MSPs detected as gamma-ray pulsars. Black dots indicate 710 radio pulsars that were (unsuccessfully) phase folded with radio ephemerides provided by the Pulsar Timing Consortium. Gray dots indicate 1,337 pulsars outside globular clusters for which phase folding was not performed. Reprinted from 2PC (Abdo et al. 2013) with permission.

error boxes) are distributed all over the sky, pointing to a relatively local origin. However, the paucity of MSPs close to the Galactic plane is probably an observational bias because the *Fermi* unidentified source radio programs preferentially select sources at high Galactic latitude, avoiding regions where gamma-ray diffuse emission is higher and source confusion more likely.

The similarities/differences between the three pulsar classes can be immediately gauged by plotting their log$N$-log$S$ distributions (see **Figure 6**). Whereas young pulsars (both radio-loud and radio-quiet) have comparable fluxes and a number-flux slope in the range of -0.7 to -0.9, number-flux distribution of MSPs is definitely steeper (slope $= -1.6$) with a much lower average source flux. According to a straightforward geometric interpretation of the log$N$-log$S$ plot, a Galactic cylindrical distribution with no boundaries would yield a -1 slope, whereas a homogeneously filled sphere would be characterized by a -1.5 slope. Thus, young pulsars seem to trace a Galactic disk-like population that lacks faint sources, probably owing to the high Galactic background, whereas MSPs point to a spherical distribution of intrinsically fainter sources in our Galactic Neighborhood.

4.2. General Population Properties: Are Radio-Loud Pulsars Different from Radio-Quiet Pulsars?

To put the 117 *Fermi* pulsars in context, it is useful to plot them on the pulsar $P - \dot{P}$ diagram. This is given in **Figure 7**, extracted from the 2PC (Abdo et al. 2013). At first glance, it is easy to note that all the detections lie above the line corresponding to $\dot{E} = 10^{33}$ erg sec^{-1}, which could

Table 1 Radio-loud isolated neutron stars

Pulsar name	Galactic latitude (deg)	Galactic longitude (deg)	P (ms)	$\dot{P}$ (s s^{-1})	$\dot{E}$ (erg s^{-1})	Luminosity (erg s^{-1})	Distance (kpc)	Age (year)	B_s (G)	B_{LC} (G)	Comments
J0205+6449	130.72	3.08	65.727	1.90×10^{-13}	2.64×10^{37}	2.44×10^{34}	1.95	5.48×10^{3}	3.58×10^{12}	1.16×10^{5}	
J0248+6021	136.90	0.70	217.107	5.50×10^{-14}	2.12×10^{35}	2.47×10^{34}	2.0	6.25×10^{4}	3.50×10^{12}	3.15×10^{3}	
J0534+2200 (B0531+21)	184.56	−5.78	33.635	4.20×10^{-13}	4.36×10^{38}	6.19×10^{35}	2.0	1.27×10^{3}	3.80×10^{12}	9.21×10^{5}	Crab
J0631+1036	201.22	0.45	287.803	1.05×10^{-13}	1.73×10^{35}	5.57×10^{33}	1.0	4.36×10^{4}	5.55×10^{12}	2.14×10^{3}	
J0659+1414 (B0656+14)	201.11	8.26	384.919	5.50×10^{-14}	3.81×10^{34}	2.35×10^{32}	0.28	1.11×10^{5}	4.66×10^{12}	7.52×10^{2}	
J0729-1448	230.39	1.42	251.691	1.14×10^{-13}	2.82×10^{35}	1.72×10^{34}	3.52	3.50×10^{4}	5.42×10^{12}	3.13×10^{3}	
J0742-2822 (B0740-28)	243.77	−2.44	166.771	1.68×10^{-14}	1.43×10^{35}	8.89×10^{33}	2.07	1.57×10^{5}	1.69×10^{12}	3.36×10^{3}	
J0835-4510 (B0833-45)	263.55	−2.79	89.365	1.25×10^{-13}	6.90×10^{36}	8.93×10^{34}	0.29	1.14×10^{4}	3.38×10^{12}	4.36×10^{4}	Vela
J0908-4913 (B0906-49)	270.27	−1.02	106.755	1.51×10^{-14}	4.90×10^{35}	3.50×10^{34}	2.57	1.12×10^{5}	1.29×10^{12}	9.72×10^{3}	
J0940-5428	277.51	−1.29	87.545	3.28×10^{-14}	1.93×10^{36}	4.22×10^{33}	2.95	4.23×10^{4}	1.7[illegible] $\times 10^{12}$	2.35×10^{4}	
J1016-5857	284.08	−1.88	107.386	8.06×10^{-14}	2.57×10^{36}	5.48×10^{34}	2.9	2.11×10^{4}	2.93×10^{12}	2.21×10^{4}	
J1019-5749	283.84	−0.68	162.506	2.01×10^{-14}	1.84×10^{35}	4.27×10^{34}	6.8	1.28×10^{5}	1.83×10^{12}	3.92×10^{3}	
J1028-5819	285.07	−0.50	91.403	1.61×10^{-14}	8.33×10^{35}	1.58×10^{35}	2.33	8.99×10^{4}	1.23×10^{12}	1.48×10^{4}	
J1048-5832 (B1046-58)	287.43	0.58	123.701	9.57×10^{-14}	2.00×10^{36}	1.76×10^{35}	2.74	2.05×10^{4}	3.48×10^{12}	1.69×10^{4}	
J1057-5226 (B1055-52)	285.98	6.65	197.114	5.83×10^{-15}	3.01×10^{34}	4.33×10^{33}	0.35	5.35×10^{5}	1.09×10^{12}	1.30×10^{3}	EGRET
J1105-6107	290.49	−0.85	63.198	1.58×10^{-14}	2.48×10^{36}	1.45×10^{35}	4.98	6.32×10^{4}	1.01×10^{12}	3.69×10^{4}	

(Continued)

Table 1 ***(Continued)***

J1112-6103	291.22	−0.46	64.962	3.15×10^{-14}	4.54×10^{36}	3.62×10^{35}	12.2	3.27×10^{4}	1.45×10^{12}	4.86×10^{4}	
J1119-6127	292.15	−0.54	408.732	4.03×10^{-12}	2.33×10^{36}	6.03×10^{35}	8.4	1.61×10^{3}	4.11×10^{13}	5.53×10^{3}	
J1124-5916	292.04	1.75	135.493	7.50×10^{-13}	1.19×10^{37}	1.70×10^{35}	4.8	2.86×10^{3}	1.02×10^{13}	3.78×10^{4}	
J1357-6429	309.92	−2.51	166.167	3.57×10^{-13}	3.07×10^{36}	2.53×10^{34}	2.5	7.37×10^{3}	7.80×10^{12}	1.56×10^{4}	
J1410-6132	312.19	−0.09	50.052	3.18×10^{-14}	1.00×10^{37}	7.66×10^{35}	15.6	2.50×10^{4}	1.28×10^{12}	9.37×10^{4}	
J1420-6048	313.54	0.23	68.202	8.29×10^{-14}	1.03×10^{37}	6.39×10^{35}	5.61	1.30×10^{4}	2.41×10^{12}	6.98×10^{4}	
J1509-5850	319.97	−0.62	88.925	9.17×10^{-15}	5.15×10^{35}	1.05×10^{35}	2.62	1.54×10^{5}	9.14×10^{11}	1.20×10^{4}	
J1513-5908 (B1509-58)	320.32	−1.16	151.578	1.53×10^{-12}	1.73×10^{37}	6.88×10^{34}	4.21	1.57×10^{3}	1.54×10^{13}	4.07×10^{4}	Comptel
J1531-5610	323.90	0.03	84.201	1.38×10^{-14}	9.12×10^{35}	1.01×10^{33}	2.09	9.67×10^{4}	1.09×10^{12}	1.68×10^{4}	
J1648-4611	339.44	−0.79	164.958	2.37×10^{-14}	2.09×10^{35}	1.60×10^{35}	4.96	1.10×10^{5}	2.00×10^{12}	4.11×10^{3}	
J1702-4128	344.74	0.12	182.153	5.23×10^{-14}	3.42×10^{35}	7.67×10^{34}	4.75	5.52×10^{4}	3.12×10^{12}	4.76×10^{3}	
J1709-4429 (B1706-44)	343.11	−2.67	102.496	9.28×10^{-14}	3.40×10^{36}	8.53×10^{35}	2.3	1.75×10^{4}	3.12×10^{12}	2.67×10^{4}	EGRET
J1718-3825	348.95	−0.43	74.675	1.32×10^{-14}	1.25×10^{36}	1.38×10^{35}	3.6	8.98×10^{4}	1.00×10^{12}	2.22×10^{4}	
J1730-3350 (B1727-33)	354.14	0.09	139.497	8.48×10^{-14}	1.23×10^{36}	3.56×10^{34}	3.55	2.61×10^{4}	3.48×10^{12}	1.18×10^{4}	
J1741-2054	6.42	4.91	413.701	1.70×10^{-14}	9.47×10^{33}	2.06×10^{33}	0.38	3.86×10^{5}	2.68×10^{12}	3.49×10^{2}	
J1747-2958	359.31	−0.84	98.827	6.13×10^{-14}	2.51×10^{36}	5.70×10^{35}	4.75	2.55×10^{4}	2.49×10^{12}	2.38×10^{4}	
J1801-2451 (B1757-24)	5.25	−0.89	124.948	1.27×10^{-13}	2.57×10^{36}	3.91×10^{34}	5.22	1.56×10^{4}	4.03×10^{12}	1.90×10^{4}	
J1833-1034	21.50	−0.89	61.888	2.02×10^{-13}	3.36×10^{37}	1.56×10^{35}	4.7	4.85×10^{3}	3.58×10^{12}	1.39×10^{5}	
J1835-1106	21.22	−1.51	165.916	2.06×10^{-14}	1.78×10^{35}	5.80×10^{33}	2.83	1.28×10^{5}	1.87×10^{12}	3.77×10^{3}	
J1952+3252 (B1951+32)	68.77	2.82	39.534	5.83×10^{-15}	3.72×10^{36}	6.61×10^{34}	2.0	1.07×10^{5}	4.86×10^{11}	7.24×10^{4}	EGRET
J2021+3651	75.22	0.11	103.742	9.56×10^{-14}	3.38×10^{36}	5.91×10^{36}	10.0	1.72×10^{4}	3.19×10^{12}	2.63×10^{4}	
J2030+3641	76.12	−1.44	200.129	6.51×10^{-15}	3.20×10^{34}	3.38×10^{34}	3.0	4.87×10^{5}	1.15×10^{12}	1.33×10^{3}	
J2032+4127	80.22	1.03	143.248	2.04×10^{-14}	2.73×10^{35}	1.69×10^{35}	3.65	1.11×10^{5}	1.73×10^{12}	5.41×10^{3}	
J2043+2740	70.61	−9.15	96.131	1.23×10^{-15}	5.46×10^{34}	3.83×10^{33}	1.8	1.24×10^{6}	3.48×10^{11}	3.60×10^{3}	
J2229+6114	106.65	2.95	51.643	7.79×10^{-14}	2.23×10^{37}	1.94×10^{34}	0.8	1.05×10^{4}	2.03×10^{12}	1.36×10^{5}	
J2240+5832	106.57	−0.11	139.941	1.52×10^{-14}	2.19×10^{35}	7.68×10^{34}	7.7	1.46×10^{5}	1.48×10^{12}	4.96×10^{3}	

Table 2 Radio-quiet isolated neutron stars

Pulsar name	Galactic latitude (deg)	Galactic longitude (deg)	P (ms)	$\dot{P}$ ($s\ s^{-1}$)	$\dot{E}$ ($erg\ s^{-1}$)	Luminosity ($erg\ s^{-1}$)	Distance (kpc)	Age (year)	B_s (G)	B_{LC} (G)	Comments
J0007+7303	119.66	10.46	315.893	3.57×10^{-13}	4.48×10^{35}	9.39×10^{34}	1.4	1.40×10^{4}	1.08×10^{13}	3.14×10^{3}	CTA-1
J0106+4855	125.47	−13.87	83.157	4.28×10^{-16}	2.94×10^{34}	2.09×10^{34}	3.01	3.08×10^{6}	1.91×10^{11}	3.06×10^{3}	
J0357+3205	162.76	−16.01	444.105	1.31×10^{-14}	5.90×10^{33}	$<5.16 \times 10^{35}$	<8.2	5.37×10^{5}	2.44×10^{12}	2.56×10^{2}	
J0622+3749	175.88	10.96	333.208	2.54×10^{-14}	2.71×10^{34}	$<1.19 \times 10^{35}$	<8.3	2.08×10^{5}	2.94×10^{12}	7.33×10^{2}	
J0633+0632	205.09	−0.93	297.397	7.96×10^{-14}	1.19×10^{35}	$<8.52 \times 10^{35}$	<8.7	5.92×10^{4}	4.92×10^{12}	1.72×10^{3}	
J0633+1746	195.13	4.27	237.104	1.10×10^{-14}	3.25×10^{34}	3.17×10^{34}	0.25	3.42×10^{5}	1.63×10^{12}	1.13×10^{3}	Geminga
J0734-1559	232.06	2.02	155.141	1.25×10^{-14}	1.32×10^{35}	$<7.05 \times 10^{35}$	<10.3	1.96×10^{5}	1.41×10^{12}	3.48×10^{3}	
J1023-5746	284.17	−0.41	111.479	3.82×10^{-13}	1.09×10^{37}	$<6.58 \times 10^{36}$	<16.8	4.62×10^{3}	6.61×10^{12}	4.39×10^{4}	
J1044-5737	286.58	1.16	139.030	5.46×10^{-14}	8.02×10^{35}	$<5.52 \times 10^{36}$	<17.2	4.03×10^{4}	2.79×10^{12}	9.55×10^{3}	
J1135-6055	293.79	0.58	114.487	7.84×10^{-14}	2.06×10^{36}	$<1.94 \times 10^{36}$	<18.4	2.31×10^{4}	3.03×10^{12}	1.86×10^{4}	
J1413-6205	312.37	−0.74	109.741	2.74×10^{-14}	8.18×10^{35}	$<8.62 \times 10^{36}$	<21.4	6.35×10^{4}	1.75×10^{12}	1.22×10^{4}	
J1418-6058	313.32	0.13	110.577	1.69×10^{-13}	4.94×10^{36}	9.24×10^{34}	1.6	1.03×10^{4}	4.38×10^{12}	2.98×10^{4}	
J1429-5911	315.26	1.30	115.844	3.05×10^{-14}	7.74×10^{35}	$<4.57 \times 10^{36}$	<21.8	6.02×10^{4}	1.90×10^{12}	1.13×10^{4}	
J1459-6053	317.89	−1.79	103.151	2.53×10^{-14}	9.09×10^{35}	$<7.62 \times 10^{36}$	<22.2	6.47×10^{4}	1.63×10^{12}	1.37×10^{4}	
J1620-4927	333.89	0.41	171.935	1.05×10^{-14}	8.15×10^{34}	$<1.08 \times 10^{37}$	<24.1	2.60×10^{5}	1.36×10^{12}	2.46×10^{3}	
J1732-3131	356.31	1.01	196.544	2.80×10^{-14}	1.46×10^{35}	8.62×10^{33}	0.61	1.11×10^{5}	2.38×10^{12}	2.88×10^{3}	
J1746-3239	356.96	−2.17	199.541	6.56×10^{-15}	3.26×10^{34}	$<5.53 \times 10^{36}$	<25.3	4.82×10^{5}	1.16×10^{12}	1.34×10^{3}	
J1803-2149	8.14	0.19	106.332	1.95×10^{-14}	6.41×10^{35}	$<7.01 \times 10^{36}$	<25.2	8.63×10^{4}	1.46×10^{12}	1.12×10^{4}	
J1809-2332	7.39	−1.99	146.789	3.44×10^{-14}	4.30×10^{35}	1.64×10^{35}	1.7	6.76×10^{4}	2.2[illegible] $\times 10^{12}$	6.62×10^{3}	Taz
J1813-1246	17.24	2.44	48.073	1.76×10^{-14}	6.24×10^{36}	$<1.84 \times 10^{37}$	<24.7	4.34×10^{4}	9.3[illegible] $\times 10^{11}$	7.70×10^{4}	
J1826-1256	18.56	−0.38	110.227	1.21×10^{-13}	3.58×10^{36}	$<2.79 \times 10^{37}$	<24.7	1.44×10^{4}	3.7[illegible] $\times 10^{12}$	2.54×10^{4}	Eel

(Continued)

Table 2 (*Continued*)

J1836+5925	88.88	25.00	173.264	1.50×10^{-15}	1.14×10^{34}	2.04×10^{34}	0.53	1.83×10^{6}	5.16×10^{11}	9.14×10^{2}	Next Geminga
J1838-0537	26.51	0.21	145.709	4.65×10^{-13}	5.93×10^{36}	$<1.30 \times 10^{37}$	<24.1	4.97×10^{3}	8.33×10^{12}	2.48×10^{4}	
J1846+0919	40.69	5.34	225.552	9.93×10^{-15}	3.42×10^{34}	$<1.40 \times 10^{36}$	<22.0	3.60×10^{5}	1.51×10^{12}	1.21×10^{3}	
J1907+0602	40.18	−0.89	106.636	8.67×10^{-14}	2.82×10^{36}	3.14×10^{35}	3.21	1.95×10^{4}	3.08×10^{12}	2.34×10^{4}	
J1954+2836	65.24	0.38	92.710	2.12×10^{-14}	1.05×10^{36}	$<4.28 \times 10^{36}$	<18.6	6.94×10^{4}	1.42×10^{12}	1.64×10^{4}	
J1957+5033	84.60	11.00	374.806	6.83×10^{-15}	5.12×10^{33}	$<6.59 \times 10^{35}$	<14.5	8.69×10^{5}	1.62×10^{12}	2.83×10^{2}	
J1958+2846	65.88	−0.35	290.397	2.12×10^{-13}	3.42×10^{35}	$<3.74 \times 10^{36}$	<18.5	2.17×10^{4}	7.94×10^{12}	2.98×10^{3}	
J2021+4026	78.23	2.09	265.320	5.42×10^{-14}	1.14×10^{35}	2.57×10^{35}	1.5	7.76×10^{4}	3.84×10^{12}	1.89×10^{3}	γ Cygni
J2028+3332	73.36	−3.01	176.707	4.86×10^{-15}	3.48×10^{34}	$<2.06 \times 10^{36}$	<17.2	5.76×10^{5}	9.37×10^{11}	1.56×10^{3}	
J2030+4415	82.34	2.89	227.070	6.49×10^{-15}	2.19×10^{34}	$<1.69 \times 10^{36}$	<15.7	5.54×10^{5}	1.23×10^{12}	9.66×10^{2}	
J2055+2539	70.69	−12.52	319.561	4.11×10^{-15}	4.97×10^{33}	$<1.50 \times 10^{36}$	<15.3	1.23×10^{6}	1.16×10^{12}	3.27×10^{2}	
J2111+4606	88.31	−1.45	157.830	1.43×10^{-13}	1.44×10^{36}	$<1.15 \times 10^{36}$	<14.8	1.75×10^{4}	4.81×10^{12}	1.13×10^{4}	
J2139+4716	92.63	−4.02	282.849	1.80×10^{-15}	3.15×10^{33}	$<5.56 \times 10^{35}$	<14.1	2.49×10^{6}	7.23×10^{11}	2.94×10^{2}	
J2238+5903	106.56	0.48	162.734	9.70×10^{-14}	8.88×10^{35}	$<1.18 \times 10^{36}$	<12.4	2.66×10^{4}	4.02×10^{12}	8.59×10^{3}	

Table 3 Millisecond pulsars

Pulsar name	Galactic latitude (deg)	Galactic longitude (deg)	P (ms)	$\dot{P}$ (s s^{-1})	$\dot{E}$ (erg s^{-1})	Luminosity (erg s^{-1})	Distance (kpc)	Age (year)	B_s (G)	B_{LC} (G)	Comments
J0023+0923	111.38	−52.85	3.050	1.09×10^{-20}	1.51×10^{34}	4.56×10^{32}	0.69	4.45×10^{9}	1.84×10^{8}	5.97×10^{4}	
J0030+0451	113.14	−57.61	4.870	1.02×10^{-20}	3.62×10^{33}	5.75×10^{32}	0.28	7.28×10^{9}	2.30×10^{8}	1.83×10^{4}	
J0034-0534	111.49	−68.07	1.880	4.98×10^{-21}	1.72×10^{34}	5.67×10^{32}	0.54	1.03×10^{10}	7.46×10^{7}	1.03×10^{5}	
J0101-6422	301.19	−52.72	2.570	4.80×10^{-21}	1.01×10^{34}	3.79×10^{32}	0.55	9.42×10^{9}	1.07×10^{8}	5.79×10^{4}	
J0102+4839	124.87	−14.17	2.960	1.17×10^{-20}	1.75×10^{34}	8.51×10^{33}	2.32	4.01×10^{9}	1.88×10^{8}	6.68×10^{4}	
J0218+4232	139.51	−17.53	2.320	7.74×10^{-20}	2.43×10^{35}	3.80×10^{34}	2.64	4.78×10^{8}	4.27×10^{8}	3.15×10^{5}	
J0340+4130	153.78	−11.02	3.300	5.90×10^{-21}	7.87×10^{33}	7.29×10^{33}	1.73	8.86×10^{9}	1.41×10^{8}	3.62×10^{4}	
J0437-4715	253.39	−41.96	5.760	5.73×10^{-20}	2.91×10^{33}	4.86×10^{31}	0.16	6.49×10^{9}	2.88×10^{8}	1.39×10^{4}	
J0610-2100	227.75	−18.18	3.860	1.23×10^{-20}	8.35×10^{32}	9.85×10^{33}	3.54	5.03×10^{10}	6.93×10^{7}	1.11×10^{4}	
J0613-0200	210.41	−9.30	3.060	9.59×10^{-21}	1.20×10^{34}	2.90×10^{33}	0.9	5.55×10^{9}	1.65×10^{8}	5.31×10^{4}	
J0614-3329	240.50	−21.83	3.150	1.78×10^{-20}	2.20×10^{34}	4.72×10^{34}	1.9	2.80×10^{9}	2.40×10^{8}	7.06×10^{4}	
J0751+1807	202.73	21.09	3.480	7.78×10^{-21}	7.21×10^{33}	2.54×10^{32}	0.4	7.16×10^{9}	1.66×10^{8}	3.62×10^{4}	
J1024-0719	251.70	40.52	5.160	1.85×10^{-20}	4.60×10^{32}	5.67×10^{31}	0.39	5.10×10^{10}	9.20×10^{7}	6.17×10^{3}	
J1124-3653	284.10	22.76	2.410	5.75×10^{-21}	1.71×10^{34}	4.29×10^{33}	1.72	6.64×10^{9}	1.19×10^{8}	7.83×10^{4}	
J1125-5825	291.89	2.60	3.100	6.09×10^{-20}	8.05×10^{34}	7.31×10^{33}	2.62	8.06×10^{8}	4.40×10^{8}	1.36×10^{5}	
J1231-1411	295.53	48.39	3.680	2.12×10^{-20}	5.15×10^{33}	2.36×10^{33}	0.43	8.98×10^{9}	1.56×10^{8}	2.89×10^{4}	
J1446-4701	322.50	11.43	2.190	9.85×10^{-21}	3.68×10^{34}	1.89×10^{33}	1.46	3.52×10^{9}	1.49×10^{8}	1.30×10^{5}	
J1514-4946	325.25	6.81	3.590	1.87×10^{-20}	1.60×10^{34}	4.76×10^{33}	0.94	3.04×10^{9}	2.62×10^{8}	5.22×10^{4}	
J1600-3053	344.09	16.45	3.600	9.50×10^{-21}	7.30×10^{33}	1.68×10^{33}	1.63	6.61×10^{9}	1.78×10^{8}	3.52×10^{4}	
J1614-2230	352.64	20.19	3.150	9.62×10^{-21}	6.33×10^{33}	1.23×10^{33}	0.65	9.96×10^{9}	1.27×10^{8}	3.74×10^{4}	
J1658-5324	334.87	−6.63	2.440	1.10×10^{-20}	3.02×10^{34}	2.99×10^{33}	0.93	3.51×10^{9}	1.66×10^{8}	1.05×10^{5}	
J1713+0747	28.75	25.22	4.570	8.53×10^{-21}	3.44×10^{33}	1.34×10^{33}	1.05	8.72×10^{9}	1.97×10^{8}	1.90×10^{4}	
J1741+1351	37.89	21.64	3.750	3.02×10^{-20}	2.18×10^{34}	3.36×10^{32}	1.08	2.04×10^{9}	3.34×10^{8}	5.83×10^{4}	
J1744-1134	14.79	9.18	4.070	8.92×10^{-21}	4.11×10^{33}	6.76×10^{32}	0.41	9.19×10^{9}	1.71×10^{8}	2.33×10^{4}	
J1747-4036	350.21	−6.41	1.650	1.33×10^{-20}	1.17×10^{35}	1.36×10^{34}	3.39	1.97×10^{9}	1.50×10^{8}	3.07×10^{5}	

(Continued)

Table 3 (*Continued*)

J1810+1744	44.64	16.81	1.660	4.63×10^{-21}	3.97×10^{34}	1.12×10^{34}	2.0	5.68×10^{9}	8.87×10^{7}	1.78×10^{5}	
J1823-3021A (B1820-30A)	2.79	−7.91	5.440	3.38×10^{-18}	8.28×10^{35}	7.39×10^{34}	7.6	2.55×10^{7}	4.34×10^{9}	2.48×10^{5}	
J1858-2216	13.58	−11.39	2.380	3.87×10^{-21}	1.13×10^{34}	7.64×10^{32}	0.94	9.74×10^{9}	9.71×10^{7}	6.63×10^{4}	
J1902-5105	345.65	−22.38	1.740	9.00×10^{-21}	6.86×10^{34}	3.59×10^{33}	1.18	3.06×10^{9}	1.27×10^{8}	2.21×10^{5}	
J1939+2134 (B1937+21)	57.51	−0.29	1.560	1.05×10^{-19}	1.10×10^{36}	1.39×10^{34}	3.56	2.34×10^{8}	4.10×10^{8}	9.95×10^{5}	
J1959+2048 (B1957+20)	59.20	−4.70	1.610	1.68×10^{-20}	7.63×10^{34}	1.26×10^{34}	2.49	3.16×10^{9}	1.15×10^{8}	2.54×10^{5}	
J2017+0603	48.62	−16.03	2.900	8.30×10^{-21}	1.30×10^{34}	9.82×10^{33}	1.57	5.54×10^{9}	1.57×10^{8}	5.93×10^{4}	
J2043+1711	61.92	−15.31	2.380	5.70×10^{-21}	1.27×10^{34}	1.00×10^{34}	1.76	8.73×10^{9}	1.03×10^{8}	7.01×10^{4}	
J2047+1053	57.05	−19.68	4.290	2.10×10^{-20}	1.05×10^{34}	3.10×10^{33}	2.05	3.24×10^{9}	3.04×10^{8}	3.54×10^{4}	
J2051-0827	39.19	−30.41	4.510	1.28×10^{-20}	5.42×10^{33}	4.37×10^{32}	1.04	5.68×10^{9}	2.41×10^{8}	2.42×10^{4}	
J2124-3358	10.93	−45.44	4.930	2.06×10^{-20}	3.67×10^{33}	3.96×10^{32}	0.3	7.01×10^{9}	2.37×10^{8}	1.82×10^{4}	
J2214+3000	86.86	−21.67	3.120	1.50×10^{-20}	1.92×10^{34}	9.29×10^{33}	1.54	3.30×10^{9}	2.19×10^{8}	6.63×10^{4}	
J2215+5135	99.87	−4.16	2.610	2.34×10^{-20}	5.19×10^{34}	1.28×10^{34}	3.01	1.77×10^{9}	2.50×10^{8}	1.29×10^{5}	
J2241-5236	337.46	−54.93	2.190	8.70×10^{-21}	2.60×10^{34}	1.05×10^{33}	0.51	3.99×10^{9}	1.40×10^{8}	1.22×10^{5}	
J2302+4442	103.40	−14.00	5.190	1.33×10^{-20}	3.82×10^{33}	6.21×10^{33}	1.19	6.18×10^{9}	2.66×10^{8}	1.75×10^{4}	

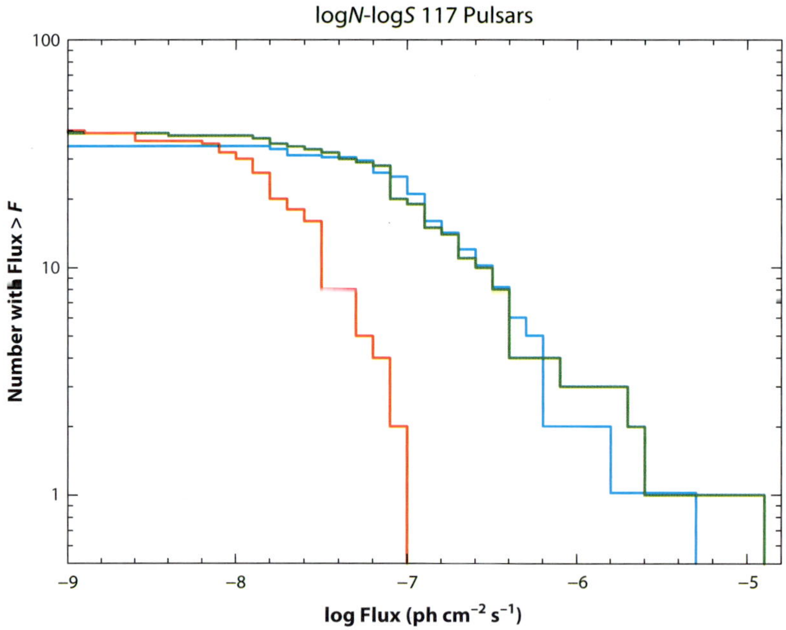

Figure 6

A log$N$-log$S$ plot for the three classes of gamma-ray pulsars: radio-loud pulsars are indicated by the green line, radio-quiet pulsars by the blue line, and millisecond pulsars by the red line.

represent either an observational bias or a true death line for gamma-ray pulsar emission. Among the young gamma-ray pulsars a "segregation" effect is seen, with radio-quiet INSs dominating the $\dot{E}$ interval ranging from 10^{33} to 10^{35} erg sec^{-1}, whereas the opposite is true for $\dot{E} > 10^{37}$ erg sec^{-1}, where only 1 of the 9 pulsars detected by *Fermi* is radio-quiet. This feature, first noted by Ravi et al. (2010), can be linked to the radio and gamma-ray beaming factors. For high-$\dot{E}$ pulsars the beams should be similar both in sky coverage and in location, and they should be rather high in the pulsar magnetosphere, whereas for low-$\dot{E}$ pulsars the radio-emitting region should migrate toward the NS surface, thus shrinking the sky coverage of the radio beam and resulting in a higher percentage of radio-quiet INSs. Such a beaming evolution is also discussed by Watters & Romani (2011) in their population synthesis simulation.

To understand the *Fermi* pulsars, one of the key parameters seems to be the magnetic field at the light cylinder. As we have already remarked, though the surface B-field of young pulsars is 10^4 higher than that inferred for old, recycled ones, the different sizes of their corotating magnetospheres result in a similar B-field at the light cylinder. This is shown in **Figure 8**, where the value of the magnetic field at the light cylinder is plotted against the characteristic age of the pulsars. Once more, the radio-loud/radio-quiet segregation is apparent with the majority of the middle-aged "young" pulsars being radio-quiet, whereas the opposite is true for very young pulsars. Such an effect could result from an observational bias hampering the detection of young, but far-away and thus faint, radio-quiet, very young pulsars. However, if true, such behavior could also be interpreted in the framework of the migration of the radio beaming described above, whereby older and less energetic pulsars are more likely to be spotted as gamma-ray emitters, rather than radio pulsars.

A similar segregation is also seen in the young pulsar spectral parameters. All 117 2PC (Abdo et al. 2013) pulsars have been fitted in a consistent way using a power law with exponential cutoff spectrum. For each pulsar, a power-law index, as well as a value for the cutoff energy, has been

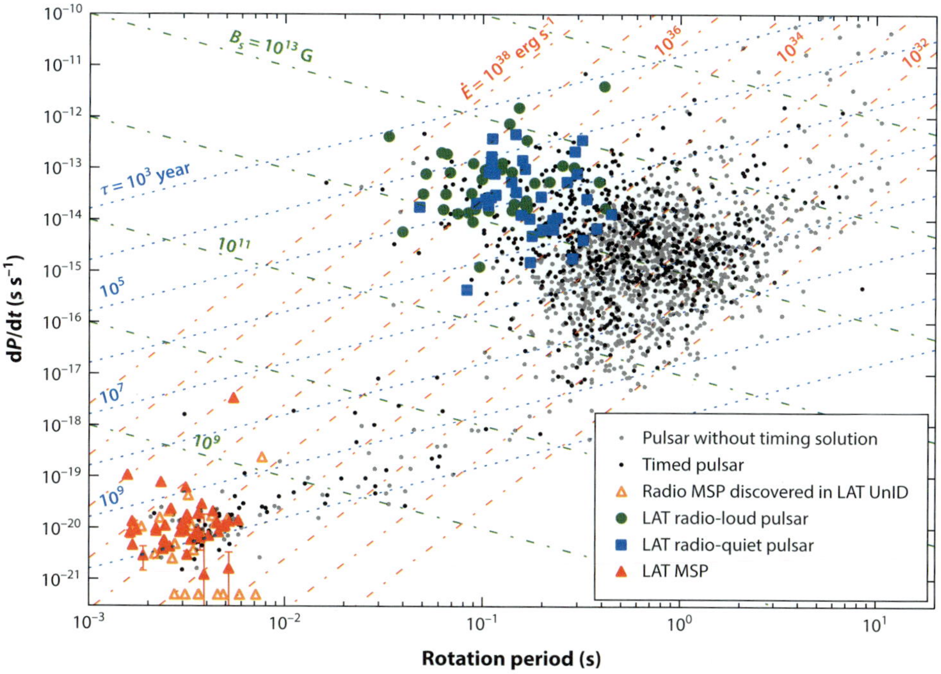

Figure 7

$P - \dot{P}$ distribution of the 117 *Fermi* pulsars [radio-loud pulsars shown as *green circles*, radio-quiet as *blue squares*, and millisecond pulsars (MSPs) as *red triangles*] plotted together with the entire radio pulsar sample known today (*black dots*, 710 in all, represent timed pulsars that were phase folded, but not detected, whereas 1,337 gray dots represent pulsars without timing solution). Lines of constant rotational energy loss ($\dot{E}_{\rm rot} = -4\pi^2 I P \dot{P}^{-3}$), characteristic age ($\tau_{\rm c} = \frac{P}{2\dot{P}}$), and surface *B* field ($B_{\rm s} = (1.5 I c^3 P \dot{P})^{1/2} / 2\pi R_{\rm NS}^3$) are also shown. Recently discovered MSPs, for which no $\dot{P}$ has been measured, are plotted at $\dot{P} = 5 \times 10^{-22}$. All the *Fermi* LAT pulsars lie above $\dot{E} = 10^{33}$ erg sec^{-1}. Reprinted from 2PC (Abdo et al. 2013) with permission.

computed. Although the power-law indices inferred for MSPs and young pulsars do show positive correlations with their rotational energy losses, when one plots the cutoff energy as a function of the magnetic field at the light cylinder (as in **Figure 9**), radio-quiet pulsars dominate in the region characterized by low cutoff and low magnetic field.

A further effort to characterize radio-loud versus radio-quiet INSs has been done through their X-ray emission. A thorough analysis of all the X-ray data available for *Fermi* pulsars has been performed [Marelli et al. 2011, Marelli 2012; later updated within the 2PC (Abdo et al. 2013)] to search for correlation (if any) between the distance independent ratio $F_\gamma/F_{\rm x}$ (gamma-ray flux over X-ray flux) and the pulsar age. As there is significant spread in the $F_\gamma/F_{\rm x}$ values that spans 3–4 orders of magnitude for objects of similar age and energetics, no correlation was found, although inspection of the X-ray flux data revealed that radio-quiet pulsars tend to be underluminous in X-rays.

4.3. The Efficiencies of Gamma-Ray Pulsars: Aged to Perfection

Plotting the gamma-ray luminosity as a function of rotational energy loss of pulsars does not yield a clear picture. Although the plot (shown in **Figure 10**) is hampered by the lack of distance estimates for the majority of radio-quiet NSs, a different trend is apparent for the young pulsars

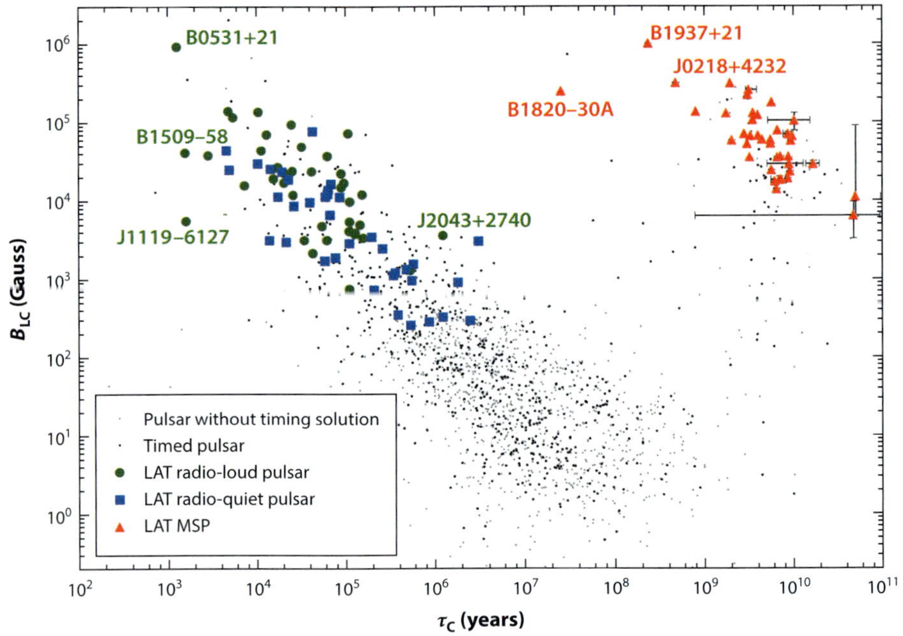

Figure 8

The magnetic field at the light cylinder $B_{\mathrm{LC}} = 4\pi^2(1.5I\dot{P})^{1/2}(c^3P^5)^{-1/2}$ is plotted against the isolated neutron star characteristic age $\tau_c = \frac{P}{2\dot{P}}$. Radio-loud and radio-quiet pulsars (*green circles* and *blue squares*, respectively) are distributed between 10^2 and 10^6 G, whereas millisecond pulsars (MSPs, *red triangles*) are more clustered in the 10^4–10^6 G region. Courtesy of David Smith, using data from 2PC (Abdo et al. 2013).

as opposed to the old, recycled ones. Though young pulsars seem to follow a $\sqrt{\dot{E}}$ trend, MSPs prefer a steeper function, pointing to a proportionality between L_γ and $\dot{E}$. However, the presence of considerable scattering should not be overlooked. It is probably due to the combination of distance uncertainty with the assumption of a common $f_\Omega = 1$. The latter assumption is generally considered acceptable for an outer-magnetosphere fan-like beam(s) sweeping the entire sky. But it has been questioned by Pierbattista et al. (2012), who found a large spread of f_Ω among the different emission models and, in more general terms, between radio-loud and radio-quiet pulsars. Clearly, an f_Ω value optimized for each pulsar using the model yielding the best fit for its light curve would be highly desirable.

However, when plotting the *Fermi* gamma-ray efficiency (i.e., the ratio between the gamma-ray luminosity and the pulsar rotational energy loss $L_\gamma/\dot{E}$) as a function of the pulsar rotational energy loss as in **Figure 11**, a clear trend appears, showing that old pulsars, which can count on a much smaller energy reservoir, are more efficient in converting their rotational energy loss into gamma rays.

4.4. Do Gamma-Quiet Pulsars Exist?

So far we have discussed the characteristics of the INSs detected by *Fermi*. What about those not seen? The point is a nontrivial one because, given the beamed nature of the pulsar emission, it is natural to expect a geometric configuration such that radio emission is detected but gamma-ray pulsation is not. According to Romani & Yadigaroglu (1995), for instance, 8% of the pulsar

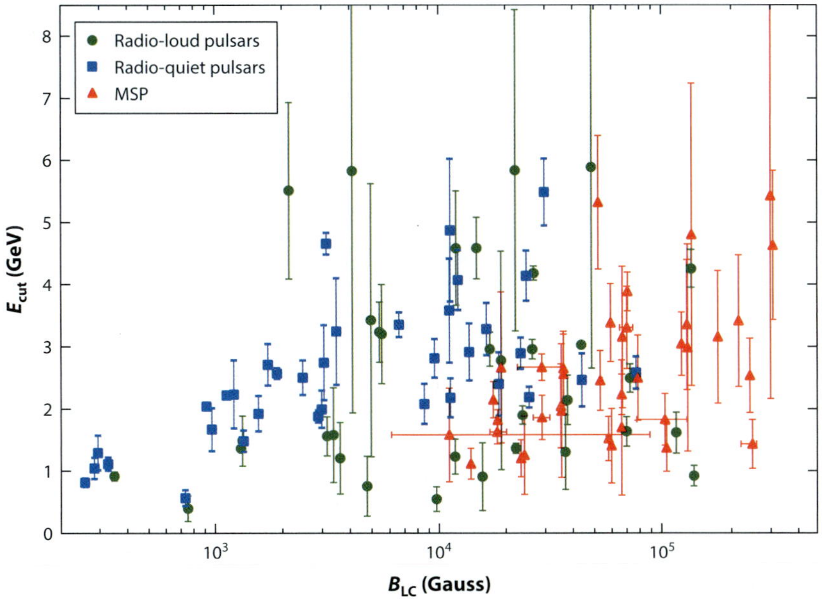

Figure 9

Cutoff energy as a function of the magnetic field at the light cylinder. Radio-loud pulsars are shown as *green circles*, radio-quiet as *blue squares*, and millisecond pulsars (MSPs) as *red triangles*. Courtesy of David Smith, using data from 2PC (Abdo et al. 2013).

population should be gamma-ray quiet. Guillemot & Tauris (2014) find that the most energetic MSPs may go undetected in gamma rays owing to unfavorable geometry.

Are there pulsars highly ranked on the basis of $\frac{\sqrt{\dot{E}}}{4\pi d^2}$ that are not detected by *Fermi*? 28 of the 64 known radio pulsars with $\dot{E} > 10^{36}$ erg sec^{-1} have not been detected and, comparing their upper limits with the "expected" flux (on the basis of their ranking), a number of interesting gamma-quiet candidates appear to be present. However, as discussed in 2PC (Abdo et al. 2013) as well as by Romani et al. (2011), distance uncertainties, confusion with nearby sources, and the possible presence of significant timing noise make it difficult to provide a clear-cut answer. A few INSs have expected gamma-ray flux values well above the current upper limits, but none is yet recognized as a fully convincing gamma-quiet radio pulsar.

4.5. Light Curves and Their Interpretation

Upon inspecting the light curves of the 117 gamma-ray pulsars, one immediately realizes that the majority of the pulsars (70% of the young pulsars and 60% of the MSPs) have two peaks (respectively, *P*1 and *P*2, where *P*1 is defined as the one soonest in phase after the radio main peak), and that the ratio *P*2/*P*1 increases with energy, indicating a harder second peak. Many double-peaked pulsars display a crescent-type light curve with significant emission between the two peaks. For the Crab and Vela, a third peak emerges from the interpeak bridge emission. However, the third peak of Vela is seen to move in phase as a function of energy (Abdo et al. 2010e), an effect not expected in the current geometric models.

For the radio pulsars, the lag between the radio signal and *P*1 is carefully evaluated because it is one of the important parameters in characterizing a pulsar's multiwavelength behavior. Indeed,

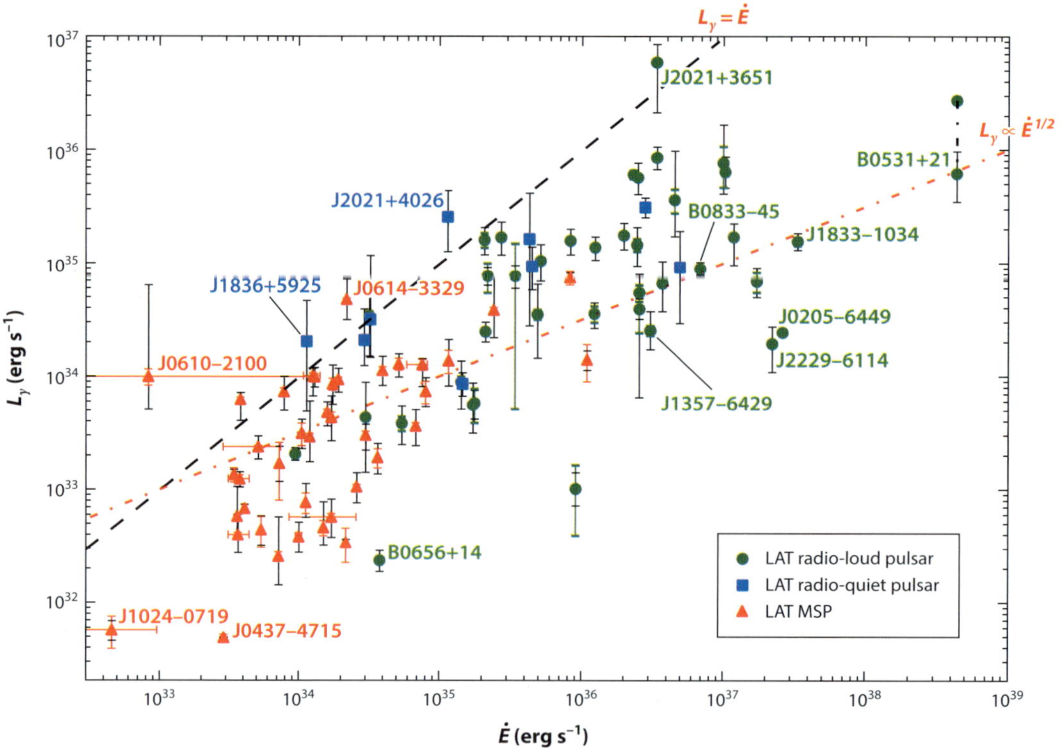

Figure 10

Gamma-ray luminosity ($L_\gamma = 4\pi d^2 F_\gamma f_\Omega$, computed assuming $f_\Omega = 1$) as a function of the rotational energy loss. The difference in slope between the red cloud of millisecond pulsars (MSPs, *triangles*) and the green cloud of radio-loud pulsars (*circles*) is evident. The paucity of radio-quiet pulsars (*squares*) is due to the lack of distance information. A dashed line represents $L_\gamma = \dot{E}$, a rather extreme case used to highlight pulsars whose distances must be significantly overestimated or that require a much narrower beaming factor. A dash-dotted line follows $\sqrt{\dot{E}}$. The Crab pulsar, PSR B0531+21, is the only pulsar detected as a very bright X-ray source. The lower point represents the gamma-ray luminosity of the Crab pulsar, whereas the upper point indicates the total luminosity including X-rays. Reprinted from 2PC, Abdo et al. (2013) with permission.

the lag between the radio peak and *P*1 is greater for MSPs than for young INSs, indicating that a smaller magnetosphere implies a stronger aberration of the radio pulses.

Such a wealth of information on pulsar light curves represents a new challenge for theoreticians who try to constrain the geometry of pulsars as well as the relevant magnetospheric physics. Starting from the location(s) of the emitting region(s), namely PC, OG, and SG, or its variation TPC, in a dipole geometry, one can build an "atlas" of predicted gamma-ray light curves to compare to observations. The computed light curves are sensitive to both the magnetic axis inclination (α) and the viewing angle (ζ), and with no a priori knowledge on such variables, all the combinations should be considered. **Figure 12** provides an example of emission pattern phase plots computed for a given magnetic inclination ($\alpha = 45°$) for all possible viewing angles. The actual pulsar light curve, obtained cutting the phase plot for a definite value of ζ, can be used to select the best fitting value of ζ on the basis of the different models. Conversely, external inputs, such as radio polarization angle or X-ray morphological study, can help to constrain the parameter space. The first gamma-ray atlas was compiled by Watters et al. (2009) for vacuum dipole field geometries. Bai & Spitkovsky (2010) considered numerically modeled force-free geometries, whereas Venter

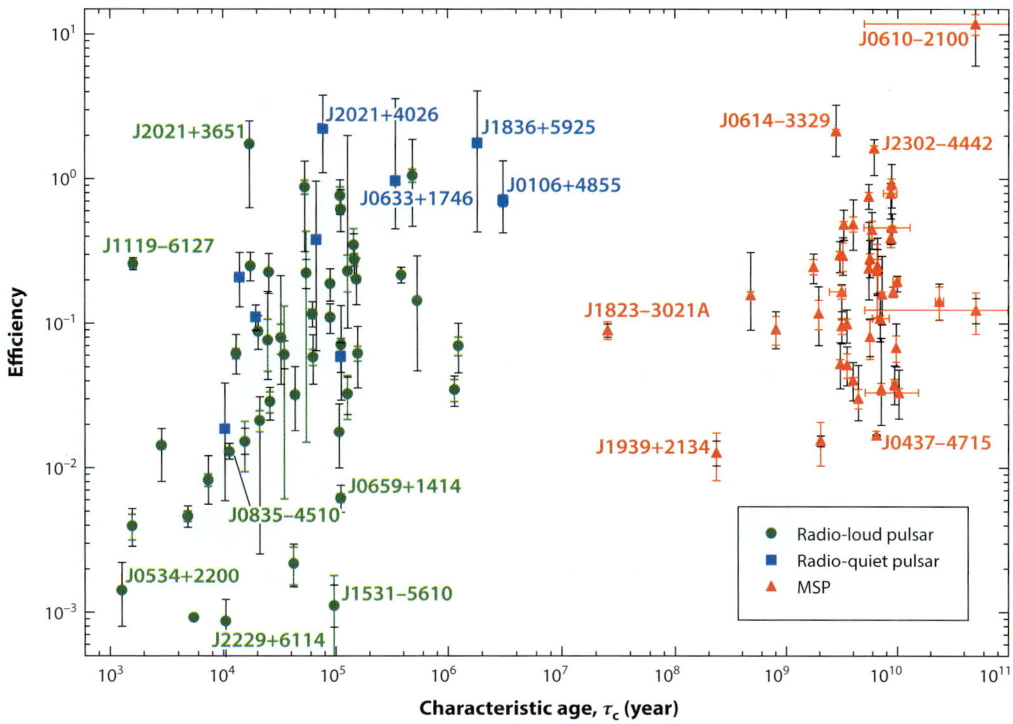

Figure 11

Gamma-ray efficiency ($L_\gamma / \dot{E}$) plotted as a function of a pulsar's characteristic age, showing that older and less energetic pulsars are more efficient in converting their rotational energy loss into high-energy gamma rays. Radio-loud pulsars are shown as *green circles*, radio-quiet pulsars as *blue squares*, and millisecond pulsars (MSPs) as *red triangles*. The number of pulsars with an efficiency near or above 1 is to be ascribed to overestimated distances as well as to a too-large beaming factor (here assumed to be 1 for all pulsars). Such a hidden beaming uncertainty is certainly responsible for the unrealistically high efficiency computed for Geminga (J0633+1746), for which the distance is well known. Courtesy of David Smith, using data from 2PC (Abdo et al. 2013).

et al. (2009) concentrated on the newly established MSPs, focusing on their higher percentage of alignment between radio and gamma-ray light curves (Venter et al. 2012). In general, OG models yield better fits (Romani & Watters 2010), but they are not able to account for all the detected pulsars. Lower-altitude emission is preferred for a sizable minority of pulsars, especially those MSPs with aligned radio and gamma-ray light curves. In parallel, emission beyond the light cylinder continues to be a viable alternative (as discussed by Petri 2011).

Pierbattista (2010) extended the atlas approach to also include radio light curves by trying to constrain the α and ζ parameters on the basis of a joint radio-gamma fit. Although promising, the procedure still needs fine-tuning in order to avoid being driven by the radio data. Once a comprehensive gamma-ray and radio light curve atlas is in hand, it can be used to synthesize a pulsar population to be compared with the *Fermi* findings (Pierbattista et al. 2012).

Not surprisingly, the OG model, with its extended beams, can easily account for the observed number of *Fermi* detections while the narrow PC beams can only account for a few pulsars, and the SG need a boost in efficiency to account for the observed set of LAT detections. The population synthesis, however, fails to reproduce the LAT results for high $\dot{E}$ pulsars. All the models predict too few high $\dot{E}$ pulsars and cannot explain the high probability of detecting energetic radio-loud pulsars.

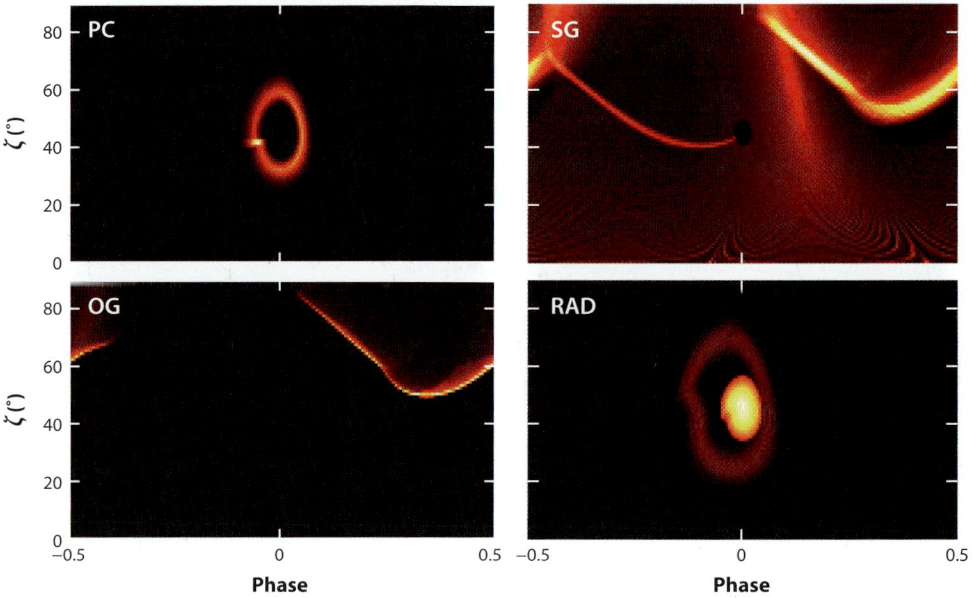

Figure 12

Emission pattern phase-plots computed, respectively, for polar cap (PC), slot gap (SG), outer gap (OG), and radio core plus cone models (RAD). For PC and radio models the phase-plots have been obtained for a magnetic field $B = 10^{12}$ Gauss and spin period of 30 ms, whereas for SG and OG models the phase-plots were computed for gap widths of 0.04 and 0.01, respectively. All plots have been obtained for a magnetic obliquity $\alpha = 45$. The emission intensity decreases from yellow to black. (Courtesy of Marco Pierbattista.)

In a nutshell, none of the models proposed so far is able to account for the phenomenology of the observed *Fermi*-LAT pulsars: OG and SG are generally better but not adequate to fit all objects. Some pulsars can be reasonably well fitted by more than one model, some by none.

5. PARADIGM LOST: WINNING THE PULSAR REVOLUTION

While the 2PC (Abdo et al. 2013) was being written, the gamma-ray pulsar family gained 15 more members that are listed in table 4 of the catalog. Of the new entries, 11 are MSPs while 4 are young radio pulsars. Since then, new detections have been announced, and as of June 2014, 147 NSs have been seen in gamma rays. Their parameters can be found at **https://confluence.slac.stanford.edu/display/GLAMCOG/Public+List+of+LAT-Detected+Gamma-Ray+Pulsars**. Thanks to the extremely successful synergy between radio and gamma-ray astronomy to find, characterize, and phase-fold MSPs, the number of MSP discoveries currently accounts for two-thirds of the new pulsars. However, one such MSP was not found through radio searches but rather through an original approach exploiting at once gamma-ray and optical astronomy. Performing blind searches for MSPs in binary systems is vastly more difficult than searching for young isolated pulsars, because the search must cover a much larger frequency interval, scanning for (at least three) additional unknown orbital parameters. This becomes an impossible task, even for the most powerful supercomputer. The source 0FGL J1311.9-3419, bright enough to be already listed in the very first *Fermi* Bright Source Catalog (Abdo et al. 2009g), however, yielded a valuable hint through its optical emission. In an observing campaign aimed at finding optical variation of MSPs' binary companions within unidentified *Fermi* sources, Romani (2012) detected a quasi-sinusoidal 93-minute modulation that was thought to arise from a black-widow system (where the MSP irradiates its low-mass companion leading, eventually, to its evaporation). Indeed, the 93-m period is the shortest known for such a binary system. By focusing on the coordinates of the variable optical source and exploiting the tight constraints placed on the orbital parameters, Pletsch et al. (2012c) were able to detect a 2.5-ms MSP, the first found through a gamma-ray blind search, an historic accomplishment. However, PSR J1311-3430 is not the long-sought radio-quiet MSP: radio astronomers went back to the source, which had been already observed to no avail and, with considerable effort, Ray et al. (2013) found the radio signal that is present only during a small fraction of the radio observations. Further optical studies allowed Romani et al. (2012) to constrain the mass of the NS to be $>2.1\,M_{\odot}$.

A similar optical strategy was applied to 0FGL J2339.8-0530. Variability pointed to a tentative orbital period (Romani & Shaw 2011) that was the driver behind a successful radio search that soon led to a gamma-ray detection listed in the 2PC (Abdo et al. 2013). Those success stories rest on new multiwavelength strategies that will certainly bring more results in the future.

5.1. 2013: Enter Citizen Science

Blind searches for gamma-ray pulsars are computationally intensive, thus any additional computer resource is most welcome. To increase (at no cost) the computer power to be used in their searches, Pletsch et al. (2013) exploited the potential of the volunteer distributed computer system Einstein@Home (E@H, Allen et al. 2013), a well-known Citizen Science project that was started in 2005 to search for GWs in data collected by the LIGO-Virgo Collaboration. By downloading the E@H software, volunteers agree to devote the power of their computer, when not in use, to perform the E@H tasks. With more than 300,000 volunteers, E@H is one of the most popular Citizen Science projects and averages a total computing power of 1 PFLOPS, comparable to the largest supercomputers. Since 2009, the system has been adapted to analyze radio telescope data, yielding several pulsar discoveries (Allen et al. 2013), and is now being used to perform blind searches on *Fermi* data. So far, four young pulsars have been discovered in unidentified 2FGL (Nolan et al. 2012) sources. Their parameters are characteristic of energetic objects relatively nearby, but none of them has yet been detected in radio. Considering the shortage of energetic radio-quiet pulsars, this result is remarkable. Two of the new pulsars have already been seen to glitch: PSR J1522-5734 glitched once, whereas PSR J1422-6138 glitched twice, proving that the system works well in spite of those timing jumps. The additional computer power provided by E@H combined with an improved search technique is a good omen for more findings, hopefully also of the long-sought radio-quiet MSP.

5.2. Breaking a Decade-Long Paradigm

The discovery of the Crab Nebula variability has been one the most astonishing results in high-energy gamma-ray astronomy. On September 2010, a sudden enhancement of the overall Crab flux was reported first by AGILE and immediately confirmed by *Fermi* (Abdo et al. 2011, Tavani et al. 2011). A quick sequence of radio, X-ray, and optical observations made it clear that the pulsar was behaving normally. Indeed, in gamma rays the pulsed flux from the Crab was also unchanged, leaving the nebula as the only suspect for the flux increase. Moreover, the short timescale of the variability pointed to a quite small region of interest, possibly next to the pulsar. To achieve the highest angular resolution to study the interior of the Crab Nebula, near the pulsar, the *Chandra X-ray Observatory* and the *Hubble Space Telescope* (HST) were immediately triggered for target-of-opportunity (ToO) observations that were performed a few days after the event. The high-resolution optical and X-ray images failed to show any dramatic change in the notoriously active Crab Nebula inner region. A search in both AGILE and *Fermi* data did prove that enhancements had been detected previously, making it clear that the Crab's September 2010 flaring episode was not a unique event but rather a recurring one. A massive ToO campaign was organized, waiting for the next flares, which were recorded in April 2011 (Striani et al. 2011, Buehler et al. 2012, Weisskopf et al. 2013) and March 2013 (Mayer et al. 2013), when both *Chandra* and HST repeatedly observed the nebula. Once again, comparing images taken before, during, and after the flare, nothing obvious was seen to change. The Crab flares shine only in gamma rays, and such events are possibly linked to sudden particle acceleration (e.g., Buehler & Blandford 2014), driven perhaps by magnetic reconnection. Hopefully the coming years may provide more flares to test the various proposed theories.

In spite of the pulsarness definition we have used so far, variability in INSs and in their surroundings may not be all that exceptional. Allafort et al. (2013) recently reported the detection of a significant flux variation; this time there was a decrease from PSR J2021+4026, a radio-quiet INS in the Cygnus region with an X-ray counterpart (Weisskopf et al. 2011). Judging from its light curve, PSR J2021+4026 is similar to Geminga and recently joined Geminga and CTA-1 in the very small club of radio-quiet gamma-ray pulsars seen to pulsate in X-rays (Lin et al. 2013). The gamma-ray flux decrease, which took place in less than a week, is associated with a 4% increase in the pulsar's spin-down rate and a change in the light curve. The timing parameters of the pulsar have changed in a way never before seen in gamma rays. A jump in $\dot{P}$ is typically followed by a recovery pattern. Moreover, flux variability has never been associated with a glitch, despite repeated analyses during major Vela glitches to search for such a behavior. Is the jump due to a shift in the magnetic field structure? If so, PSR J2021+4026 may have done it before, as AGILE has detected variability from this gamma-ray source (Chen et al. 2011), although the finding could not be linked to the pulsar. As puzzling as it may sound, these results may be heralding a new era where the pulsar steady flux paradigm is superseded.

6. NOW WHAT?

Figure 13 provides a snapshot of the current census of *Fermi* pulsars (divided between radio-loud pulsars, radio-quiet pulsars, and MSPs) together with an overall view of four decades of pulsar studies in high-energy gamma rays. While the growth during the past five years is dramatic, to say the least, it is apparent that the most recent detections have changed the family balance, making MSPs the dominant class among the gamma-ray pulsar types. This is the unexpected, surprising, and really revolutionary result from the *Fermi* mission. Predicting the detection of more MSPs is easy on the basis of the discovery rate shown in **Figure 3**. The next pulsar catalog will be dominated by MSPs, many in black-widow systems.

Together with more MSPs, fainter young radio pulsars will be found (Hou et al. 2014) owing to a major *Fermi* software update, which promises improved sensitivity at low energies (Atwood et al. 2013). The statistical significance for faint sources builds up slowly, and endurance is required to spot the less energetic pulsars. However, this effort should be handsomely rewarded as the detections of even a handful of pulsars fainter than the current death line in the $P - \dot{P}$ diagram will have important implications in pulsar physics. Moreover, lowering the $\dot{E}$ threshold for gamma-ray detection would bring many more INS candidates into play, thus increasing their contribution to the overall Galactic emission while also providing an obvious source for the overabundance of positrons detected by the Pamela Collaboration (Adriani et al. 2009), *Fermi* (Ackerman et al. 2012b), and the AMS Collaboration (Aguilar et al. 2013).

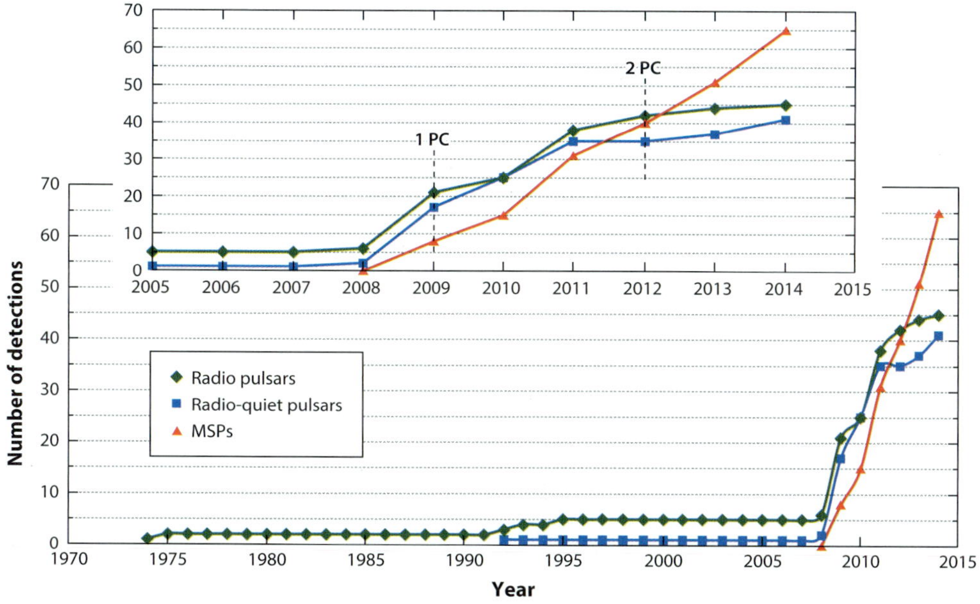

Figure 13

Pulsar census for radio-loud pulsars (*green diamonds*), radio-quiet pulsars (*blue squares*), and millisecond pulsars (MSPs, *red triangles*). The inset shows the dramatic advancements of the past five years. The pulsar accounting published in the first (1PC, Abdo et al. 2010b) and second (2PC, Abdo et al. 2013) pulsar catalogs are highlighted.

The discovery of gamma-quiet INSs detected only at radio wavelengths is also an important goal because it would provide a missing piece of the puzzle for pulsar emission geometry. So far, all the INSs we know of come in two flavors: radio-and-gamma and gamma-only. It is time to complete the picture with a few representatives of the gamma-quiet class. Detecting just unpulsed gamma-ray emission from a NS magnetosphere is also an intriguing possibility. But it will be hard to secure a convincing identification without a pulsar time signature.

By solving the riddle of dozens of previously unidentified gamma-ray sources, pulsars, be they young or recycled, prove to be the most promising candidates to account for a sizable fraction the remaining unassociated Galactic gamma-ray sources. By continuing to study unassociated sources, *Fermi* will certainly unveil many more surprising results.

Surprises may come also from the scores of INSs already detected in gamma-rays. Although variability may be more common than previously thought, the combined use of many of the MSPs unveiled by *Fermi* may provide a tool to directly detect, at last, gravitational waves.

DISCLOSURE STATEMENT

The author is not aware of any affiliations, memberships, funding, or financial holdings that might be perceived as affecting the objectivity of this review.

ACKNOWLEDGMENTS

This paper is my own mostly first-hand account of the 40-year-long struggle to detect and understand gamma-ray emission from NSs. During this long time span, I enjoyed working and discussing with many colleagues whose results and opinions I have tried to describe as accurately as possible.

I am indebted to the whole *Fermi* LAT collaboration, which did a tremendous job in changing our views of the gamma-ray sky. Working with the Galactic group has always been fun and stimulating. I have learned a lot from Alice Harding and Roger Romani, who have been the theoretical task force behind many *Fermi* papers and, quite rightly, have been awarded the 2013 Rossi prize for their contributions toward the understanding of gamma-ray pulsar emission. I am especially grateful to David Smith, who helped me to handle all the *Fermi*-related figures as well as the pulsar tables. I really appreciated his critical assessments and his clever suggestions. Elizabeth Ferrara and Pablo Saz Parkinson were kind enough to read and comment on the manuscript. To them my most sincere thanks for their time and their patience. Marco Pierbattista kindly provided **Figure 12**. I am also grateful to the Fermi Publication Board, which granted permission to use published material as well as unpublished data from the most recent pulsar census.

LITERATURE CITED

Abdo AA, Ackermann M, Ajello M, et al. (Fermi LAT Collab.) 2009a. *Astropart. Phys.* 32:193
Abdo AA, Ackermann M, Ajello M, et al. (Fermi LAT Collab.) 2009b. *Ap. J. Lett.* 699:L102
Abdo AA, Ackermann M, Ajello M, et al. (Fermi LAT Collab.) 2009c. *Science* 325:848
Abdo AA, Ackermann M, Ajello M, et al. (Fermi LAT Collab.) 2009d. *Science* 325:845
Abdo AA, Ackermann M, Ajello M, et al. (Fermi LAT Collab.) 2009e. *Science* 325:840
Abdo AA, Ackermann M, Ajello M, et al. (Fermi LAT Collab.) 2009f. *Ap. J.* 700:1059
Abdo AA, Ackermann M, Ajello M, et al. (Fermi LAT Collab.) 2009g. *Ap. J. Suppl.* 183:46
Abdo AA, Ackermann M, Ajello M, et al. (Fermi LAT Collab.) 2010a. *Astron. Astrophys.* 524:75
Abdo AA, Ackermann M, Ajello M, et al. (Fermi LAT Collab.) 2010b. *Ap. J. Suppl.* 187:460
Abdo AA, Ackermann M, Ajello M, et al. (Fermi LAT Collab.) 2010c. *Ap. J.* 708:1254
Abdo AA, Ackermann M, Ajello M, et al. (Fermi LAT Collab.) 2010d. *Ap. J.* 712:1209
Abdo AA, Ackermann M, Ajello M, et al. (Fermi LAT Collab.) 2010e. *Ap. J.* 713:154
Abdo AA, Ackermann M, Ajello M, et al. (Fermi LAT Collab.) 2010f. *Ap. J.* 720:272
Abdo AA, Ackermann M, Ajello M, et al. (Fermi LAT Collab.) 2010g. *Ap. J. Suppl.* 188:405
Abdo AA, Ackermann M, Ajello M, et al. (Fermi LAT Collab.) 2010h. *Ap. J.* 711:64
Abdo AA, Ackermann M, Ajello M, et al. (Fermi LAT Collab.) 2011. *Science* 331:739
Abdo AA, Ackermann M, Atwood WB, et al. (Fermi LAT Collab.) 2008. *Science* 322:1218
Abdo AA, Ackermann M, Atwood WB, et al. (Fermi LAT Collab.) 2009h. *Ap. J. Lett.* 695:L72
Abdo AA, Ackermann M, Atwood WB, et al. (Fermi LAT Collab.) 2009i. *Ap. J.* 696:1084
Abdo AA, Ackermann M, Atwood WB, et al. (Fermi LAT Collab.) 2009j. *Ap. J.* 699:1171
Abdo AA, Ajello M, Allafort A, et al. (Fermi LAT Collab.) 2013. *Ap. J. Suppl.* 208:17
Ackermann M, Ajello M, Allafort A, et al. (Fermi LAT Collab.) 2012a. *Ap. J.* 753:83
Ackermann M. Ajello M, Allafort A, et al. (Fermi LAT Collab.) 2012b. *Phys. Rev. Lett.* 108:011103
Adriani O, Barbarino GC, Bazilevskaya GA (Pamela Collab.) 2009. *Nature* 448:607
Aguilar M, Alberti G, Alpat B, et al. (AMS Collab.) 2013. *Phys. Rev. Lett.* 110:141102
Aleksic J, Alvarez EA, Antonelli LA, et al. (MAGIC Collab.) 2012. *Astron. Astrophys.* 540:69
Aliu E, Anderhub H, Antonelli LA, et al. (MAGIC Collab.) 2008. *Science* 322:1221
Aliu E, Arlent T, Aune T, et al. (Veritas Collab.) 2011. *Science* 334:69
Allafort A, Baldini L, Ballet J, et al. 2013. *Ap. J. Lett.* 777:L2
Allen B, Knispel B, Cordes JM, et al. 2013. *Ap. J.* 773:91
Arons J. 1983. *Ap. J.* 266:215
Arons J. 1996. *Astron. Astrophys.* 129:49
Atwood WB, Abdo AA, Ackermann M, et al. (Fermi LAT Collab.) 2009. *Ap. J.* 697:1071
Atwood WB, Baldini L, Bregeon J, et al. 2013. *Ap. J.* 774:76
Atwood WB, Ziegler M, Johnson RP, Baughman BM. 2006. *Ap. J. Lett.* 652:L49

Bai XN, Spitkovsky A. 2010. *Ap. J.* 715:1282
Baldini L, Bastieri L, Boinee D, et al. 2006. *Nuclear Phys. B Proc. Suppl.* 150:62
Baring M. 2004. *Adv. Space Res.* 33:552
Bennett K, Bignami GF, Boella G, et al. 1977. *Astron. Astrophys.* 61:279
Bertsch DL, Brazier KTS, Fichtel CE, et al. 1992. *Nature* 357:306
Bignami GF. 2008. *Science* 322:1193
Bignami GF, Bennett K, Buccheri R, et al. 1981a. *Astron. Astrophys.* 93:71
Bignami GF, Boella G, Burger JJ, et al. 1975. *Space Sci. Instrum.* 1:245
Bignami GF, Caraveo PA. 1980. *Ap. J.* 241:1161
Bignami GF, Caraveo PA. 1992. *Nature* 357:287
Bignami GF, Caraveo PA. 1996. *Ann. Rev. Astron. Astrophys.* 34:331
Bignami GF, Caraveo PA, Lamb RC. 1983. *Ap. J. Lett.* 272:L9
Bignami GF, Caraveo PA, Lamb RC, Markert TH, Paul JA. 1981b. *Ap. J. Lett.* 247:L85
Bignami GF, Hermsen W. 1983. *Annu. Rev. Astron. Astrophys.* 21:67
Brazier KTS, Kanbach G, Carramiñana A, Guichard J, Merck M. 1996. *MNRAS* 281:1033
Brazier KTS, Reimer O, Kanbach G, Carramiñana A. 1998. *MNRAS* 295:819
Buccheri R, Bennett K, Bignami GF, et al. 1983. *Astron. Astrophys.* 128:245
Buccheri R, D'Amico N, Massaro E, Scarsi L. 1978. *Nature* 274:572
Buehler R, Scargle JD, Blandford RD, et al. 2012. *Ap. J.* 749:26
Bühler R, Blandford R. 2014. *Rep. Prog. Phys.* 77:066901
Bulgarelli A, Tavani M, Caraveo P, et al. 2008. *Astron. Astrophys.* 489:L17
Camilo F, Kerr M, Ray PS, et al. 2012. *Ap. J.* 746:39
Camilo F, Ray PS, Ransom SM, et al. 2009. *Ap. J.* 705:1
Caraveo PA. 1982. *Space Sci. Rev.* 36:207
Caraveo PA. 2010. *Proc. HTRA-IV Agios Nikolaos, Crete, Greece, May 5–7*. arXiv:10092421
Caraveo PA, Bignami GF, De Luca A, et al. 2003. *Science* 301:1345
Caraveo PA, De Luca A, Marelli M, et al. 2010. *Ap. J. Lett.* 725:L6
Caraveo PA, De Luca A, Mereghetti S, Pellizzoni A, Bignami GF. 2004. *Science* 305:376
Chen AW, Piano G, Tavani M, et al. 2011. *Astron. Astrophys.* 525:33
Cheng KS, Ho C, Ruderman M. 1986. *Ap. J.* 300:500
Chiang J, Romani RW. 1992. *Ap. J.* 400:692
Chiang J, Romani RW. 1994. *Ap. J.* 436:754
Cognard I, Guillemot L, Johnson TJ, et al. 2011. *Ap. J.* 732:47
Cordes JM, Shannon RM. 2012. *Ap. J.* 750:89
Coroniti FV. 1990. *Ap. J.* 349:538
D'Amico N, Kaspi VM, Manchester RN, et al. 2001. *Ap. J. Lett.* 552:L45
Daugherty JK, Harding AK. 1982. *Ap. J.* 252:337
Daugherty JK, Harding AK. 1996. *Ap. J.* 458:278
de Jager OC, Büsching I. 2010. *Astron. Astrophys.* 517:L9
de Jager OC, Raubenheimer BC, Swanepoel JWH. 1989. *Astron. Astrophys.* 221:180
De Luca A, Caraveo PA, Mattana F, Pellizzoni A, Bignami GF. 2006. *Astron. Astrophys.* 445:L9
De Luca A, Caraveo PA, Mereghetti S, Negroni M, Bignami GF. 2005. *Ap. J.* 623:1051
Dormody M, Johnson RP, Atwood WB. 2011. *Ap. J.* 742:126
Dyks J, Rudak B. 2003. *Ap. J.* 598:1201
Fichtel CE, Hartman RC, Kniffen DA, et al. 1975. *Ap. J.* 198:163
Fierro JM, Bertsch DL, Brazier, et al. 1993. *Ap. J. Lett.* 413:L27
Freire PCC, Abdo AA, Ajello M, et al. 2011. *Science* 334:1107
Gehrels N, Chipman E, Kniffen D. 1994. *Ap. J. Suppl.* 92:351
Gehrels N, Macomb DJ, Bertsch DL, Thompson DJ, Hartman RC. 2000. *Nature* 40:363
Goldreich P, Julian WH. 1969. *Ap. J.* 157:869
Grenier I, Hermsen W, Clear J. 1988. *Astron. Astrophys.* 204:117
Guillemot L, Tauris TM. 2014. *MNRAS* 439:2033–42
Halpern JP, Camilo F, Giuliani A, et al. 2008. *Ap. J. Lett.* 688:L33

Halpern JP, Camilo F, Gotthelf EV, et al. 2001a. *Ap. J. Lett.* 552:L125
Halpern JP, Gotthelf EV, Camilo F, Helfand DJ, Ransom SM. 2004. *Ap. J.* 612:398
Halpern JP, Gotthelf EV, Leighly KM, Helfand DJ. 2001b. *Ap. J.* 547:323
Halpern JP, Gotthelf EV, Mirabal N, Camilo F. 2002. *Ap. J. Lett.* 573:L41
Halpern JP, Holt SS. 1992. *Nature* 357:222
Harding AK, Gonthier PL, Ouellette MS, O'Brien S, Berrier J. 2002. In *Neutron Stars in Supernova Remnants*, ed. PO Slane, BM Gaensler. *ASP Conf. Ser.* 271:335. San Francisco: ASP.
Harding AK, Grenier I, Gonthier PL. 2007. *Astrophys. Space Sci.* 309:221
Harding AK, Tademaru E, Esposito LW. 1978. *Ap. J.* 225:226
Harding AK, Usov V, Muslimov V, Alex G. 2005. *Ap. J.* 622:531
Hartman RC, Bertsch DL, Bloom SD, et al. 1999. *Ap. J. Suppl.* 123:79
Hirotami K, Harding AK, Shibata S. 2003. *Ap. J.* 591:334
Hou X, Smith D, Reposeur T, Rousseau R. 2014. *Astron. Nachr.* 335(3):334
Jackson MS, Halpern JP. 2005. *Ap. J.* 633:1114
Jenet FA, Hobbs GB, Lee KJ, Manchester RN. 2005. *Ap. J. Lett.* 625:L123
Kanbach G, Arzoumanian Z, Bertsch DL, et al. 1994. *Astron. Astrophys.* 289:855
Kanbach G, Bennett K, Bignami GF, et al. 1980. *Astron. Astrophys.* 90:163
Kanbach G, Bertsch DL, Fichtel CE, et al. 1988. *Space Sci. Rev.* 49:69
Kaspi VM, Lackey JR, Mattox J, et al. 2000. *Ap. J.* 528:445
Keith MJ, Johnston S, Kramer M, et al. 2008. *MNRAS* 389:1881
Keith MJ, Johnston S, Ray PS, et al. 2011. *MNRAS* 414:1292
Kerr M. 2011. *Ap. J.* 732:38
Kerr M, Camilo F, Johnson TJ, et al. 2012. *Ap. J. Lett.* 748:L2
Kong AKH, Hui CY, Cheng KS. 2010. *Ap. J. Lett.* 712:L36
Kniffen DA, Hartman RC, Thompson DJ, Bignami GF, Fichtel CE. 1974. *Nature* 251:397
Kramer M, Bell JF, Manchester RN, et al. 2003. *MNRAS* 342:1299
Kuiper L, Hermsen W, Cusumano G, et al. 2001. *Astron. Astrophys.* 378:918–35
Kuiper L, Hermsen W, Krijger JM, et al. 1999. *Astron. Astrophys.* 351:119
Kuiper L, Hermsen W, Verbunt F, et al. 2000. *Astron. Astrophys.* 359:615
Lin LCC, Hui CY, Hu CP, et al. 2013. *Ap. J . Lett.* 770:L9
Marelli M. 2012. *The X-Ray Behaviour of Fermi/LAT Pulsars.* PhD thesis, Univ. Insubria, Como, Italy. arXiv:1205.1748
Marelli M, De Luca A, Caraveo PA. 2011. *Ap. J.* 733:82
Mattox JR, Bertsch DL, Fichtel CE, et al. 1992. *Ap. J. Lett.* 401:L23
Mayer M, Buehler R, Hays E, et al. 2013. *Ap. J. Lett.* 775:L37
Mayer-Hasselwander HA, Bertsch DL, Brazier KTS, et al. 1994. *Ap. J.* 421:276
Meegan C, Lichti G, Bhat PN, et al. (GBM Collab.) 2009. *Ap. J.* 702:791
Mirabal N, Halpern JP. 2001. *Ap. J. Lett.* 547:L137
Mukherjee R, Bertsch DL, Dingus BL. 1995. *Ap. J. Lett.* 441:L61
Muslimov AG, Harding AK. 2003. *Ap. J.* 588:430
Muslimov AG, Harding AK. 2004. *Ap. J.* 606:1143
Nel HI, Arzoumanian Z, Bailes M, et al. 1996. *Ap. J.* 465:898
Ng C-Y, Roberts MSE, Romani RW. 2005. *Ap. J.* 627:904
Nolan PL, Abdo AA, Ackermann M, et al. (Fermi LAT Collab.) 2012. *Ap. J. Suppl.* 199:31
Nolan PL, Arzoumanian Z, Bertsch DL, et al. 1993. *Ap. J.* 409:697
Ogelman H, Fichtel CE, Kniffen DA, Thompson DJ. 1976. *Ap. J.* 209:584
Pavlov GG, Bhattacharyya S, Zavlin VE. 2010. *Ap. J.* 715:66
Pellizzoni A, Pilia M, Possenti A, et al. 2009a. *Ap. J.* 691:1618
Pellizzoni A, Pilia M, Possenti A, et al. 2009b. *Ap. J. Lett.* 695:L115
Pellizzoni A, Trois A, Tavani M, et al. 2010. *Science* 327:663
Petri J. 2011. *MNRAS* 412:1870
Pierbattista M. 2010. PhD thesis, Université Paris 7, Denis Diderot. arXiv:1309.5982
Pierbattista M, Grenier IA, Harding AK, Gonthier PL. 2012. *Astron. Astrophys.* 545A:42P

Pilia M, Pellizzoni A, Trois A, et al. 2010. *Ap. J.* 723:707
Pineault S, Landecker TL, Mardore B, Gaumont-Guay S. 1993. *Astron. J.* 105:1060
Pletsch HJ, Guillemot L, Allen B, et al. 2012a. *Ap. J.* 744:105
Pletsch HJ, Guillemot L, Allen B, et al. 2012b. *Ap. J. Lett.* 755:L20
Pletsch HJ, Guillemot L, Allen B, et al. 2013. *Ap. J. Lett.* 779:L11
Pletsch HJ, Guillemot L, Fehrmann HA, et al. 2012c. (Fermi LAT Collab.) *Science* 338:1314
Ramanamurthy PV, Bertsch DL, Dingus BL, et al. 1995. *Ap. J. Lett.* 447:L109
Ramanamurthy PV, Fichtel CE, Harding AK, et al. 1996. *Astron. Astrophys. Suppl.* 120:115
Ransom SM, Ray PS, Camilo F, et al. 2011. *Ap. J. Lett.* 727:L16
Ravi V, Manchester RN, Hobbs G. 2010. *Ap. J. Lett.* 716:L85
Ray PS, Abdo AA, Parent D, et al. 2012. *Proc. 2011 Fermi Symp., Rome, Italy, May 9–12*, eConf C110509. arXiv:12053089
Ray PS, Kerr M, Parent D, et al. 2011. *Ap. J. Suppl.* 194:17
Ray PS, Ransom SM, Cheung CC, et al. 2013. *Ap. J. Lett.* 763:L13
Razzano M, Harding AK, Baldini L, et al. 2009. *Astropart. Phys.* 32:1
Reimer O, Brazier KTS, Carramiñana A, et al. 2001. *MNRAS* 324:772
Roberts MSE. 2013. In *Neutron Stars and Pulsars: Challenges and Opportunities after 80 Years, Proc. IAU Symp. 291*, ed. J van Leeuwen, 8:127–32. Cambridge, UK: Cambridge Univ. Press
Roberts MSE, Brogan CL. 2008. *Ap. J.* 681:320
Roberts MSE, Hessels JWT, Ransom SM, et al. 2002. *Ap. J. Lett.* 577:L19
Roberts MSE, Romani RW, Kawai N. 2001. *Astron. Astrophy. Suppl.* 133:451
Romani RW. 1996. *Ap. J.* 470:469
Romani RW. 2012. *Ap. J. Lett.* 754:L25
Romani RW, Filippenko AV, Silverman JM, et al. 2012. *Ap. J. Lett.* 760:L36
Romani RW, Kerr M, Craig HA, et al. 2011. *Ap. J.* 738:114
Romani RW, Shaw MS. 2011. *Ap. J. Lett.* 743:L26
Romani RW, Watters KP. 2010. *Ap. J.* 714:810
Romani RW, Yadigaroglu IA. 1995. *Ap. J.* 438:314
Ruderman MA, Sutherland PG. 1975. *Ap. J.* 196:51
Saz Parkinson PM, Belfiore A, Dormody M, et al. (Fermi LAT Collab.) 2012. Proc. 2011 *Fermi Symp. Rome, Italy, May 9–12.*
Saz Parkinson PM, Dormody M, Ziegler M, et al. 2010. *Ap. J.* 725:571
Saz Parkinson PM (Fermi LAT Collab.). 2009. In *Science with the New Generation of High Energy Gamma-Ray Experiments : Bridging the Gap Between GeV and TeV*. AIP Conf. Proc., 1112:79
Smith DA, Guillemot L, Camilo F, et al. 2008. *Astron. Astrophy.* 492:923
Smith DA, Thompson DJ. 2009. In *Neutron Stars Pulsars*, ed. W Becker. *Astrophys. Space Sci. Libr.* 357:621. Heidelberg, Ger.: Springer-Verlag
Striani E, Tavani M, Piano G, et al. 2011. *Ap. J. Lett.* 741:L5
Sturrock PA. 1971. *Ap. J.* 164:529
Swanenburg BN, Bennett K, Bignami GF, et al. 1981. *Ap. J. Lett.* 243:L69
Tavani M, Barbiellini G, Argan A, et al. (Agile Collab.) 2009. *Astron. Astrophys.* 502:995
Tavani M, Bulgarelli A, Vittorini V, et al. 2011. *Science* 331:736
Thompson DJ. 2001. In *Int. Symp. High-Energy Gamma-Ray Astron. Heidelberg, Ger.*, FA Aharonian, HJ Völk. *AIP Conf. Proc.* 558:103–14. Melville, NY: AIP
Thompson DJ. 2004. In *Cosmic Gamma-Ray Sources*, ed. KS Cheng, GE Romero. *Astrophys. Space Sci. Libr.* 304149. Dordrecht, Neth.: Springer
Thompson DJ, Arzoumanian Z, Bertsch DL, et al. 1992. *Nature* 359:615
Thompson DJ, Bailes M, Bertsch DL, et al. 1996. *Ap. J.* 465:385
Thompson DJ, Bailes M, Bertsch DL, et al. 1999. *Ap. J.* 516:297
Thompson DJ, Fichtel CE, Kniffen DA, Ogelman HB. 1975. *Ap. J. Lett.* 200:L79
Vasseur JP, Parlier J, Leray B, et al. 1971. *Nature* 233:46
Vasseur JP, Parlier J, Leray B, Forichon JP, Agrinier M. 1970. *Nature* 226:534

Venter C, Harding AK, Guillemot L. 2009. *Ap. J.* 707:800
Venter C, Johnson TJ, Harding AK. 2012. *Ap. J.* 744:34
Watters KP, Romani RW. 2011. *Ap. J.* 727:123
Watters KP, Romani RW, Weltevrede P, Johnston S. 2009. *Ap. J.* 695:1289
Weisskopf MC, Romani RW, Razzano M, et al. 2011. *Ap. J.* 743:74
Weisskopf MC, Tennant AF, Arons J, et al. 2013. *Ap. J.* 765:56
Yadigaroglu I-A, Romani RW. 1995. *Ap. J.* 449:211
Ziegler M, Baughman BM, Johnson RP, Atwood WB. 2008. *Ap. J.* 680:620

Solar Dynamo Theory

Paul Charbonneau

Département de Physique, Université de Montréal, Montréal, Québec H3T 1N8, Canada; email: paulchar@astro.umontreal.ca

Annu. Rev. Astron. Astrophys. 2014. 52:251–90

First published online as a Review in Advance on June 4, 2014

The *Annual Review of Astronomy and Astrophysics* is online at astro.annualreviews.org

This article's doi:
10.1146/annurev-astro-081913-040012

Keywords

Sun, magnetohydrodynamics, hydromagnetic dynamo, solar cycle, turbulent convection, differential rotation

Abstract

The Sun's magnetic field is the engine and energy source driving all phenomena collectively defining solar activity, which in turn structures the whole heliosphere and significantly impacts Earth's atmosphere down at least to the stratosphere. The solar magnetic field is believed to originate through the action of a hydromagnetic dynamo process operating in the Sun's interior, where the strongly turbulent environment of the convection zone leads to flow-field interactions taking place on an extremely wide range of spatial and temporal scales. Following a necessarily brief observational overview of the solar magnetic field and its cycle, this review on solar dynamo theory is structured around three areas in which significant advances have been made in recent years: (*a*) global magnetohydrodynamical simulations of convection and magnetic cycles, (*b*) the turbulent electromotive force and the dynamo saturation problem, and (*c*) flux transport dynamos, and their application to model cycle fluctuations and grand minima and to carry out cycle prediction.

1. SOLAR MAGNETISM

Forget your old astronomy textbook; the Sun is a variable star, its variability is strongly imprinted across interplanetary space, there is no such thing as the solar constant, and the Sun's magnetic field is behind it all. It structures the solar atmosphere from the photosphere across the corona and into the solar wind and heliosphere, modulates the Sun's corpuscular and radiative output, and drives all geoeffective solar eruptive phenomena collectively defining solar activity. Nor is the Sun anomalous in this respect; every solar-type star observed with sufficient sensitivity shows similar signs of magnetically driven activity.

Numerous good reviews are available on the observational side of solar magnetism (e.g., Solanki et al. 2006, de Wijn et al. 2009); consequently, the following overview focuses on aspects most pertinent to solar dynamo theory. The solar magnetic field is structured and evolves over an astoundingly wide range of spatial and temporal scales (see **Figure 1**). Sunspots (**Figure 1*a***) are now understood to be the surface manifestations of emerging magnetic fields produced in the solar interior. Restricted to low heliocentric latitudes but seldom seen very near the equator, sunspots are the seats of strong magnetic fields ($\sim$0.1–0.5 T), and the larger sunspots usually appear in pairs of opposite magnetic polarities (*black* versus *white* on **Figure 1*b***). The favored physical picture is that of magnetic flux ropes rising from below and piercing the photosphere in the form of so-called Ω-loops (Parker 1955a, Caligari et al. 1995, Fan 2009 and references therein). The leading spots (with respect to the direction of solar rotation, from left to right in **Figure 1*a,b***) show the same magnetic polarity in each hemisphere, with this leading polarity reversing across hemispheres. These striking hemispheric regularities, known as Hale's Polarity law (Hale et al. 1919), reflect the presence of an internal magnetic field that is spatially well organized on the scale of the Sun as a whole and is antisymmetric about the solar equator.

Temporally extended observations (**Figure 1*c,d***) also reveal some remarkable large-scale order. The number of sunspots on the solar disk waxes and wanes on a timescale of about 11 years, with sunspot emergences occurring at midlatitudes in the beginning of this cycle and progressively closer to the equator as the cycle proceeds. The reversal of the relative magnetic polarities of sunspot pairs from one cycle to the next indicates that the underlying magnetic cycle has a period of twice that of the sunspot cycle. A well-defined dipole moment is also present, with the average polar cap magnetic field reaching a few 10^{-3} T at times of sunspot minimum and reversing its polarity at times of maximum sunspot counts, i.e., it lags the deep-seated sunspot-producing magnetic field by about half a sunspot cycle.

The sunspot record extends back to the beginning of the telescopic era in the early seventeenth century and, taken as a proxy of solar magnetism, makes it possible to trace the solar magnetic cycle over that time span (*red curve* on **Figure 1*d***; see also Hathaway 2010). The sunspot cycle shows significant variability in both its amplitude and duration, including an extended "quiet" epoch spanning the years 1645 to 1715, now known as the Maunder Minimum, during which very few sunspots were observed (Eddy 1976). Other indirect proxies, notably cosmogenic radioisotopes (e.g., Beer 2000, Usoskin 2013), make it possible to go much further back in time, although with some loss of temporal resolution. These data nonetheless show that episodes similar to the Maunder Minimum have occurred intermittently in the more distant past (*gray bands* in **Figure 1*d***; see also Usoskin et al. 2007, McCracken et al. 2013).

Although the strongest magnetic fields observed at the solar surface are found in sunspots, magnetographic observations also reveal their presence away from sunspots and active regions, in the form of small magnetic elements of sizes going down to the smallest spatial scale currently resolved by solar observing instruments (de Wijn et al. 2009). These small magnetic elements collectively define the magnetic network, and have sizes distributed as a power-law (Parnell et al.

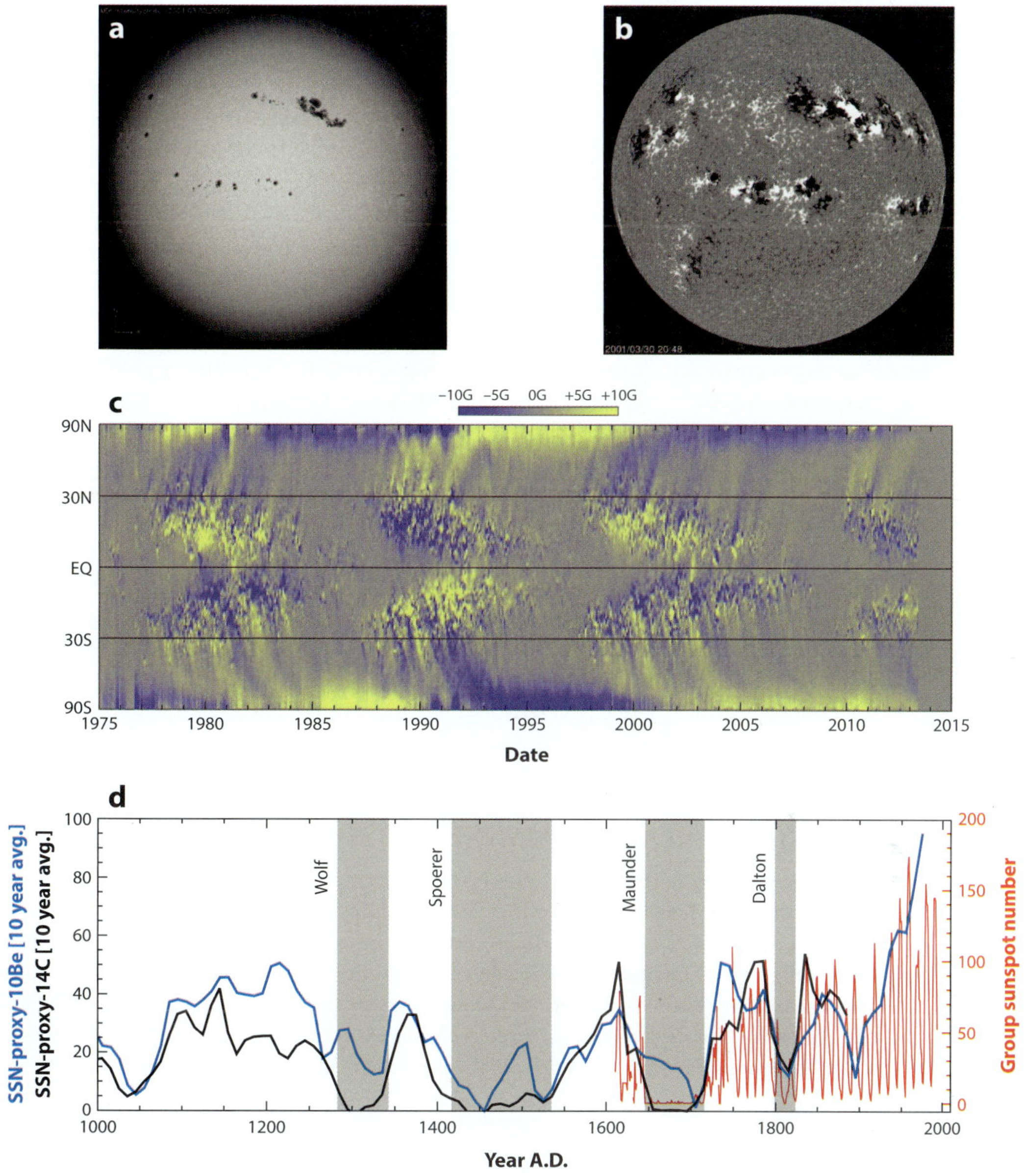

Figure 1

The solar magnetic field and its cycle. (*a*) A continuum image and (*b*) a line-of-sight magnetogram, both taken on March 30, 2001, by the MDI instrument onboard SOHO (ESA/NASA). (*c*) A synoptic magnetogram (courtesy of D. Hathaway, NASA/MSFC), constructed by zonally averaging full-disk magnetograms over successive solar rotations and stacking such averages into a time-latitude diagram. (*d*) The time series of the group sunspot number (SSN; in *red*; Hoyt & Schatten 1998) together with pseudoSSN time series constructed from two cosmogenic radioisotopes (data courtesy of I. Usoskin, Sodankylä Obs.). These provide measures of the overall activity levels at lower temporal resolution but over a much longer timespan than the sunspot record.

2009) spanning at least five orders of magnitude in size, which is indicative of scale invariance. Their lifetimes are much smaller than sunspots and associated magnetic structures, as they continuously emerge, submerge, merge or cancel with other elements of same/opposite magnetic polarity (Schrijver et al. 1997). These surface processes may actually account for the observed scale invariance (Thibault et al. 2012 and references therein). The magnetic network shows little, if any, dependence on the phase of the magnetic cycle and contributes little to the net (signed) hemispheric flux. It does, however, play an important role in driving the nonthermal photospheric radiative output, especially at wavelengths shorter than the visible.

The challenge facing dynamo theory is thus to explain the existence and observed evolution of the solar magnetic field. This is an immense topic, which has already been the subject of numerous monographs (e.g., Moffatt 1978, Parker 1979, Krause & Rädler 1980, Rüdiger & Hollerbach 2004, Charbonneau 2013). The present review focuses on a few areas where significant advances have been made in the past two decades in a manner such as to complement, rather than duplicate, other recent reviews on the topic, notably Ossendrijver (2003), Brandenburg & Subramanian (2005), Miesch & Toomre (2009), and Charbonneau (2010). General aspects of magnetohydrodynamics most pertinent to magnetic field amplification in electrically conducting fluid are first briefly reviewed in Section 2. This is followed in Section 3 by a survey of recent global magnetohydrodynamical simulations of solar convection having achieved the production of large-scale magnetic fields undergoing polarity reversals. Section 4 introduces mean-field electrodynamics, both as a tool to analyze and interpret numerical simulations as well as a means of building simplified evolutionary models of the solar magnetic cycle. Section 5 focuses on a class of mean-field-like simplified dynamo models, the so-called flux transport dynamos, which have received considerable attention in recent years as they provide an attractive framework within which to understand the origin of fluctuations in solar cycle amplitude and duration and perhaps even reliably predict these characteristics for future cycles. The review closes in Section 6 with a survey of promising future research directions and outstanding challenges in solar dynamo theory.

2. FUNDAMENTALS OF MAGNETOHYDRODYNAMICS

Physical conditions in the solar interior up to the photosphere are such that the interaction of fluid flow and magnetic field is well-described by the magnetohydrodynamical approximation (hereafter MHD; see, e.g., Davidson 2001). In its classical formulation, MHD describes the behavior of an electrically neutral mixture of electrically charged microscopic constituents in which the collision frequency largely exceeds any relevant plasma frequencies. In such a (moving) collisionally dominated plasma, Ohm's law holds in a reference frame comoving with the fluid at the macroscopic scale. For a nonrelativistic fluid flow **u**, the rest frame electric field **E** is then simply

$$\mathbf{E} = \mathbf{J}/\sigma - \mathbf{u} \times \mathbf{B}, \tag{1}$$

where **J** and **B** are, respectively, the current density and magnetic field, and σ is the electrical conductivity (SI units are used throughout). Excluding externally imposed rapid variations of **E** (e.g., turning batteries on or off), Ampère's law holds in its pre-Maxwellian form,

$$\nabla \times \mathbf{B} = \mu_0 \mathbf{J}, \tag{2}$$

where μ_0 is the magnetic permeability. Using this expression to substitute for **J** in Equation 1 and inserting the resulting expression for **E** into Faraday's law leads to the MHD induction equation,

$$\frac{\partial \mathbf{B}}{\partial t} = \nabla \times (\mathbf{u} \times \mathbf{B} - \eta \nabla \times \mathbf{B}), \tag{3}$$

where $\eta = (\mu_0\sigma)^{-1}$ is the magnetic diffusivity. The first term on the right-hand side expresses induction by the flow of electrically charged constituents across the magnetic field and the second Ohmic dissipation of the current systems supporting that same magnetic field, as per Equation 2. The relative importance of induction versus dissipation is measured by the magnetic Reynolds number,

$$\mathrm{Rm} = \frac{u_0 L}{\eta}, \tag{4}$$

where u_0 and L are characteristic values for the flow speed and length scale, respectively, the latter assumed here to adequately characterize the spatial variations of both the flow and magnetic field. For most astrophysically relevant circumstances, and all aspects of dynamo action considered in what follows, one finds $\mathrm{Rm} \gg 1$ if L is set equal to the scale of the system (e.g., the solar radius, $\mathrm{R}_\odot$), indicating that Ohmic dissipation is very inefficient on this global scale.

In the infinite conductivity ideal MHD case ($\eta = 0$), Equation 3 becomes identical to the evolution equation for a line element advected by a flow $\mathbf{u}$, implying that the magnetic field lines move with the fluid (see, e.g., Davidson 2001, his section 2.7.4), a result known as Alfvén's theorem and dubbed flux freezing. The consequent ability of the fluid flow to displace, bend, and stretch magnetic field lines is crucial to all dynamo mechanisms considered in all that follows.

Dimensional analysis of Equation 3 yields two natural timescales for magnetic field evolution: the advective timescale $\tau_u = L/u_0$ and diffusive timescale $\tau_\eta = L^2/\eta$. Using again $L \sim \mathrm{R}_\odot$ and $\eta \sim 1\ \mathrm{m}^2\ \mathrm{s}^{-1}$ for the bulk of the convection zone, one obtains $\tau_\eta \simeq 10^{10}$ year, twice the age of the Sun. From this point of view, the very existence of the solar magnetic field is not an issue; any fossil magnetic field remaining from the Sun's gravitational collapse from a (weakly magnetized) molecular cloud would still be present in the solar interior at a strength hardly differing from its zero-age main-sequence value. It also implies that we must look to the flow $\mathbf{u}$ to explain the much shorter evolutionary timescales observed, from the decadal cycle period, down to minutes for the evolution of small photospheric magnetic flux concentrations. The lower end of this range is actually commensurate with the advective time of surface convection, for which $L \sim 10^6$ m and $u_0 \sim 10^3\ \mathrm{m\ s}^{-1}$, yielding $\tau_u \sim 15$ min. The oddity lies therefore with the large-scale magnetic field and its decadal polarity reversals. Moreover, this problem with the disparity of timescales carries over to length scales. A turbulent flow can generate strong magnetic fields on its own small spatial scale provided Rm is large enough (see, e.g., Cattaneo 1999, Cattaneo et al. 2003, Stein & Nordlund 2006, Vögler & Schüssler 2007). However, it is also quite apt at enhancing the dissipation of magnetic fields having larger spatial scales. How, then, can turbulent convection, with typical scales on the order of 10^5–10^7 m and 10^3–10^5 s, induce and sustain against dissipation a spatiotemporally coherent magnetic component with scales on the order of 10^9 m and 10^8 s? This quandary is often referred to as the magnetic self-organization problem (Tobias et al. 2011).

From the MHD induction (Equation 3) one can obtain an evolution equation for the magnetic energy (E_B) integrated over the volume V of the system (here the Sun):

$$\frac{\partial E_B}{\partial t} = -\oint_{\partial V} \mathbf{S} \cdot \hat{\mathbf{n}} \mathrm{d}A - \frac{1}{\sigma} \int_V \mathbf{J}^2 \mathrm{d}V - \int_V (\mathbf{u} \cdot \mathbf{F}) \mathrm{d}V, \tag{5}$$

where $\mathbf{S} = \mu_0^{-1} \mathbf{E} \times \mathbf{B}$ is the Poynting flux at the domain boundary ∂V, and $\mathbf{F} = \mathbf{J} \times \mathbf{B}$ is the Lorentz (magnetic) force per unit volume acting on the plasma. The first term on the right-hand side thus measures the flux of electromagnetic energy in or out of the domain; for a star-like system embedded in vacuum, this is typically set to zero. The second term reflects the action of Ohmic dissipation, inexorably turning magnetic energy into heat. The third term is where the action is; it corresponds to the work done by the flow against the magnetic force. This is the channel through

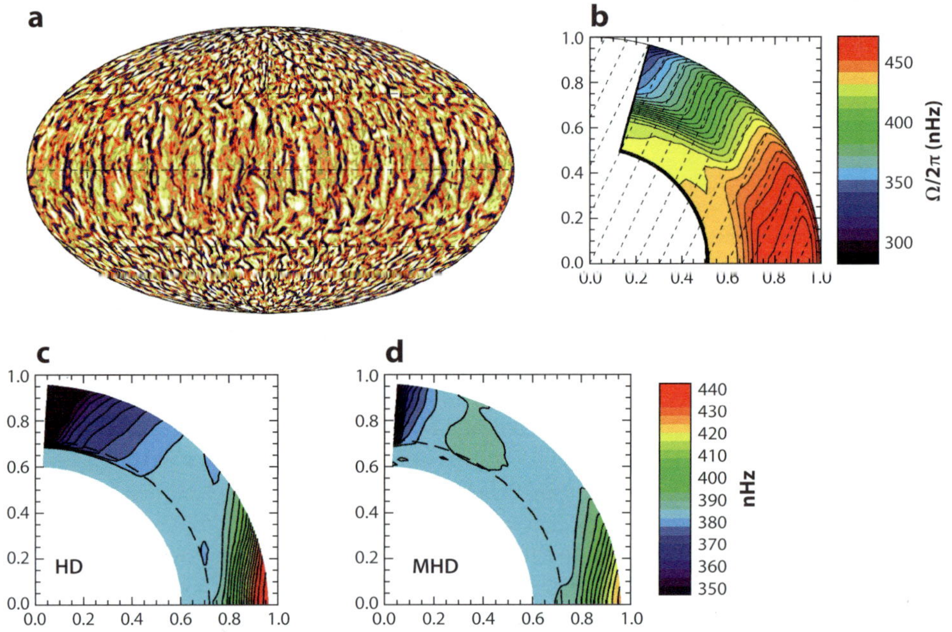

Figure 2

Plasma flows in the solar interior: (*a*) vertical flow velocities in subsurface layers, as produced by a high-resolution EULAG simulation, courtesy of Guerrero et al. (2013); (*b*) helioseismic inversion of the solar internal differential rotation, taken from Howe (2009); and differential rotation in (*c*) a purely hydrodynamical (HD) EULAG numerical simulation and (*d*) a EULAG numerical simulation obtained in the magnetohydrodynamic (MHD) regime. The dashed line indicates the base of the convection zone.

which the kinetic energy of the flow can be converted to electromagnetic energy, a process that must take place in a manner sufficiently efficient to offset losses due to Ohmic dissipation. This requirement is the very essence of any dynamo process.

In its simplest form, the dynamo problem consists of finding a flow **u** that, when inserted into Equation 3, leads to amplification of **B**. In principle, we are off to a good start, because vigorous flows abound in the Sun (see **Figure 2**). In the outer 30% of the Sun's radius, the luminosity is transported by thermally driven turbulent fluid motions known as convection. Inhomogeneous due to stratification, and lacking reflection symmetry due to the influence of rotation, convection at small spatial scales drives large-scale flows, notably differential rotation. These two flow components, energetically, are the primary contributors to **u** in Equations 3 and 5. Yet the magnetic force acting on the charged constituents of the fluid also resist the flow, potentially quenching it and, in doing so, turning off induction. The dynamo problem is inherently nonlinear, and therein lie the physical, mathematical, and computational challenges.

3. GLOBAL MAGNETOHYDRODYNAMICAL SIMULATIONS OF SOLAR CONVECTION

Global (i.e., full-sphere) numerical simulation of the MHD induction (Equation 3), together with the usual fluid equations augmented by the Lorentz force, is an obvious brute force way to tackle the solar dynamo problem. In practice, this is an extremely challenging task, in view of the aforementioned wide separation of spatial and temporal scale characterizing solar convection and magnetic field evolution. However, unlike some of the simpler modeling approaches considered in Sections 4 and 5 below, such simulations do properly incorporate the magnetic backreaction on the inductive flow at all resolved spatial and temporal scales.

The underlying physics is not at all exotic, as the full set of MHD equations simply expresses conservation of mass, momentum, energy, and magnetic induction as expressed by Equation 3; the latter is duplicated here for completeness:

$$\frac{\partial \rho}{\partial t} + \nabla \cdot (\rho \mathbf{u}) = 0, \tag{6}$$

$$\frac{\partial \mathbf{u}}{\partial t} + (\mathbf{u} \cdot \nabla)\mathbf{u} = -\frac{1}{\rho}\nabla p - 2\boldsymbol{\Omega} \times \mathbf{u} + \mathbf{g} + \frac{1}{\mu_0 \rho}(\nabla \times \mathbf{B}) \times \mathbf{B} + \frac{1}{\rho}\nabla \cdot \boldsymbol{\tau}, \tag{7}$$

$$\frac{\partial e}{\partial t} + (\gamma - 1)e\, \nabla \cdot \mathbf{u} = \frac{1}{\rho}\left\{\nabla \cdot [(\chi + \chi_r)\nabla T] + \phi_u + \phi_B\right\}, \tag{8}$$

$$\frac{\partial \mathbf{B}}{\partial t} = \nabla \times (\mathbf{u} \times \mathbf{B} - \eta \nabla \times \mathbf{B}). \tag{9}$$

Here ρ is the fluid density; e is internal energy; p is gas pressure; $\boldsymbol{\tau}$ is the viscous stress tensor; χ and χ_r are the kinetic and radiative thermal conductivities, respectively; ϕ_u and ϕ_B are the viscous and Ohmic dissipation functions, respectively; and other symbols have their usual meaning. In the solar dynamo context it is customary to write the MHD equation in a rotating reference frame; the centrifugal force is absorbed in the gravitational term so that only the Coriolis force appears on the right-hand side of Equation 7. Equations 6–9 need to be augmented by an equation of state (typically the perfect gas law), and the magnetic field needs to be subjected to the solenoidal constraint $\nabla \cdot \mathbf{B} = 0$. The specification of appropriate boundary conditions completes the mathematical specification of the problem.

The anelastic approximation, in common usage in the present context, consists of introducing a time-invariant background density profile, thus replacing Equation 6 by $\nabla \cdot (\rho \mathbf{u}) = 0$, and only retaining density variations about this reference profile induced by thermal dilation; in this way sound waves are filtered out of the system, allowing larger time steps, but the crucial buoyancy force is retained.

Equations 6–9 represent a set of strongly nonlinear partial differential equations that are to be solved in a thick rotating stratified shell of electrically conducting fluid subjected to thermal forcing driving convection. The first such simulations carried out by Gilman (1983) and Glatzmaier (1984, 1985) used strongly enhanced dissipative coefficients to maintain numerical stability, but nonetheless exhibited many encouraging features, including a rapidly rotating equatorial region, the buildup of large-scale magnetic fields (but showing strong hemispheric asymmetries), magnetic field migration (but poleward rather than equatorward), and polarity reversals (although quite irregular). As ever more powerful computing platforms became available in later years, simulations could be carried out at higher resolution and using lower diffusivities, thus reaching more strongly turbulent regimes (see, e.g., Brun et al. 2004). These simulations proved to be potent dynamos, in the sense that strong magnetic fields could be generated on the spatial scale of convection, but without generating any significant large-scale magnetic component. The presence of a convectively stable, mechanically forced underlying fluid layer proved efficient at producing an axisymmetric large-scale component (Browning et al. 2006), which, however, remained resolutely steady over the time span of these simulations. By contrast, the simulations of Brown et al. (2010) achieved the buildup of large-scale magnetic fields concentrated within the low-latitudes portions of the convecting layers, but again without reversals of magnetic polarity.

Achieving solar-like cyclic polarity reversals of a large-scale magnetic component in such simulations is a very recent, exciting development, on which the remainder of this section is focused. Before reviewing the design and characteristics of these simulations, a brief detour is first required through the dynamics of rotating, stratified convection and its associated large-scale flows.

3.1. Convection and Large-Scale Flows

The dynamics of solar/stellar convection is discussed in depth in the recent review articles by Miesch (2005) and Miesch & Toomre (2009), and by Howe (2009) on the observational (helioseismic) front. The key dynamical factors are rotation and stratification, which break the homogeneity and reflection symmetry of the turbulent fluid motions, leading to latitudinal temperature differences and Reynolds stresses that power large-scale flows. The key dimensionless parameter measuring the influence of rotation is the Rossby number (Ro), defined as the ratio of inertial to Coriolis forces on the right-hand side of Equation 7, or equivalently, of the rotation period $P_{\rm rot}$ to the convective turnover (coherence) time $\tau_{\rm c}$:

$$\mathrm{Ro} \equiv \frac{u'_{\rm rms}}{2\Omega\ell} \equiv \frac{P_{\rm rot}}{\tau_{\rm c}}, \tag{10}$$

where $u'_{\rm rms}$ and ℓ are the typical velocity and length scale of the convective eddies, respectively. With Ro $\sim$ 0.1 in the deep convection zone, from the point of view of turbulent convection the Sun is actually a fast rotator. This leads to characteristic convective patterns, with equatorial regions organized in a longitudinal stack of latitudinally elongated banana cells, extending across equatorial latitudes up to $\pm 35°$ in the Sun, at least according to global hydrodynamical simulations (see **Figure 2**; also Busse 2002).

Rotational influences also lead to a differential rotation pattern characterized by equatorial regions rotating some 30% faster than the poles, with the underlying radiative core rotating at an intermediate rate corresponding to surface midlatitudes. This is relatively well-reproduced by numerical simulations, except for rotational isocontours being too cylindrical at low to midlatitudes, and the radial shear at low latitudes too concentrated within the convection zone, as compared with helioseismic measurements (cf. **Figure 2*b*,*d***). In the Sun, this deviation from cylindrical isocontours is now understood to be caused by a latitudinal entropy gradient present throughout the convection zone, which results in a thermal wind balance that breaks the Taylor-Proudman theorem (Kitchatinov & Rudiger 1995, Miesch et al. 2006, Balbus et al. 2009). This has been reproduced in purely hydrodynamical numerical simulations including a stably stratified fluid layer underlying the convecting layers, as in **Figure 2*c*** (see also Brun et al. 2011). A thin radial shear layer is also present beneath the photosphere, and another, known as the tachocline, straddles the base of the convection zone and ensures a smooth match between the latitudinal differential rotation of the convection zone and the rigid rotation of the outer radiative core.

Meridional circulation refers to the large-scale axisymmetric flow component confined to meridional planes. It is weaker and shows more temporal variability than differential rotation. Directed poleward with speeds of $\sim$10 m s^{-1} at the surface low to midlatitudes, it is also ultimately powered by convection. Observational attempts to detect the expected equatorward return flow, carried out through helioseismology or the tracking of deeply anchored magnetic tracers, have yielded conflicting results, some inferring a return flow located deep in the convective envelope, whereas others infer a shallow return flow (cf. Braun & Fan 1998, Schou & Bogart 1998, Hathaway 2012, Zhao et al. 2013). Theory (Kitchatinov & Rüdiger 1995) and numerical simulations (Miesch & Toomre 2009, see their section 3.5) indicate that multiple flow cells stacked in radius and/or latitude are indeed likely. Both differential rotation and meridional circulation show

SUBGRID MODELS

In the numerical solution of high Reynolds number, multiscale turbulent flows, finite computing resources usually make it impossible to capture the turbulent cascade all the way down to the nominal dissipation scale. Subgrid models bypass this difficulty by displacing dissipation at the smallest scales that can be stably resolved on the computational grid. This can be carried either explicitly via modifications to the dissipative terms in the governing equations (as in the ASH simulation of **Figure 3**) or implicitly via the numerical algorithm used to solve the governing equations (as in the EULAG simulation of **Video 1**). Either way, artificially enhanced dissipation of the small scales is introduced, and the challenge is to ensure that larger scales remain unaffected. In some classes of problems, e.g., decaying turbulence, the role of the subgrid model is simply to dissipate energy at the end of the turbulent cascade; in many such cases robust subgrid models have been designed and/or validated against measurements. In the context of solar/stellar MHD convection, however, the procedure is more delicate because inverse cascades also operate, e.g., the driving of differential rotation and meridional flows by turbulent Reynolds stresses, and dual forward/inverse cascades of magnetic helicity.

low-amplitude cyclic variability phase locked to the magnetic cycle (see Howe 2009, see her section 9; Ulrich 2010).

3.2. Cycles and Polarity Reversals: A Survey

It is instructive to compare the flows and magnetic fields produced by three independent global simulations all working on, in principle, the same calculation: thermally driven MHD convection and dynamo action in a thick, stratified, thermally forced rotating shell of electrically conducting fluid. The virtue of this comparison lies with the fact that each of these simulations uses distinct physical approximations, treatment of subgrid dissipation, and computational methodologies. This allows, to some extent at least, identification of which patterns and behaviors are physically robust and which are not.

The PENCIL code (**www.nordita.org/pencil-code**) is a parallel high-order finite difference code operating as a direct numerical simulation (DNS), using fixed values for dissipative coefficients such as viscosity and thermal and magnetic diffusivity. The pseudoglobal compressible simulations in a spherical wedge discussed below are taken from Käpylä et al. (2010, 2012, 2013). The ASH (anelastic spherical harmonics; Clune et al. 1999) code is a massively parallel spectral code formulated in spherical geometry. It uses a large-eddy simulation (LES) approach, whereby stability is maintained through the use of an explicit subgrid model (see the sidebar, Subgrid Models). The simulation chosen here is taken from Brown et al. (2011) and makes use of a static eddy diffusivities formulation, as in earlier ASH MHD simulations (e.g., Brun et al. 2004, Browning et al. 2006, Brown et al. 2010). EULAG-MHD (Smolarkiewicz & Charbonneau 2013) is an MHD version of the EULAG code (EUlerian-LAGrangian, **www.mmm.ucar.edu/eulag/**; see also Prusa et al. 2008), a robust, parallel multiscale flow solver formulated in generalized curvilinear coordinates. It can operate an implicit large-eddy simulation (ILES; see Grinstein et al. 2007), in which dissipation is delegated to the numerical advection scheme, rather than being formulated explicitly. The simulations discussed below are of this type and are essentially identical to those discussed by Ghizaru et al. (2010) and Racine et al. (2011).

All three simulations generate convective patterns and large-scale flows that turn out remarkably similar. As expected, convection is organized in cells of hot, ascending fluid delineated by a network of narrow downflow lanes, as in purely hydrodynamical simulations. In equatorial

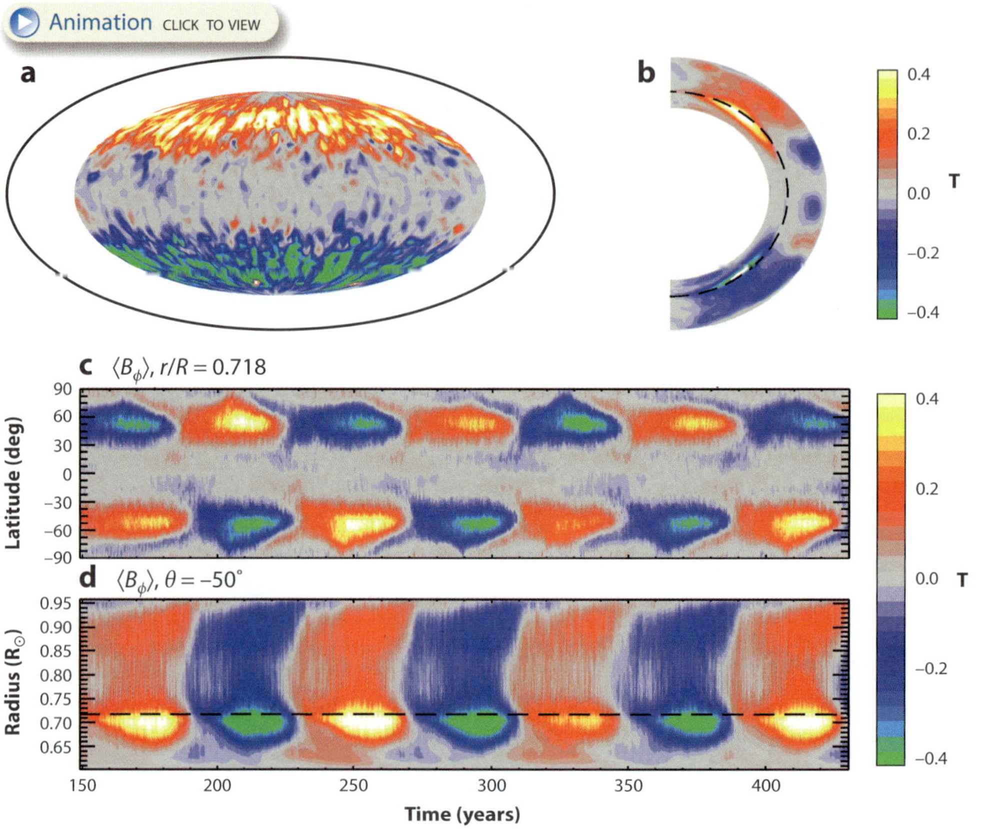

Video 1

Magnetic cycles in a global EULAG-MHD anelastic simulation, essentially identical to those by Ghizaru et al. (2010) and Racine et al. (2011). To view the video, access this article on the Annual Reviews website at **http://www.annualreviews.org**. This simulation includes a convectively stable fluid layer underlying the convecting layers. (*a*) A snapshot in Mollweide projection of the toroidal (zonal) magnetic component at depth $r/R_\odot = 0.718$; (*b*) a snapshot of the zonally averaged toroidal field in a meridional plane taken at the same time as panel *a*. (*c*) Time-latitude and (*d*) radius-latitude diagrams of the zonally averaged toroidal field, the former at depth $r/R_\odot = 0.718$ and the latter at latitude $+25°$. The dashed lines in panels *b* and *d* indicate the bottom of the convectively unstable layers. This is a moderate-resolution simulation, rotating at the solar rate but subluminous with respect to the Sun.

regions, convection again organizes itself in a longitudinal stack of elongated rolls with their axes parallel to the rotation axis (see **Figure 2*a***) (cf. Käpylä et al. 2010, their figure 2; Brown et al. 2011, their figure 3a,b; Ghizaru et al. 2010, their figure 1; and Miesch & Toomre 2009, their section 2.2.3). The mean differential rotation in all cases is characterized by equatorial acceleration with angular velocity isocontours approximately aligned with the rotational axis, but now it is a pole-to-equator angular velocity difference markedly smaller than solar (cf. Käpylä et al. 2010, their figure 1; Brown et al. 2011, their figure 3c; and Racine et al. 2011, their figure 4). Again in all three cases the mean (axisymmetric) meridional circulation is weak and highly time variable, and it shows little spatial structuring on large scales. With the axisymmetric mean subtracted, the zonally averaged flow kinetic helicity is also quite similar in all three simulations, being negative (positive) in the Northern (Southern) Hemisphere, peaking at high latitude with a secondary maximum at

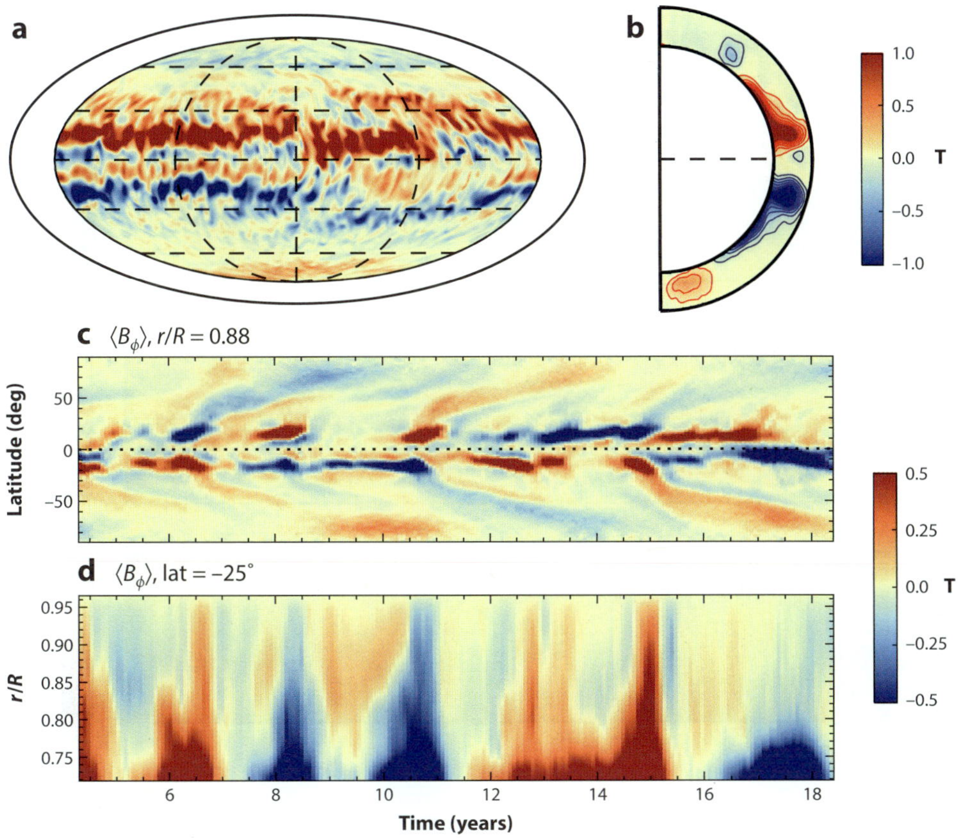

Figure 3

Identical in format to **Video 1**, but showing magnetic cycles in a global anelastic spherical harmonics simulation, specifically simulation D5 by Brown et al. (2011). This high-resolution anelastic simulation is close to the solar luminosity, rotates at five times the solar rate, and does not include a convectively stable fluid layer beneath the convecting layers. Figure provided by M. Miesch (HAO/NCAR).

low latitudes, and changing sign at the base of the convecting fluid layer (cf. Käpylä et al. 2012, their figure 2; Brown et al. 2010, their figure 8a; and Racine et al. 2011, their figure 15C).

With the flows so similar at small and large scales, one might expect that dynamo action will show comparable levels of similarity. Interestingly, this expectation is only partly borne out. **Video 1** and **Figures 3** and **4** offer different representations of the spatiotemporal evolution of a toroidal magnetic field in the three simulations considered. **Video 1** and **Figures 3*a*** and **4*a*** show snapshots in Mollweide projection of the toroidal magnetic field component at mid-depth in the convecting layers, panels *b* a zonal average in the meridional plane, and panels *c* and *d* time-latitude and radius-latitude cuts of that same zonally averaged toroidal magnetic component.

In all three simulations the magnetic field is highly turbulent, but also shows prominent banded structures fairly well aligned in the zonal direction and reaching strengths in the 0.1–1 T range. These axisymmetric toroidal field bands undergo polarity reversals that can be more or less regular and synchronized across hemispheres, depending on the specific simulation considered. All three simulations show significant accumulation of magnetic field at the base of the convecting layers, as expected of turbulent convection in a stratified environment (e.g., Tobias et al. 2001). In the

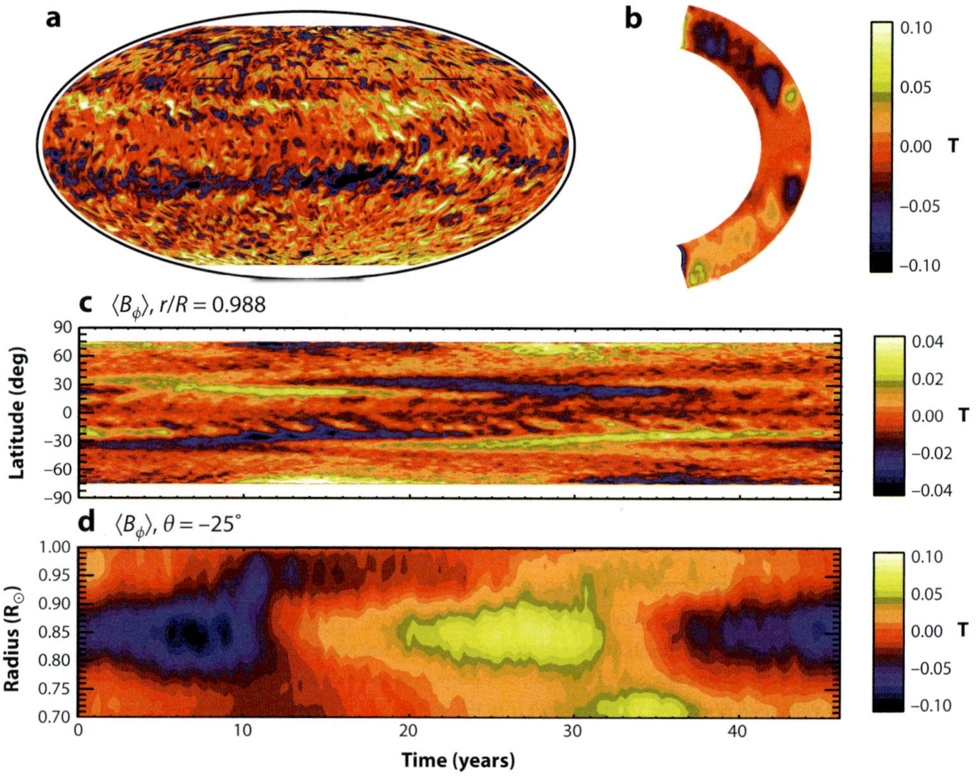

Figure 4

Identical in format to **Video 1** and **Figure 3**, but showing magnetic cycles in a PENCIL compressible simulation (case E4 in Käpylä et al. 2013). This is a high-resolution simulation in a spherical wedge excluding latitudes higher than ±15°. This simulation rotates at 4.4 times the solar rate, is superluminous as compared with the Sun, and does not include a convectively stable fluid layer beneath the convecting layers. Produced from numerical data provided by P. Käpylä (Univ. Helsinki).

EULAG-MHD simulation, the magnetic field actually peaks in the outer reaches of the stably stratified fluid layer underlying the convection zone, whereas in the ASH and PENCIL simulations strong magnetic field concentrations also build up at low latitudes within the convecting layers. The ASH and PENCIL simulations show poleward propagation at mid- to high latitudes, and the latter also exhibits solar-like equatorward propagation at low latitudes (**Figure 4*c***). The axisymmetric toroidal field in the EULAG-MHD simulation of **Video 1**, by contrast, peaks at midlatitudes and only shows a hint of equatorward propagation at its high-latitude edge. The PENCIL and EULAG-MHD simulations also show upward propagation in the outer half of the convecting layers, a feature absent in the ASH simulation. The magnetic cycle periods also differ significantly: 3–6 years in ASH, ~35 years in PENCIL, and up to ≃80 years in EULAG-MHD.

All three simulations also generate cyclic modulations of large-scale flows (see Brown et al. 2010, 2011; Passos et al. 2012). Beaudoin et al. (2013) carried out a detailed analysis of the EULAG-MHD simulation of **Video 1**, where the amplitude and phase of the torsional oscillations are found to be remarkably close to those detected in the Sun via helioseismology (cf. Howe 2009, her figures 5 and 25). They could also show that torsional oscillations developing in their simulation do not simply result from the action of the Lorentz force associated with the cyclic large-scale magnetic component, but involve magnetically mediated alteration of all other contributions to the angular momentum flux: small-scale Reynolds and Maxwell stresses, and advection of angular momentum by meridional flows.

3.3. Self-Organization of Large-Scale Magnetic Fields

All three simulations considered above are strongly turbulent, yet clearly achieve varying levels of self-organization of magnetic components characterized by spatial and temporal scales much larger/longer than those of convection. How exactly this self-organization takes place is still not understood, but various contributing factors have already been identified:

1. **Rotation**. Rotation is a powerful large-scale organizing mechanism via the action of the associated Coriolis force on convective turbulence. The Coriolis force can impart hemispheric cyclonicity on turbulence, generate latitudinal temperature gradients, and even structure convection on large scales, as in **Figure 2**. In the Sun, photospheric convection operates on timescales much shorter than the rotation period (i.e., the Rossby number $\mathrm{Ro} \gg 1$) and so is expected to be unaffected by rotation, whereas in the bulk of the convection zone one expects $\mathrm{Ro} \leq 1$, implying significant rotational influence. Indeed, otherwise identical global MHD simulations computed at varying rotation rates can exhibit very different types of magnetic self-organization; operating close to the solar luminosity, when run with the rotation lowered to three times solar, the ASH simulation of **Figure 3** develops low-latitude toroidal field bands that remain steady in time; with rotation further lowered to the solar rate, no significant large-scale field is generated. By contrast, a cousin of the EULAG-MHD simulation of **Video 1** running at close to solar luminosity requires the rotation rate to be increased to three times solar to recover reasonably regular cycles. These converging bits of empirical evidence all point to the Rossby number as a key parameter.
2. **Differential rotation**. Differential rotation powered by turbulent Reynolds stresses is a potentially powerful mechanism for spatial organization, as this flow is structured on global scales and, at least in the Sun, remains close to stationary on cycle timescales. Differential rotation is also efficient at dissipating nonaxisymmetric large-scale magnetic fields, thus also favoring the buildup of solar-like axisymmetric large-scale magnetic fields.
3. **Subgrid models**. The choice of a subgrid model to maintain nonlinear stability of the numerical solutions can have a strong impact on global characteristics of the large-scale dynamo solutions, either directly on magnetic field generation or indirectly via changes it may produce in large-scale flow contributing to magnetic field induction. For example, recent ASH simulations computed using different types of subgrid models, though otherwise identical or closely similar, achieve markedly distinct types of magnetic self-organization (cf. Brown et al. 2010, 2011; Nelson et al. 2013; K. Augustson, M.S. Miesch, & J. Toomre, manuscript in preparation).
4. **The magnetic Prandtl number**. This is the (dimensionless) ratio of viscosity to magnetic diffusivity $\mathrm{Pm} = \nu/\eta$ and thus measures the ratio of length scales at which the flow and magnetic field transit from turbulent to laminar and dissipate. In the parameter regime $\mathrm{Pm} > 1$, there exist a range of length scales in which the destructive folding of the magnetic field by the flow ceases while Ohmic dissipation is still small, which should favor dynamo action. Empirical evidence from MHD numerical simulations running in the $\mathrm{Pm} > 1$ range indicates that this is indeed the case and that the numerical value of Pm has a significant impact on magnetic self-organization. Note, however, that $\mathrm{Pm} < 1$ in the bulk of the solar convection zone.
5. **The tachocline**. Stably stratified and home to significant shear that is both radial and latitudinal (e.g., ***Figure 2b***), the tachocline is the ideal environment to amplify and store

the strong magnetic field that later will buoyantly rise through the convection zone and form sunspots upon emergence at the photosphere. Moreover, a wide variety of flow-field dynamical interactions can develop therein. Some may contribute to the self-organization of the large-scale magnetic field, and perhaps feed back, directly or indirectly, onto dynamo processes operating in the bulk of the overlying convection zone.

Mechanisms of self-organization notwithstanding, the simulations discussed in the preceding section are all dynamically consistent at all resolved spatial and temporal scales and therefore represent outstanding virtual laboratories allowing investigation of many key questions in solar dynamo theory. The remainder of this section focuses on three such questions, in which breakthroughs have been made in recent years.

3.4. Instabilities and Waves in the Tachocline

Large-scale magnetic fields imbedded in stably stratified stellar interiors are known to be sensitive to an astoundingly wide range of instabilities (see, e.g., Pitts & Tayler 1985, Spruit 1999, Braithwaite & Nordlund 2006). The tachocline is clearly a potential candidate, as it combines significant rotational shear, a strong magnetic field, weakly subadiabatic stratification, and possibly coupling to a fossil magnetic field in the deeper radiative interior (see Strugarek et al. 2011 and references therein). Perhaps most pertinent in this context is the class of magnetoshear instabilities investigated by Gilman & Fox (1997; see also Cally et al. 2003, Dikpati et al. 2003, and Miesch et al. 2007 and references therein). This joint MHD instability is inherently two-dimensional and nonaxisymmetric and operates in stably stratified environments, developing on spherical shells in the joint presence of a latitudinal angular velocity shear and a strong toroidal magnetic field. The instability has been found to operate for a wide range of toroidal field profiles, including latitudinally broad profiles similar to those produced in the stable fluid layers in the EULAG-MHD simulation of **Video 1**.

Miesch (2007) has presented thin-shell MHD numerical simulations demonstrating that in the presence of steady mechanical forcing of differential rotation and forced replenishment of a latitudinal magnetic field component, cyclic waxing and waning of this instability can take place; the toroidal field and differential rotation are first destroyed by the growth of the instability, then rebuilt through the forcing terms following nonlinear saturation. This leads to a characteristic pattern where the magnetic energies associated with the axisymmetric background toroidal field and nonaxisymmetric components resulting from the growth of the instability both oscillate with a well-defined period, the latter lagging the former by $\sim\pi/2$ in phase (see Miesch 2007, his figure 2). **Figure 5** shows equivalent time series for axisymmetric (*black*) and nonaxisymmetric (*blue*) magnetic field energies in the stable layers of the EULAG-MHD simulation of **Video 1**. The periodic behavior and $\sim\pi/2$ phase offset are both clearly apparent, even though here the mechanical and magnetic forcing occurs "naturally" through the action of turbulent Reynolds stresses and the operation of the dynamo in the overlying convecting layers. Ongoing analyses of this simulation (N. Lawson & P. Charbonneau, submitted) indicates that the preferred mode of instability has an azimuthal wavenumber $m = 3$ or sometimes $m = 4$, unlike in the simulation analyzed by Miesch (2007), where the $m = 1$ "clamshell" mode dominates, in agreement with linear theory (see Gilman et al. 2007).

The striking similarities between **Figure 5** and Miesch's (2007) figure 2 do not strictly prove that the same physical instability is at play in both cases, but nonetheless they strongly suggest that some kind of related magnetoshear instability operates in the stably stratified deep layers of the simulation. This raises a number of interesting questions: Is this instability the primary mechanism saturating the amplitude of the deep-seated dynamo-generated large-scale magnetic

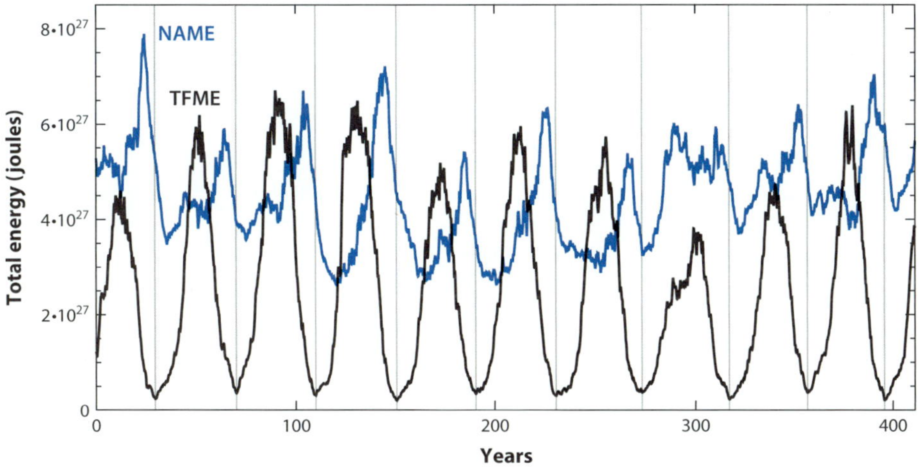

Figure 5

Magnetic energy evolution in the tachocline in the EULAG-MHD simulation of **Video 1**. The magnetic energy associated with the zonally averaged (axisymmetric) toroidal magnetic field is plotted in black and labeled TFME (toroidal field magnetic energy), and that comprising all other nonaxisymmetric contributions is plotted in blue, labeled NAME (nonaxisymmetric magnetic energy). Here the magnetic energies are volume-integrated in the Southern Hemisphere over the depth range $0.602 \leq r/R \leq 0.693$, which excludes the strongly turbulent, thin convective overshoot layer developing immediately beneath the base of the convecting layers ($r/R = 0.718$ in this simulation). Both energies oscillate with a well-defined period identical to that of the magnetic cycle and maintain a well-defined phase lag of about $\pi/2$ for most cycles. Compare with Miesch (2007, his figure 2).

field? Is it a mere by-product of a dynamo operating autonomously in the convecting layers or does it feed back into the dynamo process? Is the timescale for the growth and decay of the instability entirely set by the dynamo period imposed from above or is the growth of the instability playing a role in the global polarity reversals? Moreover, and to be discussed further below in Section 4.6, the development of some of these instabilities can drive an electromotive force, which could then power a dynamo contained entirely within the tachocline.

Another potentially important aspect of tachocline dynamics involves the excitation of internal gravity waves by convective overshoot as well as the contribution such waves might make on the dynamics of the tachocline and radiative core (e.g., Talon et al. 2002, Kim & MacGregor 2003, Rogers & Glatzmaier 2006, Rogers et al. 2008, Brun et al. 2011) and, indirectly, on dynamo action via their impact on the rotational shear. Already difficult to model in the purely hydrodynamical regime because of the extreme disparity of spatial and temporal scales between the waves and the large-scale flows, gravity wave dynamics is further complicated by the presence of magnetic fields. Geometrically simplified models (Barnes et al. 1998) and numerical simulations (Rogers & MacGregor 2011) both indicate that gravity wave propagation, absorption, and attendant angular momentum transport are all strongly altered by the presence of strong magnetic fields within the tachocline.

3.5. Coherent Structures and Sunspot Formation

A recent exciting development in global MHD numerical simulations of solar convection is the spontaneous production of strongly magnetized flux strands in the ASH simulations by Nelson et al. (2011, 2013). This process is favored by the strong turbulent intermittency characterizing these simulations, which make use of a dynamical subgrid-scale model for dissipation, as opposed

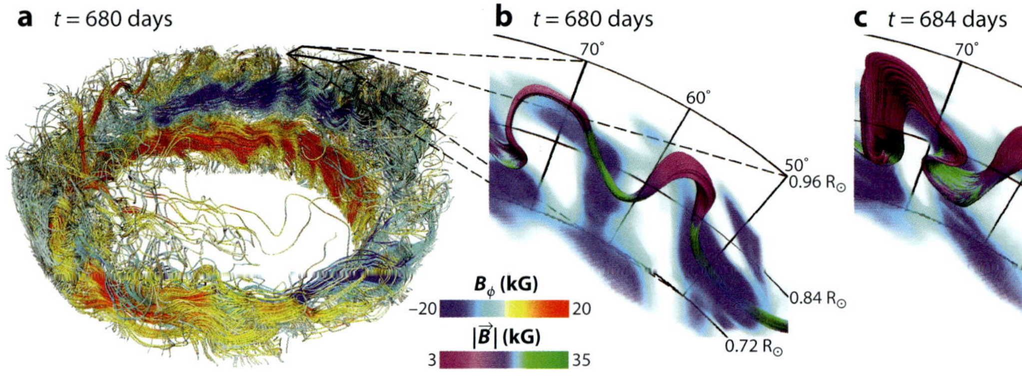

Figure 6

Formation and rise of rope-like magnetic flux systems in the toroidal wreaths produced by the ASH simulations by Nelson et al. (2013). The superequipartition-strength magnetic flux strands form as substructures within the toroidal wreaths and expand as they rise through the convecting layers. Reproduced from Nelson et al. (2013, their figure 18) with permission.

to the static eddy-diffusivity formulation characterizing most earlier ASH simulations (e.g., **Figure 3**). **Figure 6** shows an example, taken from Nelson et al. (2013, their figure 18). Many of these flux strands, forming within existing low-latitude toroidal wreaths, are pulled away by strong convective updraft but have developed enough of a density deficit in their cores for magnetic buoyancy to contribute to their rise through the convecting layers.

Further analyses of many such buoyant loop events by Nelson et al. (2014) have shown that most manage to maintain their approximately east-west orientation as they rise through the turbulent environment of the modeled convection zone, and therefore upon emergence would comply with Hale's Polarity law. This reopens the possibility that the sunspot-forming magnetic fields could be generated and stored entirely within the convection zone, challenging the currently prevalent view that the magnetic flux ropes giving rise to sunspots are stored, and perhaps even formed, in the overshoot layer beneath the base of the convection zone.

3.6. Magnetic Modulation of Convective Energy Transport

Another recent exciting development is the detection of a magnetically mediated modulation of convective energy transport in a set of EULAG-MHD simulations, including that shown in **Video 1** (Cossette et al. 2013). In the anelastic approximation the convective luminosity (enthalpy) crossing a spherical surface takes the form

$$L_c = \iint \rho c_p u_r \tilde{T} \, \mathrm{d}S, \tag{11}$$

where $\tilde{T}$ is the temperature fluctuation about the mean value on the shell. One might rightfully expect that the Lorentz force will resist the convective flow, thus reducing u_r, and therefore that the convective energy flux should be smaller at the peak of magnetic cycles than at times of polarity reversals. In fact, precisely the opposite behavior is observed in the simulation; energy transport varies in phase with the magnetic cycle, as shown in **Figure 7**. This in-phase positive correlation ($r = +0.63$ over the 1,300-year span of this simulation) can be traced to an enhancement of the stronger convective plumes, accompanied by a reduction of negative convective flux contributions by turbulent entrainment, i.e., cold fluid pushed upward, warm fluid pushed downward. Interestingly, the probability density function of vertical turbulent velocities remains essentially invariant across the cycles, indicating that the magnetic field is somehow altering the correlation

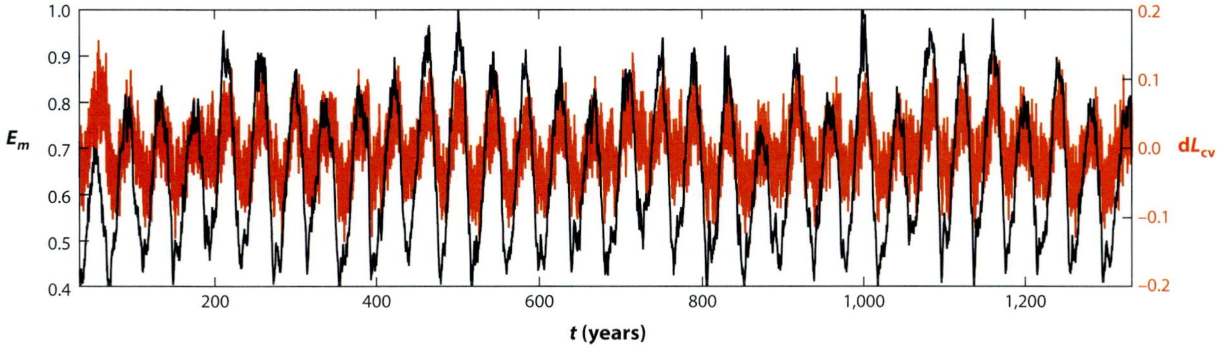

Figure 7

Time series of total magnetic energy (*black*) and convective luminosity (*red*) in the EULAG-MHD simulation of **Video 1**. The two time series vary in phase, with a linear correlation coefficient $r = +0.63$. Temporally extended version of Cossette et al. (2013, their figure 4).

between vertical velocities and temperature perturbations. How exactly this takes place is still being investigated, but the important point is that a variation of the total irradiance in phase with the magnetic cycle is also observed on the Sun (see, e.g., Fröhlich & Lean 2004 and references therein). The simulation from which the time series of **Figure 7** is extracted is actually subluminous as compared with the Sun, yet here the peak-to-peak convective luminosity variation, if scaled to the solar luminosity, amounts to $\sim$0.3%, which compares favorably with the observed decadal variation in total solar irradiance.

Although the varying coverage of magnetic structures clearly dominates irradiance variations on timescales going from minutes to weeks, the possibility that part (or even all) of the variations over timescales decadal and up could be owing to internal structural/convective changes has been a topic of great debate for over three decades now (e.g., Li & Sofia 2001, Foukal et al. 2006, Fröhlich 2009). Global MHD simulations have reached the point where the underlying physical mechanism can perhaps be pinned down, and a "smoking gun" observational signature identified.

4. MEAN-FIELD ELECTRODYNAMICS

The same wide disparity of spatial and temporal scales making direct numerical simulations so challenging can also be capitalized upon via a mathematical approach known as mean-field electrodynamics in order to design much simpler dynamo models capturing only the "slow" spatiotemporal evolution of the large-scale magnetic component. For nearly half a century, mean-field electrodynamics has served as the workhorse of solar dynamo modeling, and is consequently very well-covered in a number of recent and not-so-recent monographs and review articles (Moffatt 1978, Parker 1979, Krause & Rädler 1980, Hoyng 2003, Ossendrijver 2003, Rempel 2006a). After an overview of its theoretical underpinnings and associated mathematical formulation, the following focuses primarily on recent developments, including the use of the mean-field approach as an interpretive framework to investigate the nature of dynamo action observed in numerical simulations.

4.1. The Turbulent Electromotive Force

At the core of mean-field electrodynamics is the hypothesis of scale separation between the small-scale flow and field (here convection and the turbulent magnetic field it generates) and the large-scale flow and field (here differential rotation, meridional circulation, and the magnetic component associated with the solar cycle). The total flow and field are thus each separated into these two components:

$$\mathbf{B} = \langle \mathbf{B} \rangle + \mathbf{b}', \quad \mathbf{U} = \langle \mathbf{U} \rangle + \mathbf{u}', \tag{12}$$

where the angular brackets denote averaging over an intermediate length scale, sufficiently larger than the scales characterizing $\mathbf{u}'$ and $\mathbf{b}'$ so that $\langle \mathbf{u}' \rangle = 0$ and $\langle \mathbf{b}' \rangle = 0$. This is not a linearization, as no assumptions are being made regarding the magnitude of $\mathbf{u}'$ versus $\langle \mathbf{U} \rangle$, or $\mathbf{b}'$ versus $\langle \mathbf{B} \rangle$. Inserting Equation 12 into the MHD induction (Equation 3) and applying this generic averaging operator leads to the mean-field induction equation:

$$\frac{\partial \langle \mathbf{B} \rangle}{\partial t} = \nabla \times (\langle \mathbf{U} \rangle \times \langle \mathbf{B} \rangle + \boldsymbol{\xi} - \eta \nabla \times \langle \mathbf{B} \rangle), \tag{13}$$

where $\boldsymbol{\xi} = \langle \mathbf{u}' \times \mathbf{b}' \rangle$ is the mean electromotive force (hereafter emf) produced by correlated fluctuations of the flow and field at small scales. As we shall see presently, the key point is that this emf can act as a source term for $\langle \mathbf{B} \rangle$. If Equation 12 is inserted into the MHD induction equation, without averaging but now subtracting Equation 13, one obtains an evolutionary equation for the small-scale field:

$$\frac{\partial \mathbf{b}'}{\partial t} = \nabla \times (\langle \mathbf{U} \rangle \times \mathbf{b}' + \mathbf{u}' \times \langle \mathbf{B} \rangle + \mathbf{u}' \times \mathbf{b}' - \boldsymbol{\xi} - \eta \nabla \times \mathbf{b}'). \tag{14}$$

Now, the whole point of the mean-field approach is not to have to deal explicitly with small scales, so formally solving Equations 13 and 14 as a coupled system is out of the question. Observe, however, that if $\mathbf{u}'$ is considered to be given, Equation 14 is linear in $\mathbf{b}'$, with a source term $(\mathbf{u}' \times \langle \mathbf{B} \rangle)$ linear in $\langle \mathbf{B} \rangle$; similarly, with $\langle \mathbf{U} \rangle$ and $\mathbf{u}'$ given, Equation 13 is linear in $\langle \mathbf{B} \rangle$, with $\boldsymbol{\xi}$ providing a source term linear in $\mathbf{b}'$. It follows that the mean emf can be expressed as a linear (tensorial) development in terms of the large-scale magnetic field:

$$\mathcal{E}_i = \alpha_{ij} \langle B \rangle_j + \beta_{ijk} \frac{\partial \langle B \rangle_j}{\partial x_k} + \cdots, \tag{15}$$

where the tensors α_{ij}, β_{ijk}, etc., depend on the properties of the flow, but cannot depend on $\langle \mathbf{B} \rangle$. With closure achieved in this manner, specifying the form of these tensors becomes the crux of the matter. For homogeneous, isotropic turbulence, $\alpha_{ij} = \alpha \delta_{ij}$ and $\beta_{ijk} = \beta \epsilon_{ijk}$; truncating Equation 15 after the second term and substituting in the mean-field induction (Equation 13) then yields

$$\frac{\partial \langle \mathbf{B} \rangle}{\partial t} = \nabla \times (\langle \mathbf{U} \rangle \times \langle \mathbf{B} \rangle + \alpha \langle \mathbf{B} \rangle - (\eta + \beta) \nabla \times \langle \mathbf{B} \rangle). \tag{16}$$

The α term now emerges as a (mean) turbulent electromotive force aligned with the mean-magnetic field, in contrast to the usual motional emf $\langle \mathbf{U} \rangle \times \langle \mathbf{B} \rangle$, which is perpendicular to $\langle \mathbf{B} \rangle$. This contribution to the total turbulent emf, crucial in many dynamo models discussed below, is known as the α-effect, and is nonzero for flows lacking reflection symmetry. The β term reduces to an additive contribution to the magnetic diffusivity η and can thus be interpreted as turbulent diffusion of $\langle \mathbf{B} \rangle$. The α- and β-effects in Equation 16 embody, respectively, constructive and destructive folding of the mean-magnetic field by the small-scale flow. There is no free lunch here; turbulence may, under certain circumstances to be elucidated presently, yield a mean-electromotive force acting as a source for the mean-magnetic field, but it also inevitably provides enhanced dissipation of that same mean-magnetic field.

Because they relate to the intrinsic topological properties of the flow, α and β are in principle computable if the small-scale flow—or at least its statistical properties—are known. Even in such a case, when all is said and done, the calculation is tractable only in a few specific physical regimes (see, e.g., Ossendrijver 2003, Rempel 2006a): The first is for turbulent flows in which the correlation time τ_c is smaller than their turnover time ℓ/u'_{rms}, where ℓ and u'_{rms} are typical values for the turbulent eddies. The ratio of these two quantities is known as the Strouhal number $\mathrm{St} = \tau_c u'_{\mathrm{rms}}/\ell$, so the requirement is $\mathrm{St} < 1$. The second tractable regime is that of low magnetic Reynolds number turbulent flows, i.e., $\mathrm{Rm} \ll 1$. The third applies to situations in which the magnetic field is dominated by its large-scale component, i.e., $|\mathbf{b}'| \ll |\langle\mathbf{B}\rangle|$. For near-homogeneous, near-isotropic turbulence satisfying at least one of these criteria, the α and β coefficients in Equation 16 take these forms:

$$\alpha = -\frac{\tau_c}{3}\langle\mathbf{u}' \cdot \nabla \times \mathbf{u}'\rangle, \tag{17}$$

$$\beta = \frac{\tau_c}{3}\langle(\mathbf{u}')^2\rangle. \tag{18}$$

The β-effect is here proportional to the turbulent intensity. Within the convection zone, with $|\mathbf{u}'| \sim 10\ \mathrm{m\ s^{-1}}$, the magnitude of β is expected to exceed that of η by many orders of magnitude. The α-effect, by contrast, is proportional to the mean kinetic helicity. Cyclonic flows are thus prime candidates for driving an α-effect (Parker 1955b). In the solar/stellar context, this cyclonicity is imparted on horizontally expanding convective updrafts and converging downdrafts by the Coriolis force, thus creating a strong correlation between vertical flow speed and helicity.

The right-hand sides of Equations 17 and 18 have the benefit of being relatively straightforward to extract from numerical simulations. However, for solar interior conditions one expects $\mathrm{Rm} \gg 1$, $\mathrm{St} \sim 1$, and $|\mathbf{b}'|/|\langle\mathbf{B}\rangle| > 1$, which violates the conditions under which these expressions are physically valid. Nonetheless, empirical evidence discussed further below (Section 4.3) indicates that Equations 17 and 18 hold rather well at moderately high Rm, $\mathrm{St} \sim 1$, and $|\mathbf{b}'|/|\langle\mathbf{B}\rangle| \sim 1$.

Returning to the general tensorial formulation of the emf, it is possible to separate the α-tensor in symmetric and antisymmetric contributions:

$$\alpha_{ij} = \alpha_{ij}^S - \varepsilon_{ijk}\gamma_k, \tag{19}$$

so that the mean emf becomes

$$\mathcal{E}_i = \alpha_{ij}^S\langle\mathbf{B}\rangle_j + [\boldsymbol{\gamma} \times \langle\mathbf{B}\rangle]_i + \beta_{ijk}\frac{\partial\langle\mathbf{B}\rangle_j}{\partial x_k} + \cdots. \tag{20}$$

Substitution into the mean-field induction (Equation 13) reveals that $\boldsymbol{\gamma}$ emerges therein as an additive contribution to $\langle\mathbf{U}\rangle$, i.e., it acts on $\langle\mathbf{B}\rangle$ as would a large-scale flow. Under the same physical conditions under which Equations 17 and 18 can be expected to hold, it can be expressed as

$$\boldsymbol{\gamma} = -\frac{\tau_c}{3}\nabla\langle(\mathbf{u}')^2\rangle, \tag{21}$$

i.e., it is proportional to the negative gradient of turbulent intensity. This pseudoflow is known as general turbulent pumping and is increasingly recognized as a potentially important contributor to the evolution of the solar large-scale magnetic field. In particular, with u'_{rms} decreasing with increasing depth in the solar convection zone, downward pumping is expected all the way through the bottom of the convecting layers. This has indeed been observed in a variety of numerical simulations (e.g., Tobias et al. 2001, Ossendrijver et al. 2002, Ziegler & Rüdiger 2003, Racine et al. 2011). The off-diagonal elements of the symmetric part α_{ij}^S also represent turbulent pumping, but they are specific to individual field components and therefore cannot be subsumed into an advective pseudoflow (see discussion by Ossendrijver et al. 2002).

Whichever forms are assumed for the α and β-tensors, their specifications achieve closure, in the sense that with $\langle\mathbf{U}\rangle$ also specified, the mean-field induction (Equation 13) can be solved for $\langle\mathbf{B}\rangle$. This forms the basis of mean-field dynamo models, which offer a modeling approach much simpler than the numerical simulations reviewed in the preceding section. They allow, among other things, efficient exploration of parameter space and the tackling of questions and issues related to long-timescale behavior (e.g., **Figure 1*d***), otherwise inaccessible in practice to bona fide MHD numerical simulations.

4.2. Axisymmetric Mean-Field Dynamo Models

The large-scale magnetic field associated with the solar cycle being axisymmetric to a reasonable degree, the averaging operator of the mean-field approach can then be associated naturally with a zonal average. Working in spherical polar coordinates (r, θ, ϕ) and for the time being considering only differential rotation, the large-scale flow and field can be written as

$$\langle \mathbf{U} \rangle(r, \theta, t) = r \sin\theta\, \Omega(r, \theta, t) \hat{\mathbf{e}}_\phi, \tag{22}$$

$$\langle \mathbf{B} \rangle(r, \theta, t) = B(r, \theta, t)\hat{\mathbf{e}}_\phi + \nabla \times [A(r, \theta, t)\hat{\mathbf{e}}_\phi], \tag{23}$$

where Ω is the angular velocity of rotation, $B\hat{\mathbf{e}}_\phi$ is the toroidal magnetic field (zonally oriented), and the toroidal vector potential $A\hat{\mathbf{e}}_\phi$ defines the poloidal magnetic field (component in meridional planes). We first restrict ourselves to the kinematic approximation, in which the magnetic backreaction of $\langle \mathbf{B} \rangle$ on $\langle \mathbf{U} \rangle$ is neglected, so that Ω is considered to be given. Substituting these expressions into Equation 13 allows the separation of the mean-field induction equation into two coupled evolution equations for A and B:

$$\frac{\partial A}{\partial t} = \eta \left(\nabla^2 - \frac{1}{\varpi^2} \right) A + \alpha B, \tag{24}$$

$$\frac{\partial B}{\partial t} = \eta \left(\nabla^2 - \frac{1}{\varpi^2} \right) B + \varpi (\nabla \times A\hat{\mathbf{e}}_\phi) \cdot (\nabla \Omega) + \hat{\mathbf{e}}_\phi \cdot \nabla \times (\alpha \nabla \times A\hat{\mathbf{e}}_\phi), \tag{25}$$

with $\varpi = r \sin\theta$, and for constant η and an isotropic (diagonal) α-effect. These expressions highlight the crucial role of the α-effect. The associated turbulent emf is the sole source term appearing on the right-hand side of Equation 24, unlike in Equation 25 where shearing by differential rotation also acts as a source for B. Indeed, without the αB source term in Equation 24, Cowling's theorem ensures that an axisymmetric flow cannot sustain an axisymmetric magnetic field against Ohmic dissipation (see Moffatt 1978, his section 6.4).

Dimensional analysis allows us to measure the relative importance of the various source terms in Equations 24 and 25 through the dynamo numbers:

$$C_\alpha = \frac{\alpha_0 \mathrm{R}_\odot}{\eta_0}, \quad C_\Omega = \frac{\Omega_0 \mathrm{R}_\odot^2}{\eta_0}, \tag{26}$$

where α_0, Ω_0, and η_0 are characteristic measures of the magnitude of the α-effect, large-scale rotational shear, and net (turbulent) magnetic diffusivity, respectively. In the solar case, with $\alpha_0 \sim 1$ m s^{-1}, one finds $C_\alpha \ll C_\Omega$, implying that rotational shearing of the large-scale poloidal magnetic component by differential rotation dominates the induction of the toroidal component. On this basis, the turbulent emf contribution is often omitted from Equation 25, resulting in the $\alpha\Omega$ dynamo model, whereas retaining all components of $\boldsymbol{\xi}$ yields the $\alpha^2\Omega$ dynamo model. Dynamo action is also possible in the absence of rotational shear (i.e., $C_\Omega = 0$), which results in the α^2 dynamo model.

As shown analytically by Parker (1955b) for semi-infinite Cartesian geometry, the $\alpha\Omega$ dynamo model supports traveling-wave-like solutions and offers a simple and elegant explanation for the observed equatorward migration of sunspot belts in the course of the solar cycle (e.g., **Figure 1**). This behavior is robust, in that it carries over to spherical geometry (e.g., Yoshimura 1975, Stix

1976) and to the $\alpha^2\Omega$ regime provided shear remains significant (Choudhuri 1990, Charbonneau & MacGregor 2001). In the $\alpha\Omega$ regime, dynamo waves propagate along isocontours of angular velocity Ω, in a direction **s** given by

$$\mathbf{s} = \alpha \nabla \Omega \times \hat{\mathbf{e}}_\phi. \tag{27}$$

In the solar context, equatorward propagation at low latitude arises if the $\alpha_{\phi\phi}$ tensor component is negative in the Northern Hemisphere, and the dynamo is restricted to the equatorial portion of the tachocline, where $\nabla\Omega \simeq \mathrm{d}\Omega/\mathrm{d}r > 0$ (see **Figure 2*b***).

4.3. Measuring the α-Effect

The α-effect is a kinematic property of turbulent flows and as such can in principle be calculated under specific physical, geometrical, and statistical assumptions; for example, an important and influential early result owing to Steenbeck & Krause (1969) concerns the crucial $\alpha_{\phi\phi}$ tensor component, which can be expressed as

$$\alpha_{\phi\phi} \simeq -\frac{16}{15}\tau_c^2 (u'_{\rm rms})^2 \boldsymbol{\Omega} \cdot \nabla \ln(\rho u'_{\rm rms}). \tag{28}$$

With $u'_{\rm rms}$ increasing more slowly outward than ρ is decreasing across the convection zone (according to mixing length estimates) and Ω pointing north, $\alpha_{\phi\phi}$ turns out positive (negative) in the Northern (Southern) Hemisphere. For $u'_{\rm rms}$ depending only on depth, Equation 28 also predicts that the magnitude of $\alpha_{\phi\phi}$ should peak at the poles and vary with polar angle as $\cos\theta$. It also predicts a sign change near the base of the convecting layers, as $u'_{\rm rms}$ is expected to vanish rapidly as one moves downward into the convectively stable radiative core, leading to $\nabla(\rho u'_{\rm rms}) > 0$.

Alternately, the α-effect can be directly measured in numerical simulations of turbulent convection. This is not as straightforward as one may imagine. In principle, applying a uniform large-scale field on a statistically stationary numerical simulation of turbulent convection—or any turbulent flow for that matter—allows the measurements of the emf so generated; the original flow provides the $\mathbf{u}'$, and its inductive action on the imposed large-scale field generates the $\mathbf{b}'$, from which the mean emf $\boldsymbol{\xi} = \langle \mathbf{u}' \times \mathbf{b}' \rangle$ is computed by suitable spatiotemporal averaging. The individual components of the α-tensor are then readily computed by projecting the emf on the applied mean-field. This imposed field method has been used on a variety of local and global simulations of thermally-driven convection in stratified, rotating environments (see, e.g., Brandenburg et al. 1990; Ossendrijver et al. 2001, 2002; Ziegler & Rüdiger 2003; Cattaneo & Hughes 2006). Such calculations have allowed exploration of the dependence of the α-effect on rotation rate, stratification, etc. Particularly pertinent in the solar context are simulations of thermally-driven convection in local Cartesian boxes positioned at various latitudes set by the imposed inclination of the rotation vector $\boldsymbol{\Omega}$ (see, e.g., Brummell et al. 1996), allowing easier access to more strongly turbulent regimes than would global simulations. Surprisingly, many of these numerical experiments have corroborated at least qualitatively some key predictions of mean-field theory, including Equation 28, even though the simulations are typically run in parameter regimes under which closure expressions such as Equations 17, 18, or 21 are not expected to hold. In particular, at moderately low rotation rates, in the sense that the inverse Rossby number $\mathrm{Ro}^{-1} \equiv \Omega\tau_c < 1$, the critical (from the point of view of axisymmetric mean-field dynamos) $\alpha_{\phi\phi}$ component is positive (negative) in the Northern (Southern) Hemisphere, peaks in polar regions, decreases monotonically toward the equator, and exhibits a sign change near the base of the convecting layers (Ossendrijver et al. 2001), all of which is in general agreement with Equation 28. However, even a modest increase in rotation rate is found to shift the peak of the α-effect to low latitudes (Käpylä et al. 2006a). At most rotation rates explored, the α-tensor is full, in that its off-diagonal components have magnitudes comparable to their diagonal counterparts, which is indicative of the break in homogeneity caused by stratification and rotation. As a consequence, significant turbulent pumping is also present, downward-directed in the bulk of the convecting layers but showing in addition a significant equatorward latitudinal component at mid- to low latitudes, in the meter-per-second range for the toroidal component (Ossendrijver et al. 2002). Such calculations have also shown that the components of the α-tensor are strongly fluctuating quantities, with standard deviations about the mean comparable to the magnitude of the mean (see, e.g., Ossendrijver et al. 2001, their figure 3).

To reliably measure the α-effect of the original turbulent flow in this manner, it is essential for the applied magnetic field not to alter the flow and for the flow itself not to act as a small-scale dynamo autonomously producing its own $\mathbf{b}'$ independently of the applied large-scale field. The second of these constraints can be satisfied by restricting the analysis to magnetic Reynolds numbers smaller than those at which small-scale dynamo action is sustained. Working with very weak imposed fields can bypass the first constraint, provided the averaging is carried out on time segments over which the induced $\mathbf{b}'$ also remains sufficiently weak. Producing a time series long enough to yield stable averages may then require periodic resetting of the field to low values or the combination of statistically independent shorter simulation segments. The test-field method (Schrinner et al. 2007; see also Käpylä et al. 2009) aims at bypassing both constraints simultaneously by solving a set of evolution equations for $\mathbf{b}'$, which is essentially Equation 14 herein but with a set of imposed weak test fields replacing $\langle \mathbf{B} \rangle$ on the right-hand side with the small-scale flow $\mathbf{u}'$ taken directly from the simulation being analyzed. Even then, the results turn out to be sensitive to what one would normally have hoped to be "details," such as the size and aspect ratio of the simulation box, boundary conditions, and convective parameter regime [see the discussion by Hughes & Cattaneo (2008) and Hughes et al. (2011)].

A somewhat distinct situation arises in MHD numerical simulations that autonomously produce a large-scale magnetic field in their statistically stationary saturated state; in such cases, one can choose an averaging operator, directly measure $\langle \mathbf{B} \rangle$ (rather than impose it), define $\mathbf{u}' \equiv \mathbf{u} - \langle \mathbf{U} \rangle$ and $\mathbf{b}' \equiv \mathbf{B} - \langle \mathbf{B} \rangle$, compute $\boldsymbol{\xi} = \langle \mathbf{u}' \times \mathbf{b}' \rangle$, and then reconstruct the individual components of the α-tensor. The tensor measured in this manner is no longer a kinematic property of the flow per se, but rather a measure of the net turbulent induction in the nonlinearly saturated dynamo simulation (on this point see also Hubbard et al. 2009).

Assuming a scalar α-effect (i.e., $\boldsymbol{\xi} = \alpha \langle \mathbf{B} \rangle$), Brown et al. (2010, 2011) failed to find a satisfactory relationship between the turbulent emf and the large-scale magnetic field measured in their simulations. Working with a EULAG-MHD simulation essentially identical to that shown in **Video 1**, Racine et al. (2011) introduced a method for calculating the components of the full α-tensor based on a least-squares fit of the tensorial relation:

$$\xi_i(t) = \alpha_{ij} \langle B \rangle_j(t), \quad i, j = r, \theta, \phi, \tag{29}$$

on the time series of $\boldsymbol{\xi}$ and $\langle \mathbf{B} \rangle$ at each grid point of the meridional plane in their simulation (the zonal direction being the averaging direction). The fit itself is carried out separately for each vector component of $\boldsymbol{\xi}$ and $\langle \mathbf{B} \rangle$ using singular value decomposition (SVD). Sample results of this procedure are shown in **Figure 8*a*** for the $\alpha_{\phi\phi}$ component, and the associated standard deviation of the fit is plotted in **Figure 8*b***. In this specific instance, the fit spans 16 full magnetic cycles. Breaking the time series into segments and carrying out the fitting procedure on each segment confirms the temporal steadiness of the α-tensor, implying that the measured cyclicity of the emf is directly associated with the cyclicity of the mean-magnetic field.

Examination of **Figure 8** indicates that the $\alpha_{\phi\phi}$ component is positive in the Northern Hemisphere, peaks at the pole, and shows a sign change near the base of the convecting layer, in agreement with Equation 28 as well as with the Cartesian box calculations discussed above. A secondary peak at low latitudes is also present. Except in the equatorial region, where both $\boldsymbol{\xi}$ and

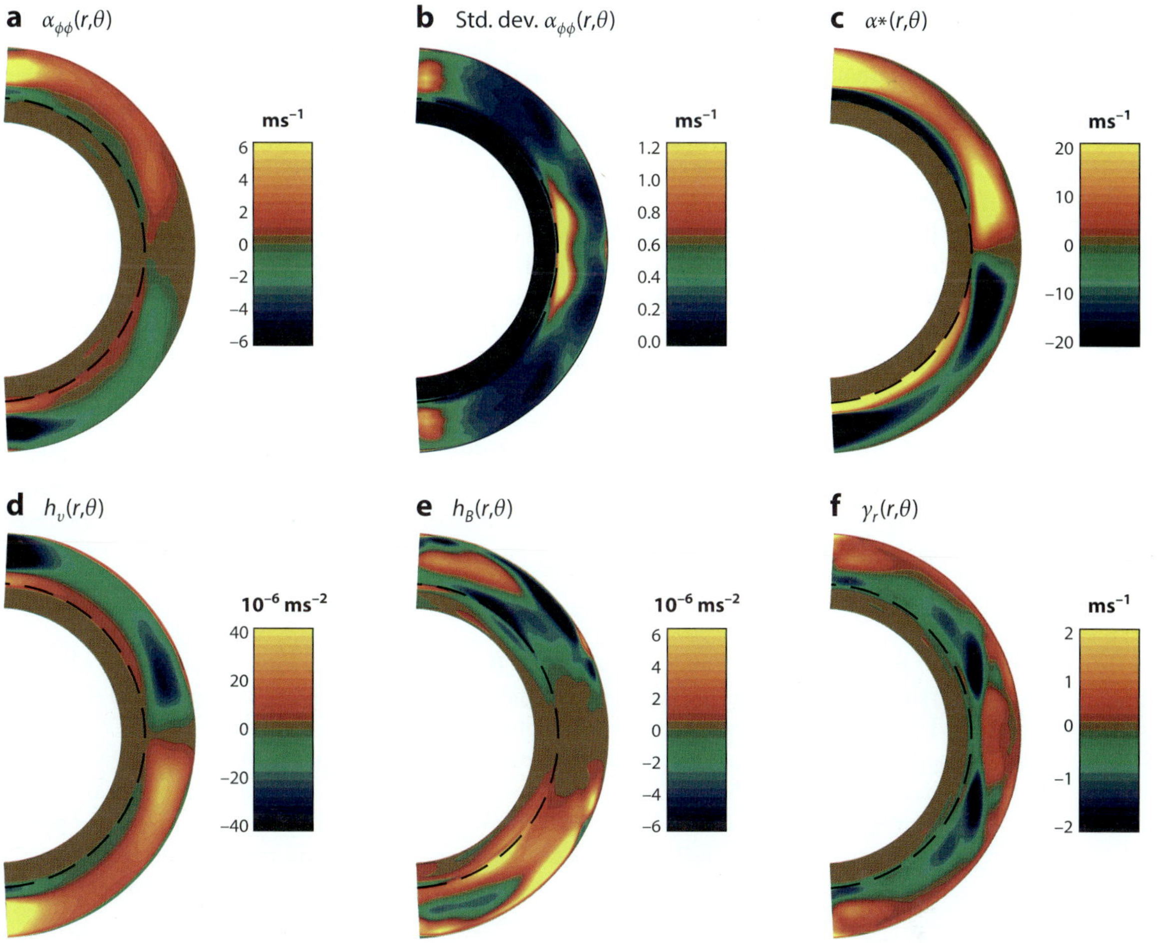

Figure 8

The α-tensor in the EULAG-MHD simulation of **Video 1**. (*a*) The $\alpha_{\phi\phi}$ component as extracted by linear least-squares fit on Equation 29, and (*b*) the associated standard deviation. (*c*) α as reconstructed via Equation 17, using the kinetic helicity profile in panel *d*, as extracted from the simulation. (*e*) The extracted current helicity profile (see Section 4.5). (*f*) The vertical turbulent pumping velocity, constructed from the full α-tensor via Equation 19.

$\langle \mathbf{B} \rangle$ are quite weak, and in the stable layer, where $\langle \mathbf{B} \rangle$ can be large but $\boldsymbol{\xi}$ is very weak, the magnitude of $\alpha_{\phi\phi}$ significantly exceeds the standard deviation returned by the SVD fit.

Figure 8*c* shows the α-coefficient computed via Equation 17; the kinetic helicity has been extracted from the same simulation (**Figure 8*d***), and the correlation time estimated as $H_\rho/u'_{\rm rms}$, where H_ρ is the density scale height. The match with the true $\alpha_{\phi\phi}$ in **Figure 8*a*** is remarkable, although the latter has a magnitude that is smaller by a factor of about five. Interestingly, if this mismatch is attributed to an overestimate of the correlation time τ_c, then a peculiar internal consistency is recovered; writing $\tau_c = H_\rho/u'_{\rm rms}$ amounts to setting the Strouhal number St $= 1$, but reducing it to St $= 0.2$ to match the magnitudes of **Figure 8*a*,*c*** brings us back to the regime St < 1, where Equation 17 is now expected to hold.

Figure 8*f* shows the radial component of the turbulent pumping velocity (e.g., Equation 19), as reconstructed from the antisymmetric component of the full α-tensor. Here again, pumping is predominantly downward in the bulk of the convecting layers and into the upper portion of the underlying stably stratified fluid layer. The upper boundary condition causes a rapid reduction of the turbulent intensity moving up across the subsurface layers, which leads to upward turbulent pumping, in agreement with Equation 21. Except in subsurface layers, equatorward latitudinal pumping is also observed at most latitudes, peaking at 2–3 m s^{-1} at $\pm 25°$ latitude.

It is quite noteworthy that very distinct numerical simulations and equally distinct means of measuring the α-tensor in rotating, stratified thermally-driven convective turbulence simulations yield α-tensors that often show a surprising degree of agreement with each other, as well as with closed-form expressions used outside of their nominal range of validity. This is not always the case with mechanically forced helical flows (see Courvoisier et al. 2006 and references therein). The distinction hinges on the varying degrees of spatial coherence characterizing each class of flows (Courvoisier et al. 2009).

4.4. Magnetic Helicity and Its Constraint on Dynamo Action

Magnetic helicity is a topological measure of linkage between magnetic flux systems (Berger & Field 1984). In a closed system, i.e., without helicity flux through its boundaries, the total helicity evolves according to

$$\frac{\mathrm{d}}{\mathrm{d}t}\int \mathbf{A}\cdot\mathbf{B}\,\mathrm{d}V = -2\mu_0\eta\int \mathbf{J}\cdot\mathbf{B}\mathrm{d}V\,, \tag{30}$$

where the vector potential $\mathbf{A}$ is such that $\mathbf{B} = \nabla\times\mathbf{A}$. In the ideal limit $\eta\to 0$, which is the relevant limit for dynamo action in the interior of the Sun and stars, the right-hand side vanishes. Equation 30 then indicates that total helicity must be conserved or at best vary on the "long" diffusive timescale τ_η. This puts a strong constraint on the amplification of any magnetic field that has a net helicity, which is certainly the case with the large-scale solar magnetic field, comprised as it is of well-defined axisymmetric poloidal and toroidal components; indeed, any nontrivial axisymmetric field defined through Equation 23 is clearly helical, as the vector potential defining the poloidal component is colinear with the toroidal magnetic component. How, then, can any field amplification be achieved?

One obvious way to bypass the constraints posed by Equation 30 is to allow helicity fluxes through the outer boundary of the dynamo region, here the solar photosphere. This idea is supported by simulations (e.g., Brandenburg & Dobler 2001) as well as photospheric measurements (e.g., Kusano et al. 2002, Démoulin & Berger 2003). In particular, coronal mass ejections are suggested to be a primary contributor to the net outward helicity flux (Low 2001). Recent numerical simulations of helically-forced turbulence akin to the spherical wedge PENCIL simulations described in Section 3 and including a simplified outer "corona" do eject helicity in intermittent bursts (Warnecke et al. 2011), and this process is found to enhance dynamo action (Warnecke et al. 2012). However, it remains unclear whether the photospheric outward helicity flux makes a significant contribution to the global internal magnetic helicity budget (Berger & Ruzmaikin 2000).

It is also possible to bypass the helicity conservation constraints by having the dynamo produce helicity of opposite signs at large and small spatial scales (Brandenburg 2001). Following the scale separation logic already introduced in Section 4.1, and because $\mathbf{J}$ and $\mathbf{A}$ are linearly related to $\mathbf{B}$, the total vector potential and electric current density can be written as $\mathbf{A} = \langle\mathbf{A}\rangle + \mathbf{a}'$ and $\mathbf{J} = \langle\mathbf{J}\rangle + \mathbf{j}'$. Substituting into Equation 30 and averaging leads to an evolution equation for the helicity associated with the large-scale field:

$$\frac{\mathrm{d}}{\mathrm{d}t}\int \langle\mathbf{A}\rangle\cdot\langle\mathbf{B}\rangle\mathrm{d}V = +2\int \boldsymbol{\xi}\cdot\langle\mathbf{B}\rangle\mathrm{d}V - 2\mu_0\eta\int \langle\mathbf{J}\rangle\cdot\langle\mathbf{B}\rangle\mathrm{d}V, \tag{31}$$

where $\boldsymbol{\xi} = \langle \mathbf{u}' \times \mathbf{b}' \rangle$ is the usual turbulent emf. Subtracting Equation 31 from the unaveraged form of Equation 30 then yields an evolution equation for the helicity of the small-scale magnetic component:

$$\frac{\mathrm{d}}{\mathrm{d}t}\int \langle \mathbf{a}' \cdot \mathbf{b}' \rangle \mathrm{d}V = -2\int \boldsymbol{\xi} \cdot \langle \mathbf{B} \rangle \mathrm{d}V - 2\mu_0 \eta \int \langle \mathbf{j}' \cdot \mathbf{b}' \rangle \mathrm{d}V. \tag{32}$$

Observe that in the ideal limit, the total helicity—now the sum of Equations 31 and 32—is still conserved, as the first term on the right-hand side of Equations 31 and 32 are identical but for their sign. This also indicates that the turbulent emf leads to the buildup of helicity of opposite signs at large and small spatial scales, corresponding to a dual cascade away from the generation scale (Brandenburg 2001). The cascade to small scales is attractive because, in principle, Ohmic dissipation can operate efficiently there, whereas it is expected to be insignificant at larger scales. Under this view the buildup of the large-scale magnetic field is controlled by the rate at which helicity can be transported and dissipated at small scales. However, this accumulation of magnetic helicity—and magnetic energy—at small scales has profound consequences for the operation of the α-effect.

4.5. The Nonlinear α-Effect

Constructive folding of the large-scale magnetic field $\langle \mathbf{B} \rangle$, as embodied in the α-effect, requires work exerted against the magnetic tension force. If $\langle \mathbf{B} \rangle$ is amplified exponentially by dynamo action, there will inevitably come a point when the flow will no longer be able to deform the magnetic field; this turning point defines the equipartition field strength (B_{eq}), at which the magnetic energy density becomes equal to the kinetic energy density of the flow, i.e., $\mu_0^{-1} B_{\mathrm{eq}}^2 = \rho (u'_{\mathrm{rms}})^2$. Near the base of the solar convection zone, where $\rho \simeq 10^2$ kg m^{-3} and $u'_{\mathrm{rms}} \sim 10$ m s^{-1}, one has $B_{\mathrm{eq}} \sim 0.1$ T, which is comparable with sunspot field strengths. This idea forms the basis of the α-quenching formalism, which consists of introducing an explicit $\langle \mathbf{B} \rangle$-dependence to the α-tensor components via an ad hoc algebraic relation of the form

$$\alpha_{ij}(\langle \mathbf{B} \rangle) \rightarrow \frac{\alpha_{ij}}{1 + (\langle \mathbf{B} \rangle^2 / B_{\mathrm{eq}}^2)}. \tag{33}$$

From a purely practical point of view, when inserted into the mean-field dynamo equations, this expression "does the right thing," in that $\alpha_{ij} \rightarrow 0$ when $\langle \mathbf{B} \rangle$ grows beyond equipartition; it remains of common usage in mean-field dynamo modeling. The situation turns out to be nowhere near as simple, however. Working with a mechanically forced "toy" flow, Cattaneo & Hughes (1996) carried out a series of numerical experiments at varying magnetic Reynolds number Rm and found that their results did not fit Equation 33, but rather the closely related form

$$\alpha_{ij}(\langle \mathbf{B} \rangle) \rightarrow \frac{\alpha_{ij}}{1 + \mathrm{Rm}(\langle \mathbf{B} \rangle^2 / B_{\mathrm{eq}}^2)} \tag{34}$$

(see also Blackman & Brandenburg 2002). This result, which came to be known as catastrophic quenching, has profound implications for astrophysical dynamos in general, because in most astrophysically relevant situations $\mathrm{Rm} \gg 1$, implying saturation of the large-scale field at a factor $\sqrt{\mathrm{Rm}}$ smaller than equipartition. In the solar interior, $\mathrm{Rm} \sim 10^{10}$ leads to a saturation at the level $|\langle \mathbf{B} \rangle| \sim 10^{-6}$ T, which is now a long way from sunspot field strengths.

A pioneering measurement of the α-effect in numerical simulations by Pouquet et al. (1976) indicates that in nonlinearly saturated helical MHD turbulence, Equation 17 picks up a magnetic contribution of sign opposite to that associated with the kinetic helicity:

$$\alpha = -\frac{\tau_c}{3}(\langle \mathbf{u}' \cdot \nabla \times \mathbf{u}' \rangle - \rho^{-1}\langle \mathbf{j}' \cdot \mathbf{b}' \rangle). \tag{35}$$

The second term on the right-hand side is the mean current helicity associated with the small-scale magnetic field and is closely related to magnetic helicity (think spectrally!). The cascade of turbulent helicity to small scales, required for the buildup of a large-scale magnetic component, thus inevitably leads to the growth of current helicity, which then quenches the α-effect. This idea has been successfully implemented in mean-field dynamo models by complementing the mean-field dynamo equation with an evolution equation for small-scale magnetic helicity driven by helicity buildup at large scales (e.g., Blackman & Brandenburg 2002, Kleeorin et al. 2003, Chatterjee et al. 2011, Hubbard & Brandenburg 2012), a procedure known as dynamical α-quenching.

Figure 8*e* shows the mean current helicity in the EULAG-MHD simulation of **Video 1**. Comparison with the mean kinetic helicity in **Figure 8*d*** reveals that both are of opposite sign over a large portion of the meridional plane, although here the kinetic helicity exceeds the current helicity by a factor of about five.

Equation 35 also allows, in principle, a sign change in the α-effect in the nonlinearly saturated regime, if the second term on the right-hand side dominates the first. Something akin to this may be at play in the simulations by Käpylä et al. (2012, 2013), where poleward propagation of the low-latitude toroidal field in the kinematic growth phase of the large-scale dynamo switches to equatorward propagation following nonlinear saturation (see Käpylä et al. 2012, their figure 3; see also **Figure 4** herein), consistent with a sign change in the α-effect.

4.6. Instabilities and Dynamo Action in the Tachocline

The aforementioned difficulties encountered with conventional mean-field dynamo models, in particular catastrophic quenching of the turbulent α-effect, have motivated the study of alternate dynamo mechanisms operating in the tachocline, where amplification and storage of strong magnetic fields is in principle easier than in the strongly turbulent overlying convection zone. The class of magnetoshear instabilities discussed in Section 3.4 has been shown to generate kinetic helicity as it develops, which could, upon acting on a pre-existing large-scale toroidal magnetic field, drive an α-effect-like poloidal field regeneration mechanism (Dikpati & Gilman 2001, Dikpati et al. 2003).

Another related potential dynamo mechanism involves the development, through the action of the Coriolis force, of helical waves along the axis of a toroidal flux rope undergoing buoyant destabilization (Ferriz-Mas et al. 1994, Ossendrijver 2000b). In either case, with tachocline differential rotation shearing this poloidal field into a toroidal component sustaining the instability, a complete dynamo loop can be produced, residing entirely within the tachocline [these dynamo models are discussed further by Charbonneau (2010, his section 4)].

4.7. Recent Progress in Mean-Field Dynamo Modeling

The design of solar cycle models based on mean-field electrodynamics remains a very active field of research. Recent progress focuses on models bypassing catastrophic α-quenching by spatially separating the α-effect and the strong toroidal fields produced by rotational shearing. These interface dynamos (Parker 1993) can achieve the desired result (Charbonneau & MacGregor 1996), but tend to be rather sensitive to modeling details and so far have not satisfactorily reproduced solar cycle characteristics (see also Tobias 1996, Petrovay & Kerekes 2004).

Another very active research thrust in mean-field dynamo modeling involves the relaxation of the kinematic approximation; solving an evolution equation for the magnetically driven, time-dependent deviation ($\langle \mathbf{U}' \rangle$) from the imposed (steady) large-scale flow ($\langle \mathbf{U} \rangle$),

$$\frac{\partial \langle \mathbf{U}' \rangle}{\partial t} = \frac{\Lambda}{\mu_0 \rho} (\nabla \times \langle \mathbf{B} \rangle) \times \langle \mathbf{B} \rangle + \mathrm{Pm} \nabla^2 \langle \mathbf{U}' \rangle, \tag{36}$$

and the total flow is now given by

$$\mathbf{u}(\mathbf{x}, t, \mathbf{B}) = \langle \mathbf{U} \rangle(\mathbf{x}) + \langle \mathbf{U}' \rangle(\langle \mathbf{B} \rangle(\mathbf{x}, t)) \tag{37}$$

(see Tobias 1997, Moss & Brooke 2000). Solving Equation 36 together with the $\alpha\Omega$ form of Equations 24 and 25 yields a nonkinematic mean-field dynamo in which the cycle amplitude develops modulations on timescales longer than the cycle via the magnetically driven modulation of differential rotation (see, e.g., Tobias 1997, Küker et al. 1999, Brooke et al. 2002, Bushby 2006 and references therein). The period of this modulation is usually found to scale with the inverse of the magnetic Prandtl number Pm. This offers a possible explanation for Maunder Minimum-like episodes of reduced activity, if the internal field strength falls below the threshold for sunspot formation, while allowing for residual magnetic activity, as during the Maunder Minimum (Beer et al. 1998).

4.8. Mean-Field Electrodynamics as a Probe on Magnetohydrodynamic Simulations

The large-scale flows and α-tensor extracted from global numerical simulations can be reinserted into the kinematic axisymmetric mean-field dynamo (Equations 24 and 25) to ascertain the degree to which the resulting large-scale magnetic field resembles what is produced by the numerical simulations. This seemingly circular exercise can actually help pin down the mode of dynamo action operating in these simulations and disentangle aspects of magnetic field evolution associated with the nonlinear backreaction on the large-scale flow (see Käpylä et al. 2006b, Simard et al. 2013, Dubé & Charbonneau 2013).

One interesting conclusion to arise from such analyses is that even though the underlying toroidal structures are often not strictly axisymmetric, the common tendency for poleward migration observed in many global MHD simulations (e.g., **Figure 3*c***; also Dubé & Charbonneau 2013, their figures 7 and 8) can be understood at some level in terms of classical dynamo waves. With the northern hemispheric $\alpha_{\phi\phi} > 0$ in the bulk of the convection zone and low-latitude differential rotation increasing outward on isocontours approximately parallel to the rotation axis (e.g., **Figure 2**), propagation away from the equatorial plane is predicted by Equation 27.

Käpylä et al. (2006b) used the α-tensor extracted from their local MHD convection simulations (Käpylä et al. 2006a) to construct $\alpha^2\Omega$ kinematic axisymmetric mean-field dynamo models that proved to exhibit a number of solar-like characteristics, including equatorward migration of the large-scale magnetic fields, although the latter was found to peak at too-high latitudes as compared with solar observations. Simard et al. (2013) carried out a similar exercise using the full α-tensor extracted from the EULAG-MHD simulation of **Video 1**, again as input into a kinematic α-quenched $\alpha^2\Omega$ model. Their mean-field dynamo solution showed many similarities with the axisymmetric magnetic component building up in the parent global MHD simulations (Simard et al. 2013, their figure 5). By selectively zeroing out various combinations of tensorial components, they could demonstrate the importance of turbulent pumping and that the complex form of the α-tensor could support cycling solutions even in the absence of differential rotation (see also Rüdiger et al. 2003).

The Simard et al. (2013) $\alpha^2\Omega$ solution also reproduced the observed tendency of the global simulation to produce "double cycles," i.e., spatially segregated dynamo modes with distinct cycle periods. This tendency can be understood upon noting that differential rotation is typically strongest at the equator and decreases toward the poles (e.g., **Figure 2**), whereas most α-tensor components, including $\alpha_{\phi\phi}$, peak at the pole and decrease in magnitude toward the equator (see **Figure 8*a***). Interestingly, hints of such double cycle behaviors have been detected for many years in indicators of solar activity (e.g., Mursula et al. 2003) and more recently in acoustic oscillation frequencies, suggesting a subphotospheric origin (Fletcher et al. 2010). The numerical experiments of Simard et al. (2013) thus support the notion that there may be two large-scale dynamos operating in the Sun (see also Benevolenskaya 1995, 1998; Mason et al. 2002; Brandenburg 2005).

The mean-field machinery can also offer some useful hints as to the manner in which magnetic self-organization takes place in the global simulations. Consider the mean-field induction equation in the limit where dissipation is inefficient at large spatial scales:

$$\frac{\partial \langle \mathbf{B} \rangle}{\partial t} = \nabla \times (\langle \mathbf{U} \rangle \times \langle \mathbf{B} \rangle + \boldsymbol{\xi}). \tag{38}$$

Now, the turbulent emf $\boldsymbol{\xi} = \langle \mathbf{u}' \times \mathbf{b}' \rangle$ can be expected to vary on the "short" advective timescale associated with $\mathbf{u}'$, which is much smaller than the cycle period of $\langle \mathbf{B} \rangle$. The large-scale flow $\langle \mathbf{U} \rangle$, if it varies significantly, will also vary on this "long" cycle timescale. The only way for $\langle \mathbf{B} \rangle$ to remain quasi-steady on the convective timescale is to have

$$\langle \mathbf{U} \rangle \times \langle \mathbf{B} \rangle + \boldsymbol{\xi} \equiv \boldsymbol{\varepsilon} \simeq 0; \tag{39}$$

the small residual emf $\boldsymbol{\varepsilon}$ (small in the sense that $|\boldsymbol{\varepsilon}| \ll |\langle \mathbf{U} \rangle \times \langle \mathbf{B} \rangle|$ and $|\boldsymbol{\varepsilon}| \ll |\boldsymbol{\xi}|$) drives the evolution of $\langle \mathbf{B} \rangle$ on the long timescale. That the small- and large-scale emfs should conspire to nearly cancel each other out in this manner is rather counterintuitive, but has actually been observed in global numerical simulations producing either steady or cycling large-scale magnetic fields (Brown et al. 2010, Racine et al. 2011). This offers a possible explanation as to why magnetic self-organization on large scales seems to be such a sensitive affair. Varying numerical treatment of small-scale dissipation through explicit or implicit subgrid models will surely affect the turbulent emf to some extent; even if the differences introduced by the choice of subgrid model were small (and available empirical evidence indicates that they are not) a small relative change in $\boldsymbol{\xi}$ can still translate into a large change in the residual $\boldsymbol{\varepsilon}$, perhaps even changing its sign. Computational "details" matter!

5. FLUX TRANSPORT DYNAMOS

With the turbulent α-effect positive (negative) in the bulk of the convection zone's Northern (Southern) Hemisphere, coupled to the primarily latitudinal rotation therein, low-latitude equatorward propagation of dynamo waves is problematic unless dynamo action is concentrated in the equatorial portion of the tachocline, where $\mathrm{d}\Omega/\mathrm{d}r > 0$, and the α-effect experiences a sign change. Unfortunately, the two instabilities discussed in Section 4.6 both yield a positive α-effect in the Northern Hemisphere, which is not surprising because in all cases it is the Coriolis force that sets the sign of the kinetic helicity. One appealing strategy to bypass this problem is to invoke a large-scale meridional flow, directed equatorward at the base of the convection zone, to bodily advect the magnetic field and thus achieve dynamo-wave-like equatorward propagation of the sunspot-forming toroidal magnetic component in the course of the cycle (e.g., Choudhuri et al. 1995, Durney 1995, Küker et al. 2001, Pipin & Kosovichev 2011). These solar cycle models are generically known as flux transport dynamos.

One specific class of such models, the so-called Babcock-Leighton dynamos, has received increasing attention in recent years because of their ability to reproduce many observed cycle fluctuation patterns, including Maunder-Minimum-like intermittent episodes of suppressed activity, and they have even been used to predict characteristics of upcoming solar cycles. Partially eclipsed for a time by the rise of mean-field electrodynamics, they offer a very attractive alternative to turbulent dynamos based on the α-effect, and the bulk of this section is consequently dedicated to these models.

5.1. Surface Flux Evolution and the Babcock-Leighton Mechanism

Large sunspots often appear in pairs of opposite magnetic polarity; the member of the pair that is farther ahead in the direction of solar rotation lies closer to the solar equator than the trailing member. A line segment connecting both members thus shows a tilt with respect to the east-west direction, and this tilt increases with increasing latitude, a pattern known as Joy's law (Hale et al. 1919). Because leading/following polarities of sunspot pairs are opposite in both solar hemispheres, in their decay phase the leading member tends to experience more cross-equatorial diffusive cancellation with the leading member of sunspot pairs located in the other hemisphere, leading to the accumulation, in each hemisphere, of magnetic flux having the polarity of the trailing member. Transport of this magnetic flux to the pole by meridional circulation and/or diffusion then leads to the reversal of the polar cap field from the previous activity cycle and subsequent buildup of a net dipole moment of polarity opposite to that present in the rising phase of the cycle (Babcock 1961; and **Figure 9**). This process shows up clearly in synoptic solar magnetograms (e.g., **Figure 1*c***).

This surface magnetic flux evolution is readily computable by solving the MHD induction equation on a spherical shell subjected to the advective and shearing action of surface differential rotation, poleward meridional flow, diffusion, and injection of tilted bipolar sunspot pairs (Wang et al. 1989, Wang & Sheeley 1991, Baumann et al. 2004, Jiang et al. 2013). One example is shown in **Figure 9*b***. The similarity with the observed synoptic magnetogram shown in **Figure 1*c*** is quite remarkable. Such simulations reveal that the bulk of the magnetic flux contained in bipolar sunspot pairs experiences resistive decay and that only a small fraction makes it to the solar poles. With a mean polar field strength of $\sim 10^{-3}$ T, the polar cap magnetic flux is $\sim 10^{14}$ Wb; this is similar to the total unsigned flux in a single large bipolar sunspot pair and three orders of magnitude lower than the total unsigned magnetic flux emerging in active regions in the course of a typical activity cycle. The conversion efficiency required of the Babcock-Leighton mechanism is therefore quite modest.

The east-west tilt of sunspot pairs is crucial to the operation of the Babcock-Leighton mechanism and originates from the action of the Coriolis force on the flow developing along the axis of a sunspot-forming magnetic flux rope during its rise through the convection zone (Fan et al. 1993, D'Silva & Choudhuri 1993, Caligari et al. 1995). The pattern of tilt compatible with Joy's law turns out to be only possible in a finite range of magnetic intensity; flux ropes with field strengths inferior to about 1 T are deflected to high latitude and emerge with randomized tilts even if they survive their rise through the turbulent convection zone. Beyond ~ 10 T, by contrast, the tilt vanishes because the ropes rise to the photosphere too rapidly for the Coriolis force to have time to act. This implies that there exist both lower and upper thresholds on the internal toroidal magnetic field for the Babcock-Leighton mechanism to operate.

5.2. Magnetic Flux Transport: Meridional Flow, Diffusion, and Turbulent Pumping

Because sunspot pairs represent the surface manifestation of a deep-seated toroidal magnetic component, globally the Babcock-Leighton mechanism produces a (surface) poloidal magnetic

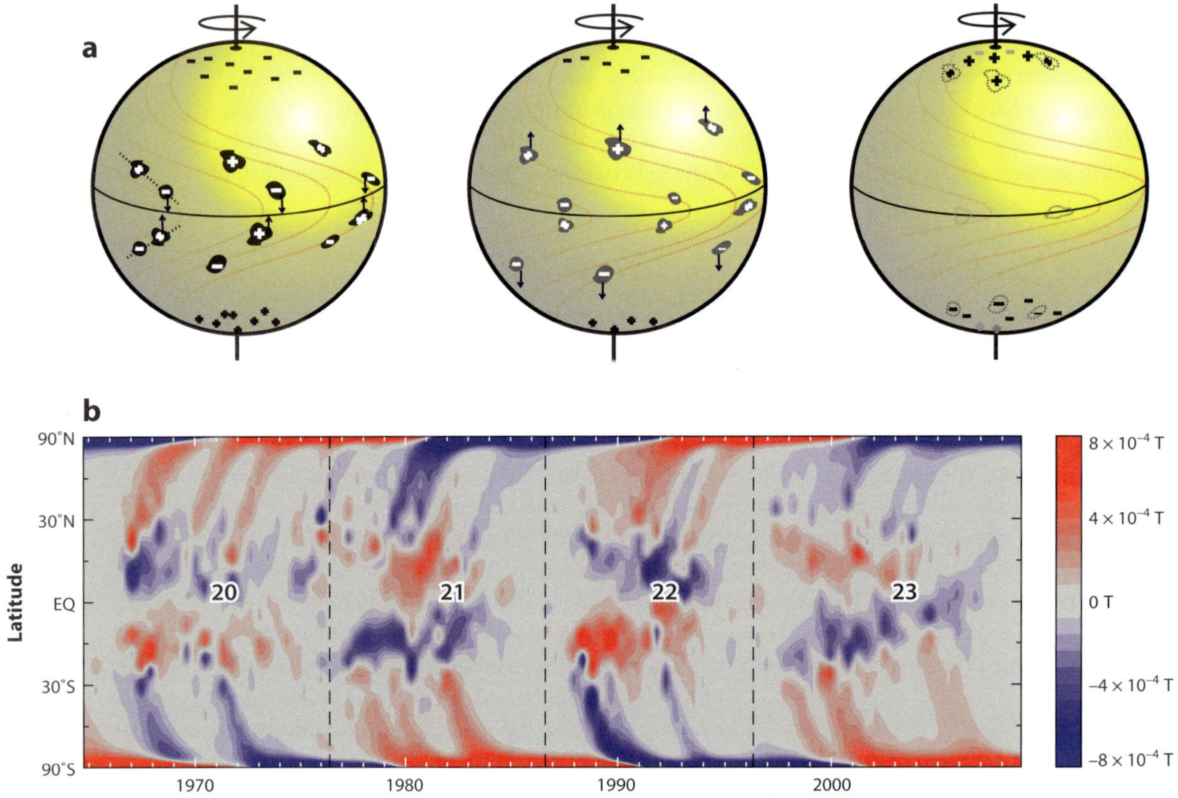

Figure 9

The Babcock-Leighton mechanism, illustrated (*a*) in diagrams and (*b*) via a 2D surface magnetic flux transport simulation covering the years 1965–2009 (activity cycles 20 to 23, numbered from one activity minimum to the next, as delineated by *vertical dashed lines*). On this latter plot, showing the variation with latitude and time of the zonally averaged surface radial magnetic field component (measured in teslas), transport to the poles shows up as inclined streaks extending poleward from low latitudes, where active region emergence takes place. Note how polarity reversal of the polar cap magnetic field occurs at about the time of activity maximum, 4–5 years after cycle onset. Compare panel *b* with the synoptic magnetogram plotted on **Figure 1*c***. Graphics in panel *a* produced by D. Passos (CENTRA/IST), simulation in panel *b* by A. Lemerle (Univ. Montréal).

component from a (deep) toroidal component. Because the trailing polarity is identical to that of the original underlying toroidal field and is that which accumulates at the pole, the polarity of the dipole moment produced is identical to that of the deep-seated toroidal field. This is conceptually similar to the action of a positive α-effect in mean-field electrodynamics. In conjunction with the shearing of this poloidal field by differential rotation, a complete dynamo loop can thus be constructed, without invoking the turbulent α-effect as a source term on the right-hand side of Equation 24. However, here the poloidal and toroidal source terms are spatially segregated, with the Babcock-Leighton mechanism operating in the surface layers, whereas toroidal field amplification and magnetic flux rope formation and storage are presumably taking place at or immediately beneath the base of the convection zone. These two source regions need to be coupled to close the dynamo loop. The first Babcock-Leighton solar cycle models (Babcock 1961, Leighton 1969) assumed the sunspot-generating toroidal magnetic field to reside in the subphotospheric layers, so this coupling issue was not deemed problematic. With the formation and storage site of magnetic flux ropes now pushed down to the tachocline, an efficient coupling mechanism becomes essential. Following Wang et al. (1989) and Wang & Sheeley (1991), large-scale meridional flows have taken up this role (see, e.g., Wang et al. 1991, Choudhuri et al. 1995, Durney 1995, Dikpati & Charbonneau 1999).

In practice, the vast majority of Babcock-Leighton solar cycle models published to date are mean-field-like in that they model the evolution of the large-scale axisymmetric solar magnetic field by solving the $\alpha\Omega$ form of Equations 24 and 25 augmented by an advection term associated with the meridional flow, with all large-scale flows considered to be given (kinematic regime):

$$\frac{\partial A}{\partial t} = \eta\left(\nabla^2 - \frac{1}{\varpi^2}\right)A - \frac{1}{\varpi}\mathbf{u}_p \cdot \nabla(\varpi A) + S(r, \theta, B), \tag{40}$$

$$\begin{aligned}\frac{\partial B}{\partial t} = \eta\left(\nabla^2 - \frac{1}{\varpi^2}\right)B + \frac{1}{\varpi}\frac{\mathrm{d}\eta}{\mathrm{d}r}\frac{\partial(\varpi B)}{\partial r} - \varpi\nabla\cdot\left(\frac{B}{\varpi}\mathbf{u}_p\right) \\ + \varpi(\nabla\times A\hat{\mathbf{e}}_\phi)\cdot(\nabla\Omega),\end{aligned} \tag{41}$$

where $\mathbf{u}_p \equiv u_r\hat{\mathbf{e}}_r + u_\theta\hat{\mathbf{e}}_\theta$ (compare with Equations 24 and 25). A mean-field-like turbulent magnetic diffusivity is implicitly invoked, now allowing for a variation with depth, and the novel (nonlinear) source term $S(r, \theta, B)$ appearing on the right-hand side of Equation 40 replaces the α-effect and operates only in the near-surface layers. It is typically formulated as an ad hoc parameterization of the Babcock-Leighton mechanism, required here because a fundamentally nonaxisymmetric process must be "forced" into a global axisymmetric model (see, e.g., Durney 1995, Dikpati & Charbonneau 1999, Nandy & Choudhuri 2001, Chatterjee et al. 2004, Charbonneau et al. 2005, Muñoz-Jaramillo et al. 2010). Most models use a steady meridional flow characterized by a single flow cell per hemisphere, with the flow speed set by the average poleward flow measured at the solar surface (e.g., Hathaway 1996, Ulrich 2010).

As with other flux transport dynamos, equatorward propagation of the internal toroidal field is achieved here through the advective effect of the meridional flow. If the advective terms on the right-hand sides of Equations 40 and 41 dominate over the diffusive terms (the "advection-dominated regime"), then the cycle period is inversely proportional to the meridional flow speed (Wang et al. 1991, Dikpati & Charbonneau 1999). The required (turbulent) magnetic diffusivity values turn out to be two orders of magnitude smaller than estimates based on, e.g., Equation 18 or mixing length theory, but even with larger magnetic diffusivity values (the diffusion-dominated regime), solar-like dynamo solutions are readily obtained (see also Muñoz-Jaramillo et al. 2011). The inclusion of turbulent pumping as an additional contribution to the large-scale flows further improves the solutions, also providing an additional coupling from the surface source region to the base of the convection zone (Guerrero & De Gouveia Dal Pino 2008, Kitchatinov & Olemskoy 2012). With surface poleward flow speeds on the order of 10 m s^{-1}, as observed, reversal of the polar field occurs around the time of cycle maximum, also as observed. However, the associated polar field strengths tend to be much larger than observed, and the deep toroidal field often develops a branch at high latitude associated with the shearing action of the radial shear in the high-latitude reaches of the tachocline. Some of these difficulties can be alleviated by changes in the imposed meridional flow profile, enhanced magnetic diffusivity in the surface layers, and/or inclusion of downward turbulent pumping in the subsurface layers (see, e.g., Dikpati et al. 2004, Jiang et al. 2009, Kitchatinov & Olemskoy 2012). Finally, these dynamo models can in principle operate in the strong-field regime—indeed, they require strong internal toroidal fields to operate.

5.3. Cycle Amplitude Fluctuations and Grand Minima

An appealing property of Babcock-Leighton solar cycle models is their response to fluctuations in the (imposed) meridional flow speed. Many observed features of the sunspot cycles can be

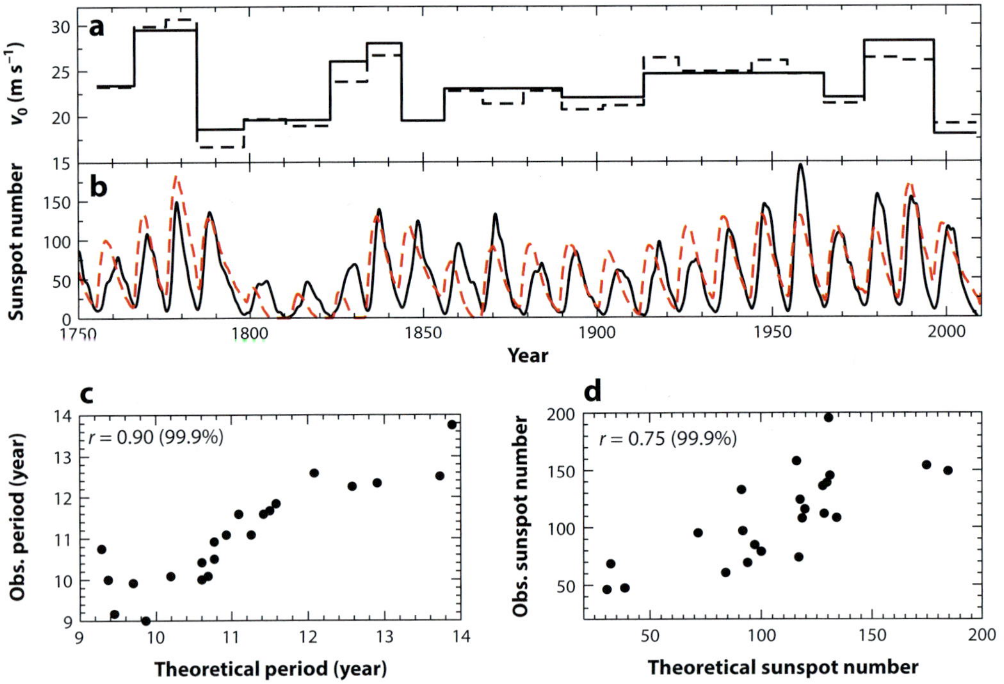

Figure 10

Impact of forced fluctuations of the meridional flow in the kinematic Babcock-Leighton dynamo model by Chatterjee et al. (2004). The (*a*) piecewise-constant imposed meridional flow variations lead to (*b*) a time series of pseudosunspot emergences (*red*) whose (*c*) amplitudes and (*d*) durations correlate quite well with the observed sunspot number time series (*black* in panel *b*). Reproduced from Karak (2010) with permission.

recovered, including the observed relationship between cycle amplitude and period (e.g., Dikpati et al. 2010, Karak & Choudhuri 2011, Nandy et al. 2011). **Figure 10**, taken from Karak (2010) offers a particularly impressive example. Working with the Chatterjee et al. (2004) kinematic model, the meridional flow speed is artificially adjusted from one sunspot cycle to the next as plotted in **Figure 10*a***, which leads to the cyclic behavior plotted as a red dashed line in **Figure 10*b***. The correlation with the observed sunspot number time series (in *black*) is excellent (e.g., **Figure 10*c*,*d***). This appears to be a robust result in that a similar behavior materializes in other formulations of Babcock-Leighton dynamos (e.g., Passos & Lopes 2008, Lopes & Passos 2009). Another sunspot feature reproduced by such models is the observed alternance between higher-than-average and lower-than-average cycle amplitude, a pattern that can arise naturally from the time delay between poloidal field production at the surface and its shearing by differential rotation upon being transported in the tachocline by the meridional flow (see Charbonneau 2001, Charbonneau et al. 2007 and references therein).

Because the Babcock-Leighton mechanism is characterized by a lower operating threshold on the internal toroidal field strength, the associated dynamo can shut off if the field falls below this threshold in response to perturbations of whatever origin. This feature is also shared by dynamos relying on the instability of magnetic flux tubes in the tachocline, mentioned in Section 4.6. Recovering "normal" cyclic behavior then requires an additional source to kick-start the dynamo once again (see, e.g., Schmitt et al. 1996, Ossendrijver 2000a, Charbonneau et al. 2004). Unlike the nonlinear amplitude modulation of mean-field dynamos discussed in Section 4.7, here the dynamo

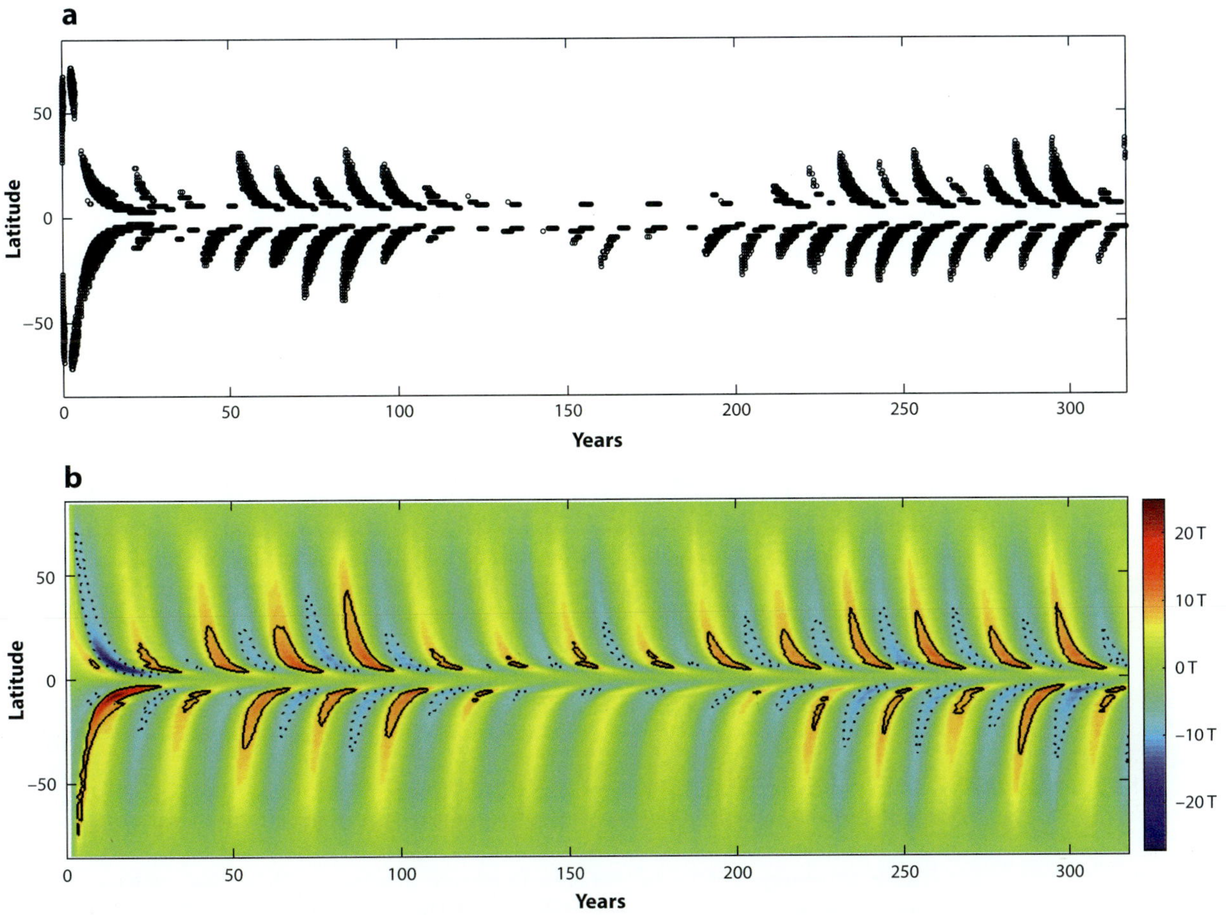

Figure 11

A grand minimum in the stochastically forced kinematic Babcock-Leighton dynamo model by Passos et al. (2014). (*a*) The time-latitude distribution of individual sunspot emergences in the model; (*b*) the time-latitude distribution of the deep toroidal magnetic field giving rise to these emergences. The first ~50 years of these simulations are spent relaxing the initial condition. Note the residual activity during the minimum epoch going from 120 to 200 years and the associated significant north-south asymmetry in emergences. Figure provided by D. Passos (CENTRA/IST).

is switching between two distinct dynamical states, a process known as intermittency. **Figure 11** shows an example taken from Passos et al. (2014). Intermittency is driven here by stochastic forcing of two source terms, one of the Babcock-Leighton type, the other a conventional α-effect operating throughout the convection zone. Similar behavior has also been obtained by simultaneous stochastic forcing of the meridional flow and Babcock-Leighton source term (Choudhuri & Karak 2012).

All of these results are predicated on a meridional flow pattern characterized by a single flow cell per hemisphere. Multiple cells stacked in depth and/or latitude can have a strong impact on the operation of these models (see, e.g., Jouve & Brun 2007). However, recent modeling work suggests that what matters primarily is the presence of an equatorward flow near the base of the convection zone and an efficient mechanism, or combination of mechanisms, to carry the surface magnetic field there (Hazra et al. 2014, Pipin & Kosovichev 2013). Much remains to be learned

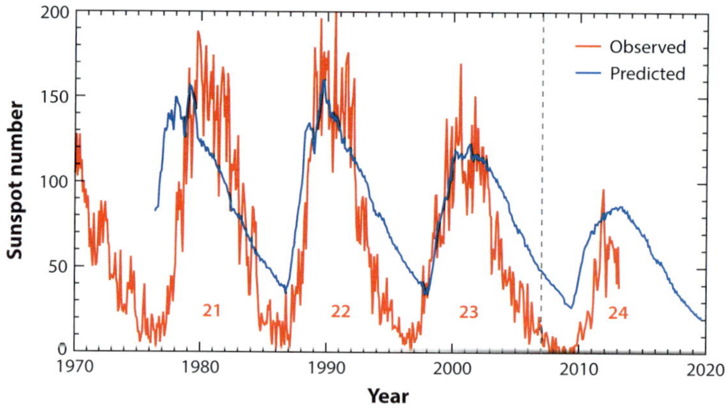

Figure 12

Prediction of the timing and amplitude of activity cycle 24 according to the dynamo-based scheme by Choudhuri et al. (2007). The observed monthly sunspot number (SSN) is plotted in red, and the blue curve shows the model's prediction after internal parameter calibration on the 1975–2007 interval. This prediction for cycle 24 was actually made in 2007 (*vertical dotted line*) in the descending phase of cycle 23. The high SSN value at cycle minima is a consequence of the excessive cycle overlap characterizing the underlying dynamo model; it affects the prediction for the timing of cycle minima, but has no impact here on prediction of cycle peak amplitude. Figure produced from numerical data provided by Arnab Rai Choudhuri (IIS, Bangalore).

also on the magnetic backreaction on the meridional flow and how it affects dynamo action in flux transport dynamos in general (Rempel 2006b, Karak & Choudhuri 2012, Passos et al. 2012).

5.4. Cycle Prediction

Because the solar magnetic cycle modulates the frequency of all geoeffective solar eruptive phenomena, prediction of its characteristics is considered an important aspect of space weather forecasting (Petrovay 2010). In Babcock-Leighton dynamo models, the evolving surface magnetic field at cycle n acts as a source of the toroidal field for cycle $n + 1$ after being transported in the interior by the meridional flow. This implies that some prediction of the characteristics of cycle $n + 1$ could be produced from magnetic field observations at cycle n. This idea forms the basis of the so-called precursor methods (see Petrovay 2010, his section 2).

Dikpati et al. (2006) have developed a novel, dynamo-based prediction scheme in which a parameterization of the zonally averaged observed surface magnetic flux is used as a surface boundary condition to a modified Babcock-Leighton model (see also Dikpati & Gilman 2006). **Figure 12** shows a cycle 24 prediction produced by the similar prediction scheme developed by Choudhuri et al. (2007; see also Jiang et al. 2007). Interestingly, these two closely related dynamo-based approaches have led to markedly distinct predictions for cycle 24, indicating that the prediction is sensitive to modeling details (see also Cameron & Schüssler 2007, Bushby & Tobias 2007). Substantial efforts are underway to understand and improve the predictive capabilities of this type of forecasting scheme (e.g., Yeates et al. 2008, Karak & Nandy 2012, Dikpati & Anderson 2012, Muñoz-Jaramillo et al. 2013).

6. CONCLUDING REMARKS

These are very exciting times for solar dynamo theory. The past two decades have witnessed a wide diversification of dynamo mechanisms and solar cycle models, propelled and constrained by the ever more-detailed view of solar internal flows provided by helioseismology and of ever finer surface magnetic field observations. The subtle role of magnetic helicity in constraining large-scale dynamo action has been recognized and by now largely clarified. In parallel, spectacular advances in computing power and algorithmic design have jointly contributed to the development of global MHD simulations of solar convection and dynamo action, which can reproduce more and more features of the real solar cycle. These represent outstanding virtual laboratories within which physical insight gained from simpler modeling approaches can be tested and validated, and these in turn can guide the design of improved global simulations. Understanding the workings of the solar dynamo remains the ultimate goal, and in this respect the words of computer scientist R.W. Hamming (1962) are as timely now as they were over fifty years ago: "The purpose of computing is insight, not numbers."

FUTURE ISSUES

1. **How critical are subgrid models?** Empirical evidence accumulated to date indicates that this seemingly secondary and highly technical computational issue is actually of great importance for magnetic self-organization. Detailed intercomparison of similar global MHD simulations using distinct subgrid formulations is essential to disentangle which aspects of magnetic self-organization are critically dependent on such computational details.
2. **What sets the dynamo period?** It is sobering to reflect upon the fact that over 150 years after the discovery of the solar cycle, the mechanism(s) setting its period has not yet been pinned down with confidence. In mean-field and mean-field-like models, the possibility to adjust parameter values not well-constrained observationally often makes it possible to obtain solar-like cycle periods no matter which dynamo modeling framework is adopted. This is where observations of magnetic activity cycles in stars other than the Sun can offer powerful discriminants. The importance of sustained monitoring of solar-type stars, such as those carried out by the Mt. Wilson program for decades, cannot be overstated.
3. **Is the tachocline important in driving the dynamo?** Various lines of evidence point to the tachocline as an important agent in achieving magnetic self-organization on large spatial scales. This is a towering computational and physical challenge given the expected strong anisotropy of turbulence therein, the interactions of waves excited by convective overshoot with large-scale flow and magnetic fields, the development of dynamical instabilities, and possible coupling to an underlying fossil magnetic field.
4. **Is the Babcock-Leighton mechanism crucial?** The Babcock-Leighton mechanism is clearly observed to operate at the solar surface; however, it remains unclear whether it represents a mere by-product or an essential element of the dynamo loop. This question may perhaps be answered through an accurate determination of the magnetic flux budget of the solar polar caps, which represents a challenging observational undertaking.
5. **What triggers grand minima?** As reviewed in Sections 4.7 and 5.3, many "simple" mean-field or mean-field-like dynamo models can produce Maunder-Minimum-like episodes of interrupted cyclic behavior, or strongly suppressed magnetic cycle amplitude, through distinct combinations of deterministic and stochastic forcing mechanisms. Most of these scenarios were developed in the context of mean-field or mean-field-like dynamo models, and the hunt is now on to produce similar behavior in dynamically consistent MHD simulations of large-scale magnetic cycles.

DISCLOSURE STATEMENT

The author is not aware of any affiliations, memberships, funding, or financial holdings that might be perceived as affecting the objectivity of this review.

ACKNOWLEDGMENTS

I wish to thank the many colleagues who kindly answered my queries, sent preprints, or provided numerical data or graphical material for inclusion in this review; I am particularly indebted to Arnab Choudhuri, Gustavo Guerrero, Petri Käpylä, Bidya Karak, Mark Miesch, and Dário Passos, as well as my graduate students Patrice Beaudoin, Caroline Dubé, Nicolas Lawson, Alexandre Lemerle and Jean-Francois Cossette for providing custom made figures "on order." The preparation of this review has been supported in part by the Natural Science and Engineering Research Council of Canada.

LITERATURE CITED

Babcock HW. 1961. *Ap. J.* 133:572–87
Balbus SA, Bonart J, Latter HN, Weiss NO. 2009. *MNRAS* 400:176–82
Barnes G, MacGregor KB, Charbonneau P. 1998. *Ap. J. Lett.* 498:L169–72
Baumann I, Schmitt D, Schüssler M, Solanki SK. 2004. *Astron. Astrophys.* 426:1075–91
Beaudoin P, Charbonneau P, Racine É, Smolarkiewicz PK. 2013. *Solar Phys.* 282:335–60
Beer J. 2000. *Space Sci. Rev.* 94:53–66
Beer J, Tobias SM, Weiss NO. 1998. *Solar Phys.* 181:237–49
Benevolenskaya EE. 1995. *Solar Phys.* 161:1–8
Benevolenskaya EE. 1998. *Ap. J. Lett.* 509:49–52
Berger MA, Field GB. 1984. *J. Fluid. Mech.* 147:133–48
Berger MA, Ruzmaikin A. 2000. *J. Geophys. Res.* 105(A5):10481–90
Blackman EG, Brandenburg A. 2002. *Ap. J.* 579:359–73
Braithwaite J, Nordlund Å. 2006. *Astron. Astrophys.* 450:1077–95
Brandenburg A. 2001. *Ap. J.* 550:824–40
Brandenburg A. 2005. *Ap. J.* 625:539–47
Brandenburg A, Dobler W. 2001. *Astron. Astrophys.* 369:329–38
Brandenburg A, Subramanian K. 2005. *Phys. Rep.* 417:1–209
Brandenburg A, Tuominen I, Nordlund Å, Pulkkinen P, Stein RF. 1990. *Astron. Astrophys.* 232:277–91
Braun DC, Fan Y. 1998. *Ap. J. Lett.* 508:L105–8
Brooke JM, Moss D, Phillips A. 2002. *Astron. Astrophys.* 395:1013–22
Brown BP, Browning MK, Brun AS, Miesch MS, Toomre J. 2010. *Ap. J.* 711:424–38
Brown BP, Miesch MS, Browning MK, Brun AS, Toomre J. 2011. *Ap. J.* 731:69
Browning MK, Miesch MS, Brun AS, Toomre J. 2006. *Ap. J. Lett.* 648:L157–60
Brummell NH, Hurlburt NE, Toomre J. 1996. *Ap. J.* 473:494–513
Brun AS, Miesch MS, Toomre J. 2004. *Ap. J.* 614:1073–98
Brun AS, Miesch MS, Toomre J. 2011. *Ap. J.* 742:79
Bushby PJ. 2006. *MNRAS* 371:772–80
Bushby PJ, Tobias SM. 2007. *Ap. J.* 661:1289–96
Busse FH. 2002. *Phys. Fluids* 14:1301–14
Caligari P, Moreno-Insertis F, Schüssler M. 1995. *Ap. J.* 441:886–902
Cally PS, Dikpati M, Gilman PA. 2003. *Ap. J.* 582:1190–205
Cameron R, Schüssler M. 2007. *Ap. J.* 659:801–11
Cattaneo F. 1999. *Ap. J. Lett.* 515:L39–42
Cattaneo F, Emonet T, Weiss NO. 2003. *Ap. J.* 515:1183–98
Cattaneo F, Hughes DW. 1996. *Phys. Rev. E* 54:4532–35

Cattaneo F, Hughes DW. 2006. *J. Fluid Mech.* 553:401–18
Charbonneau P. 2001. *Solar Phys.* 199:385–404
Charbonneau P. 2010. *Living Rev. Solar Phys.*7:lrsp-2010-3
Charbonneau P. 2013. *Solar and Stellar Dynamos*. Berlin: Springer. 237 pp.
Charbonneau P, Beaubien G, St-Jean C. 2007. *Ap. J.* 658:657–62
Charbonneau P, Blais-Laurier G, St-Jean C. 2004. *Ap. J. Lett.* 616:L183–86
Charbonneau P, MacGregor KB. 1996. *Ap. J. Lett.* 473:L59–62
Charbonneau P, MacGregor KB. 2001. *Ap. J.* 559:1094–107
Charbonneau P, St-Jean C, Zacharias P. 2005. *Ap. J.* 619:613–22
Chatterjee P, Guerrero G, Brandenburg A. 2011. *Astron. Astrophys.* 525:A5
Chatterjee P, Nandy D, Choudhuri AR. 2004. *Astron. Astrophys.* 427:1019–30
Choudhuri AR. 1990. *Ap. J.* 355:733–44
Choudhuri AR, Chatterjee P, Jiang J. 2007. *Phys. Rev. Lett.* 98:131103
Choudhuri AR, Karak BB. 2012. *Phys. Rev. Lett.* 109:171103
Choudhuri AR, Schüssler M, Dikpati M. 1995. *Astron. Astrophys.* 303:L29–32
Clune TL, Elliot JR, Glatzmaier GA, Miesch MS, Toomre J. 1999. *Parallel Comput.* 25:361
Cossette JF, Charbonneau P, Smolarkiewicz PK. 2013. *Ap. J. Lett.* 777:L29
Courvoisier A, Hughes DW, Tobias SM. 2006. *Phys. Rev. Lett.* 96(3):034503
Courvoisier A, Hughes DW, Tobias SM. 2009. *J. Fluid Mech.* 627:403–21
Davidson PA. 2001. *An Introduction to Magnetohydrodynamics*. Cambridge: Cambridge Univ. Press. 431 pp.
de Wijn AG, Stenflo JO, Solanki SK, Tsuneta S. 2009. *Space Sci. Rev.* 144:275–315
Démoulin P, Berger MA. 2003. *Solar Phys.* 215:203–15
Dikpati M, Anderson JL. 2012. *Ap. J.* 756:20
Dikpati M, Cally PS, Gilman PA. 2004. *Ap. J.* 610:597–615
Dikpati M, Charbonneau P. 1999. *Ap. J.* 518:508–20
Dikpati M, de Toma G, Gilman PA. 2006. *Geophys. Res. Lett.* 33:L05102
Dikpati M, Gilman PA. 2001. *Ap. J.* 610:597–615
Dikpati M, Gilman PA. 2006. *Ap. J.* 649:498–514
Dikpati M, Gilman PA, de Toma G, Ulrich RK. 2010. *Geophys. Res. Lett.* 37:L14107
Dikpati M, Gilman PA, Rempel M. 2003. *Ap. J.* 596:680–97
D'Silva S, Choudhuri AR. 1993. *Astron. Astrophys.* 272:621–33
Dubé C, Charbonneau P. 2013. *Ap. J.* 775:69
Durney BR. 1995. *Solar Phys.* 160:213–35
Eddy JA. 1976. *Science* 192(4245):1189–202
Fan Y. 2009. *Living Rev. Solar Phys.* 6:lrsp-2009-4
Fan Y, Fisher GH, DeLuca EE. 1993. *Ap. J.* 405:390–401
Ferriz-Mas A, Schmitt D, Schüssler M. 1994. *Astron. Astrophys.* 289:949–56
Fletcher ST, Broomhall A-M, Salabert D, Basu S, Chaplin WJ, et al. 2010. *Ap. J. Lett.* 718:L19–22
Foukal PF, Fröhlich C, Spruit H, Wigley TML. 2006. *Nature* 443(7109):61–166
Fröhlich C. 2009. *Astron. Astrophys.* 501:L27–30
Fröhlich C, Lean J. 2004. *Astron. Astrophys. Rev.* 12:273–320
Ghizaru M, Charbonneau P, Smolarkiewicz PK. 2010. *Ap. J. Lett.* 715:L133–37
Gilman PA. 1983. *Ap. J. Suppl.* 53:243–68
Gilman PA, Dikpati M, Miesch MS. 2007. *Ap. J. Suppl.* 170:203–27
Gilman PA, Fox PA. 1997. *Ap. J.* 484:439–54
Glatzmaier GA. 1984. *J. Comput. Phys.* 55:461–84
Glatzmaier GA. 1985. *Ap. J.* 291:300–307
Grinstein FF, Margolin LG, Rider WJ, eds. 2007. *Implicit Large Eddy Simulation: Computing Turbulent Fluid Dynamics*. Cambridge: Cambridge Univ. Press. 546 pp.
Guerrero G, de Gouveia Dal Pino EM. 2008. *Ap. J.* 485:267–73
Guerrero G, Smolarkiewicz PK, Kosovichev A, Mansour N. 2013. *Ap. J.* 779:176
Hale GE, Ellerman F, Nicholson SB, Joy AH. 1919. *Ap. J.* 49:153–78
Hamming RW. 1962. *Numerical Methods for Scientist and Engineers*. New York: McGraw-Hill. 721 pp.

Hathaway DH. 1996. *Ap. J.* 460:1027–33
Hathaway DH. 2010. *Living Rev. Solar Phys.* 7:lrsp-2010-1
Hathaway DH. 2012. *Ap. J.* 760:84
Hazra G, Karak BB, Choudhuri AR. 2013. *Ap. J.* 782:93
Howe R. 2009. *Living Rev. Solar Phys.* 6:lrsp-2009-1
Hoyng P. 2003. In *Advances in Nonlinear Dynamos*, ed. A Ferriz-Mas, M Nuñez, pp. 1–36. London: Taylor & Francis
Hoyt DV, Schatten K. 1998. *Solar Phys.* 179:189–219
Hubbard A, Brandenburg A. 2012. *Ap. J.* 748:51
Hubbard A, Del Sordo F, Käpylä P, Brandenburg A. 2009. *MNRAS* 398:1891–99
Hughes DW, Cattaneo F. 2008. *J. Fluid Mech.* 594:445–61
Hughes DW, Proctor MRE, Cattaneo F. 2011. *MNRAS* 414:L45–49
Jiang J, Cameron R, Schmitt D, Isik E. 2013. *Astron. Astrophys.* 553:A128
Jiang J, Cameron R, Schmitt D, Schüssler M. 2009. *Ap. J. Lett.* 693:L96–99
Jiang J, Chatterjee P, Choudhuri AR. 2007. *MNRAS* 381:1527–42
Jouve L, Brun AS. 2007. *Astron. Astrophys.* 474:239–50
Käpylä PJ, Korpi MJ, Brandenburg A. 2009. *Astron. Astrophys.* 500:633–46
Käpylä PJ, Korpi MJ, Brandenburg A, Mitra D, Tavakol R. 2010. *Astron. Nachr.* 331:73
Käpylä PJ, Korpi MJ, Ossendrijver MAJH, Stix M. 2006a. *Astron. Astrophys.* 455:401–12
Käpylä PJ, Korpi MJ, Tuominen I. 2006b. *Astron. Nachr.* 327(9):884–94
Käpylä PJ, Mantere MJ, Brandenburg A. 2012. *Ap. J. Lett.* 755:L22
Käpylä PJ, Mantere MJ, Cole E, Warnecke J, Brandenburg A. 2013. *Ap. J.* 778:41
Karak BB. 2010. *Ap. J.* 724:21021–29
Karak BB, Choudhuri AR. 2011. *MNRAS* 410:1503–12
Karak BB, Choudhuri AR. 2012. *Solar Phys.* 278:137–48
Karak BB, Nandy D. 2012. *Ap. J. Lett.* 761:L13
Kim E-J, MacGregor KB. 2003. *Ap. J.* 588:645–54
Kitchatinov LL, Olemskoy SV. 2012. *Solar Phys.* 276:3–17
Kitchatinov LL, Rüdiger G. 1995. *Astron. Astrophys.* 299:446–52
Kleeorin N, Kuzanyan K, Moss D, et al. 2003. *Astron. Astrophys.* 409:1097–105
Krause F, Rädler K-H. 1980. *Mean-Field Magnetohydrodynamics and Dynamo Theory*. Oxford, UK: Pergamon. 271 pp.
Küker M, Arlt R, Rüdiger G. 1999. *Astron. Astrophys.* 343:977–82
Küker M, Rüdiger G, Schultz M. 2001. *Astron. Astrophys.* 374:301–8
Kusano K, Maeshiro T, Yokoyama T, Sakurai T. 2002. *Ap. J.* 577:501–12
Leighton RB. 1969. *Ap. J.* 156:1–26
Li LH, Sofia S. 2001. *Ap. J.* 549:1204–11
Lopes I, Passos D. 2009. *Solar Phys.* 257:1–12
Low BC. 2001. *J. Geophys. Res.* 106(A11):25141–64
Mason J, Hughes DW, Tobias SM. 2002. *Ap. J. Lett.* 580:L89–92
McCracken KG, Beer J, Steinhilber F, Abreu J. 2013. *Solar Phys.* 286(2):609–27
Miesch MS. 2005. *Living Rev. Solar Phys.* 2:lrsp-2005-1
Miesch MS. 2007. *Ap. J. Lett.* 658:L131–34
Miesch MS, Brun AS, Toomre J. 2006. *Ap. J.* 641:618–25
Miesch MS, Gilman PA, Dikpati M. 2007. *Ap. J. Suppl.* 168:337–61
Miesch MS, Toomre J. 2009. *Annu. Rev. Fluid Mech.* 41:317–40
Moffatt HK. 1978. *Magnetic Field Generation in Electrically Conducting Fluids*. Cambridge: Cambridge Univ. Press. 343 pp.
Moss D, Brooke JM. 2000. *MNRAS* 315:521–33
Muñoz-Jaramillo A, Balmaceda LA, DeLuca EE. 2013. *Phys. Rev. Lett.* 111(4):041106
Muñoz-Jaramillo A, Nandy D, Martens PCH, Yeates AR. 2010. *Ap. J. Lett.* 720:L20–25
Muñoz-Jaramillo A, Nandy D, Martens PCH. 2011. *Ap. J. Lett.* 727:L23–26
Mursula K, Ziegler B, Vilppola JH. 2003. *Solar Phys.* 212:201–7

Nandy D, Choudhuri AR. 2001. *Ap. J.* 551:576–85
Nandy D, Muñoz-Jaramillo A, Martens PCH. 2011. *Nature* 471(7336):80–82
Nelson NJ, Brown BP, Brun AS, Miesch MS, Toomre J. 2011. *Ap. J. Lett.* 739:L38
Nelson NJ, Brown BP, Brun AS, Miesch MS, Toomre J. 2013. *Ap. J.* 762:73
Nelson NJ, Brown BP, Brun AS, Miesch MS, Toomre J. 2014. *Solar Phys.* 289:441–58
Ossendrijver MAJH. 2000a. *Astron. Astrophys.* 359:364–72
Ossendrijver MAJH. 2000b. *Astron. Astrophys.* 359:1205–10
Ossendrijver MAJH. 2003. *Astron. Astrophys. Rev.* 11:287–367
Ossendrijver MAJH, Stix M, Brandenburg A. 2001. *Astron. Astrophys.* 376:713–26
Ossendrijver MAJH, Stix M, Brandenburg A, Rüdiger G. 2002. *Astron. Astrophys.* 394:735–45
Parker EN. 1955a. *Ap. J.* 121:491–507
Parker EN. 1955b. *Ap. J.* 122:293–314
Parker EN. 1979. *Cosmical Magnetic Fields*. Oxford, UK: Clarendon. 858 pp.
Parker EN. 1993. *Ap. J.* 408:707–19
Parnell CE, DeForest CE, Hagenaar HJ, et al. 2009 *Ap. J.* 698:75–82
Passos D, Charbonneau P, Beaudoin P. 2012. *Solar Phys.* 279:1–22
Passos D, Lopes I. 2008. *Ap. J.* 686:1420–25
Passos D, Nandy D, Hazra S, Lopes I. 2014. *Astron. Astrophys.* 563:A18
Petrovay KP. 2010. *Living Rev. Solar Phys.* 7:lrsp-2010-6
Petrovay KP, Kerekes A. 2004. *MNRAS* 351:L59–62
Pipin VV, Kosovichev AG. 2011. *Ap. J.* 738:104
Pipin VV, Kosovichev AG. 2013. *Astron. Astrophys.* 776:36
Pitts E, Tayler RJ. 1985. *MNRAS* 216:139–54
Pouquet A, Frisch U, Leorat J. 1976. *J. Fluid Mech.* 77:321–54
Prusa JM, Smolarkiewicz PK, Wyszogrodzki AA. 2008. *Comp. Fluids* 37:1193–207
Racine É, Charbonneau P, Ghizaru M, Bouchat A, Smolarkiewicz PK. 2011. *Ap. J.* 735:46
Rempel M. 2006a. In *Heliophysics: Plasma Physics of the Local Cosmos*, ed. CJ Schrijver, GL Siscoe, pp. 42–74. Cambridge, UK: Cambridge Univ. Press
Rempel M. 2006b. *Ap. J.* 647:662–75
Rogers TM, Glatzmaier GA. 2006. *Ap. J.* 653:756–64
Rogers TM, MacGregor KB. 2011. *MNRAS* 410:946–62
Rogers TM, MacGregor KB, Glatzmaier GA. 2008. *MNRAS* 387:616–30
Rüdiger G, Elstner D, Ossendrijver MAJH. 2003. *Astron. Astrophys.* 406:15–21
Rüdiger G, Hollerbach R. 2004. *The Magnetic Universe: Geophysical and Astrophysical Dynamos*. Weihheim: Wiley-YCH. 343 pp.
Schmitt D, Schüssler M, Ferriz-Mas A. 1996. *Astron. Astrophys.* 311:L1–4
Schou J, Bogart RS. 1998. *Ap. J. Lett.* 504:L131–34
Schrijver CJ, Title AM, van Ballegooijen AA, Hagenaar HJ, Shine RA. 1997. *Ap. J.* 487:424–36
Schrinner M, Rädler K-H, Schmitt D, Rheinhardt M, Christensen UR. 2007. *Geophys. Astrophys. Fluid Dyn.* 101(2):81–116
Simard C, Charbonneau P, Bouchat A. 2013. *Ap. J.* 768(1):16
Smolarkiewicz PK, Charbonneau P. 2013. *J. Comput. Phys.* 236:608–23
Solanki SK, Inhester B, Schüssler M. 2006. *Rep. Prog. Phys.* 69:563–668
Spruit HC. 1999. *Astron. Astrophys.* 349:189–202
Steenbeck M, Krause F. 1969. *Astron. Nachr.* 291:49–84
Stein RF, Nordlund Å. 2006. *Ap. J.* 642:1246–55
Stix M. 1976. *Astron. Astrophys.* 47:243–54
Strugarek A, Brun AS, Zahn J-P. 2011. *Astron. Astrophys.* 532:A34
Talon S, Kumar P, Zahn J-P. 2002. *Ap. J. Lett.* 574:L175–78
Thibault K, Charbonneau P, Crouch AD. 2012. *Ap. J.* 757:187
Tobias SM. 1996. *Ap. J.* 467:870–80
Tobias SM. 1997. *Astron. Astrophys.* 322:1007–17
Tobias SM, Brummell NH, Clune TL, Toomre J. 2001. *Ap. J.* 549:1183–203

Tobias SM, Cattaneo F, Brummell NH. 2011. *Ap. J.* 728:153
Ulrich RK. 2010. *Ap. J.* 725:658–69
Usoskin IG. 2013. *Living Rev. Solar Phys.* 10:lrsp-2013-1
Usoskin IG, Solanki SK, Kovaltsov GA. 2007. *Astron. Astrophys.* 471:301–9
Vögler A, Schüssler M. 2007. *Astron. Astrophys.* 465:L43–46
Wang Y-M, Nash AG, Sheeley NR Jr. 1989. *Science* 245:712–18
Wang Y-M, Sheeley NR Jr. 1991. *Ap. J.* 375:761–70
Wang Y-M, Sheeley NR Jr, Nash AG. 1991. *Ap. J.* 383:431–42
Warnecke J, Brandenburg A, Mitra D. 2011. *Astron. Astrophys.* 534:A11
Warnecke J, Käpylä P, Mantere MJ, Brandenburg A. 2012. *Solar Phys.* 280:299–319
Yeates AR, Nandy D, Mackay DH. 2008. *Ap. J.* 673:544–56
Yoshimura H. 1975. *Ap. J.* 201:740–48
Zhao J, Bogart RS, Kosovichev AG, Duvall TL Jr, Hartlep T. 2013. *Ap. J. Lett.* 774:L29
Ziegler U, Rüdiger G. 2003. *Astron. Astrophys.* 394:735–45

The Evolution of Galaxy Structure Over Cosmic Time

Christopher J. Conselice

Center for Astronomy and Particle Theory, School of Physics and Astronomy, University of Nottingham, Nottingham, NG7 2UH, United Kingdom; email: conselice@nottingham.ac.uk

Annu. Rev. Astron. Astrophys. 2014. 52:291–337

First published online as a Review in Advance on June 16, 2014

The *Annual Review of Astronomy and Astrophysics* is online at astro.annualreviews.org

This article's doi:
10.1146/annurev-astro-081913-040037

Keywords

galaxy evolution, galaxy morphology

Abstract

I present a comprehensive review of the evolution of galaxy structure in the Universe from the first galaxies currently observable at $z \sim 6$ down to galaxies observable in the local Universe. Observed changes in galaxy structures reveal formation processes that only galaxy structural analyses can provide. This pedagogical review provides a detailed discussion of the major methods used to study galaxies morphologically and structurally, including the well-established visual method for morphology; Sérsic fitting to measure galaxy sizes and surface brightness profile shapes; and nonparametric structural methods [such as the concentration (C), asymmetry (A), clumpiness (S) (CAS) method and the Gini/M_{20} parameters, as well as newer structural indices]. These structural indices measure fundamental properties of galaxies, such as their scale, star-formation rate, and ongoing merger activity. Extensive observational results demonstrate how broad galaxy morphologies and structures change with time up to $z \sim 3$, from small, compact and peculiar systems in the distant Universe to the formation of the Hubble sequence, dominated by spirals and ellipticals. Structural methods accurately identify galaxies in mergers and allow measurements of the merger history out to $z \sim 3$. I depict properties and evolution of internal structures of galaxies, such as bulges, disks, bars, and at $z > 1$ large star-forming clumps. I describe the structure and morphologies of host galaxies of active galactic nuclei and starbursts/submillimeter galaxies, along with how morphological galaxy quenching occurs. The role of environment in producing structural changes in galaxies over cosmic time is also discussed. Galaxy sizes can also change with time, with measured sizes up to a factor of 2–5 smaller at high redshift at a given stellar mass. I conclude with a discussion of how the evolving trends, in sizes, structures, and morphologies, reveal the formation mechanisms behind galaxies and provides a new and unique way to test theories of galaxy formation.

1. INTRODUCTION

Galaxy structure is one of the fundamental ways in which galaxy properties are described and by which galaxy evolution is inferred. There is a long history of the development of this idea, which began with the earliest observations of galaxies, and continues up to the modern day as one of the major ways we study galaxies. This review gives a detailed description of the progress made up to late 2013 in using galaxy structure to understand galaxy formation and evolution. It is meant to be used as a primer for obtaining basic information from galaxy structures, including how they are measured and applied through cosmic time.

The introduction to this review first gives an outline of the basic events in the history of galaxy morphology and structure analyses, whereas the second part of the introduction describes how galaxy structure fits into the general picture of galaxy formation. I also give a detailed description of the goals of this review at the end of the introduction.

1.1. Historical Background

Galaxy morphology has a long history, one that even predates the time when we finally knew galaxies were extragalactic. When objects that today we call galaxies were first observed, what clearly distinguished them from stars was their resolved structure. Since this time, structure and morphology has remained one of the most common ways galaxies are described and studied. Initially this involved visual impressions of galaxy forms. This has now been expanded to include quantitative methods to measure galaxy structures all the way back to the earliest galaxies we can currently see.

The first published descriptions of galaxy structure and morphology predate the telescopic era. For example, the Andromeda nebula was described as a "small cloud" by the Persian astronomer Abd al-Rahman al-Sufi in the tenth century (Kepple & Sanner 1998). The study of galaxies remained descriptive until the late twentieth century, although more and more detail was resolved as technology improved. As a result, for about 150 years the science of galaxies was necessarily restricted to cataloging and general descriptions of structure, with notable achievements by Messier and William and John Herschel, who located galaxies or "nebula" by their resolved structure as seen by eye. Even before photography revolutionized the study of galaxies some observers, such as William Parsons, the third Earl of Rosse, noted that the nebulae have a spiral morphology and first used this term to describe galaxies, most notably and famously in the case of M51.

It was, however, the advent of photography that allowed astronomers to study in earnest the morphologies and structures of external galaxies. The most notable early schemes were developed by Wolf (1908) and Lundmark (1926), among others. This ultimately led to what is today called the Hubble classification, which was published in essentially its modern form by Hubble (1926), with the final "Hubble Tuning Fork" established by Hubble (1936) and Sandage (1961). The basic Hubble sequence (**Figure 1**) consists of two main types of galaxies, ellipticals and spirals, with a further division of spirals into those with bars and those without bars. Hubble, and the astronomers who followed him, could classify most nearby bright galaxies in terms of this system.

The development of morphological classification methods continued into the twentieth century, with newer methodologies based solely on visual impressions. For example, de Vaucouleurs (1959) developed a revised version of the Hubble sequence, which included criteria such as bars, rings, and other internal features that were prominent on photographic plates of galaxies. Likewise, van den Bergh (1960, 1976), and later Elmegreen & Elmegreen (1987) developed a system to classify galaxies based on the form of spiral arms and the apparent clumpiness of the light in these arms.

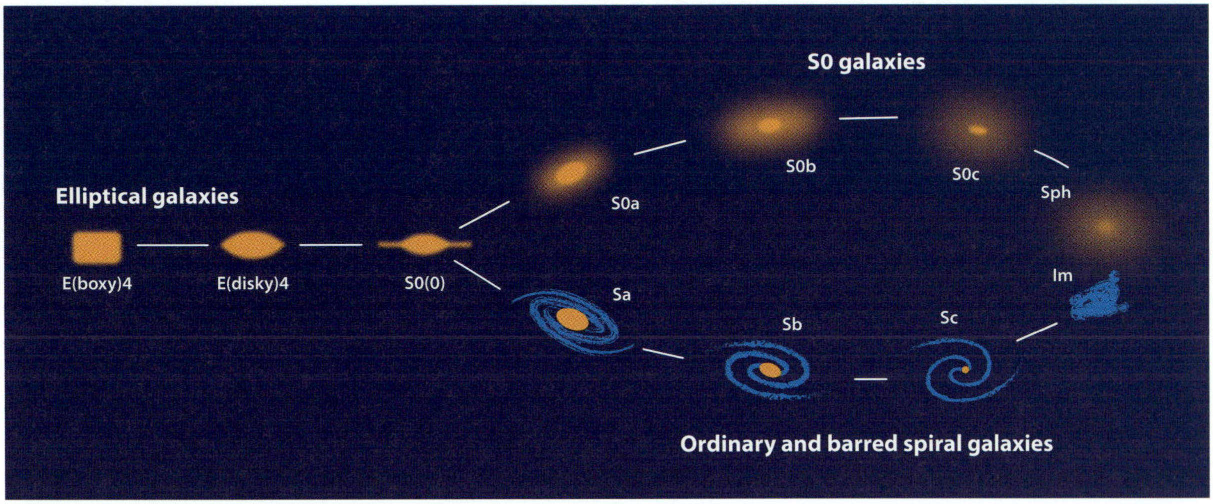

Figure 1

A modern form of the Hubble sequence showing the sequence of ellipticals and S0s and the "tuning fork" in spirals. The elliptical sequence is determined by the overall shape of the galaxy, whereas spiral classifications are divided into different types (a–c), depending on how wound-up the spiral arms are, how large the bulge relative to the disk is, and how smooth the spiral arms in the spirals arm are. The tuning fork is the differential between spirals with and without bars. Also shown is the extension of this sequence to dwarf spheroidal galaxies and irregular galaxies, both of which are lower-mass systems. (Reprinted from Kormendy & Bender 2012 with permission).

Although it is important to classify galaxies visually, and all classification systems have some use, as all features should be explained by physics, it is not obvious which structural features of galaxies are fundamental to their formation history. Ultimately morphology and structure need to be proven useful for understanding galaxies, as there is now extensive use of photometric and spectroscopic methods permitting measurements of perhaps more fundamental measures of stellar populations and dust/gas properties in galaxies. Along these lines, at roughly the same time as when progressively complicated classification systems were developed, astronomers such as Holmberg (1958) established that physical properties of nearby galaxies correlate with morphology in a broad context. Holmberg (1958) found that ellipticals are typically massive and red, and show little star formation, whereas spirals are less massive and bluer and have evidence for ongoing star formation. This quantitatively expands into other physical parameters as well (e.g., Roberts 1963, Roberts & Haynes 1994, Conselice 2006a, Allen et al. 2006). It is also well known that this segregation of morphology in the local Universe provides an important clue for understanding the physics of galaxy formation, especially as local environment is found to strongly correlate with a galaxy's morphology (e.g., Dressler 1984; see Section 4.7).

A revolution in morphological and structurally studies came about with the advent of photometric photometry, and especially the later use of charged coupled devices (CCDs), which made detailed quantitative measurements of light distributions in galaxies possible. The first major contribution from this type of work was by de Vaucouleurs (1948), who used photometry to show that the light profiles of what we would identify today as massive ellipticals all follow roughly the same fundamental light distribution, known as the de Vaucouleurs profile.

This was later expanded by others, most notably Sérsic (1963), who demonstrated that a more general form of light distribution matched galaxy-light profiles with disks having exponential light profiles, whereas the light distribution within massive ellipticals generally follows the de Vaucouleurs profile. This has led to a large industry in measuring the light profiles of galaxies in the nearby and distant Universe that continues today (Section 2.2).

During the 1970s and 1980s the study of galaxy structure expanded to include the decomposition of galaxy light into bulge and disk profiles (e.g., Kormendy 1977, Caon et al. 1993, Graham & Guzmán 2003) as well as features such as bars, rings, and lenses (e.g., Kormendy 1979, de Vaucouleurs et al. 1991). The three-dimensional structure of disk galaxies was investigated (e.g., van der Kruit & Searle 1982) as well as were detailed studies of bulges and disks in spiral systems (e.g., de Jong 1996, Peletier & Balcells 1996). We also now know there is great diversity in elliptical galaxy internal structures (e.g., Caon et al. 1993, Graham & Guzmán 2003, Kormendy et al. 2009).

Similar investigations demonstrated that secular evolution within disks can provide an explanation for how bars, rings, and lenses can form (e.g., Kormendy 1979, Combes & Sanders 1981). These effects, not driven by hierarchical galaxy formation, are also likely responsible for the formation of pseudobulges and may drive the formation of central massive black holes (e.g., Kormendy & Kennicutt 2004, Sellwood 2013).

Although there is a large amount of work done on the structures and morphologies of galaxies in the nearby Universe (e.g., see Kormendy et al. 2009, Buta 2013, Graham 2013), it is difficult to investigate more than the very basics of structure and morphology when studying distant galaxies. This is owing to the fact that current technology does not allow us to resolve these distant galaxies in the same detail as we can for closer systems. As such, this review concentrates on the features and properties of galaxy structure that we can measure in distant galaxies and how this reveals how galaxy evolution and formation occurs.

The result of this is that one of the areas where galaxy structure and morphology have made their biggest impact is their ability to measure fundamental properties of distant galaxies that we can compare with nearby galaxies to determine evolution. There are extensive methods for studying galaxy evolution such that galaxy structure analyses are becoming an essential aspect of, and providing unique information on, the history and physics of galaxy assembly, which I detail in this review.

1.2. Galaxy Structure within the Context of Galaxy Formation

We know that there is significant evolution in galaxies over time as the stellar mass density of galaxies evolves rapidly at $1 < z < 3$, with about half of all stellar mass formed by $z = 1$ (e.g., Bundy et al. 2005, Mortlock et al. 2011). We also know that there is a vast diversity of star-formation histories for individual galaxies and that the integrated star-formation rate density in the Universe's history peaks at $z \sim 2.5$ and declines at higher and lower redshifts (e.g., Shapley 2011, Madau & Dickinson 2014 in this volume). However, it is not clear from these observations what are/were the driving forces creating galaxies.

Theory offers several approaches for understanding how galaxies form that detailed studies are starting to probe. We now believe that galaxy formation can happen in a number of ways. This includes in situ star formation in a collapsed galaxy, major and minor mergers, and gas accretion from the intergalactic medium. Galaxy structure and morphology are perhaps the best ways to trace these processes, as I discuss in this review.

Another major question I address in this review is how do the structures and morphologies of galaxies change through cosmic time? Major issues that this topic addresses include the formation history of the Hubble Sequence; whether galaxies form "inside-out" or "outside-in"; how long a galaxy retains its morphology; whether morphology is an invariant quantity in a galaxy over a long cosmic time span, and furthermore what relative roles star formation and merging play in galaxy formation.

The use of the Sérsic profile to describe nearby galaxies is extensive (e.g., Caon et al. 1993, Graham & Guzmán 2003, Ferrarese & Jordán 2006, Kormendy et al. 2009, Graham 2013), and it has more recently been applied to distant galaxies, as I discuss in Section 4.2. For reviews on the use of resolved photometry through surface brightness profiles to study early-type galaxies, see e.g., Kormendy & Djorgovski (1989), Allen et al. (2006), and Simard et al. (2011).

Recently the fitting of galaxy two-dimensional profiles with various forms, such as the Sérsic, exponential, and de Vaucouleurs profiles, is done through the GALFIT code by Peng et al. (2002) as well as GIM2D by Simard et al. (2011). These are simple and quick methods for measuring the light profiles and radii of many galaxies, providing data for understanding the evolution of galaxy structure. This allows for the measurements of different light components at high z, although these codes and other similar ones have limitations, such as a constant ellipticity assumption within a given component, but are sufficient for gross measures of galaxy structure.

2.3. Nonparametric Measurements of Structure

Another more recent measurement technique involves the nonparametric method of measuring galaxy light distributions. Nonparametric methods of measuring galaxy structure began in the photographic era with attempts to quantify the light concentration in galaxies by Morgan (1962), although extensive quantitative measures were not done until the mid-1990s.

The development of methods to measure the light structures of galaxies began in earnest when the first deep images of distant galaxies were obtained with HST (Schade et al. 1995, Abraham et al. 1996), although their use for low-redshift measurements was also noted at about the same time, but in terms of a physical property rather than a descriptive quantity (e.g., Rix & Zaritsky 1995; Conselice 1997; Bershady et al. 2000; Conselice et al. 2000a,b). These early papers show that quantitative galaxy structure correlates with other parameters, such as color and peculiar features indicating mergers or galaxy interactions (e.g., Rix & Zaritsky 1995, Conselice 1997, Conselice et al. 2000a).

At present, the most common methods for measuring galaxy structure in a nonparametric way is through the concentration (C), asymmetry (A), clumpiness (S) (CAS) system (e.g., Conselice 2003; Sections 2.3.1–2.3.3) and through similar parameters (Takamiya 1999; Papovich et al. 2003, 2005; Abraham et al. 2003; Lotz et al. 2004; Law et al. 2007; Scarlata et al. 2007, Freeman et al. 2013). These parameters are designed to capture the major features of the underlying structures of these galaxies, but in a way that does not involve assumptions about the underlying form, as is done with Sérsic fitting (Section 2.2). These nonparametric parameters are also measurable out to high redshifts, making them ideal for deriving galaxy evolution over many epochs, as we discuss in Section 4.

I give a brief description for how these parameters are measured. Typically, as I discuss below, corrections must be applied to account for noise, and a reproducible radius must be used (e.g., Bershady et al. 2000, Conselice et al. 2000a). This radius issue has been addressed in detail by, e.g., Conselice et al. (2000a), Bershady et al. (2000), and Graham et al. (2005). The radius typically used in these measurements is the Petrosian radius, which is defined as the location where the ratio of surface brightness at a radius, $I(R)$, divided by the surface brightness within the radius, $<I(<R)>$, reaches some value, which is denoted by $\eta(R)$ (Petrosian 1976). The value of η changes from $\eta(0) = 1$, at the center of a galaxy, down to $\eta(\infty) = 0$, when the light from the galaxy is zero at its outer "edge."

This method of measuring the radius is much less influenced by surface brightness dimming than other methods, such as using an isophotal radius, and is therefore useful for measuring the

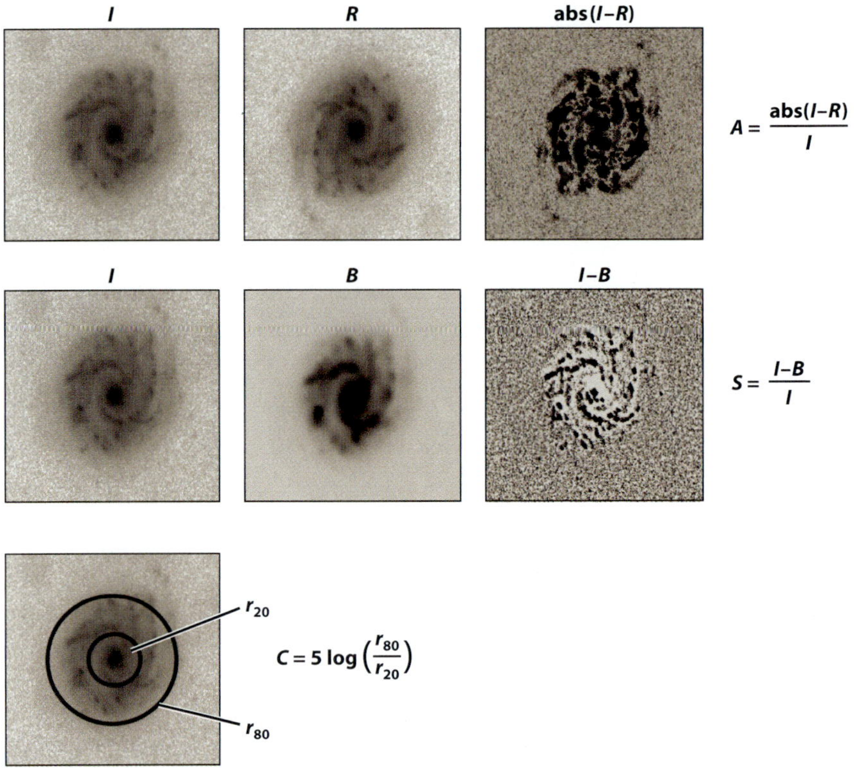

Figure 2

A graphical representation of how the concentration (*C*), asymmetry (*A*), and clumpiness (*S*) are measured on an example nearby galaxy. Within the measurements for *A* and *S*, the value "*I*" represents the original galaxy image, whereas "*R*" is this image rotated by 180°. For the clumpiness *S*, "*B*" is the image after it has been smoothed (blurred) by the factor $0.3 \times 1.5r(\eta = 0.2)$. The details of these measurements can be found in Conselice et al. (2000a) for asymmetry, *A*, Bershady et al. (2000) for concentration, *C*, and Conselice (2003) for clumpiness, *S*.

same physical parts of galaxies at different redshifts (e.g., Petrosian 1976, Bershady et al. 2000, Graham et al. 2005). The mathematical form for this radius is given by

$$\eta(R) = \frac{I(R)}{<I(<R)>}, \tag{2}$$

where most observables in nonparametric morphologies are measured at a radius that corresponds to the location where $\eta(R) = 0.2$ or a relatively small multiplicative factor of this radius (often 1.5 times) (e.g., Bershady et al. 2000, Conselice 2003, Lotz et al. 2004).

2.3.1. Asymmetry index. One of the more commonly used indices is the asymmetry index (*A*), which is a measure of how asymmetric a galaxy is after rotating along the line-of-sight center axis of the galaxy by 180° (**Figure 2**). It can be thought of as an indicator of what fraction of the light in a galaxy is in nonsymmetric components.

The basic formula for calculating the asymmetry index (*A*) is given by

$$A = \min\left(\frac{\Sigma|I_0 - I_{180}|}{\Sigma|I_0|}\right) - \min\left(\frac{\Sigma|B_0 - B_{180}|}{\Sigma|I_0|}\right), \tag{3}$$

Table 1 The average concentration (*C*), asymmetry (*A*), and clumpiness (*S*) parameters for nearby galaxies as measured in the optical R-band (Conselice 2003)

Galaxy type	Concentration (R)	Asymmetry (R)	Clumpiness (R)
Ellipticals	4.4 ± 0.3	0.02 ± 0.02	0.00 ± 0.04
Early-type disks (Sa-Sb)	3.9 ± 0.5	0.07 ± 0.04	0.08 ± 0.08
Late-type disks (Sc-Sd)	3.1 ± 0.4	0.15 ± 0.06	0.29 ± 0.13
Irregulars	2.9 ± 0.3	0.17 ± 0.10	0.40 ± 0.20
Edge-on disks	3.7 ± 0.6	0.17 ± 0.11	0.45 ± 0.20
ULIRGs	3.5 ± 0.7	0.32 ± 0.19	0.50 ± 0.40
Starbursts	2.7 ± 0.2	0.53 ± 0.22	0.74 ± 0.25
Dwarf ellipticals	2.5 ± 0.3	0.02 ± 0.03	0.00 ± 0.06

Abbreviation: ULIRGs, ultraluminous infrared galaxies.

where I_0 represents the original galaxy image and I_{180} is the image after rotating it from its center by 180°. The measurement of the asymmetry parameter, however, involves several steps beyond this simple measure. This includes carefully dealing with the background noise in the same way as is done on the galaxy itself, by using a blank background area (B_0) and finding the location for the center of rotation. The radius is usually defined as the Petrosian radius at which $\eta(R) = 0.2$, although out at a large radius the measured parameters are remarkably stable.

Operationally, the area B_0 is a blank part of the sky near the galaxy. The center of rotation is not defined a priori but is measured through an iterative process, whereby the value of the asymmetry is calculated at the initial central guess (usually the geometric center or light centroid), and then the asymmetry is calculated around this central guess using some fraction of a pixel difference. This is repeated until a global minimum is found (Conselice et al. 2000a).

Typical asymmetry values for nearby galaxies are discussed by Conselice (2003): Ellipticals have values at $A \sim 0.02 \pm 0.02$; spiral galaxies are found in the range from $A \sim 0.07-0.2$; for ultraluminous infrared galaxies (ULIRGs), which are often mergers, the average is $A \sim 0.32 \pm 0.19$; and merging starbursts are at $A \sim 0.53 \pm 0.22$ (Conselice 2003). **Table 1** lists the typical asymmetry and other CAS values (Conselice 2003). Quantitative structural values for the same galaxy can also differ significantly between wavelengths. This is important for measuring these parameters at higher redshifts, where often the rest-frame optical cannot be probed, an issue I discuss in more detail in Section 2.3.5.

2.3.2. Light concentration. The concentration of light quantifies how much light is in the center of a galaxy as opposed to its outer parts. It is a very simple index in this regard, and it is similar to, and correlates strongly with, Sérsic *n* values, which are also a measure of the light concentration in a galaxy. There are many ways of measuring the concentration, including taking ratios of radii that contain a certain fraction of light as well as measuring the ratio of the amount of light at two given radii (e.g., Bershady et al. 2000, Graham et al. 2005). These radii are often defined by the total amount of light measured within some Petrosian radius, often at the same location as used for measuring the asymmetry index.

The ratio most commonly used is that of two circular radii that contain an inner and outer fraction (20% and 80% or 30% and 70% are the most common) ($r_{\text{inner}}, r_{\text{outer}}$) of the total galaxy flux (**Figure 2**),

$$C = 5 \times \log\left(\frac{r_{\text{outer}}}{r_{\text{inner}}}\right). \tag{4}$$

A higher value of C indicates that a larger amount of light in a galaxy is contained within a central region. The concentration index, however, has to be measured very carefully, as different regions and radii used can produce very different values that systematically do not reproduce well when observed under degraded conditions (e.g., Graham et al. 2001a,b; Graham et al. 2005).

2.3.3. Clumpiness. The clumpiness (or smoothness) (S) parameter is used to describe the fraction of light in a galaxy that is contained in clumpy distributions. Clumpy galaxies have a relatively large amount of light at high spatial frequencies, whereas smooth systems, such as elliptical galaxies, contain light at low spatial frequencies. Galaxies that are undergoing star formation tend to have very clumpy structures and, thus, high S values. Clumpiness can be measured in a number of ways; the most common method used, as described by Conselice (2003), is

$$S = 10 \times \left[\left(\frac{\Sigma(I_{x,y} - I^{\sigma}_{x,y})}{\Sigma I_{x,y}} \right) - \left(\frac{\Sigma(B_{x,y} - B^{\sigma}_{x,y})}{\Sigma I_{x,y}} \right) \right], \tag{5}$$

where the original image, $I_{x,y}$, is blurred to produce a secondary image, $I^{\sigma}_{x,y}$ (**Figure 2**). This blurred image is then subtracted from the original image, leaving a residual map containing only high-frequency structures in the galaxy (Conselice 2003). The size of the smoothing kernel σ is determined by the radius of the galaxy, and the value $\sigma = 0.3 \cdot 1.5 \times r(\eta = 0.2)$ gives the best signal for nearby systems (Conselice 2003). Note that the centers of galaxies are removed when this procedure is carried out as they often contain unresolved high spatial frequency light.

Figure 3 shows a diagram for how these three CAS parameters are measured for a typical nearby spiral galaxy. Furthermore, the CAS parameters can be combined together to create a 3D space in which different galaxy types can be classified. For example, **Figure 3** shows the concentration versus asymmetry versus clumpiness diagram, demonstrating how these parameters can be used to determine morphological types of galaxies in the nearby Universe in CAS space.

2.3.4. Other coefficients. Other popular structural measurement systems are the Gini/M_{20} parameters, which are used in a fashion similar to the CAS parameters to find galaxies of broad morphological types, especially galaxies undergoing mergers (e.g., Abraham et al. 2003, Lotz et al. 2004). These parameters measure the relative distribution of light within pixels and do not involve subtraction, as is used for the asymmetry and clumpiness parameters, and therefore in principle may be less sensitive to high levels of background noise (e.g., Lotz et al. 2004).

The Gini coefficient (G) is a statistical tool originally used in economics to determine the distribution of wealth within a population; higher values indicate a very unequal distribution (Gini of 1 means all wealth/light is in one person/pixel), whereas a lower value indicates a more even distribution among the population (Gini of 0 means everyone/every pixel has an equal share). The value of G is defined by the Lorentz curve of the galaxy's light distribution, which does not take into consideration spatial position.

In the calculation of these parameters, each pixel is ordered by its brightness and counted as part of the cumulative distribution (see Lotz et al. 2004, 2008a). A galaxy in this case is considered a system with n pixels, each with a flux f_i, where i ranges from 0 to n. The Gini coefficient is then measured by

$$\mathrm{G} = \frac{1}{|\overline{f}|n(n-1)} \sum_{i}^{n} (2i - n - 1)|f_i|, \tag{6}$$

where $\overline{f}$ is the average pixel flux value.

The second-order moment parameter, M_{20}, is similar to the concentration in that it gives a value that indicates whether light is concentrated within an image. However, an M_{20} value denoting a high concentration (a very negative value) does not imply a central concentration as, in

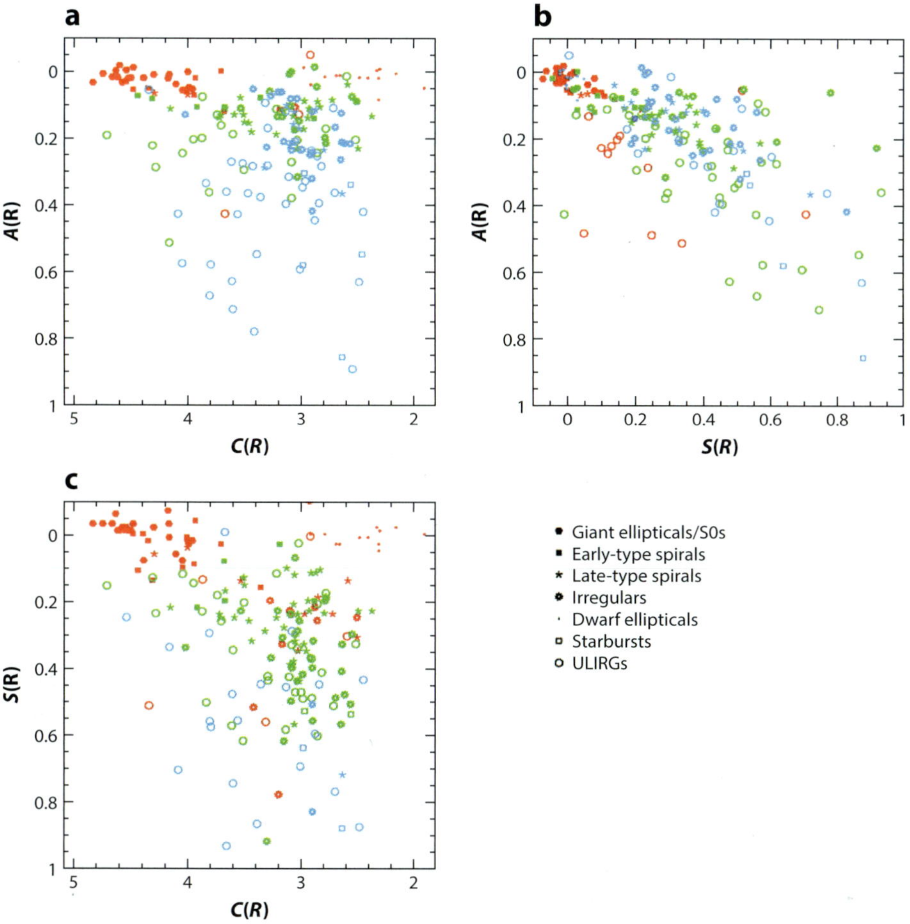

Figure 3

The different forms of the realizations of nearby galaxies of different morphologies and evolutionary states plotted together in terms of their CAS [concentration (C), asymmetry (A), clumpiness (S)] parameters. (*a*) The A-C indexes are plotted with colored points that reflect the value of the clumpiness for each galaxy. Systems that have clumpiness values $S < 0.1$ are shown in red, systems where $0.1 < S < 0.35$ are green, and systems where $S > 0.35$ are blue. (*b*) Similarly, the A-S diagram shows concentration values; systems where $C > 4$ are red, systems where $3 < C < 4$ are green, and systems where $C < 3$ are blue. (*c*) For the S-C diagram, systems having asymmetries $A < 0.1$ are red, systems where $0.1 < A < 0.35$ are green, and systems where $A > 0.35$ are blue. When using these three morphological parameters, all known nearby galaxy types can be distinctly separated and distinguished in structural space (Conselice 2003). Abbreviation: ULIRGs, ultraluminous infrared galaxies.

principle, the light could be concentrated in any location in a galaxy. The value of M_{20} is the moment of the fluxes of the brightest 20% of light in a galaxy, which is then normalized by the total light moment for all pixels (Lotz et al. 2004, 2008a). The mathematical form for the M_{20} index is

$$M_{20} = \log 10 \left(\frac{\sum_i M_i}{M_{\rm tot}} \right), \text{ while } \sum_i f_i < 0.2 f_{\rm tot}, \tag{7}$$

where the value of $M_{\rm tot}$ is

$$M_{\rm tot} = \sum_i^n M_i = \sum_i^n f_i \left[(x_i - x_c)^2 + (y_i - y_c)^2\right];$$

here x_c and y_c indicate the center of the galaxy, and in the case of M_{20} this center is defined as the location where the value of $M_{\rm tot}$ is minimized (Lotz et al. 2004). The separation for nearby ellipticals, spirals, and ULIRGs is similar to that found by the CAS parameters (see Lotz et al. 2004, 2008a).

Other popular parameters include the multiplicity index, Ψ, which is a measure of the potential energy of a light distribution (e.g., Law et al. 2007). Values of Ψ range from 0, for systems that are in the most compact forms, to $\Psi > 10$, which indicates systems that are often very irregular/peculiar (e.g., Law et al. 2012b). Another recent suite of parameters developed by Freeman et al. (2013) includes features that measure the multimode (M), intensity (I), and deviation (D) of a galaxy's light profile with the intention to locate galaxy mergers.

2.3.5. Redshift effects on structure. One of the major issues with nonparametric structural indices is that they change for more distant galaxies, due to not only any evolution but also distance effects, creating a smaller and fainter image of the same system. This must be accounted for when using galaxy structure as a measure of evolution (e.g., Conselice et al. 2000a, Conselice 2003, Lisker 2008).

There are several ways to deal with this issue, and they are similar to how corrections for point-spread functions in parametric fitting or weak lensing analyses are done. The most common correction method for nonparametric parameters is to use image simulations. These simulations are such that nearby galaxies are reduced in resolution and surface brightness to match the redshift at which the galaxy is to be simulated. These new simulated images are then placed into a background appropriate for the instrument and exposure time in which the simulation takes place (Conselice 2003). The outline for how to do these simulations is provided in papers by Giavalisco et al. (1996) and Conselice (2003), among others.

To give some idea of the difficultly in reproducing the morphologies and structures of galaxies, **Figure 4** shows simulated nearby early-type spirals as they would appear in WFC3 imaging data from the Hubble Ultra Deep Field (Conselice et al. 2011a). What can be clearly seen is that it is difficult, and sometimes even impossible, to discern features of these galaxies after they have been simulated.

Another issue when examining the structures of distant galaxies is that these systems are often observed at bluer wavelengths than that typically observed within the nearby Universe owing to the effects of redshift. For example, pictures of galaxies at $z > 1.2$ taken with WFPC2 and ACS are all imaged in the rest-frame UV. **Figure 5** shows what rest-frame wavelength various popular filters probe as a function of redshift. This shows that we must go to the near-IR to probe rest-frame optical light for galaxies at $z > 1$.

It turns out that the qualitative and quantitative morphologies and structures of galaxies can vary significantly between rest-frame UV and rest-frame optical images (e.g., Meurer et al. 1995, Hibbard & Vacca 1997, Windhorst et al. 2002, Taylor-Mager et al. 2007), though these morphologies are not significantly different for starbursting galaxies with little dust at both low and high redshift (Conselice et al. 2000c, Dickinson 2000). Although it is clear that the CAS method works better at distinguishing types at redder wavelengths (e.g., Lanyon-Foster et al. 2012), its use has also expanded into image analyses with HI and dust-emission maps from *Spitzer Space Telescope* (e.g., Bendo et al. 2007; Holwerda et al. 2011, 2012, 2013, 2014).

The process for accounting for the effects of image degradation is to measure the morphological index of interest at $z = 0$ and then to remeasure the same values at higher redshift after simulating.

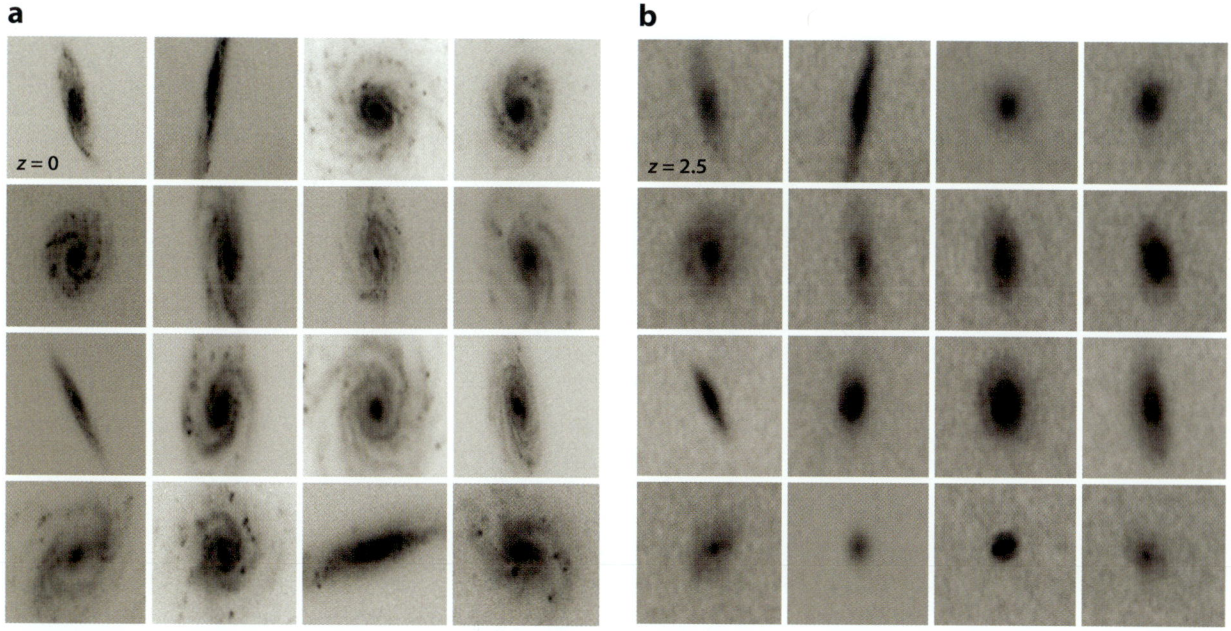

Figure 4

Nearby galaxies (*a*) originally observed at $z = 0$ in the rest-frame B-band (*b*) simulated to show how they would appear at $z = 2.5$, also observed in the rest-frame B-band, within the Hubble Ultra Deep Field WFC3 F160W H-band. These systems are classified as late-type spirals (Sc and Sd) in the nearby Universe, but can appear very different when simulated to higher redshifts as seen here and when using quantitative measures. The galaxies shown here have typical sizes of several kiloparsecs in effective radii and are at a variety of distances (see Conselice et al. 2000a, Conselice et al. 2011a). These changes in structure, both in apparent morphology and in terms of the structural indices, must be carefully considered before evolution is derived (e.g., Conselice et al. 2008, Mortlock et al. 2013).

For the morphological k-correction, the approach has been to measure the parameter of interest at different wavelengths and to determine by interpolation its value at the rest-frame wavelength of interest.

Using the asymmetry index as an example, the final measure after correcting for redshift effects is

$$A_{\text{final}} = (A_{\text{obs}} + \delta A_{\text{SB-dim}} + \delta A_{\text{k-corr}}), \tag{8}$$

where $\delta A_{\text{k-corr}}$ is the (usually negative) morphological k-correction, and $\delta A_{\text{SB-dim}}$ is the (positive) correction for image degradation effects. Other parameters can be measured in the same way. This is a necessary correction to examine the evolution of a selected population at the same effective depth, resolution, and rest-frame wavelength.

3. THE PHYSICAL NATURE OF GALAXY STRUCTURE

The ultimate goals in this review are to trace how structure evolves over cosmic time and to use this as a method to decipher galaxy evolution. In this section I present work to date that describes how physical information is derived from galaxy structural parameters.

3.1. Star Formation and Galaxy Structure

The star-formation process within galaxies is critical, as galaxies do not exist without stars in them. Star formation is also one of the major criteria for classification within the Hubble sequence. The

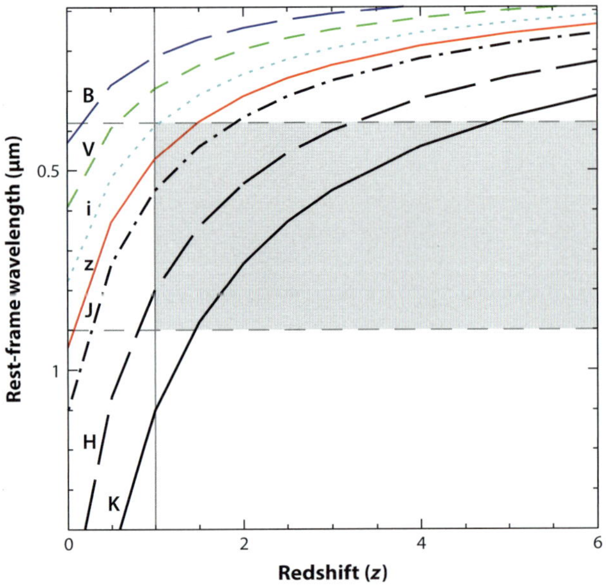

Figure 5

Plot showing the rest-frame wavelength probed by the most common filters used to image distant galaxies as a function of redshift from $z \sim 0$–6. The filters shown are the ACS B_{450}, V_{550}, i_{775}, and z_{950} filters, and the WFC3 J_{110} and H_{160} filters and a K-band filter centered at 2.2 μm. The shaded area shows the region in which the rest-frame optical light of distant galaxies can be probed from between 0.38 μm and 0.9 μm. As can be seen, the H-band allows rest-frame light up to $z \sim 3$ to be imaged, whereas the K-band can extend this out to $z \sim 4.5$.

effects of star formation have also been used to classify spiral galaxies into various classes (e.g., van den Bergh 1976, Elmegreen & Elmegreen 1987).

Star formation is an enormous topic with a great amount of work published (e.g., Kennicutt 1998, Kennicutt & Evans 2012, Madau & Dickinson 2014), and I thus limit my discussion to how star formation and galaxy structure are related. The integrated star-formation density evolution of galaxies in the Universe has been studied in detail and is now well characterized. The integrated star-formation rate increases from a low initial value at $z > 6$ to a peak at $z \sim 2$, and thereafter the rate declines. At higher redshifts $z > 1$, there is also a well-defined relation between star-formation rate and stellar mass, such that higher-mass galaxies have a higher rate of star formation (e.g., Noeske et al. 2007, Bauer et al. 2011).

This is important, as galaxies undergoing star formation can have very different morphologies and structures from passive galaxies. Examples of this include clumpy spiral arms, knots of star formation, central bright starbursts, etc. This can be seen, for example, when viewing local galaxies, whereby those with star formation appear clumpier and more asymmetric than those without star formation.

Furthermore, there are also morphological k-correction effects, such that star-forming galaxies have a smaller difference in their morphology between UV through optical and near-IR light (e.g., Windhorst et al. 2002, Taylor-Mager et al. 2007). This generally reveals that at shorter wavelengths the morphologies and quantitative structures are tracing the distribution of star formation directly. At optical wavelengths longer than the Balmer break, we are sampling a mixture of stars at different ages, with older ages dominating the spectral energy distributions (SEDs) at longer wavelengths. There are also very dusty galaxies such as submillimeter sources and ULIRGs that can have a significant fraction of their optical light absorbed (e.g., Calzetti et al. 2000).

It is apparently not just morphology that is affected by star formation but also the quantitative structure. It is well known that star-forming galaxies without significant dust are quite blue, but the effects of star formation can also be seen in their structure. Quantitative measurements of structure strongly correlate with the star-formation rate within galaxies as measured by the correlation between the clumpiness index (S) and the Hα equivalent width (Conselice 2003). This is also seen in more asymmetric and clumpy light distributions within Hα imaging of nearby galaxies, and when examining the light distribution at 24-μm imaging using *Spitzer Space Telescope* imaging (e.g., Bendo et al. 2007). Conselice (2003) shows how the clumpiness index can be used as a measure of star formation, and Conselice et al. (2000a) show that asymmetry values correlate strongly with (B-V) color for nearby galaxies.

3.2. Structure as a Merging Indicator

One of the primary physical effects that can be seen in the structures of galaxies is when two galaxies merge or interact with each other. When these dynamical events occur the structures of these systems often become very peculiar and distorted, especially when the merging galaxies contain a similar amount of mass in a major merger. (Note that a major merger throughout this review is a merger in which the ratio of the stellar masses of the progenitors are 1:4 or greater. A minor merger is one with a mass ratio of less than 1:4.) We have learned much about nearby galaxy mergers, such as ULIRGs (e.g., Joseph & Wright 1985, Sanders & Mirabel 1996) through numerical simulations (e.g., Mihos & Hernquist 1996) that have shown convincingly that peculiar galaxies are often mergers (Toomre & Toomre 1972). This demonstrates that there is a strong correlation between structure and this fundamental galaxy formation process.

Early measurements of nearby galaxies found a correlation between galaxy lopsidedness, quantified through Fourier decomposition of structure and the asymmetry parameter, and the presence of interacting or merging neighbors (e.g., Rix & Zaritsky 1995, Conselice 1997, Reichard et al. 2008). As such, galaxy structure is a powerful method for determining whether a galaxy is undergoing a recent major merger. This has been measured in many ways, from using visual estimates of mergers based on peculiar structures to examining more quantitative results.

One automatic method for finding mergers is the CAS approach (Conselice 2003), in which merging galaxies are those with a high asymmetry that is also higher than the value of the clumpiness. The simple condition

$$(A > 0.35)\ \&\ (A > S) \tag{9}$$

accounts for a large fraction, but not all, of local galaxies that are mergers—i.e., ULIRGs and starbursts in mergers (see **Figure 3**). Although the contamination from nonmergers is fairly low (at a few percent), the fraction of actual mergers that are identified is roughly 50% (Conselice 2003). This is largely due to the fact that galaxies involved in the merger process are only quantitatively asymmetric for about a third of the lifetime of the merger (see Section 3.4).

There is also the relationship found by Lotz et al. (2008a) for locating major mergers using Gini/M_{20} parameters, which is given by,

$$G > -0.14 \times M_{20} + 0.33. \tag{10}$$

More recent criteria developed by Freeman et al. (2013) use multimode (M), intensity (I), and deviation (D) statistics to quantify which galaxies are mergers. Their study shows that a higher fraction of real mergers can be found using these indices compared with CAS or Gini/M_{20}.

One ultimate result of finding these mergers is that it allows us to calculate the merger fraction within a population of galaxies. The basic merger fraction (f_m) is calculated as the number of mergers selected within a given redshift bin and stellar mass limit (or luminosity cut) (N_m) divided by the total number of galaxies within the same redshift bin and stellar mass selection (N_T). The merger fraction is thus defined as

$$f_m(M_*, z) = \frac{N_m}{N_T}. \tag{11}$$

This merger fraction is also a function of stellar mass and redshift. The CAS mergers are nearly all major mergers (Conselice 2003, 2006b; Lotz et al. 2008b), whereas Gini/M_{20} measure all types of mergers, both minor and major.

Furthermore, for structural samples, we calculate the merger fraction as opposed to the galaxy merger fraction. The difference is important when comparing with pair studies in which the two progenitors can be resolved. The difference between these two is subtle, but important. The merger fraction considers a merger as having already happened; the two galaxies that have merged now count as a single system. The galaxy merger fraction is the fraction calculated when both of these merging galaxies are considered as two separate galaxies, which they were before the final merger.

The galaxy merger fraction (f_{gm}) is thus the number of galaxies merging, where a system that has already condensed into a single galaxy is counted as two galaxies, divided by the number of galaxies in the total sample. For small merger fractions this ratio is about a factor of two larger than the merger fraction that counts only the merger remnants (Conselice 2006b). The equation to derive the galaxy merger fraction with observables through morphology, with the assumption that every merging galaxy has exactly two progenitor galaxies, is given by

$$f_{gm}(M_*, z) = \frac{2 \times N_m}{(N_T + N_m)} = \frac{2 \times f_m}{(1 + f_m)}. \tag{12}$$

This relation does not hold if a merger occurs with more than two galaxies (Conselice 2006b), although these are very rare (de Propris et al. 2007). The morphological measurement of the nearby merger fraction gives values of $f_m = 0.01$ (de Propris et al. 2007). A discussion of the measurement of this at higher redshifts is included in Section 4.3.

3.3. Galaxy-Scale Properties and Galaxy Structure

One of the interesting facts about galaxies is that many of their characteristics can be explained by an underlying property, which is likely the galaxy's halo or total mass (Caon et al. 1993, Disney et al. 2008). As an example, it was noted early on that galaxy light profile shapes of ellipticals correlated strongly with the radius or magnitude of a galaxy (e.g., Caon et al. 1993). This implies that the scale or mass of an elliptical galaxy has an influence on a galaxy's overall light profile and shape.

This can also be seen in the detailed structures of galaxies. In general it appears that on average galaxies with a higher degree of central concentration have larger total or stellar masses. This is also seen in the concentration index, which is another measure of the degree of light concentration; more massive galaxies have a higher value of concentration (e.g., Conselice 2003). This concentration also correlates with the fraction of light in bulge and disk components. This relation is such that the more concentrated a galaxy is, the less likely it contains a significant disk (e.g., Conselice 2003). In fact, it is likely that it is the fraction of bulge light that drives this correlation, and more massive systems are more likely to have significant and concentrated bulges.

Concentration also separates galaxies having different star-formation histories in the local and high-redshift Universe. In a study using the Sloan Digital Sky Survey (SDSS), Strateva (2001) showed that nonstar-forming galaxies are more concentrated than star-forming blue systems. This can also be demonstrated by other overall galaxy properties (e.g., Allen et al. 2006, Conselice 2006a). The light concentration for ellipticals also correlates with the mass of the central massive black hole (e.g., Graham et al. 2001, Savorgnan et al. 2013).

3.4. Numerical Simulations of Galaxy Structure

Simulations of galaxy formation are critical for interpreting and understanding the meaning of structural indices of galaxies. In fact, one of the first computer simulations of galaxy formation by Toomre & Toomre (1972) showed that the peculiar morphologies of galaxies seen in, e.g., the Arp (1966) atlas were due to systems undergoing major mergers rather than some other cause. Since then, numerical simulations of galaxies have proven an effective method for interpreting the structures and morphologies of galaxies in both the local Universe and at higher redshifts. In many ways this approach toward understanding galaxy morphology has just begun and promises to be a powerful and effective approach for interpreting the meaning of structure in the future.

Mihos & Hernquist (1996) provided one of the first papers to demonstrate in detail how the peculiar galaxies seen in deep HST imaging were in fact due to the merger process using the TREESPH hydrodynamical method from Hernquist & Katz (1989). Interest in comparing simulated merger results with observables was largely in terms of starbursts and SEDs of galaxies (e.g., Barnes & Hernquist 1991). However, some attempts were made even very early to use these merger N-body models as methods of interpreting HST morphologies of galaxies (Mihos 1995).

Applying quantitative structural methods described in Section 2 to these numerical simulations of structural evolution was first carried out by Conselice (2006b), who used the CAS method to calculate the timescale of the merger process on dark-matter particle simulations, finding a merger timescale of ~0.3–0.8 Gyr for galaxies having a high-enough asymmetry to be identified as an unambiguous merger (Section 3.2). This timescale is critical for interpreting galaxy merger fractions through cosmic time, as it allows us to convert merger fractions to merger rates and thus derive how mergers are driving galaxy formation. Using these results, Conselice (2006b) deduces that a typical galaxy undergoes $4.4^{+1.6}_{-0.9}$ mergers from $z \sim 3$ to $z \sim 0$.

Conselice (2006b) also shows that the time when a merger is asymmetric is distributed throughout the merger process and is not located at one particular time. Moreover, the merging systems would not always be identifiable as such when studied with the CAS parameters. In addition to giving a robust timescale, Conselice (2006b) also shows that only a fraction (about a third) of a merging galaxy's time sequence would be identified as a major merger using CAS. The timescale derived is also largely independent of the viewing angle of the merger, and an asymmetry signal is only present within major mergers with mass ratios of 1:4 or greater.

The simulations used by Conselice (2006b) are, however, simple in that they do not include the effects of star formation or dust, which are well known to produce dramatic changes in morphology (e.g., Taylor-Mager et al. 2007; and see Section 2.3.5). When star formation and dust are added to simulations of galaxy structure the quantitative structural parameters measured are similar to those seen in nearby galaxies, and the measured structure correlates with other properties, such as color, in the same way it does for nearby galaxies (e.g., Lotz et al. 2008b, Hambleton et al. 2011).

Lotz et al. (2008b) include the first measurements of CAS and Gini/M_{20} parameters on numerical simulations that include old and young stars, star formation, gas, and dust. Lotz et al. (2008b, 2010a,b) use GADGET/N-body/hydrodynamical simulations of galaxies when imaging the appearance of these galaxy mergers. Lotz et al. (2008b) utilize disk galaxies of the same total mass, whereas later they investigate mergers with a variety of mass ratios (Lotz et al. 2010a) and examine how the amount of cold gas mass in progenitor galaxies influences morphology (Lotz et al. 2010b).

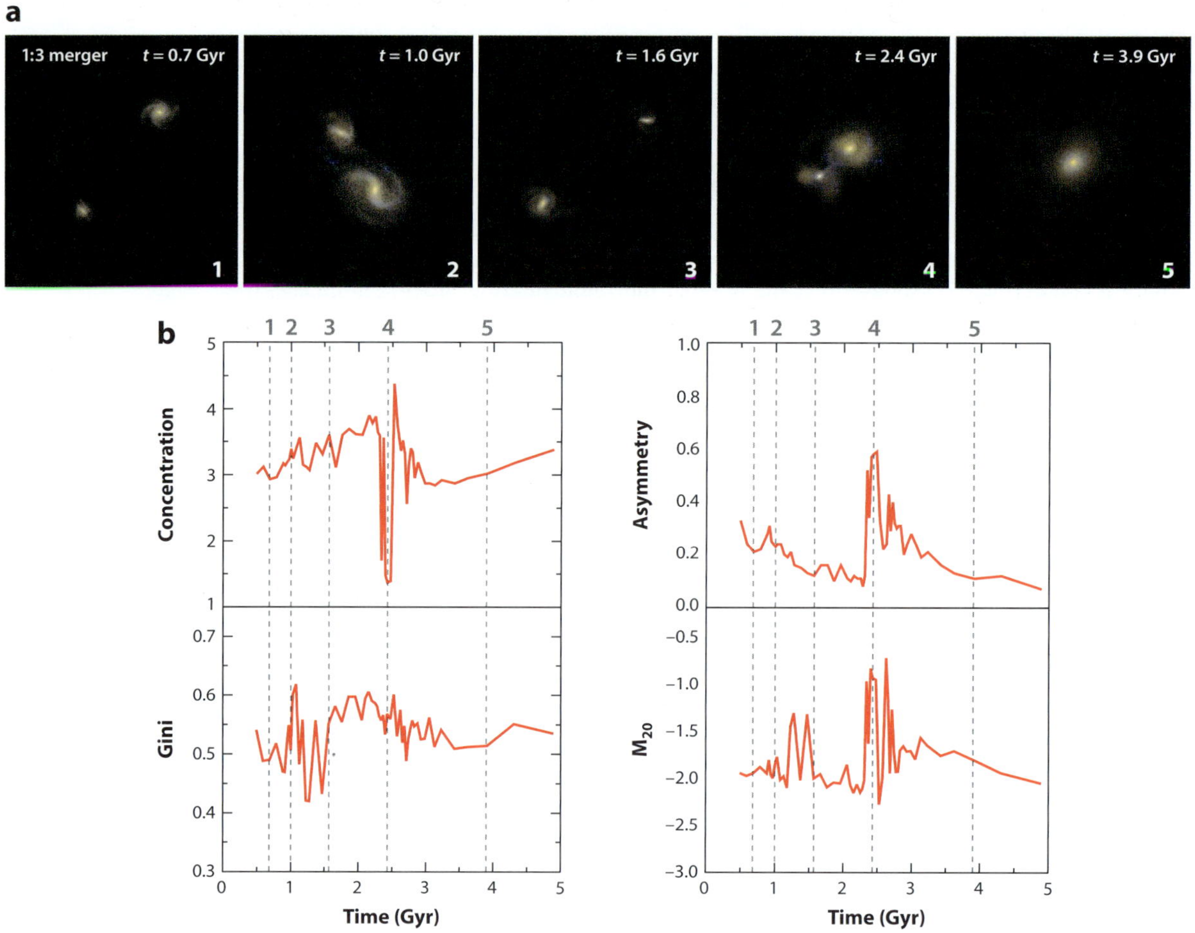

Figure 6

An N-body/hydrodynamical model from Lotz et al. (2008b, 2010a) showing two equal-sized disk galaxies merging as a function of time. (*a*) The numbers across the top of the realizations of this model show the various snapshots of time through the simulation, whereas (*b*) the bottom panels show the changes in the concentration, asymmetry, Gini, and M_{20} values for this particular simulation. This demonstrates the changing form of quantitative indices during a merger and how these systems are only identifiable within the different morphological systems as a merger at specific times. Courtesy of Jennifer Lotz.

These simulation results are passed through the SUNRISE Monte-Carlo radiative transfer code to simulate, as realistically as possible, how galaxies would appear based on the simulation output.

Lotz et al. (2008b, 2010a,b) further investigate the location in CAS and Gini/M_{20} parameter space for mergers in different scenarios and for different properties of the merging galaxies. They investigate the timescale for how long these simulated galaxies appear as a "merger" on the basis of where they fall in these nonparametric structural spaces (see **Figure 6** for an example of these simulations). These papers also investigate how the dust, viewing angle, orbital parameters, gas properties, supernova feedback, and total mass alter the structural merger timescale. Lotz et al. (2008b, 2010a) find that most properties—the total mass, supernova feedback, viewing angle, and orbital properties of mergers—have very little influence on the derived timescales. The mass ratio and gas mass fraction of the merging galaxies affect the derived merger timescales significantly, however.

Mergers are identified within both CAS and Gini/M_{20} at the first pass of the merger as well as when the systems finally coalesce to form a remnant (Lotz et al. 2008b). However, merging galaxies are not found in the merger area of the nonparametric structural parameters for the entire merger. This, however, allows the timescales for structural mergers to be calculated. Lotz et al. (2008b, 2010a) find that the asymmetry timescales for gas-rich major mergers are 0.2–0.4 Gyr and 0.06 Gyr for minor mergers (Lotz et al. 2010a). The Gini/M_{20} timescales are $\tau_{\rm m} = 0.2-0.4$ Gyr. These are relatively quick timescales and, thus, suggest that the observed merger fraction converts to a high merger rate.

This is similar to, but not exactly the same as, the dynamical friction merger timescales calculated for merging objects that have a separation change from r_i to r_f. The dynamical friction timescale is given by $t_{\rm fric}$:

$$t_{\rm fric} = 0.0014\,{\rm Gyr}\left(r_i^2 - r_f^2\right)\left(\frac{v_c}{100\,{\rm km\,s^{-1}}}\right)\left(\frac{10^{10}\,{\rm M}_\odot}{M}\right), \tag{13}$$

where v_c is the relative velocity between the two merging galaxies at a given time, M is the mean accreted mass, and the Coulomb logarithm $\ln \Lambda = 2$ (Dubinski et al. 1999). Dynamical friction calculations such as these have dominated the calculation of galaxy timescales up until simulations of mergers revealed more subtle results, although the blunt calculations from Equation 13 are often a good rough estimate for merger timescales, giving values of $\sim$0.5 Gyr for equal-mass mergers.

Lotz et al. (2008b, 2010a) also find that the asymmetry index is sensitive to major mergers of ratios of 1:4 or less, whereas the Gini/M_{20} parameters are sensitive for mergers down to 1:9, which enables probing more minor mergers. Lotz et al. (2010b), however, find that very gas-rich galaxies, such as those seen in high redshift, may have longer timescales for merging with gas-rich progenitors, which are likely to be more common at higher redshifts. This would provide a "merger" asymmetry signal for more minor mergers as long as they were more gas rich. However, it is clear that massive galaxies with $M_0 > 10^{10}\,{\rm M}_\odot$, where most measurements have been made to date at $z < 3$, have a low gas mass fraction (e.g., Erb et al. 2006, Mannuci et al. 2010, Conselice et al. 2013).

4. MEASUREMENTS OF GALAXY STRUCTURAL EVOLUTION

The above sections describe how we can measure the structures and morphologies of galaxies through various approaches as well as the meaning of these structures and morphologies. In this section, I discuss how these measurements have been applied to galaxies at all redshifts to decipher how evolution is occurring within the galaxy population.

When galaxies were first found in the distant Universe, they were not resolved enough to study their structures and morphologies, and the evolution of galaxies was observationally driven by number counting and colors (e.g., Koo & Kron 1992), and the "faint blue galaxy excess" problem at faint magnitudes dominated the field for twenty years until redshifts for these systems became available (e.g., Ellis 1997).

The problem of galaxy evolution and formation is a large one, and this review does not focus on this question, although I give a brief overview here of the important questions in understanding galaxy evolution. I also demonstrate where galaxy structure and morphology reveal information that cannot be provided by other methods. For a general review of galaxy properties at $z > 2$ see Shapley (2011) for an observational perspective and Silk & Mamon (2012) for a theoretical one. For nearby galaxies a few recent relevant reviews have been provided by Blanton & Moustakas (2009), van der Kruit & Freeman (2011), and Conroy (2013).

Galaxies are now studied up to redshifts of $z \sim 7$–10, although at the highest redshifts less information is available. The most common measures for these distant galaxies are colors, stellar masses, star-formation rates, sizes, and basic structures. From these we know that the volume integrated star-formation rate increases with time from these ultrahigh redshifts until around $z \sim 2$, when the star-formation rate begins to decline (see Madau & Dickinson 2014 in this volume). Stellar mass measurements roughly agree with this picture, such that about half of all stellar mass is formed by $z \sim 1$ (e.g., Mortlock et al. 2011). Galaxies are also much bluer in the past (Finkelstein et al. 2012), and there is some debate and uncertainty concerning the star-formation history of individual galaxies and the relevant role and commonality of very old and/or very dusty galaxies at redshifts $z > 1$.

What is largely unavailable from examining stellar masses, stellar populations, and star-formation histories is how these galaxies assembled. Clearly galaxies are fed gas or have very large gas reservoirs in them to sustain and produce star formation. How this gas gets into galaxies is a fundamental question, as is the relative role of mergers versus star formation in forming galaxies. Because the number of massive galaxies at high redshifts is a factor of ten or so less than those in existence today, clearly much evolution and formation in these systems has occurred.

Galaxy structure provides a way to examine this problem, as it permits us to determine which modes of galaxy formation are active within a galaxy. The first and by far the most common method is to study the merger history through the techniques described in Section 3.2. Another method is to simply examine the visual morphologies of galaxies to determine when the Hubble sequence is in place and combine those results with color, size, and star-formation rates to determine when spirals and ellipticals are roughly in their current form. In particular, the examination of the sizes of galaxies has provided an evolutionary puzzle, such that galaxies of similar masses are up to a factor of 2–5 times smaller than corresponding galaxies seen today (e.g., Buitrago et al. 2008, Cassata et al. 2013).

Furthermore, resolved imaging allows us to study the formation history of individual components of galaxies, such as disks, bulges, spiral arms, clumps of star formation, etc., which reveals formation information not available when examining the galaxy as a whole. This section, which is the heart of this review, provides the current observational evidence for morphological and structural evolution and what it implies for galaxy formation.

4.1. Observed Evolution of the Hubble Sequence

This section of the review examines the formation and evolution of the Hubble sequence. I discuss here only observations that answer the question of when galaxy types appear, without addressing the "how" of their formation mechanisms, which is discussed later in this review. I first describe the bulk morphological evolution in Section 4.1.1, which includes when Hubble types (spirals and ellipticals), as well as peculiar galaxies, formed. I then discuss in the next subsection (Section 4.1.2) the evolution and formation of the more detailed aspects in the Hubble sequence. Specifically I address observations of the earliest observed spiral arms, how the bar fraction in spirals evolves, and how the relative properties of bulges and disks change through time.

4.1.1. Bulk morphological evolution. The first science result I discuss is how the visual Hubble sequence evolves throughout the Universe. This can simply be restated as measuring the number density and relative fractions of galaxy types at a given selection, which are classified as ellipticals, spirals, and peculiars. As mentioned earlier, I do not consider irregulars as these are typically lower-mass galaxies that are not detected at high redshift owing to their faintness. It is also largely impossible to use finer classifications, such as Sa or Sb, on distant galaxies, as the resolution is not good enough, even with HST imaging, to resolve this type of detail. More distant galaxies also appear to be quite different from Hubble types, making these types of detailed morphologies unnecessary and undesirable (e.g., Conselice et al. 2005).

This relates to a fundamental question that has been asked since galaxies were discovered, which is whether or not a galaxy retains its morphology over a long period of cosmic time. Otherwise, if morphological transformations do occur, what processes drive this (internal or external), and how often does a galaxy change its morphology?

One of the first observations noted when examining the first deep HST images was that many of the fainter galaxies were peculiar (e.g., Driver et al. 1995, Glazebrook et al. 1995, Schade et al. 1995, Abraham et al. 1996). These early studies were limited to examining galaxy number counts, as no redshifts were known for these faint galaxies. In addition to a faint blue excess, it was clear that there was also a peculiar excess, often for the faintest galaxies. This is where the field remained until redshifts for a significant number of these peculiar galaxies were obtained.

The field of high-redshift studies changed dramatically in 1995–1997 with the advent of the Hubble Deep Fields (HDFs) (Williams et al. 1996, Ferguson et al. 2000) and the discovery of a significant population of high-redshift galaxies that could be discovered by the Lyman-Break technique (Steidel et al. 1996, Shapley 2011), now referred to as Lyman-Break Galaxies (LBGs). The HDFs, and later significant campaigns to obtain very deep Hubble imaging, such as the Hubble Deep Field South (Williams et al. 2000), the Great Observatories Origins Deep Survey (GOODS) (Giavalisco et al. 2004), the Hubble Ultra Deep Field (UDF) (Beckwith et al. 2006), the COSMOS field (Scoville et al. 2007), the Extended Groth Strip survey (EGS) (Davis et al. 2007), and most recently the CANDELS survey (Grogin et al. 2011, Koekemoer et al. 2011), have revolutionized the field of galaxy-formation studies and in particular the study of morphologies.

The critical nature of these deep fields is not simply that they are deeper than previous deep HST imaging but that they are the fulcrum of large-scale efforts to obtain photometry at the faintest levels possible at nearly all wavelengths. This allows redshifts to be measured for most galaxies using photometry through so-called photometric redshifts [e.g., Dahlen et al. (2013) for a recent discussion of the use of this method]. The availability of these redshifts allows us to measure evolution over broad redshift ranges that were not possible before, as spectroscopic samples were few and far between, even with the use of relatively large multislit spectrographs, such as LRIS on Keck (e.g., Cohen et al. 1996).

Early observations using a mixture of spectroscopic and photometric redshifts showed that the Hubble sequence was certainly not in place at high redshifts $z > 1$ (van den Bergh et al. 1996) when examined using WFC2 data. This was verified in the rest-frame optical after deep near-IR NICMOS observations of the HDF were taken in 1998 (Dickinson 2000). A deeper analysis showed that, using the rest-frame optical structures of galaxies, the Hubble sequence was nearly completely absent at $z > 2$, and only at $z \sim 1.5$ did spirals and ellipticals become as common as peculiar galaxies (Ravindranath et al. 2004, Conselice et al. 2005, Buitrago et al. 2013; **Figure 7**).

The latest results on the evolution of visual morphology, as defined solely by visual types, from the CANDELS survey, is shown in **Figure 8** (Mortlock et al. 2013) for systems with stellar masses $M_* > 10^{10}\,M_\odot$. It must be stressed that this figure only shows the visual estimates of galaxy morphology. The classification of a spiral, elliptical, or peculiar does not imply that these galaxies have a certain star-formation rate, color, mass, or size. In fact what is often seen is that these visual morphologies do not correlate well with other physical properties (e.g., Conselice et al. 2011b, Mortlock et al. 2013). What is also seen is a stellar mass difference in the formation history of the Hubble sequence, with the highest-mass galaxies appearing to form into visual ellipticals and disks before lower-mass systems (Mortlock et al. 2013).

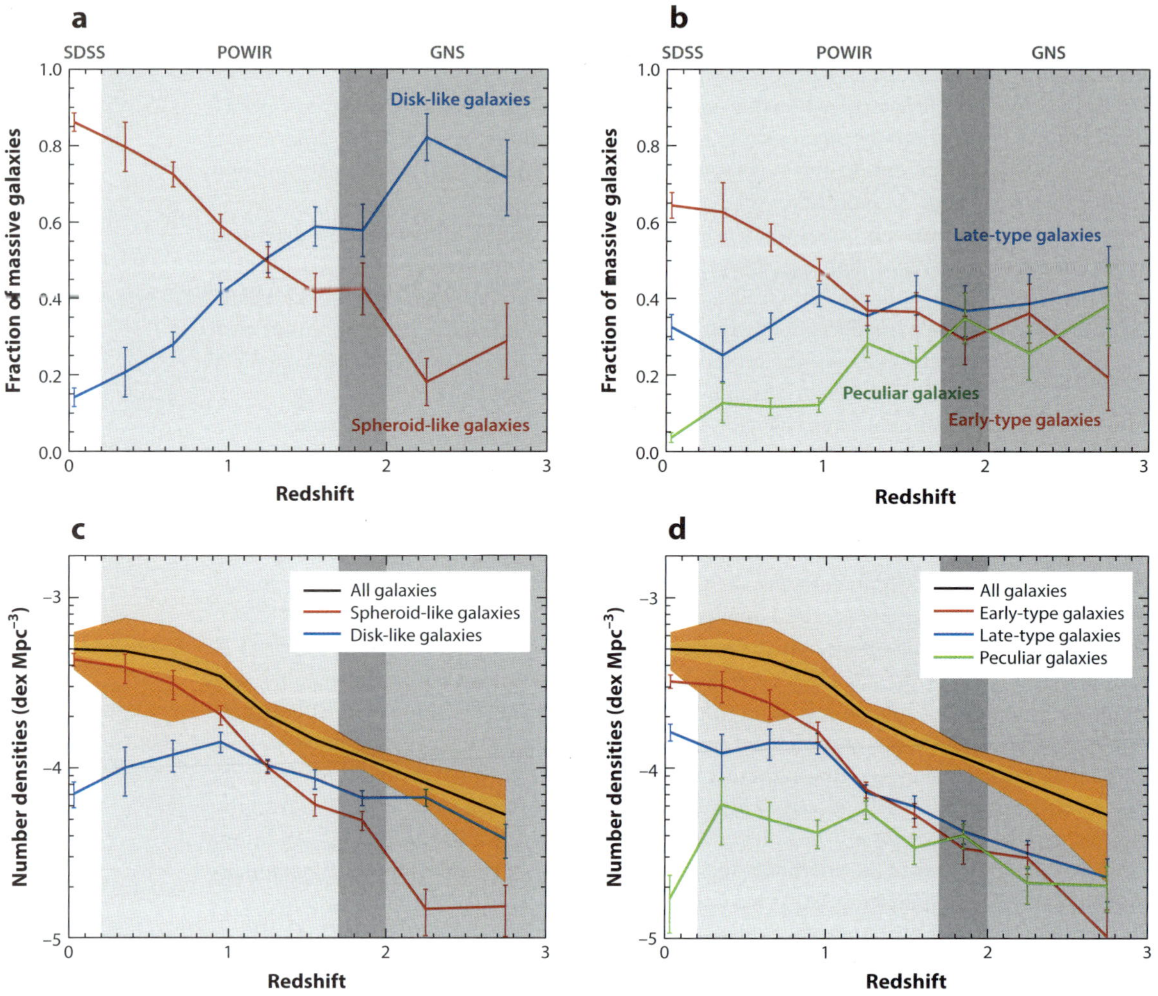

Figure 7

The evolution of apparent morphology and Sérsic-index-based classifications for massive galaxies with $M_* > 10^{11}\,M_\odot$ from Buitrago et al. (2013). Shown are the morphologies and structures derived from three different surveys—the Sloan Digital Sky Survey (SDSS) for nearby galaxies, the POWIR survey (Conselice et al. 2007) for systems up to $z \sim 2$, and the Great Observatories Origins Deep Survey NICMOS survey (GNS) for systems at $z \sim 1.8-3$. (*a*) In this plot, disk-like galaxies are those with Sérsic indices $n < 2.5$, and spheroid-like galaxies are for those with $n > 2.5$. (*b*) The morphological evolution as judged from visual estimates. Plotted here are both the fraction of types and the number density evolution. (*c*,*d*) The orange shading gives the total number density of all galaxies as a function of redshift up to $z \sim 3$. As can be seen, there is a gradual transition from galaxies that appear peculiar and "disk-like" in their Sérsic indices at high redshifts $z > 1.5$, which gradually transform into early types today.

What **Figures 7** and **8** show is that at $z > 2$ the dominant morphological type is peculiar, whereas disks and ellipticals become more common at lower redshifts (see also Conselice et al. 2005, Kriek et al. 2009, Delgado-Serrano et al. 2010, Mortlock et al. 2013). **Figures 9** and **10** show images of the most massive galaxies at both $z < 1$ and $z > 2$ in the Hubble UDF, demonstrating the stark differences between the two. This review later examines in Section 4.3 what these peculiars are potentially evolving into at lower redshifts. There is, however, a general trend such that peculiar systems and galaxies with internal features, such as blobs, have a higher star-formation rate than

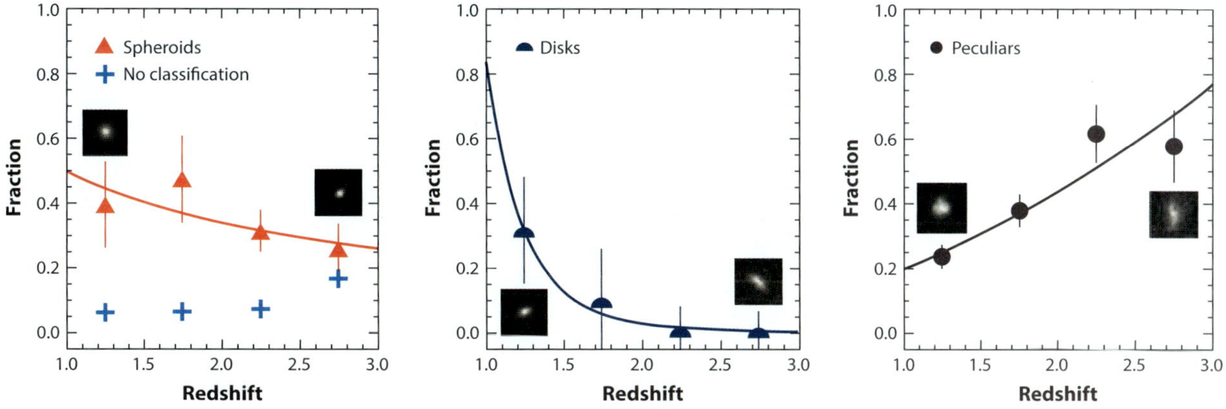

Figure 8

The latest version of the evolution of the Hubble sequence with redshift for galaxies with $M_* > 10^{10}\,M_\odot$. These classifications are corrected for image degradation such that misclassifications owing to distance are accounted for within these fractions (Mortlock et al. 2013). Examples of images of galaxies in each of these bins are shown in the observed H_{160}-band or rest-frame optical. Further analysis shows that there is a downsizing trend such that the most massive galaxies form into Hubble sequence galaxies earlier than lower-mass galaxies (Mortlock et al. 2013).

smoother galaxies (e.g., Lee et al. 2013). This shows that though the Hubble sequence itself is not formed early in the history of the Universe, there are trends that suggest it is becoming established.

Another interesting aspect to examine is not only the fraction of different galaxy types at various redshifts but also their number density evolution as a function of galaxy type. We know from galaxy stellar mass measurements that the most massive systems are largely in place at $z \sim 1$ at the same number density as at $z \sim 1$ (e.g., Bundy et al. 2005, Mortlock et al. 2011). It is the lower-mass galaxies that typically increase the most in terms of the relative number density from the epoch $z \sim 0$–2. In the nearby Universe most of the massive galaxies are ellipticals; this implies that most of the evolution in morphology is also within the lower-mass galaxies, which is what is found. In addition to a mass downsizing, we also find a morphological downsizing, such that morphological ellipticals are formed before the other galaxy types, most notably the spirals. However, whether these galaxies are inherently similar to the ellipticals and disks today is a separate question.

4.1.2. The formation of ellipticals and disks: bars, bulges, disks, and spiral arms. One of the major questions in galaxy evolution we would like to address is when modern spirals and ellipticals form. This is related to the formation of the Hubble sequence, but requires a more detailed answer because what appears as an elliptical/spiral may be quite different from systems classified this way in the local Universe. We thus are interested in determining when systems with the same morphology as galaxies we see in the local Universe (i.e., disks and ellipticals) achieve a similar physical state as measured through other properties. Although morphological fractions are similar at $z \sim 0.5$–0.8, as seen in the nearby Universe, similar physical properties of galaxies—such as colors, star-forming knots, and tidal features—have not reached the same level as locally. It is only recently at $z < 0.3$ that the galaxy population appears in most ways similar to that of today. This question can also be further divided into many subquestions, but for the purpose of this review, I examine the formation history of galaxies identifiable as disks and ellipticals in their gross morphology as well as the formation of more detailed features such as spiral arms, bulges, and bars.

One major issue is the bar fraction of galaxies and how it has evolved with time. Early studies found that the bar fraction evolves significantly (Abraham et al. 1999b), whereas later studies found

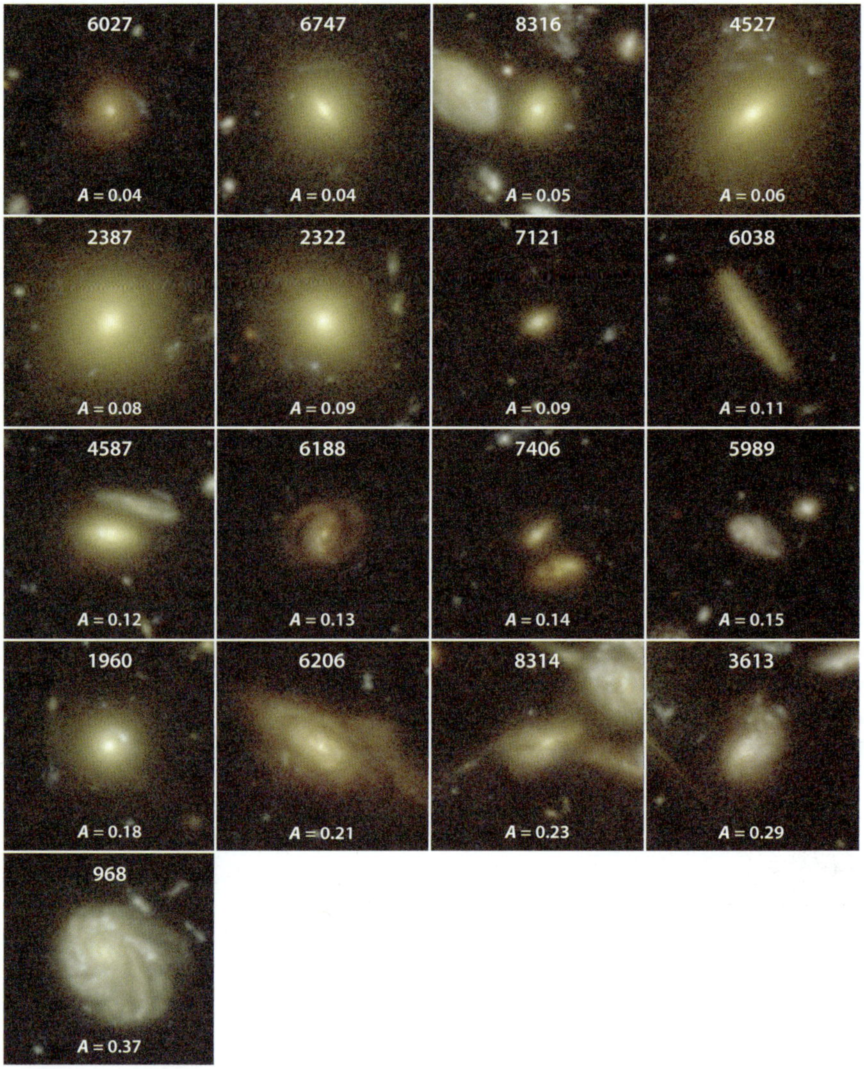

Figure 9

Galaxies in the Hubble Ultra Deep Field as imaged through the ACS camera and ordered by how asymmetric they are. These are all galaxies with redshifts $0.5 < z < 1.2$ and stellar masses $M_* > 10^{10}\,M_\odot$. The ID is the number used by Conselice et al. (2008), and the A value is the value of the asymmetry. At these redshifts most of the massive galaxies can still be classified as being on the Hubble sequence.

that bars were already in place up to $z \sim 1$ (Elmegreen et al. 2004, Jogee et al. 2004). Using the two-degree area COSMOS HST survey, Sheth et al. (2008) find that the bar fraction increases from $z = 0.84$ to $z = 0.2$, from 20% to ~60% of all disk galaxies. Sheth et al. (2008), however, find that the bar fraction is roughly constant with redshift for the most massive and red disk galaxies. In fact most of the observed evolution occurs for the lower-mass bluer disk galaxies.

The fraction of spiral galaxies with bars tells us when the disks in these galaxies become dynamically mature enough to form these structures. The fraction of bars also allows us to determine whether bars have a role in the evolution of star formation and bulge formation and in

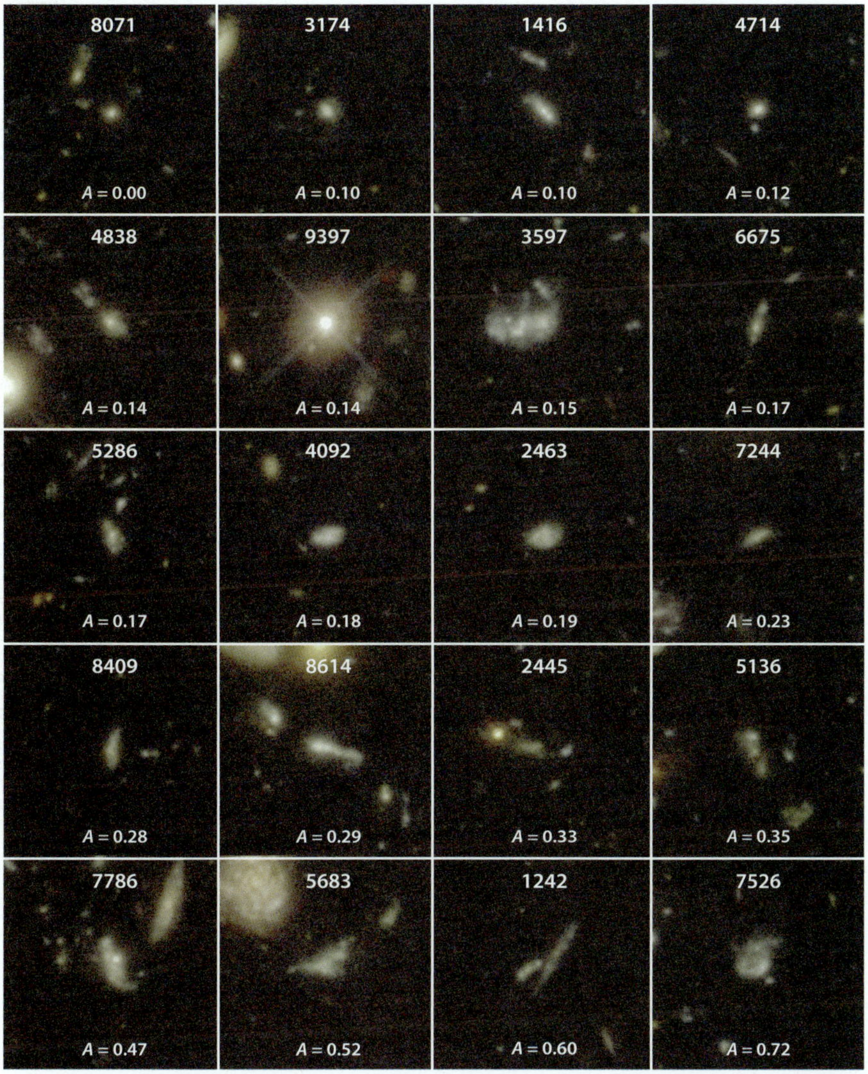

Figure 10

Massive galaxies in the Hubble Ultra Deep Field as imaged through the ACS camera and ordered by the value of their asymmetries from most symmetric to most asymmetric. Shown in this figure are systems with stellar masses $M_* > 10^{10}\,M_\odot$ at redshifts $2.2 < z < 3$. These galaxies are typically much smaller and bluer and have a higher asymmetry and inferred merger fraction than galaxies of comparable mass today (Conselice et al. 2008).

the triggering of AGNs and the formation of supermassive black holes. The observational studies above all locate bars within disks either through changes in ellipticity and position angle in the surface brightness profiles of galaxies or through visual inspection. It remains to be seen how the bar fraction holds when observed with a near-IR band such as within the CANDELS survey.

A related issue is finding the onset of spiral structure, which has remained a problem for a variety of reasons, but especially because of the difficulty of achieving an unambiguous detection owing to resolution/depth issues. Several papers published with the advent of ACS on HST showed that there are many examples of disk-like morphological systems at high redshifts. This includes the disks found by Labbe et al. (2003), the Luminous Diffuse Objects of Conselice et al. (2004), and later systems identified by Elmegreen et al. (2007) as clump-clusters. More recently, Law et al. (2012a) discovered a galaxy with a likely bona fide spiral structure at $z = 2.18$ that also contained a large internal velocity dispersion based on integral field unit (IFU) spectroscopy. Spiral arms, however, appear very rare at high redshift, and likely only a small number are formed before $z \sim 1$. This is a major problem that needs more attention and could be addressed with recently available data. However, there are many galaxies with clumpy features that are potential spiral arms and disks in formation, which I discuss in detail in Section 4.4.

Another feature of spirals and disks is the formation of bulges. Although the traditional scenario is that bulges form in mergers (Section 4.3), recent results suggest that bulges can form in multiple ways. Pseudobulges are different from classical bulges in that they are likely formed through secular processes within a disk. This can be seen in the different correlations between bulge properties and central massive black holes (e.g., Graham 2008, Hu 2008, Graham & Scott 2013, Kormendy & Ho 2013).

Although morphology itself can often be ambiguous in terms of matching with contemporary Hubble types, IFU or long-slit spectroscopy can remove some degeneracies. In particular the use of integral field spectroscopy on $z > 1$ systems reveals important clues about the nature of these high-redshift galaxies (Förster Schreiber et al. 2009, Glazebrook 2013). Currently the most influential studies have been carried out with the SINFONI IFU on the VLT along with some work done with OSIRIS on Keck (Law et al. 2012a,b).

These surveys, most notably the SINS collaboration (e.g., Förster Schreiber et al. 2009) find an equal distribution of different kinematic classes that are rotationally dominated, mergers, and very compact galaxies with high velocity dispersions (e.g., Buitrago et al. 2014 for massive galaxies). The true nature of these systems is not yet known, although in general it appears that galaxies at $z > 2$, which are clumpy, tend to have high velocity dispersions. This perhaps reveals that the formation modes of disk galaxies at high redshift results in a high velocity dispersion. The future looks promising for combining larger surveys of IFU measurements of distant galaxies with resolved morphologies to decipher evolution. Large surveys with, for example, KMOS on VLT and MOSFIRE on Keck should revolutionize this area in the coming years.

4.2. Size and Profile-Shape Evolution

One of the most important findings in galaxy evolution studies in the past decade has been the discovery that distant galaxies are more compact than systems of the same mass in the local Universe (e.g., Daddi et al. 2005; Trujillo et al. 2007; Buitrago et al. 2008; van Dokkum et al. 2008, 2010; Weinzirl et al. 2011; Barro et al. 2013; Williams et al. 2014). Change in sizes with time is now well characterized, and the evolution of galaxy sizes at a constant stellar mass selection of $M_* > 10^{11}\,M_\odot$ can be characterized by a power law of the form $R_e \sim \alpha(1+z)^\beta$. The value of the power-law slope changes with the galaxy surface brightness profile type such that the disk-like galaxies with Sérsic indices $n < 2.5$ evolve with $\beta = -0.82 \pm 0.03$, whereas spheroid-like galaxies with $n > 2.5$ have $\beta = -1.48 \pm 0.04$ (**Figure 11**). This demonstrates that there is a faster evolution in measured sizes for spheroid-like galaxies, which therefore have a more effective increase in size over cosmic time than the disk-like objects.

This size evolution is such that the effective radii of massive galaxies increases by up to a factor of five between $z = 3$ and today for galaxies at the same stellar mass (e.g., Buitrago et al. 2008, Cassata et al. 2013). The form of this evolution has been investigated to determine whether or

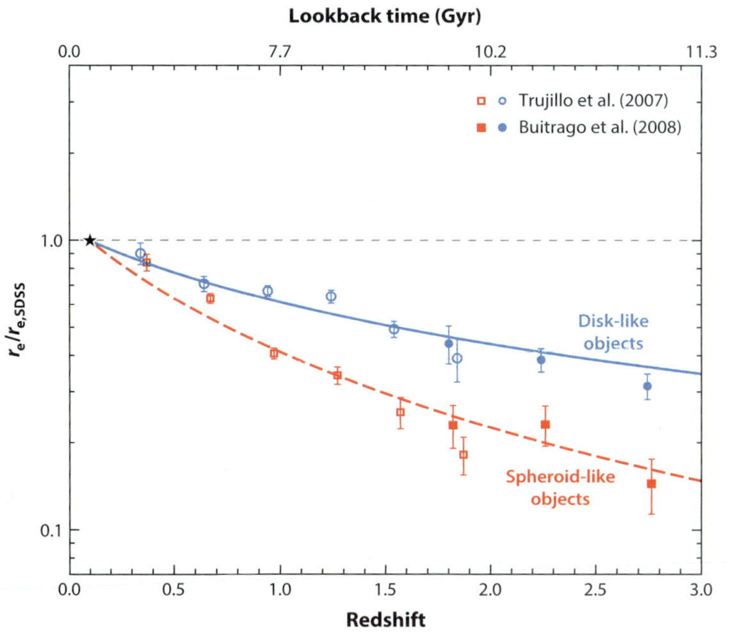

Figure 11

The average sizes of massive galaxies selected with $M_* > 10^{11}$ $M_\odot$ as imaged in the POWIR (Conselice et al. 2007) $z < 2$ data and GOODS NICMOS survey (GNS) > 1.5 images (Buitrago et al. 2008, Conselice et al. 2011b). The size evolution is divided among galaxies with elliptical-like profiles, Sérsic indices $n > 2.5$, and disk-like profiles having $n < 2.5$. The measured effective radius, R_e, is plotted as a function of the ratio with the average size of galaxies at the same stellar mass measurements with $M_* > 10^{11}$ $M_\odot$ at $z = 0$ from Shen et al. (2003).

not the increase is due to the buildup of the entire galaxy or just the inner or outer parts. The data to date show that galaxy growth through sizes is occurring in its outer parts, with the central parts in place at early times (e.g., Carrasco et al. 2010, van Dokkum et al. 2010). This indicates that the buildup of massive galaxies is an inside-out process, whereby the inner parts of massive galaxies are in place before the outer parts in galaxies with the same stellar mass density as today (e.g., Hopkins et al. 2009).

An alternative way to investigate this problem is to examine the number of compact and ultracompact galaxies at various redshifts. There is some controversy over whether or not there exist in the local Universe compact galaxies with sizes similar to those seen at high redshifts. However, what is clear is that the number densities of these ultracompact galaxies declines in relative abundance very steeply at $z < 2$ (Cassata et al. 2013).

The processes responsible for this increase in sizes at lower redshifts is not well understood and is currently a source of much debate. The most popular explanation is that this size increase is produced through minor mergers (e.g., Bluck et al. 2012, McLure et al. 2013), although other ideas such as AGNs performing work on gas is another idea (e.g., Bluck et al. 2011). However, the outer parts of nearby massive galaxies are too old to have been formed in relatively recent star formation, and the star formation observed at high redshift is not sufficient to produce the observed increase in sizes (Ownsworth et al. 2012).

The major idea for the physical mechanism producing galaxy-size evolution is through dry minor mergers, as major mergers are not able to produce the observations of increasing size without significantly increasing mass (e.g., Khochfar & Silk 2006, Naab et al. 2009, Bluck et al. 2012, Oser et al. 2012, Shankar et al. 2013). There is currently some controversy over whether or not the observed minor merger rate is high enough to provide this increase in sizes, with the most massive galaxies of $M_* > 10^{11}$ $M_\odot$ appearing to have enough minor mergers (e.g., Kaviraj et al. 2009) to produce this size evolution (Bluck et al. 2012), but this may not be the case for lower-mass systems (e.g., Newman et al. 2012). It does appear, however, that minor mergers are a significant mechanism for producing low levels of star formation in early types at $z \sim 0.8$ as well as for adding significant amounts of stellar mass to these galaxies (Kaviraj et al. 2009, 2011). Alternatively, the compact high-z galaxies may evolve into bulges of modern large spiral galaxies (e.g., Graham 2013, Dullo & Graham 2013). One of the major issues is determining not only the number of minor dry mergers but also the timescale for these mergers (Section 3.4), which more simulations would help elucidate.

Along with the evolution of galaxy sizes, there is also significant evolution in the underlying structures of galaxies at higher redshifts. One of the cleanest ways to see this is through the evolution of the Sérsic parameter, n (**Figure 7**). When examining the evolution of derived values of n as a function of redshift both for a stellar mass and at a constant number density selection, it is apparent that galaxies have lower n values at higher redshifts for the same selection (e.g., Buitrago et al. 2013). This has been interpreted by some to imply that these galaxies are more "disk-like" at high redshifts (Bruce et al. 2012), although the morphologies of these systems by visual inspection, and their internal structures and colors, are not similar to modern disks (e.g., Conselice et al. 2011a, Mortlock et al. 2013). These disk-like galaxies, while having light profiles similar to modern disks, are much smaller, have a higher stellar mass, and are often undergoing intense star formation with peculiar morphologies, making them undisk-like in all other regards. They are indeed likely a type of galaxy with no local counterpart.

4.3. The Merger History

One of the primary, if not the primary, uses of galaxy structure at high redshift at the time of writing is to measure the merger history of galaxies. This is a major issue in extragalactic astronomy, as merging is not only a method for galaxies to form but is also a potential way in which black holes, star formation, and other internal features of galaxies are assembled. Merging is also one of the key predictions of the cold dark matter model, which is the dominant idea for how galaxy evolution occurs through cosmic time (e.g., White & Rees 1978, Blumenthal et al. 1984, Cole et al. 2000).

The merger history of galaxies was first measured using galaxies in pairs—systems of at least two galaxies near enough to each other to merge in a reasonable amount of time (typically 20 or 30 kpc) and in cases of kinematic pairs with a low velocity difference of around 200 km s^{-1} (e.g., Patton et al. 1997, 2002; Le Fèvre et al. 2000; Lopez-Sanjuan et al. 2011, 2012). Simulation results show that most pairs of galaxies selected in this way will eventually merge within a timescale comparable to dynamical friction (e.g., Moreno et al. 2013). Using pairs to find galaxies that are merging is still a large industry, and interested readers are referred to the latest papers in this field (e.g., Lopez-Sanjuan et al. 2011, 2012; Tasca et al. 2014). Pairs of galaxies also provide a check on the methodology we use to find major mergers, as it is an independent measure of this formation mode. However, the merger fraction measured through pairs is more statistical in nature than structure and does not reveal for certain whether a galaxy is a merger or not.

A detailed CAS structural study of starburst, ULIRGs, and other merging galaxies shows that another route for measuring the merger history is through the use of the asymmetry index, whereby the most asymmetric galaxies are ones involved in mergers (Conselice 1997, 2003; Conselice et al. 2000a,b; Bridge et al. 2007) (Section 3.2). The merger history is also measured through other

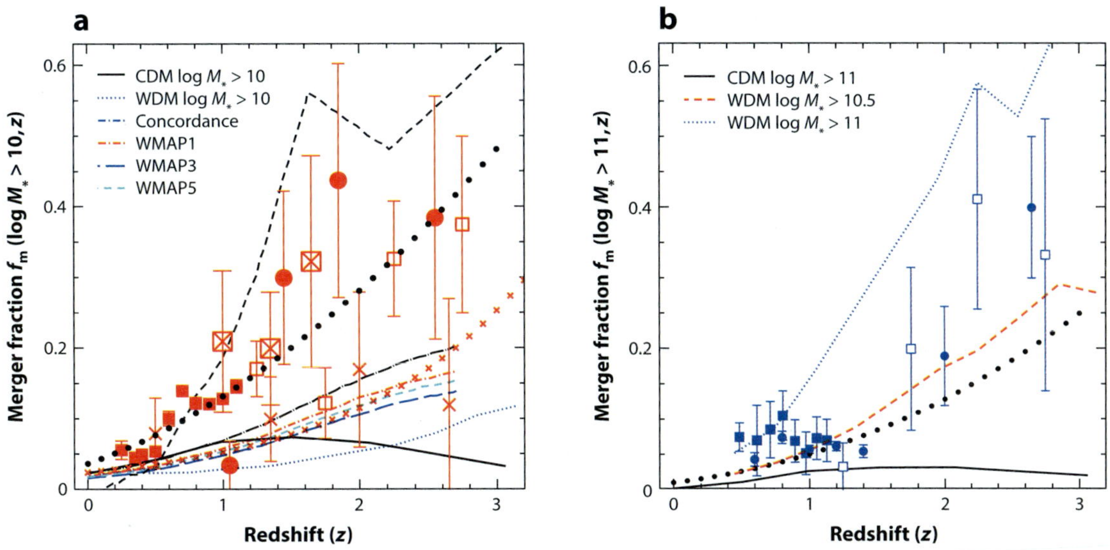

Figure 12

A compilation of the merger fraction history for galaxies selected with stellar masses $M_* > 10^{10}\,M_\odot$ and $M_* > 10^{11}\,M_\odot$. (*a*) Points shown here are for $M_* > 10^{10}\,M_\odot$ galaxies, including results from Conselice et al. (2003a) at $z > 1$ in the Hubble Deep Field (*solid circles*); from Conselice (2009) at $z < 1.2$ (*solid boxes*); and from Mortlock et al. (2013) at $z > 1$ (*open boxes*). Also shown are pair merger fractions at separations of <30 kpc: from Man et al. (2012) at $z > 1$ (*crosses*) and from Lopez-Sanjuan et al. (2010) (*large open boxes with crosses*). (*b*) The merger history for $M_* > 10^{11}\,M_\odot$ systems, including results from Conselice (2009) at $z < 1.2$ (*solid boxes*); from Bluck et al. (2009, 2012) at $z > 0.5$ (*solid circles*); and from Mortlock et al. (2013) at $z > 1$ (*open boxes*). In both panels, the line with the dark solid circles is the best fit relation for a merger fraction parameterization as $\sim(1+z)^m$. The blue dotted lines in both panels *a* and *b* and the red dashed line in panel *b* show the predicted merger history within warm dark matter (WDM) simulation predictions at the labeled respective masses. The solid black line shows the corresponding predictions from cold dark matter (CDM) simulations. For the $M_* > 10^{10}\,M_\odot$ panel I also show the merger fraction calculation from abundance matching using various WMAP and concordance cosmologies as well as abundance matching from Stewart et al. (2008) (line made up of *red crosses*).

parameters, whereby mergers occupy a unique parameter space (e.g., Lotz et al. 2004, 2008a; Freeman et al. 2013).

Although there is no perfect 1:1 relation between parameter space definitions of mergers with all mergers identifiable by eye, we still find that about half of all identifiable mergers fall within specific regions of parameter space and that there is low contamination from other galaxy types (Section 3.2). Thus, galaxy merger fractions are observations that have to be interpreted carefully with the use of galaxy merger models.

4.3.1. The merger fraction evolution. The merger fraction history is the basic observable that reveals how mergers are changing and evolving through cosmic time. It is measured at high redshift through the criteria described in Section 3.2, and the resulting merger fractions are shown in **Figure 12** (e.g., Conselice et al. 2008, 2009; Bluck et al. 2012). This figure shows that the inferred merger fraction increases with redshift. This increase is typically fit as a power law of the form

$$f_m = f_0 \times (1+z)^m, \quad (14)$$

where f_0 is the merger fraction at $z = 0$ and m is the power-law index for measuring mergers. In general, the higher the value of m the more steeply the merger history increases at higher redshifts. An alternative parameterization of the merger history is given by a combined power-law exponential (e.g., Conselice et al. 2008), whose form is

$$f_{\rm m} = \alpha \times (1+z)^{m} \exp[\beta(1+z)]. \quad (15)$$

The local $z = 0$ merger fraction in this formalism is given by $f_{\rm m}(0) = \alpha \exp(\beta)$, and the merger peak is located at $z_{\rm peak} = -(1 + m/\beta)$. This form of the merger fraction evolution appears to fit lower-mass galaxy merger fractions better than a single power law up to $z \sim 3$, as this allows for a merger fraction peak and a decline at higher redshifts. In fact, only the highest-mass galaxies with $M_* > 10^{10}\,{\rm M}_\odot$ appear to increase up to $z \sim 3$, whereas lower-mass galaxies have a merger peak around $z \sim 1.5$–2.5, which declines at higher redshifts (e.g., Conselice et al. 2008).

Although initially there were significant differences in merger histories between different studies (e.g., Lin et al. 2004), it is now clear that these are due to several effects. The first is that the value of the power-law index m can vary significantly just due to the value of the anchor redshift at $z = 0$. Secondly, when comparisons are done between galaxies that have been selected in the same way (stellar mass or absolute magnitude), and correct timescales are used for different techniques (Section 3.4), then merger rates agree within the uncertainties (e.g., Conselice 2009, Jogee et al. 2009, Lotz et al. 2011).

The merger history tends to peak at $z \sim 2.5$ for massive galaxies with $M_* > 10^{10}\,{\rm M}_\odot$ at values of $f_{\rm m} \sim 0.3-0.4$ and decline at lower redshifts. The values for m found in the literature can vary significantly, but most of this is due to different selections and different redshift ranges used as well as the use of various values for the local merger fraction. The first studies using pairs found a very steep increase up to $z \sim 1$, with $m = 2.8 \pm 0.9$ for a luminosity-selected sample up to $z \sim 0.4$ (Patton et al. 1997). The value of m was later found by Patton et al. (2002) to be $m = 2.3 \pm 0.7$ within the CNOC2 redshift survey up to $z \sim 0.55$. Le Fèvre et al. (2000) measure the merger fraction using 285 galaxies in the Canada-France Redshift Survey and the Low Dispersion Survey Spectrograph surveys up to $z \sim 1$, finding a power-law index of $m = 3.2 \pm 0.6$, although this lowers to $m = 2.7 \pm 0.6$ after considering selection effects. Other studies have found similar values, with $m = 1.5 \pm 0.7$ for brighter galaxies in the VIMOS VLT Deep Survey (de Ravel et al. 2009), and $m = 3.1 \pm 0.1$ for pairs up to $z \sim 1$ in the COSMOS field (Kartaltepe et al. 2007). These are, however, all relatively nearby galaxies at $z < 1$. For higher redshifts, Bluck et al. (2009) find a power law of $m = 3.0 \pm 0.4$ for galaxies with $M_* > 10^{11}\,{\rm M}_\odot$ using pairs from the GOODS NICMOS survey.

Within morphological studies, Conselice (2003) finds a high m index of $z \sim 4$ for massive galaxies with $M_* > 10^{10}\,{\rm M}_\odot$, with m values around $m \sim 1$–2 for lower-mass galaxies. This is similar to what is found by Le Fèvre et al. (2000) when examining the merger history of visually disturbed galaxies in Hubble imaging. This was confirmed by Conselice et al. (2008) using the same methods, but on the Hubble UDF data. Conselice (2003) also shows how the merger fraction slope m can vary significantly depending on what redshift limit is used and whether stellar mass or luminosity cuts are applied to the selected sample. In a detailed study of $z < 1$ galaxies from the COSMOS survey, Conselice (2009) finds an index of $m = 2.3 \pm 0.4$ for galaxies with stellar masses of $M_* > 10^{10}\,{\rm M}_\odot$. By contrast, Lotz et al. (2008a) find that the merger fraction does not evolve significantly with redshift in the range of $z \sim 0.2$–1.2, with a weak increase in the merger fraction with $m = 0.23 \pm 1.03$ using the Gini/M_{20} methods. Later, however, it was shown that Gini/M_{20} is very sensitive to minor mergers, and once these effects are considered the fitted parameter is $m = 2$ up to $z \sim 1.2$ (Lotz et al. 2011). Other methods of measuring the merger or interaction rate include looking for ringed galaxies (D'Onghia et al. 2008), a method that derives a merger fraction index with $m \sim 3$.

4.3.2. Galaxy merger rate evolution. The merger fraction is simply just an observational quantity, as it is the fraction of galaxies in a sample that have merged with another galaxy. This merger fraction can effectively be anything from zero to near unity depending on the mass ratio and timescales of interest. Merger fractions must therefore be carefully interpreted. The physical quantity we are interested in is the merger rate, which gives the number of mergers occurring per unit time and in some cases per unit volume per unit time.

Bluck et al. (2012), Lotz et al. (2008a), Jogee et al. (2009), and Conselice (2006b) show that CAS is only sensitive to major mergers of mass ratios of 4:1 and lower and has a particular timescale of about ~0.5 Gyr associated with the merger sensitivity (Section 3.4). This is very similar to the merger parameter sensitivity when using pairs of galaxies with mass ratios of 4:1 or lower (i.e., major pairs). Thus, by examining the merger fraction with CAS, we are likely tracing the same systems as measured through galaxies in pairs.

The merger fraction is converted into a galaxy merger rate through the use of the merger timescale $\tau_{\rm m}$, which can be derived through simulations (Lotz et al. 2010a) (see Section 3.4) or through the decline in the merger fraction at lower redshifts (Conselice 2009). The result of both approaches is that the CAS merger timescale is around 0.5 Gyr for a galaxy to remain "peculiar" in parameter space; the merger timescale in Gini/M_{20} remains at a similar level, but is more sensitive to minor mergers (Section 3.4).

Using the merger timescale, the merger rate per galaxy is thereby defined by the parameter Γ (e.g., Bluck et al. 2009, Conselice 2009),

$$\Gamma(M_*, z) = \frac{\tau_{\rm m}}{f_{\rm gm}}. \tag{16}$$

The value of Γ is in units of gigayears and gives the average amount of time between mergers for a galaxy within the given selection property, typically stellar mass, and within a given redshift range. Note that within the definition of Γ the galaxy merger fraction is used rather than the merger fraction (Section 3.2). The value of Γ as a function of redshifts for CAS merger measures and the pair selection mergers sample is shown in **Figure 13**.

Using the evolution of the galaxy merger rate, per galaxy, Γ, the number of mergers that occur between various redshifts is calculated by integrating the inverse of Γ over redshift,

$$N_{\rm merg} = \int_{t_1}^{t_2} \Gamma^{-1} dt = \int_{z_1}^{z_2} \Gamma^{-1} \frac{t_H}{(1+z)} \frac{dz}{E(z)}. \tag{17}$$

The result of this is studied in detail by Conselice (2009) and Bluck et al. (2009); initial results using simulations are discussed by Conselice (2006b). The result of these calculations shows that the number of major mergers a galaxy undergoes between $z = 3$ and $z = 1$ is 4.3 ± 0.8 major mergers at $z < 3$ (Conselice et al. 2008) for galaxies selected with stellar masses $M_* > 10^{10}\,M_\odot$. There also appears to be a limited number of mergers at higher redshifts of $z > 3$ (Conselice & Arnold 2009).

The Γ definition of the merger rate is per galaxy and thus does not take into account the overall merger rate within the Universe. Thus ultimately we are interested in the galaxy merger rate, $\Re(M_*, z)$:

$$\Re(M_*, z) = f_{\rm gm} \tau_{\rm m}^{-1} n_m, \tag{18}$$

where n_m is the number density of galaxies at a given redshift. Although this is the ultimate quantity in galaxy merger studies, it is difficult to measure, and the number density, $n_{\rm m}$, has its own associated uncertainties (e.g., Mortlock et al. 2011). In **Figure 13**, we plot the merger rate for galaxies using the CAS systems for pairs and mergers using the number densities of galaxies from Mortlock et al. (2011).

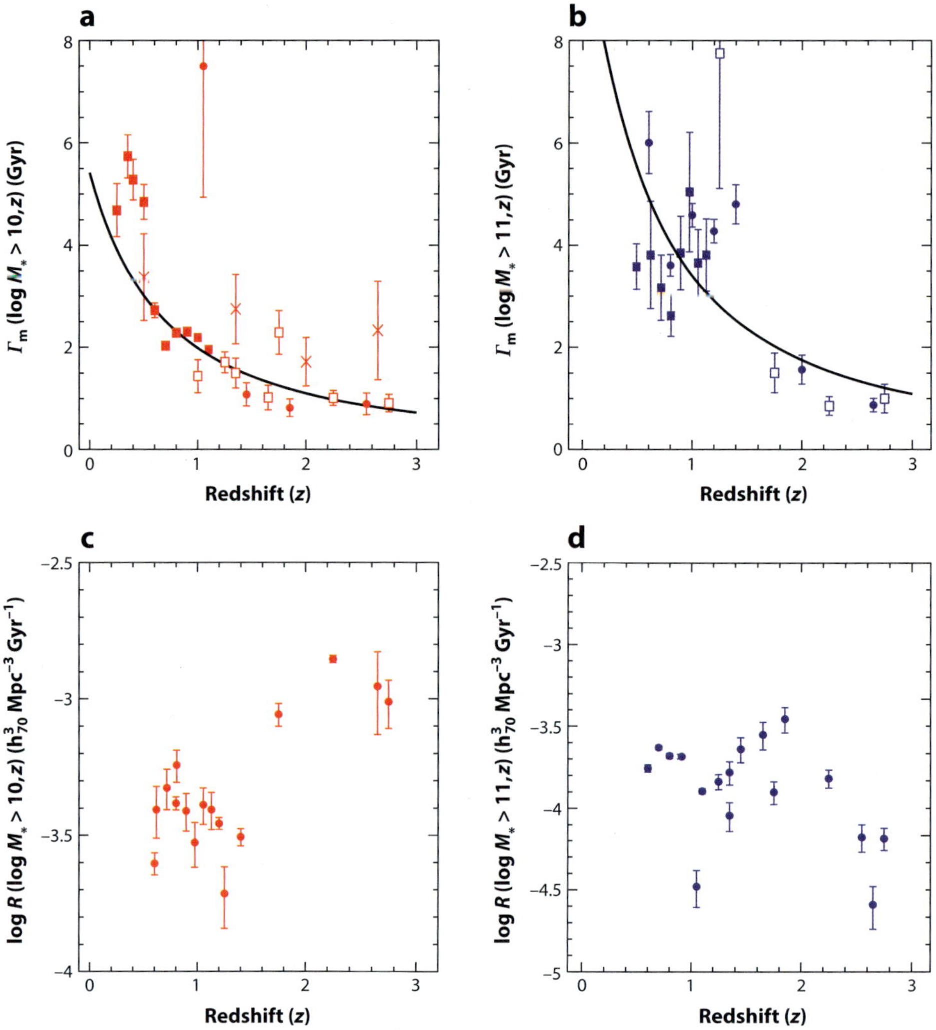

Figure 13

The merger rate for galaxies plotted in two different ways. The upper panels plot the merger rate per galaxy, (*a*) $\Gamma(z)$ at stellar mass limits of $M_* > 10^{10}\, M_\odot$ and (*b*) $M_* > 10^{11}\, M_\odot$. The units of Γ are in gigayears and represent the average time between mergers for a typical galaxy at the given mass limit. (*c,d*) The ultimate realization of the merger rate is shown at these two stellar mass limits, as the number of mergers occurring per gigayear per comoving cubic megaparsec. The point types in the merger rate per galaxy have the same meanings as the merger fraction plot in **Figure 12**.

Although the number of mergers is an interesting and fundamental quantity, we are ultimately interested in the amount of stellar mass added from mergers to galaxies over time. The total amount of stellar mass accreted into a galaxy is calculated as a double integral over the redshift range of interest (z_1 to z_2 or lookback times t_1 and t_2) and over the stellar masses range that is being probed (M_1 to M_2), and can be expressed as

$$M_{*,M} = \int_{t_1}^{t_2} \int_{M_1}^{M_2} M_* \times \frac{f'_m(z, M_*)}{\tau_m(M_*)} dM_* dz, \tag{19}$$

where $\tau_m (M_*)$ is the merger timescale, which depends on the stellar mass of the merging pair (Bluck et al. 2012). From this we calculate the total integration of the amount of mass assembled through merging for galaxies with stellar masses $M_* > 10^{11}\, M_\odot$. For a $z < 3$ mass complete sample at this limit, Conselice et al. (2013) find a value $M_{*,M}/M_*(0) = 0.56 \pm 0.15$, where $M_*(0)$ is the initial average stellar mass at $z \sim 3$, and $M_{*,M}$ is the average amount of stellar mass accreted in mergers at $z = 1$–3. This ratio is the fractional amount of stellar mass added due to both major and minor mergers for systems with stellar mass ratios down to 1:100 for an average massive galaxy after a merger adjusted for constant comoving density (Conselice et al. 2013). By observing the number of mergers, we can get an idea of how much gas is added to galaxies through the merger process. Comparing this with the star-formation rate, there is a large deficiency, which must be accounted for by gas accreted from the intergalactic medium. Using this number and the observed star-formation rate within these galaxies, the gas accretion rate from the intergalactic medium is calculated at around $100\, M_\odot$ year^{-1}, adding roughly the same amount of stellar mass to these galaxies as mergers (Conselice et al. 2013).

4.4. Resolved Morphological Formation Histories

Independent of the formation of the gross, or bulk, galaxy structure is the formation of their stars and how this correlates with structure and its assembly. This is effectively done by examining the SEDs of individual components in galaxies (i.e., disks or bulges) and/or using a newer technique of examining the SEDs of individual pixels. With resolved imaging it is therefore possible to determine to some extent the star-formation history of individual resolution elements (e.g., Abraham et al. 1999a, Zibetti et al. 2009) as well as the stellar mass within these (e.g., Lanyon-Foster et al. 2007).

Using early results from small samples of a few dozen galaxies at $z \sim 1$, Abraham et al. (1999a) demonstrated that many morphologically selected ellipticals in their sample show a diversity in their star-formation histories, with recent bursts of star formation commonly seen. Abraham et al. (1999a) also showed that bulges in these systems almost always have ages older than their disks. This was followed up with a study of 79 field spheroids by Menanteau et al. (2001), who found that a third of their morphologically regular systems have recent star formation in their centers, with so-called blue cores. It is possible that some of these systems are forming the pseudobulges that we see in the nearby Universe.

There are other studies of the bulges of distant galaxies, including that by Hathi et al. (2009) who find that bulges at $0.8 < z < 1.3$ have ages of 1–2 Gyr, with stellar masses up to $10^{10}\, M_\odot$ in the Hubble UDF. They also find that late-type bulges are younger than early types, a finding that also exists in the local Universe. However, classical bulges up to $z \sim 1$ with de Vaucouleurs profiles are found to have old stellar populations similar to giant ellipticals at the same epoch (Koo et al. 2005).

Another potentially powerful approach toward understanding the formation history of galaxies is to examine their light distribution on a pixel-by-pixel basis. The idea here is that each pixel or resolution element is independent of others and each has their own SED that can be fit by stellar population analysis methods (e.g., Bothun 1986, Abraham et al. 1999a, Lanyon-Foster et al. 2007). Local galaxies have different pixel color magnitude diagrams, depending on morphological type, with early types having a much more narrow locus of points than disk galaxies.

Galaxies also often look more symmetrical in stellar mass maps than in light (Lanyon-Foster et al. 2012, Wuyts et al. 2012; **Figure 14**). Star-forming galaxies in the CANDELS survey were recently studied in a pixel-by-pixel approach by Wuyts et al. (2012), who examined 323 systems at $0.5 < z < 1.5$ and a further 326 higher redshift systems at $1.5 < z < 2.5$. This study uses the optical ACS and near-IR WFC3 filters to construct SEDs for all pixels in these galaxies. Wuyts

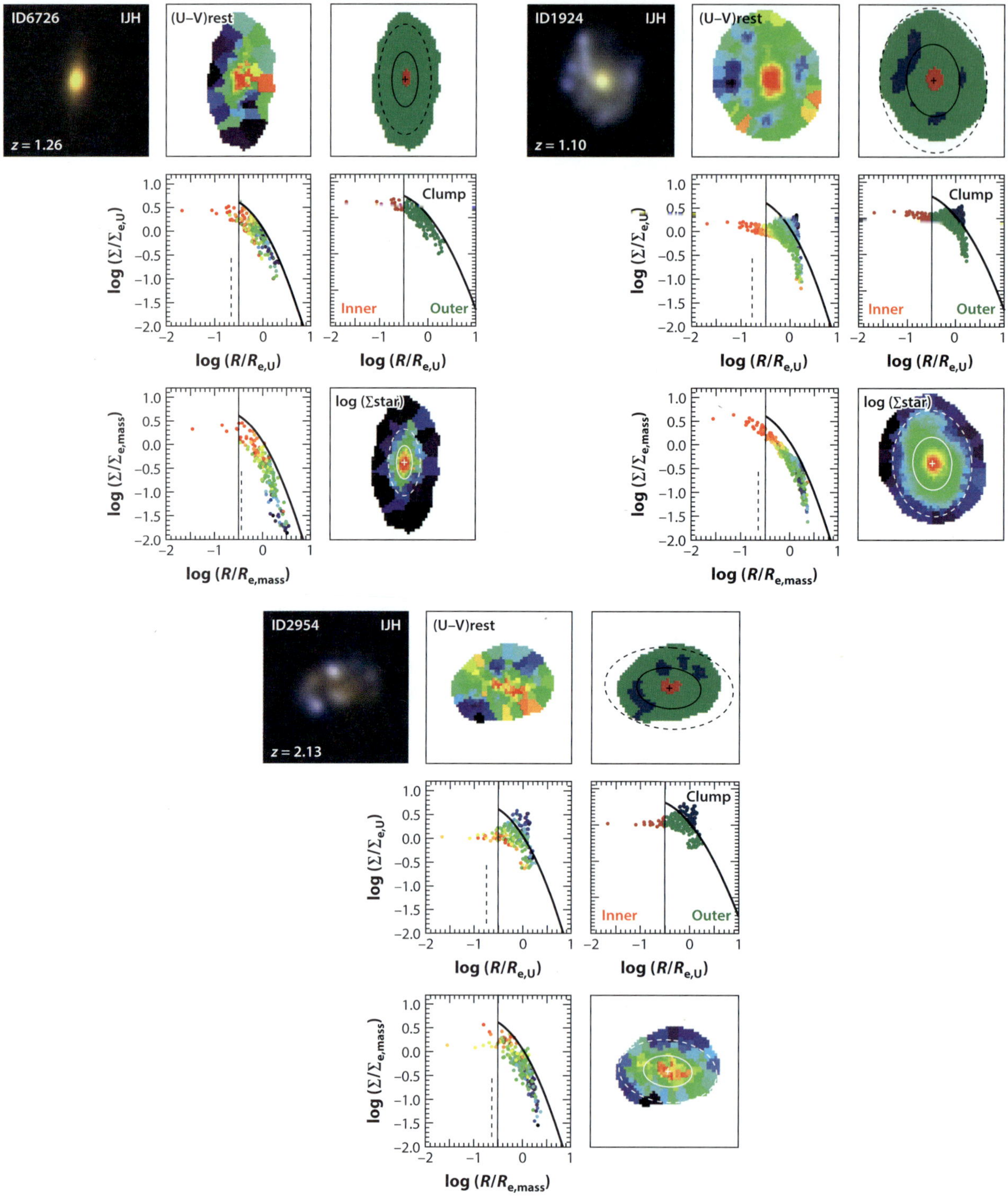
ID6726
IJH
z = 1.26
(U–V)rest
Clump
Inner
Outer
log (Σ/Σe,U)
log (R/Re,U)
log (Σ/Σe,mass)
log (R/Re,mass)
log (Σstar)
ID1924
IJH
z = 1.10
(U–V)rest
Clump
Inner
Outer
log (Σ/Σe,U)
log (R/Re,U)
log (Σ/Σe,mass)
log (R/Re,mass)
log (Σstar)
ID2954
IJH
z = 2.13
(U–V)rest
Clump
Inner
Outer
log (Σ/Σe,U)
log (R/Re,U)
log (Σ/Σe,mass)
log (R/Re,mass)

et al. (2012) find that the nuclei of star-forming galaxies are redder and have older ages than their outer parts. Clumps in these galaxies also generally occupy a smaller fraction of the total mass than the total light, demonstrating that the large clumps seen in distant star-forming galaxies are mostly star-forming regions. Clumps are also found to be central, off-central, or outer, and these clumps may play a role in forming the bulges of these galaxies through a secular process (Bournaud et al. 2007, Elmegreen et al. 2008, Genzel et al. 2008) (Section 5). This shows that there are methods beyond hierarchical clustering for the formation of structures within galaxies.

4.5. High-Redshift Active Galactic Nuclei/Starbursts and Star-Formation Quenching

One of the major ideas behind how mergers drive galaxy formation is that when galaxies merge, gas clouds collide, triggering star formation. At the same time, gas is driven to the centers of galaxies, producing AGN (e.g., Hopkins et al. 2008). Observationally we also know that galaxies that are concentrated, with for example a high Sérsic profile, are more likely to be quenched (e.g., Bell et al. 2012, Fang et al. 2013). There is therefore a strong theoretical reason to expect that structure, as influenced by mergers, should correlate with galaxy formation. There is also strong evidence that the star formation in galaxies is regulated by structure. How these relate, however, is not yet clear.

Observationally it has proven thus far difficult to correlate the presence of mergers, through either pairs or the presence of distorted or peculiar structures, with the presence of AGN or a star-formation excess at high redshift. Recent studies such as that by Cotini et al. (2013) use CAS indices to show that the fraction of nearby galaxies with AGN are roughly five times more likely than a control sample to have a distorted structure. However, studies at higher redshifts generally do not find that more asymmetric or merging galaxies have a higher AGN fraction (e.g., Grogin et al. 2005, Gabor et al. 2009, Kocevski 2012). This is also found in the deepest ACS Hubble imaging, although it does appear that AGN are found in more concentrated galaxies at higher redshifts (Grogin et al. 2005). There is also a lack of a higher fraction of peculiar/merging galaxies found in the GEMS/STAGES HST survey by Bohm et al. (2013) for AGN with luminosities of $L_X < 10^{44}$ erg s^{-1}. This may be an indication that Type 1 AGN are more likely found in early-type massive systems. It remains to be seen if obscured AGN, the Type 2s, are found in more merging systems and thereby represent an earlier phase of the merger.

Another major issue in which morphology and structure play an important role is understanding the origin of dusty star-forming galaxies, the so-called ULIRGs or submillimeter galaxies. Submillimeter galaxies are those that appear very bright at submillimeter wavelengths, typically

Figure 14

Internal pixel structures and profiles for galaxies at $z \sim 1$–2 with and without star formation. The examples are ID6726, which is a nonstar-forming galaxy, and ID1924 and ID2954, which contain clumpy star-forming knots throughout their structures. These are representative examples of commonly found galaxies at high redshifts, typically around $z \sim 2$ as seen in deep WFC3 images from surveys such as CANDELS. For each of these galaxies, the top rows demonstrate the image in the IJH *Hubble Space Telescope* bands, the rest-frame (UV) color map, and a map that indicates whether pixels belong to the inner (*red*), outer (*green*), or clump region (*dark blue*) for each galaxy. The solid black ellipses show the area that contains half of the rest-frame U-band light for each system. The middle rows for each galaxy show the U-band light profile, which is color coded on the left by the pixel type labeled in the (U–B) color map and on the right by the location of the various pixels. The lines shown in the middle row panels show the separation between the different spatial distributions. The bottom rows show the distribution in terms of stellar mass. The vertical dashed lines show the resolution limit of WFC3. The final panels show the stellar mass map for each system, with the inner solid and outer dashed lines showing the location of $R_{e,mass}$ and $2R_{e,mass}$. Reprinted from Wuyts et al. (2012) with permission and courtesy of Stijn Wuyts.

at 850 μm, and are at high redshift at $z > 1$. Morphological analyses of these submillimeter galaxies showed early on that they are involved in merger activity and have peculiar morphologies (e.g., Chapman et al. 2003, Conselice et al. 2003b). Many of these galaxies have structures and morphologies consistent with being involved in mergers (e.g., Ricciardelli et al. 2010). Kartaltepe et al. (2012) find that a sample of ULIRGs selected from the CANDELS fields are more likely than a field galaxy sample to be involved in galaxy interactions and mergers (72^{+5}_{-7}% versus 32 ± 3%). However, Swinbank et al. (2010) argue, using a large sample of submillimeter galaxies, that their morphologies are not significantly different from other star-forming field galaxies at similar redshifts. Swinbank et al. (2010) also find that their submillimeter galaxies have light profiles more similar to early types rather than disks. Although, Targett et al. (2011) find that submillimeter galaxies have disk-like profiles and conclude that these systems are more like forming disks than forming spheroids in mergers. This suggests that the morphology of submillimeter galaxies is still open for debate, and results appear to be conflicting. More detailed work, likely with adaptive optics in the K-band to avoid issues with dust and morphological k-corrections, are needed to make further progress.

Furthermore, as mentioned earlier, there is a strong observed correlation between galaxies that have steep surface brightness profiles, with $n > 2.5$, and the quenching of star formation (e.g., Bell et al. 2012). This shows that galaxy structure either produces a change in the galaxy history or more likely is a symptom of the effects that produce this quenching. In general, there are a few ways to quench the star formation seen in galaxies at high redshift. Some of these are environmental, such as ram-pressure stripping and strangulation (e.g., Peng et al. 2010). However, these processes are ineffective for the bulk of the field galaxies that we study in this review. What is more likely is that quenching of typical field galaxies is merger driven or it is driven by the stellar/halo mass in some form of feedback process (e.g., Peng et al. 2010, 2012; Carollo et al. 2013; Lilly et al. 2013). The most likely candidates are mergers that either heat gas or remove most of it in giant starbursts, preventing further star formation. The other idea explaining this feedback is that it is the result of AGNs, whereby the ongoing star formation is truncated by the existing gas in a galaxy being heated or removed by an active AGN (e.g., Croton et al. 2006). It is therefore likely that a few critical processes are ongoing to produce the Hubble sequence, specifically the red, passive and concentrated high-mass systems.

4.6. The $z \sim 3$–6 Frontier

By far the bulk of what we know of galaxy structural evolution is at $z < 3$. The reason for this is simply because this is the limit where we can observe the rest-frame optical using observations in the near-IR from the HST, which is only effective at imaging with filters bluer than the H-band (often H_{160} with WFC3 and NICMOS). However, there are some observations of galaxy structure at even higher redshifts that provide some information about the formation of these galaxies. To date most of these observations are done with a red filter using the ACS camera on HST, either the I_{814} or z_{850} band. It must be remembered that these systems are being observed in the rest-frame UV, and thus their morphologies are dominated by young stars.

There have only been a few major studies that focus on the structures of these ultrahigh-redshift galaxies. Ferguson et al. (2004) study the sizes and the axis ratios of Lyman-break galaxies (LBGs) galaxies up to $z \sim 6$ in the GOODS fields, finding that galaxies are smaller and more "disk-like" in their axis ratios at higher redshifts. This was also shown in an extensive study of 4,700 LBGs by Ravindranath et al. (2006), who find that 40% have exponential light profiles, 30% have de Vaucouleurs profiles, and the remaining 30% have multiple cores. The ellipticity distribution of these LBGs shows that these systems are skewed toward high values with $\epsilon > 0.5$, which cannot be explained by viewing disks and spheroids at various angles (Ravindranath et al. 2006). This is either an indication that these systems are mergers or that star formation is distributed in the outer parts of these galaxies, thus creating these high-ellipticity structures.

For ultrahigh-redshift galaxies, Conselice & Arnold (2009) examine the visual morphologies and pair fractions of LBGs in the Hubble UDF ACS filters from $z \sim 4$–6. These ACS data on the UDF still provide the highest resolution and deepest imaging of the most distant galaxies. Conselice & Arnold (2009) find that the fraction of $z \sim 3$–6 LBGs that are peculiar in appearance is roughly constant at ~30% throughout this redshift range. Conselice & Arnold furthermore demonstrate that many of the LBGs at these redshifts have tidal-like features—fans, shells, etc.—that resemble merger signatures seen at lower redshifts. The derived merger fraction from LBG pairs also agrees with the merger fraction based on CAS and visual estimates (Conselice & Arnold 2009, Cooke et al. 2010).

It is perhaps surprisingly easier to identify galaxies in pairs at these redshifts than at lower redshifts, as one can use the drop-out band to remove contamination, and thus ensure that two galaxies close by in the sky are at least at a similar redshift. This results in a smaller correction being needed to calculate merger fractions, and thus the merger fraction in principle can be measured more accurately (Conselice & Arnold 2009). Jiang et al. (2013) similarly examine the rest-frame UV morphologies of 51 Lyman-α galaxies and 16 LBGs and find a merger fraction for the brightest galaxies of around 50% and otherwise a diversity in morphology.

Most recently, using WFC3 data from the UDF, Oesch et al. (2010) show that $z \sim 7$–8 galaxies are very compact; these systems have a typical size of 0.7 ± 0.3 kpc with little size evolution down to $z \sim 6$. There is more development down to $z \sim 4$, with the observation of more extensive wings of light at these lower redshifts, and a corresponding increase in sizes (e.g., Ferguson et al. 2004) following a similar power law with redshift, as is found for size evolution between $z \sim 3$ and $z \sim 0$ (Buitrago et al. 2008, Mosleh et al. 2012).

4.7. Role of Environment in Structure Formation

Galaxy morphology is well known to correlate strongly with environment in the local Universe (e.g., Dressler 1984, Postman et al. 2005). It is also clear that there is a strong relationship between morphology and stellar mass such that the most massive galaxies tend to have elliptical morphologies and lack star formation. Combining this with the morphology-redshift relation shows that the structure of a galaxy depends upon its mass and local environment as well as time. Which of these is the leading cause for producing galaxy evolution is an active area of study.

The problem of galaxy morphology as a function of density is a large area of research and is outside the immediate scope of this review. However, it is relevant to discuss some of the major findings and how they relate to the evolution of galaxy structure with time. The major effect of morphology is that the type of galaxy, either elliptical or spiral, depends to a large degree in the nearby Universe on the local density of that particular galaxy's environment. This relation is such that the higher the density of the local environment, the more likely a galaxy is early type and nonstar-forming (e.g., Dressler 1984, Gomez et al. 2003, Blanton & Moustakas 2009). Disk properties are also highly environmentally driven, with few classical bulge or elliptical systems in low-density environments (Kormendy et al. 2010).

It is also the case that more massive galaxies are more likely to be early type. The question is which relationship is more fundamental, and this relates to the old issue of "nature versus nurture" for galaxy formation. The structures and morphologies of galaxies can help address this problem, especially by examining the limited number of observations we have of galaxy structure in high-redshift overdensities or (proto)clusters.

Observations of overdensities at high redshifts are just starting in earnest, but already provide some clues. Very massive clusters at high redshifts, up to $z \sim 1.2$, contain a similar pattern of morphologies and densities as local galaxies, such that the denser areas contain a higher fraction of systems that are elliptical. This tends to break down for the limited number of cluster candidates found at higher redshifts, where the galaxy population is irregular and peculiar, as is found for the general high-redshift galaxy population (e.g., Papovich et al. 2012).

Detailed studies are, however, possible up to $z \sim 1$ by both using field galaxies of various local environments as well as looking at the morphological and structural distributions of galaxies within rich clusters at various redshifts. For field galaxies, Tasca et al. (2009) examine the morphology-density relationship for 100,000 galaxies in the COSMOS survey. They find that the morphology-density relation changes slightly with redshift, becoming flatter at higher z (e.g., Grützbauch et al. 2011a,b). Above a stellar mass of about $10^{10.6}\ M_{\odot}$ the morphologies of galaxies appear to become more dominated by the stellar mass as the critical factor rather than density (e.g., Grützbauch et al. 2011b). The situation in rich clusters at $z \sim 1$, and the formation of S0s, is such that the trend with environment is not as steep as is found at $z \sim 0$. This suggests that S0s are not entirely formed yet in these distant clusters; however, the elliptical population does seem to be in place compared with the population at $z \sim 0$.

Another structural feature that can be investigated is the size evolution and how it varies with environment. The limited number of investigations of this have found that galaxies at $z > 1$ in higher-density environments show signs of a more rapid increase of galaxy size with redshift in comparison with the field (e.g., Cooper et al. 2012, Lani et al. 2013). This is one indication whereby a dense environment can facilitate a more rapid evolution in galaxies, although it is a slight effect that needs further confirmation.

5. COMPARISONS TO THEORY

Examination of galaxy morphology and structure provides a new perspective to compare against cosmologically based galaxy formation models, including those that entail extensive physics, such as star formation, AGN feedback, and supernovae, in more detailed hydrodynamical models. This review only briefly discusses this large topic and how it relates to galaxy structure. For a more detailed recent review on the theory of galaxy formation from a theoretical prospective, see Silk & Mamon (2012).

Galaxy formation models were first developed to explain the structures of galaxies, namely the bulge/disk/halo trichotomy, and the ages of the stars in these components (Eggen et al. 1962). The default initial assumption in the first galaxy formation models was that galaxies formed like stars in a relatively rapid collapse. In the 1980s the first computer simulations of structure formation showed that a Universe dominated by cold dark matter (CDM) matched observations of galaxy clustering on large scales (Davis et al. 1985) and that within this framework galaxy assembly should be hierarchical (Blumenthal et al. 1984), yet this is a fundamental prediction that is just now starting to be tested with only a few papers comparing the observations with the theoretical predictions (e.g., Bertone & Conselice 2009, Jogee et al. 2009, Hopkins et al. 2010, Lotz et al. 2011).

Today there are many simulations used to predict properties of the galaxy population and how it evolves through time. These models are largely successful when predicting basic properties of nearby galaxies, such as their luminosities, masses, colors, and star-formation rates, as well as the evolution of galaxy scaling relationships. However, problems still exist in predicting the abundances of low- and high-mass galaxies (e.g., Conselice et al. 2007, Guo et al. 2011). Within galaxy-formation models there are very famous problems such as the satellite and the CDM dark matter profile, but there are also significant issues when examining how the evolution of galaxies occurs and trying to match this with the theory. Another major problem is that there are several large disk galaxies without significant bulges in the nearby Universe that are not predicted in CDM (e.g., Kormendy et al. 2010).

One of the ways to further test these models is to investigate how well CDM models can reproduce the formation history of galaxies as seen through the merging process using the so-called semianalytical method (e.g., Bower et al. 2006, Guo et al. 2011). We show this comparison with the measured merger fractions in **Figure 12** at two different stellar mass ranges of $M_* > 10^{10}\,M_\odot$ and $M_* > 10^{11}\,M_\odot$. Plotted as the thin solid black line toward the lower part of each diagram is the prediction for the major merger fraction for galaxies from the Millennium simulation (Bertone & Conselice 2009). Also shown on these figures by the dotted blue line are the same predictions for major mergers as for warm dark matter models (e.g., Menci et al. 2012), which do a better job than CDM in matching the observed data. However, CDM better matches if minor mergers are taken into account, although the comparison merger fraction is only for major mergers based on the methodology used (Conselice et al. 2003a, Lotz et al. 2010a).

Other recent attempts to predict the merger history of observed galaxies include the abundance matching technique (e.g., Stewart et al. 2008), in which observed galaxies are matched to halos in models through their comparative abundance levels. Hopkins et al. (2010) predict, based on this abundance matching, the merger rate and fraction for galaxies. The result of this is shown in **Figure 12** for galaxies between $M_* = 10^{10-11}\,M_\odot$. Although the merger fractions from Hopkins et al. (2010) are higher than those from the CDM models, they are still lower than the observations [see also Jogee et al. (2009) and Lotz et al. (2011) for further discussions]. Similar results from Stewart et al. (2008) are also shown in **Figure 12**, who find results similar to those of Hopkins et al. (2010).

Finally, as a contrast to these, Maller et al. (2006) present cosmological hydrodynamical simulation results for similar mass galaxies of a few times $10^{10}\,M_\odot$ and find the highest merger fraction predictions of any simulation result (**Figure 12**). This shows that the predictions for merger histories are not correct or consistent with each other and that more simulation work should be focused on this critical aspect of the galaxy population. This is an area where future work is certainly needed.

There are several other types of simulations in which galaxy structure and morphology can be directly compared with observations of galaxies through cosmic time. Perhaps the most direct of these is to compare the properties and structural features of distant galaxies to hydrodynamical models of galaxy formation. Some of this work for galaxy mergers is discussed in Section 3.4. Early work in this area shows that the components of galaxies—namely bulges and disks—were the result of accretion events (e.g., Steinmetz & Navarro 2002), and these authors argued on the basis of their simulations that the Hubble type of a galaxy is not stable for long periods of cosmic time. Governato et al. (2007) show that disk galaxies having properties that match the morphological properties and kinematics of nearby disks can be simulated, although their simulation is not in a cosmological context. Overall, however, it is very difficult to predict the formation of galaxy morphology in simulations, and in a real sense this will be the ultimate test of galaxy formation models in the future.

Also, as discussed in Section 4.6, one of the most commonly seen properties in high-redshift star-forming galaxies is that they often contain large clumps of star formation within their disks. A major question is how these clumps form and evolve and how they may play a role in the formation of other galaxy components such as bulges and AGN. Bournaud et al. (2014) examined this problem computationally to determine how clumps with stellar masses of $M_* = 10^{8-9}\,M_\odot$ evolve in gaseous disks. The major question is whether these clumps dissipate within 50 Myr or so—which is the dynamical timescale of the clumps—or whether they regenerate and survive.

Bournaud et al. (2014) find that these clumps can last around 300 Myr by acquiring new gas from their disks, although some mass is lost through tidal effects. This is enough time to migrate toward the center of the galaxy, which can fuel the AGN or merge to form a bulge. This shows that these clumps may provide a significant route for galaxies to form. Thus, we have evidence for both inside-out and outside-in formation occurring in the galaxy population. What remains to be seen is whether one of these mechanisms is dominant as well as the relative role of both mechanisms in forming galaxies.

6. SUMMARY AND THE FUTURE

I present here a review of galaxy structure and morphology studies in the galaxy population through cosmic time from $z = 6$ until today. The approach taken in this review is largely observational with a limited amount of interpretation, although I do show where galaxy structure and morphology can test galaxy formation and even cosmological models in a new, largely unexplored way.

As of January 2014, the major conclusions concerning galaxy structure and its evolution, as discussed in this review, can be summarized by the following:

1. Galaxy structure and morphology are the longest-studied observational features of galaxies. In this review galaxy morphology is the apparent classification based on visual inspection, whereas structure is a way to quantify the light distributions in galaxies. In many ways, morphology is still a descriptive science, and visual efforts continue to provide useful information in the form of large-scale projects to classify many galaxies, as in the Galaxy Zoo effort. The Hubble sequence has and likely will remain the major paradigm in which we consider galaxy morphology, although this system does not "work" at high redshifts where most galaxies cannot be classified into a single Hubble type (Section 4.1).
2. Using the Hubble scheme, the evolution of three broad classes of galaxies is now classified accurately out to $z \sim 3$—namely ellipticals, spirals, and peculiars. The relative abundance of these galaxies has been measured as a function of redshift out to these early times. What we find is that the peculiar galaxies dominate the galaxy population at $z \sim 2.5$–3, and they have a relative fraction of at least 70%. Galaxies that are elliptical and spiral-like in appearance (but not necessarily in physical properties; see Section 4.1) become progressively more common at lower redshifts. The number densities of these two normal galaxy types together equal that of the peculiars by $z \sim 1.4$ (Mortlock et al. 2013).
3. Because galaxy morphology by visual estimates is limited in its ability to derive the physics behind galaxy formation and by its nature is not quantitative, the use of parametric (Section 2.2) and nonparametric (Section 2.3) methods is essential for deriving in a quantitative way how galaxies are evolving. These quantitative indices also correlate to some degree with the present and past star-formation history and properties of a galaxy. More work is needed to establish these relationships with more certainty, but it appears that the Sérsic index and concentration correlate with the scale or mass of a galaxy, the clumpiness index with the star formation, and the asymmetry parameter with ongoing merging activity (Section 3).
4. The merger history is now known from applying structural analyses to galaxy images in deep HST surveys such as the HDF (Section 4). The result is that the galaxy merger fraction increases with redshift at all stellar mass and luminosity selections. This increase can be fit well by a power law $(1+z)^m$ up to $z \sim 3$, although at higher redshifts the structurally derived merger fraction may plateau (Conselice & Arnold 2009). Using numerical/hydrodynamical simulations, the timescales for these mergers can be calculated, and thus merger fractions can be converted into merger rates (Section 4.3.2). The merger rate allows for the calculation of the number of mergers galaxies undergo at various masses as well as the amount of stellar

mass that is added to galaxies owing to the merger process. The result is that up to half of the stellar mass in modern massive galaxies was formed in mergers between $1 < z < 3$, although at $z < 1$ dry mergers are more likely responsible for the further formation of these galaxies.

5. The resolved structures of galaxies also allow us to measure the internal features of galaxies and how they are assembling. There is some controversy over the formation history of bulges, disks, and bars, although many of these are likely formed by secular processes produced internally by disk dynamical evolution. This is an area where significant progress could be made in the next few years. The most up-to-date results suggest that the bar fraction for spiral galaxies at $z < 1$ depends upon the stellar mass of the galaxy. The most massive galaxies have a similar bar fraction at $z \sim 0.8$ as they do today, yet lower-mass and bluer disk galaxies have a significantly lower bar fraction than similarly low-mass nearby disks. This mirrors the evolution of the Hubble sequence itself, where more massive galaxies settle into normal ellipticals and disks before lower-mass galaxies. Spiral structure is a difficult problem, and though some examples exist at high redshifts, even at $z > 2$, the general onset of when disks form spirals is almost totally unconstrained by observations.
6. Resolved imaging also permits us to measure the SEDs and colors of galaxy components and individual pixels of different galaxies. This is another area where more work needs to be performed, but it appears that bulges of spirals tend to be older than their disks at high redshift; there are examples of many ellipticals that have blue cores and central star formation (Section 4.4). Pixel-pixel analyses show that galaxies have a mixed star-formation history and that the inner parts of galaxies are often older than their outer parts. Pixel studies also show that the clumps seen in distant star-forming galaxies are composed of young stellar populations and thus must have recently formed or regenerated themselves.
7. Perhaps the most popular (at present) problem in galaxy structural evolution is the apparent compactness in size of galaxies at high redshifts. The observations show that massive galaxies at $z > 1$ have sizes that are a factor of 2–5 smaller than similar massive galaxies in today's Universe. This result has been studied in many different ways, and the sizes of a stellar mass–selected sample of galaxies increases gradually as a function of redshift with a power-law increase $\sim (1 + z)^{\beta}$, where β varies from -0.8 to -1.5 depending upon whether the selected samples are disk-like or elliptical-like (Section 4.2). Results to date suggest that these galaxies are building up their outer parts over time to become larger systems rather than adding mass to their centers. This process is unlikely driven by star formation, and theory suggests that this formation is produced by minor merger events (Section 4.2).

In summary, we have learned much about galaxy morphology and structure over the past 15 years. There are, however, many open questions still remaining on all aspects of using structure to determine evolution. More work needs to be done to tie galaxy structure to underlying physics, both through empirical work and in simulations. Furthermore, the timescales for structural features, such as mergers and large clump survival, are critical to achieve better understanding. Broad morphological features will remain important over the next decades as telescopes, such as the *James West Space Telescope*, *Euclid*, and the Large Synoptic Survey Telescope, and the Dark Energy Survey, among others, will all resolve many more galaxies than we can currently study, and they will do so at higher redshifts. This opens up entirely new possibilities, and with careful, thoughtful planning, a new revolution in galaxy structure may be upon us soon.

DISCLOSURE STATEMENT

The author is not aware of any affiliations, memberships, funding, or financial holdings that might be perceived as affecting the objectivity of this review.

ACKNOWLEDGMENTS

A review such as this is written with help from many people. In particular I thank Alice Mortlock, Asa Bluck, Fernando Buitrago, Jamie Ownsworth, and Ken Duncan for illuminating conversations and collaboration on these topics over the past few years. I also thank Jennifer Lotz, Fernando Buitrago, Stijn Wuyts, and Alice Mortlock for kindly providing figures. I personally thank the STFC and the Leverhulme Trust in the United Kingdom as well as the NSF and NASA in the United States of America for financial support.

LITERATURE CITED

Abraham RG, Ellis RS, Fabian AC, Tanvir NR, Glazebrook K. 1999a. *MNRAS* 303:641
Abraham RG, Merrifield MR, Ellis RS, Tanvir NR, Brichmann J. 1999b. *MNRAS* 308:569
Abraham RG, van den Bergh S, Glazebrook K, et al. 1996. *Ap. J. Suppl.* 107:1
Abraham RG, van den Bergh S, Nair P. 2003. *Ap. J.* 588:218
Akiyama M, Minowa Y, Kobayashi N, et al. 2008. *Ap. J. Suppl.* 175:1
Allen PD, Driver SP, Graham AW, et al. 2006. *MNRAS* 371:2
Arp H. 1966. *Ap. J. Suppl.* 14:1
Barnes JE, Hernquist LE. 1991. *Ap. J.* 370:65
Barro G, Faber SM, Pérez-González PG, et al. 2013. *Ap. J.* 765:104
Bauer AE, Conselice CJ, Pérez-González PG, et al. 2011. *MNRAS* 417:289
Beckwith SVW, Stiavelli M, Koekemoer AM, et al. 2006. *Astron. J.* 132:1729
Bell EF, van der Wel A, Papovich C, et al. 2012. *Ap. J.* 753:167
Bendo GJ, Calzetti D, Engelbracht CW, et al. 2007. *MNRAS* 380:1313
Bershady MA, Jangren A, Conselice CJ. 2000. *Astron. J.* 119:2645
Bertone S, Conselice CJ. 2009. *MNRAS* 396:2345
Blanton MR, Moustakas J. 2009. *Annu. Rev. Astron. Astrophys.* 47:159
Bluck AFL, Conselice CJ, Almaini O, et al. 2011. *MNRAS* 410:1174
Bluck AFL, Conselice CJ, Bouwens RJ, et al. 2009. *MNRAS* 394:L51
Bluck AFL, Conselice CJ, Buitrago F, et al. 2012. *Ap. J.* 747:34
Blumenthal GR, Faber SM, Primack JR, Rees MJ. 1984. *Nature* 311:517
Bohm A, Wisotzki L, Bell EF, et al. 2013. *Astron. Astrophys.* 549:46
Bothun GD. 1986. *Astron. J.* 91:507
Bournaud F, Elmegreen BG, Elmegreen DM. 2007. *Ap. J.* 670:237
Bournaud F, Perret V, Renaud F, et al. 2014. *Ap. J.* 780:57
Bower RG, Benson AJ, Malbon R, et al. 2006. *MNRAS* 370:645
Bridge C, Appleton PN, Conselice CJ, et al. 2007. *Ap. J.* 659:931
Bruce VA, Dunlop J, McLure R, et al. 2012. *MNRAS* 427:1666
Buitrago F, Conselice CJ, Epinat B, et al. 2014. *MNRAS* 439:1494
Buitrago F, Trujillo I, Conselice CJ, et al. 2008. *Ap. J.* 687:L61
Buitrago F, Trujillo I, Conselice CJ, et al. 2013. *MNRAS* 428:1460
Bundy K, Ellis RS, Conselice CJ. 2005. *Ap. J.* 625:621
Buta RJ. 2013. In *Extragalactic Astronomy and Cosmology*, ed. TD Oswalt, WC Keel. *Planets, Stars and Stellar Systems*, 6:1–89. Dordrecht, Neth.: Springer
Calzetti D, Armus L, Bohlin RC, et al. 2000. *Ap. J.* 533:682
Caon N, Capaccioli M, D'Onofrio M. 1993. *MNRAS* 265:1013
Carollo M, Cibinel A, Lilly SJ, et al. 2013. *Ap. J.* 776:71
Carrasco ER, Conselice CJ, Trujillo I. 2010. *MNRAS* 405:2253
Cassata P, Giavalisco M, Williams CC, et al. 2013. *Ap. J.* 775:106
Chapman SC, Windhorst R, Odewahn S, Yan H, Conselice CJ. 2003. *Ap. J.* 599:92
Cohen JG, Cowie LL, Hogg DW, et al. 1996. *Ap. J.* 471:5
Cole S, Lacey CG, Baugh CM, Frenk CS. 2000. *MNRAS* 319:168

Combes F, Sanders RH. 1981. *Astron. Astrophys.* 96:164
Conroy C. 2013. *Annu. Rev. Astron. Astrophys.* 51:393
Conselice CJ. 1997. *Publ. Astron. Soc. Pac.* 109:1251
Conselice CJ. 2003. *Ap. J. Suppl.* 147:1
Conselice CJ. 2006a. *MNRAS* 373:1389
Conselice CJ. 2006b. *Ap. J.* 638:686
Conselice CJ. 2009. *MNRAS* 399:L16
Conselice CJ, Arnold J. 2009. *MNRAS* 397:208
Conselice CJ, Bershady MA, Dickinson M, Papovich C. 2003a. *Astron. J.* 126:1183
Conselice CJ, Bershady MA, Gallagher JS III. 2000a. *Astron. Astrophys.* 354:L21
Conselice CJ, Bershady MA, Jangren A. 2000b. *Ap. J.* 529:886
Conselice CJ, Blackburne J, Papovich C. 2005. *Ap. J.* 620:564
Conselice CJ, Bluck AFL, Buitrago F, et al. 2011b. *MNRAS* 413:80
Conselice CJ, Bluck AFL, Ravindranath S, et al. 2011a. *MNRAS* 417:2770
Conselice CJ, Bundy K, Trujillo I, et al. 2007. *MNRAS* 381:962
Conselice CJ, Chapman SC, Windhorst RA. 2003b. *Ap. J.* 596:L5
Conselice CJ, Gallagher JS, Calzetti D, Homeier N, Kinney A. 2000c. *Astron. J.* 119:79
Conselice CJ, Grogin N, Jogee S, et al. 2004. *Ap. J.* 600:L139
Conselice CJ, Morlock A, Bluck AFL, Grützbauch R, Duncan K. 2013. *MNRAS* 430:1051
Conselice CJ, Rajgor S, Myers R. 2008. *MNRAS* 386:909
Conselice CJ, Yang C, Bluck AFL. 2009. *MNRAS* 394:1956
Cooke J, Berrier JC, Barton EJ, Bullock JS, Wolfe AM. 2010. *MNRAS* 403:1020
Cooper MC, Griffith RL, Newman JA, et al. 2012. *MNRAS* 419:3018
Cotini S, Ripamonti E, Caccianiga A, et al. 2013. *MNRAS* 431:2661
Croton D, Springel V, White SDM, et al. 2006. *MNRAS* 365:11
Daddi E, Renzini A, Pirzkal N, et al. 2005. *Ap. J.* 626:680
Dahlen T, Mobasher B, Faber SM, et al. 2013. *Ap. J.* 775:93
Davis M, Efstathiou G, Frenk CS, White SDM. 1985. *Ap. J.* 292:371
Davis M, Guhathakurta P, Konidaris NP, et al. 2007. *Ap. J.* 660:L1
de Jong RS. 1996. *Astron. Astrophys.* 313:45
Delgado-Serrano R, Hammer F, Yang YB, et al. 2010. *Astron. Astrophys.* 509:78
De Propris R, Conselice CJ, Liske J, et al. 2007. *Ap. J.* 666:212
de Ravel L, Le Fèvre O, Tresse L, et al. 2009. *Astron. Astrophys.* 498:379
de Vaucouleurs G. 1948. *Ann. Astrophys.* 11:247
de Vaucouleurs G. 1959. *Handb. Phys.* 53:275
de Vaucouleurs G, de Vaucouleurs A, Corwin HG, et al. 1991. *Third Reference Catalogue of Bright Galaxies.* New York: Springer
Dickinson M. 2000. *Philos. Trans. R. Soc. Lond. Ser. A* 358:2001
Disney MJ, Romano JD, Garcia-Appadoo DA, et al. 2008. *Nature* 455:1082
D'Onghia E, Mapelli M, Moore B. 2008. *MNRAS* 389:1275
Dressler A. 1984. *Annu. Rev. Astron. Astrophys.* 22:185
Dressler A, Oemler A, Butcher HR, Gunn JE. 1994. *Ap. J.* 430:107
Driver SP, Windhorst RA, Ostrander EJ, et al. 1995. *Ap. J.* 449:L23
Dubinski J, Mihos JC, Hernquist L. 1999. *Ap. J.* 526:607
Dullo BT Graham AW. 2013. *Ap. J.* 768:36
Eggen OJ, Lyden-Bell D, Sandage AR. 1962. *Ap. J.* 136:748
Ellis RE. 1997. *Annu. Rev. Astron. Astrophys.* 35:389
Elmegreen BG, Bournaud F, Elmegreen DB. 2008. *Ap. J.* 688:67
Elmegreen BG, Elmegreen DM, Hirst AC. 2004. *Ap. J.* 612:191
Elmegreen DM, Elmegreen BG. 1987. *Ap. J.* 314:3
Elmegreen DM, Elmegreen BG, Ravindranath S, Coe DA. 2007. *Ap. J.* 658:763
Erb DK, Steidel CC, Shapley AE, et al. 2006. *Ap. J.* 646:107
Fang JJ, Faber SM, Koo DC, Dekel A. 2013. *Ap. J.* 776:63

Ferguson HC, Dickinson M, Giavalisco M, et al. 2004. *Ap. J.* 600:L107
Ferguson HC, Dickinson M, Williams R. 2000. *Annu. Rev. Astron. Astrophys.* 38:667
Ferrarese LCP, Jordán A. 2006. *Ap. J. Suppl.* 164:334
Finkelstein SL, Papovich C, Salmon B, et al. 2012. *Ap. J.* 756:164
Förster Schreiber NM, Genzel R, Bouche N, et al. 2009. *Ap. J.* 706:1364
Freeman PE, Izbicki R, Lee AB, et al. 2013. *MNRAS* 434:282
Gabor JM, Impey CD, Jahnke K, et al. 2009. *Ap. J.* 691:705
Genzel R, Burkert A, Bouche N, et al. 2008. *Ap. J.* 687:59
Giavalisco M, Ferguson HC, Koekemoer AM, et al. 2004. *Ap. J.* 600:L93
Giavalisco M, Livio M, Bohlin RC, Macchetto FD, Stecher TP. 1996. *Astron. J.* 112:369
Glazebrook K. 2013. *Publ. Astron. Soc. Aust.* 30:56
Glazebrook K, Ellis R, Santiago B, Griffiths R. 1995. *MNRAS* 275:19
Gomez PL, Nichol RC, Miller CJ, et al. 2003. *Ap. J.* 584:210
Governato F, Willman B, Mayer L, et al. 2007. *MNRAS* 374:1479
Graham AW. 2008. *MNRAS* 680:143
Graham AW. 2013. In *Extragalactic Astronomy and Cosmology*, ed. TD Oswalt, WC Keel. *Planets, Stars and Stellar Systems*, 6:91–140. Dordrecht, Neth.: Springer
Graham AW, Driver SP, Petrosian V, et al. 2005. *Astron. J.* 130:1535
Graham AW, Erwin P, Caon N, Trujillo I. 2001a. *Ap. J.* 563:11
Graham AW, Guzmán R. 2003. *Ap. J.* 125:2936
Graham AW, Scott N. 2013. *Ap. J.* 764:151
Graham AW, Trujillo I, Caon N. 2001b. *Astron. J.* 122:1707
Grogin NA, Conselice CJ, Chatzichristou E, et al. 2005. *Ap. J.* 627:L97
Grogin NA, Kocevski DD, Faber SM, et al. 2011. *Ap. J. Suppl.* 197:35
Grützbauch R, Chuter RW, Conselice CJ, et al. 2011a. *MNRAS* 412:2361
Grützbauch R, Conselice CJ, Varela J, et al. 2011b. *MNRAS* 411:929
Guo Q, White SDM, Boylan-Kolchin M, et al. 2011. *MNRAS* 413:101
Hambleton KM, Gibson BK, Brook CB, et al. 2011. *MNRAS* 418:801
Hathi NP, Ferreras I, Pasquali A, et al. 2009. *Ap. J.* 690:1866
Hernquist L, Katz, N. 1989. *Ap. J. Suppl.* 70:419
Hibbard J, Vacca WD. 1997. *Astron. J.* 114:1741
Holmberg E. 1958. *Lund Medd. Astron. Obs. Ser. II* 136:1
Holwerda BW, Munoz-Mateos J-C, Comeron S, et al. 2014. *Ap. J.* 781:12
Holwerda BW, Pirzkal N, de Blok WJG, et al. 2011. *MNRAS* 416:2401
Holwerda BW, Pirzkal N, de Blok WJG, Blyth S-L. 2013. *MNRAS* 435:1020
Holwerda BW, Pirzkal N, Heiner JS. 2012. *MNRAS* 427:3159
Hopkins PF, Bundy K, Murray N, et al. 2009. *MNRAS* 398:898
Hopkins PF, Croton D, Bundy K, et al. 2010. *Ap. J.* 724:915
Hopkins PF, Hernquist L, Cox TJ, Keres D. 2008. *Ap. J. Suppl.* 175:356
Hu J. 2008. *MNRAS* 386:2242
Hubble EP. 1926. *Ap. J.* 64:321
Hubble EP. 1936. *Realm of the Nebulae*. New Haven: Yale Univ. Press
Jiang L, Egami E, Fan X, et al. 2013. *Ap. J.* 773:153
Jogee S, Barazza FD, Rix H-W, et al. 2004. *Ap. J.* 615:L105
Jogee S, Miller SH, Penner K, et al. 2009. *Ap. J.* 697:1971
Joseph RD, Wright GS. 1985. *MNRAS* 214:87
Kartaltepe JS, Dickinson M, Alexander DM, et al. 2012. *Ap. J.* 757:23
Kartaltepe JS, Mozena M, Kocevski D, et al. 2014. *Ap. J.* Submitted. arXiv:1401.2455
Kartaltepe JS, Sanders DB, Scoville NZ, et al. 2007. *Ap. J. Suppl.* 172:320
Kaviraj S, Peirani S, Khochfar S, Silk J, Kay S. 2009. *MNRAS* 394:1713
Kaviraj S, Tan K-M, Ellis RS, Silk J. 2011. *MNRAS* 411:2148
Kennicutt R. 1998. *Annu. Rev. Astron. Astrophys.* 36:189
Kennicutt R, Evans NJ. 2012. *Annu. Rev. Astron. Astrophys.* 50:531

Kepple GR, Sanner GW. 1998. *The Night Sky Observer's Guide*, Vol. 1. Richmond, VA: Willmann-Bell
Khochfar S, Silk J. 2006. *Ap. J.* 648:21
Kocevski DD. 2012. *Ap. J.* 744:148
Kocevski DD, Faber SM, Mozena M, et al. 2012. *Ap. J.* 744:148
Koekemoer AM, Faber SM, Ferguson HC, et al. 2011. *Ap. J. Suppl.* 197:36
Koo DC, Kron RG. 1992. *Annu. Rev. Astron. Astrophys.* 30:613
Koo DC, Simard L, Willmer CNA, et al. 2005. *Ap. J. Suppl.* 157:175
Kormendy J. 1977. *Ap. J.* 217:406
Kormendy J. 1979. *Ap. J.* 227:714
Kormendy J, Bender R. 2012. *Ap. J. Suppl.* 198:2
Kormendy J, Djorgovski S. 1989. *Annu. Rev. Astron. Astrophys.* 27:235
Kormendy J, Drory N, Bender R, Cornell ME. 2010. *Ap. J.* 723:54
Kormendy J, Fisher DB, Cornell ME, Bender R. 2009. *Ap. J. Suppl.* 182:216
Kormendy J, Ho LC. 2013. *Annu. Rev. Astron. Astrophys.* 51:511
Kormendy J, Kennicutt R. 2004. *Annu. Rev. Astron. Astrophys.* 42:603
Kriek M, van Dokkum P, Franx M, et al. 2009. *Ap. J.* 705:71
Labbe I, Rudnick G, Franx M, et al. 2003. *Ap. J. Lett.* 591:L95
Lani C, Almaini O, Hartley WG, et al. 2013. *MNRAS* 435:207
Lanyon-Foster MM, Conselice CJ, Merrifield MR. 2007. *MNRAS* 380:571
Lanyon-Foster MM, Conselice CJ, Merrifield MR. 2012. *MNRAS* 424:1852
Law DR, Shapley AE, Steidel CC, et al. 2012a. *Nature* 487:338
Law DR, Steidel CC, Erb DK, et al. 2007. *Ap. J.* 656:1
Law DR, Steidel CC, Shapley AE, et al. 2012b. *Ap. J.* 759:29
Lee B, Giavalisco M, Williams CC, et al. 2013. *Ap. J.* 774:47
Le Fèvre O, Abraham R, Lilly SJ, et al. 2000. *MNRAS* 311:565
Lilly S, Carollo CM, Pipino A, Renzini A, Peng Y. 2013. *Ap. J.* 772:119
Lin L, Koo DC, Willmer CNA, et al. 2004. *Ap. J.* 617:L9
Lintott C, Schawinski K, Bamford S, et al. 2011. *MNRAS* 410:166
Lisker T. 2008. *Ap. J. Suppl.* 179:319
Lopez-Sanjuan C, Balcells M, Perez-Gonzalez PG, et al. 2010. *Ap. J.* 710:1170
Lopez-Sanjuan C, Le Fèvre O, de Ravel L, et al. 2011. *Astron. Astrophys.* 530:20
Lopez-Sanjuan C, Le Fèvre O, Ilbert O, et al. 2012. *Astron. Astrophys.* 548:7
Lotz JM, Davis M, Faber SM, et al. 2008a. *Ap. J.* 672:L177
Lotz JM, Jonsson P, Cox TJ, et al. 2011. *Ap. J.* 742:103
Lotz JM, Jonsson P, Cox TJ, Primack JR. 2008b. *MNRAS* 391:1137
Lotz JM, Jonsson P, Cox TJ, Primack JR. 2010a. *MNRAS* 404:575
Lotz JM, Jonsson P, Cox TJ, Primack JR. 2010b. *MNRAS* 404:590
Lotz JM, Primack J, Madau P. 2004. *Astron. J.* 128:163
Lundmark K. 1926. *Arkiv Mat. Astron. Fys.* 19(8):1–5
Madau P, Dickinson M. 2014. *Annu. Rev. Astron. Astrophys.* 52:415–86
Maller AH, Katz N, Keres D, Dave R, Weinberg DH. 2006. *Ap. J.* 647:763
Man AWS, Toft S, Zirm AW, et al. 2012. *Ap. J.* 744:85
Mannuci F, Cresci G, Maiolino R, et al. 2010. *MNRAS* 408:2115
McLure RJ, Pearce HJ, Dunlop JS, et al. 2013. *MNRAS* 428:1088
Menanteau F, Abraham RG, Ellis RS. 2001. *MNRAS* 322:1
Menci N, Fiore F, Lamastra A. 2012. *MNRAS* 421:2384
Meurer GR, Heckman TM, Leitherer C, et al. 1995. *Astron. J.* 110:2665
Mihos JC. 1995. *Ap. J.* 438:L75
Mihos JC, Hernquist L. 1996. *Ap. J.* 464:641
Moreno J, Bluck AFL, Ellison SL, et al. 2013. *MNRAS* 436:1765
Morgan WW. 1962. *Ap. J.* 135:1
Mortlock A, Conselice CJ, Bluck AFL, et al. 2011. *MNRAS* 413:2845
Mortlock A, Conselice CJ, Hartley W, et al. 2013. *MNRAS* 433:1185

Mosleh M, Williams RJ, Franx M, et al. 2012. 756:12
Naab T, Johansson PH, Ostriker JP. 2009. *Ap. J.* 699:178
Newman AB, Ellis RS, Bundy K, Treu T. 2012. *Ap. J.* 746:162
Noeske KG, Weiner BJ, Faber SM, et al. 2007. *Ap. J.* 660:43
Oesch PA, Bouwens RJ, Carollo CM, et al. 2010. *Ap. J.* 709:21
Oser L, Naab T, Ostriker JP, Johansson PH. 2012. *Ap. J.* 744:63
Ownsworth JR, Conselice CJ, Mortlock A, Hartley W, Buitrago F. 2012. *MNRAS* 426:764
Papovich C, Bassett R, Lotz JM, et al. 2012. *Ap. J.* 750:93
Papovich C, Dickinson M, Giavalisco M, Conselice CJ, Ferguson HC. 2005. *Ap. J.* 631:101
Papovich C, Giavalisco M, Dickinson M, Conselice CJ, Ferguson HC. 2003. *Ap. J.* 598:827
Patton DR, Pritchet CJ, Carlberg RG, et al. 2002. *Ap. J.* 565:208
Patton DR, Pritchet CJ, Yee HKC, Ellingson E, Carlberg RG. 1997. *Ap. J.* 475:29
Peletier RF, Balcells M. 1996. *Astron. J.* 111:2238
Peng CY, Ho LC, Impey CD, Rix H-W. 2002. *Astron. J.* 124:266
Peng Y, Lilly S, Kovac K, et al. 2010. *Ap. J.* 721:193
Peng Y, Lilly S, Renzini A, Carollo M. 2012. *Ap. J.* 757:4
Petrosian V. 1976. *Ap. J.* 209:L1
Postman M, Franx M, Cross NJG, et al. 2005. *Ap. J.* 623:721
Ravindranath S, Ferguson HC, Conselice CJ, et al. 2004. *Ap. J.* 604:L9
Ravindranath S, Giavalisco M, Ferguson HC, et al. 2006. *Ap. J.* 652:963
Reichard TA, Heckman TM, Rudnick G, Brinchmann J, Kauffmann G. 2008. *Ap. J.* 677:186
Ricciardelli E, Trujillo I, Buitrago F, Conselice CJ. 2010. *MNRAS* 406:230
Rix H-W, Zaritsky D. 1995. *Ap. J.* 447:82
Roberts MS. 1963. *Annu. Rev. Astron. Astrophys.* 1:149
Roberts MS, Haynes MP. 1994. *Annu. Rev. Astron. Astrophys.* 32:115
Sandage A. 1961. *The Hubble Atlas of Galaxies*. Washington, DC: Carnegie Inst.
Sandage A. 1975. *Galaxies and the Universe*, ed. A Sandage, M Sandage, J Kristian, p. 1. Chicago: Univ. Chicago Press
Sanders DB, Mirabel IF. 1996. *Annu. Rev. Astron. Astrophys.* 34:749
Savorgnan G, Graham AW, Marconi A, et al. 2013. *MNRAS* 434:387
Scarlata C, Carollo CM, Lilly S, et al. 2007. *Ap. J. Suppl.* 172:406
Schade D, Lilly SJ, Crampton D, et al. 1995. *Ap. J.* 451:L1
Scoville N, Aussel H, Brusa M, et al. 2007. *Ap. J. Suppl.* 172:1
Sellwood JA. 2014. *Rev. Modern Phys.* 86:1
Sérsic JL. 1963. *BAAA* 6:41
Shankar F, Marulli F, Bernardi M, et al. 2013. *MNRAS* 428:109
Shapley AE. 2011. *Annu. Rev. Astron. Astrophys.* 49:525
Shen S, Mo HJ, White SDM, et al. 2003. *MNRAS* 343:978
Sheth K, Elmegreen DM, Elmegreen BG, et al. 2008. *Ap. J.* 675:1141
Silk J, Mamon GA. 2012. *Res. Astron. Astrophys.* 12:917
Simard L, Meden JT, Patton DR, Ellison SL, McConnachie AW. 2011. *Ap. J. Suppl.* 196:11
Steidel C, Giavalisco M, Pettini M, Dickinson M, Adelberger KL. 1996. *Ap. J.* 462:17
Steinmetz M, Navarro JF. 2002. *New Astron.* 7:155
Stewart KR, Bullock JS, Wechsler RH, et al. 2008. *Ap. J.* 683:597
Strateva I. 2001. *Astron. J.* 122:1861
Swinbank AM, Smail I, Chapman SC, et al. 2010. *MNRAS* 405:234
Takamiya M. 1999. *Ap. J. Suppl.* 122:109
Targett TA, Dunlop JS, McLure RJ, et al. 2011. *MNRAS* 412:295
Tasca LAM, Kneib J-P, Iovino A, et al. 2009. *Astron. Astrophys.* 503:379
Tasca LAM, Le Fèvre O, Lopez-Sanjuan C, et al. 2014. *Astron. Astrophys.* 565:10
Taylor-Mager V, Conselice CJ, Windhorst R, Jansen R. 2007. *Ap. J.* 659:162
Toomre A, Toomre J. 1972. *Ap. J.* 178:623
Trujillo I, Conselice CJ, Bundy K, et al. 2007. *MNRAS* 382:109

van den Bergh S. 1960. *Ap. J.* 131:215
van den Bergh S. 1976. *Ap. J.* 206:883
van den Bergh S, Abraham RG, Ellis RS, et al. 1996. *Astron. J.* 112:359
van der Kruit PC, Freeman K. 2011. *Annu. Rev. Astron. Astrophys.* 49:301
van der Kruit PC, Searle L. 1982. *Astron. Astrophys.* 110:61
van Dokkum PG, Franx M, Kriek M, et al. 2008. *Ap. J.* 677:5
van Dokkum PG, Whitaker KE, Brammer G, et al. 2010. *Ap. J.* 709:1018
Weinzirl T, Jogee S, Conselice CJ, et al. 2011. *Ap. J.* 743:87
White SDM, Rees MJ. 1978. *MNRAS* 183:341
Williams CC, Giavalisco M, Cassata P, et al. 2014. *Ap. J.* 780:1
Williams RE, Baum S, Bergeron LE, et al. 2000. *Astron. J.* 120:2735
Williams RE, Blacker B, Dickinson M, et al. 1996. *Astron. J.* 112:1335
Windhorst RA, Taylor VA, Janse RA, et al. 2002. *Ap. J. Suppl.* 143:113
Wolf M. 1908. *Publ. Astrophys. Inst. Koenigstuhl-Heidelb.* 3:109
Wuyts S, Förster Schreiber NM, Genzel R, et al. 2012. *Ap. J.* 753:114
Zibetti S, Charlot S, Rix H-W. 2009. *MNRAS* 400:1181

Microarcsecond Radio Astrometry

M.J. Reid[1] and M. Honma[2]

[1]Harvard-Smithsonian Center for Astrophysics, Cambridge, Massachusetts 02138; email: reid@cfa.harvard.edu

[2]Mizusawa VLBI Observatory, National Astronomical Observatory of Japan & Department of Astronomical Science, The Graduate University for Advanced Study, Mitaka 181-8588, Japan

Annu. Rev. Astron. Astrophys. 2014. 52:339–72

First published online as a Review in Advance on June 2, 2014

The *Annual Review of Astronomy and Astrophysics* is online at astro.annualreviews.org

This article's doi:
10.1146/annurev-astro-081913-040006

Keywords

distance, parallax, proper motion, VLBI, galactic structure, star formation, evolved stars, pulsars, Hubble constant

Abstract

Astrometry provides the foundation for astrophysics. Accurate positions are required for the association of sources detected at different times or wavelengths, and distances are essential to estimate the size, luminosity, mass, and ages of most objects. Very long baseline interferometry at radio wavelengths, with diffraction-limited imaging at submilliarcsecond resolution, has long held the promise of microarcsecond astrometry. However, only in the past decade has this been routinely achieved. Currently, parallaxes for sources across the Milky Way are being measured with ~10 μas accuracy, and proper motions of galaxies are being determined with accuracies of ~1 μas year^{-1}. The astrophysical applications of these measurements cover many fields, including star formation, evolved stars, stellar and supermassive black holes, Galactic structure, the history and fate of the Local Group, the Hubble constant, and tests of general relativity. This review summarizes the methods used and the astrophysical applications of microarcsecond radio astrometry.

1. INTRODUCTION

1.1. History

For centuries astrometry was the primary focus of astronomy. Before the seventeenth century, astronomers charted the locations of naked-eye stars with an accuracy of a fraction of an arcminute. With this level of accuracy, all but a handful of the highest proper motion stars remain "fixed" for an astronomer's lifetime, and it was not until Galileo's introduction of the telescope in astronomy that arcsecond precision became a possibility.

For millennia two great questions remained unanswered: Was Earth at the center of the Universe, and how large was the Universe? Both questions could be answered through astrometry by measuring the parallax of stars. If the Earth revolved around the Sun, stars would exhibit yearly shifts in apparent position (annual parallax) and, if measured, these would indicate their distances (at that time equivalent to the "size of the Universe"). This scientific opportunity led most scientists of the period to attempt astrometric measurements to detect parallax (Hirshfeld 2001).

Early attempts to measure parallax by Robert Hooke, and later James Bradley, compared the tilt of a near-vertical telescope to that of a plumb-line as the star γ Draconis transited directly overhead in London. Bradley, after discovering and accounting for aberration of light (a yearly ± 20 arcsec effect caused by Earth's orbital motion), was only able to place an upper limit of 0.5 arcsec on the annual parallax of the star. It was not until 1838 that Friedrich Wilhelm Bessel presented the first convincing measurement of stellar parallax: 0.314 arcsec for the star 61 Cygni (chosen as a candidate because of its large proper motion of ≈ 5 arcsec year^{-1}).

Bessel and others followed a suggestion attributed to Galileo to measure relative parallax, the differential position shift of a nearby target star relative to a distant background star, rather than absolute position shifts. This approach still forms the basis of the highest accuracy measurements made today. Currently astrometric accuracy using very long baseline interferometry (VLBI) at centimeter wavelengths is approaching the $\sim$1-µas level. This review documents the state of the art in radio astrometry, focusing both on the techniques and the scientific results.

1.2. Radio Astrometry

Karl Jansky made the first astronomical observations at radio wavelengths in the 1920s. Although searching for the source of interference in transatlantic telephone calls (mostly from lightning in the tropics), he noted strong emission localized within a few degrees in the constellation of Sagittarius. The peak of the signal drifted in time at a sidereal rate, indicating an origin outside the Solar System, and was ultimately traced to energetic (synchrotron emitting) electrons throughout the Milky Way, but concentrated toward the center.

The development of radar during WWII gave a boost to radio astronomy in the 1940s and 1950s. Astrometric precision with a single radio telescope was generally limited to some fraction of the diffraction limit, $\theta_d \approx \lambda / D$, where λ is the observing wavelength and D is the antenna diameter. Even though radio antennas were an order of magnitude larger than optical telescopes, the roughly four-orders-of-magnitude-longer wavelength seemed to doom radio astrometry. However, that changed with the development of radio interferometry. In the 1960s, interferometers with baselines of $\sim$1 km localized the positions of quasi-stellar objects (QSOs), leading to the discovery of highly redshifted optical emission. Over the years, with increasing baseline length, positional accuracy with (connected-element) interferometers improved from $\sim$1 to $\sim$0.03 arcsec (Wade & Johnston 1977).

In the late 1960s, radio interferometry was greatly extended by removing the need for a direct connection (either with cables or microwave links) between antennas. Signals were accurately time tagged using independent atomic clocks and recorded on magnetic tape, allowing separations of interferometer elements across Earth. This technique, called VLBI, led to many discoveries, including superluminal motion in jets from active galactic nuclei (AGNs) and upper limits of ~1 pc on the size of the emitting regions. Both results provided strong evidence for supermassive black holes as the engines for AGNs (Reid 2009). Early VLBI observations with intrinsic angular resolution better than 1 mas offered absolute position accuracy of ~0.3 mas using group-delay observables (Clark et al. 1976, Ma et al. 1986) and relative astrometric accuracy between fortuitously close pairings of QSOs of ~10 μas using phase-delay information (Marcaide et al. 1985).

Starting in the 2000s, calibration techniques improved to the point where relative positional accuracy of ~10 μas could be routinely achieved for most bright targets, relative to a detectable QSO usually within a couple of degrees on the sky. This changed the game dramatically. Such astrometric accuracy is unsurpassed in astronomy and is comparable to, or better than, the target accuracy of the next European astrometric space mission: *Gaia* (Bourda et al. 2011). Because radio waves are not absorbed significantly by interstellar dust, the entire Milky Way is available for observation. Also, because QSOs are used as the position reference, absolute parallax and proper motions are directly measurable. Such measurements for over 100 star-forming regions have now been made for sources as distant as 11 kpc. Significant results (see Section 2) include a resolution of the Hipparcos Pleiades distance controversy (C. Melis, M.J. Reid, A.J. Meouduszewski, J.R. Stauffer, G.C. Bower, et al., submitted), three-dimensional "imaging" of nearby star-forming regions (Loinard et al. 2007), and the most accurate measurements to date of the distance to the Galactic center, R_0, and the circular rotation speed of the local standard of rest (LSR), Θ_0 (Reid et al. 2014).

2. ASTROPHYSICAL APPLICATIONS

Because distance is fundamental to astrophysical understanding, it should not be surprising that radio astrometry and parallax measurements are critical for characterizing a wide variety of phenomena and classes of sources. Here we briefly discuss some of the more important astrophysical applications of radio astrometry.

2.1. Galactic Structure

Surprisingly, we know little of the spiral structure of the Milky Way; there is considerable debate over the number of spiral arms, the nature of the central bar, and the values of the fundamental parameters R_0 (distance to Galactic center) and Θ_0 (circular rotation speed of the LSR). Major projects are underway to map the Milky Way with the VLBI Exploration of Radio Astrometry (VERA) array and the Very Long Baseline Array (VLBA), through the Bar and Spiral Structure Legacy (BeSSeL) survey. With typical parallax accuracy of ±20 μas, and best accuracy of ±5 μas, one can measure distances of 5 and 20 kpc, respectively, with ±10% accuracy. For example, parallax data for the most distant source measured to date, the massive star-forming region W49N, are shown in **Figure 1**. Over 100 distances to high-mass star-forming regions have been measured using the astronomical "gold standard" technique of trigonometric parallax [see Reid et al. (2009c, 2014), Honma et al. (2012) and references therein].

Attempts to decode the nature of the spiral structure of the Milky Way have long relied on kinematic distance estimates (by comparing the Doppler velocity of a source and that expected from a model of Galactic rotation with distance as a free parameter). One of the first high-precision VLBA parallaxes was for the massive star-forming region W3(OH); Xu et al. (2006) found a distance of 1.95 ±0.04 kpc, and this was confirmed by Hachisuka et al. (2006). This distance was more than a factor of two less than its kinematic distance, demonstrating how unreliable

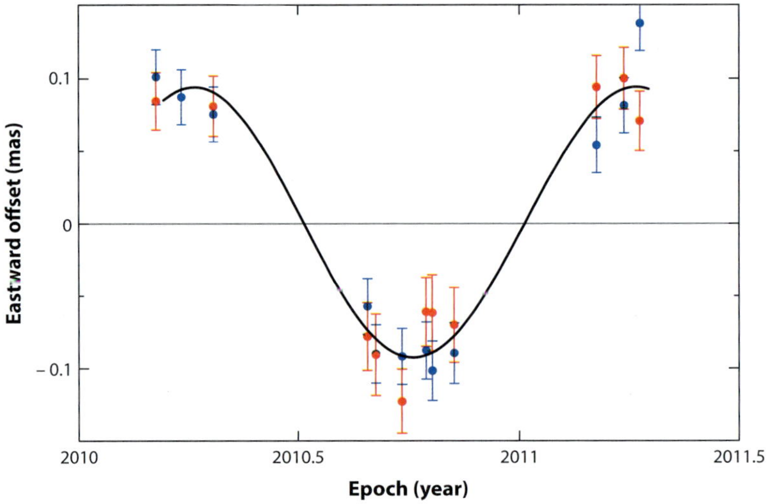

Figure 1

Parallax data for W49N measured with the Very Long Baseline Array, adapted from Zhang et al. (2013). Plotted are eastward position offsets versus time for H_2O maser spots at a local-standard-of-rest (LSR) velocity of 8.3 km s^{-1} relative to the background quasar J1905+0952 (shown in *red*) and a maser spot at an LSR velocity of 4.9 km s^{-1} relative to the background quasar J1922+0841 (shown in *blue*). The best-fit proper motion has been removed, allowing the data to be overlaid and effects of parallax to be more clearly seen. This source has a parallax of 0.090 ±0.006 mas, corresponding to a distance of 11.1 ±0.8 kpc.

kinematic distances can be. Now, with gold standard trigonometric parallaxes, the major spiral features of the Milky Way are, for the first time, being accurately located and spiral arm pitch angles measured (see **Figure 2**). Recent surprises include that the Local (Orion) arm, thought to be a minor structure, is longer and has far more ongoing star formation than previously thought and rivals the Perseus spiral arm in the second and third Galactic quadrants (Xu et al. 2013).

With measured source coordinates, distance, proper motion, and radial velocity, one has full phase-space (three-dimensional position and velocity) information. These data can be modeled to yield an estimate of R_0, Θ_0, and the slope of the Galaxy's rotation curve. The latest modeling indicates that $R_0 = 8.34 \pm 0.16$ kpc and $\Theta_0 = 240 \pm 8$ km s^{-1} (Reid et al. 2014). Also, the rotation curve between 4 and 13 kpc from the Galactic center is very flat (Honma et al. 2007, Reid et al. 2014). Such a large value for Θ_0 [about 10% larger than the International Astronomical Union (IAU)–recommended value of 220 km s^{-1}], if confirmed by other measurements, would have widespread impact in astrophysics, including increasing the total mass (including dark matter halo) of the Milky Way by about 30%, revising Local Group dynamics by changing the Milky Way's mass and velocities of Group members (when transforming from Heliocentric to Galactocentric coordinate systems), and increasing the expected signal from dark matter annihilation radiation.

2.2. Star Formation

Gould's Belt, a flattened structure of star-forming regions of radius ≈1 kpc and centered ≈0.1 kpc from the Sun (toward the Galactic anticenter), contains most of the sites of current star formation near the Sun (e.g., the Ophiuchus, Lupus, Taurus, Orion, Aquila Rift, and Serpens star-forming regions). Most of our knowledge about the formation of stars like the Sun comes from in-depth

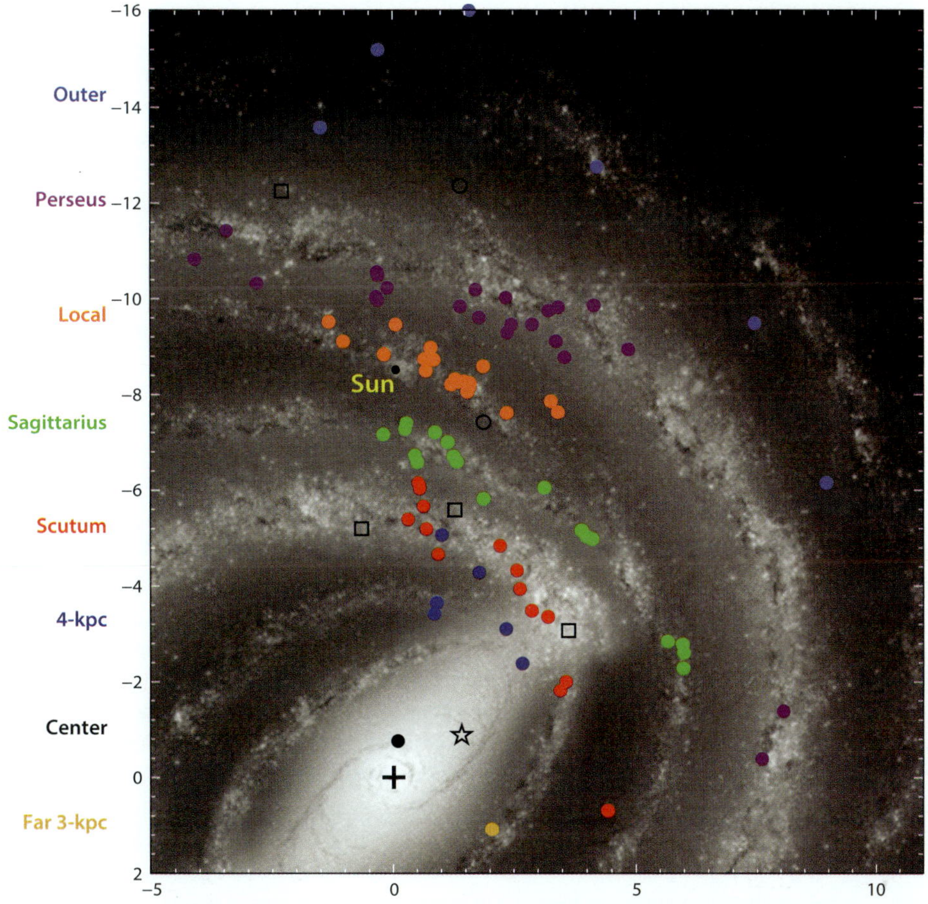

Figure 2

Plan view of the Milky Way. The background is an artist conception, guided by very long baseline interferometry (VLBI) astrometry and *Spitzer Space Telescope* photometry (R. Hurt: IPAC). Colored dots are the locations of newly formed OB-type stars determined from trigonometric parallaxes using associated maser emission. The parallaxes were determined with the Very Long Baseline Array, the VLBI Exploration of Radio Astrometry array, and the European VLBI Network. Assignment to spiral arms (indicated by *dot color*) has been made by comparison with large-scale emission of carbon monoxide in Galactic longitude–Doppler velocity space, independent of distance. The Galactic center is denoted by the plus (+) sign at (0,0) kpc; the Sun is labeled in yellow at (0,8.4) kpc. In this view, the Milky Way rotates clockwise.

studies of these regions, for example, from the *Spitzer* c2d survey (Evans et al. 2009), the *XMM-Newton* Extended Survey of Taurus (Güdel et al. 2008), and the *Herschel* Gould Belt Survey (André et al. 2010). Of course, knowledge of distance is critical for quantitative measures of cloud and young stellar object (YSO) sizes, masses, luminosities, and ages. For such deeply embedded sources, which are optically invisible, typically distances had not been estimated to an accuracy of better than ±30%. Because some physical parameters depend on distance squared or cubed, these parameters can be in error by factors of ≈2.

Trigonometric parallax measurements with VLBI techniques can be made by observing gyro-synchrotron emission from T Tauri objects, which is usually confined to a region of a few stellar radii or from H_2O maser emission often associated with Herbig-Haro outflows. Currently, about a dozen YSOs' parallaxes have been obtained, with up to 200 planned in this decade.

In the Ophiuchus cloud, Loinard et al. (2008), using the VLBA, found that sources S1 and DoAr21 are at a mean distance of 120.0 ± 4.5 pc; Imai et al. (2007) using VERA found a consistent distance of 178^{+18}_{-37} pc, albeit with larger uncertainty, for the H_2O masers toward IRAS 16293-2422. The Taurus molecular cloud has been well studied, and parallaxes for five YSOs have been reported from VLBA observations in a series of papers (Loinard et al. 2005, 2007; Torres et al. 2007, 2009, 2012). The authors locate three YSOs associated with the L 1495 dark cloud at a distance of 131.4 ± 1.4 pc and, interestingly, T Tauri Sb at 146.7 ± 0.6 pc and HP Tau/G3 at 161.9 ± 0.9 pc, respectively. Clearly, these observations trace the three-dimensional structure of the cloud, which is extended over about 30 pc along the line of sight, consistent with its extent on the sky of 10 deg (≈25 pc).

VERA observations have yielded parallaxes to YSOs in the Perseus molecular cloud. Observing H_2O masers, Hirota et al. (2008) measured parallax distances of 235 ± 18 pc for the SVS 13 toward NGC 1333 and 232 ± 18 pc for L 1448 C (Hirota et al. 2011). These measurements clarified the distance to the Perseus cloud, which was highly uncertain—estimated at between 220 pc (Cernis 1990) and 350 pc (Herbig & Jones 1983). The Serpens cloud provides another example where previous optical-based distance estimates varied considerably from 250 to 700 pc, with recent convergence toward the low end at ~230 ± 20 pc, as summarized by Eiroa, Djupvik & Casali (2008). However, Dzib et al. (2010) used the VLBA to obtain a trigonometric parallax distance for EC 59 of 414.9 ± 4.4 pc and suggested that faulty identification of dusty clouds in the foreground Aquila Rift might account for low optical distance estimate.

The Orion Nebula is perhaps the most widely studied region of star formation. Prior to 1981, distance estimates for the Orion Nebula ranged from about 380 to 520 pc as summarized by Genzel et al. (1981). By comparing radial and proper motions (measured with VLBI observations) of H_2O masers toward the Kleinman-Low (KL) Nebula, an active region of star formation within the Orion Nebula, Genzel et al. estimated the distance to be 480 ± 80 pc. Since then, four independent trigonometric parallax measurements have been performed. Sandstrom et al. (2007) observed gyro-synchrotron emission from a YSO over a period of nearly two years and obtained a parallax distance of 389 ± 24 pc using the VLBA. Hirota et al. (2007) using VERA measured H_2O masers and determined a distance of 437 ± 19 pc for the KL region. Recently, two very accurate parallaxes have been published: Menten et al. (2007) used the VLBA and observed continuum emission from three YSOs and determined a distance of 414 ± 7 pc; Kim et al. (2008) using VERA and observing SiO masers estimated a distance of 418 ± 6 pc. The latter two measurements from different groups, using different VLBI arrays, different target sources, and different correlators and software are in excellent agreement. Together they indicate that the KL nebula/Trapezium region of the Orion Nebula is at a distance of 416 ± 5 pc, nearly a 1% accurate distance! Because the Orion Nebula is the subject of large surveys, from X-rays to radio waves, having such a gold-standard distance will enable precise estimates of sizes, luminosities, masses, and ages.

A large number (>100) of parallaxes have been measured using the VLBA and European VLBI Network (EVN) arrays (Bartkiewicz et al. 2008; Brunthaler et al. 2009; Hachisuka et al. 2009; Moellenbrock, Claussen & Goss 2009; Sanna et al. 2009; Xu et al. 2009; Zhang et al. 2009; Rygl et al. 2010; Sato et al. 2010b; Moscadelli et al. 2011; Xu et al. 2011; Sanna et al. 2012; Immer et al. 2013) and the VERA array (Sato et al. 2008, 2010a; Oh et al. 2010; Ando et al. 2011; Honma et al. 2011; Kurayama et al. 2011; Matsumoto et al. 2011; Motogi et al. 2011; Nagayama et al. 2011a,b; Niinuma et al. 2011; Shiozaki et al. 2011; Sakai et al. 2012) for maser sources in high-mass star-forming regions. The focus of these observations has been to better understand Galactic structure, and we discuss these in Section 2.1.

2.3. Asymptotic Giant Branch Stars

Asymptotic giant branch (AGB) stars are in late stages of stellar evolution and have large convective envelopes and high mass-loss rates. They often exhibit maser emission from circumstellar OH, H_2O, and SiO molecules, providing good targets for radio astrometry with VLBI. Accurate distances to AGB stars are necessary to constrain their physical parameters, such as size and luminosity, and these are crucial to test theories of late stages of stellar evolution. And, of course, distances are needed to calibrate the period-luminosity (P-L) relation of Mira variables, which can be used as standard candles (e.g., Whitelock, Feast & Van Leeuwen 2008). Giant stars are not good astrometric targets at optical wavelengths, and the best optical parallaxes have accuracies poorer than a few milliarcseconds. However, radio observations of circumstellar masers have yielded more than a factor of ten better parallax accuracy, allowing improved calibration of the Mira P-L relation.

Early astrometric observations with the VLBA for red-giant OH masers demonstrated the potential for parallax measurements, achieving between 0.3 and 2 mas uncertainties (Langevelde et al. 2000, Vlemmings et al. 2003). For example, parallax distances for S CrB (418^{+21}_{-18} pc) and U Her (266^{+32}_{-28} pc) calculated by Vlemmings & Langevelde (2007) are considerably more accurate than those measured by the *Hipparcos* satellite. Astrometry of OH masers at the relatively low observing frequency of 1.6 GHz is limited by uncompensated ionospheric effects, but these effects can be minimized by observations near solar minimum and using in-beam calibrators very close to the targets.

The first VLBI parallax of an AGB star using H_2O masers at 22 GHz revealed a distance for UX Cyg, a long period Mira variable, of $1.85^{+0.25}_{-0.19}$ kpc (Kurayama, Sasao & Kobayashi 2005). Since then, the VERA array has been used to obtain parallaxes for many Mira and semiregular variables, including S Crt (Nakagawa et al. 2008), SY Scl (Nyu et al. 2011), and RX Boo (Kamezaki et al. 2012). Of particular interest are the astrometric measurements for the symbiotic system R Aqr (Kamohara et al. 2010) and the parallax distance of 218^{+12}_{-11} pc (Min et al. 2014). Future observations may enable us to trace the orbital motion of binaries.

Red supergiants are rare objects, typically at kiloparsec distances. At these distances, and owing to their very large sizes and irregular photospheres, they are beyond the reach of optical parallax measurements. Choi et al. (2008) observed H_2O masers toward VY CMa, one of the best-studied red supergiants, and found a parallax distance of $1.14^{+0.11}_{-0.09}$ kpc using VERA. This distance was later confirmed by Zhang et al. (2012), who used the VLBA and found a distance of $1.20^{+0.13}_{-0.10}$ kpc. Generally, the distance had been assumed to be 1.5 kpc, requiring an extraordinary luminosity of 5×10^5 $L_\odot$; the parallax distances reduce the luminosity estimate by 40% to a more reasonable value. VLBI parallaxes accurate to $\pm 10\%$ have been obtained for several other red supergiants, including S Per (Asaki et al. 2010), NML Cyg (Zhang et al. 2012), and IRAS 22480+6002 (Imai et al. 2012).

Protoplanetary nebulae are in the final stage of evolution of intermediate-mass stars, linking the AGB and planetary nebula phases. These sources are known to exhibit bipolar outflows with velocities exceeding $\sim$100 km s^{-1}; they can have strong H_2O masers and are often called water-fountains. Parallaxes have been measured for IRAS 19134+2131 (Imai, Sahai & Morris 2007), IRAS 19312+1950 (Imai et al. 2011), IRAS 18286-0959 (Imai et al. 2013), and K 3-35 (Tafoya et al. 2011).

2.4. X-Ray Binaries

The first trigonometric parallax for a black hole candidate was for the X-ray binary V404 Cyg (Miller-Jones et al. 2009), indicating a distance of 2.39 $\pm$ 0.14 kpc. This value was significantly lower than previously estimated and indicated that its 1989 outburst was not super-Eddington. Its peculiar velocity, derived from the radio proper motion, is only about 40 km s^{-1}, suggesting that it did not receive a large natal "kick" from an asymmetric supernova explosion. This differs from many pulsars, which often have peculiar motions that are larger by an order of magnitude (see Section 2.5).

The long-standing uncertainty over the distance to Cyg X-1 (see, e.g., Caballero-Nieves et al. 2009) limited understanding of this famous binary, including whether or not the unseen companion was a black hole. Recently a trigonometric parallax measurement from VLBA data yielded a distance of 1.86 ± 0.12 kpc (Reid et al. 2011). Knowing the distance to the binary removed the mass–distance degeneracy that limited the modeling of optical/IR data (light and velocity curves) and revealed that the unseen companion in Cyg X-1 has a mass of $14.8 \pm 1.0\,M_\odot$ (Orosz et al. 2011). This mass confidently exceeds the limit for a neutron star and firmly established it as a black hole. Once the masses of the two stars were accurately determined, X-ray data could be well modeled. This revealed that the black hole spin is near maximal and, because the binary is too young for accretion to have appreciably spun up the black hole, most of the spin angular momentum is probably natal (Gou et al. 2011). Finally, the three-dimensional space-motion of the binary, from the radio astrometry and optical Doppler shifts, confirms Cyg X-1 as a member of the Cyg OB3 association, and the lack of evidence for a supernova explosion in this region suggests that the black hole may have formed via prompt collapse without an explosion (Mirabel & Rodrigues 2003).

Recently, using the EVN and the VLBA, Miller-Jones et al. (2013) obtained a parallax distance of 114 ± 2 pc for SS Cyg, a binary system composed of both a white and a red dwarf. This distance is significantly less than an optical parallax distance of 159 ± 12 pc measured using the *Hubble Space Telescope* (Harrison et al. 1999). The larger optical distance required the source to be significantly more luminous and proved difficult to reconcile with accretion disk theory. However, the smaller distance from radio astrometry seems to resolve this problem.

2.5. Pulsars

Pulsar radio emissions are extremely compact and relatively bright, which make them suitable for VLBI observations, and radio astrometry of pulsars dates back to the 1980s. The first interferometric parallax measurements were made by Gwinn et al. (1986), who reported parallaxes for two pulsars using a VLBI array that included the Arecibo 305-m telescope. Later, Bailes et al. (1990a) measured a parallax for PSR 1451-68 of 2.2 ± 0.3 mas, using the Parkes-Tidbinbilla Interferometer in Australia, and determined a line-of-sight average interstellar electron density of 0.019 ± 0.003 cm^{-3} by combining the distance and dispersion measures, suggesting that the ISM in the Solar Neighborhood is typical of that over larger scales in the Galaxy. Pulsar proper motions for six pulsars conducted with the Parkes-Tidbinbilla Interferometer revealed that motions away from the Galactic plane were between 70 and 600 km s^{-1} (Bailes et al. 1990b). These early studies provided distances and confirmed the expectation that pulsars are high-velocity objects, most likely due to kicks associated with their parent supernova explosions.

Most recent pulsar astrometry uses "in-beam" calibrators to greatly improve positional accuracy. Other improvements include gating the pulsar signal during correlation to improve signal-to-noise ratios and performing ionospheric corrections based on multifrequency phase fitting. With these improvements and using the VLBA, a parallax with better than ±10% accuracy was obtained for PSR B0950+08 at a distance of 280 pc (Brisken et al. 2000), and nine other pulsar parallaxes with distances between 160 and 1,400 pc have been reported (Brisken et al. 2002).

By observing at higher frequencies (5 GHz instead of 1.6 GHz), in order to reduce the effects of the ionosphere, higher accuracy pulsar parallaxes have been obtained, for example, for PSR B0355+54 (0.91 $\pm$ 0.16 mas) and PSR B1929+10 (2.77 $\pm$ 0.07 mas) (Chatterjee et al. 2004). More recently, Chatterjee et al. (2009) obtained results for 14 pulsars using the VLBA, including a parallax for the most distant pulsar yet measured: PSR B1514+09 with a parallax of 0.13 $\pm$ 0.02 mas, corresponding to a distance of $7.2^{+1.3}_{-1.2}$ kpc. With this sample, it is clear that most pulsars are moving away from the Galactic plane with speeds of hundreds of kilometers per second.

Astrometry also provides a unique opportunity to constrain pulsar birthplaces through velocity and distance measurements. For instance, Campbell et al. (1996) used a global VLBI array to measure the parallax and proper motion for PSR B2021+51, which ruled out the supernova remnant HB 21 as the origin of the pulsar, as it implied an improbably low age of only 700 years. Similar studies have been performed by others (Dodson et al. 2003, Ng et al. 2007, Chatterjee et al. 2009, Bietenholz et al. 2013).

Astrometric observations also provide information on the physical properties of pulsars. For example, Brisken et al. (2003) combined a VLBA parallax for PSR B0656+14 with a thermal X-ray emission model to constrain the stellar radius between 13 and 20 km. Deller et al. (2012a) measured a parallax for the transitional millisecond pulsar J1023+0038 using the VLBA. Their distance of 1368^{+42}_{-39} pc was twice that predicted by the standard interstellar plasma model. When combined with timing and optical observations of this binary system, the new distance indicated a mass of $M \sim 1.71 \pm 0.1\,M_{\odot}$, suggesting that it is a recycled pulsar. Deller et al. (2009b) conducted astrometric observations of seven pulsars in the Southern Hemisphere using the Australian Long Baseline Array. Their new distance to PSR J0630-2834 required the efficiency of conversion of spin-down energy to X-rays to be less than 1%, an order of magnitude lower than previous estimates using a less accurate distance.

Magnetars, pulsars having extremely strong magnetic fields, are also interesting astrometric targets. Helfand et al. (2007) observed XTE J1810-197 with the VLBA and obtained a proper motion of 212 km s^{-1}, suggesting that this magnetar has a lower space motion than theoretically predicted (Duncan & Thompson 1992). Recently, Deller et al. (2012b) found a similarly low space motion for the magnetar J1550-5418.

Finally, pulsar astrometry can be critical for testing the constancy of physical parameters. Using the VLBA, Deller et al. (2008) observed J0437-4715, a millisecond binary pulsar with a white dwarf companion. They obtained a very precise parallax distance of 156.3 $\pm$ 1.3 pc. Comparing this to a kinematic distance from pulsar timing gives strong limits on unmodeled accelerations, which provide a limit on the constancy of the gravitational constant of $\dot{G}/G = (-5 \pm 26) \times 10^{-13}$ year^{-1}. This constraint is consistent with those obtained from lunar laser ranging as well as gravitational wave backgrounds.

2.6. Radio Stars

The Algol system is an eclipsing binary (with a period of 2.9 days, consisting of B8 V primary and K0 IV secondary) and a distant companion with an orbital period of 1.86 years. VLBI astrometry by Lestrade et al. (1993) revealed that the radio emission originates from the K0 subgiant, and they traced the orbital motion of the eclipsing binary, indicating that the orbit of the binary is nearly orthogonal to that of the tertiary companion. Observations of another hierarchical triple (Algol-like) system, UX Ari, by Peterson et al. (2011) detected the acceleration of the tight binary caused by the tertiary star. This acceleration measurement dynamically constrains the mass of the tertiary to be $\approx 0.75\,M_{\odot}$, a value consistent with a spectroscopic identification of a K1 main-sequence star.

Astrometric observations of radio stars allow one to accurately tie together the fundamental radio and the optical reference frames. A comparison between the radio and preliminary *Hipparcos* frames by Lestrade et al. (1995) revealed systematic discrepancies that could be removed by a global rotation (and its time derivative). Further VLBI observations by Lestrade et al. (1999) achieved parallax accuracies of $\approx$0.25 mas for nine sources, whose distances ranged from 20 to 150 pc.

Dzib et al. (2013) monitored radio emission from colliding winds in Cyg OB2#5 and obtained a marginal parallax of 0.61 $\pm$ 0.22 mas, consistent with other distance measurements of Cyg X regions (Rygl et al. 2012, Zhang et al. 2012). These observations also revealed a high radio-brightness temperature ($\gtrsim 10^7$ K), providing information for modeling the stellar winds.

2.7. Star Clusters

The Hyades and Pleiades clusters play pivotal roles in quantitative astrophysics, serving as pillars of the astronomical distance ladder. Recently, the *Hipparcos* space mission, which measured $\approx$100,000 stellar parallaxes with typical accuracies of $\pm$1 mas, presented a (revised) parallax distance of 120.2 $\pm$ 1.9 pc for the Pleiades (van Leeuwen 2009). This result has been quite controversial, because a variety of other techniques, including main-sequence fitting, generally give distances between 131 and 135 pc. Using the HST fine guidance sensor, Soderblom et al. (2005) measured relative trigonometric parallaxes for three Pleiads, which, after making corrections to absolute parallaxes, average to a distance of 134.6 $\pm$ 3.1 pc.

In an attempt to provide a totally independent and straightforward distance to the Pleiades, Melis and colleagues (C. Melis, M.J. Reid, A.J. Meouduszewski, J.R. Stauffer, G.C. Bower, et al., submitted) used the VLBA with the Green Bank and Arecibo telescopes to measure absolute parallaxes to several Pleiads that display compact radio emission. Results indicate a cluster distance of 136.2 $\pm$ 1.2 pc, where the uncertainty includes measurement error and cluster depth effects. This result seems to rule out the *Hipparcos* value, and it may even be in some tension with the ensemble of astrophysical-based distance indicators. Because the source of error for the *Hipparcos* parallax for the Pleiades has not been convincingly established, there could be concern for the *Gaia* mission, which is targeting a parallax accuracy of $\pm$20 μas, because *Gaia* might inherit some unknown systematics from *Hipparcos*. Intercomparison of high-accuracy VLBA parallaxes with those from *Gaia* will provide critical cross checking.

2.8. Sgr A*

Sgr A*, the candidate supermassive black hole (SMBH) at the center of the Galaxy, is a strong radio source. It is precluded from optical view by >20 mag of visual extinction, but can sometimes be detected when flaring at 2.2 μm wavelength (through $\lesssim$3 mag of extinction at this wavelength). Astrometric observations in the IR of stars orbiting an unseen mass have provided compelling evidence of a huge mass concentration, almost surely a black hole (Ghez et al. 2008, Gillessen et al. 2009). These observations require a grid of sources with accurate positions relative to Sgr A* to calibrate the IR plate scale, rotation, and low-order distortion terms. The calibration sources have been provided by the Jansky Very Large Array (JVLA) and VLBA astrometric observations, relative to the radio bright Sgr A*, of SiO and H_2O masers in the circumstellar envelopes of red giant and supergiant stars within the central parsec (Reid et al. 2003). This allowed researchers to locate the position of Sgr A* on IR images, which matched that of the gravitational focal position of the stellar orbits to $\approx$1 mas accuracy, as well as confirm Sgr A*'s extremely low luminosity (Menten et al. 1997).

Astrometric observations of Sgr A* at 7-mm wavelength, relative to background quasars with the VLBA, yielded the angular motion of the Sun about the Galactic center and placed extremely stringent limits on the mass density of the SMBH candidate. Although stars orbiting Sgr A* have been observed to move with speeds exceeding 5,000 km s^{-1} (Schödel et al. 2002), the velocity component perpendicular to the Galactic plane of Sgr A* is less than $\approx$1 km s^{-1}, requiring that most of the 4×10^6 $M_\odot$ indicated by the stellar orbits is tied to the radiative source Sgr A* (Reid & Brunthaler 2004). Because the millimeter wavelength emission from Sgr A* is confined to a $\sim$1 Schwarzschild radii region (Doeleman et al. 2008), the implied mass density approaches that theoretically expected for a black hole (Reid 2009).

2.9. Megamasers and the Hubble Constant

The Hubble constant, H_0, is a critical cosmological parameter, not only for determining the extragalactic distance scale but also for the flatness of the Universe and the nature of dark energy. Some AGNs with thin, edge-on accretion disks surrounding their central SMBH exhibit H_2O megamaser emission, with bright maser spots coming from clouds in Keplerian orbit about the SMBH. Astrometric observations can be used to map the positions and velocities of these clouds. Coupled with time monitoring of the maser spectra, which allows direct measurement of cloud accelerations (by tracking the velocity drift over time of maser features), one can obtain a geometric (angular-diameter) distance estimate for the galaxy.

Observations of the archetypal megamaser galaxy NGC 4258 demonstrated the power of radio astrometric observations for better understanding of AGN accretion disks (Herrnstein et al. 2005) and yielded an accurate distance of 7.2 ± 0.5 Mpc to the galaxy (Herrnstein et al. 1999). Because NGC 4258 is nearby and its (unknown) peculiar motion is likely to be a large fraction of its cosmological recessional velocity ($\approx$500 km s^{-1}), one cannot directly measure H_0 by dividing velocity by distance. However, because Cepheid variables have been observed in NGC 4258, one can use the radio distance to recalibrate the zero point of the Cepheid P-L relation (Macri et al. 2006) and then revise estimates of H_0 (Riess, Fliri & Valls-Gabaud 2012, Riess et al. 2012). Recently, Humphreys et al. (2013) analyzed a decade of observations of NGC 4258 and, via more detailed modeling of disk kinematics, obtained an extremely accurate distance of 7.60 ± 0.23 Mpc, which constrains $H_0 = 72.0 \pm 3.0$ km s^{-1} Mpc^{-1}.

By observing H_2O masers in galaxies like NGC 4258, but more distant and well into the "Hubble flow," the Megamaser Cosmology Project is obtaining direct estimates of H_0. For galaxies in the Hubble flow, unknown galaxy peculiar motions are a small source of systematic uncertainty ($\lesssim 5\%$). Results for the megamaser galaxy UGC 3789 (Reid et al. 2009a, Braatz et al. 2010) have yielded $H_0 = 68.9 \pm 7.1$ km s^{-1} Mpc^{-1} (Reid et al. 2013) and for NGC 6264 $H_0 = 68 \pm 9$ km s^{-1} Mpc^{-1} (Kuo et al. 2013). On the basis of these results, and preliminary results for Mrk 1419, Braatz et al. (2013) find a combined result of $H_0 = 68.0 \pm 4.8$ km s^{-1} Mpc^{-1}. The goal of the Megamaser Cosmology Project is measurement of $\approx$10 megamaser galaxies, each with an accuracy near $\pm 10\%$, which should yield a combined estimate of H_0 with $\pm 3\%$ accuracy. Although, formally, a $\pm 3\%$ uncertainty may be slightly larger than claimed by other techniques, the megamaser method is direct (not dependent on standard candles) and totally independent.

2.10. Extragalactic Proper Motions

In the concordance ΛCDM cosmological model, galaxies grow hierarchically by accreting smaller galaxies. Nearby examples of galaxy interactions are found in the environments of the Milky Way and the Andromeda Galaxy, the dominant galaxies in the Local Group. In the past, only radial velocities for Local Group galaxies were known, and statistical approaches had to be used to model the system. Although the radial velocity of Andromeda indicates that it is approaching the Milky

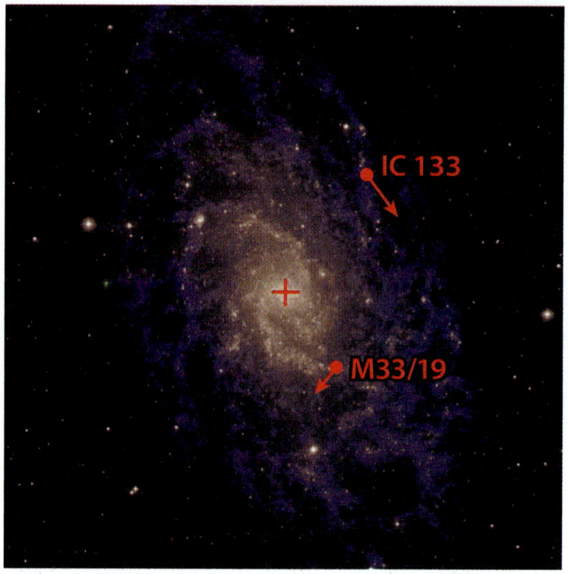

Figure 3

Image of M33, a satellite of the Andromeda galaxy, with the locations and measured proper motions of H_2O masers in two regions of massive star formation [adapted from Brunthaler et al. (2005) with permission]. Both the relative motions (i.e., the van Maanen experiment) and the absolute motions with respect to a background quasar have been measured. The relative motions give a rotational parallax distance, and the absolute motion can be used to constrain the mass of the Andromeda galaxy.

Way, without knowledge of its proper motion one cannot know, for example, if the two galaxies are on a collision course or if they are in a relatively stable orbit.

Astrometric VLBI observations have yielded both the internal angular rotation and the absolute proper motion of M33, a satellite of Andromeda (Brunthaler et al. 2005). The angular rotation of M33 was the focus of the van Maanen-Hubble debate in the 1920s (van Maanen 1923, Hubble 1926), with van Maanen claiming an angular rotation of 20 ± 1 mas year^{-1}. Such a large angular rotation required it to be nearby and part of the Milky Way in order to avoid implausibly large rotation speeds. Indeed, Shapley forwarded this as evidence that spiral nebulae were Galactic objects in the famous Shapley-Curtis debate. The angular rotation measured by Brunthaler et al. is $\sim$1,000 times smaller than van Maanen claimed, and, of course, consistent with an external galaxy (see **Figure 3**). Coupled with knowledge of the HI rotation speed and inclination of M33, the angular rotation rate yields a direct estimate of distance (rotational parallax) of 730 ± 168 kpc, consistent with standard candle estimates. Significant improvement in the accuracy of this distance estimate can come from better HI data and a longer time baseline for the proper motion measurements.

Absolute proper motions (with respect to background quasars) have been measured for two Andromeda satellites: M33 and IC 10. Assuming these galaxies are gravitationally bound to Andromeda, their three-dimensional motions yield a lower mass limit for Andromeda of $>7.5 \times 10^{11}\,M_\odot$ (Brunthaler et al. 2007). Of course, knowledge of the proper motion of Andromeda would refine this estimate and is key to unlocking the history and fate of the Local Group. Observations with the upgraded bandwidth of the VLBA and the Green Bank and Effelsberg 100-m telescopes are underway to measure the proper motion of M31* (the weak AGN at the center of Andromeda) and the motions of a handful of H_2O masers recently discovered in the galaxy (Darling 2011). Awaiting a direct measurement of the absolute proper motion of Andromeda, one can use the fact that M33 has not been tidally disrupted by a close encounter with Andromeda in the past to place constraints on the motion of Andromeda. In this way, Loeb et al. (2005) showed that Andromeda likely has a proper motion of $\sim$100 km s^{-1}.

2.11. Tests of General Relativity

The dominant uncertainty in modeling the Hulse-Taylor binary pulsar and measuring the effects of gravitational radiation on the binary orbit comes from uncertainty in the accelerations of the Sun (Θ_0^2/R_0) and the pulsar ($\Theta(R)^2/R$) as they orbit the center of the Galaxy (Damour & Taylor 1991). Using recently improved values for the fundamental Galactic parameters reduces the uncertainty in the binary orbital decay expected from gravitational radiation by nearly a factor of four compared with using the IAU recommended values (Reid et al. 2014). The main uncertainty in the general relativistic test parameter is now dominated by the uncertainty in the pulsar distance, and a VLBI parallax for the binary accurate to $\pm$14% would bring the contribution from distance uncertainty down to that of Galactic parameter uncertainty.

Deller, Bailes & Tingay (2009) obtained an accurate parallax distance of $1.15^{+0.22}_{-0.16}$ kpc for the pulsar binary J0737-3039 A/B. Given that these pulsars are only a kiloparsec distant (not $\sim$10 kpc as is the Hulse-Taylor pulsar), with this parallax accuracy uncertainties in the correction for the effects of Galactic accelerations are an order of magnitude smaller than for the Hulse-Taylor system. Thus, with perhaps another decade of pulsar timing, one might achieve a test of the effects of gravitational radiation predicted by general relativity at 0.01% level.

Pulsars orbiting the supermassive black hole at the Galactic center, Sgr A*, may show general relativistic effects that can be measured and tested on the basis of high-accuracy astrometric observations. Pulsars near the Galactic center are strongly affected by interstellar scattering, which makes it difficult to observe these pulsars at $\lesssim$10 GHz. For instance, the recently discovered transient magnetar PSR J1745-2900 (Mori et al. 2013), located only 3 arcsec from Sgr A*, is highly likely to be in the Galactic center region. If another pulsar is found closer to Sgr A*, it could prove to be an excellent target for testing general relativity with unprecedented accuracy.

Gravity Probe B (GP-B) was a satellite mission to test general relativistic frame dragging (Lense-Thirring effect) caused by the spin of Earth. The satellite was equipped with four ultrastable gyroscopes, allowing satellite altitude variation to be measured with unprecedented accuracy. On the basis of data from the four gyroscopes spanning $\sim$1 year, Everitt et al. (2011) reported a geodetic drift at $-6{,}601.8 \pm 18.3$ mas year^{-1} and a frame-dragging drift at -37.2 ± 7.2 mas year^{-1}. These results are fully consistent with the predictions of Einstein's theory of general relativity ($-6{,}606.1$ mas year^{-1} and -39.2 mas year^{-1}, respectively). VLBI astrometry played a fundamental role in the calibration of the GP-B data. IM Peg (HR 8703), an RS CVn type binary system, was used as the reference star. Its parallax and proper motion were accurately measured using VLBA observations spanning 5 years [Shapiro et al. (2012) and references therein]. On the basis of astrometry of IM Peg relative to 3C454.3, the absolute proper motion of IM Peg was determined with an accuracy better than 0.1 mas year^{-1} (Ratner et al. 2012), which provided the fundamental basis for measuring the geodetic and frame-dragging effect by GP-B.

The classical test of general relativity, measuring gravitational bending of star light by the Sun (Dyson, Eddington & Davidson 1919), has been repeated with radio interferometry by observing background radio sources and has provided one of the most accurate tests of general relativity. The most recent analysis of VLBI data reported that the post-Newtonian parameter γ is consistent with unity to an accuracy of one part in 10^{-4} (Lambert & Le Poncin-Lafitte 2009, 2011; Fomalont et al. 2009). Additionally, Fomalont & Kopeikin (2003) detected the deflection of radio waves by

Jupiter, including the retarded delay caused by Jupiter's motion, and claimed that this constrained the speed of gravitational waves, although this claim is controversial (see Asada 2002, Will 2003, Fomalont & Kopeikin 2009).

2.12. Extrasolar Planets and Brown Dwarfs

A radio astrometric search for extrasolar planets was conducted by Lestrade et al. (1996) toward σ^2 CrB over a 7.5-year period. After removing the effects of parallax and proper motion, their postfit residuals of <0.3 mas excluded a Jupiter-mass planet orbiting at a radius of 4 AU from the central star. Guirado et al. (1997) observed the active K0 star AB Dor with the Australian LBA and, combined with *Hipparcos* data, inferred a companion with a mass of $\sim$0.1 $M_\odot$. These studies demonstrate the potential of radio astrometry to detect low-mass companions, including brown dwarfs and planets.

Low-mass stars ($<1\,M_\odot$) could have habitable planets that are difficult to detect with optical radial-velocity techniques, because the stars are faint and activity can distort their line profiles, reducing the accuracy of velocity measurements. Thus, radio astrometric planet searches are complementary to other approaches. The Radio Interferometric Planet (RIPL) Search (Bower et al. 2009) and the Radio Interferometric Survey of Active Red Dwarfs (Gawroński, Goździewski & Katarzyński 2013) seek to detect the effects of planets around nearby low-mass stars, specifically active M dwarfs. For RIPL, the demonstrated astrometric accuracy is 260 μas, which for GJ 897A limits a planetary companion at 2 AU radius to $<0.15\,M_J$ (Bower et al. 2011).

2.13. Active Galactic Nuclei Cores

AGNs often show very strong and compact radio emission, with brightness temperatures up to and exceeding 10^{12} K. Bartel et al. (1986) measured the relative positions of a radio bright pair of QSOs, NRAO 512 and 3C 345, with a global VLBI network, revealing that the core positions of the sources were stable to within ±20 μas year^{-1}. This stability corresponds to an upper limit of $\sim 1c$ for the core-motions, which is significantly lower than observed in the jets of superluminal sources. Marcaide, Elosegui & Shapiro (1994), Rioja et al. (1997), Guirado et al. (1995), and Marti-Vidal et al. (2008) reported little if any relative proper motion between other QSO pairs (at levels of $\sim$10 μas year^{-1}), which is consistent with contamination of stationary cores with some expanding plasma. Such studies point out a source of limiting "noise" when using radio loud QSOs for establishing fundamental reference frames.

In addition to measuring relative positions between two QSOs, one can precisely measure positions of jet components within a source at different frequencies. In this manner, Hada et al. (2011) measured the theoretically predicted core-shift as a function of observing frequency (Blandford & Königl 1979) for the supermassive black hole of M87 (see **Figure 4**) and located the black hole relative to the jet emission with an accuracy of $\sim 10R_{\rm sch}$. Kovalev et al. (2008) and Sokolovsky et al. (2011) have confirmed that frequency-dependent core-shifts, such as observed in M87, are common in large samples of AGNs with prominent jets. Bietenholz, Bartel & Rupen (2004) used SN 1993J in M81 as a position reference for multifrequency imaging of M81* (the AGN in that galaxy) to accurately locate the supermassive black hole. Later, Marti-Vidal et al. (2011) found that the M81* core position appears variable at lower frequencies, and they suggest this is a result of jet precession.

Porcas (2009) considered the effects of an AGN core-shift on group-delay astrometry, used for measuring antenna locations for geodetic VLBI and for establishing reference frames (e.g., the International Celestial Reference Frame, ICRF). Usually group-delay measurements are conducted

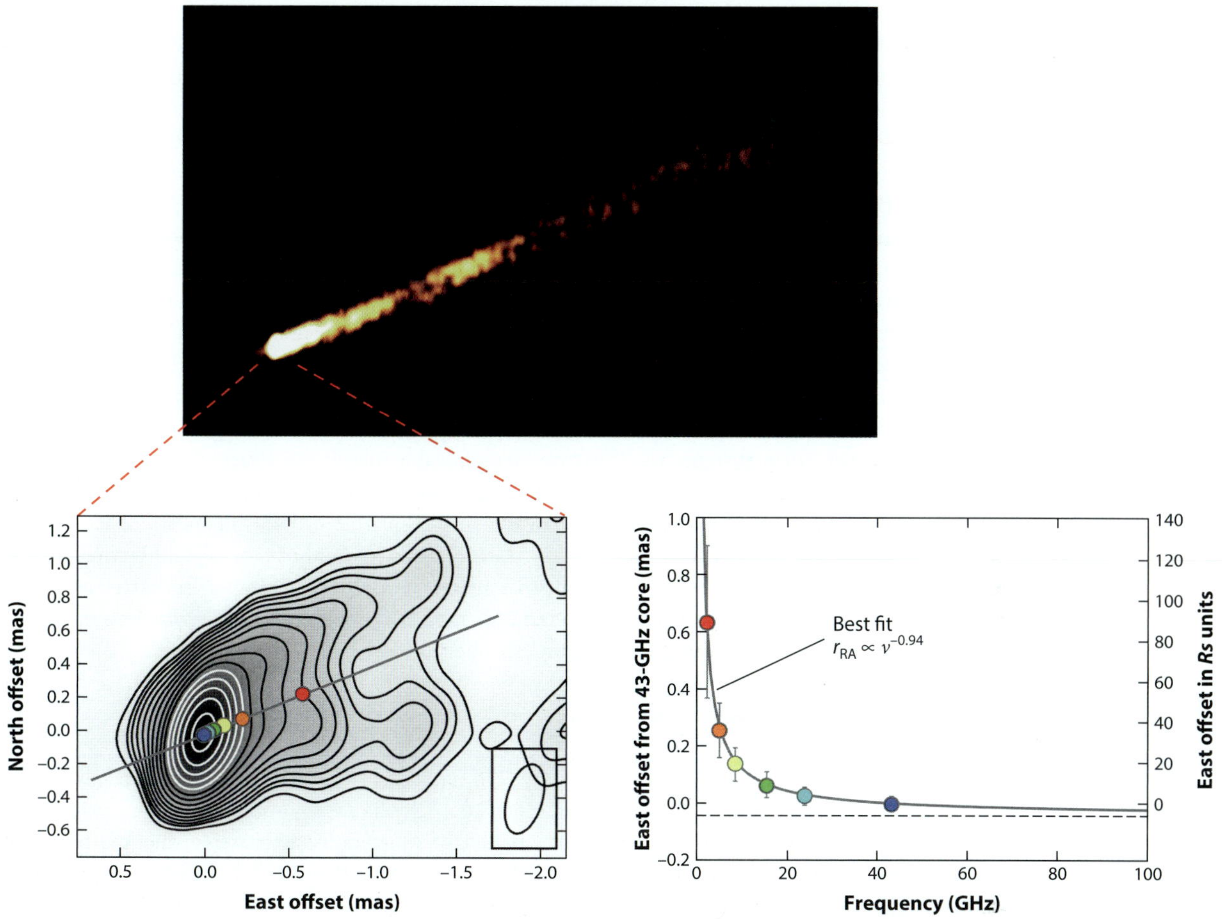

Figure 4

The apparent shift in the core of M87 as a function of observing frequency. This effect can be explained by frequency-dependent optical depths and allows the position of the supermassive black hole to be accurately located on the images. The bottom right panel gives offsets in angular units (mas) and in units of Schwarzschild radii (*Rs*). Abbreviation: RA, right ascension. Adapted from Hada et al. (2012) with permission.

at 8.4 GHz and, to remove propagation delays in the ionosphere, simultaneous observations are also made at 2.3 GHz. When removing the dispersive component of delay to correct for ionospheric propagation delay, generally one assumes that AGN core positions are the same at the two observing frequencies. However, the core-shift effect on dual-frequency geodetic VLBI observations can be comparable to a position error of ~170 μas and should be measured and compensated for in the future, for example, when tying the radio reference frame with a new optical frame constructed by *Gaia* (Bourda et al. 2011).

2.14. Satellite Tracking

The technique of high-accuracy VLBI astrometry can be applied to locate spacecraft using telemetry signals as radio beacons. In fact, astrometric observations of radio beacons placed on the moon by Apollo missions were among the earliest applications of phase-referencing VLBI. The Apollo Lunar Surface Experiments Package and Lunar Roving Vehicle radio transmitters from various

Apollo missions spanned hundreds of kilometers across the Lunar surface. These transmitters were observed by Counselman et al. (1973) and Salzberg (1973), who measured their separations with an accuracy of ~1 m (~0.5 mas). Later, Slade et al. (1977) were able to locate these transmitters with respect to the ICRF with a positional accuracy of ~1 mas. King, Counselman & Shapiro (1976) combined VLBI observations of Lunar transmitters with ranging data to better estimate the mass of the Earth-Moon system and the moment-of-inertia ratios of the Moon. [See Lanyi, Bagri & Border (2007) for a recent review.]

Recently, high-accuracy radio astrometry played a part in generating a precise map of the lunar gravity field in conjunction with the *Seleneological and Engineering Exporer* (SELENE) mission (Goossens et al. 2011). This was made possible by the high relative-position accuracy possible with in-beam astrometry (Kikuchi et al. 2009). This method was originally developed in the 1990s for observing artificial radio sources near Venus (Border et al. 1992, Folkner et al. 1993) and made possible measurements of wind speed as a function of altitude by monitoring the trajectories of probes released into the Venusian atmosphere (Counselman et al. 1979, Sagdeyev et al. 1992).

In the outer Solar System, VLBI astrometry played a role in the *Phobos*-2 (Hildebrand et al. 1994) and Mars *Odyssey* missions (Antreasian et al. 2005). The motion of the Huygens probe as it descended in Titan's atmosphere revealed the vertical profile of its wind speed (Witasse et al. 2006). The *Cassini* satellite itself was used to trace the gravity field of Saturn, and Jacobson et al. (2006) combined VLBI data with optical observations and Doppler ranging to better constrain the mass and potential of the Saturnian system. Also, VLBA observations of *Cassini* have measured the center of mass of the Saturnian system with an accuracy of 2 km (0.3 mas) with respect to the ICRF (Jones et al. 2011). The potential of radio astrometry for improving future space missions has been reviewed by Duev et al. (2012).

3. ARRAYS FOR RADIO ASTROMETRY

There are four major VLBI arrays that are regularly doing radio astrometry: the VLBA in the United States, the EVN in Europe, the VERA array in Japan, and the LBA in Australia. Basic parameters of these arrays are summarized in **Table 1**.

The VLBA (see **Figure 5**) consists of ten 25-m-diameter telescopes spread over US territory from Hawaii in the west to New Hampshire and St. Croix in the east. With a maximum baseline length of ≈8,000 km and frequency coverage from 300 MHz to 86 GHz, the VLBA is a flexible and sensitive array. The homogeneity of the array, with all the antennas and instruments identical, is advantageous as it makes calibration straightforward. Although the VLBA is a general-purpose imaging array, recently phase-referencing astrometry has occupied a major portion of its observing time. Currently there are several large astrometric projects being conducted using the VLBA. The BeSSeL Survey is measuring parallaxes and proper motions of hundreds of Galactic masers associated with high-mass star-forming regions. The Gould's Belt Distances Survey aims to measure distances to and provide a detailed view of star formation in the Solar Neighborhood. RIPL is an astrometric search for planets around nearby low-mass stars. The Pleiades Distance Project seeks absolute parallaxes for up to ten cluster stars in order to resolve the current cluster distance controversy and provide a solid foundation for many aspects of stellar astrophysics. Finally, the Megamaser Cosmology Project seeks to measure H_0 directly for ten H_2O megamaser galaxies well into the Hubble flow in order to independently constrain H_0 with ±3% accuracy.

The VERA array consists of four 20-m-diameter telescopes located across Japan with a maximum baseline length of 2,300 km (see **Figure 6**). Each antenna is equipped with two receiver systems that are independently steerable in the focal plane. As such, VERA can observe target and reference sources simultaneously to effectively cancel tropospheric fluctuations. VERA is also

Table 1 Major radio astrometric interferometers

Array	Country/ region	Antenna diameters and number in array	Maximum baseline	Operating frequencies	Beam size (mas)	Comments
VLBA	USA	25 m × 10	8,600 km	0.3–86 GHz	0.17 at 43 GHz	Homogeneous and best imaging capability
VERA	Japan	20 m × 4	2,300 km	6.7–43 GHz	0.63 at 43 GHz	Dedicated to astrometry with dual-beam system
EVN	Europe	14 m–100 m, × ∼10	3,000–10,000 km	1.6–22 GHz	0.30 at 22 GHz	High sensitivity with large dishes
LBA	Australia	22 m–70 m, × ∼10	1,700 km	1.4–22 GHz	1.7 at 22 GHz	The only VLBI array in the Southern Hemisphere
JVLA	USA	25 m × 27	36 km	0.07–50 GHz	40 at 43 GHz	Connected array with high sensitivity and excellent imaging
ALMA	Chile	12 m × 54 + 7 m × 12	16 km	43–900 GHz	4.5 at 850 GHz	Large connected array at millimeter and submillimeter wavelengths
SKA	Australia/ South Africa	∼1 km^2 aperture	∼3,000 km	0.1–22 GHz	0.7 at 22 GHz	Future large centimeter-wavelength array

Figure 5

Schematic portrayal of the locations of the Very Long Baseline Array (VLBA) antennas. The VLBA has ten 25-m-diameter antennas spanning the globe from Hawaii to New Hampshire and St. Croix.

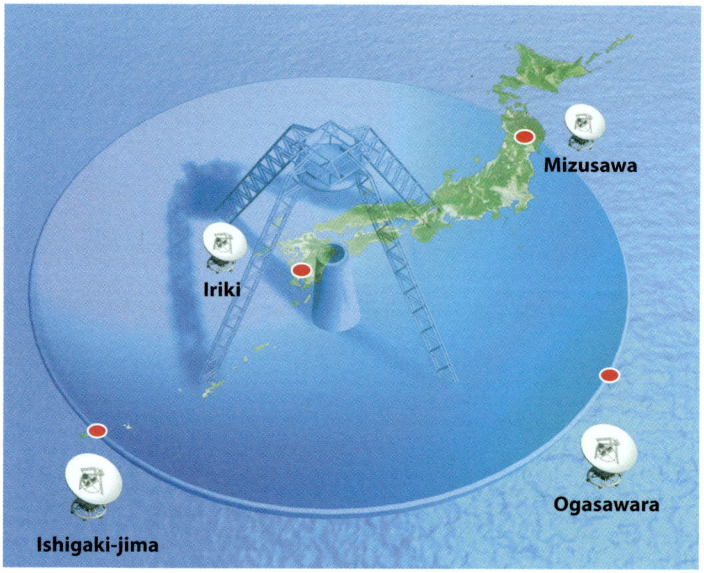

Figure 6

Schematic portrayal of the locations of the VLBI Exploration of Radio Astrometry (VERA) antennas. The VERA array has four 20-m-diameter antennas spanning Japan and is dedicated to astrometric observations.

the only array dedicated full-time to phase-referencing astrometry. Most of the observing time is spent on parallax measurements of maser sources tracing spiral structure in the Milky Way and of red giant stars.

The EVN array has antennas distributed across Europe as well as in other countries, including China, South Africa, and the United States. The EVN is most sensitive at frequencies <10 GHz, owing to some large antennas: the Effelsberg 100-m, the Jodrell Bank 76-m, and soon the Sardinia 64-m telescope. The array has been used for astrometric measurements of OH masers at 1.6 GHz, methanol masers at 6.7 GHz, and active stars.

The LBA in Australia is the only VLBI array regularly operating in the Southern Hemisphere. It has high sensitivity when the Parkes 64-m and Tidbinbilla 70-m telescopes are included. The LBA has provided astrometric measurements for southern pulsars, and hopefully it will soon be used for maser parallaxes. This is necessary in order to trace the three-dimensional structure of the roughly one-third of the Milky Way that cannot be observed from the north.

4. FUNDAMENTALS OF RADIO ASTROMETRY

The fundamental observable for a radio interferometer is the arrival time difference of wavefronts between antennas, owing to the finite propagation speed of electro-magnetic waves (the speed of light c). For an ideal interferometer, the arrival time difference is determined by the locations of the antennas with respect to the line of sight to the source and is referred to as the geometric delay. The basic equation of a radio interferometer, which relates the geometric delay, τ_g, to the source unit vector, s, and the baseline vector, $\vec{B}$, is given by

$$\tau_g = \frac{\vec{s} \cdot \vec{B}}{c}. \tag{1}$$

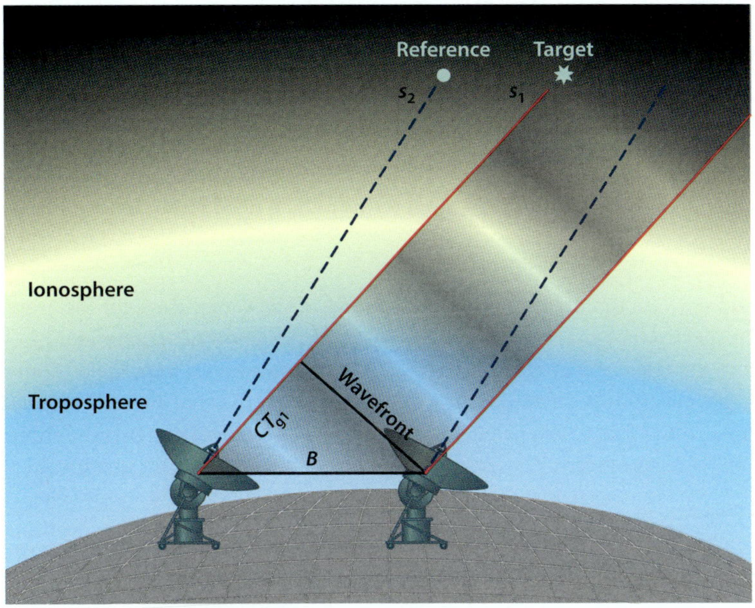

Figure 7

Schematic view of a delay measurement in phase-referencing astrometry. Here, for simplicity, an array with two stations is shown. The baseline vector and source directions are indicated by lines. In relative astrometry, the target and adjacent reference sources are observed (nearly) at the same time, thus delay errors can be effectively canceled in the relative measurement. For relative radio astrometry, the dominant error sources are generally uncompensated propagation delays in troposphere and/or ionosphere.

In **Figure 7**, we sketch the geometry of these vectors. Note that while the source vector $\vec{s}$ has two parameters (e.g., right ascension and declination), the geometric delay is a scalar, and so multiple measurements of geometric delays are required to solve for a source position. Such measurements can be made, even with a single baseline, by utilizing Earth's rotation, which changes the orientation of the baseline vector with respect to a celestial frame.

We now consider the effects of observational errors on astrometric accuracy. Given the uncertainty in a delay measurement, which we denote as $\Delta\tau$, from Equation 1 one can roughly estimate the astrometric error, Δs:

$$\Delta s \approx \frac{c\,\Delta\tau}{|B|}. \tag{2}$$

As seen in Equation 2, for a given $\Delta\tau$, astrometric accuracy improves with longer baselines. This is the fundamental reason why VLBI, which utilizes continental and/or intercontinental baselines, can achieve the highest accuracy astrometry. For example, for an 8,000-km baseline and a typical delay error of $\sim$2 cm (converted to path length by $c\,\Delta\tau$), one expects an astrometric error of $\sim$0.5 mas. This is a typical error for absolute astrometry, such as measuring QSO positions in the ICRF using broadband delay measurements.

For Galaxy-scale parallaxes, however, this accuracy is insufficient because parallaxes can be $\sim$0.1 mas, corresponding to distances of $\sim$10 kpc. At such distances, one needs astrometric accuracy of $\pm$10 μas to achieve $\pm$10% uncertainty. In order to obtain μas-level accuracy, one must substantially cancel systematic delay errors through relative astrometry, i.e., a position measurement with respect to a nearby reference source. Historically, the concept of relative astrometry has been attributed to Galileo, who proposed measuring the parallax of a nearby star with respect to a more distant star located close by on the sky.

4.1. Relative Astrometry

The delay measured with an interferometer can be considered as the sum of the geometric delay (which we would like to know) and additional terms (which need to be removed from the data):

$$\tau_{\rm obs} = \tau_{\rm g} + \tau_{\rm tropo} + \tau_{\rm iono} + \tau_{\rm ant} + \tau_{\rm inst} + \tau_{\rm struc} + \tau_{\rm therm}. \tag{3}$$

Here $\tau_{\rm tropo}$ and $\tau_{\rm iono}$ are the delays in the propagation of the signal through the troposphere and ionosphere, respectively; $\tau_{\rm ant}$ is caused by an error in the location of an antenna; $\tau_{\rm inst}$ is an instrumental delay in the telescope or electronics; $\tau_{\rm struc}$ is the delay due to unmodeled source structure; and $\tau_{\rm therm}$ is the uncertainty in measuring delay caused by thermal noise. Note that generally throughout this review the terms delay and phase (or phase-delay) can be used interchangeably: for instance, the interferometric phase caused by the geometric delay in Equation 1 can be written as

$$\phi_{\rm g} = 2\pi\nu\tau_{\rm g}, \tag{4}$$

where ν is the observing frequency.

For relative astrometry, one observes two sources, the target and reference, at nearly the same time and nearly the same position on the sky, and differences the observed delays (phases) between the pair of sources. This type of observation is often referred to as phase referencing, as the observed phase of the reference source is used to correct the phase of the target. Applying Equation 3 for the difference between two sources, we obtain the delay difference observable:

$$\begin{aligned}\Delta\tau_{\rm obs} = &(\tau_{\rm geo,1} - \tau_{\rm geo,2})\\ &+ (\tau_{\rm tropo,1} - \tau_{\rm tropo,2}) + (\tau_{\rm iono,1} - \tau_{\rm iono,2})\\ &+ (\tau_{\rm ant,1} - \tau_{\rm ant,2}) + (\tau_{\rm inst,1} - \tau_{\rm inst,2})\\ &+ (\tau_{\rm struc,1} - \tau_{\rm struc,2}) + (\tau_{\rm therm,1} - \tau_{\rm therm,2}).\end{aligned} \tag{5}$$

Here, subscripts 1 and 2 denote the quantities for the target and reference sources, respectively. The first term is the difference in geometric delay between the source and reference, which corresponds to the relative position of the target source with respect to the reference source. The additional terms in Equation 5 may be reduced if they are similar for the two lines of sight. As is discussed in detail below, four terms ($\tau_{\rm trop}$, $\tau_{\rm iono}$, $\tau_{\rm star}$, and $\tau_{\rm inst}$) are antenna-based quantities (delays originated at each antenna) and generally are similar for the target and reference lines of sight, for small source separations.

Antenna-based terms can be effectively reduced by phase referencing. For example, the delay error generated by an antenna position offset, $\Delta\vec{B}$, is given by

$$\tau_{\rm ant} = \frac{\vec{s}\cdot\Delta\vec{B}}{c} \sim \frac{|\Delta B|}{c}. \tag{6}$$

The approximation used to obtain the right-hand side of Equation 6 is that the trigonometric terms in the vector dot product are generally $\lesssim 1$. Differencing the antenna delays for two adjacent sources yields

$$\Delta\tau_{\rm ant} = (\tau_{\rm ant,1} - \tau_{\rm ant,2}) = \frac{(\vec{s}_1 - \vec{s}_2)\cdot\Delta\vec{B}}{c} \sim \theta_{\rm sep}\frac{|\Delta B|}{c}, \tag{7}$$

where $\theta_{\rm sep}$ is the separation angle between the sources. Comparison of Equations 6 and 7 shows that the delay error caused by an antenna position error is reduced by a factor of $\theta_{\rm sep}$ in the differenced delay. This reduction can significantly improve relative over absolute astrometric accuracy. For a separation of $\theta_{sep} = 1°$, the error reduction factor (θ_{sep}) is about 0.02 (radians).

As we discuss below in more detail, the dominant error source in radio astrometry is uncompensated propagation delays, generally tropospheric for observing frequencies $\gtrsim$10 GHz and ionospheric for lower frequencies. These terms are antenna-based, just as are the antenna location errors described above. A rough estimate of astrometric accuracy achievable in the presence of propagation delay errors is given by

$$\Delta s_{rel} \approx \theta_{sep} \frac{c\,\Delta\tau}{|B|}. \tag{8}$$

The reduction in position uncertainty in going from Equation 2 to Equation 8 is, of course, the addition of the canceling term, θ_{sep}. Using the same example values previously used for absolute astrometry, ($|B| = 8{,}000$ km and a delay error of $c\,\Delta t \sim 2$ cm), relative astrometric accuracy for a 1-deg source separation is ~10 μas.

4.2. Sources of Delay Error

Uncompensated delay differences between antennas in an interferometric array usually limit radio astrometric accuracy. In this subsection, we discuss sources of delay errors individually.

4.2.1. Troposphere. The propagation speed of electromagnetic waves in the troposphere is slower than in vacuum, causing an extra delay (τ_{tropo}) to be added to the geometric delay. The tropospheric delay is almost entirely nondispersive (frequency independent) at radio frequencies. It is convenient to separate the tropospheric delay into two components on the basis of the timescales of fluctuation: a slowly (~ hours) and a rapidly (~ minutes) varying term.

The rapidly varying term is associated with the passage of small "clouds" of water vapor flowing at wind speeds of ~10 m s^{-1} at a characteristic height of ~1 km over each antenna. This can be the main cause of coherence loss in interferometric observations. The characteristic timescales for interferometer phase fluctuations induced by tropospheric water vapor can be described by the Allan standard deviation (σ_A) and is typically ~0.5 to 1×10^{-13} over timescales of 1 to 100 s (Thompson, Moran & Swenson 2001; Honma et al. 2003). An interferometer coherence time, τ_{coh}, can be defined as the time interval over which phase fluctuation differences between two antennas accumulate one radian. At an observing frequency ν, this implies

$$2\pi\sigma_A \nu \tau_{coh} \sim 1. \tag{9}$$

For $\sigma_A = 0.7 \times 10^{-13}$ and $\nu = 22$ GHz, this implies a coherence time of ~100 s. One must measure and remove the phase variations on a timescale shorter than the coherence time. Rapid switching between the target and calibrator sources can usually accomplish this, provided the antennas can slew and settle on sources rapidly (see Section 4.3).

The rapid variations from tropospheric delays can be completely removed if one can simultaneously observe the target and calibration sources. This is done using the VERA antennas, which have dual-beam observing systems. Alternatively, this can also be accomplished if one is fortunate in finding a source pair having a very small angular separation, so that both fit within the primary beam of the antennas (in-beam calibration). Note that with either method, this still leaves a spatial variation component, but that term is generally smaller than the temporal term for sources separated by a few degrees.

The slowly varying delay term is generally associated with the "dry" part of the troposphere, although the slow term can also contain a small, relatively stable contribution from water vapor.

At sea level, a typical path delay for the dry component is 230 cm at the zenith, and water vapor can contribute up to several tens of centimeters for very humid locations. The total dry delay can be estimated from antenna latitude and elevation and, if necessary, using local temperature and pressure values. Because the total dry delay is considerable, it is important to remove its effects during correlation, or early in postcorrelation processing, by carefully accounting for the exact slant path through the atmosphere (air mass), which will be different for the target and calibrator. The slowly varying wet component cannot reliably be estimated from surface weather parameters and needs to be directly measured and removed to achieve microarcsecond astrometry (see Sections 5.1.1 and 5.1.2 for details).

4.2.2. Ionospheric delay. The ionosphere is a partially ionized layer located between ~50 and ~500 km altitude. An electromagnetic wave propagating through this plasma experiences phase and group delays that are frequency dependent (i.e., dispersive delays). The phase and group delay are given by

$$\tau_{\text{iono}} \equiv \frac{1}{2\pi}\frac{\phi}{\nu} = -\frac{c r_0}{2\pi\nu^2} I_e, \tag{10}$$

and

$$\tau_{\text{grp,iono}} \equiv \frac{1}{2\pi}\frac{\partial\phi}{\partial\nu} = \frac{c r_0}{2\pi\nu^2} I_e = -\tau_{\text{iono}}, \tag{11}$$

where r_0 is the classical electron radius and $I_e = \int n_e dl$ is the electron column density along the line of sight or total electron content (TEC). Owing to the dispersive nature of plasma, delays are larger at lower frequency because of the ν^{-2} dependence. (Note that the ionospheric phase delay has the opposite sign of the group delay; therefore correction of interferometer phase for ionospheric delays is different than for tropospheric delays.)

The ionization of the ionosphere is predominantly from solar radiation, and thus there is a strong diurnal variation, as well as from the 11-year solar cycle. Owing to the diurnal variation, the ionospheric delay needs to be modeled (or measured) and calibrated on hour scales. Modeling of the ionospheric delays is done by a combination of a vertical TEC (VTEC) and a mapping (air mass like) function. Typical values for VTEC range from a few to 100 TECU, where 1 TECU corresponds to an electron column density of 10^{16} m^{-2}. Once the TEC toward a source is known, the ionospheric delay (in path length units) can be calculated by the following relation:

$$|c\,\tau_{\text{iono}}| = 40.3 \left(\frac{I_e}{\text{TECU}}\right)\left(\frac{\nu}{\text{GHz}}\right)^{-2} \text{ (cm)}. \tag{12}$$

At a frequency of 22 GHz, a 50-TECU column density causes a delay equivalent to an extra path length of ~4 cm, but it reaches ~400 cm at a frequency of 2 GHz.

Because of the dispersive nature of the ionospheric delay, dual-frequency observations are effective for calibration. Most geodetic measurements involve simultaneous observations at 2.3 and 8.4 GHz in order to calculate and then remove ionospheric delays. Another way to measure ionospheric TEC values is to use global positioning system (GPS) data, which provide dual-frequency signals at 1.23 and 1.58 GHz. Several services accumulate GPS data from stations around Earth and provide global TEC models approximately every two hours. Generally, radio interferometric data are calibrated for ionospheric delays using these models.

4.2.3. Instrumental delay. Radio propagation in antenna structures, feeds, and electronics causes additional delays, which we lump together as "instrumental delays" (τ_{inst}). In addition, there are electronic phase offsets associated with the generation of the local oscillators used to mix the observed frequencies to intermediate frequencies (IFs; ~1 GHz) prior to correlation. The data recorded at each antenna are time-tagged using a clock tied to the fundamental frequency standard at each station (usually a hydrogen maser atomic oscillator), and owing to slight frequency offsets (that do not affect local oscillator stability) these clocks generally gain or lose time at a rate of one part in $\sim 10^{-14}$ (only ~ 1 ns per day!). However, a delay error of 1 ns causes a phase slope of π radians across a 500-MHz observational bandwidth, and it is critical to correct for this. Calibration observations using "geodetic blocks" (see Section 5.1.1) can remove clock errors to an accuracy of $\lesssim 0.1$ ns. Other instrumental delay and phase offsets are generally relatively slowly varying, for well-designed systems and temperature-controlled electronics, and can be calibrated and removed by observations of strong sources a few times a day.

Any residual instrumental delay and phase offsets will be mostly canceled when switching between two sources, provided the electronics are not changed. Note that if one of the sources is a spectral line and the other a continuum emitter, one should try to place the line near the center of the continuum band to ensure better calibration. In contrast, if the two sources are observed with different receivers, which is the case for dual-beam receiving with VERA, there is a need for additional calibration of the instrumental delay and phase, such as by using an artificial noise source (see Section 4.3).

4.2.4. Antenna position. The antenna position is usually defined as the intersection of azimuth and elevation axes and can be measured to an accuracy of $\lesssim 3$ mm with regular geodetic observations. Earth's rotation rate varies (measured as a time correction, UT1–UTC) as does the location of an antenna on Earth's crust with respect to its spin axis (polar motion). Together these are called Earth Orientation Parameters (EOPs). EOP values are determined on a daily basis by the International Earth Rotation and Reference Systems Service by utilizing global GPS data and geodetic VLBI observations. A typical error in EOP values is ≈ 0.1 mas, provided the final (not preliminary or extrapolated) values are used. With high-accuracy antenna locations and EOP corrections, the total uncertainty in the position of an antenna contributes less to the astrometric error budget than uncertainty in atmospheric propagation delays.

4.2.5. Source structure. If the observed sources are not point like, interferometric delay/phase can be affected by source structure. Because structure phase shifts are independent for the target and reference sources, their effects do not cancel when differencing the phases in relative astrometry. Source structure phases are baseline-based quantities. Because there are $\sim N^2$ baselines for N antennas in an array, structure phase can be separated from station-based quantities and estimated using self-calibration (closure techniques and hybrid mapping). Once a source image is obtained, phase shifts caused by source structure can be calculated and subtracted from observed interferometric data. Note that because self-calibration loses absolute position information, if one self-calibrates the reference source data, one must apply the same calibrations to both the reference and target source to preserve relative astrometric accuracy. Generally it is not a good idea to self-calibrate the target source data but instead measure the structure in the phase-referenced images.

4.2.6. Thermal noise. Thermal (usually dominated by receiver) noise is random and cannot be calibrated and removed. However, a random process does not cause systematic offsets, and by integrating over many samples its effects can be reduced. Thermal noise in an interferometric image leads to position measurement uncertainty given by equation 1 of Reid et al. (1988):

$$\Delta\theta_{\text{therm}} \approx 0.5\frac{\theta_{\text{beam}}}{SNR} \approx 0.5\frac{\lambda}{B}\frac{1}{SNR}. \tag{13}$$

For a baseline length $B = 8{,}000$ km and observing wavelength $\lambda = 1.3$ cm, the synthesized beam (FWHM) size is $\theta_{\rm beam} \approx 0.3$ mas. Thus, for a source with an image signal-to-noise ratio $SNR \sim 30$, the expected position error due to thermal noise is only ≈5 μas. Therefore, thermal noise often is not a major source of radio astrometric error.

4.3. Observing Methods

Several methods of phase referencing are commonly used: source-switching, dual-beam, and in-beam observations. Source switching or nodding is the most commonly used observing method, because it does not require a special radio telescope and can be used for all source pairs. The only requirement is that slewing/settling times are short enough to track tropospheric phase fluctuations. As discussed in Section 4.2.1, coherence times at an observing frequency of 22 GHz are ~100 s. Therefore switching cycles should be shorter than this time, e.g., a cycle of 60 s obtained by integrating for 20 s on the reference, 10 s for slewing to the target, integrating for 20 s on the target, then 10 s for slewing back to the reference. Because of the nondispersive nature of tropospheric delays, the interferometer coherence time scales linearly with wavelength, i.e., $\tau_{\rm coh} \propto \lambda$. Hence at short wavelengths, the coherence time can be problematic. For an observing wavelength of $\lambda = 1.3$ mm ($\nu = 230$ GHz), the coherence time is likely to be ~10 s. This makes it practically impossible to conduct switched VLBI observations at $\lesssim$1 mm wavelength.

In dual-beam observations, two sources are observed simultaneously using independent feeds and receivers. This can be done using multifeed systems on a single antenna, as with VERA, or with multiple antennas at each site (Rioja et al. 2009; Broderick, Loeb & Reid 2011). There is no gap between the observations of the target and reference sources, and hence there is no coherence loss owing to temporal phase fluctuations. The antennas of the VERA array achieve this with two receivers installed at the Cassegrain focus. These receivers are independently steerable with a Steward-mount platform, and one can observe a source pair with separation angles between 0.3 deg to 2.2 deg. In dual-beam radio astrometry, special care needs to be taken to calibrate the instrumental delay, because it is not common to the target and reference source and thus will not cancel when phase referencing.

In-beam observations can be regarded as a special case of phase referencing, in which the target and reference sources are so closely located that the two sources can be observed simultaneously with a single feed (within the primary beam of each antenna). In such a case, calibration error can be very effectively reduced, because the observations are done at the same time for the target and reference and, because the separation angle is small, most systematic errors are largely canceled. However, finding a sufficiently strong reference source may be difficult; as such, in-beam observations are more common at lower observing frequencies, as the primary beam becomes larger and the reference sources stronger.

5. ADVANCED TECHNIQUES

For parameters typical of centimeter-wavelength VLBI observations seeking to measure relative positions between two sources separated by ≈1 deg with ≈10 μas accuracy, phase-reference sources should have coordinates known to $\lesssim$5 mas, antenna locations known to $\lesssim$1 cm, and electronic, clock, and propagation delays known to $\lesssim$0.05 ns.

5.1. Tropospheric Delay Calibration

Tropospheric (nondispersive) delays are usually calibrated by one of two methods: (*a*) using geodetic-like observing blocks or (*b*) using GPS data.

5.1.1. Geodetic block calibration. One can use observations of radio sources spread over the sky to measure broadband (group) delays. For sources with positions known to better than $\approx$1 mas, group delay residuals are generally dominated by the effects of tropospheric (and at low frequencies ionospheric) mismodeling, provided the geometric model for the array is accurate to better than $\pm$ 1 cm (including antenna locations, EOPs, and solid Earth tides). By observing $\approx$10 sources over a range of source azimuths and, most importantly, elevations in rapid succession ($\lesssim$30 min), one can estimate residual "zenith" tropospheric delays at each telescope. These observing blocks are similar to those used for geodetic VLBI observations to determine source and telescope positions, as well as Earth's orientation parameters and, hence, are called geodetic blocks (Reid et al. 2009b).

The observing sequence can be determined by Monte Carlo simulations of large numbers of blocks and choosing the block with the lowest expected zenith-delay uncertainties. Operationally, it may be best to choose sources above an elevation of $\approx$8 deg and available at a minimum of about ~60% of the telescopes in the array. Geodetic blocks should be placed before the start, during (roughly every 2 hours), and at the end of phase-reference observations. This allows one to monitor slow changes in the total tropospheric delay for each telescope. In order to measure group delays accurately, the observations should span the maximum IF bandwidth, spaced in a "minimum redundancy pattern" to uniformly sample, as best as possible, all frequency differences. For a system with 8 IF bands, this can be accomplished with bands spaced by 0, 1, 4, 9, 15, 22, 32, and 34 units. If the recording system limited to 500 MHz IF bandwidth, the unit separation would be $\approx$14.7 MHz. With such a setup, estimated zenith delays are generally accurate to $\approx$0.03 ns ($\approx$1 cm of propagation path delay). Comparisons of the geodetic block technique with those using GPS data and an image-optimization approach confirm this accuracy (Honma, Tamura & Reid 2008).

Residual multiband (group) delays and fringe rates are modeled as owing to a vertical tropospheric delay and delay rate, as well as a clock offset and clock drift rate, at each antenna. Note that the geodetic blocks need not be observed at the same frequency as the phase-referencing observations. If one has independent calibration of ionospheric (dispersive) delays (see Section 5.2), then it is simplest to observe the geodetic blocks at a frequency (ν) above $\approx$20 GHz to avoid contamination of the group delays by residual ionospheric effects, which decline approximately as $1/\nu^2$. However, if one is limited to one (wide-band) receiver, then for observing frequencies below about $\approx$10 GHz, it is important to remove unmodeled ionospheric contributions to the geodetic group delays, because correction of interferometer phases has a different sign for dispersive compared with nondispersive delays. This can be accomplished with dual-frequency geodetic blocks as outlined in Section 5.2.

5.1.2. Global positioning system tropospheric delay calibration. The GPS is a navigation system used to determine an accurate three-dimensional position for an observer on Earth. The system consists of more than 20 satellites constituting an artificial constellation of reference sources. Accurate positions of the satellites are broadcast by radio transmission to receivers on Earth. The basic principles for position determination using GPS signals are similar to geodetic VLBI: The receiver position can be accurately determined based on delay measurements from multiple satellites. Because the broadcasts are at low frequencies (1.2 to 1.6 GHz), propagation delays in the ionosphere strongly affect the data. Dual frequencies allow removal of the ionospheric (dispersive) delay, and by observing several GPS satellites at different slant-paths through the atmosphere simultaneously, one can solve for the tropospheric delay as well as the receiver position. As such, one can use GPS data to calibrate tropospheric delays. Steigenberger et al. (2007) and Honma, Tamura & Reid (2008) have conducted detailed comparisons of tropospheric delay measurements by GPS and geodetic-mode VLBI (see Section 5.1.1) observations; both studies conclude that the difference between the two methods are small, $\lesssim$2 cm, with no systematic differences.

5.2. Ionospheric Delay Calibration

In principle, the effects of ionospheric (dispersive) delays can be largely removed by using global models of the total electron content (Walker & Chatterjee 2000), by direct use of GPS data at each antenna, or from dual-frequency geodetic block observations. Because the ionosphere is confined to a shell far above Earth's surface, rays from the source to a telescope can penetrate the ionosphere far from the telescope location on Earth's surface. This generally favors using global models, based on smoothed results from a grid of GPS stations, over measurements made only at the VLBI telescopes. However, this is an area of current development and the methods of dispersive delay calibration may improve.

If one conducts observations at frequencies below $\approx$10 GHz, one can in principle obtain both dispersive and nondispersive delays simultaneously from wide-band geodetic blocks. For example, with receivers covering 4 to 8 GHz, one can generate two separate sets of geodetic block data spanning 500 MHz centered at the low end of the band ($\nu_L = 4.25$ GHz) and at the high end of the band ($\nu_H = 7.75$ GHz). This gives observables $\tau(\nu_L)$ and $\tau(\nu_H)$, which contain both dispersive and nondispersive delays (e.g., tropospheric delays and clock terms). Differencing $\tau(\nu_L)$ and $\tau(\nu_H)$ values for all sources and baselines yields the dispersive contribution. The dispersive contribution can be used to model the ionospheric contribution and, importantly, can be scaled and subtracted from the $\tau(\nu_H)$ delays to produce pure nondispersive delays. The nondispersive delays can then be modeled as described in Section 4.2.1 to solve for pure tropospheric and clock terms. Note that if one uses observations of a strong continuum source (fringe finder) to remove electronic delay differences among IF sub-bands (i.e., zeroing the delay on the fringe finder scan), this must be taken into account when modeling the dispersive delays.

An alternative to direct measurement is to use global maps of total ionospheric electron content generated from GPS data. Global ionosphere maps (GIMs) are produced on a regular basis by several groups, such as NASA, EAS, CODE, and UCP, who provide GIMs every two hours. Basic trends in these global maps are generally consistent with each other, although there are differences in scale for TEC values. Also note that currently GIMs have a typical angular resolution of 2.5 deg, and thus small-scale fluctuations in the ionosphere may not be well traced. Hernández-Pajares et al. (2009) compared TEC values obtained by GPS and direct measurements (such as with JASON and TOPEX) and found that vertical TEC values among the methods differ at the $\approx$20% level. For a 50-TECU zenith column and an error of $\pm$20%, the uncertainty in the ionospheric delay correction would be $\sim$1 cm at an observing frequency of 22 GHz, which is slightly smaller than the delay error caused by the troposphere. However, at frequencies below 10 GHz the ionosphere is the dominant contributor to the delay error budget, even after corrections with GIMs are applied.

5.3. Dual-Beam Systems

For a dual-beam receiving system, an additional calibration is required, because the propagation paths through the antenna and electronics are independent for the target and reference sources. At VERA, the horn-on-dish method is utilized (Honma et al. 2008), in which an artificial noise source is mounted on an antenna's main reflector and a common signal is injected into both receivers. During observations, cross correlation of the noise source between the two receivers is monitored in real time, so that one can trace the time variation of the instrumental delay difference. The measured delay difference between the two receivers can be applied in postprocessing. Honma et al. (2008) have conducted tests of the horn-on-dish calibration and found that the VERA system can measure instrumental (path) delay differences to an accuracy of $\pm$ 0.1 mm, which is more than adequate for 10 μas relative astrometry.

5.4. Reference Source Position

The error in the position of the source used as the phase reference causes an error in its delay/phase measurements, and these errors are then propagated to the target source. The effect of a reference source position offset differs from other errors previously discussed. The first-order effect is that the position offset of the reference source is transferred to the target source position. If the reference source has negligible proper motion and parallax, this only introduces a constant term, which is absorbed when fitting for parallax and proper motion. This is why we do not include such a term in Equation 3.

However, if the reference source position offset becomes large ($\gtrsim$10 mas), there are additional second-order effects that come from the fact that the target and reference sources are separated on the sky. Because they are at different positions, the interferometer phase response to a position offset at one position on the sky does not exactly mimic the response at the other position. These second-order effects lead to a small position error and to degraded image quality. For an interferometric fringe spacing of 1 mas and a pair separation of 1 deg, one radian of second-order phase-shift results from a calibrator position offset of $\sim$10 mas [this offset, δ_θ, can be estimated by $2\pi\theta_{\text{sep}}(\delta_\theta/\theta_{\text{beam}}) \sim 1$]. In general, for phase-referencing astrometry, one needs to know the calibrator position with an accuracy of $\lesssim$10 mas. For a finer beam size or larger separation angle, the required positional accuracy is even higher.

There are several methods for determining the absolute position of a phase reference source. If one can find a compact ICRF radio source (with $\delta_\theta \approx 1$ mas) close to the target source, one can transfer the position accuracy to the phase-reference source. However, ICRF sources are sparsely distributed on the sky, and some are heavily resolved on long interferometer baselines. In such cases, one should include an ICRF source in the phase-referenced observations, even if offset by up to about $\approx$5 deg from the other sources used for differential astrometry, in order to determine their positions. Even if using a distant ICRF source as a reference produces poor images for other sources, one can usually obtain $\sim$1-mas accurate positions (and then discard the ICRF source data).

Alternatively, one can do preparatory observations to measure the position of a phase-reference source. Usually connected-element interferometers (e.g., JVLA, eMERLIN, or the ATCA), which are limited in baseline length to $\lesssim$100 km, can yield absolute positions with accuracies between 0.01 and 1 arcsec. Although this is usually adequate for VLBI data correlation, it is not good enough for high-accuracy VLBI astrometry. Instead, VLBI observations using broadband group-delay observables (as done for ICRF campaigns) are preferable.

If a maser source is to serve as the phase reference, then broadband group-delays cannot be measured, because spectral lines are intrinsically very narrow, leading to large uncertainty in group-delay estimates. In this case, it is best to use an ICRF source to transfer positional accuracy via phase referencing. As a last resort, if no qualified source is available, one can attempt to synthesize a maser spot image without phase referencing. If the VLBI data have been correlated (or later corrected) with a model accurate to a few wavelengths of path delay, then such an image will resemble optical speckles (caused by short-term phase fluctuations induced by tropospheric water fluctuations and instabilities in the frequency standard). However, the centroid of these speckles can often be determined to $\sim$10 mas, which may be sufficient for the position of the phase reference.

5.5. Source Elevation Limits

Often the dominant source of systematic error is uncompensated tropospheric delays. At any telescope, if one misestimates the zenith (i.e., $z = 0$) tropospheric delay by $\delta\tau_0$, this error is magnified at larger zenith angles by a factor of $\approx$sec z. The differential effect between the target and a background source, with a zenith angle difference of Δz, is then given by $\Delta\tau(z) = \delta\tau_0 \frac{\partial \sec z}{\partial z} \Delta z = \delta\tau_0 \sec z \tan z \Delta z$. Note that $\Delta\tau(z)$ increases dramatically at large zenith angles where both sec z and tan z tend to diverge. For example, the ratio of differential delay error $\Delta\tau(z)$ at $z = 70°$ to $z = 45°$ is a factor of 8.0. This suggests that astrometric observations should be limited, whenever possible, to small zenith angles. Of course, when restricting observations to small zenith angles, one must balance the loss of interferometer (u, v) coverage, which affects both sensitivity and image fidelity, against increasing systematic errors that come with large zenith angle observations. In practice, it may prove valuable to fully simulate interferometric observations in order to estimate an optimum zenith angle limit for the declination of a given target source and the orientation and separation on the sky of the background source.

5.6. Measuring Positions

Perhaps the simplest method of measuring a position from radio interferometric data is to work in the image domain. This involves placing the visibility (u, v) data on a grid, performing a two-dimensional Fourier transformation to sky coordinates, $(u, v) \rightarrow (x, y)$, deconvolving the resulting "dirty" map using the "dirty" beam (point source response) with the CLEAN algorithm, and fitting a two-dimensional Gaussian brightness distribution to compact emission components. This provides offsets, $(\Delta x, \Delta y)$, from the map center, whose absolute position is defined by the position used when cross correlating the interferometer data and any subsequent shifts applied during calibration.

Care must be taken in both shifting positions and interpreting the measured offsets from the map center to properly account for the effects of precession, nutation, aberration, and even general relativistic ray bending near the Sun. For example, position shifts applied to (u, v) data are best done by calculating the full interferometric delay and phase for the desired source position and subtracting those values for the original (correlator) position. Simply calculating a differential correction can lead to significant astrometric errors. Similarly, one should not simply add map offsets $(\Delta x, \Delta y)$—which "know nothing" of physical effects—to the map-center coordinates without correcting for differential precession, nutation, and aberration unless the map offsets are very small. For example, over time-scales of less than a few years, neglecting differential effects generally leads to position errors of $\sim 10^{-4}$ $(\Delta x, \Delta y)$. So, to ensure astrometric accuracies of ~1 μas, $(\Delta x, \Delta y)$ values should be $\lesssim 10$ mas. Therefore, if one measures large $(\Delta x, \Delta y)$ values in a map, one should (accurately) shift positions in the (u, v) data and then reimage and remeasure the offsets.

6. FUTURE

6.1. Space Very Long Baseline Interferometry

Space VLBI involves using a radio antenna in orbit around Earth to obtain baselines lengths greater than an Earth diameter. As is clear from Equation 2, for fixed delay uncertainty one gains astrometric accuracy as the baseline length increases. For space VLBI to improve on Earth-based astrometry, the satellite position (orbit determination) must be determined to ~1-cm accuracy, which is very challenging. However, the complementary application of using VLBI to accurately track spacecraft holds great promise (Duev et al. 2012).

VSOP/HALCA was a space-VLBI mission launched in 1997. Using VSOP, Guirado et al. (2001) conducted space-VLBI astrometry of the radio QSO pair B1342+662/B1342+663 (which are only 5 arcmin apart) and demonstrated that the satellite position error was ~3 meters and that the useful astrometry could be accomplished only for very close pairs. Because of the poor performance of the VSOP 22-GHz receiving system, maser astrometry was not attempted.

Currently the Russian space-VLBI satellite, RadioAstron, is in orbit (with a maximum interferometric baseline of ~300,000 km, which is comparable to the Earth-Moon separation). Fringes from space baselines have been obtained (Kardashev et al. 2013) and, perhaps, high-accuracy space-VLBI astrometry can be realized.

6.2. The Event Horizon Telescope

At millimeter and submillimeter wavelengths, VLBI can achieve an angular resolution sufficient to resolve event-horizon scales for nearby supermassive black holes. A prediction of general relativity is that at this scale one should see the "shadow" of the black hole (Falcke, Melia & Agol 2000). Impressive results have been achieved using ad hoc arrays of antennas that can observe at the short wavelengths required to "see through" a screen of electrons that blurs the image of Sgr A*, the supermassive black hole at the center of the Milky Way (Doeleman et al. 2008).

The Event Horizon Telescope is a worldwide collaboration to realize a powerful VLBI array operating at 1-mm or shorter wavelengths, anchored by the phased-ALMA (acting as a single telescope with great collecting area). In addition to imaging event horizon scales for Sgr A* and M87, one should be able to explore the detailed structure of accretion disks and jet launching points in the vicinity of these supermassive black holes using multifrequency astrometry. Broderick, Loeb & Reid (2011) investigated the possibility of such observations by using the subarray mode of ALMA, SMA, CARMA, and other telescopes and demonstrated that astrometry at the ~3-μas level is possible. With such accuracy one can trace positional variations of the black holes owing to perturbations from surrounding stars and/or a black hole companion. In addition, one should be able to monitor structural variations occurring on dynamical timescales of the innermost stable circular orbit, which for Sgr A* would be ~10 min.

6.3. Square Kilometer Array

The Square Kilometer Array (SKA) is a worldwide project to build and operate the next generation of radio interferometers with an aggregate collecting area of ~1 km^2. Achieving the SKA will likely require three independent arrays, using different technologies to cover the frequency range of ~100 MHz to ~20 GHz. Microarcsecond astrometry can be achieved at frequencies above ~3 GHz, provided baselines of several thousand kilometers are an integral part of the design. For astrometric observations, the SKA holds the promise of significantly increased relative positional accuracy, because its great sensitivity allows the use of weak calibrators much closer in angle on the sky to the astrometric target. Compared with the VLBA, for example, the increased sensitivity of the full SKA should allow the use of background compact radio sources at least an order of magnitude closer to the target, resulting in astrometric accuracy better than ±1 μas! In addition, if multibeam feeds are used on individual antennas, one could simultaneously observe many sources, greatly increasing astrometric survey speed.

DISCLOSURE STATEMENT

The authors are not aware of any affiliations, memberships, funding, or financial holdings that might be perceived as affecting the objectivity of this review.

ACKNOWLEDGMENTS

M.H. would like to thank Hikaru Chida, Kazuhiro Hada, Osamu Kameya, Fuyuhiko Kikuchi, Naoko Matsumoto, Yuka Oizumi, and Nobuyuki Sakai for their support in preparing the manuscript. M.H. also acknowledges financial support by the MEXT/JSPS KAKENHI Grant Numbers 24540242 and 25120007.

LITERATURE CITED

Ando K, Nagayama T, Omodaka T, Handa T, Imai H, et al. 2011. *Publ. Astron. Soc. Jpn.* 63:45–51
André P, Men'schikov A, Bontemps S, Könyves V, Motte F, et al. 2010. *Astron. Astrophys.* 518:L102
Antreasian PG, Baird DT, Border JS, Burckhart PD, Graat EJ, et al. 2005. *J. Spacecr. Rockets* 42:394–405
Asada H. 2002. *Ap. J. Lett.* 574:L69–70
Asaki Y, Deguchi S, Imai H, Hachisuka K, Miyoshi M, Honma M. 2010. *Ap. J.* 721:267–77
Bailes M, Manchester R, Kesteven M, Norris RP, Reynolds JE. 1990a. *Nature* 343:240–41
Bailes M, Manchester R, Kesteven MJ, Norris RP, Revnolds JE. 1990b. *MNRAS* 247:322
Bartel N, Herring TA, Ratner MI, Shapiro II, Corey BE. 1986. *Nature* 319:733–83
Bartkiewicz A, Brunthaler A, Szymczak M, van Langevelde HJ, Reid MJ. 2008. *Astron. Astrophys.* 490:787–92
Bietenholz MF, Bartel N, Rupen MP. 2004. *Ap. J.* 615:173–80
Bietenholz MF, Kondratiev V, Ransom S, Slane P, Bartel N, Buchner S. 2013. *MNRAS* 431:2590–98
Blandford RD, Königl A. 1979. *Ap. J.* 232:34–48
Border JS, Folkner WM, Kahn RD, Zukor KS. 1992. *Telecommun. Data Acquis. Prog. Rep.* 42:110
Bourda G, Collioud A, Charlot P, Porcas R, Garrington S. 2011. *Astron. Astrophys.* 526:102
Bower GC, Bolatto A, Ford EB, Fries A, Kalas P, et al. 2011. *Astron. J.* 740:32
Bower GC, Bolatto A, Ford EB, Kalas P. 2009. *Astron. J.* 701:1922–39
Braatz JA, Reid MJ, Humphreys EML, Henkel C, Condon JJ, Lo KY. 2010. *Ap. J.* 718:657–65
Braatz J, Reid M, Kuo CY, Impellizzeri V, Condon J, et al. 2013. In *Advancing the Physics of Cosmic Distances, Proc. IAU Symp. 289*, ed. R de Grijs, G Bono, 8:255–61. Cambridge, UK: Cambridge Univ. Press
Brisken WF, Benson JM, Beasley AJ, Fomalont EB, Goss WM, Thorsett SE. 2000. *Ap. J.* 541:959–62
Brisken WF, Benson JM, Goss WM, Thorsett SE. 2002. *Ap. J.* 571:906–17
Brisken WF, Thorsett S, Golden A, Goss WM. 2003. *Ap. J.* 593:89–92
Broderick AE, Loeb A, Reid MJ. 2011. *Ap. J.* 735:57
Brunthaler A, Reid MJ, Falcke H, Greenhill LJ, Henkel C. 2005. *Science* 307:1440–43
Brunthaler A, Reid MJ, Falcke H, Henkel C, Menten KM. 2007. *Astron. Astrophys.* 462:101–6
Brunthaler A, Reid MJ, Menten KM, Zheng XW, Moscadelli L, Xu Y. 2009. *Ap. J.* 693:424–29
Caballero-Nieves SM, Gies DR, Bolton CT, Hadrava P, Herrero A, et al. 2009. *Ap. J.* 701:1895–905
Campbell R, Bartel N, Shapiro II, Ratner MI, Cappallo RJ, et al. 1996. *Ap. J. Lett.* 461:95
Cernis K. 1990. *Ap. Space Sci.* 166:315–30
Chatterjee S, Brisken WF, Vlemmings WHT, Goss WM, Lazio TJW. 2009. *Ap. J.* 698:250–65
Chatterjee S, Cordes JM, Vlemmings WHT, Arzoumanian Z, Goss WM, Lazio TJW. 2004. *Ap. J.* 604:339–45
Choi Y, Hirota T, Honma M, Kobayashi H, Bushimata T, et al. 2008. *Publ. Astron. Soc. Jpn.* 60:1007–12
Clark TA, Hutton LK, Marandino GE, Counselman CC, Robertson DS, et al. 1976. *Astron. J.* 81:599–603
Counselman CC III, Gourevitch SA, King RW, Loriot GB, Prinn RG. 1979. *Science* 205:85–86
Counselman CC III, Hinteregger HF, King RW, Shapiro II. 1973. *Science* 181:772–4
Damour T, Taylor JH. 1991. *Ap. J.* 366:501–11
Darling J. 2011. *Ap. J.* 732:2
Deller AT, Archibald AM, Brisken WF, Chatterjee S, Janssen GH, et al. 2012a. *Ap. J. Lett.* 756:L25
Deller AT, Bailes M, Tingay SJ. 2009. *Science* 323:1327–29
Deller AT, Camilo F, Reynolds JE, Halpern JP. 2012b. *Ap. J. Lett.* 748:L1
Deller AT, Tingay SJ, Bailes M, Reynolds JE. 2009. *Ap. J.* 701:1243–57
Deller AT, Verbiest J, Tingay SJ, Bailes M. 2008. *Ap. J. Lett.* 685:L67–70
Dodson R, Legge D, Revnolds JE, McCulloch PM. 2003. *Ap. J.* 596:1137–41

Doeleman SS, Weintroub J, Rogers AEE, Plambeck R, Freund R, et al. 2008. *Nature* 455:78–80
Duev DA, Calves GM, Pogrebenko SV, Gurvits LI, Cimó G, Bocanegra BT. 2012. *Astron. Astrophys.* 541:43
Duncan RC, Thompson C. 1992. *Ap. J. Lett.* 392:L9–12
Dyson FW, Eddington AS, Davidson C. 1919. *Ap. J.* 220:291–333
Dzib S, Loinard L, Mioduszewski AJ, Boden AF, Rodríguez LF, Torres RM. 2010. *Ap. J.* 718:610–9
Dzib SA, Rodriguez LF, Loinard L, Mioduszewski AJ, Ortiz-Leon GN, Araudo AT. 2013. *Ap. J.* 763:139
Eiroa C, Djupvik AA, Casali MM. 2008. In *Handbook of Star Forming Regions, Vol. II*, ed. B Reipurth. *ASP Monogr. Publ.* 5:693–734. San Francisco: ASP
Evans NJ II, Dunham MM, Jørgensen JK, Enoch ML, Merín B, et al. 2009. *Ap. J. Suppl.* 181:321–50
Everitt C, DeBra D, Parkinson B, Turneaure J, Conklin J, et al. 2011. *Phys. Rev. Lett.* 106:1
Falcke H, Melia F, Agol E. 2000. *Ap. J. Lett.* 528:L13–6
Folkner WM, Border JS, Nandi S, Zukor KS. 1993. *Telecommun. Data Acquis. Prog. Rep.* 42–113:22–36
Fomalont EB, Kopeikin S. 2003. *Ap. J.* 598:704
Fomalont EB, Kopeikin S. 2009. In *A Giant Step: From Milli- to Micro- Arcsecond Astrometry, Proc. IAU Symp. 248*, ed. W Jin, I Platais, MAC Perryman, 3:383–86. Cambridge, UK: Cambridge Univ. Press
Fomalont EB, Kopeikin S, Lanyi G, Benson J. 2009. *Ap. J.* 699:1395
Gawronski MP, Gozdziewski K, Katarzynski K. 2013. *Project RISARD.* Paper presented at EVN Symp., 11th, Oct. 9–12, 2012, Bordeaux, France. arXiv:1309:4639
Genzel R, Reid MJ, Moran JM, Downes D. 1981. *Ap. J.* 244:884–902
Ghez AM, Salim S, Weinberg NN, Lu JR, Do T, et al. 2008. *Ap. J.* 689:1044–62
Gillessen S, Eisenhauer F, Trippe S, Alexander T, Genzel R, et al. 2009. *Ap. J.* 692:1074–109
Goossens S, Matsumoto K, Liu Q, Kikuchi F, Sato K, et al. 2011. *J. Geod.* 85:205–28
Gou L, McClintock JE, Reid MJ, Orosz JA, Steiner JF, et al. 2011. *Ap. J.* 742:85
Güdel M, Arzner K, Audard M, Bouvier J, Briggs K, et al. 2008. *Astron. Soc. Pac. Conf. Ser.* 384:65–69
Guirado JC, Marcaide JM, Alberdi A, Elosegui P, Ratner MI, et al. 1995. *Astron. J.* 110:2586
Guirado JC, Reynolds JE, Lestrade JF, Preston RA, Jauncey DL, et al. 1997. *Ap. J.* 490:835–9
Guirado JC, Ros E, Jones DL, Lestrade JF, Marcaide JM, et al. 2001. *Astron. Astrophys.* 371:766–70
Gwinn C, Taylor J, Weisberg JM, Rawley LA. 1986. *Astron. J.* 91:338–42
Hachisuka K, Brunthaler A, Menten KM, Reid MJ, Hagiwara Y, Mochizuki N. 2009. *Ap. J.* 696:1981–6
Hachisuka K, Brunthaler A, Menten KM, Reid MJ, Imai H, et al. 2006. *Ap. J.* 645:337–44
Hada K, Doi A, Kino M, Nagai H, Hagiwara Y, Kawaguchi N. 2011. *Nature* 477:185–7
Hada K, Kino M, Nagai H, Doi A, Hagiwara Y, et al. 2012. *Ap. J.* 760:52
Harrison TE, McNamara BJ, Szkody P, McArthur BE, Benedict GF, et al. 1999. *Ap. J. Lett.* 515:L93–96
Helfand D, Chatterjee S, Brisken WF, Camilo F, Reynolds J, et al. 2007. *Ap. J.* 662:1198–203
Herbig GH, Jones BF. 1983. *Astron. J.* 88:1040–52
Hernández-Pajares M, Juan JM, Sanz J, Orus R, Garcia-Rigo A, et al. 2009. *J. Geod.* 83:265–75
Herrnstein JR, Moran JM, Greenhill LJ, Diamond PJ, Inoue M, et al. 1999. *Nature* 400:539–41
Herrnstein JR, Moran JM, Greenhill LJ, Trotter AS. 2005. *Ap. J.* 629:719–38
Hildebrand CE, Iijima BA, Kroger PM, Folkner WM, Edwards CD. 1994. *Telecommun. Data Acquis. Prog. Rep.* 119:46–82
Hirota T, Bushimata T, Choi YK, Honma M, Imai H, et al. 2007. *Publ. Astron. Soc. Jpn.* 59(5):897–903
Hirota T, Bushimata T, Choi YK, Honma M, Imai H, et al. 2008. *Publ. Astron. Soc. Jpn.* 60(1):37–44
Hirota T, Honma M, Imai H. 2011. *Publ. Astron. Soc. Jpn.* 63:1–8
Hirshfeld A. 2001. *Parallax: The Race to Measure the Cosmos.* New York: Freeman
Honma M, Bushimata T, Choi YK, Hirota T, Imai H, et al. 2007. *Publ. Astron. Soc. Jpn.* 59:889–95
Honma M, Fujii T, Hirota T, Horiai K, Iwadate K, et al. 2003. *Publ. Astron. Soc. Jpn.* 55:L57–60
Honma M, Hirota T, Kan-Ya Y, Kawaguchi N, Kobayashi H, et al. 2011. *Publ. Astron. Soc. Jpn.* 63:17–22
Honma M, Kijima M, Suda H, Kawaguchi N, Kobayashi H, et al. 2008. *Publ. Astron. Soc. Jpn.* 60:935–50
Honma M, Nagayama T, Ando K, Bushimata T, Choi YK, et al. 2012. *Publ. Astron. Soc. Jpn.* 64:136
Honma M, Tamura Y, Reid MJ. 2008. *Publ. Astron. Soc. Jpn.* 60:951–60
Hubble EP. 1926. *Ap. J.* 63:236–74
Humphreys EML, Reid MJ, Moran JM, Greenhill LJ, Argon AL. 2013. *Ap. J.* 775:13
Imai H, Kurayama T, Honma M, Miyaji T. 2013. *Publ. Astron. Soc. Jpn.* 65:28

Imai H, Nakashima K, Bushimata T, Choi YK, Hirota T, et al. 2007. *Publ. Astron. Soc. Jpn.* 59:1107–13
Imai H, Sahai R, Morris M. 2007. *Ap. J.* 669:424–34
Imai H, Sakai N, Nakanishi H, Sakanoue H, Honma M, Miyaji T. 2012. *Publ. Astron. Soc. Jpn.* 64:142
Imai H, Tafoya D, Honma M, Hirota T, Miyaji T. 2011. *Publ. Astron. Soc. Jpn.* 63:81–87
Immer K, Reid MJ, Menten KM, Brunthaler A, Dame TM. 2013. *Astron. Astrophys.* 553:117
Jacobson RA, Antreasian PG, Bordi JJ, Criddle KE, Ionasescu R, et al. 2006. *Astron. J.* 132:2520–6
Jones DL, Fomalont EB, Dhawan V, Romney J, Folkner WM, et al. 2011. *Astron. J.* 141:29
Kamezaki T, Nakagawa A, Omodaka T, Kurayama T, Imai H, et al. 2012. *Publ. Astron. Soc. Jpn.* 64:7
Kamohara R, Bujarrabal V, Honma M, Nakagawa A, Matsumoto N, et al. 2010. *Astron. Astrophys.* 510:A69
Kardashev N, Khartov VV, Abramov VV, Avdeev VY, Alakoz AV, et al. 2013. *Astron. Rep.* 57:153–94
Kikuchi F, Liu Q, Hanada H, Kawano N, Matsumoto K, et al. 2009. *Radio Sci.* 44:2008–14
Kim MK, Hirota T, Honma M, Kobayashi H, Bushimata T, et al. 2008. *Publ. Astron. Soc. Jpn.* 60:991–99
King RW, Counselman CC III, Shapiro II. 1976. *J. Geophys. Res.* 81:6251–56
Kovalev YY, Lobanov AP, Pushkarev AB, Zensus JA. 2008. *Astron. Astrophys.* 483:759–86
Kuo CY, Braatz JA, Reid MJ, Lo KY, Condon JJ, et al. 2013. *Ap. J.* 767:155
Kurayama T, Nakagawa A, Sawada-Satoh S, Sato K, Honma M, et al. 2011. *Publ. Astron. Soc. Jpn.* 63:513–25
Kurayama T, Sasao T, Kobayashi H. 2005. *Ap. J. Lett.* 627:L49–52
Lambert SB, Le Poncin-Lafitte C. 2009. *Astron. Astrophys.* 499:331–35
Lambert SB, Le Poncin-Lafitte C. 2011. *Astron. Astrophys.* 529:70
Lanyi G, Bagri DS, Border J. 2007. *Proc. IEEE* 95:2193–201
Lestrade JF, Jones DL, Preston RA, Phillips RB, Titus MA, et al. 1995. *Astron. Astrophys.* 304:182–88
Lestrade JF, Phillips RB, Hodges MW, Preston RA. 1993. *Ap. J.* 410:808–14
Lestrade JF, Phillips RB, Jones DL, Preston RA. 1996. *J. Geophys. Res. Planets* 101(E6):14837–42
Lestrade JF, Preston RA, Jones DL, Phillips RB, Rogers AEE, et al. 1999. *Astron. Astrophys.* 344:1014–26
Loeb A, Reid MJ, Brunthaler A, Falcke H. 2005. *Ap. J.* 633:894–98
Loinard L, Mioduszewski AJ, Rodríguez LF, González RA, Rodríguez MI, Torres RM. 2005. *Ap. J. Lett.* 619:L179–82
Loinard L, Torres RM, Mioduszewski AJ, Rodríguez LF. 2008. *Ap. J. Lett.* 675:L29–32
Loinard L, Torres RM, Mioduszewski AJ, Rodríguez LF, González-Lópezlira RA, et al. 2007. *Ap. J.* 671:546–54
Ma C, Clark TA, Ryan JW, Herring TA, Shapiro II, et al. 1986. *Astron. J.* 92:1020–29
Macri LM, Stanek KZ, Bersier D, Greenhill LJ, Reid MJ. 2006. *Ap. J.* 652:1113–49
Marcaide JM, Elosegui P, Shapiro II. 1994. *Astron. J.* 108:368–73
Marcaide JM, Shapiro II, Corey BE, Cotton WD, Gorenstein MV, et al. 1985. *Astron. Astrophys.* 142:71–84
Marti-Vidal I, Marcaide JM, Alberdi A, Pérez-Torres MA, Ros E, Guirado JC. 2011. *Astron. Astrophys.* 533:A111
Marti-Vidal I, Marcaide JM, Guirado JC, Pérez-Torres MA, Ros E. 2008. *Astron. Astrophys.* 478:267–75
Matsumoto N, Honma M, Isono Y, Ujihara H, Kimura K, et al. 2011. *Publ. Astron. Soc. Jpn.* 63:1345–56
Menten KM, Reid MJ, Eckart A, Genzel R. 1997. *Ap. J. Lett.* 475:L111–14
Menten KM, Reid MJ, Forbrich J, Brunthaler A. 2007. *Astron. Astrophys.* 474:515–20
Miller-Jones JCA, Jonker PG, Dhawan V, Brisken W, Rupen MP, et al. 2009. *Ap. J. Lett.* 706:L230–34
Miller-Jones JCA, Sivakoff GR, Knigge C, Körding EG, Templeton M, Waagen EO. 2013. *Science* 340:950–52
Min C, Matsumoto N, Kim MK, Hirota T, Shibata KM, et al. 2014. *Publ. Astron. Soc. Jpn.* 66:38
Mirabel FI, Rodrigues I. 2003. *Science* 300:1119–20
Moellenbrock GA, Claussen MJ, Goss WM. 2009. *Ap. J.* 694:192–204
Mori K, Gotthelf EV, Zhang S, An H, Baganoff FK, et al. 2013. *Ap. J. Lett.* 770:L23
Moscadelli L, Cesaroni R, Rioja MJ, Dodson R, Reid MJ. 2011. *Astron. Astrophys.* 526:66
Motogi K, Sorai K, Habe A, Honma M, Kobayashi H, et al. 2011. *Publ. Astron. Soc. Jpn.* 63:31–44
Nagayama T, Omodaka T, Handa T, Honma M, Kobayashi H, et al. 2011a. *Publ. Astron. Soc. Jpn.* 63:719–25
Nagayama T, Omodaka T, Nakagawa A, Handa T, Honma M, et al. 2011b. *Publ. Astron. Soc. Jpn.* 63:23–30
Nakagawa A, Tsushima M, Ando K, Bushimata T, Choi YK, et al. 2008. *Publ. Astron. Soc. Jpn.* 60:1013–21
Ng CY, Romani RW, Brisken WF, Chatterjee S, Kramer M. 2007. *Ap. J.* 654:487–93
Niinuma K, Nagayama T, Hirota T, Honma M, Motogi K, et al. 2011. *Publ. Astron. Soc. Jpn.* 63:9–16

Nyu D, Nakagawa A, Matsui M, Imai H, Sofue Y, et al. 2011. *Publ. Astron. Soc. Jpn.* 63:63–70
Oh CS, Kobayashi H, Honma M, Hirota T, Sato K, et al. 2010. *Publ. Astron. Soc. Jpn.* 62:101–14
Orosz JA, McClintock JE, Aufdenberg JP, Remillard RA, Reid MJ, et al. 2011. *Ap. J.* 742:84
Peterson WM, Mutel RL, Lestrade JF, Güdel M, Goss WM. 2011. *Astron. J.* 737:104
Porcas RW. 2009. *Astron. Astrophys.* 505:L1–4
Ratner MI, Bartel N, Bietenholz MF, Lebach DE, Lestrade JF, et al. 2012. *Ap. J. Suppl.* 201:5
Reid MJ. 2009. *Int. J. Mod. Phys. D* 18:889–910
Reid MJ, Braatz JA, Condon JJ, Greenhill LJ, Henkel C, Lo KY. 2009a. *Ap. J.* 695:287–91
Reid MJ, Braatz JA, Condon JJ, Lo KY, Kuo CY, et al. 2013. *Ap. J.* 767:154
Reid MJ, Brunthaler A. 2004. *Ap. J.* 616:872–84
Reid MJ, McClintock JE, Narayan R, Gou L, Remillard RA, Orosz JA. 2011. *Ap. J.* 742:83
Reid MJ, Menten KM, Brunthaler A, Zheng XW, Dame TM, et al. 2014. *Ap. J.* 783:130
Reid MJ, Menten KM, Brunthaler A, Zheng XW, Moscadelli L, Xu Y. 2009b. *Ap. J.* 693:397–405
Reid MJ, Menten KM, Genzel R, Ott T, Schödel R, Eckart A. 2003. *Ap. J.* 587:208–20
Reid MJ, Menten KM, Zheng XW, Brunthaler A, Moscadelli L, et al. 2009c. *Ap. J.* 700:137–48
Reid MJ, Schneps MH, Moran JM, Gwinn CR, Genzel R, et al. 1988. *Ap. J.* 330:809–16
Riess AG, Fliri J, Valls-Gabaud D. 2012. *Ap. J.* 745:156
Riess AG, Macri L, Casterno S, Lampeitl H, Ferguson HC, et al. 2012. *Ap. J.* 752:76
Rioja M, Dodson R, Porcas RW, Ferris D, Reynolds J, et al. 2009. *Proc. 8th Int. e-VLBI Workshop*, Madrid, Spain, June 22–26, PoS(EXPReS09)014 **http://pos.sissa.it/archive/conferences/082/014/EXPReS09_014.pdf**
Rioja MJ, Marcaide JM, Elosegui P, Shapiro II. 1997. *Astron. Astrophys.* 325:383–90
Rygl KLJ, Brunthaler A, Reid MJ, Menten KM, van Langevelde HJ, Xu Y. 2010. *Astron. Astrophys.* 511:2
Rygl KLJ, Brunthaler A, Sanna A, Menten KM, Reid MJ, et al. 2012. *Astron. Astrophys.* 539:A79
Sagdeyev RZ, Kerzhanovitch VV, Kogan LR, Kostenko VI, Linkin VM, et al. 1992. *Astron. Astrophys.* 254:387
Salzberg IM. 1973. *Proc. IEEE* 61:1233–36
Sandstrom KM, Peek JEG, Bower GC, Bolatto AD, Plambeck RL. 2007. *Ap. J.* 667:1161–69
Sanna A, Reid MJ, Dame TM, Menten KM, Brunthaler A, et al. 2012. *Ap. J.* 745:82
Sanna A, Reid MJ, Moscadelli L, Dame TM, Menten KM, et al. 2009. *Ap. J.* 706:464–70
Sakai N, Honma M, Nakanishi H, Sakanoue H, Kurayama T, et al. 2012. *Publ. Astron. Soc. Jpn.* 64:108
Sato M, Hirota T, Honma M, Kobayashi H, Sasao T, et al. 2008. *Publ. Astron. Soc. Jpn.* 60:975–89
Sato M, Hirota T, Reid MJ, Honma M, Kobayashi H, et al. 2010a. *Publ. Astron. Soc. Jpn.* 62:287–99
Sato M, Reid MJ, Brunthaler A, Menten KM. 2010b. *Ap. J.* 720:1055–65
Schödel R, Ott T, Genzel R, Hofmann R, Lehnert M, et al. 2002. *Nature* 419:694–96
Shapiro II, Bartel N, Bietenholz MF, Lebach DE, Lestrade JF, et al. 2012. *Ap. J. Suppl.* 201:1
Shiozaki S, Imai H, Tafoya D, Omodaka T, Hirota T, et al. 2011. *Publ. Astron. Soc. Jpn.* 63:1219–28
Slade MA, Preston RA, Harris AW, Skjerve LJ, Spitzmesser DJ. 1977. *Moon* 17:133–47
Soderblom DR, Nelan E, Benedict GF, McArthur B, Ramirez I, et al. 2005. *Astron. J.* 129:1616–24
Sokolovsky KV, Kovalev YY, Pushkarev AB, Lobanov AP. 2011. *Astron. Astrophys.* 532:A38
Steigenberger P, Tesmer V, Krügel M, Thaller D, Schmid R, et al. 2007. *J. Geod.* 81:503–14
Tafoya D, Imai H, Gomez Y, Torrelles JM, Patel NA, et al. 2011. *Publ. Astron. Soc. Jpn.* 63:71–80
Thompson AR, Moran JM, Swenson GW Jr. 2001. *Interferometry and Synthesis in Radio Astronomy*. New York: Wiley. 2nd ed.
Torres RM, Loinard L, Mioduszewski AJ, Boden AF, Franco-Hernández R, et al. 2012. *Ap. J.* 747:18
Torres RM, Loinard L, Mioduszewski AJ, Rodríguez LF. 2007. *Ap. J.* 671:1813–19
Torres RM, Loinard L, Mioduszewski AJ, Rodríguez LF. 2009. *Ap. J.* 698:242–49
van Langevelde HJ, Vlemmings WHT, Diamond PJ, Baudry A, Beasley AJ. 2000. *Astron. Astrophys.* 357:945–50
van Leeuwen F. 2009. *Astron. Astrophys.* 497:209–42
van Maanen A. 1923. *Ap. J.* 57:264–78
Vlemmings WHT, van Langevelde HJ. 2007. *Astron. Astrophys.* 472:547–53
Vlemmings WHT, van Langevelde HJ, Diamond PJ, Habing HJ, Schilizzi RT. 2003. *Astron. Astrophys.* 407:213–24
Wade C, Johnston KJ. 1977. *Astron. J.* 82:791–95

Walker RC, Chatterjee S. 2000. *VLBA scientific memo 23*. Natl. Radio Astron. Obs., Charlottesville, VA. **https://science.nrao.edu/facilities/vlba/other/publications/memos/sci/gps_ion/gps_ion**
Whitelock PA, Feast MW, Van Leeuwen F. 2008. *MNRAS* 386:313–23
Will CM. 2003. *Ap. J.* 590:683–90
Witasse O, Lebreton JP, Bird MK, Dutta-Roy R, Folkner WM, et al. 2006. *J. Geophys. Res.* 111:E07S01
Xu Y, Li JJ, Reid MJ, Menten KM, Zheng XW, et al. 2013. *Ap. J.* 769:15
Xu Y, Moscadelli L, Reid MJ, Menten KM, Zhang B, et al. 2011. *Ap. J.* 733:25
Xu Y, Reid MJ, Menten KM, Brunthaler A, Zheng XW, Moscadelli L. 2009. *Ap. J.* 693:413–18
Xu Y, Reid MJ, Menten KM, Zheng XW. 2006. *Science* 311:54–57
Zhang B, Reid MJ, Menten KM, Zheng XW, Brunthaler A. 2012. *Astron. Astrophys.* 544:A42
Zhang B, Reid MJ, Menten KM, Zheng XW, Brunthaler A, et al. 2013. *Ap. J.* 775:79
Zhang B, Zheng XW, Reid MJ, Menten KM, Xu Y, et al. 2009. *Ap. J.* 693:419–23

Far-Infrared Surveys of Galaxy Evolution

Dieter Lutz

Max-Planck-Institut für extraterrestrische Physik, 85741 Garching, Germany;
email: lutz@mpe.mpg.de

Annu. Rev. Astron. Astrophys. 2014. 52:373–414

First published online as a Review in Advance on June 18, 2014

The *Annual Review of Astronomy and Astrophysics* is online at astro.annualreviews.org

This article's doi:
10.1146/annurev-astro-081913-035953

Keywords

infrared galaxies, active galaxies, star formation, cosmic infrared background, interstellar medium, high-redshift galaxies

Abstract

Roughly half of the radiation from evolving galaxies in the early Universe reaches us in the far-infrared and submillimeter wavelength ranges. Recent major advances in observing capabilities, in particular the launch of the *Herschel Space Observatory* in 2009, have dramatically enhanced our ability to use this information in the context of multiwavelength studies of galaxy evolution. Near its peak, three-quarters of the cosmic infrared background is now resolved into individually detected sources. The use of far-infrared diagnostics of dust-obscured star formation and of interstellar medium conditions has expanded from extreme and rare extreme high-redshift galaxies to more typical main-sequence galaxies and hosts of active galactic nuclei out to $z \gtrsim 2$. These studies shed light on the evolving role of steady equilibrium processes and of brief starbursts at and since the peak of cosmic star formation and black hole accretion. This review presents a selection of recent far-infrared studies of galaxy evolution with an emphasis on *Herschel* results.

1. THE ROAD TO *HERSCHEL* SURVEYS

The very first steps to use far-infrared emission as a tool to study galaxy evolution date back to the pioneering *Infrared Astronomical Satellite* (IRAS) mission, in which 60-μm source counts in the ecliptic pole region were found to exceed the counts predicted by no-evolution models (Hacking et al. 1987). The *Infrared Space Observatory* (ISO) obtained the first deep surveys at both mid- and far-infrared wavelengths; these surveys detected strong evolution of the luminosity function out to $z \sim 1$ and supported by plausible extrapolation that such dusty galaxies constitute the cosmic infrared background (CIB). ISO also pioneered the application of the rich mid-infrared spectra as a tool for studying energy sources and physical conditions in dusty galaxies (for a review, see Genzel & Cesarsky 2000). The *Spitzer Space Telescope* revolutionized mid-infrared surveys and opened a window to direct mid-infrared spectroscopy of faint high-z galaxies (for a review, see Soifer et al. 2008). The *AKARI* mission (Murakami et al. 2007) provided uniquely detailed mid-infrared photometric coverage. Independently and at a similar time, ground-based (sub)millimeter surveys at wavelengths from 850 μm to 1.2 mm detected luminous Submillimetre Common-User Bolometer Array (SCUBA) galaxies (Blain et al. 2002; see also Casey et al. 2014), which are among the most actively star-forming systems at redshifts $z \sim 2$.

For most redshifts of interest, both spaceborne mid-infrared and ground-based submillimeter surveys miss the rest frame far-infrared peak that dominates the spectral energy distributions (SEDs) of most galaxies. Extrapolation via template SEDs is then needed to characterize the energy budget of samples that were selected at these wavelengths. Also, such samples can be significantly biased compared with a truly calorimetric selection by infrared luminosity. When aiming for the crucial deep far-infrared surveys, fully cryogenic space missions with small 60–80-cm diameter primary mirrors, such as ISO, *Spitzer*, and *AKARI*, were strongly limited by source confusion. At 250–500 μm, results from the 2-m Balloon-borne Large Aperture Submillimeter Telescope (BLAST) (Pascale et al. 2008) provided a first deep glimpse at the extragalactic sky but are now superseded by *Herschel* data. The European Space Agency's *Herschel Space Observatory* (Pilbratt et al. 2010), in operation 2009–2013, for the first time met the need for a combination of sensitivity and large aperture (3.5 m) over the full far-infrared and submillimeter wavelength range, substantially reducing confusion levels. At 350 μm and longer wavelengths, low-spatial-resolution *Planck* all-sky maps and catalogs (Planck Collab. et al. 2013a) support studies of galaxy evolution.

Much of this review is based on *Herschel* photometric surveys and pointed observations using the camera modes of the Photodetector Array Camera and Spectrometer (PACS) (70, 100, and 160 μm; Poglitsch et al. 2010) and Spectral and Photometric Imaging Receiver (SPIRE) (250, 350, and 500 μm; Griffin et al. 2010) instruments. We refer to this range as the far-infrared, leaving the (sub)millimeter terminology to ground-based surveys typically conducted at 850 μm or longer wavelengths. Similarly, submillimeter galaxy (SMG) refers to the type of galaxy detected in these surveys, whereas equivalent *Herschel*-selected sources are called dusty star-forming galaxies (DSFGs) where appropriate.

Traditionally, studies in the local Universe make use of a terminology of luminous infrared galaxies (LIRGs), defined by their total 8–1,000 μm infrared luminosity $L_{\mathrm{IR}} > 10^{11}\,\mathrm{L}_{\odot}$, and their ultra- and hyperluminous (ULIRG and HYLIRG) equivalents greater than 10^{12} and $10^{13}\,\mathrm{L}_{\odot}$, respectively (e.g., Sanders & Mirabel 1996). These are handy acronyms, but for the purpose of galaxy evolution studies, it is important to recall that connotations of these classifications that were carefully calibrated in the local Universe may not apply at high redshift. For example, local ULIRGs are found to be major mergers with an unusually dense and warm interstellar medium (ISM). The same cannot necessarily be assumed for their higher-redshift equivalents at the same infrared luminosity. Where used in this review, the (U)LIRG acronyms should be seen as pure infrared luminosity classifications without further implications for the nature of the galaxy under study.

The *Herschel* general purpose observatory served a wide range of science goals by using capabilities that included both photometric imaging and spectroscopy over the 55–672 μm wavelength range. Approximately 15% of *Herschel*'s observing time, including both guaranteed and open time, was devoted to imaging surveys of galaxy evolution, not counting pointed imaging or spectroscopic studies of individual high-redshift sources. The list below summarizes some of the major efforts from the initial guaranteed time and open time *Herschel* key programs as well as from similar-scale programs that were allocated after later calls for proposals. In addition, a significant fraction of the science discussed below is based on smaller-scale imaging projects as well as pointed imaging and spectroscopic studies that are not listed explicitly. *Herschel* extragalactic surveys exceed 1,000 deg^2 in total but do not reach the full sky coverage that *Planck* provides at lower spatial resolution for $\lambda \geq 350$ μm and *AKARI* provides at lower resolution and sensitivity for $\lambda \leq 160$ μm. Extragalactic confusion limits and instrument design imply a focus for PACS on small-area deep surveys, whereas SPIRE surveys typically emphasize large area. Below we list some major surveys, and **Figure 1** gives a graphical overview of the area and depth of fields covered.

- Herschel Multi-tiered Extragalactic Survey (HerMES) (Oliver et al. 2012; **http://hermes.sussex.ac.uk/content/hermes-project**) is a wedding cake–type multipurpose survey of blank and cluster fields primarily using SPIRE and covering 70 deg^2 in its shallowest tier. A shallow 270-deg^2 extension has been obtained in the HerMES Large Mode Survey (HeLMS) project (PI M. Viero).
- PACS Evolutionary Probe (PEP) (Lutz et al. 2011; **http://www.mpe.mpg.de/ir/Research/PEP/index.php**) is a PACS survey of popular multiwavelength fields, such as Great Observatories Origins Deep Survey (GOODS), Cosmic Evolution Survey (COSMOS), Extended Groth Strip (EGS), Extended Chandra Deep Field South (ECDFS), Lockman Hole-X-ray Multi-Mirror Mission (LH-XMM), and cluster fields, that is coordinated with SPIRE observations of the same fields from HerMES.
- GOODS-Herschel (Elbaz et al. 2011; **http://hedam.oamp.fr/GOODS-Herschel/**) provides deep PACS and SPIRE observations of GOODS-North and ultradeep PACS coverage of part of GOODS-South (GOODS-S). A combined PEP/GOODS-Herschel data set including all PACS data for the two GOODS fields was published by Magnelli et al. (2013).
- Herschel Astrophysical Terahertz Large Area Survey (H-ATLAS) (Eales et al. 2010a; **http://www.h-atlas.org**) is the largest-area, shallow-extragalactic *Herschel* survey; it covers approximately 570 deg^2 using SPIRE and PACS.
- The Herschel Lensing Survey (HLS) (Egami et al. 2010; **http://herschel.as.arizona.edu/hls/hls.html**) provides deep PACS and SPIRE data on 44 X-ray luminous clusters as well as SPIRE snapshots of another 527 clusters. HLS aims at both lensed background objects and cluster members.
- Herschel Cosmic Assembly Near-Infrared Deep Extragalactic Legacy Survey (H-CANDELS) (PI M. Dickinson) provides deep *Herschel* data of the CANDELS (**http://candels.ucolick.org/index.html**) subregions of the COSMOS and of the United Kingdom Infrared Telescope Infrared Deep Sky Survey Ultra-Deep Survey (UKIDSS-UDS) fields that were not covered at equivalent depth by the projects mentioned above.

Figure 2 shows an example of the post-*Herschel* status of deep surveys over the full mid-infrared to submillimeter wavelength range. Surveys with beams of width 5 arcsec to 30 arcsec are now available over the full range. In the deepest fields, they reach the confusion limit for all wavelengths except 70 μm, at which a fully cryogenic 3-m class telescope such as the Space Infrared Telescope for Cosmology and Astrophysics (SPICA) project will be needed.

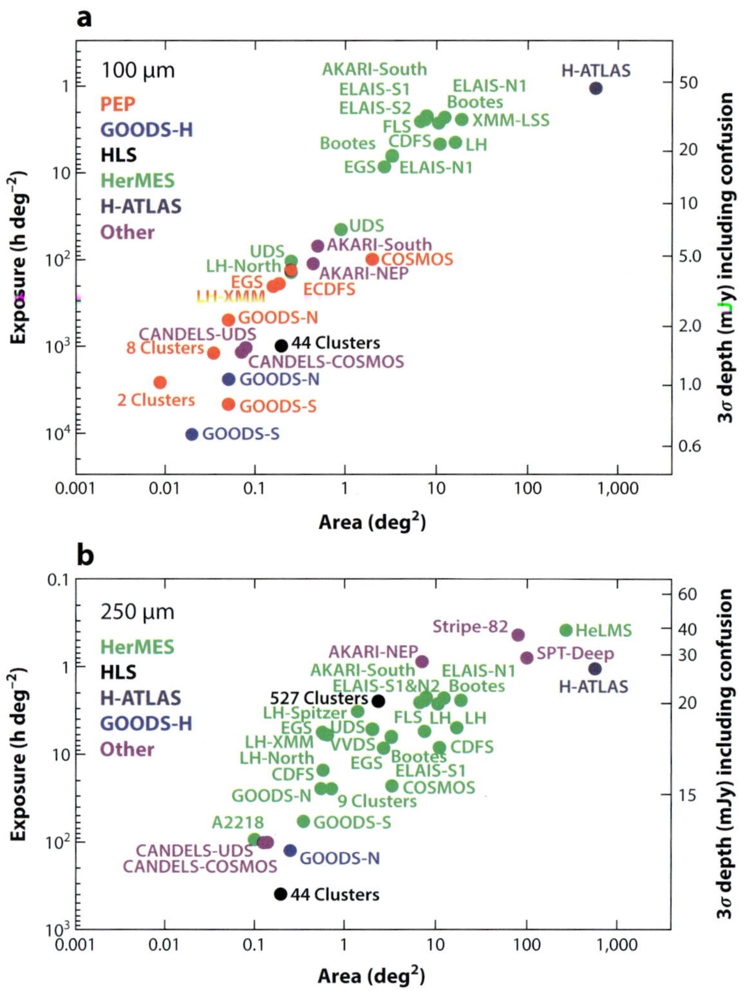

Figure 1

Survey area and point source depth reached by some extragalactic surveys with (*a*) *Herschel* Photodetector Array Camera and Spectrometer (PACS) (100-µm data are shown, but 160-µm data are available for the same fields) and (*b*) SPIRE (250-µm data are shown, but 350- and 500-µm data are available). Exposure is computed from total observing time (including overheads) and survey area. The point source depths (including confusion noise) shown on the right axis should hence be seen as indicative only, because no attempt was made to capture detailed effects of different observing layouts in the various projects. Some sets of cluster observations are included; the area shown is the sum of the areas for all clusters. The surveys shown are from the projects listed in Section 1.1 as well as other projects using the *AKARI* North Ecliptic Pole (NEP) (PI S. Serjeant), *AKARI* Deep Field South (PI T. Takagi), South Pole Telescope (SPT) deep field (PI J. Carlstrom), and Sloan Digital Sky Survey (SDSS) stripe 82 (Viero et al. 2014). Abbreviations: CANDELS, Cosmic Assembly Near-Infrared Deep Extragalactic Legacy Survey; COSMOS, Cosmic Evolution Survey; EGS, Extended Groth Strip; ELAIS, European Large Area ISO (*Infrared Space Observatory*) Survey; (E)CDFS, (Extended) Chandra Deep Field South; FLS, *Spitzer* First Look Survey; GOODS-H, Great Observatories Origins Deep Survey Herschel; H-ATLAS, Herschel Astrophysical Terahertz Large Area Survey; HeLMS, HerMES Large Mode Survey; HerMES, Herschel Multi-tiered Extragalactic Survey; HLS, Herschel Lensing Survey; LH, Lockman Hole; N, north; PEP, PACS Evolutionary Probe; S, south; UDS, Ultra-Deep Survey; VVDS, Visible Imaging Multi-Object Spectrograph (VIMOS) Very Large Telescope (VLT) Deep Survey; XMM-LSS, X-ray Multi-Mirror Mission Large Scale Structure Survey.

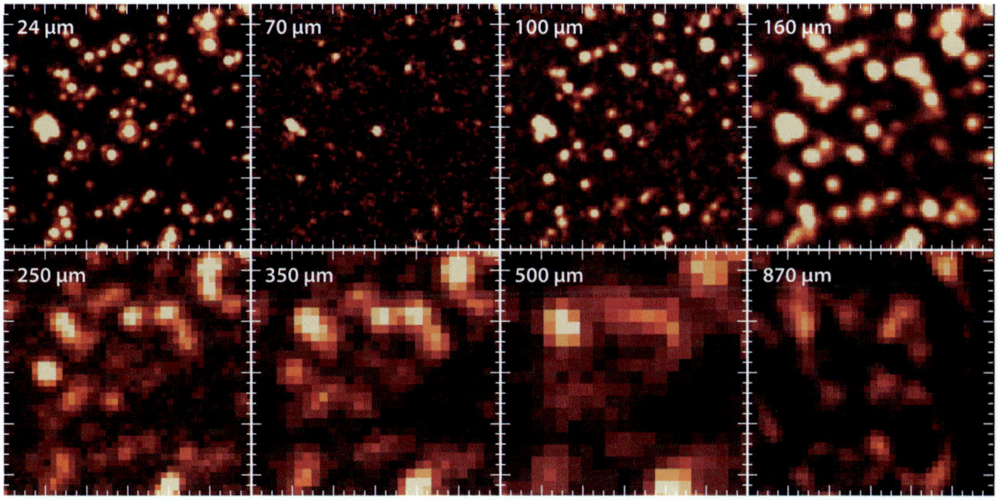

Figure 2

Current status of the deepest 24–870 μm infrared surveys, shown by 4 arcmin × 4 arcmin cutouts in the Hubble Ultra Deep Field (HUDF) region. Data are from the Great Observatories Origins Deep Survey (GOODS) project (24 μm), Photodetector Array Camera and Spectrometer (PACS) Evolutionary Probe (PEP), and the combined PEP and GOODS-Herschel data (70–160 μm); Herschel Multi-tiered Extragalactic Survey (HerMES) (250–500 μm); and the ground-based Large Apex Bolometer Camera (LABOCA) Extended Chandra Deep Field South (ECDFS) Submillimeter Survey (LESS) (870 μm; Weiß et al. 2009).

2. A RESOLVED VIEW OF THE COSMIC FAR-INFRARED BACKGROUND

The detection of an isotropic CIB by the Cosmic Background Explorer (COBE) satellite (for reviews, see Hauser & Dwek 2001, Lagache et al. 2005) was immediately interpreted in terms of radiation from galaxies in the early Universe that was absorbed and re-emitted by their interstellar dust. Strong observational support was provided by retrieving the major part of the CIB when stacking *Spitzer* 70- and 160-μm data at the positions of mid-infrared-detected galaxies (Dole et al. 2006) and by resolving the majority of its 850-μm long wavelength tail using the deepest SCUBA counts, which was assisted by gravitationally lensing clusters of galaxies (e.g., Knudsen et al. 2008, Zemcov et al. 2010). Closer to the peak of the CIB, source confusion mostly restricted the smaller cryogenic far-infrared telescopes ISO and *Spitzer* to statistical methods, a limitation that *Herschel*'s larger primary mirror and improved wavelength coverage significantly relaxes.

2.1. Number Counts and Resolved Fraction of the Far-Infrared Background

The enhanced *Herschel* capabilities were rapidly used for number count studies based on several analysis techniques, namely, direct resolved detection, stacking analysis, and probability of deflection [P(D)] analysis. At 70–160 μm, Berta et al. (2010, 2011) and Magnelli et al. (2013) provide counts of increasing depth from individually resolved detections based on a variety of deep blank fields and consistent with the analysis of the lensing cluster Abell 2218 by Altieri et al. (2010). Rigby et al. (2011) extend the 100- and 160-μm counts to a larger area, and Sibthorpe et al. (2013) reconfirm that cosmic variance is a minor concern for these results. Stacking and P(D) analyses at these wavelengths have been reported by Berta et al. (2011) but provide only a modest extension compared with the very deep resolved PACS counts. At 250–500 μm, Oliver et al.

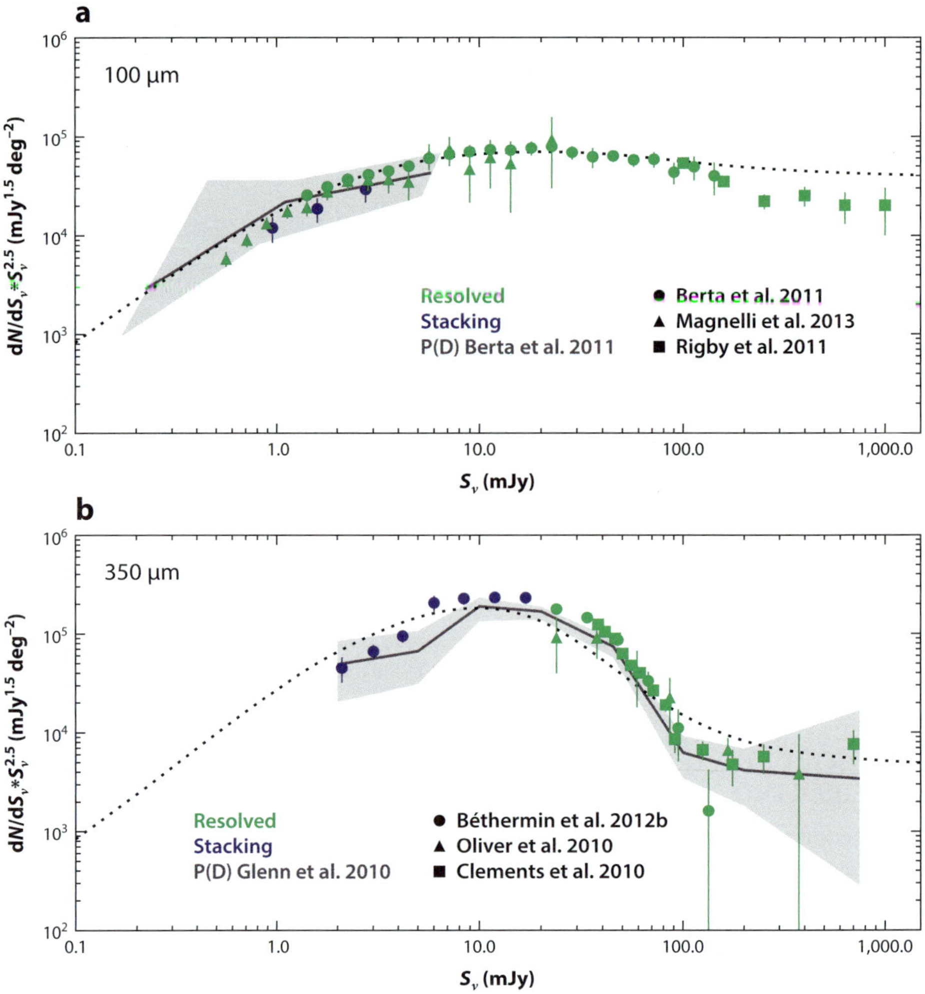

Figure 3

Euclidean normalized differential number counts for (*a*) 100 μm (*Herschel*-PACS) and (*b*) 350 μm (SPIRE). The references in the figure also provide counts at other *Herschel* wavelengths. The dotted line shows an example for a number count model (Béthermin et al. 2012a).

(2010), Clements et al. (2010), and Béthermin et al. (2012b) provide resolved counts. Because of the higher SPIRE confusion limits, statistical methods are more important and have been applied by Glenn et al. [2010; P(D)] and Béthermin et al. (2012b; stacking). At all *Herschel* wavelengths, measurements now reach below the knee in integral number counts (i.e., the peak in the Euclidean normalized differential number counts). An unusual feature of the bright end of the counts at long *Herschel* wavelengths is the significant contribution by gravitationally lensed sources (see Section 4.1).

Figure 3 summarizes number count results for the central wavelengths of the two *Herschel* cameras, again highlighting the importance of statistical methods at the longer wavelengths because of increasing source confusion. Different measures of source confusion are in use and applicable for different shapes of the number counts and for location of the detection limit in the counts (see, e.g., the discussion in Dole et al. 2004). For 500, 350, and 250 μm, Nguyen et al. (2010) report confusion noise of 6.8, 6.3, and 5.8 mJy, respectively. For 160 and 100 μm, confusion noise has been measured by Berta et al. (2011) and, using the deepest PACS data, by Magnelli et al. (2013), who find 0.68 and 0.15 mJy, respectively. No *Herschel* deep survey reaches the extragalactic confusion limit at 70 μm for a 3.5-m telescope (Berta et al. 2011).

In addition to the star-formation rates (SFRs) and black hole accretion rates of galaxies, direct predictions of far-infrared counts from semianalytic or hydrodynamic models of galaxy evolution are sensitive to the amount and geometry of obscuring dust. The amount of dust and in particular its poorly predictable distribution affect the energy and shape of the far-infrared SED. For that reason, among others, infrared counts have been traditionally interpreted in terms of backward evolution models that adopt locally calibrated empirical SEDs or families of SEDs. These models then predict counts on the basis of an assumed evolution of luminosity and number densities of the luminosity functions for the populations that are represented by these SEDs. Before the launch of *Herschel* (e.g., Valiante et al. 2009), assumptions about the evolution of SEDs at a given infrared luminosity started to be included as well. As noted in many of the papers presenting observed *Herschel* counts, pre-*Herschel* models often gave decent representations of the observed counts in some individual bands but typically did less well in providing a consistent picture for all available far-infrared bands, and in particular for the redshift distributions. It is inevitable that future, more successful models will include the *Herschel* evidence on evolution of SEDs (see Section 3). Béthermin et al. (2012a) have already attempted this; their model implements SEDs for galaxies on the evolving main sequence of star-forming galaxies and for starbursts above this sequence. More physical forward semianalytic or hydrodynamic models will face the same constraints on adopted SEDs as simple backward models do.

Figure 4 summarizes the 70–850 μm CIB as directly measured by COBE and the contributions of galaxies as determined by *Spitzer*, *Herschel*, and ground-based submillimeter surveys. The extrapolation of measured number counts is consistent with the direct CIB measurements and provides smaller uncertainties at some wavelengths near the peak of the CIB. Combining extrapolated *Spitzer*, *Herschel*, and 850-μm results, the total 8–1,000 μm background is 27^{+7}_{-3} nW m^{-2} sr^{-1} (Béthermin et al. 2012b). There is no evidence for a diffuse extragalactic infrared background that galaxies do not account for. The CIB fraction accounted for by individually resolved galaxies peaks at 75% at 100 and 160 μm (Magnelli et al. 2013). The resolved fraction is lower at 70 μm, where *Herschel* could not integrate to its confusion limit (Berta et al. 2011), and at 250–500 μm; 15% to 6% of the CIB is resolved at these wavelengths owing to increasing source confusion (Oliver et al. 2010, Béthermin et al. 2012b). At 250–500 μm, stacking analysis again attributes 73% to 59% of the CIB to known galaxies (Béthermin et al. 2012b). As expected, the typical redshift of sources contributing to the CIB increases with wavelength (**Figure 4**, *lower*). Of the major CIB fraction that is directly resolved or retrieved by stacking, half of the emission at wavelengths 100, 160, 250, 350, and 500 μm is contributed by sources above $z = 0.75$, 1.00, 1.04, 1.20, and 1.25, respectively (Béthermin et al. 2012b, Magnelli et al. 2013). Median redshifts of sources with fluxes above faint limits are naturally somewhat higher (e.g., $z_{\rm Med} = 1.37$ above $S_{100} = 1.5$ mJy; $z_{\rm Med} = 1.22$ above $S_{160} = 2.5$ mJy).

2.2. Extinction-Insensitive Star-Formation Rates and Comparison with Other Tracers

Far-infrared emission is an established tool to determine the SFRs of galaxies. The total infrared luminosity provides an observationally straightforward calorimetric measure that can be interpreted as total SFR if three aspects, described below, are taken into consideration.

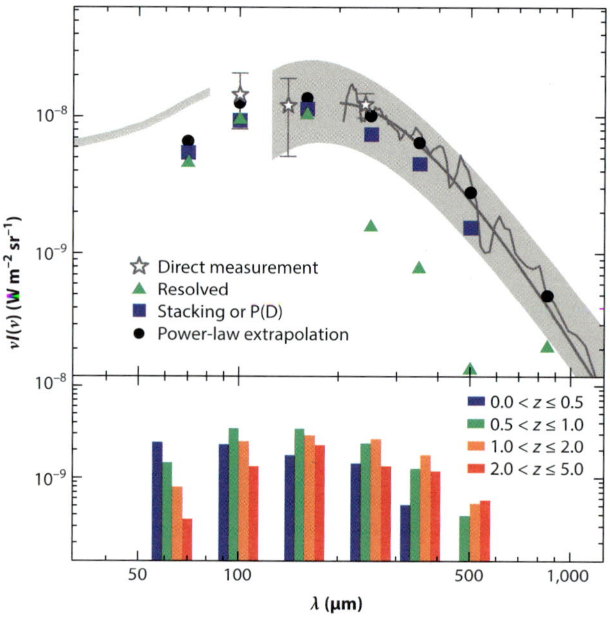

Figure 4

(*Upper*) The cosmic infrared background (CIB) as seen by direct measurements and as resolved by *Herschel*. Direct measurements (*gray*) include the γ-ray-based limits of Mazin & Raue (2007), Cosmic Background Explorer (COBE)-Diffuse Infrared Background Experiment (DIRBE) results as presented in Dole et al. (2006) (*stars*), the COBE-Far Infrared Absolute Spectrophotometer (FIRAS) $\lambda > 200$ µm spectrum of Lagache et al. (1999), and the modified blackbody fit of Fixsen et al. (1998). CIB contributions by resolved sources are from Berta et al. (2011) and Magnelli et al. (2013) (70–160 µm), Béthermin et al. (2012b) (200–500 µm), and Zemcov et al. (2010) (850 µm). Stacking results are from Béthermin et al. (2010) (70 µm), Berta et al. (2011) (100 µm, 160 µm), and Béthermin et al. (2012b) (250–500 µm); power-law count extrapolations come from the same studies and Zemcov et al. (2010) (850 µm). (*Lower*) Contributions of different redshift slices to the part of the CIB that is contained in resolved sources (PACS) and covered by stacking (SPIRE) (Berta et al. 2011, Béthermin et al. 2012b, Magnelli et al. 2013).

First, the infrared traces only the obscured part of the radiation emitted by the ensemble of young stars in a galaxy, the part that is absorbed by dust. A short-wavelength tracer must be used to quantify the escaping fraction to either ensure dominance of the obscured star formation or revert to a composite star-formation indicator that sums the obscured and unobscured star formation. In massive high-z star-forming galaxies, dust-obscured star formation is typically approximately 5–10 times stronger than unobscured star formation (Section 4.2).

Second, the contribution to the infrared luminosity by accretion onto an active galactic nucleus (AGN) must be low for star formation to dominate. This is discussed in Section 5, but in short, the generally elevated star formation in high-redshift galaxies implies that for moderate-luminosity AGNs from typical surveys, host stellar heated emission will dominate over AGN-heated emission in the far- and sometimes even mid-infrared. Elimination of known AGNs on the basis of X-ray data and of rest frame near-infrared hot dust power laws can further reduce this source of uncertainty.

Third, dust can also be heated by older stars, which in particular can dominate dust heating in regions such as the bulges of quiescent local spirals. This makes the conversion from total infrared (8–1,000 µm) luminosity to current SFR more dependent on the star-formation history than the conversion is for tracers of O stars, such as the Hα line that originates only in the ionized gas.

Compared to the assumption of constant star formation over 100 Myr, as in the widely adopted conversion of Kennicutt (1998) (note update in Kennicutt & Evans 2012), constant star formation over 10 Gyr will increase the ratio of infrared luminosity and current SFR by a factor of almost 2, an effect that can increase further for decaying star-formation histories. Further aspects of multiwavelength methods to determine SFRs, including updated calibrations, are covered in the recent review by Kennicutt & Evans (2012).

2.2.1. Rest frame far-infrared star-formation rates. Deriving total 8–1,000 μm infrared luminosities and SFRs by fitting SED templates is straightforward for fully sampled *Herschel* SEDs, but sensitivity and confusion limits often restrict detections to one or two photometric bands only. Several studies confirm that, as long as one of these detected photometric points is near the far-infrared SED peak, such monochromatic-based infrared luminosities are good to ±0.1–0.2 dex even if considering a wide range of possible SED shapes (Elbaz et al. 2011, Nordon et al. 2012, Berta et al. 2013b, Magnelli et al. 2013). For the fitting, SEDs derived from or validated on *Herschel* data should be used, because lack of observational constraints caused some earlier SED families to exhibit fluxes that were too low on the long-wavelength side of the far-infrared peak (Elbaz et al. 2010).

For the high-redshift star-forming galaxies that are individually detected in *Herschel* surveys, the ratio of infrared and ultraviolet (UV) luminosity $L_{IR}/\nu L_\nu(UV)$ [also called the infrared excess (IRX)] is typically of order 10, sometimes higher (e.g., Buat et al. 2010, 2012; Nordon et al. 2013; see also Section 4.2). These *Herschel* detections are mostly galaxies near and above main-sequence star-forming galaxies, the narrow locus of most star-forming galaxies in the SFR-mass plane (e.g., Brinchmann et al. 2004, Noeske et al. 2007, Elbaz et al. 2007, Daddi et al. 2007; see also Section 3.2). Only rarely do galaxies with individual *Herschel* detections fall short of $IRX \sim 2$ [e.g., in the UV-selected sample of Oteo et al. (2012b)]. Converting from the ratio of luminosities to SFR ratios, obscured star formation thus dominates safely, and infrared-based SFRs are useful even without adding the unobscured UV, as is of course still recommended if UV data are available. More care in this respect is needed for lower SFR and lower obscuration samples for which stacked *Herschel* detections can still be obtained [e.g., the samples of Reddy et al. (2012) and Lee et al. (2012), in which $IRX \sim 7$ and ~3–4, respectively).

Studies have found good agreement between infrared-based SFRs and the scarcer Hα SFRs (extinction corrected using the Balmer decrement). This applies both to a sample of mostly near-main-sequence star-forming galaxies with *Herschel* and Hα detection at $z \sim 0.27$, $SFR \sim 6\,M_\odot$ year^{-1} (Domínguez Sánchez et al. 2012) and to a sample of $z \sim 1.4$, $SFR \sim 100\,M_\odot$ year^{-1} SPIRE-selected galaxies near and above the main sequence (Roseboom et al. 2012). The Balmer decrement suggests these samples are moderately dusty [$E(B-V) \sim 0.5$]. These findings suggest dust heating by old stars is of minor importance in these star-forming objects. Also, despite sizeable SFRs, these galaxies cannot have a large contribution of heavily obscured star formation that is not traced by Hα. More care in interpreting the far-infrared luminosity is needed for passive objects well below the star-forming sequence. Given realistic survey depths, this occurs for individual detections mostly within a subset of $z < 0.5$ detections as well as for stacks of other passive samples.

2.2.2. Star-formation rates from mid-infrared emission. Given the superb *Spitzer* mapping speed at 24 μm, mid-infrared surveys remain a key resource, including for many sources that are too faint for *Herschel* detection. Deducing SFRs for high-z galaxies from 24-μm fluxes requires an extrapolation to L_{IR} that is more uncertain, given the diverse factors that shape this spectral range, namely, a continuum caused by warm or transiently heated dust in HII regions, aromatic polycyclic aromatic hydrocarbon (PAH) emission features, possibly AGN-heated emission, and silicate absorption.

Traditionally, this extrapolation was done using locally calibrated, luminosity-dependent spectral templates (e.g., Chary & Elbaz 2001). These encode the physical properties of local infrared galaxies. Specifically, the ratio of 8-μm PAH emission to total infrared is lower for local ULIRGs than for lower-luminosity galaxies because of the compact star-forming regions and intense radiation fields of these galaxy merger ULIRGs. As early as during the *Spitzer* era, observations suggested that at $z \sim 2$, application of these templates leads to overpredicted infrared luminosities (Papovich et al. 2007, Daddi et al. 2007). This mid-IRX was ascribed either to relatively stronger PAH emission in $z \sim 2$ galaxies or to a strong AGN mid-infrared continuum.

Herschel observations have quantified this mismatch for large samples and for individual detections. There is a factor of ~5 overprediction of $z \sim 2$ SFRs if using 24-μm photometry and typical locally calibrated template families (Nordon et al. 2010; Elbaz et al. 2010, 2011; Nordon et al. 2012). The effect conspicuously sets in close to $z \sim 2$ when the strongest PAH feature enters the Multiband Imaging Photometer (MIPS) 24-μm band (Elbaz et al. 2010, 2011; see also **Figure 8**), suggesting that it is due to enhanced PAH emission rather than to an AGN continuum. This is unambiguously confirmed (Nordon et al. 2012) by fully reproducing the photometric trends with trends in the PAH emission in the ultradeep low-resolution *Spitzer* spectra of $z \sim 1$ and $z \sim 2$ galaxies by Fadda et al. (2010). Section 3 discusses the relation of these findings to the evolving star-forming main sequence and to changes in redshift in the ISM conditions of galaxies of a given infrared luminosity.

Several procedures are in use to avoid the overprediction that arises when applying local luminosity-dependent templates to MIPS data from high-z sources. The main-sequence SED template of Elbaz et al. (2011) encodes the proper 8 μm/infrared ratio that is valid for MIPS galaxies except for the minor fraction of starbursts above the main sequence. Similar arguments apply to the single template of Wuyts et al. (2008). Nordon et al. (2012) provide a procedure to choose the right template as a continuous function of specific SFR ($SSFR = SFR/M_*$) offset from the main sequence, whereas Rujopakarn et al. (2013) derive $z \sim 2$ relations by treating high-z galaxies as main-sequence objects and linking their SEDs to local objects of the same infrared surface density. If sufficient UV to mid-infrared SED information is available, it is also possible to derive infrared luminosities by rescaling templates from libraries that span SED shapes without a fixed link between shape and infrared luminosity. These include the library of Polletta et al. (2007) and the derivation by Berta et al. (2013b) that also considers *Herschel* data. Nordon et al. (2012) and Berta et al. (2013b) compare some of these revised prescriptions. Given that methods that trace changes in SEDs with SSFR offset from the main sequence will amplify noise in the 24-μm photometry and internal spread in the SEDs at given $L_{\rm IR}$, there is a trade-off between simple single templates (low scatter but residual systematics in the comparison of $L_{\rm IR}$ predicted from 24-μm data and of true $L_{\rm IR}$) and more complex approaches (larger scatter but lower systematics).

The methods discussed above have been tested using MIPS and *Herschel* detections, i.e., effectively for the main sequence above $\sim 10^{10}(10^{10.5})\,M_\odot$ at $z \sim 1$ (2) as well as for more actively star-forming galaxies. Application to lower mass and lower SFR main-sequence objects may need further validation: The stacked $L_{\rm IR}/L_{8\,\mu\rm m} = 8.9 \pm 1.3$ reported by Reddy et al. (2012) for fainter UV-selected galaxies is approximately 80% larger than the main-sequence value, possibly due to a weakening of PAH for more compact star formation and/or lower metallicity.

2.2.3. Radio emission and the radio–far-infrared correlation. Thirty years after its discovery, the tight correlation between the radio emission and far-infrared emission of star-forming galaxies is still a challenge for models (e.g., Lacki et al. 2010, Lacki & Thompson 2010, and references therein). Empirically, the radio–far-infrared correlation enables interferometric radio continuum surveys to provide a convenient high-spatial-resolution and extinction-insensitive access to high-z SFRs. This is true as long as no AGN-related excess radio emission is present and changes in the radio–far-infrared correlation with redshift are absent or calibrated. Supplementary radio interferometric data have hence played a key role, for example, in identifying and characterizing the SMG population detected in ground-based surveys (see also Section 4.3).

Again, verification of the relation at high redshift, with minimal extrapolation uncertainties, requires samples of star-forming galaxies with SED detections near the rest frame far-infrared peak. Selection effects in both radio and far-infrared wavelength ranges are of paramount consideration for current studies. The first results using early *Herschel* data sets in combination with Very Large Array (VLA) surveys give no compelling evidence for an evolution of the radio–far-infrared relation out to $z \sim 2$ (Ivison et al. 2010, Jarvis et al. 2010) but are consistent with modest evolution toward a lower ratio of infrared to radio emission. Combining full *Herschel* data with upcoming Jansky VLA (JVLA) radio surveys should significantly enhance the power of this approach and guide the further use of deep radio surveys to measure SFRs.

2.2.4. Comparison to extinction-corrected ultraviolet emission and spectral energy distribution fitting. Using a deep GOODS-S data set out to $z \sim 3$, Wuyts et al. (2011a) compare SFRs from rest UV to near-infrared SED fitting with total SFRs that add unobscured SFR_{UV} and re-emitted SFR_{IR} (directly measured by *Herschel* or *Herschel*-recalibrated 24 μm). SFRs from SED fitting and total SFRs are reasonably consistent at low and intermediate SFRs, but SED fitting underpredicts the total SFR_{UV+IR} for galaxies with the largest SFR_{IR}/SFR_{UV} ratio, which are typically found at the highest $SFR_{UV+IR} > 100\,M_{\odot}\,year^{-1}$ and at $z \gtrsim 2.5$.

A mixed picture emerges from studies comparing *Herschel* far-infrared SFRs with SFRs estimated from UV emission and slope via the IRX-β relation (e.g., Meurer et al. 1999). Although some populations match this relation on average (Reddy et al. 2012), the scatter is large (see also Section 4.2). Specifically, the most infrared-luminous galaxies often have much larger IRX than this relation would indicate, reminiscent of the situation for local ULIRGs (Goldader et al. 2002). This offset correlates with the offset from the star-forming main sequence (Nordon et al. 2013).

2.2.5. Summary. Given these constraints, studies of star-forming galaxies over a wide range of redshifts and SFRs may often benefit from a combined approach that, in order of decreasing SFR, uses:

- Far-infrared SFRs for galaxies with at least one *Herschel* detection near the SED peak,
- Mid-infrared SFRs calculated using recipes that have been validated against *Herschel* data, and
- SFRs from rest UV to near-infrared SED fitting.

Specific limits between these steps and the need to invoke all or just some of these steps also depend on survey practicalities. Studies utilizing such an approach include Rodighiero et al. (2011), Wuyts et al. (2011b), Wang et al. (2013), and Magnelli et al. (2014).

2.3. Evolution of the Infrared Luminosity Function and the Star-Formation Rate Density

Evolution of the infrared luminosity function and the infrared luminosity density encode the cosmic history of dusty star formation and are an essential part of measuring the total cosmic star-formation history; see also the review by Madau & Dickinson (2014). Previous determinations

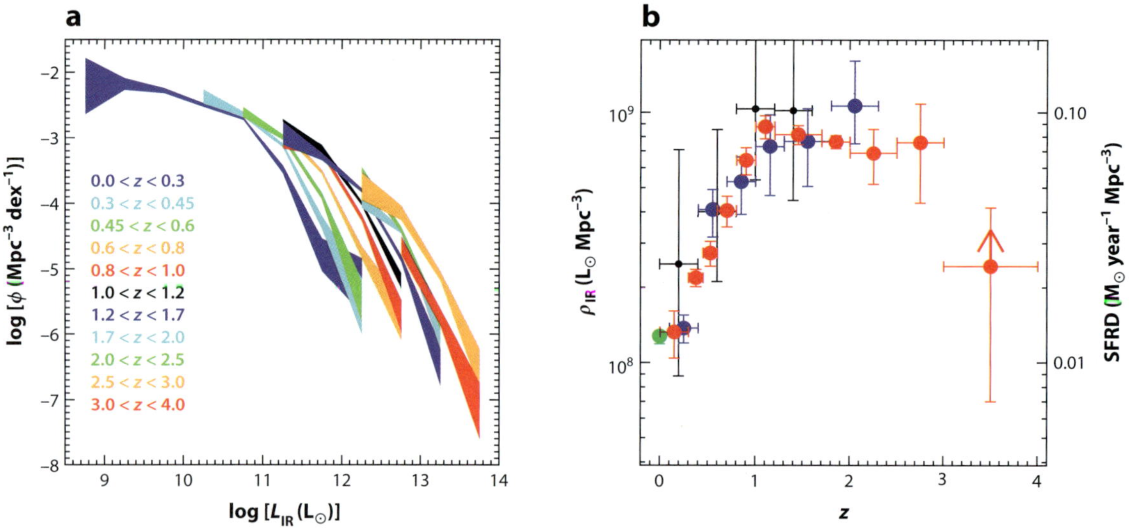

Figure 5

(*a*) Evolution of the total infrared (8–1,000 μm) luminosity function. Results are from Gruppioni et al. (2013); the width of the band indicates the $\pm 1\sigma$ poissonian error. (*b*) Evolution of the infrared luminosity density. Data are from Gruppioni et al. (2013) (*red*), Magnelli et al. (2013) (*blue*), Casey et al. (2012) (*black*), and the local results of Sanders et al. (2003) (*green*). As argued by Gruppioni et al. (2013), the $z \sim 3.5$ point may be a mild lower limit due to selection at $\lambda < 200$ μm. Adopted cosmology and initial mass function are as in Gruppioni et al. (2013). The right axis [star-formation rate density (SFRD)] represents only the dust-obscured part of star formation.

were based mostly on ground-based submillimeter data (requiring significant extrapolation to the total infrared emission and covering only a small luminosity range) or *Spitzer* mid-infrared data (also requiring significant extrapolation to the total infrared; see Section 2.2).

Early *Herschel* results clearly demonstrate strong evolution of the total and monochromatic infrared luminosity functions (e.g., Gruppioni et al. 2010, Vaccari et al. 2010, Dye et al. 2010, Eales et al. 2010b, Lapi et al. 2011) and, at $z \sim 0$, dependence on environment, with the Virgo cluster showing a peaked luminosity function (Davies et al. 2010, 2012). Perhaps the most complete analyses of the evolving total infrared luminosity function are provided by Gruppioni et al. (2013) from a large combined PACS and SPIRE data set at $z \lesssim 4$ and by Magnelli et al. (2013) at $z \lesssim 2$ using the deepest PACS data. **Figure 5** shows evolving total infrared luminosity functions and infrared luminosity densities from these and other studies. Given the depth limits of the surveys used, the faint end slope of the luminosity function is only weakly constrained at all but the lowest redshifts (**Figure 5**), and assumptions vary. For example, at $1.0 < z < 1.2$ and in the $L_{\mathrm{IR}} < 10^{11.2}\,\mathrm{L}_{\odot}$ range that is not constrained by *Herschel*, Gruppioni et al. (2013) adopt a slope of the luminosity function $\Phi = dN/[d(\log L_{\mathrm{IR}})dV] \propto L_{\mathrm{IR}}^{-0.2}$ Mpc^{-3} dex^{-1}, guided by their lowest redshift bin, whereas Magnelli et al. (2013) adopt $\Phi \propto L_{\mathrm{IR}}^{-0.6}$. The deepest *Spitzer* 24-μm data provide some guidance at these luminosities (Magnelli et al. 2013) in a regime where extrapolations from mid-infrared to total infrared are rather certain, but these data cannot fully resolve the ambiguities in slope. The effect of these faint-end slope differences on the integrated infrared luminosity density is small, however.

In summary, infrared luminosity functions and infrared luminosity density are now determined by direct rest far-infrared measurements out to $z > 3$. The contribution of LIRG and ULIRG luminosity bins to the luminosity density increases from 5% and 0.4% in the local Universe to 31% and 50%, respectively, at $z \sim 2$ (Magnelli et al. 2013). This does not necessarily imply that these important $z \sim 2$ luminous objects physically resemble local (U)LIRGs. Most of the $z = 1$–2 infrared luminosity density arises in near-main-sequence objects rather than starbursting outliers, which dominate only the upper end of the luminosity function (Gruppioni et al. 2013). Considering stellar mass bins, the infrared luminosity density is dominated by the range $10^{10} < M\,[M_\odot] < 10^{11}$ at all redshifts studied by Gruppioni et al. (2013).

Burgarella et al. (2013) combine the results of Gruppioni et al. (2013) with recent UV luminosity functions and compare the luminosity densities $LD_{\rm IR}$ and $LD_{\rm UV}$ obtained by integrating the respective luminosity functions. The ratio of infrared to UV luminosity densities traces the average obscuration of the galaxy population and is well known to rise steeply from the local Universe to $z \sim 1$. Burgarella et al. (2013) report that the local ratio $LD_{\rm IR}/LD_{\rm UV} = 4.25 \pm 0.35$ rises by a factor of $\sim$3 to reach a peak of 13.44 ± 4.68 at $z \sim 1.2$, and then it falls again, reaching values similar to local ones at their last data point at $z \sim 3.6$. This is in accordance with recent evidence from rest frame UV data for decreasing obscuration of the galaxy population at high redshifts.

Because previous mid-infrared-based determinations of the infrared luminosity function and infrared luminosity density used a range of recipes to extrapolate to total infrared, no single correction can be applied to them. Comparisons in the references cited above indicate that the new *Herschel*-based luminosity functions are consistent with or slightly lower than typical *Spitzer*-based results. Several past studies reported a discrepancy between cosmic star-formation history and cosmic mass assembly: The cosmic stellar mass densities inferred by integrating the SFR densities were a factor of $\sim$2 larger than those from stellar mass measurements (e.g., Hopkins & Beacom 2006, Pérez-González et al. 2008). Burgarella et al. (2013) argue that these two measurements are now in good agreement due to a combination of the somewhat reduced *Herschel*-based SFR densities; the evidence for a decreased average UV obscuration at $z \gtrsim 1.2$, which lowers UV-based SFRs at high redshift; and proper consideration of the stellar mass that is returned to the ISM during stellar evolution in the comparison of stellar mass densities and integrated SFR densities (see also Behroozi et al. 2013). The total stellar mass inferred from infrared/UV SFRs clearly depends on the assumed stellar initial mass function (IMF) on the lower main sequence. The improved agreement between inferred and measured total stellar mass thus reduces the need to invoke bottom-light IMFs at high z (e.g., Davé 2008).

2.4. Clustering of Infrared Galaxies and the Role of Environment

Large survey areas and sample sizes as well as improved systematics of current far-infrared surveys provide new opportunities to explore the clustering properties of DSFGs and to investigate the impact of environment on their properties.

Magliocchetti et al. (2011, 2013) present observed and deprojected two-point correlation functions for individually detected PACS sources; comoving correlation lengths r_0 increase toward higher redshift and fainter sources. Sources selected at rest frame 60 μm with similar $SFR \gtrsim 100\,M_\odot$ year^{-1} locally reside in low-density environments but at $z \sim 2$ are found in massive $M_{\rm min} \sim 10^{13.5}\,M_\odot$ halos. For SPIRE-selected $z < 0.3$ samples, Guo et al. (2011) and van Kampen et al. (2012) report $r_0 \sim 5$ Mpc, similar to $z < 1$ PACS sources. Exploiting the gravitational magnification of background Lyman break galaxies (LBGs), Hildebrandt et al. (2013) suggest that luminous $S_{250} > 15$ mJy SPIRE sources reside in massive $\log_{10}(M) = 13.2$ halos, which is in broad agreement with the $z > 1$ PACS sources and $z = 1$–3 SMGs (Hickox et al. 2012).

Given significant confusion limits but large-area high-quality data sets, much of the related analysis using SPIRE as well as *Planck* has focused on power spectrum analyses of the CIB fluctuations

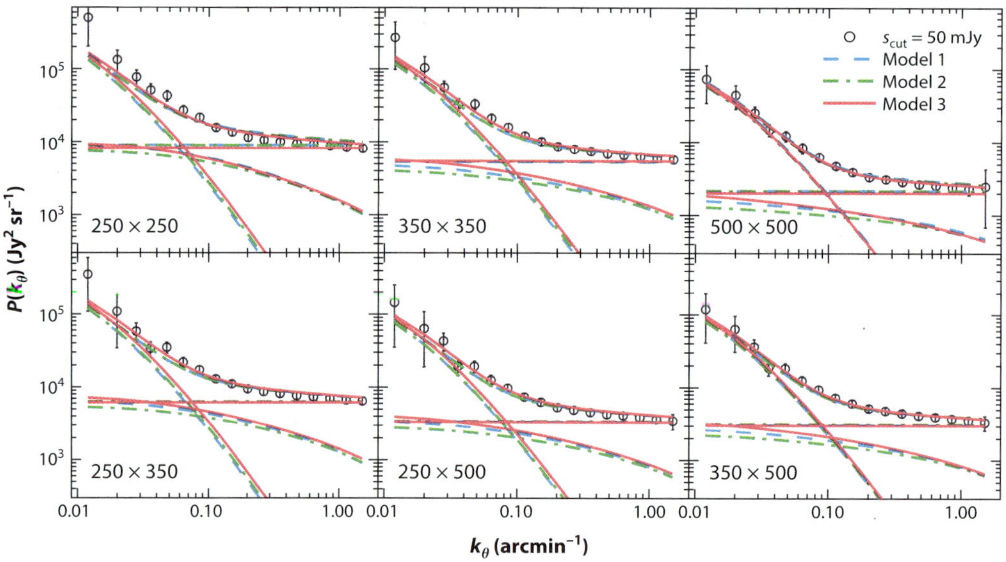

Figure 6

SPIRE auto- and cross-frequency power spectra with sources at fluxes greater than $s_{\mathrm{cut}} = 50$ mJy masked and cirrus subtracted. Three halo models (see Viero et al. 2013) are fit to these power spectra as well as to the number counts of Glenn et al. (2010). Spectra are fit with three terms: Poisson (horizontal lines), 2-halo (steep lines and dominant at low k_θ), and 1-halo (less steep). Figure reproduced from Viero et al. (2013, figure 11) with permission.

(e.g., Amblard et al. 2011; Planck Collab. et al. 2011, 2013b; Thacker et al. 2013); Viero et al. (2013) present the currently most complete SPIRE analysis. Power spectra and cross-frequency power spectra from SPIRE are available over scales $k_\Theta = 0.01$–2 arcmin^{-1} and clearly show the 1-halo and 2-halo clustering terms (**Figure 6**).

Several halo and abundance matching models are under development to match these and other far-infrared data (e.g., Béthermin et al. 2013, Planck Collab. et al. 2013b, Viero et al. 2013, Wang et al. 2013). Results to date indicate that $M \sim 10^{12}$ M$_\odot$ halos generally dominate star formation at all redshifts. These halos represent the progenitors of structures of increasingly larger present-day mass as one goes back in time. Variations still exist in the prescriptions used by the various models to match halo mass and SFR, as well as in the adopted or fitted infrared SEDs, and are worth further scrutiny.

Local LIRGs and ULIRGs preferentially reside in underdense regions rather than clusters. *Herschel* studies at $z < 0.5$ confirm this (Dariush et al. 2011, Burton et al. 2013) and suggest that the difference is mostly mediated by a change in the ratio of blue star-forming to red passive galaxies rather than by a change in the intrinsic properties of the star-forming galaxies. However, Rawle et al. (2012b) present evidence for a population of unusually warm galaxies in $z \sim 0.3$ massive clusters; they suggest these galaxies to be objects in which cold dust has been stripped from the periphery. Comparing the total SFR per halo mass ($\Sigma(SFR)/M$) in the field, groups, and clusters out to $z > 1$, Popesso et al. (2012) find that groups have low $\Sigma(SFR)/M$ locally but rapidly catch up with the field $\Sigma(SFR)/M$ as redshift increases. The cluster $\Sigma(SFR)/M$ increases with redshift but stays below the values for the field and for groups out to $z \sim 1$. It is currently less clear to what extent these changes in the relation of the SFR, as traced by infrared emission, to environment, as well as the general increase of star-forming activity with redshift, express themselves in a change (or not) in the local relation between galaxy (S)SFR and density (Popesso et al. 2011, Ziparo et al.

2014) or in the level of enhancement of infrared emission in galaxy pairs compared with isolated galaxies (Hwang et al. 2011, Xu et al. 2012).

Rawle et al. (2012a) give an overview of SFRs in the brightest cluster galaxies. Use of *Herschel* fluxes avoids potential problems with AGN contamination that could affect previous results that are based on mid-infrared observations. The SFRs are typically low, 1–150 $M_\odot$ year^{-1} even in the far-infrared-detected subset of the sample, and comparison with Hα-based SFRs suggests that star formation is little obscured in these galaxies and likely spatially very extended. SFRs are clearly anticorrelated with the cooling times of the X-ray gas for cool-core clusters but fall short of nominal X-ray-based cooling rates. This is in line with mounting evidence for the importance of AGN feedback in such systems. Only in rare cases does the SFR of the brightest galaxy reach an appreciable fraction of the cooling rate that is suggested by X-ray data (McDonald et al. 2012).

3. GALAXIES AND INFRARED SPECTRAL ENERGY DISTRIBUTIONS IN THE CONTEXT OF THE EVOLVING STAR-FORMING SEQUENCE

The power of far-infrared emission as a star-formation indicator simply relies on the total infrared emission, but the detailed shape of the SED of a galaxy encodes the conditions in the ISM in which the emitting dust grains reside. A dust grain will equilibrate to a temperature that increases with the intensity of the local radiation field. Because the grain's re-emission scales with $T_{\rm dust}^{4+\beta}$ and the dust emissivity index is $\beta = 1.5-2$, variations of a few Kelvin around dust temperatures of typically 30 K already reflect significant variations in the average radiation field. Dust heating is a local process that is dependent on the spatial distribution of the heating sources and the dust (acting both as heated and obscuring material), but strongly simplified assumptions can give insights for possible scalings. Compact high-pressure and high–radiation field intensity star-formation regions in the nuclear regions of galaxy mergers have warmer dust. Such a link between size and temperature cannot be extended to the total size of a galaxy, however, if temperature is driven by the conditions in localized star-forming regions; more or fewer of these regions may be present in a galaxy's disk, but each will have similar conditions. Another simple assumption is to consider the dust as a calorimeter that re-emits the total radiation from young stars. The temperature then reflects the ratio of SFR to dust mass if the ratio of SFR and gas mass is fixed (which is plausible in regimes with a near-linear star-formation law); lower metallicity galaxies with lower dust-to-gas ratios will then have warmer dust.

The dust temperature also depends on each grain's optical properties as driven by chemical composition, size, and structure. The grain's emission at this temperature again depends on optical properties via the wavelength-dependent emissivity. Small grains are transiently heated by individual photons to higher than equilibrium temperature. Such transient heating is important for the continuum on the short wavelength (Wien) side of the far-infrared SED peak, particularly for the mid-infrared PAH emission features that reflect the aromatic nature of their carriers. An example of a current model of interstellar dust reflecting these considerations is Draine & Li (2007). Warmer far-infrared SEDs or colors thus reflect a more intense radiation field via a long chain involving the properties of the mixed-grain population. An additional complication arises because the dust emission may not be optically thin, as implicitly assumed above and valid for normal galaxies, but rather optically thick into the far-infrared. Comparing infrared luminosities, SED temperatures, and sizes to the Stefan-Boltzmann law for blackbodies suggests this is an issue for the most compact local ULIRGs. In contrast, attempts to infer the wavelength of optical thickness from fits to high-redshift SEDs should be viewed with some skepticism, given that the photometric coverage and signal-to-noise ratio (S/N) are typically limited and that the assumptions made about the intrinsic SED are often simplified.

Caution is warranted when comparing dust temperatures from different references because various conventions are used in assigning a single temperature to a complex SED. Most often, dust temperatures are derived by fitting the photometric data with a variant of the modified blackbody function $S_\nu \propto B_\nu(T_{\rm dust})(1 - e^{-\tau_\nu})$, which incorporates a wavelength-dependent optical depth $\tau_\nu = \tau(\nu_0)(\nu/\nu_0)^\beta$. For a given set of fluxes, the derived dust temperature $T_{\rm dust}$ depends on the adopted dust emissivity index β (typically assumed to be 1.5–2) and on adoption of an optically thin model rather than a model that is optically thick below some frequency ν_0 (which may be fitted to the data or adopted as a fixed value). Finally, the rest wavelength range included will influence $T_{\rm dust}$ because galaxy SEDs deviate from the single modified blackbody, especially on the Wien side of the peak, due to hotter dust and/or transiently heated dust grains. Differences in nominal $T_{\rm dust}$ of order 10 K can emerge from different conventions. This is significant for typical $T_{\rm dust}$ = 30 K and relevant given the good quality and rest frame coverage of current data. Further modifications in use include (*a*) fits with modified blackbodies at several discrete temperatures or with a power law distribution of temperatures, (*b*) fits with a modified blackbody joined with a power law on the Wien side, and (*c*) use of fitted peak wavelengths or frequencies of the SED, which are converted to a temperature via Wien's displacement law.

In the local Universe, an extensively studied relation exists between the infrared luminosity of a galaxy and its far-infrared color or dust temperature: Local infrared luminous galaxies with high SFRs arising from often small regions show warmer dust (e.g., Chapin et al. 2009, Hwang et al. 2010 and references therein). *Herschel* data permit researchers to test whether this $L_{\rm IR}$–$T_{\rm dust}$ relation evolves with redshift and how the findings relate to the current picture of high-z galaxies in the context of the evolving star-forming sequence.

3.1. Evolution of Infrared Spectral Energy Distributions with Redshift

As discussed in the context of the use of mid-infrared emission as a star-formation indicator (Section 2.2), there is an evolution with redshift in the ratio of rest frame 8-μm PAH emission and total infrared emission. At a given infrared luminosity, high-z galaxies have relatively stronger PAH than local ULIRG mergers. This indicates that they have less intense local radiation fields (e.g., Tielens 2008).

3.1.1. Evolution of dust temperature. The other main line of studies addresses the dust temperature that characterizes the rest far-infrared peak. Care is needed due to the common presence of selection effects. In particular, SMGs from ground-based 850-μm surveys were long expected to be biased toward cold objects. This has been confirmed by *Herschel* (Section 4.3): Warmer objects at the same z and $L_{\rm IR}$ escape the SMG selection. Selection effects certainly also contribute to lower dust temperature and larger dust mass when comparing $z < 0.5$, 250-μm-selected galaxies and local IRAS-selected samples (Smith et al. 2012).

Herschel selection near the SED peak is less biased. In a first comparison of $z > 0.5$ *Herschel* detections with *AKARI* data on local galaxies that have SED coverage out to 160 μm, Hwang et al. (2010) conclude that the median $T_{\rm dust}$ of $L_{\rm IR} > 5 \times 10^{10}$ L$_\odot$ *Herschel* sources is 2–5 K colder than that of local galaxies with similar luminosity and that the dispersion in dust temperature is larger in the *Herschel* sample. Such a shift to colder SEDs at higher redshift but equivalent infrared luminosity is also reported by Rex et al. (2010). Symeonidis et al. (2013) address this issue using a very large SPIRE and PACS data set. Based on comparison with a wide grid of model SEDs, they restrict the luminosity and redshift range studied to carefully ensure the absence of selection effects on the dust temperature. At the $z \gtrsim 1$ $L_{IR} \gtrsim 10^{12}$ L$_\odot$ upper end of the range studied, SEDs are 5–10 K colder than for local sources of equivalent luminosity (**Figure 7**). High-redshift

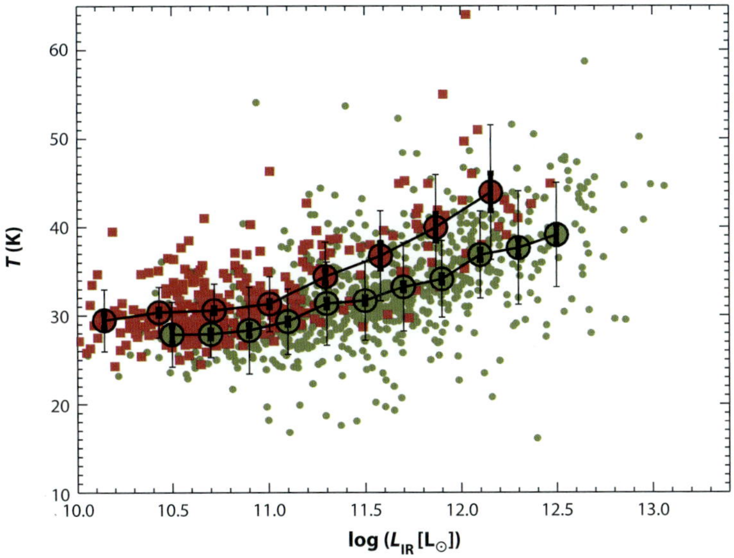

Figure 7

L_{IR}–T_{dust} relation for local $z < 0.1$ galaxies (*red squares*) and for a *Herschel* sample (*green circles*), representing redshifts from $z \sim 0.2$ to $z \sim 1.5$ as infrared luminosity increases. Figure reproduced from Symeonidis et al. (2013, figure 18) with permission.

galaxies have lower average interstellar radiation field intensities than their same-luminosity local analogs, reflecting more extended star formation.

Placing the infrared SED in the context of the basic parameters of stellar mass, SFR, and redshift, Magnelli et al. (2014) derive average dust temperatures from stacked SEDs in bins gridding the SFR–M_* plane for six redshift slices out to $z \sim 2$. They derive scaling relations for the dust temperature as a function of these parameters. At all redshifts, T_{dust} smoothly increases with infrared luminosity, with SSFR, and with SSFR offset from the star-forming main sequence. The classical relation of dust temperature to infrared luminosity is less tight than the other two relations, however, as also evident in the sloped isotemperature lines in the SFR–M_* planes for the various redshift slices (example in **Figure 8*d***). Dust temperature is better expressed in relation to the evolving main sequence than in relation to absolute infrared luminosity. The scalings derived by Magnelli et al. (2014) for massive $z \lesssim 2$ star-forming galaxies as a function of redshift and SSFR and as a function of redshift and SSFR offset from the main sequence are

$$T_{dust} = 98 \times (1 + z)^{-0.065} + 6.9 \times \log(SSFR)$$

and

$$T_{dust} = 26.5 \times (1 + z)^{0.18} + 6.5 \times [\log(SSFR) - \log(SSFR_{MS})],$$

respectively, with the two scalings having similar quality. For normal star-forming galaxies on the star-forming main sequence, both Magdis et al. (2012a) and Magnelli et al. (2014) report a slow increase in dust temperature (radiation field intensity) with redshift. In a simple scenario in which dust temperature is directly linked to the ratio of SFR to dust mass as an approximation for radiation field intensity, for main-sequence galaxies at increasing redshift this increase may

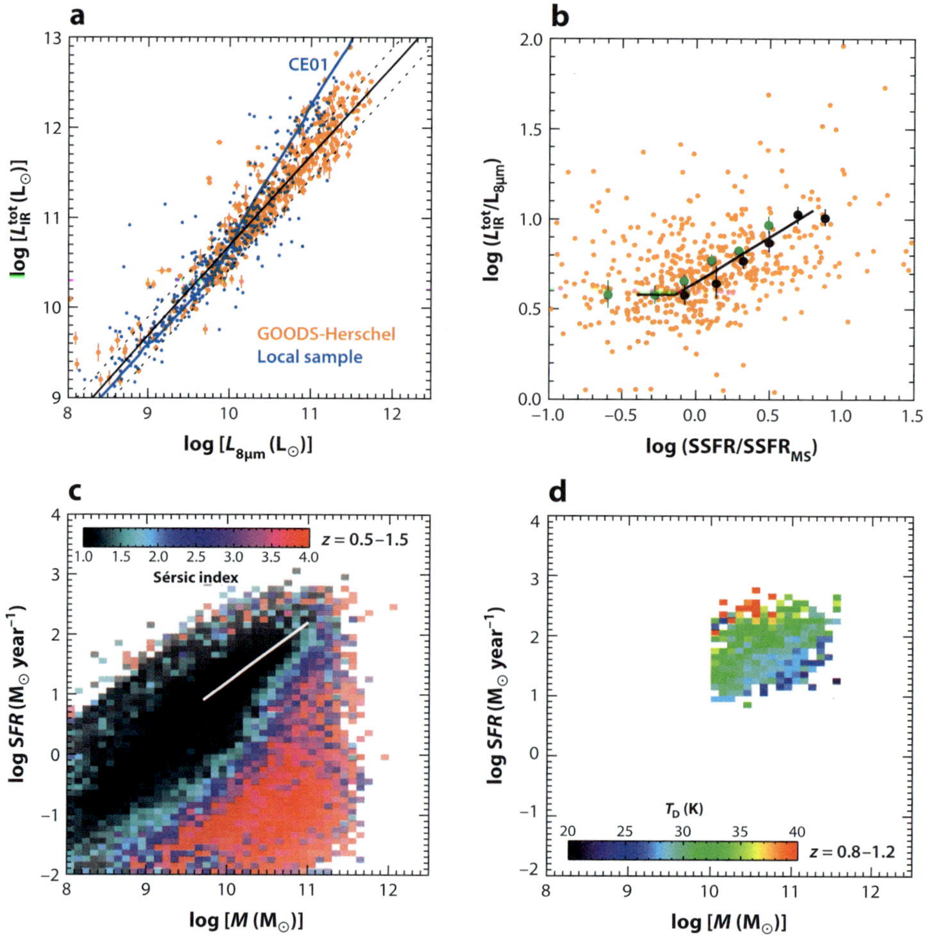

Figure 8

Changes in infrared spectral energy distributions (SEDs) of galaxies and their relation to the evolving star-forming sequence. (*a*) Total infrared versus 8-μm luminosity for a local sample (*blue dots*) and a $z \lesssim 2.5$ GOODS-*Herschel* sample (*orange dots*; *diagonal black solid and dashed lines* visualize median ratio and dispersion for this sample). The blue line shows the locus traced by the Chary & Elbaz (2001) local SED library. Panel adapted from Elbaz et al. (2011) with permission. (*b*) Relation between the offset in specific star-forming rates (SSFRs) from the evolving main sequence (MS) (starburstiness) and the ratio of total infrared and 8-μm luminosity. The high-z sample of Elbaz et al. (2011) is shown in orange. Average values from Nordon et al. (2012) are plotted for $z \sim 1$ (*green dots*) and $z \sim 2$ (*black dots*) along with their proposed redshift-independent relation (*black line*; their equation 3). (*c*) Median Sérsic indices for galaxies in the $z \sim 1$ SFR–stellar mass plane. Low values indicating a preference for disk morphologies are found near the main sequence, which is qualitatively located by the white line [panel adapted from Wuyts et al. (2011b) with permission]. (*d*) Mean temperature of dust in the $z \sim 1$ SFR–stellar mass plane [panel adapted from Magnelli et al. (2014) with permission].

be linked to a combination of a decrease in molecular gas depletion times $M_{\mathrm{Mol}}/SFR \propto (1+z)^{-1}$ (Tacconi et al. 2013) and a decrease in metallicity. Such trends should then continue to regimes that are not covered by *Herschel* blank field surveys. Indeed, for an optically selected sample of gravitationally lensed, $z = 1.4$–3.1, $M_* \sim 10^{10}\,\mathrm{M}_\odot$, near-main-sequence galaxies that would be too faint for an SED study in blank fields, Saintonge et al. (2013) report rather warm dust temperatures of order 50 K, which are plausibly linked to such galaxies' short gas depletion times M_{Mol}/SFR and low metallicity.

Magdis et al. (2011b) explore whether deep *Herschel* surveys uncover a qualitatively new population of galaxies that is missing in even very deep mid-infrared *Spitzer* data. Indeed, a small ~2% fraction of GOODS-S field PACS sources are undetected even in the deepest 24-μm data. According to spectroscopic or optical/near-infrared photometric redshifts, these are found to cluster around redshifts $z \sim 0.4$ and $z \sim 1.3$, where the 24-μm band can drop out due to strong silicate absorption. These sources are similar to other compact and obscured starbursts that are detected at both mid- and far-infrared wavelengths.

3.1.2. Far-infrared photometric redshifts for large samples. Identifications and redshifts are relatively easily secured for $z \lesssim 2.5$ *Herschel* galaxies in the best-studied fields with deep multiwavelength imaging, in particular from *Spitzer*, and with rich sets of spectroscopic and photometric redshifts. Spectroscopic completeness of ~70% and essentially complete optical/near-infrared photometric redshifts are in hand for PACS sources in the best fields, such as GOODS-S. For higher redshifts and in the largest fields with less complete multiwavelength data, it is of interest to directly estimate redshifts for large samples, without a need to identify counterparts in deep optical/near-infrared imaging. This is analogous to earlier radio–submillimeter–infrared attempts to estimate redshifts for SMGs (e.g., Carilli & Yun 2000, Aretxaga et al. 2007, Daddi et al. 2009). The observed SED peak wavelength can be well measured for large numbers of SPIRE/PACS-detected sources, but conversion into a redshift estimate is subject to the well-known T_{dust}–z degeneracy. Given the scatter and evolution with redshift of dust temperature even at a given infrared luminosity, this is a significant limitation. Specifically, in $z \lesssim 1$ flux-limited samples, selection effects can conspire to leave very little evolution in the median observed peak wavelength, as Casey et al. (2012) showed on the basis of a spectroscopic redshift survey of SPIRE sources.

Several different approaches to *Herschel* photometric redshifts based on rest frame far-infrared data have been reported (e.g., Amblard et al. 2010, Dannerbauer et al. 2010, Lapi et al. 2011); in the best reported case, the average accuracy in $(1 + z_{\mathrm{phot}})/(1 + z_{\mathrm{spec}})$ is about 10% (Chakrabarti et al. 2013). Calibration data sets are often biased toward sources with previous ground-based (sub)millimeter detections and redshifts; in some cases the method encodes the L_{IR}–T_{dust} relation of SMGs, which is biased toward low T_{dust} for low infrared luminosities (Roseboom et al. 2012). Further validation is clearly needed. Although *Herschel* SEDs can provide first redshift estimates at $z \gtrsim 1$ or help resolve significant ambiguities in photometric redshifts that are based on other wavelengths, their use in any analysis that needs high-quality photometric redshifts should be viewed with caution.

An independent route toward a redshift distribution of *Herschel* sources has been explored by Mitchell-Wynne et al. (2012). Spatially cross-correlating galaxy samples selected at 250 to 500 μm with larger samples selected at optical and near-infrared wavelengths and with known redshift distributions, they infer redshift distributions for the *Herschel* sources with mean $<z> = 1.8 \ldots 1.9$ that are broadly consistent with other evidence. Uncertainties are currently large but may improve with inclusion of a larger fraction of the existing *Herschel* data.

3.2. Infrared Galaxies in Relation to the Star-Forming Main Sequence

In recent years, star-forming galaxies out to at least $z \sim 2$ have been shown to follow a relation between SFR and stellar mass dubbed the star-forming sequence or main sequence of star formation (e.g., Brinchmann et al. 2004, Noeske et al. 2007, Elbaz et al. 2007, Daddi et al. 2007). Starburst galaxies rise to higher SFRs above this sequence, and a population of passive galaxies is found at low SFRs, in analogy to the division of galaxies into red sequence and blue cloud in color-magnitude diagrams. The star-forming sequence is rather tight, as it has a dispersion of ~0.3 dex, and SFR increases with stellar mass according to the power law $SFR \propto M_*^p$ with slope $p \sim 0.6$–1. Most importantly, SFRs of galaxies on this sequence increase with redshift according to $(S)SFR \propto (1+z)^{2.5-3}$ to $z \sim 2$, with a possible flattening at higher redshift (e.g., González et al. 2010, but note ongoing discussion). The existence and tightness of the sequence argue for rather steady star-formation histories. This has contributed to the ongoing reassessment of the roles in the driving of star formation that are ascribed to steadier processes, such as gas inflow from the cosmic web and minor mergers rather than dramatic major mergers. Here, a close link exists to the finding that many $z \sim 1$–2 massive star-forming galaxies are (clumpy) disk galaxies (e.g., Förster Schreiber et al. 2009) with large molecular gas fractions (e.g., Daddi et al. 2010; Tacconi et al. 2010, 2013).

The SFR limits of large-area *Herschel* surveys reach the main sequence only at its $\gtrsim 10^{11}\ M_\odot$ tip and create samples dominated by starbursts above the sequence. Although such surveys provide important statistics for luminous objects, attempts to identify the main sequence from such data alone fail (e.g., Lee et al. 2013). Studying main-sequence objects at lower masses requires the deepest *Herschel* data, e.g., from the GOODS fields; combination with other star-formation indicators; or *Herschel* stacking of star-formation samples selected at other wavelengths.

Most star formation happens on the star-forming sequence. Combining *Herschel* SFRs for strongly star-forming galaxies with optical/near-infrared data for BzK galaxies with lower SFRs, Rodighiero et al. (2011) estimate that only 10% of the $z \sim 2$ SFR density is due to objects with SSFRs more than a factor of 4 above the main sequence (**Figure 9**). In numbers, only 2% of $z \sim 2$ star-forming galaxies are found above this threshold. With different parameterization (two overlapping Gaussian components rather than a threshold separating main sequence and starbursts), the outlying starbursts may contribute ~20% of the star-formation density (Sargent et al. 2012). Such a predominant role for the star-forming sequence is similar to what observations in the local Universe suggest (e.g., Brinchmann et al. 2004).

Elbaz et al. (2011) find that at a given infrared luminosity, the ratio $\nu L_\nu(8\ \mu\mathrm{m})/L_{\mathrm{IR}}$ increases with redshift [i.e., its inverse, dubbed IR8 by Elbaz et al. (2011), decreases], causing the mid-IRX that affected previous mid-infrared-based SFR estimates (Section 2.2). Elbaz et al. (2011) systematically study this ratio for both a $0 < z < 2.5$ *Herschel* sample and a local reference sample of star-forming galaxies (**Figure 8*a***). In both samples, the distribution of IR8 is characterized by a prominent Gaussian-like peak centered near IR8 ~ 4 and a tail toward larger values (weaker PAH emission). The peak can be identified with the main sequence of star-forming galaxies. In particular, for the well-studied local sample, the tail toward larger IR8 represents compact starbursts of high infrared surface brightness and high SSFR. Changes in IR8 can be related to the main sequence in a way that is consistent over $0 < z \lesssim 2.5$. This can be done as two fixed ratios for main-sequence objects and for starbursts above the main sequence, which are separated at IR8 = 8 and have IR8 = 5 and IR8 = 11 for the derived main sequence and starburst spectral templates, respectively (Elbaz et al. 2011). Alternatively, $\nu L_\nu(8\ \mu\mathrm{m})/L_{\mathrm{IR}}$ can be expressed as a steady function of the SSFR offset from the main sequence (Nordon et al. 2012),

$$\log[\nu L_\nu(8\ \mu\mathrm{m})/L_{\mathrm{IR}}] = \begin{cases} -0.58_{\pm 0.04} & : \Delta_{\mathrm{MS}} < -0.14 \\ -0.65_{\pm 0.03} - 0.50_{\pm 0.06} \times \Delta_{\mathrm{MS}} & : \Delta_{\mathrm{MS}} \geq -0.14 \end{cases}, \tag{1}$$

where $\Delta_{\mathrm{MS}} = \log(SSFR) - \log(SSFR_{\mathrm{MS}})$ is the SSFR offset from the main sequence (Gyr^{-1}) (**Figure 8*b***). Galaxies with a similar ratio $\nu L_\nu(8\ \mu\mathrm{m})/L_{IR}$ tend to lie along lines of constant logarithmic SSFR offset from the main sequence in the $z \sim 1$ and $z \sim 2$ samples of Nordon et al. (2012).

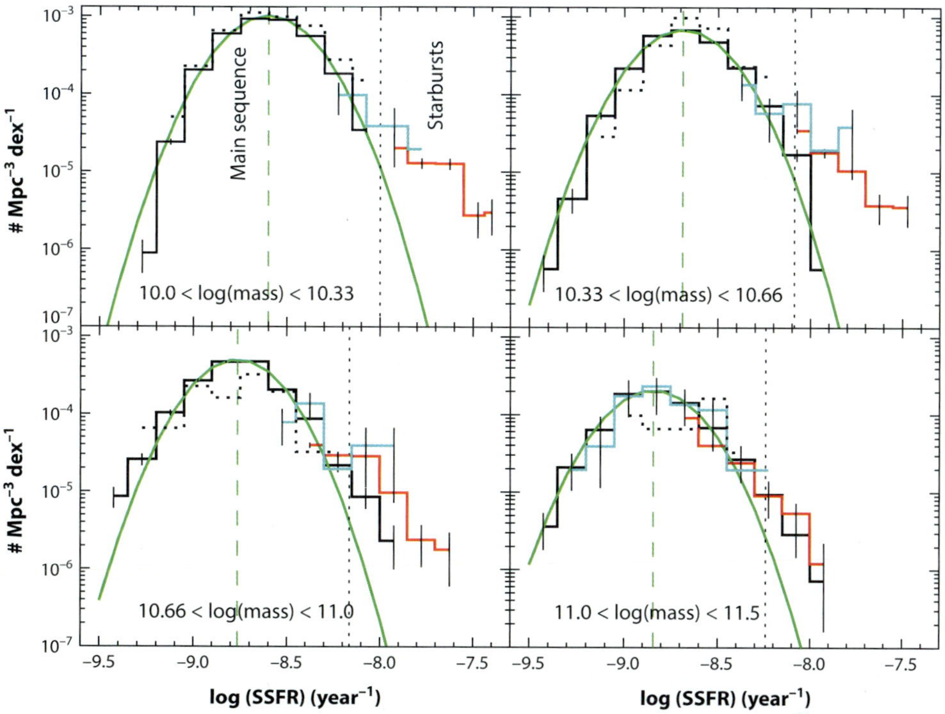

Figure 9

Distribution of specific star-formation rates (SSFRs) for several mass bins in a sample of star-forming galaxies at $1.5 < z < 2.5$. SFRs are from *Herschel* for high-SFR sources in COSMOS (*red*) and GOODS-S (*cyan*) and from dust-corrected UV emission for BzK galaxies in GOODS-S (*dotted black*) and COSMOS (*solid black*). The green curves show the Gaussian fits to the black solid histograms; the main-sequence location at the vertex is indicated by the green dashed line. The black dashed lines are 0.6 dex above the main sequence and show the adopted separation between the main sequence and starbursts that are off the sequence. Figure reproduced from Rodighiero et al. (2011, figure 2) with permission.

The link between these findings and size and morphology is emphasized by Wuyts et al. (2011b), who use a hierarchy of SFRs from *Herschel*, *Spitzer*, and UV-optical-near-infrared SED fitting to place large samples of $z \sim 0.1$ to $z \sim 2.5$ galaxies on the SFR-mass plane and then study their Hubble Space Telescope (HST) morphologies. In a median sense, main-sequence galaxies show the largest sizes at a given stellar mass, as well as disk-like Sérsic indices, whereas quiescent galaxies as well as starbursts in the upper envelope of the main sequence prefer higher Sérsic indices. The starbursts may be in the process of building up central mass concentrations. Interestingly, disk-like morphologies provide a tracing of the main sequence (**Figure 8*c***) that is independent of the original identification of the star-forming main sequence via the correlation of star formation and stellar mass for star-forming galaxies. The morphology trends reported by Wuyts et al. (2011b) refer to the median in a certain SFR, M_*, z bin. This leaves room for a mix of disk and merger morphologies within such a bin. Reliable quantification of such fractions is affected by the well-known intricacies of accurately distinguishing clumpy single objects from mergers if using broadband morphological information only.

Regarding the rest frame far-infrared peak, high-redshift starbursts above the main sequence exhibit increased dust temperatures (Elbaz et al. 2011, Magdis et al. 2012b, Magnelli et al. 2014).

This is analogous to the difference between warm (U)LIRGs and cold lower-luminosity disk galaxies in the local Universe. As discussed above, when looking at a broader redshift range, dust temperature can be expressed as a function of main-sequence offset with less scatter and less redshift dependence rather than expressing temperature as a function of infrared luminosity only.

Another key ISM diagnostic is the rest frame far-infrared fine structure lines. The [CII] 158-μm line, the main ISM cooling line under many conditions, has been known since early airborne detections and the ISO mission to show a deficit (decreased [CII]/$L_{\rm IR}$ ratio) in local ULIRGs (e.g., Malhotra et al. 2001, Luhman et al. 2003 and references therein). Decreased photoelectric heating efficiency in intense radiation fields and dust-bounded HII regions likely play a role in shaping this trend. *Herschel* high-redshift detections of this line are difficult, but large ground-based submillimeter telescopes are stepping in to provide detections. This field is about to expand vastly with the Atacama Large Millimeter Array (ALMA) beginning operations. At a given infrared luminosity, high-redshift galaxies typically show larger [CII]/$L_{\rm IR}$ than local objects (e.g., Maiolino et al. 2009, Stacey et al. 2010, Rigopoulou et al. 2014). Although many of these targets cannot yet be located with respect to the main sequence, owing to lack of reliable stellar masses, Graciá-Carpio et al. (2011) combine such literature high-z [CII] detections with local *Herschel* measurements. They find that the difference between the [CII] deficit at low and high z largely disappears if the deficit is plotted as a function of $L_{\rm IR}/M_{\rm Mol}$ rather than $L_{\rm IR}$. This ratio of infrared luminosity and molecular gas mass is a proxy to offset from the main sequence. Graciá-Carpio et al. (2011) find that sources showing a deficit in [CII] and also in other far-infrared fine structure lines are located at $L_{\rm FIR}/M_{\rm H_2} \gtrsim 80\,{\rm L_\odot\,M_\odot^{-1}}$.

To summarize these findings for massive $M_* \gtrsim 10^{10}\,{\rm M_\odot}$ star-forming galaxies: For a wide set of ISM diagnostics (PAH emission, far-infrared dust temperature, fine structure line emission) as well as morphology, which are all known to locally follow well-characterized trends with infrared luminosity from disk galaxies to merger ULIRGs, the main-sequence paradigm seems to provide a better and more consistent reference to express properties in a redshift-independent way. Star formation in main-sequence galaxies appears to occur in a similar mode at all redshifts, despite a variation in (S)SFR of more than a factor of 20 for main-sequence galaxies over the redshift range studied. Compression of the gas content, for example, in major mergers, changes the ISM conditions and enhances the specific SFRs of starbursts above the main sequence.

4. FAR-INFRARED EMISSION OF HIGH-REDSHIFT GALAXY POPULATIONS

4.1. Gravitational Lenses as Detailed Probes of the Interstellar Medium at High Redshift

Gravitational lensing magnification by individual foreground galaxies or by clusters of galaxies permits detailed studies that are impossible for equivalent sources in blank fields due to sensitivity limitations. Starting with the detection of IRAS F10214+4724 in a redshift survey of the IRAS catalog (Rowan-Robinson et al. 1991), a handful of lensed $z > 2$ sources have become accessible for in-depth infrared to millimeter studies. The first few such objects were detected through a variety of techniques that often selected systems hosting powerful AGNs. Early on, several studies realized that large-area surveys at submillimeter and millimeter wavelengths provide an efficient and less biased method for detecting lensed dusty galaxies at high redshift (e.g., Blain 1996, Negrello et al. 2007). At these wavelengths, the combination of the steep decay of unlensed source counts toward bright fluxes and the approximate constancy of source flux with redshift due to the negative

K-correction on the Rayleigh-Jeans side of the far-infrared SED ensures that a significant fraction of bright sources in a large-area survey will be lensed high-redshift galaxies rather than local objects. This differs from optical and X-ray wavelengths, for which lensed objects are relatively faint compared with the multitude of foreground sources due to positive K-corrections.

Herschel-SPIRE surveys and ground-based South Pole Telescope (SPT) millimeter surveys have now covered approximately 1,000 and 2,500 deg^2, respectively, sufficient for application of these methods. Flux cuts at $S_{500\,\mu m} = 100$ mJy or $S_{1.4\,mm} = 20$ mJy produce samples that contain candidate lensed dusty galaxies at the tens of percent level. Separating these from radio galaxies, low-redshift galaxies, and galactic dust clouds at the same flux level is relatively straightforward on the basis of (sub)millimeter colors as well as readily available radio and optical data (Vieira et al. 2010, Negrello et al. 2010); see also Marsden et al. (2014) for a search based on Atacama Cosmology Telescope data. Hundreds of lensed galaxies are expected to be present at these levels in the existing surveys, given surface densities of candidate lensed galaxies of 0.14 to 0.26 ± 0.04 deg^{-2} at $S_{500\,\mu m} > 100$ mJy (Bussmann et al. 2013, Wardlow et al. 2013) and $\sim$0.1 deg^{-2} at $S_{1.4\,mm} > 20$ mJy (Vieira et al. 2010). Refined selection methods including color information may push the number of candidates to more than 1,000 (González-Nuevo et al. 2012). Related searches for somewhat fainter sources with extremely red SPIRE colors (Dowell et al. 2014) have the potential to uncover the highest redshift DSFGs (e.g., Riechers et al. 2013), likely with a smaller fraction of lensed objects. Firming up the census of $z > 4$ sources from these searches may tighten the constraints that very high SFR high-z sources place on models of galaxy evolution, beyond those known from the $z < 4$ SMG and *Herschel* populations.

Most redshifts of these lensed DSFGs have been obtained from direct CO detections. CO lines (or alternatively rest frame mid-infrared features observable with space telescopes) are quite closely linked to the rest frame far-infrared emission that defines these targets. Their direct detection largely avoids the effort, misidentification risks, and biases that are inherent to the traditional way of obtaining first a radio- or millimeter-interferometric accurate position, then identification in the optical or near-infrared, and finally an optical spectroscopic redshift of the counterpart. The high flux of these lensed objects and the recent availability of wideband spectrometers at both single-dish telescopes and interferometers have contributed to a rather rapid buildup of redshift information (e.g., Negrello et al. 2010, Cox et al. 2011, Frayer et al. 2011, Scott et al. 2011, Combes et al. 2012, Harris et al. 2012, Lupu et al. 2012, Bussmann et al. 2013, Weiß et al. 2013). Spectroscopic redshift completeness well exceeds 50% for the best-studied lensed samples.

Herschel- and SPT-detected systems include some of the highest redshift $z \sim 6$ dusty star-forming objects that are currently known and do not host a quasi-stellar object (QSO) (Riechers et al. 2013, Weiß et al. 2013), and they have already been followed up using a variety of other tracers of the warm and dense ISM (e.g., Cox et al. 2011; Omont et al. 2011, 2013; Bothwell et al. 2013; Rawle et al. 2014). Corrected (where known) for the lensing magnification, these objects typically show very high SFRs ranging from several hundred to a few thousand solar masses per year (e.g., Negrello et al. 2014), including extreme cases that reach $\gtrsim 2{,}000\,M_\odot$ $year^{-1}$ and are single galaxies or physically associated pairs according to interferometric follow-up (e.g., Fu et al. 2012, Fu et al. 2013, Ivison et al. 2013, Riechers et al. 2013). These last cases have passed the detection thresholds simply by their huge luminosities; the detailed follow-up suggests only weak amplification (e.g., **Figure 10**). In their intrinsically large luminosities ($L_{IR} \sim 1.5 \times 10^{13}\,L_\odot$), compact sizes ($R_{half,880\,\mu m} \sim 1.5$ kpc), and warm dust temperatures ($T_{dust} \sim 39$ K), the lensed sources studied by Bussmann et al. (2013) resemble high-luminosity blank field SMGs; the effect of lensing biases is still under study.

Placing these sources in the context of other galaxy evolution studies is slowed down by the need to construct reliable lensing models on the basis of $<1''$ resolution data either from (sub)millimeter

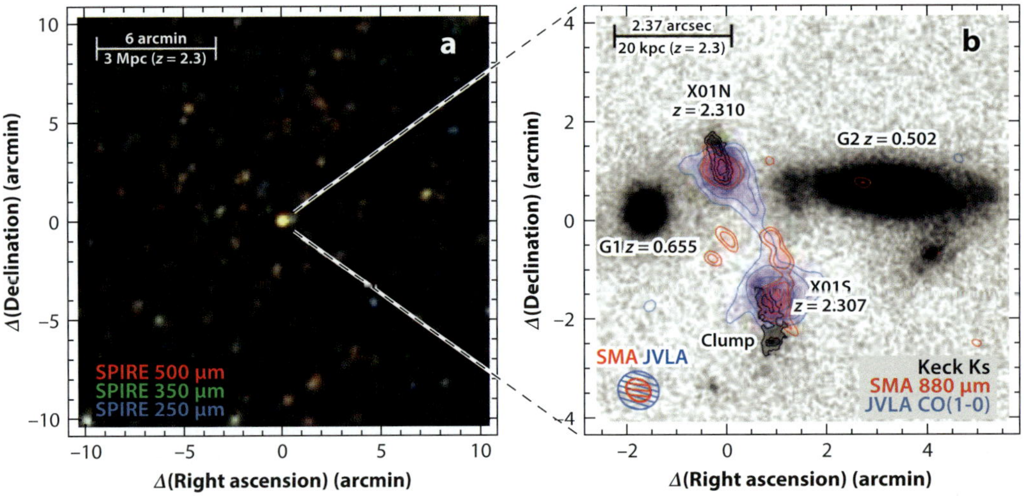

Figure 10

Follow-up of bright *Herschel* 500-µm or South Pole Telescope (SPT) millimeter sources uncovers a large fraction of lensed high-redshift galaxies. Properties range from only weakly magnified, intrinsically extremely luminous targets, as in the example of the weakly amplified galaxy pair HXMM01, to strongly magnified targets. (*a*) A SPIRE three-color image of HXMM01. (*b*) High-resolution follow-up in the near-infrared stellar continuum, submillimeter dust continuum, and Jansky Very Large Array (JVLA) CO emission. Figure reproduced from Fu et al. (2013) with permission. Abbreviation: SMA, submillimeter array.

interferometry or from near-infrared imaging from HST or adaptive optics, even if methods for crude magnification estimates on the basis of line widths have also been proposed (Harris et al. 2012). Of the ~30 published lensing magnifications, many are based on ~0.5″ resolution data, which are coarser than desirable given the likely presence of differential lensing effects, especially for objects that are strongly amplified by galaxy lenses or near the caustics of cluster lenses. In a statistical sense, lenses from flux-limited wide-area submillimeter surveys prefer intrinsically compact sources (Hezaveh et al. 2012). Comparison of the expected magnification distributions to observations is still ongoing (Bussmann et al. 2013). This size bias will also affect the ISM conditions. The difficulty of separating lens and lensed object at optical/near-infrared wavelengths severely limits the number of lensed DSFGs from wide-area searches for which key host parameters such as the stellar mass are currently known. Cases with published stellar mass (Fu et al. 2012, 2013; Negrello et al. 2014; see also Swinbank et al. 2010) mostly lie above the main sequence; they resemble bright blank field SMGs in that respect as well.

4.2. Ultraviolet-Selected Galaxies

Over the past two decades, galaxies selected in the rest frame UV either by broadband photometry and the Lyman break dropout technique (for LBGs) or by detection as Lyman α emitters (LAEs) in narrowband searches have played a pivotal role in our understanding of $z \gtrsim 3$ galaxy evolution. An important uncertainty concerns the obscured fraction of their SFR. One typical way of deducing obscuration (often expressed as $IRX = L_{\mathrm{IR}}/L_{\mathrm{UV}}$) is from the slope β of the rest frame UV continuum ($f_\lambda \propto \lambda^\beta$) by applying locally calibrated relations (e.g., Meurer et al. 1999). Alternatively, SED fitting techniques are applied to the same plus longer wavelength photometric information. Direct detection of rest frame far-infrared or submillimeter emission from these systems turned out to be difficult (Chapman et al. 2000). For typical members of the population, such detections have been restricted mostly to a few lensed systems (e.g., Baker et al. 2001, Siana et al. 2009).

4.2.1. Lyman break galaxies. The direct detection of typical blank field LBGs remains out of reach of *Herschel* data. In large samples, a rather small fraction is individually detected but biased toward LBGs that are more massive and have redder rest frame UV colors. This is true both for Galaxy Evolution Explorer (GALEX)-selected $0.7 < z < 2.0$ LBGs (Burgarella et al. 2011, Oteo et al. 2013b) and for $z \sim 3$ LBGs (Oteo et al. 2013a). These *Herschel*-detected LBGs have large infrared luminosities $L_{IR} > 10^{11} \ldots 10^{12}\, L_{\odot}$. For larger $z \gtrsim 3$ LBG samples (sometimes preselected for larger expected SFR), even stacking analyses are just able to sometimes achieve detections in *Herschel* or ground-based (sub)millimeter data (Magdis et al. 2010, Rigopoulou et al. 2010, Lee et al. 2012, Davies et al. 2013). The SED information that can be extracted is limited given the modest S/N of the stacks, but in some cases (unless affected by sample contamination) it suggests a moderately cold dust temperature ($T_{dust} \sim$ 30–40 K) (Lee et al. 2012, Davies et al. 2013). Perhaps the best SED constraints for $z \lesssim 3$ LBG-like galaxies come from small numbers of gravitationally lensed systems, for which Saintonge et al. (2013) find warm dust temperatures reaching $T_{dust} \sim 50$ K (see also Sklias et al. 2014). There remains uncertainty concerning the rest frame far-infrared SEDs of $z > 3$ LBGs. This implies limitations for interpreting detections that ALMA will obtain on the long-wavelength tail of the SED, in terms of the balance of obscured and unobscured star formation.

4.2.2. The infrared excess-β relation at high redshift. In the absence of infrared data, obscured star formation must be quantified from the rest frame UV emission. Discussing and testing at high redshift the derivation via the IRX-β relation mentioned above is also of relevance for more complex SED fitting techniques that are making use of the same photometric information and similar underlying concepts. An empirical IRX-β relation, such as the classical one of Meurer et al. (1999) for local UV-selected starbursts (and many revisions since), encodes the properties of the unobscured stellar population, such as star-formation history and metallicity, and an attenuation law. The attenuation law is sensitive not only to the properties of the obscuring dust that determine the classical extinction curve but also to the geometric configuration of stars and obscuring dust. Inevitably, stellar emission and dust obscuration in a galaxy are spatially mixed to some extent. In this approach, however, attenuation is formally treated as a foreground screen.

Several papers have used *Herschel* data to test the IRX-β relation outside the local Universe. A main finding is the wide spread in the IRX-β plane, which also depends on the specific sample studied and thus makes it impossible to distill a unique and reliable recipe to estimate the obscured star formation. A sample of $z < 0.3$, 250-μm-selected galaxies (Buat et al. 2010) spreads from roughly the IRX-β relation for local starbursts to that of local normal star-forming galaxies, with IRX varying by at least an order of magnitude at a given β. A similarly large spread is observed for $1 < z < 2.5$ PACS detections (Buat et al. 2012, Nordon et al. 2013) and in several other studies (Wijesinghe et al. 2011; Oteo et al. 2012b, 2013a, 2014).

Analyzing *Herschel* stacks of individually undetected $z \sim 2$ and $z \sim 1.5$, UV-selected galaxies, Reddy et al. (2012) and Heinis et al. (2013) find that the IRX for such stacks increases with β, similar to the trends for local galaxies. There is, however, a noticeable factor >2 difference between these two studies in the absolute IRX level at given β, which is reflected in a better match to local starbursts (Reddy et al. 2012) versus a better match to local normal star-forming galaxies (Heinis et al. 2013). A similar spread may be present in the $3.3 \lesssim z \lesssim 4.3$ stacking study of Lee et al. (2012).

Buat et al. (2010) report a decrease in IRX for $z \sim 1$ galaxies compared with local galaxies of the same infrared luminosity. Nordon et al. (2013) observe that the offset of $1 < z < 2.5$ PACS galaxies from a mean IRX-β relation correlates strongly with the offset from the star-forming main sequence; IRX is larger above the main sequence. This finding likely relates to the offset from a standard IRX-β that is found for local dusty ULIRGs (Goldader et al. 2002) as well to the connection that has been found locally between the IRX-β relation and the stellar birthrate parameter (Kong et al. 2004). UV obscuration properties may be closely connected to the position of a galaxy with respect to the evolving main sequence, in analogy to the infrared SEDs (Section 3).

In their detailed study combining short wavelength and *Herschel* data, Buat et al. (2012) report detection of a 2,175 Å UV bump in the attenuation curves for 20% of their sample as well as a diversity of attenuation-law slopes. Both findings might relate to variations between the galaxies in dust geometry as well as dust properties.

Obscuration of star-forming galaxies (IRX) increases with stellar mass (e.g., Wuyts et al. 2011b, Buat et al. 2012, Heinis et al. 2014). Grouping galaxies detected in both infrared and UV by UV luminosity yields a trend in the sense that the most far-UV luminous galaxies avoid large IRX (Burgarella et al. 2011, Buat et al. 2012, Oteo et al. 2013b). Selection as well as evolution may play a role here, and no obscuration trend is obvious in stacks of $z \lesssim 2$ UV-selected galaxies that are grouped by L_{UV} (Reddy et al. 2012, Heinis et al. 2013). Some decrease of IRX toward high L_{UV} may be present for stacks of $z \sim 3$–4 UV-selected galaxies (Heinis et al. 2014).

In summary, there is a large spread in the IRX-β properties of *Herschel*-detected galaxies. Combined with the tendency for underpredicting the SFRs of the most star-forming galaxies when using rest UV-optical SED fitting (Wuyts et al. 2011a) or a local IRX-β relation (e.g., Oteo et al. 2013a, 2014), this finding supports the use of a hierarchy of SFR indicators in which SFR is based on *Herschel* or mid-infrared (plus unobscured UV where available) for the most heavily star-forming systems, and SFR is based on rest frame UV to near-infrared SED fitting for lower SFR systems (Section 2.2).

4.2.3. Lyman α emitters. Detection of dust continuum emission from $z \gtrsim 5$ LAEs is firmly in the regime requiring ALMA, although an initial upper limit indicates a complex picture (Ouchi et al. 2013). At lower redshifts, insights can be gained from single-dish submillimeter telescopes or *Herschel*. A small subset of $z \sim 2$–3.5 LAEs is detected by *Herschel* (e.g., Bongiovanni et al. 2010, Oteo et al. 2012a); in another $z \sim 2.8$–4 study there are no individual detections (Wardlow et al. 2014). Finding a few LAEs to be infrared-bright is in line with the earlier realization of significant Lyman α emission from many SMGs (Chapman et al. 2005). These individual identifications with DSFGs form a biased subset of LAEs, though, calling for stacking analyses or study of local analogs. At $z \sim 0.3$, *Herschel* detection rates of GALEX-selected LAEs are more favorable (Oteo et al. 2011), and LAEs are found to be biased to lower obscuration compared with similarly UV-luminous systems that are lacking Lyman α emission (Oteo et al. 2012b). In all properties, classical LAEs selected from narrowband Lyman α searches should be contrasted with the extremely infrared-luminous dust-obscured population that was recently retrieved from a Wide-field Infrared Survey Explorer (WISE) selection by Bridge et al. (2013), which also exhibits prominent and extended Lyman α emission.

4.3. Submillimeter Galaxies from Ground-Based Surveys

Ground-based 850-μm and millimeter observations provided the first detections of luminous $z \sim 2$ infrared galaxies. These SMGs continue to be of interest, as ever larger samples are being obtained with improving detector technology at ground-based single-dish telescopes. Reproducing such a significant high-redshift population with SFRs reaching the regime of 1,000 $M_\odot$ year^{-1} has long been known to be a challenge to theoretical models of galaxy evolution (e.g., Baugh et al. 2005, Davé et al. 2010). Because SMG luminosities traditionally had to be extrapolated from the submillimeter and radio or from the mid-infrared using locally calibrated relations, confirmation of these large luminosities is important.

For the difficult task of identifying short-wavelength counterparts to SMGs, *Herschel* photometric surveys can provide better spectral continuity between the detection wavelength and its mid- and near-infrared counterparts (Dannerbauer et al. 2010). But, given *Herschel* beam sizes, such surveys cannot replace radio and in particular (sub)millimeter interferometry for reliable identification and for characterizing the fraction of single-dish detections that breaks up into several sources, which are either physically associated and possibly interacting galaxies or chance projections (e.g., Hayward et al. 2011, 2013; Barger et al. 2012; Smolcić et al. 2012; Karim et al. 2013).

PACS and SPIRE studies sample the SED peaks of bright $z \lesssim 4$ SMGs with good S/N (Magnelli et al. 2010, 2012a; Chapman et al. 2010; Swinbank et al. 2014). These results strongly confirm the dust temperature selection effects that ground-based SMG samples suffer at given infrared luminosity and redshift. In particular for $z \sim 1$, $L_{\mathrm{IR}} \gtrsim 10^{12}$ $L_{\odot}$ galaxies, an object has to be colder than representative for bolometrically selected samples to be detectable in a typical ground-based submillimeter survey. These selection effects are less severe for the most luminous $z \sim 2$ SMGs. *Herschel* data also confirmed that many previously known optically faint radio galaxies with radio continuums indicative of strong star formation but nondetection in submillimeter surveys are indeed strongly star-forming objects but have warmer dust temperatures (Magnelli et al. 2010, Chapman et al. 2010). A further confirmation of the dust temperature biases in the SMG population stems from the comparison with galaxies selected solely via the stellar rest frame 1.6 μm bump. These extend to warmer dust temperatures than SMGs at the same redshift and infrared luminosity (Magdis et al. 2010).

The huge luminosities of SMGs previously extrapolated from other wavelengths are broadly confirmed by *Herschel*. As for other high-redshift galaxies (Section 2.2), infrared luminosities estimated from 24-μm emission via locally calibrated luminosity-dependent templates are too high and show large scatter (Magnelli et al. 2012a). Luminosity estimates based on radio and submillimeter flux, redshift, and adoption of the local radio–far-infrared correlation do better, but they also require downward revision for the SMG sample with spectroscopic redshifts that was studied with *Herschel* by Magnelli et al. (2012a). It would, however, be premature to interpret this as evidence for evolution of the radio–far-infrared correlation because these particular SMGs are somewhat biased to high radio fluxes owing to the role radio interferometry plays in the traditional SMG identification and redshift determination process. In line with this, the small SMG sample of Barger et al. (2012) with full submillimeter and radio identification is consistent with the local radio–far-infrared correlation.

A picture emerges in which traditional SMG samples that are selected over just a small $S_{850} \sim$ 3–10 mJy range are in fact rather heterogeneous (e.g., Magnelli et al. 2012a). Such samples cover two orders of magnitude in total infrared luminosity, range from objects near the star-forming main sequence to those more than an order of magnitude above it (but note discussion on SMG stellar masses, e.g., Michałowski et al. 2012), and cover almost a factor of 3 in dust temperature. In studying the detailed nature of SMGs and their ISM, it is hence key to differentiate these groups rather than globally refer to SMGs. A $z \gtrsim 1$, $L_{\mathrm{IR}} \sim 10^{12}$ $L_{\odot}$ sample [as, e.g., studied in [OI] 63 μm by Coppin et al. (2012)] will almost certainly differ in its properties from ten times more luminous galaxies at $z \sim 2.5$, despite sharing detection in a ground-based SMG survey. This caveat will become ever more important with improving single-dish submillimeter surveys and with ALMA detections of significantly fainter objects at similar wavelengths. The analysis of Hayward et al. (2013) further emphasizes the heterogeneity of the SMG population. They use a semiempirical model based on evolving stellar mass functions and scalings for SFR and gas fractions, combined with hydrodynamic and radiative transfer simulations for isolated disks and mergers, to model the SMG population. Number counts are fit without a need to invoke a top-heavy IMF. At faint fluxes of $S_{850} \sim 1$ mJy, isolated normal star-forming disks and pairs that are blended by the large single-dish beam dominate somewhat expectedly. But isolated disks and blended pairs still do so at $S_{850} \sim 3$ mJy, and even at the $S_{850} \sim 10$ mJy bright end, only two-thirds of the SMGs are suggested to be luminous mergers. In going from single-dish to interferometric studies, number counts should be reduced by as much as a factor of 2 as blended pairs break up, as also suggested by recent observational studies (Barger et al. 2012, Smolcić et al. 2012, Karim et al. 2013).

4.4. Dust-Obscured Galaxies

A considerable portion of *Spitzer* observations and follow-up has been devoted to a population of $z \sim 2$ dust-obscured galaxies (DOGs), which are defined by a high ratio of mid-infrared to rest UV (observed R band) flux density $S_{24\,\mu m}/S_R \gtrsim 1{,}000$ (e.g., Dey et al. 2008). At the $S_{24\,\mu m} \gtrsim 1$ mJy bright end, these samples include a large fraction of obscured AGNs, whereas at fainter levels obscured star formation is often suggested by the presence of a rest near-infrared stellar bump rather than an AGN power-law continuum.

Several studies address the far-infrared properties of this population and are in broad agreement with this general picture (for a related discussion of a bright 24-μm selected sample, see Sajina et al. 2012). *Herschel* detects half of the $S_{24\,\mu m} > 100$ μJy, $z \sim 2$ DOGs at average infrared luminosity 2.8×10^{12} $L_\odot$ (Calanog et al. 2013). Sources with bump rest frame near-infrared SED require a star formation–dominated overall SED, whereas sources with power-law near-infrared SEDs are better fit with SEDs, such as that of Mrk 231, which include a strong AGN component (Melbourne et al. 2012). For a sample of DOGs that emphasizes the faint, mostly star-forming end of the population (median $S_{24\,\mu m} = 161$ μJy), Penner et al. (2012) analyze the far-infrared to mid-infrared flux ratio. This ratio is found to be similar to those of other DSFGs at these redshifts, suggesting that the large mid-infrared to rest UV ratio is not due to an extra mid-infrared component but to atypically large obscuration of the star-forming regions in the rest frame UV.

The WISE color-selected population with strong Lyman α emission studied by Bridge et al. (2013) appears to form a yet more extreme end of the dust-obscured population; it has extreme $L_{IR} > 10^{13}$–10^{14} $L_\odot$ luminosities and warm dust temperatures and likely a dominant AGN.

5. STAR FORMATION IN HOSTS OF ACTIVE GALACTIC NUCLEI

Ideally, coevolution of supermassive black holes and their host galaxies should be studied in samples with measurements for all of the four key quantities that capture the status and growth of the supermassive black hole (black hole mass and accretion rate) as well as the host galaxy (assembled stellar mass and SFR). Concerning host properties, many of the popular star-formation indicators (rest frame UV-optical SED fitting, optical emission lines, mid-infrared flux) are significantly affected by AGN-driven emission, in particular for type 1 AGNs. Because in the rest frame far-infrared the contrast is the largest between the typical SED of a star-forming galaxy and an AGN SED, many researchers have exploited *Herschel* surveys to study star formation in the hosts of high-redshift AGNs. One goal of these surveys is to search for correlations between phases of intense host and black hole growth that may occur in the merger scenario (e.g., Sanders et al. 1988, Di Matteo et al. 2005) in which major galaxy mergers play a strong role in shaping the local scaling relations. Another goal is to test to what extent AGNs are hosted by galaxies with already quenched star formation. See Heckman & Best (2014) for a review of galaxy–black hole coevolution studies in the local Universe.

Even in the far-infrared part of the SED, host dominance is not assured. However, to explore the conditions under which the host is dominant, theoretical models of AGN-heated dust are of somewhat limited value. The freedom to distribute dust outside the sublimation radius in different ways leads to a large variety of warm to cold SEDs, not all of which are actually present around real AGNs. Two different empirical approaches have been pursued instead. Netzer et al. (2007), Mullaney et al. (2011), and Mor & Netzer (2012) have decomposed the total AGN + host SEDs of local systems using *Spitzer* Infrared Spectrograph (IRS) spectra and spectral templates of star-forming galaxies. The resulting intrinsic AGN SEDs drop steeply from the rest mid- to far-infrared. Hatziminaoglou et al. (2010) and Rosario et al. (2012) have compared the infrared colors of high-redshift AGN hosts with inactive reference samples, finding the mid-infrared flux is often boosted by AGN emission but rest frame far-infrared colors are similar in AGN hosts and reference samples. Both approaches suggest that for $z \lesssim 3$ AGNs with luminosity similar to typical $L_{2-10\,\mathrm{keV}} \lesssim 10^{45}$ X-ray deep field sources, the rest frame far-infrared emission is, on average, dominated by the host. This may not be the case for some of the most luminous AGNs or for AGNs in hosts with particularly low SFRs.

In practice, the mid- to far-infrared intrinsic AGN SEDs mentioned above, scaled according to AGN luminosity, are often used to assess whether a given study is still in a regime in which host dominance in the rest frame far-infrared can be confidently assumed or not. This is illustrated in **Figure 11**, which shows a summary of average far-infrared luminosities (tracing host SFRs) for AGNs of different luminosities and redshifts. Pre-*Herschel* studies of samples of powerful AGNs indicated correlations between SFR and AGN luminosity that are suggestive of merger-induced correlations (e.g., Lutz et al. 2008, Netzer 2009). The much better statistics and the breaking of redshift-luminosity degeneracies provided by the *Herschel* samples provide a more differentiated picture. Host SFRs at a given AGN luminosity rise steeply with redshift. At $z < 0.8$, dependence of SFR on AGN luminosity is not significant for moderate-luminosity AGNs but appears to be present for the most luminous AGNs (bolometric $L_{\mathrm{AGN}} \sim 10^{45}$ erg s^{-1}). At $0.8 < z < 2.5$, no significant SFR variations are seen over several orders of magnitude in AGN luminosity. Typical SFRs of optically selected QSOs at similar redshifts, as derived from stacking or maximum-likelihood analysis, also find a clear increase in host SFRs with redshift as well as typically weak dependence on AGN luminosity (Serjeant et al. 2010, Bonfield et al. 2011). Powerful $z > 2$ optically selected QSOs are found to be hosted, on average, in highly star-forming systems (Serjeant et al. 2010, Mor et al. 2012, Leipski et al. 2013, Netzer et al. 2013). These new measurements, which are probing closer to the peak of the star-forming SED component, reinforce previous ground-based (sub)millimeter studies of powerful high-z QSOs. The AGN luminosity range covered by these high-z QSOs is too small to probe for trends with luminosity in the same way as for X-ray AGNs at $z \lesssim 2.5$, but some correlation of host star formation and AGN luminosity among very luminous AGNs is suggested (e.g., Netzer et al. 2013).

This absence or weakness of trends in average SFR with AGN luminosity among moderate-luminosity ($L_{\mathrm{AGN}} < 10^{45}$ erg s^{-1}) X-ray-selected AGNs is reported by several studies with partly overlapping deep X-ray and *Herschel* data sets (Shao et al. 2010, Mullaney et al. 2012b, Rosario et al. 2012, Rovilos et al. 2012). The increase in AGN host SFRs with redshift (**Figure 11**) is similar to the increase in SFR with redshift of main-sequence star-forming galaxies, a link that is strongly reinforced by studies that compare active and inactive galaxies of the same stellar mass and redshift. AGN host SFRs and SSFRs are enhanced with respect to mass-matched galaxies if both star-forming and passive galaxies are included in the matched reference sample (Santini et al. 2012), but they are close to those of mass-matched star-forming main-sequence galaxies only (Mullaney et al. 2012b, Santini et al. 2012, Rosario et al. 2013a). In other words, the hosts of moderate-luminosity $z \lesssim 2$ AGNs resemble typical massive main-sequence star-forming galaxies at these redshifts and are less likely to be quenched (Mullaney et al. 2012b, Rosario et al. 2013a). The similarity between AGN hosts and star-forming galaxies of the same stellar mass extends to

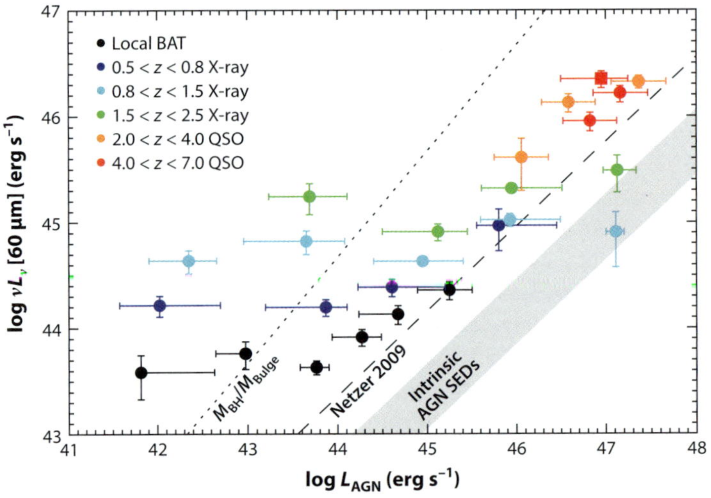

Figure 11

Growth of galaxies and their black holes. Average rest frame far-infrared emission of active galactic nucleus (AGN) hosts expressing the star-formation rate is plotted as a function of bolometric AGN luminosity in different redshift bins. Data for local Burst Alert Telescope (BAT) AGNs and for $z < 2.5$ X-ray-selected AGNs are reproduced from Rosario et al. (2012). Data for optically selected high-redshift quasi-stellar objects (QSOs) are from Serjeant et al. (2010), S. Serjeant (personal communication), and the $z \sim 4.8$ sample of Mor et al. (2012) (*square*). Far-infrared luminosities are mean values that include direct detections as well as stacked nondetections. They include all AGNs in a bin and are plotted at the median AGN luminosity with horizontal error bars showing the range including 80% of the bin's sources. The dotted line indicates the proportionality for a continuous host and black hole growth that would produce the local Universe ratio of black hole mass to bulge mass (Häring & Rix 2004) for an assumed black hole accretion efficiency of 0.1. The diagonal dashed line is the correlation for local AGN-dominated sources as proposed by Netzer (2009), and the diagonal gray band is the approximate 1σ range exhibited by empirical pure AGN intrinsic spectral energy distributions (SEDs) (for details, see Rosario et al. 2012).

the distribution of rest frame $U-V$ colors (Rosario et al. 2013a). Because of dust, those colors are poor tracers of star formation, however. AGNs populate the green valley of a color-mass diagram in a way similar to massive star-forming galaxies and far-infrared-based SSFR locates many green valley AGNs on the main sequence. Location on the green valley then does not correspond to ongoing quenching of star formation, and the green optical colors are indicative of dust reddening.

The link between the hosts of moderate-luminosity AGNs and normal star-forming galaxies provided by these *Herschel* studies is fully consistent with the lack of evidence for enhanced merger fractions in morphological studies of AGN hosts at these redshifts (e.g., Cisternas et al. 2011, Kocevski et al. 2012 and references therein). It is also consistent with the absence of some correlations that are expected in at least some versions of the merger scenario: Obscured AGNs as defined either by a high X-ray obscuring column or by optical type 2 are not more star-forming than unobscured AGNs at the same redshift (Rosario et al. 2012, Rovilos et al. 2012, Merloni et al. 2014), there is no evidence for enhanced star formation in the hosts of those AGNs accreting with the highest Eddington ratios (Rosario et al. 2013b), and high ionization broad absorption line (HiBal) and non-broad absorption line QSOs do not differ significantly in their far-infrared properties (Cao Orjales et al. 2012).

Focusing on more luminous ($L_{2-10\,\mathrm{keV}} > 10^{44}$ erg s^{-1}) $1 < z < 3$ AGNs, Page et al. (2012) reported reduced *Herschel* detection rates and a drop in average SFRs in comparison with moderate-luminosity AGNs. They interpret this in terms of star-formation suppression by the AGN. This finding is in tension with the results from the much larger sample studied by Rosario et al. (2012) (**Figure 11**), which finds no such drop, and the significant *Herschel* detection rates of luminous AGNs at similar redshift in other studies (e.g., Hatziminaoglou et al. 2010, Dai et al. 2012). It is also not reproduced in a follow-up with a larger sample (Harrison et al. 2012) and methodology closely matching that of Page et al. (2012). The differences between these studies are probably caused by the combination of small number statistics and a considerable dispersion in the individual SFRs that compose the average SFRs for a certain $L_{2-10\,\mathrm{keV}}$, z bin. There likely is a fraction of hosts of luminous $z \sim 2$ AGNs that are quenched (i.e., have very low SFR) even if the average host SFR of such luminous AGNs is not lower than the average SFR for lower-luminosity AGNs.

Evidence for suppressed star formation has been reported in massive radio galaxies at $z < 0.8$ and $L_K > 1.5\,\mathrm{L}_\odot$, in comparison with magnitude and color-matched inactive galaxies (Virdee et al. 2013). Similarly, radio excess AGNs at $0.2 \lesssim z \lesssim 3$ show lower specific SFRs than X-ray AGNs at similar redshifts (Del Moro et al. 2013). These two results may currently present the strongest *Herschel*-based evidence for quenching in a subset of the high-redshift AGN population. Of course, not all radio galaxies are passive. Strong star formation in some powerful high-z radio galaxies has been implied by previous ground-based submillimeter observations and is being studied with detailed *Herschel* SEDs (e.g., Seymour et al. 2011, Barthel et al. 2012, Wylezalek et al. 2013).

Quenching by an AGN may also be responsible for the fast (up to $\sim$1,000 km s^{-1}), massive (outflow rates up to $\sim$ 1,200 $\mathrm{M}_\odot$ year^{-1}) molecular outflows found in local ULIRGs via both CO emission and *Herschel* OH absorptions (e.g., Feruglio et al. 2010, Fischer et al. 2010, Sturm et al. 2011, Veilleux et al. 2013, Cicone et al. 2014). These could indicate a phase in which the cold gas reservoir of these galaxies is expelled on short timescales within $\sim$10^6–10^8 years.

The studies discussed above address the SFR of various AGN populations. Taking a reverse view, several recent papers indicate a roughly proportional increase of mean black hole accretion rate with host SFR at $z < 1$ (Rafferty et al. 2011, Chen et al. 2013). Mullaney et al. (2012a) group $z \sim 1$ and $z \sim 2$ star-forming galaxies by stellar mass. Averaging over the samples in these stellar mass bins, they derive mean accretion and mean SFRs that increase with host stellar mass at both redshifts. Absolute growth rate of the black hole and star formation are larger at $z \sim 2$, but ratios of black hole growth to star formation are similar at both redshifts and all stellar masses (**Figure 12**). In all three studies, the mean ratios of black hole to host growth are close to, but somewhat lower than, the local ratio of black hole to bulge mass. A variety of factors may contribute to such a mismatch; for example, obscured AGNs may be missing from the samples studied. Also, it would be necessary to compare black hole accretion with only the fraction of the SFR that corresponds to bulge growth rather than with disk growth.

These results support an average tracking of star formation and black hole accretion in galaxy evolution since $z \sim 2$. They plausibly fit the known trends of AGN fractions and star formation with stellar mass as well as the similarity of the cosmic black hole accretion and star-formation histories. However, in all these studies, the use of sample means implies that black hole accretion is effectively averaged over the $\sim$10$^{7\ldots8}$-year timescales in which SFRs may vary or over even longer galaxy evolution timescales. At any given time, most of the rapidly variable black hole accretion actually happens in an AGN subset of the galaxies. Whereas the mean relation of star formation and black hole accretion may be plausibly linked to gas that is supplied to both, the specific processes of star-formation feeding and black hole accretion, which occur on hugely different spatial scales, remain little constrained by this tracking in the long-term average.

Returning to **Figure 11**, we conclude that the lack of correlation between SFR and AGN luminosity for $z < 2.5$ moderate-luminosity AGNs and the similarity of their hosts to main-sequence star-forming galaxies imply that they largely are part of the same population. The classification as

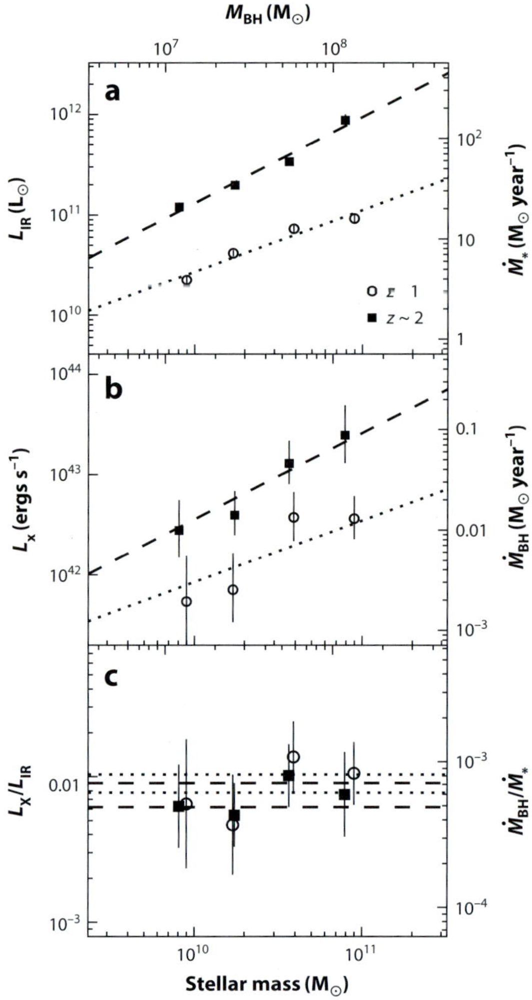

Figure 12

(*a*) Average star-formation rates (SFRs) (right-hand axis) versus stellar mass for $z \sim 1$ (*open circles*) and $z \sim 2$ (*filled squares*) samples of star-forming galaxies (left-hand axis gives equivalent infrared luminosity for illustrative purposes only). Lines are least squares fits to the data. (*b*) Average X-ray luminosities of these star-forming galaxies (same symbols as panel *a*) after accounting for any host-galaxy contribution. Lines have the same gradients as in panel *a* but are normalized to fit the inferred $\dot{M}_{\rm BH}$, which is indicated on the right-hand axis. (*c*) Average supermassive black hole accretion rate to SFR ratio for the two redshift samples. The uncertainties on these points are consistent with a flat $\dot{M}_{\rm BH}$/SFR ratio with respect to M_* for both samples, as indicated by the dotted and dashed lines, respectively. Figure reproduced from Mullaney et al. (2012a, figure 1) with permission.

AGN or as inactive as well as AGN luminosity vary rapidly through processes that do not strongly relate to the global SFR.

Merger-induced effects may still play a role in shaping the increase in SFR with AGN luminosity for low-z and high-luminosity AGNs. This is in qualitative agreement with a visualization into this plane of recent merger hydrodynamic models (Rosario et al. 2012) showing that these models spend considerable time in the region near the correlation line proposed by Netzer (2009). Fast

AGN variability very likely plays a role in shaping this diagram, but different mechanisms deserve further scrutiny. Scenarios qualitatively reproducing **Figure 11** range from one in which accretion tracks constant star formation in the long-term average but with a time-variable Schechter-like accretion rate distribution (Hickox et al. 2014), to one in which accretion events and short starbursts mediated by (minor) mergers are placed on top of a slowly varying background star formation (Neistein & Netzer 2014).

In summary, *Herschel* studies suggest that $z \lesssim 2.5$ AGNs are hosted mostly in massive main-sequence star-forming galaxies with properties that are typical for their redshift. Special events such as major mergers likely play a less important role. Black hole accretion and star formation seem connected by the common gas reservoir and supply on a long galaxy evolution timescale. Given observational limitations, such as the widespread use of averaging over subsamples, and the theoretical uncertainties, the detailed mechanisms of feeding the AGN are not strongly constrained.

6. DUST AS AN INTERSTELLAR MEDIUM TRACER IN DISTANT GALAXIES

Measuring dust masses to study their scalings with other galaxy properties requires submillimeter detections that in the past were typically available only for modestly sized samples preselected by either IRAS far-infrared emission or optical emission. This is now changing with unbiased large surveys and large local samples.

Cortese et al. (2012) use SPIRE data to study dust scaling relations for a magnitude- and volume-limited sample of approximately 300 nearby galaxies that include star-forming as well as passive objects. The ratio of dust mass to stellar mass decreases with stellar mass, stellar mass surface density, and NUV-r color, and it is significantly lower for early-type galaxies. Such scalings are reminiscent of the equivalent scalings for the neutral hydrogen (HI) component (Catinella et al. 2010) and molecular component (Saintonge et al. 2011) of the ISM in local galaxies. Dust stripping by environmental effects is indicated for the Cortese et al. (2012) sample, but at a lower level than for HI, consistent with the fact that dust traces both extended atomic and more compact molecular material. Bourne et al. (2012) stack SPIRE data for a large $z < 0.35$ optically selected spectroscopic sample. Again, dust masses increase with stellar mass while the $M_{\rm dust}/M_*$ ratio decreases, and there are clear indications of an increase in dust masses with redshift in the redshift range studied. Evolution of the dust mass function is very prominent in the $z < 0.5$ study of Dunne et al. (2011). Dust masses increase by a factor of $\sim$5 from local to $z \sim 0.5$ galaxies, and a similar increase is reported for the ratio of dust mass and stellar mass. Santini et al. (2014) bin the redshift–stellar mass–SFR space out to $z = 2.5$ and investigate scalings between dust mass, stellar mass, and SFR from stacked *Herschel* data. Dust masses increase with stellar mass and with redshift and strongly correlate with SFR. There is less of a trend with redshift for dust mass at given SFR and stellar mass.

These observed dust scalings plausibly fit the context in which SFR and stellar mass are linked via the evolving star-forming sequence and the increase in the gas fraction of normal galaxies with redshift (e.g., Daddi et al. 2010, Tacconi et al. 2010). Given rather short gas depletion times for high-redshift galaxies, models of the cosmic evolution of dust must increasingly move from closed box assumptions to considering gas inflow as well as outflow to and recycling from the intergalactic medium, all of which are related to current equilibrium attempts to represent the interplay of gas flows and stellar mass buildup in galaxy evolution (e.g., Lilly et al. 2013). Another interesting question is to what extent it is observationally possible in this context to quantify the cold ISM of high-redshift galaxies via the dust mass rather than direct tracers of the gas.

The cold gas content of evolving high-redshift galaxies is a key measure for characterizing the interplay of gas inflow, gas consumption by star formation, metal enrichment, and outflows or feedback during steady evolution and while influenced by merging and other external effects. Traditionally, low-lying rotational CO transitions are used to measure the molecular gas content. Even at the beginning of the ALMA era, $z > 1$ CO detections remain hard to obtain, however. Roughly 200 detections have been reported in total (Carilli & Walter 2013). Of these, only approximately 60 refer to galaxies near the star-forming main sequence (Tacconi et al. 2013). In addition, use of CO as a molecular gas tracer relies on the CO (1-0) luminosity to molecular gas mass conversion factor that depends on the physical conditions of the source and on metallicity (for a recent review, see Bolatto et al. 2013). Often excitation corrections are applied to measurements of CO lines originating in higher rotational states rather than measuring the CO (1-0) line for which the conversion to gas mass is calibrated.

Herschel has obtained a large number of dust continuum detections of high-redshift galaxies to which ALMA has begun adding longer-wavelength submillimeter detections of unprecedented depth (e.g., Scoville et al. 2014). These resources have spawned great interest in using dust as a proxy for cold gas content, applying locally calibrated conversions from observed rest frame far-infrared and submillimeter fluxes to dust mass, and employing a metallicity-dependent conversion from dust to cold gas mass. Several assumptions and steps are necessary for such methods.

Conversions from dust emission SEDs to dust mass are calibrated in the Milky Way or nearby galaxies and implicitly assume dust properties similar to the ISM of these galaxies. ISM dust has a variety of sources including asymptotic giant branch star mass loss and supernova explosions. The relative weight of these can vary, especially toward the highest redshifts, depending on the histories of star formation and outflow from the galaxy. In addition, dust destruction and reformation in the ISM proper may play a significant role (e.g., Jones et al. 1996), depending on physical conditions. Observational evidence against homogeneous dust properties includes long-known variations in the optical/UV extinction curves, variations of the submillimeter emissivity index with dust temperature (e.g., Paradis et al. 2010; note discussion in the literature on influence of fitting techniques), presence of crystalline silicate dust in some local ULIRGs but not normal galaxies (Spoon et al. 2006), unusual silicate properties in some AGN spectra (Sturm et al. 2005, Markwick-Kemper et al. 2007), and the rest frame $\gtrsim$500 μm submillimeter excess often appearing in the SEDs of low-metallicity galaxies (e.g., Rémy-Ruyer et al. 2013 and references therein).

Dust masses are most easily measured on the Rayleigh-Jeans tail of the SED, where the dependence on dust temperature is only linear and optically thick emission is of almost no concern. This wavelength condition is often not met for pure *Herschel* data, as the 500-μm end of the SPIRE range and in particular source confusion severely limit the accessible rest wavelength range for high-redshift galaxies, except for extremely luminous objects. High S/N color information is then needed instead to constrain either the dust temperature or equivalent parameters, such as radiation field intensity, in physical dust emission models. In addition, such measurements are insensitive to any very cold dust. Draine et al. (2007) demonstrate that for a local disk galaxy sample and the physical dust model of Draine & Li (2007), coverage out to rest wavelength 160 μm is sufficient to obtain dust masses with an extra error due to this wavelength constraint of less than a factor of 2.2, and often less than a factor of 1.5, for appropriately constrained model parameters. Similarly, Dale et al. (2012) compare dust masses of 61 local Key Insights on Nearby Galaxies: a Far-Infrared Survey with *Herschel* (KINGFISH) galaxies from fitting Draine & Li (2007) models to either the *Spitzer*-only data out to 160 μm or combined *Spitzer* and *Herschel* data out to 500 μm. Masses from these two approaches agree well on average, with only 0.2 dex dispersion, but the ratio has trends with both dust temperature and metallicity. Such use of $\lambda_{rest} \leq 160$ μm fits to derive dust masses may be transferable to high-redshift galaxies as long as the ISM conditions are similar to the local calibration sample and dust emission is not optically thick near the far-infrared peak.

Conversion from dust mass to (atomic plus molecular) gas mass $\delta_{GDR} M_{dust} = M_{H_2} + M_{HI}$ must assume a metallicity dependence of the gas-to-dust ratio δ_{GDR}. This is typically assumed to be close to inverse linear (e.g., Leroy et al. 2011), but steeper relations have also been argued for (Muñoz-Mateos et al. 2009) or constancy assumed (Scoville 2012). For a $z \sim 2$ lensed sample with measured metallicities, *Herschel* dust masses, and CO measurements, Saintonge et al. (2013) calculate a relation of gas-to-dust ratio to metallicity that is larger by a factor of 1.7 with respect to the local one of Leroy et al. (2011). Irrespective of an origin in changed dust properties or changed CO conversion, this offset suggests caution about the transfer of local calibrations. Scatter around such relations is significant (Draine & Li 2007, Rémy-Ruyer et al. 2014), and several high-redshift SMGs have been reported to have much smaller gas-to-dust ratios than expected from their nebular metallicity (Santini et al. 2010). Typically, however, individual metallicities are unavailable for high-z galaxies, requiring researchers to adopt scalings such as redshift-dependent mass-metallicity relations (e.g., Zahid et al. 2013) or the fundamental metallicity relation of Mannucci et al. (2010), which links metallicity to stellar mass and SFR in a redshift-independent way. Differences between such scalings and the largely unconstrained scatter of high-redshift galaxies around each scaling can lead to uncertainties of a factor of a few in metallicity alone.

These data practicalities and systematic effects suggest that dust-based methods provide a largely independent second view of the cold gas content of high-z galaxies in addition to CO measurements, but with significant uncertainty.

For a sample of ten normal spiral galaxies in the local Universe with spatially resolved high-quality PACS and SPIRE photometry as well as CO and HI data, Eales et al. (2012) assumed homogeneous dust and gas properties and solved for minimal dispersion between dust-based and CO + HI–based estimates of the ISM mass. They verified that this leads to a consistent picture with CO conversion factor and dust emissivity close to commonly adopted values for the Milky Way.

At higher redshift, Magdis et al. (2011a, 2012a) study small samples of galaxies with *Herschel* and (sub)millimeter detections and low-J CO data, applying the Draine & Li (2007) dust model. Like Magnelli et al. (2012b), they also find dust-based ISM masses in support of a decrease in the CO conversion factor α when moving from main-sequence $z \sim 1$–2 galaxies to (merger) starbursts. Resorting to stacking because of the requirements that even the Draine & Li (2007) model puts on photometric S/N and rest wavelength coverage, Magdis et al. (2012a) then study the properties of larger $z \lesssim 2.3$ samples near the main sequence. They find that SSFR trends across and along the main sequence are related mostly to varying gas fractions. Detected variations with redshift are a significant increase in gas fraction and a mild decrease in gas depletion time M_{Gas}/SFR, in agreement with CO-based results (e.g., Tacconi et al. 2013). Santini et al. (2014) use similar stacking methods for mostly near-main-sequence bins in the redshift–stellar mass–SFR space out to $z = 2.5$, finding a steeper decrease of gas depletion time and suggesting a fit by a redshift-independent fundamental relation between gas fraction, stellar mass, and SFR.

Following a different approach, Nordon et al. (2013) derive a method for estimating the molecular gas content of near-main-sequence galaxies solely from more readily available integrated far-infrared luminosities and rest UV obscuration properties. Based on a study of the UV properties of *Herschel* galaxies and comparison with CO, they derive a quantity that encodes the attenuation contributed by the molecular gas mass per young star. These estimates thus include a next order correction rather than the average gas depletion time of the calibration sample with CO data. Berta et al. (2013a) use this method as well as gas depletion time scalings from Tacconi et al. (2013) to derive the molecular gas mass function since $z \sim 3$.

Dust emission has been successfully applied as an ISM tracer for high-redshift galaxies with results in satisfactory agreement with and partly calibrated on CO-based studies. Moving from current samples to higher redshifts, lower stellar masses and lower metallicities will require more sensitive (and difficult) observations with ALMA as well as larger extrapolation from the ISM conditions at which current dust-based methods are calibrated.

7. OUTLOOK

Productive mining of the current generation of far-infrared surveys will continue for years. Because infrared surveys are now firmly integrated into both the science goals and field choice of multiwavelength surveys of galaxy evolution, their legacy value will only be strengthened as other facilities continue to expand our knowledge of the galaxy populations and survey fields under study.

Fruitful links between *Herschel* and the current generation of (sub)millimeter interferometers, such as the Institut de Radioastronomie Millimétrique Plateau de Bure Interferometer (IRAM-PdBI), indicate the unique role that such interferometers and in particular ALMA will assume in the detailed characterization of far-infrared-detected galaxies and in strengthening our knowledge of their gas content, ISM conditions, and kinematics.

Concerning photometric surveys, large steps in the submillimeter regime can be expected by bringing improved detector technology to large telescopes at the best ground-based sites, as the Cerro Chajnantor Atacama Telescope (CCAT) project plans to do. In space, the planned Japanese SPICA large cryogenic space telescope will for the first time provide far-infrared spectroscopic diagnostics, which for high-z galaxies were effectively out of reach of the passively cooled *Herschel* telescope. With fast and deep mapping in the mid-infrared and in the far-infrared out to $\sim$200 μm, SPICA will also open a new regime in size and quality of SED information for galaxy evolution studies.

SUMMARY POINTS

New observing capabilities from *Herschel* and other facilities have enhanced the power of far-infrared studies of galaxy evolution for redshifts up to $z \sim 2$ and beyond.

1. Near its peak, three-quarters of the cosmic infrared background is resolved into individually detected galaxies.
2. Far-infrared calorimetric star-formation rates are now available for massive normal star-forming galaxies at high redshift and help calibrate other indicators. Direct far-infrared luminosity functions and star-formation rate densities have been obtained out to $z \gtrsim 3$.
3. The interstellar medium conditions of high-z massive star-forming galaxies, as expressed in the far-infrared spectral energy distribution and in the ratio of mid- and far-infrared emission, are better described in relation to the evolving main sequence of star-forming galaxies than by absolute infrared luminosity. Most star formation happens near the main sequence.
4. The far-infrared emission shows that active galactic nuclei hosts out to $z \sim 2$ typically are normal massive star-forming galaxies. The role of major mergers is less important.
5. Steps have been made toward using dust emission as a tracer of the total interstellar medium mass of high-z galaxies.

DISCLOSURE STATEMENT

The author is not aware of any affiliations, memberships, funding, or financial holdings that might be perceived as affecting the objectivity of this review.

ACKNOWLEDGMENTS

The author would like to thank all who provided comments, insights, discussions, data, and figures, in particular Stefano Berta, David Elbaz, Sandy Faber, Reinhard Genzel, Hagai Netzer, Benjamin Magnelli, Emma Rigby, David Rosario, Stephen Serjeant, Toshinobu Takagi, and Stijn Wuyts.

LITERATURE CITED

Altieri B, Berta S, Lutz D, et al. 2010. *Astron. Astrophys.* 518:L17
Amblard A, Cooray A, Serra P, et al. 2010. *Astron. Astrophys.* 518:L9
Amblard A, Cooray A, Serra P, et al. 2011. *Nature* 470:510–12
Aretxaga I, Hughes DH, Coppin K, et al. 2007. *MNRAS* 379:1571–88
Baker AJ, Lutz D, Genzel R, Tacconi LJ, Lehnert MD. 2001. *Astron. Astrophys.* 372:L37–40
Barger AJ, Wang WH, Cowie LL, et al. 2012. *Ap. J.* 761:89
Barthel P, Haas M, Leipski C, Wilkes B. 2012. *Ap. J. Lett.* 757:L26
Baugh CM, Lacey CG, Frenk CS, et al. 2005. *MNRAS* 356:1191–200
Behroozi PS, Wechsler RH, Conroy C. 2013. *Ap. J.* 770:57
Berta S, Lutz D, Nordon R, et al. 2013a. *Astron. Astrophys.* 555:L8
Berta S, Lutz D, Santini P, et al. 2013b. *Astron. Astrophys.* 551:A100
Berta S, Magnelli B, Lutz D, et al. 2010. *Astron. Astrophys.* 518:L30
Berta S, Magnelli B, Nordon R, et al. 2011. *Astron. Astrophys.* 532:A49
Béthermin M, Daddi E, Magdis G, et al. 2012a. *Ap. J. Lett.* 757:L23
Béthermin M, Dole H, Beelen A, Aussel H. 2010. *Astron. Astrophys.* 512:A78
Béthermin M, Le Floc'h E, Ilbert O, et al. 2012b. *Astron. Astrophys.* 542:A58
Béthermin M, Wang L, Doré O, et al. 2013. *Astron. Astrophys.* 557:A66
Blain AW. 1996. *MNRAS* 283:1340–48
Blain AW, Smail I, Ivison RJ, Kneib JP, Frayer DT. 2002. *Phys. Rep.* 369:111–76
Bolatto AD, Wolfire M, Leroy AK. 2013. *Annu. Rev. Astron. Astrophys.* 51:207–68
Bonfield DG, Jarvis MJ, Hardcastle MJ, et al. 2011. *MNRAS* 416:13–21
Bongiovanni A, Oteo I, Cepa J, et al. 2010. *Astron. Astrophys.* 519:L4
Bothwell MS, Aguirre JE, Chapman SC, et al. 2013. *Ap. J.* 779:67
Bourne N, Maddox SJ, Dunne L, et al. 2012. *MNRAS* 421:3027–59
Bridge CR, Blain A, Borys CJK, et al. 2013. *Ap. J.* 769:91
Brinchmann J, Charlot S, White SDM, et al. 2004. *MNRAS* 351:1151–79
Buat V, Giovannoli E, Burgarella D, et al. 2010. *MNRAS* 409:L1–6
Buat V, Noll S, Burgarella D, et al. 2012. *Astron. Astrophys.* 545:A141
Burgarella D, Buat V, Gruppioni C, et al. 2013. *Astron. Astrophys.* 554:A70
Burgarella D, Heinis S, Magdis G, et al. 2011. *Ap. J. Lett.* 734:L12
Burton CS, Jarvis MJ, Smith DJB, et al. 2013. *MNRAS* 433:771–86
Bussmann RS, Pérez-Fournon I, Amber S, et al. 2013. *Ap. J.* 779:25
Calanog JA, Wardlow J, Fu H, et al. 2013. *Ap. J.* 775:61
Cao Orjales JM, Stevens JA, Jarvis MJ, et al. 2012. *MNRAS* 427:1209–18
Carilli CL, Walter F. 2013. *Annu. Rev. Astron. Astrophys.* 51:105–61
Carilli CL, Yun MS. 2000. *Ap. J.* 530:618–24
Casey C, Narayanan D, Cooray A. 2014. *Phys. Rep.* In press (arXiv 1402.1456)
Casey CM, Berta S, Béthermin M, et al. 2012. *Ap. J.* 761:140
Catinella B, Schiminovich D, Kauffmann G, et al. 2010. *MNRAS* 403:683–708

Chakrabarti S, Magnelli B, McKee CF, et al. 2013. *Ap. J.* 773:113
Chapin EL, Hughes DH, Aretxaga I. 2009. *MNRAS* 393:653–62
Chapman SC, Blain AW, Smail I, Ivison RJ. 2005. *Ap. J.* 622:772–96
Chapman SC, Ivison RJ, Roseboom IG, et al. 2010. *MNRAS* 409:L13–18
Chapman SC, Scott D, Steidel CC, et al. 2000. *MNRAS* 319:318–30
Chary R, Elbaz D. 2001. *Ap. J.* 556:562–81
Chen CTJ, Hickox RC, Alberts S, et al. 2013. *Ap. J.* 773:3
Cicone C, Maiolino R, Sturm E, et al. 2014. *Astron. Astrophys.* 562:A21
Cisternas M, Jahnke K, Inskip KJ, et al. 2011. *Ap. J.* 726:57
Clements DL, Rigby E, Maddox S, et al. 2010. *Astron. Astrophys.* 518:L8
Combes F, Rex M, Rawle TD, et al. 2012. *Astron. Astrophys.* 538:L4
Coppin KEK, Danielson ALR, Geach JE, et al. 2012. *MNRAS* 427:520–32
Cortese L, Ciesla L, Boselli A, et al. 2012. *Astron. Astrophys.* 540:A52
Cox P, Krips M, Neri R, et al. 2011. *Ap. J.* 740:63
Daddi E, Bournaud F, Walter F, et al. 2010. *Ap. J.* 713:686–707
Daddi E, Dannerbauer H, Stern D, et al. 2009. *Ap. J.* 694:1517–38
Daddi E, Dickinson M, Morrison G, et al. 2007. *Ap. J.* 670:156–72
Dai YS, Bergeron J, Elvis M, et al. 2012. *Ap. J.* 753:33
Dale DA, Aniano G, Engelbracht CW, et al. 2012. *Ap. J.* 745:95
Dannerbauer H, Daddi E, Morrison GE, et al. 2010. *Ap. J. Lett.* 720:L144–48
Dariush A, Cortese L, Eales S, et al. 2011. *MNRAS* 418:64–73
Davé R. 2008. *MNRAS* 385:147–60
Davé R, Finlator K, Oppenheimer BD, et al. 2010. *MNRAS* 404:1355–68
Davies JI, Baes M, Bendo GJ, et al. 2010. *Astron. Astrophys.* 518:L48
Davies JI, Bianchi S, Cortese L, et al. 2012. *MNRAS* 419:3505–20
Davies LJM, Bremer MN, Stanway ER, Lehnert MD. 2013. *MNRAS* 433:2588–603
Del Moro A, Alexander DM, Mullaney JR, et al. 2013. *Astron. Astrophys.* 549:A59
Dey A, Soifer BT, Desai V, et al. 2008. *Ap. J.* 677:943–56
Di Matteo T, Springel V, Hernquist L. 2005. *Nature* 433:604–7
Dole H, Lagache G, Puget JL, et al. 2006. *Astron. Astrophys.* 451:417–29
Dole H, Rieke GH, Lagache G, et al. 2004. *Ap. J. Suppl.* 154:93–96
Domínguez Sánchez H, Mignoli M, Pozzi F, et al. 2012. *MNRAS* 426:330–41
Dowell CD, Conley A, Glenn J, et al. 2014. *Ap. J.* 780:75
Draine BT, Dale DA, Bendo G, et al. 2007. *Ap. J.* 663:866–94
Draine BT, Li A. 2007. *Ap. J.* 657:810–37
Dunne L, Gomez HL, da Cunha E, et al. 2011. *MNRAS* 417:1510–33
Dye S, Dunne L, Eales S, et al. 2010. *Astron. Astrophys.* 518:L10
Eales S, Dunne L, Clements D, et al. 2010a. *Publ. Astron. Soc. Pac.* 122:499–515
Eales S, Smith MWL, Auld R, et al. 2012. *Ap. J.* 761:168
Eales SA, Raymond G, Roseboom IG, et al. 2010b. *Astron. Astrophys.* 518:L23
Egami E, Rex M, Rawle TD, et al. 2010. *Astron. Astrophys.* 518:L12
Elbaz D, Daddi E, Le Borgne D, et al. 2007. *Astron. Astrophys.* 468:33–48
Elbaz D, Dickinson M, Hwang HS, et al. 2011. *Astron. Astrophys.* 533:A119
Elbaz D, Hwang HS, Magnelli B, et al. 2010. *Astron. Astrophys.* 518:L29
Fadda D, Yan L, Lagache G, et al. 2010. *Ap. J.* 719:425–50
Feruglio C, Maiolino R, Piconcelli E, et al. 2010. *Astron. Astrophys.* 518:L155
Fischer J, Sturm E, González-Alfonso E, et al. 2010. *Astron. Astrophys.* 518:L41
Fixsen DJ, Dwek E, Mather JC, Bennett CL, Shafer RA. 1998. *Ap. J.* 508:123–28
Förster Schreiber NM, Genzel R, Bouché N, et al. 2009. *Ap. J.* 706:1364–428
Frayer DT, Harris AI, Baker AJ, et al. 2011. *Ap. J. Lett.* 726:L22
Fu H, Cooray A, Feruglio C, et al. 2013. *Nature* 498:338–41
Fu H, Jullo E, Cooray A, et al. 2012. *Ap. J.* 753:134
Genzel R, Cesarsky CJ. 2000. *Annu. Rev. Astron. Astrophys.* 38:761–814

Glenn J, Conley A, Béthermin M, et al. 2010. *MNRAS* 409:109–21
Goldader JD, Meurer G, Heckman TM, et al. 2002. *Ap. J.* 568:651–78
González V, Labbé I, Bouwens RJ, et al. 2010. *Ap. J.* 713:115–30
González-Nuevo J, Lapi A, Fleuren S, et al. 2012. *Ap. J.* 749:65
Graciá-Carpio J, Sturm E, Hailey-Dunsheath S, et al. 2011. *Ap. J. Lett.* 728:L7
Griffin MJ, Abergel A, Abreu A, et al. 2010. *Astron. Astrophys.* 518:L3
Gruppioni C, Pozzi F, Andreani P, et al. 2010. *Astron. Astrophys.* 518:L27
Gruppioni C, Pozzi F, Rodighiero G, et al. 2013. *MNRAS* 432:23–52
Guo Q, Cole S, Lacey CG, et al. 2011. *MNRAS* 412:2277–85
Hacking P, Condon JJ, Houck JR. 1987. *Ap. J. Lett.* 316:L15–19
Häring N, Rix HW. 2004. *Ap. J. Lett.* 604:L89–92
Harris AI, Baker AJ, Frayer DT, et al. 2012. *Ap. J.* 752:152
Harrison CM, Alexander DM, Mullaney JR, et al. 2012. *Ap. J. Lett.* 760:L15
Hatziminaoglou E, Omont A, Stevens JA, et al. 2010. *Astron. Astrophys.* 518:L33
Hauser MG, Dwek E. 2001. *Annu. Rev. Astron. Astrophys.* 39:249–307
Hayward CC, Keres D, Jonsson P, et al. 2011. *Ap. J.* 743:159
Hayward CC, Narayanan D, Keres D, et al. 2013. *MNRAS* 428:2529–47
Heckman TM, Best PN. 2014. *Annu. Rev. Astron. Astrophys.* 52:589–660
Heinis S, Buat V, Béthermin M, et al. 2013. *MNRAS* 429:1113–32
Heinis S, Buat V, Béthermin M, et al. 2014. *MNRAS* 437:1268–83
Hezaveh YD, Marrone DP, Holder GP. 2012. *Ap. J.* 761:20
Hickox RC, Mullaney JR, Alexander DM, et al. 2014. *Ap. J.* 782:9
Hickox RC, Wardlow JL, Smail I, et al. 2012. *MNRAS* 421:284–95
Hildebrandt H, van Waerbeke L, Scott D, et al. 2013. *MNRAS* 429:3230–37
Hopkins AM, Beacom JF. 2006. *Ap. J.* 651:142–54
Hwang HS, Elbaz D, Dickinson M, et al. 2011. *Astron. Astrophys.* 535:A60
Hwang HS, Elbaz D, Magdis G, et al. 2010. *MNRAS* 409:75–82
Ivison RJ, Magnelli B, Ibar E, et al. 2010. *Astron. Astrophys.* 518:L31
Ivison RJ, Swinbank AM, Smail I, et al. 2013. *Ap. J.* 772:137
Jarvis MJ, Smith DJB, Bonfield DG, et al. 2010. *MNRAS* 409:92–101
Jones AP, Tielens AGGM, Hollenbach DJ. 1996. *Ap. J.* 469:740
Karim A, Swinbank AM, Hodge JA, et al. 2013. *MNRAS* 432:2–9
Kennicutt RC Jr. 1998. *Annu. Rev. Astron. Astrophys.* 36:189–232
Kennicutt RC Jr, Evans NJ. 2012. *Annu. Rev. Astron. Astrophys.* 50:531–608
Knudsen KK, van der Werf PP, Kneib JP. 2008. *MNRAS* 384:1611–26
Kocevski DD, Faber SM, Mozena M, et al. 2012. *Ap. J.* 744:148
Kong X, Charlot S, Brinchmann J, Fall SM. 2004. *MNRAS* 349:769–78
Lacki BC, Thompson TA. 2010. *Ap. J.* 717:196–208
Lacki BC, Thompson TA, Quataert E. 2010. *Ap. J.* 717:1–28
Lagache G, Abergel A, Boulanger F, Désert FX, Puget JL. 1999. *Astron. Astrophys.* 344:322–32
Lagache G, Puget JL, Dole H. 2005. *Annu. Rev. Astron. Astrophys.* 43:727–68
Lapi A, González-Nuevo J, Fan L, et al. 2011. *Ap. J.* 742:24
Lee KS, Alberts S, Atlee D, et al. 2012. *Ap. J. Lett.* 758:L31
Lee N, Sanders DB, Casey CM, et al. 2013. *Ap. J.* 778:131
Leipski C, Meisenheimer K, Walter F, et al. 2013. *Ap. J.* 772:103
Leroy AK, Bolatto A, Gordon K, et al. 2011. *Ap. J.* 737:12
Lilly SJ, Carollo CM, Pipino A, Renzini A, Peng Y. 2013. *Ap. J.* 772:119
Luhman ML, Satyapal S, Fischer J, et al. 2003. *Ap. J.* 594:758–75
Lupu RE, Scott KS, Aguirre JE, et al. 2012. *Ap. J.* 757:135
Lutz D, Poglitsch A, Altieri B, et al. 2011. *Astron. Astrophys.* 532:A90
Lutz D, Sturm E, Tacconi LJ, et al. 2008. *Ap. J.* 684:853–61
Madau P, Dickinson M. 2014. *Annu. Rev. Astron. Astrophys.* 52:415–86
Magdis GE, Daddi E, Béthermin M, et al. 2012a. *Ap. J.* 760:6

Magdis GE, Daddi E, Elbaz D, et al. 2011a. *Ap. J. Lett.* 740:L15
Magdis GE, Daddi E, Sargent M, et al. 2012b. *Ap. J. Lett.* 758:L9
Magdis GE, Elbaz D, Dickinson M, et al. 2011b. *Astron. Astrophys.* 534:A15
Magdis GE, Elbaz D, Hwang HS, et al. 2010. *MNRAS* 409:22–28
Magliocchetti M, Popesso P, Rosario D, et al. 2013. *MNRAS* 433:127–37
Magliocchetti M, Santini P, Rodighiero G, et al. 2011. *MNRAS* 416:1105–17
Magnelli B, Lutz D, Berta S, et al. 2010. *Astron. Astrophys.* 518:L28
Magnelli B, Lutz D, Saintonge A, et al. 2014. *Astron. Astrophys.* 561:A86
Magnelli B, Lutz D, Santini P, et al. 2012a. *Astron. Astrophys.* 539:A155
Magnelli B, Popesso P, Berta S, et al. 2013. *Astron. Astrophys.* 553:A132
Magnelli B, Saintonge A, Lutz D, et al. 2012b. *Astron. Astrophys.* 548:A22
Maiolino R, Caselli P, Nagao T, et al. 2009. *Astron. Astrophys.* 500:L1–L4
Malhotra S, Kaufman MJ, Hollenbach D, et al. 2001. *Ap. J.* 561:766–86
Mannucci F, Cresci G, Maiolino R, Marconi A, Gnerucci A. 2010. *MNRAS* 408:2115–27
Markwick-Kemper F, Gallagher SC, Hines DC, Bouwman J. 2007. *Ap. J. Lett.* 668:L107–10
Marsden D, Gralla M, Marriage TA, et al. 2014. *MNRAS* 439:1556–74
Mazin D, Raue M. 2007. *Astron. Astrophys.* 471:439–52
McDonald M, Bayliss M, Benson BA, et al. 2012. *Nature* 488:349–52
Melbourne J, Soifer BT, Desai V, et al. 2012. *Astron. J.* 143:125
Merloni A, Bongiorno A, Brusa M, et al. 2014. *MNRAS* 437:3550–67
Meurer GR, Heckman TM, Calzetti D. 1999. *Ap. J.* 521:64–80
Michałowski MJ, Dunlop JS, Cirasuolo M, et al. 2012. *Astron. Astrophys.* 541:A85
Mitchell-Wynne K, Cooray A, Gong Y, et al. 2012. *Ap. J.* 753:23
Mor R, Netzer H. 2012. *MNRAS* 420:526–541
Mor R, Netzer H, Trakhtenbrot B, Shemmer O, Lira P. 2012. *Ap. J. Lett.* 749:L25
Muñoz-Mateos JC, Gil de Paz A, Boissier S, et al. 2009. *Ap. J.* 701:1965–91
Mullaney JR, Alexander DM, Goulding AD, Hickox RC. 2011. *MNRAS* 414:1082–110
Mullaney JR, Daddi E, Béthermin M, et al. 2012a. *Ap. J. Lett.* 753:L30
Mullaney JR, Pannella M, Daddi E, et al. 2012b. *MNRAS* 419:95–115
Murakami H, Baba H, Barthel P, et al. 2007. *Publ. Astron. Soc. Jpn.* 59:369
Negrello M, Hopwood R, De Zotti G, et al. 2010. *Science* 330:800–4
Negrello M, Hopwood R, Dye S, et al. 2014. *MNRAS* 440:1999–2012
Negrello M, Perrotta F, González-Nuevo J, et al. 2007. *MNRAS* 377:1557–68
Neistein E, Netzer H. 2014. *MNRAS* 437:3373–84
Netzer H. 2009. *MNRAS* 399:1907–20
Netzer H, Lutz D, Schweitzer M, et al. 2007. *Ap. J.* 666:806–16
Netzer H, Mor R, Trakhtenbrot B, Shemmer O, Lira P. 2013. arXiv:1308.0012
Nguyen HT, Schulz B, Levenson L, et al. 2010. *Astron. Astrophys.* 518:L5
Noeske KG, Weiner BJ, Faber SM, et al. 2007. *Ap. J. Lett.* 660:L43–46
Nordon R, Lutz D, Genzel R, et al. 2012. *Ap. J.* 745:182
Nordon R, Lutz D, Saintonge A, et al. 2013. *Ap. J.* 762:125
Nordon R, Lutz D, Shao L, et al. 2010. *Astron. Astrophys.* 518:L24
Oliver SJ, Bock J, Altieri B, et al. 2012. *MNRAS* 424:1614–35
Oliver SJ, Wang L, Smith AJ, et al. 2010. *Astron. Astrophys.* 518:L21
Omont A, Neri R, Cox P, et al. 2011. *Astron. Astrophys.* 530:L3
Omont A, Yang C, Cox P, et al. 2013. *Astron. Astrophys.* 551:A115
Oteo I, Bongiovanni Á, Magdis G, et al. 2014. *MNRAS* 439:1337–63
Oteo I, Bongiovanni A, Pérez García AM, et al. 2011. *Ap. J. Lett.* 735:L15
Oteo I, Bongiovanni A, Pérez García AM, et al. 2012a. *Astron. Astrophys.* 541:A65
Oteo I, Bongiovanni A, Pérez García AM, et al. 2012b. *Ap. J.* 751:139
Oteo I, Cepa J, Bongiovanni Á, et al. 2013a. *Astron. Astrophys.* 554:L3
Oteo I, Magdis G, Bongiovanni Á, et al. 2013b. *MNRAS* 435:158–86
Ouchi M, Ellis R, Ono Y, et al. 2013. *Ap. J.* 778:102

Page MJ, Symeonidis M, Vieira JD, et al. 2012. *Nature* 485:213–16
Papovich C, Rudnick G, Le Floc'h E, et al. 2007. *Ap. J.* 668:45–61
Paradis D, Veneziani M, Noriega-Crespo A, et al. 2010. *Astron. Astrophys.* 520:L8
Pascale E, Ade PAR, Bock JJ, et al. 2008. *Ap. J.* 681:400–14
Penner K, Dickinson M, Pope A, et al. 2012. *Ap. J.* 759:28
Pérez-González PG, Rieke GH, Villar V, et al. 2008. *Ap. J.* 675:234–61
Pilbratt GL, Riedinger JR, Passvogel T, et al. 2010. *Astron. Astrophys.* 518:L1
Planck Collab, Ade PAR, Aghanim N, et al. 2011. *Astron. Astrophys.* 536:A18
Planck Collab, Ade PAR, Aghanim N, et al. 2013a. arXiv:1303.5062
Planck Collab, Ade PAR, Aghanim N, et al. 2013b. arXiv:1309.0382
Poglitsch A, Waelkens C, Geis N, et al. 2010. *Astron. Astrophys.* 518:L2
Polletta M, Tajer M, Maraschi L, et al. 2007. *Ap. J.* 663:81–102
Popesso P, Biviano A, Rodighiero G, et al. 2012. *Astron. Astrophys.* 537:A58
Popesso P, Rodighiero G, Saintonge A, et al. 2011. *Astron. Astrophys.* 532:A145
Rafferty DA, Brandt WN, Alexander DM, et al. 2011. *Ap. J.* 742:3
Rawle TD, Edge AC, Egami E, et al. 2012a. *Ap. J.* 747:29
Rawle TD, Egami E, Bussmann RS, et al. 2014. *Ap. J.* 783:59
Rawle TD, Rex M, Egami E, et al. 2012b. *Ap. J.* 756:106
Reddy N, Dickinson M, Elbaz D, et al. 2012. *Ap. J.* 744:154
Rémy-Ruyer A, Madden SC, Galliano F, et al. 2013. *Astron. Astrophys.* 557:A95
Rémy-Ruyer A, Madden SC, Galliano F, et al. 2014. *Astron. Astrophys.* 563:A31
Rex M, Rawle TD, Egami E, et al. 2010. *Astron. Astrophys.* 518:L13
Riechers DA, Bradford CM, Clements DL, et al. 2013. *Nature* 496:329–33
Rigby EE, Maddox SJ, Dunne L, et al. 2011. *MNRAS* 415:2336–48
Rigopoulou D, Hopwood R, Magdis GE, et al. 2014. *Ap. J.* 781:L15
Rigopoulou D, Magdis G, Ivison RJ, et al. 2010. *MNRAS* 409:L7–12
Rodighiero G, Daddi E, Baronchelli I, et al. 2011. *Ap. J. Lett.* 739:L40
Rosario DJ, Santini P, Lutz D, et al. 2012. *Astron. Astrophys.* 545:A45
Rosario DJ, Santini P, Lutz D, et al. 2013a. *Ap. J.* 771:63
Rosario DJ, Trakhtenbrot B, Lutz D, et al. 2013b. *Astron. Astrophys.* 560:A72
Roseboom IG, Ivison RJ, Greve TR, et al. 2012. *MNRAS* 419:2758–73
Rovilos E, Comastri A, Gilli R, et al. 2012. *Astron. Astrophys.* 546:A58
Rowan-Robinson M, Broadhurst T, Oliver SJ, et al. 1991. *Nature* 351:719–21
Rujopakarn W, Rieke GH, Weiner BJ, et al. 2013. *Ap. J.* 767:73
Saintonge A, Kauffmann G, Kramer C, et al. 2011. *MNRAS* 415:32–60
Saintonge A, Lutz D, Genzel R, et al. 2013. *Ap. J.* 778:2
Sajina A, Yan L, Fadda D, Dasyra K, Huynh M. 2012. *Ap. J.* 757:13
Sanders DB, Mazzarella JM, Kim DC, Surace JA, Soifer BT. 2003. *Astron. J.* 126:1607–64
Sanders DB, Mirabel IF. 1996. *Annu. Rev. Astron. Astrophys.* 34:749–92
Sanders DB, Soifer BT, Elias JH, et al. 1988. *Ap. J.* 325:74–91
Santini P, Maiolino R, Magnelli B, et al. 2010. *Astron. Astrophys.* 518:L154
Santini P, Maiolino R, Magnelli B, et al. 2014. *Astron. Astrophys.* 562:A30
Santini P, Rosario DJ, Shao L, et al. 2012. *Astron. Astrophys.* 540:A109
Sargent MT, Béthermin M, Daddi E, Elbaz D. 2012. *Ap. J. Lett.* 747:L31
Scott KS, Lupu RE, Aguirre JE, et al. 2011. *Ap. J.* 733:29
Scoville N, Aussel H, Sheth K, et al. 2014. *Ap. J.* 783:84
Scoville NZ. 2013. In *Secular Evolution of Galaxies*, ed. J Falcón-Barroso, JH Knapen, pp. 491–554. Cambridge, UK: Cambridge Univ. Press
Serjeant S, Bertoldi F, Blain AW, et al. 2010. *Astron. Astrophys.* 518:L7
Seymour N, Symeonidis M, Page MJ, et al. 2011. *MNRAS* 413:1777–86
Shao L, Lutz D, Nordon R, et al. 2010. *Astron. Astrophys.* 518:L26
Siana B, Smail I, Swinbank AM, et al. 2009. *Ap. J.* 698:1273–81
Sibthorpe B, Ivison RJ, Massey RJ, et al. 2013. *MNRAS* 428:L6–10

Sklias P, Zamojski M, Schaerer D, et al. 2014. *Astron. Astrophys.* 561:A149
Smith DJB, Dunne L, da Cunha E, et al. 2012. *MNRAS* 427:703–27
Smolcić V, Aravena M, Navarrete F, et al. 2012. *Astron. Astrophys.* 548:A4
Soifer BT, Helou G, Werner M. 2008. *Annu. Rev. Astron. Astrophys.* 46:201–40
Spoon HWW, Tielens AGGM, Armus L, et al. 2006. *Ap. J.* 638:759–65
Stacey GJ, Hailey-Dunsheath S, Ferkinhoff C, et al. 2010. *Ap. J.* 724:957–74
Sturm E, González-Alfonso E, Veilleux S, et al. 2011. *Ap. J. Lett.* 733:L16
Sturm E, Schweitzer M, Lutz D, et al. 2005. *Ap. J. Lett.* 629:L21–23
Swinbank AM, Simpson J, Smail I, et al. 2014. *MNRAS* 438:1267–87
Swinbank AM, Smail I, Longmore S, et al. 2010. *Nature* 464:733–36
Symeonidis M, Vaccari M, Berta S, et al. 2013. *MNRAS* 431:2317–40
Tacconi LJ, Genzel R, Neri R, et al. 2010. *Nature* 463:781–84
Tacconi LJ, Neri R, Genzel R, et al. 2013. *Ap. J.* 768:74
Thacker C, Cooray A, Smidt J, et al. 2013. *Ap. J.* 768:58
Tielens AGGM. 2008. *Annu. Rev. Astron. Astrophys.* 46:289–337
Vaccari M, Marchetti L, Franceschini A, et al. 2010. *Astron. Astrophys.* 518:L20
Valiante E, Lutz D, Sturm E, Genzel R, Chapin EL. 2009. *Ap. J.* 701:1814–38
van Kampen E, Smith DJB, Maddox S, et al. 2012. *MNRAS* 426:3455–63
Veilleux S, Meléndez M, Sturm E, et al. 2013. *Ap. J.* 776:27
Vieira JD, Crawford TM, Switzer ER, et al. 2010. *Ap. J.* 719:763–83
Viero MP, Asboth V, Roseboom IG, et al. 2014. *Ap. J.* 210:22
Viero MP, Wang L, Zemcov M, et al. 2013. *Ap. J.* 772:77
Virdee JS, Hardcastle MJ, Rawlings S, et al. 2013. *MNRAS* 432:609–25
Wang L, Farrah D, Oliver SJ, et al. 2013. *MNRAS* 431:648–61
Wardlow JL, Cooray A, De Bernardis F, et al. 2013. *Ap. J.* 762:59
Wardlow JL, Malhotra S, Zheng Z, et al. 2014. *Ap. J.* 787:9
Weiß A, De Breuck C, Marrone DP, et al. 2013. *Ap. J.* 767:88
Weiß A, Kovács A, Coppin K, et al. 2009. *Ap. J.* 707:1201–16
Wijesinghe DB, da Cunha E, Hopkins AM, et al. 2011. *MNRAS* 415:1002–12
Wuyts S, Förster Schreiber NM, Lutz D, et al. 2011a. *Ap. J.* 738:106
Wuyts S, Förster Schreiber NM, van der Wel A, et al. 2011b. *Ap. J.* 742:96
Wuyts S, Labbé I, Schreiber NMF, et al. 2008. *Ap. J.* 682:985–1003
Wylezalek D, Vernet J, De Breuck C, et al. 2013. *MNRAS* 428:3206–19
Xu CK, Shupe DL, Béthermin M, et al. 2012. *Ap. J.* 760:72
Zahid HJ, Geller MJ, Kewley LJ, et al. 2013. *Ap. J. Lett.* 771:L19
Zemcov M, Blain A, Halpern M, Levenson L. 2010. *Ap. J.* 721:424–30
Ziparo F, Popesso P, Finoguenov A, et al. 2014. *MNRAS* 437:458–74

Cosmic Star-Formation History

Piero Madau[1] and Mark Dickinson[2]

[1]Department of Astronomy and Astrophysics, University of California, Santa Cruz, California 95064; email: pmadau@ucolick.org

[2]National Optical Astronomy Observatory, Tucson, Arizona 85719; email: med@noao.edu

Annu. Rev. Astron. Astrophys. 2014. 52:415–86

First published online as a Review in Advance on June 18, 2014

The *Annual Review of Astronomy and Astrophysics* is online at astro.annualreviews.org

This article's doi:
10.1146/annurev-astro-081811-125615

Keywords

cosmology, galaxy formation, evolution, star formation, stellar populations

Abstract

Over the past two decades, an avalanche of new data from multiwavelength imaging and spectroscopic surveys has revolutionized our view of galaxy formation and evolution. Here we review the range of complementary techniques and theoretical tools that allow astronomers to map the cosmic history of star formation, heavy element production, and reionization of the Universe from the cosmic "dark ages" to the present epoch. A consistent picture is emerging, whereby the star-formation rate density peaked approximately 3.5 Gyr after the Big Bang, at $z \approx 1.9$, and declined exponentially at later times, with an e-folding timescale of 3.9 Gyr. Half of the stellar mass observed today was formed before a redshift $z = 1.3$. About 25% formed before the peak of the cosmic star-formation rate density, and another 25% formed after $z = 0.7$. Less than $\sim$1% of today's stars formed during the epoch of reionization. Under the assumption of a universal initial mass function, the global stellar mass density inferred at any epoch matches reasonably well the time integral of all the preceding star-formation activity. The comoving rates of star formation and central black hole accretion follow a similar rise and fall, offering evidence for coevolution of black holes and their host galaxies. The rise of the mean metallicity of the Universe to about 0.001 solar by $z = 6$, one Gyr after the Big Bang, appears to have been accompanied by the production of fewer than ten hydrogen Lyman-continuum photons per baryon, a rather tight budget for cosmological reionization.

1. INTRODUCTION

The origin and evolution of galaxies are among the most intriguing and complex chapters in the formation of cosmic structure, and observations in this field have accumulated at an astonishing pace. Multiwavelength imaging surveys with the *Hubble* (HST) and *Spitzer* space telescopes and ground-based facilities, together with spectroscopic follow-up with 8-m-class telescopes, have led to the discovery of galaxies with confirmed redshifts as large as $z = 7.5$ (Finkelstein et al. 2013) as well as compelling photometric candidates as far back as $z \approx 11$ (Coe et al. 2013) when the Universe was only 3% of its current age. Following the seminal work of Steidel et al. (1996), color-selection criteria that are sensitive to the presence of intergalactic HI absorption features in the spectral energy distribution (SED) of distant sources have been used to build increasingly large samples of star-forming galaxies at $2.5 \lesssim z \lesssim 9$ (e.g., Madau et al. 1996, Steidel et al. 2003, Giavalisco et al. 2004a, Bouwens et al. 2011b). Infrared (IR)-optical color selection criteria efficiently isolate both actively star-forming and passively evolving galaxies at $z \approx 2$ (Franx et al. 2003, Daddi et al. 2004). Photometric redshifts have become an unavoidable tool for placing faint galaxies onto a cosmic timeline. *Spitzer*, *Herschel*, and submillimeter telescopes have revealed that dusty galaxies with star-formation rates (SFRs) of order $100\,M_\odot$ year^{-1} or more were abundant when the Universe was only 2–3 Gyr old (Barger et al. 1998, Daddi et al. 2005, Gruppioni et al. 2013). Deep near-infrared (NIR) observations are now commonly used to select galaxies on the basis of their optical rest-frame light and to chart the evolution of the global stellar mass density (SMD) at $0 < z < 3$ (Dickinson et al. 2003). The *Galaxy Evolution Explorer* (GALEX) satellite has quantified the UV galaxy luminosity function (LF) of galaxies in the local Universe and its evolution at $z \lesssim 1$. Ground-based observations and, subsequently, UV and IR data from GALEX and *Spitzer* have confirmed that star-formation activity was significantly higher in the past (Lilly et al. 1996, Le Floc'h et al. 2005, Schiminovich et al. 2005). In the local Universe, various galaxy properties (colors, surface mass densities, and concentrations) have been observed by the Sloan Digital Sky Survey (SDSS) to be "bimodal" around a transitional stellar mass of $3 \times 10^{10}\,M_\odot$ (Kauffmann et al. 2003), showing a clear division between faint, blue, active galaxies and bright, red, passive systems. The number and total stellar mass of blue galaxies appear to have remained nearly constant since $z \sim 1$, whereas those of red galaxies (around L_*) have been rising (Faber et al. 2007). At redshifts $0 < z < 2$ at least, and perhaps earlier, most star-forming galaxies are observed to obey a relatively tight "main-sequence" correlation between their SFRs and stellar masses (Brinchmann et al. 2004, Daddi et al. 2007, Elbaz et al. 2007, Noeske et al. 2007). A minority of starburst galaxies have elevated SFRs above this main sequence as well as a growing population of quiescent galaxies that fall below it.

With the avalanche of new data, galaxy taxonomy has been enriched by the addition of new acronyms such as LBGs, LAEs, EROs, BzKs, DRGs, DOGs, LIRGs, ULIRGs, and SMGs. Making sense of it all and fitting it together into a coherent picture remain one of astronomy's great challenges, in part because of the observational difficulty of tracking continuously transforming galaxy subpopulations across cosmic time and in part because theory provides only a partial interpretative framework. The key idea of standard cosmological scenarios is that primordial density fluctuations grow by gravitational instability driven by cold, collisionless dark matter, leading to a "bottom-up" ΛCDM (cold dark matter) scenario of structure formation (Peebles 1982). Galaxies form hierarchically: Low-mass objects ("halos") collapse earlier and merge to form increasingly larger systems over time—from ultrafaint dwarfs to clusters of galaxies (Blumenthal et al. 1984). Ordinary matter in the Universe follows the dynamics dictated by the dark matter until radiative, hydrodynamic, and star-formation processes take over (White & Rees 1978). The "dark side" of galaxy formation can be modeled with high accuracy and has been explored in detail through N-body numerical simulations of increasing resolution and size (e.g., Davis et al. 1985; Dubinski & Carlberg 1991; Moore et al. 1999; Springel et al. 2005, 2008; Diemand et al. 2008; Stadel et al. 2009; Klypin et al. 2011). However, the same does not hold for the baryons. Several complex processes are still poorly understood, for example, baryonic dissipation inside evolving CDM halos, the transformation of cold gas into stars, the formation of disks and spheroids, the chemical enrichment of gaseous material on galactic and intergalactic scales, and the role played by "feedback" [the effect of the energy input from stars, supernovae (SNe), and massive black holes on their environment] in regulating star formation and generating galactic outflows. The purely phenomenological treatment of complex physical processes that is at the core of semianalytic schemes of galaxy formation (e.g., White & Frenk 1991, Kauffmann et al. 1993, Somerville & Primack 1999, Cole et al. 2000) and—at a much higher level of realism—the "subgrid modeling" of star formation and stellar feedback that must be implemented even in the more accurate cosmological hydrodynamic simulations (e.g., Katz et al. 1996, Yepes et al. 1997, Navarro & Steinmetz 2000, Springel & Hernquist 2003, Keres et al. 2005, Ocvirk et al. 2008, Governato et al. 2010, Guedes et al. 2011, Hopkins et al. 2012, Kuhlen et al. 2012, Zemp et al. 2012, Agertz et al. 2013) are sensitive to poorly determined parameters and suffer from various degeneracies, a weakness that has traditionally prevented robust predictions to be made in advance of specific observations.

Ideally, an in-depth understanding of galaxy evolution would encompass the full sequence of events that led from the formation of the first stars after the end of the cosmic dark ages to the present-day diversity of forms, sizes, masses, colors, luminosities, metallicities, and clustering properties of galaxies. This is a daunting task, and it is perhaps not surprising that an alternative way to look at and interpret the bewildering variety of galaxy data has become very popular in the past two decades. The method focuses on the emission properties of the galaxy population as a whole, traces the evolution with cosmic time of the galaxy luminosity density from the far-UV (FUV) to the far-infrared (FIR), and offers the prospect of an empirical determination of the global history of star formation and heavy element production of the Universe, independently of the complex evolutionary phases of individual galaxy subpopulations. The modern version of this technique relies on some basic properties of stellar populations and dusty starburst galaxies:

1. The UV-continuum emission in all but the oldest galaxies is dominated by short-lived massive stars. Therefore, for a given stellar initial mass function (IMF) and dust content, it is a direct measure of the instantaneous star-formation rate density (SFRD).
2. The rest-frame NIR light is dominated by near-solar-mass-evolved stars that make up the bulk of a galaxy's stellar mass and can then be used as a tracer of the total SMD.
3. Interstellar dust preferentially absorbs UV light and reradiates it in the thermal IR, so that the FIR emission of dusty starburst galaxies can be a sensitive tracer of young stellar populations and the SFRD.

By modeling the emission history of all stars in the Universe at UV, optical, and IR wavelengths from the present epoch to $z \approx 8$ and beyond, one can then shed light on some key questions in galaxy formation and evolution studies: Is there a characteristic cosmic epoch of the formation of stars and heavy elements in galaxies? What fraction of the luminous baryons observed today were already locked into galaxies at early times? Are the data consistent with a universal IMF? Do galaxies reionize the Universe at a redshift greater than 6? Can we account for all the metals produced by the global star-formation activity from the Big Bang to the present? How does the cosmic history of star formation compare with the history of mass accretion onto massive black holes as traced by luminous quasars?

This review focuses on the range of observations, methods, and theoretical tools that are allowing astronomers to map the rate of transformation of gas into stars in the Universe, from the cosmic dark ages to the present epoch. Given the limited space available, it is impossible to provide a thorough survey of such a huge community effort without leaving out significant contributions or whole subfields. We have therefore tried to refer only briefly to earlier findings and present recent observations in more detail, limiting the number of studies cited and highlighting key research areas. In doing so, we hope to provide a manageable overview of how the field has developed and matured in line with new technological advances and theoretical insights and of the questions with which astronomers still struggle nowadays.

The remainder of this review is organized as follows: The equations of cosmic chemical evolution that govern the consumption of gas into stars and the formation and dispersal of heavy elements in the Universe as a whole are given in Section 2. We turn to the topic of measuring mass from light and draw attention to areas of uncertainty in Section 3. Large surveys, key data sets, and the analyses thereof are highlighted in Section 4. An up-to-date determination of the star-formation history (SFH) of the Universe is provided and its main implications are discussed in Section 5. Finally, we summarize our conclusions in Section 6. Unless otherwise stated, all results presented here will assume a "cosmic concordance cosmology" with parameters $(\Omega_M, \Omega_\Lambda, \Omega_b, h) = (0.3, 0.7, 0.045, 0.7)$.

2. THE EQUATIONS OF COSMIC CHEMICAL EVOLUTION

To pursue and cast light into a quantitative form of the idea of a history of cosmic star formation and metal enrichment—not of any particular type of galaxy but of the Universe as a whole—it is useful to start by generalizing the standard equations of galaxy evolution (Tinsley 1980) over all galaxies and intergalactic gas in the Universe. In a representative cosmological comoving volume with density ρ_* in long-lived stars and stellar remnants (white dwarfs, neutron stars, black holes) and gas density ρ_g and in which new stars are formed at the rate ψ, the equations of cosmic chemical evolution can be written as

$$\begin{aligned} \frac{d\rho_*}{dt} &= (1-R)\psi \\ \frac{d\rho_g}{dt} &= -\frac{d\rho_*}{dt} \\ \rho_g\frac{dZ}{dt} &= y(1-R)\psi. \end{aligned} \tag{1}$$

Here, Z is the metallicity in the gas and newly born stars, R is the "return fraction" or the mass fraction of each generation of stars that is put back into the interstellar medium (ISM) and intergalactic medium (IGM), and y is the net metal yield or the mass of new heavy elements created and ejected into the ISM/IGM by each generation of stars per unit mass locked into stars. The above equations govern the formation, destruction, and distribution of heavy elements as they cycle through stars and are ultimately dispersed into the ISM/IGM. By treating all galaxies as a single stellar system and all baryons in the ISM/IGM as its gas reservoir, their solution enables the mean trends of galaxy populations to be calculated with the fewest number of free parameters. The equations state that, for every new mass element locked forever into long-lived stars and stellar remnants, $\Delta\rho_*$, the metallicity of the ISM/IGM increases as $\Delta Z = y\Delta\rho_*/\rho_g$, whereas the mass of heavy elements in the ISM/IGM changes as $\Delta(Z\rho_g) = (y - Z)\Delta\rho_*$. The latter expression is a consequence of metals being released into the gas from mass loss during post-main-sequence stellar evolution as well as being removed from the ISM/IGM when new stars condense out.

However, compared with the source term, the metal sink term can be neglected at early epochs when $Z \ll y$.

At redshift z, Equation 1 can be integrated to give the following:

1. The total mass density of long-lived stars and stellar remnants accumulated from earlier episodes of star formation,

$$\rho_*(z) = (1 - R) \int_0^{t(z)} \psi \, \mathrm{d}t = (1 - R) \int_z^{\infty} \psi \frac{\mathrm{d}z'}{H(z')(1 + z')},$$

$$\rho_*(z) = (1 - R) \int_0^{t(z)} \psi \, \mathrm{d}t = (1 - R) \int_z^{\infty} \psi \frac{\mathrm{d}z'}{H(z')(1 + z')}, \quad (2)$$

   where $H(z') = H_0[\Omega_M(1 + z')^3 + \Omega_\Lambda]^{1/2}$ is the Hubble parameter in a flat cosmology.

2. The total mass density of gas,

$$\rho_g(z) = \rho_{g,\infty} - \rho_*(z), \quad (3)$$

   where $\rho_{g,\infty}$ is the comoving density of gas at some suitable high redshift where there are no stars or heavy elements.

3. The total mass density of heavy elements in the ISM/IGM,

$$Z(z)\rho_g(z) = y(1 - R) \int_0^{t(z)} \psi \, \mathrm{d}t - (1 - R) \int_0^{t(z)} Z\psi \, \mathrm{d}t$$

$$\equiv [y - \langle Z_*(z) \rangle]\rho_*(z), \quad (4)$$

   where the term $\langle Z_* \rangle \rho_*$ is the total metal content of stars and remnants at that redshift. Note that the instantaneous total metal ejection rate, E_Z, is the sum of a recycle term and a creation term (Maeder 1992),

$$E_Z = ZR\psi + y(1 - R)\psi, \quad (5)$$

   where the first term is the amount of heavy elements initially lost from the ISM when stars formed that are now being rereleased, and the second represents the new metals synthesized by stars and released during mass loss.

For a given universal stellar IMF, the quantities R and y can be derived using the following formulas:

$$R = \int_{m_0}^{m_u} (m - w_m)\phi(m)\mathrm{d}m \quad (6)$$

$$y(1 - R) = \int_{m_0}^{m_u} m y_m \phi(m)\mathrm{d}m, \quad (7)$$

where m is mass of a star, w_m is its remnant mass, $\phi(m)$ is the IMF [normalized so that $\int_{m_l}^{m_u} m\phi(m)\mathrm{d}m = 1$], and y_m is the stellar yield, i.e., the fraction of mass m that is converted to metals and ejected. (In this review, the term "yield" generally indicates the net yield y of a stellar population as defined in Equation 7; instead, we explicitly speak of "stellar yields" to indicate the y_m resulting from nucleosynthesis calculations.) The above equations have been written under the simplifying assumptions of "instantaneous recycling" (where the release and mixing of the products of nucleosynthesis by all stars more massive than m_0 occur on a timescale that is much shorter than the Hubble time, whereas stars with $m < m_0$ live forever), "one zone" (where the heavy elements are well mixed at all times within the volume under consideration), "closed box" (flows of gas in and out of the chosen volume are negligible), and "constant IMF and metal yield."

Recall now that the main-sequence timescale is shorter than 0.6 Gyr (the age of the Universe at $z = 8.5$) for stars more massive than $2.5\,\mathrm{M}_\odot$, whereas stars less massive than $0.9\,\mathrm{M}_\odot$ never evolve off the main sequence. [Stellar evolutionary models by Schaller et al. (1992) show that, for $m < 7\,\mathrm{M}_\odot$, solar-metallicity stars have longer lifetimes than their metal-poor counterparts, whereas the opposite is true for $m > 9\,\mathrm{M}_\odot$.] So over the redshift range of interest here, the instantaneous recycling approximation may break down in the limited mass range $0.9 < m < 2.5\,\mathrm{M}_\odot$. For illustrative purposes, in the following we adopt the initial-final mass values for white dwarfs tabulated by Weidemann (2000), which can be fit to few-percent accuracy over the interval $1\,\mathrm{M}_\odot < m < 7\,\mathrm{M}_\odot$ as $w_m = 0.444 + 0.084m$. We also assume that all stars with $8\,\mathrm{M}_\odot < m < m_{\mathrm{BH}} = 40\,\mathrm{M}_\odot$ return all but a $w_m = 1.4\,\mathrm{M}_\odot$ remnant, and stars above m_{BH} collapse to black holes without ejecting material into space, i.e., $w_m = m$. Few stars form with masses above $40\,\mathrm{M}_\odot$, so the impact of the latter simplifying assumption on chemical evolution is minimal. Thus, taking $m_0 = 1\,\mathrm{M}_\odot$ as the dividing stellar mass for instantaneous recycling and a Salpeter (1955) IMF with $\phi(m) \propto m^{-2.35}$ in the range $m_l = 0.1\,\mathrm{M}_\odot < m < m_u = 100\,\mathrm{M}_\odot$, one derives a return fraction of $R = 0.27$. Under the same assumptions, a Chabrier (2003) IMF,

$$\phi(m) \propto \begin{cases} e^{-(\log m - \log m_c)^2/2\sigma^2}/m & (m < 1\,\mathrm{M}_\odot) \\ m^{-2.3} & (m > 1\,\mathrm{M}_\odot) \end{cases} \tag{8}$$

(with $m_c = 0.08\,\mathrm{M}_\odot$ and $\sigma = 0.69$) is more weighted toward short-lived massive stars and yields a larger return fraction, $R = 0.41$. In the instantaneous recycling approximation, the fraction of "dark" stellar remnants formed in each generation is

$$D = \int_{m_0}^{m_u} w_m \phi(m) \mathrm{d}m. \tag{9}$$

The two IMFs produce a dark remnant mass fraction of $D = 0.12$ and $D = 0.19$, respectively. The stellar nucleosynthetic yields depend on metallicity, rotation, and the mass limit for black hole formation m_{BH}. By integrating over the IMF the subsolar metallicity stellar yields (where the effect of mass loss is negligible) tabulated by Maeder (1992) from $10\,\mathrm{M}_\odot$ to $m_{\mathrm{BH}} = 40\,\mathrm{M}_\odot$, we obtain $y = 0.016$ for a Salpeter and $y = 0.032$ for a Chabrier IMF. When integrated to $m_{\mathrm{BH}} = 60\,\mathrm{M}_\odot$, the same tabulation implies $y = 0.023$ (with $R = 0.29$) and $y = 0.048$ (with $R = 0.44$) for a Salpeter and Chabrier IMF, respectively. Notice that some of the uncertainties associated with the IMF and the mass cutoff m_{BH} become smaller when computing the term $y(1 - R)$ in the equations (Equation 1). For massive stars at solar metallicities, stellar winds eject large amounts of helium and carbon into the ISM before these are processed into heavier elements, but the effect on the integrated metal yields (10–$40\,\mathrm{M}_\odot$) is weak (Maeder 1992). Total stellar yields (including the wind and pre-SN contributions) obtained from rotating stellar models at solar metallicity have been presented by Hirschi et al. (2005). Over the same range 10–$40\,\mathrm{M}_\odot$, in this case we derive $y = 0.019$ for Salpeter and $y = 0.038$ for Chabrier. For comparison, the zero-metallicity stellar yields of Chieffi & Limongi (2004) imply $y = 0.015$ for Salpeter and $y = 0.030$ for Chabrier.

Although disfavored by many observations, a Salpeter IMF in the mass range 0.1–$100\,\mathrm{M}_\odot$ is used as a reference throughout the rest of this review. Similarly, for consistency with prior work, we assume the canonical metallicity scale where solar metallicity is $\mathrm{Z}_\odot = 0.02$, rather than the revised value $\mathrm{Z}_\odot = 0.014$ of Asplund et al. (2009).

3. MEASURING MASS FROM LIGHT

Fundamentally, deriving the history of star formation in galaxies involves inferring mass from light. We observe the emission from galaxies at various wavelengths, and from those measurements, we try to infer either the rates at which the galaxies are forming stars or their integrated stellar masses. **Figure 1** illustrates the sensitivity of today's premier multiwavelength surveys to the SFRs and stellar masses of galaxies at high redshift. Rest-frame UV, IR, submillimeter and radio emission, as well as nebular lines such as Hα are all used to measure SFRs and are discussed in this section. In the absence of extinction, UV measurements are more sensitive than current IR or radio data by orders of magnitude, but in practice, dust attenuation is often severe. Long-wavelength data are essential to gain a comprehensive picture of cosmic star formation but are limited by current instrumental sensitivities, although ALMA (Atacama Large Millimeter Array) enables dramatic improvements at submillimeter wavelengths that are particularly valuable at higher redshifts. NIR to mid-infrared (MIR) measurements are critical for deriving stellar masses. Their sensitivity to stellar mass depends critically on the mass-to-light ratio of the stellar population in a distant galaxy, hence on its age, SFH, and extinction. **Figure 1*b*** illustrates two limiting cases: a maximum-M/L model defined as a passively evolving stellar population as old as the Universe and a minimum-M/L model defined as a very young, unreddened, actively star-forming galaxy. In principle, surveys should be mass-complete to the maximum-M/L limits. Much less massive galaxies with young, low-M/L stellar populations can easily be detected, but observations would miss dusty or evolved galaxies with lower masses. The HST WFC3 camera has significantly improved NIR sensitivity compared with most ground-based imaging, but it samples only optical rest-frame light at $z < 3$. *Spitzer*'s IRAC remains the premier resource for deriving stellar masses at higher redshifts, and *James Webb Space Telescope* (JWST) will provide a major advance.

The conversions from light to mass are derived or calibrated using stellar population synthesis models, which encode our knowledge of stellar evolution and of the SEDs of stars, and compute the emergent spectrum for a galaxy with given properties. This knowledge is imperfect, although astronomers have made great progress developing population synthesis models and improving the libraries of empirical and theoretical stellar spectra that they use (for a recent review, see Conroy 2013).

A galaxy (or the Universe as a whole) consists of stars that span a wide range of masses, ages, and metal abundances. The light from those stars may be attenuated by dust before it emerges from the galaxy; the dust dims and generally reddens the galaxy spectrum, and the heated dust re-emits energy in the IR. A galaxy spectrum arises from a composite stellar population whose true distribution of properties is generally unknown. For nearby galaxies, resolved color-magnitude diagrams can reveal the actual distributions of stellar properties, but for most galaxies, we can observe only their integrated light and properties of the emergent spectrum (particularly broadband colors) are often degenerate to different intrinsic properties. An often-noted example is the degeneracy between age, metallicity, and dust attenuation, all of which can redden the spectrum of a galaxy. Observations at higher spectral resolution, for example, of individual spectral lines, can help to resolve some degeneracies (e.g., to constrain stellar metallicities, population ages from absorption line strengths, or reddening from emission line ratios), but never all: The inherently composite nature of stellar populations requires that we make simplifying assumptions when interpreting the light, assumptions that generally cannot be uniquely tested for individual galaxies. Examples of such assumptions include the form of the IMF, the stellar metallicity distribution, the wavelength dependence of dust attenuation, or the precise SFH of the galaxy. The hope is that these assumptions can be made as reasonably as possible, that their impact on derived masses or SFRs can be estimated, and that ultimately they may be tested or constrained by observations in various ways.

The IMF underlies the relation between mass, light, and stellar population age. It controls the ratio of hot, bright stars that dominate the light to cool, faint stars that usually dominate the mass. It regulates the luminosity and color evolution of the integrated stellar population, as stars with different masses evolve at different rates. It also affects the time evolution of the integrated stellar mass, which changes as more massive stars lose gas to the ISM via winds or detonate as SNe.

It is essentially impossible to constrain the IMF from photometric measurements of the integrated light from galaxies: The color of a galaxy does not uniquely reveal its underlying IMF, as there are too many degeneracies to permit useful constraints. Even detailed spectroscopy does not usually offer strong constraints on the IMF overall, although certain spectral features can be useful diagnostics of the number of stars in a given mass range (e.g., Leitherer et al. 1999). The most direct constraints on the IMF come from counting stars as a function of mass in resolved, nearby stellar populations, but they must be very nearby (within our Galaxy and its satellites) to detect subsolar dwarf stars that dominate the mass of a stellar system. The next-best constraints come from integrated measurements of the mass-to-light ratio for star clusters or galaxies, using kinematics (velocity dispersions or rotation curves) to derive a mass for comparison to the luminosity. However, these measurements are difficult to make for faint galaxies at high redshift and require careful modeling to account for the role of dark matter and many other effects.

For lack of better information, astronomers often assume that the IMF is universal, with the same shape at all times and in all galaxies. Although the IMF of various stellar populations within the Milky Way appears to be invariant (for a review, see Bastian et al. 2010), recent studies suggest that the low-mass IMF slope may be a function of the global galactic potential, becoming increasingly shallow (bottom light) with decreasing galaxy velocity dispersion (Conroy & van Dokkum 2012, Geha et al. 2013). It is still unknown, however, how galaxy to galaxy variations may affect the "cosmic" volume-averaged IMF as a function of redshift. In Section 5, we see how a universal IMF can provide a reasonably consistent picture of the global SFH. The exact shape of the IMF at low stellar masses is fairly unimportant for deriving relative stellar masses or SFRs for galaxies. Low-mass stars contribute most of the mass but almost none of the light and do not evolve over a

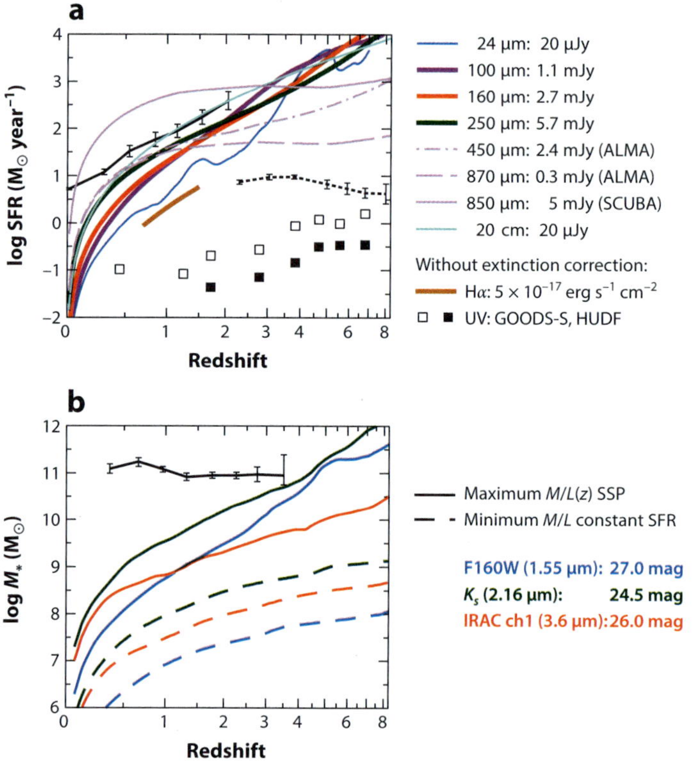

Hubble time. Therefore, changing the low-mass IMF mainly rescales the mass-to-light ratio M/L and, hence, affects both stellar masses and SFRs derived from photometry to a similar degree. Changes to the intermediate- and high-mass region of the IMF, however, can have significant effects on the luminosity, color evolution, and the galaxy properties derived from photometry. It is quite common to adopt the simple power-law IMF of Salpeter (1955), truncated over a finite mass range (generally, 0.1 to 100 $M_\odot$, as adopted in this review). However, most observations show that the actual IMF turns over from the Salpeter slope at masses $<1\ M_\odot$, resulting in smaller M/L ratios than those predicted by the Salpeter IMF. Some common versions of such an IMF are the broken power-law representation used by Kroupa (2001) and the log-normal turnover suggested by Chabrier (2003).

Dust extinction is another important effect that must often be assumed or inferred, rather than directly measured. The shape of the extinction law depends on the properties of the dust grains causing the extinction. For observations of a single star, photons may be absorbed by dust or scattered out of the observed sightline. However, galaxies are 3D structures with mixed and varying distributions of stars and dust. Photons may be scattered both into and out of the sightline, and the optical depth of dust along the line of sight to the observer will be different for every star in the galaxy. These effects are generally lumped together into the simplifying assumption of a net dust

Figure 1

(*a*) Sensitivity to star formation versus redshift for deep survey data at various wavelengths. The key at right indicates the wavelengths and the corresponding flux limits. Most of these limits were computed using data available in the GOODS fields or are simply chosen to be representative values for very deep surveys. For ALMA, we use projected sensitivities of the completed interferometer for modest 10-min integration times, noting that the small ALMA primary beam at 870 μm and 450 μm requires approximately 50 and 170 pointings, respectively, to map 1 square arcminute with uniform sensitivity. All conversions to star-formation rate (SFR) assume a Salpeter IMF from 0.1 to 100 $M_\odot$. The limits for mid-IR (*Spitzer*), far-IR (*Herschel*), submillimeter, and radio data use bolometric corrections from the observed wavelength based on spectral energy distribution templates by Magdis et al. (2012). Open square points show rest-frame 1,500-Å sensitivities assuming no extinction; in practice, dust attenuation can significantly increase these limits. The open squares are based on GOODS data from GALEX, ground-based *U*-band (Nonino et al. 2009), and HST ACS and WFC3 (CANDELS), whereas the filled square points show HST ACS and WFC3 limits for the HUDF (Bouwens et al. 2011b). The Hα curve shown here represents the typical limiting sensitivity of the 3D-HST IR grism spectroscopic survey (Brammer et al. 2012), again assuming no extinction, with conversion to SFR from Kennicutt (1998). The data points (shown as error bars) connected by a solid line show the SFR corresponding to the characteristic infrared luminosity L_*(IR) from Magnelli et al. (2013), and the points connected by a dotted line show the SFR corresponding to the characteristic FUV luminosity L_*(FUV) for Lyman break galaxies at $2 < z < 8$, uncorrected for extinction (Reddy & Steidel 2009; Bouwens et al. 2012b). (*b*) Sensitivity to stellar mass versus redshift, for flux limits at several wavelengths, as indicated at right. The solid curves assume a passively evolving simple stellar population with the age of the Universe, solar metallicity, a Salpeter IMF, and no extinction, computed using the models of Conroy et al. (2009). This approximates the maximum mass-to-light ratio potentially visible at any redshift and, hence, provides an upper bound to the mass completeness limit for a given survey. Less massive galaxies can easily be detected, however, if they have young stellar populations. The dashed curves show sensitivities for an unreddened galaxy with a constant SFR and age of 10^7 years, when M/L at these wavelengths reaches a minimum value. The IR sensitivity limits (given in AB magnitudes) are chosen to be representative of deep surveys such as GOODS and CANDELS [e.g., K_s data from Retzlaff et al. (2010) or Wang et al. (2010)], but are not specific to a particular data set. The data points (shown as error bars) connected by a solid line show the characteristic stellar mass M_* at redshifts $0.2 < z < 4$ (Ilbert et al. 2013). Abbreviations: 3D, three-dimensional; ACS, Advanced Camera for Surveys; ALMA, Atacama Large Millimeter Array; GALEX, *Galaxy Evolution Explorer*; GOODS, Great Observatories Origins Deep Survey; HST, *Hubble Space Telescope*; HUDF, Hubble Ultradeep Field; IMF, initial mass function; IR, infrared; SFR, star-formation rate; SSP, simple stellar population.

attenuation curve, and such relations have been derived for local galaxy samples both empirically (e.g., Calzetti et al. 2000) and using theoretical modeling (Charlot & Fall 2000). However, all galaxies are not equal, and no net attenuation law is equally appropriate for all galaxies. There can always be stars that are completely obscured behind optically thick dust such that little or none of their light emerges directly from the galaxy, except reradiated as dust emission. Although this may not be a significant factor for many galaxies, there are certainly some starburst galaxies in which huge and bolometrically dominant star-formation activity takes place in regions screened by hundreds of magnitudes of dust extinction. UV/optical measurements will never detect this light, but the star formation can be detected and measured at other wavelengths, e.g., with FIR or radio data.

To derive star formation rates or stellar masses for galaxies using stellar population synthesis models, astronomers typically assume relatively simple, parameterized SFHs. However, the SFHs of individual galaxies are unlikely to be smooth and simple; they may vary on both long and short timescales. The fact that young stars are more luminous than older stars leads to the problem of "outshining" (e.g., Papovich et al. 2001, Maraston et al. 2010)—the light from older stars can be lost in the glare of more recent star formation and contributes relatively little to the observed photometry from a galaxy, even if those stars contribute significantly to its mass. SED model fits to galaxies with recent star formation tend to be driven largely by the younger, brighter starlight and may not constrain the mass (or other properties) of older stars that may be present.

For the Universe as a whole, there is one "cosmic" IMF that represents the global average at a given time or redshift, regardless of whether the IMF varies from one galaxy to another. Similarly, there is a "cosmic" distribution of metallicities, a "cosmic" net attenuation of starlight by dust at a given wavelength, and the Universe as a whole obeys one "cosmic" SFH that, moreover, was probably relatively smooth over time—i.e., any stochasticity or "burstiness" averages out when considered for the Universe as a whole. In principle, these facts can simplify the determination of the cosmic SFH, particularly when it is derived from measurements of integrated light averaging over all galaxies. In practice, however, astronomers often derive SFRs and stellar masses for individual galaxies in their deep surveys and then sum them to derive comoving volume averages. In that case, some of the advantages of the "cosmic averaging" are reduced.

3.1. Star-Formation Rates

There are many ways in which to infer SFRs from observations of the integrated light from galaxies. Kennicutt (1998) and Kennicutt & Evans (2012) have presented extensive reviews of this topic, and here we recap only points that are especially relevant for measurements of the global SFH, particularly at high redshift. Virtually all observational tracers of star formation fundamentally measure the rate of massive star formation, because massive stars emit most of the energy from a young stellar population. However, different observational tracers are sensitive to different ranges of stellar masses; hence, they respond differently as a function of stellar population age. For example, Hα emission arises primarily from HII regions photoionized by O stars with lifetimes shorter than 20 Myr, whereas the UV continuum is produced by stars with a broader mass range and with longer lifetimes. The time dependence of different indicators can complicate efforts to derive accurate SFRs for individual galaxies, especially if their SFRs may be rapidly changing (e.g., during a starburst event), but they should average out when summing over a whole population of galaxies.

3.1.1. UV light. Newly formed stellar populations emit radiation over a broad spectrum. For a normal IMF, low-mass stars dominate the mass integrated over the whole stellar population, but at young ages, the luminosity is dominated by UV emission from massive stars. These stars have

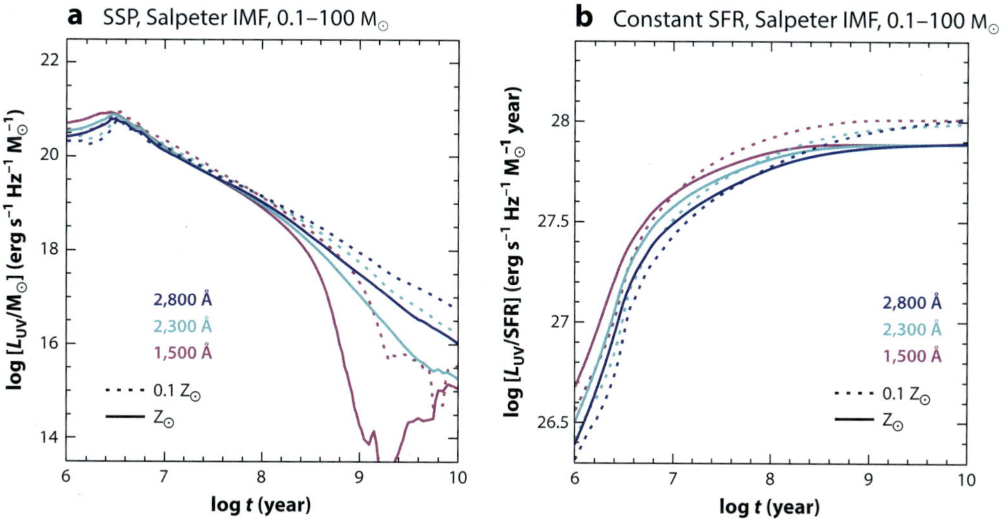

Figure 2

(*a*) Time dependence of the UV luminosity of an SSP with initial mass 1 M⊙, formed with a Salpeter IMF in the range 0.1–100 M⊙, measured at three wavelengths, 1,500 Å, 2,300 Å, and 2,800 Å, and computed with the stellar population synthesis models of Conroy et al. (2009): (*solid lines*) solar metallicity ($Z = 0.02$), (*dashed lines*) 1/10 solar metallicity ($Z_* = 0.002$). (*b*) Time dependence of the UV luminosity per unit SFR for a model with a constant SFR, shown for the same UV wavelengths and metallicities as in the left panel. Abbreviations: IMF, initial mass function; SFR, star-formation rate; SSP, simple stellar population.

short lifetimes, so the UV emission fades quickly. For a Salpeter IMF, the 1,500-Å luminosity from an evolving simple stellar population (SSP) (i.e., an ensemble of stars formed instantaneously and evolving together) with solar metallicity fades by a factor of 100 after 10^8 years and by factors of 10^3 to 10^6 after 10^9 years, depending on metallicity (**Figure 2**). Bolometrically, at least half of the luminous energy that an SSP produces over a 10-Gyr cosmic lifetime emerges in the first 100 Myr, mostly in the UV, making this a natural wavelength from which to infer SFRs.

For a galaxy forming stars at a constant rate, the 1,500-Å luminosity stabilizes once O stars start to evolve off the main sequence. For solar metallicity, by an age of $10^{7.5}$ years, the 1,500-Å luminosity has reached 75% of its asymptotic value, although convergence is somewhat slower at lower metallicity (**Figure 2**). For these reasons, the UV luminosity at wavelengths of ∼1,500 Å (wavelengths from 1,400 Å to 1,700 Å have been used in the literature for both local and high-redshift studies) is regarded as a good tracer of the formation rate of massive stars, provided that the timescale for significant fluctuations in the SFR is longer than a few 10^7 years. For shorter bursts or dips in the SFR, changes in the UV continuum flux may lag those in the SFR and smooth over such variations.

Although the 1,500-Å rest frame is readily accessible with ground-based optical observations of galaxies at redshifts $z \gtrsim 1.4$, measurements at lower redshifts require space-based UV data (e.g., from GALEX or HST) or are limited to longer UV wavelengths. Mid-UV reference wavelengths that have been used in the literature include 2,300 Å (the approximate central wavelength of the GALEX near-UV passband) and 2,800 Å (used, e.g., by Lilly et al. 1996). The mid-UV emission from a galaxy can have a larger contribution from longer-lived, lower-mass stars, particularly at later ages, and the time evolution of the luminosity is more gradual. This is particularly true after ∼250 Myr, when the 1,500-Å luminosity of an SSP drops off sharply, whereas the 2,800-Å luminosity continues to fade at an approximately exponential rate (**Figure 2**). For a constant

SFR, the UV spectral slope reddens moderately with time, as the 1,500-Å luminosity reaches a steady-state level earlier, whereas longer-lived (B and A) stars continue to build up and contribute to the 2,800-Å luminosity. This complicates the conversion from luminosity to SFR as well as any correction for dust extinction based on the UV spectral slope. Still, for young ages, both shorter and longer UV wavelengths usefully trace the SFR and have been used extensively in the literature. Moreover, longer UV wavelengths are subject to somewhat lesser dust attenuation. Wavelengths shorter than that of Lyα (1,216 Å) are rarely used to estimate SFRs, particularly at high redshift where absorption from neutral hydrogen in the IGM is strong.

The UV luminosity output by a stellar population also depends on its metallicity, which affects stellar temperatures and line blanketing. Generally speaking, less-metal-rich stars produce more UV light. The amplitude of this effect is not insignificant and depends on the details of the SFH. From a Salpeter IMF and constant SFR, the range of FUV luminosity per unit SFR for stars spanning a factor of 100 in metallicity (from $Z = 0.0003$ to 0.03) is less than 0.24 dex, or 70%. These variations are larger at higher metallicities and older ages; therefore, we may expect significant evolution in the $L_{\rm FUV}$ to the SFR conversion factor as the global metallicity of galaxies evolves.

We express the conversion factor between the intrinsic FUV-specific luminosity $L_\nu(\rm FUV)$ (before extinction or corrected for extinction) and the ongoing SFR as

$$\mathrm{SFR} = \mathcal{K}_{\rm FUV} \times L_\nu(\mathrm{FUV}), \tag{10}$$

where $L_\nu(\rm FUV)$ is expressed in units of erg s^{-1} Hz^{-1} and SFR in units of M$_\odot$ year^{-1}. The precise value of the conversion factor $\mathcal{K}_{\rm FUV}$ is sensitive to the recent SFH and metal-enrichment history as well as to the choice of IMF. It is relatively insensitive to the exact FUV wavelength, as the UV spectrum of a galaxy with a constant SFR is quite flat in f_ν units, at least for ages much longer than 10^7 years. Generally in this review, we use FUV to refer to 1,500-Å emission or are explicit when we refer to other UV wavelengths. For a Salpeter IMF in the mass range $0.1-100\,\mathrm{M}_\odot$ and constant SFR, the flexible stellar population synthesis (FSPS) models of Conroy et al. (2009) yield $\mathcal{K}_{\rm FUV} = (1.55, 1.3, 1.1, 1.0) \times 10^{-28}$ for $\log Z_*/Z_\odot = (+0.2, 0, -0.5, -1.0)$ at age $\gtrsim$300 Myr. The GALAXEV models of Bruzual & Charlot (2003) yield values of $\mathcal{K}_{\rm FUV}$ that are ∼5% smaller.

Figure 3 illustrates the combined effects of the evolution of the global SFR and metal density on the global mean UV-to-SFR conversion factor as a function of redshift on the basis of the FSPS models. Concentrating on the FUV behavior at 1,500 Å for constant $\psi(z)$, the conversion factor is nearly constant, though slightly elevated at the highest redshifts as the cosmic age gets young, particularly for lower-metallicity models. A SFH that increases with time from $z = 12$ to 1.7, so that the UV-emitting population is on average younger over that redshift range, leads to a more gradually declining trend in $\mathcal{K}_{\rm FUV}(z)$ with time. The dashed line incorporates one scenario for a global change in the metallicity of the star-forming population, evolving as $Z_* = Z_\odot 10^{-0.15z}$ (Kewley & Kobulnicky 2007). This particular evolution is only moderately well constrained at lower redshifts and should be taken as illustrative only, but we may certainly expect metallicities to be lower on average at higher redshift. The effects of metallicity dominate over those of age in this scenario, but the two counterbalance each other to a certain degree, so that $\mathcal{K}_{\rm FUV}(z)$ changes by less than 20%. At 2,800 Å, the redshift dependence of $\mathcal{K}_{\rm FUV}(z)$ is stronger, particularly at $z < 2$ as the global SFRD declines with time, although this may be partially canceled by the effects of metallicity evolution. This is an example of why shorter FUV wavelengths should be preferred for deriving galaxy SFRs.

In this review, we adopt a constant FUV conversion factor $\mathcal{K}_{\rm FUV} = 1.15 \times 10^{-28}$ M$_\odot$ year^{-1} erg^{-1} s Hz (we typically leave out the units) as a compromise value based on the evolutionary scenario from **Figure 3**. The widely used value from Kennicutt (1998) (and based on the calibration by Madau et al. 1998b), $\mathcal{K}_{\rm FUV} = 1.4 \times 10^{-28}$, is 20% larger than our calibration. Other recent

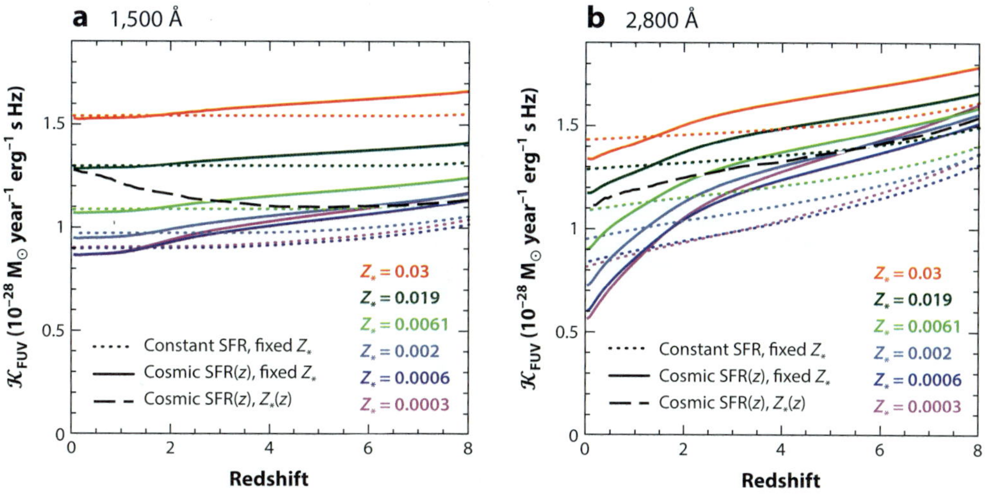

Figure 3

Age and metallicity dependence of the SFR to UV luminosity ratio ($= \mathcal{K}$) (Equation 10) for a stellar population with a Salpeter IMF in the range $0.1-100\,M_\odot$, using the spectral population synthesis models of Conroy et al. (2009): (*dotted lines*) $\mathcal{K}_{FUV}(z)$ assuming constant SFR (starting at $z = 12$) and fixed stellar metallicity; (*solid lines*) same assuming the functional form for cosmic SFR density given in Equation 15 (again starting at $z = 12$) and fixed stellar metallicity; (*dashed lines*) same assuming the functional form for cosmic SFR density given in Equation 15 (again starting at $z = 12$) and stellar metallicity evolving as $Z_* = Z_\odot 10^{-0.15z}$ (Kewley & Kobulnicky 2007). This relation is shown at (*a*) 1,500 Å and (*b*) 2,800 Å, respectively. Abbreviations: IMF, initial mass function; SFR, star-formation rate.

analyses based on the GALAXEV libraries have also found lower mean conversion factors, for both low- and high-redshift galaxy populations (e.g., Salim et al. 2007, Haardt & Madau 2012). The FUV conversion tabulated in Kennicutt & Evans (2012) (from Murphy et al. 2011), if rescaled from the Kroupa to Salpeter IMF, is very close to the $z = 0$ solar metallicity value of $\mathcal{K}_{FUV}(z)$ in **Figure 3**, but our somewhat smaller value should be more representative for the peak era of cosmic star formation at high redshift. Some authors express FUV luminosity as $L_{FUV} = \nu L_\nu$ in solar units. In that case, $\mathcal{K}_{FUV} = 2.2 \times 10^{-10}\,M_\odot\,year^{-1}\,L_\odot^{-1}$ at 1,500 Å, and the conversion factor will depend on the wavelength.

Figure 4 shows the ratio of $\mathcal{K}_{FUV}$ for the Chabrier or Kroupa IMFs to that for the Salpeter IMF as a function of age for a constant SFR calculated using FSPS. This ratio is nearly constant, varying by only 5% with age and 3% over a factor of 100 in metallicity. Where necessary to convert SFRs from the literature from Chabrier or Kroupa IMFs to the Salpeter IMF, we divide by constant factors of 0.63 (Chabrier) or 0.67 (Kroupa). Similarly, **Figure 4** examines mass-to-light ratios for an SSP as a function of age in various bandpasses, comparing values for the Chabrier or Kroupa IMFs to the Salpeter IMF. Again, these ratios are fairly constant with age and have very little dependence on the bandpass. In other words, the color evolution for an SSP with Chabrier or Kroupa IMFs is very similar to that for the Salpeter IMF, showing a roughly constant offset in M/L. [The similar time dependence of M/L for the "bottom-light" Chabrier or Kroupa IMFs to that for the Salpeter IMF is something of a coincidence (or a conspiracy). The rate of luminosity evolution for an SSP depends on the logarithmic IMF slope at masses greater than $1\,M_\odot$ and is faster for the flatter $x = 1.3$ (Kroupa or Chabrier) than for the Salpeter value $x = 1.35$. However, the evolution of the recycled mass fraction is also faster for the Kroupa and Chabrier IMFs because their low-mass turnovers give them smaller mass fractions of long-lived stars. These two effects roughly cancel for $x = 1.3$, resulting in a time dependence for M/L that

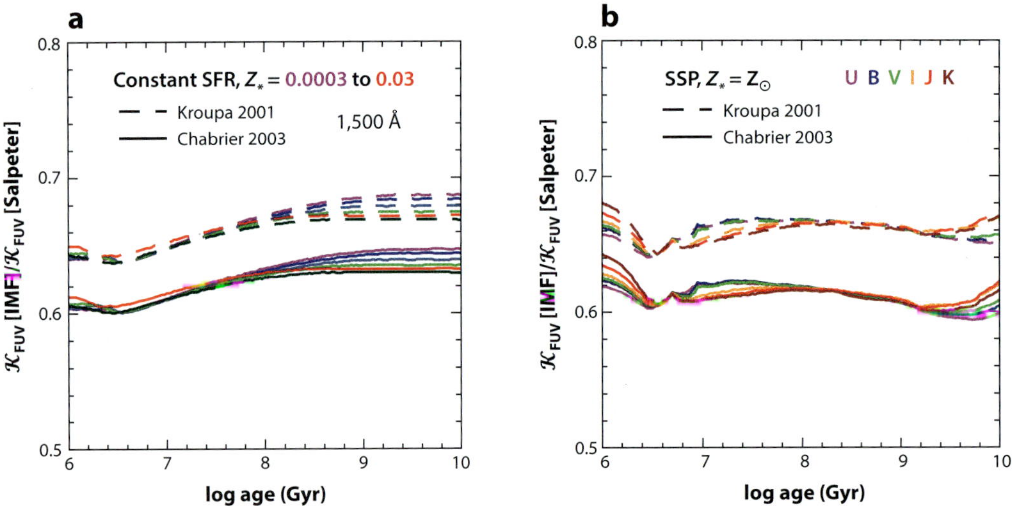

Figure 4

(*a*) The ratio of SFR/L_{FUV} conversion factors $\mathcal{K}_{FUV}$ (Equation 10) for Chabrier or Kroupa IMFs to that for a Salpeter IMF (0.1 to 100 $M_\odot$), for a constant SFR and for various metallicities computed with the FSPS models of Conroy et al. (2009). The difference in this conversion factor varies only slightly with age or metallicity over the ranges shown here. (*b*) Mass-to-light ratio for a simple stellar population as a function of age, for various bandpasses from the near UV through near IR, comparing values for Chabrier and Kroupa IMFs to Salpeter. This factor also has only a small dependence on age, bandpass, or metallicity (the last not shown here). Abbreviations: FSPS, flexible star population synthesis; IMF, initial mass function; SFR, star-formation rate.

is nearly the same as that for the Salpeter IMF. For an SSP with a "bottom-light" IMF with a Salpeter slope $x = 1.35$, the ratio of M/L compared with that for a Salpeter IMF would decrease by ~16% over ~5 Gyr, and a constant IMF rescaling factor for derived stellar masses would be inappropriate.] The dependence on metallicity (not shown) is very weak. To rescale stellar masses from Chabrier or Kroupa to Salpeter IMF, we divide by constant factors 0.61 and 0.66, respectively.

The greatest drawback for UV measurements of star formation is the obscuring effect of dust. Extinction is strong in the UV, so even modest amounts of dust can dramatically suppress the emerging UV flux. Dust re-emits the absorbed energy in the IR, which we discuss in the next section. A reliable measurement of SFRs from UV light must either correct for the effects of dust absorption or measure the absorbed energy directly through IR emission. We will return to the relation between UV dust attenuation and IR emission in Section 3.1.3 below.

3.1.2. Infrared emission. The energy that dust absorbs from the UV is reradiated at MIR and FIR wavelengths, making IR observations another important tool for measuring SFRs. The effect of dust extinction at FIR wavelengths is generally regarded as negligible, although in the MIR extinction can still be relevant for the most deeply buried star formation and active galactic nucleus (AGN). The total IR luminosity (L_{IR}, usually defined as being integrated over the wavelength range 8–1,000 μm) is a measurement of the energy that was absorbed by dust mainly at UV wavelengths. Because most UV emission comes from star formation, the IR luminosity is often interpreted as being directly proportional to the absorbed fraction of the energy from star formation. However, active nuclei can also produce strong UV emission, often in dusty environments, and may contribute to IR emission by heating dust in the torus and clouds surrounding the AGN. Older stellar populations can also heat dust that is present in the ISM of a galaxy, contributing to the FIR emission. This is important particularly for "mature" galaxies with low current SFRs in the nearby Universe: For a galaxy such as our Milky Way, perhaps half of the FIR emission comes from dust heated by older stars, not from young star-forming regions (Lonsdale Persson & Helou 1987). However, for very actively star-forming galaxies without AGN, it is generally assumed that most of the IR emission arises from new star formation. Ideally, a galaxy's total IR luminosity would be measured by fitting a dust emission model to observations at several wavelengths, hopefully spanning the peak of dust emission. In practice, however, such multiwavelength data are often unavailable, and astronomers frequently use an SED template that is often derived from observations of local galaxies to extrapolate from a single observed flux density at some MIR or FIR wavelength, not necessarily close to the dust emission peak, to a total L_{IR}. Thus, variations in the dust emission properties from galaxy to galaxy can lead to significant uncertainties in not only this bolometric correction but also the estimation of SFRs.

Arising from various components heated to different temperatures, the spectrum of dust emission is fairly complex. Most of the dust mass in a galaxy is usually in the form of relatively cold dust (15–60 K) that contributes strongly to the emission at FIR and submillimeter wavelengths (30–1,000 μm). Dust at several different temperatures may be present, including both colder grains in the ambient ISM and warmer grains in star-forming regions. Emission from still hotter, small-grain dust in star-forming regions, usually transiently heated by single photons and not in thermal equilibrium, can dominate the MIR continuum ($\lambda < 30$ μm) and may serve as a useful SFR indicator (e.g., Calzetti et al. 2007). The MIR spectral region (3–20 μm) is both spectrally and physically complex: It has strong emission bands from polycyclic aromatic hydrocarbons and absorption bands primarily from silicates. The strength of emission from polycyclic aromatic hydrocarbons can depend strongly on ISM metallicity and radiation field intensity (e.g., Engelbracht et al. 2005, 2008; Smith et al. 2007). Strong silicate absorption features are seen when the column density of dust and gas is particularly large toward obscured AGN and perhaps even nuclear starburst regions. AGN may contribute strong continuum emission from warm dust and can dominate over star formation at MIR wavelengths. By contrast, in the FIR, their role is less prominent.

The *Infrared Space Observatory* (ISO) and the *Spitzer Space Telescope* were the first telescopes with MIR sensitivities sufficient to detect galaxies at cosmological redshifts. In particular, *Spitzer* observations at 24 μm with the MIPS instrument are very sensitive and capable of detecting "normal" star-forming galaxies out to $z \approx 2$ in modest integration times. *Spitzer* is also very efficient for mapping large sky areas. It has a 24-μm beam size that is small enough (5.7 arcsec) to reliably identify faint galaxy counterparts to the IR emission. However, only a fraction of the total IR luminosity emerges in the MIR. As noted above, it is a complicated spectral region that leads to large and potentially quite uncertain bolometric corrections from the observed MIR flux to the total IR luminosity. At $z \approx 2$, where 24-μm observations sample rest-frame wavelengths around 8 μm, where the strongest polycyclic aromatic hydrocarbon bands are found, spectral templates based on local galaxies span more than an order of magnitude in the ratio $L_{IR}/L_{8\ \mu m}$ (e.g., Chary & Elbaz 2001, Dale & Helou 2002, Dale et al. 2005). More information about the type of galaxy being observed is needed to choose with confidence an appropriate template to convert the observed MIR luminosity to L_{IR} or an SFR.

The FIR thermal emission is a simpler and more direct measurement of star-formation energy. Partly owing to their large beam sizes that resulted in significant confusion and blending of sources and in difficulty localizing galaxy counterparts, ISO and *Spitzer* offer only relatively limited FIR sensitivity for deep observations. The *Herschel Space Observatory* dramatically improved such observations: Its 3.5-m mirror diameter provided a point spread function FWHM (full width half maximum) small enough to minimize confusion and to identify source counterparts in observations from 70 to 250 μm. However, at the longest wavelengths of the *Herschel* SPIRE instrument, 350 and 500 μm, confusion becomes severe. *Herschel* observations can directly detect galaxies near the peak of their FIR dust emission: Dust SEDs typically peak at 60–100 μm in the rest frame, within the range of *Herschel* observations out to $z < 4$. Temperature variations in galaxies lead to variations in the bolometric corrections for observations at a single wavelength, but these differences are much smaller than for MIR data, generally less then factors of 2.

Despite *Herschel*'s FIR sensitivity, deep *Spitzer* 24-μm observations, in general, still detect more high-z sources down to lower limiting IR luminosities or SFRs. At $z \approx 2$, the deepest *Herschel* observations only barely reach to roughly $L^*_{\rm IR}$ [the characteristic luminosity of the "knee" of the infrared luminosity function (IRLF)], leaving a large fraction of the total cosmic SFRD undetected, at least for individual sources, although stacking can be used to probe to fainter levels. Deep *Spitzer* 24-μm observations detect galaxies with SFRs several times lower, and many fields were surveyed to faint limiting fluxes at 24 μm during *Spitzer*'s cryogenic lifetime. Therefore, there is still value in trying to understand and calibrate ways to measure star formation from deep MIR data, despite the large and potentially uncertain bolometric corrections.

In practice, observations of IR-luminous galaxies detected at high redshift with both *Spitzer* and *Herschel* have demonstrated that the IR SEDs for many galaxies are well behaved and that variations can be understood at least in part. Several pre-*Herschel* studies (Daddi et al. 2007; Papovich et al. 2007; Magnelli et al. 2009, 2011) compared 24-μm observations of distant galaxies with those of other SFR tracers, including *Spitzer* FIR measurements (either individual detections or stacked averages) and radio emission. On average, the MIR to FIR flux ratios for galaxies at $z \lesssim 1.3$ match those predicted by local IR SED templates such as those of Chary & Elbaz (2001), implying that 24-μm-derived SFRs should be reliable. However, at higher redshift, $1.3 < z < 2.5$, the 24-μm fluxes were brighter than expected relative to the FIR or radio data, i.e., SFRs derived from 24-μm data using local SED templates may be systematically overestimated at $z \approx 2$. This result was upheld by early *Herschel* studies (Elbaz et al. 2010, Nordon et al. 2010). In a joint analysis of the IR SED properties of both nearby and high-redshift IR-luminous galaxies, Elbaz et al. (2011) provided an explanatory framework for these observations in terms of the distinction between a majority population of galaxies obeying a "main-sequence" correlation between their SFRs and stellar masses and a minority "starburst" population with substantially higher SFRs per unit mass (or sSFR). Locally, starburst galaxies have more compact, high surface density star-forming regions, whereas normal disk galaxies on the star-forming main sequence have star formation distributed on larger scales with lower surface density. Starbursts also have warmer average dust temperatures and a significantly larger ratio between their FIR and 8-μm rest-frame luminosities than those of the main-sequence disk galaxies. Locally, most luminous and ultraluminous infrared galaxies (LIRGs and ULIRGs, with $L_{\rm IR} > 10^{11}\ L_\odot$ and $>10^{12}\ L_\odot$, respectively) are merger-driven starbursts, but at $z \approx 2$ where the SFRs and sSFRs of galaxies are globally much larger, the majority of LIRGs and ULIRGs are "normal" main-sequence galaxies. Their IR SEDs are more similar to those of ordinary, local star-forming spiral galaxies and have smaller bolometric corrections from observed 24-μm data (rest frame $\lambda \approx 8$ μm) than those predicted by SED templates designed to match local LIRGs and ULIRGs. Elbaz et al. (2011) constructed a "universal" main-sequence SED from the ensemble of high-z *Spitzer* and *Herschel* photometry for galaxies in the Great Observatories Origins Deep Survey (GOODS) fields at $0.3 < z < 2.5$. This SED leads to consistent total IR luminosities for the large majority of galaxies over that redshift range. Although no single template can be used to accurately derive $L_{\rm IR}$ or SFR from MIR observations for all galaxies, we now have a better understanding of how this can be done on average, which may be sufficient for deriving the global redshift evolution of the IR luminosity density or its corresponding SFRD. Rodighiero et al. (2011) (see also Sargent et al. 2012) showed that starbursts (whose IR SEDs deviate significantly from those of the main-sequence population) account for only 10% of the global SFRD at $z \approx 2$. With the data now available from *Herschel* and *Spitzer*, a broad understanding of the evolving IRLF and IR luminosity density, at least at $0 < z < 2.5$, seems within reach.

MIR and FIR observations require space-based telescopes, but at submillimeter and millimeter wavelengths, observations can once again be made from the ground within certain atmospheric transmission windows. The advent of submillimeter bolometer array cameras such as SCUBA on the JCMT revolutionized the field and led to the first detections of a large population of ULIRGs at high redshift (e.g., Smail et al. 1997, Hughes et al. 1998, Barger et al. 1998). Until recently, only the most luminous high-z objects could be readily detected, but the new ALMA interferometer will improve detection sensitivities by more than an order of magnitude, albeit over small fields of view. As noted above, submillimeter observations measure emission beyond the peak of dust emission, where flux is declining steeply with wavelength in the Rayleigh-Jeans part of the SED. This leads to a negative K correction so strong that it cancels out the effects of distance: A galaxy with a given IR luminosity will have roughly constant submillimeter flux if it is observed at any redshift $1 < z < 10$. By contrast, the bolometric corrections from the observed submillimeter wavelengths to the total IR luminosities are large and depend strongly on dust temperature. This can lead to significant uncertainties interpreting submillimeter fluxes from high-redshift sources and a bias toward detecting galaxies with the coldest dust emission.

By analogy with Equation 10, we express the conversion from IR luminosity ($L_{\rm IR}$) to ongoing SFR as

$$\mathrm{SFR_{IR}} = \mathcal{K}_{\rm IR} \times L_{\rm IR}, \tag{11}$$

where $L_{\rm IR}$ is the IR luminosity integrated over the wavelength range from 8 to 1,000 μm. Here, it is assumed that the IR emission is entirely due to recent star formation, but in practice, AGN and older stars can contribute to dust heating. Furthermore, if the net dust opacity to young star-forming regions in a galaxy is not large and if a significant amount of UV radiation emerges, then the SFR derived from the IR luminosity will represent only a fraction of the total. Hence, we write $\mathrm{SFR_{IR}}$ in Equation 11 to indicate that this is only the dust-obscured component of the SFR. For this reason, some authors advocate summing the SFRs derived from the observed IR and UV luminosity densities, the latter uncorrected for extinction. Once again, we calibrate the conversion factor $\mathcal{K}_{\rm IR}$ using the FSPS models of Conroy et al. (2009), which also incorporate dust attenuation and re-emission. We assume simple foreground-screen dust attenuation from Calzetti et al. (2000), although the details of the dust absorption model matter relatively little. The luminosity integrated from 8 to 1000 μm depends only mildly on the detailed dust emission parameters (essentially, the dust temperature distribution) for a broad range of reasonable values. Because the dust luminosity is primarily reprocessed UV emission from young star formation, the conversion factor $\mathcal{K}_{\rm IR}$ also depends on the details of the SFH and on metallicity. In practice, we may expect that galaxies with substantial extinction and bolometrically dominant dust emission are unlikely to have low metallicities; here, we assume solar metallicity for our calibration. We modify Equation 11 to account for both the FUV and FIR components of star formation:

$$\mathrm{SFR_{tot}} = \mathcal{K}_{\rm FUV} L_{\rm FUV} + \mathcal{K}_{\rm IR} L_{\rm IR}, \tag{12}$$

where $L_{\rm FUV}$ is the observed FUV luminosity at 1,500 Å with no correction for extinction. We use FSPS models with a Salpeter IMF, solar metallicity, and constant SFR to compute $L_{\rm FUV}$ and $L_{\rm IR}$ as a function of age for various levels of dust attenuation; we then solve for $\mathcal{K}_{\rm IR}$. **Figure 5** shows the result of this calculation: SFR is expressed in units of $\mathrm{M_\odot}$ year^{-1}, and both the FUV and IR luminosities are expressed in solar units (with $L_{\rm FUV} = \nu L_\nu$) to display both on the same

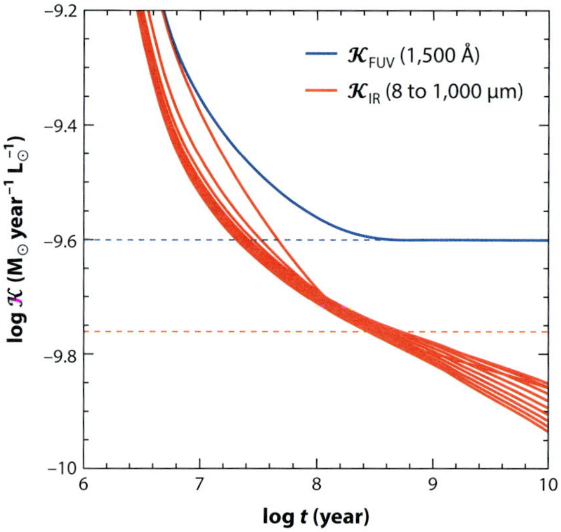

Figure 5

The SFR to luminosity ratio $\mathcal{K}$ for dusty galaxies, in solar luminosity units, calibrated with the FSPS models of Conroy et al. (2009), assuming a constant SFR, a Salpeter IMF, and solar metallicity. The blue curve shows the FUV conversion factor $\mathcal{K}_{\rm FUV}$ (see Section 3.1.1), and the blue dotted line indicates its asymptotic value. The red curves show the FIR conversion factor $\mathcal{K}_{\rm IR}$ computed as per Equation 12, for 1,500-Å extinction ranging from 0.055 to 5.5 magnitudes. The dotted red line indicates the value of $\mathcal{K}_{\rm IR}$ from Kennicutt (1998), which we also adopt here. Abbreviations: FSPS, flexible star population synthesis; FIR, far-infrared; FUV, far-UV; IMF, initial mass function; SFR, star-formation rate.

scale. As shown in Section 3.1.1, the FUV emission reaches a steady state after ~300 Myr, and for this calculation, we use the asymptotic value $\mathcal{K}_{\rm FUV} = 2.5 \times 10^{-10}\,{\rm M}_\odot$ year^{-1} ${\rm L}_\odot^{-1}$ (equivalently, $\mathcal{K}_{\rm FUV} = 1.3 \times 10^{-28}\,{\rm M}_\odot$ year^{-1} erg^{-1} s Hz). Instead, $L_{\rm IR}$ increases slowly (hence, $\mathcal{K}_{\rm IR}$ decreases) as the optical rest-frame luminosity of longer-lived stars continues to build, some fraction of which is then absorbed by dust and re-emitted. This model with constant SFR and constant dust attenuation results in a modest effect of ~0.1 dex in log $\mathcal{K}_{\rm IR}$ per dex in log t. However, in practice, older stars will likely have lower dust extinction than younger stars, thus further reducing this trend. At ages of a few 10^8 years, $\mathcal{K}_{\rm IR}$ depends very little on the total extinction. Kennicutt (1998) proposed a calibration factor $\mathcal{K}_{\rm IR} = 1.73 \times 10^{-10}\,{\rm M}_\odot$ year^{-1} ${\rm L}_\odot^{-1}$, which is fully consistent with the models shown in **Figure 5** for an age of 300 Myr. We adopt that value for this review. For luminosities measured in cgs units, we can write $\mathcal{K}_{\rm IR} = 4.5 \times 10^{-44}\,{\rm M}_\odot$ year^{-1} erg^{-1} s.

3.1.3. UV extinction and IR emission. As noted above, dust can substantially attenuate UV emission, not only compromising its utility for measuring SFRs, but also producing IR emission, which is a valuable tracer of star-formation activity. Considerable effort has been invested in understanding the physics and phenomenology of extinction in galaxies (for a review, see Calzetti 2001). In principle, the best way to account for the effect of dust attenuation is to directly measure the energy emitted at both UV and IR wavelengths, i.e., both the luminosity that escapes the galaxy directly and that which is absorbed and reradiated by dust. This provides a "bolometric" approach to measuring SFRs. In practice, however, data sensitive enough to measure FIR luminosities of high-redshift galaxies are often unavailable. *Herschel* greatly advanced these sorts of observations, but its sensitivity, although impressive, was sufficient to detect only galaxies with high SFRs > $100\,{\rm M}_\odot$ year^{-1}, at $z > 2$.

For star-forming galaxies with moderate extinction at $z > 1$, optical photometry measuring rest-frame UV light is obtained much more easily than are suitably deep FIR, submillimeter, or radio data. Current observations of UV light are also typically much more sensitive to star formation than are those at other wavelengths (**Figure 1**). As a result, trying to infer SFRs from rest-frame UV observations alone is tempting, but this requires reliable estimates of dust extinction corrections. For example, Lyman break galaxies (LBGs) are a UV-selected population of star-forming high-redshift galaxies. Their selection would favor relatively low extinction, but even LBGs are quite dusty: Reddy et al. (2012) used *Herschel* observations to determine that, on average, 80% of the FUV emission from typical ($\sim L^*_{\rm FUV}$) LBGs at $z \approx 2$ is absorbed by dust and reradiated in the FIR. Many more massive galaxies with high SFRs have greater extinction. So-called dust-obscured galaxies (Dey et al. 2008) have MIR to UV flux density ratios >1,000 (typically corresponding to $L_{\rm IR}/L_{\rm FUV} > 100$) (Penner et al. 2012) and are quite common, contributing 5–10% of the SFRD at $z \approx 2$ (Pope et al. 2008); many of these are nearly or entirely invisible in deep optical images.

Nevertheless, the widespread availability of rest-frame UV data for high-redshift galaxies encourages their use for measuring the cosmic SFH. Presently, at $z \gg 2$, there is little alternative: Even the deepest *Spitzer*, *Herschel*, radio, or submillimeter surveys can detect only the rarest and most ultraluminous galaxies at such redshifts. By contrast, deep optical and NIR surveys have now identified samples of thousands of UV-selected star-forming galaxies out to $z \approx 7$ and beyond.

Attempts to measure and correct for dust extinction in high-z galaxies have generally used the UV spectral slope (designated β) as a measure of UV reddening and have adopted empirical correlations between UV reddening and UV extinction. Calzetti et al. (1994, 2000) used ultraviolet and optical spectroscopy to derive an empirical, average dust attenuation curve for a sample of local UV-bright star-forming galaxies. Meurer et al. (1999) (later updated by Overzier et al. 2011) used UV and FIR data for a similar local sample to empirically calibrate the relation between UV reddening (β) and UV extinction (IRX $\equiv L_{\rm IR}/L_{\rm FUV}$, which can be directly related to $A_{\rm FUV}$). The reasonably tight IRX-β relation obeyed by the local UV-bright galaxies is broadly consistent with the Calzetti attenuation law, hence reinforcing its popularity. However, other local studies showed clearly that some galaxies deviate from these relations. Goldader et al. (2002) found that nearby ULIRGs deviate strongly from the Meurer IRX-β relation; these ULIRGs have very large values of IRX but often with relatively blue UV spectral index β. This was interpreted to mean that the observed UV light from local ULIRGs is relatively unreddened star formation in the host galaxy that is unrelated to the bolometrically dominant star formation, which is entirely obscured from view at UV-optical wavelengths and detected only in the FIR. Instead, observations of ordinary spiral galaxies (Kong et al. 2004, Buat et al. 2005) measured redder values of β for a given IRX. This is generally taken as evidence that light from older and less massive stars contributes significantly to the near-UV emission, leading to redder UV colors for reasons unrelated to extinction. In general, different relative distributions of stars and dust can lead to different net attenuation properties. Extinction can easily be patchy: Winds from star-forming regions can blow away dust on certain timescales, whereas other regions that are younger or more deeply embedded in the galaxy's ISM remain more heavily obscured. Dust heating also depends on geometry, leading to different distributions of dust temperatures and different emission spectra at IR and submillimeter wavelengths.

At high redshift, there are only relatively limited tests of the relation between UV reddening and extinction. Reddy & Steidel (2004) and Reddy et al. (2006, 2010, 2012) have compared various SFR tracers (including radio, *Spitzer* 24-μm, and *Herschel* 100–160-μm emission) to show that Calzetti/Meurer UV extinction laws are broadly appropriate for the majority of L_* LBGs at $z \approx 2$. However, they found evidence for systematic deviations for galaxies with the largest SFRs (>100 $M_\odot$ year^{-1}), which, similar to local ULIRGs, show "grayer" effective attenuation (i.e., less

UV reddening for their net UV extinction). They also found evidence for systematic deviations for the youngest galaxies, which show stronger reddening for their net FUV extinction, perhaps because of the metallicity or geometric effects that steepen the wavelength dependence of the UV reddening function compared with results from the Calzetti law. Assuming Calzetti attenuation, Daddi et al. (2007) and Magdis et al. (2010) also found broad consistency between UV-based and IR- or radio-based SFR measurements for samples at $z \approx 2$–3 (although, see Carilli et al. 2008). However, studies that have selected galaxies primarily on the basis of their IR emission have tended to find significant deviations from Meurer/Calzetti attenuation. In general, these deviations indicate that UV-based SFRs using Meurer/Calzetti UV slope corrections significantly underestimate total SFRs (e.g., Chapman et al. 2005, Papovich et al. 2007). Such studies have also found that differently selected populations may obey systematically different net dust attenuation relations depending on the properties of the galaxies (Buat et al. 2012, Penner et al. 2012).

Therefore, we must remain cautious about SFRs derived from UV data alone, even when estimates of UV reddening are available. Current evidence suggests that these may work well on average for UV-bright LBGs with relatively low reddening but may fail for other galaxies including the most IR-luminous objects that dominate the most rapidly star-forming galaxy population. Star formation that is obscured by too much dust, e.g., in compact starburst regions, will be unrecorded by UV observations and can be measured directly only with deep IR, submillimeter, or radio measurements.

3.1.4. Other indicators: nebular line, radio, and X-ray emission. Star formation also produces nebular line emission from excited and ionized gas in HII regions. Recombination lines of hydrogen such as Hα and Lyα are often used to measure SFRs, because they have a close relation to photoionization rates that are mainly due to intense UV radiation from OB stars. Hence, they trace massive star formation quite directly, although the presence of AGN can also contribute to these lines. Other lines from heavier elements such as [OII] 3,727 Å or [OIII] 5,007 Å have been used, but they tend to have more complex dependence on ISM conditions such as metallicity or excitation. Emission lines are also subject to absorption by dust in the star-forming regions. This is particularly true for Lyα, which is a resonance line, scattered by encounters with neutral hydrogen atoms. Such encounters can greatly increase the path length of travel for Lyα and hence increase the likelihood that it may encounter a dust grain and be absorbed. Overall, Hα is regarded as the most reliable among the easily accessible nebular SFR tracers (e.g., Moustakas et al. 2006). Weaker but less extinguished hydrogen lines in the NIR, such as Paschen α, can be very useful for measuring SFRs in dusty galaxies, but they are generally accessible only at very low redshift, although the JWST will open the possibility for measuring these for significant numbers of galaxies at cosmological distances.

Radio emission is also correlated with star formation, as SN-accelerated electrons emit nonthermal radiation at centimeter wavelengths; thermal (free-free) emission from electrons in HII regions can also contribute, particularly at higher frequencies (>5 GHz). The physics is somewhat complicated and not entirely understood, but a remarkably tight correlation is observed between radio emission and FIR emission in local galaxies spanning many orders of magnitude in luminosity (e.g., Condon 1992, Yun et al. 2001). Radio emission is free from dust extinction and thus offers a relatively unbiased tracer of star formation. However, it is difficult to obtain radio observations deep enough to detect ordinary star-forming galaxies at high redshift, although recent upgrades to the Karl G. Jansky VLA have significantly improved its sensitivity. AGN can also contribute to radio emission, occasionally dominating for radio-loud AGN (which are a minority population). Radio emission should also be suppressed at earlier cosmic epochs, as electrons should lose energy by inverse Compton scattering off microwave background photons whose energy density increases at high redshift. Recent studies have found little evidence for redshift evolution in the FIR to radio correlation (Appleton et al. 2004; Ivison et al. 2010a,b; Sargent et al. 2010a,b; Mao et al. 2011).

Even X-rays have been used to trace SFRs. X-rays are typically regarded as a quintessential signature of AGN activity in galaxies, but they are also produced by young stellar populations, notably by X-ray binaries. In the absence of an AGN, X-ray emission may be measured from individual star-forming galaxies out to $z \approx 1$ in the deepest *Chandra* fields, and stacking measurements have been used to reach fainter fluxes in studies of UV-selected galaxies, with detections at $1 < z < 4$ and upper limits at higher redshifts (Reddy & Steidel 2004; Laird et al. 2005, 2006; Lehmer et al. 2005; Basu-Zych et al. 2013). However, the proportionality between X-ray luminosity and SFR may vary with stellar population age and other parameters that could affect the mix of low- and high-mass X-ray binaries present in a galaxy; various calibrations that differ significantly have been published (e.g., Ranalli et al. 2003, Persic et al. 2004). Overall, because most of the cosmic X-ray background arises from AGN (for a review, see Brandt & Hasinger 2005), the value of using X-rays to measure the cosmic SFH seems limited (we do not discuss this method further).

3.2. "Weighing" Stellar Mass

Whereas hot young stars emit most of their energy at UV wavelengths, the cooler low-mass stars that dominate the stellar mass of a galaxy emit most of their light at red optical and NIR wavelengths. If we examine the SED of an evolving SSP from ages older than 10^9 years, the bulk of the luminosity (in λf_λ energy units) is emitted in a broad plateau between 0.4 and 2.5 μm, peaking at ~1 μm for ages >2 Gyr. (In f_ν flux density or AB magnitude units, the SED peak is at approximately 1.6 μm, where there is a minimum in the H^- opacity of cool stellar atmospheres.) The effects of dust extinction are also greatly reduced at NIR wavelengths: For Calzetti attenuation, the extinction (in magnitudes) A in the K-band is 10 times smaller than that in the V-band and 25 times smaller than that at 1,600 Å.

The luminosity, and hence the mass-to-light ratio, of a stellar population evolves very steeply with time at UV and blue wavelengths: Young stars evolve quickly off the main sequence but more slowly at red and NIR wavelengths. Therefore, observations in the NIR rest frame more closely trace the integrated stellar mass of a galaxy, but we cannot neglect the effects of evolution: The flux at 1 μm still changes by more than an order of magnitude as a stellar population ages from 0.1 to 10 Gyr (see, e.g., figure 9 from Bruzual & Charlot 2003). Therefore, we need to do more than simply measure the NIR luminosity to infer a stellar mass.

In effect, astronomers use the colors or SED of a galaxy to infer the expected mass-to-light ratio at some wavelength (preferably in the red or NIR) and then multiply the observed luminosity by M/L to estimate the stellar mass (M_*). The most common method is to fit spectral templates generated by stellar population synthesis models to broadband photometry in whatever bands are available that span rest-frame UV to NIR wavelengths, where stellar photospheric emission dominates the galaxy light. Generally speaking, researchers generate a large suite of models that span a wide range of stellar population parameters, including the past SFH, age, metallicity, and dust attenuation. The IMF is typically fixed, because there is almost no photometric signature that can usefully constrain it. The suite of models is redshifted to match a galaxy of interest. The models are then convolved by the filter bandpasses to generate synthetic broadband fluxes that are fit to the photometry, allowing the luminosity normalization to vary and minimizing χ^2 or some other likelihood parameter. The unnormalized models have a specified unit mass; therefore, normalization of the best-fitting model provides the best estimate of the galaxy's stellar mass, given the range of input parameters that were allowed.

In principle, this method can be used to constrain other stellar population parameters such as the galaxy's age, SFR, or the degree of extinction that is present. In practice, the fitting results for various parameters are often quite degenerate. For example, age, extinction, and metallicity all affect the integrated colors of a galaxy. As a result, the derived values of these parameters tend to be highly covariant: A galaxy may be red because it is old, dusty, or very metal rich. With very good photometry, particularly spanning a large range of wavelength and with many bandpasses that can more accurately sample the detailed spectral shape (e.g., measuring relatively sharp age-sensitive features such as the Balmer or 4,000-Å breaks), these constraints can be improved, but it is hard to avoid significant degeneracies. Careful practitioners may consider joint probability distributions for models that fit with acceptable likelihood. The stellar mass tends to be the best-constrained parameter, largely because the degeneracies in other parameters all tend to affect the net M/L of the model in similar ways. Redder colors from age, dust, or metallicity all tend to affect M/L to a similar (but not identical) degree. Whereas parameters such as age or reddening may be individually uncertain, the net M/L of acceptable models does not vary so much. Thus, the total mass is well constrained. Many papers have discussed stellar population modeling uncertainties in estimating galaxy masses; these are thoroughly reviewed by Conroy (2013).

Other than the choice of the IMF, the largest uncertainty that affects the derived stellar mass is usually the necessarily imperfect knowledge of the galaxy's past SFH. Fundamentally, more recently formed stars can easily outshine older stars and dominate the observed light, even at red wavelengths. The observed photometry may be dominated by the younger starlight, even though the actual galaxy mass may be dominated by older stars that are lost in the glare of the younger stars ("outshining," e.g., Papovich et al. 2001, Maraston et al. 2010) and thus have little impact on the choice of the best-fitting models. Therefore, the model fitting often underestimates the age of the galaxy or the potential contribution of older stars to the mass, and it may also underestimate the mass. If the actual SFH were well known (which is almost never the case in practice), this might not be a problem. For example, the models used to fit the photometry are often assumed to have smoothly varying SFRs, but the actual SFHs of real galaxies can be complex and nonmonotonic, fluctuating with time and perhaps punctuated by short-duration bursts. Even if very large suites of models with complex SFHs are considered, outshining tends to ensure that recently formed stars drive the model fitting, whereas the mass in older stars is poorly constrained. This effect generally leads to underestimation of galaxy stellar masses (Pforr et al. 2012). Realistically constraining the distribution of allowable past SFHs for real galaxies, especially at high redshift, remains a basic limitation when deriving stellar masses.

In practice, these SFH degeneracies are largest for galaxies with recent star formation. For galaxies that have not formed stars in a long while (say >1 Gyr) or for which the current SFR is small compared with the stellar mass (often quantified by the sSFR), the outshining is small and, thus, so is the resulting systematic uncertainty on M/L. Therefore, stellar masses for present-day elliptical galaxies, which are old with little or no ongoing star formation, or for ordinary spiral galaxies such as the Milky Way tend to be reasonably well-constrained, whereas those for very actively star-forming galaxies are less certain. As an example, Papovich et al. (2001) fit models to HST WFPC2 and NICMOS photometry for faint LBGs at $z \approx 2.5$ in the Hubble Deep Field North (HDF-N). When using models with smoothly varying SFHs, they found stellar mass uncertainties to be $\sigma(\log M_*) < 0.5$ dex, with typical uncertainties of 0.25 dex, i.e., less than a factor of 2. However, if they considered "maximal M/L" models, which allowed for as much older stellar mass as possible within the χ^2 fitting constraints formed at $z = \infty$, the masses could in principle be as much as 3 to 8 times larger. In practice, such extreme models seem unlikely. Moreover, the early work of Papovich et al. (2001) used photometry only out to the K-band or rest-frame wavelengths $\sim$6,000 Å at $z = 2.5$. Today, deep *Spitzer* IRAC photometry routinely measures fluxes for high-z galaxies at redder rest-frame wavelengths and can significantly improve stellar mass constraints. Nevertheless, even with the best data from *Spitzer* (or the JWST in the future), the effects of outshining fundamentally limit our certainty about stellar mass estimates for individual objects. These effects can be reduced only if reasonable prior assumptions can more tightly constrain the range of allowable SFHs.

Interestingly, at very high redshifts some of these SFH uncertainties are reduced, simply because the Universe is much younger. At $z > 6$, the Universe is less than 1 Gyr old, and the oldest stars in the galaxies must be younger than that; this sets a cap on M/L for a hypothetical unseen old population and thus on its possible contribution to the total stellar mass. Curtis-Lake et al. (2013) provided a recent and detailed discussion of stellar population modeling uncertainties for galaxies at $z \sim 6$.

Additionally, the practitioners who create stellar population models have not reached complete consensus: Questions regarding evolutionary tracks, the contributions of certain stellar subpopulations, and the behavior of stellar populations at low and high metallicities remain topics of debate or are poorly calibrated by observations. One widely recognized example of such uncertainties was highlighted by Maraston (2005), whose models featured significantly greater contributions of emission from thermally pulsating asymptotic giant branch (TP-AGB) stars to the red and NIR rest-frame light at SSP ages between a few hundred million years and ~2 Gyr. The enhanced red luminosity led to lower M/L at these wavelengths and redder colors, with potentially quite significant effects (factors of 2 or more) in derived stellar masses for galaxies dominated by stars in this age range. Although such populations may not dominate in most present-day galaxies, at $z \approx 2-4$ when the Universe was only a few billion years old their role must be accurately modeled to ensure proper estimates of stellar masses. Maraston et al. (2006) found that this could reduce derived stellar masses by ~60% on average for K-band-selected star-forming galaxies at $z \approx 2$ compared with results computed using the popular models of Bruzual & Charlot (2003). Although Bruzual et al. (2013) released new models in 2007 that featured enhanced TP-AGB emission, they have argued in recent conference presentations for weaker TP-AGB emission more similar to that in the older models. Given the lack of a completely satisfactory way to compute this contribution on theoretical principles, a lot hangs on the sparseness of data available to empirically calibrate the emission and evolution of TP-AGB stars.

4. TRACING THE GALAXY EMISSION HISTORY WITH LARGE SURVEYS

Over the past 18 years, a sea of published measurements of the cosmic SFRD or SMD at many different redshifts have used different data sets and methods. Much of the observational work has been carried out in deep survey fields that have accumulated outstanding multiwavelength data for this purpose and cover different angular scales to different depths (**Figure 6**). We do not attempt a comprehensive review of this literature. Instead, we highlight key data sets and analyses from roughly the first decade of cosmic SFR measurements (1996–2006) and then focus on the most important or extensive recent measurements that we use in our analysis in Section 5. Hopkins (2004) and Hopkins & Beacom (2006) provided an extensive compilation of SFRD measurements up to 2006, whereas Wilkins et al. (2008a) summarized SMD derivations through 2007. Other authors have also compiled these data more recently (e.g., Behroozi et al. 2013).

The number of papers that present measurements of the cosmic SFR history vastly exceeds the number of different data sets that have been used for this purpose because certain well-trodden surveys such as the HDF-N, the Hubble Ultradeep Field (HUDF), GOODS, and COSMOS have been used repeatedly by many groups or by the same groups who continue to refine their analyses

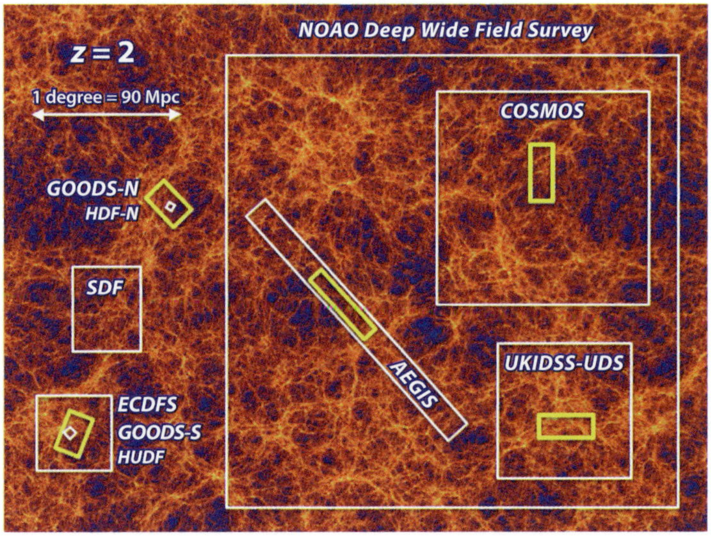

Figure 6

Relative sizes of the regions on the sky observed in several important surveys of the distant Universe. The two Great Observatories Origins Deep Survey (GOODS) fields, the Subaru Deep Field (SDF) and the Extended Chandra Deep Field South (ECDFS) are shown on the left. Very-deep surveys such as the Hubble Deep Field North (HDF-N) and the Hubble Ultradeep Field (HUDF) [Advanced Camera for Surveys (ACS) area shown], which are embedded within the GOODS fields, can detect fainter galaxies, but cover only very tiny regions on the sky. Other surveys such as the Cosmic Evolution Survey (COSMOS), the UK Infrared Deep Sky Survey (UKIDSS), the Ultradeep Survey (UDS), the All-Wavelength Extended Groth Strip International Survey (AEGIS), and the National Optical Astronomy Observatory (NOAO) Deep Wide Field Survey cover wider regions of the sky, usually to shallower depths, i.e., with less sensitivity to very faint galaxies. However, they encompass larger and perhaps more statistically representative volumes of the Universe. The yellow boxes indicate the five fields from the Cosmic Assembly Near-Infrared Deep Extragalactic Legacy Survey (CANDELS), each of which is embedded within another famous survey area. The image in the background shows a cosmological *N*-body simulation performed within the MultiDark project (see **http://www.multidark.org/MultiDark/**), viewed at $z = 2$, more than 10 Gyr ago. The colors represent the matter density distribution in a slice 43-Mpc thick, or $\Delta(z) = 0.03$ at that redshift, and all lengths are given in comoving units for $h = 0.7$. Small surveys may sample under- or overdense regions, whereas larger surveys can average over density variations, but may not be sensitive to the ordinary, relatively faint galaxies that are most numerous in the Universe. Averaging over redshift intervals that are greater than that shown in the background figure will smooth over density variations, but for any redshift binsize cosmic variance will be smaller for wider surveys or when a survey is divided into fields sampling multiple, independent sightlines.

or add new observational information. One should thus be cautious compiling results from many different studies: Although the analyses are independent, the data used and the actual galaxies measured may not be different. For example, the SFHs of GOODS-South and COSMOS are particularly well studied, but true cosmic variance due to clustering in those fields will not cancel out from one analysis to another.

4.1. UV Surveys

The largest number of analyses of the cosmic SFRD have used rest-frame UV continuum measurements, largely because the method is quite sensitive (see **Figure 1**) and can be applied over a very broad range of redshifts. Rest-frame FUV (1,500 Å) SFRs at $1.4 < z < 6$ can be measured using optical photometry that is (relatively) easily obtained with ground-based or HST imaging.

The heavy use of photometric redshifts in recent years has led to much work based on imaging data alone, with little or no spectroscopy. However, in popular deep survey fields such as GOODS or COSMOS, the photo-z values are typically well calibrated thanks to the widespread availability of thousands of spectroscopic measurements. At $z < 1$, rest-frame UV measurements ideally require space-based telescopes such as GALEX or HST to reach rest-frame wavelengths near 1,500 Å, but several studies have used blue or U-band imaging to sample mid-UV wavelengths (e.g., 2,800 Å) at $z < 1$ instead.

The modern era of SFRD measurements arguably began with the analysis of Lilly et al. (1996), who were the first to combine a large and deep (for its time) spectroscopic redshift survey with multiwavelength photometry and to derive LFs and luminosity densities at several different rest-frame wavelengths, including the rest-frame UV. The Canada-France Redshift Survey (CFRS) was carried out using the 4-m *Canada-France-Hawaii Telescope* and mainly surveyed the Universe out to $z < 1$. The available *BVIK*-band photometry permitted direct measurement of 2,800-Å rest-frame luminosities at $z > 0.5$ and down to $z \approx 0.3$ with modest spectral extrapolation. Lilly et al. (1996) found that the 2,800-Å luminosity density declined by approximately one order of magnitude from $z = 1$ to the present, which they interpreted as a steep decline in the SFRD.

Madau et al. (1996) used the then-new HDF observations to extend this analysis to much higher redshift. They employed color-selected LBG samples at $\langle z \rangle = 2.75$ and 4. The deep HST WFPC2 photometry allowed luminosities to be measured at 1,500 Å in the rest frame, reaching fainter than contemporaneous ground-based LBG data at $z \approx 3$ from Steidel et al. (1996) and thus integrating further down the LF. Madau et al. (1996) quoted only lower limits for the SFRD, without extrapolation to fainter luminosities (relatively small, given the depth of the HDF imaging) and without correction for dust absorption (significant, but at the time little known). Later analyses (e.g., Sawicki et al. 1997, Madau et al. 1998b, Steidel et al. 1999) fit Schechter LFs to the photometric samples to extrapolate to total UV luminosity densities. Connolly et al. (1997) and Pascarelle et al. (1998) combined the optical HST imaging of the HDF with ground-based NIR data to improve photometric redshift analyses in the redshift "desert" at $1 < z < 2$, between the regime of Lilly et al. (1996) and that of Madau et al. (1996). Taken together, the HDF measurements at $z > 1$ and the CFRS measurements at $z < 1$ suggested a general "rise and fall" picture of a UV luminosity density and, by inference, the cosmic SFH that peaked somewhere between $z \approx 1$ and 2.

Various surveys subsequently extended this finding using other data sets or different analyses. Several groups reanalyzed the HDF (and later the HDF-South) data in various ways or made use of deeper spectroscopic surveys with the *Keck* telescope. Cowie et al. (1999) and Wilson et al. (2002) combined *Keck* spectroscopy in several fields with deep U-band imaging to measure shorter rest-frame UV wavelengths (2,000–2,500 Å) at $z < 1$ than were probed in the CFRS analysis of Lilly et al. (1996) and derived a shallower rate of decline in the SFRD. Wolf et al. (2005) used 17-filter intermediate and broadband imaging to obtain high-quality photometric redshifts at $z < 1.25$ in the Extended Chandra Deep Field South and analyzed the 2,800-Å luminosity density evolution.

Ideally, UV rest-frame observations at $z < 1$ should be done from space telescopes to sample shorter UV wavelengths than those used in the ground-based studies by Lilly, Wilson, Wolf, and others. In early work, Treyer et al. (1998) and Sullivan et al. (2000) used the FOCA balloon-borne UV telescope to measure 2,000-Å luminosity densities at $z \approx 0.15$. GALEX has since provided vastly more FUV data, including both wide-area and deeper surveys. Wyder et al. (2005) combined GALEX all-sky imaging survey data with distances from the 2dF Galaxy Redshift Survey (2dFGRS) over 56 deg^2 to measure local ($z \approx 0.055$) LFs at 1,500 Å and 2,300 Å. Budavári et al. (2005) analyzed a similar total sky area using somewhat deeper GALEX data and SDSS-derived photometric redshifts to compute LFs and densities at $z < 0.25$. Salim et al. (2007) and

Robotham & Driver (2011) have since analyzed much larger GALEX data sets with SDSS data to cover as much as 830 deg^2. Robotham & Driver (2011) presented a straightforward derivation of the UV luminosity function (UVLF) and luminosity density, whereas Salim et al. (2007) used GALEX photometry as one ingredient to derive SFRs and the total SFRD.

Arnouts et al. (2005) combined much deeper GALEX observations with spectroscopy from the VIMOS VLT Deep Survey (VVDS) and derived 1,500-Å rest-frame LFs at $0.2 < z < 1.2$. Schiminovich et al. (2005) integrated these LFs to determine the 1,500-Å luminosity density, measuring evolution consistent with $\rho_{\rm FUV} \propto (1+z)^{2.5}$. This remains the most direct and frequently cited GALEX study of FUV luminosity densities at $z < 1$, which is somewhat surprising, as there are many more deep GALEX observations in fields with extensive spectroscopy (COSMOS, AEGIS, GOODS, etc.). Yet, to our knowledge, there have been no other published LFs. The Arnouts/Schiminovich analysis is admirable, but it used only ~1,000 galaxies with spectroscopic redshifts over the whole range $\Delta z = 1$ in a single field covering 0.5 deg^2. Hence, it may be subject to cosmic variance issues. This area is ripe for further exploitation of existing archival data.

HST is the only other modern space telescope with UV capabilities, particularly with the UVIS channel of the WFC3 camera. Until recently, only one 50-arcmin^2 field in GOODS-South had been surveyed to interesting depths to study distant galaxies (Hathi et al. 2010, Oesch et al. 2010). At $z < 1.4$, where these data measure FUV rest-frame emission, the survey volume and counting statistics are poor. These data have also been used for Lyman break color selection at $z \approx 1.5$–2. The analyses generally support relatively steep UVLFs with $\alpha < -1.5$ with large uncertainties. Expanded WFC3 UVIS observations in the HUDF and GOODS-North field have been recently completed and should improve the existing measurements. A recent WFC3 UVIS survey of gravitationally lensed galaxies behind the massive cluster Abell 1689 (Alavi et al. 2013) has been used for Lyman break color selection to unprecedentedly faint FUV luminosities at $z \approx 2$ to $M_{AB} \approx -13$, or ~1,000 times fainter than L_* at that redshift. Alavi et al. (2013) found no turnover to the LF down to those limits and measured a faint-end slope $\alpha = -1.56 \pm 0.13$.

Cucciati et al. (2012) analyzed a larger, deeper and more complete spectroscopic sample in the same VVDS survey field studied by Arnouts et al. (2005) and Schiminovich et al. (2005). They fit SED models to multiband photometry from the *U*- to the *K*-bands to extrapolate flux density measurements to FUV rest-frame wavelengths at $z < 1.4$. They demonstrated consistency with the GALEX luminosity densities from Schiminovich et al. (2005) but did not make direct use of the GALEX data. Although the largest and deepest spectroscopic sample used to derive UVLFs at $0 < z < 2$, it is based on only one sightline and requires SED extrapolation to rest frame 1,500 Å at lower redshifts. Tresse et al. (2007) presented a similar, earlier analysis using shallower spectroscopy in two VVDS fields.

Over many years, Steidel and collaborators have carried out an extensive campaign of *Keck* spectroscopy for Lyman break–selected galaxies, especially at $z \approx 2$ and 3 (Steidel et al. 2003, 2004). Their survey covers many widely spread sightlines and provides excellent control over cosmic variance. Among several LF analyses from these data, the most recent and definitive are those of Reddy et al. (2008) and Reddy & Steidel (2009). These still rely on deep photometric color-selected samples to probe the faint end of the LF, but with a degree of spectroscopic confirmation and calibration for brighter galaxies that is unmatched by any other survey. The use of UV rest-frame selection means that any LBG-based study will miss heavily dust-obscured star formation at these redshifts, but as a measure of the evolving UV luminosity density, the LBG surveys have provided the most robust method to date.

At higher redshifts, $4 < z < 7$, deep HST observations (discussed below) have dominated surveys for LBGs in recent years, but several ground-based imaging programs have made significant contributions, particularly surveying wider areas at (relatively) shallower depths to constrain the bright end of the LF. The *Subaru* telescope and its SuPrime Cam imager have been particularly important in this respect, although deep IR imaging from UKIRT and the VLT has also been used. Notable examples (not an exhaustive list) include Ouchi et al. (2004), Yoshida et al. (2006), Iwata et al. (2007), and McLure et al. (2009) at $z = 4$ to 6 and Ouchi et al. (2009), Castellano et al. (2010a,b), Bowler et al. (2012), and Tilvi et al. (2013) at $z \approx 7$.

The installation of the Advanced Camera for Surveys (ACS) enabled substantially more efficient optical HST imaging that covers fields much wider than the original HDF. ACS also offered significant gains in sensitivity at the reddest wavelengths, making Lyman break selection practical out to $z \approx 6$. Two major ACS surveys led to new derivations of the cosmic SFRD. GOODS (Giavalisco et al. 2004b) observed two independent fields with combined area >60 times larger than the HDF through four filters. This provided a sample of several thousand Lyman break candidates at $z \approx 4$ and of order 1,000 at $z \approx 5$, reaching significantly fainter than L_* and permitting robust characterization of the luminosity density. LBG selection at $z \approx 6$ was less secure from GOODS ACS data alone, as it was based on a single color $(i - z)$ and could sample only relatively bright galaxies. The HUDF (Beckwith et al. 2006) observed a single ACS pointing ($\sim$11 arcmin^2) located within the GOODS-South region with very long exposure times and reaching fainter than the original HDF and with better sensitivity at higher redshifts. Both GOODS and the HUDF have been repeatedly revisited with new observations from HST over the years to add deeper optical imaging as well as NIR data, first from NICMOS (Thompson et al. 2006, Conselice et al. 2011) and later with WFC3 in the HUDF09 and HUDF12 programs (Bouwens et al. 2011b, Ellis et al. 2013) and the Cosmic Assembly Near-Infrared Deep Extragalactic Legacy Survey (CANDELS) of several premier deep survey fields including GOODS (Grogin et al. 2011, Koekemoer et al. 2011). These IR observations make Lyman break selection at $z \approx 6$ far more robust and extend the method out to $z \approx 8$, with a handful of unconfirmed candidates identified out to $z \approx 12$.

Early analyses of the GOODS data (Giavalisco et al. 2004a) found evidence for relatively mild evolution of the UV luminosity density from $2 < z < 5$ and clear evidence that there were fewer high-luminosity galaxies at $z \approx 6$ (Stanway et al. 2003, Dickinson et al. 2004). Subsequent studies have repeatedly mined the combined GOODS+HUDF observations, using deeper data and more rigorous analyses. At least 20 papers about high-redshift LFs using WFC3 data in the HUDF and GOODS/CANDELS have been published since 2010, augmenting at least a dozen others pre-WFC3. We cannot attempt to review them all here, but they have convincingly demonstrated that the UVLF evolves significantly at $z > 4$. The current consensus is that this is primarily luminosity evolution, at least at $4 < z < 7$, with L_* brightening over time (e.g., Bouwens et al. 2007, 2012b; Grazian et al. 2011). As a result, the number density of bright LBGs increases rapidly with time, more quickly than does the integrated luminosity density. Analyses using the CANDELS and HUDF09+12 NIR imaging point to continued evolution out to $z = 8$ and perhaps beyond (Oesch et al. 2012, Yan et al. 2012, Lorenzoni et al. 2013, Schenker et al. 2013), although it would be prudent to recall that only a handful of galaxies at $z \approx 7$ have reasonably secure spectroscopic confirmation and none at $z > 7.5$. There is broad agreement that UV spectral slopes for LBGs are bluer at $z > 4$ than at lower redshifts (Bouwens et al. 2012a, Finkelstein et al. 2012a, Dunlop et al. 2013), which has implications for their dust extinction and total SFRD. Most of these studies have examined the faint-end slope of the LF, measuring very steep values, in some cases approaching or exceeding the divergent value $\alpha = -2$ (Bouwens et al. 2012b, McLure et al. 2013). Several studies have also extended SFRD analyses to $9 < z < 12$, using data from the HUDF (Bouwens et al. 2011a, Ellis et al. 2013) or from lensing cluster studies (Coe et al. 2013). Although still in flux as better data accumulate, these measurements have considerable significance for the earliest phases of galaxy evolution and for the reionization of the IGM (e.g.,

Robertson et al. 2013), but relatively little impact on the global star-formation budget of the Universe. According to current estimates, only ~1% of the cosmic SMD present today was formed at $z > 6$.

4.2. Infrared Surveys

IRAS enabled the first measurements of the local far-infrared luminosity function (FIRLF) (Lawrence et al. 1986, Soifer et al. 1987, Saunders et al. 1990, Sanders et al. 2003). These were typically measured either at 60-μm observed wavelength or using FIR luminosities integrated over a broader wavelength range. They were also generally extrapolated from the measured IRAS fluxes using fitting formulas. Although various representations of FIR luminosity have been adopted, here we consider L_{IR} as the luminosity integrated over the range 8–1,000 μm, which encompasses most of the bolometric luminosity of dust emission from nearly all sources of interest. The longest wavelength IRAS band was at 100 μm, but AKARI extended all-sky FIR measurements out to 160 μm to provide more reliable measurements of the bolometric luminosity and reduced bias against galaxies with cold dust temperatures. Goto et al. (2011a,b) reanalyzed the local IRLF incorporating AKARI data. Despite differences in detail, the results are largely consistent with previous IRAS measurements in the luminosity range of overlap. LFs have also been measured for IRAS samples selected at 12 and 25 μm (Rush et al. 1993, Shupe et al. 1998). LFs at longer IR wavelengths were measured for (rather small) local samples with ISO [90 μm (Serjeant et al. 2004); 170 μm (Takeuchi et al. 2006)], *Herschel* (250–500 μm) (Dey et al. 2010, Vaccari et al. 2010), and with ground-based (sub)millimeter observations generally for IRAS-selected samples [1.2 mm (Franceschini et al. 1998); 850 μm (Dunne et al. 2000)]. New, large-area measurements using the largest *Herschel* surveys (e.g., H-ATLAS, covering 550 deg^2 with observations from 100 to 500 μm) have not yet appeared in the literature.

Nearly all studies found that the bright end of the IRLF cuts off less sharply than does the exponential used in the Schechter function. This has typically been modeled either as a double power law (e.g., Lawrence et al. 1986, Sanders et al. 2003) or as a combined log-normal and power law (e.g., Saunders et al. 1990). Locally, the bright end of the IRLF is dominated by galaxies with warmer dust temperatures, which tend to be starburst galaxies and dusty AGN (Saunders et al. 1990). Several studies have measured a steep faint-end slope α ($\mathrm{d}N/\mathrm{d}L \propto L^{\alpha}$), e.g., $\alpha = -1.6$ (Sanders et al. 2003), $\alpha = -1.8$ (Goto et al. 2011a), and $\alpha = -1.99$ (Goto et al. 2011b). However, other studies have found flatter distributions, e.g., $\alpha = -1.2$ to -1.0 (Saunders et al. 1990, Takeuchi et al. 2003, Vaccari et al. 2010). In practice, the faint end has not been well sampled locally except in a few of the IRAS surveys. Future analysis of the widest *Herschel* surveys may help resolve this.

In local, relatively quiescent spiral galaxies such as the Milky Way, more than half of the FIR luminosity is believed to arise, not from dust in active star-forming regions, but from dust in the general ISM that is heated by ambient starlight from intermediate- and older-age stellar populations (Lonsdale Persson & Helou 1987, Sodriski et al. 1997). The luminosity of the Milky Way is typical ($L_{\mathrm{IR}} \approx 10^{10}\,\mathrm{L}_{\odot}$) (Sodriski et al. 1997) compared with the knee of the local IRLF at $L^{*}_{\mathrm{IR}} = 10^{10.5}\,\mathrm{L}_{\odot}$ (Sanders et al. 2003). This implies that a significant fraction of the local IR luminosity density is not the direct result of young star formation. Thus, it may not be a good measure of the global SFR today. At higher redshift when the specific SFR of typical galaxies was much larger and the net dust extinction to star-forming regions was, on average, larger (see Section 5.1), we may expect the IRLF and its integral to more reliably trace the total SFRD. However, Salim et al. (2009) suggested that, even at higher redshift ($z \approx 0.7$), intermediate-age stars may significantly contribute to MIR dust emission observed by *Spitzer* at 24 μm.

The deepest surveys with ISO at 15 μm detected a few hundred galaxies, mainly at $z \leq 1$, in the HDF and a few other deep survey regions where spectroscopic and photometric redshifts were available (Rowan-Robinson et al. 1997, Aussel et al. 1999, Flores et al. 1999). Analyses of these generally agreed that the emission from dusty star formation increased steeply with redshift, although statistics were generally too poor to construct redshift-dependent LFs. Chary & Elbaz (2001) used measurements from ISO and SCUBA as well as constraints from the FIR background as measured by the COBE satellite to constrain a model for the evolution of the cosmic SFH. Their model exhibited a sharp decline in the SFRD by a factor of 10 or more from $z \approx 0.8$ to the present, with a plateau of nearly constant star formation at $0.8 < z < 2$. At higher redshifts, the SFRD was more poorly constrained. Submillimeter sources placed a rough lower bound, whereas the cosmic infrared background (CIRB) set an upper limit. Acceptable solutions ranged from flat evolution to an increase by a factor of ~10 from $z = 4.5$ to 2.

Spitzer greatly enhanced the sensitivity and mapping efficiency for deep IR observations, particularly at 24 μm where the beam size (FWHM ≈ 5.7 arcsec) was small enough to enable relatively straightforward association with optical counterparts. *Spitzer* also observed in an era when very large spectroscopic redshift surveys were available or under way and when photometric redshift techniques were well established. Le Floc'h et al. (2005) produced an early, seminal analysis of 24-μm sources at $0.3 < z < 1.2$ in the Extended Chandra Deep Field South. Integrating over derived IRLFs, they inferred an evolution of the IR luminosity density proportional to $(1+z)^{3.9\pm0.4}$, significantly steeper than the evolving UV luminosity density, $\rho_{\mathrm{FUV}} \propto (1+z)^{2.5}$ (Schiminovich et al. 2005). With strong luminosity evolution, the fraction of the IR luminosity density produced by LIGs and ULIGs evolved even more steeply: Le Floc'h et al. (2005) found that galaxies with $L_{\mathrm{IR}} > 10^{11}\ \mathrm{L}_\odot$ produced 70% ± 15% of the IR luminosity density at $z \approx 1$, compared with ~5–15% today, depending on the adopted local LF. Several analyses of LFs, mainly at $z < 1$, using shallower *Spitzer* data covering significantly wider areas have also been published: These include Babbedge et al. (2006) and Rujopakarn et al. (2010) at 24 μm and Patel et al. (2013) at 70 μm and 160 μm. The work by Rujopakarn et al. (2010) is particularly notable for its combination of (relatively) wide area (9 deg^2), extensive spectroscopic redshifts (4,047 galaxies with $z \leq 0.65$), and 24-μm sensitivity (0.27 mJy, sufficient to reach $\sim L_*$ out to $z = 0.65$), making it arguably the best bridge study to date between local (IRAS and AKARI) measurements and deep-field studies at $z \geq 1$.

Several studies extended *Spitzer* 24-μm-based LF measurements to higher redshifts, $z \approx 2$ to 2.5 (Pérez-González et al. 2005, Caputi et al. 2007, Rodighiero et al. 2010). Such studies primarily use deeper 24-μm data and fainter spectroscopic and photometric redshifts available in the two GOODS fields. [Pérez-González et al. (2005) used shallower 24-μm data, whereas Caputi et al. (2007) and Rodighiero et al. (2010) employed deeper data from the GOODS team. Rodighiero et al. (2010) also incorporated relatively shallow *Spitzer* 24-μm data covering 0.85 deg^2 in one of the VVDS redshift survey fields.] Depending on the analysis, these studies all found flatter IR luminosity density evolution at higher redshifts with modestly lower or higher $\rho(L_{\mathrm{IR}})$ at $z = 2$ than at $z = 1$. All three studies also found that the characteristic IR luminosity L^*_{IR} brightened further at $z > 1$, such that ULIRGs emitted either close to 50% of the total energy density at $z \approx 2$ (Caputi et al. 2007, Rodighiero et al. 2010) or the majority of it (Pérez-González et al. 2005).

Depending on the data, methodology, and assumptions that are used, IRLFs estimated at high redshift have differed at both the faint and bright ends. At the faint end, the available data rarely constrain the slope of the LF at high redshift. Indeed, as noted above, there are significant differences in the faint-end slopes that have been measured at $z \approx 0$. Data with a limiting 24-μm flux density of 80 μJy (as found for most of the earlier *Spitzer* studies described above) reach only a factor of a few fainter than typical estimates of L^*_{IR} at $z \approx 1$, and they barely reach L^*_{IR} at $z \approx 2$. Thus, most analyses are forced to assume a faint-end slope based on measurements at lower redshifts, making them subject to large (typically >100%) and uncertain extrapolations to total IR luminosity densities.

Moreover, depending on the SED templates that are adopted, there are significant differences in the (large) extrapolations from observed MIR rest-frame measurements (e.g., 8-μm rest frame at $z = 2$) to the bolometric IR luminosity and SFR. Indeed, we expect such variations among real galaxies. Compared with several other studies, Pérez-González et al. (2005) found many more galaxies with $L_{IR} > 10^{12}\ L_{\odot}$ at $z \approx 2$, in part owing to different assumptions about these bolometric corrections. Also, different procedures to account for AGN emission (which can be particularly strong in the MIR) as well as heavy reliance on photometric redshifts may contribute to systematic issues in the IRLF, particularly at the bright end.

As we noted above (Section 3.1.2), several studies (Papovich et al. 2007, Daddi et al. 2007) stacked data at longer FIR wavelengths (e.g., 70 μm) or in radio and submillimeter data and found that standard SED templates such as those of Chary & Elbaz (2001) tend to overestimate typical bolometric corrections from observed 24-μm data for galaxies at $z \approx 2$. This suggested that true FIR measurements were needed to reliably determine luminosities and SFRs at high redshift. Huynh et al. (2007) made early measurements of the IRLF at $z < 1$ using the deepest available *Spitzer* 70-μm data in GOODS-North. The sample of detected sources was very small, but it was generally consistent with the earlier 24-μm work by Le Floc'h et al. (2005).

Magnelli et al. (2009, 2011) used comparably deep 70-μm data over a much wider area from the Far-Infrared Deep Extragalactic Legacy Survey to extend this work. In addition to counting detected sources (mostly at $z < 1.3$, given the depth of the 70-μm data), Magnelli et al. (2009, 2011) stacked 70-μm data in bins of 24-μm flux and redshift to measure empirically the average conversion between observed MIR and FIR luminosities. Compared with previous studies, they also used significantly deeper 24-μm catalogs extending down to 30 μJy in the GOODS fields. At $z < 1.3$, Magnelli et al. (2009) found that the average FIR over MIR flux ratios closely matched predictions from the template library of Chary & Elbaz (2001). They also measured LFs that were similar to previous measurements, but that extended to fainter luminosities significantly below the bend in the IRLF at $z \approx 1$ and with better statistics. At $z > 1.3$, however, Magnelli et al. (2011) confirmed previous suggestions that the average 70-μm to 24-μm flux ratios deviated systematically from the predictions of standard local SED templates. They extrapolated from the averaged 70-μm fluxes to the bolometric IRLF and found only a mild increase in L_* and the luminosity contribution of ULIRGs from $z \approx 1$ to 2. At lower redshifts, the faint-end slope was consistent with $\alpha = -1.6$ as measured for local IRAS galaxies by Sanders et al. (2003). By $z = 2$, the data reach only slightly fainter than the IRLF knee, and the slope is not constrained. However, extrapolating with a fixed slope $\alpha = -1.6$, Magnelli et al. (2011) found that the faint IRLF at $z = 2$ would be quite similar to that predicted by Reddy et al. (2008) on the basis of UV-selected galaxies and the dust absorption predicted from their UV spectral slopes.

The *Herschel Space Observatory* significantly improved sensitivity and reduced the beam size for FIR (70–500 μm) observations, and several large programs were dedicated to surveys of the most important multiwavelength-deep fields. Even the deepest *Herschel* surveys do not detect as many sources per square arcminute as are found in the deepest *Spitzer* 24-μm observations, but direct access to FIR wavelengths is invaluable for reliably estimating bolometric luminosities and SFRs at high redshift. Several analyses presented preliminary LFs out to $z = 2$ to 3 using data sets obtained early in the *Herschel* mission (Eales et al. 2010, Gruppioni et al. 2010, Lapi et al. 2011).

More extensive analyses of larger *Herschel* data sets have recently appeared in the literature. Gruppioni et al. (2013) used sources selected in *Herschel* PACS observations at 70, 100, and 160 μm in the two GOODS fields as well as shallower but wider observations of the Extended Chandra Deep Field South and COSMOS. They fit customized IR SED templates to photometry from both the PACS and SPIRE (250–500 μm) instruments and computed IR luminosities based on spectroscopic and photometric redshifts. They derived LFs out to $z < 4.2$. However, at the highest redshifts ($3 < z < 4.2$), the data are sensitive only to the rarest hyperluminous sources. Limiting their analysis to $z < 2.3$, Magnelli et al. (2013) restricted their study to the deepest 70–160-μm data available in the GOODS fields and extracted FIR fluxes to still-fainter limits at positions of 24-μm-detected sources. Note that, as is often the case, the fields analyzed and the data used in these studies overlap considerably; even if the methods of analysis are different, they cannot be considered to be fully independent. That said, in their range of overlap, the two analyses are generally consistent: They find somewhat stronger luminosity evolution at $z > 1$ than in the *Spitzer* analysis of Magnelli et al. (2011) and, hence, demonstrate a larger contribution of ULIRGs to the total IR luminosity density at $z \approx 2$ [but not as large as in some earlier *Spitzer* studies, e.g., by Pérez-González et al. (2005)]. Magnelli et al. (2011) concluded that this difference (compared with their own very similar *Spitzer* analysis) is mainly due to better determination of the total IR luminosities of galaxies using the improved *Herschel* FIR measurements. Gruppioni et al. (2013) found that the characteristic luminosity $L^*_{\rm IR}$ continued to brighten at $z > 2$, albeit at a slower rate. Neither survey reliably measured the faint-end slope of the IRLF at high redshift, and both fixed it to values derived locally. Each study adopted distinct values: Magnelli et al. (2011) used $\alpha = -1.6$, whereas Gruppioni et al. (2013) used $\alpha = -1.2$. Given these different slopes, it is striking and perhaps surprising that these two analyses derive similar values for the total IR luminosity density at redshifts $0 < z < 2$. Broadly speaking, both studies find evolution by a factor of ~6 between $z = 1.1$ and today and comparatively flat evolution at higher redshift to $\langle z \rangle = 2.05$ (Magnelli et al. 2011) and $\langle z \rangle = 2.75$ (Gruppioni et al. 2013), albeit with an increasing range of values allowed within the measurement uncertainties.

Figure 7 shows recent determinations of the IRLFs and UVLFs at $0 < z < 4$. The UVLFs in the figure show the observed luminosities uncorrected for extinction and are presented in units of solar luminosities for more direct comparison with the IRLFs. This figure illustrates several points that indicate low extinction for galaxies with lower SFRs and a significant contribution from low-luminosity galaxies to the global SFRD at high redshift.

1. Compared with the UVLFs, the IRLFs cut off less steeply at high luminosities.
2. The IRLFs extend to much higher luminosities than the UVLFs at the same redshifts, as the most actively star-forming galaxies tend to be strongly obscured by dust.
3. There is strong luminosity evolution, particularly for the IRLFs, but also in the UV, with more modest density evolution.
4. The UVLFs shown in **Figure 7** (from Cucciati et al. 2012) exhibit a trend toward steeper faint-end slopes at higher redshifts, especially for $z > 2$. Although this point lacks universal agreement, most studies do measure quite steep UVLFs at $z > 2$ and a trend toward bluer UV colors at faint luminosities.

Before *Spitzer* and *Herschel*, ground-based submillimeter bolometer arrays, especially SCUBA at JCMT, provided an essential glimpse at dusty star formation at very high redshifts. As is frequently noted, the negative *K* correction at submillimeter wavelengths approximately cancels luminosity distance dimming at $z > 1$. The flux limits of most submillimeter surveys constrain individual source detections to ultra- and hyperluminous galaxies, so that only the tip of the IRLF is sampled. In practice, the greatest limitation for deriving LFs or SFRD is identifying galaxy counterparts to submillimeter sources and measuring their redshifts. This limitation is due mainly to the large beam size of single-dish submillimeter observations, but it is also due to the fact that the optical counterparts are often very faint and sometimes invisible. Another consequence of the negative *K* correction is that substantial redshift uncertainties translate to only relatively small uncertainties in the bolometric luminosity. Hence, using radio-identified counterparts and

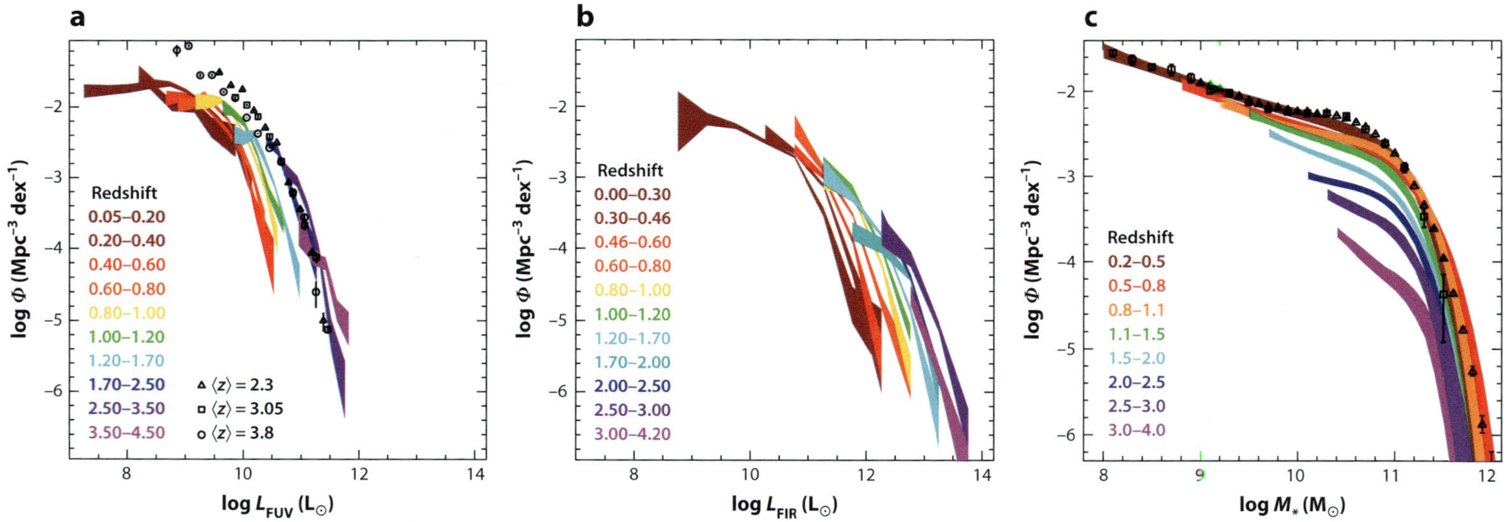

Figure 7

(*a*) Redshift evolution of the FUV luminosity function at $0 < z < 4$. The colored bands indicate the 68% confidence intervals on the space densities over the observed luminosities (uncorrected for dust attenuation), in different redshift ranges as indicated by the legend, from Cucciati et al. (2012). Data points, coded by shape, also as indicated in the legend, show the FUV luminosity functions for Lyman break galaxies at mean redshifts 2.3 and 3.05 from Reddy & Steidel (2009) and 3.8 from Bouwens et al. (2007). These luminosity functions use color selection techniques to extend the measurements to much fainter luminosities than those measured in the purely spectroscopic samples from Cucciati et al. (2012). The FUV luminosity functions at $2.3 < z < 3.8$ are observed to be quite similar. (*b*) Redshift evolution of the FIR luminosity function at $0 < z < 4$ from Gruppioni et al. (2013). The bands indicate the 68% confidence intervals at each redshift, as indicated by the color coding.
(*c*) Galaxy stellar mass function at $0 < z < 4$ for a large, deep ($K_s < 24$) sample of 220,000 galaxies, from Ilbert et al. (2013). Once again, the bands correspond to the 68% confidence intervals at each redshift, including estimated uncertainties in the derived stellar masses. The open triangles and squares correspond to the local estimates by Moustakas et al. (2013) and Baldry et al. (2012), respectively. Abbreviations: FIR, far-infrared; FUV, far-UV. Panel *c* adapted from Ilbert et al. (2013).

very rough radio-millimetric redshift estimates, Barger et al. (2000) were able to make plausible estimates of the SFRD from submillimeter sources in broad redshift bins. Barger et al. (2012) recently updated their findings using a complete submillimeter galaxy (SMG) sample with accurate interferometric positions from the submillimeter array (SMA) and with more extensive redshift information, made possible in part thanks to recent advances in millimetric CO spectroscopy. Chapman et al. (2005) measured optical spectroscopic redshifts for a sample of radio-identified SMGs and derived the first SMG LFs at $\langle z \rangle = 0.9$ and $\langle z \rangle = 2.5$. These estimates were recently updated by Wardlow et al. (2011) using (mainly) photometric redshifts for another well-defined SMG sample. These analyses demonstrated a significant contribution of dusty SMGs to the cosmic SFRD at high redshift, mainly limited to $z \lesssim 4$. However, recent discoveries of substantial numbers of SMGs out to $z = 6.3$ (e.g., Riechers et al. 2013) suggest that the dusty ultraluminous population may be important even in the first few billion years of the cosmic SFH.

Although direct FIR detection of individual sources at $z \gg 2$ is limited to the most extremely luminous objects (**Figure 1**), the CIRB provides additional constraints on dusty star formation at the highest redshifts (Pei et al. 1999, Gispert et al. 2000). FIR source counts and the CIRB were reviewed by Lagache et al. (2005) and recent results from *Herschel* are reviewed by Lutz (2014) in this volume, so we only briefly discuss this topic here. Sources directly detected by *Herschel* in the deepest observations resolve ~75% of the CIRB at 100 and 160 μm (Berta et al. 2011, Magnelli et al. 2013), albeit with significant uncertainties at 100 μm. Stacking *Herschel* data at the position of sources detected by Spitzer at 24 μm detects an even larger fraction that, with modest extrapolation, can account for the entire CIRB at these wavelengths. At 250 to 500 μm, where confusion is more severe in *Herschel* SPIRE data, the directly resolved fractions are smaller (from 15% to 5% at 250 to 500 μm), whereas stacking detects 73% to 55%. Again, an extrapolation is consistent with resolving the entire background (Béthermin et al. 2012). Because the deepest *Herschel* observations were carried out in fields such as GOODS and COSMOS with exceptional ancillary data, it is possible to stack in bins of photometric redshift to constrain the redshift distribution of the CIRB emission. At higher redshifts, the peak of dust emission from galaxies shifts to longer FIR and submillimeter wavelengths, and the fractional contribution of more-distant galaxies increases with the wavelength of the bandpass analyzed. Combining data from *Spitzer* through *Herschel* to ground-based submillimeter observations, Béthermin et al. (2012) estimated that 4.7 ± 2.0 nW m^{-2} sr^{-1}, or 17^{+11}_{-9}% of the integrated CIRB, is produced by galaxies at $z > 2$. Although the fraction is small, it still allows for a significant amount of dusty star formation to take place at $z > 2$, beyond what is seen in directly detected sources. However, the 24-μm sources used for these stacking analyses are also subject to strong *k*-correction dimming at $z > 2$. Thus, a larger fraction of the high-redshift CIRB may have been missed.

4.3. Emission Line Surveys

Among the nebular emission lines that have been most frequently used to quantify SFRs at high redshift, Hα is arguably the most useful (see Section 3.1.4). Lines from other elements, most commonly [OII] 3,727 Å, have also been used to measure the cosmic SFR (e.g., Hogg et al. 1998), but their more complex dependence on metallicity and ISM conditions as well as their larger extinction make them problematic. Narrow-band Lyα surveys are popular at high redshift, but Lyα is so strongly subject to resonant scattering and extinction that it must always be regarded as setting only a lower limit on the true SFR.

Hα is accessible with optical observations only at $z < 0.5$, and NIR observations are needed to follow it out to higher redshifts ($z \lesssim 2.5$). Recent technological developments in IR instrumentation have significantly increased the potential for such measurements. A new generation of wide-field imagers using mosaics of IR arrays is now operating on 4-m- and 8-m-class telescopes (e.g., UKIRT WFCAM, CFHT WIRCAM, NOAO NEWFIRM, VISTA VIRCAM, VLT HAWK-I), thereby significantly increasing the comoving volumes accessible for deep narrow-band imaging. NIR multiobject spectrographs are now becoming mature and efficient (e.g., *Subaru* MOIRCS and FMOS, *Keck* MOSFIRE, VLT KMOS, LBT LUCI). Slitless grism spectroscopy with WFC3 on the HST can measure faint Hα lines out to $z < 1.5$ for all objects within its field of view. Each method has advantages and disadvantages. Narrow-band imaging surveys are tuned to specific, narrow redshift ranges and are strongly subject to density variations due to line-of-sight clustering effects. Flux calibration for objects whose redshifts place emission lines in the wings of the narrow bandpasses can also be problematic. Only statistical corrections can be made for the flux contribution from [NII] or for stellar absorption. Multislit spectroscopy is subject to slit losses that complicate measurements of integrated line fluxes, and atmospheric absorption and emission significantly limit the accessible redshifts and can introduce complicated selection effects. (The new VLT KMOS spectrograph uses multiple deployable integral field units, thus eliminating slit-loss concerns, and may prove to be a valuable tool for Hα surveys.) HST WFC3 slitless spectroscopy avoids concerns about the atmosphere and slit losses, but deep observations covering adequately large solid angles are time intensive. As of this writing, results on Hα LFs from the largest ongoing programs [WISPS (Atek et al. 2010) and 3D-HST (Brammer et al. 2012)] have not appeared in the literature to supersede earlier HST NICMOS results (Yan et al. 1999, Hopkins et al. 2000, Shim et al. 2009). With all methods, reliable extinction corrections depend on the measurement of the Balmer decrement (the ratio of Hα to Hβ line fluxes). This is rarely available for narrow-band surveys, but it is sometimes available for spectroscopic samples. However, in such cases, the sample selection is often limited by the weaker Hβ line. Hence, statistical corrections are often adopted. AGN and LINERs can also contribute significantly to the samples of emission line galaxies and can be reliably taken into account only using high-quality spectroscopic data to measure line-excitation diagnostics. Most careful studies of the local galaxy population using spectroscopy from SDSS or GAMA (see below) have done this, but it is rarely possible at higher redshifts. Thus, most studies have resorted to statistical corrections or none at all.

Much literature discusses LF and SFRD measurements from Hα, [OII] and [OIII], and we note only selected works here. Gunawardhana et al. (2013) included a thorough and up-to-date compilation of these measurements at low and high redshifts. Using objective prism photographic data, Gallego et al. (1995) presented an important early study of the local Hα LF. Subsequently, the SDSS provided a vast number of spectroscopic redshifts and line-flux measurements, although the small aperture size of the spectroscopic fibers requires careful and inevitably uncertain corrections to the total emission line flux for each galaxy. SDSS spectroscopy covers both Hα and Hβ and can provide a measurement of extinction via the Balmer decrement, although care is needed to account for stellar absorption, flux limits, and selection effects. Brinchmann et al. (2004) conducted a widely cited study of local star formation from SDSS optical spectroscopy and photometry. They used a full analysis of the emission and absorption line spectroscopy. Thus, their study was not strictly based on Hα alone, although the Balmer lines carry significant weight in the SFR determinations. As noted above, Salim et al. (2007) carried out an independent SDSS analysis based mainly on photometry including GALEX UV measurements but with extensive cross comparison to the Hα data. By using photometry, the Salim analysis bypasses uncertainties inherent in the spectroscopic Hα aperture corrections. The local SFRDs (normalized to $z = 0.1$) from the studies by Brinchmann et al. (2004) and Salim et al. (2007) agree extremely well. Most recently, Gunawardhana et al. (2013) combined data from the SDSS with significantly deeper (but also much smaller) fiber spectroscopy from the GAMA survey to probe fainter down the LF and extend the analysis to higher redshifts ($z < 0.35$).

Optical spectroscopic surveys have measured Hα SFRDs typically out to $z < 0.4$ (e.g., Tresse & Maddox 1998, Sullivan et al. 2000, Westra et al. 2010, Gunawardhana et al. 2013). Until recently, NIR spectroscopy was measured only for small samples, object by object, typically at $0.5 < z < 1.1$ (Glazebrook et al. 1999, Tresse et al. 2002, Doherty et al. 2006). In general, nebular line extinction is not directly measured and can only be assumed. The same is true for corrections for stellar absorption. Larger-scale and deeper spectroscopic surveys from a new generation of multiobject IR spectrographs on 8–10-m telescopes should be forthcoming.

Several studies have exploited slitless grism spectroscopy of Hα at $0.7 < z < 1.9$ from the HST NICMOS instrument (Yan et al. 1999, Hopkins et al. 2000, Shim et al. 2009). These studies are of small but relatively unbiased samples that are observed with relatively uniform sensitivity over a broad redshift range and without concerns for flux losses due to a finite spectrograph slit size. Once again, direct measurements of extinction are generally unavailable, and the low-resolution slitless spectroscopy inextricably blends Hα with [NII], thus requiring purely statistical corrections. New, more sensitive grism surveys with HST WFC3 that cover larger solid angles are under way.

The largest number of high-redshift measurements has come from narrow-band imaging surveys using wide-field CCD cameras out to $z \approx 0.4$ and with IR arrays at higher redshifts. These are generally carried out using filters that fit into spectral windows relatively unaffected by atmospheric absorption or emission and that correspond to specific redshifts ($z = 0.24$, 0.40, 0.84, 1.47, and 2.23 are all common). Using a new generation of wide-field imagers, the current state-of-the-art Hα surveys include those of Ly et al. (2007, 2011), Hayes et al. (2010), and Sobral et al. (2013). The latter survey (HiZELS) combines data from *Subaru* Suprime-Cam ($z = 0.40$) and UKIRT WFCAM ($z = 0.84$, 1.47 and 2.23): It covers $\sim$2 deg^2 in two survey fields, with deeper but narrower (0.03 deg^2) observations at $z = 2.23$ from the VLT and its HAWK-I imager. Between 500 and 1750 Hα emitters are detected in each redshift window, thereby providing good statistics. These authors measured a steady increase in the characteristic luminosity $L_*(\mathrm{H}\alpha)$ with redshift and a faint-end slope $\alpha \approx 1.6$ that is constant with redshift and is also similar to that found in most of the UV continuum surveys.

4.4. Radio Surveys

Centimeter-wavelength radio continuum emission from star-forming galaxies arises from a combination of flat-spectrum free-free emission, which is prominent at high frequencies, and steep spectrum synchrotron emission, which dominates at lower frequencies. Although the free-free emission should be a direct tracer of star-formation rates, it has been impractical to observe this at high redshift. However, the improved high-frequency sensitivity of the Jansky VLA should open this capability in the future. The lower-frequency emission arises mainly as a consequence of SNe. As such, it is also related to the SFR. In practice, its calibration as an SFR measurement is primarily empirical, e.g., based on the tight correlation observed between radio and far-IR emission for IRAS-selected galaxies in the local Universe (e.g., Yun et al. 2001). This correlation together with insensitivity to dust extinction makes radio emission an attractive wavelength for studying star formation.

Problematically, AGN can also produce powerful radio emission. Locally, radio sources with $L_\nu(1.4\,\mathrm{GHz}) > 10^{23}$ W Hz^{-1} are predominantly AGN. This radio luminosity corresponds to an SFR $> 94\,M_\odot$ year^{-1}, using the conversion factor from Murphy et al. (2011) scaled to a Salpeter IMF. Such galaxies are extremely rare locally, so very powerful radio sources are commonly excluded as likely AGN. However, at $z > 1$, galaxies with these SFRs (corresponding to ULIRGs in terms of their IR luminosities) are common, even "normal" (see Section 3.1.2). Thus, it is problematic to disregard them simply on the basis of their radio emission. Other considerations are needed to distinguish AGN from star-forming radio sources, such as radio spectral index, radio morphology, or radio/IR flux ratios, but these are all more difficult to measure, especially for very faint sources.

The local radio LF has been used to estimate the current SFRD in many studies (for recent examples, see Machalski & Godlowski 2000, Condon et al. 2002, Sadler et al. 2002, Serjeant et al. 2002, Mauch & Sadler 2007). In these studies, the authors carefully employed various criteria to separate AGN from star-forming galaxies. Data were often compared with IRAS IR measurements and excluded objects with a radio excess relative to their IR luminosities.

At $z > 1$, even the deepest VLA surveys have been able to detect galaxies with SFRs only $\gtrsim 100\,M_\odot$ year^{-1} (**Figure 1**) and $\gtrsim 1{,}000\,M_\odot$ year^{-1} at $z > 3$, although the Jansky VLA is now pushing to fainter sensitivities. Haarsma et al. (2000) were among the first to combine very deep VLA observations with spectroscopic and photometric redshift information to derive radio LFs and the cosmic SFRD at high redshift, primarily at $0.3 < z < 0.8$. However, they also considered one very broad bin from $1 < z < 4.4$. Based on individually detected radio sources, this work was updated in later studies that took advantage of more extensive multiwavelength data to distinguish AGN from star-forming galaxies. Seymour et al. (2008) identified AGN based on radio spectral indexes, radio morphology, and radio to NIR and MIR flux ratios. Smolčić et al. (2009) used optical SED criteria to identify star-forming galaxies detected in VLA 1.4 GHz data for COSMOS at more modest redshifts $z < 1.3$. Seymour et al. (2008) assumed pure luminosity evolution for the LF, consistent with the measurements by Smolčić et al. (2009) (and earlier by Haarsma et al. 2000). All these studies found broad agreement between the radio SFRD evolution and optical and IR surveys, noting a steep decline from $z = 1$ to 0; Seymour et al. (2008) measured a peak SFRD at $z \approx 1.5$, albeit with large uncertainties.

Other studies have used radio stacking to probe to fainter luminosities below the detection limits for individual sources—particularly, stacking for NIR samples that approximate stellar mass selection. Here, the assumption is that, as in the local Universe, radio sources at fainter luminosities will primarily trace star formation with relatively little AGN contribution. Dunne et al. (2009) stacked VLA 1.4 GHz and GMRT 610 MHz data for *K*-band-selected sources in bins of redshift and *K*-band luminosity. They found a linear (but redshift-dependent) relation between radio and NIR luminosity. Using a measurement of the evolving *K*-band LF from the same data set, they then used this radio/NIR ratio to extrapolate the observations to the total radio luminosity density and SFRD. Karim et al. (2011) used a large 1.4-GHz survey of COSMOS and a *Spitzer* 3.6-μm-selected sample to carry out the most extensive study of this sort to date. Stacking in bins of stellar mass and photometric redshift and converting the mean radio fluxes to SFRs, they extensively analyzed the SFR-M_* correlation and used this and the evolving stellar mass function (SMF) (from Ilbert et al. 2010) to compute the integrated SFRD. Dunne et al. (2009) measured a steady increase in the SFRD from $z \approx 0$ to a peak at $z \approx 1.5$ that declined at higher redshift to $z \approx 4$. By contrast, Karim et al. (2011) found a monotonic decline in the SFRD from $z = 3$ to today.

4.5. Stellar Mass Density Surveys

As the technology of NIR detectors advanced, so did surveys that used NIR photometry to better sample galaxies by stellar mass at both low and high redshifts (e.g., Cowie et al. 1996, Gavazzi et al. 1996). However, it was not until the turn of the millennium that authors started to routinely use stellar population synthesis models to convert photometry and redshift information to stellar masses for large samples of galaxies, especially at high redshift. Before the era of large, modern redshift surveys such as the SDSS and the 2dFGRS, several authors made estimates of the local baryonic densities and SMDs (Persic & Salucci 1992; Fukugita et al. 1998; Salucci & Persic 1999).

This effort accelerated, however, as new spectroscopic surveys mapped the local Universe. Cole et al. (2001) used 2dFGRS redshifts and NIR photometry from 2MASS to measure the local K-band LF more accurately than had previously been possible. Then, using stellar population modeling, they inferred stellar masses from the galaxy colors and luminosities and derived the local galaxy stellar mass function (GSMF) and the local comoving SMD. Bell et al. (2003) did the same using SDSS and 2MASS. Many studies have subsequently derived the GSMF from incrementally improving SDSS data releases, using additional ancillary data and a variety of methods for stellar population modeling; mass-to-light ratios are sometimes derived from photometry and from the SDSS spectra (Panter et al. 2007, Baldry et al. 2008, Li & White 2009). Moustakas et al. (2013) incorporated photometry from the UV (GALEX) to the MIR (WISE) in their analysis of the SDSS GSMF. Baldry et al. (2012) analyzed a sample from GAMA, a wide-area spectroscopic survey extending significantly fainter than the SDSS, modeling M/L from optical SDSS photometry. As the samples have grown larger, more elaborate functional forms have been fit to the GSMF, including double and even triple Schechter functions, and there is some evidence that the GSMF slope at low masses ($<10^9\ M_\odot$) may be steeper than was previously believed (e.g., $\alpha = -1.47$) (Baldry et al. 2012). Other authors have modeled stellar populations for ensembles of galaxies rather than deriving individual galaxy stellar masses, e.g., by fitting the integrated local luminosity densities for the entire local volume from UV to NIR wavelengths (Baldry & Glazebrook 2003) or by modeling coadded SDSS spectra in bins of luminosity, velocity dispersion, and 4,000-Å break strength (Gallazzi et al. 2008).

At higher redshifts, Brinchmann & Ellis (2000) derived stellar masses for galaxies at $z < 1$ in fields with both HST and NIR imaging. As expected given the declining cosmic SFRD at those redshifts, they found relatively little evolution in the integrated mass density at $0.4 < z < 0.9$ but a marked evolution in the mass breakdown by morphological type. Cohen (2002) similarly found no significant evolution in the SMD at $0.25 < z < 1.05$. Because the cosmic SFRD declines steeply with time at $z < 1$, relatively little new stellar mass accumulates in the late stages of cosmic history. Moustakas et al. (2013) recently broke new ground with a low-resolution prism spectroscopic survey measuring ~40,000 redshifts for galaxies at $0.2 < z < 1$ in five fields with *Spitzer* IRAC photometry covering 5.5 deg^2. Even with such outstanding data, the evolution in the overall SMF at $0 < z < 1$ is nearly imperceptible, but the migration of galaxies from the star forming to the quiescent population is confirmed with exquisite detail.

Sawicki & Yee (1998), Giallongo et al. (1998), Papovich et al. (2001), and Shapley et al. (2001) pioneered the use of stellar population models to derive stellar masses for LBGs at $z \approx 2$–3. Giallongo et al. (1998) computed comoving SMDs out to $z \approx 4$ for galaxies from a relatively bright, optically selected sample, and measured steep evolution, but did not attempt to correct for unobserved galaxies fainter than the limits of their data. Subsequently, Dickinson et al. (2003), Fontana et al. (2003), and Rudnick et al. (2003) used the deepest NIR imaging then available for the two HDFs, HST NICMOS (HDF-N) and VLT ISAAC (HDF-S), together with extensive photometric and spectroscopic redshifts to derive the comoving SMD in several redshift bins out to $z \approx 3$. Strong evolution was found over that longer redshift baseline; the SMD at $z \approx 3$ measured in the range 5% to 15% of its present-day value, although a somewhat broader range of values would be permitted if systematic assumptions about the galaxy SFHs or stellar metallicities were pushed well beyond the range of models used for standard analysis. The SMD reached half its present-day value somewhere between $z = 2$ and 1. Strong cosmic variance in these small fields was also evident: At $z > 2$, red galaxies with high mass-to-light ratios were nearly absent in the HDF-N but were found in moderate abundance in the HDF-S. By contrast, the corresponding SMDs differed substantially. This indicated the importance of surveying larger fields and more sightlines, but obtaining IR imaging to satisfactory depth over these larger regions of sky has proven to be very challenging and has required another decade of effort.

The launch of *Spitzer* and the impressive performance of its IRAC camera for imaging at 3.6 to 8 μm made it possible to measure rest-frame NIR photometry for galaxies at high redshift, and major public survey imaging campaigns such as GOODS, S-COSMOS, and SWIRE produced widely accessible and heavily used imaging data sets spanning a wide range of area-depth parameter space, ideal for deriving SMFs and densities at high redshift. Indeed, NIR imaging has struggled to catch up with IRAC in terms of comparable area-depth coverage. Despite the vast swaths of telescope time that have been invested in obtaining NIR data on popular fields such as GOODS and COSMOS, imaging at 1 to 2.5 μm still tends to fall short of *Spitzer*'s sensitivity at 3.6 and 4.5 μm. At $z > 4.5$, the ground-based *K*-band samples rest-frame UV wavelengths, and IRAC offers the only viable way to measure optical rest-frame light to constrain stellar masses.

From 2006 onward, most (although not all) papers on SMFs and densities at high redshift have made use of IRAC data—often in the same survey fields that are repeatedly analyzed. Among many other papers, Fontana et al. (2006), Pérez-González et al. (2008), Kajisawa et al. (2009), and Marchesini et al. (2009) analyzed stellar masses in the GOODS fields (sometimes in combination with other data sets) for galaxies out to $z \approx 4$. Arnouts et al. (2007), Pozzetti et al. (2010), Ilbert et al. (2010), and Brammer et al. (2011) analyzed wider-area but shallower surveys (e.g., COSMOS, VVDS-SWIRE, NMBS), generally focusing on redshifts $z \leq 2$. Despite differences in their methodologies, the conclusions of these papers painted a remarkably consistent picture of the evolution of the SMF at $0 < z < 3$, with very little change in its shape, characteristic mass M_*, or faint-end slope, but with steady evolution in the characteristic density ϕ^*. There are indications that the faint-end slope of the mass function may steepen at higher redshifts (e.g., Kajisawa et al. 2009, Santini et al. 2012). The integrated SMDs measured in the different analyses generally agreed within factors of 2 at most redshifts, and a consistent picture of mass build-up emerged.

Some of the most recent additions to this literature have taken advantage of deeper, wider NIR imaging from the largest-format cameras on 4-m-class telescopes (Bielby et al. 2012, Ilbert et al. 2013, Muzzin et al. 2013) to map relatively wide survey areas such as COSMOS or the CFHT Legacy Survey fields to depths previously reserved for small, deep surveys such as GOODS. Together with ever-growing spectroscopic surveys and increasingly excellent photometric redshifts, these have yielded the most statistically robust measurements of the SMFs at $z < 2.5$. However, several of these surveys repeat analysis in COSMOS. Some even use essentially the same imaging data sets, so these analyses are not always robust against cosmic variance.

Spitzer IRAC has been essential for deriving stellar masses at $z > 4$, and very deep observations are necessary to detect typical galaxies at those redshifts. For this reason, nearly all analyses of SMDs at $z > 4$ have been carried out in GOODS and the HUDF [in a departure from the GOODS-dominated norm, McLure et al. (2009) stacked relatively shallow IRAC data for LBGs at $z = 5$ and 6 in the UKIDSS Ultra Deep Survey to measure average SEDs and mass-to-light ratios and, hence, to estimate the SMD], and nearly all have studied UV-selected LBGs, for which there are abundant samples. Early analyses of small samples of galaxies at $z = 5$ and 6, including estimates of the integrated SMD, were presented by Yan et al. (2006), Eyles et al. (2007), Stark et al. (2007), Verma et al. (2007), and Yabe et al. (2009). These were followed by larger and more systematic analyses of LBG samples at $4 \geq z \geq 7$ (Stark et al. 2009, González et al. 2011, Lee et al. 2012), all of which used similar procedures and found broadly similar results. In particular, the derived LBG SMFs have somewhat shallower low-mass slopes than do the UVLFs, because M/L_{UV} decreases at fainter UV luminosities, at least at $z = 4$ and 5 where this could be measured with some robustness from galaxies with individual IRAC detections (e.g., Lee et al. 2012). Using

HST WFC3-selected samples in the HUDF and GOODS/CANDELS fields, Labbé et al. (2013) recently extended this analysis to $z \approx 8$.

Exceptionally, some studies have used IRAC selection to avoid the potential bias of UV selection against older or dustier galaxies. Caputi et al. (2011) analyzed an IRAC 4.5-μm-selected sample in the UKIDSS Ultradeep Survey, thereby computing SMDs at $3 \leq z \leq 5$. The depth of their IRAC data limited direct detections to a stellar mass completeness limit $\gtrsim 10^{10.4}\,M_\odot$. Their extrapolated mass densities based on Schechter function fits fall somewhat below those from other LBG-based studies (e.g., González et al. 2011, Lee et al. 2012), but this is likely due to uncertainties in the large extrapolation required. Several other authors have analyzed partially or wholly IRAC-selected candidates for massive galaxies at $z \gtrsim 3.5$ (Wiklind et al. 2008, Mancini et al. 2009, Marchesini et al. 2010, Caputi et al. 2012). In some cases, they have estimated comoving SMDs, although generally without fitting SMFs and often without rigorous analysis of sample completeness.

Several studies have suggested that LBGs at $z \gtrsim 4$ have much stronger optical rest-frame nebular line emission (particularly Hα and [OIII]) than do similar UV-selected galaxies at lower redshifts (Chary et al. 2005, Raiter et al. 2010, Shim et al. 2011, Labbé et al. 2013, Stark et al. 2013). In most cases, this has been inferred on the basis of *Spitzer* IRAC colors that would be unusual for pure stellar populations but that can be explained if strong line emission boosts the IRAC fluxes. This line emission, if not taken into account, can significantly affect derived stellar population parameters and generally leads to overestimated stellar masses. For LBG samples at $z = 4$ to 8, Stark et al. (2013) and Labbé et al. (2013) estimated that average stellar masses derived from models without nebular lines should be reduced by factors from 10% to 70%, with the effect increasing at higher redshifts. Although the photometric evidence for this strong nebular emission is compelling, it will be vitally important for JWST spectroscopy to confirm and quantify its effects.

4.6. The State of the Art, and What's Wrong with It

Galaxy evolution surveys have amassed a remarkable wealth of new data in the past two decades. Newer, larger, and deeper surveys have supplanted earlier work and have provided more statistically robust measurements of the SFR and stellar mass densities at many different redshifts. Nevertheless, there are still shortcomings in even the best data or in our use of those data, which may significantly limit the accuracy or reliability of measurements of the cosmic SFH. Here we consider the strengths and weaknesses of the best measurements to date, from the local Universe out to the highest redshifts.

4.6.1. Local measurements. To be statistically meaningful, measurements of the current SFR or SMD require surveys covering a large fraction of the sky. Salim et al. (2007) and Robotham & Driver (2011) used most or all of the GALEX Medium Imaging Survey data, covering $\sim$1,000 deg^2 overlapping the SDSS and 2dFGRS spectroscopic surveys, and there is little prospect for improving the UV data in the near future. Only refinements in the analysis can be expected, e.g., incorporating improved photometric data at optical or NIR wavelengths, or further joint analysis with spectroscopic stellar population measurements. As discussed below (Section 5.1) (**Figure 8**), there is significant disagreement in the literature about the net FUV extinction correction at $z \approx 0$. Wide-area spectroscopic emission line surveys (e.g., from SDSS or GAMA) (Brinchmann et al. 2004, Gunawardhana et al. 2013) are limited by uncertain aperture corrections to line fluxes, whereas narrow-band imaging surveys have yet to cover enough galaxies over a wide enough area and are usually limited by the absence of direct measurements of extinction from the Balmer decrement, the [NII] contribution to Hα measurements, or the contribution of AGN emission. There is still room for progress in combined narrow-band plus spectroscopic data for large local

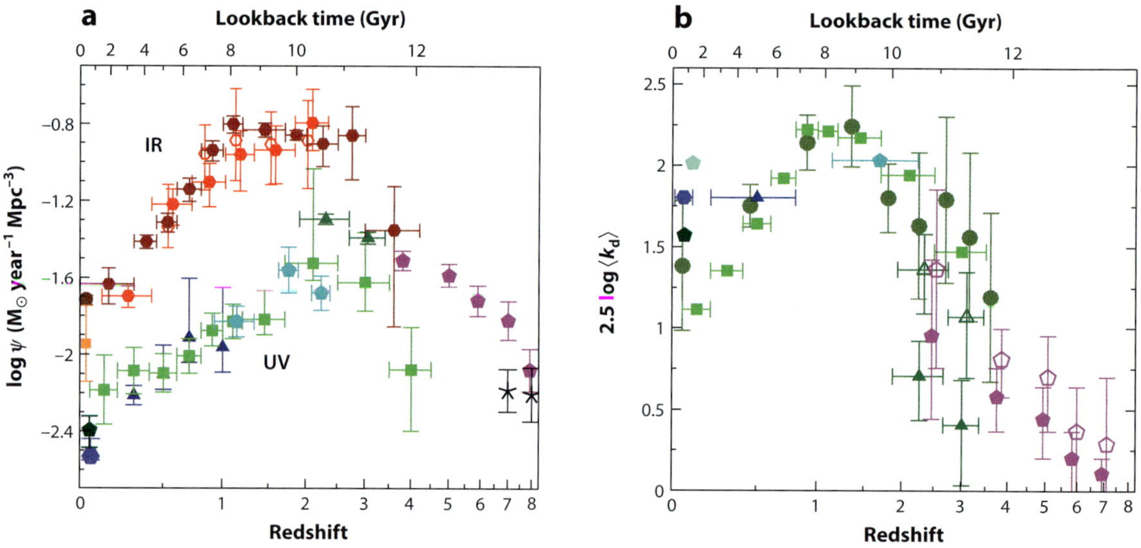

Figure 8

(*a*) SFR densities in the FUV (uncorrected for dust attenuation) and in the FIR. The data points with symbols are given in **Table 1**. All UV and IR luminosities have been converted to instantaneous SFR densities using the factors $\mathcal{K}_{\rm FUV} = 1.15 \times 10^{-28}$ and $\mathcal{K}_{\rm IR} = 4.5 \times 10^{-44}$ (cgs units) valid for a Salpeter IMF. (*b*) Mean dust attenuation in magnitudes as a function of redshift. Most of the data points shown are based on UV spectral slopes or stellar population model fitting. The symbol shapes and colors correspond to the data sets cited in **Table 1**, with the addition of Salim et al. (2007) (*cyan pentagon*). Two versions of the attenuation factors are shown for UV-selected galaxies at $2 < z < 7$ (Reddy & Steidel 2009, Bouwens et al. 2012a) (offset slightly in the redshift axis for clarity): one integrated over the observed population (*open symbols*), the other extrapolated down to $L_{\rm FUV} = 0$ (*filled symbols*). Data points from Burgarella et al. (2013) (*olive green dots*) are calculated by comparing the integrated FIR and FUV luminosity densities in redshift bins, rather than from the UV slopes or UV-optical spectral energy distributions. Abbreviations: FIR, far-infrared; FUV, far-UV; IMF, initial mass function; IR, infrared; SFR, star-formation rate.

samples. The local FIRLF has not been drastically revised since the final IRAS analyses (Sanders et al. 2003, Takeuchi et al. 2003); additional AKARI data did not drastically change earlier results (Goto et al. 2011a,b; Sedgwick et al. 2011). The biggest remaining uncertainties pertain to the faint-end slope, where measurements vary significantly from $\alpha = -1.2$ to -1.8 (or, somewhat implausibly, even -2.0) (e.g., Goto et al. 2011b). Analysis of the widest-area FIR surveys from *Herschel*, such as H-ATLAS (570 deg^2) (Eales et al. 2010), may help with this. The present uncertainties lead to a difference of a factor of at least 2 to 3 in the local FIR luminosity density. Nevertheless, as previously noted, in today's relatively "dead" epoch of cosmic star formation, a significant fraction of the FIR emission from ordinary spiral galaxies may arise from dust heated by intermediate-age and older stellar populations, not newly formed OB stars. Hence, it is not necessarily the best measure of the SFR. At higher redshifts, when the cosmic-specific SFR was much larger, new star formation should dominate dust heating, making the IR emission a more robust global tracer.

Local measurements of the SMD have relied mainly on purely optical data (e.g., SDSS photometry and spectroscopy) or on relatively shallow NIR data from 2MASS. There may still be concerns about missing light, surface brightness biases, etc., in the 2MASS data (e.g., Bell et al. 2003), and deeper very-wide-field NIR data would be helpful. All-sky MIR data from WISE may be valuable and have been used by Moustakas et al. (2013), but without extensive analysis specifically focused on this topic. Deeper NIR data covering a significant fraction of the sky, either from new wide-field ground-based NIR telescopes with large apertures or from space-based surveys with EUCLID or WFIRST, would make a valuable new contribution. Otherwise, as for UV SFRs, the most likely improvements will come from refined stellar population analyses, rather than from new data.

4.6.2. $0 < z < 1$. During the decline and fall of cosmic star formation, from $z \approx 1$ to 0, the greatest weakness of current measurements is that they have generally covered small sky areas and small co-moving volumes over few independent sightlines. Hence, they may be subject to significant cosmic variance uncertainties. Fields such as GOODS, which have been analyzed many times, are too small to offer precision measurements in fine redshift bins at $z < 1$. Even the 2 deg^2 COSMOS field subtends less than 100 Mpc at $z < 1$ and can be spanned by large-scale structure; as a single sightline, it is subject to density fluctuations. Although very good data for measuring the SFRD or SMD at $z < 1$ exist in many fields, relatively little information has been analyzed thoroughly, in part because intensive effort on spectroscopic (or even photometric) redshifts has been applied to only the few, most famous fields. Sometimes, even fields such as AEGIS, which has outstanding spectroscopy and deep GALEX, *Spitzer*, and *Herschel* data, have not been fully exploited for this purpose.

For example, very deep GALEX data exist for several of the most famous survey fields, but the one widely cited analysis of the UV luminosity density at $z < 1$ (Schiminovich et al. 2005) uses only ~1,000 sources with redshifts in a single 0.5-deg^2 field. Expanded analysis of comparably deep GALEX data in COSMOS, AEGIS, and several other survey fields with existing, extensive spectroscopy is long overdue. There are no opportunities to collect more GALEX data, but deep *U*-band imaging measuring somewhat longer rest-frame UV wavelengths may be quite adequate for many purposes. Such data exist or could be obtained with wide-field imagers, but the best analysis to date (Cucciati et al. 2012), using ~11,000 spectroscopic redshifts, is limited to a single 0.6-deg^2 sightline [indeed, the same field analyzed by Schiminovich et al. (2005) with GALEX]. Much more work can be done to improve this situation, with relatively limited new observational effort and often using data that already exist.

Similarly, most analyses of MIR and FIR data from *Spitzer* and *Herschel* data have used data from, at most, three independent sightlines (e.g., Magnelli et al. 2009, Rodighiero et al. 2010, Gruppioni et al. 2013), nearly always combining the two GOODS fields with one shallower but wider data set (e.g., COSMOS or the VVDS-SWIRE field). The widest-area analysis to date is that of Rujopakarn et al. (2010), which used 24-μm-selected sources at $z < 0.65$ from the 9-deg^2 Boötes survey. In practice, more data over more sightlines exist. Large consortium surveys such as the PEP and HerMES Herschel Guaranteed Time programs have mapped many fields, often with an elegant hierarchy of different areas and depths, but these have not yet been exploited and combined into a single, definitive analysis. Typically, this is because the ancillary imaging and spectroscopy needed to identify IR source counterparts and to determine their redshifts is available or adequate in only a few fields (hence, the repeated analyses of GOODS and COSMOS).

Many of these same comments apply to SMD estimates at $z < 1$. In this case, the state of the art has recently been improved by Moustakas et al. (2013), who analyzed five independent fields with a combined solid angle of 5.5 deg^2 with (relatively shallow) IRAC photometry and (low-resolution) spectroscopic redshifts. This is the best combination of area, depth, number of sightlines, and redshift quality for any study of the SMD at $z < 1$ and is also superior to any data used to date for SFRD studies at similar redshifts. Deeper ground-based NIR data were used by Bielby et al. (2012), who analyzed four fields covering 2.1 deg^2, and by several studies of the single 2-deg^2 COSMOS field (e.g., Ilbert et al. 2013). New wide-area surveys such as VISTA VIDEO (Jarvis 2012) (ground-based NIR, three fields, 12 deg^2) and SERVS (Mauduit et al. 2012) (*Spitzer* IRAC, five fields, 18 deg^2, overlapping VIDEO) will provide excellent new data to improve mass function estimates, but only if adequate redshift information and supporting optical photometry are available.

4.6.3. $1 < z < 4$. At $z > 1$, deep surveys are needed to probe typical (L_* and fainter) luminosities and to directly detect the majority of cosmic star formations. In principle, many fields have suitable data. However, in practice, a few survey fields have been reanalyzed many times, in part because they have the best spectroscopic and photometric redshift measurements and in part because they have the richest multiwavelength data. As such, they are magnets for studies of all kinds. In the UV, the 1,500-Å rest frame is easily probed at $z > 1.5$ using ground-based or HST optical imaging. At $z \approx 2$ and 3, the surveys of LBGs by Steidel and collaborators (e.g., Reddy & Steidel 2009) cover many independent sightlines. They offer excellent control of cosmic variance and have outstanding spectroscopic calibration. UV selection is biased against dusty star formation, but for pure measurements of the UVLF, their surveys are arguably definitive. The best direct measurements of dust emission from *Spitzer* and *Herschel* are limited to fewer fields, and still fewer have the very deep data needed to probe galaxies near L_*. As described above, most analysis has revolved around GOODS and COSMOS, and even in GOODS, the *Herschel* data only barely reach L_* at $z = 2$. The faint-end slope of the IRLF is not directly constrained by individually detected sources at $z > 2$, leading to potentially large uncertainties in any extrapolation to the bolometric IR luminosity density. With no more sensitive FIR space missions on the horizon, there is little prospect for wide-area surveys to fainter flux limits. ALMA can reach fainter submillimeter continuum limits and detect dust continuum from ordinary galaxies at very high redshifts, but only for very small solid angles. Mosaics of hundreds or even thousands of pointings would be needed to survey a field the size of GOODS. Well-designed observations targeting intelligently selected galaxy samples, rather than panoramic mapping, may be required.

Stellar masses below the characteristic mass M_* can be probed using the best ground-based NIR data and using *Spitzer* IRAC even with modest integration times. Thus, more fields have been analyzed, although GOODS and COSMOS still tend to dominate the literature. The extended *Spitzer* warm mission generated a wealth of valuable data that has yet to be fully exploited. In practice, there is uncannily good agreement between most determinations of the SMF and density at $1 < z < 4$, and it is not clear that new surveys are needed, rather than more sophisticated analysis of the existing data. However, the CANDELS HST WFC3 survey reaches fainter multiband NIR fluxes in fields that already have the deepest IRAC data and, hence, provides an important opportunity for measuring photometric redshifts and stellar masses for galaxies fainter than the limits of most studies to date. This should provide a better constraint on the slope of the SMF at low masses.

4.6.4. $z > 4$. At $z \geq 4$, the large majority of UV-based SFR and SMD measurements have been derived from HST data, mainly in the GOODS fields and the HUDF plus its deep parallel fields (also located in and around GOODS-South). Although ground-based imaging can select galaxies at $z \geq 4$, in practice the HST surveys have gone deeper, especially at the very red optical wavelengths (*I*- and *z*-bands) and the NIR wavelengths needed to select galaxies at the highest redshifts; only with these data can LF measurements probe significantly fainter than L_*. GOODS also has the deepest *Spitzer* IRAC data, essential for deriving stellar masses at $z > 4$, where even the *K*-band samples rest-frame UV wavelengths. Although the comoving volumes in the GOODS fields are significant at these large redshifts (nearly 10^6 comoving Mpc^3 at $z = 4$ for the combined GOODS fields), one may still worry about clustering and cosmic variance, given only two sightlines. The CANDELS program (Grogin et al. 2011, Koekemoer et al. 2011) is obtaining multiband optical and NIR HST imaging over five fields, each similar in size, including the two GOODS fields. This program is supplemented by very deep IRAC imaging from SEDS (Ashby et al. 2013) and

S-CANDELS (G. Fazio, research in progress) and will help with cosmic variance control and improved statistics.

At $6 < z < 8$, the HUDF data become essential to get any handle on the faint-end slope of the LF, and one must worry about the very small field size. At $z > 8$, the HUDF (or lensing clusters) are needed to detect convincing candidates. Hence, nearly all the literature on the SFRD or stellar masses at these redshifts consists of serial and parallel analyses of the same HUDF data sets as well as recent work from CLASH (Zheng et al. 2012, Coe et al. 2013). Here, the new Hubble Frontier Fields should be important. This program, to be executed from 2013 to 2016, will obtain multiband optical and NIR HST imaging with near-HUDF depth along with ultradeep *Spitzer* IRAC data for six massive galaxy clusters. Its lensing potential will magnify the faintest high-redshift background galaxies. In addition, there will be six new "blank" parallel fields that will improve statistics for unlensed sources and provide cosmic variance control for analyses that now depend on the HUDF (and its satellites) alone.

At $z > 4$, galaxies detected in ground-based NIR data (and at $z > 3$ for HST observations in the reddest WFC3-IR filters) are observed at rest-frame UV wavelengths. Hence, even with CANDELS, the HUDF, and the Frontier Fields, HST selection will be biased against quiescent or dusty galaxies. Massive galaxies with huge SFRs that are detected at submillimeter wavelengths but invisible even out to the *K*-band have been detected. Redshifts $z > 4$ have been confirmed from CO measurements (e.g., Daddi et al. 2009), but it is unknown how much they might contribute to the SFR or SMD. IRAC selection should, in principle, provide a less biased census (e.g., Mancini et al. 2009; Huang et al. 2011; Caputi et al. 2011, 2012), but spectroscopic confirmation or even photometric redshift estimates for this population may prove to be very difficult before the launch of the JWST.

5. FROM OBSERVATIONS TO GENERAL PRINCIPLES

Equation 1 was first used by Lanzetta et al. (1995) to study the chemical evolution of the damped Lyα absorption systems, where one infers the comoving rate of star formation from the observed cosmological mass density of HI; Pei & Fall (1995) then generalized it to models with inflows and outflows. Madau et al. (1996, 1998b) and Lilly et al. (1996) developed a different method where data from galaxy surveys were used to infer the SFRD $\psi(t)$ directly. This new approach relies on coupling the equations of chemical evolution to the spectrophotometric properties of the cosmic volume under consideration. The specific luminosity density at time t of a "cosmic stellar population" characterized by an SFRD $\psi(t)$ and a metal-enrichment law $Z_*(t)$ is given by the convolution integral

$$\rho_\nu(t) = \int_0^t \psi(t-\tau)\mathcal{L}_\nu[\tau, Z_*(t-\tau)]\mathrm{d}\tau, \tag{13}$$

where $\mathcal{L}_\nu[\tau, Z_*(t-\tau)]$ is the specific luminosity density radiated per unit initial stellar mass by a SSP of age τ and metallicity $Z_*(t-\tau)$. The theoretical calculation of $\mathcal{L}_\nu$ requires stellar evolutionary tracks, isochrones, and stellar atmosphere libraries. As an illustrative example of this technique, we provide in this section a current determination of the SFH of the Universe and discuss a number of possible implications.

5.1. Star-Formation Density

Rather than trying to be exhaustive, we base our modeling below on a limited number of contemporary (mostly post-2006) galaxy surveys (see **Table 1**). For the present purpose, we consider only surveys that have measured SFRs from rest-frame FUV (generally 1,500 Å) or MIR

Table 1 Determinations of the cosmic star-formation rate density from UV data (*top group*) and infrared data (*bottom group*) used in this review

Reference(s)	Redshift range	A_{FUV}[a] [mag]	log ψ[b] [$M_{\odot}$ year^{-1} Mpc^{-3}]	Symbols used in Figure 9
Wyder et al. (2005)	0.01–0.1	1.80	$-1.82^{+0.09}_{-0.02}$	Blue-gray hexagon
Schiminovich et al. (2005)	0.2–0.4	1.80	$-1.50^{+0.05}_{-0.05}$	Blue triangles
	0.4–0.6	1.80	$-1.39^{+0.15}_{-0.08}$	
	0.6–0.8	1.80	$-1.20^{+0.31}_{-0.13}$	
	0.8–1.2	1.80	$-1.25^{+0.31}_{-0.13}$	
Robotham & Driver (2011)	0.05	1.57	$-1.77^{+0.08}_{-0.09}$	Dark green pentagon
Cucciati et al. (2012)	0.05–0.2	1.11	$-1.75^{+0.18}_{-0.18}$	Green squares
	0.2–0.4	1.35	$-1.55^{+0.12}_{-0.12}$	
	0.4–0.6	1.64	$-1.44^{+0.10}_{-0.10}$	
	0.6–0.8	1.92	$-1.24^{+0.10}_{-0.10}$	
	0.8–1.0	2.22	$-0.99^{+0.09}_{-0.08}$	
	1.0–1.2	2.21	$-0.94^{+0.09}_{-0.09}$	
	1.2–1.7	2.17	$-0.95^{+0.15}_{-0.08}$	
	1.7–2.5	1.94	$-0.75^{+0.49}_{-0.09}$	
	2.5–3.5	1.47	$-1.04^{+0.26}_{-0.15}$	
	3.5–4.5	0.97	$-1.69^{+0.22}_{-0.32}$	
Dahlen et al. (2007)	0.92–1.33	2.03	$-1.02^{+0.08}_{-0.08}$	Turquoise pentagons
	1.62–1.88	2.03	$-0.75^{+0.12}_{-0.12}$	
	2.08–2.37	2.03	$-0.87^{+0.09}_{-0.09}$	
Reddy & Steidel (2009)	1.9–2.7	1.36	$-0.75^{+0.09}_{-0.11}$	Dark green triangles
	2.7–3.4	1.07	$-0.97^{+0.11}_{-0.15}$	
Bouwens et al. (2012a,b)	3.8	0.58	$-1.29^{+0.05}_{-0.05}$	Magenta pentagons
	4.9	0.44	$-1.42^{+0.06}_{-0.06}$	
	5.9	0.20	$-1.65^{+0.08}_{-0.08}$	
	7.0	0.10	$-1.79^{+0.10}_{-0.10}$	
	7.9	0.0	$-2.09^{+0.11}_{-0.11}$	
Schenker et al. (2013)	7.0	0.10	$-2.00^{+0.10}_{-0.11}$	Black crosses
Sanders et al. (2003)	0.03	NA	$-1.72^{+0.02}_{-0.03}$	Brown circle
Takeuchi et al. (2003)	0.03	NA	$-1.95^{+0.20}_{-0.20}$	Dark orange square
Magnelli et al. (2011)	0.40–0.70	NA	$-1.34^{+0.22}_{-0.11}$	Red open hexagons
	0.70–1.00	NA	$-0.96^{+0.15}_{-0.19}$	
	1.00–1.30	NA	$-0.89^{+0.27}_{-0.21}$	
	1.30–1.80	NA	$-0.91^{+0.17}_{-0.21}$	
	1.80–2.30	NA	$-0.89^{+0.21}_{-0.25}$	
Magnelli et al. (2013)	0.40–0.70	NA	$-1.22^{+0.08}_{-0.11}$	Red filled hexagons
	0.70–1.00	NA	$-1.10^{+0.10}_{-0.13}$	
	1.00–1.30	NA	$-0.96^{+0.13}_{-0.20}$	
	1.30–1.80	NA	$-0.94^{+0.13}_{-0.18}$	
	1.80–2.30	NA	$-0.80^{+0.18}_{-0.15}$	

(Continued)

Table 1 (*Continued*)

Reference(s)	Redshift range	A_{FUV}[a] [mag]	log ψ[b] [$M_\odot$ year^{-1} Mpc^{-3}]	Symbols used in Figure 9
Gruppioni et al. (2013)	0.00–0.30	NA	$-1.64^{+0.09}_{-0.11}$	Dark red filled hexagons
	0.30–0.45	NA	$-1.42^{+0.03}_{-0.04}$	
	0.45–0.60	NA	$-1.32^{+0.05}_{-0.05}$	
	0.60–0.80	NA	$-1.14^{+0.06}_{-0.06}$	
	0.80–1.00	NA	$-0.94^{+0.05}_{-0.06}$	
	1.00–1.20	NA	$-0.81^{+0.04}_{-0.05}$	
	1.20–1.70	NA	$-0.84^{+0.04}_{-0.04}$	
	1.70–2.00	NA	$-0.86^{+0.02}_{-0.03}$	
	2.00–2.50	NA	$-0.91^{+0.09}_{-0.12}$	
	2.50–3.00	NA	$-0.86^{+0.15}_{-0.23}$	
	3.00–4.20	NA	$-1.36^{+0.23}_{-0.50}$	

[a]In our notation, $A_{FUV} \equiv -2.5\log_{10}\langle k_d \rangle$.
[b]All our star-formation rate densities are based on the integration of the best-fit luminosity function parameters down to the same relative limiting luminosity, in units of the characteristic luminosity L_*, of $L_{min} = 0.03L_*$. A Salpeter initial mass function has been assumed.
Abbreviation: NA, not applicable.

and FIR measurements. Other surveys of nebular line or radio emission are also important, but they provide more limited or indirect information as discussed in previous sections (Sections 4.3 and 4.4). For the IR measurements, we emphasize surveys that make use of FIR data from *Spitzer* or *Herschel*, rather than relying on MIR (e.g., *Spitzer* 24-μm) measurements alone, owing to the complexity and lingering uncertainty over the best conversions from MIR luminosity to SFR, particularly at high redshift or high luminosity. In a few cases, we include older measurements when they are the best available, particularly for local luminosity densities from IRAS or GALEX, or GALEX-based measurements at higher redshifts that have not been updated since 2005.

For rest-frame FUV data, we use local GALEX measurements by Wyder et al. (2005) and Robotham & Driver (2011) and also include the 1,500 Å GALEX measurements at $z < 1$ from Schiminovich et al. (2005). We use the FUV luminosity densities of Cucciati et al. (2012) at $0.1 < z < 4$, noting that for $z < 1$ these are extrapolations from photometry at longer UV rest-frame wavelengths. At $1 \lesssim z \lesssim 3$ we also use FUV luminosity densities from Dahlen et al. (2007) and Reddy & Steidel (2009). At redshifts $4 \leq z \leq 8$, there are a plethora of HST-based studies, with some groups of authors repeatedly reanalyzing samples in GOODS and the HUDF as new and improved data have accumulated. We restrict our choices to a few of the most recent analyses, taking best-fit Schechter parameters (ϕ^*, L_*, α) from Bouwens et al. (2012b) and Schenker et al. (2013). For the present analysis, we stop at $z = 8$ and do not consider estimates at higher redshifts.

For local IR estimates of the SFRD, we use IRAS LFs from Sanders et al. (2003) and Takeuchi et al. (2003). At $0.4 < z < 2.3$, we include data from Magnelli et al. (2009, 2011), who used stacked *Spitzer* 70-μm measurements for 24-μm-selected sources. We also use the *Herschel* FIRLFs of Gruppioni et al. (2013) and Magnelli et al. (2013). Although both groups analyze data from the GOODS fields, Gruppioni et al. (2013) incorporate wider/shallower data from COSMOS. By contrast, Magnelli et al. (2013) include the deepest 100- and 160-μm data from GOODS-*Herschel*, extracting sources down to the faintest limits using 24-μm prior positions.

All the surveys used here provide best-fit LF parameters—generally Schechter functions for the UV data, but other functions for the IR measurements, such as double power laws or the function by Saunders et al. (1990). These allow us to integrate the LF down to the same relative limiting luminosity in units of the characteristic luminosity L_*. We adopt an integration limit $L_{\rm min} = 0.03L_*$ when computing the luminosity density $\rho_{\rm FUV}$ or $\rho_{\rm IR}$. For the case of a Schechter function, this integral is

$$\rho_{\rm FUV}(z) = \int_{0.03L_*}^{\infty} L\phi(L, z){\rm d}L = \Gamma(2 + \alpha, 0.03)\phi^* L_*. \quad (14)$$

Here α denotes the faint-end slope of the Schechter parameterization, and Γ is the incomplete gamma function. The integrated luminosity density has a strong dependence on $L_{\rm min}$ at high redshift, where the faint-end slope is measured to be very steep, i.e., $\alpha = -2.01 \pm 0.21$ at $z \sim 7$ and $\alpha = -1.91 \pm 0.32$ at $z \sim 8$ (Bouwens et al. 2011b). Slopes of $\alpha \lesssim -2$ lead to formally divergent luminosity densities. Our choice of a limiting luminosity that is 3.8 magnitudes fainter than L_*, although it samples a significant portion of the faint-end of the FUV LF, requires only a mild extrapolation (1.3 mag) from the deepest HST WFC3/IR observations (~2.5 mag beyond L_* at $z \sim 5$–8) of the HUDF (Bouwens et al. 2011b). For the IR data, we use analytic or numerical integrations depending on the LF form adopted by each reference, but the same faint-end slope considerations apply. (Note, however, that some authors use logarithmic slopes for IRLFs, which differ from the linear form used in the standard Schechter formula by $\Delta\alpha = +1$.)

Multiplying the integrated FUV and IR comoving luminosity densities by the conversion factors $\mathcal{K}_{\rm FUV}$ (Section 3.1.1) and $\mathcal{K}_{\rm IR}$ (Section 3.1.2), we obtain measurements of the "observed" UV and IR SFRDs (shown in **Figure 8**). Here, the FUV measurements are uncorrected for dust attenuation. This illustrates the now well-known result that most of the energy from star-forming galaxies at $0 < z < 2$ is absorbed and reradiated by dust; only a minority fraction emerges directly from galaxies as UV light. The gap between the UV and IR measurements increases with redshift out to at least $z \approx 1$ and then may narrow from $z = 1$ to 2. Robust measurements of the FIR luminosity density are not yet available at $z > 2.5$.

Clearly, a robust determination of dust attenuation is essential to transform FUV luminosity densities into total SFRDs. **Figure 8** shows measurements of the effective dust extinction, $\langle k_d \rangle$, as a function of redshift. This is the multiplicative factor needed to correct the observed FUV luminosity density to the intrinsic value before extinction or, equivalently, $\langle k_d \rangle = \rho_{\rm IR}/\rho_{\rm FUV} + 1$ (e.g., Meurer et al. 1999). For most of the data shown in **Figure 8**, the attenuation has been estimated from the UV spectral slopes of star-forming galaxies using the attenuation-reddening relations from Meurer et al. (1999) or Calzetti et al. (2000) or occasionally from stellar population model fitting to the full UV-optical SEDs of galaxies integrated over the observed population (e.g., Salim et al. 2007, Cucciati et al. 2012). Robotham & Driver (2011) used the empirical attenuation correction of Driver et al. (2008). We note that the estimates of UV attenuation in the local Universe span a broad range, suggesting that more work needs to be done to firmly pin down this quantity (and perhaps implying that we should be cautious about the estimates at higher redshift). Several studies of UV-selected galaxies at $z \geq 2$ (Reddy & Steidel 2009, Bouwens et al. 2012a, Finkelstein et al. 2012b) have noted strong trends for less luminous galaxies as having bluer UV spectral slopes and, hence, lower inferred dust attenuation. Because the faint-end slope of the far-UV luminosity function (FUVLF) is so steep at high redshift, a large fraction of the reddened FUV luminosity density is emitted by galaxies much fainter than L_*; this extinction-luminosity trend also implies that the net extinction for the entire population will be a function of the faint integration limit for the sample. In **Figure 8**, the points from Reddy & Steidel (2009) (at $z = 2.3$ and 3.05) and from Bouwens et al. (2012a) (at $2.5 \leq z \leq 7$) are shown for two faint-end integration limits: These are roughly down to the observed faint limit of the data, $M_{\rm FUV} < -17.5$ to -17.7 for the different redshift subsamples and extrapolated to $L_{\rm FUV} = 0$. The net attenuation for the brighter limit, which more closely represents the sample of galaxies actually observed in the study, is significantly larger than for the extrapolation—nearly two times larger for the Reddy & Steidel (2009) samples and by a lesser factor for the more distant objects from Bouwens et al. (2012a). In our analysis of the SFRDs, we have adopted the mean extinction factors inferred by each survey to correct the corresponding FUV luminosity densities.

Adopting a different approach, Burgarella et al. (2013) measured total UV attenuation from the ratio of FIR to observed (uncorrected) FUV luminosity densities (**Figure 8**) as a function of redshift, using FUVLFs from Cucciati et al. (2012) and *Herschel* FIRLFs from Gruppioni et al. (2013). At $z < 2$, these estimates agree reasonably well with the measurements inferred from the UV slope or from SED fitting. At $z > 2$, the FIR/FUV estimates have large uncertainties owing to the similarly large uncertainties required to extrapolate the observed FIRLFs to a total luminosity density. The values are larger than those for the UV-selected surveys, particularly when compared with the UV values extrapolated to very faint luminosities. Although galaxies with lower SFRs may have reduced extinction, purely UV-selected samples at high redshift may also be biased against dusty star-forming galaxies. As we noted above, a robust census for star-forming galaxies at $z \gg 2$ selected on the basis of dust emission alone does not exist, owing to the sensitivity limits of past and present FIR and submillimeter observatories. Accordingly, the total amount of star formation that is missed from UV surveys at such high redshifts remains uncertain.

Figure 9 shows the cosmic SFH from UV and IR data following the above prescriptions as well as the best-fitting function

$$\psi(z) = 0.015\frac{(1+z)^{2.7}}{1+[(1+z)/2.9]^{5.6}}\,\mathrm{M_\odot\ year^{-1}\ Mpc^{-3}}. \tag{15}$$

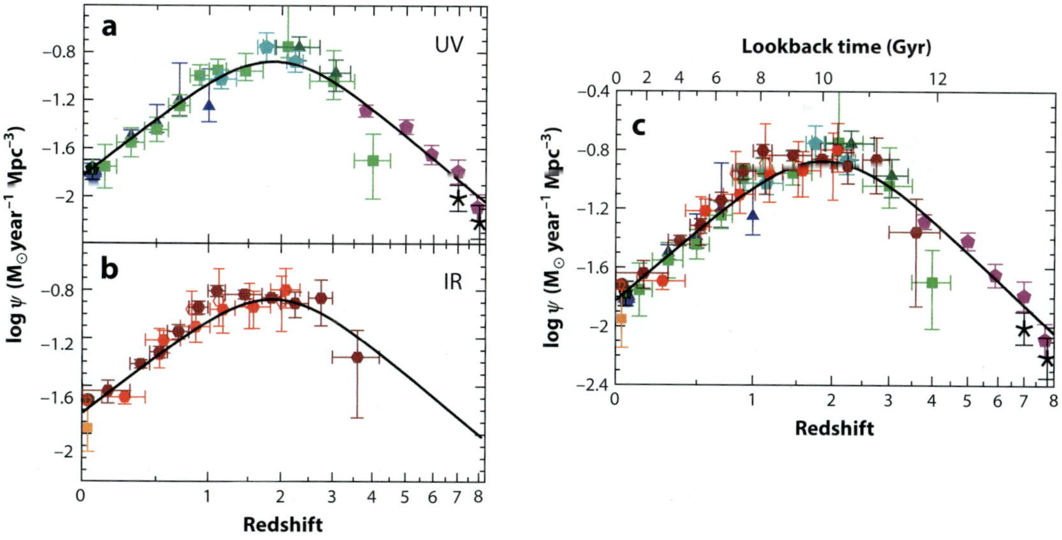

Figure 9

The history of cosmic star formation from (*a*) FUV, (*b*) IR, and (*c*) FUV+IR rest-frame measurements. The data points with symbols are given in **Table 1**. All UV luminosities have been converted to instantaneous SFR densities using the factor $\mathcal{K}_{\rm FUV} = 1.15 \times 10^{-28}$ (see Equation 10), valid for a Salpeter IMF. FIR luminosities (8–1,000 μm) have been converted to instantaneous SFRs using the factor $\mathcal{K}_{\rm IR} = 4.5 \times 10^{-44}$ (see Equation 11), also valid for a Salpeter IMF. The solid curve in the three panels plots the best-fit SFR density in Equation 15. Abbreviations: FIR, far-infrared; FUV, far-UV; IMF, initial mass function; IR, infrared; SFR, star-formation rate.

These state-of-the-art surveys provide a remarkably consistent picture of the cosmic SFH: a rising phase, scaling as $\psi(z) \propto (1+z)^{-2.9}$ at $3 \lesssim z \lesssim 8$, slowing and peaking at some point probably between $z = 2$ and 1.5, when the Universe was ~3.5 Gyr old, followed by a gradual decline to the present day, roughly as $\psi(z) \propto (1+z)^{2.7}$. The comoving SFRD at redshift 7 was approximately the same as that measured locally. The increase in $\psi(z)$ from $z \approx 8$ to 3 appears to have been steady, with no sharp drop at the highest redshifts, although there is now active debate in the literature about whether that trend continues or breaks at redshifts 9 and beyond (Coe et al. 2013, Ellis et al. 2013, Oesch et al. 2013). Although we have adopted a fitting function that is a double power law in $(1+z)$, we note that the SFRD data at $z < 1$ can also be fit quite well by an exponential decline with cosmic time and an e-folding timescale of 3.9 Gyr. Compared with the recent empirical fit to the SFRD by Behroozi et al. (2013), the function in Equation 15 reaches its peak at a slightly higher redshift, with a lower maximum value of ψ and with slightly shallower rates of change at both lower and higher redshift, and produces 20% fewer stars by $z = 0$.

We also note that each published measurement has its own approach to computing uncertainties on the SFRD and takes different random and systematic factors into account, and we have made no attempt to rationalize these here. Moreover, the published studies integrate their measurements down to different luminosity limits. We have instead adopted a fixed threshold of $0.03L_*$ to integrate the published LFs, and given the covariance on the measurements and uncertainties of LF parameters, there is no simple way for us to correct the published uncertainties to be appropriate for our adopted integration limit. Therefore, we have simply retained the fractional errors on the SFRD measurements published by each author without modification to provide an indication of the relative inaccuracy derived by each study. These should not be taken too literally, especially when there is significant difference in the faint-end slopes of LFs reported in different studies, which can lead to large differences in the integrated luminosity density. Uncertainties in the faint-end slope and the resulting extrapolations are not always fully taken into account in published error analyses, especially when LFs are fit at high redshift by fixing the slope to some value measured only at lower redshift.

5.2. Core-Collapse Supernova Rate

Because core-collapse supernovae (CC SNe) (i.e., Type II and Ibc SNe) originate from massive, short-lived stars, the rates of these events should reflect ongoing star formation and offer an independent determination of the cosmic star formation and metal production rates at different cosmological epochs (e.g., Madau et al. 1998a, Dahlen et al. 2004). Although poor statistics and dust obscuration are major liming factors for using CC SNe as a tracer of the SFH of the Universe, most derived rates are consistent with each other and increase with lookback time between $z =$ 0 and $z \sim 1$ (see **Figure 10**). The comoving volumetric SN rate is determined by multiplying Equation 15 by the efficiency of forming CC SNe

$$R_{\rm CC}(z) = \psi(z) \times \frac{\int_{m_{\rm min}}^{m_{\rm max}} \phi(m){\rm d}m}{\int_{m_l}^{m_u} m\phi(m){\rm d}m} \equiv \psi(z) \times k_{\rm CC}, \tag{16}$$

where the number of stars that explode as SNe per unit mass is $k_{\rm CC} = 0.0068\ {\rm M}_\odot^{-1}$ for a Salpeter IMF, $m_{\rm min} = 8\,{\rm M}_\odot$ and $m_{\rm max} = 40\,{\rm M}_\odot$. The predicted cosmic SN rate is shown in **Figure 10** and appears to be in good agreement with the data. The IMF dependence in Equation 16 is largely canceled out by the IMF dependence of the derived SFRD $\psi(z)$, as the stellar mass range probed by SFR indicators is comparable to the mass range of stars exploding as CC SNe. Recent comparisons between SFRs and CC SN rates have suggested a discrepancy between the two rates:

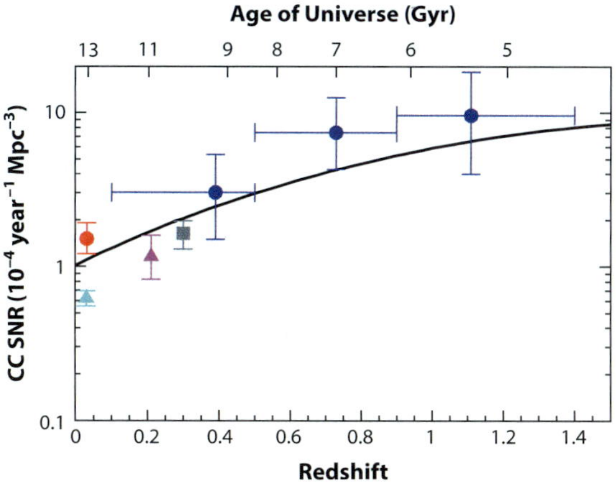

Figure 10

The cosmic core-collapse supernova rate (CC SNR). The data points are taken from Li et al. (2011) (*cyan triangle*), Mattila et al. (2012) (*red dot*), Botticella et al. (2008) (*magenta triangle*), Bazin et al. (2009) (*gray square*), and Dahlen et al. (2012) (*blue dots*). The solid line shows the rates predicted from our fit to the cosmic star-formation history. The local overdensity in star formation may boost the local rate within 10–15 Mpc of Mattila et al. (2012).

The numbers of CC SNe detected are too low by a factor of approximately 2 (Horiuchi et al. 2011). Our revised cosmic SFH does not appear to show such systematic discrepancy (see also Dahlen et al. 2012).

Observations show that at least some long-duration gamma-ray bursts (GRBs) happen simultaneously with CC SNe, but neither all SNe nor even all SNe of Type Ibc produce GRBs (for a review, see Woosley & Bloom 2006). In principle, the rate of GRBs of this class could provide a complementary estimate of the SFRD (e.g., Porciani & Madau 2001), but it is only a small fraction (<1% after correction for beaming) of the CC SN rate (Gal-Yam et al. 2006), suggesting that GRBs are an uncommon chapter in the evolution of massive stars requiring special conditions that are difficult to model. Recent studies of the GRB-SFR connection have claimed that GRBs do not trace the SFR in an unbiased way and are more frequent per unit stellar mass formed at early times (Kistler et al. 2009, Robertson & Ellis 2012, Trenti et al. 2012).

5.3. Stellar Mass Density

Figure 11 shows a compilation (see also **Table 2**) of recent (mostly post 2006) measurements of the SMD as a function of redshift (for a compilation of older data, see Wilkins et al. 2008a). We show local SDSS-based SMDs from Gallazzi et al. (2008), Li & White (2009), and Moustakas et al. (2013). Moustakas et al. (2013) also measured SMFs at $0.2 < z < 1$. However, at $z > 0.5$, their mass completeness limit is larger than $10^{9.5}$ $M_{\odot}$, so we have used their points only below that redshift. At higher redshifts (as in Moustakas et al. 2013), nearly all the modern estimates incorporate *Spitzer* IRAC photometry; we include only one recent analysis (Bielby et al. 2012) that does not but that otherwise uses excellent deep, wide-field NIR data in four independent sightlines. We also include measurements at $0.1 < z \lesssim 4$ from Arnouts et al. (2007), Pérez-González et al. (2008), Kajisawa et al. (2009), Marchesini et al. (2009), Pozzetti et al. (2010), Reddy & Steidel (2009), Ilbert et al. (2013), and Muzzin et al. (2013). We show measurements for the IRAC-selected sample of Caputi et al. (2011) at $3 \leq z \leq 5$ and for UV-selected LBG samples at $4 < z < 8$ by Yabe et al. (2009), González et al. (2011), Lee et al. (2012), and Labbé et al. (2013).

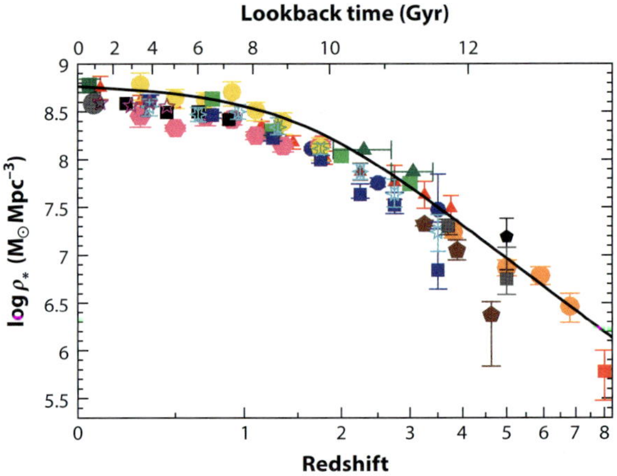

Figure 11

The evolution of the stellar mass density. The data points with symbols are given in **Table 2**. The solid line shows the global stellar mass density obtained by integrating the best-fit instantaneous star-formation rate density $\psi(z)$ (Equations 2 and 15) with a return fraction $R = 0.27$.

When needed, we have scaled from a Chabrier IMF to a Salpeter IMF by multiplying the stellar masses by a factor of 1.64 (**Figure 4**). At high redshift, authors often extrapolate their SMFs beyond the observed range by fitting a Schechter function. Stellar mass completeness at any given redshift is rarely as well defined as luminosity completeness, given the broad range of M/L values that galaxies can exhibit. Unlike the LFs used for the SFRD calculations where we have tried to impose a consistent faint luminosity limit (relative to L_*) for integration, in most cases we have simply accepted whatever low-mass limits or integral values that the various authors reported. Many authors found that the characteristic mass M_* appears to change little for $0 < z < 3$ (e.g., Fontana et al. 2006, Ilbert et al. 2013) and is roughly 10^{11} M$_\odot$ (Salpeter). Therefore, a low-mass integration limit similar to that which we used for the LFs ($L_{\rm min} = 0.03\, L_*$) would correspond to $M_{\rm min} \approx 10^{9.5}$ M$_\odot$ in that redshift range. A common but by no means universal low-mass integration limit used in the literature is $10^8\, M_\odot$. Generally, SMFs have flatter low-mass slopes than do UVLFs (and sometimes IRLFs), so the lower-mass limit makes less difference to the SMDs than it does to the SFRDs.

Our model predicts an SMD that is somewhat high (~0.2 dex on average, or 60%) compared with many, but not all, of the data at $0 < z \lesssim 3$. At $0.2 < z < 2$, our model matches the SMD measurements for the *Spitzer* IRAC-selected sample of Arnouts et al. (2007), but several other modern measurements in this redshift range from COSMOS (Pozzetti et al. 2010, Ilbert et al. 2013, Muzzin et al. 2013) fall below our curve. Carried down to $z = 0$, our model is somewhat high compared with several, but not all (Gallazzi et al. 2008), local estimates of the SMD (e.g., Cole et al. 2001, Baldry et al. 2008, Li & White 2009).

Several previous analyses (Hopkins & Beacom 2006, but see the erratum; Hopkins & Beacom 2008; Wilkins et al. 2008a) have found that the instantaneous SFH overpredicts the SMD by larger factors, up to 0.6 dex at redshift 3. We find little evidence for such significant discrepancies, although there does appear to be a net offset over a broad redshift range. Although smaller, a ~60% effect should not be disregarded. One can imagine several possible causes for this discrepancy; we consider several of them here.

SFRs may be overestimated, particularly at high redshift during the peak era of galaxy growth. For UV-based measurements, a likely culprit may be the luminosity-weighted dust corrections,

Table 2 Determinations of the cosmic stellar mass density used in this review

Reference	Redshift range	$\log \rho_*$[a] [$M_\odot$ Mpc^{-3}]	Symbols used in Figure 11
Li & White (2009)	0.07	$8.59^{+0.01}_{-0.01}$	Gray dot
Gallazzi et al. (2008)	0.005–0.22	$8.78^{+0.07}_{-0.08}$	Dark green square
Moustakas et al. (2013)	0.0–0.2	$8.59^{+0.05}_{-0.05}$	Magenta stars
	0.2–0.3	$8.56^{+0.09}_{-0.09}$	
	0.3–0.4	$8.59^{+0.06}_{-0.06}$	
	0.4–0.5	$8.55^{+0.08}_{-0.08}$	
Bielby et al. (2012)[b]	0.2–0.4	$8.46^{+0.09}_{-0.12}$	Pink filled hexagons
	0.4–0.6	$8.33^{+0.03}_{-0.03}$	
	0.6–0.8	$8.45^{+0.08}_{-0.1}$	
	0.8–1.0	$8.42^{+0.05}_{-0.06}$	
	1.0–1.2	$8.25^{+0.04}_{-0.04}$	
	1.2–1.5	$8.14^{+0.06}_{-0.06}$	
	1.5–2.0	$8.16^{+0.32}_{-0.03}$	
Pérez-González et al. (2008)	0.0–0.2	$8.75^{+0.12}_{-0.12}$	Red triangles
	0.2–0.4	$8.61^{+0.06}_{-0.06}$	
	0.4–0.6	$8.57^{+0.04}_{-0.04}$	
	0.6–0.8	$8.52^{+0.05}_{-0.05}$	
	0.8–1.0	$8.44^{+0.05}_{-0.05}$	
	1.0–1.3	$8.35^{+0.05}_{-0.05}$	
	1.3–1.6	$8.18^{+0.07}_{-0.07}$	
	1.6–2.0	$8.02^{+0.07}_{-0.07}$	
	2.0–2.5	$7.87^{+0.09}_{-0.09}$	
	2.5–3.0	$7.76^{+0.18}_{-0.18}$	
	3.0–3.5	$7.63^{+0.14}_{-0.14}$	
	3.5–4.0	$7.49^{+0.13}_{-0.13}$	
Ilbert et al. (2013)	0.2–0.5	$8.55^{+0.08}_{-0.09}$	Cyan stars
	0.5–0.8	$8.47^{+0.07}_{-0.08}$	
	0.8–1.1	$8.50^{+0.08}_{-0.08}$	
	1.1–1.5	$8.34^{+0.10}_{-0.07}$	
	1.5–2.0	$8.11^{+0.05}_{-0.06}$	
	2.0–2.5	$7.87^{+0.08}_{-0.08}$	
	2.5–3.0	$7.64^{+0.15}_{-0.14}$	
	3.0–4.0	$7.24^{+0.18}_{-0.20}$	
Muzzin et al. (2013)	0.2–0.5	$8.61^{+0.06}_{-0.06}$	Blue squares
	0.5–1.0	$8.46^{+0.03}_{-0.03}$	
	1.0–1.5	$8.22^{+0.03}_{-0.03}$	
	1.5–2.0	$7.99^{+0.05}_{-0.03}$	
	2.0–2.5	$7.63^{+0.11}_{-0.04}$	
	2.5–3.0	$7.52^{+0.13}_{-0.09}$	
	3.0–4.0	$6.84^{+0.43}_{-0.20}$	

(Continued)

Table 2 *(Continued)*

Reference	Redshift range	log ρ_*[a] $[M_\odot\ Mpc^{-3}]$	Symbols used in Figure 11
Arnouts et al. (2007)	0.3	$8.78^{+0.12}_{-0.16}$	Yellow dots
	0.5	$8.64^{+0.09}_{-0.11}$	
	0.7	$8.62^{+0.08}_{-0.10}$	
	0.9	$8.70^{+0.11}_{-0.15}$	
	1.1	$8.51^{+0.08}_{-0.11}$	
	1.35	$8.39^{+0.10}_{-0.13}$	
	1.75	$8.13^{+0.10}_{-0.13}$	
Pozzetti et al. (2010)	0.1–0.35	8.58	Black squares
	0.35–0.55	8.49	
	0.55–0.75	8.50	
	0.75–1.00	8.42	
Kajisawa et al. (2009)	0.5–1.0	8.63	Green squares
	1.0–1.5	8.30	
	1.5–2.5	8.04	
	2.5–3.5	7.74	
Marchesini et al. (2009)	1.3–2.0	$8.11^{+0.02}_{-0.02}$	Blue dots
	2.0–3.0	$7.75^{+0.05}_{-0.04}$	
	3.0–4.0	$7.47^{+0.37}_{-0.13}$	
Reddy et al. (2012)	1.9–2.7	$8.10^{+0.03}_{-0.03}$	Dark green triangles
	2.7–3.4	$7.87^{+0.03}_{-0.03}$	
Caputi et al. (2011)	3.0–3.5	$7.32^{+0.04}_{-0.02}$	Brown pentagons
	3.5–4.25	$7.05^{+0.11}_{-0.10}$	
	4.25–5.0	$6.37^{+0.14}_{-0.54}$	
González et al. (2011)[c]	3.8	$7.24^{+0.06}_{-0.06}$	Orange dots
	5.0	$6.87^{+0.08}_{-0.09}$	
	5.9	$6.79^{+0.09}_{-0.09}$	
	6.8	$6.46^{+0.14}_{-0.17}$	
Lee et al. (2012)[c]	3.7	$7.30^{+0.07}_{-0.09}$	Gray squares
	5.0	$6.75^{+0.33}_{-0.16}$	
Yabe et al. (2009)	5.0	$7.19^{+0.19}_{-0.35}$	Small black pentagon
Labbé et al. (2013)	8.0	$5.78^{+0.22}_{-0.30}$	Red square

[a]All the stellar mass densities have been derived assuming a Salpeter initial mass function.
[b]Stellar mass densities were computed by averaging over the four fields studied by Bielby et al. (2012).
[c]Following Stark et al. (2013), the mass densities of González et al. (2011) and Lee et al. (2012) at $z \simeq 4$, 5, 6, and 7 have been reduced by the factor 1.1, 1.3, 1.6, and 2.4, respectively, to account for contamination by nebular emission lines.

which could be too large, although it is often asserted that UV data are likely to underestimate SFRs in very dusty, luminous galaxies. IR-based SFRs may be overestimated and indeed were overestimated for some high-redshift galaxies in earlier *Spitzer* studies, although this seems less likely now in the era of deep *Herschel* FIR measurements. It seems more plausible that the SFRs inferred for individual galaxies may be correct on average but that the LF extrapolations could be too large. Many authors adopt fairly steep ($\alpha \geq -1.6$) faint-end slopes to both the UVLFs and IRLFs for distant galaxies. For the UVLFs, the best modern data constrain these slopes quite well, but in the IR current measurements are not deep enough to do so. However, although these extrapolations may be uncertain, the good agreement between the current best estimates of the UV- and IR-based SFRDs at $0 < z < 2.5$ (**Figure 9**) does not clearly point to a problem in either one.

Instead, stellar masses or their integrated SMD may be systematically underestimated. This is not implausible, particularly for star-forming galaxies, where the problem of recent star formation "outshining" older high-M/L stars is well known (see Section 5.1). By analyzing mock catalogs of galaxies drawn from simulations with realistic (and complex) SFHs, Pforr et al. (2012) found that the simplifying assumptions that are typically made when modeling stellar masses for real surveys generally lead to systematically underestimated stellar masses at all redshifts. That said, other systematic effects can work in the opposite direction and lead to mass overestimates, e.g., the effects of TP-AGB stars on the red and NIR light if these are not correctly modeled (Maraston 2005). A steeper low-mass slope to the GSMF could also increase the total SMD. This has been suggested even at $z = 0$, where mass functions have been measured with seemingly great precision and dynamic range (e.g., Baldry et al. 2008). At high redshift, most studies to date have found relatively flat low-mass SMF slopes, but galaxy samples may be incomplete (and photometric redshift estimates poor) for very faint, red, high-M/L galaxies if they exist in significant numbers. Some recent SMF determinations using very deep HST WFC3 observations have found steeper SMF slopes at $z > 1.5$ (e.g., Santini et al. 2012), and new measurements from extremely deep NIR surveys such as CANDELS are eagerly anticipated. That said, it seems unlikely that the SMF slope at low redshift has been underestimated enough to account for a difference of 0.2 dex in the SMD.

Recent evidence has suggested that strong nebular line emission can significantly affect broadband photometry for galaxies at high redshift, particularly $z > 3.8$, where Hα (and, at $z > 5.3$, [OIII]) enter the *Spitzer* IRAC bands (Shim et al. 2011). Therefore, following Stark et al. (2013), we have divided the SMD of González et al. (2011) at $z \simeq 4$, 5, 6, and 7 in **Figure 11** by the factor 1.1, 1.3, 1.6, and 2.4, respectively, to account for this effect. Although considerable uncertainties in these corrections remain, such downward revision to the inferred early SMD improves consistency with expectations from the time-integrated SFRD.

Alternatively, some authors have considered how changing the IMF may help reconcile SMD(z) with the time-integrated SFRD(z) (e.g., Wilkins et al. 2008b). Generally, a more top-heavy or bottom-light IMF will lead to larger luminosities per unit SFR, hence smaller SFR/L conversion factors $\mathcal{K}$ (Section 3.1). Mass-to-light ratios for older stellar populations will also tend to be smaller, but not necessarily by the same factor. Although we have used a Salpeter IMF for reference in this review, an IMF with a low-mass turnover (e.g., Chabrier or Kroupa) will yield a larger mass return fraction R and proportionately lower final stellar masses for a given integrated past SFH by the factor $(1 - R_1)/(1 - R_2) = 0.81$, where R_1 and $R_2 = 0.41$ and 0.27 for the Chabrier and Salpeter IMFs, respectively (Section 2). The apparent offset between the SMD data and our integrated model $\psi(z)$ can be reduced further to only ~0.1 dex without invoking a particularly unusual IMF. Given the remaining potential for systemic uncertainties in the measurements of SFRDs and SMDs, it seems premature to tinker further with the IMF, although if discrepancies remain after further improvements in the measurements and modeling then this topic may be worth revisiting.

5.4. Fossil Cosmology

In concordance with estimates from the cosmic SFH, the measurements of the SMD discussed above imply that galaxies formed the bulk ($\gtrsim$75%) of their stellar mass at $z < 2$. Stars formed in galaxies before 11.5 Gyr are predicted to contribute only 8% of the total stellar mass today. An important consistency check for all these determinations could come from studies of the past

SFH of the Universe from its present contents. This "fossil cosmology" approach has benefited from large spectroscopic surveys in the local Universe such as the 2dFGRS (Colless et al. 2001) and the SDSS (York et al. 2000), which provide detailed spectral information for hundreds of thousands of galaxies. Using a sample of 1.7×10^5 galaxies drawn from the SDSS DR2 and comparing the spectrum of each galaxy to a library of templates by Bruzual & Charlot (2003) (the comparison was based on five spectral absorption features, namely D4,000, Hβ, and Hδ_A + Hγ_A as age-sensitive indexes and [Mg_2Fe] and [MgFe]$'$ as metal-sensitive indexes), Gallazzi et al. (2008) have constructed a distribution of stellar mass as a function of age (for a similar analysis on the SDSS DR3 sample, see also Panter et al. 2007). In **Figure 12**, we compare this distribution with the one predicted by our best-fit cosmic SFH. The latter implies a mass-weighted mean stellar age,

$$\langle t_{\rm age} \rangle = t_0 - \int_0^{t_0} t\psi(t){\rm d}t \left[\int_0^{t_0} \psi(t){\rm d}t\right]^{-1}, \tag{17}$$

equal to $\langle t_{\rm age} \rangle = 8.3$ Gyr. Both distributions have a peak at 8–10 Gyr and decline rapidly at younger ages, with the peak age corresponding to the formation redshift, $z \sim 2$, where the cosmic star-formation density reaches a maximum. The SDSS distribution, however, appears to be skewed toward younger ages. This is partly caused by a bias toward younger populations in the SDSS "archaeological" approach, where individual galaxies are assigned only an average (weighted by mass or light) age that is closer to the last significant episode of star formation. Such bias appears to be reflected in **Figure 12** where the mass assembly history predicted by our model SFH is compared with that inferred by translating the characteristic age of the stellar populations measured by Gallazzi et al. (2008) into a characteristic redshift of formation. The agreement is generally good, although the SDSS distribution would predict later star formation and more rapid SMD growth at $z < 2$ and correspondingly less stellar mass formed at $z > 2$. The present-day total SMD derived by Gallazzi et al. (2008) is $(6.0 \pm 1.0) \times 10^8$ M$_\odot$ Mpc^{-3} (scaled up from a Chabrier to a Salpeter IMF), in excellent agreement with $\rho_* = 5.8 \times 10^8$ M$_\odot$ Mpc^{-3} predicted by our model SFH. This stellar density corresponds to a stellar baryon fraction of only 9% (5% for a Chabrier IMF).

5.5. The Global Specific Star-Formation Rate

In recent years, there has been considerable interest in the sSFR (sSFR $\equiv$ SFR$/M_*$) of galaxies with different masses at different times in the history of the Universe. The sSFR describes the fractional growth rate of stellar mass in a galaxy or, equivalently, the ratio of current to past star formation. The inverse of the sSFR is the characteristic stellar mass doubling time (Guzman et al. 1997). At $0 < z < 2$ and quite possibly at higher redshifts, most star-forming galaxies follow a reasonably tight relation between SFR and M_*, whose normalization (e.g., the mean sSFR at some fiducial mass) decreases steadily with cosmic time (decreasing redshift) at least from $z = 2$ to the present (Brinchmann et al. 2004, Daddi et al. 2007, Elbaz et al. 2007, Noeske et al. 2007). A minority population of starburst galaxies exhibits elevated sSFRs, whereas quiescent or passive galaxies lie below the SFR-M_* correlation. For the "main sequence" of star-forming galaxies, most studies find that the average sSFR is a mildly declining function of stellar mass (e.g., Karim et al. 2011). This implies that more massive galaxies completed the bulk of their star formation earlier than that did lower-mass galaxies (Brinchmann & Ellis 2000, Juneau et al. 2005), a "downsizing" picture first introduced by Cowie et al. (1996). Dwarf galaxies continue to undergo major episodes of activity. The tightness of this SFR-M_* correlation has important implications for how star formation is regulated within galaxies and perhaps for the cosmic SFH. Starburst galaxies, whose SFRs are significantly elevated above the main-sequence correlation, contribute only a small fraction of the global SFRD at $z \leq 2$ (Rodighiero et al. 2011, Sargent et al. 2012).

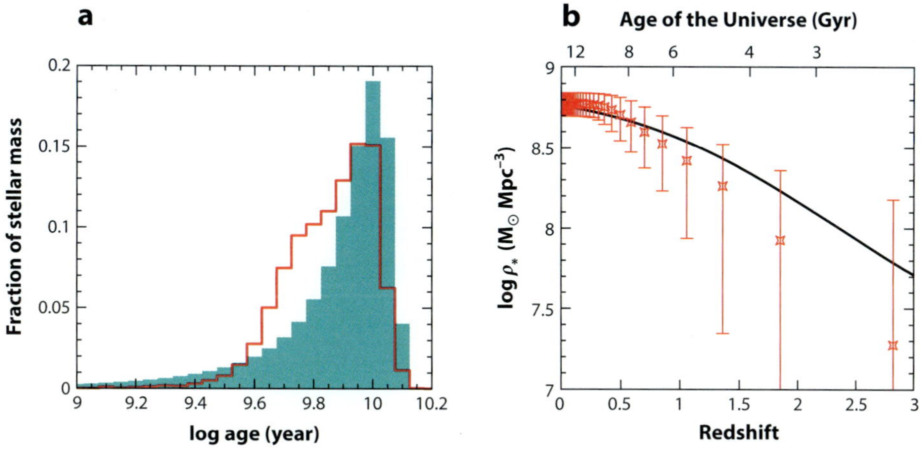

Figure 12

"Stellar archaeology" with the SDSS. (*a*) Normalized distribution of stellar mass in the local Universe as a function of age. The red histogram shows estimates from the SDSS (Gallazzi et al. 2008). The measured (mass-weighted) ages have been corrected by adding the lookback time corresponding to the redshift at which the galaxy is observed. The turquoise shaded histogram shows the distribution generated by integrating the instantaneous star-formation rate density in Equation 2. (*b*) Evolution of the stellar mass density with redshift. The points, shown with systematic error bars, are derived from the analysis of SDSS data by Gallazzi et al. (2008), assuming a Salpeter IMF in the range $0.1-100\,M_\odot$. The solid line shows the mass assembly history predicted by integrating our best-fit star-formation history. Abbreviations: IMF, initial mass function; SDSS, Sloan Digital Sky Survey.

Instead, the evolution of the cosmic SFR is primarily due to the steadily evolving properties of main-sequence disk galaxies.

Figure 13 compares the sSFR (in Gyr^{-1}) for star-forming galaxies with estimated stellar masses in the range $10^{9.4}-10^{10}\,M_\odot$ from a recent compilation by González et al. (2014) with the predictions from our best-fit SFH. At $z < 2$, the globally averaged sSFR ($\equiv \psi/\rho_*$) declines more steeply than does that for the star-forming population, as star formation is "quenched" for an increasingly large fraction of the galaxy population. These passive galaxies are represented in the global sSFR, but not in the sSFR of the star-forming "main sequence." Previous derivations showed a nearly constant sSFR of $\sim 2\ \text{Gyr}^{-1}$ for galaxies in the redshift range $2 < z < 7$, suggesting relatively inefficient early star formation and exponential growth in SFRs and stellar masses with cosmic time. Recent estimates of reduced stellar masses were derived after correcting for nebular emission in broadband photometry and appear to require some evolution in the high-z sSFR (Stark et al. 2013, González et al. 2014). At these epochs, the global sSFR decreases with increasing cosmic time t as sSFR $\sim 4/t\ \text{Gyr}^{-1}$, a consequence of the power-law scaling of our SFRD, $\psi(t) \propto t^{1.9}$.

5.6. Cosmic Metallicity

According to Equation 4, the sum of the heavy elements stored in stars and in the gas phase at any given time, $Z\rho_g + \langle Z_* \rangle \rho_*$, is equal to the total mass of metals produced over cosmic history, $y\rho_*$. It can be useful to express this quantity relative to the baryon density,

$$Z_b(z) \equiv \frac{y\rho_*(z)}{\rho_b}, \tag{18}$$

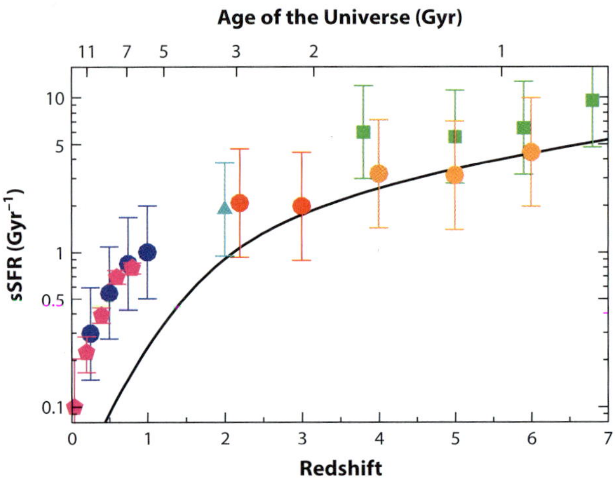

Figure 13

The mean specific star-formation rate (sSFR $\equiv$ SFR$/M_*$) for galaxies with estimated stellar masses in the range $10^{9.4}$–10^{10} $M_\odot$. The values are from Daddi et al. (2007) (*cyan triangle*), Noeske et al. (2007) (*blue dots*), Damen et al. (2009) (*magenta pentagons*), Reddy & Steidel (2009) (*red dots*), González et al. (2014) (*orange dots*), and Stark et al. (2013) (*green squares*). The error bars correspond to systematic uncertainties. The high-redshift points from González et al. (2014) and Stark et al. (2013) have been corrected upward owing to the effect of optical emission lines on the derived stellar masses, using the "RSF with emission lines" model of González et al. (2014) and the "fixed Hα EW" model of Stark et al. (2013). The curve shows the predictions from our best-fit star-formation history.

where $\rho_b = 2.77 \times 10^{11} \Omega_b h^2$ $M_\odot$ Mpc^{-3}. The evolution of the mean metallicity of the Universe, Z_b, predicted by our model SFH is plotted in **Figure 14**. The global metallicity is $Z_b \simeq 0.09$ $(y/Z_\odot)$ solar at the present epoch (note that this is the same value derived by Madau et al. 1998b). It drops to $Z_b \simeq 0.01(y/Z_\odot)$ solar at $z = 2.5$, i.e., the star-formation activity we believe to have taken place between the Big Bang and $z = 2.5$ (2.5 Gyr later) was sufficient to enrich the Universe as a whole to a metallicity of ∼1% solar (for $y \simeq Z_\odot$). Note that the metal production term $y\rho_*$ (and therefore Z_b) depends only weakly on the IMF (at a fixed luminosity density): Salpeter-based mass-to-light ratios are 1.64 times higher than those based on Chabrier. This is counterbalanced by Salpeter-based net metal yields that are a factor of ∼2 lower than those based on Chabrier (see Section 2).

Figure 14 also shows the metallicity of a variety of astrophysical objects at different epochs. The mass-weighted average stellar metallicity in the local Universe, $\langle Z_*(0) \rangle = (1.04 \pm 0.14)\,Z_\odot$ (Gallazzi et al. 2008), is plotted together with the metallicity of three different gaseous components of the distant Universe: (*a*) galaxy clusters, the largest bound objects for which chemical enrichment can be thoroughly studied and perhaps the best example in nature of a "closed box"; (*b*) the damped Lyα absorption systems that originate in galaxies and dominate the neutral-gas content of the Universe; and (*c*) the highly ionized circumgalactic and intergalactic gas that participates in the cycle of baryons in and out of galaxies in the early Universe.

The iron mass in clusters is several times larger than could have been produced by CC SNe if stars formed with a standard IMF, a discrepancy that may indicate an IMF in clusters that is skewed toward high-mass stars (e.g., Portinari et al. 2004) and/or to enhanced iron production by Type Ia SNe (e.g., Maoz & Gal-Yam 2004). The damped Lyα absorption systems are detected in absorption (i.e., have no luminosity bias), and their large optical depths at the Lyman limit eliminate the need for uncertain ionization corrections to deduce metal abundances. Their metallicity is

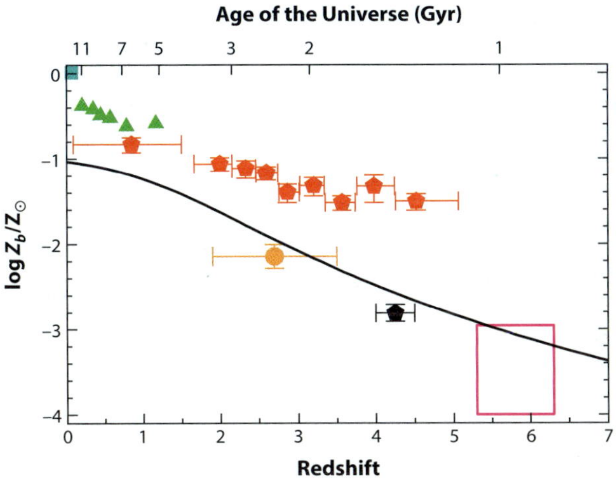

Figure 14

Mean metallicity of the Universe (in solar units): (*solid curve*) mass of heavy elements ever produced per cosmic baryon from our model SFH, for an assumed IMF-averaged yield of $y = 0.02$; (*turquoise square*) mass-weighted stellar metallicity in the nearby Universe from the SDSS (Gallazzi et al. 2008); (*green triangles*) mean iron abundances in the central regions of galaxy clusters (Balestra et al. 2007); (*red pentagons*) column density-weighted metallicities of the damped Lyα absorption systems (Rafelski et al. 2012); (*orange dot*) metallicity of the IGM as probed by OVI absorption in the Lyα forest (Aguirre et al. 2008); (*black pentagon*) metallicity of the IGM as probed by CIV absorption (Simcoe 2011); (*magenta rectangle*) metallicity of the IGM as probed by CIV and CII absorption (Ryan-Weber et al. 2009, Becker et al. 2011, Simcoe et al. 2011). Abbreviations: IGM, intergalactic medium; IMF, initial mass function; SDSS, Sloan Digital Sky Survey; SFH, star-formation history.

determined with the highest confidence from elements such as O, S, Si, Zn, and Fe and decreases with increasing redshift down to ≈1/600 solar to $z \sim 5$ (e.g., Rafelski et al. 2012). The enrichment of the circumgalactic medium, as probed in absorption by CIII, CIV, SiIII, SiIV, OVI, and other transitions, provides us with a record of past star formation and of the impact of galactic winds on their surroundings. **Figure 14** shows that the ionization corrected metal abundances from OVI absorption at $z \sim 3$ (Aguirre et al. 2008) and CIV absorption at $z \sim 4$ (Simcoe 2011) track well the predicted mean metallicity of the Universe, i.e., that these systems are an unbiased probe of the cosmic baryon cycle.

The Universe at redshift 6 remains one of the most challenging observational frontiers, as the high opacity of the Lyα forest inhibits detailed studies of hydrogen absorption along the line-of-sight to distant quasars. In this regime, the metal lines that fall longwards of the Lyα emission take on a special significance as the only tool at our disposal to recognize individual absorption systems, whether in galaxies or the IGM, and probe cosmic enrichment following the earliest episodes of star formation. Here, we use recent surveys of high- and low-ionization intergalactic absorption to estimate the metallicity of the Universe at these extreme redshifts. According to Simcoe et al. (2011) (see also Ryan-Weber et al. 2009), the comoving mass density of triply ionized carbon over the redshift range 5.3–6.4 is (expressed as a fraction of the critical density) $\Omega_{\rm CIV} = (0.46 \pm 0.20) \times 10^{-8}$. Over a similar redshift range, CII absorption yields $\Omega_{\rm CII} = 0.9 \times 10^{-8}$ (Becker et al. 2011). The total carbon metallicity by mass, $Z_{\rm C}$, at $\langle z \rangle = 5.8$ implied by these measurements is

$$Z_{\rm C} = \frac{\Omega_{\rm CIV} + \Omega_{\rm CII}}{\Omega_b} \times \frac{\rm C}{\rm C_{II} + C_{IV}} \simeq 3 \times 10^{-7} \frac{\rm C}{\rm C_{II} + C_{IV}}, \quad (19)$$

where (C II + C IV)/C is the fraction of carbon that is either singly or triply ionized. In **Figure 14**, we plot Z_C in units of the mass fraction of carbon in the Sun, $Z_{C\odot} = 0.003$ (Asplund et al. 2009). The lower bound to the rectangle centered at redshift 5.8 assumes no ionization correction, i.e., (C II + C IV)/C = 1. To derive the upper bound, we adopt the conservative limit (C II + C IV)/C ≥ 0.1; this is the minimum fractional abundance reached by C II + C IV under the most favorable photoionization balance conditions at redshift 6. [To obtain this estimate, we have computed photoionization models based on the CLOUDY code (Ferland et al. 1998) assuming the UV radiation background at $z = 6$ of Haardt & Madau (2012) and a range of gas overdensities $0 < \log \delta < 3$.] If the ionization state of the early metal-bearing IGM is such that most of the C is either singly or triply ionized, then most of the heavy elements at these epochs appear to be "missing" compared with the expectations based on the integral of the cosmic SFH (Pettini 2006, Ryan-Weber et al. 2009). Conversely, if C II and C IV are only trace ion stages of carbon, then the majority of the heavy elements produced by stars 1 Gyr after the Big Bang ($z = 6$) may have been detected already.

A simple argument can also be made against the possibility that our best-fit SFH significantly overpredicts the cosmic metallicity at these early epochs. The massive stars that explode as Type II SNe and seed the IGM with metals are also the sources of nearly all of the Lyman-continuum (LyC) photons produced by a burst of star formation. It is then relatively straightforward to link a given IGM metallicity to the minimum number of LyC photons that must have been produced up until that time. The close correspondence between the sources of metals and photons makes the conversion from one to the other largely independent of the details of the stellar IMF (Madau & Shull 1996). Specifically, the energy emitted in hydrogen ionizing radiation per baryon, $E_{\rm ion}$, is related to the average cosmic metallicity by

$$E_{\rm ion} = \eta m_p c^2 Z_b, \qquad 20.$$

where $m_p c^2 = 938$ MeV is the rest mass of the proton and η is the efficiency of conversion of the heavy element rest mass into LyC radiation. For stars with $Z_* = Z_\odot/50$, one derives $\eta = 0.014$ (Schaerer 2002, Venkatesan & Truran 2003). An average energy of 22 eV per LyC photon together with our prediction of $Z_b = 7 \times 10^{-4} Z_\odot$ (assuming a solar yield) at redshift 6 implies that approximately 8 LyC photons per baryon were emitted by early galaxies prior to this epoch. Although at least one photon per baryon is needed for reionization to occur, this is a reasonable value because the effect of hydrogen recombinations in the IGM and within individual halos will likely boost the number of photons required. A global metallicity at $z = 6$ that was much lower than our predicted value would effectively create a deficit of UV radiation and leave the reionization of the IGM unexplained. In Section 5.8 below, we link the production of LyC photons to stellar mass, show that the efficiency of LyC production decreases with increasing Z_*, and discuss early star formation and the epoch of reionization in more detail.

5.7. Black Hole Accretion History

Direct dynamical measurements show that most local massive galaxies host a quiescent massive black hole in their nuclei. Their masses correlate tightly with the stellar velocity dispersion of the host stellar bulge, as manifested in the $M_{\rm BH} - \sigma_*$ relation of spheroids (Ferrarese & Merritt 2000, Gebhardt et al. 2000). It is not yet understood whether such scaling relations were set in primordial structures and maintained throughout cosmic time with a small dispersion or which physical processes established such correlations in the first place. Nor it is understood whether the energy released during the luminous quasar phase has a global impact on the host, generating

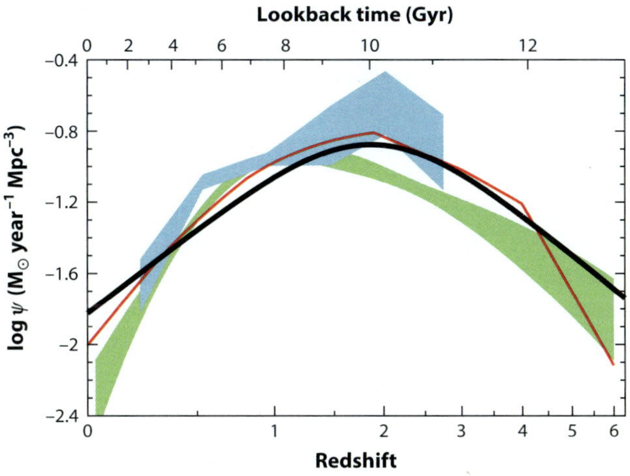

Figure 15

Comparison of the best-fit star-formation history (*thick solid curve*) with the massive black hole accretion history from X-ray [*red curve* (Shankar et al. 2009); *light green shading* (Aird et al. 2010)] and infrared (*light blue shading*) (Delvecchio et al. 2014) data. The shading indicates the $\pm 1\sigma$ uncertainty range on the total bolometric luminosity density. The radiative efficiency has been set to $\epsilon = 0.1$. The comoving rates of black hole accretion have been scaled up by a factor of 3,300 to facilitate visual comparison to the star-formation history.

large-scale galactic outflows and quenching star formation (Di Matteo et al. 2005), or just modifies gas dynamics in the galactic nucleus (Debuhr et al. 2010).

Here, we consider a different perspective on the link between the assembly of the stellar component of galaxies and the growth of their central black holes. The cosmic mass accretion history of massive black holes can be inferred using Soltan's argument (Soltan 1982), which relates the quasar bolometric luminosity density to the rate at which mass accumulates into black holes,

$$\dot{\rho}_{\rm BH}(z) = \frac{1-\epsilon}{\epsilon c^2} \int L\phi(L, z)\mathrm{d}L, \qquad (21)$$

where ϵ is the efficiency of conversion of rest-mass energy into radiation. In practice, bolometric luminosities are typically derived from observations of the AGN emission at X-ray, optical or IR wavelengths, scaled by a bolometric correction. In **Figure 15**, several recent determinations of the massive black hole mass growth rate are compared with the cosmic SFRD (Equation 15). Also shown is the accretion history derived from the hard X-ray LF of Aird et al. (2010), assuming a radiative efficiency $\epsilon = 0.1$ and a constant bolometric correction of 40 for the observed 2–10 KeV X-ray luminosities. This accretion rate peaks at lower redshift than does the SFRD and declines more rapidly from $z \approx 1$ to 0. However, several authors have discussed the need for luminosity-dependent bolometric corrections, which in turn can affect the derived accretion history (e.g., Marconi et al. 2004, Hopkins et al. 2007, Shankar et al. 2009). Moreover, although the hard X-ray LF includes unobscured as well as moderately obscured sources that may not be identified as AGN at optical wavelengths, it can miss Compton thick AGN, which may be identified in other ways, particularly using IR data. Delvecchio et al. (2014) have used deep *Herschel* and *Spitzer* survey data in GOODS-S and COSMOS to identify AGN by SED fitting. This is a potentially powerful method but depends on reliable decomposition of the IR emission from AGN and star formation.

Black hole mass growth rates derived from the bolometric AGN LFs of Shankar et al. (2009) and Delvecchio et al. (2014) are also shown in **Figure 15**. These more closely track the cosmic

SFH, peaking at $z \approx 2$, and suggest that star formation and black hole growth are closely linked at all redshifts (Boyle & Terlevich 1998, Silverman et al. 2008). However, the differences between accretion histories published in the recent literature would caution that it is premature to consider this comparison to be definitive.

5.8. First Light and Cosmic Reionization

Fundamental to our understanding of how the Universe evolved to its present state is the epoch of "first light," the first billion years after the Big Bang when the collapse of the earliest baryonic objects—the elementary building blocks for the more massive systems that formed later—determined the "initial conditions" of the cosmological structure formation process. The reionization in the all-pervading IGM—the transformation of neutral hydrogen into an ionized state—is a landmark event in the history of the early Universe. Studies of Lyα absorption in the spectra of distant quasars show that the IGM is highly photoionized out to redshift $z \gtrsim 6$ (for a review, see Fan et al. 2006), whereas polarization data from the *Wilkinson Microwave Anisotropy Probe* constrain the redshift of any sudden reionization event to be significantly higher, $z = 10.5 \pm 1.2$ (Jarosik et al. 2011). It is generally thought that the IGM is kept ionized by the integrated UV emission from AGN and star-forming galaxies, but the relative contributions of these sources as a function of epoch are poorly known (e.g., Madau et al. 1999, Faucher-Giguère et al. 2008, Haardt & Madau 2012, Robertson et al. 2013).

Establishing whether massive stars in young star-forming galaxies were responsible for cosmic reionization requires a determination of the early history of star formation, of the LyC flux emitted by a stellar population, and of the fraction of hydrogen-ionizing photons that can escape from the dense sites of star formation into the low-density IGM. We can use stellar population synthesis to estimate the comoving volumetric rate at which photons above 1 Ryd are emitted from star-forming galaxies as

$$\dot{n}_{\rm ion} = I_{\rm ion} \times \psi(t), \tag{22}$$

where $\dot{n}_{\rm ion}$ is expressed in units of s^{-1} Mpc^{-3} and ψ in units of M$_\odot$ year^{-1} Mpc^{-3}. The LyC photon yield $I_{\rm ion}$ is plotted in **Figure 16** for our reference Salpeter IMF and for a wide range of metallicities spanning from extremely metal-poor to metal-rich stars (Schaerer 2003). The yield increases with decreasing metallicity by more than a factor of 3. Integrating Equation 22 over time and dividing by the mean comoving hydrogen density $\langle n_{\rm H}\rangle = 1.9 \times 10^{-7}$ cm^{-3}, we can write the total number of stellar LyC photons emitted per hydrogen atom since the Big Bang as

$$\mathcal{N}_{\rm ion}(z) \equiv \frac{n_{\rm ion}}{\langle n_{\rm H}\rangle} = I_{\rm ion} \times 7.8 \times 10^{-60} \times \rho_*(z), \tag{23}$$

where $\rho_*(z)$ is the cosmic SMD. [Within purely stellar radiation or energetic X-ray photons, either the total number of ionizing photons produced or the total radiated energy, respectively, is what matters for reionization. This is because, in a largely neutral medium, each photoionization produces a host of secondary collisional ionizations, with approximately one hydrogen secondary ionization for every 37 eV of energy in the primary photoelectron (Shull & van Steenberg 1985). As the medium becomes more ionized, however, an increasing fraction of this energy is deposited as heat.] **Figure 16** depicts the quantity $\mathcal{N}_{\rm ion}$ at $z > 6$ according to our best-fit SFH for the range of stellar metallicities $0 < Z_* < Z_\odot$. Cosmological reionization requires at least one LyC photon per hydrogen atom escaping into the intergalactic space, i.e., $\mathcal{N}_{\rm ion}\langle f_{\rm esc}\rangle > 1$. The exact number depends on the rate of radiative recombinations in a clumpy IGM. Here, the escape fraction $\langle f_{\rm esc}\rangle$ is the angle-averaged, absorption cross-section-weighted, and luminosity-weighted

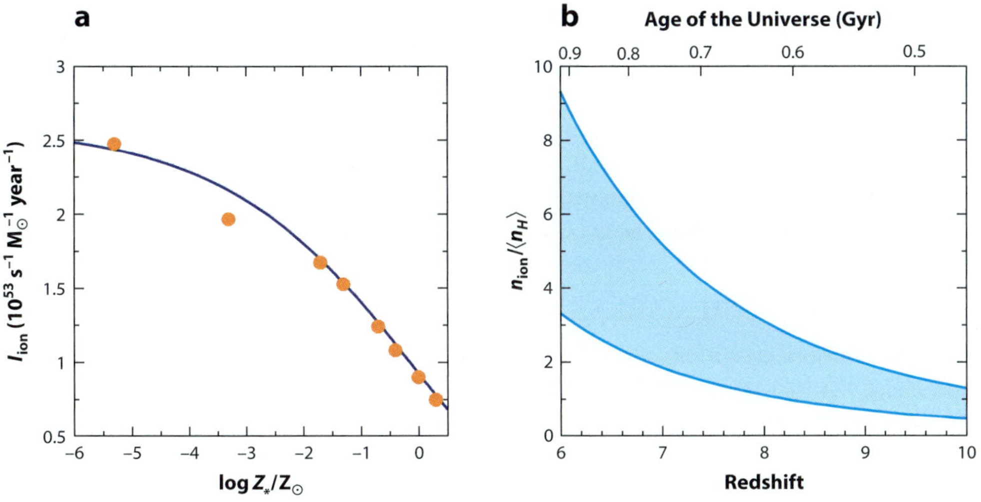

Figure 16

(*a*) Metallicity dependence of the ionizing photon yield, $I_{\rm ion}$, for a stellar population with a Salpeter IMF (initial mass function) and constant star-formation rate. The points show the values given in table 4 of Schaerer (2003), computed for a Salpeter IMF in the $1-100\,M_\odot$ range, and divided by a mass conversion factor of 2.55 to rescale to a mass range $0.1-100\,M_\odot$. The solid curve shows the best-fitting function, $\log I_{\rm ion} = [(0.00038 Z_*^{0.227} + 0.01858)]^{-1} - \log(2.55)$. (*b*) Number of H-ionizing photons emitted per hydrogen atom since the Big Bang, $n_{\rm ion}/\langle n_{\rm H}\rangle$, according to our best-fit star-formation history. The shaded area is delimited by stellar populations with metallicities $Z_* = Z_\odot$ (lower boundary) and $Z_* = 0$ (upper boundary).

fraction of ionizing photons that leaks into the IGM from the dense star-forming regions within galaxies.

Figure 16 shows the well-known result that, even if the emission of LyC photons at early cosmic times was dominated by extremely low-metallicity stars, escape fractions in excess of 20% would be required to reionize the Universe by redshift 6–7 (e.g., Bolton & Haehnelt 2007, Ouchi et al. 2009, Bunker et al. 2010, Finkelstein et al. 2012b, Haardt & Madau 2012). Although the mechanisms regulating the escape fraction of ionizing radiation from galaxies and its dependence on cosmic time and luminosity are unknown, these leakage values are higher than typically inferred from observations of LBGs at $z \sim 3$ (e.g., Nestor et al. 2013, Vanzella et al. 2012, and references therein).

The "reionization budget" is made even tighter by two facts. First, the volume-averaged hydrogen recombination timescale in the IGM,

$$\langle t_{\rm rec}\rangle = (\chi \langle n_{\rm H}\rangle \alpha_B\, C_{\rm IGM})^{-1} \simeq 3.2 \text{ Gyr} \left(\frac{1+z}{7}\right)^{-3} C_{\rm IGM}^{-1}, \qquad (24)$$

is ~60% of the expansion timescale $H^{-1} = H_0^{-1}[\Omega_M(1+z)^3 + \Omega_\Lambda]^{-1/2}$ at $z = 10$, i.e., close to two ionizing photons per baryon are needed to keep the IGM ionized. Here, α_B is the recombination coefficient to the excited states of hydrogen, $\chi = 1.08$ accounts for the presence of photoelectrons from singly ionized helium, and $C_{\rm IGM}$ is the clumping factor of ionized hydrogen. The above estimate assumes a gas temperature of 2×10^4 K and a clumping factor for the IGM, $C_{\rm IGM} = 1 + 43\, z^{-1.71}$, that is equal to the clumpiness of gas below a threshold overdensity of 100 found at $z \geq 6$ in a suite of cosmological hydrodynamical simulations (Pawlik et al. 2009; see also Shull et al. 2012). Second, the rest-frame UV continuum properties of very high redshift galaxies appear to show little evidence for the "exotic" stellar populations (e.g., extremely subsolar metallicities or top-heavy IMFs) that would significantly boost the LyC photon yield (e.g., Finkelstein et al. 2012a, Dunlop et al. 2013). Although improved knowledge of the SFRD beyond redshift 9 may help to better characterize this photon shortfall, the prospects of a direct observational determination of the escape fraction of ionizing photons leaking into the IGM at these early epochs look rather bleak, as the Universe becomes opaque to LyC radiation above redshift 4.

6. CONCLUDING REMARKS

The cosmic history of star formation is one of the most fundamental observables in astrophysical cosmology. We have reviewed the range of complementary techniques and theoretical tools that are allowing astronomers to map the transformation of gas into stars, the production of heavy elements, and the reionization of the Universe from the cosmic dark ages to the present epoch. Under the simple assumption of a universal IMF, there is reasonable agreement between the global SMD inferred at any particular time and the time integral of all the preceding instantaneous star-formation activity, although modest offsets may still point toward systematic uncertainties. A consistent picture is emerging, whereby the SFRD peaked ~3.5 Gyr after the Big Bang and dropped exponentially at $z < 1$ with an e-folding timescale of 3.9 Gyr. The Universe was a much more active place in the past: Stars formed at a peak rate approximately nine times higher than is seen today. Approximately 25% of the present-day SMD formed at $z > 2$, before the peak of the SFRD, and another 25% formed since $z = 0.7$, i.e., roughly over the last half of the Universe's age. From the peak of the SFRD at $z \approx 2$ to the present day, and perhaps earlier as well, most stars formed in galaxies that obey a relatively tight SFR-M_* correlation, and only a small fraction formed in starbursts with significantly elevated specific SFRs. The smooth evolution of this dominant main-sequence galaxy population suggests that the evolution of the cosmic SFH is primarily determined by a balance between gas accretion and feedback processes, both closely related to galaxy mass, and that stochastic events such as merger-driven starbursts play a relatively minor role. The growth histories of the stellar component of galaxies and their central black holes are similar in shape, suggesting broad coevolution of black holes and their host galaxies. The rise of the mean metallicity of the Universe to 0.001 solar by redshift six, 1 Gyr after the Big Bang, appears to have been accompanied by the production of fewer than 10 hydrogen LyC photons per baryon, indicating a rather tight budget for cosmological reionization. The SFRD at $z \approx 7$ was approximately the same as that of today, at $z \approx 0$, but only 1% of today's SMD was formed during the epoch of reionization.

As far as the observations and data are concerned, there is still room for improvement in both SFRD and SMD measurements at virtually every redshift, from the local Universe to the epoch of reionization (Section 4.6). That said, it would be somewhat surprising if new measurements changed the picture dramatically at $z < 1$; it is more likely that stellar population modeling, e.g., for deriving stellar masses or SFRs, could still change the details of the picture during the decline and fall of the cosmic SFH. Indeed, at all redshifts, limitations of our methods for interpreting light as mass may play a significant, even dominant, role in the error budget for the analyses described in this review. The peak era of cosmic star formation has been extensively mapped, and yet even with the current data (**Figure 9**), it is still hard to accurately pinpoint the redshift of maximum SFRD within a range $\Delta z = 1$. Our fitting function (Equation 15) places this peak at $z \approx 1.85$, which is plausible but still uncertain. Uncertainties in the faint-end slope of the IRLF and in extinction corrections for the UVLF still dominate at this peak era of cosmic star formation.

Although evidence seems to point clearly to a steady increase in the SFRD from $z = 8$ to $z \approx 2$, our direct knowledge of dust-obscured star formation at these redshifts is, for the most part, limited to the rarest and most ultraluminous galaxies, leaving considerable uncertainty about how much SFRD we may be missing in the UV census of that early phase of galaxy evolution. At $z > 4$, our galaxy surveys have been strongly biased toward UV-bright galaxies and may underestimate both SFRDs and SMDs. Even for UV-selected galaxies, the measurements at $z \geq 8$ are very new and likely uncertain, unsupported by spectroscopic confirmation to date. In addition to measuring redshifts, spectroscopy from the JWST will help clarify basic issues about nebular line emission and the degree to which it has affected the photometric analyses that have been carried out to date.

Painstaking though all this vast effort has been, it does miss a crucial point. It says little about the inner workings of galaxies, i.e., their "metabolism" and the basic process of ingestion (gas infall and cooling), digestion (star formation), and excretion (outflows). Ultimately, it also says little about the mapping from dark matter halos to their baryonic components. Its roots are in optical-IR astronomy, statistics, stellar populations, and phenomenology, rather than in the physics of the ISM, self-regulated accretion and star formation, stellar feedback, and SN-driven galactic winds. It provides a benchmark against which to compare semianalytic modeling and hydrodynamical simulations of galaxy formation, but it offers little guidance in identifying the smaller-scale basic mechanisms that determine the rate of conversion of gas into stars and lead to the grandiose events in the history of the Universe described in this review.

A variety of physical processes are thought to shape the observed distribution of galaxy properties, ranging from those responsible for galaxy growth (e.g., star formation and galaxy merging) to those that regulate such growth (e.g., energetic feedback from SNe, AGN, and the UV radiation background). However, many of these processes likely depend primarily on the mass of a galaxy's dark matter halo. Relating the stellar masses and SFRs of galaxies to the masses and assembly histories of their parent halos is a crucial piece of the galaxy formation and evolution puzzle. With the accumulation of data from large surveys and from cosmological numerical simulations, several statistical methods have been developed over the past decade to link the properties of galaxies to the underlying dark matter structures (e.g., Berlind & Weinberg 2002, Yang et al. 2003, Vale & Ostriker 2004). One of them, the "abundance matching" technique, assumes in its simplest form a unique and monotonic relation between galaxy light and halo mass, and it reproduces galaxy clustering as a function of luminosity over a wide range in redshift (e.g., Conroy et al. 2006, Guo et al. 2010, Moster et al. 2010). Modern versions of this approach (Behroozi et al. 2013, Moster et al. 2013) have shown that (*a*) halos of mass $\sim 10^{12}\,M_\odot$ are the most efficient at forming stars at every epoch, with baryon conversion efficiencies of 20–40% that fall rapidly at both higher and lower masses; (*b*) in halos similar to that of the Milky Way, approximately half of the central stellar mass is assembled after redshift 0.7; and (*c*) in low-mass halos, the accretion of satellites contributes little to the assembly of their central galaxies, whereas in massive halos more than half of the central stellar mass is formed "ex situ." These studies represent promising advances, albeit with serious potential shortcomings (e.g., Guo & White 2013, Zentner et al. 2013). The assumption of a monotonic relation between stellar mass and the mass of the host halo is likely incorrect in detail, and it predicts only numerically converged properties on scales that are well resolved in simulations. The matching procedure requires minimal assumptions and avoids an explicit treatment of the physics of galaxy formation. As such, it provides relatively little new insight into this physics. In the version of this technique by Behroozi et al. (2013), for example, the cosmic SFH is reproduced by construction.

As of this writing, a solid interpretation of the cosmic SFH from first principles is still missing (for a recent review, see Mac Low 2013). Generically, one expects that star formation may be limited at early times by the build-up of dark matter halos and quenched at low redshift as densities decline from Hubble expansion to the point where gas cooling is inhibited. These two regimes could then lead to a peak in the SFH at intermediate redshifts (Hernquist & Springel 2003). A decade ago, hydrodynamical simulations predicted that the peak in star-formation activity should occur at a much higher redshift, $z \gtrsim 5$, than is actually observed (Springel & Hernquist 2003, Nagamine et al. 2004). Theoretical modeling has been unable to correctly forecast the evolution of the SFRD because of the large range of galaxy masses that contribute significantly to cosmic star formation and the difficulty in following the feedback of energy into the ISM and circumgalactic medium from stellar radiation, SN explosions, and accreting massive black holes. Gas cooling in an expanding Universe is an intrinsically unstable process because cooling acts to increase the density of the gas, which in turn increases the cooling rate. Systems collapsing at low redshift have low mean densities and long cooling times, whereas systems collapsing at higher redshifts have higher mean densities and cool catastrophically. Without feedback processes that transfer energy to the ISM and reheat it, one is faced with the classical overcooling problem—the unphysical cooling of hot gas in the poorly resolved inner regions of galaxies—and with the consequent overproduction of stars at early times. And yet, a completely satisfactory treatment of feedback in hydrodynamical simulations that capture large cosmological volumes remains elusive, as these mechanisms operate on scales too small to be resolved and must therefore be incorporated via ad hoc recipes that are too simplistic to capture the complex subgrid physics involved (e.g., Schaye et al. 2009).

In-depth knowledge of the mechanisms responsible for suppressing star formation in small halos (e.g., Governato et al. 2010, Krumholz & Dekel 2012, Kuhlen et al. 2012), more powerful supercomputers, better algorithms, as well as more robust numerical implementations of stellar feedback (e.g., Agertz et al. 2013) all now appear as crucial prerequisites for predicting more realistic SFHs. Newer and deeper observations from the ground and space should improve our measurements of the galaxy population and its integrated properties, especially at and beyond the current redshift frontier where data remain sparse. It seems likely, however, that the most important contribution of new surveys and better modeling will be toward a detailed understanding of the physics of galaxy evolution, not simply its demographics.

DISCLOSURE STATEMENT

The authors are not aware of any affiliations, memberships, funding, or financial holdings that might be perceived as affecting the objectivity of this review.

ACKNOWLEDGMENTS

This review has benefited from many discussions with and the help of J. Aird, R. Chary, C. Conroy, O. Cucciati, D. Elbaz, S. Faber, H. Ferguson, A. Gallazzi, V. González, A. Klypin, K.-S. Lee, D. Maoz, P. Oesch, M. Pettini, J. Pforr, L. Pozzetti, J. Primack, J.X. Prochaska, M. Rafelski, A. Renzini, B. Robertson, M. Schenker, and D. Stark. Support for this work was provided by the National Science Foundation (NSF) through grant OIA-1124453, by NASA through grant NNX12AF87G (P.M.), and by NOAO, which is operated by the Association of Universities for Research in Astronomy, under a cooperative agreement with the NSF (M.D.).

LITERATURE CITED

Agertz O, Kravtsov AV, Leitner SN, Gnedin NY. 2013. *Ap. J.* 770:25
Aguirre A, Dow-Hygelund C, Schaye J, Theuns T. 2008. *Ap. J.* 689:851

Aird J, Nandra K, Laird ES, et al. 2010. *MNRAS* 401:2531
Alavi A, Siana B, Richard J, et al. 2013. *Ap. J.* 780:143
Appleton PN, Fadda DT, Marleau FR, et al. 2004. *Ap. J. Suppl.* 154:147
Arnouts S, Schiminovich D, Ilbert O, et al. 2005. *Ap. J. Lett.* 619:L43
Arnouts S, Walcher CJ, Le Fevre O, et al. 2007. *Astron. Astrophys.* 476:137
Ashby MLN, Willner SP, Fazio GG, et al. 2013. *Ap. J.* 679:80
Asplund M, Grevesse N, Sauval AJ, Scott P. 2009. *Annu. Rev. Astron. Astrophys.* 47:481
Atek H, Malkan M, McCarthy P. 2010. *Ap. J.* 723:104
Aussel H, Cesarsky CJ, Elbaz D, Stark JL. 1999. *Astron. Astrophys.* 342:313
Babbedge TSR, Rowan-Robinson M, Vaccari M, et al. 2006. *MNRAS* 370:1159
Baldry IK, Driver SP, Loveday J, et al. 2012. *MNRAS* 421:621
Baldry IK, Glazebrook K. 2003. *Ap. J.* 593:258
Baldry IK, Glazebrook K, Driver SP. 2008. *MNRAS* 388:945
Balestra I, Tozzi P, Ettori S, et al. 2007. *Astron. Astrophys.* 462:429
Barger A, Cowie LL, Richards EA. 2000. *Astron. J.* 119:2092
Barger A, Cowie LL, Sanders DB, et al. 1998. *Nature* 394:248
Barger A, Wang W-H, Cowie LL, et al. 2012. *Ap. J.* 761:89
Bastian N, Covey KR, Meyer MR. 2010. *Annu. Rev. Astron. Astrophys.* 48:339
Basu-Zych AR, Lehmer BD, Hornschemeier AE, et al. 2013. *Ap. J.* 762:45
Bazin G, Palanque-Delabrouille N, Rich J, et al. 2009. *Astron. Astrophys.* 499:653
Becker GD, Sargent WLW, Rauch M, Calverley AP. 2011. *Ap. J.* 735:93
Beckwith SVW, Stiavelli M, Koekemoer AM, et al. 2006. *Astron. J.* 132:1729
Behroozi PS, Wechsler RH, Conroy C. 2013. *Ap. J.* 770:57
Bell EF, MacIntosh DH, Katz N, Weinberg MD. 2003. *Ap. J. Suppl.* 149:289
Berlind AA, Weinberg DH. 2002. *Ap. J.* 575:587
Berta S, Magnelli B, Nordon R, et al. 2011. *Astron. Astrophys.* 532:A49
Béthermin M, Le Floc'h E, Olbert O, et al. 2012. *Astron. Astrophys.* 542:A57
Bielby R, Hudelot P, McCracken HJ, et al. 2012. *Astron. Astrophys.* 545:23
Blumenthal GR, Faber SM, Primack JR, Rees MJ. 1984. *Nature* 311:517
Bolton JS, Haehnelt MG. 2007. *MNRAS* 382:325
Botticella MT, Riello M, Cappellaro E, et al. 2008. *Astron. Astrophys.* 479:49
Bouwens RJ, Illingworth GD, Franx M, Ford H. 2007. *Ap. J.* 670:928
Bouwens RJ, Illingworth GD, Labbé I, et al. 2011a. *Nature* 469:504
Bouwens RJ, Illingworth GD, Oesch PA, et al. 2011b. *Ap. J.* 737:90
Bouwens RJ, Illingworth GD, Oesch PA, et al. 2012a. *Ap. J.* 754:83
Bouwens RJ, Illingworth GD, Oesch PA, et al. 2012b. *Ap. J. Lett.* 752:L5
Bowler RAA, Dunlop JS, McLure RJ, et al. 2012. *MNRAS* 426:2772
Boyle BJ, Terlevich RJ. 1998. *MNRAS* 293:L49
Brammer GB, van Dokkum PG, Franx M, et al. 2012. *Ap. J. Suppl.* 200:13
Brammer GB, Whitaker KE, van Dokkum PG, et al. 2011. *Ap. J.* 739:24
Brandt WN, Hasinger G. 2005. *Annu. Rev. Astron. Astrophys.* 43:827
Brinchmann J, Charlot S, White SDM, et al. 2004. *MNRAS* 351:1151
Brinchmann J, Ellis RS. 2000. *Ap. J. Lett.* 536:L77
Bruzual G, Charlot S. 2003. *MNRAS* 344:1000
Bruzual G, Charlot S, González Lópezlira R, et al. 2013. *Proc. Int. Astron. Union* 295:435
Buat V, Iglesias-Páramo J, Seibert M, et al. 2005. *Ap. J. Lett.* 619:L51
Buat V, Noll S, Burgarella D, et al. 2012. *Astron. Astrophys.* 545:A141
Budavári T, Szalay AS, Charlot S, et al. 2005. *Ap. J. Lett.* 619:L31
Bunker AJ, Wilkins S, Ellis RS, et al. 2010. *MNRAS* 409:855
Burgarella D, Buat V, Gruppioni C, et al. 2013. *Astron. Astrophys.* 554:A70
Calzetti D. 2001. *Publ. Astron. Soc. Pac.* 113:1449
Calzetti D, Armus L, Bohlin RC, et al. 2000. *Ap. J.* 533:682
Calzetti D, Kennicutt RC, Engelbracht CW, et al. 2007. *Ap. J.* 666:870

Calzetti D, Kinney AL, Storchi-Bergmann T. 1994. *Ap. J.* 429:582
Caputi KI, Cirasuolo M, Dunlop JS, et al. 2011. *MNRAS* 413:162
Caputi KI, Dunlop JS, McLure RJ, et al. 2012. *Ap. J.* 750:20
Caputi KI, Lagache G, Yan L, et al. 2007. *Ap. J.* 660:97
Carilli CL, Lee N, Capak P, et al. 2008. *Ap. J.* 689:883
Castellano M, Fontana A, Boutsia K, et al. 2010a. *Astron. Astrophys.* 511:A20
Castellano M, Fontana A, Paris D, et al. 2010b. *Astron. Astrophys.* 524:A28
Chabrier G. 2003. *Publ. Astron. Soc. Pac.* 115:763
Chapman SC, Blain AW, Smail I, Ivison RJ. 2005. *Ap. J.* 622:772
Charlot S, Fall SM. 2000. *Ap. J.* 539:718
Chary R-R, Elbaz D. 2001. *Ap. J.* 556:562
Chary R-R, Stern D, Eisenhardt P. 2005. *Ap. J. Lett.* 635:L5
Chieffi A, Limongi M. 2004. *Ap. J.* 608:405
Coe D, Zitrin A, Carrasco M, et al. 2013. *Ap. J.* 762:32
Cohen JG. 2002. *Ap. J.* 567:672
Cole S, Lacey CG, Baugh CM, Frenk CS. 2000. *MNRAS* 319:168
Cole S, Nordberg P, Baugh CM, et al. 2001. *MNRAS* 326:255
Colless M, Dalton G, Maddox S, et al. 2001. *MNRAS* 328:1039
Condon JJ. 1992. *Annu. Rev. Astron. Astrophys.* 30:575
Condon JJ, Cotton WD, Broderick JJ. 2002. *Astron. J.* 124:675
Connolly AJ, Szalay AS, Dickinson M, Subbarao MU, Brunner RJ. 1997. *Ap. J. Lett.* 486:L11
Conroy C. 2013. *Annu. Rev. Astron. Astrophys.* 51:393
Conroy C, Gunn JE, White M. 2009. *Ap. J.* 699:486
Conroy C, van Dokkum PG. 2012. *Ap. J.* 760:71
Conroy C, Wechsler RH, Kravtson AV. 2006. *Ap. J.* 647:201
Conselice CJ, Bluck AF, Buitrago F, et al. 2011. *MNRAS* 413:80
Cowie LL, Songaila A, Barger AJ. 1999. *Astron. J.* 118:603
Cowie LL, Songaila A, Hu EM, Cohen JG. 1996. *Astron. J.* 112:839
Cucciati O, Tresse L, Ilbert O, et al. 2012. *Astron. Astrophys.* 539:31
Curtis-Lake E, McLure RJ, Dunlop JS, et al. 2013. *MNRAS* 429:302
Daddi E, Cimatti A, Renzini A, et al. 2004. *Ap. J.* 617:746
Daddi E, Dannerbauer H, Krips M, et al. 2009. *Ap. J. Lett.* 695:L176
Daddi E, Dickinson M, Chary R, et al. 2005. *Ap. J. Lett.* 631:L13
Daddi E, Dickinson M, Morrison G, et al. 2007. *Ap. J.* 670:156
Dahlen T, Mobasher B, Dickinson M, et al. 2007. *Ap. J.* 654:172
Dahlen T, Strolger L-G, Riess AG, et al. 2004. *Ap. J.* 613:189
Dahlen T, Strolger L-G, Riess AG, et al. 2012. *Ap. J.* 757:70
Dale DA, Bendo GJ, Engelbracht CW, et al. 2005. *Ap. J.* 633:857
Dale DA, Helou G. 2002. *Ap. J.* 576:159
Damen M, Labbé I, Franx M, et al. 2009. *Ap. J.* 690:937
Davis M, Efstathiou G, Frenk CS, White SDM. 1985. *Ap. J.* 292:371
Debuhr J, Quataert E, Ma C-P, Hopkins P. 2010. *MNRAS* 406:L55
Delvecchio I, Gruppioni C, Pozzi F, et al. 2014. *MNRAS* 439:2736
Dey A, Soifer BT, Desai V, et al. 2008. *Ap. J.* 677:943
Di Matteo T, Springel V, Hernquist L. 2005. *Nature* 433:604
Dickinson M, Papovich C, Ferguson HC, Budavári T. 2003. *Ap. J.* 587:25
Dickinson M, Stern D, Giavalisco M, et al. 2004. *Ap. J. Lett.* 600:L99
Diemand J, Kuhlen M, Madau P, et al. 2008. *Nature* 454:735
Doherty M, Bunker A, Sharp R, et al. 2006. *MNRAS* 370:331
Driver SP, Popescu CC, Tuffs RJ, et al. 2008. *Ap. J. Lett.* 678:L101
Dubinski J, Carlberg RG. 1991. *Ap. J.* 378:496
Dunlop JS, Rogers AB, McLure RJ, et al. 2013. *MNRAS* 432:3520
Dunne L, Eales S, Edmunds M, et al. 2000. *MNRAS* 315:115

Dunne L, Ivison RJ, Maddox S, et al. 2009. *MNRAS* 394:3
Dyle S, Dunne L, Eales S, et al. 2010. *Astron. Astrophys.* 518:L10
Eales SA, Raymond G, Roseboom IG, et al. 2010. *Astron. Astrophys.* 518:L23
Elbaz D, Daddi E, Le Borgne D, et al. 2007. *Astron. Astrophys.* 468:33
Elbaz D, Dickinson M, Hwang HS, et al. 2011. *Astron. Astrophys.* 533:119
Elbaz D, Hwang HS, Magnelli B, et al. 2010. *Astron. Astrophys.* 518:L29
Ellis RS, McLure RJ, Dunlop JS, et al. 2013. *Ap. J. Lett.* 763:L7
Engelbracht CW, Gordon KD, Rieke GH, et al. 2005. *Ap. J. Lett.* 628:L29
Engelbracht CW, Rieke GH, Gordon KD, et al. 2008. *Ap. J.* 678:804
Eyles LP, Bunker AJ, Ellis RS, et al. 2007. *MNRAS* 374:910
Faber SM, Willmer CNA, Wolf C, et al. 2007. *Ap. J.* 665:265
Fan X, Carilli CL, Keating B. 2006. *Annu. Rev. Astron. Astrophys.* 44:415
Faucher-Giguère C-A, Prochaska JX, Lidz A, Hernquist L, Zaldarriaga M. 2008. *Ap. J.* 681:831
Ferland GJ, Korista KT, Verner DA, et al. 1998. *Publ. Astron. Soc. Pac.* 110:761
Ferrarese L, Merritt D. 2000. *Ap. J. Lett.* 539:L9
Finkelstein SL, Papovich C, Dickinson M, et al. 2013. *Nature* 502:524
Finkelstein SL, Papovich C, Ryan RE, et al. 2012a. *Ap. J.* 758:93
Finkelstein SL, Papovich C, Salmon B, et al. 2012b. *Ap. J.* 756:164
Flores H, Hammaer F, Thuan TX, et al. 1999. *Ap. J.* 517:148
Fontana A, Donnarumma I, Vanzella E, et al. 2003. *Ap. J. Lett.* 594:L9
Fontana A, Salimbeni S, Grazian A, et al. 2006. *Astron. Astrophys.* 459:745
Franceschini A, Andreani P, Danese L. 1998. *MNRAS* 296:709
Franx M, Labbé I, Rudnick G, et al. 2003. *Ap. J. Lett.* 587:L79
Fukugita M, Hogan CJ, Peebles PJE. 1998. *Ap. J.* 503:518
Gallazzi A, Brinchmann J, Charlot S, White SDM. 2008. *MNRAS* 383:1439
Gallego J, Zamorano J, Aragon-Salamanca A, Rego M. 1995. *Ap. J. Lett.* 455:L1
Gal-Yam A, Ofek EO, Poznanski D, et al. 2006. *Ap. J.* 639:331
Gavazzi G, Pierini D, Boselli A. 1996. *Astron. Astrophys.* 312:397
Gebhardt K, Bender R, Bower G, et al. 2000. *Ap. J. Lett.* 539:L13
Geha M, Brown TM, Tumlinson J, et al. 2013. *Ap. J.* 771:29
Giallongo E, D'Odorico S, Fontana A, et al. 1998. *Ap. J.* 115:2169
Giavalisco M, Dickinson M, Ferguson HC, et al. 2004a. *Ap. J. Lett.* 600:L103
Giavalisco M, Ferguson HC, Koekemoer, et al. 2004b. *Ap. J. Lett.* 600:L93
Gispert R, Lagache G, Puget JL. 2000. *Astron. Astrophys.* 360:1
Glazebrook K, Blake C, Economou F, Lilly S, Colless M. 1999. *MNRAS* 306:843
Goldader JD, Meurer G, Heckman TM, et al. 2002. *Ap. J.* 568:651
González V, Bouwens R, Illingworth G, et al. 2014. *Ap. J.* 781:34
González V, Labbé I, Bouwens RJ, et al. 2011. *Ap. J. Lett.* 735:L34
Goto T, Arnouts S, Inami H, et al. 2011a. *MNRAS* 410:573
Goto T, Arnouts S, Malkan M, et al. 2011b. *MNRAS* 414:1903
Governato F, Brook C, Mayer L, et al. 2010. *Nature* 463:203
Grazian A, Cstellano M, Koekemoer AM, et al. 2011. *Astron. Astrophys.* 532:33
Grogin NA, Kocevski DD, Faber SM, et al. 2011. *Ap. J. Suppl.* 197:35
Gruppioni C, Pozzi F, Andreani P, et al. 2010. *Astron. Astrophys.* 518:L27
Gruppioni C, Pozzi F, Rodighiero G, et al. 2013. *MNRAS* 432:23
Guedes J, Callegari S, Madau P, Mayer L. 2011. *Ap. J.* 742:76
Gunawardhana MLP, Hopkins AM, Bland-Hawthorn J, et al. 2013. *MNRAS* 433:2764
Guo Q, White S. 2013. *MNRAS* 437:3228
Guo Q, White S, Li C, Boylan-Kolchin M. 2010. *MNRAS* 404:1111
Guzman R, Gallego J, Koo DC, et al. 1997. *Ap. J.* 489:559
Haardt F, Madau P. 2012. *Ap. J.* 746:125
Haarsma DB, Partridge RB, Windhorst RA, Richards EA. 2000. *Ap. J.* 544:641
Hathi NP, Ryan RE, Cohen SH, et al. 2010. *Ap. J.* 720:1708

Hayes M, Schaerer D, Östlin G. 2010. *Astron. Astrophys.* 509:L5
Hernquist L, Springel V. 2003. *MNRAS* 341:1253
Hirschi R, Meynet G, Maeder A. 2005. *Astron. Astrophys.* 433:1013
Hogg DW, Cohen JG, Blandford R, Pahre MA. 1998. *Ap. J.* 504:622
Hopkins AM. 2004. *Ap. J.* 615:209
Hopkins AM, Beacom JF. 2006. *Ap. J.* 651:142
Hopkins AM, Beacom JF. 2008. *Ap. J.* 682:1486
Hopkins AM, Connolly AJ, Szalay AS. 2000. *Astron. J.* 120:2843
Hopkins PF, Quataert E, Murray N. 2012. *MNRAS* 421:3522
Hopkins PF, Richards GT, Hernquist L. 2007. *Ap. J.* 654:731
Horiuchi S, Beacom JF, Kochanek CS, et al. 2011. *Ap. J.* 738:154
Huang J-S, Zheng XZ, Rigopoulou D, et al. 2011. *Ap. J.* 742:L13
Hughes DH, Serjeant S, Dunlop J, et al. 1998. *Nature* 394:241
Huynh MT, Frayer DT, Mobasher M, et al. 2007. *Ap. J. Lett.* 667:L9
Ilbert O, McCracken HJ, Le Fèvre O, et al. 2013. *Astron. Astrophys.* 556:A55
Ilbert O, Salvato M, Le Floc'h E, et al. 2010. *Ap. J.* 709:644
Iwata I, Ohta K, Tamura N. 2007. *MNRAS* 376:1557
Ivison RJ, Alexander DM, Biggs AD, et al. 2010a. *MNRAS* 402:245
Ivison RJ, Magnelli B, Ibar E, et al. 2010b. *Astron. Astrophys.* 518:L31
Jarosik N, Bennett CL, Dunkley J, et al. 2011. *Ap. J. Suppl.* 192:14
Jarvis M. 2012. *The VISTA Deep Extragalactic Observations* (*VIDEO*) *Survey*. Presented at Sci. Next Gen. Imaging Spectrosc. Surv., ESO Garching, 15–18 Oct.
Juneau S, Glazebrook K, Crampton D, et al. 2005. *Ap. J. Lett.* 619:L135
Kajisawa M, Ichikawa T, Tanaka I, et al. 2009. *Ap. J.* 702:1393
Karim A, Schinnerer E, Martínez-Sansigre A, et al. 2011. *Ap. J.* 730:61
Katz N, Weinberg DH, Hernquist L. 1996. *Ap. J. Suppl.* 105:19
Kauffmann G, Heckman TM, White SDM, et al. 2003. *MNRAS* 341:33
Kauffmann G, White SDM, Guiderdoni B. 1993. *MNRAS* 264:201
Kennicutt RC. 1998. *Annu. Rev. Astron. Astrophys.* 36:189
Kennicutt RC, Evans NJ. 2012. *Annu. Rev. Astron. Astrophys.* 50:531
Keres D, Katz N, Weinberg DH, Dave R. 2005. *MNRAS* 363:2
Kewley L, Kobulnicky HA. 2007. *Island Universes: Structure and Evolution of Disk Galaxies*, ed. RS de Jong. Dordrecht: Springer
Kistler M, Yüksel H, Beacom JF, Hopkins AM, Wyithe JSB. 2009. *Ap. J.* 705:104
Klypin A, Trujillo-Gomez S, Primack J. 2011. *Ap. J.* 740:102
Koekemoer AM, Faber SM, Ferguson HC. 2011. *Ap. J. Suppl.* 197:36
Kong X, Charlot S, Brinchmann J, Fall SM. 2004. *MNRAS* 349:769
Kroupa P. 2001. *MNRAS* 322:231
Krumholz MR, Dekel A. 2012. *Ap. J.* 753:16
Kuhlen M, Krumholz MR, Madau P, Smith B, Wise J. 2012. *Ap. J.* 749:36
Labbé I, Oesch PA, Bouwens RJ, et al. 2013. *Ap. J. Lett.* 777:L19
Lagache G, Puget J-L, Dole H. 2005. *Annu. Rev. Astron. Astrophys.* 43:727
Laird ES, Nandra K, Adelberger KL, et al. 2005. *MNRAS* 359:47
Laird ES, Nandra K, Hobbs A, et al. 2006. *MNRAS* 373:217
Lanzetta KM, Wolfe AM, Turnshek DA. 1995. *Ap. J.* 440:435
Lapi A, González-Nuevo J, Fan L, et al. 2011. *Ap. J.* 742:24
Lawrence A, Walker D, Rowan-Robinson M, Leech KJ, Penston MV. 1986. *MNRAS* 219:687
Le Floc'h E, Papovich C, Dole H, et al. 2005. *Ap. J.* 632:169
Lee K-S, Ferguson HC, Wiklind T, et al. 2012. *Ap. J.* 752:66
Lehmer BD, Brandt WN, Alexander DM, et al. 2005. *Astron. J.* 129:1
Leitherer C, Schaerer D, Goldader JD, et al. 1999. *Ap. J. Suppl.* 123:3
Li C, White SDM. 2009. *MNRAS* 398:2177
Li W, Chornock R, Leaman J, et al. 2011. *MNRAS* 412:1473

Lilly SJ, Le Fevre O, Hammer F, Crampton D. 1996. *Ap. J. Lett.* 460:L1
Lonsdale Persson CJ, Helou G. 1987. *Ap. J.* 314:513
Lorenzoni S, Bunker AJ, Wilkins SM, et al. 2013. *MNRAS* 429:150
Lutz D. 2014. *Annu. Rev. Astron. Astrophys.* 52:373–414
Ly C, Lee JC, Dale DA, et al. 2011. *Ap. J.* 726:109
Ly C, Malkan MA, Kashikawa N, et al. 2007. *Ap. J.* 657:738
Mac Low M-M. 2013. *Science* 340:6140
Machalski J, Godlowski W. 2000. *Astron. Astrophys.* 360:463
Madau P, Della Valle M, Panagia N. 1998a. *MNRAS* 297:L17
Madau P, Ferguson HC, Dickinson ME, et al. 1996. *MNRAS* 283:1388
Madau P, Haardt F, Rees MJ. 1999. *Ap. J.* 514:648
Madau P, Pozzetti L, Dickinson M. 1998b. *Ap. J.* 498:106
Madau P, Shull JM. 1996. *Ap. J.* 457:551
Maeder A. 1992. *Astron. Astrophys.* 264:105
Magdis GE, Daddi E, Béthermin M, et al. 2012. *Ap. J.* 760:6
Magdis GE, Elbaz D, Daddi E, et al. 2010. *Ap. J.* 714:1740
Magnelli B, Elbaz D, Chary RR, et al. 2009. *Astron. Astrophys.* 496:57
Magnelli B, Elbaz D, Chary RR, et al. 2011. *Astron. Astrophys.* 528:A35
Magnelli B, Popesso P, Berta S, et al. 2013. *Astron. Astrophys.* 553:A132
Mancini C, Matute I, Cimatti A, et al. 2009. *Astron. Astrophys.* 500:705
Mao MY, Huynh MT, Norris RP, et al. 2011. *Ap. J.* 731:79
Maoz D, Gal-Yam A. 2004. *MNRAS* 347:951
Maraston C. 2005. *MNRAS* 362:799
Maraston C, Daddi E, Renzini A, et al. 2006. *Ap. J.* 652:85
Maraston C, Pforr J, Renzini A, et al. 2010. *MNRAS* 407:830
Marchesini D, van Dokkum PG, Forster Schreiber NM, et al. 2009. *Ap. J.* 701:1765
Marchesini D, Whitaker KE, Brammer G, et al. 2010. *Ap. J.* 725:1277
Marconi A, Risaliti G, Gilli R, et al. 2004. *MNRAS* 351:169
Mattila S, Dahlen T, Efstathiou A, et al. 2012. *Ap. J.* 756:111
Mauch T, Sadler EM. 2007. *MNRAS* 375:931
Mauduit J-C, Lacy M, Farrah D, et al. 2012. *Publ. Astron. Soc. Pac.* 124:714
McLure RJ, Cirasuolo M, Dunlop JS, et al. 2009. *MNRAS* 395:2196
McLure RJ, Dunlop JS, Bowler RAA, et al. 2013. *MNRAS* 432:2696
Meurer GR, Heckman TM, Calzetti D. 1999. *Ap. J.* 521:64
Moore B, Ghigna S, Governato F, et al. 1999. *Ap. J. Lett.* 524:L19
Moster BP, Naab T, White SDM. 2013. *MNRAS* 428:3121
Moster BP, Somerville RS, Maulbetsch C, et al. 2010. *Ap. J.* 710:903
Moustakas J, Coil A, Aird J, et al. 2013. *Ap. J.* 767:50
Moustakas J, Kennicutt RC, Tremonti CA. 2006. *Ap. J.* 642:775
Murphy EJ, Condon JJ, Schinnerer E, et al. 2011. *Ap. J.* 737:67
Muzzin A, Marchesini D, Stefanon M, et al. 2013. *Ap. J.* 777:18
Nagamine K, Cen R, Hernquist L, Ostriker JP, Springel V. 2004. *Ap. J.* 610:45
Navarro JF, Steinmetz M. 2000. *Ap. J.* 538:477
Nestor DB, Shapley AE, Kornei KA, Steidel CC, Siana B. 2013. *Ap. J.* 765:47
Noeske KG, Faber SM, Weiner BJ, et al. 2007. *Ap. J. Lett.* 660:L47
Nonino M, Dickinson M, Rosati P, et al. 2009. *Ap. J. Suppl.* 183:244
Nordon R, Lutz D, Shao L, et al. 2010. *Astron. Astrophys.* 518:L24
Ocvirk P, Pichon C, Teyssier R. 2008. *MNRAS* 390:1326
Oesch PA, Bouwens RJ, Carollo CM, et al. 2010. *Ap. J. Lett.* 725:L150
Oesch PA, Bouwens RJ, Illingworth GD, et al. 2012. *Ap. J.* 759:135
Oesch PA, Bouwens RJ, Illingworth GD, et al. 2013. *Ap. J.* 773:75
Ouchi M, Mobasher B, Shimasaku K, et al. 2009. *Ap. J.* 706:1136
Ouchi M, Shimasaku K, Okamura S, et al. 2004. *Ap. J.* 611:660

Overzier RA, Heckman TM, Wang J, et al. 2011. *Ap. J. Lett.* 726:L7
Panter B, Jimenez R, Heavens AF, Charlot S. 2007. *MNRAS* 378:1550
Papovich C, Dickinson M, Ferguson HC. 2001. *Ap. J.* 559:620
Papovich C, Rudnick G, Le Floc'h E, et al. 2007. *Ap. J.* 668:45
Pascarelle SM, Lanzetta KM, Fernández-Soto A. 1998. *Ap. J.* 508:L1
Patel H, Clements DL, Vaccari M, et al. 2013. *MNRAS* 428:291
Pawlik AH, Schaye J, van Scherpenzeel E. 2009. *MNRAS* 394:4
Peebles PJE. 1982. *Ap. J. Lett.* 263:L1
Pei YC, Fall SM. 1995. *Ap. J.* 454:69
Pei YC, Fall SM, Hauser MG. 1999. *Ap. J.* 522:604
Penner K, Dickinson M, Pope A, et al. 2012. *Ap. J.* 759:28
Pérez-González PG, Rieke GH, Egami E, et al. 2005. *Ap. J.* 630:82
Pérez-González PG, Rieke GH, Villar V, et al. 2008. *Ap. J.* 675:234
Persic M, Rephaeli Y, Braito V, et al. 2004. *Astron. Astrophys.* 419:849
Persic M, Salucci P. 1992. *MNRAS* 258:14
Pettini M. 2006. In *The Fabulous Destiny of Galaxies: Bridging Past and Present*, ed. V LeBrun, A Mazure, S Arnouts, D Burgarella, pp. 319–335. Paris: Frontiers Group
Pforr J, Maraston C, Tonini C. 2012. *MNRAS* 422:3285
Pope A, Bussmann RS, Dey A, et al. 2008. *Ap. J.* 689:127
Porciani C, Madau P. 2001. *Ap. J.* 548:522
Portinari L, Moretti A, Chiosi C, Sommer-Larsen J. 2004. *Ap. J.* 604:579
Pozzetti L, Bolzonella M, Zucca E, et al. 2010. *Astron. Astrophys.* 523:A13
Rafelski M, Wolfe AM, Prochaska JX, Neeleman M, Mendez AJ. 2012. *Ap. J.* 755:89
Raiter A, Fosbury RAE, Teimoorinia H. 2010. *Astron. Astrophys.* 510:A109
Ranalli P, Comastri A, Setti G. 2003. *Astron. Astrophys.* 399:39
Reddy NA, Erb DK, Pettini M, et al. 2010. *Ap. J.* 712:1070
Reddy NA, Dickinson M, Elbaz D, et al. 2012. *Ap. J.* 744:154
Reddy NA, Steidel CC. 2004. *Ap. J. Lett.* 603:L13
Reddy NA, Steidel CC. 2009. *Ap. J.* 692:778
Reddy NA, Steidel CC, Fadda D, et al. 2006. *Ap. J.* 744:792
Reddy NA, Steidel CC, Pettini M, et al. 2008. *Ap. J. Suppl.* 175:48
Retzlaff J, Rosati P, Dickinson M, et al. 2010. *Astron. Astrophys.* 511:A50
Riechers DA, Bradford CM, Clements DL, et al. 2013. *Nature* 496:329
Robertson BE, Ellis RS. 2012. *Ap. J.* 744:95
Robertson BE, Furlanetto SR, Schneider E, et al. 2013. *Ap. J.* 768:71
Robotham ASG, Driver SP. 2011. *MNRAS* 413:2570
Rodighiero G, Cimatti A, Gruppioni C, et al. 2010. *Astron. Astrophys.* 518:25
Rodighiero G, Daddi E, Baronchelli I, et al. 2011. *Ap. J. Lett.* 739:L40
Rowan-Robinson M, Mann RG, Oliver SJ, et al. 1997. *MNRAS* 289:490
Rudnick G, Rix H-W, Franx M, et al. 2003. *Ap. J.* 599:847
Rujopakarn W, Eisenstein DJ, Rieke GH, et al. 2010. *Ap. J.* 718:1171
Rush B, Malkan MA, Spinoglio L. 1993. *Ap. J. Suppl.* 89:1
Ryan-Weber EV, Pettini M, Madau P, Zych B. 2009. *MNRAS* 395:1476
Sadler EM, Jackson CA, Cannon RD. 2002. *MNRAS* 329:227
Salim S, Dickinson M, Rich RM, et al. 2009. *Ap. J.* 700:161
Salim S, Rich RM, Charlot S, et al. 2007. *Ap. J. Suppl.* 173:267
Salpeter EE. 1955. *Ap. J.* 121:161
Salucci P, Persic M. 1999. *MNRAS* 309:923
Sanders DB, Mazzarella JM, Kim DC, Surace JA, Soifer BT. 2003. *Astron. J.* 126:1607
Santini P, Fontana A, Grazian A, et al. 2012. *Astron. Astrophys.* 538:A33
Sargent MT, Béthermin M, Daddi E, Elbaz D. 2012. *Ap. J. Lett.* 747:L31
Sargent MT, Schinnerer E, Murphy E, et al. 2010a. *Ap. J. Suppl.* 186:341
Sargent MT, Schinnerer E, Murphy E, et al. 2010b. *Ap. J. Lett.* 714:L190

Saunders W, Rowan-Robinson M, Lawrence A, et al. 1990. *MNRAS* 242:318
Sawicki M, Yee HKC. 1998. *Ap. J.* 115:1329
Sawicki MJ, Lin H, Yee HKC. 1997. *Astron. J.* 113:1
Schaerer D. 2002. *Astron. Astrophys.* 382:28
Schaerer D. 2003. *Astron. Astrophys.* 397:527
Schaller G, Schaerer D, Meynet G, Maeder A. 1992. *Astron. Astrophys.* 96:269
Schaye J, Dalla Vecchia C, Booth CM, et al. 2010. *MNRAS* 402:1536
Schenker MA, Robertson BE, Ellis RS, et al. 2013. *Ap. J.* 768:196
Schiminovich D, Ilbert O, Arnouts S, et al. 2005. *Ap. J. Lett.* 619:L47
Sedgwick C, Serjeant S, Pearson C, et al. 2011. *MNRAS* 416:1862
Serjeant S, Carramiñana A, Gonzáles-Solares E, et al. 2004. *MNRAS* 355:813
Serjeant S, Gruppioni C, Oliver S. 2002. *MNRAS* 330:621
Seymour N, Dwelly T, Moss D, et al. 2008. *MNRAS* 386:1695
Shankar F, Weinberg DH, Miralda-Escude J. 2009. *Ap. J.* 690:20
Shapley A, Steidel CC, Adelberger KL. 2001. *Ap. J.* 562:95
Shim H, Chary RR, Dickinson M, et al. 2011. *Ap. J.* 738:69
Shim H, Colbert J, Teplitz H, et al. 2009. *Ap. J.* 696:785
Shull JM, Harness A, Trenti M, Smith BD. 2012. *Ap. J.* 747:100
Shull JM, van Steenberg ME. 1985. *Ap. J.* 298:268
Shupe DL, Fang F, Hacking PB, Huchra JP. 1998. *Ap. J.* 501:597
Silverman JD, Green PJ, Barkhouse WA, et al. 2008. *Ap. J.* 679:118
Simcoe RA. 2011. *Ap. J.* 738:159
Simcoe RA, Cooksey KL, Matejek M, et al. 2011. *Ap. J.* 743:21
Smail I, Ivison RJ, Blain AW. 1997. *Ap. J. Lett.* 490:L5
Smith JDT, Draine BT, Dale DA, et al. 2007. *Ap. J.* 656:770
Smolčić V, Schinnerer E, Zamorani G, et al. 2009. *Ap. J.* 690:610
Sobral D, Best PN, Smail I, et al. 2013. *MNRAS* 428:1128
Sodriski TJ, Odegard N, Arendt RG, et al. 1997. *Ap. J.* 480:173
Soifer BT, Sanders DB, Madore BF, et al. 1987. *Ap. J.* 320:238
Soltan A. 1982. *MNRAS* 200:115
Somerville RS, Primack JR. 1999. *MNRAS* 310:1087
Springel V, Hernquist L. 2003. *MNRAS* 339:289
Springel V, Wang J, Vogelsberger M, et al. 2008. *MNRAS* 391:1685
Springel V, White SDM, Jenkins A, et al. 2005. *Nature* 435:629
Stadel J, Potter D, Moore B, et al. 2009. *MNRAS* 398:L21
Stanway ER, Bunker AJ, McMahon RG. 2003. *MNRAS* 342:439
Stark DP, Bunker A, Ellis RS, et al. 2007. *Ap. J.* 659:84
Stark DP, Ellis RS, Bunker A, et al. 2009. *Ap. J.* 697:1493
Stark DP, Schenker MA, Ellis RS, et al. 2013. *Ap. J.* 763:129
Steidel CC, Adelberger KL, Giavalisco M, Dickinson M, Pettini M. 1999. *Ap. J.* 519:1
Steidel CC, Adelberger KL, Shapley AE, Pettini M, Dickinson M. 2003. *Ap. J.* 592:728
Steidel CC, Giavalisco M, Pettini M, Dickinson M, Adelberger KL. 1996. *Ap. J. Lett.* 462:L17
Steidel CC, Pettini M, Hamilton D. 1995. *Astron. J.* 110:2519
Steidel CC, Shapley AE, Pettini M, et al. 2004. *Ap. J.* 604:534
Sullivan M, Treyer MA, Ellis RS, et al. 2000. *MNRAS* 312:442
Takeuchi TT, Ishii TT, Dole H, et al. 2006. *Astron. Astrophys.* 448:525
Takeuchi TT, Yoshikawa K, Ishii TT. 2003. *Ap. J. Lett.* 587:L89
Thompson RI, Eisenstein D, Fan X, et al. 2006. *Ap. J.* 647:787
Tilvi V, Papovich C, Tran KVH, et al. 2013. *Ap. J.* 768:56
Tinsley BM. 1980. *Fundam. Cosmic Phys.* 5:287
Trenti M, Perna R, Levesque EM, Shull JM, Stocke JT. 2012. *Ap. J. Lett.* 749:L38
Tresse L, Ilbert O, Zucca E, et al. 2007. *Astron. Astrophys.* 472:403
Tresse L, Maddox SJ. 1998. *Ap. J.* 495:691

Tresse L, Maddox SJ, Le Fevre O, Cuby J-G. 2002. *MNRAS* 337:369
Treyer M, Ellis RS, Milliard B, Donas J, Bridges TJ. 1998. *MNRAS* 300:303
Vaccari M, Marchetti L, Franceschini A, et al. 2010. *Astron. Astrophys.* 518:L20
Vale A, Ostriker JP. 2004. *MNRAS* 353:189
Vanzella E, Guo Y, Giavalisco M, et al. 2012. *Ap. J.* 751:70
Venkatesan A, Truran JW. 2003. *Ap. J. Lett.* 594:L1
Verma A, Lehnert MD, Förster Schreiber NM, et al. 2007. *MNRAS* 377:1024
Wang W-H, Cowie LL, Barger AJ, et al. 2010. *Ap. J. Suppl.* 187:251
Wardlow JL, Smail I, Coppin KEK, et al. 2011. *MNRAS* 415:1479
Weidemann V. 2000. *Astron. Astrophys.* 363:647
Westra E, Geller MJ, Kurtz MJ, Fabricant DG, Dell'Antonio I. 2010. *Ap. J.* 708:534
White SDM, Frenk CS. 1991. *Ap. J.* 379:52
White SDM, Rees MJ. 1978. *MNRAS* 183:341
Wiklind T, Dickinson M, Ferguson HC, et al. 2008. *Ap. J.* 676:781
Wilkins SM, Hopkins AM, Trentham N, Tojeiro R. 2008b. *MNRAS* 391:363
Wilkins SM, Trentham N, Hopkins AM. 2008a. *MNRAS* 385:687
Wilson G, Cowie LL, Barger AJ, Burke DJ. 2002. *Astron. J.* 124:1258
Wolf C, Bell EF, McIntosh DH, et al. 2005. *Ap. J.* 630:771
Woosley SE, Bloom JS. 2006. *Annu. Rev. Astron. Astrophys.* 44:507
Wyder TK, Treyer MA, Milliard B, et al. 2005. *Ap. J. Lett.* 619:L15
Yabe K, Ohta K, Iwata I, et al. 2009. *Ap. J.* 693:507
Yan H, Dickinson M, Giavalisco M, et al. 2006. *Ap. J.* 651:24
Yan H, Finkelstein SL, Kuang K-H, et al. 2012. *Ap. J.* 761:177
Yan L, McCarthy PJ, Freudling W, et al. 1999. *Ap. J. Lett.* 519:L47
Yang X, Mo HJ, van den Bosch FC. 2003. *MNRAS* 339:1057
Yepes G, Kates R, Khokhlov A, Klypin A. 1997. *MNRAS* 284:235
York DG, Adelman J, Anderson JE, et al. 2000. *Astron. J.* 120:1579
Yoshida M, Shimasaku K, Kashikawa N, et al. 2006. *Ap. J.* 653:988
Yun MS, Reddy NA, Condon JJ. 2001. *Ap. J.* 554:803
Zemp M, Gnedin OY, Gnedin NY, Kravtsov AV. 2012. *Ap. J.* 748:54
Zentner AR, Hearin AP, van den Bosch FC. 2013. *MNRAS.* Submitted. arXiv:1311.1818
Zheng W, Postman M, Zitrin A, et al. 2012. *Nature* 489:406

Mass Loss: Its Effect on the Evolution and Fate of High-Mass Stars

Nathan Smith

Steward Observatory, University of Arizona, Tucson, Arizona 85721;
email: nathans@as.arizona.edu

Annu. Rev. Astron. Astrophys. 2014. 52:487–528

First published online as a Review in Advance on June 16, 2014

The *Annual Review of Astronomy and Astrophysics* is online at astro.annualreviews.org

This article's doi:
10.1146/annurev-astro-081913-040025

Keywords

binary stars, massive stars, stellar evolution, stellar winds, supernovae, variables

Abstract

Our understanding of massive star evolution is in flux due to recent upheavals in our view of mass loss and observations of a high binary fraction among O-type stars. Mass-loss rates for standard metallicity-dependent winds of hot stars are lower by a factor of 2–3 compared with rates adopted in modern stellar evolution codes, due to the influence of clumping on observed diagnostics. Weaker hot star winds shift the burden of H-envelope removal to the winds, pulsations, and eruptions of evolved supergiants, as well as binary mass transfer. Studies of stripped-envelope supernovae, in particular, require binary mass transfer. Dramatic examples of eruptive mass loss are seen in Type IIn supernovae, which have massive shells ejected just a few years earlier. These eruptions are a prelude to core collapse, and may signify severe instabilities in the latest nuclear burning phases. We encounter the predicament that the most important modes of mass loss are also the most uncertain, undermining the predictive power of single-star evolution models. Moreover, the influence of winds and rotation has been evaluated by testing single-star models against observed statistics that, it turns out, are heavily influenced by binary evolution. Altogether, this may alter our view about the most basic outcomes of massive-star mass loss—are Wolf-Rayet stars and Type Ibc supernovae the products of winds, or are they mostly the result of binary evolution and eruptive mass loss? This is not fully settled, but mounting evidence points toward the latter. This paradigm shift impacts other areas of astronomy, because it changes predictions for ionizing radiation and wind feedback from stellar populations, it may alter conclusions about star-formation rates and initial mass functions, it affects the origin of compact stellar remnants, and it influences how we use supernovae as probes of stellar evolution across cosmic time.

1. INTRODUCTION

Massive star: star with an initial mass above $M_{ZAMS} = 8M_{\odot}$ that experiences a core collapse or other SN at death

Supernova (SN): the collapse of the iron core in massive stars, but could also result from an electron-capture SN ($8-10\,M_{\odot}$) or pair instability explosion

Metallicity (*Z*): relative abundance of elements heavier than He; important for radiatively driven winds and for the opacity in parts of a star's envelope

Because of the scaling of luminosity with initial mass, relatively rare stars born with masses above $\sim$10–20 $M_{\odot}$ vastly outshine the much larger number of lower-mass stars. The radiative output of massive stars is so intense that photon momentum can drive strong winds, and the energy transport through the stellar interior may jeopardize the stability of the star itself. Mass loss from massive stars has a *deterministic* influence on the structure and evolution of those stars, which in turn have tremendous impact on other areas of astronomy.

Massive stars are the cosmic engines that provide most of the luminosity in star-forming galaxies. Because massive stars have short lifetimes, their UV radiation (reprocessed in various forms) is the most observable tracer of star formation and is used to calculate star-formation rates (Kennicutt 1998, Kennicutt & Evans 2012). Feedback in the form of UV radiation, stellar winds, and supernovae (SNe) stirs interstellar gas, driving turbulence and perhaps triggering new generations of stars by sweeping gas into dense filaments (Elmegreen & Lada 1977). Through this feedback, massive stars have a profound impact on the evolution of disk galaxies (van der Kruit & Freeman 2011). Massive-star feedback may also terminate star formation locally, blowing giant bubbles (Cox 2005) and leaking hot processed gas into the Galactic Halo (Putman et al. 2012). Elemental yields from nuclear burning in the cores of massive stars and explosive nucleosynthesis in SNe provide the elements in the periodic table that pollute the ISM, driving galactic chemical evolution and the metallicity (*Z*) evolution of the Universe. Through their violent deaths as SNe, massive stars provide brilliant displays that permit us to dissect individual stellar interiors at distances of many megaparsecs while the star's inner layers are peeled away for us to see (Filippenko 1997). They leave behind exotic corpses such as black holes, neutron stars, pulsars, magnetars, and all the bizarre high-energy phenomena that occur when compact objects remain bound in a binary system (e.g., Remillard & McClintock 2006). Shock fronts in their beautifully complex SN remnants echo through the ISM for thousands of years after the bright SN display has faded (Reynolds 2008).

Mass loss affects a star's luminosity, burning lifetime, and apparent temperature; the hardness of its emitted radiation field; and its He core mass; it will also profoundly impact the end fate of a star. As a consequence, changes in estimates of mass-loss rates can alter expectations for their collective ionizing radiation, UV luminosity, winds, and SNe. Thus, understanding massive stars and their mass loss also remains important as we push ever farther to the distant reaches of the Universe. Inferences about the initial mass function (IMF; Bastian et al. 2010) and variation of the star-formation rate through cosmic time (Madau et al. 1998, Hopkins & Beacom 2006) hinge upon converting reprocessed massive-star UV luminosity to a collective star-formation rate. Gamma ray bursts (GRBs; Woosley & Bloom 2006, Gehrels et al. 2009) and SNe also provide a probe of very distant populations and stellar evolution in early environments, but this requires that we understand the connection between stars of various types, their mass loss, and the eventual type of SN seen. The first stars in *Z*-poor environments are expected to be very massive (Bromm & Larson 2004), and hence the scaling of mass loss with *Z* affects the detailed abundances of the lowest *Z* stars (Beers & Christlieb 2005, Bromm & Yoshida 2011) polluted by a small number of early SNe. Because dense stellar winds can absorb portions of the Lyman continuum emitted by a hot star's photosphere (Najarro et al. 1996), estimates of mass-loss rates and their scaling with metallicity can profoundly impact topics as remote as reionization of the Universe and the interpretation of spectra from galaxies at the highest redshifts currently being detected (Loeb & Barkana 2001, Fan et al. 2006, Morales & Wyithe 2010).

Although convenient recipes and simple scaling relations to account for the collective effects of mass loss and feedback will always be available, researchers working in these other branches of astronomy should not believe that such recipes are reliable to better than order-of-magnitude levels. Even in the very local Universe, where we have excellent multiwavelength observations, there is still tremendous uncertainty in derived mass-loss rates for massive stars, and so there is large uncertainty in their influence on evolution. This is exacerbated by the predicament that these uncertainties are largest for the most massive and most luminous stars, but these stars also tend to be the most influential. Extrapolating to the early Universe is still quite risky and should always raise eyebrows. This review provides a broad overview of the current understanding of mass loss and its influence on the evolution of massive stars and raises a flag of caution about the uncertainties involved.

Main sequence (MS): core-H burning lifetime of massive stars, when they appear as O-type stars or early B-type stars; also includes some WNH stars

ZAMS: zero-age main sequence

Non-LTE: radiative transfer calculations not invoking local thermodynamic equilibrium

1.1. The Importance of Mass Loss in Massive-Star Evolution

For low- and intermediate-mass stars ($M_{\rm ZAMS} < 8\,\rm M_{\odot}$; zero-age main sequence, ZAMS), wind mass loss is relatively unimportant for evolution until the final stages as the asymptotic-giant-branch (AGB) star transitions to a protoplanetary nebula. For massive stars, however, mass loss cannot be ignored. For most of their lives—even on the main sequence (MS)—massive stars above $\sim 20\,\rm M_{\odot}$ shed mass in fast winds that affect their subsequent evolution (i.e., $\int \dot{M} dt$ can be a significant fraction of the stellar mass), and in post-MS phases the mass loss becomes critical in determining the type of resulting SN explosion.

Thus, mass loss is inexorably linked to evolution for massive stars. This article will not provide a detailed review of our understanding of stellar evolution from a theoretical perspective, because this has already been done in a number of excellent reviews. Although somewhat dated, Chiosi & Maeder (1986) provide a good description of how models that incorporate mass loss compare with stellar evolution models without mass loss. Maeder & Meynet (2000) review how the inclusion of rotation can influence single-star evolution models, whereas Langer (2012) reviews more recent advances as well as important aspects of how close binarity may dramatically change the evolutionary paths of stars. Finally, Woosley et al. (2002) discuss stellar evolution models with particular emphasis on the late pre-SN burning phases and their connection to core collapse.

Instead, this review concentrates on observational estimates of mass loss and their impact on stellar evolution, because this is where the largest uncertainty currently resides. Stellar evolution calculations must adopt prescriptions for mass-loss rates as input to their code, but these assumed rates determine the outcome of evolution. Depending on precarious assumptions about wind strength, a red supergiant (RSG) can be made to evolve to the blue on the Hertzsprung-Russell (HR) diagram or not. A massive star can be driven to the luminous blue variable (LBV) phase or it can avoid this phase altogether. Moreover, the mass-loss rates typically used in stellar evolution models are time-averaged, even though eruptive and explosive mass-loss events are observed. Hence, there is great uncertainty in the predictions of all evolutionary models. Model tracks on the HR diagram are not plotted with error bars that reflect these uncertain assumptions.

Owing to the need for time-averaged prescriptions in stellar evolution calculations, the steady line-driven winds of hot stars have consumed the majority of effort in understanding mass loss in the massive-star community for the past three decades, both theoretically and observationally. There have been great leaps in quantitative non-LTE (local thermodynamic equilibrium) modeling of spectra influenced by winds. A previous review by Kudritzki & Puls (2000) dealt with line-driven winds of hot stars, focusing on the theory of the driving mechanism as well as common scaling relations used to convert observables to rates. A more recent review by Puls et al. (2008) also concentrated on the relatively steady line-driven winds of hot stars, providing a thorough discussion of wind theory and its connection to a wide array of observational diagnostics. Those reviews did not focus on more extreme winds of evolved massive stars or mass loss in binaries, and they did not discuss highly time-dependent mass loss associated with transient events (eruptions, explosions, mergers). This review therefore emphasizes these latter topics, but it also discusses more recent developments in steady winds, in particular the reduction in mass-loss rates and the consequent paradigm shift in stellar evolution.

Roche-lobe overflow (RLOF): phase when either star of a binary system has a radius exceeding the inner Lagrange point and begins to transfer mass to the other star

1.2. Historical Perspective and Paradigm Shift

Our present understanding of mass loss has undergone dramatic changes, due in large part to major shifts in our quantitative estimates of what mass-loss rates actually are. The historical path to our current understanding has had some interesting swings.

Very early observations of transient events like Tycho's SN, P Cygni's 1600 AD eruption, and many nova eruptions, as well as the broad emission lines in Wolf-Rayet (WR) stars (Wolf & Rayet 1867) and the broad blueshifted absorption seen in spectra of many objects, indicated the presence of outflowing material. However, these were seen as rare stars or brief eruptive events, and their connection to the lives of normal stars was unclear. Later, estimates for the solar wind (Parker 1958) powered by hot gas pressure had such low mass-loss rates that winds would not matter much in stellar evolution. There were some important early indications and expectations of mass loss from hot stars (Sobolev 1960), but the mass-loss rates were very uncertain. Therefore, the dominant paradigm (e.g., Paczynski 1966, 1967, 1971) was that, as in low-mass stars, binary Roche-lobe overflow (RLOF) played a major role in making massive stripped-envelope stars like the He-rich WR stars.

The birth of UV astronomy triggered a revolution in our understanding of massive stars by providing the decisive evidence that all luminous hot stars have strong winds, indicated by deep P Cyg absorption in their UV resonance lines (Morton 1967). This led to a new paradigm wherein line-driven winds dominate the stripping of the H envelope, rather than binaries, leading to WR stars and Type Ibc SN progenitors via single-star evolution (i.e., the so-called Conti scenario; Conti 1976). As with UV observations, the development of IR detectors led to a similar recognition of the importance of dust-driven winds in RSGs (Gehrz & Woolf 1971). With the assumption that mass loss via steady winds is the dominant form of mass loss, theorists could adopt simple prescriptions for that mass loss and calculate single-star evolutionary tracks on the HR diagram (see reviews by Chiosi & Maeder 1986, Maeder & Meynet 2000). These single-star models were able to provide a plausible explanation for observed distributions of stars, including the relative amounts of time spent in different evolutionary phases as O-type stars, WR stars, and RSGs. Massey (2003) reviewed how these single-star models have been compared to observations and how observed statistics have been used to test and refine the single-star models. This single-star paradigm then permitted one to extend even further, to compute models for the collective radiative output of entire stellar populations as a function of age (e.g., codes like STARBURST99; Leitherer et al. 1999).

In the past decade this paradigm has shifted yet again because of two important realizations: (*a*) Due to the effects of clumping (see below), empirical mass-loss rates for line-driven winds are lower than previously thought, and (*b*) other unsteady modes of mass loss are more important than previously recognized, owing to increased estimates for mass ejected in episodic mass-loss events (Smith & Owocki 2006) and the very high binary fraction among massive stars (Sana et al. 2012). These changes to our view of mass loss are discussed in detail below. This may alter a huge number of predictions for the evolutionary tracks of stars and their variation with *Z*. The pendulum appears to be swinging back to binaries as the dominant agent in massive-star evolution, although not everyone is on the bandwagon. Justifiably, this topic occupies much of the current debate and excitement in the field.

TYPES OF H-BURNING MASSIVE STARS

OB dwarfs: Hot massive MS stars with luminosity class V, early in core-H burning. *Examples:* θ^1 *Ori C*, ζ *Oph*

Of, Oe, Ofpe/WN9 (etc.) supergiants: O stars with evidence for strong winds in their spectra (i.e., emission lines). These are thought to be more evolved than O dwarfs, either in the later phases of core-H burning or in transition to He burning. *Examples:* ζ Puppis, Hen 3–519, S61.

WNH: H-rich stars with WR signatures in their spectra; more colloquially known as "O-stars on steroids." If single, these stars are in the late stages of core-H burning MS evolution of the most massive stars. If in binary systems, these are the possible products of mergers, like more massive analogs of blue stragglers. These are the most massive stars in young massive star clusters and the most massive stars measured in binaries. *Examples:* WR25, WR20a.

2. THE DIMINISHED ROLE OF STEADY LINE-DRIVEN WINDS

Massive stars spend the majority of their lives as hot stars (mostly as OB types, and more briefly as WR stars), and during these phases the dominant mode of mass loss is through line-driven winds (see the sidebar, Types of H-Burning Massive Stars). Again, see Kudritzki & Puls (2000) and Puls et al. (2008) for reviews concerning steady, line-driven winds from hot stars. Here we focus on the treatment of winds in evolution, as well as important updates.

2.1. The Standard View

In the wind of a hot star, momentum is transferred from the outwardly propagating radiation to the gas through absorption and scattering by UV metal lines (Lucy & Solomon 1970, Castor et al. 1975). Thus, the rate at which mass is lifted from the star by this mechanism depends on the UV luminosity of the star, the temperature (and ionization), and the metallicity Z (Mokiem et al. 2007, Puls et al. 2008). Most stellar evolution models therefore adopt simple prescriptions for the mass-loss rates that scale smoothly with the stellar luminosity, temperature, and metallicity. For the generation of models calculated throughout the 1990s, the most commonly adopted mass-loss rates were those of de Jager et al. (1988) and Nieuwenhuijzen & de Jager (1990) for O-type stars and RSGs, and the rates of Nugis & Lamers (2000) were adopted for WR stars. A representative example of the application of this technique to infer the evolution and fate of massive stars as a function of metallicity is discussed by Heger et al. (2003), where Z-dependent mass loss ($\propto \sqrt{Z}$) is adopted [many other stellar evolution models are reviewed by Langer (2012)]. Hence, it is expected that the total mass lost by a star increases smoothly with luminosity and Z, yielding trends with M_{ZAMS} and Z such as that shown in **Figure 1** (Heger et al. 2003). This basic picture has been widely regarded as the "standard view" of single-star evolution at high initial mass, although details of the implementation differ from one model to the next.

2.2. Observational Diagnostics and Clumping

The mass-loss rate, $\dot{M} = 4\pi r^2 \rho(r) v_\infty$, depends on the average mass density $\rho(r)$ at a particular radius at which the wind has reached its terminal velocity v_∞. Connecting these ideal values to observations is nontrivial and requires detailed models including realistic opacities, because the radius from which the emissivity or absorption originates is wavelength dependent. The wind terminal speed can be deduced from resolved profiles of P Cygni absorption in UV resonance

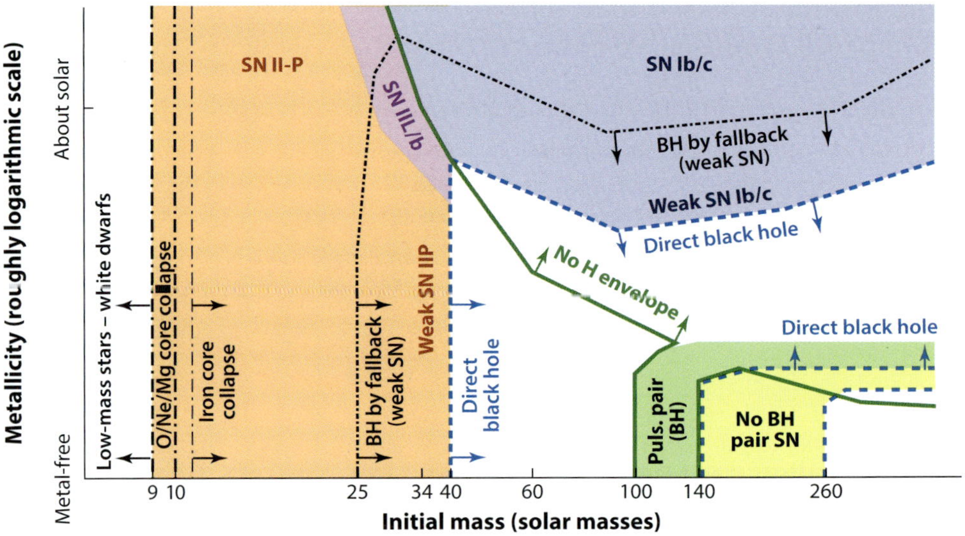

Figure 1

Example of the standard expectations for the fates of massive stars as a function of initial mass and metallicity. Adapted from Heger et al. (2003) with permission. Abbreviations: black hole, BH; supernova, SN.

lines. There are several diagnostics of the wind density, the most common being the strength of wind free-free emission in the IR or radio (Wright & Barlow 1975) and recombination emission lines like Hα or others, as well as the strength of blueshifted P Cygni absorption features in unsaturated UV resonance lines (see Puls et al. 2008).

Radio/IR free-free continuum excess and emission lines like Hα, HeI, and HeII are recombination processes, so their emissivity varies as ρ^2, whereas P Cyg absorption varies linearly with ρ. The quadratic density dependence of recombination emissivity raises a problem—if small-scale inhomogeneities (i.e., "clumps") permeate the wind, then recombination emission arising in dense clumps is stronger than emission from the same amount of mass distributed uniformly throughout the wind (in other words, $\langle\rho^2\rangle > \langle\rho\rangle^2$). It is now well established that winds are in fact clumpy (see below), so when mass-loss rates are derived from Hα or free-free excess using the assumption of a homogeneous wind, the mass-loss rates are overestimated. This is the case for the frequently used "standard" mass-loss rates of de Jager et al. (1988) and Nieuwenhuijzen & de Jager (1990), which assumed a smooth wind. The factor by which mass-loss rates are overestimated is $\sqrt{f_{\rm cl}}$, where it is standard practice to define the "clumping factor" as $f_{\rm cl} = \langle\rho^2\rangle/\langle\rho\rangle^2$. (This assumes that the gas is optically thin.) Constraining the value of $f_{\rm cl}$ observationally is paramount for understanding stellar evolution.

Significant wind clumping is expected on theoretical grounds (Owocki & Rybicki 1984, Owocki et al. 1988, Feldmeier 1995, Owocki & Puls 1999, Dessart & Owocki 2005, Sundqvist & Owocki 2013), due mostly to the line-driven instability (this arises because the force of line driving is velocity dependent; gas parcels that absorb line photons are accelerated and are therefore Doppler shifted out of the line to absorb adjacent photons). Clumps are expected on a size scale comparable to the Sobolov length, given by the thermal velocity divided by the radial velocity gradient in the wind (dv/dr), which means that clumping should exist on a size scale smaller than the stellar radius. Clumping may also be induced at the base of the wind because of subsurface convection

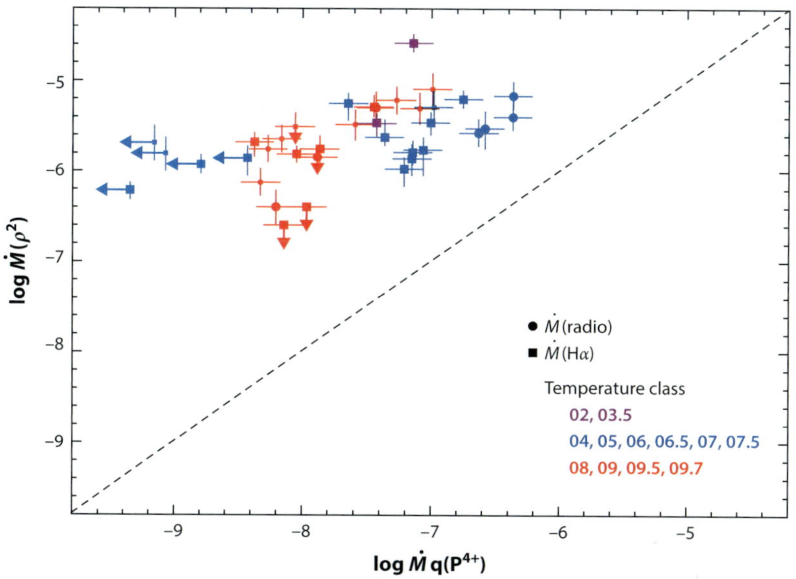

Figure 2

A comparison of the mass-loss rates derived from diagnostics that are linearly proportional to wind density, like UV P Cyg absorption, and that are proportional to ρ^2, like free-free and Hα emission. Reproduced from Fullerton et al. (2006) with permission. Although there is still discussion about the P V lines used for this study and whether they overestimate the reductions in $\dot{M}$ that are applied, this research nevertheless forced an important discussion in massive-star research by highlighting the potential influence of clumping.

driven by the Fe opacity bump (Cantiello et al. 2009). In fact, it has long been known that hot-star winds are probably clumpy on small scales (Hillier 1991, Drew et al. 1994, Moffat & Robert 1994) and inhomogeneous on larger scales; large scales include such complexities as the time-variable discrete absorption components (Howarth et al. 1995, Massa et al. 1995, Cranmer & Owocki 1996, Fullerton et al. 1997) or axisymmetric winds due to rapid rotation (Owocki et al. 1996), and a wide array of magnetically induced inhomogeneities (Townsend et al. 2005, ud-Doula & Owocki 2002). However, it was only recently that the severity of the problem for the global mass-loss rate was quantified.

One can check the influence of clumping on ρ^2 diagnostics (and thereby measure $f_{\rm cl}$) by studying the same winds using diagnostics that are linearly proportional to density, like P Cyg absorption in UV resonance lines and the strengths of electron-scattering wings. Using UV resonance absorption lines in O-type stars, Fullerton et al. (2006) proposed a reduction by a factor of 10 or more from traditional mass-loss rates ($f_{\rm cl}$ values of 100 or more), whereas Bouret et al. (2005) require reductions by factors of ~3 (see **Figure 2**). For the Milky Way, Large Magellanic Cloud (LMC), and Small Magellanic Cloud (SMC), various studies using modern non-LTE analysis find $f_{\rm cl} \simeq 10$ (Crowther et al. 2002, Figer et al. 2002, Hillier et al. 2003, Massa et al. 2003, Evans et al. 2004, Puls et al. 2006), corresponding to $\dot{M}$ reductions by a factor of ~3 with typical uncertainties of ~30%. Based on unifying Hα measurements with the theoretical wind momentum relation, both Repolust et al. (2004) and Markova et al. (2004) found $f_{\rm cl} = 5$ or a mass-loss reduction by 2.3. [Note that long ago, small-scale clumping with $f_{\rm cl}$ values of 4–20 (a reduction in mass-loss rates by factors of 2–4) was required to fit both the emission cores ($\propto \rho^2$) and electron-scattering wings ($\propto \rho$) in WR stars (Hillier 1991, Moffat & Robert 1994). Line wings of O-type stars are too weak for this analysis.] In addition, polarization variations in WR+O eclipsing binaries (St. Louis et al. 1993) implied $f_{cl} \simeq 10$.

Puls et al. (2006, 2008) and others have discussed that the larger mass-loss rate reduction of 10 found by Fullerton et al. (2006) may be an overestimate because of how ionization can affect the optically thin P V lines that the large clumping factor is based upon as well as because of possible mediating effects of porosity in the wind. These topics are not completely settled [the interested reader can consult the proceedings volume of a recent conference on this topic for more detailed information (Hamann et al. 2008)], but nevertheless, most observational studies agree that for mid/early O-type stars, clumping is significant enough to warrant mass-loss rates reduced by at least a factor of 2 to 3 relative to the standard rates from Hα and radio flux that assume homogeneous winds (de Jager et al. 1988, Nieuwenhuijzen & de Jager 1990). Reductions by factors of 2–4 are confirmed by X-ray observations (Kramer et al. 2003; Cohen et al. 2010, 2011). Thus, reductions in mass-loss rates for normal O-type stars by a factor of 3 ($\pm 30\%$) are a good guide.

2.3. Recent Developments and Modifications

In current generations of stellar evolution models (see the recent review by Langer 2012), the most commonly used prescription for the mass-loss rates of hot stars with line-driven winds is from Vink et al. (2001). These are theoretical mass-loss rates based on the expected radiative acceleration of a wind, calculated by loss of photon energy using a Monte-Carlo method. These $\dot{M}$ values are comparable to the values of the old "standard" rates derived observationally from ρ^2 diagnostics assuming homogeneous winds. The prescriptions for O-star mass-loss rates taken directly from de Jager et al. (1988) and Vink et al. (2001) are compared in **Figure 3** (this figure also includes $\dot{M}$ for other types of stars discussed in this review). For comparison, **Figure 3** also plots the de Jager et al. (1988) rates divided by factors of 3 (the favored reduction) and 10 (possibly an overestimate) to account for the way clumping affects the observationally derived rates. These $\dot{M}$ prescriptions are for O-type MS stars over a range of luminosities (using different M_{ZAMS} and T_{eff} values from stellar evolution models) at $Z_{\odot}$. All these values of $\dot{M}$ increase with both increasing L and decreasing T_{eff}, so $\dot{M}$ climbs by a factor of 3–4 as an O-type star evolves along the MS, giving rise to observed properties in Of and WNH stars. The most luminous WNH stars are generally assumed to be in the late phases of core-H burning, rather than He burning like traditional WR stars. Some other WN stars with H in their spectra are indeed thought to be transition objects and possibly related to LBVs. Binary evolution may of course play a role in creating some of these WN stars with H, and the connections between terminology and evolutionary state are often complicated.

An important point to recognize is that the theoretical Vink et al. (2001) prescription for $\dot{M}$ is almost the same as the empirical de Jager et al. (1988) prescription (in fact, for much of the range of O-star luminosities, the Vink et al. mass-loss rates are higher). This is for physical parameters from model ZAMS stars from Ekström et al. (2012); note that the Vink et al. recipe depends on L, T_{eff}, M, and v_{∞}, whereas the de Jager prescription uses only L and T_{eff}. The two prescriptions therefore change differently as a star evolves. In any case, it appears that stellar evolution calculations are still using mass-loss rates that are too high by a factor of $\sim$3 during the MS lifetimes of massive stars. [The Vink et al. (2001) mass-loss rates are indeed a factor of $\sim$2 lower than rates used in some massive-star evolution models. In particular, older models by Meynet et al. (1994) adopted mass-loss rates that were artificially a factor of 2 higher than the de Jager et al. (1988) mass-loss rates, because this enhanced mass loss did a better job of accounting for the observed statistics of WR stars. Moreover, these models with enhanced mass-loss rates are still often employed in

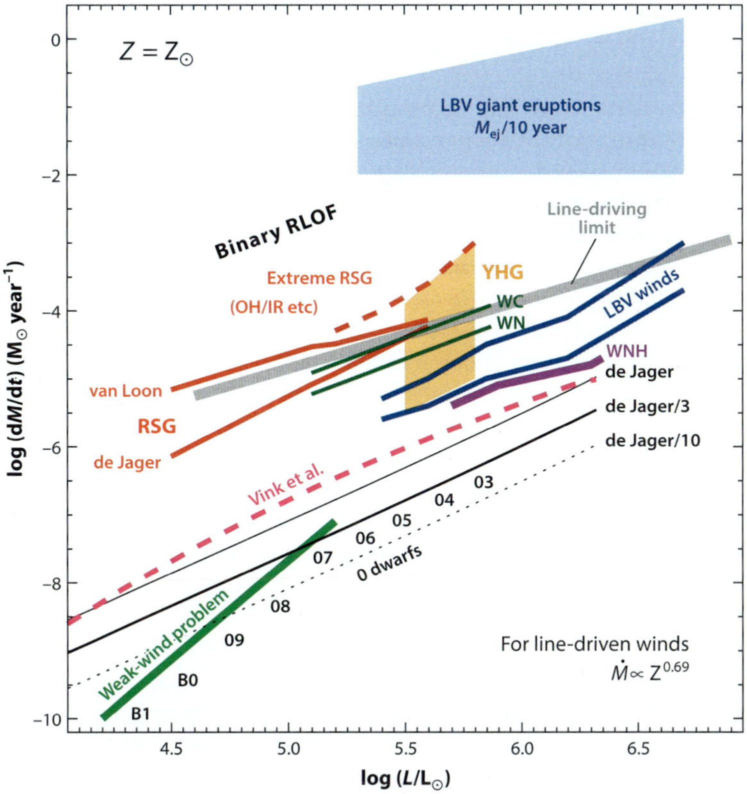

Figure 3

A number of different prescriptions for wind mass loss used in models, as well as typical observed ranges of mass-loss rates for a number of different types of stars. For O-type stars, the theoretical rates from the prescription by Vink et al. (2001) are shown, along with "standard" observational rates using the prescription by de Jager et al. (1988), as well as these same prescriptions divided by factors of 3 and 10 for comparison. The green line labeled "weak-wind problem" refers to lower mass-loss rates for late O-type and early B-type MS stars. Rates for nitrogen-sequence Wolf-Rayet (WN) and carbon-sequence Wolf-Rayet (WC) stars are from Crowther (2007). Red supergiant (RSG) mass-loss prescriptions are from de Jager et al. (1988) and van Loon et al. (2005), as indicated. For yellow hypergiants (YHGs), see de Jager (1998). For $\dot{M}$ corresponding to normal winds of luminous blue variables (LBVs), values were compiled from a number of studies (Hillier et al. 2001, Vink & de Koter 2002, Smith et al. 2004, Groh et al. 2009). For LBV eruptions, the "rates" shown are calculated from total masses observed in LBV circumstellar shells (Smith & Owocki 2006) divided by a nominal eruption duration of 10 years (see **Figure 5**). For "binary RLOF," an order-of-magnitude value for the strongest mass-transfer rates expected in brief RLOF (Roche-lobe overflow) phases is noted, although the mass-transfer or mass-loss rate can be much less for slow mass transfer or possibly more for dynamical common-envelope ejection events; see references in the text, especially the review by Langer (2012).

stellar population synthesis models.] It is likely that adopting the reduced mass-loss rates will have a profound impact on the outcome of single-star evolution calculations, but a meaningful comparison with observed properties of stars cannot be attempted until this is updated.

Modifications to standard mass-loss rates have also occurred at the high and low ends of the luminosity range. As the most massive O-type stars (spectral types of O3 and O2) evolve toward the terminal-age MS, their luminosities go up and they move close to $\Gamma = 1$, where $\Gamma = \kappa L/4\pi GMc$ is the Eddington ratio. High Γ values in hot stars can substantially affect the winds and increase mass-loss rates (Gräfener et al. 2011, Vink et al. 2011). Thus, it may be possible for stars with initial masses of 80–100 $M_\odot$ or more to lose mass so fast in a WNH phase that they avoid the LBV phase altogether, although this effect has not yet been included in stellar evolution calculations. For these very massive stars it is likely that steady winds could have a more significant impact on evolution than for the majority of SN progenitors with lower initial masses.

Eddington limit ($\Gamma = 1$): point at which a star's luminosity is so strong that the radiation force balances gravity

Although the most massive H-burning stars may have winds that are stronger than the standard Vink et al. (2001) prescriptions, it has been found that later O-type and early B-type stars have surprisingly weak winds for their luminosity compared to theoretical expectations. Below $\log(L/L_\odot) = 5.2$ (spectral types of O7 and later), observed wind momenta and mass-loss rates are much lower than theoretical predictions (see **Figure 3**). This is known as the weak-wind problem for O dwarfs (Puls et al. 2008, Muijres et al. 2012) and may indicate inefficient line driving in hot dwarfs. Analysis of the bow shock around the O9.5 V runaway star ζ Oph suggests that the weak-wind problem may not be as bad as indicated by UV absorption (factor of 100 lower), but that the mass-loss rates are still a factor of 6–7 lower (Gvaramadze et al. 2012). Similarly, Huenemoerder et al. (2012) find from considering X-ray diagnostics that the $\dot{M}$ reduction may not be as severe as suggested from UV diagnostics, although their favored rate of $\dot{M} = 2 \times 10^{-9}\, M_\odot\, \text{year}^{-1}$ for the O9.5 V star μ Col lies on the green line for weak winds in **Figure 3**. Note that similar considerations of UV diagnostics were noted earlier as well (Drew et al. 1994; Cohen et al. 1997, 2008).

2.4. Wolf-Rayet Winds

The strong winds of WR stars yield spectacular spectra with extremely strong and broad emission lines, as well as strong excess in the IR and radio from free-free emission. Crowther (2007) has recently reviewed the properties of WR stars, and typical mass-loss rates for WN and WC (carbon-sequence Wolf-Rayet) stars are plotted in **Figure 3**. Much of the discussion of clumping and mass-loss rates of WR stars echoes that of O-type stars, except for the fact that the effects of clumping were already known more than a decade ago based on the relative strengths of electron-scattering wings and emission-line cores, as noted earlier (Hillier 1991). The mass-loss rates of WR stars, like O-type stars, are caused by line-driven winds and are Z dependent. Vink & de Koter (2005) find that Fe dominates the driving in WN stars and that they have a metallicity dependence similar to O-type stars, whereas WC stars have a somewhat shallower dependence on Z because intermediate-mass elements from self-enrichment contribute more to the driving as Z drops.

2.5. Implications of Lower O-Star Mass-Loss Rates

It would appear that the net result of decades of detailed study of line-driven winds is that, at least concerning stellar evolution, they don't matter as much as was previously believed (except perhaps for the most massive stars where proximity to the Eddington limit enhances the winds; see above). It is still commonly stated in the literature that a massive star of $M_{\text{ZAMS}} = 60\, M_\odot$ will shed half its mass on the MS, but if clumping requires us to reduce mass-loss rates by a factor of 3, then such statements are no longer true. For example, a 60-$M_\odot$ star will begin the MS with $\log(L/L_\odot) = 5.7$ and $\log(T_{\text{eff}}) = 4.7$ according to standard evolutionary models, and $\dot{M} = 10^{-5.86}$ according to the de Jager et al. (1988) prescription. The mass-loss rate will climb throughout the MS because the luminosity goes up, and so the average $\dot{M}$ is about a factor of 2 higher. However, with $\dot{M}$ reduced by a factor of 3 as a standard clumping correction, a 60-$M_\odot$ star would ultimately lose only a few solar masses during the entire 3.5 Myr of MS evolution. This raises the important question of where WR stars come from in light of lower mass-loss rates. No H-free WR stars are known with masses above roughly 20–25 $M_\odot$ (see Crowther 2007, Smith & Conti 2008), so the lion's share of mass loss is yet to come. Stars with $M_{ZAMS} = 40-60\,M_\odot$ don't become RSGs, and the observed $\dot{M}$ values of steady winds of post-MS stars at these luminosities [i.e., blue supergiants (BSGs) and quiescent LBVs; see **Figure 3**] are not high enough when combined with the short duration of these post-MS phases envisioned in single-star evolution models. There are only a couple options (Smith & Owocki 2006): (*a*) eruptive LBV mass loss makes up the difference or (*b*) single stars with $M_{ZAMS} = 40-60\,M_\odot$ don't fully shed their H envelopes before core collapse, and binaries are instead responsible for most of the observed WR stars that may come from these initial masses. For stars of lower initial mass $M_{ZAMS} = 40-60\,M_\odot$, the problem is worse. Below $M_{ZAMS} \simeq 35\,M_\odot$, single stars should go through a RSG phase and this may help shed the H envelope to make WR stars. RSG mass-loss rates are highly uncertain as well, however (see below).

At higher initial masses above 80–100 $M_\odot$, stars pass through a WNH phase with very strong winds enhanced by high Γ values (see above). Although the mass loss here may be strong enough to evaporate the H envelope, stars of such high initial mass would yield He cores that are more massive than any observed H-poor WR stars. Because the most massive stars are rare, it is uncertain whether this is a show stopper. As discussed later, however, it seems easy for known binaries to account for WR stars, and indeed binary evolution (mass accretion, mergers) may factor prominently in producing some N-rich WR stars that may still be H burning.

Even with $\dot{M}$ values reduced by a factor of 3, however, line-driven winds operating over the entire MS lifetime of O-type stars may still be quite important in angular momentum loss and the rotational evolution of massive stars, especially with the possible aid of magnetic fields. This important aspect of rotational evolution is discussed by Langer (2012).

2.6. Metallicity Dependence and Implications for Feedback

Although the winds of OB-type MS stars may be less important for a star's evolution than previously thought, having relatively precise (better than a factor of ~2) estimates of $\dot{M}$ is still desirable to assess the role of wind feedback in clustered star-forming regions, starbursts, and disk galaxy evolution. This is because massive stars spend most of their lifetimes as H-burning O-type stars, and the youngest ages are when the natal ISM of the star-forming environment is still close to the star and susceptible to the direct impact of radiation pressure and winds. Although binary RLOF, eruptive LBV mass loss, and RSG winds remove more mass from a typical O-type star than line-driven winds, this mass loss is generally slow and/or cold, it usually happens on a very short timescale, and it usually occurs late in a star's life, so that the energy and momentum injection into the surrounding ISM integrated over the lifetime of the star is far less. The eventual SNe tend to explode in a large cavity and may be less influential as well. Thus, for assessing local mechanical feedback from massive stars, line-driven winds are still an important consideration.

The good news is that, modulo the uncertainty in $\dot{M}$ caused by clumping, the simple Z-dependent scaling of line-driven winds (e.g., Vink et al. 2001, Mokiem et al. 2007) is probably reliable enough to estimate the global contribution of feedback of MS O-type stars from solar to mildly subsolar metallicities. At a given temperature and luminosity, one expects mass loss to scale as $\dot{M} \propto Z^m$. Theoretically, Vink et al. (2001) predict $m = 0.69 \pm 0.10$ for O-type stars, whereas observations suggest $m = 0.83 \pm 0.16$ (Mokiem et al. 2007). These are in reasonable agreement, although note that both are steeper than the $Z^{0.5}$ scaling given by Kudritzki & Puls (2000) and adopted by Heger et al. (2003) (**Figure 1**) owing to the Z dependence of wind speed (Vink et al. 2001). These relations have only been tested to roughly $1/5\,Z_\odot$, so extrapolating to hyper-metal-poor environments is still uncertain.

EVOLVED MASSIVE STARS I: COOL TYPES

Red supergiants (RSGs): Coolest evolved massive stars. Strongest mass-loss phase for single stars with initial masses below about 30 $M_{\odot}$. *Example:* Betelgeuse.

Extreme RSGs (OH/IR stars): The most luminous RSGs whose mass loss is so strong that they are self-obscured at visual wavelengths, with strong IR excess from dust and maser emission from OH, SiO, and H_2O. *Examples:* VY CMa, NML Cygni.

Yellow supergiants (YSGs): Rare stars that appear in the middle of the HR diagram, possibly as post-RSGs, and usually with strong mass loss (Drout et al. 2012). This is a short-lived phase in both binary and single-star models. *Example:* progenitor of SN 2011dh.

Yellow hypergiants (YHGs): The most luminous YSGs, usually with extreme mass loss. Often designated with a luminosity class of Ia^+ (de Jager 1998). *Examples:* IRC+10420, HR 5171A, ρ Cas.

It should be noted that lower mass-loss rates are also important for assessing the collective radiative feedback from a cluster or starburst, because lower values of $\dot{M}$ allow more UV radiation to escape the wind. Moreover, the lower mass-loss rates that result from clumping may indirectly but substantially influence the global feedback of SNe from a stellar population, because weaker winds during H burning could modify the burning lifetime, core size, and end fate of the star and, hence, the characteristics of the resulting SN explosion.

3. DENSE WINDS FROM COOL SUPERGIANTS

Winds of cool supergiants (see the sidebar, Evolved Massive Stars I: Cool Types) have received less attention from the massive-star community than hot-star winds, even though RSG winds are much stronger (**Figure 3**) and far more important for the evolution of 8–35 $M_{\odot}$ stars (i.e., the vast majority of SN progenitors). Willson (2000) reviewed mass loss from cool stars, although that paper focused on lower masses ($M_{ZAMS} = 1$–$9\ M_{\odot}$). The basic physical picture of the mechanism by which more massive RSG stars lose mass is similar; pulsations lift gas to a few R_*, where the equilibrium temperature becomes low enough (1,000–1,500 K) for substantial dust condensation to occur. Radiation pressure on newly formed dust (coupled to the gas by collisions in these dense winds) then takes over and pushes the wind to escape the star's gravity. Willson (2000) also reviewed observational diagnostics of mass-loss rates in these cool stars. For lower mass-loss rates in cool supergiants with coronal winds, UV spectra and radio emission can be used to investigate the mass-loss rate and other wind properties (see also Bennett 2010). For stronger winds, the primary methods used to measure the mass loss are by thermal-IR excess from hot dust and molecular emission. For extreme mass-loss rates, one can also use masers (Habing 1996) and spatially resolved circumstellar material using a variety of techniques like IR interferometry (Monnier et al. 2004).

The most common prescriptions for RSG mass loss in modern stellar evolution codes are from the same sources as the old rates for hot stars (de Jager et al. 1988, Nieuwenhuijzen & de Jager 1990), and they come with comparably large uncertainty. This uncertainty must be kept in mind when evaluating the predictions of single-star evolution models that pass through the RSG phase. Some models adopt the significantly higher empirical RSG mass-loss prescription of van Loon et al. (2005), which is intended for dust-enshrouded RSGs. Both these relations are shown in **Figure 3**. Depending on which RSG mass-loss recipe is chosen for a model, RSGs can

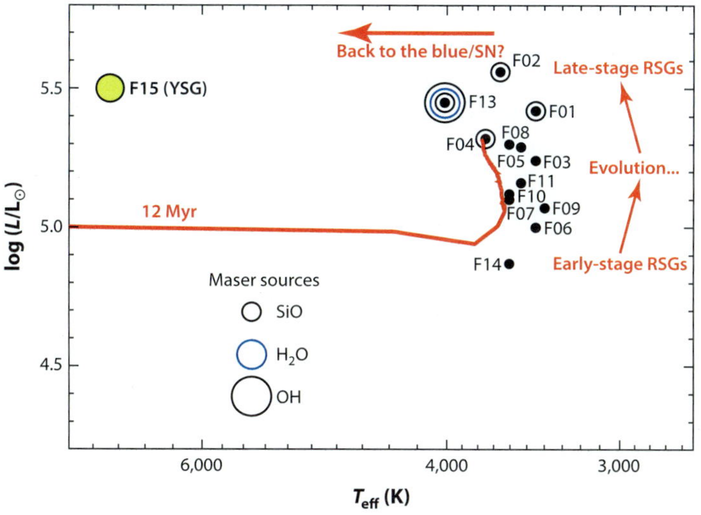

Figure 4

A Hertzsprung-Russell diagram of the cluster RSGC1, with red supergiants (RSGs) and a yellow supergiant (YSG). Sources that are circled include maser emission from dense envelopes, indicating especially strong mass loss. Adapted from Davies et al. (2008) with permission.

stay in the red (lower $\dot{M}$) or be driven to hotter temperatures on the HR diagram. This general behavior has been known for a long time and discussed in detail with regard to models for the blue progenitor of SN 1987A (Arnett et al. 1989). However, after Smartt (2009) discussed that the most massive RSGs apparently do not explode as Type II-P SNe, new evolutionary tracks that included blueward evolution for RSGs with initial masses above $\sim 20\,M_\odot$ became more common (Ekström et al. 2012). Thus, the model outcome of the RSG phase is sensitively dependent on an uncertain input mass-loss prescription. Unfortunately, there is, as yet, no well-established quantitative theoretical prediction for the mass-loss rates of RSG winds or the detailed physics driving them (including pulsations) like there is for hot stars. RSG mass loss is time dependent, and there appears to be a wide dispersion even for a given luminosity and temperature (Willson 2000). Moreover, the mass loss may not obey any single prescription throughout the whole RSG evolution of an individual star, and in fact there is evidence suggestive of this.

Observations of massive clusters with numerous RSGs have shed some important light on this topic. These provide excellent probes of massive-star evolution in general—and RSGs in particular—because massive clusters sample a relatively coeval group of stars, whereas different clusters sample different ages and initial masses of stars that have reached the RSG phase (see Davies et al. 2012 and references therein). An example from the cluster RSGC1 is shown in **Figure 4** (Davies et al. 2008). This compares the locations of several RSGs in the HR diagram with a 12-Myr isochrone (Maeder & Meynet 2000), indicating that these RSGs probably all have initial masses close to $18\,M_\odot$. It is interesting that the RSGs with the strongest mass-loss indicators (traced by H_2O, SiO, and OH maser emission; encircled in **Figure 4**) are found only at the top of the RSG branch. This seems to indicate that the highest mass-loss rates that lead to self-obscuration and maser emission (the sources for which the van Loon et al. mass-loss rates are appropriate) may be concentrated toward the very end of the RSG phase when turbulence and pulsations are most vigorous. Moreover, this cluster also includes a yellow supergiant (YSG) at a luminosity comparable to the most luminous RSGs with masers in the same cluster. This provides compelling evidence that at around $18\,M_\odot$ or above, enhanced RSG mass loss can indeed drive these stars toward warmer temperatures at the ends of their lives. This, in turn, has implications for connections between progenitor stars and the types of SNe that they make, as well as their circumstellar environments into which the SNe explode.

Some of the most luminous RSGs tend to have extremely high mass-loss rates that cause self-obscuration by dust and strong maser emission. The best-studied Galactic object in this category is VY Canis Majoris, which has an extended dust-scattering nebula seen in *Hubble Space Telescope* (HST) images (Smith et al. 2001). In the case of VY CMa, the density of its nebula drops off sharply at ~8,000 AU from the star, indicating that its high average mass-loss rate of $\sim 10^{-3}\,M_\odot$ year^{-1} has been limited to the previous 10^3 years or so (Smith et al. 2001, 2009; Decin et al. 2006). The nebula around VY CMa is quite similar to that around the yellow hypergiant (YHG) IRC+10420, which is regarded as the prototypical post-RSG because of its maser shell surviving around such a warm star and its observed fast blueward evolution (Humphreys et al. 1997). Cases like this provide additional evidence that strong mass loss among more luminous RSGs will drive them on blueward evolutionary tracks. Some models predict that this very strong mass-loss phase is short-lived and enhanced before core collapse caused by increasingly violent pulsations during C burning due to a high *L/M* ratio (Heger et al. 1997, Yoon & Cantiello 2010, Arnett & Meakin 2011).

4. SUPER-EDDINGTON WINDS, ERUPTIVE MASS LOSS, AND TRANSIENTS

Stellar evolution models demonstrate the critical impact of mass loss on stellar evolution, and we see results of mass loss in the existence of WR stars and the diversity of SN types. The uncertainty gets large as we move toward post-MS phases, as we have seen in the case of RSG mass-loss rates, and the probably dominant role of binaries (Section 5). The uncertainty is worst at the highest stellar luminosities, where mass loss is strongest and where systems observed in detail are few. Proximity to the Eddington limit can enhance the strength of a steady wind, or worse, it can make the star unstable, potentially leading to violent eruptive or explosive mass loss associated with transient events that can dramatically change the star in a short time. For extreme cases in very massive stars, a single eruptive event can remove more mass in a few years than is shed during its MS evolution. It is therefore sobering to recognize that none of these effects are included in current generations of stellar evolution models.

In Section 2.3, we briefly mentioned the role of the Eddington ratio, Γ, in enhancing the relatively steady mass-loss rates of very luminous H-burning stars, like the WNH stars. Here we focus on the more extreme cases in which the high Γ leads to instability in the star and highly time-dependent mass loss or in which advanced nuclear burning stages may play a role (Arnett & Meakin 2011). When this variability is observed, the stars are designated as LBVs, or they are designated as LBV candidates (see the sidebar, Evolved Massive Stars II: Hot Types) if they are suspected to be dormant versions of the same stars.

4.1. Luminous Blue Variables: History and Phenomenology

The most dramatic instability arising in post-MS evolution is the class of objects known as LBVs. These were recognized early as the brightest blue irregular variables in nearby galaxies (Hubble & Sandage 1953, Tammann & Sandage 1968), and these classic examples were referred to as the "Hubble-Sandage variables." Famous Galactic objects like P Cygni and η Carinae had spectacular outbursts in the seventeenth and nineteenth centuries, respectively, but their connection to other eruptive massive stars was unclear. Conti (1984) recognized that many different classes of hot,

EVOLVED MASSIVE STARS II: HOT TYPES

Wolf-Rayet star (WR): He-burning massive stars with very strong emission lines of He in their spectra, caused by very strong winds. WN (WR with N) and WC (with C lines) are exposed He cores of massive stars that have lost their H envelopes through prior mass loss. *Examples:* γ^2 *Vel, EZ CMa.*

Luminous blue variable (LBV): A group of evolved massive stars that exhibit eruptive mass loss or irregular variability. A union of various subtypes, including giant eruptions (η Car variables), S Dor variables, α Cyg variables, P Cygni stars, Hubble-Sandage variables, etc. Most have strong winds and strong emission-line spectra. Candidate LBVs are stars that have similar spectra and/or dust shells, but have not yet been seen to exhibit variability. *Examples:* η *Car, P Cygni, AG Car, S Dor, HR Car.*

Blue supergiant (BSG): Post-MS massive stars with B spectral types. The relative number of BSGs in observed HR diagrams of stellar populations is not well understood. *Examples: Sk-69 202, Sher 25, SBW1.*

Be and B[e] stars: B-type stars with strong and usually time-variable emission lines, often showing evidence for disk-like circumstellar material. Be stars are rapid rotators, possibly resulting from increased angular momentum through mass accretion in binaries. The B[e] stars have strong forbidden line emission and IR excess from dust that are thought to arise in a circumstellar disk or torus. Some are high-luminosity evolved supergiants similar to LBVs. *Examples:* γ *Cas (Be), R4 in the SMC (B[e]).*

irregular variable stars in the Milky Way and Magellanic Clouds were probably related to the Hubble-Sandage variables and to η Car and P Cygni, so he grouped them together as "LBVs." The LBVs are a rather diverse class, consisting of a wide range of irregular variable phenomena (Humphreys & Davidson 1994; van Genderen 2001; Smith et al. 2004, 2011d; Clark et al. 2005; Van Dyk & Matheson 2012). Their initial masses are uncertain, but comparing their luminosities with single-star evolution tracks suggests initial masses greater than 25 $M_\odot$ (Smith et al. 2004). LBVs are defined by their irregular eruptive variability. There are, however, stars that spectroscopically resemble LBVs in their quiescent state, but which have not (yet) been observed to show the signature variability of LBVs; these are often called LBV candidates, and they are usually of spectral type Ofpe/WN9 or early B supergiants. It is not known whether LBVs pass through long dormant periods, and if they do, the duty cycle is unknown. As noted below, the detection of a dense circumstellar shell is often taken to indicate a prior giant outburst.

4.1.1. S Doradus phases. S Dor outbursts are seen as a visual brightening that occurs when the peak of the star's energy distribution shifts from the UV to visual wavelengths. The increase in visual brightness (i.e., 1–2 mag, typically) corresponds roughly to the bolometric correction, so that hotter stars exhibit larger amplitudes. In their quiescent states, LBVs have apparent temperatures that increase with increasing luminosity: They often appear as Ofpe/WN9 stars at high luminosity or as early/mid B supergiants at the lower luminosity end (Humphreys & Davidson 1994, Smith et al. 2004). Visual maximum occurs at a constant temperature of ~8,000 K, causing the star to resemble a late F supergiant. S Dor events were originally proposed to occur at constant bolometric luminosity (Humphreys & Davidson 1994), but quantitative studies do reveal variations in L_{Bol} (Groh et al. 2009). The traditional explanation for the apparent temperature change was that the star dramatically increased its mass-loss rate, driving the wind to very high optical depth and causing a pseudo photosphere (Davidson 1987, Humphreys & Davidson 1994). However, quantitative spectroscopy revealed that the measured mass-loss rates in outbursts do not increase enough to cause a pseudo photosphere (de Koter et al. 1996) and that the increasing photospheric radius is therefore more akin to a pulsation. A possible cause of this inflation of the star's outer layers may be near-Eddington luminosities in the subsurface Fe opacity bump (Gräfener et al. 2012, Guzik & Lovekin 2012). S Dor eruptions of LBVs are therefore not major mass-loss events. However, the average mass-loss rate in a wind throughout the LBV phase (quiescent or not) is about an order of magnitude higher than for O-type stars of comparable luminosity (**Figure 3**).

4.1.2. Luminous blue variable giant eruptions. The most pronounced variability attributed to LBVs is their so-called giant eruptions, in which stars are observed to increase their bolometric luminosity for months to years, accompanied by extreme mass loss (Humphreys et al. 1999). The best-studied example is the Galactic object η Carinae, which has provided us with its historically observed light curve (Smith & Frew 2011) as well as its complex ejecta that contain 10–20 $M_\odot$ and $\sim 10^{50}$ ergs of kinetic energy (Smith et al. 2003, Smith 2006). Light echoes from the Great Eruption of η Carinae have just recently been discovered (Rest et al. 2012), and their continued study through the use of spectroscopy may modify long-held ideas about LBVs. A less well-documented case is P Cygni's 1600 AD eruption, for which a much smaller ejecta mass of 0.1 $M_\odot$ has been measured (Smith & Hartigan 2006). P Cyg's nebula has an expansion speed of $\sim$140 km s^{-1} (Barlow et al. 1994, Smith & Hartigan 2006), with an implied total kinetic energy of a few 10^{46} ergs. P Cyg and η Car are the only two cases of observed LBV giant eruptions in which the ejected mass has actually been measured, because they have spatially resolved shell nebulae ejected in the events. With decade-long durations, the implied mass-loss rates are at least 0.01 $M_\odot$ $year^{-1}$ and 1 $M_\odot$ $year^{-1}$ for P Cyg and η Car, respectively. These rates are too high to be driven by traditional stellar winds because the material is opaque (Owocki et al. 2004, Smith & Owocki 2006). Dust formation in LBV eruptions also points toward eruptive mass loss (Kochanek 2011).

4.1.3. Extragalactic supernova impostors. LBV giant eruptions are rare, so our only other observed examples are a few dozen found in nearby galaxies (Smith et al. 2011d, Van Dyk & Matheson 2012). Owing to their serendipitous discovery in SN searches, they are sometimes called SN impostors. Other names include Type V SNe, η Car analogs, and various permutations of intermediate luminosity transients. These have peak absolute magnitudes of -11 to -15 mag (Smith et al. 2011d, Van Dyk & Matheson 2012). Typical expansion speeds observed in outburst spectra are 100–1,000 km s^{-1} (Smith et al. 2011d), although lower speeds can be seen along the line of sight if the ejection speed is latitude dependent (Smith 2006). A realization in the past decade is that there is wide diversity among the SN impostors and their progenitors; some events that resemble LBV eruptions may actually arise in lower-mass progenitor stars (Prieto et al. 2008a, Thompson et al. 2009). A review of the lower-mass analogs of LBV giant eruptions is beyond the scope of this review, but the fact that lower-mass stars may experience similar transient events casts doubt on the long-held belief that these eruptions result from high luminosities near the Eddington limit (see Section 4.2.1 below).

4.1.4. Luminous blue variable winds. Most LBVs exhibit strong emission lines in their visual-wavelength spectra, similar to WR stars but with narrower widths and stronger H lines. Wind speeds are typically 100–600 km s^{-1}, reflecting the lower escape speed of BSG stars as compared with 1,000–2,000 km s^{-1} in more compact O and WR stars. The wind mass-loss rates implied by quantitative models of the spectra typically range from 10^{-5} to 10^{-4} $M_\odot$ $year^{-1}$ (de Koter et al. 1996, Vink & de Koter 2002, Smith et al. 2004, Groh et al. 2009) or even 10^{-3} $M_\odot$ $year^{-1}$ in the extreme case of η Car (Hillier et al. 2001). These LBV wind mass-loss rates are indicated in **Figure 3**. LBV winds are strong enough to play an important role in the evolution of the star if the LBV phase lasts more than 10^5 years, and eruptions further enhance (or actually dominate)

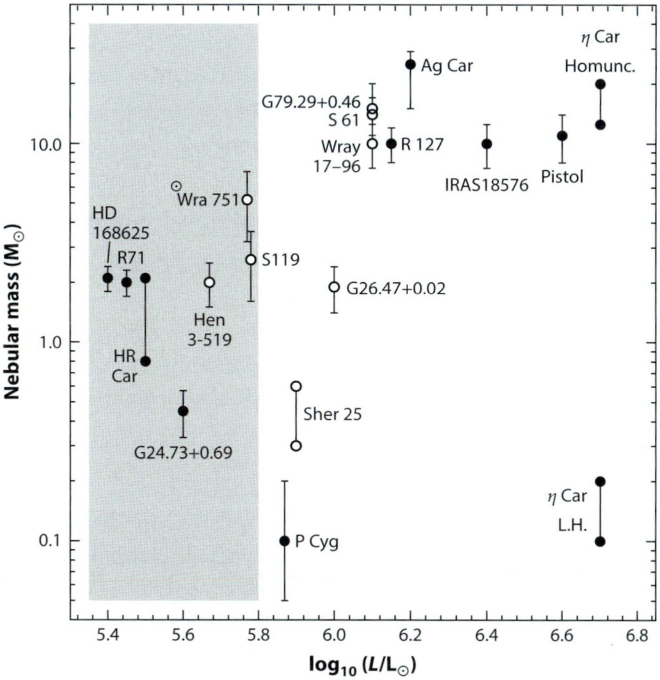

Figure 5

Masses of circumstellar shells around luminous blue variables (LBVs) and LBV-like stars, as a function of luminosity, modified from Smith & Owocki (2006). The left side of the plot (*gray box*) corresponds to stars below $\log(L/L_{\odot}) = 5.8$, so these LBVs could be post-RSGs (red supergiants) and the nebular mass could have been ejected in the RSG phase and swept up into a shell. Objects on the right must have ejected their massive shells in giant LBV eruptions. There may be many lower-mass shells that are hard to detect around the very bright central stars.

the mass loss. Other stars that exhibit similar spectra but are not necessarily LBVs include WNH stars, Ofpe/WN9 stars, and B[e] supergiants, which overlap on the HR diagram.

4.1.5. Circumstellar shells. Many LBVs have spatially resolved circumstellar shells that are fossils of previous eruptions. Stars that resemble LBVs spectroscopically and have massive shells, but have not been observed to exhibit LBV variability, are called LBV candidates, as noted earlier. LBV circumstellar shells are extremely important, as they provide the only reliable way to estimate the amount of mass ejected in an LBV giant eruption. A large number of LBVs and candidates in the Milky Way and Magellanic Clouds are surrounded by massive shell nebulae (Clark et al. 2005, Smith & Owocki 2006, Gvaramadze et al. 2010, Wachter et al. 2010). Thus, eruptive LBV mass loss is important in the late evolution of massive stars. Masses of LBV nebulae occupy a very large range from $\sim$20 $M_{\odot}$ at the upper end down to 0.1 $M_{\odot}$, although even smaller masses become difficult to detect around bright central stars. In some cases a very large range in mass is seen in multiple shells around the same star, as for η Car (Smith 2005), so there is no clear one-to-one correlation of shell mass and stellar luminosity, although there may be such a relation for the most massive shell a star can eject. Masses for a collection of LBV shells are shown in **Figure 5**, compiled by Smith & Owocki (2006). To compare with steady winds, LBV giant eruption mass-loss "rates" in **Figure 3** are shown by dividing LBV shell masses by the $\sim$10 year duration of the eruptions of

η Car and P Cyg, although the true instantaneous $\dot{M}$ may be higher. Dynamical ages of the shells around LBVs/candidates range from 10^2 years to several thousand years. It is, however, difficult to use this age as an indication of the duration of the LBV phase, because expanding shells can decelerate. The duration of the LBV phase may be further complicated, because it may vary with initial mass and may depend on details of binary evolution.

4.2. Physics of Eruptive Mass Loss

Although the violent variability of LBVs has been known for a century or more, the search for a physical theory of LBV eruptions is still in the early stages. Most work so far has concentrated on how to lift material off the star or on tighter observational constraints on the mass, speed, and energy of outbursts. In terms of driving the mass loss, two broad classes of models have developed: super-Eddington winds and explosions. Both may operate at some level, but neither of these addresses the deeper question of what initiates LBV eruptions in the first place. Ideas for the underlying trigger are still speculative.

4.2.1. Super-Eddington winds. Traditionally, LBV giant eruptions have been discussed as super-Eddington (SE) winds driven by a sudden unexplained increase in the star's bolometric luminosity (Humphreys & Davidson 1994, Humphreys et al. 1999, Shaviv 2000, Owocki et al. 2004, Smith & Owocki 2006). This is motivated mostly by the fact that η Car's Great Eruption had an observed luminosity that indicated $\Gamma \simeq 5$ for about a decade or more and that extragalactic SN impostors show similar high luminosities. With Γ values substantially above unity, one naturally expects strong mass loss, but the detailed physical picture of such winds is not obvious. SE winds are expected to be very dense but porous, allowing the star's atmosphere to remain in steady state while exceeding the classical Eddington limit (Shaviv 2000, Owocki et al. 2004). An important point is that the mass loss is so strong that the high density in the wind causes UV absorption lines to be saturated. Therefore, SE winds are driven by photon momentum transferred to gas through electron scattering opacity and not line opacity (Owocki et al. 2004). This makes SE wind mass loss essentially independent of metallicity, which may allow this mode of mass loss to operate in Pop III stars (Smith & Owocki 2006).

Numerical simulations of continuum-driven SE winds show complex structure with both infall and outflow (van Marle et al. 2008, 2009), confirming expectations that these winds should be highly inhomogeneous (Shaviv 2000, Owocki et al. 2004). SE winds may also account for the bipolar shape of nebulae around LBVs like η Car if the star is a rapid rotator, because equatorial gravity darkening leads to a higher $\dot{M}$ and faster speed in the polar wind (Owocki et al. 1996, Dwarkadas & Owocki 2002).

Open questions surrounding SE winds are whether the mechanism can supply the mass loss in the most extreme observed cases and what initiates the SE phase. Current estimates of the mass lost in η Car's nineteenth-century eruption are of order 15 $M_\odot$ or more (Smith et al. 2003, Smith & Ferland 2007). SE winds can in principle cause that much mass loss averaged over 20 years (Owocki et al. 2004), but those same models predict relatively slow outflow speeds. This makes it hard to explain the nebula around η Car, with most of the mass moving at speeds of 500–600 km s^{-1}. Moreover, η Car also shows a smaller mass of extremely fast material moving at 5,000 km s^{-1} or more (Smith 2008), which is hard for a SE wind model to achieve while driving such a large amount of mass. SE winds are still viable for most LBVs, which are less extreme than η Car's nineteenth-century eruption. As for the extra radiative energy output that initiates the SE wind, its source is not known. Lastly, SE winds are assumed to be launched from the surface of the star, but it is also not yet clear whether the star's interior can remain stable at $\Gamma = 5$ for more than a decade.

4.2.2. Explosions. There is growing observational evidence that some giant LBV eruptions may be nonterminal hydrodynamic explosions. Part of the motivation for this is based on detailed study of η Carinae, which has shown several signs that the 1840s eruption had a shock-powered component to it. This includes estimates of the ratio of total ejecta kinetic energy to the integrated radiated energy of $E_k/E_{\rm rad} > 3$, which is hard for a radiation-driven wind to achieve (although perhaps not impossible with extreme photon-tiring; Owocki et al. 2004). The very thin walls of the nebula indicate a small range of expansion speed (Smith 2006), which is easiest to achieve from compression in a shock. Lastly, observations show extremely high-speed ejecta moving at 5,000 km s^{-1}, which seems impossible to achieve without a strong blast wave (Smith 2008). A number of extragalactic LBV-like eruptions show spectra that closely resemble shock-powered Type IIn SNe and also show evidence for extremely fast ejecta that may signify a shock-powered event, such as the precursor outbursts of SN2009ip (Smith et al. 2010b, Foley et al. 2011).

CSM: circumstellar material or circumstellar medium

One normally expects sudden, hydrodynamic events to be brief (i.e., a dynamical time), which at first may seem incompatible with the decade-long Great Eruption of η Car. However, an explosion followed by circumstellar material (CSM) interaction can generate a high sustained luminosity, as in core-collapse SNe IIn. Smith (2013a) showed that a shock-powered event with CSM interaction could account for the 1845–1860 light curve of η Car using a SN IIn-type model but with lower explosion energy. The resulting slower shock speed from a subenergetic explosion (10^{50} instead of 10^{51} ergs) produces a lower CSM interaction luminosity compared with a core-collapse SN IIn and takes much longer to expand through the CSM. The duration of the event is determined by the outer extent of the dense CSM—in principle, a CSM-interaction-powered LBV eruption might continue for several decades or it could last only 100 days (van Marle et al. 2010), depending on the extent of the dense pre-explosion wind. Because shock/CSM interaction is such an efficient way to convert explosion kinetic energy into luminosity, it is plausible that many of the SN impostors with narrow emission lines may be powered in this way. The shock model would help explain the wide observed diversity of SN impostors (Smith et al. 2011d). The catch is that even this model requires something else to create the dense CSM into which the shock expands, which may be where SE winds or binary interaction play an important role.

4.2.3. Eruption triggers The reasons for the onset of an LBV eruption along with its power source remain unanswered for either mechanism. In the SE wind model, even if the wind can be driven at the rates required, we have no underlying physical explanation for why the star's bolometric luminosity suddenly increases by factors of 5–10, and we don't know how the star's envelope would process that high energy flux. In the explosion model, the underlying trigger for an explosive event is unknown. In either case, something must inject a large amount of extra energy (10^{48}–10^{50} ergs) into the star's interior in a highly time-dependent way. There have been a number of physical mechanisms discussed in connection with LBVs, recently reviewed in detail by Smith et al. (2011d). In brief, there is no clearly favored explanation, but some ideas can be ruled out based on the required energy and mass budgets. These can be thought of in two broad categories: instability and energy deposition.

4.2.3.1. Envelope instability. LBVs are massive stars that are in close proximity to the Eddington limit. Consequently, their loosely bound envelopes may be susceptible to strange-mode instabilities (Glatzel & Kiriakidis 1993, Glatzel et al. 1999), runaway mass loss (a.k.a. the "Geyser" model; Maeder 1992, Humphreys & Davidson 1994), or the critical rotation limit (a.k.a. the "Ω limit"; Langer 1998, 2012). Although these may help explain some of the irregular variability seen in S Dor outbursts of LBVs, they are unsatisfactory explanations for giant LBV eruptions. This is because the total mass ejected can be much more than the small mass in the outer H envelope where the relevant instabilities reside, and LBV eruptions can have substantially more kinetic energy than the total thermal energy in the star's envelope.

PPI: pulsational pair instability

4.2.3.2. Energy deposition. A large amount of extra energy can be deposited deep in a massive star's envelope by a number of suggested mechanisms, including unsteady burning (Smith & Arnett 2014), the pulsational pair instability (PPI; Woosley et al. 2002, 2007), other explosive shell burning instabilities (Dessart et al. 2010, Smith & Arnett 2014), wave-driven mass loss (Meakin & Arnett 2007, Quataert & Shiode 2012, Shiode & Quataert 2014), and stellar collisions or mergers in a binary system (Podsiadlowski et al. 2010, Smith 2011, Smith & Arnett 2014). [Soker and collaborators discussed a model to power the luminosity in LBV eruptions using accretion onto a companion star (Kashi & Soker 2009), but this invokes an eruption to provide the mass that is then accreted; it does not explain what initiates the mass loss from the primary in the first place.] Although any of these provide a plausible result, the main criticism for explaining LBV eruptions is that the mechanisms that are related to late nuclear burning instabilities are expected to occur only in the few years preceding core collapse. However, many LBVs with massive shells appear to have survived for 10^2–10^4 years after a giant eruption. Such an objection turns into an advantage, however, in the case of the violent pre-SN eruptions needed for SNe IIn (see Section 6.3 below).

Research on LBV eruptions and pre-SN eruptions is actively ongoing, and it is a major unsolved problem in astrophysics. Observations demonstrate that these events do occur, and the mass budget involved can dominate or significantly contribute to the total mass lost by a massive star.

4.3. The Role of Luminous Blue Variable Mass Loss in Stellar Evolution

If this section were to review the influence of LBV eruptions when they are included in stellar evolution models, it would be a very short section. Without exception, no current stellar evolution models account for LBV giant eruptions. This is partly because we don't know how to include them properly. Observationally, we don't have reliable estimates of the typical mass ejected as a function of the star's initial mass, how many repeating eruptions occur for a given star, or the duty cycle of eruptions. Theoretically, we don't know the underlying physical mechanism(s), when they occur during the evolution of a star, or how they should vary with $M_{\rm ZAMS}$.

The traditional view of LBVs, which emerged in the 1980s and 1990s, is that they correspond to a very brief transitional phase of evolution, when the massive star moves from core-H burning to core-He burning (Humphreys & Davidson 1994). A typical monotonic evolutionary scheme is

$$100\,M_\odot : \mathrm{O\,star} \rightarrow \mathrm{Of/WNH} \rightarrow \mathrm{LBV} \rightarrow \mathrm{WN} \rightarrow \mathrm{WC} \rightarrow \mathrm{SN\,Ibc}.$$

In this scenario, the strong mass loss experienced by LBVs is important for removing what is left of the star's H envelope after the MS, leaving a WR star following the LBV phase. The motivation for a very brief phase comes from the fact that LBVs are extremely rare: The duration of the LBV phase is presumed to be only a few 10^4 years (Humphreys & Davidson 1994).

However, a number of inconsistencies have arisen with this standard view. The very short inferred LBV lifetime depends on the assumption that the observed LBVs occupy the whole transitional phase. In fact, there is a much larger number of blue supergiant stars that are not seen in eruption—these are the LBV candidates. Examining populations in nearby galaxies, Massey et al. (2007) find that there are greater than an order of magnitude more spectroscopically similar LBV candidates than there are LBVs confirmed by their variability. If LBV candidates are included, then the average LBV phase rises from a few 10^4 years to several 10^5 years. This is comparable with the whole He burning lifetime of very massive stars, making it impossible for LBVs to be mere transitional objects. Massey (2006) has pointed to the case of P Cygni as a salient example: Its 1600 AD giant LBV eruption was observed and so we call it an LBV, but it has shown no eruptive LBV-like behavior since then. If the observational record had started in 1700, then we would have no idea that P Cygni was an LBV.

Another major issue is that we have growing evidence that LBVs or something like them (massive H-rich stars with high mass loss, N enrichment, slow 100–500 km s^{-1} winds, massive shells) are exploding as core-collapse SNe while still in an LBV-like phase (see Section 6.3 below). This could not be true if LBVs are only in a brief transition to the WR phase, which should last another 0.5–1 Myr.

A prolonged LBV phase, and indeed any very massive stars above $M_{ZAMS} = 40\,M_{\odot}$ making it to core collapse with H envelopes still intact at $Z_{\odot}$, is in direct conflict with the single-star evolution models. The evolutionary state and basic nature of LBVs is therefore still quite uncertain. This is problematic for our understanding of massive-star evolution, in which mass loss is a key ingredient, because LBVs have the highest known mass-loss rates of any stars (**Figure 3**).

4.4. Low Metallicity

Although we don't yet know the root cause of LBV eruptions, we do know that the huge observed mass-loss rates demand that the mechanism imparting momentum to the ejecta is not a Z-dependent wind, because the outflowing material is very optically thick and lines are saturated. LBV eruptions must be either continuum-driven SE winds or hydrodynamic events (violent pulsations, explosions), and both of these are relatively insensitive to Z (Smith & Owocki 2006). Because this mode of eruptive mass loss may actually dominate the total mass shed by a massive star at $Z_{\odot}$, there is not yet any reason to think it won't also work in the early Universe. This may be important for Pop III stars, because they are argued to have been preferentially very massive. There may, of course, be some unrecognized way that metallicity creeps into the problem (e.g., Fe opacity bumps, some Z dependence of explosive burning, etc.), but this has not yet been investigated. LBVs have been identified in nearby low-Z dwarf galaxies (Izotov & Thuan 2009, Isotov et al. 2011), in addition to the very nearby case of HD 5980 in the SMC, and the LBV-like eruptions that precede SNe IIn (see Section 6.3 below) often occur in dwarf galaxies. Thus, there is clear empirical evidence that low metallicity does not inhibit eruptive mass loss.

5. BINARY MASS TRANSFER, MERGERS, AND MASS LOSS

At massive-star conferences, a common refrain heard upon completion of a talk about single-star evolutionary models is, "What about binaries?" This is often followed by an uncomfortable silence, shrugging of shoulders, nervous laughter by the speaker, or intervention by the session chair to move to a serious question.

Actually, this is a serious problem. Exclusion of complicated binary effects and a focus on single-star evolution models is valid, in principle, because indeed we must start by having a foundation in understanding single stars. However, comparing single-star models to the observed statistical properties of stars—and then using these diagnostics to inform the validity of assumptions in those single-star models—can lead to serious errors if binaries make a significant contribution to observed distributions. The most critical influence of binaries on the observed distribution of massive stars comes through the process of mass loss and mass transfer via RLOF, which exceeds the influence of stellar winds and rotation for most (and possibly all) initial masses.

5.1. Massive Stars Are Mostly in Binaries

Based on work in the past decade, we now have secure evidence that the binary fraction is not only high among massive stars, but that the large majority of massive stars (roughly 2/3–3/4) reside in binary systems that have orbital periods short enough that the stars will interact and exchange mass (or merge) during their lives (Mason et al. 2009, Sana & Evans 2011, Sana et al. 2012). There have been several monitoring campaigns to measure the observed spectroscopic binary fraction among massive stars in clusters using radial velocities; such campaigns typically find an observed binary fraction of 20–60% (Gies 1987; Garcia & Mermilliod 2001; De Becker et al. 2006; Evans et al. 2006; Kobulnicky & Fryer 2007; Sana et al. 2008, 2009, 2012; Mahy et al. 2009; Chini et al. 2012; Kiminki & Kobulnicky 2012; Kiminki et al. 2012). However, these observed binary fractions are only a lower limit, because they must be corrected for the spectroscopic binary systems that are missed because of low inclination, periods that are too long compared to the observational cadence or have high eccentricity, or low-mass companions that are more difficult to detect. Some fraction of these have orbital separations small enough that the stars will actually interact and exchange mass; whether or not they will interact is typically determined by the maximum radius of an RSG (roughly 5 AU) or LBVs in outburst (a few AU) depending on the initial masses of the stars (Kiminki & Kobulnicky 2012).

Recent estimates including the results from several young clusters suggest that, when corrected for observational bias, the fraction of massive stars in binary systems whose orbital period is so short that the stars must exchange mass or merge is something like 3/4 (Kobulnicky & Fryer 2007, Kiminki & Kobulnicky 2012, Sana et al. 2012). Of the total population of massive stars, Sana et al. (2012) estimate that ~25% will merge, ~33% will have their H envelopes stripped before death, and ~14% will be spun up by accretion, whereas only about 25% of massive stars are actually effectively single (including stars in wide binaries). This means that binary RLOF is not just a factor that we should perhaps consider but that it must dominate the observed effects of mass loss and mixing seen in massive stars. It is also likely that the vast majority of the most rapidly rotating stars, including potentially all Be stars, result from interaction with a companion star (de Mink et al. 2013) rather than being single stars born with a very high rotation rate. This, in turn, suggests that the influence of rotation and rotational mixing in current single-star models may be overestimated (Vanbeveren 2009, de Mink et al. 2013).

5.2. Physics of Mass Loss in Binaries

From the ZAMS until the H envelope is substantially stripped or removed completely, a massive star tends to increase its radius owing to interior evolution. This begins on the MS as stars steadily become more luminous and slightly cooler, and the radius then expands more severely in post-MS phases. In a binary system, significant mass transfer begins when the primary star expands to a size where its photosphere crosses the inner Lagrange point (L1).

The onset of RLOF therefore depends sensitively on the initial orbital separation; it may occur while the primary is still on the MS (orbital periods of a few days) or when it expands to a much larger size in post-MS supergiant phases (orbital periods of tens of days or longer). RLOF on the MS is referred to as Case A, whereas Case B is for RLOF during H-shell burning and Case C for core-He burning (Podsiadlowski et al. 1992, Petrovic et al. 2005, Langer 2012). Mass transfer becomes increasingly unstable and rapid from Case A to C, in some cases involving dynamical events, common envelopes, or inspiral to a merger. The mass-transfer rate also depends sensitively on the mass ratio $q = M_2/M_1$ ($q \leq 1$), with increasing rates for lower values of q. For Case A, the orbit widens, and the mass-transfer rate drops as q approaches unity.

Mass-transfer rates (and hence, mass-loss rates from the primary) can be very high and can far exceed any mass-loss rate for a line-driven wind (see **Figure 3**). The detailed physics of mass transfer in RLOF is quite complicated and time-dependent. There is much remaining uncertainty about the proper treatment of contact systems, as well as of their mass and angular momentum loss. A common prescription for the strongest phases of RLOF mass transfer is to assume that the mass transfer is limited by the thermal timescale of a massive star with a radiative envelope:

$$\dot{M} = (M_0 - M_{\mathrm{WR}})/\tau_{\mathrm{KH}},$$

where τ_{KH} is the thermal (Kelvin-Helmholtz) timescale of the envelope, M_0 is the initial mass, and M_{WR} is the mass of the resulting WR star. Mass-transfer rates can therefore be the highest for more massive and more luminous stars, which have short thermal timescales. Resulting mass-transfer rates can be extremely high; during fast Case A or Case B and C phases, mass-loss rates can be of order $10^{-3}\,\mathrm{M}_\odot$ year^{-1} or higher (Taam & Sandquist 2000, Langer 2012), although there is a wide range (only order-of-magnitude values are represented in **Figure 3**). With thermal timescales of order 10^4 years, binary RLOF is therefore capable of quickly removing almost the entire H envelope of a massive star and leaving behind a WR star (Petrovic et al. 2005). When RLOF ends, the star's radius shrinks and there will be a small residual H layer on the star.

The physics and observed phenomenology of the RLOF phase and common envelopes can be very complicated (Taam & Sandquist 2000) and could probably fill several review articles. The main uncertainties are the degree to which mass transfer is conservative ("conservative" mass transfer means that no significant mass is lost from the binary system in RLOF) (Cantiello et al. 2007, de Mink et al. 2007), the related question of how the mass gainer responds to the added angular momentum that may lead to critical rotation (de Mink et al. 2013), and the consequent orbital evolution. The most important point for our purpose here is to concentrate on the net result: Large amounts of mass are "instantly" (relative to nuclear burning timescales) stripped from one star, and much or all of this is accreted by the other. The envelope stripping of the primary is similarly efficient to the also "instantaneous" removal of $\sim$10 $\mathrm{M}_\odot$ in giant eruptions of LBVs. Indeed, their mass-loss rates are suspiciously similar in **Figure 3**, and it remains possible that some LBV giant eruptions are extreme mass-transfer or merger events (many LBVs and LBV candidates have only a single massive CSM shell). Podsiadlowski et al. (2010) have discussed extreme cases in which mixing of fresh fuel into deeper layers during a binary merger might lead to explosive burning and removal of the H envelope, reminiscent of some ideas for explosive LBV giant eruptions.

There are few observational constraints on mass loss and mass-transfer rates in binaries undergoing RLOF. Because the most active phases are very brief ($\sim 10^4$ years; less than 1% of a massive star's life), systems undergoing strong RLOF at any given time are rare. (The longer-lasting systems in slow Case A RLOF are more common, but have lower mass-transfer rates; see de Mink et al. 2007.) One well-studied example of a short-period system caught in the phase of fast Cas A mass transfer is the 11-day eclipsing binary RY Scuti. It has component masses of roughly $8\,\mathrm{M}_\odot$ and $30\,\mathrm{M}_\odot$, with a mass gainer surrounded by an opaque disk (Grundstrom et al. 2007) and a spatially resolved toroidal nebula (Smith et al. 2011a). A noteworthy object for wider orbital separations (probably Case C) is the famous YHG star HR 5171A, which Chesneau et al. (2014) recently resolved as a mass-transfer binary using interferometry. This system hints that some of the YHGs may actually be wide binary systems, where RLOF truncates further redward evolution of the primary, rather than products of single-star RSG mass loss. Because the companion was not discovered until it was resolved with interferometry, this also demonstrates that these binaries may be easily hidden. Interestingly, Prieto et al. (2008b) report the discovery of a rare extragalactic YSG eclipsing binary, which may provide a similar indication. Aside from these rare cases, much of our empirical understanding of RLOF therefore comes from studying the more common post-RLOF binaries (WR+OB systems). Their inferred histories depend on a number of assumptions, however (Hellings 1984).

5.3. Binary Evolution Models and Population Synthesis

The idea that binary RLOF strips a star's H envelope to dominate the production of WR stars and SNe Ibc is an old one (Paczynski 1967). As noted in the introduction, this view fell out of favor owing to the estimated strength of radiatively driven winds, but is now experiencing a resurgence due to lower wind mass-loss rates and very high observed binary fractions. As such, it is now clear that the complexity of RLOF and its many varying parameters are a necessary evil to consider. Some advocates for the importance of binaries may point out that this was true all along (e.g., Vanbeveren et al. 1998).

Before and after binary stars interact, they behave largely as single stars, and so binary stellar evolution models come with all the assumptions and uncertainties that go into single-star models, like the treatment of convection, assumptions about convective overshoot, the importance of rotational mixing and angular momentum diffusion, and of course wind mass loss (Maeder & Meynet 2000, Woosley et al. 2002). Binary evolution introduces additional parameters (Langer 2012). The total mass lost or transferred in RLOF, its transfer rate, the amount of angular momentum lost/transferred, and the time dependence of these are influenced by several initial conditions: (*a*) the primary/secondary initial mass ratio q, (*b*) the primary star's initial mass, (*c*) the initial orbital separation, (*d*) the orbit eccentricity, and (*e*) the relative wind mass-loss rates of both stars. It may also matter that a system has additional multiplicity, but this is usually ignored. Calculating grids of detailed binary stellar evolution that explore this parameter space and also provide a detailed treatment of rotation and mixing in the stars would require a large fraction of the computing power on Earth. Therefore, present state-of-the-art binary models and population synthesis must make simplifying assumptions (Langer 2012). Binary population synthesis models have demonstrated that RLOF can naturally account for many of the observed statistical distributions of stellar types as well as the relative rates of various types of SNe (Podsiadlowski et al. 1992; Vanbeveren et al. 1998, 2007; Petrovic et al. 2005; Eldridge et al. 2008; Eldridge & Stanway 2009; Yoon et al. 2010; Dessart et al. 2011; Sana et al. 2012). Earlier studies made inferences about what the interacting binary fraction would need to be in order to account for observations. If new estimates of the very high binary fraction are correct, it now seems unavoidable that binary evolution will dominate the observed populations of WR stars and SNe Ibc (Sections 6.1 and 6.2).

The fact that binary population synthesis can naturally explain the observed statistical properties of massive stars (like the observed ratio of WR to OB stars, relative numbers of various SN types, etc.) means that single-star models (which represent a minority of stars) should not do so. The point is that by including efficient rotational mixing and stellar winds that are too strong, single-star models mimic outcomes that are in fact dominated to a large extent by binary RLOF. This raises concerns about the correctness of physical ingredients adopted in these single-star models, including the treatment of rotation and turbulent convection in addition to the overestimated mass-loss rates.

5.4. Low Metallicity

The physics of RLOF is governed by the gravitational interaction of two stars and is insensitive to the metallicity of the gas being transferred. The insensitivity to Z for the extreme mass loss induced by RLOF is therefore similar, in principle, to the continuum-driven winds and explosions of LBVs (Smith & Owocki 2006). This should have strong implications for populations of evolved stars and SNe at low Z, although this aspect has not been much explored in the literature.

Observations do indicate evolution with metallicity, such as the WC/WN ratio (Massey 2003) and the relative rates of Type Ibc to Type II SNe (Prantzos & Boissier 2003, Prieto et al. 2008a, Boissier & Prantzos 2009). However, we should be cautious that even binary RLOF may have some dependence on metallicity, because metallicity affects the opacity in the star's envelope and, hence, the hydrostatic stellar radius. With lower opacity, low-Z stars are more compact (Maeder & Meynet 2000, Heger et al. 2003). The onset of RLOF depends on the primary star's radius, so a low-Z stellar population might be less affected by RLOF on average than at $Z_\odot$. It might be easy to mistakenly attribute such apparent Z dependence entirely to line-driven winds, so the Z dependence (or not) of RLOF therefore deserves additional study. Of course, extrapolating RLOF to low Z also requires detailed knowledge of binary star-formation physics and the resulting period distribution as a function of Z, which are not readily available, but may be quite important. This impacts a number of issues in astrophysics, but most obvious is the progenitors of GRBs.

Nevertheless, even if we were to adopt zero Z dependence for RLOF, observed trends with Z of WC/WN stars and SNe Ibc/II do not contradict the dominant role of binary RLOF. An important point for interpreting WR subtypes and SN progenitors is that even when stripping of the H envelope is done by RLOF, the subsequent evolution from WN to WC (and SN Types IIb to Ib to Ic) is still determined largely by Z-dependent line-driven winds. This is discussed next.

6. END RESULTS OF MASS LOSS AND IMPLICATIONS

6.1. Outcomes I: Wolf-Rayet Stars as the Product of Mass Loss

One of the most fundamental tenets of massive-star evolution is that strong mass loss through winds will strip off a star's H envelope and leave a bare He core that we observe as a luminous WR star. (For the purpose of discussion here, we exclude WNH stars.) The agent that dominates that stripping of the H envelope and how it varies with metallicity is a long-standing unsolved issue. Does every massive O-type star evolve to become a WR star, or do only those in interacting binaries or in certain initial mass ranges become WR stars? The answer to this question has important implications for relative nuclear burning timescales in various phases, SN progenitors, and many other issues.

In a single-star framework, only the most massive stars are luminous enough to have radiation-driven winds that can remove the massive H envelope, so one expects a Z-dependent minimum initial mass that can yield an H-poor WR star, $M_{\rm WR}(Z)$. Standard single-star models predict $M_{\rm WR} \simeq 35\,M_\odot$ at $Z_\odot$ (Heger et al. 2003, Georgy et al. 2012), increasing to about 45 $M_\odot$ and 70 $M_\odot$ at $Z_{\rm LMC}$ and $Z_{\rm SMC}$, respectively (Heger et al. 2003). $M_{\rm WR}$ can be lowered by adopting substantially enhanced mass-loss rates in models (Meynet et al. 1994, Ekström et al. 2012) or by including the effects of relatively rapid rotation (Georgy et al. 2012). [Vanbeveren et al. (2007) criticize the application of these rotating models to observed trends, pointing out that the initial 300-km-s^{-1} rotation speeds in models are not representative of most massive stars. Correcting observed rotation speeds for a distribution of inclination angles, they argue that most O-type stars rotate more slowly at 100–120 km s^{-1}.] At first glance, models and observations would seem to be reasonably well aligned (Massey 2003): $M_{\rm WR}$ is inferred to be about 25 $M_\odot$ at $Z_\odot$ (Crowther 2007), increasing to about 30 $M_\odot$ and 70 $M_\odot$ at $Z_{\rm LMC}$ and $Z_{\rm SMC}$, respectively (Massey et al. 2000). However, theoretical expectations for $M_{\rm WR}$ are based on models that incorporate mass-loss rates that are known to be a factor of $\sim$3 too high. (They also do not include the weak-wind problem discussed earlier, which is important in this mass range and which affects a large fraction of SN progenitors.) With the $Z^{0.69}$ scaling of mass loss in line-driven winds (Vink et al. 2001), this means that the appropriate mass-loss rates for $Z_\odot$ are actually similar to those currently adopted in SMC models. Grids of models with appropriate mass-loss rates have not been published, but we can infer that the net effect will be to move the predicted $M_{\rm WR}$ upward significantly in single-star models for each Z range to a point that probably cannot be reconciled with observed $M_{\rm WR}$.

In a binary evolution paradigm, by contrast, He stars stripped of their H envelope can occur over a wider range of initial mass (Vanbeveren et al. 2007, Claeys et al. 2011). In that case, the dominant factors controlling observationally inferred values of $M_{\rm WR}$ are detectability and classification, as well as the wind strength that removes whatever residual H layer may be left at the end of RLOF. Even if the H envelope is removed (by any mechanism), classifying an object as a WR star observationally requires a strong wind to produce strong emission lines, which favors sources of higher L and Z. It may be hard to detect exposed He cores resulting from binary RLOF in $M_{\rm ZAMS} = 10-25\,{\rm M}_{\odot}$ stars, and their low luminosity and weak emission lines would prevent them from being classified as WR stars. The brighter, cooler companion star may be overluminous because it has just accreted its companion's H envelope. Such stripped-envelope stars should be the most common SN Ibc progenitors (see Section 6.2 below).

Two other key considerations are the observed WC/WN ratio that increases with Z (Massey 2003) and the fact that even early-type WR stars in the SMC tend to have some small amount of H present in their atmospheres (Foellmi et al. 2003); both of these are generally attributed to the important role of Z-dependent, line-driven winds. Again, it is important to recognize the influence of the stellar wind after the H envelope is stripped in binary RLOF. No matter what mechanism removes the H envelope (binary RLOF, LBV eruptions, RSG winds, or hot-star winds), the subsequent evolution is dominated by a line-driven wind (Vanbeveren et al. 2007, Claeys et al. 2011). Binary RLOF usually leaves a thin H layer, and stars at lower Z or lower L have a harder time removing it. Similarly, the evolution from a WN to a WC star (if this progression is monotonic) is harder at lower Z or lower L, even if the H envelope was stripped by binary interaction. Therefore, comparing models to the observed WC/WN ratio and showing that it increases with Z, for example, is not indicative of the importance of line-driven winds in any earlier phases of evolution, because any scenario should predict an increasing WC/WN ratio with Z.

Altogether, it is difficult to rule out the hypothesis that binary RLOF is the dominant agent responsible for producing most or all WR stars, and it remains unclear under what ranges of L and Z (if any) a single-star can become a WR star via its own wind mass loss. Unfortunately, finding examples of apparently single WR stars does not provide a conclusive answer, because a companion star may have exploded already. Moreover, the possible importance of RSG mass loss in producing WR stars is still not well understood. It is suspicious that most H-poor WR stars have luminosities of $\log(L/{\rm L}_{\odot}) = 5.5$ to 5.8 (Hamann et al. 2006, Crowther 2007), which also corresponds to the strongest RSG mass loss (**Figure 3**). The creation of WR stars is, of course, also closely related to the end fate in a SN and the rates of various SN subtypes that are seen, which is discussed next.

6.2. Outcomes II: The Main Supernova Subtypes

Core-collapse SNe exhibit a wide diversity of properties, summarized briefly in the sidebar titled Supernova Subtypes. Different SN types are the direct product of different amounts of mass loss from massive stars, and they provide key constraints that inform our understanding of stellar evolution. Here we discuss the main types of SNe II and Ibc with "normal" SN atmospheres. Types IIn and Ibn with narrow emission lines from dense CSM that indicate eruptive pre-SN mass loss are discussed separately in the next section. **Table 1** includes a map of SN type to progenitor star properties based on current prevailing ideas.

For understanding the main population of SNe, the relative rates of various subtypes are critical, because they must match the relative fractions of different progenitor stars for a given initial mass function (IMF). Volume-limited rates of the various core-collapse SN subtypes (excluding SNe Ia) measured in a controlled sample are now available (Smith et al. 2011b); these come from the Lick Observatory SN Search, which targeted mainly large galaxies representative of $\sim Z_{\odot}$ (Li et al.

SUPERNOVA SUBTYPES

SN explosions exhibit a wide diversity of observed properties. Proliferating classifications are based on their spectra and light curves, not a physical mechanism. Types I and II are determined by the presence (II) or absence (I) of H lines in the spectrum. The main categories of SNe are listed here (see also Filippenko 1997).

Type II-P: They are the most common type of core-collapse SN, exhibiting broad H lines in the spectrum and showing a plateau of typically ~100 days in the visual-wavelength light curve. They arise primarily from RSGs with initial masses of $8-20\,M_\odot$ (Smartt 2009) and mass-loss rates of $10^{-6}-10^{-5}\,M_\odot\ year^{-1}$.

Type II-L: Spectroscopically these are very similar to SNe II-P, but their light curves show a linear decay. The faster decline may result from a lower-mass H envelope, indicating heavier pre-SN mass loss by the progenitor.

Type II-pec: Similar to SNe II-P spectroscopically, but with a light curve that rises slowly from an initially faint state due to a more compact BSG progenitor, with SN 1987A being the prototype.

Type IIn: SNe with prominent narrow H lines in their spectra. The narrow lines arise in nearby (a few 10^{15} cm) CSM that is photoionized or shock-heated by the SN. These SNe require strong mass loss immediately (a few years) preceding the SN and exhibit wide diversity.

Type IIn-P: A new subclass of SNe IIn with plateau-shaped light curves, which show a pronounced (3–6 mag) drop in flux at times around 120 days (Mauerhan et al. 2013b).

Type IIb: These are very similar to Type Ib, except that they show transient broad H lines in their early-time spectra. This is due to a low-mass ($0.01\,M_\odot$ or less) residual H envelope remaining on the outer layers of the progenitor.

Type Ib: SNe that do not show H in their spectra, but which have strong broad He lines.

Type Ibn: Analogous to SNe IIn caused by strong CSM interaction, but with strong narrow lines of He instead of H.

Type Ic: SNe that show no H and little or no He lines in their spectra, requiring them to be the most extreme examples of stripped-envelope progenitors.

Type Ic-BL/GRB: These are SNe Ic with very broad (20,000–30,000 km s^{-1}) lines in their spectra. This is the only type of SN observed to be associated with GRB explosions.

Superluminous SN (SLSN): In principle, this class would include any of the above with a peak absolute magnitude more luminous than about −20 or −21 mag (i.e., brighter than the brightest SNe Ia). In practice, only spectral types IIn, II-L, and Ic have been seen in this class so far.

Type Ia: Thermonuclear SNe from white dwarf progenitor stars.

Type Ia/IIn (Ia-CSM): SNe Ia with narrow H lines in their spectra, indicating dense CSM. When strong CSM interaction veils underlying Ia spectral features, it is difficult to distinguish these from core-collapse SNe IIn.

2011). The most common core-collapse SNe are Type II-P (48.2 ± 6%), marking the explosions of relatively low-mass RSGs. Types II-L and IIn contribute 6.4 ± 3% and 9 ± 3%, respectively. The remainder (36.5 ± 6%) are stripped envelope SNe of Types IIb (10.6 ± 3.6%) and Ibc (26 ± 5%, including a few peculiar cases like SNe Ic-BL and Ibn that are rare at $Z_\odot$).

The observed fraction of stripped-envelope SNe is a key constraint on mass loss for the majority of massive stars. Smith et al. (2011b) pointed out that the observed fraction of stripped envelope SNe (IIb + Ib + Ic) of 36.5% is far too high to be reconciled with predictions of single-star evolution. If SN Ibc progenitors are assumed to be WR stars, then the observed fraction of SNe Ibc (not including SNe IIb) would require $M_{WR} = 22\,M_\odot$ (Smith et al. 2011b). Although this is not

Table 1 Mapping of supernova (SN) types to their likely progenitor star properties

SN	Progenitor Star[a]	M_{ZAMS} ($M_\odot$)[b]	$\dot{M}$ ($M_\odot$ $year^{-1}$)[c]	V_∞ (km s^{-1})
II-P	RSG	8–20	10^{-6}–10^{-5}	10–20
II-L	RSG/YSG	20–30 (?)	10^{-5}–10^{-4}	20–40
II-pec	BSG (b)	15–25	10^{-6}–10^{-4}	100–300
IIb	YSG (b)	10–25	10^{-5}–10^{-4}	20–100
Ib	He star (b)	15–25 (?)	10^{-7}–10^{-4}	100–1,000
Ic	He star (b)/WR	25–?	10^{-7}–10^{-4}	1,000
Ic-BL	He star (b)/WR	25–?	10^{-6}–10^{-5}	1,000
IIn (SL)	LBV	30–?	(1–10)	50–600
IIn	LBV/B[e] (b)	25–?	(0.01–1)	50–600
IIn	RSG/YHG	25–40	10^{-4}–10^{-3}	30–100
IIn-P	Super-AGB	8–10	0.01–1	10–600
Ibn	WR/LBV	40–?	10^{-3}–0.1	1,000
Ia/IIn	WD (b)	5–8 (?)	0.01–1	50–100

[a]Most likely progenitor star type. "(b)" indicates that a binary channel is probably key. Note that stars that shed envelopes in binary Roche-lobe overflow are likely to have a slow (10 km s^{-1}) equatorial outflow, in addition to the wind speed of the star.
[b]Masses with "(?)" indicate high uncertainty. Mass ranges with "?" as the upper end of the range indicate that these types might extend to a high and highly uncertain upper mass limit.
[c]Mass-loss rates for pre-SN eruptions are listed in parentheses and correspond roughly to the total mass ejected in the few years immediately preceding core collapse. The mass-loss rates may be lower, but still substantial, at larger radii traced by the expanding SN shock at late times.
Abbreviations: blue supergiant, BSG; luminous blue variable, LBV; red supergiant, RSG; Wolf-Rayet, WR; yellow hypergiant, YHG; yellow supergiant, YSG.

much lower than the lowest-mass WR stars observed in the Milky Way (Crowther 2007), it is much lower than can be explained by standard single-star evolution models—especially if we recognize that "standard" single-star models all adopt mass-loss rates that are too high. The observed SN statistics strongly favor the interpretation that most stripped-envelope SNe (including SNe IIb) come from lower-mass stars (10–25 $M_\odot$) that lose their H envelope in binaries. Again, ~36% is the observed SN IIb + Ib + Ic fraction; compare this to 33%, which is the fraction of massive stars that Sana et al. (2012) expect to have their H envelopes stripped in a binary system, given the observed binary fraction of O-type stars. One infers that binary RLOF can account for the observed statistics. (Recall that ~80% of SNe come from initial masses $< 25\,M_\odot$, assuming a Salpeter IMF in which every star with initial mass above 8.5 $M_\odot$ explodes as a SN.) Preference for the binary channel agrees with relatively low ejecta masses and H/He mass fractions inferred from detailed radiative transfer models of stripped-envelope SNe (Yoon et al. 2010, Dessart et al. 2011, Hachinger et al. 2012), which seem to rule out the idea that SNe IIb and Ib can come from progenitors much more massive than progenitors of SNe II-P, on average.

A dominant binary channel for stripped-envelope SNe is also consistent with available direct detections and upper limits of SN progenitors. Smartt (2009) summarized progress up until 2008, but there have been several important additions since then. Observations to date are consistent with stars having initial masses of roughly 8–20 $M_\odot$ dying as SNe II-P [the upper bound of this range is uncertain and higher than that found by Smartt (2009) if one accounts for progenitor reddening (Walmswell & Eldridge 2012)], but this does not necessarily mean that all stars in this mass range die that way; stars in the same mass range could die as stripped-envelope SNe if they are in a binary system, and in fact, there are currently three direct progenitor detections for SNe IIb (SN 1993J, 2011dh, and 2013df) that are thought to be YSGs in binary systems with inferred initial masses of 13–17 $M_\odot$ (Maund & Smartt 2009; Van Dyk et al. 2013, 2014). Two other SNe IIb, SN 2001ig and SN 2008ax, also show possible indications of a companion star (Ryder et al. 2006, Crockett et al. 2008). Moreover, there are as yet no detections of progenitors of SNe Ibc. Their nondetection seems unlikely if luminous WR stars are their progenitors (Smartt 2009), but the hot temperatures of WR stars make it hard to definitively rule them out.

What about stripped-envelope SN fractions at lower Z? The SN Ibc/II ratio decreases at lower metallicity (Prantzos & Boissier 2003, Prieto et al. 2008a, Boissier & Prantzos 2009), as noted earlier. However, Smith et al. (2011b) pointed out that the relevant ratio really is (IIb + Ibc)/(II-P + II-L + IIn), because SNe IIb are almost identical to SNe Ib except for $<0.1\,M_\odot$ of H in their outermost layer (Dessart et al. 2011, Hachinger et al. 2012). [Many studies group SNe IIb and II-L together into a transitional class between SNe II-P and Ib in a single-star framework (e.g., **Figure 1**). However, it is important to note that SNe IIb and II-L are actually quite different, and they are not part of the same continuum in decreasing H envelope mass. SNe IIb really are almost identical to SNe Ib, whereas SNe II-L have more in common with normal SNe IIn and SNe II-P.] The recent study by Arcavi et al. (2010) finds that, compared with giant galaxies at high Z, there is a much larger fraction of SNe IIb and a lower fraction of normal SNe Ibc in lower-Z dwarf galaxies. The lower fraction of SNe Ibc is expected in single-star models, but the higher fraction of SNe IIb is not. Because the removal of the H envelope itself would be greatly hindered, we would expect a much larger fraction of SNe II-P and II-L at lower Z rather than SNe IIb. This result is, however, expected in binary evolution, because low-Z line-driven winds have a harder time removing the residual H layer that remains after RLOF (Claeys et al. 2011). Interestingly, Arcavi et al. (2010) also find a larger relative fraction of SN Ic-BL in low-Z dwarf galaxies; this is not yet explained, partly because we don't have a good understanding of what physical mechanism makes some SNe Ic have such broad lines. It follows the trend that GRBs and their associated SNe Ic-BL seem to prefer low Z (Modjaz et al. 2008).

Altogether, current evidence strongly suggests that it is no longer true that single WR stars are the preferred progenitors of most stripped-envelope SNe. Massive WR stars might yield some of the SNe Ibc, of course, especially in the smaller category of SNe Ic and GRBs that may favor high-mass progenitors, but many SNe Ibc must come from lower initial masses in which the H envelope is stripped in a binary.

Using radio and X-ray observations, one can probe the density in the wind of the SN progenitor directly. Assuming wind speeds of $\sim$1,000 km s^{-1}, observations suggest a very wide range of mass-loss rates for SN Ibc progenitors, from $10^{-7}\,M_\odot$ year^{-1} up to values near the line driving limit, with an average around $10^{-5}\,M_\odot$ year^{-1} (Wellons et al. 2012). The examples near the lower end of this range are inconsistent with WR winds, whereas those near the upper end of the range are consistent (see **Figure 3**). This is further evidence that SNe Ibc may arise from a large range of initial masses, including relatively low-luminosity stars with weak winds that must have lost their envelopes in binary RLOF. Interestingly, the SN Ibc near the top of this range tend to show density modulation in their winds, which may indicate slow (note that slower outflows would reduce the value of $\dot{M}$ inferred from radio observations) and dense outflows from RLOF or interacting winds in binaries (Podsiadlowski et al. 1992), pre-SN eruptive mass loss (Smith & Arnett 2014), or S Doradus (LBV-like) variability (Kotak & Vink 2006). This pre-SN variability is not yet understood. Much more extreme cases of eruptive pre-SN mass loss are also seen, and this is discussed next.

6.3. Outcomes III: Enhanced Pre-Supernova Mass Loss and Luminous Supernovae

One of the more exciting new developments in massive star and SN research in the past decade is the recognition that a subset of massive stars undergo violent eruptive mass loss immediately preceding core collapse, and that this may yield some of the most luminous SNe in the Universe.

A SN blast wave expands outward into the CSM, and the ensuing collision, referred to as CSM interaction, is commonly observed in the form of X-ray or radio emission (Chevalier & Fransson 1994) for normal winds (**Figure 3**). In 8–9% of core-collapse SNe (Smith et al. 2011b), however, the CSM is so dense that the shock interaction gives rise to strong narrow emission lines in the visual-wavelength SN spectrum. When the CSM is very dense, it can substantially decelerate the fast SN ejecta and convert a large fraction of the kinetic energy (10–50% or more) into radiation. When these strong narrow emission lines are observed, we refer to the SN as Type IIn (narrow H lines) or Ibn (narrow He lines). In general, $\dot{M}$ values of at least 10^{-3} to $10^{-2}\ M_{\odot}\ year^{-1}$ are required for the narrow emission lines to compete with the luminosity of the normal SN photosphere. There is a huge diversity among SNe with strong CSM interaction, which can be understood in a few different regimes:

- Superluminous SNe (SLSNe) with Type IIn spectra represent the most extreme cases of eruptive pre-SN mass loss; luminosities are ~10 times higher than a normal bright SN Ia. Smith & McCray (2007) proposed that these extremely high luminosities could be achieved with normal energy core-collapse SNe (a few 10^{51} ergs) if the fast SN ejecta crash into a very massive 10–20$M_{\odot}$ CSM shell. This large mass comes from the basic physical requirement that the CSM must have enough inertia to substantially decelerate the fast SN ejecta and convert its expansion kinetic energy into thermal energy that can be radiated away. Diverse SLSN light curve shapes are possible, depending on the distribution of the mass and explosion properties (van Marle et al. 2010, Chevalier & Irwin 2011, Chatzopoulos et al. 2013, Moriya et al. 2013). SN 2006gy was the first observed event that instigated these ideas of massive CSM shell collisions (Ofek et al. 2007; Smith & McCray 2007; Smith et al. 2007, 2010a; Woosley et al. 2007), but a number of very luminous SNe IIn have been studied in detail since then, including objects like SN 2006tf (Smith et al. 2008), SN 2003ma (Rest et al. 2011), and SN 2008fz (Drake et al. 2010) (see also the sidebar, Superluminous Supernovae without Narrow Lines).
- SNe IIn with moderate luminosity represent less extreme cases than SLSNe, but they still require strong CSM interaction that indicates eruptive or episodic pre-SN mass-loss events. Instead of pre-SN ejections of 10–20 $M_{\odot}$, more typical luminosities require less massive shells of order 0.1–1 $M_{\odot}$. The lower luminosity could result from lower explosion energy (see SNe IIn-P below), but for normal SNe IIn it is more likely attributed to lower density CSM or asymmetric CSM that only intercepts a portion of the explosion solid angle. Some well-studied SNe IIn like SN1998S (Leonard et al. 2000) and SN 2009ip (Levesque et al. 2014, Smith et al. 2014) are consistent with very asymmetric or even disk-like CSM. The

SUPERLUMINOUS SUPERNOVAE WITHOUT NARROW LINES

The most luminous SNe (i.e., SLSNe) seen so far are of Types Ic or II-L, without narrow lines. These SLSNe Ic (Quimby et al. 2011, Gal-Yam 2012) and II-L (Gezari et al. 2009, Miller et al. 2009) might be explained by CSM interaction with a dense shell in special cases in which the shell has a relatively sharp outer boundary. After shock breakout from the wind, radiation diffuses out from the accelerated CSM, yielding no narrow lines (Smith & McCray 2007, Chevalier & Irwin 2011). However, these SNe may also be explained by magnetar birth (Kasen & Bildsten 2010, Woosley 2010), so they are not necessarily the product of eruptive pre-SN mass loss. SLSNe Ic do, of course, require a stripped H envelope like other SNe Ic, and they are interestingly similar to SNe Ic-BL and GRBs in that they seem to prefer low-*Z* host galaxies (Neill et al. 2011). This may also suggest a binary origin for these stars.

mass-loss rates indicated by the CSM suggest that viable progenitors could be LBVs (Gal-Yam et al. 2007, Taddia et al. 2013) as well as extreme RSGs or YHGs (Smith et al. 2009) (see **Figure 3**). Although the immediate pre-SN eruptive mass loss is less extreme than required for SLSNe IIn, some SNe IIn show strong CSM interaction that continues for years or decades (like SN 1988Z; Aretxaga et al. 1999) as well as long-lasting IR echoes from distant dust shells illuminated by the SN (Gerardy et al. 2002, Fox et al. 2011)—both of these indicate a considerable amount of mass lost by the progenitor star for centuries before core collapse, either in previous eruptions or very strong dusty winds.

- The subclass of SNe IIn with plateau light curves, SNe IIn-P, was proposed recently (Mauerhan et al. 2013b), represented by SNe like SN 1994W, SN 2009kr, and SN 2011ht. Among the wide diversity of SNe IIn, this subset is surprisingly homogeneous, with nearly identical spectral evolution and very similar light curves that all plummet sharply after ~120 days (other SNe IIn have smoothly declining or very slowly declining light curves). These may be relatively low-energy (10^{50} erg) electron-capture SNe from $M_{ZAMS} = 8-10\ M_{\odot}$ super-AGB stars that achieve luminosities of normal SNe through intense CSM interaction (Smith 2013b).

Electron-capture SN: collapse of degenerate O-Ne-Mg core due to electron capture in lower-mass (8–10 $M_{\odot}$) SN progenitors

Explosions classified as SNe Ibn are very similar to SNe IIn, except that instead of narrow H lines they exhibit narrow He lines. Their peculiar spectra arise from the same basic scenario as SNe IIn, with an SN shock interacting with dense CSM, but here the CSM is H poor. There are also a few reported transitional cases between SNe Ibn and IIn, including 2005la and SN 2011hw (Pastorello et al. 2008, Smith et al. 2012), suggesting a possible continuum in progenitor H-envelope stripping, similar to SNe II-P, IIb, and Ib. The best-studied Type Ibn is SN 2006jc (Foley et al. 2007, Pastorello et al. 2007), for which an LBV-like outburst was detected 2 years prior to the eventual SN. If SNe Ibn come from single stars, they are likely to be very massive WR stars; however, lower initial masses (10–20 $M_{\odot}$) can yield H-depleted progenitors through binary RLOF, and such stars may have dense H-poor CSM. There are not yet any detections of a quiescent SN Ibn progenitor. Interestingly, Sanders et al. (2013) report the detection of a SN Ibn in a giant elliptical galaxy with no evidence for ongoing star formation at the explosion site, which challenges the idea that these arise exclusively from very massive progenitor stars. This may echo the observation that some SNe IIn appear to be caused by an underlying thermonuclear Type Ia interacting with dense CSM (Silverman et al. 2013).

Until recently, pre-SN eruptive mass loss and connections to LBVs were mostly hypothetical, limited to (reasonable) conjectures supported by the circumstantial evidence that something must deposit a large mass of outflowing H-rich CSM so close to the star (Chugai et al. 2004; Smith et al. 2008, 2010a). However, we now have a handful of directly detected progenitor stars that seem consistent with LBVs, as well as a few examples of SN explosions in which an outburst was actually detected photometrically in the few years before a SN. In all cases the SN had narrow emission lines indicative of CSM interaction, but for each it is also difficult to prove conclusively that the event was a true core-collapse SN. So far all these objects have continued to fade and there is no clear evidence that the stars survived.

6.3.1. SN 1961V. This is Zwicky's prototype Type V SN, later associated with LBV-like giant eruptions, but it may have been a true core-collapse SN IIn that was not recognized at that time because the Type IIn class hadn't been defined yet. A luminous (−12.2-mag absolute at blue wavelengths) progenitor was detected at several epochs for ~20 years preceding the SN 1961V event (Humphreys et al. 1999), and this source has now faded. The pre-SN detections include small (~0.5 mag) fluctuations that resemble S Dor-like episodes, and in the year before the SN there is one detection at an absolute magnitude of roughly −14.5, similar to LBV giant eruptions.

Currently, a source at the same position is about 6 mag fainter than the progenitor and still shows narrow Hα emission (Van Dyk & Matheson 2012) that could represent ongoing CSM interaction. The progenitor source has not been discussed in the context of SN progenitors because the 1961 event was considered an LBV eruption (a "super η Car–like event"), not a true SN (see Van Dyk & Matheson 2012). Recently it was argued that SN 1961V was actually a true core-collapse SN IIn (Kochanek et al. 2011, Smith et al. 2011d), which would provide evidence for a $\sim$100–$M_\odot$ LBV-like progenitor that experienced eruptive mass loss before a SN IIn.

6.3.2. SN 2005gl. This was a normal SN IIn in pre-explosion HST images that showed a source at the SN position, which then faded below detection limits after the SN had faded (Gal-Yam et al. 2007, Gal-Yam & Leonard 2009). Its high luminosity suggested that the progenitor was a massive LBV, similar to P Cygni, with an initial mass of order 60 $M_\odot$ and a mass-loss rate before core collapse of $\sim 0.01\,M_\odot$ year^{-1}.

6.3.3. SN 2006jc. A precursor eruption was discovered in 2004 and noted as a possible LBV or SN impostor. It had a peak luminosity similar to that of η Car (Pastorello et al. 2007). No spectra were obtained, but the coincident SN explosion two years later was of Type Ibn with strong narrow HeI emission lines (Foley et al. 2007, Pastorello et al. 2007). There is no detection of the quiescent progenitor.

6.3.4. SN 2009ip. This source was initially discovered and studied in detail as an LBV-like outburst in 2009 before finally exploding as a much brighter SN in 2012. A quiescent progenitor star was detected in archival HST data, indicating a very massive $50-80\,M_\odot$ progenitor (Smith et al. 2010b, Foley et al. 2011). It showed slow variability consistent with an S Dor LBV-like episode (Smith et al. 2010b), followed by a series of brief LBV-like giant eruptions (Smith et al. 2010b, Mauerhan et al. 2013a, Pastorello et al. 2013). SN 2009ip is so far unique among SN progenitor detections: Not only did it have a detection of the quiescent progenitor and multiple pre-SN eruptions but, unlike any other object, we also have detailed, high-quality spectra of the pre-SN eruptions (Smith et al. 2010b, Foley et al. 2011). The presumably final SN explosion of SN 2009ip in 2012 looked like a normal SN IIn, as the fast ejecta crashed into the slow material ejected 1–3 years earlier (Mauerhan et al. 2013a, Smith et al. 2014). A number of detailed studies of the bright 2012 transient have now been published, although there has been some controversy about whether the 2012 event was a core-collapse SN (Mauerhan et al. 2013a; Ofek et al. 2013a,b; Prieto et al. 2013; Smith et al. 2014) or some type of extremely bright nonterminal event (Fraser et al. 2013a, Pastorello et al. 2013, Margutti et al. 2014). More recently, Smith et al. (2014) showed that the object continues to fade, and its late-time emission is consistent with late-time CSM interaction in normal SNe IIn. If SN 2009ip was indeed a SN, it provides the strongest case that very massive stars above 30 $M_\odot$ do in fact experience core collapse and LBV-like stars are linked to SNe IIn.

6.3.5. SN 2010mc. Ofek et al. (2013b) reported the detection of a precursor event $\sim$40 days before the peak of the Type IIn SN 2010mc. It is unclear whether this was a pre-SN outburst or the SN itself. Smith et al. (2014) showed that the double-peaked light curve of SN 2010mc was nearly identical to that of SN 2009ip, for which it has been suggested that the -40-day precursor was actually the faint SN explosion of a BSG, and the later rise to peak was caused by CSM interaction.

6.3.6. SN 2010jl. This was an SLSN IIn, with a peak absolute magnitude brighter than -20 mag. Smith et al. (2011c) identified a source at the location of the SN in pre-explosion HST images that suggested either an extremely massive progenitor star or a very young massive-star cluster; in either case it seems likely that the progenitor had an initial mass above 30 $M_\odot$.

These direct detections of LBV-like progenitors and of pre-SN outbursts provide unambiguous evidence for violent eruptive mass loss associated with the latest phases in a massive star's life. This only occurs in ~9% of core-collapse SNe (Smith et al. 2011b). The extremely short timescale of only a few years probably hints at severe instability in the final nuclear burning sequences, especially Ne and O burning (Arnett & Meakin 2011, Quataert & Shiode 2012, Shiode & Quataert 2014, Smith & Arnett 2014), each of which lasts about 1 year. These instabilities may be exacerbated in the most massive stars, although much theoretical work remains to be done. The increased instability at very high initial masses is extreme in cases in which pre-SN eruptions result from the PPI (Woosley et al. 2002, 2007), but eruptions may extend to other nuclear burning instabilities as well (Arnett & Meakin 2011, Smith & Arnett 2014). Although the events listed above are just a few lucky cases, they may also be the tip of the iceberg. Undoubtedly, continued work on the flood of new transient discoveries will reveal more of these cases. [Indeed, as this review went to press, Fraser et al. (2013b) reported the recovery in archival data of an outburst one year before the SN IIn-P 2011ht.] The limitation will be the existence of high-quality archival data over long timescales of years before the SNe, but these sorts of archives are becoming more populated and improved as time passes. When the Large Synoptic Survey Telescope arrives, it may become routine to detect pre-SN outbursts, although constraining their physical nature with spectra will still be a challenge.

Not all SNe with strong CSM interaction necessarily require LBV-like progenitors, but many of them do, and most of the detected progenitors or progenitor outbursts are consistent with this hypothesis. If the progenitors of SNe IIn are not actually LBVs, they do a very good impersonation of the eruption energy, luminosity, spectral morphology, ejecta mass, H composition, and outflow speeds observed in LBV giant eruptions. In any case, the basic observation that very massive stars are reaching core collapse with massive H envelopes still intact is in direct conflict with single-star evolution models. Once again, this may be related to the issue of overestimated mass-loss rates adopted in these models and the neglect of binary evolution.

6.4. Extrapolating to Low-*Z* and Population III Stars

Stellar evolution is complicated and fraught with tremendous uncertainties about mass loss. Faced with such road blocks, one can sympathize with the desire to divert one's attention to Population III stars, where theorists are unencumbered by observational data. More seriously, though, there is indeed great interest in thinking about what the earliest stars in the Universe might have been like, because they had a strong impact on their environment and on galaxy evolution in general. These stars may have been very massive. They made the first SNe, the first stellar mass black holes, and the first dust. They ejected the first metals back into the ISM, they enhanced or dominated reionization, and they probably triggered the formation of the earliest generations of low-mass stars (Heger et al. 2003, Bromm & Larson 2004). However, as we attempt to extrapolate from the huge uncertainty associated with local stars to a regime in which there is no data, we must be cautious.

In studying stars at very low Z, a standard approach is to extrapolate the smooth Z dependence of line-driven winds to very low or even zero Z, essentially assuming that massive stars at the lowest Z will have no mass loss (Heger et al. 2003). This leads to interesting fates for very massive stars, such as pair instability SNe (Woosley et al. 2002, Heger et al. 2003). In contrast, more recent results suggest that Z-dependent winds are weaker than we used to think, even at $Z_{\odot}$. However, binary RLOF and eruptive LBV-like mass loss are much more important than was appreciated in the past, and they are relatively insensitive to Z. Thus, it is not safe to assume that massive stars at very low Z suffer much less mass loss on average.

As noted above, LBVs are seen in nearby low-Z dwarf galaxies. Pre-SN mass loss also provides some clues in this regard. Many SNe IIn and SLSNe occur in low-Z dwarf galaxies (Neill et al. 2011, Stoll et al. 2011), demonstrating that low Z does not inhibit strong eruptive mass loss. The influence of these pre-SN eruptions may have an interesting effect on the predicted outcomes of SN explosions at low Z (Couch & Ott 2013, Smith & Arnett 2014).

GRBs, of course, present one of the most puzzling observable mysteries about mass loss and massive-star evolution at low Z. The fact that GRB progenitors have shed most of their envelopes while still retaining a great deal of angular momentum (MacFadyen & Woosley 1999) would seem to strongly favor their origin in binary evolution with mass transfer to spin up the star or even a merger. GRBs are quite rare, so attributing them to a special circumstance like a late-phase merger is plausible.

7. SUMMARY AND PERSPECTIVE

7.1. Take-Home Points

We now have a fairly firm understanding of stellar winds of hot O-type stars (Puls et al. 2008), relevant for most of their lives on the H-burning MS phase. There remain substantial issues in understanding the physics of wind driving, magnetic fields, and angular momentum loss, and it remains challenging to understand how these correspond to various observational diagnostics—but in terms of the total mass-loss rates and their impact on stellar evolution, we know that wind mass loss is weaker than we used to think owing to the effect of clumping. The consensus seems to be that mass-loss rates need to be reduced by a factor of at least 2–3 compared with rates derived observationally from standard ρ^2 diagnostics assuming homogeneous winds (de Jager et al. 1988, Nieuwenhuijzen & de Jager 1990). Reduction by a factor of $\sim$10 is probably an overestimate for most O-type stars, but not for later O-types and early B-type MS stars subject to the weak-wind problem.

Astronomers often scoff at a factor of 2 in any individual measurement, but we must recognize that systematically lowering $\dot{M}$ by even a factor of 2 in models is very significant, and a factor of 3 is huge (current debate generally centers around a factor of 2 or 3). A factor of only 2 lower $\dot{M}$ would be like replacing mass-loss rates at $Z_\odot$ with mass-loss rates currently used for $0.37\,Z_\odot$ (i.e., lower than the LMC), whereas a factor of 3 would correspond to $0.2\,Z_\odot$, which is similar to models for SMC stars (using $\dot{M} \propto Z^{0.69}$; Vink et al. 2001). With moderately weaker mass loss, we know that SMC models exhibit significantly different evolution than Milky Way stars. Most importantly, they do not produce such a large population of H-free WR stars. However, it is over this same metallicity range (Massey 2003) that we are testing the validity of the physical ingredients in stellar evolution models. This isn't good.

For post-MS supergiant phases and binary RLOF, the uncertainty in mass loss is far worse—but these are also times when mass loss is strongest! Given that single-star models neglect eruptive mass loss and binaries, but match many properties of the observed distributions anyway, predictions beyond the end of H burning are not unique. Extrapolations to low Z and prescriptions for feedback from early stellar populations are therefore highly uncertain. Researchers working in other fields where massive-star feedback is relevant—galaxy evolution, reionization, chemical evolution, low-metallicity stars, etc.—should be cognizant of this.

7.2. Perspectives and Directions

With such a high level of uncertainty in the most important phases of mass loss for massive stars, astronomers working on mass loss and massive-star evolution have a few choices on how to proceed: (*a*) restrict their attention to MS stars, (*b*) work on Population III stars (see above), or (*c*) confront the complicated effects of binaries and time-dependent mass loss as well as their influence on observed populations.

A major hurdle is disentangling the observed effects of single-star mass loss and binary RLOF. On the one hand, Occam's razor encourages us to keep things simple, motivating us to understand single stars before we can hope to tackle the "free parameter heaven" of binaries. On the other hand, the massive stars we observe are mostly binaries: A wise man once said that we should "Make things as simple as possible, but not simpler." The message here is that ignoring binaries, but testing single-star models against observed statistical properties of massive stars (which are mostly binaries), leads us to misinterpret single stars too because we are relying upon strong winds and rotation to compensate for actual outcomes of binary RLOF.

Because the observed statistics of massive-star populations are unavoidably contaminated by the effects of binary stars, one alternative approach for testing single-star models would be to place less emphasis on matching these observed statistics and more emphasis on matching individual observed stars with well-constrained physical parameters. Masses for "effectively single" stars can be measured in wide binary systems in which the stars will not interact (or O-type stars that have not yet interacted on the MS), and for these a detailed modern quantitative analysis can yield estimates of $\dot{M}$, $T_{\rm eff}$, L, rotational speed, abundances, etc. The distance must be known, but the Magellanic Clouds and clusters with known distances can help, and *Gaia* will soon lower distance uncertainties for the nearest massive stars in the Milky Way. Doing this for a number of stars at various evolutionary stages and luminosities will build up a "grid" of observational constraints that models can aim to fit. Much of this observational work has already been done.

Given the huge level of uncertainty in post-MS mass loss (for both RSGs and LBVs), another general approach for theorists could be to use stellar evolution codes as "toy models" to investigate the final outcome for a wide range of possible mass-loss prescriptions that include episodic mass loss. For example, one could calculate evolutionary tracks for very massive stars that have lower wind $\dot{M}$ during H burning and then simply invoke one, two, or multiple sudden ejections of $10\,M_\odot$ in post-MS phases to mimic LBV eruptions in order to learn how the star's further evolution responds. This approach may seem artificial, but the goal would be to reverse-engineer stars, not to provide uniquely correct evolutionary tracks. In any case, much additional work on constraining post-MS mass loss of all forms is needed. Several specific possible future directions for theory and observations are listed below.

FUTURE ISSUES: FOR THEORISTS

1. A major task is to calculate stellar evolution models with lower wind mass-loss rates for MS phases, including rates appropriate for the weak-wind problem in later O-type and early B-type stars. These can be compared with precise measurements of physical parameters for well-studied individual stars rather than populations, as noted above. For post-MS phases, mass loss is simply too uncertain to predict unique outcomes. Until we have a firmer understanding of the mass loss in post-MS stages, end states of models are mostly hypothetical. Toy models with various prescriptions for unsteady $\dot{M}$ can help constrain the possible outcomes and their influence on SNe. Perhaps new open-source codes like MESA (Paxton et al. 2011) will facilitate this.
2. For episodic mass loss in the latest pre-SN evolutionary phases, further work on stellar evolution with hydrodynamics and advanced nuclear burning is essential. Pre-SN eruptions constitute a new and very important unsolved problem in astrophysics, which may influence the outcome of core collapse itself.

3. Theoretical investigations are needed to probe the underlying cause of LBV eruptions and their impact on the subsequent stellar evolution. Predictions for observed consequences of envelope instabilities, mergers/collisions, and explosive burning events are needed to link various possible physical mechanisms to observables for transient sources.
4. The hydrodynamics of RLOF and its dependence (or lack thereof) on metallicity is an important issue. A better understanding of mass and angular momentum transfer versus mass loss from binary systems is needed to determine what conditions lead to mergers. In principle, mass transfer itself should be insensitive to Z, but in practice, Z affects opacity and the stellar radius, and thus the onset of RLOF. In the end, we may find that observed trends with Z hinge upon the formation of massive binaries (Kratter et al. 2010) and its dependence on fragmentation and cooling as a function of Z.
5. We need additional theoretical investigations of RSG mass loss and its scaling with initial mass, Z, and as a function of evolutionary time. How does the global $\dot{M}$ depend on the interplay between pulsational instability and dust formation?

FUTURE ISSUES: FOR OBSERVERS

1. To help constrain single-star evolution, we need precise estimates of physical parameters for a number of nearby "effectively single" stars at various evolutionary stages that can serve as anchors for stellar evolution models. Relatively wide binaries that have not yet interacted are good targets for this.
2. To better understand how binarity influences observed trends, we need continued work on the binary fraction and orbital parameters as a function of environment: clusters versus field and Z. Orbits are often assumed to be circular in models, but the eccentricity distribution may also be critical for some problems.
3. We need better constraints on the episodic mass loss of LBVs, including the lifetime of the LBV phase, the total mass lost per eruption, the duty cycle of eruptions, and how these vary with Z and the initial masses and binarity of central stars. The large number of shell nebulae recently discovered by IR surveys (Gvaramadze et al. 2010, Wachter et al. 2010) may be very helpful in improving statistics and associations with the central stars. These improved constraints are needed to guide prescriptions for including episodic mass loss in models.
4. Regarding observational constraints on the total mass lost in binary RLOF as compared with mass-transfer rates, studies of RLOF binaries or post-RLOF systems can help constrain under what conditions mass transfer is conservative or the degree to which it is conservative.
5. Ongoing studies of RSGs in clusters may help provide better constraints on measured RSG $\dot{M}$ as a function of evolutionary stage for a range of Z. For how long and for which stars is the strongest mass loss (self-obscured, masers) at work? Connections to pulsation amplitudes and stars on blueward tracks are also important.

6. SN progenitors and SNe with CSM provide critical clues about the mass loss of massive stars, and well-constrained individual cases are limited to relatively nearby SNe. This provides important links between stars and their end fates and has opened a new window for observing episodic mass loss in the very final stages of evolution. Also, among nearby stars, we need better constraints on the properties of the faint He-rich stars in binaries that have lost their H envelope to a companion in RLOF, because they may be the dominant stripped envelope SN progenitors.
7. Stellar evolution codes adopt time-averaged mass-loss prescriptions, but increasing evidence suggests that brief disruptive events are very important. Observational work on transient phenomena in general is still a relatively new topic, providing important constraints on physical parameters and rates for these events. Are there observational signatures that will tell us if they are binaries?
8. Studies of SN and transient environments, including their surrounding stellar populations, can help disentangle the relative evolutionary phases of their progenitors (ages, burning stages, etc.). This may help to identify types of stars that don't fit with the expected age of their surrounding populations, perhaps flagging stars that are mainly binaries.

DISCLOSURE STATEMENT

The author is not aware of any affiliations, memberships, funding, or financial holdings that might be perceived as affecting the objectivity of this review.

ACKNOWLEDGMENTS

I am grateful for numerous discussions with many people, but especially Stan Owocki, Jo Puls, Dave Arnett, Paul Crowther, Jorick Vink, Jose Prieto, Todd Thompson, Norbert Langer, Selma de Mink, Luc Dessart, John Hillier, Ben Davies, John Eldridge, Phil Podsiadlowski, and Matteo Cantiello.

LITERATURE CITED

Arcavi I, Gal-Yam A, Kasliwal MM, et al. 2010. *Ap. J.* 721:777
Aretxaga I, Benetti S, Terlevich RJ, et al. 1999. *MNRAS* 309:343
Arnett WD, Bahcall JN, Kirshner RP, Woosley SE. 1989. *Annu. Rev. Astron. Astrophys.* 27:629
Arnett WD, Meakin C. 2011. *Ap. J.* 741:33
Barlow J, Drew JE, Meaburn J, Massey RM. 1994. *MNRAS* 268:L29
Bastian N, Covey KR, Meyer MR. 2010. *Annu. Rev. Astron. Astrophys.* 48:339
Beers TC, Christlieb N. 2005. *Annu. Rev. Astron. Astrophys.* 43:531
Bennett PD. 2010. In *Hot and Cool: Bridging Gaps in Massive Star Evolution*, ed. C Leitherer, P Bennett, P Morris, J van Loon. *ASP Conf. Ser.* 425:181. San Francisco: ASP
Boissier S, Prantzos N. 2009. *Astron. Astrophys.* 503:137
Bouret JC, Lanz T, Hillier DJ. 2005. *Astron. Astrophys.* 438:301
Bromm V, Larson RB. 2004. *Annu. Rev. Astron. Astrophys.* 42:79
Bromm V, Yoshida N. 2011. *Annu. Rev. Astron. Astrophys.* 49:373
Cantiello M, Langer N, Brott I, et al. 2009. *Astron. Astrophys.* 499:279
Cantiello M, Yoon SC, Langer N, Livio M. 2007. *Astron. Astrophys.* 465:L29

Castor JI, Abbott DC, Klein RI. 1975. *Ap. J.* 195:157
Chatzopoulos E, Wheeler JC, Vinko J, Horvath ZL, Nagy A. 2013. *Ap. J.* 773:76
Chesneau O, Meilland A, Chapellier E, et al. 2014. *Astron. Astrophys.* 563:A71
Chevalier RA, Fransson C. 1994. *Ap. J.* 420:268
Chevalier RA, Irwin 2011. *Ap. J.* 729:L6
Chini R, Hoffmeister VH, Nasseri A, Stahl O, Zinnecker H. 2012. *MNRAS* 424:1925
Chiosi C, Maeder A. 1986. *Annu. Rev. Astron. Astrophys.* 24:329
Chugai NN, Blinnikov SI, Cumming RJ, et al. 2004. *MNRAS* 352:1213
Claeys JSW, de Mink SE, Pols OR, Eldridge JJ, Baes M. 2011. *Astron. Astrophys.* 528:A131
Clark JS, Larionov VM, Arkharov A. 2005. *Astron. Astrophys.* 435:239
Cohen DH, Cassinelli JP, Macfarlane JJ. 1997. *Ap. J.* 487:867
Cohen DH, Gagne M, Leutenegger MA, et al. 2011. *MNRAS* 415:3354
Cohen DH, Kuhn MA, Gagne M, Jensen ELN, Miller NA. 2008. *MNRAS* 306:1855
Cohen DH, Leutenegger MA, Wollman EE, et al. 2010. *MNRAS* 405:2391
Conti PS. 1976. *Mem. Soc. R. Sci. Lie'ge* 9:193
Conti PS. 1984. In *Observational Tests of the Stellar Evolution Theory, IAU Symp. 105*, ed. A Maeder, A Renzini, p. 233. Dordrecht, Neth.: Reidel
Couch S, Ott C. 2013. *Ap. J. Lett.* 778:7
Cox DP. 2005. *Annu. Rev. Astron. Astrophys.* 43:337
Cranmer SR, Owocki SP. 1996. *Ap. J.* 462:469
Crockett RM, Eldridge JJ, Smartt SJ, et al. 2008. *MNRAS* 391:L5
Crowther PA. 2007. *Annu. Rev. Astron. Astrophys.* 45:177
Crowther PA, Hillier DJ, Evans CJ, et al. 2002. *Ap. J.* 579:774
Davidson K. 1987. *Ap. J.* 317:760
Davies B, de la Fuente D, Najarro F, et al. 2012. *MNRAS* 419:1860
Davies B, Figer DF, Law CJ, et al. 2008. *Ap. J.* 676:1016
De Becker M, Rauw G, Manfroid J, Eenens P. 2006. *Astron. Astrophys.* 456:1121
de Jager C. 1998. *Astron. Astrophys. Rev.* 8:145
de Jager C, Nieuwenhuijzen H, van der Hucht KA. 1988. *Astron. Astrophys. Suppl.* 72:259
de Koter A, Lamers HJGLM, Schmutz W. 1996. *Astron. Astrophys.* 306:501
de Mink SE, Langer N, Izzard RG, Sana H, de Koter A. 2013. *Ap. J.* 764:166
de Mink SE, Pols OR, Hilditch RW. 2007. *Astron. Astrophys.* 467:1181
Decin L, Hony S, de Koter A, et al. 2006. *Astron. Astrophys.* 456:549
Dessart L, Hillier DJ, Livne E, et al. 2011. *MNRAS* 414:2985
Dessart L, Livne E, Waldman R. 2010. *MNRAS* 405:2113
Dessart L, Owocki SP. 2005. *Astron. Astrophys.* 437:657
Drake AJ, Djorgovski SG, Prieto JL, et al. 2010. *Ap. J.* 718:L127
Drew JE, Hoare MG, Denby M. 1994. *MNRAS* 266:917
Drout M, Massey P, Maynet G. 2012. *Ap. J.* 750:97
Dwarkadas VV, Owocki SP. 2002. *Ap. J.* 581:1337
Ekström S, Georgy C, Eggenberger P, et al. 2012. *Astron. Astrophys.* 537:A146
Eldridge JJ, Izzard RG, Tout CA. 2008. *MNRAS* 384:1109
Eldridge JJ, Stanway ER. 2009. *MNRAS* 400:1019
Elmegreen BG, Lada CJ. 1977. *Ap. J.* 214:725
Evans CJ, Crowther PA, Fullerton AW, Hillier DJ. 2004. *Ap. J.* 610:1021
Evans CJ, Lennon DJ, Smartt SJ, Trundle C. 2006. *Astron. Astrophys.* 456:623
Fan X, Carilli CL, Keating B. 2006. *Annu. Rev. Astron. Astrophys.* 44:415
Feldmeier A. 1995. *Astron. Astrophys.* 299:523
Figer DF, Najarro F, Gilmore D, et al. 2002. *Ap. J.* 581:258
Filippenko AV. 1997. *Annu. Rev. Astron. Astrophys.* 35:309
Foellmi C, Moffat AFJ, Guerrero MA. 2003. *MNRAS* 338:360
Foley RJ, Berger E, Fox O, et al. 2011. *Ap. J.* 732:32
Foley RJ, Smith N, Ganeshalingam M, et al. 2007. *Ap. J.* 657:L105

Fox O, Chevalier RA, Skrutskie MF, et al. 2011. *Ap. J.* 741:7
Fraser M, Inserra C, Jerkstrand A, et al. 2013a. *MNRAS* 433:1312
Fraser M, Magee M, Kotak R, et al. 2013b. *Ap. J.* 779:L8
Fullerton AW, Massa DL, Prinja RK. 2006. *Ap. J.* 637:1025
Fullerton AW, Massa D, Prinja RK, et al. 1997. *Astron. Astrophys.* 327:699
Gal-Yam A. 2012. *Science* 337:927
Gal-Yam A, Leonard DC. 2009. *Nature* 458:865
Gal-Yam A, Leonard DC, Fox DB, et al. 2007. *Ap. J.* 656:372
Garcia B, Mermilliod J. 2001. *Astron. Astrophys.* 368:122
Gehrels N, Ramirez-Ruiz E, Fox DB. 2009. *Annu. Rev. Astron. Astrophys.* 47:567
Gehrz RD, Woolf NJ. 1971. *Ap. J.* 165:285
Georgy C, Ekstrom S, Meynet G, et al. 2012. *Astron. Astrophys.* 542:A29
Gerardy CL, Fesen RA, Nomoto K, et al. 2002. *Ap. J.* 575:1007
Gezari S, Halpern JP, Grupe D, et al. 2009. *Ap. J.* 690:1313
Gies DR. 1987. *Ap. J. Suppl.* 64:545
Glatzel W, Kiriakidis M. 1993. *MNRAS* 263:375
Glatzel W, Kiriakidis M, Chernigovskij S, Fricke KJ. 1999. *MNRAS* 303:116
Gräfener G, Owocki SP, Vink JS. 2012. *Astron. Astrophys.* 538:A40
Gräfener G, Vink JS, de Koter A, Langer N. 2011. *Astron. Astrophys.* 535:A56
Groh JH, Hillier DJ, Damineli A, et al. 2009. *Ap. J.* 698:1698
Grundstrom ED, Gies DR, Hillwig TC, et al. 2007. *Ap. J.* 667:505
Guzik JA, Lovekin CC. 2012. *Astron. Rev.* 7:c13
Gvaramadze VV, Kniazev AY, Fabrika S. 2010. *MNRAS* 405:520
Gvaramadze VV, Langer N, Mackey J. 2012. *MNRAS* 427:L50
Habing HJ. 1996. *Astron. Astrophys. Rev.* 7:97
Hachinger S, Mazzali PA, Tautenberger S, et al. 2012. *MNRAS* 422:70
Hamann W-R, Feldmeier A, Oskinova LM, eds. 2008. *Clumping in Hot-Star Winds*. Potsdam: Potsdam Univ. Press
Hamann W-R, Gräfener G, Liermann A. 2006. *Astron. Astrophys.* 457:1015
Heger A, Fryer CL, Woosley SE, Langer N, Hartmann DH. 2003. *Ap. J.* 591:288
Heger A, Jeannin L, Langer N, Baraffe I. 1997. *Astron. Astrophys.* 327:224
Hellings P. 1984. *Ap. Space Sci.* 104:83
Hillier DJ. 1991. *Astron. Astrophys.* 247:455
Hillier DJ, Davidson K, Ishibashi K, Gull T. 2001. *Ap. J.* 553:837
Hillier DJ, Lanz T, Heap SR, et al. 2003. *Ap. J.* 588:1039
Hopkins AM, Beacom JF. 2006. *Ap. J.* 651:142
Howarth ID, Prinja RK, Massa D. 1995. *Ap. J.* 452:L65
Hubble E, Sandage A. 1953. *Ap. J.* 118:353
Huenemoerder DP, Oskinova LM, Ignace R, et al. 2012. *Ap. J.* 756:L34
Humphreys RM, Davidson K. 1994. *Publ. Astron. Soc. Pac.* 106:1025
Humphreys RM, Davidson K, Smith N. 1999. *Publ. Astron. Soc. Pac.* 111:1124
Humphreys RM, Smith N, Davidson K, et al. 1997. *Astron. J.* 114:2778
Izotov YI, Guseva NG, Fricke KJ, Henkel C. 2011. *Astron. Astrophys.* 533:A25
Izotov YI, Thuan TX. 2009. *Ap. J.* 690:1797
Kasen D, Bildsten L. 2010. *Ap. J.* 717:245
Kashi A, Soker N. 2009. *New Astron.* 14:11
Kennicutt RC. 1998. *Annu. Rev. Astron. Astrophys.* 36:189
Kennicutt RC, Evans NJ. 2012. *Annu. Rev. Astron. Astrophys.* 50:531
Kiminki DC, Kobulnicky HA. 2012. *Ap. J.* 751:4
Kiminki DC, Kobulnicky HA, Erwig I, et al. 2012. *Ap. J.* 747:41
Kobulnicky HA, Fryer CL. 2007. *Ap. J.* 670:747
Kochanek CS. 2011. *Ap. J.* 743:73
Kochanek CS, Szczygiel DM, Stanek KZ. 2011. *Ap. J.* 737:76

Kotak R, Vink JS. 2006. *Astron. Astrophys.* 460:L5
Kramer RH, Cohen DH, Owocki SP. 2003. *Ap. J.* 592:532
Kratter KM, Matzner CD, Krumholz MR, Klein RI. 2010. *Ap. J.* 708:1585
Kudritzki RP, Puls J. 2000. *Annu. Rev. Astron. Astrophys.* 38:613
Langer N. 1998. *Astron. Astrophys.* 329:551
Langer N. 2012. *Annu. Rev. Astron. Astrophys.* 50:107
Leitherer C, Schaerer D, Goldader JD, et al. 1999. *Ap. J. Suppl.* 123:3
Leonard DC, Filippenko AV, Barth AJ, Matheson T. 2000. *Ap. J.* 536:239
Levesque E, Stringfellow GS, Ginsburg AG, Bally J, Keeney BA. 2014. *Astron. J.* 147:23
Li W, Leaman J, Chornock R, et al. 2011. *MNRAS* 412:1441
Loeb A, Barkana R. 2001. *Annu. Rev. Astron. Astrophys.* 39:19
Lucy LB, Solomon PM. 1970. *Ap. J.* 159:879
MacFadyen AI, Woosley SE. 1999. *Ap. J.* 524:262
Madau P, Pozzetti L, Dickenson M. 1998. *Ap. J.* 498:106
Maeder A. 1992. In *Instabilities in Evolved Super and Hypergiants*, ed. C de Jager, H Nieuwenhuijzen, p. 138. Amsterdam: North-Holland
Maeder A, Meynet G. 2000. *Annu. Rev. Astron. Astrophys.* 38:143
Mahy L, Naze Y, Rauw G, et al. 2009. *Astron. Astrophys.* 502:937
Margutti R, Milisavljevic D, Soderberg AM, et al. 2014. *Ap. J.* 780:21
Markova N, Puls J, Repolust T, Markhov H. 2004. *Astron. Astrophys.* 413:693
Mason BD, Hartkopf WI, Geis DR, Henry TJ, Helsel JY. 2009. *Astron. J.* 110:2715
Massa D, Fullerton AW, Nichols JS, et al. 1995. *Ap. J.* 452:L53
Massa D, Fullerton AW, Sonneborn G, Hutchings JB. 2003. *Ap. J.* 586:996
Massey P. 2003. *Annu. Rev. Astron. Astrophys.* 41:15
Massey P. 2006. *Ap. J. Lett.* 683:93
Massey P. 2007. *Astron. J.* 134:2474
Massey P, Waterhouse E, DeGioia-Eastwood K. 2000. *Astron. J.* 119:2214
Mauerhan JC, Smith N, Filippenko AV, et al. 2013a. *MNRAS* 430:1801
Mauerhan JC, Smith N, Silverman JM, et al. 2013b. *MNRAS* 431:2599
Maund JR, Smartt SJ. 2009. *Science* 324:486
Meakin C, Arnett WD. 2007. *Ap. J.* 667:448
Meynet G, Maeder A, Schaller G, Shaerer D, Charbonnel C. 1994. *Astron. Astrophys. Suppl.* 103:97
Miller AA, Chornock R, Perley DA, et al. 2009. *Ap. J.* 690:1303
Modjaz M, Kewley L, Kirshner RP, et al. 2008. *Ap. J.* 702:226
Moffat AFJ, Robert C. 1994. *Ap. J.* 421:310
Mokiem MR, de Koter A, Vink JS, et al. 2007. *Astron. Astrophys.* 473:603
Monnier JD, Millan-Gabet R, Tuthill PG, et al. 2004. *Ap. J.* 605:436
Morales MF, Wyithe JSB. 2010. *Annu. Rev. Astron. Astrophys.* 48:127
Moriya TJ, Blinnikov SI, Baklanov PV, Sorokina EI, Dolgov AP. 2013. *MNRAS* 430:1402
Morton DC. 1967. *Ap. J.* 147:1017
Muijres LE, Vink JS, de Koter A, Müller PE, Langer N. 2012. *Astron. Astrophys.* 537:A37
Najarro F, Kudritzki RP, Cassinelli JP, Stahl O, Hillier DJ. 1996. *Astron. Astrophys.* 306:892
Neill JD, Sullivan M, Gal-Yam A, et al. 2011. *Ap. J.* 727:15
Nieuwenhuijzen H, de Jager C. 1990. *Astron. Astrophys.* 231:134
Nugis T, Lamers HJGLM. 2000. *Astron. Astrophys.* 360:227
Ofek EO, Cameron PB, Kasliwal MM, et al. 2007. *Ap. J.* 659:L13
Ofek EO, Lin L, Kouveliotou C, et al. 2013a. *Ap. J.* 768:47
Ofek EO, Sullivan M, Cenko SB, et al. 2013b. *Nature* 494:65
Owocki SP, Castor JI, Rybicki GB. 1988. *Ap. J.* 335:914
Owocki SP, Cranmer SR, Gayey KG. 1996. *Ap. J.* 472:L115
Owocki SP, Gayey KG, Shaviv NJ. 2004. *Ap. J.* 616:525
Owocki SP, Puls J. 1999. *Ap. J.* 510:355
Owocki SP, Rybicki GB. 1984. *Ap. J.* 284:337

Paczynski B. 1966. *Acta Astron.* 16:231
Paczynski B. 1967. *Acta Astron.* 17:355
Paczynski B. 1971. *Annu. Rev. Astron. Astrophys.* 9:183
Parker EN. 1958. *Ap. J.* 128:664
Pastorello A, Cappaellaro E, Inserra C, et al. 2013. *Ap. J.* 767:1
Pastorello A, Quimby RM, Smartt SJ, et al. 2008. *MNRAS* 389:131
Pastorello A, Smartt SJ, Mattila S, et al. 2007. *Nature* 447:829
Paxton B, Bildsten L, Dutter A, et al. 2011. *Ap. J. Suppl.* 192:3
Petrovic J, Langer N, van der Hucht KA. 2005. *Astron. Astrophys.* 435:1013
Podsiadlowski P, Ivanova N, Justham S, Pappaport S. 2010. *MNRAS* 406:840
Podsiadlowski P, Joss PC, Hsu JJL. 1992. *Ap. J.* 391:246
Prantzos N, Boissier S. 2003. *Astron. Astrophys.* 406:259
Prieto JL, Brimacombe J, Drake AJ, et al. 2013. *Ap. J.* 763:L27
Prieto JL, Stanek KZ, Beacom JF. 2008a. *Ap. J.* 673:999
Prieto JL, Stanek KZ, Kochanek CS, et al. 2008b. *Ap. J.* 673:L59
Puls J, Markova N, Scuderi S, et al. 2006. *Astron. Astrophys.* 454:625
Puls J, Vink JS, Najarro F. 2008. *Astron. Astrophys. Rev.* 16:209
Putman ME, Peek JEG, Joung MR. 2012. *Annu. Rev. Astron. Astrophys.* 50:491
Quataert E, Shiode J. 2012. *MNRAS* 423:L92
Quimby RM, Kulkarni SR, Kasliwal MM, et al. 2011. *Nature* 474:487
Remillard RA, McClintock JE. 2006. *Annu. Rev. Astron. Astrophys.* 44:49
Repolust T, Puls J, Herrero A. 2004. *Astron. Astrophys.* 415:349
Rest A, Foley RJ, Gezari S, et al. 2011. *Ap. J.* 729:88
Rest A, Prieto JL, Walborn NR, et al. 2012. *Nature* 482:375
Reynolds SP. 2008. *Annu. Rev. Astron. Astrophys.* 46:89
Ryder SD, Murrowood CE, Stathakis RA. 2006. *MNRAS* 369:L32
Sana H, de Mink SE, de Koter A, et al. 2012. *Science* 337:444
Sana H, Evans CJ. 2011. In *Active OB Stars: Structure, Evolution, Mass Loss, and Critical Limits, IAU Symp. 272*, ed. C Neiner, G Wade, G Meynet, G Peters, pp. 474–85. Cambridge, UK: Cambridge Univ. Press
Sana H, Gosset S, Evans CJ. 2009. *MNRAS* 400:1479
Sana H, Gosset S, Naze Y, Rauw G, Linder N. 2008. *MNRAS* 386:447
Sanders NE, Soderberg AM, Foley RJ, et al. 2013. *Ap. J.* 769:39
Shaviv NJ. 2000. *Ap. J.* 532:L137
Shiode JH, Quataert E. 2014. *Ap. J.* 780:96
Silverman JM, Nugent PE, Gal-Yam A, et al. 2013. *Ap. J. Suppl.* 207:3
Smartt SJ. 2009. *Annu. Rev. Astron. Astrophys.* 47:63
Smith N. 2005. *MNRAS* 357:1330
Smith N. 2006. *Ap. J.* 644:1151
Smith N. 2008. *Nature* 455:201
Smith N. 2011. *MNRAS* 415:2020
Smith N. 2013a. *MNRAS* 429:2366
Smith N. 2013b. *MNRAS* 434:102
Smith N, Arnett WD. 2014. *Ap. J.* 785:82
Smith N, Chornock R, Li W, et al. 2008. *Ap. J.* 686:467
Smith N, Chornock R, Silverman JM, Filippenko AV, Foley RJ. 2010a. *Ap. J.* 709:856
Smith N, Conti PS. 2008. *Ap. J.* 679:1467
Smith N, Ferland GJ. 2007. *Ap. J.* 655:911
Smith N, Frew D. 2011. *MNRAS* 415:2009
Smith N, Gehrz RD, Campbell R, et al. 2011a. *MNRAS* 418:1959
Smith N, Gehrz RD, Hinz PM, et al. 2003. *Astron. J.* 125:1458
Smith N, Hartigan P. 2006. *Ap. J.* 638:1045
Smith N, Hinkle KH, Ryde N. 2009. *Astron. J.* 137:3558
Smith N, Humphreys RM, Davidson K, et al. 2001. *Astron. J.* 121:1111

Smith N, Li W, Filippenko AV, Chornock R. 2011b. *MNRAS* 412:1522
Smith N, Li W, Foley RJ, et al. 2007. *Ap. J.* 666:1116
Smith N, Li W, Miller AA, et al. 2011c. *Ap. J.* 732:63
Smith N, Li W, Silverman JM, Ganeshalingam M, Filippenko AV. 2011d. *MNRAS* 415:773
Smith N, Mauerhan JC, Prieto JL. 2014. *MNRAS* 438:1191
Smith N, Mauerhan JC, Silverman JM, et al. 2012. *MNRAS* 426:1905
Smith N, McCray R. 2007. *Ap. J.* 671:L17
Smith N, Miller AA, Li W, et al. 2010b. *Astron. J.* 139:1451
Smith N, Owocki SP. 2006. *Ap. J.* 645:L45
Smith N, Vink JS, de Koter A. 2004. *Ap. J.* 615:475
Sobolev VV. 1960. *Moving Envelopes of Stars*. Cambridge, MA: Harvard Univ. Press
St. Louis N, Moffatt AFJ, Lapointe L, et al. 1993. *Ap. J.* 410:342
Stoll R, Prieto JL, Stanek KZ, et al. 2011. *Ap. J.* 730:34
Sundqvist JO, Owocki SP. 2013. *MNRAS* 428:1837
Taam RE, Sandquist EL. 2000. *Annu. Rev. Astron. Astrophys.* 38:113
Taddia F, Stritzinger MD, Sollerman J, et al. 2013. *Astron. Astrophys.* 555:A10
Tammann GA, Sandage A. 1968. *Ap. J.* 151:825
Thompson TA, Prieto JL, Stanek KZ, et al. 2009. *Ap. J.* 705:1364
Townsend RHD, Owocki SP, Groote D. 2005. *Ap. J.* 630:L81
ud-Doula A, Owocki SP. 2002. *Ap. J.* 576:413
van der Kruit PC, Freeman KC. 2011. *Annu. Rev. Astron. Astrophys.* 49:301
Van Dyk SD, Matheson T. 2012. *Ap. J.* 746:179
Van Dyk SD, Zheng WK, Clubb KI, et al. 2013. *Ap. J.* 772:L32
Van Dyk SD, Zheng WK, Fox OD, et al. 2014. *Astron. J.* 147:37
van Genderen A. 2001. *Astron. Astrophys.* 366:508
van Loon JT, Cioni MRL, Zijlstra AA, Loup C. 2005. *Astron. Astrophys.* 438:273
van Marle AJ, Owocki SP, Shaviv NJ. 2008. *MNRAS* 389:1353
van Marle AJ, Owocki SP, Shaviv NJ. 2009. *MNRAS* 394:595
van Marle AJ, Smith N, Owocki SP, van Veelen B. 2010. *MNRAS* 407:2305
Vanbeveren D. 2009. *New Astron. Rev.* 53:27
Vanbeveren D, De Loore C, Van Rensbergen W. 1998. *Astron. Astrophys. Rev.* 9:63
Vanbeveren D, Van Bever J, Belkus H. 2007. *Ap. J.* 662:L107
Vink JS, de Koter A. 2002. *Astron. Astrophys.* 393:543
Vink JS, de Koter A. 2005. *Astron. Astrophys.* 442:587
Vink JS, de Koter A, Lamers HJGLM. 2001. *Astron. Astrophys.* 369:574
Vink JS, Muijres LE, Anthonisse D, et al. 2011. *Astron. Astrophys.* 531:A132
Wachter S, Mauerhan JC, Van Dyk SD, et al. 2010. *Astron. J.* 139:2330
Walmswell JJ, Eldridge JJ. 2012. *MNRAS* 419:2054
Wellons S, Soderberg AM, Chevalier RA. 2012. *Ap. J.* 752:17
Willson LA. 2000. *Annu. Rev. Astron. Astrophys.* 38:573
Wolf CJE, Rayet GAP. 1867. *Comptes Rendus* 65:291
Woosley SE. 2010. *Ap. J.* 719:L204
Woosley SE, Blinnikov S, Heger A. 2007. *Nature* 450:390
Woosley SE, Bloom JS. 2006. *Annu. Rev. Astron. Astrophys.* 44:507
Woosley SE, Heger A, Weaver TA. 2002. *Rev. Mod. Phys.* 74:1015
Wright AE, Barlow MJ. 1975. *MNRAS* 170:41
Yoon SC, Cantiello M. 2010. *Ap. J.* 717:L62
Yoon SC, Woosley SE, Langer N. 2010. *Ap. J.* 725:940

Hot Accretion Flows Around Black Holes

Feng Yuan[1] and Ramesh Narayan[2]

[1]Shanghai Astronomical Observatory, Chinese Academy of Sciences, Shanghai 200030, China; email: fyuan@shao.ac.cn

[2]Harvard-Smithsonian Center for Astrophysics, Cambridge, Massachusetts 02138; email: rnarayan@cfa.harvard.edu

Annu. Rev. Astron. Astrophys. 2014. 52:529–88

First published online as a Review in Advance on June 18, 2014

The *Annual Review of Astronomy and Astrophysics* is online at astro.annualreviews.org

This article's doi:
10.1146/annurev-astro-082812-141003

Keywords

accretion disks, active galactic nuclei, active galactic feedback, black holes, black hole X-ray binaries, jet, outflow

Abstract

Black hole accretion flows can be divided into two broad classes: cold and hot. Whereas cold accretion flows consist of cool optically thick gas and are found at relatively high mass accretion rates, hot accretion flows, the topic of this review, are virially hot and optically thin, and occur at lower mass accretion rates. They are described by accretion solutions such as the advection-dominated accretion flow and luminous hot accretion flow. Because of energy advection, the radiative efficiency of these flows is in general lower than that of a standard thin accretion disk. Moreover, the efficiency decreases with decreasing mass accretion rate. Observations show that hot accretion flows are associated with jets. In addition, theoretical arguments suggest that hot flows should produce strong winds. Hot accretion flows are believed to be present in low-luminosity active galactic nuclei and in black hole X-ray binaries in the hard and quiescent states. The prototype is Sgr A*, the ultralow-luminosity supermassive black hole at our Galactic center. The jet, wind, and radiation from a supermassive black hole with a hot accretion flow can interact with the external interstellar medium and modify the evolution of the host galaxy.

1. INTRODUCTION

Black hole accretion is a fundamental physical process in the universe and is the primary power source behind active galactic nuclei (AGNs), black hole binaries (BHBs), and, possibly, gamma-ray bursts. The first genuine model of an accretion disk—by which we mean a rotating flow with viscous transport of angular momentum—is the celebrated thin disk model developed in the early 1970s (Novikov & Thorne 1973, Shakura & Sunyaev 1973, Lynden-Bell & Pringle 1974; see reviews by Pringle 1981, Frank et al. 2002, Kato et al. 2008, Abramowicz & Fragile 2013, Blaes 2013). Depending on the mass of the central black hole, the gas temperature in this model lies in the range 10^4–10^7 K, which is quite cold relative to the virial temperature. The disk is geometrically thin, whereas the gas is optically thick and radiates thermal blackbody-like radiation. Many accreting black hole sources have been successfully modeled as thin disks, e.g., luminous AGNs (see reviews above; but also Koratkar & Blaes 1999) and BHBs in the thermal state (Remillard & McClintock 2006, McClintock et al. 2013).

The thin disk model applies whenever the disk luminosity L is somewhat below the Eddington luminosity $L_{\rm Edd}$ or, equivalently, when the mass accretion rate $\dot{M}$ is below the Eddington rate: $\dot{M}_{\rm Edd} \equiv 10L_{\rm Edd}/c^2 = 1.39 \times 10^{18}(M/{\rm M}_\odot)$ g s^{-1}, where M is the mass of the black hole. [Although everyone agrees on the definition of the Eddington luminosity, e.g., $L_{\rm Edd} = 4\pi GMc/\kappa_{\rm es}$, where $\kappa_{\rm es}$ is the electron scattering opacity, usually taken to be 0.4 cm^2 g^{-1}, many definitions are used for $\dot{M}_{\rm Edd}$. In this article we use a definition that assumes the accretion disk has a nominal radiative efficiency of 10%; hence, $L_{\rm Edd} = 0.1\dot{M}_{\rm Edd}c^2$. Some authors use $L_{\rm Edd} = \dot{M}_{\rm Edd}c^2$, others use $L_{\rm Edd} = \dot{M}_{\rm Edd}c^2/12$ or $\dot{M}_{\rm Edd}c^2/16$, and yet others use $L_{\rm Edd} = \eta(a_*)\dot{M}_{\rm Edd}c^2$, where $\eta(a_*)$ is the relativistic radiative efficiency of a thin disk around a black hole with dimensionless spin parameter $a_* \equiv a/M$.] When $\dot{M}$ approaches or exceeds $\dot{M}_{\rm Edd}$, the accreting gas becomes optically too thick to radiate all the dissipated energy locally (a key requirement of the thin disk model). Radiation is then trapped and advected inward with the accretion flow. Consequently, the radiative efficiency becomes lower, and L becomes progressively smaller than $0.1\dot{M}c^2$. The disk solution that describes such a system is called the slim disk (Katz 1977, Begelman 1979, Begelman & Meier 1982, Abramowicz et al. 1988, Chen & Taam 1993, Ohsuga et al. 2005). The slim disk model has been applied to narrow-line Seyfert galaxies (Mineshige et al. 2000), SS433 (Fabrika 2004), and ultraluminous X-ray sources (Watarai et al. 2001).

The thin disk and slim disk both belong to the class of cold accretion flows. Both consist of optically thick gas. In contrast to these disks is the hot accretion flow model, which was first described by Shapiro et al. (1976; hereafter SLE). The temperature of the gas in the SLE solution is much higher, approaching virial, and the gas is optically thin. A key innovation of the SLE model is the introduction of a two-temperature accreting plasma, where the ions are much hotter than the electrons. The main success of the SLE solution, indeed its motivation, is that, for the first time, it was able to explain the hard X-ray emission seen in some black hole sources. Unfortunately, soon after the SLE model was introduced, it was shown to be thermally unstable, so the model as originally developed is unlikely to be realized in nature.

The important role of advection in hot accretion flows was first emphasized by Ichimaru (1977), who pointed out that in certain regimes the viscously dissipated accretion energy can go into heating the accretion flow rather than being radiated away. This is the most important feature of the general class of advection-dominated accretion flows (ADAFs), one of the hot accretion solutions we discuss in this review. Ichimaru further argued that, because of the inclusion of advection, his hot accretion solution should be thermally stable. Independently, Rees et al. (1982) described similar ideas in their two-temperature "ion torus" model. However, they did not emphasize the relation between their model and those of SLE and Ichimaru, nor did they discuss stability. Sadly, these pioneering studies were not followed up for many years.

With the rediscovery of the ADAF solution in the mid-1990s (Narayan & Yi 1994, 1995a,b; Abramowicz et al. 1995; Chen et al. 1995) and the subsequent detailed study of its properties, hot accretion flow models finally became established in the accretion literature. [Note that advection-dominated accretion is not synonymous with hot accretion. For instance, the slim disk is advection dominated, although for a very different reason (long radiative diffusion time) compared with a hot ADAF (long cooling time). In this review, we classify accretion solutions as hot or cold and focus our attention on the hot solutions. Although our classification is somewhat arbitrary, at least in certain respects, the two solution branches are clearly distinct.] The dynamical and radiative properties of the ADAF solution have been studied in significant detail, and the model has been applied to various black hole systems, including the supermassive black hole in our Galactic Center, Sagittarius A* (Sgr A*), low-luminosity AGNs (LLAGNs), and BHBs in the hard and quiescent states. Because this article focuses on current developments in our understanding of hot accretion flows, readers are referred to earlier reviews for a more detailed discussion of topics not covered here (e.g., Narayan et al. 1998b, Lasota 1999, Quataert 2001, Ho 2008, Narayan & McClintock 2008, Abramowicz & Fragile 2013, Blaes 2013).

Before proceeding further, we briefly explain some terminology. The most popular and widely used term for a hot accretion flow is ADAF. Two variants of ADAF are adiabatic inflow-outflow solution (ADIOS) and convection-dominated accretion flow (CDAF), which emphasize the roles of two distinct physical phenomena in hot accretion flows: outflows and convection (Section 3.4). Hot accretion flows are usually radiatively inefficient. Perhaps to emphasize that the low efficiency is not just because of advection, but may also be the result of other effects such as outflows and convection, some authors use the term radiatively inefficient accretion flow. However, as discussed in Section 2, the radiative efficiency of a hot accretion flow increases with increasing mass accretion rate. In fact, the efficiency can even be comparable with that of a standard thin disk. This is especially the case with a luminous hot accretion flow (LHAF) (Section 2.6), an extension of an ADAF to accretion rates above the original range of validity of the ADAF solution. Because the common feature of all these accretion solutions is that the gas is very hot, we use the generic term hot accretion flow. Nevertheless, because of its popularity, we sometimes also use the term ADAF.

2. ONE-DIMENSIONAL DYNAMICS AND RADIATION

2.1. One-Dimensional Equations and Self-Similar Solutions

Consider a steady axisymmetric accretion flow, and focus for now only on the dynamics. Conservation of mass, radial momentum, angular momentum, and energy are described by the following height-integrated differential equations (e.g., Abramowicz et al. 1988, Narayan & Yi 1994, Narayan et al. 1998b):

$$\frac{\mathrm{d}}{\mathrm{d}R}(\rho R H v) = 0, \tag{1}$$

$$v\frac{\mathrm{d}v}{\mathrm{d}R} - \Omega^2 R = -\Omega_K^2 R - \frac{1}{\rho}\frac{\mathrm{d}}{\mathrm{d}R}(\rho c_s^2), \tag{2}$$

$$v\frac{\mathrm{d}(\Omega R^2)}{\mathrm{d}R} = \frac{1}{\rho R H}\frac{\mathrm{d}}{\mathrm{d}R}\left(\nu\rho R^3 H\frac{\mathrm{d}\Omega}{\mathrm{d}R}\right), \tag{3}$$

$$\rho v\left(\frac{\mathrm{d}e}{\mathrm{d}R}-\frac{p}{\rho^2}\frac{\mathrm{d}\rho}{\mathrm{d}R}\right)=\rho\nu R^2\left(\frac{\mathrm{d}\Omega}{\mathrm{d}R}\right)^2-q^-, \tag{4}$$

where ρ is the mid-plane density of the gas, R is the radius, $H \approx c_s/\Omega_K$ is the vertical scale height, v is the radial velocity, Ω is the angular velocity, Ω_K is the Keplerian angular velocity, $c_s \equiv \sqrt{p/\rho}$ is the isothermal sound speed, p is the pressure, e is the specific internal energy, and q^- is the radiative cooling rate per unit volume. The kinematic viscosity coefficient ν may be parameterized via the Shakura & Sunyaev (1973) prescription,

$$\nu \equiv \alpha c_s H = \alpha \frac{c_s^2}{\Omega_K}, \tag{5}$$

where the dimensionless parameter α is generally assumed to be a constant.

Equations 1–4 are quite general in the sense that they encompass all accretion models, including the thin disk, slim disk, and ADAF (Section 2.7). Note, however, that Equation 1 implies the mass accretion rate $\dot{M} = 4\pi\rho RH|v|$ is independent of the radius. Although this may be a reasonable approximation for a thin disk, numerical simulations of hot accretion flows indicate that outflows are almost inevitable (Section 3.4), causing the mass accretion rate to decrease with decreasing radius. Therefore, assuming a power-law variation for simplicity, it is useful to generalize Equation 1 to (Blandford & Begelman 1999)

$$\dot{M}(R)=4\pi\rho RH|v|=\dot{M}_{\rm BH}\left(\frac{R}{R_S}\right)^s, \quad R_S \le R \le R_{\rm out}, \tag{6}$$

where $R_S = 2GM/c^2$ is the Schwarzschild radius of the black hole, $\dot{M}_{\rm BH}$ is the mass accretion rate at this radius, and $R_{\rm out}$ is the outer radius of the accretion flow. The index s is a measure of the strength of the outflow; s cannot exceed 1 for energetic reasons (Blandford & Begelman 1999), whereas $s = 0$ corresponds to a constant mass accretion rate (no outflow). Equations 2–4 should also be modified when there is mass outflow (e.g., Poutanen et al. 2007, Xie & Yuan 2008). However, the main effect of an outflow is probably through the density profile. Hence, simply replacing Equation 1 with Equation 6 and retaining Equations 2–4 as written yield a reasonable first approximation. Another caveat is that the power-law variation of $\dot{M}$ with R is not likely to continue all the way down to R_S as written above. It probably ceases at some inner radius $R_{\rm in}$ of order ten (or even tens of) R_S (Section 3.4).

Equation 4 needs more discussion. The two terms on the left-hand side represent the rate of change of the internal energy per unit volume and the work done by compression; we call the latter q^c. Together, the two terms represent energy advection, which we write compactly as $q^{\rm adv}$. More precisely, $q^{\rm adv}$ corresponds to $\rho v T \mathrm{d}s/\mathrm{d}R$, where T and s are the temperature and specific entropy of the gas; $q^{\rm adv}$ is thus the radial rate of advection of entropy. The first term on the right-hand side of Equation 4 is the heating rate per unit volume or, more precisely, the rate at which entropy is added to the gas via viscous dissipation. Calling this term q^+, Equation 4 takes the simple form

$$\rho v\frac{\mathrm{d}e}{\mathrm{d}R}-q^c \equiv q^{\rm adv}=q^+-q^- \equiv fq^+, \tag{7}$$

where the parameter $f \equiv q^{\rm adv}/q^+$ measures the relative importance of advection. Out of the total heat energy q^+ released by viscous dissipation per unit volume per unit time, a fraction f is advected and the rest $(1-f)$ is radiated. The standard thin disk and SLE models assume $q^+ = q^-$ and thus correspond to $f = 0$, i.e., vanishing energy advection. The slim disk and various hot accretion flows have nonzero f. Quite often, e.g., when $\dot{M}_{\rm BH} \gg \dot{M}_{\rm Edd}$ (slim disk) or $\dot{M}_{\rm BH} \ll \dot{M}_{\rm Edd}$ (hot accretion flow), one finds $q^+ \gg q^-$, $f \to 1$. These accretion flows are then strongly advection dominated.

Assuming a Newtonian gravitational potential and taking the advection parameter f to be independent of radius, Narayan & Yi (1994, 1995b) showed that Equations 1–4 have a self-similar solution [also obtained by Spruit et al. (1987), but in a different context]. Their solution corresponds to a constant mass accretion rate without outflows ($s = 0$). Including mass outflow via Equation 6 and making the reasonable assumption that the only important change is in the density profile (Xie & Yuan 2008), the self-similar solution of Narayan & Yi (1994, 1995b) becomes approximately (Yuan et al. 2012b)

$$v \approx -1.1 \times 10^{10} \alpha r^{-1/2} \text{ cm s}^{-1}, \tag{8}$$

$$\Omega \approx 2.9 \times 10^{4} m^{-1} r^{-3/2} \text{ s}^{-1}, \tag{9}$$

$$c_s^2 \approx 1.4 \times 10^{20} r^{-1} \text{ cm}^2 \text{ s}^{-2}, \tag{10}$$

$$n_e \approx 6.3 \times 10^{19} \alpha^{-1} m^{-1} \dot{m}_{\rm BH}\, r^{-3/2+s} \text{ cm}^{-3}, \tag{11}$$

$$B \approx 6.5 \times 10^{8} (1+\beta)^{-1/2} \alpha^{-1/2} m^{-1/2} \dot{m}_{\rm BH}^{1/2}\, r^{-5/4+s/2} \text{ G}, \tag{12}$$

$$p \approx 1.7 \times 10^{16} \alpha^{-1} m^{-1} \dot{m}_{\rm BH}\, r^{-5/2+s} \text{ g cm}^{-1} \text{ s}^{-2}, \tag{13}$$

where the black hole mass M, the mass accretion rate $\dot{M}$, and the radius R have been scaled to solar, Eddington, and Schwarzschild units, respectively:

$$m \equiv \frac{M}{\mathrm{M}_\odot}, \quad \dot{m} \equiv \frac{\dot{M}}{\dot{M}_{\rm Edd}}, \quad r \equiv \frac{R}{R_S}. \tag{14}$$

Correspondingly, $\dot{m}_{\rm BH} = \dot{M}_{\rm BH}/\dot{M}_{\rm Edd}$, where $\dot{M}_{\rm BH}$ is defined in Equation 6. The parameter β is a measure of the strength of the magnetic field:

$$\beta \equiv \frac{p_{\rm gas}}{p_{\rm mag}}, \tag{15}$$

where $p_{\rm gas}$ is the gas pressure and $p_{\rm mag} \equiv B^2/8\pi$ is the magnetic pressure. [This is the standard definition of β as used in plasma physics; however, following Narayan & Yi (1995b), much of the ADAF literature uses a different $\beta_{\rm ADAF} \equiv p_{\rm gas}/(p_{\rm gas}+p_{\rm mag})$, which is confusingly also called β. The two β are related by $\beta_{\rm ADAF} = \beta/(\beta+1)$.] Numerical magnetohydrodynamic (MHD) simulations usually give $\beta \simeq 10$ (Section 3.2).

The advection parameter f is generally a function of radius r and, more importantly, the mass accretion rate $\dot{m}$. When $\dot{m}$ is significantly smaller than unity (say, $\lesssim 10^{-4}$), f is nearly equal to unity and the flow is well-described as a true ADAF. As $\dot{m}$ becomes larger, radiation plays an increasingly important role and f becomes smaller or even negative in some regimes (LHAF). Section 2.6 discusses the energetics of various kinds of hot accretion flows.

Apart from being convenient for estimating gas properties in hot accretion flows, the self-similar solution reveals several distinct features of these solutions, which distinguish hot flows from a standard (cool) thin disk, for example:

- The temperature of a hot accretion flow is almost virial:

$$T \simeq GMm_p/6kR \sim (10^{12}/r) \text{ K}, \tag{16}$$

which is much larger than the temperature of a thin disk. Because of the near-virial temperature, the accretion flow is geometrically quite thick: $H/R \sim 0.5$. Nevertheless, the height-integrated equations used in the one-dimensional (1D) analysis appear to be reasonably accurate (Narayan & Yi 1995a).

- The radial velocity is much larger than in a thin disk because accretion theory predicts $v \sim \alpha c_s H/R$ (e.g., Equation 3) and both c_s and H/R are much larger in a hot accretion flow.
- The angular velocity is sub-Keplerian because the pressure is much larger than in a thin disk (higher temperature), so gravity is partially balanced by the radial pressure gradient (right-hand side of Equation 2).
- The large radial velocity and the low mass accretion rate generally cause the optical depth to be less than unity. Therefore, the emitted radiation is almost never blackbody, but instead is dominated by processes such as synchrotron, bremsstrahlung, and inverse Compton scattering. In addition, as we discuss in Section 2.5, the radiative efficiency,

$$\epsilon \equiv \frac{L}{\dot{M}_{\rm BH}c^2}, \tag{17}$$

where L is the luminosity of the accretion flow, is much lower than the fiducial 10% efficiency of a standard thin accretion disk, especially when $\dot{m}$ is small.

- In the low radiative efficiency limit, the entropy increases with decreasing radius because the gas is heated but hardly cools. Hot accretion flows are therefore potentially unstable to convection. Rotation can stabilize a system against convection even if the entropy gradient is unstable. The role of magnetic fields is less clear (Section 3.4).
- Finally, the self-similar solution implies that the Bernoulli parameter, *Be*, of the flow is positive, which suggests that hot accretion flows should have strong outflows and jets (Narayan & Yi 1994, 1995a; Blandford & Begelman 1999). Global solutions (Section 2.4) indicate that *Be* may be either positive or negative, depending on outer boundary conditions (Nakamura 1998, Yuan 1999).

Whereas much work on hot accretion flows has focused on the time-steady self-similar solution described above, Ogilvie (1999) has derived a beautiful similarity solution that describes the radiatively inefficient evolution of an initially narrow ring of viscous orbiting fluid. This solution confirms several of the features discussed above. In addition, it avoids an annoying singularity that is present in the time-steady self-similar solution when the gas adiabatic index approaches 5/3 (Narayan & Yi 1994, Blandford & Begelman 1999, Quataert & Narayan 1999a).

2.2. Two-Temperature Flow: Thermal Properties

In the discussion so far, we have focused on dynamics. When dealing with the thermodynamics of a hot accretion flow, it is customary to follow the pioneering work of SLE and to allow the ions and electrons to have different temperatures. For such two-temperature plasmas, the energy equation (Equation 4 or 7) is replaced by two coupled equations (e.g., Nakamura et al. 1997, Quataert & Narayan 1999b):

$$q^{\mathrm{adv},i} \equiv \rho v \left(\frac{\mathrm{d}e_i}{\mathrm{d}R} - \frac{p_i}{\rho^2}\frac{\mathrm{d}\rho}{\mathrm{d}R} \right) \equiv \rho v \frac{\mathrm{d}e_i}{\mathrm{d}R} - q^{i,c} = (1-\delta)q^+ - q^{\mathrm{ie}}, \tag{18}$$

$$q^{\mathrm{adv,e}} \equiv \rho v \left(\frac{\mathrm{d}e_e}{\mathrm{d}R} - \frac{p_e}{\rho^2}\frac{\mathrm{d}\rho}{\mathrm{d}R} \right) \equiv \rho v \frac{\mathrm{d}e_e}{\mathrm{d}R} - q^{e,c} = \delta q^+ + q^{\mathrm{ie}} - q^-. \tag{19}$$

Here $e_i \equiv kT_i/[(\gamma_i - 1)\mu_i m_p]$ and $e_e \equiv kT_e/[(\gamma_e - 1)\mu_e m_p]$ are the internal energies of ions and electrons per unit mass of the gas. Similarly, γ_i and γ_e are the respective adiabatic indices,

p_i and p_e are the respective pressures, and $q^{i,c}$ and $q^{e,c}$ are the respective compression work done per unit volume. The quantity $q^{\rm ie}$ is the rate of transfer of thermal energy from ions to electrons via Coulomb collisions. The parameter δ denotes the fraction of the viscously dissipated energy that directly heats electrons; the remainder $(1-\delta)$ goes into the ions. There have been attempts to estimate this important parameter from first principles (Section 2.3), but δ is often treated as a free parameter. The above energy equations are further modified when the contribution of the magnetic field is included (Quataert & Narayan 1999a), but we ignore this complication here.

It is important to note that the two-temperature nature of the gas in a hot accretion flow is not simply an assumption, but rather a generic consequence of the physics of these solutions. First, electrons radiate much more efficiently than do ions (which is why we include a cooling term q^- only in Equation 19) and thus have a tendency to be cooler. Second, the primary channel by which ions cool is by transferring their energy to the electrons. Coupling via Coulomb collisions is inefficient at the low densities found in hot flows, thus Coulomb equilibration of temperatures is suppressed. Third, we see from the energy equation that gravitational energy is transformed into the thermal energy of the gas via two comparably important channels: viscous heating (q^+) and compressional heating ($q^{i,c}$, $q^{e,c}$). As we discuss in Section 2.3, viscous heating probably deposits comparable amounts of energy in the ions and electrons, with electrons perhaps receiving a somewhat smaller share ($\delta \sim 0.1-0.5$). Under adiabatic conditions, compressional heating causes the temperature to scale as $T \propto \rho^{\gamma-1}$. Because the ions remain nonrelativistic throughout the accretion flow (even at $T_i \sim 10^{12}$ K), they have $\gamma_i \sim 5/3$. However, in the inner regions of the accretion flow, the electrons become relativistic, $kT_e > m_e c^2$, and $\gamma_e \rightarrow 4/3$. Therefore, whereas ions heat up by compression as $T_i \sim \rho^{2/3}$, electrons heat up only as $T_e \sim \rho^{1/3}$. This drives the gas to a two-temperature state at radii $r \lesssim 10^3$.

Despite all these arguments, the gas would still be single temperature if there were efficient modes of energy transfer (over and above Coulomb collisions) from ions to electrons. Only one mechanism has been discussed in the literature (Begelman & Chiueh 1988), and its importance in situations of interest is unclear (Narayan & Yi 1995b). Plasmas are complicated, so there may well be some as-yet unidentified mechanism that succeeds in maintaining the gas at a single temperature. Nevertheless, the two-temperature nature of hot accretion flows seems to be supported by observations (Yuan et al. 2006). Furthermore, the plasma in the solar wind is both two temperature and anisotropic (Marsch 2012), and the plasma behind shocks in supernova remnants is also two temperature (Rakowski 2005). So, nature certainly does not abhor a two-temperature plasma.

2.3. Two-Temperature Flow: Heating and Acceleration of Electrons and Ions

Early work on the two-temperature ADAF model assumed that most of the turbulent viscous energy goes into the ions (Ichimaru 1977, Rees et al. 1982, Narayan & Yi 1995b) and that only a small fraction $\delta < 10^{-2}$ goes into the electrons. However, the existence of neither a two-temperature plasma (Section 2.2) nor a radiatively inefficient flow (Section 2.6) requires such a small value of δ. What is essential is that $\dot{m}$ needs to be low.

A few attempts have been made to estimate δ from microphysics, by considering magnetic reconnection (Bisnovatyi-Kogan & Lovelace 1997; Quataert & Gruzinov 1999; Ding et al. 2010; Hoshino 2012, 2013), MHD turbulence (Quataert 1998, Blackman 1999, Quataert & Gruzinov 1999, Medvedev 2000, Lehe et al. 2009), or dissipation of pressure anisotropy in a collisionless plasma (Sharma et al. 2007a). There is no consensus at the moment, but work so far generally suggests that $\delta \gg 10^{-2}$.

By modeling astrophysical observations of hot accretion flows, researchers have obtained weak constraints on the value of δ. In the case of Sgr A*, where we may have the most detailed observations of a hot accretion flow, Yuan et al. (2003) (see Section 4.1 for details) estimated $\delta \approx 0.5$. However, from models of black hole sources at higher luminosities, $\delta \sim 0.1$ (Yu et al. 2011, Liu & Wu 2013). At present, the best we can say is that δ probably lies in the range 0.1–0.5. This revision in the value of δ means that our understanding of hot accretion flows has evolved significantly since the early work reviewed in Narayan et al. (1998b), which was based entirely on models with $\delta < 0.01$. Accordingly, the radiative efficiency of a hot accretion flow is not as low as previously imagined, even when $\dot{m}$ is small (Section 2.6).

Is the energy distribution of the hot electrons thermal or nonthermal? The answer depends on the details of energy dissipation, particle acceleration, and thermalization. Processes such as magnetic reconnection, weak shocks, and turbulent dissipation are all likely to accelerate a fraction of the ions and electrons into a nonthermal power-law distribution (e.g., Ding et al. 2010, Hoshino 2013). How rapidly are the distributions then thermalized?

Mahadevan & Quataert (1997) showed that Coulomb collisions are far too inefficient to thermalize the ions, so ions retain whatever energy distribution they acquire through viscous dissipation and heating. Coulomb coupling between ions and electrons is also inefficient (though less so), which is why hot accretion flows develop a two-temperature structure (Section 2.2). By contrast, electrons can exchange energy quite efficiently through Coulomb collisions as well as through the emission and absorption of synchrotron photons. Thus, for accretion rates $\dot{m} > 10^{-3}$, the electrons are expected to have a more or less thermal distribution throughout the accretion flow. However, very-high-energy electrons are not easily thermalized and could, in principle, retain a power-law distribution even at these high accretion rates. The electron energy distribution may thus be Maxwellian for the bulk of the electrons, but it may follow a power law for a small population of electrons at higher energies.

At lower accretion rates, thermalization is less efficient and the electron distribution function is expected to retain a stronger memory of the heating/acceleration process. Thus, a hybrid thermal-nonthermal energy distribution should form readily. Observationally, nonthermal electrons are needed to explain the quiescent low-frequency radio emission in Sgr A* (Mahadevan 1998, Özel et al. 2000, Yuan et al. 2003) (see Section 4.1) and other LLAGNs (Liu & Wu 2013) as well as the X-ray emission in flares in Sgr A* (Yuan et al. 2004).

2.4. Global Solutions

The great virtue of the self-similar solution presented in Section 2.1 is that it is analytic, thereby providing a transparent way of understanding the key properties of an ADAF. However, because the self-similar solution is scale free, it cannot describe the flow near the inner or outer boundary. Especially when calculating the radiation spectrum, a global solution is required, as most of the radiation comes from the region close to the inner boundary where the self-similar solution is invalid.

A global solution refers to a numerical solution obtained by solving directly the differential equations of the problem, e.g., Equations 1–4. Usually, an integrated version of the angular momentum Equation 3 is used:

$$\frac{d\Omega}{dR} = \frac{v\Omega_K(\Omega R^2 - j)}{\alpha R^2 c_s^2}, \tag{20}$$

where the integration constant j is the angular momentum per unit mass accreted by the central mass. This constant is an eigenvalue of the problem and is obtained as part of the numerical solution. If the model under consideration includes mass loss in a wind, then (at the simplest level) Equation 1 is simply replaced by Equation 6 with the chosen value of s [which is assumed to be independent of the radius; numerical simulations suggest that mass loss begins only at radii greater than ten or tens of R_S (Section 3.4), so it is an oversimplification to assume a constant s all the way down to R_S (Equation 6). Presumably, though not confirmed, the error introduced is not large.]. To study the thermodynamics of the two-temperature gas consistently, Equation 4 is replaced by Equations 18 and 19. To go beyond the assumption of a constant value of f, the radiative cooling term q^- is kept in the energy equation, with contributions from relevant radiative processes (Section 2.5). In many studies, the pseudo-Newtonian gravitational potential of Paczyński & Wiita (1980) is adopted to mimic the effective potential of a Schwarzschild black hole.

Mathematically, obtaining a global solution involves a three-point boundary value problem. Whereas the radial velocity of the accreting gas at large radius is highly subsonic, the gas falls into the black hole horizon at the speed of light. Thus, there has to be an intermediate sonic radius $R_{\rm sonic}$ where the radial velocity equals the sound speed. The global solution must satisfy two boundary conditions at this radius, one of which is $v = c_s$. In addition, because the black hole cannot support a shear stress, the viscous torque must be zero at the horizon. This boundary condition is not always applied at the horizon; sometimes it is transferred to the sonic radius. Finally, at the outer edge of the solution ($R = R_{\rm out}$), the flow should match the properties of the gas flowing in from the outside.

The above boundary value problem is usually solved by one of two numerical methods (for details, see Press et al. 1992, 2002): relaxation (Chen et al. 1997, Esin et al. 1997, Narayan et al. 1997c) or shooting (Nakamura et al. 1996, 1997; Manmoto et al. 1997; Yuan 1999, 2001; Yuan et al. 2003). The main parameters are black hole mass M, mass accretion rate $\dot{M}_{\rm BH}$, viscosity parameter α, magnetization parameter β, wind parameter s, and electron heating parameter δ. Among these, M is usually found through observations; δ is discussed in Section 2.3; and rough values of α, β, and s may be obtained from numerical simulations, though s, in particular, is somewhat uncertain (Section 3.4). In addition, $\dot{M}_{\rm BH}$ is a free parameter that either is allowed to range over many values if a parameter study is being conducted (e.g., **Figure 1**) or is fitted to observations such as the luminosity and spectrum of a source. The global solution then gives the radial distributions of v, Ω, c_s, ρ, T_i, T_e, and B, together with the eigenvalue j and the sonic radius $R_{\rm sonic}$. Away from the boundaries, global solutions generally agree well with the self-similar solution (Chen et al. 1997, Narayan et al. 1997c), confirming the validity and usefulness of the latter.

The relativistic global problem, where the Newtonian equations discussed here are replaced by their general relativistic versions corresponding to the Kerr metric, has been solved by several authors (Abramowicz et al. 1996, Peitz & Appl 1997, Gammie & Popham 1998, Popham & Gammie 1998, Manmoto 2000). Solutions of the relativistic equations are similar to those of the Newtonian problem for radii $R \gtrsim 10 R_S$, but they differ significantly at smaller radii. In addition, the black hole's spin has a substantial effect at small radii, which can impact the observed spectrum (Jaroszynski & Kurpiewski 1997).

2.5. Radiation Processes, Spectrum, and Radiative Efficiency

Because gas close to a black hole in a hot accretion flow has a very high temperature and is also optically thin and magnetized, the relevant radiation processes are synchrotron emission and bremsstrahlung, modified by Comptonization. The radiative cooling rate, the shape of the spectrum, the different components in the spectrum, and how all these scale with parameters are described in various papers (e.g., Narayan & Yi 1995b, Narayan 1996, Esin et al. 1997, Mahadevan 1997, Manmoto et al. 1997, Nakamura et al. 1997, Narayan et al. 1998b, Quataert & Narayan 1999b, Yuan et al. 2003).

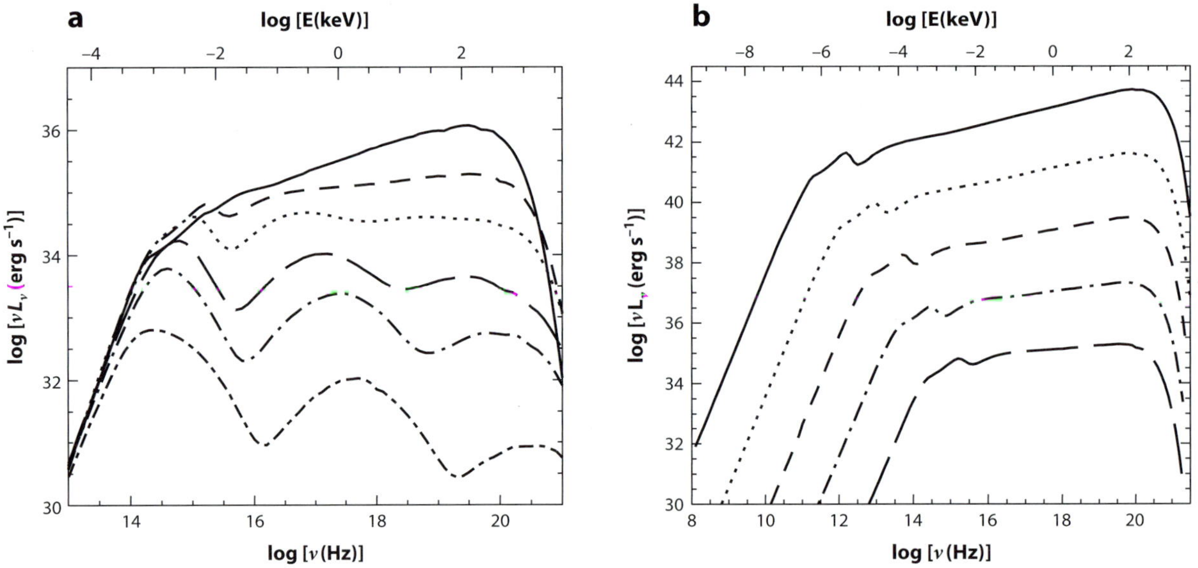

Figure 1

Model spectra of hot accretion flows for the following parameters: viscosity parameter $\alpha = 0.1$, magnetization parameter $\beta = 9$, electron heating parameter $\delta = 0.5$, and wind parameter $s = 0.4$ (Sections 2.1 and 2.2). (*a*) Spectra corresponding to a 10-$M_\odot$ black hole accreting with a mass accretion rate, from bottom to top, $\dot{m}_{\rm BH} = 8 \times 10^{-6}$, 5×10^{-5}, 1.6×10^{-4}, 8×10^{-4}, 2.4×10^{-3}, and 5×10^{-3}. (*b*) Spectra corresponding to $\dot{m}_{\rm BH} = 2.4 \times 10^{-3}$ for black hole masses, from bottom to top, $M/M_\odot = 10, 10^3, 10^5, 10^7$, and 10^9. The model spectra are shown for hot thermal accretion flows. When there is a cool outer disk beyond a transition radius, the spectrum has an additional thermal blackbody-like component (see **Figure 8**). If nonthermal electrons are present, there is enhanced emission at radio and infrared wavelengths. In addition, the prominent inverse Compton bumps shown here at low mass accretion rates are smoothed out to some degree (**Figures 6** and **8**). Figure adapted from Narayan (1996), but using modern parameters.

Figure 1 shows model spectra for hot accretion flows with different mass accretion rates $\dot{m}$ and black hole masses m. The results can be understood as follows (based on Mahadevan 1997). At photon energies below and up to the first peak in the spectrum, the radiation is primarily due to synchrotron emission from the thermal electrons. The emission is highly self-absorbed and is very sensitive to the electron temperature ($\nu L_\nu \propto T_e^7$). The emission at the peak comes from gas near the black hole, whereas the radiation at lower frequencies comes from larger radii. The peak frequency scales roughly as $\nu_{\rm peak} \propto m^{-1/2}\dot{m}^{1/2}$. Synchrotron photons are Compton upscattered by the hot electrons and produce hard radiation extending up to around the electron temperature: $kT_e \gtrsim 100$ keV for typical two-temperature models. The importance of this Compton component depends on $\dot{m}$. At high values of $\dot{m}$, it dominates the spectrum, becoming even stronger than the primary synchrotron peak. As $\dot{m}$ decreases, the Compton component is softer and becomes weaker (bolometrically) than the synchrotron component. At a sufficiently low $\dot{m}$, Comptonization is so weak that the X-ray spectrum is dominated by bremsstrahlung emission, which again cuts off at $h\nu \sim kT_e$.

The above discussion pertains to a pure hot accretion flow. If the hot flow is surrounded by a standard thin disk at larger radii (Section 4.2.2), there will be an additional multicolor blackbody component in the spectrum from the thermal disk. The Compton component will also be modified because, in addition to synchrotron photons, there is a second source of soft photons from the outer disk. The importance of the latter depends on where the transition radius $R_{\rm tr}$ between the hot flow and the thin disk is located.

In addition to thermal radiation from hot electrons, proton-proton collisions in a hot accretion flow can create pions, whose decay will yield gamma rays (Mahadevan et al. 1997). The same collisions will also produce a population of relativistic nonthermal electrons whose synchrotron radiation may explain the excess radio emission observed in Sgr A* (Mahadevan 1998, 1999) (this process is less important in current models, which use higher values of δ than in the past and thus have lower mass accretion rates), although other processes can also produce such nonthermal electrons (Section 2.3). Interestingly, although the electrons in a hot accretion flow reach relativistic temperatures, pair processes are generally unimportant (Björnsson et al. 1996, Kusunose & Mineshige 1996, Esin 1999, Mościbrodzka et al. 2011) because the low opacity and low radiation energy density mean that there are very few pair-producing interactions in the medium.

At radii $\gtrsim 10^4 R_S$, the gas in a hot accretion flow is cool enough that heavier atomic species, especially iron-peak elements, are able to retain one or two electrons. As a result, the X-ray emission from these regions is expected to show emission lines on top of the inverse Compton and bremsstrahlung continuum (Narayan & Raymond 1999). The utility of these lines lies in their ability to constrain the run of gas density with radius and, thus, to provide an observational estimate of the outflow parameter s (Perna et al. 2000, Xu et al. 2006, Wang et al. 2013).

Figure 2 shows the radiative efficiency of a hot accretion flow as a function of the mass accretion rate for various values of the electron heating parameter δ. Mass loss has been included via Equation 6 with $s = 0.4$. As can be seen, the efficiency depends strongly on the assumed value of δ. In addition, for a given δ, the efficiency increases steeply with increasing mass accretion rate. Indeed, near the upper end, the efficiency of a hot accretion flow approaches the efficiency $\epsilon_{SSD} \approx 10\%$ of a standard Shakura-Sunyaev disk (SSD). Xie & Yuan (2012) give piecewise power-law fitting formulae for the dependence of the radiative efficiency on $\dot{m}$ and δ.

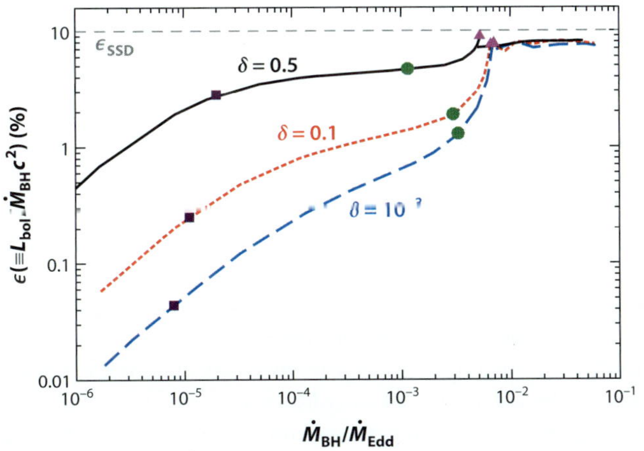

Figure 2

Radiative efficiency (Equation 17) of a luminous hot accretion flow as a function of the mass accretion rate at the black hole $\dot{M}_{BH}$ for three values of the electron heating parameter δ. Model parameters: $\alpha = 0.1$, $\beta = 9$, and $s = 0.4$. The nominal radiative efficiency of a standard thin disk, $\epsilon_{SSD} = 10\%$, is indicated by the horizontal dashed line at the top. When δ is large, the efficiency of a hot accretion flow is not very different from ϵ_{SSD} for a wide range of $\dot{M}_{BH}$ down to $\sim 10^{-5}\ \dot{M}_{Edd}$. In contrast, when δ is small, the efficiency drops precipitously for $\dot{M}_{BH} \lesssim 10^{-2} \dot{M}_{Edd}$. Squares, filled circles, and triangles indicate $\dot{M}_{eADAF}$, $\dot{M}_{crit,ADAF}$, and $\dot{M}_{crit,LHAF}$, respectively, for each value of δ (Section 2.6 defines these quantities). The horizontal extensions of the curves above $\sim 7 \times 10^{-3}\ \dot{M}_{Edd}$ show approximate radiative efficiencies assuming a two-phase accretion flow. Abbreviations: ADAF, advection-dominated accretion flow; BH, black hole; Edd, Eddington; LHAF, luminous hot accretion flow; SSD, Shakura-Sunyaev disk. Figure adapted from Xie & Yuan (2012).

2.6. Energetics: Electron ADAF, ADAF, LHAF, and Beyond

We now consider the energy equation of a hot accretion flow and discuss the role of the various terms that appear in it: viscous heating, compressional heating, energy advection, Coulomb energy transfer, and radiative cooling. For simplicity, we begin with the energy Equation 7, which corresponds to a single-temperature flow.

When $\dot{M}_{\rm BH}$ is very low, the gas density ρ is also low and the radiative cooling rate q^- (which decreases rapidly with decreasing ρ) becomes negligibly small (Section 2.5). The viscous heating rate is then balanced primarily by energy advection rather than cooling. Hence we have

$$q^+ \approx q^{\rm adv} \gg q^-, \quad f \approx 1. \tag{21}$$

That is, most of the viscous heat energy is stored in the flow and advected into the black hole rather than being radiated away. This is the classic regime of an ADAF. In the terminology used in this field, advection plays a "cooling" role.

With increasing $\dot{M}_{\rm BH}$, the radiative cooling q^- increases faster than $q^{\rm adv}$ and advective cooling becomes progressively less dominant. At a critical accretion rate $\dot{M}_{\rm crit,ADAF}$, the condition

$$q^+ = q^-, \quad f \approx 0 \tag{22}$$

is satisfied. An ADAF is allowed only for $\dot{M}_{\rm BH} \leq \dot{M}_{\rm crit,ADAF}$.

When $\dot{M}_{\rm BH} > \dot{M}_{\rm crit,ADAF}$, $q^+ < q^-$, i.e., radiative cooling is stronger than the rate of heating by viscosity ($f < 0$). Yuan (2001) showed that hot accretion flows are still permitted in this regime up to a second critical accretion rate $\dot{M}_{\rm crit,LHAF}$, which is determined by the condition

$$q^c + q^+ = q^-. \tag{23}$$

Solutions over the range $\dot{M}_{\rm crit,ADAF} < \dot{M}_{\rm BH} < \dot{M}_{\rm crit,LHAF}$ are called LHAFs: They are hot, but unlike ADAFs, they are radiatively efficient and luminous. The gas in these solutions remains hot, despite the strong cooling, because of the action of compressional heating q^c. Even though the entropy of the gas decreases with decreasing radius, the quantity $\rho v(de/dR) = q^+ + q^c - q^-$ is still positive. Thus the gas temperature continues to increase inward, and the flow remains hot (provided it starts out hot at a large radius). Over the entire LHAF branch, we have

$$q^c + q^+ > q^- > q^+. \tag{24}$$

Thus, $q^{\rm adv} = q^+ - q^- < 0$ and $f < 0$, so energy advection plays a "heating" role. In other words, the extra energy to heat the gas is supplied, not by viscous dissipation, but by the entropy already stored in the gas at large radius. Because of the high radiative efficiency and relatively large $\dot{M}_{\rm BH}$, LHAFs are expected to be much more luminous than ADAFs.

Consider now the more realistic case of a two-temperature hot accretion flow, where Equation 7 is replaced by Equations 18 and 19. In the early literature on ADAFs, this case was treated in an approximate fashion by considering only the energy Equation 18 of the ions. Neglect of the electron energy Equation 19 is valid whenever δ is small, as was the case in early studies that assumed $\delta \approx 0$–0.01. In this limit, almost all the viscous heat goes into the ions. Moreover, the critical bottleneck that prevents gas from radiating is the rate of transfer of energy from ions to electrons, $q^{\rm ie}$. Whenever $q^{\rm ie}$ is substantial (as happens at larger values of $\dot{M}_{\rm BH}$), the electrons have no trouble radiating whatever energy they receive from the ions, i.e., $q^{\rm ie} \approx q^-$. Thus, the approximation is self-consistent, though it does require very small values of δ. Numerically, in this regime, $\dot{M}_{\rm crit,ADAF} \approx 0.4\alpha^2 \dot{M}_{\rm Edd}$ and $\dot{M}_{\rm crit,LHAF} \approx \alpha^2 \dot{M}_{\rm Edd}$ (Narayan 1996, Esin et al. 1997, Yuan 2001).

As discussed in Section 2.3, the current consensus is that hot accretion flows have a larger value of $\delta \sim 0.1$–0.5. Viscous heating of electrons is then no longer negligible, nor is the Coulomb energy transfer rate q^{ie} the sole bottleneck. Consequently, it is now necessary to consider both the ion and electron energy equations (for early works on electron advection, see Mahadevan & Quataert 1997, Nakamura et al. 1997, Yuan 2001). As before, the two critical mass accretion rates, $\dot{M}_{\rm crit,ADAF}$ and $\dot{M}_{\rm crit,LHAF}$, may still be defined by the conditions given in Equations 22 and 23. [Xie & Yuan (2012) adopt $(1-\delta)q^+ = q^{ie}$ to define $\dot{M}_{\rm crit,ADAF}$. However, the definition (Equation 23) is more physical, and there is little difference in the numerical results.] However, a third critical accretion rate, $\dot{M}_{\rm eADAF}$, appears, which is explained below. By computing global models for δ in the range 0.1–0.5, Xie & Yuan (2012) obtain the following rough estimates for the three critical accretion rates (measured at the black hole):

$$\dot{M}_{\rm eADAF} \approx 0.001\, \alpha^2 \dot{M}_{\rm Edd}, \tag{25}$$

$$\dot{M}_{\rm crit,ADAF} \approx (0.1\text{--}0.3)\, \alpha^2 \dot{M}_{\rm Edd}, \tag{26}$$

$$\dot{M}_{\rm crit,LHAF} \approx (0.06\text{--}0.08)\, \alpha \dot{M}_{\rm Edd}. \tag{27}$$

In **Figure 2**, these critical rates are indicated by squares, filled circles, and triangles, respectively. Note that $\dot{M}_{\rm crit,ADAF}$ is smaller than the value $(0.4\alpha^2 \dot{M}_{\rm Edd})$ mentioned above for $\delta \approx 0 - 0.01$. This is because electrons now receive more energy directly via viscous heating and, hence, radiate more efficiently. [Two caveats should be mentioned. First, all quantities, such as $q^{\rm adv}$, q^c, q^+, and q^-, are functions of the radius. So we should, in principle, define a "local" critical accretion rate as a function of the radius (e.g., see the review by Narayan et al. 1998b). Here we adopt a simpler and more "global" definition where we check if the relevant condition is satisfied at any radius within the range of interest. For example, we call a solution an LHAF whenever the condition $q^+ < q^-$ is satisfied at any radius. Second, the results quoted here are from Xie & Yuan (2012), who assume $s = 0.4$ and $\beta = 9$. The results are likely to change for other choices of the parameters.]

The three critical mass accretion rates listed above separate different regimes of hot accretion as follows:

- $\dot{M}_{\rm BH} < \dot{M}_{\rm eADAF}$: Here both ions and electrons are radiatively inefficient. In particular, the electrons are unable to radiate either the viscous heat they acquire directly (δq^+) or the small amount of energy they receive from ions via Coulomb collisions (q^{ie}). Systems in this regime are truly radiatively inefficient because even the electrons are advection dominated; we call this regime an electron ADAF (eADAF). These systems correspond to the dimmest black hole accretion sources known, e.g., Sgr A* at the Galactic Center and quiescent BHBs.
- $\dot{M}_{\rm eADAF} < \dot{M}_{\rm BH} < \dot{M}_{\rm crit,ADAF}$: Here electrons radiate efficiently their own viscous energy (δq^+) as well as any energy they receive from the ions (q^{ie}). However, Coulomb collisions are inefficient. So the ions transfer only a small fraction of their energy to the electrons and therefore remain advection dominated. Systems in this regime are expected to be fairly radiatively efficient, with an efficiency of the order of 1% or more (depending on the value of δ). They are thus substantially brighter than classic ADAF models (see **Figure 2**). However, the flows are hot, geometrically thick, and still ADAFs in the sense that $q^+ > q^-$.
- $\dot{M}_{\rm crit,ADAF} < \dot{M}_{\rm BH} < \dot{M}_{\rm crit,LHAF}$: Here we have an LHAF with $q^+ < q^-$. All the energy terms in the ion energy equation are roughly comparable in magnitude. The entropy decreases as the gas flows in, but the gas remains hot because of compressional heating. The radiative efficiency increases rapidly with increasing $\dot{M}_{\rm BH}$, as shown in **Figure 2**. This plot further shows that, when $\delta \ll 1$, the LHAF branch is restricted to quite a narrow range of $\dot{M}_{\rm BH}$. However, for currently accepted values of $\delta \sim 0.1\text{--}0.5$, the LHAF solution extends over a factor of several in $\dot{M}_{\rm BH}$.

- $\dot{M}_{\rm BH} > \dot{M}_{\rm crit,LHAF}$: In this regime, the 1D global equations have no hot accretion flow solution. Radiative cooling is too strong, and even compressional heating is insufficient to keep the gas hot. In the traditional view, the accretion flow transitions to a standard thin accretion disk. However, there are large uncertainties, and Yuan (2003) speculates that the gas may transition to a two-phase medium with cold dense clumps embedded in hot gas. Alternatively, Oda et al. (2010) propose that a magnetically dominated accretion flow may form (see Section 3.2.3).

2.7. Stability and Relationship to Other Accretion Solutions

Hot optically thin gas generally tends to be thermally unstable. Therefore, all hot accretion flows are potentially unstable. What saves them is the fact that the accretion timescale is shorter than the instability growth time.

Narayan & Yi (1995b) and Abramowicz et al. (1995) showed that ADAFs are stable to long wavelength perturbations. For small-scale perturbations, however, the results are somewhat subtle. Wu & Li (1996) and Wu (1997) showed that ADAFs are stable under most reasonable conditions, whereas Kato et al. (1996, 1997) showed that ADAFs are potentially unstable at short wavelengths. Using a time-dependent analysis, Manmoto et al. (1996) showed that small-scale density perturbations in a one-temperature ADAF grow as the gas flows in, but they do not do so sufficiently quickly to affect the global viability of the solution. All these results were derived assuming that advection dominates. Hence, they do not apply to the SLE or LHAF solutions.

In the case of the SLE solution, Piran (1978) showed that the model is thermally unstable. Yuan (2003) studied the thermal stability of LHAFs and concluded that these flows are thermally unstable. However, if the accretion rate is below $\dot{M}_{\rm crit,LHAF}$, the growth timescale of the instability remains longer than the accretion timescale and the solution can survive. Above this accretion rate, however, the instability will grow quickly. It is possible that the instability will not destroy the solution but will lead instead to a two-phase medium in which cold dense blobs are intermixed with hot gas (Section 2.6). Dynamically, the hot phase would behave like an LHAF with radiative cooling stronger than viscous heating. Xie & Yuan (2012) estimated the luminosity of such a two-phase accretion flow and found that the radiative efficiency is expected to be approximately 10%, as indicated by the horizontal extensions in **Figure 2**.

The above discussion deals with thermal stability. What about viscous stability? The most convenient way to investigate this is by plotting the locus of accretion solutions in the two-dimensional (2D) plane of accretion rate $\dot{M}$ and surface density $\Sigma \equiv 2\rho H$ (e.g., Frank et al. 2002). If the solution track has a positive slope, the solution is viscously stable, and vice versa. By including all solutions (both hot and cold) in such a diagram, one can appreciate the relationship between the various solutions.

Figure 3 shows an example of such a plot (taken from Yuan 2003; see also Abramowicz et al. 1995, Chen et al. 1995). The various solution tracks shown have been obtained by solving simple equations, such as Equations 1–4, or more complex versions of these that include radiative transfer, two-temperature plasma, etc. Usually, approximations are needed, e.g., assuming $\Omega = \Omega_K$ and $q^{\rm adv} = \xi(\dot{M}c_s^2/2\pi R^2 H)$ with a constant ξ (see Abramowicz et al. 1995). Then, for a given set of parameters, α, M, R, ξ, one can solve for $\dot{M}$ as a function of Σ.

The black solid lines in **Figure 3** show all the standard solution branches: The ADAF and SLE solutions belong to the sequence of hot solutions, and the SSD and slim disk solutions belong to the sequence of cold solutions. All four of these solution branches are viscously stable because each track has a positive slope. At a given Σ, if there are multiple solutions, the

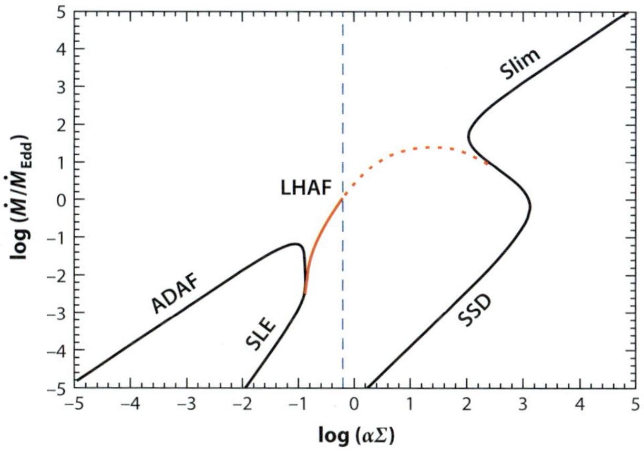

Figure 3

Thermal equilibrium curves of various accretion solutions for the following parameter values: $M = 10\,M_{\odot}$, $\alpha = 0.1$, and $r = 5$. The accretion rate is normalized to $\dot{M}_{\rm Edd}$, and the surface density is in units of grams per centimeter squared. The black solid lines correspond to the classic solution branches, e.g., the hot branch consisting of ADAF and SLE as well as the cold branch consisting of a slim disk and SSD. The blue vertical dashed line separates optically thin solutions on the left from optically thick solutions on the right. The red line corresponds to the LHAF solution. Although the LHAF branch appears to go all the way across from the hot to the cold branch, global models indicate that this solution is self-consistent only to the left of the vertical line (*solid red segment*). Abbreviations: ADAF, advection-dominated accretion flow; LHAF, luminous hot accretion flow; SLE, Shapiro-Lightman-Eardley; SSD, Shakura-Sunyaev disk. Figure adapted from Yuan (2003) (for similar plots without the LHAF solution branch, also see Abramowicz et al. 1995, Chen et al. 1995).

uppermost (highest $\dot{M}$) solution is thermally stable to long wavelength perturbations, the next one below is unstable, and the next is stable. Therefore, the ADAF, slim disk, and SSD solutions are thermally stable. However, the SLE solution is thermally unstable, as is the segment between the SSD and slim disk. [Recently, there has been considerable interest in the thermal stability of the latter solution branch, which corresponds to a radiation pressure–dominated thin disk. Using numerical radiation MHD simulations, Hirose et al. (2009) concluded that a thin disk in this regime is thermally stable. However, later work by Jiang et al. (2013) showed that the disk is, in fact, thermally unstable. They also discussed the reason for the discrepancy.] The branch between the SSD and slim disk in **Figure 3** is also viscously unstable (Lightman & Eardley 1974) because it has negative slope. In terms of the advection parameter f, $f \approx 0$ for the SSD and SLE, and $f \approx 1$ for the ADAF and slim disk. The red line in **Figure 3** corresponds to the LHAF solution, where ξ is negative, advection plays a heating role (Section 2.6), and $f < 0$.

The results presented in **Figure 3** are approximate because they are based on a single-temperature model (however, for equivalent results for a two-temperature model, see Chen et al. 1995, their figure 1c) and are, moreover, based on a local rather than a global analysis. Nevertheless, these plots are believed to be qualitatively correct and therefore raise an important question: Given that for certain ranges of the mass accretion rate, both hot and cold solutions are available and both are thermally stable, which solution does nature pick? Narayan & Yi (1995b) discussed two options.

The first option is that, if the accreting gas is hot at the outer feeding radius where mass first enters the accretion disk, and if a hot accretion solution is permitted at that radius, then the gas will start off in the hot mode of accretion and will remain in the hot accretion state all the way down to the black hole. By contrast, if the gas starts out on the cool SSD branch on the outside, then it will remain in that branch down to the black hole unless the disk enters the viscous instability zone of Lightman & Eardley (1974) (where the gas would become a hot accretion flow, the only stable solution remaining).

The second, more revolutionary, option is that the accretion flow will switch to the hot accretion branch whenever the latter solution is allowed, i.e., whenever $\dot{M} < \dot{M}_{\rm crit,LHAF}(R)$ corresponding to the local radius R. In other words, accretion occurs via the SSD solution only if it is the sole stable solution available. This so-called strong ADAF principle appears to be generally consistent with observations (Section 1.2.2).

3. NUMERICAL SIMULATIONS

The 1D solutions considered so far are easy to calculate and often capture the important physics. However, hot accretion flows are geometrically thick, so one cannot be sure that the vertically integrated equations from which 1D solutions are derived are valid. In particular, height-integration eliminates multidimensional structures such as outflows.

Analytical 2D solutions have been obtained by a number of authors over the years (e.g., Begelman & Meier 1982, Narayan & Yi 1995b, Xu & Chen 1997, Blandford & Begelman 2004, Xue & Wang 2005, Tanaka & Menou 2006, Jiao & Wu 2011, Begelman 2012). However, these models make simplifying assumptions such as self-similarity and therefore have limited applicability. If we wish to understand the multidimensional structure of hot accretion flows, numerical simulations are the only way.

3.1. Hydrodynamic Simulations

Although angular momentum transport in hot accretion flows occurs via MHD turbulence driven by the magnetorotational instability (Section 3.2.1), early computer simulations were carried out without magnetic fields, using 2D hydrodynamic (HD) codes and an α-like prescription for the viscous stress (Igumenshchev et al. 1996, Igumenshchev & Abramowicz 1999, Stone et al. 1999, Igumenshchev & Abramowicz 2000, Igumenshchev et al. 2000, De Villiers & Hawley 2002, Fragile & Anninos 2005). There were some differences in the adopted form of the shear stress. For example, Stone et al. (1999) assumed that only the azimuthal component of the shear stress tensor is present, whereas Igumenshchev & Abramowicz (1999) also included poloidal components.

HD simulations of hot accretion flows reveal rich and complicated time-dependent structures. In particular, there are convective motions (Igumenshchev & Abramowicz 1999, Stone et al. 1999, Igumenshchev et al. 2000), confirming an early prediction of Narayan & Yi (1994) (Section 2.1). The level of convective turbulence depends on details; for example, convection becomes weaker if a larger value of α is used or if poloidal components of the shear stress are included (Igumenshchev & Abramowicz 1999, Stone et al. 1999, Yuan & Bu 2010). The radial dynamic range of simulations is usually fairly limited [even more so for the three-dimensional (3D) MHD simulations discussed below], but Yuan et al. (2012b) recently achieved an unprecedented four decades of dynamic range in an HD simulation using a "two-zone" approach.

The time-averaged "steady-state" flow in HD simulations is usually well described by a radial power-law distribution of various quantities, with the power-law indices depending on the specific form of the adopted shear stress. For the usual α-prescription of Shakura & Sunayev (1973), the radial scalings are consistent with the self-similar solution (Equations 8–13) (see Stone et al. 1999). By contrast, the initial conditions used in the simulations appear to affect some results such as the streamline structure and the Bernoulli parameter *Be* (Yuan et al. 2012a).

3.2. Magnetohydrodynamic Simulations

For obvious reasons, MHD simulations are much more realistic than HD simulations. Most simulations performed in recent years have been carried out with MHD codes.

3.2.1. Magnetorotational instability. It is now widely accepted that the mechanism of angular momentum transport in ionized accretion flows is the magnetorotational instability (MRI) (Balbus & Hawley 1991, 1998). This instability takes a seed magnetic field in the accreting gas and amplifies it exponentially until the system becomes nonlinear and develops MHD turbulence. The Maxwell and Reynolds stresses in the turbulent state transport angular momentum outward, causing gas to accrete inward.

Whereas the basic MRI is a linear instability and can be understood analytically, the nonlinear turbulent state relevant for disk accretion can be studied only with numerical simulations. A number of codes have been used, notably ZEUS (Stone & Norman 1992a,b), HARM (Gammie et al. 2003), the general relativistic MHD (GRMHD) code of De Villiers & Hawley (2003a), COSMOS++ (Anninos et al. 2005), and ATHENA (Stone et al. 2008). Many studies have been done in the limit of a local shearing box, which permits high spatial resolution (e.g., Hawley & Balbus 1991; Brandenburg et al. 1995; Hawley et al. 1995, 1996; Matsumoto & Tajima 1995; Stone et al. 1996). These studies show that MHD turbulence is inevitable so long as the gas and the magnetic field are well coupled and that the Maxwell stress dominates over the Reynolds stress by a factor of several.

When data from different published 3D shearing box simulations are combined, a tight correlation is seen between the parameter β (Equation 15) and the viscosity parameter α, e.g., $\alpha\beta \sim 0.5$ (Blackman et al. 2008, Guan et al. 2009, Sorathia et al. 2012; see also Hawley & Balbus 1996). However, the individual values of β and α vary substantially from one numerical experiment to the next. For example, Hawley et al. (2011) obtained β values in the range 10–200, corresponding to $\alpha \sim 0.01-0.003$, and other authors have found even larger variations. The value of α is thus not constrained. It seems to depend on the magnitude of the net initial magnetic field (Hawley et al. 1995, 1996; Pessah et al. 2007), a dependence that is also confirmed in localized regions of global simulations (Sorathia et al. 2010). Some shearing box simulations even find α values larger than unity (e.g., Bai & Stone 2013). Numerical resolution also plays a role. Generally, better resolution gives a larger α (up to some saturation value). However, when the net magnetic flux is zero, increasing the resolution actually causes α to decrease (Fromang & Papaloizou 2007, Fromang et al. 2007), with α going to zero in the limit of infinite resolution.

Interestingly, the uncertainty in the value of α is reduced if shearing box simulations include vertical stratification to mimic the effect of vertical disk gravity (Davis et al. 2010, Bai & Stone 2013). Perhaps because of this, global disk simulations, which automatically include vertical gravity, show less variation in the effective value of α. These simulations generally evolve to steady state with $\alpha \sim 0.05-0.2$ (Hawley & Balbus 2002, Penna et al. 2013b).

3.2.2. Global simulations: general results. Global MHD simulations of hot accretion flows are more realistic than global HD simulations, as they self-consistently generate shear stress through MRI-induced MHD turbulence. By contrast, HD simulations must include an ad hoc viscosity. Early pioneers in global MHD simulations include Matsumoto & Shibata (1997), Armitage (1998), Hawley (2000), Machida et al. (2000), Hawley & Krolik (2001), Stone & Pringle (2001), and Igumenshchev & Narayan (2002).

Compared with shearing box simulations, global simulations enable the MRI to sample much larger radial and azimuthal wavelengths. The largest radial dynamic ranges are achieved in 2D.

[With current computer resources, 3D simulations can reach inflow equilibrium over, at best, only 2 orders of magnitude in radius (e.g., Pang et al. 2011, McKinney et al. 2012, Narayan et al. 2012b, Sadowski et al. 2013a), a substantial part of which is dominated by either inner or outer boundary conditions. Two-dimensional simulations can achieve a factor of several larger dynamic range.] However, 2D simulations do not treat the MRI accurately because of Cowling's antidynamo theorem (Cowling 1933), which limits the growth of the poloidal magnetic field and causes turbulence to die away. Thus, no steady accretion is possible in 2D, and one has to carefully select a period of time after the disk has become turbulent but before the turbulence dies out. There has been no systematic study of how well the properties of this intermediate period in 2D simulations agree with those of 3D simulations with sustained turbulence. Qualitatively, it appears that the differences are not large.

Hawley (2000) compared the results of local shearing box and global simulations using two different initial configurations of the magnetic field: toroidal and vertical. In terms of growth of the MRI and transition to MHD turbulence, he found global and shearing box simulations to behave similarly. The magnetic shear stress $T_{R\phi}$ is directly proportional to the magnetic pressure, $2\langle B_R B_\phi \rangle \approx (0.4\text{–}0.5)\langle B^2 \rangle$; this is equivalent to $\alpha\beta \approx 0.5$ mentioned earlier. Depending on the value of β, the resulting $\alpha \approx 0.05\text{–}0.2$ in the interior of the disk (e.g., Penna et al. 2013b). The Maxwell (magnetic) stress is always larger than the Reynolds stress by a factor of several. Also, the toroidal component of the field is significantly larger than the radial component, which is somewhat larger than the vertical component. Although most global simulations start with a weak magnetic field (initial $\beta \sim 100$), Machida et al. (2000) used a strong initial toroidal field with $\beta = 1$. There was no MRI in their simulation, but they found the Parker instability, which led to the formation of a magnetized corona.

Global 3D MHD simulations have been run by various groups, and the results are fairly similar. Early work assumed Newtonian dynamics and modeled the black hole at the center via a pseudo-Newtonian potential (Armitage 1998; Hawley 2000; Machida et al. 2000, 2004; Hawley & Krolik 2001; De Villiers & Hawley 2003b; Igumenshchev et al. 2003). GRMHD codes were later developed (Koide et al. 1999, De Villiers & Hawley 2003a, Gammie et al. 2003, Fragile et al. 2007). Much of the recent work in this field is based on the latter codes, which provide a more realistic description of phenomena close to the black hole. Nevertheless, pseudo-Newtonian simulations are still useful for studying large-scale properties of the accretion flow.

Representative results from 3D GRMHD simulations can be found in the series of early papers by De Villiers and collaborators (De Villiers et al. 2003, 2005; Hirose et al. 2004; Krolik et al. 2005) and by Gammie and collaborators (Gammie et al. 2004, McKinney & Gammie 2004, McKinney 2006). The simulations are initialized with a rotating torus in hydrostatic equilibrium and embedded with a weak poloidal magnetic field. Accretion occurs self-consistently as a result of MHD turbulence generated by the MRI, and the accretion flow separates into three qualitatively different regions (**Figure 4**): disk body, corona, and axial funnel.

The disk body is turbulent and dense and has a roughly constant value of H/R, consistent with 1D self-similar and global solutions (Section 2). The magnetic field within the disk is subequipartition ($\beta \sim 10\text{–}100$), and both the magnetic field and velocity streamlines are chaotic, as expected for turbulence. Inside the innermost stable circular orbit (ISCO) is the plunging region. Here, the flow spirals in rapidly toward the black hole horizon, and the motion is almost laminar. Although the ISCO is roughly where the turbulent gas in the disk transitions to laminar inflow, there is no other specific signature in the flow dynamics associated with the ISCO. By contrast, in thin disks, the flow changes dramatically across the ISCO (Reynolds & Fabian 2008, Shafee et al. 2008, Penna et al. 2010; but see Noble et al. 2010).

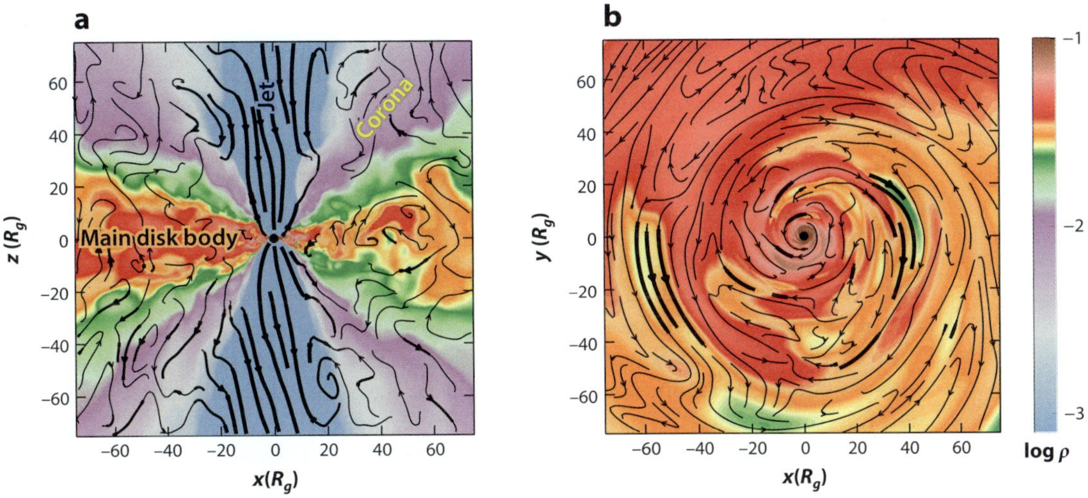

Figure 4

Instantaneous slice at a fixed azimuthal angle (*a*) and a slice through the equatorial plane (*b*) from a three-dimensional general relativistic magnetohydrodynamic (GRMHD) simulation of a hot accretion flow around a spinning black hole ($a_* \equiv a/M = 0.7$). The black hole is at (0,0), and lengths are in units of $R_g = GM/c^2$. Three regions of the flow are identified: main disk body, corona, and jet. Color indicates density, with fluctuations caused by turbulence evident. Lines trace the magnetic field in the two image planes; the out-of-plane component is ignored. Arrows show the direction of the magnetic field, and line thickness indicates magnetic energy density relative to other energy densities: The thickest lines correspond to regions with comoving $\frac{B^2}{4\pi} > \rho c^2$ (found primarily in the region of the jet), intermediate thickness lines indicate regions with (see Equation 15) $\beta < 4$ (mostly in the corona), and thin lines correspond to regions with the weakest magnetization ($\beta > 4$, primarily in the main disk body). Figure courtesy of A. Tchekhovskoy; data taken from Sadowski et al. (2013a).

Above and below the main disk is the corona, where the gas density is much lower. The magnetic field here is more regular than in the disk body and tends to be toroidal. The magnetic and gas pressure are roughly comparable ($\beta \sim 1$). The value of β decreases with increasing distance away from the midplane, with $\beta \sim 0.1$ above approximately two density scale heights (De Villiers et al. 2003, 2005). The corona is the launchpad for the disk wind (Section 3.4).

The axial funnel is a magnetically dominated region in which the gas is very tenuous. It is also the location of the jet (Section 3.3). The magnetic field is predominantly radial close to the black hole, where the jet extracts rotational energy from the black hole spin (Section 3.3.1). Far from the black hole the field becomes mostly toroidal and carries the jet power in the form of a Poynting flux. The boundary of the funnel (the funnel wall) corresponds to the centrifugal barrier associated with material originating from the innermost region of the disk. The jet here is less relativistic and is powered at least partly by the rotation of the accretion flow (Section 3.3.2). Hence, it is less sensitive to the black hole spin. Overall, there is a smooth variation of properties, going from highly magnetically dominated conditions at the axis to progressively larger gas content with increasing distance from the axis. Past the funnel wall, the jet merges with the corona where the disk wind is launched (Section 3.4).

The effect of different initial magnetic field geometries has been investigated by several authors (e.g., Hawley & Krolik 2002 and Igumenshchev et al. 2003, using a pseudo-Newtonian potential; Beckwith et al. 2008 and McKinney & Blandford 2009, using GRMHD). Models with a purely toroidal initial field evolve much more slowly than those with a poloidal initial field, because the former have neither an initial vertical field, which is needed for the linear MRI, nor a radial field, which is needed for field amplification via shear. Inflow begins only after the MRI has produced turbulence of sufficient amplitude (Hawley & Krolik 2002), which happens much later when the initial field is toroidal. Generally, once saturation has been reached, the disk properties do not depend much on the initial field topology. By contrast, jet properties are very sensitive (Section 3.3).

For geometrically thin accretion disks, Lense-Thirring precession may cause a tilted disk to align with the spin axis of the black hole out to a fairly large radius (Bardeen & Petterson 1975, Scheuer & Feiler 1996, Lodato & Pringle 2006). Fragile and collaborators have carried out a number of numerical simulations of tilted hot accretion flows (Fragile et al. 2007, 2009; Fragile 2009; Dexter & Fragile 2011, 2013). They found that the disk does not align with the black hole, in agreement with theoretical predictions for a geometrically thick disk (Papaloizou & Lin 1995). Instead, the disk precesses as a whole out to some radius. Alignment does happen when accretion occurs in the MAD regime (Section 3.2.3) and may have observational implications for relativistic jets (McKinney et al. 2013).

The precession of a tilted disk will lead to time-dependence in the observed radiation from a hot accretion flow, and could potentially explain some low frequency quasiperiodic oscillations (QPOs) seen in BHBs (Ingram et al. 2009, Ingram & Done 2011, Veledina et al. 2013). It is unlikely that high-frequency QPOs can be explained in a similar fashion (Dexter & Fragile 2011). Several authors (e.g., Bursa et al. 2004; Blaes et al. 2006; Abramowicz et al. 2006) have explored oscillation modes of tori in this connection. Recently, high-frequency QPOs have been reported in numerical GRMHD simulations of geometrically thick hot accretion flows (Dolence et al. 2012) and magnetically arrested disks (MADs) (Section 3.2.3) (McKinney et al. 2012; Shcherbakov & McKinney 2013).

Although beyond the scope of this review, a few authors have carried out simulations of magnetized spherical accretion (Igumenshchev & Narayan 2002, Igumenshchev 2006) as well as accretion of low-angular momentum gas (Proga & Begelman 2003, Mościbrodzka et al. 2007, Janiuk et al. 2009).

We conclude with some general remarks on numerical accuracy. Because energy advection plays a key role in hot accretion flows, it is important to ensure that numerical codes conserve energy accurately. Early numerical simulations were based on codes that evolve the internal energy of the gas. Such codes do not conserve total energy and can introduce an effective numerical cooling that is hard to quantify. An alternative Godunov-based approach enforces strict mass, momentum, and total energy conservation, as exemplified by the pioneering relativistic MHD code of Komissarov (2001), the GRMHD codes HARM (Gammie et al. 2003) and COSMOS++ (Anninos et al. 2005), and the Newtonian MHD code ATHENA (Stone et al. 2008). Direct comparison of simulations using both techniques shows that accretion flows simulated with nonenergy-conserving codes tend to be geometrically thinner than they should be.

The effect of numerical resolution on global simulation results has been investigated recently (Sorathia et al. 2010; Hawley et al. 2011, 2013; Shiokawa et al. 2012). These studies achieved numerical convergence in terms of shell-averaged quantities, azimuthal correlation length of fluid variables, and synthetic spectra. They indicate that most previous global simulations may have been somewhat underresolved.

3.2.3. Magnetically arrested disk. Magnetically dominated hot accretion flows have recently become more prominent, thanks to the advent of numerical MHD simulations. A version of these models called a magnetically arrested disk (MAD) (Igumenshchev et al. 2003, Narayan et al. 2003, Igumenshchev 2008), or a magnetically choked accretion flow (McKinney et al. 2012), is based on an idea originally proposed by Bisnovatyi-Kogan & Ruzmaikin (1974), in which a strong vertical bipolar magnetic field is pushed into the central black hole by the thermal and ram pressure of the accreting gas. A significant amount of magnetic flux threads the horizon. As a result, the field outside the black hole becomes so strong that it disrupts the axisymmetric accretion flow, forcing the gas to move inward via streams and blobs through an interchange instability. This behavior was first noted by Igumenshchev et al. (2003) in 3D Newtonian MHD simulations and was later confirmed in 3D GRMHD simulations (Tchekhovskoy et al. 2011, 2012; McKinney, Tchekhovskoy & Blandford 2012). Current interest in MAD accretion is driven by the discovery that it leads to very powerful relativistic jets (Section 3.3).

All magnetized accretion flows cause a certain amount of magnetic flux to thread the black hole. The MAD state is special in that the flux threading the hole is at its maximum saturation value for the given mass accretion rate $\dot{M}_{\rm BH}$. This saturation flux is approximately (Tchekhovskoy et al. 2011, 2012)

$$\Phi_{\rm MAD} \approx 50\ \dot{M}_{\rm BH}^{1/2} R_g c^{1/2} = 1.5 \times 10^{21}\ m^{3/2} \dot{m}_{\rm BH}^{1/2}\ {\rm G\ cm}^2, \tag{28}$$

where m and $\dot{m}$ are defined in Equation 14 and $R_g = GM/c^2 = R_S/2$ (half the Schwarzschild radius) is the gravitational radius of the black hole. [$\Phi_{\rm MAD}$ has a weak dependence on the black hole spin as well as the disk thickness (see Tchekhovskoy et al. 2014), but we ignore this complication for clarity.] The corresponding field strength at the horizon is roughly

$$B_{\rm MAD} \approx \frac{\Phi_{\rm MAD}}{2\pi R_g^2} = 10^{10}\ m^{-1/2} \dot{m}_{\rm BH}^{1/2}\ {\rm G} \tag{29}$$

(compare with Equation 12). [Simulation results are often given in Heaviside-Lorentz units, whereas numerical estimates in this article are in Gaussian units. The two differ by a factor of $\sqrt{4\pi}$. For instance, the magnetic pressure is $B^2/8\pi$ in Gaussian units but $B^2/2$ in Heaviside-Lorentz units. Note also that the magnetic field strength is frame-dependent. For instance, when evaluating the magnetic pressure, especially for computing the value of β (Equation 15 and see **Figure 4**), one must consider the field strength in the comoving fluid frame, i.e., $B^2/4\pi \rightarrow b^2$ in the notation of Komissarov (1999) and Gammie et al. (2003). However, $B_{\rm MAD}$ and $\Phi_{\rm MAD}$ in Equation 29 are evaluated in the stationary coordinate frame or "lab frame".] Systems that have not reached the MAD limit have been referred to as SANE (standard and normal evolution) (Narayan et al. 2012b). They span a one-parameter family of models extending from $\Phi = 0$ up to a magnetic flux just below $\Phi_{\rm MAD}$. Structural differences, most notably in the jet, are evident between MAD and SANE models (Narayan et al. 2012b, Sadowski et al. 2013a).

Unlike small-scale fields, a large-scale vertical magnetic field cannot be dissipated locally because the plasma has negligible resistivity. It also cannot be absorbed by the central black hole; for example, even when field lines thread the horizon, the external magnetic flux is unaffected (see Igumenshchev et al. 2003). But how does vertical field even reach the center? Presumably the field is advected in from whatever external mass reservoir feeds the accretion flow. Such advection happens quite efficiently in numerical simulations, especially in the case of geometrically thick hot accretion flows. However, most simulations are limited to radii relatively close to the black hole, and it is not clear whether the same physics will necessarily operate at larger radii. If outward diffusion of the magnetic field via reconnection is inefficient (Spruit & Uzdensky 2005, Bisnovatyi-Kogan & Lovelace 2007, Rothstein & Lovelace 2008, Cao 2011), as seems especially likely for geometrically thick accretion flows (Livio et al. 1999; Guilet & Ogilvie 2012, 2013), the magnetic field should be readily advected in from large radii. The magnetic field can also be brought in efficiently via the corona (Beckwith et al. 2009). At least in the case of supermassive black holes accreting from an external medium, plenty of magnetic flux is available in the mass reservoir (e.g., Narayan et al. 2003). Therefore, all supermassive black holes with hot accretion flows should quickly approach the MAD limit, provided flux is advected efficiently. The situation is less clear in the case of BHBs, as the supply of net magnetic flux depends on the properties of the companion star and the details of the mass transfer.

Two additional magnetically dominated accretion models have been discussed. Meier (2005; see also Fragile & Meier 2009) proposed that the inner regions ($R \lesssim 50 R_S$) of a hot accretion flow may be converted into a magnetically dominated magnetosphere-like phase in which a strong, well-ordered field is present rather than the weak, turbulent field usually seen in a hot flow. Stimulated by MHD simulations described in Machida et al. (2006), Oda et al. (2010) proposed a different kind of magnetically supported accretion flow model, which exists only when the accretion rate is relatively high, well above $\dot{M}_{\rm crit,ADAF}$. Both of these models have a magnetic field geometry dominated by radial and toroidal fields, which is different from the vertical poloidal field envisaged in the MAD model.

3.3. Jets in Hot Accretion Flows

Early papers (Narayan & Yi 1994, 1995a; Blandford & Begelman 1999) conjectured that hot accretion flows should have strong winds and, by extension, jets. Observational evidence for such an association has accumulated in recent years, with the recognition that essentially all LLAGNs are radio loud (Falcke et al. 2000, Nagar et al. 2000, Ho 2002) and the parallel discovery that virtually every BHB in a hard state has radio emission (Corbel et al. 2000; Fender 2001, 2006; Fender et al. 2004). Because all these systems may have hot accretion flows (Section 4.2), a direct causal connection between hot flows and radio-emitting jets seems likely. In contrast, jets are much weaker and are often not seen in systems that have cool geometrically thin disks. Although there is no definitive explanation for this dichotomy, three effects likely play a role: (*a*) Compared with thin disks, geometrically thick disks more easily advect the magnetic field to the black hole (Livio et al. 1999, Guilet & Ogilvie 2012); (*b*) the Bernoulli parameter of the gas in hot accretion flows is larger, hence enhancing winds in these systems (Narayan & Yi 1994, 1995a; Blandford & Begelman 1999); and (*c*) strong winds help to collimate and stabilize the jet (Appl & Camenzind 1992, 1993; Beskin & Malyshkin 2000).

Although many jet models have been proposed over the years, the current consensus is that jets arise from a combination of magnetic fields and rotation. Especially influential in this field are the Blandford-Znajek (BZ) model (Blandford & Znajek 1977; see also Ruffini & Wilson 1975, Lovelace 1976, MacDonald & Thorne 1982, Phinney 1983, Thorne et al. 1986, Punsly & Coroniti 1989, Komissarov & McKinney 2007, Tchekhovskoy et al. 2011, Penna et al. 2013a) and the Blandford-Payne (BP) model (Blandford & Payne 1982; see also Pudritz & Norman 1983, Heyvaerts & Norman 1989, Li et al. 1992, Contopoulos & Lovelace 1994, Ostriker 1997, Vlahakis & Konigl 2003). The primary distinction between the two models is the energy source of the jet. In the BZ model, the source is the rotational energy of the black hole, whereas in the BP model, it is the rotational energy of the accretion flow. Numerical simulations suggest that truly relativistic jets are produced primarily by the BZ mechanism, whereas quasi-relativistic jets and nonrelativistic winds may be driven by a combination of the BP and other mechanisms. In the following discussion, we use the term "BZ jet" for the truly relativistic BZ-powered outflow and refer to the quasi-relativistic outflow from the inner region of the accretion flow as the "disk jet."

3.3.1. Relativistic BZ jet. In the BZ model, a large-scale poloidal magnetic field passes through the ergosphere of the rotating black hole and threads the horizon (cf. Section 3.2.3). Frame dragging by the rotating hole leads to the creation of a toroidal field and a Poynting flux. The key to the BZ process, which goes back to the influential work of Penrose (1969) and subsequently

Ruffini & Wilson (1975), is the possibility of having within the ergosphere a negative inward electromagnetic energy flux as measured at infinity. This negative energy flux enters the horizon, thereby reducing the mass-energy and angular momentum of the hole. Correspondingly, there is an outgoing Poynting-dominated jet that carries positive energy and angular momentum. At its most basic, the outflowing power in the BZ model is given by (Blandford & Znajek 1977; for more accurate approximations, see Tchekhovskoy et al. 2010)

$$P_{\rm BZ} = \frac{\kappa}{4\pi c}\Phi^2\Omega_{\rm H}^2, \tag{30}$$

where Φ is the magnetic flux threading the horizon, $\Omega_{\rm H} = a_* c/2R_{\rm H}$ is the angular velocity of the horizon, $a_* \equiv a/M$ is the dimensionless spin parameter of the black hole, and $R_{\rm H} = R_g(1+\sqrt{1-a_*^2})$ is the radius of the horizon. The numerical coefficient κ depends weakly on the magnetic field geometry and is approximately $\approx$0.05. The above formula highlights the fact that the BZ mechanism requires two key ingredients: an ordered magnetic flux at the horizon (Φ) and rotation of the black hole ($\Omega_{\rm H}$).

Many MHD simulations of hot accretion flows have been performed to study jet formation (e.g., Kudoh et al. 1998; Koide et al. 1999, 2000; Kuwabara et al. 2000; Hawley & Balbus 2002; Koide 2003; McKinney & Gammie 2004; Kato et al. 2004a,b; De Villiers et al. 2005; McKinney 2005, 2006; Hawley & Krolik 2006; Komissarov et al. 2007; McKinney & Blandford 2009; Tchekhovskoy et al. 2011; McKinney et al. 2012; Sadowski et al. 2013a). The simulations are typically initialized with a gas torus threaded with a weak magnetic field. As the simulation progresses, a large-scale magnetic field forms self-consistently at the black hole horizon, as required by the BZ model, even though such a field is not present in the initial state (e.g., De Villiers et al. 2003, 2005; Igumenshchev et al. 2003; McKinney 2006; Tchekhovskoy et al. 2011). Also, the magnetic flux is trapped within a funnel, causing the outgoing power to be collimated in a relativistic jet. However, a powerful jet forms only if the initial field in the simulation has a favorable poloidal configuration. A dipolar field is ideal. If a quadrupolar initial field is adopted, the field in the funnel is much weaker, and if a toroidal field is adopted, no funnel field develops (Igumenshchev et al. 2003, De Villiers et al. 2005, Beckwith et al. 2008, McKinney & Blandford 2009). In the latter case, there is no BZ jet.

The jet power measured in simulations shows good agreement with the predictions of the BZ model (Equation 30), with only modest changes in the coefficient due to the presence of the surrounding thick disk. A rough estimate of the BZ jet power is obtained by combining Equations 28 and 30:

$$P_{\rm jet} \approx 2.5\left(\frac{a_*}{1+\sqrt{1-a_*^2}}\right)^2\left(\frac{\Phi}{\Phi_{\rm MAD}}\right)^2 \dot{M}_{\rm BH}c^2, \tag{31}$$

where $\Phi_{\rm MAD}$ is the limiting magnetic flux defined in Equation 28 and Φ is the actual flux threading the black hole horizon. This formula slightly underestimates the jet power for slow spins and overestimates the power for rapid spins. A better approximation is $P_{\rm jet} \approx 0.65a_*^2(1+0.85a_*^2)(\Phi/\Phi_{\rm MAD})^2\dot{M}_{\rm BH}c^2$ (A. Tchekhovskoy, private communication). As is evident from Equation 31, the most favorable situation is when the magnetic flux has reached the MAD limit ($\Phi \to \Phi_{\rm MAD}$) and the black hole spin is maximum ($a_* \to 1$). In this limit, the BZ jet power can exceed the total accretion energy budget of $\dot{M}_{\rm BH}c^2$ (Tchekhovskoy et al. 2011, 2012; Tchekhovskoy & McKinney 2012). Although at first sight this may appear to violate energy conservation, there is no inconsistency because most of the jet energy comes directly from the spinning black hole via a generalization of the Penrose (1969) process.

Over the years, many published studies have provided estimates of the jet power as a function of black hole spin (e.g., Hawley & Balbus 2002; De Villiers et al. 2005; McKinney 2005, 2006;

Hawley & Krolik 2006). Even though the numerical values given do not always agree, the results are consistent with the above BZ-derived relation, once the dependence on magnetic flux is taken into account. Modest differences are seen between prograde and retrograde disks, with the former producing somewhat stronger jets (Tchekhovskoy & McKinney 2012). In addition, the physics near the horizon in the simulations matches very closely the physics of the BZ mechanism (Penna et al. 2013a) as described in the membrane paradigm (Thorne et al. 1986). It also satisfies all the requirements to be viewed as a form of generalized Penrose process (Lasota et al. 2014).

Another quantity of interest is the asymptotic Lorentz factor of the jet, $\gamma_{\rm jet}$. Unfortunately, the value of $\gamma_{\rm jet}$ depends on how much mass is loaded on magnetic field lines, the physics of which is very poorly understood. The current best guess is that mass loading occurs via pair creation through the breakdown of a vacuum gap (Beskin et al. 1992, Hirotani & Okamoto 1998). However, in simulations, mass loading is treated in an entirely ad hoc fashion by applying a minimum "floor" value for the gas density. The resulting jet Lorentz factor tends to be large on the axis and to decrease outward, but the values obtained do not mean much without a physical model of mass loading. Mass loss in the jet $\dot{M}_{\rm jet}$ and jet Lorentz factor $\gamma_{\rm jet}$ are related by $P_{\rm jet} \approx \gamma_{\rm jet}\dot{M}_{\rm jet}c^2$ (including the rest mass energy of the ejected gas). Whereas $P_{\rm jet}$ can be calculated with reasonable confidence from simulations, neither $\dot{M}_{\rm jet}$ nor $\gamma_{\rm jet}$ can be estimated reliably.

3.3.2. Quasi-relativistic disk jet. Surrounding the relativistic BZ jet discussed in Section 3.3.1 is a quasi-relativistic disk jet. The disk jet is distinct in two key ways: (*a*) It is matter dominated, not Poynting flux dominated, and (*b*) it is powered by the inner regions of the accretion disk, not directly by the black hole [the recent detection of Doppler-shifted X-ray emission lines in the candidate BHB 4U1630–47 (Diaz Trigo et al. 2013) suggests that at the time of the observations this system may have had a baryon-loaded jet; thus, a disk jet may have been observed and perhaps also an episodic jet (Section 3.3.3)]. There is no unambiguous way to demarcate the two regions, and different authors have used different prescriptions to identify the boundary (e.g., Hawley & Krolik 2006, Tchekhovskoy et al. 2011). The outflow power of the disk jet is typically $< 0.1\dot{M}_{\rm BH}c^2$, and it varies only modestly with the parameters Φ and a_*, in contrast to the steep dependence of the BZ jet (Equation 31). As a result, when $\Phi \ll \Phi_{\rm MAD}$ (extreme SANE limit), the power in the disk jet can be larger than that in the BZ jet, even though the black hole may be spinning rapidly (De Villiers et al. 2005, Sadowski et al. 2013a). Observationally, the disk jet will produce radio emission and will behave in many respects like a relativistic jet. However, its Lorentz factor is usually modest.

The quasi-relativistic disk jet receives energy from the disk via magnetic fields anchored in the accretion flow. In the BP model, if the field lines are angled outward sufficiently with respect to the disk rotation axis, there is a net outward centrifugal force on matter threading the field. As gas is accelerated outward along the rotating field lines, its angular momentum increases, thereby causing further acceleration. The relevance of the BP mechanism to quasi-relativistic outflows in GRMHD simulations has not been explored systematically.

Lynden-Bell (2003) has proposed an alternative magnetic tower mechanism in which a strong toroidal magnetic field is produced by the differential rotation of the accretion flow and the resulting magnetic pressure gradient causes gas to be accelerated away from the disk surface. Structures analogous to a magnetic tower have been seen in some MHD simulations (e.g., Shibata & Uchida 1985, 1986; Kato et al. 2004b). In addition, there is a suggestion (Hawley & Krolik 2006) that acceleration of the disk jet is caused, not by centrifugal force (BP), but by the gradient of magnetic and gas pressure (magnetic tower).

3.3.3. Episodic jet. Observations show that BHBs have two distinct kinds of jets (Fender & Belloni 2004): steady jets and episodic (or ballistic) jets. Episodic jets are most obviously observed in BHBs during the transition from the hard to the soft state (Remillard & McClintock 2006, McClintock & Remillard 2006), often at luminosities close to Eddington. However, there is a hint that these jets can also occur at lower luminosities (Section 4.1.3), e.g., Cyg X-1 (Fender et al. 2006) and Sgr A* (Yusef-Zadeh et al. 2006) (other examples are reviewed in Yuan et al. 2009a). The most distinctive difference between the two jets is that the episodic jet is transient and is in the form of discrete, isolated blobs, whereas the steady jet behaves like a continuous outflow. Other differences have been noted regarding polarization, spectrum, and power (Fender & Belloni 2004).

It is unclear whether the models reviewed in Sections 3.3.1 and 3.3.2 are applicable to episodic jets because the simulations discussed there generally give quasi-steady jets. Numerical MHD simulations of accretion disks in other contexts do produce episodic ejections of magnetized blobs from the disk surface (e.g., Romanova et al. 1998, Kudoh et al. 2002, Kato et al. 2004a, Dyda et al. 2013), although the underlying physics has not been clearly identified. For example, episodic ejections have been found in MHD simulations of accretion disks around young stellar objects, where they may be better understood (e.g., Hayashi et al. 1996, Goodson et al. 1999, Goodson & Winglee 1999).

By analogy with coronal mass ejections in the Sun, which is another example of episodic mass ejection, Yuan et al. (2009a) proposed an MHD model for the formation of episodic jets (for other interesting similarities between coronal mass ejections and AGN jets, see Massi & Poletto 2011). In this model, a flux rope is first formed in the corona as a result of the twisting of magnetic loops emerging from the disk body via the Parker instability. The flux rope is initially in force equilibrium between magnetic tension and magnetic pressure. However, with further twisting of the field lines, a threshold energy is reached. The flux rope jumps upward, causing a reconnection event to occur. This results in substantial enhancement of the magnetic pressure force and weakening of the tension force, causing the flux rope to be ejected. This model is similar to the magnetic tower model (Lynden-Bell 2003) discussed above, except that here it is time dependent and involves a flux rope.

3.4. Disk Wind from Hot Accretion Flows

Outside the quasi-relativistic disk jet and above the main disk body lies the disk wind, which is the bulk of the mass outflow from the disk (Narayan et al. 2012b; Yuan et al. 2012a; Sadowski et al. 2013a). In contrast to the BZ jet and the disk jet, the disk wind is nonrelativistic and moves slowly. However, it occupies a much larger solid angle. The mass loss rate is also quite high, although the rate of outflow of energy is small compared with the power in the BZ jet or the disk jet (Yuan et al. 2012a; Sadowski et al. 2013a). As with the case of the boundary between the BZ jet and the disk jet, there is no unambiguous way to identify the boundary between the disk jet and the disk wind.

Narayan & Yi (1994, 1995a) noted the likelihood that ADAFs will have strong winds, but these authors were unable to derive a quantitative estimate of the amount of mass loss in the wind. Blandford & Begelman (1999) described a family of self-similar solutions with a wide range of assumed outflow efficiencies, again emphasizing the inability of analytical models to say anything definite about the magnitude of mass and energy loss in disk winds. This uncertainty has been a serious bottleneck in the development of 1D models of hot accretion flows and is an important motivation for doing numerical simulations. From a practical standpoint, it is essential to understand the nature of disk winds in hot accretion flows because (*a*) mass loss can strongly affect the dynamics of the accreting gas and (*b*) disk winds can be powerful contributors to AGN feedback (Section 5).

In an important pioneering study of winds from hot accretion flows, Stone et al. (1999) carried out numerical 2D HD simulations and calculated the mass inflow, outflow, and net accretion rates via the following integrals:

$$\dot{M}_{\rm in}(r) = -2\pi r^2 \left\langle \int_0^\pi \rho \min(v_r, 0) \sin\theta \mathrm{d}\theta \right\rangle_{t\phi}, \tag{32}$$

$$\dot{M}_{\rm out}(r) = 2\pi r^2 \left\langle \int_0^\pi \rho \max(v_r, 0) \sin\theta \mathrm{d}\theta \right\rangle_{t\phi}, \tag{33}$$

$$\dot{M}_{\rm net} = \dot{M}_{\rm in} - \dot{M}_{\rm out}, \tag{34}$$

where the angle brackets represent an average over time (and also azimuthal angle ϕ in the case of 3D simulations). The quantity $\dot{M}_{\rm net}$ is the net mass accretion rate; in steady state, it is equal to the accretion rate on the black hole $\dot{M}_{\rm BH}$. Stone et al. (1999) found, as has been confirmed in many later simulations (see references below), that both $\dot{M}_{\rm in}$ and $\dot{M}_{\rm out}$ decrease inward, roughly following a power-law behavior (see Equation 6),

$$\dot{M}_{\rm in}(r) = \dot{M}_{\rm in}(r_{\rm out}) \left(\frac{r}{r_{\rm out}}\right)^s, \quad s > 0 \tag{35}$$

(see **Figure 5**). Correspondingly, the radial profile of density becomes flatter than in a self-similar ADAF solution: $\rho(r) \propto r^{-p}$ with $p < 1.5$. These statements apply to hydro simulations (Stone et al. 1999, Yuan & Bu 2010, Yuan et al. 2012b, Li et al. 2013a) as well as MHD simulations (Hawley et al. 2001, Machida et al. 2001, Stone & Pringle 2001, Hawley & Balbus 2002, Igumenshchev et al. 2003, Pen et al. 2003, Kato et al. 2004b, Pang et al. 2011, Yuan et al. 2012a).

The values of s and p in various simulations are summarized in Yuan et al. (2012b): $s = 0.4$–0.8, $p = 0.5$–1. The variations between different simulations seem to be due to differences in the value of the viscosity parameter α (in the case of hydro simulations), choice of Newtonian gravity versus general relativity, initial configuration of the magnetic field (toroidal or poloidal), and the strength of the initial field (weak or strong). Observationally, there is support for a value of $s \sim 0.3$, $p \sim 1$ (see the discussion of Sgr A* in Section 4.1). The apparent discrepancy between theory and observations may be due to the low angular momentum of the accretion flow in Sgr A* (Bu et al. 2013) or the dynamical importance of thermal conduction (Section 3.6) (Johnson & Quataert 2007).

Competing models have been proposed to explain the radially varying inflow and outflow rates seen in numerical simulations. In the ADIOS model (Blandford & Begelman 1999, 2004; Begelman 2012), the inward decrease of $\dot{M}_{\rm in}$ is due to a genuine mass loss in a wind. What drives the wind is unspecified in the model. Assuming merely that a mechanism exists for draining energy from the interior of the accretion flow to launch a wind, Blandford & Begelman (1999, 2004) constructed 1D and 2D self-similar solutions. In their models, the index s is left as a free parameter, limited only by the condition $0 \le s \le 1$. In the most recent version of the ADIOS model, however, Begelman (2012) considered the inflow and outflow zones on an equal footing and, using a conserved outward energy flux, found that s should be close to unity. This is somewhat larger than the range of values seen in numerical simulations.

An alternative scenario is the CDAF model (Narayan et al. 2000, Quataert & Gruzinov 2000, Abramowicz et al. 2002a, Igumenshchev 2002), which is based on the assumption that a hot accretion flow is convectively unstable (Narayan & Yi 1994). In this model, inward angular momentum transport by convection and outward transport by viscous stresses almost cancel each other. A convective envelope then forms that has a conserved (outward) convective luminosity and a flat density profile. The gas constantly moves in and out in turbulent convective eddies, and this motion gives the impression that there are large fluxes of inflowing and outflowing matter. However, none of the outgoing gas really escapes, and the net accretion rate is quite small. Whether the

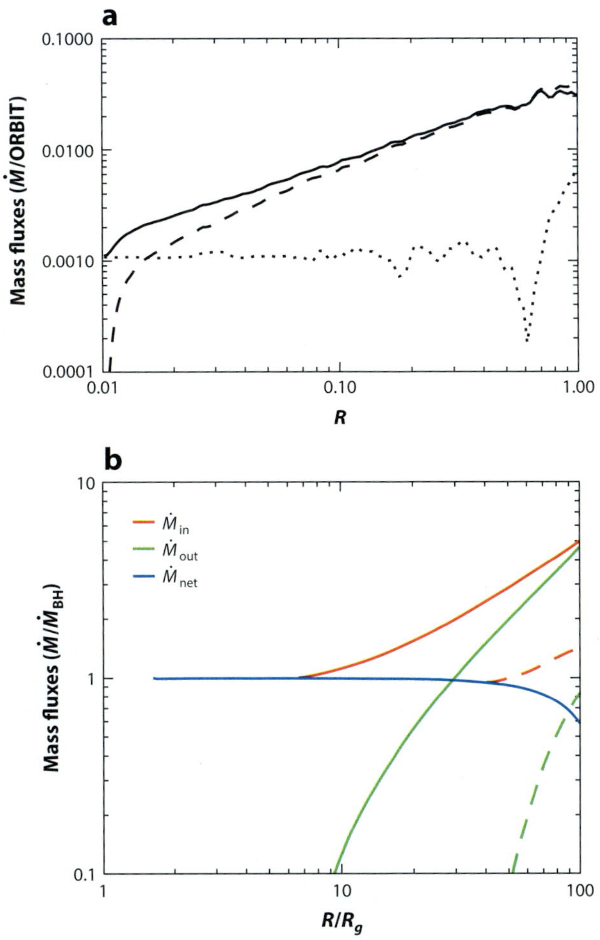

Figure 5

Radial profiles of mass inflow rate $\dot{M}_{\rm in}$, mass outflow rate $\dot{M}_{\rm out}$, and net mass accretion rate $\dot{M}_{\rm net}$ (Equations 32–34). (*a*) Results from a two-dimensional Newtonian HD simulation of a hot accretion flow (Stone et al. 1999). Solid, dashed, and dotted lines correspond to $\dot{M}_{\rm in}$, $\dot{M}_{\rm out}$, and $\dot{M}_{\rm net}$, respectively. Note the rapid increase of $\dot{M}_{\rm out}$ with increasing radius. (*b*) Solid lines indicate equivalent results from a three-dimensional GRMHD simulation of a hot accretion flow around a nonspinning black hole ($a_* = 0$) (Narayan et al. 2012b). Mass outflow becomes important only beyond a radius $\sim 30R_g$, though the slope outside this radius is similar to that in panel *a*. Dashed lines indicate results for a different kind of time averaging, as described in the text. Here the estimated mass outflow rate is significantly smaller (see also Yuan et al. 2012a, their figure 14, top-left panel). The true mass outflow rate is likely to fall in between the solid and dashed green lines. Abbreviations: BH, black hole; GRMHD, general relativistic magnetohydrodynamic; HD, hydrodynamic.

CDAF model can be applied to MHD accretion flows remains unresolved (Stone & Pringle 2001, Hawley & Balbus 2002, Narayan et al. 2002).

The ADIOS and CDAF scenarios are very different from each other, so numerical simulations ought to be able to discriminate between them. In this context, the key question is, How strong is the "real" wind in a simulated hot accretion disk? This is not easy to answer. Returning to Equations 32 and 33, note that the integrals are computed at each instant of time using instantaneous velocities, and the integrals are then averaged over t and ϕ to obtain $\dot{M}_{\rm in}(r)$ and $\dot{M}_{\rm out}(r)$.

This procedure gives undue importance to turbulent motions. Especially at large radii, where a given turbulent eddy will consist of roughly half the gas moving in and half moving out, one is likely to overestimate both the inflow rate and outflow rate. A parcel of gas that is moving out at a particular time will likely soon turn around and begin to flow in. Thus, the inflow and outflow rates estimated via Equations 32 and 33 will both be overestimates of the true values.

An alternative approach is to move the $t\phi$ average inside the integrals, i.e., to integrate $\min(\langle \rho v_r \rangle_{t\phi}, 0)$ and $\max(\langle \rho v_r \rangle_{t\phi}, 0)$ (see Narayan et al. 2012b). This eliminates contributions from the to and fro motion due to turbulence and, not surprisingly, produces substantially lower estimates for the mass outflow rate (see **Figure 5**). However, this procedure also is problematic: It could underestimate the real outflow rate if some genuine outflowing streams wander around in 3D space (Yuan et al. 2012a). Therefore, the estimated mass outflow rate obtained by this method is a lower limit.

Yuan et al. (2012a) presented an alternative way to roughly estimate the strength of the wind. They calculated and compared the various properties of inflow and outflow, such as angular momentum and temperature, and found that the properties are quite different. However, if the inflowing and outflowing motion were dominated by turbulence, the properties would be roughly similar. Therefore, they concluded that systematic outflow must exist and the rate of real outflow should be a significant fraction of that indicated by Equation 33. [In the case of convection (the HD case in Yuan et al. 2012), some differences are expected between inflow and outflow, but this might not affect the final conclusion.] Yuan et al. (2012a) investigated the production mechanism of the wind. On the basis of the much larger angular momentum of outflow compared with that of inflow, they argued that the magnetocentrifugal force must play an important role. They also briefly discussed the velocity of the wind (see also Li et al. 2013a).

An influential concept in theoretical discussions of outflows is the Bernoulli parameter *Be*, which is the sum of the kinetic energy, potential energy, and enthalpy. It measures the ratio of energy flux to mass flux. For a steady inviscid HD flow, *Be* is conserved along streamlines. Therefore, any parcel of gas with a positive *Be* can escape to infinity, whereas a parcel with negative *Be* cannot. One-dimensional hot accretion flow models often have $Be > 0$ (Narayan & Yi 1994, 1995a; Blandford & Begelman 1999), which is interpreted as a strong clue that these flows should experience heavy mass loss in winds. However, note that *Be* is not conserved if the flow is either viscous or nonsteady. Therefore, *Be* is not a useful parameter for describing gas in the interior of a turbulent disk. The situation is somewhat better in the case of outflows, which tend to be more laminar and quasi-steady.

In the area of numerical simulations of nonradiative accretion flows, Igumenshchev & Abramowicz (2000) were among the first to explore the connection between *Be* and outflows. They found that HD simulations with a large value of the viscosity parameter $\alpha \gtrsim 0.3$ have well-defined outflows with $Be > 0$, whereas simulations with a smaller $\alpha \lesssim 0.1$ have outflowing gas with $Be < 0$, i.e., the outward-moving gas in the latter models is gravitationally bound and cannot escape to infinity. More recently, Yuan et al. (2012a) carried out a detailed study of *Be* in HD and MHD simulations. They found that in the HD case the value of *Be* of outflowing gas is always larger than that of inflowing gas.

In MHD flows, the definition of *Be* must be modified to include the contribution of the magnetic field. The necessary expression is well-known in the theory of relativistic hydromagnetic winds, e.g., the "total energy-to-mass flux ratio" μ in Vlahakis & Konigl (2003; also J in Lovelace et al. 1986). Sadowski et al. (2013a) analyzed 3D GRMHD simulations using a general relativistic version of μ. They found that gas with $\mu > 0$ has an outward-pointing velocity (outflow), whereas gas with $\mu < 0$ has an inward-pointing velocity (inflow). This result appears to confirm the usefulness of *Be* as a diagnostic for MHD winds in hot accretion flows. Note that the analysis was carried out using time-averaged quantities in quasi-steady state, where *Be* is expected to be particularly well conserved.

In addition to directly estimating the strength of mass outflows in simulated hot accretion flows, the convective stability of the gas in MHD simulations may be analyzed using the Hoiland criteria (Narayan et al. 2012b; Yuan et al. 2012a). This reveals that gas is convectively stable over most of the accretion flow, in contrast to HD accretion simulations, which are convectively unstable (Section 3.1). Thus, the magnetic field in MHD simulations appears to stabilize the gas against convection. Pen et al. (2003) termed this state of affairs "frustrated convection."

On the whole, current results seem to favor the ADIOS model over the CDAF model, but not overwhelmingly. The truth likely involves some combination of the two models. More work is required to clarify this issue. In particular, simulations covering a significantly larger dynamic range in radius than currently possible will be required before we can hope to obtain unambiguous results.

Three-dimensional simulations with a large dynamic range are especially important for estimating two critical parameters: (*a*) the mass loss index s and (*b*) the radius $R_{\rm in}$ inside which mass loss is unimportant. Knowing the values of these parameters is vital to calculate quantitative global models of hot accretion flows (Section 2.4) and to apply these models to real systems (Section 4).

3.5. Effect of Radiation

Radiation is nearly always ignored in hydro and MHD simulations of hot accretion flows. However, a few studies have considered optically thin radiative cooling. Yuan & Bu (2010) included bremsstrahlung radiation in the energy equation in their HD simulations and recovered the ADAF and LHAF solutions when they varied the mass accretion rate. Surprisingly, their simulated LHAF was convectively unstable, whereas 1D models predict that the entropy gradient should be stable (Section 2.6). Apparently, the 2D structure of the flow permits an unstable entropy gradient to survive, although this behavior is not understood. Li et al. (2013a) also included bremsstrahlung cooling and showed that, by changing the mass supply rate outside the Bondi radius, they could successfully reproduce both a cool thin disk at high $\dot{M}_{\rm BH}$ and a hot accretion flow at lower $\dot{M}_{\rm BH}$.

Ohsuga and collaborators (e.g., Ohsuga et al. 2009, Ohsuga & Mineshige 2011) have carried out simulations with full radiative transfer and have studied a wide range of accretion rates. They recovered both the cold and hot accretion flow solutions at appropriate values of $\dot{M}_{\rm BH}$. Radiation generally does not appear to have a significant effect on the dynamics of their hot solutions. However, in a recent study, Dibi et al. (2012) included optically thin cooling and found that, when $\dot{M}_{\rm BH} \gtrsim 10^{-7}\dot{M}_{\rm Edd}$, radiative cooling can significantly affect the density and temperature. Their result is likely to be sensitive to the particular prescription they used to fix the electron temperature in the two-temperature plasma. Nevertheless, their study highlights the fact that, above some accretion rate, numerical simulations need to include radiation self-consistently.

3.6. Effect of Low Collisionality

Most studies of hot accretion flows are based on a fluid approximation, specifically MHD. However, the density of the accreting gas is often so low that the flow is macroscopically collisionless. As a result, a kinetic analysis is required to determine whether MHD simulations can capture the relevant physics.

For the MRI, a kinetic treatment is, in principle, required whenever the wavelength of the fastest growing mode is smaller than the collisional mean free path. Quataert et al. (2002) found that, whereas the MRI instability criterion is the same in kinetic theory as in MHD, the growth rates of the modes are different. The nonlinear development of the kinetic MRI has been studied using numerical simulations based on a fluid model with kinetic effects added (Sharma et al. 2006) as well as with the more precise particle-in-cell technique (Riquelme et al. 2012). The nonlinear evolution of the axisymmetric kinetic MRI is qualitatively similar to that of the standard MHD MRI.

The low collisionality of hot accretion flows also has an effect on thermal conduction. For a magnetized collisionless accretion flow, the collisional mean free path of electrons is larger than the electron Larmor radius, and thermal conduction is the dominant mode of heat transport. Conduction tends to occur primarily along the magnetic field lines, although cross-field diffusion is not as small as often imagined (Narayan & Medvedev 2001). Anisotropic conduction modifies the convective stability criterion, as shown by Balbus (2001). The instability in this case is usually referred to as the magnetothermal instability (MTI). Local MHD simulations with anisotropic electron thermal conduction have demonstrated that the MTI amplifies the magnetic field and causes a substantial convective heat flux (Parrish & Stone 2007). Sharma et al. (2008) investigated the effects of the MTI on nonrotating accretion flows and confirmed the main results of local simulations. Bu et al. (2011) extended this study to a rotating accretion flow and found that the MTI and MRI operate independently and can cooperatively amplify the magnetic field. Thermal conduction in a hot collisionless accretion flow can also directly affect the dynamics by flattening the temperature profile (e.g., Quataert 2004, Tanaka & Menou 2006, Johnson & Quataert 2007). This mechanism has been invoked as an alternative explanation for the very low mass accretion rate in Sgr A* (Section 4.1) (Shcherbakov & Baganoff 2010).

4. APPLICATIONS

4.1. Galactic Center Black Hole: Sgr A*

Sgr A*, the compact radio source at the center of our Galaxy, is a unique laboratory for studying black hole accretion. Observations of O and B stars orbiting the Galactic Center (Schödel et al. 2002; Ghez et al. 2003, 2008; Gillessen et al. 2009a,b; Meyer et al. 2012) provide very strong evidence for the presence of a dark compact object of mass $(4.1 \pm 0.4) \times 10^6\,M_\odot$. Measurements of the size of Sgr A* (Bower et al. 2004, Shen et al. 2005, Doeleman et al. 2008) leave little doubt that this dark object must be a supermassive black hole. Because Sgr A* is relatively nearby, there are abundant data to constrain the nature of the accretion flow (for details on the observations, see Genzel et al. 2010 and references therein).

Sgr A* spends most of its time in a steady low-luminosity state, usually referred to as the "quiescent state" (Section 4.1.1). A few times each day, strong variations are seen in the infrared and X-ray bands and sometimes also in other wavebands. These fluctuations are referred to as "flares" (Section 4.1.3). The ADAF model and its variants explain the main features of the quiescent state. However, despite important recent progress, the nature of the flares is still poorly understood.

4.1.1. Observational constraints on the quiescent state. The outer boundary of the accretion flow around Sgr A* is generally assumed to be located at the Bondi (1952) radius, $R_B \sim 10^5 R_S \approx 0.04$ pc $\approx 1''$, where the thermal energy of the external ambient gas is equal to its potential energy in the gravitational field of the black hole. Because of the high spatial resolution of the *Chandra X-ray Observatory*, the density and temperature of gas near R_B can be measured using X-ray observations (Baganoff et al. 2003), thereby enabling us to estimate the Bondi mass accretion rate: $\dot{M}_B \sim 10^{-5}\,M_\odot$ year$^{-1} \sim 10^{-4}\,\dot{M}_{\rm Edd}$. Independently, 3D numerical simulations of stellar winds accreting onto Sgr A* (Cuadra et al. 2008; see also Quataert 2004, Shcherbakov & Baganoff 2010)

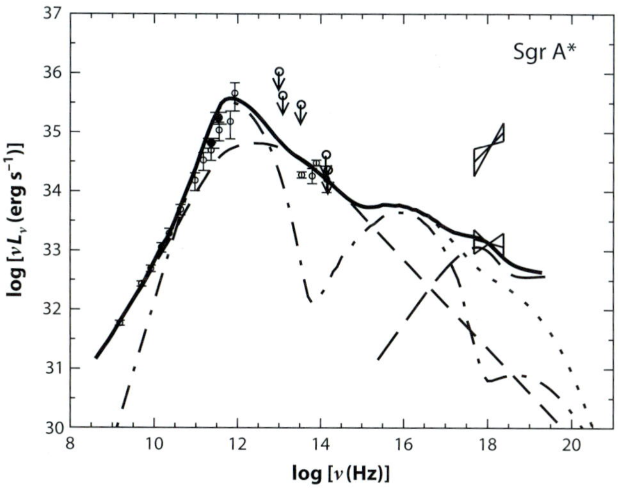

Figure 6

Spectrum corresponding to an advection-dominated accretion flow model of the quiescent state of Sagittarius A* (Sgr A*), the supermassive black hole at the Galactic Center (from Yuan et al. 2003). Circles with error bars show measurements at radio and millimeter wavelengths, circles with arrows correspond to infrared upper limits, and the two "bowties" show X-ray data in the quiescent state (*below*) and during a bright flare (*above*). Additional data are shown in the infrared waveband (from Schödel et al. 2011), which were not available when the model was originally developed. The model spectrum (*thick solid line*) is the sum of three components: synchrotron emission and its Compton humps from thermal electrons (*dot-dashed line*), synchrotron emission from a population of nonthermal electrons (*short-dashed line*), and bremsstrahlung emission from electrons near the Bondi radius (*long-dashed line*). The dotted line indicates the total synchrotron and inverse Compton emission.

predict $\dot{M}_B \sim$ few $\times 10^{-6}\, M_\odot$ year^{-1}, consistent with the above estimate. The simulations indicate that the gas has a reasonable amount of angular momentum at the Bondi radius, corresponding to a circularization radius $\sim 10^4 R_s$. Thus, gas cannot fall directly into the black hole, as in the classic Bondi model, but must accrete viscously via a hot accretion flow. The measured density and temperature of gas at the Bondi radius, and its estimated specific angular momentum, constitute outer boundary conditions that a successful accretion model must satisfy.

The spectral energy distribution of Sgr A* is shown in **Figure 6**. The radio spectrum has two components: Below ~50 GHz, the spectrum consists of a power law, and above this frequency, there is a "submillimeter bump." The X-ray emission in the quiescent state is spatially resolved (~1.4″) (Baganoff et al. 2003), consistent with the size of the Bondi radius. The bolometric luminosity of Sgr A* is extremely low. $L_{\mathrm{bol}} \sim 10^{36}\, \mathrm{erg\, s^{-1}} \sim 2 \times 10^{-9}\, L_{\mathrm{Edd}}$. If gas accretes on the black hole at the Bondi accretion rate $\dot{M}_B$ estimated above, the radiative efficiency of the accretion flow must be extremely low: $\epsilon_{\mathrm{Sgr\,A^*}} \sim 10^{-6}$, instead of the traditional $\epsilon_{\mathrm{SSD}} \approx 10\%$. This ultralow efficiency is the strongest argument for invoking an ADAF or other hot accretion flow model instead of a standard thin accretion disk. Another argument is that the observed spectrum does not look anything like the multitemperature blackbody spectrum expected for a standard thin disk. Note, however, that strong extinction in the optical and ultraviolet (UV) could hide much of the emission from a thin disk, so constraints come mainly from infrared observations (Falcke & Melia 1997).

The radio emission at submillimeter and millimeter wavelengths is linearly polarized at a level of 2–9% (Aitken et al. 2000; Bower et al. 2003; Marrone et al. 2006, 2007), and the mean rotation measure between 227 and 343 GHz is $-5.6 \pm 0.7 \times 10^5$ rad m^{-2}. The latter measurement limits the mass accretion rate at the central black hole (Agol 2000, Quataert & Gruzinov 2000). Current constraints are $\dot{M}_{BH} < 2 \times 10^{-7} M_\odot$ year^{-1} if the magnetic field is near equipartition, ordered, and largely radial, and $\dot{M}_{BH} > 2 \times 10^{-9} M_\odot$ year^{-1} if the field is subequipartition, disordered, or toroidal (Marrone et al. 2007). Because these estimates of $\dot{M}_{BH}$ are significantly smaller than $\dot{M}_B$, most of the gas available at R_B apparently does not fall into the black hole. Recently, emission lines from relatively low ionization species were detected (Wang et al. 2013) during long *Chandra* observations. The H-like Fe Kα line was extremely weak, indicating a flat radial density profile $\rho(r) \propto r^{-0.5}$ near the Bondi radius, rather than $r^{-3/2}$ as expected for the classic ADAF model. The flat density profile seems to confirm that the mass accretion rate decreases with decreasing radius (Section 3.4), consistent with millimeter-wave polarization results. Both observations suggest that the accretion flow has a significant outflow that causes $\dot{M}$ to decrease with decreasing radius (Section 3.4).

4.1.2. ADAF model for the quiescent state of Sgr A*.

Narayan et al. (1995; see also Manmoto et al. 1997; Narayan et al. 1998a) applied the ADAF model to Sgr A* and showed that the model explains the main features of the source, e.g., an ultralow radiative efficiency and an unusual (nonthin disk) spectrum. In these early studies, the accretion rate was taken to be independent of radius, and viscous dissipation was assumed to heat only ions ($\delta \ll 1$) (Section 2.3). The most serious defect of this model is that it predicts a rotation measure orders of magnitude larger than that observed (Section 4.1.1).

In the years following this early work, three separate developments led to a new paradigm: (*a*) Numerical simulations demonstrated that outflows are important (Section 3.4), (*b*) electron heating was recognized to be more efficient than previously assumed ($\delta \sim 0.1$–0.5, Section 2.3), and (*c*) Faraday rotation measurements indicated that $\dot{M}_{BH} \ll \dot{M}_B$ (Section 4.1.1). Yuan et al. (2003, 2004) presented an updated ADAF model of Sgr A* (sometimes called a radiatively inefficient accretion flow model to distinguish it from the "old" ADAF model) that allowed for an outflow and assumed more efficient electron heating. The model spectrum is shown in **Figure 6**. The submillimeter bump is produced by synchrotron emission from thermal electrons in the ADAF (the additional bumps at higher frequencies are due to inverse Compton scattering), whereas the low-frequency power-law radio spectrum is produced by synchrotron radiation from a small population of nonthermal electrons (following earlier suggestions by Mahadevan 1998 and Özel et al. 2000). The nonthermal electrons are usually introduced into the model in an ad hoc fashion, but they are thought to be the result of electron-positron production via proton-proton collisions (Mahadevan 1998, 1999), magnetic reconnection, or weak shocks (e.g., Machida & Matsumoto 2003). Bremsstrahlung emission of thermal gas near the Bondi radius is responsible for the X-ray emission (Quataert 2002). Sazonov et al. (2012) recently proposed that the extended X-ray emission may be dominated by coronal radiation from a population of low-mass stars. However, Wang et al. (2013) did not detect the predicted level of Fe Kα emission.

The net radiative efficiency of the model described by Yuan et al. (2003) is very low: $L_{\rm bol}/[\dot{M}_{\rm in}(R_B)c^2] \approx 2 \times 10^{-5}$. The low efficiency is the result of two effects: (*a*) Mass loss in an outflow gives an effective $\epsilon_{\rm outflow} \equiv \dot{M}_{\rm BH}/[\dot{M}_{\rm in}(R_B)] \sim 4 \times 10^{-2}$, and (*b*) energy advection gives an additional "real efficiency" (Equation 17) $\epsilon \equiv L_{\rm bol}/\dot{M}_{\rm BH}c^2 \sim 4 \times 10^{-4}$. The inclusion of an outflow in the model is consistent with independent modeling of X-ray emission lines (Wang et al. 2013) as well as numerical MHD simulations (Section 3.4). However, there is no direct observational evidence for any outflowing gas. Overall, the success of this (by no means unique) model provides strong evidence for the presence of a hot ADAF around Sgr A*. It has also been used to argue for the existence of an event horizon in this object (Narayan et al. 1998a, Broderick & Narayan 2006, Narayan & McClintock 2008, Broderick et al. 2009; for related arguments in the case of quiescent BHBs, see also Narayan et al. 1997b, Menou et al. 1999a, Garcia et al. 2001, McClintock et al. 2003; for a counterargument, see Abramowicz et al. 2002b).

More constraints on model parameters such as the mass accretion rate, the relative importance of the disk versus the jet (Section 4.1.4), as well as the orientation and magnitude of the black hole spin and the magnetic field could be obtained by considering additional millimeter-VLBI (very long baseline interferometry) observations and polarization data (Section 4.1.5). This has been done using the semianalytical model of Yuan et al. (2003) (Huang et al. 2007, Broderick et al. 2011a) and in numerous studies based on MHD simulations (Goldston et al. 2005; Noble et al. 2007; Sharma et al. 2007b, 2008; Chan et al. 2009; Kato et al. 2009; Mościbrodzka et al. 2009; Dexter et al. 2010; Shcherbakov et al. 2012; Dexter & Fragile 2013). In most of the latter studies, the simulations do not include radiation, so the emergent spectrum is calculated separately by postprocessing the simulation output.

4.1.3. Multiwaveband flares. Flares have been observed in Sgr A* in many wavebands, but they are strongest in X-rays (Baganoff et al. 2001) and infrared (Genzel et al. 2003, Ghez et al. 2004, Gillessen et al. 2006), where the flux can increase by up to a factor of 100 and 5, respectively. The variability timescale ranges from several minutes to 3 h, indicating that the flares must be produced close to the black hole. For comparison, the light-crossing time of the black hole is $2R_S/c \approx 30$ s, and the orbital period at the ISCO for a nonrotating black hole is $2\pi/\sqrt{GM/R_{\mathrm{ISCO}}^3} \approx 2{,}000$ s. Many multiwaveband campaigns (e.g., Eckart et al. 2004, 2006; Yusef-Zadeh et al. 2006, 2009; Dodds-Eden et al. 2009; Trap et al. 2011) have provided valuable information on flare spectra, polarization, time lags between different wavebands, and occurrence rates. The observations suggest that flares in different wavebands are likely physically related.

So far, most theoretical flare models are phenomenological and have focused mainly on interpreting the observed spectrum (e.g., Markoff et al. 2001; Yuan et al. 2004; Dodds-Eden et al. 2009, 2010). While the infrared flare is generally believed to be due to pure synchrotron emission, there is still some debate on whether the X-ray flare is due to synchrotron (Yuan et al. 2003, 2004; Dodds-Eden et al. 2009), synchrotron self-Compton, or inverse Compton of external radiation (Markoff et al. 2001; Eckart et al. 2004, 2006; Yusef-Zadeh et al. 2009). Yusef-Zadeh et al. (2006) found that, in the radio band, the peak flare emission at 43 GHz leads that at 22 GHz by ~20–40 min. They interpret this in terms of a van der Laan (1966)-like expanding plasma blob model. The blob is ejected from the accretion flow and becomes optically thin as it expands. The maximum emission at any given frequency occurs when the blob transitions from optically thick to thin at that frequency. Thus, the peak naturally occurs later at longer wavelengths. A similar process may be happening in GRS 1915+105 and other black hole sources (Fender, Belloni & Gallo 2004; see references in Yuan et al. 2009a), which strongly suggests episodic jet ejections (Section 3.3.3). By analogy with coronal mass ejections associated with solar flares, the ejections in Sgr A* may be caused by magnetic reconnection in a corona, and the same process may also be responsible for the flares (Yuan et al. 2009a). More multiwaveband observations will clarify many remaining puzzles. Theoretical models that combine detailed gas dynamics with radiative processes are also needed.

4.1.4. Alternative models of Sgr A*. Two alternative models of Sgr A* have been discussed in the literature: the jet model (Falcke & Markoff 2000, Markoff et al. 2001, Yuan et al. 2002a, Markoff et al. 2007) and the spherical accretion model (Melia 1992, Melia et al. 2001). One version of the jet model (Falcke & Markoff 2000, Markoff et al. 2007) proposes that all the emission from radio to X-rays is produced by the jet. In an alternate version, called the jet-ADAF model (Yuan et al. 2002a), only the radio spectrum below ~50 GHz is produced by the jet, whereas the rest of the emission is assumed to come from the ADAF. No radio jet has been convincingly detected in Sgr A* (but see Li et al. 2013b), even though this supermassive black hole is in our Galaxy and has been observed in the radio with both high sensitivity and very high spatial resolution.

The spherical accretion model (Melia 1992) is similar to the ADAF model in that the accretion flow is assumed to be very hot, with the temperature being nearly virial. However, in contrast to the ADAF and other hot accretion flow models, the gas in the spherical accretion model is one temperature. In addition, the angular momentum of the gas is assumed to be extremely small, with a circularization radius of only $\sim 5-10R_s$, which is rather extreme. If gas at the Bondi radius has any reasonable angular momentum, as seems likely (e.g., the numerical work of Cuadra et al. 2008), accretion cannot take place spherically but must proceed via a viscous rotating flow.

4.1.5. Future observations. As the nearest supermassive black hole, Sgr A* has always been a favorite target for observational campaigns. Two near-term opportunities may provide significant new information on the nature of the accretion flow around this black hole. Gillessen et al. (2012) discovered a dense cloud of gas called G2 on a highly eccentric orbit around Sgr A*. Pericentric passage occurred in early 2014 at a distance of $\sim 2000R_S$ from the black hole (Phifer et al. 2013, Gillessen et al. 2013). The interaction of G2 with the hot accretion flow could potentially be observed in radio, infrared, or X-rays (Narayan et al. 2012a, Crumley & Kumar 2013, Sadowski et al. 2013b, Saitoh et al. 2014, Shcherbakov 2014, Yusef-Zadeh & Wardle 2013). If a signal is detected, it will provide information on the properties (density, temperature, magnetic field) of the accretion flow in a region ($R \gtrsim 10^3 R_S$) where we have hitherto had no observational constraint. In addition, gas stripped from G2 is expected to accrete on the black hole on a viscous time, causing the quiescent radio and millimeter emission of Sgr A* to increase. Detection of this increase will provide direct measurement of the viscous time at the orbit pericenter, which is currently poorly constrained. A number of HD and GRMHD simulations of the G2 encounter have already been carried out (e.g., Anninos et al. 2012; Burkert et al. 2012; Moscibrodzka et al. 2012; Schartmann et al. 2012, 2013; Abarca et al. 2014; Sadowski et al. 2013b). Coupled with future observations, such work should provide new information on the nature of the accretion flow in Sgr A* at intermediate radii.

In recent years, ultrahigh angular resolution, millimeter-wave interferometry has become a reality, and the first detections have been made of event-horizon-scale structure in Sgr A* (Doeleman et al. 2008, Fish et al. 2011) and M87 (Doeleman et al. 2012). From these observations, some inferences have been made on the spins of the black hole and the nature of the accretion flow and jet near the horizon (e.g., Fish et al. 2009, 2011; Broderick et al. 2009, 2011a,b). Future plans are focused on commissioning the Event Horizon Telescope (Doeleman et al. 2010), which will carry out long baseline millimeter-wave interferometry using a network of up to eight telescopes spread all around the world. With this array, images of Sgr A* will be measured with unprecedented sensitivity and angular resolution, and information will be obtained both on polarization and time variability. Thus, the Event Horizon Telescope may finally provide direct information on the physics of the hot accretion flow in the vicinity of Sgr A*'s horizon. Questions such as the relative importance of the disk versus the jet; the density, temperature, and optical depth (at millimeter wavelengths) of the accreting plasma; and the direction and topology of the magnetic field will hopefully be answered by means of direct observations. In anticipation, numerical codes are being developed, and some numerical simulations have been carried out (e.g., Mościbrodzka et al. 2009, Dexter et al. 2010, Shcherbakov et al. 2012, Chan et al. 2013, Dexter & Fragile 2013). Much more work is anticipated in the future.

4.2. Other Low-Luminosity Sources

The hot accretion flow solution is essentially independent of black hole mass. If $\dot{M}_{\rm BH}$ is scaled by the Eddington rate and radius is scaled by the Schwarzschild radius, many properties of the solution, notably the gas temperature and the Eddington-scaled luminosity, are independent of mass (see Section 2.1). Accordingly, hot accretion flow systems, such as LLAGNs and BHBs in hard and quiescent states, show similar properties, despite the large disparity in their black hole masses. We begin by introducing LLAGNs and BHBs and then describe the role of hot accretion flows in determining their properties.

4.2.1. LLAGNs and BHBs. Although virtually every galaxy with a bulge has a supermassive black hole in its nucleus, at any given time only a small fraction of these black holes have luminosities close to Eddington. The vast majority are LLAGNs with luminosities spanning the range $L_{\rm bol}/L_{\rm Edd} \approx 10^{-9}$–$10^{-1}$ (Ho 2008, 2009). As with Sgr A*, most of these black holes have considerable gas available for accretion close to their Bondi radii (Fabian & Canizares 1988). The fact that the black holes are dim thus suggests that they must be accreting via a radiatively inefficient mode, i.e., a hot accretion flow (Fabian & Rees 1995, Di Matteo et al. 2000, Ho 2009, Russell et al. 2013b). Other distinctive features of LLAGNs confirm this suspicion.

The "big blue bump," a characteristic spectral feature associated with a thin accretion disk around a supermassive black hole, is absent in LLAGNs (Ho 1999, 2008; Chiaberge et al. 2006; Eracleous et al. 2010; Younes et al. 2012). In the language of $\alpha_{\rm ox}$, defined as the two-point spectral index between 2,500 Å and 2 keV, LLAGNs have $\alpha_{\rm ox} \gtrsim -1$, whereas quasars and Seyferts have $\alpha_{\rm ox} \approx -1.4$ and -1.2, respectively. The optical-UV slope is also exceptionally steep (Ho 2008). These observations strongly suggest that LLAGNs do not have a thin disk in their inner regions; however, a disk may be present at larger radii, as indicated by a "big red bump" in their spectra (Lasota et al. 1996a, Gammie et al. 1999, Quataert et al. 1999, Yuan et al. 2002b, Chiang & Blaes 2003, Ptak et al. 2004, Yuan & Narayan 2004, Nemmen et al. 2006, Wu et al. 2007, Yu et al. 2011). [Given that LLAGNs lie on the low-luminosity extrapolation of the well-known relation between $\alpha_{\rm ox}$ and luminosity, Maoz (2007) argued that LLAGNs do not differ appreciably from luminous AGNs and, hence, that LLAGN accretion disks are similar to disks in luminous AGNs. However, he does not provide a physical explanation for the correlation. By contrast, Yu et al. (2011) showed that the trend is naturally explained in the framework of hot accretion flows.] As further confirmation, the iron Kα line, which is commonly attributed to X-ray fluorescence off a cold accretion disk extending close to the black hole, is weak or absent (e.g., Fabbiano et al. 2003, Ptak et al. 2004, Binder et al. 2009, Younes et al. 2011, Kawamuro et al. 2013).

BHBs have a number of distinct spectral states (for reviews, see Zdziarski & Gierliński 2004, Done et al. 2007, McClintock & Remillard 2006, Zhang 2013, Poutanen & Veledina 2014). The most notable among these, in order of decreasing luminosity, are the soft or thermal state, the hard state, and the quiescent state. (There is also a very high state or steep power-law state which usually, but not always, occurs at an even higher luminosity than the thermal state. This state is poorly understood.) The soft state is found at luminosities down to $\sim$1.5% $L_{\rm Edd}$ (Kalemci et al. 2013). It is characterized by a strong blackbody or thermal spectrum and is well described by the standard thin disk model. The hard state is found at lower luminosities and differs dramatically. Its spectrum has only a weak blackbody component and is dominated by a strong hard power law with a cutoff at $\sim$100 keV. Whereas the soft state has little time variability, the hard state is highly variable and often exhibits QPOs (Remillard & McClintock 2006). In addition, the hard state has a continuous, steady jet, whereas the soft state almost never has jets (Fender 2006). All these differences indicate that the hard state must correspond to a very different regime of accretion compared with the thin disk.

Many studies have shown that the power-law component in the hard state must be produced by thermal Comptonization in a hot plasma with a temperature $kT \sim 100$ keV and optical depth $\tau \sim 1$ (e.g., Zdziarski et al. 1998). The ADAF model has the correct density, electron temperature, and optical depth needed to reproduce the observed spectrum. Moreover, the luminosity at which a fading thermal state source switches to the hard state ($\sim 1.5\% L_{\rm Edd}$) is reasonably consistent with the maximum mass accretion rate of a hot accretion flow (**Figure 2**). At luminosities below $\sim 10^{-3} L_{\rm Edd}$, the hard state merges smoothly with the quiescent state, which then extends down to as low as $L \sim 10^{-9} L_{\rm Edd}$. There is no clear boundary between the hard and quiescent states, so it is likely that the same accretion physics operates in both.

BHBs show a very interesting hysteresis effect in their transitions between the soft and hard state. Whereas with decreasing luminosity the soft-to-hard transition in most sources happens at roughly the same luminosity $\sim 1.5\% L_{\rm Edd}$, with increasing luminosity the hard-to-soft transition can occur at any of a wide range of luminosities extending up to $L \sim 30\% L_{\rm Edd}$ (Zdziarski & Gierliński 2004, Done et al. 2007, Yu & Yan 2009). There is as yet no convincing physical explanation for this hysteresis phenomenon, though Meyer-Hofmeister et al. (2005) and Liu et al. (2005) showed that their disk evaporation model can reproduce the observations when Compton cooling is included. Done et al. (2007) have an alternative explanation, arguing that it is simply a matter of timescales. As $\dot{M}_{\rm BH}$ increases, it takes a viscous time for the truncated thin disk (see Section 4.2.2 below) to move down to the ISCO, and during this time the luminosity continues to increase and goes well above the threshold value before the transition is completed. This plausible scenario needs to be confirmed with detailed models.

4.2.2. Accretion geometry: truncated thin disk plus hot accretion flow. Most BHBs in the hard and quiescent state, as well as many LLAGNs, are deduced to have a two-zone accretion flow consisting of a cool thin disk at large radii and a hot accretion flow at small radii. This configuration is illustrated in **Figure 7**. (In the case of Sgr A* and some elliptical galaxies such as M87, the flow starts out hot at the Bondi radius and remains hot throughout, so there is apparently no thin disk at large radii.) The main parameter that controls the transition radius $R_{\rm tr}$ between the two zones is the mass accretion rate $\dot{M}_{\rm BH}$. When $\dot{M}_{\rm BH}$ is above the maximum allowed for the hot accretion flow solution (Section 2.7), only the thin disk solution is available, so the thin disk extends all the way down to the ISCO. This corresponds to the soft state. When $\dot{M}_{\rm BH}$ becomes smaller, the source first enters an intermediate state and then, with decreasing $\dot{M}_{\rm BH}$, progresses to the hard state and finally the quiescent state. In these latter states, the thin disk is truncated at a radius $R_{\rm tr} > R_{\rm ISCO}$, and the region inside $R_{\rm tr}$ is occupied by a hot accretion flow. The radial extent of the hot zone increases with decreasing $\dot{M}_{\rm BH}$. This "truncated thin disk plus hot inner accretion flow" configuration was first proposed by Shapiro et al. (1976) to explain the hard state of Cyg X-1; however, their model was based on the unstable SLE solution. A similar concept, but using the ADAF solution, was later developed by Narayan (1996; see also Esin et al. 1997, Poutanen et al. 1997) to explain the various spectral states of BHBs.

Despite some understanding of the dynamics of the transition from the outer thin disk to the inner hot accretion flow (e.g., Abramowicz et al. 1998, Manmoto et al. 2000), the physical reason why cold gas in the outer disk is converted into hot gas on the inside is not fully understood. Some combination of the following models is likely responsible: the evaporation model (Meyer & Meyer-Hofmeister 1994; Liu et al. 1999, 2011; Meyer et al. 2000, 2007; Różańska & Czerny 2000; Spruit & Deufel 2002; Mayer & Pringle 2007; Taam et al. 2012); the turbulent diffusion model (Honma 1996, Manmoto & Kato 2000, Manmoto et al. 2000); and the large viscosity model (Gu & Lu

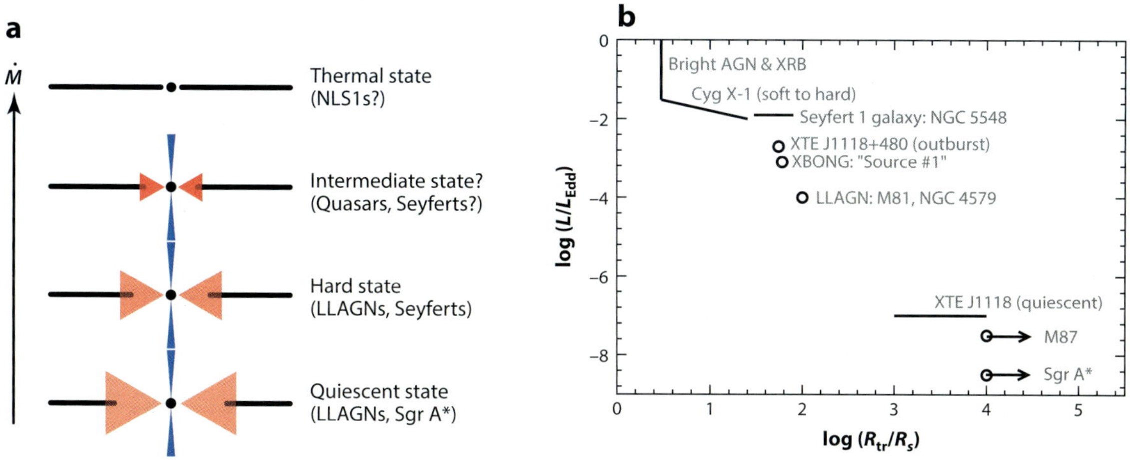

Figure 7

(*a*) Schematic diagram showing the configuration of the accretion flow in different spectral states of BHBs as a function of the mass accretion rate $\dot{M}_{\rm BH}$ (panel adapted from Esin et al. 1997, Narayan & McClintock 2008). Possibly equivalent AGN classes are indicated in parentheses. Red triangles indicate the hot accretion flow, whereas thick black horizontal lines represent the standard thin disk. The transition radius $R_{\rm tr}$ where the thin disk is truncated becomes smaller with increasing $\dot{M}_{\rm BH}$. In the thermal state, the disk is not truncated, and its inner edge is located at the ISCO. (*b*) Plot of the Eddington-scaled accretion luminosity $L/L_{\rm Edd}$ versus $R_{\rm tr}$ as deduced from observations. The transition radii were estimated by modeling spectra of individual LLAGNs and BHBs (panel adapted from Yuan & Narayan 2004). Abbreviations: AGN, active galactic nuclei; BHB, black hole binary; ISCO, innermost stable circular orbit; LLAGN, low-luminosity active galactic nuclei.

2000, Lu et al. 2004). Of these, the evaporation model has been studied most extensively, though all three models predict that $R_{\rm tr}$ should increase with decreasing $\dot{M}_{\rm BH}$. As shown in **Figure 7**, this prediction is in agreement with empirical data based on modeling spectra of individual sources (Yuan & Narayan 2004). [Not all systems follow the trend shown in **Figure 7**. For BHBs with short orbital periods, the mass transfer stream from the companion star circularizes at a fairly small radius $R_{\rm circ}$. When these systems go into the quiescent state, the continued supply of cold gas at $R_{\rm circ}$ ensures that the transition radius is pinned close to $R_{\rm circ}$ (Menou et al. 1999b). A similar situation is possible even in very low luminosity AGNs. If an AGN is fueled from the external medium by cold gas clouds with low angular momentum, the gas clouds will first circularize and form a thin disk before "evaporating" into a hot accretion flow (Inogamov & Sunyaev 2010). The value of $R_{\rm tr}$ in XTE J1118+480 shown in **Figure 7** is significantly smaller than that obtained in later work by Yuan et al. (2005) (see Section 4.2.4 for details). Yuan et al. (2005) explain this discrepancy, which does not affect the main conclusion.] Das & Sharma (2013) have used numerical simulations to explore the effect of a time-varying $\dot{M}_{\rm BH}$ on the accretion geometry and state transitions.

4.2.3. Modeling observations of LLAGNs. Following the initial application of the ADAF model to Sgr A* (Narayan et al. 1995), various authors used similar ideas to explain a variety of observations of other LLAGNs: nearby elliptical galaxies (Fabian & Rees 1995; Reynolds et al. 1996; Di Matteo et al. 2000, 2001; Loewenstein et al. 2001; Ho et al. 2003; Fabbiano et al. 2003), LINERs (Lasota et al. 1996a, Gammie et al. 1999, Quataert et al. 1999, Yuan et al. 2002b, Pellegrini et al. 2003, Ptak et al. 2004, Nemmen et al. 2006, Xu & Cao 2009, Liu & Wu 2013, Nemmen et al. 2014), BL Lac objects (Maraschi & Tavecchio 2003), FR I sources (Reynolds et al. 1996, Begelman & Celotti 2004, Wu et al. 2007, Yuan et al. 2009c, Yu et al. 2011), X-ray bright,

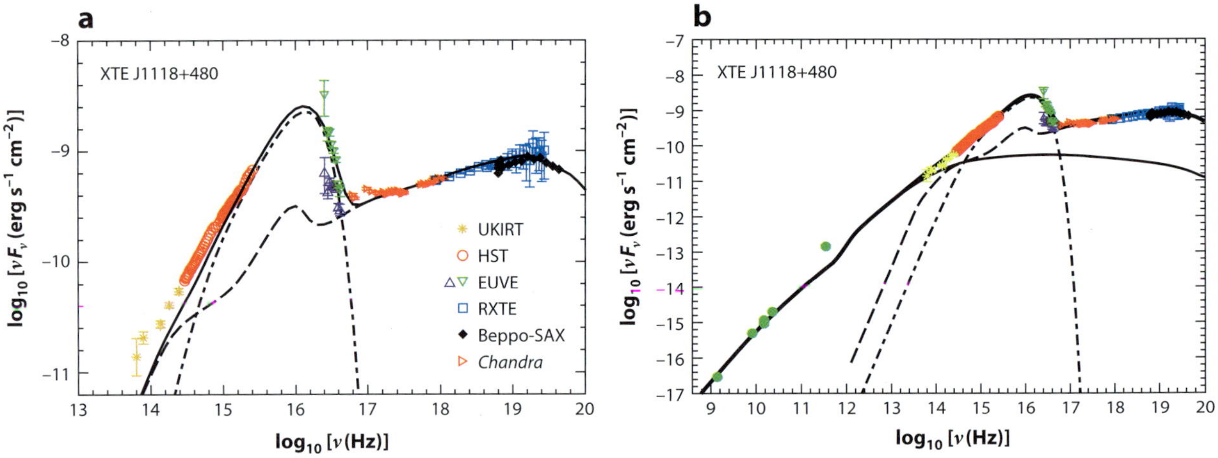

Figure 8

Modeling the spectrum of the BHB XTE J1118+480 in the hard state (from Yuan et al. 2005). (*a*) Calculated spectrum for a model consisting of an ADAF (*dashed line*) plus a truncated thin disk (*dot-dashed line*). The thin disk is essential to fit the optical-EUV part of the spectrum, whereas the hot ADAF is needed to fit the power-law X-ray emission. The model fails to explain the infrared and radio data. (*b*) A model that includes an additional jet component (*thin solid line*) and that explains all the observations. Abbreviations: ADAF, advection-dominated accretion flow; BHB, black hole binary; EUV, extreme ultraviolet.

optically normal galaxies (Yuan & Narayan 2004), and even Seyferts (Chiang & Blaes 2003, Yuan & Zdziarski 2004).

NGC 1097, a famous LINER, is an interesting case. It is the first and best-studied LLAGN to display broad, double-peaked Hα and Hβ emission lines (e.g., Storchi-Bergmann et al. 1997). Such double-peaked lines are believed to be the result of irradiation of a truncated thin disk, most likely by radiation from the inner hot accretion flow (Chen & Halpern 1989). For NGC 1097, the Hα line profile requires the transition radius to be at $R_{tr} \approx 225\, R_S$ (Storchi-Bergmann et al. 1997). Using the truncated thin disk plus hot accretion flow model, Nemmen et al. (2006) successfully modeled the optical-to-X-ray continuum spectrum of the source. In their model, the truncated thin disk dominates the optical-UV emission, whereas the inner hot accretion flow produces the X-ray emission. Impressively, the R_{tr} they require to fit the continuum spectrum agrees very well with the R_{tr} estimated from fitting the double-peaked Hα line.

4.2.4. Modeling observations of BHBs. In the field of BHBs, the truncated thin disk plus hot accretion flow scenario was first applied to the quiescent state (Narayan et al. 1996, 1997a; Menou et al. 1999a,b) and soon after to the hard state (Esin et al. 1997, 1998, 2001). The source XTE J1118+480 is a spectacular example where there is (*a*) good extreme UV (EUV) data, which are crucial for constraining the radius of the inner edge of the thin disk (McClintock et al. 2003), and (*b*) good timing information, including a low-frequency QPO (~0.1 Hz) and measurements of time lags between different wavebands (see review in Yuan et al. 2005). Spectral fitting of the EUV data indicates that the thin disk must be truncated (Esin et al. 2001, Chaty et al. 2003). **Figure 8** shows a comprehensive model (Yuan et al. 2005) that uses a more modern hot accretion flow solution, including a jet. In this model, the radio/infrared emission is dominated by the jet and the optical-EUV by the truncated thin disk, and the X-ray region of the spectrum is produced by the hot accretion flow. The transition radius is constrained to be $R_{tr} \approx 300 R_S$. The same truncated thin disk scenario can also explain the low-frequency QPO. In the model proposed by Giannios

& Spruit (2004; see also Rezzolla et al. 2003 for a similar idea), the QPO arises from a global p-mode oscillation of the hot ADAF. (As mentioned in Section 3.2.2, another QPO model invokes the precession of the ADAF, with the frequency again determined by the size of the ADAF.) The QPO frequency is roughly determined by the Keplerian frequency at R_{tr} and agrees well with the observed frequency for $R_{tr} \approx 300R_s$. In addition, the model also qualitatively explains other timing features such as time-lags between different wavebands (Yuan et al. 2005; for a similar model with similar conclusions, see also Malzac et al. 2004).

A number of observations have confirmed the basic features of the truncated thin disk scenario in BHBs (see reviews by Zdziarski & Gierliński 2004, Done et al. 2007, McClintock & Remillard 2006, Poutanen & Veledina 2014):

1. A truncated thin disk is required to model the thermal spectral component observed in BHBs in the quiescent state (Narayan et al. 1996, 1997a; Yuan & Cui 2005) and the hard state (Esin et al. 2001, Di Salvo et al. 2001, Chaty et al. 2003, Yuan et al. 2005, Cabanac et al. 2009, Tomsick et al. 2009).
2. The observed transient behavior of BHBs requires a truncated disk (Lasota et al. 1996b, Menou et al. 2000, Dubus et al. 2001), as does the time delay between the optical and X-ray outbursts (Hameury et al. 1997).
3. A reflection component is seen in the X-ray continuum spectrum in the hard state. When the X-ray spectrum steepens, both the solid angle subtended by the reflection material and the amount of relativistic smearing increase, consistent with the truncation radius moving in and thereby increasing the flux of soft photons irradiating the inner hot accretion flow (Gilfanov et al. 1999, Zdziarski et al. 2003).
4. The truncated disk model explains the correlation between the luminosity and the photon index of the X-ray spectrum in the hard state (Gardner & Done 2013, Qiao & Liu 2013).
5. The model naturally explains why the QPO frequency increases with increasing X-ray luminosity (Cui et al. 1999, Ingram & Done 2011).
6. Across both the soft-to-hard and hard-to-soft state transitions, there is a sharp change in observational features such as variability (Kalemci et al. 2013 and references therein), photon index and X-ray flux (see Zdziarski & Gierliński 2004, their figure 6), and high-energy cutoff of the X-ray spectrum (Belloni et al. 2006), all of which suggest that there must be a substantial qualitative change in the mode of accretion during this transition.

Most investigations of hot accretion flows tend to focus on thermal electrons. However, it is quite plausible that some nonthermal electrons will also be present (Section 2.3). Compared with a pure thermal model, a hybrid thermal-nonthermal model can explain a wider range of observations, with the nonthermal electrons playing a role similar to the jet component in disk-jet models. For instance, the hybrid model explains the MeV tail in the hard state spectrum of some BHBs (Poutanen & Vurm 2009), the power-law-like optical/infrared spectrum, and the concave shape of the X-ray spectrum (Poutanen & Veledina 2014).

The hard state often reaches luminosities $\sim 10\% L_{\rm Edd}$ during the hard-to-soft state transition. This is a factor of several too high for an ADAF (Section 2). Yuan & Zdziarski (2004) suggested that such luminous hard state systems, and also some Seyfert galaxies, may be explained by the LHAF model. This has been confirmed in detailed modeling of XTE J1550-564 (Yuan et al. 2007), where the X-ray spectrum is naturally explained by the LHAF model. Features such as the slope of the X-ray spectrum, the cutoff energy, and the normalization are reproduced well. The agreement in the cutoff energy indicates that the predicted electron temperature is consistent with that required by observations (e.g., Yuan et al. 2007, their figure 3), although in some cases the ADAF model may be too hot and a hybrid thermal-nonthermal electron distribution may be required to reconcile the model with observations (e.g., Poutanen & Veledina 2014). Some

BHBs achieve even higher luminosities of up to $\sim 30\% L_{\rm Edd}$ in the "bright hard state" (Gierliński & Newton 2006). Oda et al. (2010) proposed that a magnetically supported accretion flow model may explain these systems.

4.2.5. Assessment of arguments against disk truncation. The hard state sometimes shows a dim blackbody-like thermal component as well as a broad iron Kα line at X-ray luminosities $L_{0.5-10\,\rm keV} > 10^{-3} L_{\rm Edd}$. These observations have been used to argue that the thin disk is not truncated but extends down to the ISCO (e.g., Miller et al. 2006, Ramadevi & Seetha 2007, Rykoff et al. 2007, Reis et al. 2010). However, the soft component in the spectrum typically has only 10% of the total observed luminosity. It is hard to understand how a radiatively efficient thin disk can extend down to the ISCO and yet contribute so little to the emitted luminosity. In fact, other groups have obtained different estimates of the disk inner radius using the same data (e.g., Done & Gierliński 2006, Done et al. 2007, Gierliński et al. 2008). Cabanac et al. (2009) carried out a detailed analysis of systematic uncertainties in spectral fitting and concluded that when $L_{0.5-10\,\rm keV} \gtrsim 0.01 L_{\rm Edd}$, $R_{\rm tr} \lesssim 10 R_g$, whereas when $10^{-3} L_{\rm Edd} \lesssim L_{0.5-10\,\rm keV} \lesssim 10^{-2} L_{\rm Edd}$, the disk inner edge recedes well away from the ISCO.

Alternate explanations have been proposed for the weak thermal spectral component in the hard state. These do not require the thin disk to extend down to the ISCO. One possibility is that the thermal component may originate near the inner edge of the truncated thin disk that is illuminated by hard X-rays from the hot flow (D'Angelo et al. 2008). Another possibility is that the emission is from cold clumps in the hot accretion flow (Chiang et al. 2010). The same clumps may also explain timing features in the X-ray emission such as the power spectrum of rapid aperiodic variability (Böttcher & Liang 1999). Note that, at accretion rates approaching the upper limit of the hot accretion flow solution, cold clumps are expected to form naturally as a result of thermal instability in the hot gas (Yuan 2003) (see Section 2.7) or condensation of the hot flow (Różańska & Czerny 2000; Liu et al. 2007, 2011; Mayer & Pringle 2007; Meyer et al. 2007; Meyer-Hofmeister et al. 2009).

Disk inner radii derived from iron line profiles are even more controversial. Hartnoll & Blackman (2001) showed that iron lines can be readily produced in a two-phase (hot gas plus cold clumps) accretion flow. Provided the clumps can survive long enough, the line profiles are similar to those produced by a thin disk extending down to the ISCO.

Done & Diaz Trigo (2010) reanalyzed MOS iron line data in GX 339−4 in the hard state. Miller et al. 2006 (see also Reis et al. 2008) had previously claimed that the data indicated an extremely broad iron line. However, Done & Diaz Trigo (2010) showed that the line shape is strongly affected by pile-up. Furthermore, using the simultaneous PN timing-mode data, which should not be affected by pile-up, they obtained a significantly narrower line, which is easily consistent with a truncated disk. Recently, Plant et al. (2013) carried out a systematic study of the iron line in GX 339−4 and tracked the evolution of the thin disk inner radius over a range of two orders of magnitude in luminosity. They found that the data are consistent with the thin disk being truncated throughout the hard state and with the truncation radius moving closer to the black hole as the luminosity increases. In another recent study, Kolehmainen et al. (2014) used data on both the weak thermal component and the iron line to constrain the inner radius of the thin disk. They also found that the data are consistent with a truncated disk.

The question of whether the disk is truncated in the hard state is crucially important for efforts to measure black hole spin using X-ray reflection spectroscopy (for a current review, see Reynolds 2013). A key assumption of this method is that reflection occurs from a cool disk that extends down to the ISCO. Effectively, the measured profile of the iron Kα emission line is used to fit for the radius of the inner edge of the disk. Then, assuming that the disk edge is located at the ISCO, the black hole spin is estimated. However, almost all applications of the reflection method to stellar mass black holes have been carried out on X-ray data in the hard state. If the cool disk in BHBs in the hard state is truncated outside the ISCO, as the preponderance of evidence suggests (in the authors' view), most spin measurements by the reflection method will be affected. Measurements of the spins of supermassive black holes [where the reflection method originated and where considerable work has been done over many years (Reynolds 2013)] are not affected by this criticism unless those systems also have truncated disks.

4.2.6. Radio/X-ray correlation and the role of jet radiation. Corbel et al. (2003) and Gallo et al. (2003) discovered a remarkable correlation between the radio luminosity L_R and X-ray luminosity L_X of BHBs in the hard state. Soon after, Merloni et al. (2003) and Falcke et al. (2004) considered the effect of black hole mass M and showed that the correlation extends also to LLAGNs (for subsequent work, see Körding et al. 2006, Wang et al. 2006, Li et al. 2008, Gültekin et al. 2009, Yuan et al. 2009c, de Gasperin et al. 2011, Younes et al. 2012). Their generalized correlation (**Figure 9*a***) takes the form (Gültekin et al. 2009)

$$\log\left(\frac{L_R}{\mathrm{erg\ s^{-1}}}\right) = (0.7 \pm 0.1)\log\left(\frac{L_X}{\mathrm{erg\ s^{-1}}}\right) + (0.8 \pm 0.3)\log\left(\frac{M}{\mathrm{M}_\odot}\right) + (4.8 \pm 0.2) \qquad (36)$$

and is referred to as the "fundamental plane of black hole activity."

Using a simple model in which thermal gas in a hot accretion flow is responsible for the X-ray emission and relativistic electrons in a jet produce the radio emission, Heinz & Sunyaev (2003) showed that the fundamental plane can be naturally explained (see also Merloni et al. 2003, Heinz 2004, Yuan & Cui 2005, Yuan et al. 2005, Li et al. 2008). An alternative explanation has also been advanced in which the radio and X-ray emission are both produced by the jet (Markoff et al. 2003). [Heinz (2004) noted that, if the "cooling break" is properly taken into account for the electron energy distribution, the predicted correlation is different from that obtained by Markoff et al. (2003). This is confirmed by Yuan & Cui (2005), who found a steeper correlation when radiation from the jet dominates both the radio and X-ray emission (Equation 37). Plotkin et al. (2012) therefore suggested that supermassive black holes should be excluded from the sample because the cooling break "is a concern" for these objects. However, Zdziarski et al. (2012) found that the cooling break is independent of black hole mass, so the cooling break should be equally important for both LLAGNs and BHBs.]

Within the sample of Merloni et al. (2003), most sources are relatively luminous and correspond to the upper range of allowed $\dot{M}_{\mathrm{BH}}$ for the hot accretion flow solution. Yuan & Cui (2005) extrapolated the coupled ADAF-jet model of Yuan et al. (2005) to lower luminosities and predicted that, below a critical luminosity $L_{\mathrm{X,crit}}$ given by $\log(L_{\mathrm{X,crit}}/L_{\mathrm{Edd}}) \approx -5.36 - 0.17\ \log(M/\mathrm{M}_\odot)$, the correlation should steepen to

$$\log\left(\frac{L_R}{\mathrm{erg\ s^{-1}}}\right) = 1.23\log\left(\frac{L_X}{\mathrm{erg\ s^{-1}}}\right) + 0.25\log\left(\frac{M}{\mathrm{M}_\odot}\right) - 13.45 \qquad (37)$$

because radiation from a hot accretion flow at low $\dot{M}_{\mathrm{BH}}$ is roughly $\propto \dot{M}_{\mathrm{BH}}^2$ (Section 2.5), whereas that from the jet is $\propto \dot{M}_{\mathrm{BH}}$ (Heinz 2004, Yuan & Cui 2005) (also see **Figure 9**). Therefore, at a low enough $\dot{M}_{\mathrm{BH}}$, the X-ray emission from the jet will dominate. This can explain some otherwise puzzling observations of quiescent black holes (Yuan & Cui 2005). Early models of M87, for example, assumed that the X-ray emission is produced by the ADAF (e.g., Reynolds et al. 1996, Di Matteo et al. 2003). However, *Chandra* observations suggest that the emission is dominated by the jet (Wilson & Yang 2002), which is consistent because M87 has $L_X < L_{\mathrm{X,crit}}$.

On the whole, there are not enough data on quiescent BHBs to verify Equation 37. The two most promising sources are V404 Cyg (Corbel et al. 2008) and A0620-00 (Gallo et al. 2006).

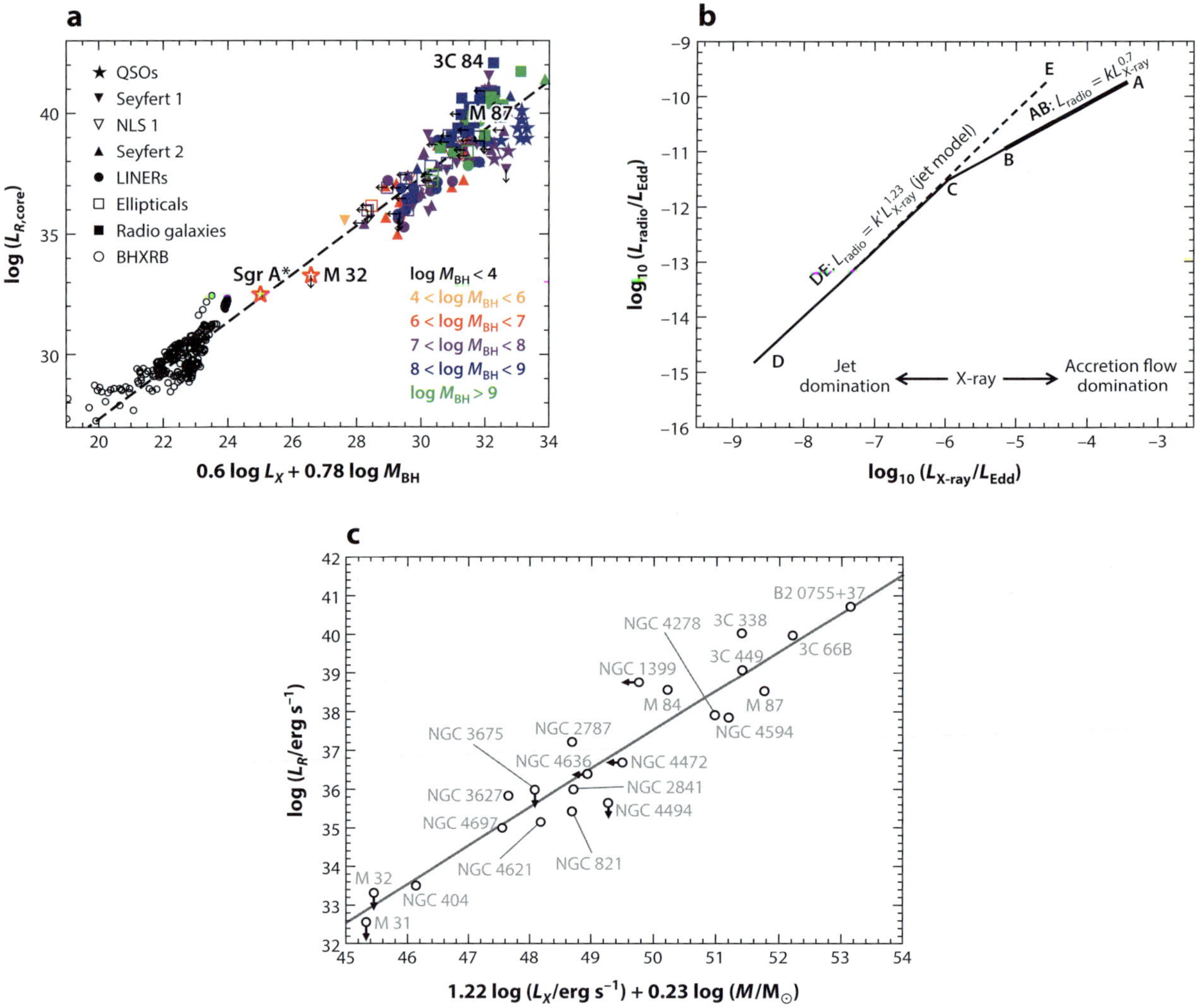

Figure 9

(*a*) The "fundamental plane of black hole activity" (from Merloni et al. 2003). Note that the correlation extends over many decades of black hole mass and accretion luminosity and includes many different source types. Courtesy of A. Merloni. (*b*) The predicted radio/X-ray correlation according to the advection-dominated accretion flow–jet model (from Yuan & Cui 2005). The segment AB, which represents luminous sources, has the same slope (~0.6) as that in panel *a*. The segment CD corresponds to objects at lower luminosities, where the jet is expected to dominate both the radio and X-ray emission. (*c*) Observational data on 22 low-luminosity active galactic nuclei with luminosities corresponding to CD in panel *b* (from Yuan et al. 2009c). The measured slope (~1.22) agrees well with the theoretically predicted slope (~1.23).

However, the former is not dim enough to explore the low-luminosity end of the correlation, and the latter has data in only the quiescent state, with none in the hard state. The situation is much better in the case of LLAGNs, where recent work has confirmed the change of slope of the $L_R - L_X$ correlation at low luminosities (Pellegrini et al. 2007, Wu et al. 2007, Wrobel et al. 2008, Yuan et al. 2009c, de Gasperin et al. 2011, Younes et al. 2012). Yuan et al. (2009c) considered 22 LLAGNs with $L_X < L_{X,\mathrm{crit}}$ and found a correlation slope ~1.22 (**Figure 9*c***), which is in excellent agreement with Equation 37.

4.2.7. Jets and black hole spin. The fundamental plane of black hole activity (**Figure 9**) (Equation 36) involves only the mass accretion rate $\dot{M}_{\mathrm{BH}}$ (through the luminosities L_R and L_X) and the black hole mass M; it does not involve the black hole spin. Van Velzen & Falcke (2013) recently obtained a very homogeneous sample of radio quasars in which they claim that most of the remaining scatter around the fundamental plane is due to the environment, leaving little room for additional scatter due to variations in the black hole spin. If jets are powered by black hole spin (BZ jet) (Section 3.3.1), we expect a strong dependence of jet power on the angular velocity of the black hole Ω_H (Equations 30 and 31). Why do the data not show this?

One possibility is that the range of black hole spins is not very large. Because the data are usually shown in a log-log plot, any modest variations due to spin could be hidden. Even the data of van Velzen & Falcke (2013) are potentially consistent with Equation 31, provided the black hole population does not span the full range of spin values but is restricted to a smaller range, say, from 0.3 to 1. Alternatively, hot accretion flows in nature may not approach anywhere near the MAD limit discussed in Section 3.2.3. If the flux Φ around the black hole is sufficiently below Φ_{MAD}, the BZ jet mechanism will be subdominant. The quasi-relativistic disk jet will then take over, and we will no longer expect a strong correlation between jet power and spin (Section 3.3.2). Note that jets in BHBs in the hard state do not appear to be relativistic (Fender 2006) and may well be disk jets rather than BZ jets. Even in LLAGNs, only a few sources [e.g., M87 (Hada et al. 2011] are known to have relativistic jets.

As discussed in Section 3.3.3, BHBs have two distinct kinds of jets: (*a*) steady quasi-relativistic jets, which are associated with the hard state, and (*b*) episodic jets, which are most obviously seen when black holes undergo state transitions from hot to cold mode accretion. Thus, a third possibility is that the sample of van Velzen & Falcke (2013), which consists of radio-loud quasars, is dominated by episodic jets (e.g., Merloni et al. 2003, Nipoti et al. 2005). If episodic jets are powered by the disk mechanism proposed by Yuan et al. (2009a) (see Section 3.3.3 for details), no strong correlation between jet power and black hole spin is expected.

Thanks to recent progress in measuring the spins of black holes (McClintock et al. 2011, 2013), we can now check directly for a correlation between spin and jet power in BHBs. A strong correlation has been seen in episodic jets (Narayan & McClintock 2012, McClintock et al. 2013, Steiner et al. 2013; but see Fender et al. 2010, Russell et al. 2013a, who question the reality of the correlation). However, the episodic jets in question are all associated with the transition from a hard to a soft state (i.e., from a hot accretion flow to a cold disk), rather than with a pure hard state. Hence, the simulation results discussed in Section 3.3.1 may not be relevant. Because the jets were produced when the source luminosities were close to Eddington (Steiner et al. 2013), additional physical effects, e.g., radiation, may also play an important role. Investigations of the effect of radiation on the BZ mechanism are just becoming possible (Sadowski et al. 2014).

4.2.8. Alternative models for the hard state. An alternative to the hot accretion flow model of hard state BHBs and LLAGNs is the "disk-corona" model. Here the thin disk extends down to the ISCO, and the hard radiation is produced above the disk in a hot corona that may be heated by magnetic reconnection (Liang & Price 1977; Haardt & Maraschi 1991, 1993). Because of the abundance of soft photons from the disk, coupled with the strong reprocessing of the coronal emission in the disk, the "slab" geometry of disk-corona models generally gives relatively soft X-ray spectra (photon indices $\Gamma \gtrsim 2$). This is not consistent with observed spectra (Stern et al. 1995, Zdziarski et al. 2003, Done et al. 2007). One way to resolve this is to postulate that the corona forms the base of an outward-moving jet (Beloborodov 1999, Malzac et al. 2001, Merloni & Fabian 2002). If so, relativistic beaming reduces the amount of reprocessing in the cold disk, making it possible to obtain hard spectra. A patchy corona is another possibility (e.g., Malzac et al. 2001). In a recent study, Schnittman et al. (2013) calculated spectra for the disk-corona model using MHD simulation data. They claimed to obtain relatively hard X-ray spectral slopes, although their spectra are inconsistent with observations in certain other respects. Apart from the spectral slope, the corona model also has trouble explaining the observed correlations between the spectral slope, iron line width, and characteristic variability frequencies (Poutanen & Veledina 2014).

Another model for the hard state is the "jet model." As in the standard hot accretion flow models, the thin disk here is truncated. The main difference is that the jet produces not only the radio emission, but also the X-ray emission. Early work proposed that the X-ray emission is due to synchrotron radiation (Markoff et al. 1999, 2003). However, a pure synchrotron model cannot reproduce the observed shape of the high-energy cutoff of the X-ray spectrum (Zdziarski et al. 2003). Later versions of the jet model invoke synchrotron self-Compton radiation from the base of the jet to explain the X-ray emission (Markoff et al. 2005). The model is then similar to the standard hot accretion flow model, except that the required electron temperature is much higher, approximately several megaelectronvolts. Because a jet involves outflowing gas, one way to distinguish between the models is by measuring the velocity of the X-ray-emitting gas. However, the outflowing gas at the base of the jet is only weakly relativistic. Therefore, although there may be some modest beaming [as in the model of Beloborodov (1999) cited above], it is not expected to be a dominant effect. Observationally, there is no evidence that X-ray emission in the hard state has any dependence on the inclination of the system (Fender et al. 2004, Narayan & McClintock 2005). Other problems with the jet model are pointed out in Malzac et al. (2009) and Poutanen & Veledina (2014).

5. HOT ACCRETION AND ACTIVE GALACTIC NUCLEI FEEDBACK

There is considerable observational evidence that AGN feedback plays an important role in the evolution of galaxies and galaxy clusters (Fabian 2012, Kormendy & Ho 2013). Arguments usually mentioned include the following:

1. The famous correlation between the black hole mass and the luminosity of the host galaxy or the velocity dispersion of the galaxy bulge (Magorrian et al. 1998, Ferrarese & Merritt 2000, Gebhardt et al. 2000, Kormendy & Ho 2013);
2. The observed exponential cutoff in the number density of galaxies at the high mass/luminosity end (Schechter 1976), even though there is no cutoff at the same mass scale in the distribution of dark matter halos;
3. The "downsizing" puzzle, where the most massive galaxies and black holes are the oldest (Cowie et al. 1996, Kriek et al. 2007, Babić et al. 2007, Fanidakis et al. 2012); and
4. The "cooling flow problem" in galaxy clusters, where the lack of any significant cooling in cluster cores (Peterson et al. 2001), despite a short cooling time, suggests that a central AGN serves as an extra source of energy (Pedlar et al. 1990; Churazov et al. 2000, 2002; Ciotti & Ostriker 2001; Brüggen & Kaiser 2002).

Two major modes of AGN feedback have been identified (Fabian 2012, Kormendy & Ho 2013): (*a*) a radiative mode, also known as quasar mode, which operates when the black hole accretes at a good fraction ($\gtrsim 0.1$) of the Eddington rate, and (*b*) kinetic mode, also known as radio mode or maintenance mode, which typically operates when $\dot{M}_{\rm BH}$ is low, i.e., when the AGN is fed by a hot accretion flow. Maintenance-mode feedback has been considered in the context of the cooling flow problem in galaxy clusters (Churazov et al. 2002, Ruszkowski & Begelman 2002): It is included in semiempirical models of galaxy formation (Croton et al. 2006, Hopkins et al. 2006, Somerville et al. 2008) and is incorporated in HD simulations of galaxy formation and evolution (Ciotti et al. 2010, Novak et al. 2011, Gaspari et al. 2012). Below, we review some of the physics of maintenance-mode feedback, highlighting simplifications in current models that could be improved using our current knowledge of hot accretion flows.

5.1. Feedback from Jets and Outflows

Because the majority of supermassive black holes accrete at highly sub-Eddington rates via hot accretion flows, maintenance-mode feedback, as opposed to quasar-mode feedback, is much more prevalent in the universe. In particular, most nearby galaxies contain LLAGNs (Section 4.2), and any feedback activity in these systems occurs via the maintenance mode. Hot accretion flows tend to be radiatively inefficient, so the maintenance mode is believed to be dominated by mechanical feedback via jets and winds rather than by radiation from the accretion disk (but see Section 5.2). Bubbles and cavities in X-ray and radio images (e.g., Fabian 2012, Morganti et al. 2013) provide direct observational evidence for interaction between jets and the interstellar or intracluster medium. The two *Fermi* bubbles detected above and below the Galactic Center (Su et al. 2010) are also thought to have been inflated by a jet or wind from Sgr A* during a period of activity in the past few million years (Zubovas et al. 2011, Guo & Mathews 2012).

Most studies of maintenance-mode feedback focus on the role of a collimated jet rather than a more isotropic nonrelativistic disk wind. In large part, this is driven by the fact that jets are easily observed in the radio and their power can be estimated directly from observations. In contrast, there is virtually no direct evidence for uncollimated winds in LLAGNs (but see Crenshaw & Kraemer 2012). In addition, jets have been investigated via simulations for a number of years (Section 3.3), whereas the study of winds from hot accretion flows has only just begun (Section 3.4).

The relevant quantities that determine the effectiveness of feedback are the rate of injection of energy and momentum into the external medium and the degree of collimation of the jet or wind. Numerical simulations of hot accretion flows can provide some useful information. Relatively better information is known for jets (Section 3.3) than is known for winds (Section 3.4), but rapid progress is expected on both fronts. One important parameter for feedback studies is "feedback efficiency" ϵ, defined as the ratio of the kinetic power in the jet or wind to the accretion power $\dot{M}_{\rm BH}c^2$. In almost all current cosmological models of feedback, ϵ is regarded as a free parameter, but it could in principle be estimated via simulations. Typically, $\epsilon \sim 10^{-3}$–10^{-4} is required to explain observations (e.g., Di Matteo et al. 2005, Springel et al. 2005, Ciotti et al. 2009, Ostriker et al. 2010, Gaspari et al. 2012). Many studies of AGN feedback on galaxies do not discriminate between quasar and maintenance mode feedback and adopt a single value of ϵ for both.

Both jets and winds carry with them substantial fluxes of energy and momentum. Typically, the jet dominates the energy output, whereas the wind dominates the momentum output (Yuan et al. 2012a, Sadowski et al. 2013a). By and large, studies of AGN feedback have tended to focus on energy feedback (e.g., Di Matteo et al. 2005; Springel et al. 2005). However, momentum feedback can more effectively push the surrounding gas and may be equally important for controlling the growth of the black hole and switching off star formation (King 2003, 2005, 2010; Ostriker et al. 2010; Debuhr et al. 2010; Silk 2013).

The jet and wind differ significantly in their degree of collimation. Even though the jet dominates the energy flux, it is not clear that this energy couples very well to the interstellar medium of the host galaxy. The jet may simply drill through the surrounding gas, depositing little energy within the galaxy. Indeed, some simulations indicate that the jet is ineffective even on the scale of galaxy clusters (Vernaleo & Reynolds 2006), though the problem is much alleviated when shear, rotation, and large-scale flows in the intracluster medium are included (Heinz et al. 2006). Models that invoke efficient jet heating generally require a relatively slow (subrelativistic) jet and a mass loss rate in the jet as large as the Eddington rate, which seems unlikely (e.g., Omma et al. 2004). In contrast, the less collimated disk wind, despite its lower energy budget, may be more important for galaxy-scale feedback.

One other difference between jets and winds is the dependence of feedback efficiency on the parameters of the system. Apart from the obvious dependence on mass accretion rate $\dot{M}_{\rm BH}$, jet power is strongly affected by the black hole spin and the magnetic flux around the hole (at least in numerical simulations) (see Section 3.3). Thus, realistic incorporation of jet feedback in cosmological simulations requires keeping track of the spins of supermassive black holes. It is also necessary to decide whether the MAD configuration is viable, which depends on whether advection of magnetic field into the black hole is efficient (Section 3.2.3). In contrast, the energy and momentum flux in winds is relatively insensitive to the black hole spin and magnetic flux and is primarily determined by $\dot{M}_{\rm BH}$. Thus, wind feedback ought to be simpler to model. Unfortunately, estimating the mass accretion rate involves large uncertainties (Section 3.4).

5.2. Feedback from Radiation

The most obvious output of black hole accretion is radiation, which can impart energy and momentum to the surrounding interstellar medium via electron scattering, photoionization, atomic resonance scattering, and absorption by dust grains. Some semianalytical models and numerical simulations (e.g., Wyithe & Loeb 2003, Di Matteo et al. 2005, Croton et al. 2006) assume that a small and constant fraction ($\sim$0.05) of the radiated luminosity from an AGN couples thermodynamically to the surrounding gas. Other studies employ a more elaborate calculation to estimate the heating rate using the Compton temperature $T_{\rm C}$ of the radiation field, which measures the frequency-weighted average energy of the emitted photons. Usually $T_{\rm C} \sim 10^7$ K is adopted, as appropriate for typical spectra of quasars (Sazonov et al. 2005, Ciotti et al. 2010, Novak et al. 2011).

Whereas in quasar mode there is no doubt that feedback from radiation and radiatively driven winds (Proga 2007, Proga et al. 2008, Liu et al. 2013) are very important, in the maintenance mode radiative feedback is usually assumed to be negligible compared with mechanical feedback. Two reasons are invoked (e.g., Churazov et al. 2005): (*a*) The kinetic power of the outflow is larger than the radiative output of the disk, and (*b*) the efficiency of radiative heating is low. However, the radiative luminosity may actually be larger than the kinetic outflow power for luminosities $\gtrsim 10^{-4} L_{\rm Edd}$ (Fender et al. 2003). In addition, the efficiency of radiative heating is not as low as usually imagined, because the efficiency is proportional to T_c. The spectrum of a hot accretion flow is much harder than that of a quasar (Section 4.2). Hence, the Compton temperature can be as high as $T_c \sim 10^9$ K (Yuan et al. 2009b), indicating that the radiative heating efficiency is much greater than in the quasar mode. [The value of $T_c \sim 10^9$ K is obtained for a pure hot accretion flow; when the contribution from a truncated thin disk is included (Section 4.2), the value will be somewhat lower.]

Although the effect of a larger T_c has not been included in studies of galaxy-wide feedback, Ostriker and collaborators (e.g., Park & Ostriker 2001, 2007; Yuan et al. 2009b) have for many years investigated its role on the scale of the accretion flow and have demonstrated its importance in that context. When the accretion rate is relatively high, nonlocal radiative feedback via Compton heating is dynamically important and can change the temperature profile of the accretion flow. Moreover, if $L \gtrsim 2\% L_{\rm Edd}$, radiative heating at and beyond the Bondi radius ($R \gtrsim 10^5 R_S$) can be so strong that the gas is heated above the virial temperature and wants to flow out rather than in.

In this case, no steady accretion solution can be found, and the accretion flow oscillates between active and inactive phases (Cowie et al. 1978; Ciotti & Ostriker 1997, 2001, 2007; Yuan et al. 2009b). This "small-scale AGN radiative feedback" effect has also been invoked to explain the intermittent activity of compact young radio sources (Yuan & Li 2011).

5.3. Estimating the Mass Accretion Rate

Both mechanical and radiative feedback depend strongly on the mass accretion rate $\dot{M}_{\rm BH}$ of the black hole. Various approaches have been adopted in the literature for estimating $\dot{M}_{\rm BH}$. These include assuming that $\dot{M}_{\rm BH}$ is equal to the Eddington rate, the Bondi rate, or some variant of these (e.g., Springel et al. 2005; Debuhr et al. 2010). The Bondi accretion rate is more relevant for maintenance-mode feedback. However, whether the Bondi model is a reasonable proxy for a hot accretion flow remains highly debatable (see Narayan & Fabian 2011 and references therein).

Given the density and temperature of the external gas at the Bondi radius $R_{\rm B} \sim 10^5 - 10^6 R_S$, it is straightforward to calculate the Bondi accretion rate $\dot{M}_{\rm B}$ at that radius, but how much of this gas reaches the black hole is highly uncertain. If the accretion rate declines with decreasing radius as r^s, with $s \sim 0.5$ (Section 3.4), then as little as 0.1–0.3% of $\dot{M}_{\rm B}$ will reach the black hole. However, some systems with powerful jets seem to require much more gas to reach the black hole to power the observed jets (Allen et al. 2006, Rafferty et al. 2006, McNamara et al. 2009, Russell et al. 2013b). A possible solution is that the value of s depends on boundary conditions. Perhaps s is effectively lower whenever the accreting gas has very low angular momentum (Narayan & Fabian 2011, Bu et al. 2013) or when accretion occurs via the MAD mode (Narayan et al. 2012b, Sadowski et al. 2013a). Unfortunately, without a better understanding of the mapping between $\dot{M}_{\rm B}$ and $\dot{M}_{\rm BH}$, it is hard to imagine any kind of quantitative modeling of maintenance-mode feedback.

Another complication is that the mass accretion rate may be dominated by cold gas from the external medium rather than the hot gas usually considered in hot accretion flow models (Pizzolato & Soker 2005, Rafferty et al. 2006). This could potentially boost the accretion rate by up to two orders of magnitude compared with the calculated Bondi rate based purely on hot gas (Gaspari et al. 2013). The cold gas would presumably first form a thin accretion disk and then evaporate to become a hot accretion flow closer to the black hole. Modeling this mode of accretion would require an understanding of the multiphase nature of the interstellar medium as well as the specific angular momentum of the external cold clouds (Inogamov & Sunyaev 2010).

6. PROSPECTS AND REMAINING OPEN QUESTIONS

The discovery of the self-similar ADAF solution 20 years ago (Narayan & Yi 1994), and the subsequent development of the ADAF model of hot accretion flows (Abramowicz et al. 1995, Narayan & Yi 1995b, Chen et al. 1995), triggered a flurry of activity that has contributed greatly to our understanding of the dynamics and thermodynamics of hot accretion flows as well as the recognition that these flows are relevant for numerous astrophysical objects: Sgr A*, low luminosity AGNs, and BHBs.

Despite the impressive progress of the past two decades, there are presently more questions than answers in this field. Below, we list what we consider are some of the more important questions:

- How are electrons and ions heated in a hot accretion flow? What particle energy distributions do these processes generate? What role do nonthermal particles play in the dynamical and radiative properties of the system?
- Are there processes in addition to Coulomb collisions that transfer energy from ions to electrons, and how do they influence the temperatures of the two species?

- How strong are mass outflows from hot accretion flows, and how does the mass accretion rate at the black hole ($\dot{M}_{BH}$) depend on boundary conditions at large radius? What if gas is supplied from an external two-phase (or even multiphase) medium?
- Why do hot accretion flows produce jets, whereas cool thin disks apparently do not? What role does the black hole, especially its spin, play in determining the properties of the jet? What fraction of the observed radiation comes from the jet versus the hot accretion flow?
- How efficiently do hot accretion flows advect large-scale ordered magnetic fields toward the center, and how often do accreting black holes approach the MAD limit?
- Why and how do state transitions in BHBs occur? What are the physical processes responsible for converting cold optically thick gas into hot optically thin gas in a "truncated thin disk and hot inner accretion flow" configuration, and how do they relate to the hysteresis phenomenon? How do the same processes behave in supermassive black holes?
- What is the thermal state of a hot accretion flow when $\dot{M}_{BH}$ is close to the upper limit for a hot solution? Does the accreting gas become a two-phase medium, and what observational signatures do the hot and cold phases produce?
- What determines whether a hot accretion flow produces a steady jet or an episodic jet, and why are the latter often associated with the hard-to-soft state transition in BHBs? How does this map to supermassive black holes, AGN jets, and the radio loud/quiet dichotomy?
- What is the angular distribution of mass, momentum, and energy outflow from a hot accretion flow around a supermassive black hole, and how do they determine the efficiency of feedback processes?
- How does the relative importance of mechanical versus radiative feedback depend on $\dot{M}_{BH}$ and other parameters of an accretion flow?
- What are the properties of hot accretion flows around compact objects with a surface? (An object with a surface introduces two important modifications compared with the black hole flows considered in this review. First, because gas comes to rest at the stellar surface, the inner boundary condition on the dynamical equations is very different and will result in vastly different densities, velocities, pressures, etc., at small radii. Second, radiation from the surface will Compton-cool the hot accreting gas and modify its temperature. There have been only a few applications of the ADAF model to accreting neutron stars and white dwarfs. For lack of space, we have not reviewed this work here. Much more research could be done in this area.)

DISCLOSURE STATEMENT

The authors are not aware of any affiliations, memberships, funding, or financial holdings that might be perceived as affecting the objectivity of this review.

ACKNOWLEDGMENTS

The authors are grateful to M.C. Begelman, C. Done, J.-P. Lasota, J.E. McClintock, J.P. Ostriker, J. Poutanen, E. Quataert, J. Stone, A. Tchekhovskoy, and A. Zdziarski for helpful comments on the manuscript. This work was supported by grants 11133005 and 11121062 from the National Natural Science Foundation of China (F.Y.) and AST1312651 from the National Science Foundation (R.N.).

LITERATURE CITED

Abarca D, Sadowski A, Sironi L. 2014. *MNRAS.* 440:1125–37

Abramowicz MA, Blaes O, Horák J, Kluźniak W, Rebusco P. 2006. *Class. Quant. Grav.* 23:1689–96

Abramowicz MA, Chen X, Granath M, Lasota JP. 1996. *Ap. J.* 471:762–73
Abramowicz MA, Chen X, Kato S, Lasota, JP, Regev O. 1995. *Ap. J. Lett.* 438:L37–39
Abramowicz MA, Czerny B, Lasota JP, Szuszkiewicz E. 1988. *Ap. J.* 332:646–58
Abramowicz MA, Fragile PC. 2013. *Living Rev. Relativ.* 16:1–88
Abramowicz MA, Igumenshchev IV, Lasota JP. 1998. *MNRAS* 293:443–46
Abramowicz MA, Igumenshchev IV, Quataert E, Narayan R. 2002a. *Ap. J.* 565:1101–6
Abramowicz MA, Kluzniak W, Lasota JP. 2002b. *Astron. Astrophys.* 396:L31–34
Agol E. 2000. *Ap. J. Lett.* 538:L121–24
Aitken DK, Greaves J, Chrysostomou A, et al. 2000. *Ap. J. Lett.* 534:L173–76
Allen SW, Dunn RJH, Fabian AC, Taylor GB, Reynolds CS. 2006. *MNRAS* 372:21–30
Anninos P, Fragile PC, Salmonson JD. 2005. *Ap. J.* 635:723–40
Anninos P, Fragile PC, Wilson J, Murray SD. 2012. *Ap. J.* 759:132
Appl S, Camenzind M. 1992. *Astron. Astrophys.* 256:354–70
Appl S, Camenzind M. 1993. *Astron. Astrophys.* 270:71–82
Armitage P. 1998. *Ap. J. Lett.* 501:L189–92
Babić A, Miller L, Jarvis MJ, et al. 2007. *Astron. Astrophys.* 474:755–62
Baganoff FK, Bautz MW, Brandt WN, et al. 2001. *Nature* 413:45–48
Baganoff FK, Maeda Y, Morris M, et al. 2003. *Ap. J.* 591:891–915
Bai X, Stone JM. 2013. *Ap. J.* 767:30
Balbus SA. 2001. *Ap. J.* 562:909–17
Balbus SA, Hawley JF. 1991. *Ap. J.* 376:214–33
Balbus SA, Hawley JF. 1998. *Rev. Mod. Phys.* 70:1–53
Bardeen JM, Petterson JA. 1975. *Ap. J. Lett.* 195:L65–67
Beckwith K, Hawley J, Krolik JH. 2008. *Ap. J.* 678:1180–99
Beckwith K, Hawley J, Krolik JH. 2009. *Ap. J.* 707:428–45
Begelman MC. 1979. *MNRAS* 187:237–51
Begelman MC. 2012. *MNRAS* 420:2912–23
Begelman MC, Celotti A. 2004. *MNRAS* 352:L45–48
Begelman MC, Chiueh T. 1988. *MNRAS* 332:872–90
Begelman MC, Meier DL. 1982. *Ap. J.* 253:873–96
Belloni T, Parolin I, Del Santo M, et al. 2006. *MNRAS* 367:1113–20
Beloborodov AM. 1999. *Ap. J. Lett.* 510:L123–26
Beskin VS, Istomin YN, Parev VI. 1992. *Astron. Z.* 69:1254–74
Beskin VS, Malyshkin LM. 2000. *Astron. Lett.* 26:208–18
Binder B, Markowitz A, Rothschild RE. 2009. *Ap. J.* 691:431–40
Bisnovatyi-Kogan G, Lovelace RVE. 1997. *Ap. J. Lett.* 486:L43–46
Bisnovatyi-Kogan G, Lovelace RVE. 2007. *Ap. J. Lett.* 667:L167–69
Bisnovatyi-Kogan GS, Ruzmaikin AA. 1974. *Ap. Space Sci.* 28:45–59
Björnsson G, Abramowicz MA, Chen X, Lasota JP. 1996. *Ap. J.* 467:99–104
Blackman EG. 1999. *MNRAS* 302:723–30
Blackman EG, Penna RF, Varniére P. 2008. *New Astron.* 13:244–51
Blaes O. 2013. *Space Sci. Rev.* In press. doi: 10.1007/s11214-013-9985-6
Blaes OM, Arras P, Fragile PC. 2006. *Ap. J.* 369:1235–52
Blandford RD, Begelman MC. 1999. *MNRAS* 303:L1–5
Blandford RD, Begelman MC. 2004. *MNRAS* 349:66–86
Blandford RD, Payne DG. 1982. *MNRAS* 199:883–903
Blandford RD, Znajek RL. 1977. *MNRAS* 179:433–56
Bondi H. 1952. *MNRAS* 112:195–204
Böttcher M, Liang EP. 1999. *Ap. J. Lett.* 511:L37–40
Bower GC, Falcke H, Herrnstein RM, et al. 2004. *Science* 304:704–8
Bower GC, Wright MCH, Falcke H, Backer DC. 2003. *Ap. J.* 588:331–37
Brandenburg A, Nordlund A, Stein RF, Torkelsson U. 1995. *Ap. J.* 446:741–54
Broderick AE, Fish VL, Doeleman SS, Loeb A. 2011a. *Ap. J.* 735:110

Broderick AE, Fish VL, Doeleman SS, Loeb A. 2011b. *Ap. J.* 738:38
Broderick AE, Loeb A, Narayan R. 2009. *Ap. J.* 701:1357–66
Broderick AE, Narayan R. 2006. *Ap. J. Lett.* 638:L21–24
Brüggen M, Kaiser CR. 2002. *Nature* 418:301–3
Bu DF, Yuan F, Stone JM. 2011. *MNRAS* 413:2808–14
Bu DF, Yuan F, Wu M, Cuadra J. 2013. *MNRAS* 434:1691–701
Burkert A, Schartmann M, Alig C, et al. 2012. *Ap. J.* 750:58
Bursa M, Abramowicz MA, Karas V, Kluźniak W. 2004. *Ap. J. Lett.* 617:L45–48
Cabanac C, Fender RP, Dunn RJH, Körding EG. 2009. *MNRAS* 396:1415–40
Cao X. 2011. *Ap. J.* 737;94
Chan CK, Liu S, Fryer CL, et al. 2009. *Ap. J.* 701:521–34
Chan CK, Psaltis D, Özel F. 2013. *Ap. J.* 777:13
Chaty S, Haswell CA, Malzac J, et al. 2003. *MNRAS* 346:689–703
Chen K, Halpern JP. 1989. *Ap. J.* 344:115–24
Chen X, Abramowicz MA, Lasota JP. 1997. *Ap. J.* 476:61–69
Chen X, Abramowicz MA, Lasota JP, Narayan R, Yi I. 1995. *Ap. J. Lett.* 443:L61–64
Chen X, Taam RE. 1993. *Ap. J.* 412:254–66
Chiaberge M, Gilli R, Macchetto FD, Sparks WB. 2006. *Ap. J.* 651:728–34
Chiang CY, Done C, Still M, Godet O. 2010. *MNRAS* 403:1102–12
Chiang J, Blaes O. 2003. *Ap. J.* 586:97–111
Churazov E, Forman W, Jones C, Böhringer H. 2000. *Astron. Astrophys.* 356:788–94
Churazov E, Sazonov S, Sunyaev R, et al. 2005. *MNRAS* 363:L91–95
Churazov E, Sunyaev R, Forman W, Böhringer H. 2002. *MNRAS* 332:729–34
Ciotti L, Ostriker JP. 1997. *Ap. J. Lett.* 487:L105–8
Ciotti L, Ostriker JP. 2001. *Ap. J.* 551:131–52
Ciotti L, Ostriker JP. 2007. *Ap. J.* 665:1038–56
Ciotti L, Ostriker JP, Proga D. 2009. *Ap. J.* 699:89–104
Ciotti L, Ostriker JP, Proga D. 2010. *Ap. J.* 717:708–23
Contopoulos J, Lovelace RVE. 1994. *Ap. J.* 429:139–52
Corbel S, Fender RP, Tzioumis AK, et al. 2000. *Astron. Astrophys.* 359:251–68
Corbel S, Koerding E, Kaaret P. 2008. *MNRAS* 389:1697–702
Corbel S, Nowak MA, Fender RP, Tzioumis AK, Markoff S. 2003. *Astron. Astrophys.* 400:1007–12
Cowie LL, Ostriker JP, Stark AA. 1978. *Ap. J.* 226:1041–62
Cowie LL, Songaila A, Hu E, Cohen JG. 1996. *Astron. J.* 112:839–64
Cowling TG. 1933. *MNRAS* 94:39–48
Crenshaw DM, Kraemer SB. 2012. *Ap. J.* 753:75
Croton DJ, Springel V, White SD, et al. 2006. *MNRAS* 365:11–28
Crumley P, Kumar P. 2013. *MNRAS* 436:1955–60
Cuadra J, Nayakshin S, Martins F. 2008. *MNRAS* 383:458–66
Cui W, Zhang SN, Chen W, Morgan EH. 1999. *Ap. J. Lett.* 512:L43–46
D'Angelo C, Giannios D, Dullemond C, Spruit H. 2008. *Astron. Astrophys.* 488:441–50
Das U, Sharma P. 2013. *MNRAS* 435:2431–44
Davis SW, Stone JM, Pessah ME. 2010. *Ap. J.* 713:52–65
de Gasperin F, Merloni A, Sell P, et al. 2011. *MNRAS* 415:2910–19
De Villiers JP, Hawley JF. 2002. *Ap. J.* 577:866–79
De Villiers JP, Hawley JF. 2003a. *Ap. J.* 589:458–80
De Villiers JP, Hawley JF. 2003b. *Ap. J.* 592:1060–77
De Villiers JP, Hawley JF, Krolik JH. 2003. *Ap. J.* 599:1238–53
De Villiers JP, Hawley JF, Krolik JH, Hirose S. 2005. *Ap. J.* 620:878–88
Debuhr J, Quataert E, Ma CP, Hopkins P. 2010. *MNRAS* 406:L55–59
Dexter J, Agol E, Fragile PC, McKinney JC. 2010. *Ap. J.* 717:1092–104
Dexter J, Fragile PC. 2011. *Ap. J.* 730:36
Dexter J, Fragile PC. 2013. *MNRAS* 432:2252–72

Di Matteo T, Allen SW, Fabian AC, Wilson AS, Young AJ. 2003. *Ap. J.* 582:133–40
Di Matteo T, Johnstone RM, Allen SW, Fabian AC. 2001. *Ap. J. Lett.* 550:L19–23
Di Matteo T, Quataert E, Allen SW, Narayan R, Fabian AC. 2000. *MNRAS* 311:507–21
Di Matteo T, Springel V, Hernquist L. 2005. *Nature* 433:604–7
Di Salvo T, Done C, Zycki PT, Burderi L, Robba NR. 2001. *Ap. J.* 547:1024–33
Diaz Trigo M, Miller-Jones JCA, Migliari S, Broderick JW, Tzioumis T. 2013. *Nature* 504:260–62
Dibi S, Drappeau S, Fragile PC, Markoff S, Dexter J. 2012. *MNRAS* 426:1928–39
Ding J, Yuan F, Liang E. 2010. *Ap. J.* 708:1545–50
Dodds-Eden K, Porquet D, Trap G, et al. 2009. *Ap. J.* 698:676–92
Dodds-Eden K, Sharma P, Quataert E, et al. 2010. *Ap. J.* 725:450–65
Doeleman SS, Agol E, Backer D, et al. 2010. In *Astro2010: The* Astronomy and Astrophysics Decadal Survey, *Science White Papers, Cosmology and Fundamental Physics No. 25*, Natl. Acad., Washington, DC **http://sites.nationalacademies.org/BPA/BPA_050603**
Doeleman SS, Fish VL, Schenck DE, et al. 2012. *Science* 338:355–58
Doeleman SS, Weintroub J, Rogers AEE, et al. 2008. *Nature* 455:78–80
Dolence JC, Gammie CF, Shiokawa H, Noble SC. 2012. *Ap. J. Lett.* 746:L10
Done C, Diaz Trigo M. 2010. *MNRAS* 407:2287–96
Done C, Gierliński M. 2006. *MNRAS* 367:659–68
Done C, Gierliński M, Kubota A. 2007. *Astron. Astrophys. Rev.* 15:1–66
Dubus G, Hameury JM, Lasota JP. 2001. *Astron. Astrophys.* 373:251–71
Dyda S, Lovelace RVE, Ustyugova GV, et al. 2013. *MNRAS* 432:127–37
Eckart A, Baganoff FK, Morris M, et al. 2004. *Astron. Astrophys.* 427:1–11
Eckart A, Baganoff FK, Schödel R, et al. 2006. *Astron. Astrophys.* 450:535–55
Eracleous M, Hwang JA, Flohic HMLG. 2010. *Ap. J. Suppl.* 187:135–48
Esin AA. 1999. *Ap. J.* 517:381–95
Esin AA, McClintock JE, Drake JJ, et al. 2001. *Ap. J.* 555:483–88
Esin AA, McClintock JE, Narayan R. 1997. *Ap. J.* 489:865–89. Erratum. 1998. *Ap. J.* 500:523
Esin AA, Narayan R, Cui W, Grove JE, Zhang SN. 1998. *Ap. J.* 505:854–68
Fabbiano G, Elvis M, Markoff S, et al. 2003. *Ap. J.* 588:175–85
Fabian AC. 2012. *Annu. Rev. Astron. Astrophys.* 50:455–89
Fabian AC, Canizares CR. 1988. *Nature* 333:829–31
Fabian AC, Rees MJ. 1995. *MNRAS* 277:L55–58
Fabrika S. 2004. *Ap. Space Phys. Rev.* 12:1–152
Falcke H, Körding E, Markoff S. 2004. *Astron. Astrophys.* 414:895–903
Falcke H, Markoff S. 2000. *Astron. Astrophys.* 362:113–18
Falcke H, Melia F. 1997. *Ap. J.* 479:740–51
Falcke H, Nagar NM, Wilson AS, Ulvestad JS. 2000. *Ap. J.* 542:197–200
Fanidakis N, Baugh CM, Benson AJ, et al. 2012. *MNRAS* 419:2797–820
Fender R, Stirling AM, Spencer RE, et al. 2006. *MNRAS* 369:603–7
Fender RP. 2001. *MNRAS* 322:31–42
Fender RP. 2006. See Lewin & van der Klis 2006, pp. 381–419
Fender R, Belloni T. 2004. *Annu. Rev. Astron. Astrophys.* 42:317–64
Fender RP, Belloni TM, Gallo E. 2004. *MNRAS* 355:1105–18
Fender RP, Gallo E, Jonker PG. 2003. *MNRAS* 343:L99–103
Fender RP, Gallo E, Russell D. 2010. *MNRAS* 406:1425–34
Ferrarese L, Merritt D. 2000. *Ap. J. Lett.* 539:L9–12
Fish VL, Broderick AE, Doeleman SS, Loeb A. 2009. *Ap. J. Lett.* 692:L14–18
Fish VL, Doeleman SS, Beaudoin C, et al. 2011. *Ap. J. Lett.* 727:L36
Fragile PC. 2009. *Ap. J. Lett.* 706:L246–50
Fragile PC, Anninos P. 2005. *Ap. J.* 623:347–61
Fragile PC, Blaes OM, Anninos P, Salmonson JD. 2007. *Ap. J.* 668:417–29
Fragile PC, Lindner CC, Anninos P, Salmonson JD. 2009. *Ap. J.* 691:482–94
Fragile PC, Meier DL. 2009. *Ap. J.* 693:771–83

Frank J, King A, Raine DJ. 2002. *Accretion Power in Astrophysics*. Cambridge, UK: Cambridge Univ. Press
Fromang S, Papaloizou J. 2007. *Astron. Astrophys.* 476:1113–22
Fromang S, Papaloizou J, Lesur G, Heinemann T. 2007. *Astron. Astrophys.* 476:1123–32
Gallo E, Fender RP, Miller-Jones JCA, et al. 2006. *MNRAS* 370:1351–60
Gallo E, Fender RP, Pooley GG. 2003. *MNRAS* 344:60–72
Gammie CF, McKinney JC, Tóth G. 2003. *Ap. J.* 589:444–57
Gammie CF, Narayan R, Blandford R. 1999. *Ap. J.* 516:177–86
Gammie CF, Popham R. 1998. *Ap. J.* 498:313–26
Gammie CF, Shapiro SL, McKinney JC. 2004. *Ap. J.* 602:312–19
Garcia MR, McClintock JE, Narayan R, et al. 2001. *Ap. J. Lett.* 553:L47–50
Gardner E, Done C. 2013. *MNRAS* 434:3454–62
Gaspari M, Brighenti F, Temi P. 2012. *MNRAS* 424:190–209
Gaspari M, Ruszkowski M, Peng Oh S. 2013. *MNRAS* 432:3401–22
Gebhardt K, Bender R, Bower G, et al. 2000. *Ap. J. Lett.* 539:L13–16
Genzel R, Eisenhauer F, Gillessen S. 2010. *Rev. Mod. Phys.* 82:3121–95
Genzel R, Schödel R, Ott T, et al. 2003. *Nature* 425:934–37
Ghez AM, Duchene G, Matthews K, et al. 2003. *Ap. J. Lett.* 586:L127–31
Ghez AM, Salim S, Weinberg NN, et al. 2008. *Ap. J.* 689:1044–62
Ghez AM, Wright SA, Matthews K, et al. 2004. *Ap. J. Lett.* 601:L159–62
Giannios D, Spruit HC. 2004. *Astron. Astrophys.* 427:251–61
Gierliński M, Done C, Page K. 2008. *MNRAS* 388:753–60
Gierliński M, Newton J. 2006. *MNRAS* 370:837–44
Gilfanov M, Churazov E, Revnivtsev M. 1999. *Astron. Astrophys.* 352:182–88
Gillessen S, Eisenhauer F, Fritz TK, et al. 2009a. *Ap. J. Lett.* 707:L114–17
Gillessen S, Eisenhauer F, Quataert E, et al. 2006. *Ap. J. Lett.* 640:L163–66
Gillessen S, Eisenhauer F, Trippe S, et al. 2009b. *Ap. J.* 692:1075–109
Gillessen S, Genzel R, Fritz TK, et al. 2012. *Nature* 481:51–54
Gillessen S, Genzel R, Fritz TK, et al. 2013. *Ap. J.* 774:44
Goldston JE, Quataert E, Igumenshchev IV. 2005. *Ap. J.* 621:785–92
Goodson AP, Bohm KH, Winglee RM. 1999. *Ap. J.* 524:142–58
Goodson AP, Winglee RM. 1999. *Ap. J.* 524:159–68
Gu WM, Lu JF. 2000. *Ap. J. Lett.* 540:L33–36
Guan X, Gammie CF, Simon JB, Johnson BM. 2009. *Ap. J.* 694:1010–18
Guilet J, Ogilvie GI. 2012. *MNRAS* 424:2097–117
Guilet J, Ogilvie GI. 2013. *MNRAS* 430:822–35
Gültekin K, Cackett EM, Miller JM, et al. 2009. *Ap. J.* 706:404–16
Guo F, Mathews WG. 2012. *Ap. J.* 756:181
Haardt F, Maraschi L. 1993. *Ap. J.* 413:507–17
Hada K, Doi A, Kino M, et al. 2011. *Nature* 477:185–87
Hameury JM, Lasota JP, McClintock JE, Narayan R. 1997. *Ap. J.* 489:234–43
Hartnoll SA, Blackman EG. 2001. *MNRAS* 324:257–66
Hawley JF. 2000. *Ap. J.* 528:462–79
Hawley JF, Balbus SA. 1991. *Ap. J.* 376:223–33
Hawley JF, Balbus SA. 1996. In *Physics of Accretion Disks*, ed. S Kato, et al., pp. 273–78. New York: Gordon & Breach
Hawley JF, Balbus SA. 2002. *Ap. J.* 573:738–48
Hawley JF, Balbus SA, Stone JM. 2001. *Ap. J. Lett.* 554:L49–52
Hawley JF, Gammie CF, Balbus SA. 1995. *Ap. J.* 440:742
Hawley JF, Gammie CF, Balbus SA. 1996. *Ap. J.* 464:690
Hawley JF, Guan X, Krolik JH. 2011. *Ap. J.* 738:84
Hawley JF, Krolik JH. 2001. *Ap. J.* 548:348–67
Hawley JF, Krolik JH. 2002. *Ap. J.* 566:164–80
Hawley JF, Krolik JH. 2006. *Ap. J.* 641:103–16

Hawley JF, Richers SA, Guan X, Krolik JH. 2013. *Ap. J.* 722:102
Hayashi MR, Shibata K, Matsumoto R. 1996. *Ap. J. Lett.* 468:L37–40
Heinz S. 2004. *MNRAS* 355:835–44
Heinz S, Brüggen M, Young A, Levesque E. 2006. *MNRAS* 373:L65–69
Heinz S, Sunyaev RA. 2003. *MNRAS* 343:L59–64
Heyvaerts J, Norman C. 1989. *Ap. J.* 347:1055–81
Hirose S, Krolik JH, Blaes O. 2009. *Ap. J.* 691:16–31
Hirose S, Krolik JH, De Villiers JP, Hawley JH. 2004. *Ap. J.* 606:1083–97
Hirotani K, Okamoto I. 1998. *Ap. J.* 497:563
Ho LC. 1999. *Ap. J.* 516:672–82
Ho LC. 2002. *Ap. J.* 564:120–32
Ho LC. 2008. *Annu. Rev. Astron. Astrophys.* 46:475–539
Ho LC. 2009. *Ap. J.* 699:626–37
Ho LC, Terashima Y, Ulvestad JS. 2003. *Ap. J.* 589:783–89
Honma F. 1996. *Publ. Astron. Soc. Jpn.* 48:77–87
Hopkins PF, Hernquist L, Cox TJ, et al. 2006. *Ap. J. Suppl.* 163:1–49
Hoshino M. 2012. *Phys. Rev. Lett.* 108:135003
Hoshino M. 2013. *Ap. J.* 773:118
Huang L, Cai M, Shen ZQ, Yuan F. 2007. *MNRAS* 379:833–40
Ichimaru S. 1977. *Ap. J.* 214:840–55
Igumenshchev IV. 2002. *Ap. J. Lett.* 577:L31–34
Igumenshchev IV. 2006. *Ap. J.* 649:361–72
Igumenshchev IV. 2008. *Ap. J.* 677:317–26
Igumenshchev IV, Abramowicz MA. 1999. *Ap. J. Lett.* 537:L27
Igumenshchev IV, Abramowicz MA. 2000. *Ap. J. Suppl.* 130:463–84
Igumenshchev IV, Abramowicz MA, Narayan R. 2000. *Ap. J. Lett.* 537:L27–30
Igumenshchev IV, Chen XM, Abramowicz M. 1996. *MNRAS* 278:236–50
Igumenshchev IV, Narayan R. 2002. *Ap. J.* 566:137–47
Igumenshchev IV, Narayan R, Abramowicz MA. 2003. *Ap. J.* 592:1042–59
Ingram A, Done C. 2011. *MNRAS* 415:2323–35
Ingram A, Done C, Fragile PC. 2009. *MNRAS* 397:L101–5
Inogamov NA, Sunyaev RA. 2010. *Astron. Lett.* 36:835–47
Janiuk A, Sznajder M, Mościbrodzka M, Proga D. 2009. *Ap. J.* 705:1503–21
Jaroszynski M, Kurpiewski A. 1997. *Astron. Astrophys.* 326:419–26
Jiang YF, Stone JM, Davis SW. 2013. *Ap. J.* 778:65–76
Jiao CL, Wu XB. 2011. *Ap. J.* 733:112
Johnson BM, Quataert E. 2007. *Ap. J.* 660:1273–81
Kalemci E, Dincer T, Tomsick JA, et al. 2013. *Ap. J.* 779:95
Kato S, Abramowicz MA, Chen X. 1996. *Publ. Astron. Soc. Jpn.* 48:67–75
Kato S, Fukue J, Mineshige S. 2008. *Black-Hole Accretion Disks: Towards a New Paradigm*. Kyoto: Kyoto Univ. Press
Kato S, Yamasaki T, Abramowicz MA, Chen X. 1997. *Publ. Astron. Soc. Jpn.* 49:221–25
Kato Y, Hayashi MR, Matsumoto R. 2004a. *Ap. J.* 600:338–42
Kato Y, Mineshige S, Shibata K. 2004b. *Ap. J.* 605:307–20
Kato Y, Umemura M, Ohsuga K. 2009. *MNRAS* 400:1742–48
Katz J. 1977. *Ap. J.* 215:265–75
Kawamuro T, Ueda Y, Tazaki F, Terashima Y. 2013. *Ap. J.* 770:157
King A. 2003. *Ap. J. Lett.* 596:L27–29
King A. 2005. *Ap. J. Lett.* 635:L121–23
King AR. 2010. *MNRAS* 402:1516–22
Koide S. 2003. *Phys. Rev. D* 67:104010
Koide S, Meier DL, Shibata K, Kudoh T. 2000. *Ap. J.* 536:668–74
Koide S, Shibata K, Kudoh T. 1999. *Ap. J.* 522:727–52

Kolehmainen M, Done C, Diaz Trigo M. 2014. *MNRAS* 437:316–26
Komissarov SS. 1999. *MNRAS* 303:343–66
Komissarov SS. 2001. *MNRAS* 326:L41–44
Komissarov SS, Barkov MV, Vlahakis N, Konigl A. 2007. *MNRAS* 380:51–70
Komissarov SS, McKinney JC. 2007. *MNRAS* 377:L49–53
Koratkar A, Blaes O. 1999. *Publ. Astron. Soc. Jpn.* 755:1–30
Körding E, Falcke H, Corbel S. 2006. *Astron. Astrophys.* 456:439–50
Kormendy J, Ho LC. 2013. *Annu. Rev. Astron. Astrophys.* 51:511–653
Kriek M, van Dokkum PG, Franx M, et al. 2007. *Ap. J.* 669:776–90
Krolik JH, Hawley JF, Hirose S. 2005. *Ap. J.* 622:1008–23
Kudoh T, Matsumoto R, Shibata K. 1998. *Ap. J.* 508:186–99
Kudoh T, Matsumoto R, Shibata K. 2002. *Publ. Astron. Soc. Jpn.* 54:267–74
Kusunose M, Mineshige S. 1996. *Ap. J.* 468:330–37
Kuwabara T, Shibata K, Kudoh T, Matsumoto R. 2000. *Publ. Astron. Soc. Jpn.* 52:1109–24
Lasota JP. 1999. *Phys. Rep.* 311:247–58
Lasota JP, Abramowicz MA, Chen X, et al. 1996a. *Ap. J.* 462:142–46
Lasota JP, Gourgoulhon E, Abramowicz M, Tchekhovskoy A, Narayan R. 2014. *Phys. Rev. D.* 89:024041
Lasota JP, Narayan R, Yi I. 1996b. *Astron. Astrophys.* 314:813–20
Lehe R, Parrish IJ, Quataert E. 2009. *Ap. J.* 707:404–19
Lewin W, van der Klis M, eds. 2006. *Compact Stellar X-Ray Sources*. Camb. Astrophys. Ser. Cambridge, UK: Cambridge Univ. Press
Li J, Ostriker J, Sunyaev R. 2013a. *Ap. J.* 767:105
Li Z, Morris MR, Baganoff FK. 2013b. *Ap. J.* 779:154–62
Li Z, Wu XB, Wang R. 2008. *Ap. J.* 688:826–36
Li ZY, Chiueh T, Begelman MC. 1992. *Ap. J.* 394:459–71
Liang EPT, Price RH. 1977. *Ap. J.* 218:247–52
Lightman AP, Eardley DM. 1974. *Ap. J. Lett.* 187:L1–4
Liu BF, Done C, Taam RE. 2011. *Ap. J.* 726:10
Liu BF, Meyer F, Meyer-Hofmeister E. 2005. *Astron. Astrophys.* 442:555–62
Liu BF, Taam RE, Meyer-Hofmeister E, Meyer F. 2007. *Ap. J.* 671:695–705
Liu BF, Yuan W, Meyer F, Meyer-Hofmeister E. 1999. *Ap. J. Lett.* 527:L17–20
Liu C, Yuan F, Ostriker JP, Gan Z, Yang X. 2013. *MNRAS* 434:1721–35
Liu H, Wu Q. 2013. *Ap. J.* 764:17
Livio M, Ogilvie GI, Pringle JE. 1999. *Ap. J.* 512:100–4
Lodato G, Pringle JE. 2006. *MNRAS* 368:1196–208
Loewenstein M, Mushotzky RF, Angelini L, Arnaud KA, Quataert E. 2001. *Ap. J. Lett.* 555:L21–24
Lovelace RVE. 1976. *Nature* 262:649–52
Lovelace RVE, Mehanian CM, Mobarry CM, Sulkanen ME. 1986. *Ap. J. Suppl.* 62:1–37
Lu JF, Lin YQ, Gu WM 2004. *Ap. J. Lett.* 602:L37
Lynden-Bell D. 2003. *MNRAS* 341:1360–72
Lynden-Bell D, Pringle JE. 1974. *MNRAS* 168:603–37
MacDonald D, Thorne KS. 1982. *MNRAS* 198:345–82
Machida M, Hayashi MR, Matsumoto R. 2000. *Ap. J. Lett.* 532:L67–70
Machida M, Matsumoto R. 2003. *Ap. J.* 585:429–42
Machida M, Matsumoto R, Mineshige S. 2001. *Publ. Astron. Soc. Jpn.* 53:L1–4
Machida M, Nakamura KE, Matsumoto R. 2004. *Publ. Astron. Soc. Jpn.* 56:671–79
Machida M, Nakamura KE, Matsumoto R. 2006. *Publ. Astron. Soc. Jpn.* 58:193–202
Magorrian J, Tremaine S, Richstone D, et al. 1998. *Astron. J.* 115:2285–305
Mahadevan R. 1997. *Ap. J.* 477:585–601
Mahadevan R. 1998. *Nature* 394:651–53
Mahadevan R. 1999. *MNRAS* 304:501–11
Mahadevan R, Narayan R, Krolik J. 1997. *Ap. J.* 486:268–75
Mahadevan R, Quataert E. 1997. *Ap. J.* 490:605–18

Malzac J, Belmont R, Fabian AC. 2009. *MNRAS* 400:1512–20
Malzac J, Beloborodov AM, Poutanen J. 2001. *MNRAS* 326:417–27
Malzac J, Merloni A, Fabian AC. 2004. *MNRAS* 351:253–64
Manmoto T. 2000. *Ap. J.* 534:734–46
Manmoto T, Kato S. 2000. *Ap. J.* 538:295–306
Manmoto T, Kato S, Nakamura KE, Narayan R. 2000. *Ap. J.* 529:127–37
Manmoto T, Mineshige S, Kusunose M. 1997. *Ap. J.* 489:791–803
Manmoto T, Takeuchi M, Mineshige S, Matsumoto R, Negoro H. 1996. *Ap. J. Lett.* 464:L135–38
Maoz D. 2007. *MNRAS* 377:1696–710
Maraschi L, Tavecchio F. 2003. *Ap. J.* 593:667–75
Markoff S, Bower GC, Falcke H. 2007. *MNRAS* 379:1519–32
Markoff S, Falcke H, Fender R. 1999. *Astron. Astrophys.* 372:L25–28
Markoff S, Falcke H, Yuan F, Biermann P. 2001. *Astron. Astrophys.* 379:L13–16
Markoff S, Nowak M, Corbel S, Fender R, et al. 2003. *Astron. Astrophys.* 397:645
Markoff S, Nowak MA, Wilms J. 2005. *Ap. J.* 635:1203–16
Marrone DP, Moran JM, Zhao JH, Rao R. 2006. *Ap. J.* 640:308–18
Marrone DP, Moran JM, Zhao JH, Rao R. 2007. *Ap. J. Lett.* 654:L57–60
Marsch E. 2012. *Space Sci. Rev.* 172:23–39
Massi M, Poletto G. 2011. *Mem. Soc. Astron. Ital.* 82:145
Matsumoto R, Shibata K. 1997. In *Accretion Phenomena and Related Outflows, IAU Colloq. 163*, ed. D Wickramsinghe, G Bicknell, L Ferrario. ASP Conf. Ser. 121:443–47. San Francisco: ASP
Matsumoto R, Tajima T. 1995. *Ap. J.* 445:767–79
Mayer M, Pringle JE. 2007. *MNRAS* 376:435–56
McClintock JE, Narayan R, Davis SW, et al. 2011. *Class. Quantum Gravity* 28:114009
McClintock JE, Narayan R, Garcia MR, et al. 2003. *Ap. J.* 593:435–51
McClintock JE, Narayan R, Steiner JF. 2013. *Space Sci. Rev.* In press. doi: 10.1007/s11214-013-0003-9
McClintock JE, Remillard RA. 2006. See Lewin & van der Klis 2006, pp. 157–213
McKinney JC. 2005. *Ap. J. Lett.* 630:L5–8
McKinney JC. 2006. *MNRAS* 368:1561–82
McKinney JC, Blandford RD. 2009. *MNRAS* 394:L126–30
McKinney JC, Gammie CF. 2004. *Ap. J.* 611:977–95
McKinney JC, Tchekhovskoy A, Blandford RD. 2012. *MNRAS* 423:3083–117
McKinney JC, Tchekhovskoy A, Blandford RD. 2013. *Science* 339:49–52
McNamara BR, Kazemzadeh F, Rafferty DA, et al. 2009. *Ap. J.* 698:594–605
Medvedev MV. 2000. *Ap. J.* 541:811–20
Meier DL. 2005. *Ap. Space. Sci.* 300:55–65
Melia F. 1992. *Ap. J. Lett.* 387:L25–28
Melia F, Liu S, Coker R. 2001. *Ap. J.* 553:146–57
Menou K, Esin AA, Narayan R, et al. 1999a. *Ap. J.* 520:276–91
Menou K, Hameury JM, Lasota JP, Narayan R. 2000. *MNRAS* 314:498–510
Menou K, Narayan R, Lasota JP. 1999b. *Ap. J.* 513:811–26
Merloni A, Fabian AC. 2002. *MNRAS* 332:165–75
Merloni A, Heinz S, Di Matteo T. 2003. *MNRAS* 345:1057–76
Meyer F, Liu BF, Meyer-Hofmeister E. 2000. *Astron. Astrophys.* 361:175–88
Meyer F, Liu BF, Meyer-Hofmeister E. 2007. *Astron. Astrophys.* 463:1–9
Meyer F, Meyer-Hofmeister E. 1994. *Astron. Astrophys.* 288:175–82
Meyer L, Ghez AM, Schödel R, et al. 2012. *Science* 338:84–87
Meyer-Hofmeister E, Liu BF, Meyer F. 2005. *Astron. Astrophys.* 432:181–87
Meyer-Hofmeister E, Liu BF, Meyer F. 2009. *Astron. Astrophys.* 508:329–37
Miller JM, Homan J, Steeghs D, et al. 2006. *Ap. J.* 653:525–35
Mineshige S, Kawaguchi T, Takeuchi M, Hayashida K. 2000. *Publ. Astron. Soc. Jpn.* 52:499–508
Morganti R, Fogasy J, Paragi Z, Oosterloo T, Orienti M. 2013. *Science* 341:1082–85
Mościbrodzka M, Gammie CF, Dolence JC, Shiokawa H. 2011. *Ap. J.* 735:9

Mościbrodzka M, Gammie CF, Dolence JC, Shiokawa H, Leung PK. 2009. *Ap. J.* 706:497–507
Mościbrodzka M, Proga D, Czerny B, Siemiginowska A. 2007. *Astron. Astrophys.* 474:1–13
Mościbrodzka M, Shiokawa H, Gammie CF, Dolence JC. 2012. *Ap. J. Lett.* 752:L1
Nagar NM, Falcke H, Wilson AS, Ho LC. 2000. *Ap. J.* 542:186–96
Nakamura KE. 1998. *Publ. Astron. Soc. Jpn.* 50:L11–14
Nakamura KE, Kusunose M, Matsumoto R, Kato S. 1997. *Publ. Astron. Soc. Jpn.* 49:503–12
Nakamura KE, Matsumoto R, Kusunose M, Kato S. 1996. *Publ. Astron. Soc. Jpn.* 48:761–69
Narayan R. 1996. *Ap. J.* 462:136–41
Narayan R, Barret D, McClintock JE. 1997a. *Ap. J.* 482:448–64
Narayan R, Fabian A. 2011. *MNRAS* 415:3721–30
Narayan R, Garcia MR, McClintock JE. 1997b. *Ap. J. Lett.* 478:L79–82
Narayan R, Igumenshchev IV, Abramowicz MA. 2000. *Ap. J.* 539:798–808
Narayan R, Igumenshchev IV, Abramowicz MA. 2003. *Publ. Astron. Soc. Jpn.* 55:L69–72
Narayan R, Kato S, Honma F. 1997c. *Ap. J.* 476:49–60
Narayan R, Mahadevan R, Grindlay JE, Popham RG, Gammie C. 1998a. *Ap. J.* 492:554–68
Narayan R, Mahadevan R, Quataert E. 1998b. In *Theory of Black Hole Accretion Disks*, ed. MA Abramowicz, G Bjornsson, JE Pringle, pp. 148. Cambridge, UK: Cambridge Univ. Press
Narayan R, McClintock JE. 2005. *Ap. J.* 623:1017–25
Narayan R, McClintock JE. 2008. *New Astron. Rev.* 51:733–51
Narayan R, McClintock JE. 2012. *MNRAS* 419:L69–73
Narayan R, McClintock JE, Yi I. 1996. *Ap. J.* 457:821–33
Narayan R, Medvedev MV. 2001. *Ap. J. Lett.* 562:L129–32
Narayan R, Özel F, Sironi L. 2012a. *Ap. J. Lett.* 757:L20
Narayan R, Quataert E, Igumenshchev IV, Abramowicz MA. 2002. *Ap. J.* 577:295–301
Narayan R, Raymond J. 1999. *Ap. J. Lett.* 515:L69–72
Narayan R, Sädowski A, Penna RF, Kulkarni AK. 2012b. *MNRAS* 426:3241–59
Narayan R, Yi I. 1994. *Ap. J. Lett.* 428:L13–16
Narayan R, Yi I. 1995a. *Ap. J.* 444:231–43
Narayan R, Yi I. 1995b. *Ap. J.* 452:710–35
Narayan R, Yi I, Mahadevan R. 1995. *Nature* 374:623–25
Nemmen RS, Storchi-Bergmann T, Eracleous M. 2014. *MNRAS* 438:2804–27
Nemmen RS, Storchi-Bergmann T, Yuan F, et al. 2006. *Ap. J.* 643:652–59
Nipoti C, Blundell KM, Binney J. 2005. *MNRAS* 361:633–37
Noble SC, Krolik JH, Hawley JF. 2010. *Ap. J.* 711:959–73
Noble SC, Leung PK, Gammie CF, Book LG. 2007. *Class. Quantum Gravity* 24:S259–74
Novak GS, Ostriker JP, Ciotti L. 2011. *Ap. J.* 737:26
Novikov ID, Thorne KS. 1973. In *Black Holes* (*Les Astres Occlus*), ed. C DeWitt, BS DeWitt, pp. 343–450. New York: Gordon & Breach
Oda H, Machida M, Nakamura KE, Matsumoto R. 2010. *Ap. J.* 712:639–52
Ogilvie GI. 1999. *MNRAS* 306:L9–13
Ohsuga K, Mineshige S. 2011. *Ap. J.* 736:2
Ohsuga K, Mineshige S, Mori M, Kato Y. 2009. *Publ. Astron. Soc. Jpn.* 61:L7–11
Ohsuga K, Mori M, Nakamoto T, Mineshige S. 2005. *Ap. J.* 628:368–81
Omma H, Binney J, Bryan G, Slyz A. 2004. *MNRAS* 348:1105–19
Ostriker E. 1997. *Ap. J.* 486:291–306
Ostriker JP, Choi E, Ciotti L, Novak GS, Proga D. 2010. *Ap. J.* 722:642–52
Özel F, Psaltis D, Narayan R. 2000. *Ap. J.* 541:234–49
Paczyński B, Wiita PJ. 1980. *Astron. Astrophys.* 88:23–31
Pang B, Pen UL, Matzner CD, Green SR, Liebendörfer M. 2011. *MNRAS* 415:1228–39
Papaloizou JCB, Lin DNC. 1995. *Ap. J.* 438:841–51
Park M, Ostriker JP. 2001. *Ap. J.* 549:100–17
Park M, Ostriker JP. 2007. *Ap. J.* 655:88–97
Parrish IJ, Stone JM. 2007. *Ap. J.* 664:135–48

Pedlar A, Ghataure HS, Davies RD, et al. 1990. *MNRAS* 246:477–89
Peitz J, Appl S. 1997. *MNRAS* 286:681–95
Pellegrini S, Baldi A, Fabbiano G, Kim DM. 2003. *Ap. J.* 597:175–85
Pellegrini S, Siemiginowska A, Fabbiano G, et al. 2007. *Ap. J.* 667:749–59
Pen UL, Matzner CD, Wong S. 2003. *Ap. J. Lett.* 596:L207–10
Penna RF, McKinney JC, Narayan R, et al. 2010. *MNRAS* 408:752–82
Penna RF, Narayan R, Sadowski A. 2013a. *MNRAS* 436:3741–58
Penna RF, Sadowski A, Kulkarni AK, Narayan R. 2013b. *MNRAS* 428:2255–74
Penrose R. 1969. *Riv. Nuovo Cim.* 1:252–76
Perna R, Raymond J, Narayan R. 2000. *Ap. J.* 541:898–907
Pessah ME, Chan CK, Psaltis D. 2007. *Ap. J. Lett.* 668:L51–54
Peterson JR, Paerels FBS, Kaastra JS, et al. 2001. *Astron. Astrophys.* 365:L104–9
Phifer K, Do T, Meyer L, et al. 2013. *Ap. J. Lett.* 773:L13
Phinney ES. 1983. In *Astrophysical Jets*, ed. A Ferrari, AG Pacholczyk, 103:201–12. Dordrecht, Ger.: Reidel
Piran T. 1978. *Ap. J.* 221:652–60
Pizzolato F, Soker N. 2005. *Ap. J.* 632:821–30
Plant DS, Fender RP, Ponti G, Munoz-Darias T, Coriat M. 2013. *MNRAS*. Submitted. arXiv:1309.4781
Plotkin RM, Markoff S, Kelly BC, Körding E, Anderson SF. 2012. *MNRAS* 419:267–86
Popham R, Gammie CF. 1998. *Ap. J.* 504:419–30
Poutanen J, Krolik JH, Ryde F. 1997. *MNRAS* 292:L21–25
Poutanen J, Lipunova G, Fabrika S, Butkevich AG, Abolmasov P. 2007. *MNRAS* 377:1187–94
Poutanen J, Veledina A. 2014. *Space Sci. Rev.* In press. doi: 10.1007/s11214-013-0033-3
Poutanen J, Vurm I. 2009. *Ap. J. Lett.* 690:L97–100
Press WH, Teukolsky SA, Vetterling WT, Flannery BP. 1992. *Numerical Recipes in FORTRAN. The Art of Scientific Computing*. Cambridge, UK: Cambridge Univ. Press
Press WH, Teukolsky SA, Vetterling WT, Flannery BP. 2002. *Numerical Recipes in C++: The Art of Scientific Computing*. Cambridge, UK: Cambridge Univ. Press
Pringle JE. 1981. *Annu. Rev. Astron. Astrophys.* 19:137–62
Proga D. 2007. *Ap. J.* 661:693–701
Proga D, Begelman MC. 2003. *Ap. J.* 582:69–81
Proga D, Ostriker JP, Kurosawa R. 2008. *Ap. J.* 676:101–12
Ptak A, Terashima Y, Ho LC, Quataert E. 2004. *Ap. J.* 606:173–84
Pudritz RE, Norman CA. 1983. *Ap. J.* 274:677–97
Punsly B, Coroniti FV. 1989. *Phys. Rev. D.* 40:3834–57
Qiao E, Liu BF. 2013. *Ap. J.* 764:2
Quataert E. 1998. *Ap. J.* 500:978–91
Quataert E. 2001. In *Probing the Physics of Active Galactic Nuclei*, ed. BM Peterson, RW Pogge, RS Polidan, pp. 71. San Francisco: ASP
Quataert E. 2002. *Ap. J.* 575:855–59
Quataert E. 2004. *Ap. J.* 613:322–25
Quataert E, Di Matteo T, Narayan R, Ho LC. 1999. *Ap. J. Lett.* 525:L89–92
Quataert E, Dorland W, Hammett GW. 2002. *Ap. J.* 577:524–33
Quataert E, Gruzinov A. 1999. *Ap. J.* 520:248–55
Quataert E, Gruzinov A. 2000. *Ap. J.* 539:809–14
Quataert E, Narayan R. 1999a. *Ap. J.* 516:399–410
Quataert E, Narayan R. 1999b. *Ap. J.* 520:298–315
Rafferty DA, McNamara BR, Nulsen PEJ, Wise MW. 2006. *Ap. J.* 652:216–31
Rakowski CE. 2005. *Adv. Space Res.* 35:1017–26
Ramadevi MC, Seetha S. 2007. *MNRAS* 378:182–88
Rees MJ, Begelman MC, Blandford RD, Phinney ES. 1982. *Nature* 295:17–21
Reis RC, Fabian AC, Miller JM. 2010. *MNRAS* 402:836–54
Reis RC, Fabian AC, Ross RR, et al. 2008. *MNRAS* 387:1489–98
Remillard RA, McClintock JE. 2006. *Annu. Rev. Astron. Astrophys.* 44:49–92

Reynolds CS. 2013. *Space Sci. Rev.* In press. doi: 10.1007/s11214-013-0006-6
Reynolds CS, Di Matteo T, Fabian AC, Hwang U, Canizares CR. 1996. *MNRAS* 283:L111–16
Reynolds CS, Fabian AC. 2008. *Ap. J.* 675:1048–56
Rezzolla L, Yoshida S, Maccarone TJ, Zanotti O. 2003. *MNRAS* 344:L37–41
Riquelme MA, Quataert E, Sharma P, Spitkovsky A. 2012. *Ap. J.* 755:50
Romanova MM, Ustyugova GV, Koldoba AV, Chechetkin VM, Lovelace RVE. 1998. *Ap. J.* 500:703–13
Rothstein DM, Lovelace RVE. 2008. *Ap. J.* 677:1221–32
Różańska A, Czerny B. 2000. *Astron. Astrophys.* 360:1170–86
Ruffini R, Wilson JR. 1975. *Phys. Rev. D* 12:2959–62
Russell DM, Gallo E, Fender RP. 2013a. *MNRAS* 431:405–14
Russell HR, McNamara BR, Edge AC, et al. 2013b. *MNRAS* 432:530–53
Ruszkowski M, Begelman MC. 2002. *Ap. J.* 581:223–28
Rykoff E, Miller JM, Steeghs D, Torres MAP. 2007. *Ap. J.* 666:1129–39
Sadowski A, Narayan R, McKinney JC, Tchekhovskoy A. 2014. *MNRAS* 439:503–20
Sadowski A, Narayan R, Penna R, Zhu Y. 2013a. *MNRAS* 436:3856–74
Sadowski A, Sironi L, Abarca D, et al. 2013b. *MNRAS* 432:478–91
Saitoh TR, Makino J, Asaki Y, et al. 2014. *Publ. Astron. Soc. Jpn.* 66:1–9
Sazonov S, Sunyaev R, Revnivtsev M. 2012. *MNRAS* 420:388–404
Sazonov SY, Ostriker JP, Ciotti L, Sunyaev RA. 2005. *MNRAS* 358:168–80
Schartmann M, Burkert A, Alig C, et al. 2012. *Ap. J.* 755:155
Schartmann M, Burkert A, Alig C, et al. 2013. Presented at Nuclei of Seyfert galaxies and QSOs - central engine and conditions of star formation. Proc. Sci., Max-Planck-Inst. Radioastron., Nov. 6–8, Bonn, Ger.
Schechter P. 1976. *Ap. J.* 203:297–306
Scheuer PAG, Feiler R. 1996. *MNRAS* 282:291–94
Schnittman JD, Krolik JH, Noblem SC. 2013. *Ap. J.* 769:156
Schödel R, Morris MR, Muzic K, et al. 2011. *Astron. Astrophys.* 532:83
Schödel R, Ott T, Genzel R, et al. 2002. *Nature* 419:694
Shafee R, McKinney JC, Narayan R, et al. 2008. *Ap. J. Lett.* 687:L25–28
Shakura NI, Sunyaev RA. 1973. *Astron. Astrophys.* 24:337
Shapiro SL, Lightman AP, Eardley DM. 1976. *Ap. J.* 204:187–99
Sharma P, Hammett GW, Quataert E, Stone JM. 2006. *Ap. J.* 637:952–67
Sharma P, Quataert E, Hammett GW, Stone JM. 2007a. *Ap. J.* 667:714–23
Sharma P, Quataert E, Stone JM. 2007b. *Ap. J.* 671:1696–707
Sharma P, Quataert E, Stone JM. 2008. *MNRAS* 389:1815–27
Shen ZQ, Lo KY, Liang MC, Ho PTP, Zhao JH. 2005. *Nature* 438:62–64
Shcherbakov RV. 2014. *Ap. J.* 783:31
Shcherbakov RV, Baganoff FK. 2010. *Ap. J.* 716:504–9
Shcherbakov RV, McKinney JC. 2013. *Ap. J. Lett.* 774:L22
Shcherbakov RV, Penna RF, McKinney JC. 2012. *Ap. J.* 755:133
Shibata K, Uchida Y. 1985. *Publ. Astron. Soc. Jpn.* 37:31–46
Shibata K, Uchida Y. 1986. *Publ. Astron. Soc. Jpn.* 38:631–60
Shiokawa H, Dolence JC, Gammie CF, Noble SC. 2012. *Ap. J.* 744:187
Silk J. 2013. *Ap. J.* 772:112
Somerville RS, Hopkins PF, Cox TJ, Robertson BE, Hernquist L. 2008. *MNRAS* 391:481–506
Sorathia KA, Reynolds CS, Armitage PJ. 2010. *Ap. J.* 712:1241–47
Sorathia KA, Reynolds CS, Stone JM, Beckwith K. 2012. *Ap. J.* 749:189
Spruit HC, Deufel B. 2002. *Astron. Astrophys.* 387:918–30
Spruit HC, Matsuda T, Inoue M, Sawada K. 1987. *MNRAS* 229:517–27
Spruit HC, Uzdensky DA. 2005. *Ap. J.* 629:960–68
Springel V, Di Matteo T, Hernquist L. 2005. *MNRAS* 361:776–94
Steiner JF, McClintock JE, Narayan R. 2013. *Ap. J.* 762:104
Stern BE, Poutanen J, Svensson R, Sikora M, Begelman MC. 1995. *Ap. J. Lett.* 449:L13–16
Stone JM, Gardiner TA, Teuben P, Hawley JF, Simon JB. 2008. *Ap. J.* 178:137–77

Stone JM, Hawley JF, Gammie CF, Balbus SA. 1996. *Ap. J.* 463:656–73
Stone JM, Norman ML. 1992a. *Ap. J. Suppl.* 80:753–90
Stone JM, Norman ML. 1992b. *Ap. J. Suppl.* 80:791–818
Stone JM, Pringle JE. 2001. *MNRAS* 322:461–72
Stone JM, Pringle JE, Begelman MC. 1999. *MNRAS* 310:1002–16
Storchi-Bergmann T, Eracleous M, Ruiz MT, et al. 1997. *Ap. J.* 489:87–93
Su M, Slatyer TR, Finkbeiner DP. 2010. *Ap. J.* 724:1044–82
Taam RE, Liu B, Yuan W, Qiao E. 2012. *Ap. J.* 759:65
Tanaka T, Menou K. 2006. *Ap. J.* 649:345–60
Tchekhovskoy A, McKinney JC. 2012. *MNRAS* 423:L55–59
Tchekhovskoy A, McKinney JC, Narayan R. 2012. *J. Phys. Chem. Sci.* 372:012040
Tchekhovskoy A, Metzger BD, Giannios D, Kelley LZ. 2014. *MNRAS* 437:2744–60
Tchekhovskoy A, Narayan R, McKinney JC. 2010. *Ap. J.* 711:50–63
Tchekhovskoy A, Narayan R, McKinney JC. 2011. *MNRAS* 418:L79–83
Thorne KS, Price RH, MacDonald DA. 1986. *Black Holes: The Membrane Paradigm*. New Haven, CT: Yale Univ. Press
Tomsick JA, Yamaoka K, Corbel S. 2009. *Ap. J. Lett.* 707:L87–91
Trap G, Goldwurm A, Dodds-Eden K, et al. 2011. *Astron. Astrophys.* 528:140
van der Laan H. 1966. *Nature* 211:1131–33
van Velzen S, Falcke H. 2013. *Astron. Astrophys.* 557:L7–10
Veledina A, Poutanen J, Ingram A. 2013. *Ap. J.* 773:165
Vernaleo JC, Reynolds C. 2006. *Ap. J.* 645:83–94
Vlahakis N, Konigl A. 2003. *Ap. J.* 596:1080–103
Wang QD, Nowak MA, Markoff SB, et al. 2013. *Science* 341:981–83
Wang R, Wu XB, Kong MZ. 2006. *Ap. J.* 645:890–99
Watarai K, Mizuno T, Mineshige S. 2001. *Ap. J. Lett.* 549:L77–88
Wilson AS, Yang Y. 2002. *Ap. J.* 568:133–40
Wrobel JM, Terashima Y, Ho L. 2008. *Ap. J.* 675:1041–47
Wu Q, Yuan F, Cao X. 2007. *Ap. J.* 669:96–105
Wu XB. 1997. *MNRAS* 292:113–19
Wu XB, Li Q. 1996. *Ap. J.* 469:776–83
Wyithe JSB, Loeb A. 2003. *Ap. J.* 595:614–23
Xie FG, Yuan F. 2008. *Ap. J.* 681:499–505
Xie FG, Yuan F. 2012. *MNRAS* 427:1580–86
Xu G, Chen X. 1997. *Ap. J. Lett.* 489:L29–32
Xu YD, Cao XW. 2009. *Res. Astron. Astrophys.* 9:401–8
Xu YD, Narayan R, Quataert E, Yuan F, Baganoff FK. 2006. *Ap. J.* 640:319–26
Xue L, Wang JC. 2005. *Ap. J.* 623:372–82
Younes G, Porquet D, Sabra B, Reeves JN. 2011. *Astron. Astrophys.* 530:149
Younes G, Porquet D, Sabra B, Reeves JN, Grosso N. 2012. *Astron. Astrophys.* 539:104
Yu W, Yan Z. 2009. *Ap. J.* 701:1940–57
Yu Z, Yuan F, Ho LC. 2011. *Ap. J.* 726:87
Yuan F. 1999. *Ap. J. Lett.* 521:L55–58
Yuan F. 2001. *MNRAS* 324:119–27
Yuan F. 2003. *Ap. J. Lett.* 594:L99–102
Yuan F, Bu D. 2010. *MNRAS* 408:1051–60
Yuan F, Bu D, Wu M. 2012a. *Ap. J.* 761:130
Yuan F, Cui W. 2005. *Ap. J.* 629:408–13
Yuan F, Cui W, Narayan R. 2005. *Ap. J.* 620:905–14
Yuan F, Li M. 2011. *Ap. J.* 737:23
Yuan F, Lin J, Wu K, Ho LC. 2009a. *MNRAS* 395:2183–88
Yuan F, Markoff S, Falcke H. 2002a. *Astron. Astrophys.* 383:854–63
Yuan F, Markoff S, Falcke H, Biermann PL. 2002b. *Astron. Astrophys.* 391:139–48

Yuan F, Narayan R. 2004. *Ap. J.* 612:724–28
Yuan F, Quataert E, Narayan R. 2003. *Ap. J.* 598:301–12
Yuan F, Quataert E, Narayan R. 2004. *Ap. J.* 606:894–99
Yuan F, Taam RE, Xue R, Cui W. 2006. *Ap. J.* 636:46–55
Yuan F, Wu M, Bu D. 2012b. *Ap. J.* 761:129
Yuan F, Xie F, Ostriker JP. 2009b. *Ap. J.* 691:98–104
Yuan F, Yu Z, Ho L. 2009c. *Ap. J.* 703:1034–43
Yuan F, Zdziarski A. 2004. *MNRAS* 354:953–60
Yuan F, Zdziarski A, Xue Y, Wu XB. 2007. *Ap. J.* 659:541–48
Yusef-Zadeh F, Wardle M. 2013. *Ap. J. Lett.* 770:L21
Yusef-Zadeh F, Bushouse H, Wardle M, et al. 2009. *Ap. J.* 706:348–75
Yusef-Zadeh F, Roberts D, Wardle M, Heinke CO, Bower GC. 2006. *Ap. J.* 650:189–94
Zdziarski AA, Gierliński M. 2004. *Prog. Theor. Phys. Suppl.* 155:99–119
Zdziarski AA, Lubiński P, Gilfanov M, Revnivtsev M. 2003. *MNRAS* 342:355–72
Zdziarski AA, Lubiński P, Sikora M. 2012. *MNRAS* 423:663–75
Zdziarski AA, Poutanen J, Mikolajewska J, et al. 1998. *MNRAS* 301:435–50
Zhang SN. 2013. *Front. Phys.* 8:630–60
Zubovas K, King AR, Nayakshin S. 2011. *MNRAS* 415:L21–25

The Coevolution of Galaxies and Supermassive Black Holes: Insights from Surveys of the Contemporary Universe

Timothy M. Heckman[1] and Philip N. Best[2]

[1]Center for Astrophysical Sciences, Department of Physics & Astronomy, The Johns Hopkins University, Baltimore, Maryland 21218; email: heckman@pha.jhu.edu

[2]Institute for Astronomy, Royal Observatory Edinburgh, Blackford Hill, Edinburgh EH9 3HJ, United Kingdom; email: pnb@roe.ac.uk

Annu. Rev. Astron. Astrophys. 2014. 52:589–660

First published online as a Review in Advance on June 16, 2014

The *Annual Review of Astronomy and Astrophysics* is online at astro.annualreviews.org

This article's doi:
10.1146/annurev-astro-081913-035722

Keywords

active galactic nuclei, Seyfert galaxies, quasi-stellar objects, QSOs, radio galaxies

Abstract

We summarize what large surveys of the contemporary Universe have taught us about the physics and phenomenology of the processes that link the formation and evolution of galaxies with their central supermassive black holes. We present a picture in which the population of active galactic nuclei (AGNs) can be divided into two distinct populations. The radiative-mode AGNs are associated with black holes (BHs) that produce radiant energy powered by accretion at rates in excess of ~1% of the Eddington limit. They are primarily associated with less massive BHs growing in high-density pseudobulges at a rate sufficient to produce the total mass budget in these BHs in ~10 Gyr. The circumnuclear environment contains high-density cold gas and associated star formation. Major mergers are not the primary mechanism for transporting this gas inward; secular processes appear dominant. Stellar feedback is generic in these objects, and strong AGN feedback is seen only in the most powerful AGNs. In jet-mode AGNs the bulk of energetic output takes the form of collimated outflows (jets). These AGNs are associated with the more massive BHs in more massive (classical) bulges and elliptical galaxies. Neither the accretion onto these BHs nor star formation in their host bulge is significant today. These AGNs are probably fueled by the accretion of slowly cooling hot gas that is limited by the feedback/heating provided by AGN radio sources. Surveys of the high-redshift Universe paint a similar picture. Noting that the volume-averaged ratio of star formation to BH growth has remained broadly constant over the past 10 Gyrs, we argue that the processes that linked the cosmic evolution of galaxies and BHs are still at play today.

1. INTRODUCTION

1.1. A Brief History

It has been nearly a century since Hubble (1925) decisively demonstrated that the Andromeda nebula is a vast island universe of stars similar to our own Milky Way galaxy. His discovery soon thereafter of an expanding cosmos filled with galaxies opened up a vista of an immense and seemingly serene Universe. However, this picture of serenity was deceptive. As we write this review, it is the 50-year anniversary of Maarten Schmidt's (1963) realization that the radio source 3C273 was associated with an optically unresolved object with the (then) enormous redshift of 0.158. Subsequent identifications of a whole population of quasi-stellar radio sources at even higher redshifts were soon followed by the discovery of a more numerous population of otherwise-similar but radio-quiet quasi-stellar galaxies (Sandage 1965). [It is fascinating in hindsight that the discovery of the cosmic X-ray background (Giacconi et al. 1962) and of the Kerr metric (Kerr 1963) occurred at the same time.] The concept of the Violent Universe was born.

It is beyond the scope of this review to examine the twists and turns that finally led to the consensus that all these quasi-stellar objects (QSOs) were the extremely luminous active nuclei of distant galaxies. Qualitatively similar objects had been found much earlier in the nuclei of some relatively nearby galaxies by Seyfert (1943). It is sobering that the significance of this discovery was long unrecognized: The first investigation into the nature of Seyfert galaxies was published 16 years later by Burbidge et al. (1959), who concluded that NGC 1068 was explosively ejecting ionized gas from its nucleus. Likewise, although other radio sources had been identified with distant galaxies before the discovery of QSOs—e.g., Cygnus A by Baade & Minkowski (1954)—there was great confusion as to the origin of this very large-scale radio emission, which had no obvious connection to the galaxy nucleus.

For most of the past five decades the communities that studied galaxies and active galactic nuclei (AGNs) remained largely disconnected. AGNs were studied primarily as laboratories in which to probe exotic high-energy processes. There was some effort to understand the role that the environment might play in triggering or fueling the AGN—for example, see the ancient review in this journal by Balick & Heckman (1982)—but there was virtually no concept that AGNs played any significant role in the evolution of typical galaxies. Of course, things could hardly be more different today. The notion of the coevolution of galaxies and AGNs has become inextricably ingrained in our current cosmogony. Indeed, this review represents the third article in this journal in consecutive years that deals with some aspect of this coevolution (Fabian 2012, Kormendy & Ho 2013).

The reasons for this change are easy to recognize. First came the realization that powerful AGNs (as represented by QSOs) were only the tip of the iceberg. Extensive surveys in the radio, optical, and X-ray domains revealed local populations of Seyfert and radio galaxies and established that signs of lower-level activity were commonplace in the nuclei of early-type galaxies. This strongly suggested that the AGN phenomenon—rather than being simply a rare spectacle—is a part of the lifecycle of typical galaxies. The demography and physical properties of these low-power AGNs have been comprehensively reviewed in this journal by Ho (2008).

A second reason followed from the documentation of the overall buildup over most of cosmic time of the populations of galaxies (as traced via star formation) and of supermassive black holes (SMBHs, whose growth is traced by AGNs). The evolution of the two populations is strikingly similar: a steep rise in both the star-formation rate (SFR) and SMBH growth rate by about a factor of 10 from redshift $z = 0$ to 1, a broad maximum in both rates at $z \sim 2$ to 3, and then a relatively steep decline at higher redshifts (see **Figure 1**) (Shankar et al. 2009 and references

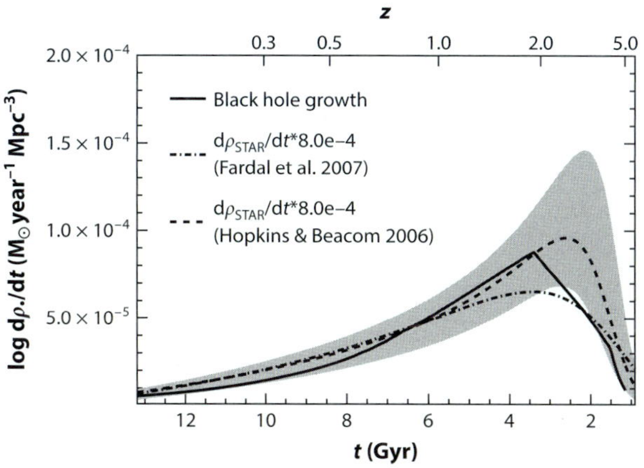

Figure 1

The cosmic history of black hole growth and stellar mass growth. The average black hole accretion rate is compared to the star-formation rate as a function of redshift, where the latter is given by Hopkins & Beacom (2006) and Fardal et al. (2007), scaled by the factor 0.8×10^{-3}. The shaded gray area shows the 3σ uncertainty region from Hopkins & Beacom (2006). Reprinted from Shankar et al. (2009) with permission.

therein). For at least the last ~11 Gyr of cosmic history the ratio of these two growth rates has remained roughly constant with a value of order 10^3. Thus, at least in a volume-averaged sense, the growth of galaxies and SMBHs has been synchronized somehow.

Finally, and even more remarkably, observations over the past decade have not only established that SMBHs exist in the nuclei of (probably) all galactic bulges, they have shown that the properties of these present-day SMBHs and the galaxies in which they live are linked on a galaxy-by-galaxy basis. This impressive fossil record of coevolution has been exhaustively reviewed by Kormendy & Ho (2013). In addition to the accumulating observational evidence for the coevolution of galaxies and SMBHs, a considerable theoretical motivation to invoke this linkage has developed as well. Without some form of feedback from AGNs, neither current semianalytic models nor numerical simulations can successfully reproduce the properties of massive galaxies. This fascinating subject has been reviewed in this journal by Fabian (2012).

In conclusion, though the evidence to date remains indirect, it is hard not to infer that the cosmic evolution of galaxies and that of SMBHs have seemingly been driven by a suite of interlinked physical processes.

1.2. Our Perspective: Large Surveys of the Contemporary Universe

Our goal here is to review the evidence for the coevolution of galaxies and SMBHs as derived from large surveys of the contemporary (low-redshift) Universe. The majority of the growth of SMBHs and of the stellar components in galaxies occurred between redshifts of roughly 0.5 and 2, and the present-day growth rates of both populations are over an order of magnitude smaller than during the peak epoch. A reasonable question is then whether one can learn anything very useful by studying the contemporary Universe. We hope that by the end of the review the reader will agree with us that one can actually learn a great deal.

We believe that there are a number of reasons why this is the case. The contemporary Universe contains a rich fossil record that (once successfully decoded) reveals the processes that produced this record. Moreover, the basic physical processes that operated during the peak epoch of SMBH/galaxy growth are still in place and can be studied in greater detail due to their relative proximity. However, the primary reason to study the contemporary Universe is that it is the only place where it has been possible to carry out the surveys of galaxies and SMBHs that are of sufficiently large size and whose data are of sufficiently high quality to have finally allowed us to be able to fully explore the complex inter-relationship between these two populations in a statistically robust way. This has led to a much clearer picture of both the fossil record of galaxy/SMBH coevolution and the processes by which the coevolution continues to play out.

The particular importance of high-quality optical spectroscopy cannot be overstated: These data form the interpretational backbone of the whole structure defined by extensive multiband imaging and photometric surveys. Of most relevance to the specific topic of our review, studies of the lower-luminosity objects that dominate the AGN population in the local Universe benefitted immensely from galaxy spectra obtained as part of the main galaxy sample of the Sloan Digital Sky Survey (SDSS; Strauss et al. 2002). The high quality of the spectra enabled them to be used to characterize both the AGNs and the stellar populations in the host galaxies. Finally, the uniformity and completeness of the SDSS main galaxy sample rendered it ideal for statistical studies of the multiparameter characteristics of the galaxy population in the contemporary Universe, including their AGN properties.

The uniformity and wide sky coverage of the 2dF Galaxy Redshift Survey (2dFGRS; Colless et al. 2001) and SDSS galaxy redshift survey make them an ideal starting point for studies of AGNs and their host galaxies in regions of the electromagnetic spectrum other than the optical. This encompasses the study of radio-selected AGNs using surveys such as the 1.4-GHz National Radio Astronomy Observatory (NRAO) VLA Sky Survey (NVSS; Condon et al. 1998) by Sadler et al. (2002) and Mauch & Sadler (2007) and the Faint Images of the Radio Sky at Twenty centimeters (FIRST; Becker et al. 1995) survey by Best et al. (2005a) and Best & Heckman (2012). Recently, samples of tens of thousands of nearby mid-IR-detected AGNs covering large areas of the sky have been constructed by cross-correlating sources detected by the *Wide-Field Infrared Survey Explorer* (WISE; Wright et al. 2010) with the SDSS (Donoso et al. 2012, Shao et al. 2013). X-ray data having the depth and wide-field sky coverage to fully exploit the SDSS galaxy sample do not exist. However, to date the *Swift/Burst Alert Telescope* (BAT) survey has detected over 700 (mostly local) AGNs at energies above 15 keV (Baumgartner et al. 2013). Near-IR data from WISE and vacuum-UV data from the *Galaxy Evolution Explorer* (GALEX; Martin et al. 2005) provide important additional information about AGN host galaxies. In subsequent sections, we summarize the new scientific insights that resulted from all these large surveys.

1.3. The Landscape of the Galaxy Population and How It Got That Way

To help frame the main issues addressed in this review it is helpful to briefly summarize the basic properties of the population of galaxies in the contemporary Universe and the current thinking about how these galaxies were built. The reader is referred to the review by Madau & Dickinson (2014) in this volume for all the details.

Results from the SDSS have shown that the galaxy population in the contemporary Universe occupies a very small part of the parameter space defined by the structure, stellar content, and chemical composition of a galaxy. In particular, the existence of clear bimodality in the galaxy population was revealed (Kauffmann et al. 2003b, Blanton et al. 2003, Baldry et al. 2004). One population (blue, for short) consists of galaxies with significant ongoing star formation, small stellar masses (M_*), low stellar surface mass densities ($\mu_* = 0.5\, M_*/(\pi R_{50}^2)$, where R_{50} is the radius containing 50% of the light; see Section 2.4), and small concentrations ($C = R_{90}/R_{50}$) of their light (late Hubble type). The other (red) consists of galaxies with little ongoing star

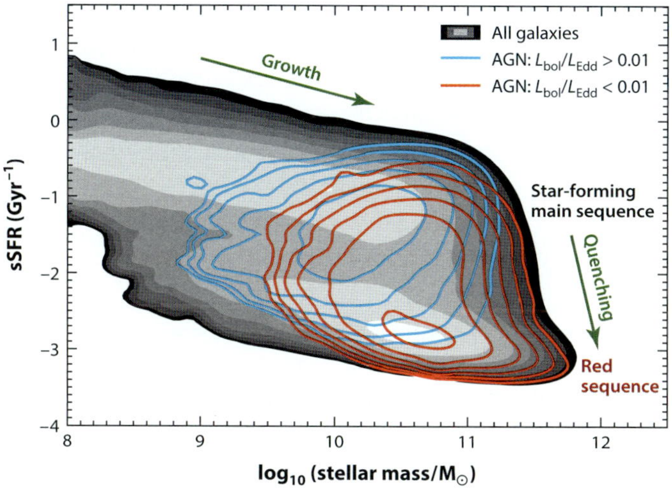

Figure 2

The distribution of galaxies in the SDSS main galaxy sample on the plane of stellar mass versus specific star-formation rate (sSFR $= SFR/M_*$). The grayscale indicates the volume-weighted distribution of all galaxies, with each lighter color band indicating a factor of two increase. Galaxies predominantly fall within two regions: a main sequence of star-forming galaxies and a red sequence of quenched galaxies. The blue and red contours show the volume-weighted distributions of high (>1%; mostly radiative-mode) and low (<1%; mostly jet-mode) Eddington-fraction AGNs, with contours spaced by a factor of two.

formation, large M_*, high μ_*, and large C (early Hubble type). The characteristic parameter values that mark the transition between populations are $M_* \sim 10^{10.5}\,M_\odot$, $\mu_* \sim 10^{8.5}\,M_\odot\,\mathrm{kpc}^{-2}$, and $C \sim 2.6$. Subsequent work showed that the blue population is characterized by a tight, almost linear relationship between the SFR and M_* (Brinchmann et al. 2004, Schiminovich et al. 2007). This has come to be called the star-forming main sequence. **Figure 2** shows the relationship between M_* and the sSFR (sSFR $= SFR/M_*$) in the contemporary Universe.

Over the past few years, deep surveys have established that the qualitative distribution shown in **Figure 2** is characteristic of the galaxy population out to at least a redshift of 2 (e.g., Whitaker et al. 2012). The most significant difference is that the actual values of the sSFR on the star-forming main sequence evolve extremely rapidly with time: Elbaz et al. (2011) find as the age of the Universe ($t_{\rm cos}$) increased from 2.2 to 13.6 Gyr ($z = 3$ to 0), the characteristic value of the sSFR declined with time as $t_{\rm cos}^{-2.2}$.

The simplest picture for the evolution of a typical galaxy (e.g., Lilly et al. 2013) is that it evolves along the (strongly evolving) blue star-forming main sequence, increasing in mass through the accretion of cold gas from the cosmic web and (secondarily) through mergers with other galaxies. As it approaches a critical mass, its supply of cold gas is shut off, the star formation is quenched, and the galaxy then evolves into the red population. It can continue to increase in mass through subsequent mergers with other galaxies. It is presently not clear whether the mass scale at which quenching occurs pertains to the stellar mass or the dark matter halo mass.

The physical process(es) that quench the galaxy are unclear. In part, quenching may be due to a change in the nature of accretion: rapid accretion of cold streams of infalling gas at low mass transitioning to slow accretion of hot gas in hydrostatic equilibrium at high mass (e.g., Dekel et al. 2009). In principle these processes are included in numerical and semianalytic models of galaxy evolution. Nevertheless, these models require some additional process to be at play in order to reproduce the observed properties of massive galaxies. Heating and/or the ejection of surrounding gas by an AGN-driven outflow to suppress the cold accretion is a popular idea. In addition the models also require AGN feedback to keep galaxies that arrive in the red/dead population from forming too many stars from the slow accretion of their hot halo gas. Thus, in the current paradigm, AGNs play a crucial role in the evolution of massive galaxies.

2. BASIC METHODOLOGY

2.1. Overview of the Local AGN Population

The fundamental property that we consider to define an AGN is that its power source involves extracting energy from the relativistically-deep potential well of an SMBH at or near the center of a galaxy.

In this review we present strong empirical evidence that the low-redshift population of AGNs can be divided into two main categories [we refer the reader to Antonucci (2012) for a complementary review]. The first category consists of objects whose dominant energetic output is in the form of electromagnetic radiation produced by the efficient conversion of the potential energy of the gas accreted by the SMBH. Historically, these objects have been called either Seyfert galaxies or QSOs depending upon rather vague and arbitrary criteria involving luminosity and/or redshift. In the rest of this review we refer to these as radiative-mode AGNs. The second category consists of objects that produce relatively little radiation and whose primary energetic output takes the form of the bulk kinetic energy transported in two-sided collimated outflows (jets). These jets may be ultimately powered by the accretion of gas or by tapping the spin energy of the SMBH. Historically, this AGN population has been called (low-excitation) radio galaxies. In this review we refer to them as jet-mode AGNs. As we show in this review, these two populations are virtually disjoint in terms of the basic properties of their SMBHs and host galaxies.

Let us then begin by describing the basic building blocks for these two types of AGNs (see **Figure 3** for schematic diagrams). We refer readers to the texts of Krolik (1999), Netzer (2013), Osterbrock & Ferland (2005), Peterson (1997), and the review by Yuan & Narayan (2014) in this volume for the gory details. In the first category (radiative-mode AGNs), the SMBH is surrounded by a geometrically thin, optically thick accretion disk through which an inflow occurs. The accretion disk has a radial temperature gradient, and the resulting total thermal continuum emission emerges from the extreme UV through to the visible portion of the electromagnetic spectrum. The accretion disk is surrounded by a hot corona that Compton-up-scatters the soft seed photons from the disk into the X-ray regime. As the X-rays impact the accretion disk their spectral energy distribution is modified through fluorescence and reflection off the accretion disk. The ionizing radiation from the disk and corona heats and photoionizes a population of dense gas clouds located on scales of light-days to light-years from the SMBH leading to the production of UV-, optical-, and near-IR-permitted emission lines. The velocity dispersion of the population of clouds is typically several thousand kilometers per second, leading to its designation as the broad-line region on the basis of the resulting emission-line spectrum.

On larger scales, the SMBH and accretion disk are surrounded by a region of dusty molecular gas (which we refer to as the obscuring structure). Its inner radius is set by the sublimation temperature of the most refractory dust grains and is hence larger in more luminous AGNs. In this region some of the incident UV/visible photons from the accretion disk and the soft X-rays from the corona are absorbed by the dust, and this absorbed energy emerges as thermal IR emission. The total column density of the obscuring structure spans a range in inferred column densities from roughly 10^{23} to 10^{25} cm^{-2}. The highest column densities are sufficient to absorb even hard X-rays (these cases are Compton-thick). As ionizing radiation escapes along the polar

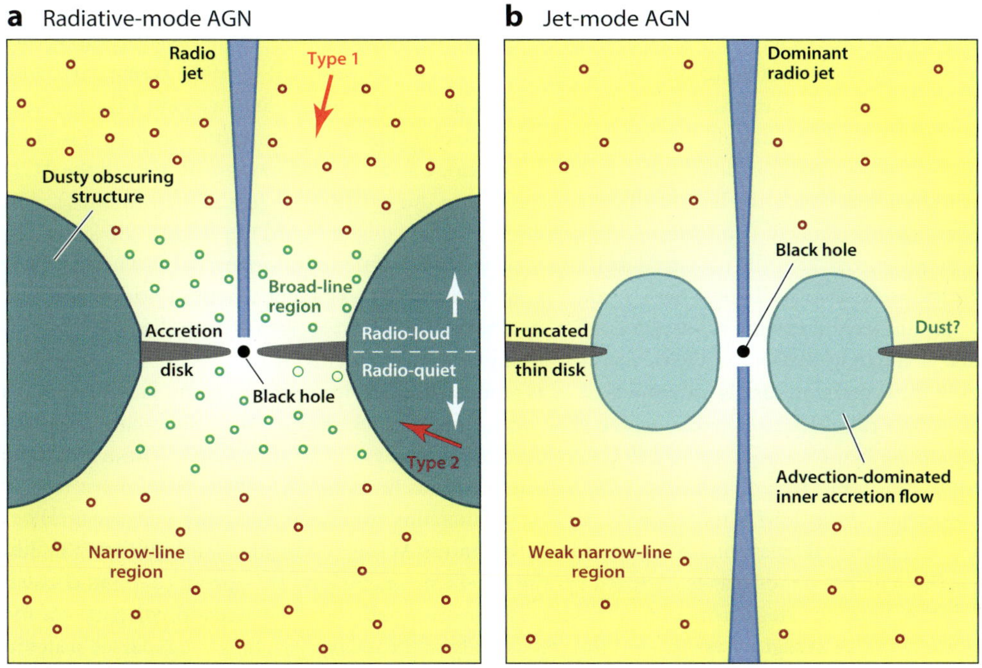

Figure 3

Schematic drawings of the central engines of radiative-mode and jet-mode AGNs (not to scale). (*a*) Radiative-mode AGNs possess a geometrically thin, optically thick accretion disk, reaching into the radius of the innermost stable orbit around the central supermassive black hole. Luminous UV radiation from this accretion disk illuminates the broad-line and narrow-line emission regions. An obscuring structure of dusty molecular gas prohibits direct view of the accretion disk and broad-line regions from certain lines of sight (Type 2 AGN), whereas they are visible from others (Type 1 AGN). In a small proportion of sources (predominantly toward the high end of the range of black hole masses), powerful radio jets can also be produced. (*b*) In jet-mode AGNs the thin accretion disk is replaced in the inner regions by a geometrically thick advection-dominated accretion flow. At larger radii (beyond a few tens of Schwarzschild radii, the precise value depending upon properties of the accretion flow, such as the Eddington-scaled accretion rate), there may be a transition to an outer (truncated) thin disk. The majority of the energetic output of these sources is released in bulk kinetic form through radio jets. Radiative emission is less powerful, but can ionize weak, low-ionization narrow-line regions, especially where the truncation radius of the thin disk is relatively low.

axis of the obscuring structure, it photoionizes gas on circumnuclear scales (a few hundred to a few thousand parsecs). This more quiescent and lower-density population of clouds produces UV-, optical-, and IR-forbidden and -permitted emission lines, Doppler-broadened by several hundred kilometers per second, and is hence called the narrow-line region (NLR).

Observing an AGN from a sight line nearer the polar axis of the obscuring structure yields a clear direct view of the SMBH, the disk/corona, and broad-line region (BLR). These are called Type 1 (or unobscured) AGNs. When observing an AGN from a sight line nearer the equatorial plane of the obscuring structure, this central region is hidden and these are called Type 2 (or obscured) AGNs. This is the basis for the standard unified model for radiative-mode AGNs (e.g., Antonucci 1993), which asserts that the Type 1 and 2 populations differ only in the viewing angle from which the AGN is observed. The presence of AGNs can still be inferred in the Type 2 objects from the thermal IR emission from the obscuring structure, from hard X-rays transmitted through the structure (when it is Compton-thin), and from the emission lines with tell-tale line ratios from the NLR.

In some cases, the obscuring material can be the larger-scale dusty ISM of the host galaxy. This is particularly relevant when the host galaxy's disk is viewed at a large inclination or the galaxy is in the throes of an ongoing major merger with a strong central concentration of dusty gas. This material is sometimes sufficient to obscure the optical, UV, and soft X-ray emission from the AGN accretion disk and BLR, but insufficient to attenuate the hard X-rays (e.g., Gelbord 2003). The effect of this is nicely demonstrated by Lagos et al. (2011), who use SDSS data to show that the host galaxies of optically obscured AGNs are skewed toward edge-on orientations, whereas those optically classified as Type 1 have a much higher probability of having a face-on orientation.

In the second category (jet-mode AGNs), a distinct mode of accretion onto the SMBH exists that is apparently associated with low accretion rates and is radiatively inefficient. The geometrically thin accretion disk is either absent or is truncated in the inner regions, and is replaced by a geometrically thick structure in which the inflow time is much shorter than the radiative cooling time (e.g., Narayan & Yi 1994, 1995; Quataert 2001; Narayan 2005; Ho 2008). These are called advection-dominated or radiatively inefficient accretion flows (ADAFs/RIAFs). A characteristic property of these flows is that they are capable of launching two-sided jets. Note that powerful jets are also launched by a small fraction of radiative-mode AGNs (e.g., radio-loud QSOs). The jets are (far and away) most easily detected via the synchrotron emission they produce at radio wavelengths. This can extend from optically thick (synchrotron self-absorbed) emission on parsec scales all the way out to regions far beyond the stellar body of the galaxy (reaching megaparsec scales in extreme cases). In typical local radio galaxies the jets travel at relativistic velocities (Lorentz gammas of several) when launched, but appear to rapidly decelerate and destabilize as they interact with the gaseous halo of the host galaxy and transition to subsonic turbulent plumes. At the highest radio luminosities (most commonly found in the radio-loud radiative-mode AGNs), the jets survive as highly collimated structures until they terminate as bright shocks (hot spots) at the interface with the circumgalactic or intergalactic medium. The survival or disruption of the jets leads to the Fanaroff & Riley (1974) morphological classification of radio galaxies (see Urry & Padovani 1995 for further discussion).

Missing from the above description are the objects that may trace the lowest-luminosity portion of the AGN population. This population has been reviewed in this journal by Ho (2008). Since his review, evidence has grown that at least some of the objects previously classified as the lowest-luminosity AGNs (the low-ionization nuclear emission-line regions, or LINERs, with the very weakest optical emission lines) are not bona fide AGNs (e.g., Cid Fernandes et al. 2011, Yan & Blanton 2012). However, the remaining more powerful LINERs are very likely to be members of the jet-mode AGN population (e.g., Nagar et al. 2005; Ho 2008; P.N. Best & T.M. Heckman, in preparation), albeit with rather modest radio luminosities.

Figure 4 summarizes this description of the local AGN population and how the common AGN classes fit within it. The blue text within that figure describes the properties of a typical example of each category of AGNs. As the review develops, we describe the evidence for this and discuss the range of properties seen for each AGN class.

2.2. Finding AGNs

This review is focused on insights into the coevolution of SMBHs and galaxies that have been derived from large surveys of the local Universe. For such investigations of the radiative-mode AGNs it is the obscured (Type 2) AGNs that are far and away the more valuable. In these objects the blinding glare of the UV and optical continuum emission from the central accretion disk has

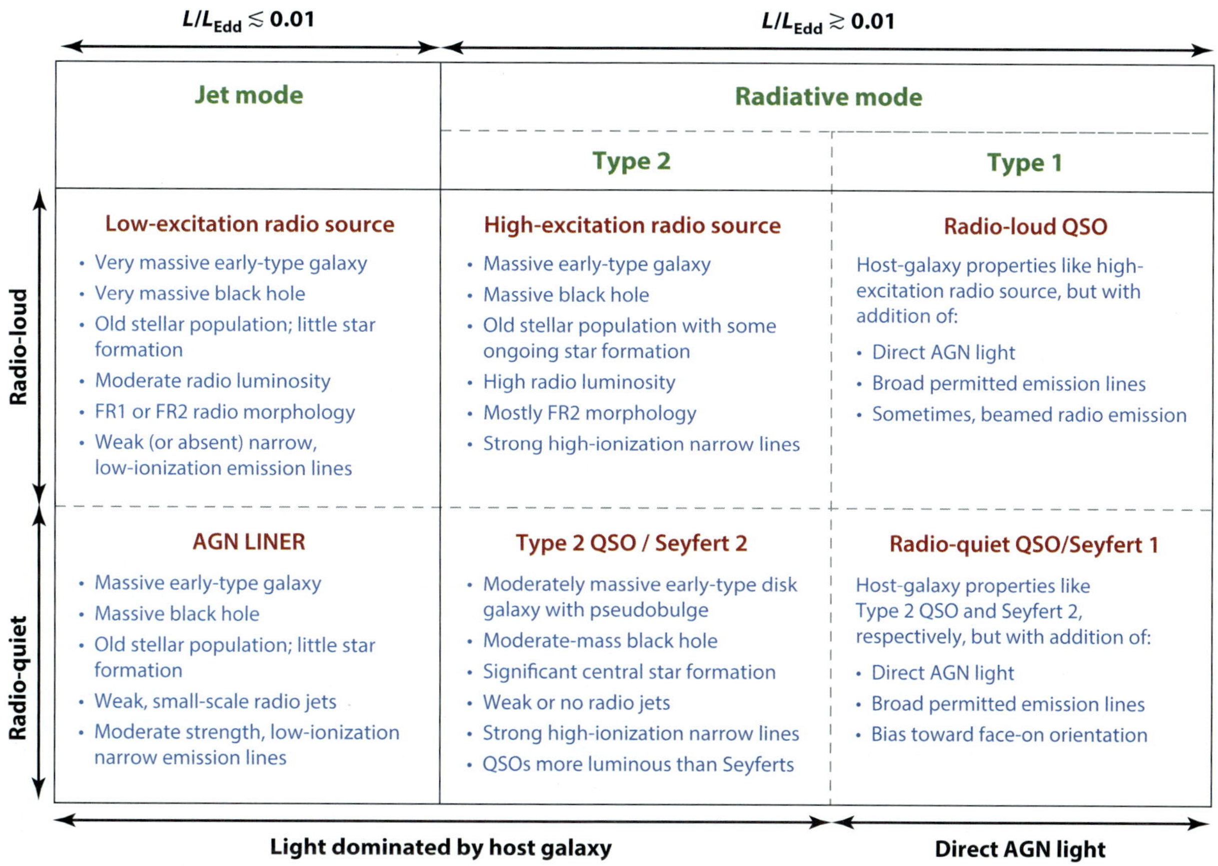

Figure 4

The categorization of the local AGN population adopted throughout this review. The blue text describes typical properties of each AGN class. These, together with the spread of properties for each class, are justified throughout the review. Abbreviations: FR1/FR2, Faranoff & Riley (1974) Class 1/2; LINER, low-ionization nuclear emission-line region; QSO, quasi-stellar object.

been blocked by the natural coronagraph created by the dusty obscuring structure. The remaining UV and optical continuum is generally dominated by the galaxy's stellar component (Kauffmann et al. 2003a), which can then be readily characterized. In the sections to follow we therefore restrict our discussion of radiative-mode AGNs to techniques that can recognize Type 2 AGNs. For the jet-mode AGNs the intrinsic UV and optical emission from the AGN is generally weak or absent unless the observer is looking directly down the jet axis (e.g., Urry & Padovani 1995). Thus, the host-galaxy properties can be easily studied without contamination.

The relevant techniques that have been used for very large surveys of the local Universe are based on (*a*) optical spectroscopy to identify emission from the AGN-powered NLR, (*b*) mid-IR photometry to identify the warm dust emission from the obscuring structure, (*c*) hard X-ray imaging spectroscopy to detect emission from the accretion disk corona, and (*d*) nonthermal radio continuum emission to locate jets and jet-powered structures.

For the radiative-mode population, one important caveat to keep in mind is the degree to which Type 2 AGNs and their hosts are the same as those of Type 1. In the most simplistic version of the AGN unified model described above, this complete overlap would be the case. However, in reality the fraction of the sky covered by obscuring material (as seen from the SMBH) varies significantly from AGN to AGN. In this case, the AGNs observed as Type 1 (2) will be systematically drawn from the portion of the AGN parent population with a smaller (larger) covering factor for the obscuring material. It would therefore not be surprising if the properties of Type 1 and Type 2 AGNs and their hosts differed in some systematic fashion.

2.2.1. Optical emission-line surveys. Large-scale spectroscopic surveys of the contemporary Universe (particularly 2dFGRS and SDSS) have revolutionized the study of emission-line AGNs, by providing high-quality spectra for over a million galaxies, with well-understood selection functions. The broad emission lines and strong nonstellar continuum emission in Type 1 radiative-mode AGNs offer easily detectable and unambiguous evidence of the presence of an AGN origin [detailed discussions of the identification of Type 1 AGNs and of the criteria used to discriminate between Type 1 and Type 2 AGNs in the SDSS are discussed by Richards et al. (2002) and Hao et al. (2005b)]. In contrast, the optical signature of Type 2 radiative-mode AGNs, and of jet-mode AGNs, is the presence of narrow emission lines. In general, such narrow emission lines can also be produced by ordinary populations of O stars associated with ongoing star formation. However, the relative intensities of the strong forbidden and permitted emission lines resulting from photoionization by the relatively hard continuum from an AGN differ systematically from those produced by the much softer continuum of O stars. This approach to recognizing AGNs was first systematized by Baldwin et al. (1981; BPT) and later refined by Veilleux & Osterbrock (1987). The diagnostic diagrams (commonly referred to as BPT diagrams) compare pairs of emission-line flux ratios. Primary among these is the flux ratio pair [OIII] 5007/Hβ versus [NII]6584/Hα, but the [OI] 6300/Hα and [SII] 6716,6731/Hα flux ratios provide diagnostics that are complementary to [NII]/Hα (e.g., Kewley et al. 2006; see **Figure 5**). The same diagnostic diagrams work for Type 1 AGNs provided that only the narrow components of the Balmer lines are considered (e.g., Stern & Laor 2012a).

We show a set of diagnostic diagrams for the SDSS main galaxy sample in **Figure 5** (taken from Kewley et al. 2006). The basic morphology of these diagrams reveals the presence of two

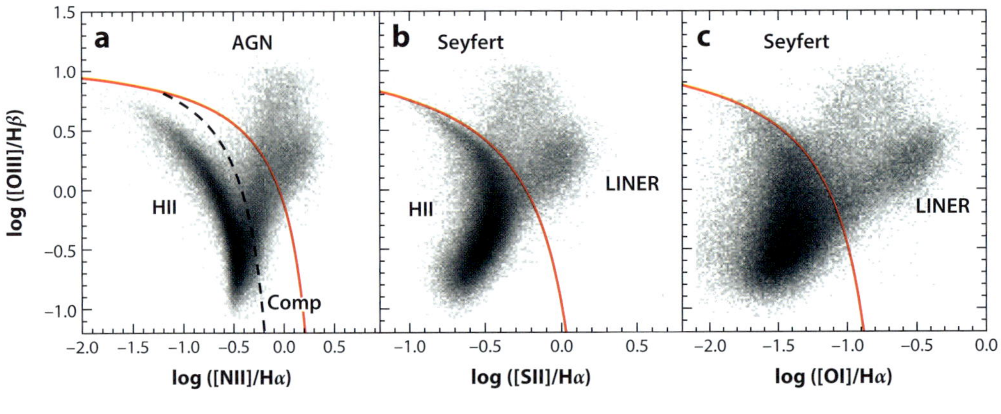

Figure 5

A set of diagnostic diagrams for the Sloan Digital Sky Survey main galaxy sample taken from Kewley et al. (2006). The red line is the maximal starburst line from Kewley et al. (2001), whereas the dashed line in panel *a* shows the more stringent dividing line between pure star-forming galaxies and AGNs adopted by Kauffmann et al. (2003a). The separation between star-forming galaxies and AGNs is most cleanly done using panel *a*, whereas panels *b* and *c* can be used to separate Seyferts from LINERs (low-ionization nuclear emission-line regions).

populations. The first is that of purely star-forming galaxies whose locus runs from the upper left to lower middle of the plots in a sequence of downwardly increasing metallicity (e.g., Pettini & Pagel 2004). The second is the AGN population, which is joined to the star-forming sequence at the high-metallicity end and then extends upward and to the right.

Physically, the observed difference in the emission from high-metallicity gas photoionized by an AGN reflects the increased rate of photoelectric heating by the hard AGN radiation field. This leads to a higher gas temperature and an increased strength of the collisionally excited forbidden lines relative to the recombination lines of hydrogen. Because real galaxies contain both AGNs and regions of star formation within the observed region, the extension of the AGN locus away from the star-forming sequence can be viewed as a mixing line along which the contribution to the total emission-line spectrum by the AGN increases relative to that from the star-forming regions as the points move further away from the star-forming locus. This inference is confirmed in a statistical sense by the result that the relative distribution of points along this sequence migrates downward toward the star-forming sequence as the redshift of the galaxy population at fixed stellar mass increases and the SDSS fiber encompasses a progressively larger region of the host galaxy (Kauffmann et al. 2003a, Kewley et al. 2006, LaMassa et al. 2013).

Kewley et al. (2001) used photoionization and stellar population synthesis models to define an upper boundary to the possible location of star-forming galaxies in the emission-line diagnostic diagrams, above which a clean sample of AGNs can be selected. Kauffmann et al. (2003a) used empirical methods, based on the tight locus of star-forming galaxies within the [NII]/Hα BPT diagram, to produce an alternative demarcation that should provide a more complete census of the AGN population that is less biased against AGNs in strongly star-forming galaxies. These demarcation lines are shown in **Figure 5**. Other authors have proposed further refinements of this division (e.g., Kewley et al. 2006, Stasinska et al. 2006, Cid Fernandes et al. 2011). This diagram has often been used to separate emission-line galaxies into three classes: star-formation-dominated (below and to the left of the Kauffmann et al. line), AGN-dominated (lying above and to the right of the Kewley line), and composite (between the Kewley and Kauffmann lines). In this review we do not distinguish between the composite and AGN-dominated objects (we regard them all as AGNs). It is important to emphasize here that though the optical emission lines used to classify AGNs and star-formation-dominated galaxies can be substantially affected by dust absorption, these optical classifications agree with mid-IR-based classifications in the great majority of cases (LaMassa et al. 2012).

In 1980 a separate class of optical emission-line AGNs was proposed, known as LINERs (Heckman 1980). LINERs are distinguished from Seyfert nuclei by the relative strength of their low-ionization emission lines; the original definition was based on the forbidden lines of oxygen, with LINERs having emission-line flux ratios [OII] 3727/[OIII] 5007 > 1 and [OI] 6300/[OIII] 5007 > 0.32. Subsequent classifications have built upon the diagnostic diagrams separating the AGN population into Seyferts and LINERs (see **Figure 5**). Although LINERs typically have lower nuclear luminosities than Seyferts, many of the fundamental characteristics of the objects close to the demarcation boundary are very similar to those of the low-luminosity end of the distribution of Seyferts (e.g., Kewley et al. 2006, Ho 2008, Netzer 2009).

Stasinska et al. (2008), Sarzi et al. (2010), Cid Fernandes et al. (2011), Capetti & Baldi (2011), and Yan & Blanton (2012) have all presented evidence that those LINERs with the weakest emission lines (which are located in galaxies with predominantly old stars) are not powered by an AGN. Instead, most of these papers argue that they are produced by photoionization of neutral atomic gas by a population of post-asymptotic giant branch (PAGB) stars. A problem with this idea is that *Hubble Space Telescope* (HST) UV images of the bulge of M31 (the site of a LINER) do not reveal the predicted population of these PAGB stars (Brown et al. 1998, 2008; Rosenfield et al.

2012). For these weak LINERs to be photoionized by old stars, a population that is less luminous and more numerous than the PAGB stars would be needed.

In this review we are conservative and consider the possibility that weak LINERs are not members of the AGN population. Here we consider weak LINERs to be objects with [OIII] 5007 equivalent widths smaller than ~1 Å (e.g., Capetti & Baldi 2011). It is important to note that although these objects constitute the majority of LINERs in the SDSS, they contribute a negligible amount to the overall AGN emissivity of the contemporary Universe. It is also important to note that objects classified as LINERs in the SDSS span over two orders of magnitude in [OIII] luminosity ($L_{[OIII]}$) and roughly three orders of magnitude in $L_{[OIII]}/M_{BH}$ [Kewley et al. 2006, Netzer 2009; where M_{BH} is the black hole (BH) mass in solar units]. Although a stellar origin for the LINER emission is likely at the lowest luminosities, the more powerful LINERs are most likely to be AGNs (P.N. Best & T.M. Heckman, in preparation).

2.2.2. X-ray and infrared surveys. X-ray surveys have played a critical role in defining the radiative-mode AGN population and delineating its evolution over cosmic time [see Brandt & Hasinger (2005) for a thorough review]. In our review we are most interested in the population of Type 2 obscured AGNs in the contemporary Universe. The *Rosat* all-sky survey (Voges et al. 1999) has the requisite wide-field coverage, but the survey is very shallow compared with the SDSS in terms of finding AGNs, and its 0.1–2.4-keV band is too soft to recover the bulk of the obscured AGN population (e.g., Anderson et al. 2003, Shen et al. 2006). Hard X-ray observations are much more useful, and the current state of the art is represented by the survey undertaken with the BAT onboard the *Swift* gamma-ray burst observatory. As of late 2012, BAT observations have yielded a catalog of 1,171 X-ray sources detected in the 14–195-keV band, drawn from a survey with nearly uniform sensitivity over the entire sky on the basis of 70 months of observations (Baumgartner et al. 2013). Of these, over 700 are AGNs. At these high X-ray energies only Compton-thick obscured AGNs are under-represented, and indeed the 70-month catalog contains roughly equal numbers of Type 1 and Type 2 Seyfert galaxies.

The signature of an obscured AGN in the IR is unusually strong mid-IR (~3- to 30-μm) emission produced by the dusty obscuring structure. The *Infrared Astronomical Satellite* (IRAS) pioneered the detection and characterization of the local AGN populations in the mid- and far-IR (e.g., Miley et al. 1985, Spinoglio & Malkan 1989). Unfortunately, the sensitivity of IRAS was not well matched to the bulk of the optically selected local AGNs in the SDSS (e.g., Pasquali et al. 2005). The *Spitzer Space Telescope* was considerably more sensitive than IRAS, but it did not undertake wide-field surveys optimized to the local AGN population. Nonetheless, a number of different techniques based on near- and mid-IR color selection were developed to find AGNs using *Spitzer* data (e.g., Lacy et al. 2004, Stern et al. 2005, Richards et al. 2006, Donley et al. 2008). The situation has been revolutionized with the completion of the mission of NASA's WISE, which has detected near- and mid-IR emission from roughly 30,000 optically classified AGNs in the SDSS main galaxy sample (Donoso et al. 2012, Shao et al. 2013) and many other AGNs as well (e.g., Stern et al. 2012, Rosario et al. 2013). It is important to note here that though samples of AGNs selected by their optical emission lines, their mid-IR emission, or their hard X-ray emission are roughly similar to one another in terms of both the properties of the AGNs and the host galaxies (e.g., Kauffmann et al. 2003a, LaMassa et al. 2010, Winter et al. 2010, Shao et al. 2013), they may not be identical (e.g., Hickox et al. 2009, Koss et al. 2011, Juneau et al. 2013).

2.2.3. Radio continuum surveys. Radio continuum surveys offer an important means to find AGNs, particularly the jet-mode AGNs for which jets represent the bulk of the AGN's energetic output. For radiative-mode AGNs, the dusty obscuring structure is optically thin to radio continuum emission, so the detection of radio-emitting Type 2 AGNs is also reasonably straightforward. Deep radio sky surveys such as NVSS, FIRST, the Westerbork Northern Sky Survey (WENSS; Rengelink et al. 1997), and the Sydney University Molonglo Sky Survey (SUMSS; Bock et al. 1999) cover large sky areas with sufficient sensitivity to probe the majority of the radio-AGNs in the contemporary Universe, reaching down to at least $P_{1.4\,\mathrm{GHz}} \sim 10^{23}$ W Hz^{-1} out to $z \sim 0.1$. More problematic issues than simple detection, however, are the accurate association of observed radio sources with their optical host galaxies and the confirmation that the observed radio emission arises from AGN activity rather than being associated with star formation within the host galaxy.

The NVSS was the first large-area radio survey of sufficiently high angular resolution (45 arcsec) to permit automated cross-correlation with optical surveys (cf. Machalski & Condon 1999). However, the positional accuracy of NVSS sources is only 1–2 arcsec at 10-mJy flux levels, increasing to 4–5 arcsec or more at the 2.5-mJy flux limit of the survey (Condon et al. 1998). This leads to significant uncertainties in cross-identifying the radio sources with their optical host galaxies, which are further exacerbated by the extended and multicomponent nature of some radio sources: Even though more than 95% of $z < 0.1$ radio sources are contained within a single NVSS component, the radio centroid of asymmetric extended sources may not match the host-galaxy position. This leads to a trade-off between the reliability and the completeness of the matched sample. For example, Sadler et al. (2002) cross-matched the NVSS with galaxies from the 2dFGRS, accepting matches within 10 arcsec of the optical galaxy; they calculated that the catalog would be 90% complete at this level, but that 5–10% of the matches were expected to be false identifications.

Higher reliability samples can be derived using radio surveys with better angular resolution, such as the 5-arcsec resolution of FIRST (cf. the the SDSS. cross-matching by Ivezić et al. 2002). However, these surveys can miss extended radio emission, leading to the radio luminosity of sources larger than a few arcseconds being systematically low (Becker et al. 1995) and also to a much larger fraction of extended radio sources being split into multiple components. These multicomponent sources can usually be associated with host galaxies by visual inspection, which is practical for moderate sample sizes (e.g., Mauch & Sadler 2007), but not for the very large samples produced from the SDSS. Automated cross-matching routines therefore need to be used, which take account of the possible multicomponent nature of radio sources. So-called collapsing algorithms aim to identify multicomponent radio sources by using the angular separations, flux densities and flux-density ratios of close source pairs to determine whether these are likely to be associated (e.g., Magliocchetti et al. 1998); for optical cross-matching, the relative separations and position angles from the host-galaxy position offer further information (e.g., McMahon et al. 2002, de Vries et al. 2006). Best et al. (2005a) developed a combined approach, using data from both NVSS (for reliable luminosities) and FIRST (for accurate cross-matching), along with all of the above information and incorporating visual analysis for the most extended radio sources. They were able to develop samples with ~95% completeness and ~99% reliability. Adaptations of this method have since been widely used (e.g., Sadler et al. 2007, Donoso et al. 2009, Lin et al. 2010, Best & Heckman 2012).

The derived radio source populations comprise a mixture of AGNs and star-forming galaxies. A variety of different methods have been adopted in the literature for separating radio-loud AGNs from star-forming galaxies. Perhaps the cleanest of these uses the ratio of far-IR to radio luminosity: Star-forming galaxies show a very tight correlation between these properties (see review by Condon 1992), whereas radio-loud AGNs are offset to brighter radio luminosities. This technique has been widely used at higher redshifts (e.g., Ibar et al. 2008, Sargent et al. 2010, Padovani et al. 2011, Simpson et al. 2012), but the shallow depth of wide-area far-IR surveys has limited its use locally (cf. Sadler et al. 2002). For local galaxies, the information provided by the large spectroscopic surveys can be used instead. The standard emission-line ratio diagnostic diagrams offer an approach consistent with that used to classify emission-line-selected AGNs, although a large fraction of the radio population lack the required emission-line detections (Sadler et al. 1999, Best et al. 2005a). Radio-loud AGNs can also be identified as being offset from the tight correlation that star-forming galaxies display between Hα emission-line luminosity and radio luminosity (Kauffmann et al. 2008, Vitale et al. 2012). An alternative method developed by Best et al. (2005a) considers the 4,000-Å break strength (a measure of stellar population age; see Section 2.4.2) against the ratio of radio power per stellar mass on the basis that star-forming galaxies with a wide range of star-formation histories occupy a similar locus in this plane.

These different AGN/star-formation classification schemes agree on the classification of most sources, but also produce significant differences (see the comparison by Best & Heckman 2012). Primary among these differences is the way that radio quiet AGNs are classified. These AGNs often display compact radio cores and weak jet-like structures, but those which lack these features lie on the same far-IR to radio correlation as star-forming galaxies (Roy et al. 1998). This suggests that the bulk of their radio emission is powered by star formation. The emission-line diagnostic diagrams classify these sources as AGNs, because the AGN drives the emission-line ratios away from the star-forming locus, but in contrast the sources are generally classified as star forming using the far-IR–to–radio or emission line–to–radio luminosity ratio diagnostics or the 4,000-Å break strength. Best & Heckman (2012) combined these different classification methods to provide a sample of 18,286 radio-loud AGNs from the SDSS, excluding radio-quiet AGNs (see **Figure 6**). This is the sample used to construct relevant plots throughout this review.

2.3. Determining AGN Physical Properties

2.3.1. Proxies for bolometric luminosity in radiative-mode AGNs. In this section we discuss how best to estimate the bolometric luminosity of radiative-mode AGNs. Type 2 AGNs are (by definition) obscured and this complicates the estimation of perhaps the most basic physical property of an AGN: its total (bolometric) luminosity. We cannot directly observe the great bulk of this radiant energy and must therefore rely on attempting to calibrate various proxies for the bolometric luminosity on the basis of the radiation that does escape. Let us then assess these various proxies and begin by briefly listing their pros and cons.

2.3.1.1. Proxy 1: The [OIII] 5007 emission line. The proxy that we primarily use later in this review is the luminosity of the [OIII] 5007 narrow emission line. The [OIII] line is produced in AGNs by ionizing radiation that escapes along the polar axis of the dusty obscuring structure where it photoionizes and heats ambient gas clouds on radial scales of order 0.1 to 1 kpc from the SMBH (in the NLR).

Pros: (1) This line is easily observed and can be measured (if present) in the existing SDSS spectra of a few million galaxies. (2) Because the NLR lies outside the very high column density in the obscuring structure, its emission suffers far less attenuation owing to dust near the AGN than the optical emission from the accretion disk. Modest amounts of dust extinction can be corrected using the ratio of the fluxes of the narrow Hα and Hβ emission lines (the Balmer decrement). Typical implied corrections are 1 to 2 mag (e.g., Kewley et al. 2006).

Cons: (1) Significant amounts of more widespread dust extinction due to the galaxy ISM may be present, and it may not be possible to reliably correct for this extinction if it is too large. (2) The [OIII] line can be contaminated by a contribution owing to star formation (which will be significant in the composite objects shown in **Figure 5**). Examples of the empirical techniques for correcting for contamination by stellar ionization are described by Kauffmann & Heckman (2009) and Wild et al. (2010). (3) The luminosity of the line depends on the uncertain fraction of the AGN ionizing luminosity that is intercepted by clouds in the NLR.

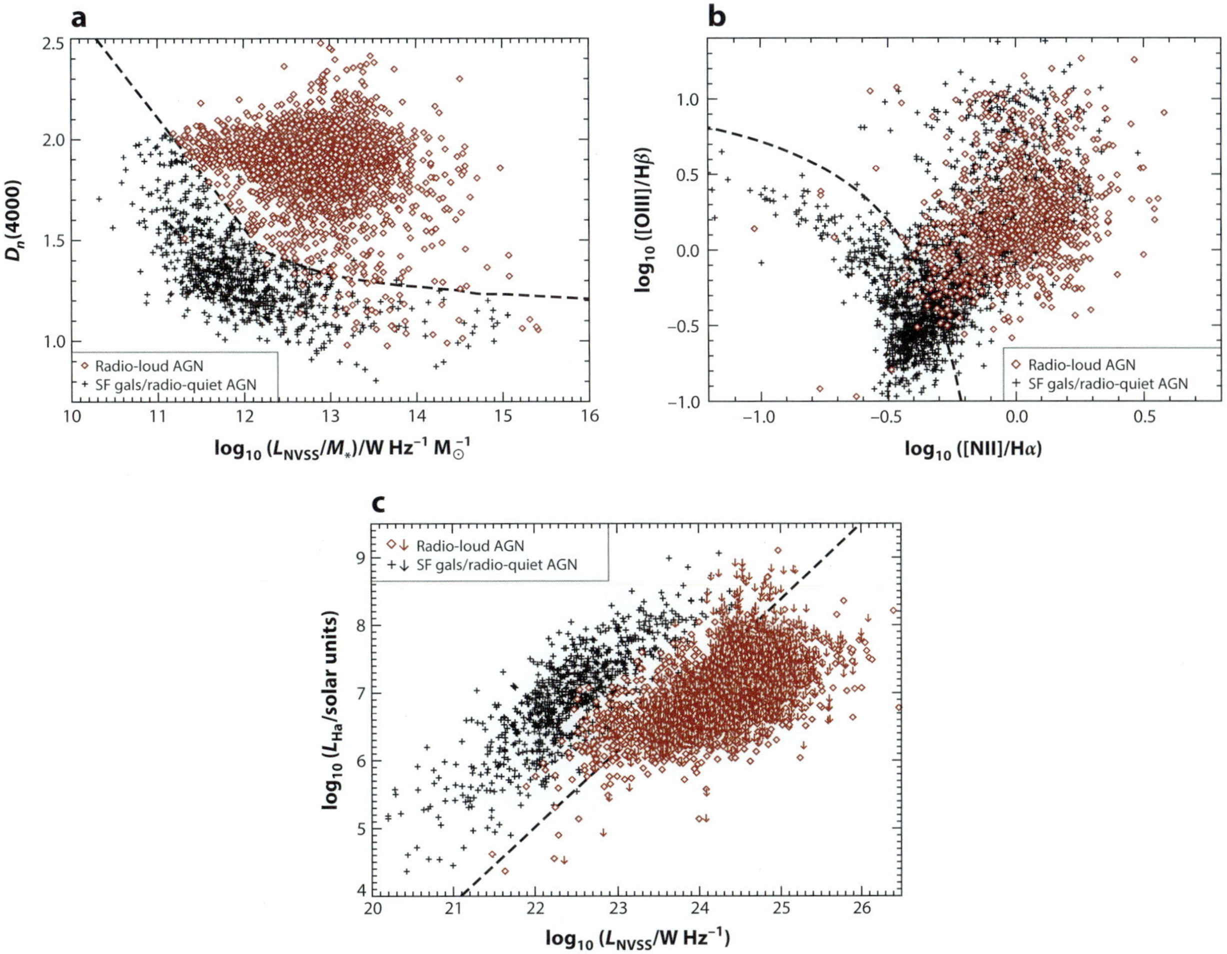

Figure 6

Diagnostic diagrams showing the galaxies within the Sloan Digital Sky Survey main galaxy sample that are detected in the radio waveband above $S_{1.4\,\mathrm{GHz}} = 5$ mJy. These diagrams are used to separate genuine radio-loud AGNs from other radio-detected galaxies in which the radio emission is powered by star formation (this includes both radio-quiet AGNs and star-forming galaxies). (*a*) The "D(4000) versus $L_{1.4\,\mathrm{GHz}}/M$" method, originally developed by Best et al. (2005a). (*b*) The widely used "BPT" (Baldwin-Phillips-Terlevich) emission-line ratio diagnostic. (*c*) The relationship between Hα and radio luminosity (*arrows* indicate upper limits to the Hα luminosity). In all plots, the dotted lines indicate the division used by Best & Heckman (2012) for that classification method. For the BPT diagnostic diagram, this division separates AGNs from star-forming galaxies in terms of what powers the emission lines, rather than what powers the radio emission. The division line in the $L_{H\alpha}$ versus $L_{1.4\,\mathrm{GHz}}$ plot was defined to ensure a clean sample of genuine radio-loud AGNs below the division. Through combinations of the locations of galaxies in the three diagrams [see Best & Heckman (2012) for details], sources can be given an overall classification as radio-loud AGNs (*red diamonds*) or star-forming (SF) galaxies/radio-quiet AGNs (*black crosses*). Adapted from Best & Heckman (2012) with permission.

2.3.1.2. Proxy 2: The [OIV] 25.9-micron and [NeV] 14.3,24.3-micron emission lines. These proxies have the same physical origin as the [OIII] 5007 line, arising from photoionized gas in the NLR.

Pros: (1) They arise well outside the region of the heaviest obscuration. (2) As mid-IR lines they are much less affected by dust extinction than the [OIII] 5007 line. (3) They have ionization potentials above the HeII edge at 54.4 eV (where a population of massive stars produces very little radiation). They therefore suffer much less contamination from stellar photoionization than [OIII]. This is especially true for [NeV]. This advantage relative to [OIII] has been exploited by Satyapal et al. (2008, 2009) and Goulding & Alexander (2009) to identify examples of low-luminosity AGNs in the nuclei of nearby late-type galaxies that were not recognized as AGNs based on optical spectra that were dominated by regions of star formation.

Cons: (1) They are expensive to observe (having been detected in hundreds of objects rather than hundreds of thousands). (2) The luminosity depends on the uncertain fraction of the AGN ionizing luminosity that is intercepted by clouds in the NLR.

2.3.1.3. Proxy 3: The mid-infrared continuum. This is produced by warm dust in the obscuring structure and as such represents the reprocessing of the primary UV/visible emission from the accretion disk and the soft X-rays from its corona.

Pros: (1) Thanks to WISE, we now have measurements of the near- to mid-IR continuum in a very large sample of galaxies in the nearby Universe (comparable in size with the SDSS main galaxy sample). (2) It represents a substantial fraction of the bolometric luminosity (roughly 25% in Type 1 AGNs where we directly observe nearly all the emission).

Cons: (1) The mid-IR light from the obscuring structure may not be emitted in an isotropic way and correction for this effect is uncertain at the level of factors of several (Buchanan et al. 2006; but see Gandhi et al. 2009). This effect is largest at the shortest mid-IR wavelengths. (2) The net observed mid-IR emission is the sum of the emission from the AGN and dust heated by hot stars. A decomposition of these two emission processes is uncertain and model dependent. This effect is worse at longer IR wavelengths.

2.3.1.4. Proxy 4: The hard X-ray continuum. In a Type 2 AGN the observed hard X-ray emission is a combination of whatever radiation succeeds in traversing the obscuring structure plus emission from the central AGN that has been scattered or reprocessed into our line of sight (e.g., Murphy & Yaqoob 2009).

Pros: (1) The hard X-ray emission probes regions very close to the SMBH (the accretion disk corona), but suffers far less obscuration than optical or UV observations of the accretion disk. (2) This hard X-ray emission is a rather unambiguous indicator of an AGN (rather than a starburst) at luminosities greater than about 10^{42} erg s^{-1}.

Cons: (1) Hard X-rays are poor probes of the AGN emission at the high column densities often inferred for the obscuring structure. This is true for column densities above 10^{24} cm^{-2} in the 2–10 keV band and above 10^{25} cm^{-2} even in the *Swift*/BAT band. These latter are the Compton-thick AGNs. (2) The *Swift*/BAT wide all-sky survey is quite shallow relative to the SDSS in its ability to detect AGNs.

2.3.1.5. Estimating the bolometric luminosity. Having briefly described the various proxies, we now summarize the evidence as to how well they actually do in estimating the AGN bolometric luminosity. The problem with this is that we do not know the ground truth needed to make this evaluation. The best approach is to intercompare these different proxies among the Type 2 AGNs and also to evaluate their behavior relative to the Type 1 AGNs (where obscuration is minimized) (see **Figure 7**).

LaMassa et al. (2010) examined a wide range of proxies in two complete samples of Type 2 AGNs. One was selected on the basis of the flux of the [OIII] 5007 emission line from the SDSS and the other was selected on the basis of the mid-IR (12-μm) flux measured with IRAS

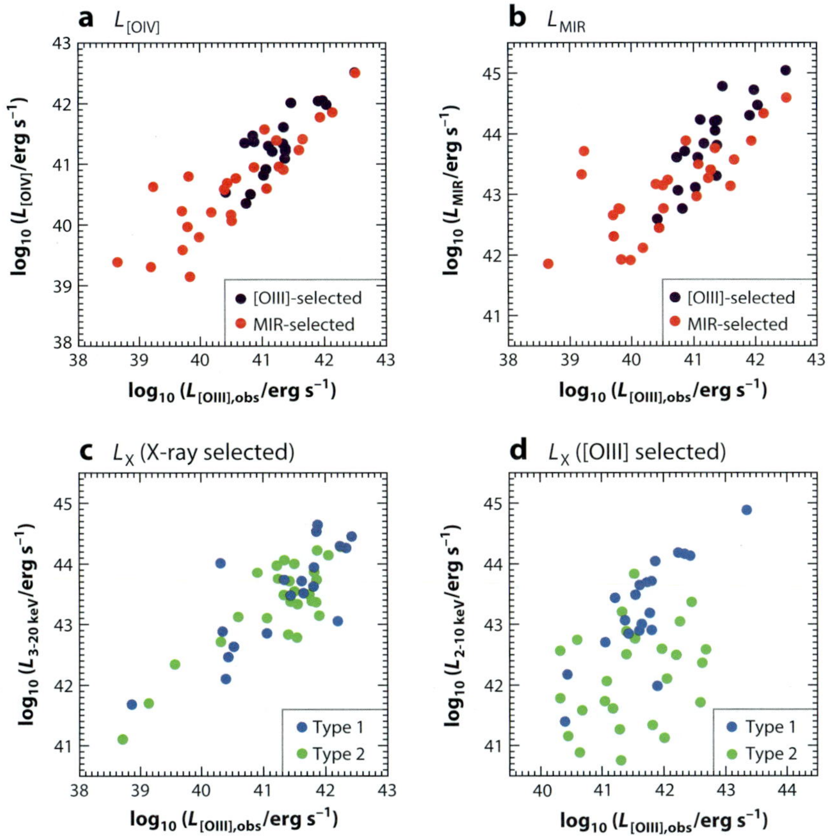

Figure 7

Relationships between different estimators of bolometric luminosity for radiative-mode AGNs. A comparison of the [OIII] emission-line luminosity with (*a*) the [OIV] line luminosity and (*b*) mid-IR (MIR) continuum luminosity, for samples of galaxies selected from their [OIII] emission line and their 12-μm mid-IR flux (data from LaMassa et al. 2010). The hard X-ray versus [OIII] luminosity for (*c*) X-ray-selected and (*d*) emission-line-selected samples of AGNs (data from Heckman et al. 2005). Type 2 AGNs in the emission-line-selected samples do not show a good correlation due to X-ray absorption.

(hereafter the SDSS-OIII and IRAS-MIR samples). The proxies they considered for the bolometric luminosity were the luminosities of the [OIII] 5007 line, the [OIV] 25.9-micron line, the mid-IR (12-μm) continuum, the hard X-ray *Swift*/BAT continuum, and the 8.4-GHz nuclear radio continuum. They examined the correlations between all possible pairs of the proxies and compared the distributions of the flux ratios of all proxy pairs with those of samples of unobscured Type 1 AGNs (cf. **Figure 7**). They concluded that the luminosity of the [OIV] line is the single best proxy and that the radio and hard X-ray luminosities are the worst. The problem with the hard X-ray luminosity is that many of the Type 2 AGNs in the SDSS-OIII and IRAS-MIR samples are Compton-thick and significantly obscured even in the *Swift*/BAT band (14–195 keV) (see also Heckman et al. 2005 and Winter et al. 2010). Similar conclusions were drawn by Rigby et al. (2009) on the basis of comparison of the flux ratio of the [OIV] 25.9-micron line and hard X-ray continuum (above 10 keV) in optically selected Type 1 and Type 2 AGNs drawn from the Revised Shapley-Ames Catalog (hereafter the RSA sample).

An analysis of multiple proxies in the RSA sample was undertaken by Diamond-Stanic et al. (2009). They found that the ratio of the flux of the 2–10-keV continuum and the [OIV] emission line was smaller on average in the Type 2 AGNs than in Type 1 AGNs and varied by about three orders of magnitude. This implies a wide range in the amount of attenuation of the 2–10-keV X-ray emission in these Type 2 AGNs (see **Figure 7**). They also found that the flux ratio of the [OIII] 5007/[OIV] 25.9-micron lines was systematically smaller in Type 2 than in Type 1 AGNs and showed more scatter in the RSA sample. They argued that this was a result of a greater amount of dust extinction in Type 2 AGNs (by typically 1 to 2 mag). LaMassa et al. (2010) found a much smaller offset in this ratio in the SDSS-OIII and IRAS-MIR samples (implying a mean excess extinction of [OIII] by only about 0.5 mag in the Type 2 AGNs). These results do underscore the importance of correcting the [OIII] flux for extinction (when possible).

Shao et al. (2013) have analyzed WISE observations of a sample of 30,000 Type 2 AGNs identified in SDSS spectra. They find an excellent overall correlation between the luminosities of the (extinction-corrected) [OIII] 5007 emission line and the (starlight-subtracted) 4.6-micron emission from hot AGN-heated dust (rms scatter of about 0.3 dex). Moreover, the partitioning of the AGN emission as a function of host-galaxy parameters is the same based on the [OIII] and 4.6-micron emission (once LINERs were excluded). Finally they show that the existing color-color techniques that have been used to identify AGNs from *Spitzer* and WISE data recover only the most luminous AGNs (objects in which the AGN significantly outshines its host galaxy in the near- and mid-IR). This misses most of the optically identified Type 2 AGNs in the SDSS, which have lower luminosities.

A simple summary of the above is that there is no single perfect proxy for the bolometric luminosity in obscured AGNs. Even relatively good proxies produce estimates of the bolometric luminosity of any individual obscured AGN that will be uncertain at the level of a factor of several (in a root-mean-square sense). Given the huge number of AGNs with measured [OIII] 5007 luminosities (e.g., Hao et al. 2005b, Kauffmann & Heckman 2009), we adopt this as our default proxy for the optically selected Type 2 AGNs. Based on an analysis of multiwaveband data for two samples of Type 1 AGNs, Heckman et al. (2004) calculated a mean bolometric correction to the [OIII] 5007 luminosity of a factor of 3,500. This refers to [OIII] luminosities derived from the raw [OIII] flux with no correction for dust extinction and assumes that Type 1 and Type 2 AGNs have the same bolometric correction. Kauffmann & Heckman (2009) adopted a mean bolometric correction of 600 for the extinction-corrected [OIII] 5007 luminosity. This was based on the multiwaveband data subsequently presented by LaMassa et al. (2010) and is consistent with the average value for the extinction of the [OIII] emission as measured in Type 2 Seyferts by Kewley et al. (2006). We adopt this factor 600 bolometric correction for calculations within this review.

Netzer (2009) finds similar mean bolometric corrections for the observed and extinction-corrected [OIII] 5007 luminosities. He argues, however, that the bolometric correction for a given AGN will depend on the ionization state of the gas in the NLR and that a combination of the luminosities of the [OIII] 5007 and [OI] 6300 lines is a more robust estimator of the bolometric luminosity. We agree with this point, but note that the [OI] 6300 line is significantly weaker than the [OIII] 5007 line (by a factor of typically 3 in LINERs and 10 in Seyfert nuclei). Requiring the [OI] line to be well detected diminishes the ability to probe AGNs at lower luminosities and/or over larger volumes in the SDSS main galaxy sample. The difference between Netzer's method and ours is most significant in the case of LINERs. As we have discussed above, a number of recent papers have concluded that the majority of LINERs in the SDSS (those with the weakest emission lines) are not bona fide AGNs. However, for the minority population of LINERs that are clearly AGNs we agree that Netzer's methodology is better.

Finally, it is important to note that these bolometric corrections can be luminosity dependent. This dependence is well established in the hard X-ray regime where the bolometric correction increases systematically with luminosity (e.g., Marconi et al. 2004). There have been claims and counter claims about a luminosity dependence for the bolometric correction to the [OIII] luminosity. Most recently, Stern & Laor (2012b) have analyzed SDSS spectra of a large sample of Type 1 AGNs and find that in log-log plots the [OIII] luminosity has a sublinear dependence on the UV continuum luminosity and on the broad Hα luminosity, but a linear dependence on the hard X-ray luminosity. They derive a bolometric correction to the [OIII] luminosity that increases by an order of magnitude over the range between $L_{[\mathrm{OIII}]} \sim 10^{6.5}$ and $10^9\,\mathrm{L}_\odot$. However, Shao et al. (2013) find a linear dependence of the [OIII] luminosity on the mid-IR luminosity (from WISE) for Type 2 SDSS AGNs over the same range in [OIII] luminosity. LaMassa et al. (2010) found a superlinear dependence of $L_{[\mathrm{OIII}]}$ on the mid-IR continuum luminosity for Type 2 AGNs. Hainline et al. (2013) confirm this relationship, extending it to the yet higher luminosities of Type 2 QSOs. The implied bolometric correction to $L_{[\mathrm{OIII}]}$ decreases by about an order of magnitude between $L_{[\mathrm{OIII}]} \sim 10^6$ and $10^{9.5}\,\mathrm{L}_\odot$ (i.e., the opposite of the result Stern & Laor found for Type 1 AGNs). The reason for this disagreement is not clear. Hainline et al. also find a possible saturation in $L_{[\mathrm{OIII}]}$ at MIR luminosities above about $10^{12}\,\mathrm{L}_\odot$ and attribute this to a transition to matter-bounded conditions where the AGN is effectively photoionizing the entire galaxy ISM (see also Netzer et al. 2006). We emphasize in closing that most of the results reviewed below (and our own new analyses) have assumed a luminosity-independent bolometric correction for $L_{[\mathrm{OIII}]}$ in Type 2 AGNs.

2.3.2. Mechanical energy outflow rates in radio jets. We now turn our attention to estimating the energy outflow rate in radio jets. In jet-mode AGNs, this can greatly exceed the radiative bolometric luminosity. It can also make a significant contribution to the total energetics in the small population of radio-loud radiative-mode AGNs.

The AGN jets are observable through their radio synchrotron emission. Monochromatic radio luminosity represents only a small fraction of the energy transport: The mechanical (kinetic) power of the jets is estimated to be about 2 orders of magnitude larger (Scheuer 1974). The jet mechanical power (P_{mech}) can be estimated from the synchrotron emission using the minimum energy condition to estimate the energy stored in the radio lobes. This assumes that the magnetic field has the strength that minimizes the combined energy content of particles and magnetic fields needed to produce the observed synchrotron emission (see Miley 1980). This is then combined with the age of the radio source (from synchrotron spectral aging techniques or radio source growth models; cf. Willott et al. 1999) and an efficiency factor that accounts for losses due to work done by the expanding radio source (Rawlings & Saunders 1991).

Our incomplete knowledge of the physics of radio sources leads to significant uncertainties in this estimation. Primary among these are the composition of the radio jet plasma and the low-energy cutoff of the electron energy distribution. The observed synchrotron emission traces only the highly relativistic electron population, and there are indications that the energy density in protons may be up to an order of magnitude higher (e.g., Bell 1978). Similarly, extrapolating the electron energy distribution down from the observed limit of $\gamma > 100$ to $\gamma \sim 1$ can increase the energy estimates by a factor of a few (depending upon the spectral index). Willott et al. (1999) combined all of the uncertainties into a single factor, f_{W}, that through observational constraints they determined to lie between ~1 and 20. Subsequent work favored values toward the upper end of that range (e.g., Blundell & Rawlings 2000, Hardcastle et al. 2007). The mechanical jet power derived by Willott et al., expressed in terms of the 1.4-GHz radio luminosity, corresponds to

$$P_{\mathrm{mech,sync}} = 4 \times 10^{35}(f_W)^{3/2}(L_{1.4\,\mathrm{GHz}}/10^{25}\ \mathrm{W\,Hz^{-1}})^{0.86}\ \mathrm{W}. \tag{1}$$

Note that weak dependencies on other radio source properties such as environment and source size also exist (see also Shabala & Godfrey 2013), but can be ignored to first order. O'Dea et al. (2009) used deep multifrequency radio data for 31 powerful radio galaxies to observationally estimate the pressure, volume, and spectral age of the lobes, and hence to derive the jet power. Daly et al. (2012) combined these with the radio luminosities to derive a relation between P_{mech} and L_{radio} and found a slope of 0.84, almost exactly the same as the Willott et al. theoretical prediction. Their normalization is consistent with Equation 1 for $f_W \sim 5$.

The expanding radio sources inflate lobes of relativistic plasma, which in X-rays can be observed as bubbles, or cavities, in the surrounding hot gas (Böhringer et al. 1993). As X-ray facilities developed the sensitivity and angular resolution to accurately measure the extent of these cavities and the pressure of the surrounding intergalactic medium, an alternative method of estimating the radio jet power became available. Bîrzan et al. (2004) derived the pV energy associated with the cavities in a sample of groups and clusters and used the buoyancy timescale (e.g., Churazov et al. 2001) to estimate their ages. They combined these cavity powers with the monochromatic 1.4-GHz radio luminosities to show that the two were well correlated, with a scatter of ~0.7 dex. Further cavity power measurements in systems ranging from giant elliptical galaxies to the most massive clusters have refined this correlation (e.g., Dunn et al. 2005, Rafferty et al. 2006, Bîrzan et al. 2008, Cavagnolo et al. 2010; see **Figure 8**). The largest uncertainty in this method is the determination of the cavity energy from the measured pressure and volume: $E_{\mathrm{cav}} = f_{\mathrm{cav}}\,pV$. The work done inflating the cavity implies $f_{\mathrm{cav}} > 1$, whereas for the relativistic plasma of the radio lobes the enthalpy of the cavity is $4pV$. Additional heating arising directly from the radio jet shocks (e.g., Forman et al. 2007) could imply higher values of f_{cav}, with some authors arguing that f_{cav}

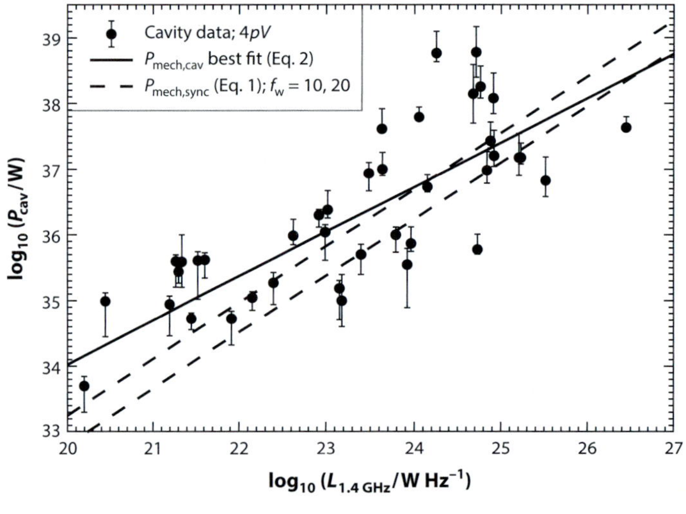

Figure 8

The jet mechanical energy of radio sources, estimated as $4pV$ from cavities and bubbles in the X-ray gas, versus the monochromatic 1.4-GHz radio luminosity. Data are primarily sourced from Cavagnolo et al. (2010), Bîrzan et al. (2008), and Rafferty et al. (2006). The solid line indicates the best-fit linear relation given in Equation 2 [$P_{\mathrm{mech,\,cav}} = 7 \times 10^{36} f_{\mathrm{cav}}(L_{1.4\,\mathrm{GHz}}/10^{25}\ \mathrm{W\,Hz^{-1}})^{0.68}$ W]. For comparison, the dotted lines show the synchrotron estimate of Equation 1 [$P_{\mathrm{mech,sync}} = 4 \times 10^{35}(f_W)^{3/2}(L_{1.4\,\mathrm{GHz}}/10^{25}\ \mathrm{W\,Hz^{-1}})^{0.86}$ W] for values of $f_W = 10$ and 20.

can be as high as ~10 (e.g., Nusser et al. 2006). The value of $f_{cav} = 4$ is the one usually adopted (e.g., Cavagnolo et al. 2010) and is consistent with a broad balance between AGN heating and radiative cooling in massive clusters (see Section 6).

Figure 8 shows a compilation of jet mechanical energy (assuming $4pV$) and radio luminosity measurements, primarily drawn from the analysis of Cavagnolo et al. (2010), recast into the same format as Equation 1. From this we derive a best-fit linear relation of

$$P_{\rm mech,\,cav} = 7 \times 10^{36} f_{\rm cav} (L_{1.4\,\rm GHz}/10^{25}\,\rm W\,Hz^{-1})^{0.68}\ W. \tag{2}$$

Equations 1 and 2 show very similar slopes, and for factors of $f_{cav} = 4$ and $f_W = 15$, the normalizations also agree for typical sources of $L_{1.4\,\rm GHz} \sim 10^{25}$ W Hz^{-1} (see **Figure 8**). Although discrepancies would exist for lower values of f_W (cf. Daly et al. 2012), the degree of agreement between these two independent estimators is encouraging. Hereafter in this review, Equation 2 with $f_{cav} = 4$ is adopted to calculate jet mechanical power from the radio luminosity.

2.3.3. Black hole masses, Eddington ratios, and accretion rates. The masses for SMBHs have been determined in two types of situations. In the case of Type 1 AGNs, the time-dependent response of the gas in the BLR to changes in the ionizing continuum flux can be used to infer a size for the BLR (from light-travel-time arguments). This size, together with the measured velocity dispersion in the BLR, can be used to estimate M_{BH} under the assumption that gas is acting solely under the dynamical influence of the SMBH. This technique is called reverberation mapping and has been recently reviewed by Peterson (2013). The subsample of SMBHs with masses determined by reverberation mapping can be used to calibrate a relation between M_{BH} and a combination of the widths of the broad emission lines and the optical continuum luminosity in Type 1 AGNs (e.g., Kaspi et al. 2000, Bentz et al. 2013).

In this review we are focused on Type 2 and jet-mode AGNs and their relationship to the more numerous population of inactive galaxies. In these cases, M_{BH} has been measured directly through the dynamical modeling of the spatially resolved kinematics of stars and/or gas. This subject has been recently reviewed by Kormendy & Ho (2013). There are still fewer than 100 galaxies in which M_{BH} has been determined directly. In order to estimate M_{BH} in large surveys like the SDSS, indirect (secondary) techniques must be used. The most widely used secondary technique is based on a fit to the relationship between M_{BH} and the stellar velocity dispersion (σ) of the surrounding galactic bulge (the M-σ relation—Ferrarese & Merritt 2000, Gebhardt et al. 2000; see **Figure 9**). This technique is particularly useful for the SDSS main galaxy sample. The SDSS fiber typically covers the region within a radius of a few kiloparsecs of the nucleus. The typical hosts for the SDSS AGNs are early-type galaxies (see Section 4 below), and thus the measured stellar velocity dispersion is representative of the bulge.

Most of the results in the literature that used the M-σ relation were based on the fit given by Tremaine et al. (2002): $\log(M_{BH}/M_\odot) = 8.13 + 4.02 \log[\sigma/(200\,\rm km\,s^{-1})]$. More recent work has found a steeper slope and a different normalization than the Tremaine et al. fit (see the Kormendy & Ho 2013 review). More specifically, the recent analyses by Graham & Scott (2013), McConnell & Ma (2013), and Woo et al. (2013) find respective slopes of 6.18, 5.64, and 5.48 and zero points of 8.15, 8.32, and 8.33. Woo et al. (2013) also conclude that the relationship between BH mass and σ is consistent between samples of AGNs and quiescent galaxies (see **Figure 9**), and this conclusion was reinforced by an investigation of powerful AGNs with more massive BHs by Grier et al. (2013). In terms of applicability to the SDSS, note that the SDSS spectra can only reliably measure stellar velocity dispersions above $\sigma \sim 70$ km s^{-1}, corresponding to BH masses larger than of order 10^6 M$_\odot$ (depending on which version of M-σ is adopted). A complementary probe of lower-mass SMBHs in the SDSS is provided by using mass estimates based on the broad

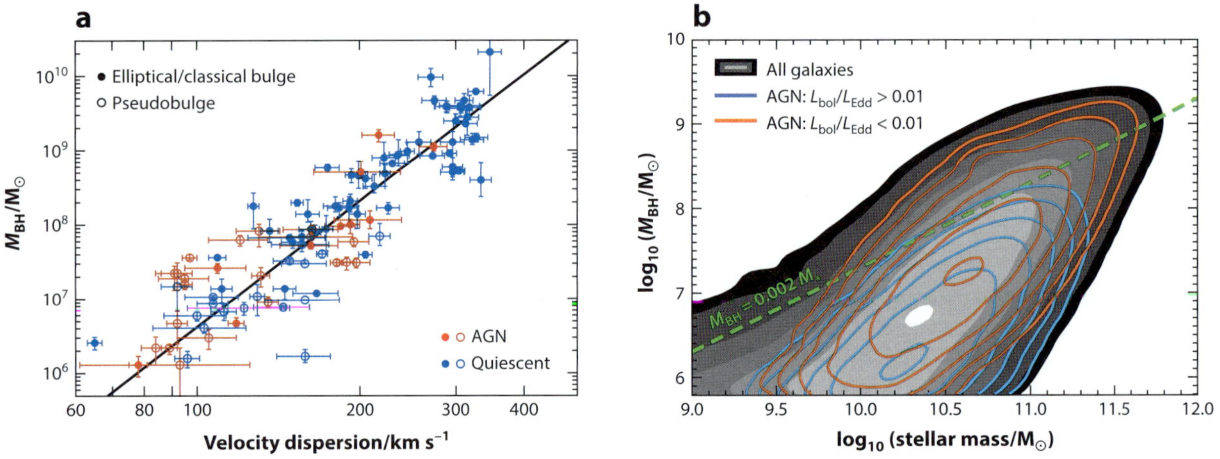

Figure 9

(*a*) The M_{BH}-σ relation of local galaxies with direct black hole mass measurements (data from Woo et al. 2013 and references therein). Both AGNs (the color-coding includes both radiative-mode and jet-mode AGNs) and quiescent galaxies are consistent with the McConnell & Ma (2013) relation, shown by the solid line. (*b*) The distribution of galaxies in the Sloan Digital Sky Survey main galaxy sample on the stellar mass versus black hole mass plane using black hole masses derived from the M_{BH}-σ relation. The grayscale indicates the volume-weighted distribution of all galaxies, with each lighter color band indicating a factor of two increase. The black hole mass is not a fixed fraction of the total stellar mass. This is also true for AGNs: The blue and red contours show the volume-weighted distributions of high (>1%; mostly radiative-mode) and low (<1%; mostly jet-mode) Eddington-fraction AGNs, with contours spaced by a factor of two.

emission-line width for Type 1 AGNs (Greene & Ho 2007b, Dong et al. 2012, Kormendy & Ho 2013).

Another secondary technique uses the correlation between M_{BH} and the stellar mass of the bulge (e.g., Marconi & Hunt 2003, Haring & Rix 2004). Although this is straightforward to apply to elliptical galaxies, the typical AGN host galaxies in the SDSS sample have significant bulge and disk components (Section 4). The SDSS images are shallow and do not well-resolve the bulge component in typical cases. This makes the task of robustly determining the bulge/disk mass ratio rather expensive in terms of computational time and the need for human intervention (e.g., Gadotti 2009, Lackner & Gunn 2012). Thus, this technique for estimating M_{BH} is less readily applicable to the SDSS AGN sample. Moreover, the new data compiled by Kormendy & Ho (2013) and the analysis by Graham & Scott (2013) both imply that the correlation of M_{BH} is significantly poorer with stellar bulge mass than with σ.

In this review we derive values for M_{BH} using the simple power-law M-σ relation used by McConnell & Ma (2013), specifically,

$$\log(M_{\mathrm{BH}}/\mathrm{M}_{\odot}) = 8.32 + 5.64 \log[\sigma/(200\,\mathrm{km\,s^{-1}})]. \tag{3}$$

It is important to emphasize here that it is not a reasonable approximation to assume that the BH mass is some fixed fraction of the total stellar mass. To illustrate this, we show in **Figure 9*b*** a plot of the total galaxy stellar mass versus the BH mass as estimated using our adopted M-σ relation for the SDSS main galaxy sample. The ratio of estimated BH–to–galaxy mass drops from an average of about $10^{-2.5}$ at $M_* \sim 10^{11.5}\,\mathrm{M}_{\odot}$ to about 10^{-4} at $M_* \sim 10^{10}\,\mathrm{M}_{\odot}$. As also shown in this figure, this is true for galaxies in general and for the hosts of both radiative-mode and jet-mode AGNs.

With an estimate of the AGN bolometric luminosity and SMBH mass in hand it is useful to then recast things in terms of the Eddington ratio for the AGN ($L_{\mathrm{bol}}/L_{\mathrm{Edd}}$). Here L_{Edd} is the luminosity of the classical Eddington limit:

$$L_{\rm Edd} = (4\pi G m_p c/\sigma_T) M_{\rm BH} = 3.3 \times 10^4 M_{\rm BH}, \tag{4}$$

where for the numerical form of the relation the luminosity and mass are in solar units. More uncertain still is an estimate of the accretion rate onto the SMBH. This is typically computed assuming a fixed efficiency ϵ for the conversion of accreted mass into radiant energy:

$$L_{\rm bol} = \epsilon \dot{M} c^2. \tag{5}$$

It is typically assumed that $\epsilon \sim 0.1$. This relatively high efficiency is well motivated theoretically for a BH radiating at a significant fraction of the Eddington limit (i.e., the radiative-mode AGNs), but may substantially underestimate the accretion rate of AGNs with very small Eddington ratios (the jet-mode AGNs). We discuss this more quantitatively in Section 5 below.

2.4. Determining Basic Galaxy Parameters

2.4.1. From imaging and photometry. With the advent of digital surveys of large regions of the sky, astronomers were motivated to design automated algorithms for measuring the principle properties of galaxies (e.g., Lupton et al. 2002). The size of a galaxy is most often parameterized by the radius that encloses a fixed fraction (often 50%; written as R_{50}) of the total light. The shape of the radial surface-brightness profile provides a simple way to estimate the prominence of the bulge with respect to the disk: Galaxies with concentration index $C = R_{90}/R_{50} > 2.6$ are predominantly bulge-dominated systems, with Hubble classification of Type Sa or earlier, whereas galaxies with $C < 2.6$ are mostly disk-dominated systems (Shimasaku et al. 2001). Gadotti (2009) performed 2D bulge/bar/disk decompositions of a representative sample of galaxies from the SDSS and found that the concentration index correlated more tightly with the bulge-to-disk ratio than the Sérsic index n obtained from 1D profile fitting. Lackner & Gunn (2012) reached a similar conclusion from analysis of a larger SDSS sample. Recently, the power of citizen science has been exploited to create visual classifications of roughly 900,000 SDSS galaxies (Lintott et al. 2011). These classifications have been used to document the nature of the host galaxies of AGNs (e.g., Schawinski et al. 2010).

The stellar mass-to-light ratio (and hence the stellar mass) of a galaxy can be measured using multiband optical and near-IR photometry to an accuracy of ~10%. Although this spectral energy distribution is sensitive to both the (luminosity-weighted) age of the stellar population and the amount of dust reddening, these two effects operate in such a way that the relation between optical color and *M/L* is linear with little scatter if galaxies have undergone relatively smooth recent star-formation histories (Bell & De Jong 2001, Salim et al. 2005). These estimates assume a fixed stellar initial mass function. The effective surface mass density μ_* is then defined as $0.5 M_*/(\pi R_{50}^2)$.

The SFR of a galaxy can also be estimated from fits to the UV through near-IR multiband photometry (e.g., Salim et al. 2005). UV photometry from GALEX is particularly important as it probes the young stellar population in star-forming galaxies. We note, however, that in the absence of far-IR photometry, which allows the thermal emission from dust to be measured directly, SFRs estimated from UV/optical photometry can be significantly biased if the range of dust extinction values in the spectral energy distribution (SED) model library is incorrect (Saintonge et al. 2011, Wuyts et al. 2011a). Most of this dust emission powered by the reprocessing of UV light from massive stars emerges in the far-IR. Unfortunately, the depth of the all-sky IRAS survey is not well matched to the SDSS main galaxy sample, with only about 2% of these galaxies being reliably detected by IRAS at 60 and 100 μm (Pasquali et al. 2005). The situation has been improved in some respects by the WISE survey, which detects about 25% of the SDSS main galaxies in all four WISE bands (3.4, 4.6, 12, and 22 μm). The disadvantage of the WISE data for estimating the SFR is that larger and more uncertain corrections need to be made to estimate the total IR luminosity than in the case of IRAS. In the case of AGNs, the WISE data suffer from the further problem of potential contamination by AGN-powered emission from warm dust in the obscuring structure, even at the longest-wavelength WISE band (22 μm).

Finally, radio continuum emission can also be used to estimate the SFR because it shows a strong correlation with the far-IR emission in star-forming galaxies (Condon 1992). Given the depth of the FIRST survey, the sensitivity of this technique over the SDSS volume is rather similar to that of the IRAS Faint Source Catalog, but—for our purposes—it has the disadvantage of being more susceptible to contamination by AGN emission than the far-IR.

2.4.2. From spectra. The spectra available for galaxies from the SDSS are taken through 3-arcsec diameter fibers. At the median redshift of the main galaxy sample ($z \sim 0.1$), only about 25–30% of the total light from the galaxy is sampled. As a result, estimates of the global SFRs of SDSS galaxies cannot be made using only the spectra (Brinchmann et al. 2004). The spectra do, however, provide a very useful probe of the central stellar populations of normal galaxies and AGN hosts.

The break occurring at 4,000 Å is the strongest discontinuity in the optical spectrum of a galaxy. The main contribution to the opacity comes from low-ionization metals. In hot stars, the elements are multiply ionized and the opacity decreases, so the 4,000-Å break will be small for young stellar populations and large for old, metal-rich galaxies. A break index, D(4000), was defined initially by Bruzual (1983). A definition using narrower and more closely spaced continuum bands was introduced by Balogh et al. (1999) and has the advantage of being considerably less sensitive to reddening effects.

Strong Balmer absorption lines arise in galaxies that experienced a burst of star formation that ended $\sim 10^8$ to 10^9 years ago, when A stars are the dominant contributors to the optical continuum (e.g., Dressler & Gunn 1983). Worthey & Ottaviani (1997) defined an Hδ_A index using a central bandpass bracketed by two pseudocontinuum bandpasses. Kauffmann et al. (2003b) used stellar population models (Bruzual & Charlot 2003) to show that the locus of galaxies in the D(4000)-Hδ_A plane is a powerful diagnostic of whether galaxies have been forming stars continuously or in bursts over the past 1–2 Gyr. In addition, these two stellar indices are largely insensitive to the dust attenuation effects that complicate the interpretation of broad-band colors. These indices are very useful diagnostics of the recent star-formation history of Type 2 and jet-mode AGNs.

Measurements of the Hδ_A index that have a precision that is good enough to provide useful constraints require relatively high signal-to-noise data. However, the Hδ line itself is only one of a handful of other high-order Balmer lines in the blue/near-UV portion of the spectrum. The situation can be significantly improved by utilizing the combined information in these lines. One technique for doing so uses principal component analysis. This approach has been successfully applied to SDSS spectra by Wild et al. (2007) and Chen et al. (2013b). More examples of the use of principal component analysis and other sophisticated techniques for extraction of information about the stellar content and star-formation history from the SDSS galaxy spectra are demonstrated by Panter et al. (2003, 2007), Cid Fernandes et al. (2005), and Yip et al. (2004).

3. BLACK HOLE MASS AND AGN LUMINOSITY FUNCTIONS

3.1. The Radiative-Mode Population

The luminosity functions for the radiative-mode AGNs (Seyferts and QSOs) can be constructed using a variety of different indicators. Some indicators, such as broad-band optical luminosities, are effective at probing unobscured BHs accreting near their Eddington limit at high redshifts.

However, when the accretion rate is low, these broad-band optical luminosities are dominated by stellar emission from the host galaxy. Moreover, the optical/UV continuum will not recover the obscured (Type 2) AGN population. Other indicators, such as high-ionization nebular emission-line luminosities, mid-IR continuum emission, and X-ray emission, have been used to probe lower accretion rate and obscured systems. Nice syntheses of these multiwaveband data have been undertaken by Hopkins et al. (2007b) and Shankar et al. (2009), who use them to produce bolometric luminosity functions for the radiative-mode AGNs that span the redshift range from ~0 to 6.

At a given redshift, the AGN bolometric luminosity function can be approximated as a double power law:

$$d\Phi/d\log(L) = \phi_*/\left[(L/L_*)^{\gamma_1} + (L/L_*)^{\gamma_2}\right]. \quad (6)$$

This is characterized by a faint-end slope (γ_1), a bright end slope (γ_2), the characteristic luminosity of the break in the function (L_*), and the comoving space density of AGNs with that luminosity (ϕ_*). The evolution of the bright end of the luminosity function from redshifts ~0 to 3 can be well described by a so-called pure luminosity evolution (PLE) model in which L_* evolves strongly with redshift while γ_2 and ϕ_* do not (Hopkins et al. 2007b). Fiore et al. (2012) show that this behavior of the bright AGN population may extend to even higher redshifts. The value for L_* (bolometric) increases from about $10^{11.5}$ $L_\odot$ at $z \sim 0$ to 10^{13} $L_\odot$ at $z \sim 2$ to 3, and then declines at higher redshift (e.g., Hopkins et al. 2007b, Shankar et al. 2009).

The first clear indications for a departure from the PLE model came from X-ray surveys of AGNs using the *Chandra X-Ray Observatory*, which generated complete samples of Type 1 AGNs down to significantly lower luminosities than was possible with optical surveys (see Brandt & Hasinger 2005). These surveys demonstrated that the space density of lower-luminosity AGNs peaks at lower redshift than that of higher-luminosity AGNs, a phenomenon that is often referred to as AGN downsizing. Similar results were later obtained by optical QSO surveys that pushed down to fainter flux levels. The QSO luminosity function of the completed 2dF-SDSS optically selected QSO survey (Croom et al. 2009) agrees extremely well in both amplitude and shape with the X-ray-selected QSO luminosity functions of Hasinger et al. (2005).

In terms of the luminosity function, this departure from a PLE model is primarily manifested as an evolution in the faint-end slope with γ_1 increasing from ~0.3 at $z = 3$ to ~0.9 at $z = 0$ (Hopkins et al. 2007b). The steep faint-end slope at low z is seen for both Type 1 and Type 2 AGNs and in both the [OIII] 5007 (Hao et al. 2005a,b) and hard X-ray (Ueda et al. 2011 and references therein) luminosity functions. Note that a faint-end slope of 1 corresponds to equal radiant output per bin in log L. This has an important implication: while at high redshifts the bulk of the radiant output occurs in AGNs near L_*, this is not the case in the contemporary Universe where it is spread rather evenly across a broad range in luminosity (and only about 20% is produced by AGNs with $L_{\rm bol} > L_*$).

Heckman et al. (2004) used data from the SDSS to compute the integrated [OIII] luminosity emitted by (Type 2 and jet-mode) emission-line AGNs per unit BH mass for nearby BHs spanning the mass range $3 \times 10^6 - 10^9$ $M_\odot$. They adopted a bolometric correction $L_{\rm bol}/L_{\rm [OIII]} = 3{,}500$ (see Section 2.3.1) and a mass-to-energy conversion efficiency of $\epsilon = 0.1$ to transform these ratios into estimates of the time taken for the entire population of BHs to have grown to their present-day mass, given their present-day volume-averaged accretion rate. They defined the mass-dependent volume-averaged growth time for the local population of BHs to be the total mass of the BH population per unit volume in a given BH mass bin divided by the total growth rate of the BH population per unit volume in the same BH mass bin. They then found that this volume-averaged growth time of the population was of order the Hubble time for BHs with masses less than

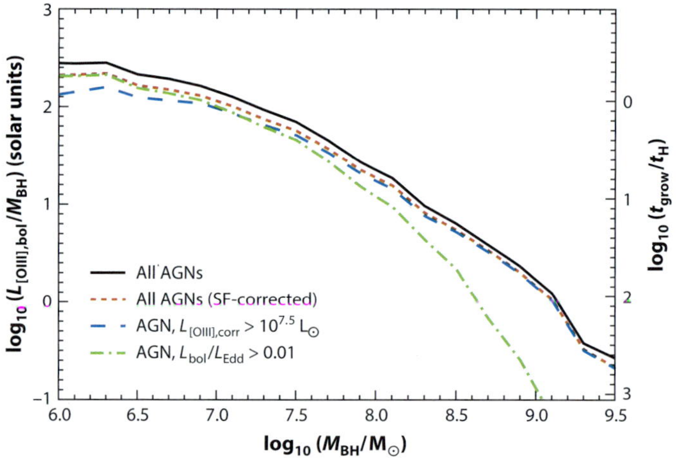

Figure 10

Logarithm of the ratio of the total bolometric radiated luminosity (as calculated from the [OIII] emission line; see Section 2.3.1) per unit volume due to emission-line AGNs of a given black hole mass to the total mass per unit volume in black holes at that black hole mass (both in solar units) in the contemporary Universe (see Heckman et al. 2004). The right axis shows the corresponding growth time (mass-doubling time) for the population of black holes in units of the Hubble time. The solid black line shows the result obtained by integrating over all AGNs in the sample. The dotted red line shows the effect of correcting the [OIII] luminosities for the contribution from star formation in composite galaxies. The dashed blue line shows the result obtained if only AGNs [reddening-corrected and star-formation (SF)-corrected] with [OIII] luminosities above $10^{7.5}\ L_\odot$ are included. The dash-dot green line includes only AGNs with bolometric luminosities above 1% of the Eddington rate (this cut effectively reduces the sample to only radiative-mode AGNs).

$10^{7.5}\ M_\odot$, but that this growth time increased sharply for higher mass BHs, implying that most of their mass was accreted at high redshifts (**Figure 10**). Greene and colleagues (Greene & Ho 2007a, Greene et al. 2009) reached similar conclusions for the local Type 1 AGN population in the SDSS.

Heckman et al. (2004) also studied the fraction of BHs of mass $M_{\rm BH}$ with accretion rates above a given value. They showed that these cumulative accretion rate functions cut off sharply at $\dot{M}_{\rm BH}/\dot{M}_{\rm Edd} \simeq 1$. Many more low-mass BHs than high-mass BHs had accretion rates near the Eddington limit (0.5% of $10^7\ M_\odot$ BHs are accreting above a tenth of the Eddington limit, as compared with 0.01% of BHs with masses $3 \times 10^8\ M_\odot$). In a later paper, Kauffmann & Heckman (2009) analyzed Eddington ratio distribution functions in narrow bins of BH mass. They found that the distribution functions can be decomposed into a log-normal distribution of accretion rates, plus a power-law function that dominates at low Eddington ratio. We describe the dependence of this distribution function on the age of the central stellar population in Section 4.2 below.

3.2. The Jet-Mode AGNs

The broad nature of the radio luminosity function of local galaxies has been well-established for many years (e.g., Toffolatti et al. 1987, Condon 1989), but the large spectroscopic surveys have enabled both improved statistics and an accurate separation of the star-forming and radio-AGN populations. Radio luminosity functions derived from the Las Campanas redshift survey (Machalski & Godlowski 2000), the 2dFGRS (Sadler et al. 2002), the SDSS (Best et al. 2005a,

Best & Heckman 2012), and the 6-degree-field galaxy survey (6dFGS; Mauch & Sadler 2007) are all in excellent agreement and show that the local radio luminosity function of AGNs is well fitted by a double power-law model similar in form to Equation 6:

$$\rho = \rho_0 / \left[(P/P_0)^{\alpha} + (P/P_0)^{\beta} \right]. \tag{7}$$

At 1.4 GHz, the best-fitting values from combining all four spectroscopic surveys above give $\rho_0 = 10^{-(5.33\pm0.12)}$ Mpc^{-3} $\log_{10}(P)^{-1}$, $P_0 = 10^{24.95\pm0.14}$ W Hz^{-1}, $\alpha = 0.42 \pm 0.04$, and $\beta = 1.66 \pm 0.21$. Mauch & Sadler (2007) show that, at the faint end, the radio luminosity function of radio-loud AGNs continues with the same slope down to at least $P_{1.4\,\mathrm{GHz}} \sim 10^{20}$ W Hz^{-1}. They argue that the luminosity function must begin to turn over at around $P_{1.4\,\mathrm{GHz}} \sim 10^{19.5}$ W Hz^{-1}, or the total space density of radio sources would exceed the space density of galaxies brighter than L_*, which form the typical hosts. Cattaneo & Best (2009) similarly argue for a turnover no fainter than $P_{1.4\,\mathrm{GHz}} \sim 10^{19.2}$ W Hz^{-1} by comparison with the local space density of SMBHs above 10^6 M$_\odot$. Thus, the observed radio luminosity functions now broadly constrain the entire distribution of AGN radio luminosities.

The AGN population as revealed by radio surveys is composed of two distinct categories of sources: a population with strong QSO/Seyfert-like emission lines (historically referred to as high-excitation sources) and a population with weak LINER-like emission lines (referred to as low-excitation sources; e.g., Hine & Longair 1979, Laing et al. 1994). In the language of the present review these are radiative-mode and jet-mode AGNs, where the high excitation sources form a small subpopulation of radiative-mode AGNs that produce powerful jets. Best & Heckman (2012) separated the radio luminosity function into these two radio source populations and showed that both are found over the full range of radio powers they observed (see **Figure 11**). The jet-mode sources constitute about 95% of the radio-AGN population at all radio luminosities below $P_{1.4\,\mathrm{GHz}} = 10^{25}$ W Hz^{-1}, but the radiative-mode population becomes dominant above 10^{26} W Hz^{-1}. In the

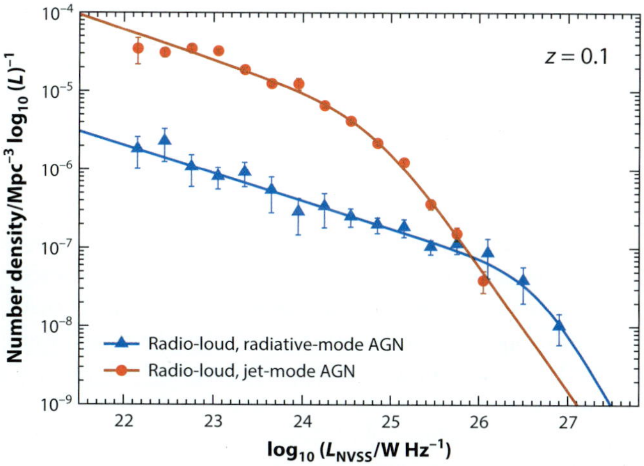

Figure 11

The local radio luminosity function of radio-loud AGNs split into radiative-mode and jet-mode sources. The data are largely drawn from the results of Best & Heckman (2012), who split these two populations and derived luminosity functions using radio-selected AGNs in the Sloan Digital Sky Survey main galaxy sample. However, as shown by Gendre et al. (2013), these results underestimate the luminosity function of the radiative-mode AGNs above 10^{26} W Hz^{-1}, where Type 1 AGNs dominate. Therefore the results of Gendre et al. are adopted for the three highest luminosity points of the radiative-mode population. The local radio luminosity functions are well fitted by broken power-law models.

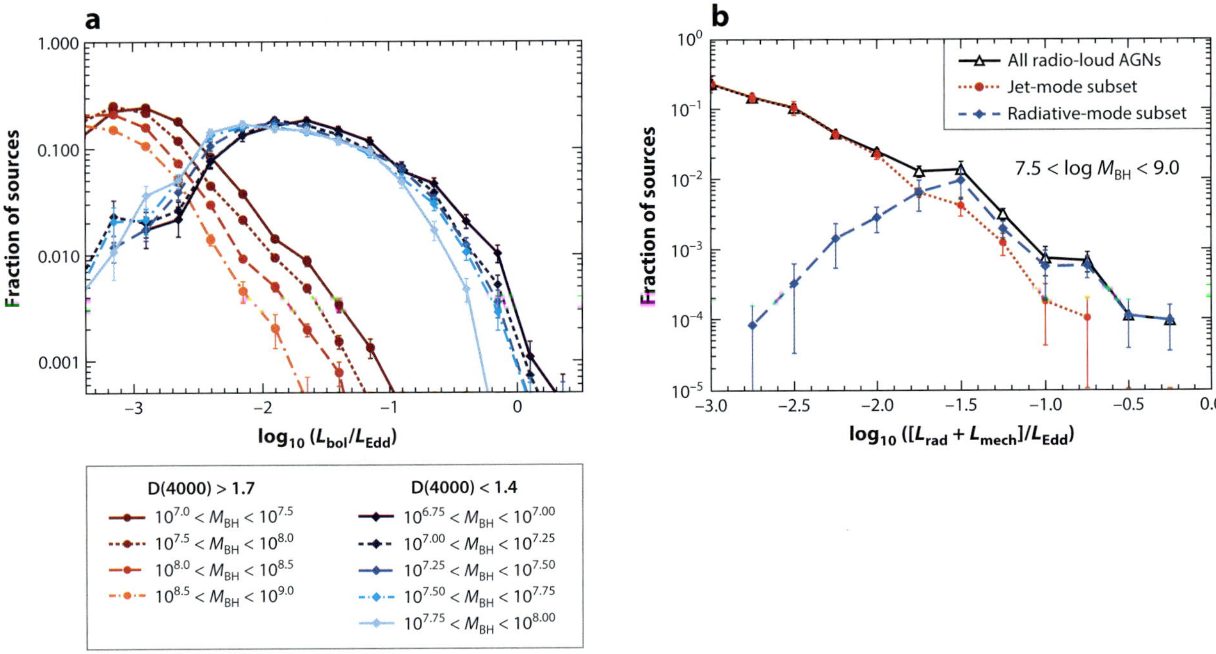

Figure 12

(*a*) The distribution of Eddington-scaled accretion ratios for emission-line-selected AGNs from the Sloan Digital Sky Survey (SDSS) main galaxy sample split by star-formation activity. Following Kauffmann & Heckman (2009), the distributions are shown separately for actively star-forming host galaxies [D(4000) < 1.4] and for passive host galaxies [D(4000) > 1.7], each for a range of black hole masses. Fractions plotted are within each 0.25-dex wide bin in the Eddington ratio. As found by Kauffmann & Heckman, the passive galaxies show a power-law distribution of Eddington fractions, the amplitude of which decreases with increasing black hole mass. In contrast, the Eddington ratios of the star-forming hosts show a log-normal distribution peaking at a few percent of Eddington. This distribution is largely independent of black hole mass (though this depends slightly upon the version of the M-σ relation adopted). (*b*) The Eddington-scaled accretion ratios of radio-loud AGNs from the SDSS split into jet-mode and radiative-mode sources; again, fractions plotted are within each 0.25-dex wide bin. Here the bolometric luminosity is calculated by summing the radiative and mechanical contributions. The two classes show distributions remarkably similar to the populations seen for emission-line AGNs in panel *a*.

radio samples derived from large spectroscopic surveys of the contemporary Universe, therefore, the radio-AGN population is predominantly composed of the jet-mode (low-excitation) sources.

The local radio-AGN population has historically been split by Fanaroff & Riley (FR; 1974) radio morphology, where FR2 sources are "edge-brightened" with jets that remain collimated until they end in bright hotspots, and FR1 sources are "edge-darkened" with jets that flare and gradually fade. The FR1 and FR2 populations are also found over a wide range of overlapping radio powers, with the switch in the dominant local population occurring at around $P_{1.4\,\mathrm{GHz}} \sim 10^{25}$ W Hz^{-1} (cf. Best 2009, Gendre et al. 2010). There is considerable overlap between the jet-mode AGN population and those radio sources morphologically classified as FR1s, and also between the radiative-mode AGNs with strong radio emission and the FR2s. However, the radio morphological split and the division in the nature of the accretion flow are not the same thing. The large overlap arises because the population switch occurs at similar radio powers, but there exists a large cross-population of jet-mode FR2s (e.g., Laing et al. 1994) and a smaller population of radiative-mode FR1s (e.g., Heywood et al. 2007), as quantified by Gendre et al. (2013).

Best et al. (2005b) considered the distribution of radio emission associated with AGN activity across galaxies of different BH masses. They showed that the prevalence of radio-AGN activity is a very strong function of BH mass, rising as $f_{\rm rad} \propto M_{\rm BH}^{1.6}$. This offered a proper quantification of earlier widespread statements in the literature that radio-loud AGNs are found only in galaxies at the top end of the black hole mass function (BHMF). Best et al. (2005b) also considered the bivariate distribution of radio luminosity and BH mass. They found that the shape of the radio luminosity function showed only weak dependence on BH mass.

The Eddington ratio distribution function of radio-loud AGNs can be calculated by summing the jet mechanical luminosity and the bolometric radiative luminosity (cf. Section 2.3) to provide the total energetic output of the AGNs. Best & Heckman (2012) showed that the radiative-mode and jet-mode radio sources show distinct distributions of Eddington ratios, with the latter typically accreting below 1% of their Eddington rate, whereas the former have typical accretion rates between 1% and 10% of Eddington. Considering sources with BH masses between $10^{7.5}$ and $10^9\ M_\odot$, **Figure 12*b*** shows the distribution of Eddington ratios of radio-loud AGNs in the SDSS main galaxy sample, above 10^{-3} of the Eddington limit (incompleteness prohibits analysis below this level). The jet-mode sources dominate at low Eddington ratios and show a roughly power-law distribution with a slope of about -1.5. In contrast, the radiative-mode population shows a distribution peaked at a few percent of the Eddington limit.

3.3. The Black Hole Mass Function

The supermassive BHMF is defined as the comoving number density of BHs per bin in log mass at a given redshift. The techniques used to determine the BHMF have been reviewed recently by Kelly & Merloni (2012). Here we briefly summarize the properties and implications of the BHMF.

In the contemporary Universe the BHMF can be measured in an empirical model–independent fashion by adopting the measured scaling relation between BH mass and the bulge mass or velocity dispersion (described in Section 2.3.3). Combined with the measured volume density of bulges as a function of velocity dispersion or stellar mass from the SDSS, the BHMF can then be computed. Shankar et al. (2009) have compared the BHMF derived in this way by different groups (**Figure 13**).

The two main conclusions to be drawn from **Figure 13** are that the distribution of mass in BHs today peaks in BHs with masses of around $10^{8.5}\ M_\odot$ and that the mass distribution is actually quite narrow (half the total mass in BHs resides in the mass range from $\sim 10^{7.8}$ to $10^{8.8}\ M_\odot$). The integral of this distribution represents the total mass per unit volume in the contemporary Universe ($\rho \sim 10^{5.7}\ M_\odot\ {\rm Mpc}^{-3}$). This mass density is directly connected to the total energetic output of AGNs over the history of the Universe (Soltan 1982). Consistency with the cosmic evolution of the AGN bolometric luminosity function (summarized above) implies a mean accretion efficiency ($\epsilon = L_{\rm bol}/\dot{M}c^2$) of order 10% (Yu & Tremaine 2002, Marconi et al. 2004, Hopkins et al. 2007b, Shankar et al. 2009).

At higher redshifts the BHMF has usually been computed in a model-dependent fashion. These computations take the present-day BHMF as a boundary condition and use a continuity equation to relate the evolution of the BHMF to the accretion history as inferred from the evolution of the AGN luminosity function. To compute BHMF(z), an assumption must be made about the mean AGN light curve [$L(M_{\rm BH}, t)$] and the accretion efficiency (ϵ). Examples of this approach are shown by Marconi et al. (2004), Shankar et al. (2009, 2013) and Cao (2010). An alternative technique for measuring BHMF(z) uses the mass function of actively accreting BHs and the so-called black hole fundamental plane (Merloni & Heinz 2008).

All these analyses reach qualitatively similar conclusions: The BHMF grows antihierarchically, with the population of more massive BHs being produced at higher redshifts. The population of BHs with masses $>10^9\ M_\odot$ was largely in place by $z \sim 2$. Since $z \sim 1$, only the population of

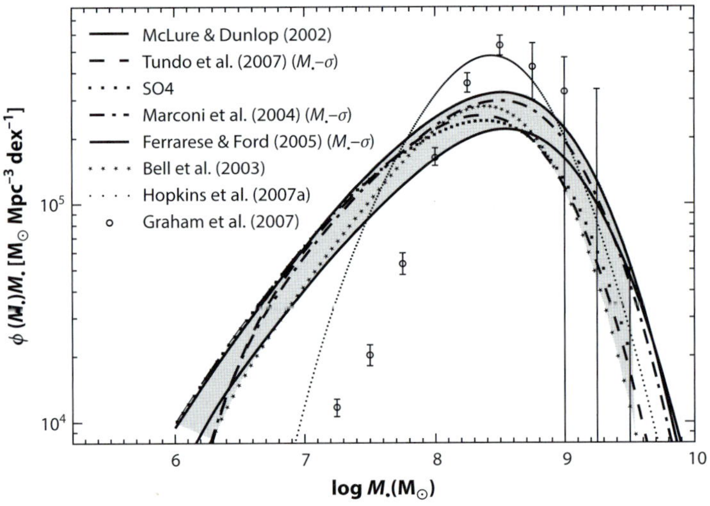

Figure 13

A collection of estimates of the distribution of mass across the local black hole population (i.e., the local black hole mass function scaled by black hole mass) taken from Shankar et al. (2009). The shaded gray area represents the range of estimates based on local scaling relations of $M_{\rm BH}$ with σ, $M_{\rm Bulge}$, and $L_{\rm Bulge}$. Readers are referred to Shankar et al. (2009) for details.

BHs with masses below about $10^8\,M_\odot$ has been growing significantly. As we have summarized above, this inference is directly confirmed by measurements of the mass-dependent growth rates of SMBHs in the contemporary Universe (Heckman et al. 2004, Kauffmann & Heckman 2009).

4. HOST-GALAXY PROPERTIES

4.1. Overview

In the introduction to this review we summarized the overall landscape defined by the population of galaxies in the contemporary Universe and described the bimodal galaxy population [lower-mass, lower-density, star-forming (blue) versus higher-mass, higher-density, quiescent (red) galaxies]. Here we want to see where the population of growing BHs today lives within this landscape. Both the structural properties of the galaxies and their stellar populations turn out to be crucial in this regard. We begin with an overview of both the radiative-mode and jet-mode AGNs in this section, and then follow with detailed discussions of the two populations in the two sections to follow.

To fully appreciate the interplay between the structural properties and the stellar population in AGN host galaxies, it is worth taking a guided tour of the pair of multipanel plots shown in **Figures 14** and **15**, derived from the SDSS. **Figure 14** plots the volume-weighted values of various key parameters in the 2D space of stellar mass (M_*) and D(4000). The top four panels show the distributions of M_*, BH mass ($M_{\rm BH}$), extinction-corrected [OIII] luminosity from AGNs ($L_{\rm [OIII]}$; tracing bolometric radiative output), and the production of mechanical energy in radio jets ($L_{\rm mech}$). The bottom four panels show the ratios of $L_{\rm [OIII]}/M_*$, $L_{\rm [OIII]}/M_{\rm BH}$, $L_{\rm mech}/M_*$, and $L_{\rm mech}/M_{\rm BH}$.

The plot weighted by M_* shows the basic layout of the bimodal galaxy population. The plot weighted by $M_{\rm BH}$ shows a shift toward higher values of M_* (consistent with the steep dependence of $M_{\rm BH}$ on M_* shown in **Figure 9**). The plot weighted by $L_{\rm [OIII]}$ traces the growth of BHs through accretion (this is essentially tracing the radiative-mode population of AGNs). It is immediately

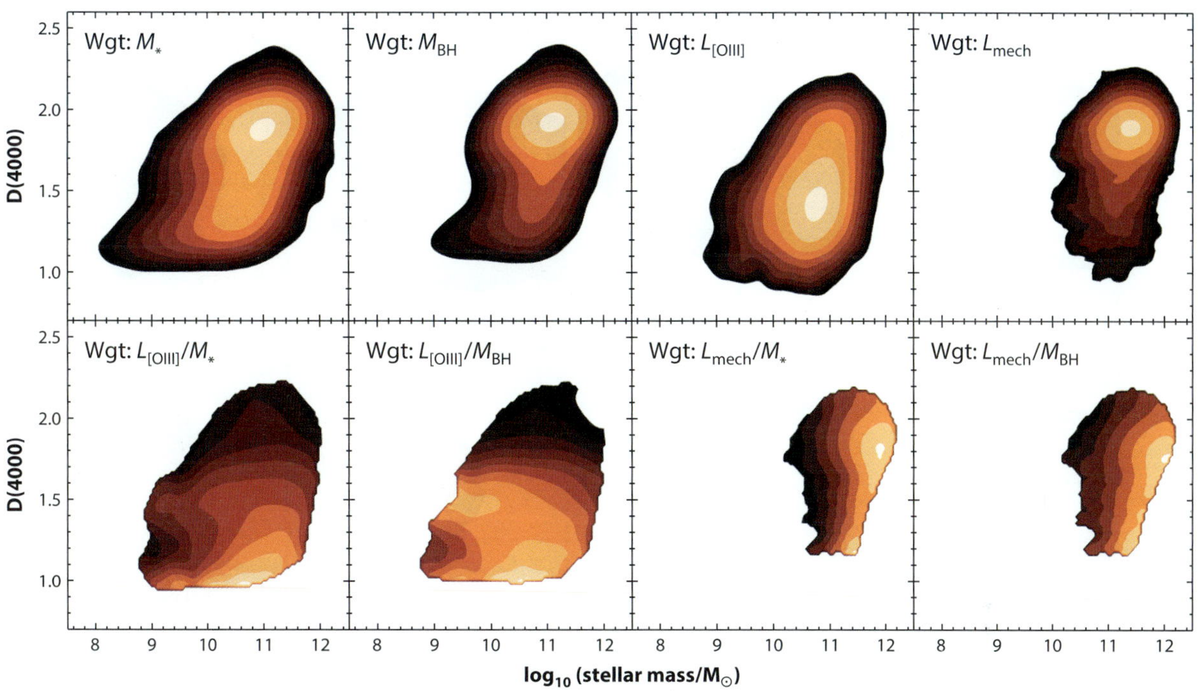

Figure 14

(*Top*) The distribution of galaxies from the Sloan Digital Sky Survey main galaxy sample in the plane of stellar mass versus 4,000-Å break strength, weighted according to (from *left* to *right*) stellar mass, black hole mass, reddening-corrected [OIII] line luminosity (i.e., bolometric radiative luminosity), and jet mechanical luminosity. Each color band corresponds to a factor of two difference, increasing from dark to light. (*Bottom*) Ratios of upper panels (same color coding) to illustrate the distributions of (from *left* to *right*) bolometric radiative luminosity per stellar mass, bolometric radiative luminosity per black hole mass, jet mechanical luminosity per stellar mass, and jet mechanical luminosity per black hole mass.

clear that BH growth is occurring primarily in moderately massive galaxies ($M_* \sim 10^{10}$ to a few times 10^{11} $M_\odot$) with a young central stellar population (see Kauffmann et al. 2003a; also see **Figure 2**). This is further quantified by the panels below. The plot of $L_{[OIII]}/M_*$ shows a strong dependence on the age of the stellar population and only a weak dependence on M_* over the range from $M_* \sim 10^{10}$ to $10^{11.5}$ $M_\odot$. The plot of $L_{[OIII]}/M_{BH}$ (a proxy for the Eddington ratio) tells a similar story, but here the dependence is almost entirely in stellar age with essentially no dependence on stellar mass for a given value of D(4000).

A complementary set of plots is shown in **Figure 15**, which shows the same set of distributions in the 2D space of M_{BH} versus D(4000). These plots reinforce the conclusions above. The growth of BHs as traced by [OIII] emission primarily occurs in the lower mass BHs ($\sim 10^{6.5}$ to 10^8 $M_\odot$; Heckman et al. 2004), more specifically in the subset of those BHs that live in galaxies undergoing significant central star formation. The plots of $L_{[OIII]}/M_*$ and $L_{[OIII]}/M_{BH}$ both illustrate the fact that the strong dependences of these ratios are primarily on the age of the central stellar population.

Turning our attention now to the panels involving L_{mech}, we see that the jet-mode population of AGNs is almost disjoint from the radiative-mode AGNs in the properties of the host galaxies and their BHs (see also **Figure 2**). The bulk of the jet mechanical energy is produced by the most massive BHs ($M_{BH} \sim 10^8$ to $10^{9.5}$ $M_\odot$) living in the most massive galaxies ($M_* \sim 10^{11}$ to 10^{12} $M_\odot$)

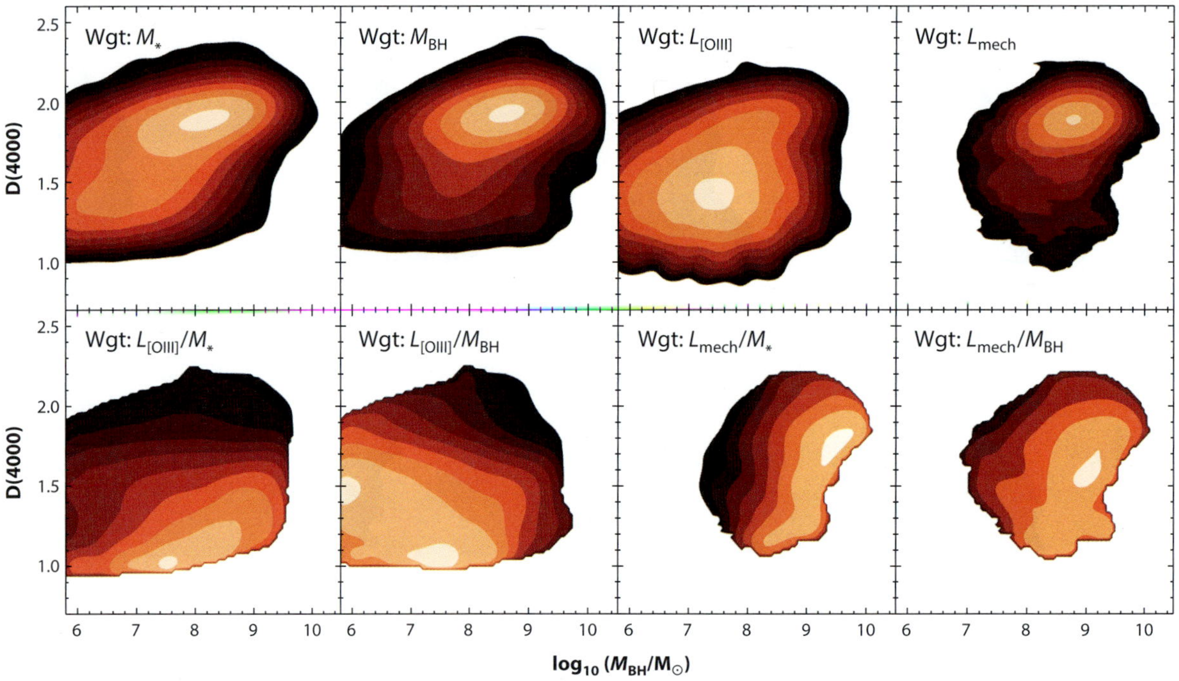

Figure 15

As **Figure 14**, but now showing the distributions in the black hole mass versus 4,000-Å break strength plane.

and generally having old stellar populations (although there is a minority population of massive galaxies with younger stellar populations). In direct contrast to the radiative-mode AGNs, the plots of L_{mech}/M_* and L_{mech}/M_{BH} show strong positive correlations with both M_* and M_{BH} (see Best et al. 2005b), but only weak dependences on the age of the stellar population.

The main dependences revealed in these 2D plots are summarized in a simpler form in **Figure 16** in which the mass-normalized AGN outputs are plotted as 1D functions of M_*, μ_*, and D(4000). Three key conclusions are evident.

First, there is only a slow rise of about 0.5 dex in L_{rad}/M_* on M_* over the range from 10^9 to $10^{10.7}$ $M_\odot$, but then there is a steeper drop at higher masses. This can be contrasted with the very steep increase in L_{mech}/M_* with M_*. This shows that the predominant form of energy produced by BHs switches from radiation to jet energy at the highest stellar masses (>$10^{11.5}$ $M_\odot$). The crossover corresponds to $M_{BH} > 10^9$ $M_\odot$. It is intriguing that the sum of the mass-normalized L_{rad} and L_{mech} remains relatively constant over the full stellar mass range plotted. Because the ratio of BH to galaxy stellar mass rises strongly with increasing mass (**Figure 9**), this near-constancy would not apply to AGN output normalized to BH mass (e.g., **Figure 10**).

Second, both modes of BH output plummet precipitously below the characteristic value $\mu_* \sim 10^{8.5}$ $M_\odot$ kpc^{-2} that divides the bimodal galaxy population described in Section 1.3. The effective growth of BHs evidently requires conditions of high central density.

Finally, the strong overall increase in L_{rad}/M_* moving from old [D(4000) > 2] to intermediate-age [D(4000) $\sim$ 1.5] stellar populations, followed by saturation at still younger ages, is very clear. This was shown most explicitly by Kauffmann & Heckman (2009) and is consistent with **Figure 12*a***. In contrast, the jet output is only significant when compared with the radiative output in the galaxies having the oldest stellar populations (see Best et al. 2005b).

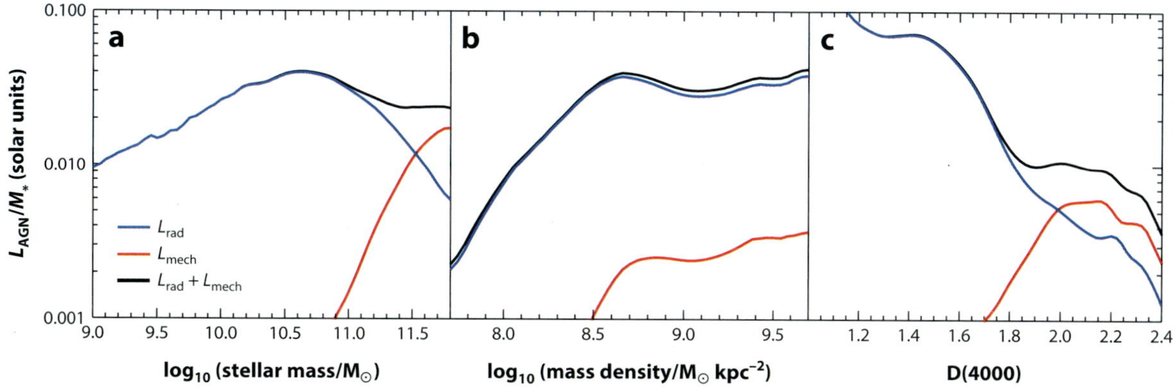

Figure 16

The distribution of AGN luminosity per unit stellar mass among galaxies from the Sloan Digital Sky Survey main galaxy sample as a function of (*a*) galaxy stellar mass, (*b*) stellar surface mass density, and (*c*) 4,000-Å break strength. In each plot the blue line indicates the radiative luminosity output (estimated from the [OIII] emission line), the red line shows the mechanical luminosity output in the radio jet, and the black line shows the sum of the two.

Taken together these figures clearly support the division of the AGN population into two separate modes, as we have stressed throughout our review. The broad conclusions on host-galaxy properties of the different AGN populations are summarized in **Figure 4**.

4.2. The Hosts of Radiative-Mode AGNs

The above plots for the radiative-mode AGNs lead to a simple and physically satisfying conclusion: The accretion-driven growth of BHs in the contemporary Universe primarily occurs in centers of galaxies where two criteria are met: (*a*) there is an abundant supply of cold gas (whose presence is implied by recent or ongoing star formation) and (*b*) the stellar surface mass density is high. Putting these two criteria together, it is natural to conclude that there is really one overarching requirement, namely an abundant central supply of cold and dense gas (e.g., Larson 2010). This is shown in an explicit and compact way in **Figure 17**. In this section, we expand on this by considering different aspects of the relationship between BH growth and star formation in some detail.

4.2.1. The relationship to the star-formation rate. So far we have discussed the relationship between AGNs and star formation only in a qualitative sense. Here we summarize what has been learned about the quantitative relationship. Let us consider the correlation between SFR and AGN luminosity. There are two complementary approaches to this. One is to select a sample of AGNs spanning a range of luminosity and then measure the SFRs for each. This is conceptually simple, but ignores the fact that many galaxies do not host detectable AGNs, and these galaxies are ignored in the above approach. Thus, the second approach is to examine the distribution of AGN luminosity as a function of SFR for a complete sample of galaxies selected without regard to the presence of an AGN.

As an example of the first approach, Kauffmann et al. (2003a) used the SDSS to measure the distribution of the strength of the 4,000-Å break for AGNs as a function of raw [OIII] luminosity. Based on subsequent calibration (by Brinchmann et al. 2004) of the 4,000-Å break as a function of sSFR (sSFR = SFR/M_*), there is a clear but shallow dependence: As the [OIII] luminosity increases from 10^6 to 10^9 $L_\odot$, the median sSFR only increases from $10^{-11.3}$ to 10^{-10} year^{-1}. This is consistent

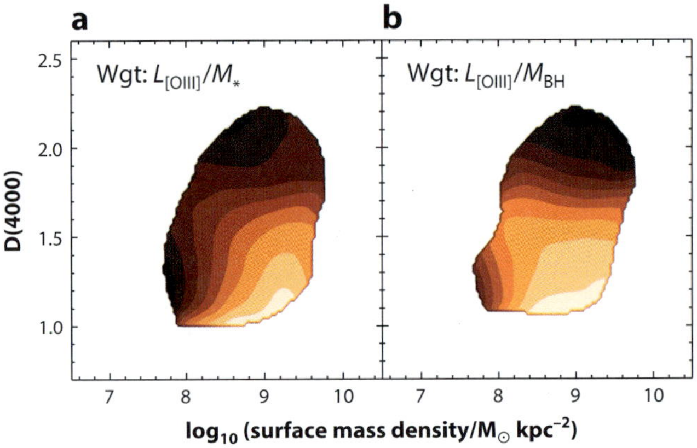

Figure 17

(*a*) The distributions of AGN bolometric radiative luminosity per unit galaxy stellar mass and (*b*) bolometric radiative luminosity per black hole mass in the stellar surface mass density versus 4,000-Å break strength plane. (Color banding as in **Figure 14**.) Radiative-mode AGN activity clearly requires both an abundant supply of cold gas (indicated by recent or ongoing star formation) and a high stellar surface mass density.

with the SDSS-based results of LaMassa et al. (2013), who find that the mean sSFR in Type 2 Seyferts (within a radius of about 2 kpc) only increases from $\sim 10^{-11}$ to 10^{-10} year^{-1} as the ratio of $L/L_{\rm Edd}$ increases from $\sim 10^{-3}$ to ~ 1. Similarly, they found that the SFR in this central region only increases from 0.1 to 1 $M_\odot$ year^{-1} as the extinction-corrected [OIII] luminosity increases from $10^{6.5}$ to 10^9 $L_\odot$ (i.e., SFR $\propto L_{\rm AGN}^{0.4}$). Diamond-Stanic & Rieke (2012) and Netzer (2009) find steeper but still sublinear relations between SFR and AGN luminosity (SFR $\propto L_{\rm AGN}^{0.6-0.8}$). Rosario et al. (2012) find no correlation (at $z \sim 0$) between the global (galaxy-wide) SFR and AGN luminosity for AGN bolometric luminosities below about 10^{44} erg s^{-1} and SFR $\propto L_{\rm AGN}^{0.5}$ at higher luminosity.

The second approach was taken by Kauffmann & Heckman (2009). They looked at the distribution of the Eddington ratio ($L_{\rm bol}/L_{\rm Edd}$) as a function of the amplitude of the 4,000-Å break for a complete SDSS-based sample of galaxies (see **Figure 12*a***). They found that for galaxies with predominantly old central stellar populations, the probability of finding an AGN decreases with increasing Eddington ratio as a power law. Using the calibration of Brinchmann et al. (2004), the results of Kauffmann & Heckman (2009) imply that the normalization of this power law increases roughly linearly with the central sSFR over the range in the latter from $\sim 10^{-12}$ to $10^{-10.5}$ year^{-1}. However, at higher values of central sSFR, the Eddington ratio distribution transitions to a log-normal distribution that peaks at an Eddington ratio of about 3%. In this regime, the Eddington ratio distribution is independent of the central sSFR. It is also largely independent of the mass of the BH (though the precise details depend upon the version of the M-σ relation adopted). Thus, Kauffmann & Heckman found no simple monotonic relation between central star formation and BH growth rates. Once the central region of the galaxy has a mass-doubling time (M_*/SFR) that is comparable with the current Hubble time, the SMBH growth rate saturates and is not boosted by additional star formation. We discuss some possible interpretations of this in Sections 5 and 6 below.

The combination of power-law plus narrow-peaked distribution for the Eddington ratio distribution found for emission-line AGNs bears close resemblance to the results of the radiative-mode versus jet-mode decomposition of Eddington ratio distributions for radio-AGNs (Section 3.2; **Figure 12*b***). This fits with the picture discussed above, that radiative-mode AGNs are generally found in star-forming host galaxies, whereas jet-mode AGNs are mostly in passive galaxies. It also indicates that the main properties of the Eddington ratio distribution are largely independent of whether or not an AGN is classified as radio-loud. This further justifies our approach throughout this review of considering just two fundamental AGN modes (cf. **Figure 4**).

4.2.2. The radial distribution of star formation and cold gas. The physical connection between the fueling of the SMBH and star formation should be clearer on smaller scales near the SMBH. Kauffmann et al. (2007) compared this relationship determined on two scales. They found that the luminosity of the AGNs (relative to the Eddington limit) correlates much more tightly with the youth of the central stellar population of the galaxy as probed with the SDSS spectra than with its global stellar population as probed with SDSS and GALEX optical/UV imaging. They concluded that ongoing star formation in the global galaxy disk is a necessary but not sufficient condition for ongoing star formation in the center and the concurrent growth of the SMBH: young bulges and their growing SMBHs are only found in galaxies with star-forming disks, but many galaxies with star-forming disks have dead bulges and quiescent SMBHs.

A similar result was found by Diamond-Stanic & Rieke (2012) based on an analysis of *Spitzer* mid-IR imaging and spectroscopy of a sample of very nearby Seyfert galaxies. They found that the correlation between the SFR and the rate of SMBH growth (measured using the luminosity of the [OIV] 25.9-μm line) was significantly stronger for the SFR measured in the inner-most kiloparsec than for the entire galaxy. The radial distribution of the star formation as a function of the AGN luminosity was mapped in a statistical sense by LaMassa et al. (2013) for a sample of 28,000 Type 2 Seyfert galaxies drawn from the SDSS. They examined the SFR within the SDSS fiber for subsamples matched in both host-galaxy properties and AGN luminosity, but spanning a range in mean redshift from $z \sim 0.05$ to 0.15 (so that the fiber mapped the star formation over different radial scales in each subsample). They found that as the AGN luminosity increased, the total star formation increased and became more and more centrally concentrated (located within about 2 kpc of the nucleus).

On smaller radial scales, the stellar populations in nearby Type 2 Seyfert nuclei have been investigated by Cid Fernandes et al. (2001) using optical spectra that probe typical radial scales of only a few hundred parsecs. By computing $L_{\rm bol}/L_{\rm Edd}$ for this sample, we find that a young stellar population can be directly detected surrounding about 75% of the Seyfert nuclei with luminosities exceeding 10% of the Eddington limit, but only in about 25% of those with lower luminosities. This investigation was extended to lower AGN luminosities and still smaller scales using ground-based and HST spectra by Cid Fernandes et al. (2004a,b) and González Delgado et al. (2004), respectively. They found that the nuclear stellar populations were usually old, except in cases whose optical emission-line properties were consistent with an ionizing contribution from both an AGN and young stars.

A complementary picture emerges from consideration of the radial distribution of the cold gas associated with star formation. By comparing the hosts of radio-quiet AGNs with control samples of non-AGNs matched in stellar mass, structural properties, and mean stellar age, Fabello et al. (2011) found that the global atomic gas (HI) content of a galaxy had no effect on the accretion rate on the central BH. This agrees with earlier studies of the HI content of smaller samples of Seyfert galaxies (Heckman et al. 1978, Mirabel & Wilson 1984, Ho et al. 2008), which did not show systematic differences in HI content in Seyferts compared with normal galaxies. Atomic gas in a galaxy is generally spread out in a large-scale disk, and the surface density of the HI exhibits a dip or even a hole in the center of the galaxy. If gas is present in the vicinity of the BH, it is likely to be at high enough densities that it would exist primarily in molecular form.

Surveys of the global molecular content of AGN hosts using the millimeter wave $J = 1$–0 CO emission line show only small (Heckman et al. 1989) or no (Maiolino et al. 1997, Saintonge et al. 2012) differences in the molecular content of AGN hosts compared with normal galaxies (in terms of CO luminosity normalized by galaxy optical luminosity, HI mass, or SFR). On smaller radial scales, thick, clumpy disks of molecular gas with radii ~30 pc have been found at the centers of nearby Seyfert galaxies (e.g., Hickox et al. 2009). More recently, Hicks et al. (2013) examined molecular gas on 50-pc to 1-kpc scales in a matched sample of Seyfert and quiescent galaxies and showed that the Seyferts had more centrally concentrated H_2 surface-brightness profiles.

The presence of dust can serve as a proxy for cold gas. Kauffmann et al. (2007) showed that the central regions of the host galaxies of more rapidly growing BHs were more heavily dust-reddened than were the same regions in the hosts of low-power AGNs. Simoes Lopes et al. (2007) used HST imaging to show that all of the 65 AGN host galaxies they studied have circumnuclear dust structures on radial scales of ~0.1 to 1 kpc (see also González Delgado et al. 2008). This includes 34 early-type AGN hosts. In a carefully constructed control sample of normal galaxies only a minority (26%) of the early-type galaxies had circumnuclear dust (though all the late-type control galaxies did). Subsequent *Spitzer* observations (Martini et al. 2013) imply dust masses of order $10^6\ M_\odot$ or associated gas masses of order $10^8\ M_\odot$. In a complementary study, Lauer et al. (2005) used HST images to discover a strong connection between the AGN emission-line activity and the presence, quantity, and structure of central dust in a sample of 77 early-type galaxies. All in all, it seems reasonable to conclude that radiative-mode AGNs reside in galaxies where the central (kiloparsec-scale) supply of cold dense gas is substantial.

4.2.3. The starburst-AGN connection. We have seen above that the connection between SMBH growth and star formation is clearer when the central kiloparsec-scale region of the galaxy is probed. Such central star formation can sometimes occur in a strong episodic burst, presumably related to the short dynamical times near the nucleus. The existence of a direct connection between the growth of an SMBH and intense star formation in the centers of galaxies has long been a matter of speculation. In its most extreme form a direct starburst-AGN connection was posited: AGNs were simply the evolved descendants of starbursts powered by a collection of supernovae and supernova remnants rather than by an SMBH (Terlevich & Melnick 1985, Terlevich et al. 1992). Many other authors have proposed less iconoclastic models in which the delivery of a fresh supply of cold gas to the center of a galaxy leads to both a strong starburst and the efficient fueling of the SMBH (e.g., Sanders et al. 1988a,b; Hopkins et al. 2006, 2008; Di Matteo et al. 2008).

Investigations of the starburst-AGN connection using SDSS spectra have shown that there is indeed a connection, but one that is more subtle than might have been expected. We have described above how SDSS spectra can be used to identify starbursts and poststarbursts using the combination of the 4,000-Å break and the strength of the high-order Balmer absorption lines. Based on this methodology, Kauffmann et al. (2003a) showed that high-luminosity Type 2 AGNs in the SDSS were more likely to have undergone a starburst within the last 10^9 years compared with low-luminosity AGNs or normal galaxies of similar stellar mass (see also Goto 2006 and Yan et al. 2006). This issue was examined in more detail by Wild et al. (2007). Based on a sample of 34,000 early-type galaxies in the SDSS, Wild et al. found that the poststarburst systems had the highest mean AGN luminosity. However, because these are relatively rare, they represent only about 15% of the total amount of SMBH growth in the contemporary Universe. They found that galaxies with ongoing (but less episodic) central star formation are the sites of the majority of SMBH growth.

In a subsequent paper, Wild et al. (2010) examined the time history of SMBH growth during and after an intense burst of central star formation (involving roughly 10% of the galaxy stellar mass). They found that the growth of the SMBH was delayed by about 200 Myr relative to the start of the starburst (**Figure 18**). These results pertain to the region within a radius of ~1 to

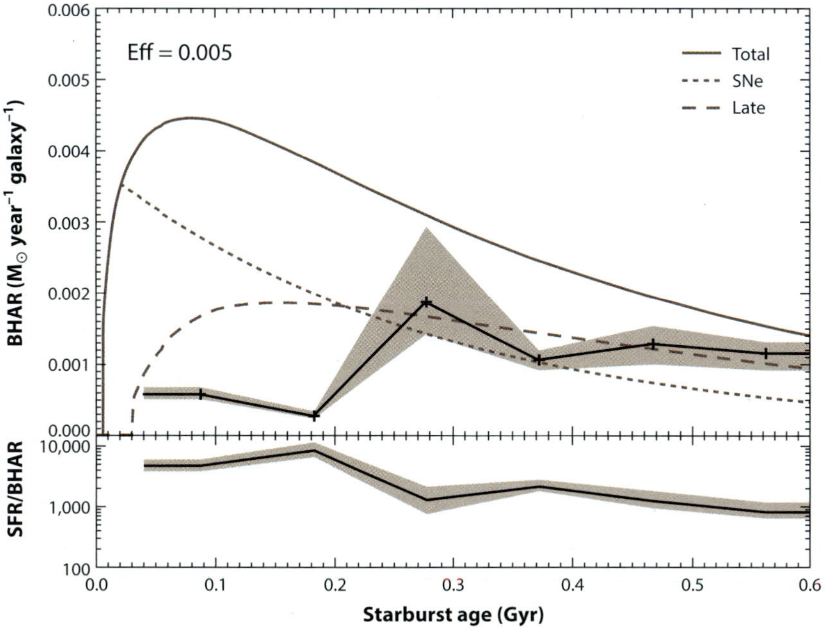

Figure 18

(*Top*) The mean black hole accretion rate (BHAR) averaged over all galaxies in a sample of starburst and poststarburst galaxies (*black line*). The gray-shaded area shows the typical 10th and 90th percentile range on the total BHAR estimated from bootstrap resampling of the data. The model prediction for the average gas mass-loss rate from the stars formed during the starburst (*solid gray line*) assumes an accretion efficiency onto the black hole of 0.5 percent of the ejecta from intermediate- and low-mass stars between the age of 250 and 600 Myr. Stars return mass to the ISM through supernova explosions and fast winds from massive stars (*dotted line*) and planetary nebula ejections and stellar winds from lower-mass stars (*dashed line*). (*Bottom*) The ratio of star-formation rate (SFR) (from Hα) to BHAR, with mean value and errors calculated as for the top panel. The BHAR lags the SFR and becomes significant only after several hundred million years. Reprinted from Wild et al. (2010) with permission.

2 kpc. Qualitatively similar results from the SDSS were found by Schawinski et al. (2007) for galaxies selected morphologically to be ellipticals. On much smaller scales (10–100 pc) and for a much smaller sample, Davies et al. (2007) also find evidence for a delay of order 100 Myr between a starburst and powerful AGN activity. We discuss the implications of these results in Section 6.1 below.

4.3. The Hosts of Jet-Mode AGNs

The results in Section 4.1 above show that the host galaxies of the BHs that dominate the production of jet mechanical energy are nearly disjoint from the host galaxies where BHs are actively growing. These jets are traced by their radio emission. Detailed investigations of the structure and stellar content of the radio-selected AGN population in the SDSS were first undertaken by Best et al. (2005b). They confirmed the widely held view that nearly all such AGNs are hosted by massive elliptical galaxies with old stellar populations: The radio source hosts are highly bulge-dominated (their BH mass versus stellar-mass relation lies on the established BH mass versus bulge mass relation), with 4,000-Å break strengths clustered around 2.0, concentration indices typically between 2.9 and 3.4, and effective stellar surface mass densities around $10^9\ M_\odot\ \mathrm{kpc}^{-2}$. This is entirely consistent with the results summarized in Section 4.1.

More interestingly, Best et al. (2005b) were able to quantify the prevalence of radio-AGN activity as a function of stellar mass and BH mass with much greater statistical robustness than previous studies. They found that the fraction of galaxies selected via radio emission scaled as $M_*^{2.5}$ and as $M_{BH}^{1.6}$, with these results being broadly independent of the radio luminosity limit. This is strikingly different from the mass dependence of the prevalence of emission-line AGN activity (see **Figure 14**). Best et al. (2005b) did not separate their sample into radiative-mode and jet-mode sources, but their sample was dominated by the latter, and it is these sources that drive the strong mass dependence. Janssen et al. (2012) considered the mass dependences of the two populations separately. They showed that the prevalence of radiative-mode radio sources has a much shallower dependence on mass, scaling as $M_*^{1.5}$, albeit that this remains steeper than that found for radio-quiet radiative-mode AGNs selected by their optical emission lines.

For jet-mode sources, Best et al. found no evidence for any strong relationship between the [OIII] and radio luminosities; indeed the radio emitting fraction was found to be largely independent whether or not the optical galaxy is classified as an emission-line AGN. This result breaks down at higher radio luminosities, when the radiative-mode radio sources begin to dominate. These have been long known to show a strong correlation between radio- and emission-line luminosities (e.g., Rawlings & Saunders 1991, Xu et al. 1999). Kauffmann et al. (2008) investigated the subset of radio sources within the SDSS that display emission lines, and confirmed this correlation, finding also that it became tighter if Eddington-scaled quantities (i.e., $L_{1.4\,\mathrm{GHz}}/M_{\mathrm{BH}}$ and $L_{[\mathrm{OIII}]}/M_{\mathrm{BH}}$) were considered.

Lin et al. (2010) compared the host galaxies of radio sources separated into different radio morphological types. As previously discussed, these two radio morphological classes show a broad overlap with the jet-mode and radiative-mode classes, with most FR1s being jet mode and most FR2s being radiative mode; however, there are substantial differences between the segregation methods, and a significant population of jet-mode (low-excitation) FR2 sources exists. Lin et al. (2010) found that the most significant differences between FR1s and FR2s occurred for the subset of the FR2 population that displayed both strong high-excitation optical emission lines and hot spots close to the end of the radio lobes. They found this FR2 subset to be hosted by lower-mass galaxies, with bluer colors, living in poorer environments and having higher accretion rates than the rest of the radio source population. Similarly Kauffmann et al. (2008) found that their radio-AGNs with emission lines showed lower stellar masses, lower velocity dispersions, lower 4,000-Å break strengths, and stronger Balmer absorption features than those radio-AGNs without emission lines.

Best & Heckman (2012) did an explicit separation of jet-mode and radiative-mode radio sources and confirmed these results. They also constructed matched samples of the two radio source classes to investigate whether observed differences in host-galaxy colors and morphologies were just secondary effects driven by, for example, the different stellar masses of the two samples. They found that the hosts of radiative-mode radio-AGNs have bluer colors, smaller half-light radii, lower concentration indices, and lower 4,000-Å break strengths than the hosts of jet-mode radio-AGNs of the same stellar mass, BH mass, and radio luminosity. These results imply that the radiative-mode radio sources are associated with ongoing star-formation activity, as is also seen in the radio-quiet AGN samples. Further evidence in support of this comes from the work by Sadler et al. (2014), who investigated differences in the mid-IR colors of the two radio source classes using WISE and concluded that star-formation activity was present in the radiative-mode AGNs, but not the jet-mode AGNs (see also Gürkan et al. 2014).

The larger scale lengths and higher concentration indices of the hosts of the jet-mode AGNs can be explained if these sources are preferentially located toward the centers of groups and clusters (as suggested by their clustering and environmental properties; see Section 4.4), where host galaxies are typically larger and may have more extended (cD-type) light profiles. Mannering et al. (2011) compared the scale lengths of radio-AGNs at both red and blue wavelengths and found that the ratio of the two was higher than that for control galaxy samples, suggesting the possible presence of extended red envelopes (although their control samples were matched only in redshift and magnitude, so this result might be driven by morphology differences between the radio-AGNs and control samples).

Janssen et al. (2012) considered the relationship between AGN classification, radio luminosity, and galaxy color. Red (passive) galaxies show a higher probability of hosting jet-mode radio-AGNs than blue (star-forming) galaxies of the same stellar mass by a factor of a few. Blue galaxies show a higher probability of hosting radiative-mode radio-AGNs at all radio luminosities. For blue galaxies, the likelihood of hosting a radio-AGN of either class is a strong positive function of the SFR. This suggests that the presence of cold, star-forming gas in a galaxy enhances the probability of its central BH becoming a radio-loud AGN. This means that jet-mode AGN activity, especially at high radio luminosities, is not solely related to hot halo gas accretion (see Section 5 below).

4.4. Clustering, Environments, and Halo Masses of AGNs

Since early in the study of Seyfert galaxies, QSOs, and powerful radio sources, attempts have been made to categorize the typical environments in which these sources are found (e.g., Gisler 1978, Longair & Seldner 1979, Yee & Green 1984). Analyses mostly relied on relatively small fields around the AGNs and either considered simple galaxy counts out to some projected radius or divided those counts into annuli to enable a cross-correlation analysis. The majority of analyses indicated that AGNs are found in a wide range of environments, with powerful QSOs and radio sources being often located in group or poor cluster environments, but Seyfert galaxies generally avoiding the densest environments. However, AGN samples were relatively small, and contradictory results were obtained in different studies. Furthermore, AGN activity depends strongly on host-galaxy properties, such as stellar mass (especially for radio-selected AGNs; see previous sections); therefore, to reliably isolate any trends of AGN activity that are explicitly caused by environment, large and homogeneous data sets like that offered by the SDSS are required so that the AGN samples can be compared against control samples carefully matched in other galaxy properties. Even so, there has remained much debate as to the nature of these trends, largely because of the different definitions adopted both for AGN samples (all emission-line AGNs, just Seyfert sources, radio-AGNs, etc.) and for environment (local galaxy density; membership of, or distance from center of, a group or cluster).

Once stellar mass effects are properly accounted for, the incidence of optical emission-line AGNs is generally found to be lower in denser environments and in groups and clusters rather than in the field (Miller et al. 2003, Kauffmann et al. 2004, von der Linden et al. 2007): The fraction of AGNs among galaxies of given stellar mass decreases by a factor of two from the field toward the cluster/group center. This is matched by a decrease in star-formation activity of these galaxies. Hwang et al. (2012) investigate these trends for early and late-type galaxies separately and find that the prevalence of AGN activity in early-type galaxies already begins to decrease at the virial radius of the cluster, whereas in late-type galaxies it only decreases as the cluster center is approached. *Chandra* has revealed significant samples of X-ray-detected AGNs in nearby galaxy clusters (e.g., Martini et al. 2006), the majority of which do not show emission lines and so are not identified as AGNs in optical observations. These may be obscured AGNs or more likely may be jet-mode AGNs. Haines et al. (2012) argue that luminous X-ray AGNs in massive clusters are mostly an infalling population. These results are all generally consistent with a picture in which there is a gradual decline in emission-line AGN activity as the denser environment decreases the cold gas supply for an AGN.

In contrast to emission-line AGNs, radio-selected AGNs are found to be preferentially located in group and cluster environments (e.g., Best et al. 2005b, Sabater et al. 2013), confirming the results of many previous studies of small radio-loud AGN samples. This is particularly true of the jet-mode subset of the radio-AGN population (Best 2004, Reviglio & Hefland 2006, Sabater et al. 2013), again suggesting a different mechanism for the triggering of AGN activity in these sources than for radiative-mode AGNs. The prevalence of radio-AGN activity among brightest cluster galaxies (BCGs) is especially high (Best et al. 2007, Croft et al. 2007; cf. Burns 1990), whereas that of optical AGNs is lower than in non-BCGs of the same mass (von der Linden et al. 2007).

An alternative view of the environments of AGNs can arise from looking at the two-point correlation function. This describes the excess probability of finding a galaxy at a given distance R from another galaxy due to clustering. In the widely adopted halo-model description of large-scale structure, on small scales (<1 Mpc) the signal in the correlation function is dominated by pairs of galaxies within the same dark matter halo (known as the one-halo term; e.g., Seljak 2000). This therefore measures the prevalence of a particular galaxy type among satellite galaxies in halos. Li et al. (2006) computed the cross-correlation function between Seyfert galaxies and the SDSS spectroscopic main galaxy sample and compared this against the results for a non-AGN control sample that was matched in redshift, stellar mass, BH mass (from velocity dispersion), concentration index, and 4,000-Å break strength. They found that on scales between 100 kpc and 1 Mpc the AGNs showed weaker clustering than the control sample. By comparison with simulations, they argued that this could be understood if AGN activity was dis-favored within satellite galaxies. This interpretation is consistent with the results arising from the study of AGN activity as a function of cluster membership or local galaxy density, discussed above.

On much larger scales (10–20 Mpc) the two-point correlation function measures the correlation between different dark matter halos (the two-halo term), which depends upon the properties of the parent halo for a given galaxy type (see the review by Cooray & Sheth 2002). If the luminosity function and two-point correlation function of a set of objects are known, then (for a given cosmological model) it is possible to estimate the mean mass of the parent dark matter halo (e.g., Sheth et al. 2001). Mandelbaum et al. (2009) used this method to estimate dark matter halo masses for ($z < 0.3$) Seyfert galaxies from the SDSS and confirmed the reliability of their measurements by using an alternative estimate from weak gravitational lensing of background galaxies. They found a mean halo mass of $\sim 10^{12}$ $M_{\odot}$. They compared these results against a control sample of non-AGNs and found that the halo mass versus stellar mass relation was independent of whether or not Seyfert activity was present.

Mandelbaum et al. (2009) also analyzed radio-loud AGNs in the same redshift range. They found for the radio-loud AGNs an average halo mass of 10^{13} solar masses, nearly an order of magnitude higher than that of optically selected AGNs. Much of this increase is because radio-loud AGNs are hosted by more massive galaxies, but Mandelbaum et al. also found that the dark matter haloes of radio-loud AGNs are about twice as massive as those of control galaxies of the same stellar mass. A similar enhancement in the clustering strength of radio-loud AGNs relative to a carefully matched control sample of massive galaxies was found by Donoso et al. (2010) at redshifts $0.4 < z < 0.8$ using the SDSS Luminous Red Galaxy sample. Donoso et al. also showed that the clustering strength of the radio galaxy population is luminosity dependent. At low radio luminosities, where the jet-mode AGNs dominate the radio-AGN samples (these also dominated the Mandelbaum et al. sample), the clustering strength is enhanced relative to radio-loud QSOs. At higher radio luminosities the radio galaxy population is dominated by radiative-mode sources, and the clustering amplitude is more in line with that of the radio-loud QSOs, as expected from the standard AGN unified model.

The 2dFGRS and SDSS QSO redshift surveys have been used to study the dark matter haloes of powerful AGNs to even higher redshifts (Porciani et al. 2004, Croom et al. 2005, Ross et al. 2009). Intriguingly, the mean mass of haloes hosting QSOs remains roughly constant (at around $2 \times 10^{12}\,M_{\odot}$) over the entire redshift range $z < 2.5$ and shows only weak dependence on QSO properties like luminosity, color, or BH mass (Croom et al. 2005, Shen et al. 2009). Only at the highest QSO luminosities is the clustering amplitude seen to increase. Shen et al. (2009) find that radio-loud QSOs reside in halos almost an order of magnitude larger than those of radio-quiet QSOs [though note that Donoso et al. (2010) do not find this strong signal among their $z \sim 0.6$ sample]. The result is in line with the local results of Mandelbaum et al.: A large part of this is likely to be driven by the higher typical stellar mass of radio-loud QSO hosts, but radio-AGN activity also seems to be preferentially found in denser environments.

In summary, radiative-mode AGNs are located in dark matter haloes with a mean mass of $10^{12}\,M_{\odot}$, which is typical for their stellar masses. Once stellar mass biases are accounted for, the prevalence of radiative-mode AGNs is lower in denser environments than in the field, because they are disfavored in satellite galaxies; this mirrors the decrease in star-formation activity in denser environments, with both trends being driven by the reduced cold gas supply. Radiative-mode AGNs that are found in denser environments are more likely to be radio-loud. This partly reflects the trend for radio-loud radiative-mode AGNs to be hosted by more massive galaxies (see **Figure 4**), but may also be enhanced by a boosting of the radio luminosity in dense environments, where confinement of the radio lobes reduces adiabatic losses (e.g., Barthel & Arnaud 1996). In contrast to radiative-mode AGNs, jet-mode AGNs have an increased prevalence in denser environments and, especially, in the central galaxies of groups and clusters. This indicates a different fueling mechanism.

5. BLACK HOLE FUELING

5.1. The Nature of the Accretion Flow

There are three main features that we need to consider when examining the fueling of AGNs: the origin of the fueling gas, the mechanism by which it is transported to the vicinity of the BH (e.g., Hopkins & Quataert 2011), and the nature of the accretion flow onto the BH. With regard to the last of these, the most important determining properties are the BH mass, its spin, and the nature and rate of the inflow. Note that even if the AGN is powered by tapping the BH spin via the Blandford & Znajek (1977) mechanism, accretion is required for this to occur.

Before embarking upon a discussion of the effects of different modes of accretion, it is worth summarizing the lessons that have been learned from studying accretion onto stellar-mass BHs in galactic X-ray binary systems. These are systems consisting of a BH of between roughly $4\,M_{\odot}$ and $20\,M_{\odot}$ in a close binary system with a normal star. At least in terms of the BH properties, these represent analogs of accreting SMBHs, scaled down in both luminosity and the accretion timescales. They are not precise equivalents of AGNs because the temperature of the accretion disk around a stellar mass BH is typically two orders of magnitude higher than that around an AGN, which can lead to differences in processes at atomic levels, such as ionization-dependent accretion instabilities. Nevertheless, stellar mass BHs offer the considerable advantage that variations in the accretion flow can be directly tracked; changes in the accretion happen on timescales of less than days. The observed accretion rates vary between the Eddington rate and values several orders of magnitude below this.

X-ray binaries have been shown to display two fundamentally different X-ray spectral states (see the review by Remillard & McClintock 2006 in this journal). In the low-hard state, the source spectrum is characterized by low luminosity and hard (predominantly >10 keV) X-ray emission, and low-power radio jets are ubiquitous. The power of the radio jets is correlated (albeit nonlinearly) with the X-ray luminosity (Gallo et al. 2003). In the high-soft state, sources are dominated by a high-luminosity thermal X-ray component, peaking at lower energies and with the characteristics of a standard geometrically thin, optically thick accretion disk. The primary difference between these two accretion modes is observed to be the accretion rate onto the BH relative to the Eddington rate: The low-hard state is associated with accretion rates below a few percent of the Eddington rate, whereas the high-soft state arises when the accretion rate exceeds that value (Maccarone 2003). X-ray binaries are observed to traverse from one state to another. Within the low-hard state, there are indications that the transition radius between an inner ADAF and an outer thin disk (cf. **Figure 3**) decreases with increasing luminosity (Narayan 2005). As the accretion rate increases, the luminosity increases in all bands until they reach a transition or intermediate phase where they undergo a short-lived but strong radio outburst in which both a thermal accretion disk and powerful and highly relativistic radio jets are seen (e.g., Fender et al. 2004). The radio emission then ceases as the source moves into the high-soft state. A further important property of X-ray binaries is that winds and jets appear to anticorrelate (Miller et al. 2008, Neilsen & Lee 2009, Ponti et al. 2012): Winds occur in the high-soft state when jets are not observed, but they are not observed in the low-hard state in which jets are ubiquitous.

These states of X-ray binary activity each have an analog among the AGN population. The low-hard state X-ray binaries are analogs of the jet-mode radiatively inefficient AGNs. Indeed these AGNs and low/hard X-ray binaries fall on the same fundamental plane of BH accretion: a relationship between the radio luminosity, X-ray luminosity, and black-hole mass of BHs (Merloni et al. 2003, Falcke et al. 2004). The high-soft state X-ray binaries are a direct analog of the radiative-mode AGNs (Seyferts and QSOs). The transition-phase objects have properties that make them a broad analog of the radio-loud radiative-mode AGNs, although in AGNs there is no observational evidence for these sources being a transition phase that pre-empts radio-quiet radiative-mode activity.

The critical accretion rate for the switch between X-ray binary states (a few percent of the Eddington rate) also appears to be broadly comparable in AGNs. As shown in Section 4.2 (**Figure 12*a***) the emission-line AGNs (predominantly Seyferts) in gas-rich (star-forming) galaxies in the SDSS show a log-normal distribution of Eddington rates peaking at a few percent of Eddington. This result is also confirmed at higher redshifts with QSOs seen to have lower limits to their accretion rates at around 1% of Eddington (e.g., Kollmeier et al. 2006, Trump et al. 2009). Radiative-mode (Seyfert/QSO-like) radio-AGNs show a similar Eddington ratio distribution (Best & Heckman 2012; see Section 3.2 and **Figure 12*b***). In contrast to this, the jet-mode radio-AGNs have accretion rates distributed below a few percent of Eddington (Best & Heckman 2012; see Section 3.2). If powered by the Blandford & Znajek mechanism then these accretion rates would be lower still. This last result is also supported at higher redshift, because BL Lac objects (believed to be beamed versions of the jet-mode sources) are also found to have luminosities below about 1% of Eddington, whereas Russell et al. (2013b) find a transition from jet-mode to radiative-mode AGNs at a similar Eddington ratio among BCGs (see also Churazov et al. 2005, Merloni & Heinz 2007).

As in the case of X-ray binaries, the spectra of the AGN subclass of LINERs (jet-mode AGNs) are best explained if the inner ADAF is supplemented by a truncated thin disk at larger radii, with the transition between the two being dependent upon the Eddington-scaled accretion rate (Yuan & Narayan 2004, Ho 2005). Sadler et al. (2014) found a weak signature of mid-IR AGN emission at the highest radio luminosities in jet-mode radio-AGNs, supporting this idea.

All of these results are in line with theoretical predictions of ADAF, which predict a change in the nature of the inner accretion flow below a critical fraction of the Eddington rate [e.g., see the review by Narayan (2002) or by Yuan & Narayan (2014) in this volume]; the precise value of that fraction depends strongly upon the viscosity of the accretion disk (e.g., Mahadevan 1997, Qiao & Liu 2009). This is consistent with the picture we have advocated in this review: The high-Eddington-rate (radiative-mode) AGNs and the low-Eddington-rate (jet-mode) sources are two fundamentally distinct populations. We include the small population of radio-loud radiative-mode AGNs together with the (radio-quiet) rest of the radiative-mode population: The results in the above sections have shown that their host-galaxy properties and Eddington-scaled accretion rates are broadly similar, albeit that the radio-loud QSOs are predominantly located at the upper end of the range of stellar and BH masses seen in the radio-quiet population (see **Figure 4**).

5.2. The Fueling of Radiative-Mode AGNs

5.2.1. Mergers and interactions. The idea that AGNs are largely triggered through the process of galaxy mergers or strong tidal interactions has long been a key component of most theoretical models (e.g., Kauffmann & Haehnelt 2000; Hopkins et al. 2006, 2008). Nevertheless, it has proven difficult to establish this idea observationally, with many claims and counter-claims in the literature. As we explain below, this is a clear case in which arriving at the correct answer requires superb control of the systematics, and this is made possible by a very large and homogeneous data set like that provided by the SDSS. We consider two types of observational tests for a link between interactions/merger and AGN fueling. The first is to measure the incidence of close companion galaxies and the second is to determine the morphological structure of the host galaxy.

The amplitude of the cross-correlation function on scales less than $\sim$100 kpc probes interacting pairs of galaxies where tidal forces begin to influence the internal structure of the pair. In two companion papers, Li et al. (2008a,b) used the amplitude of the cross-correlation function on scales of tens of kiloparsecs as a probe of the effects of galaxy interactions on star formation and AGN activity in galaxies. They first showed that on scales less than 100 kpc the amplitude of the cross-correlation function exhibits a strong dependence on the sSFR of the galaxy, reflecting the well-known result that galaxy interactions and mergers trigger starbursts in galaxies (Larson & Tinsley 1978). As the separation between the two galaxies in the pair decreased, their sSFR also increased relative to the mean value for field galaxies of the same mass. In a follow-up paper, Li et al. (2008b) applied the same techniques to AGNs in the survey, showing that interactions do not lead to enhancement of nuclear activity over and above that expected as a result of the boost in the central SFRs. In other words, the relation between central star formation and AGN activity in interacting galaxies is no different from that observed for noninteracting galaxies.

Mergers can be identified quantitatively through measures of the global asymmetry (lopsidedness) of the light distribution of the galaxy (e.g., Conselice et al. 2000). An analysis of the SDSS population of normal and AGN host galaxies by Wild et al. (2007) showed that the majority of BH growth in the contemporary Universe occurs in galaxies with relatively quiescent recent star-formation histories that have not experienced recent mergers or strong interactions. In subsequent work, Reichard et al. (2009) followed up on the Li et al. (2008a,b) analysis by showing that AGN activity is not enhanced in lopsided galaxies, compared with symmetric galaxies of the same central SFR. They found that even among AGNs radiating near the Eddington limit, only a small minority exhibited the highly lopsided morphologies of a major merger.

There are many papers reaching the seemingly contrary conclusions that AGNs are triggered by galaxy interactions and/or mergers. Recent examples are by Koss et al. (2010, 2011, 2012), Alonso et al. (2007), Ellison et al. (2011), Hwang et al. (2012), Liu et al. (2012), and Sabater et al.

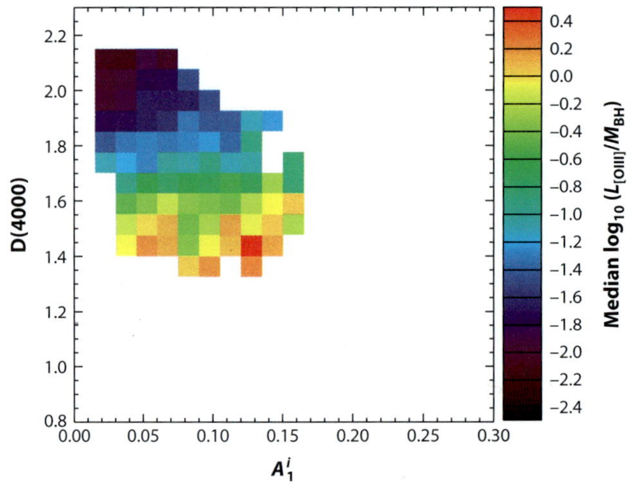

Figure 19

The relationship between $L_{[OIII]}/M_{BH}$, D(4000), and galaxy lopsidedness (A_1^i). The color coding of these two-dimensional histograms indicates the median $L_{[OIII]}/M_{BH}$. The primary correlation is between stellar age and $L_{[OIII]}/M_{BH}$, as is apparent from the horizontal nature of the color contours. There is no independent correlation between $L_{[OIII]}/M_{BH}$ and lopsidedness in these data. Reprinted from Reichard et al. (2009) with permission.

(2013). We stress that these results are actually not at odds with the results above. As Reichard et al. (2009) showed explicitly, the hosts of strong AGNs are in fact more lopsided on average than normal galaxies with the same stellar mass, radius, and concentration. However, they showed that this link is a secondary unphysical one induced by much stronger correlations between lopsidedness and central star formation and between central star formation and the presence of a strong AGN (**Figure 19**). They interpreted this as implying that the fueling of the SMBH is enhanced by the presence of circumnuclear star formation (and its associated cold gas), but this SMBH fueling process does not depend upon the way in which this central star formation arises.

In summary, contrary to the assumption of many theoretical models, most of the AGN activity in the contemporary Universe appears to occur without the need for major mergers or strong tidal interactions. All that is required is an abundant supply of cold, dense gas in the central regions of the galaxy (regardless of its origin).

5.2.2. Fueling via secular evolution. The inward radial transport of gas in a disk galaxy requires torque on the gas that allows angular momentum to be transferred from the gas to the stars or dark matter. Although this can be accomplished in a sudden and spectacular way in a major merger or strong tidal interaction, a slower but significant inflow can be driven by nonaxisymmetric perturbations in the underlying mass distribution that arise through internal dynamical processes in the disk. Examples of such nonaxisymmetric perturbations are bars, oval distortions, and even spiral arms (e.g., Kormendy & Kennicutt 2004, Athanassoula 2008, Sellwood 2014).

Kormendy & Kennicutt (2004) argued that the consequence of this secular evolution for the inner regions of disk galaxies is the creation of a pseudobulge (perhaps more descriptively called a disky bulge by Athanassoula 2005, 2008). These are defined observationally as central regions of enhanced surface brightness relative to an inward extrapolation of the radial surface-brightness profile of the galaxy disk. Unlike classical bulges, they are dynamically cold structures as revealed by the presence of spiral arms, rings, and bars. Unlike typical classical bulges, they are the sites of significant ongoing star formation (e.g., Gadotti 2009). They have exponentially declining radial surface-brightness profiles whose functional form is more similar to a classic exponential Freeman disk than the $r^{1/4}$ de Vaucouleurs law for elliptical galaxies. The boundary between classical and pseudobulges is typically set at a value for the Sérsic index of $n \sim 2$ for the bulge component (e.g., Kormendy & Kennicutt 2004, Athanassoula 2008, Gadotti 2009, Fisher & Drory 2011).

If the growth of SMBHs in the contemporary Universe is largely fueled by secular processes that transport gas inward, we would therefore expect rapidly growing SMBHs to be found primarily in pseudobulges. We have summarized above the general link between the presence of a young central stellar population (one of the defining characteristics of a pseudobulge) and the rapid growth of the SMBH. Moreover, Kormendy & Ho (2013) show in their figure 17 that the transition from classical bulges to pseudobulges occurs below a bulge velocity dispersion of $\sigma \sim 150$ km s^{-1} and below an SMBH mass $M_{\rm BH} \sim$ few $\times 10^7$ M$_\odot$ (cf. **Figure 9**; and see also Fisher & Drory 2011). This is just the regime where the SDSS shows the bulk of SMBH growth to be occurring today (Heckman et al. 2004, Kauffmann & Heckman 2009).

Is there other more direct evidence that growing SBMHs are preferentially found in pseudobulges? Intriguingly, the compilation of SMBHs by Kormendy & Ho (2013) shows that Seyfert nuclei are present in only 1 of the 45 cases classified as an elliptical galaxy (the powerful radio source Cygnus A) and in only 1 of the 20 cases classified as classical bulges (NGC 4258), but in 11 of the 22 cases classified as pseudobulges. The preference of Seyfert nuclei for pseudobulges is confirmed by the HST imaging survey of nearby active galaxies by Malkan et al. (1998). Simple visual inspection of these images shows that spiral arms, rings, dust lanes, and bars are present in the majority of the AGN hosts on radial scales on the order of a kiloparsec or less from the nucleus (see **Figure 20**). Malkan et al. assigned an inner-Hubble type for each galaxy on the basis of these images, and only 20 of the 107 (19%) of the Type 2 Seyferts had inner Hubble types of E or S0 as expected for a classical bulge. Jiang et al. (2011) reached similar conclusions for an SDSS sample of Type 1 AGNs associated with BHs with estimated masses below 10^6 M$_\odot$.

If AGNs are typically fueled via secular processes, one might naturally expect that they would be preferentially found in barred galaxies (because bars are one of the major mechanisms for driving radial gas flows; e.g., Shlosman et al. 1990). As in the case of establishing a link between interactions and AGNs, there have been many studies over the years that have reached contradictory conclusions on this matter. The most recent studies based on the SDSS do not show evidence for a link once a proper control sample has been identified (Lee et al. 2012, Cisternas et al. 2013 and references therein). Finally we note that there are rare cases of AGNs in galaxies without any detectable bulge at all (Filippenko et al. 1993, Satyapal et al. 2009, Simmons et al. 2013). The fueling mechanism for these systems is mysterious.

5.3. Fueling of Jet-Mode AGNs

The idea that SMBHs can be fed through the accretion of gas shed by evolved stars in the central region of the galaxy is an old one (e.g., Bailey 1980, David et al. 1987, Norman & Scoville 1988, Ciotti & Ostriker 1997). As discussed in Section 4.2, Kauffmann & Heckman (2009) found evidence for two distinct regimes of BH growth in nearby galaxies. We highlight here the regime in which the Eddington ratio distribution is a steep power law. This regime dominates for the more massive BHs living in the older bulges and typically for those accreting at the low Eddington rates associated with jet-mode AGNs (see **Figure 12**). For this population, Kauffmann & Heckman found that at fixed 4,000-Å break strength (broadly, fixed mean stellar age), the time-averaged accretion rate on to the BH scales in proportion with the stellar mass of the galaxy bulge, and the overall rate is higher in galaxies with younger mean stellar ages. They argued that this is consistent with the BH being fueled from the recycled hot gas arising from stellar mass loss.

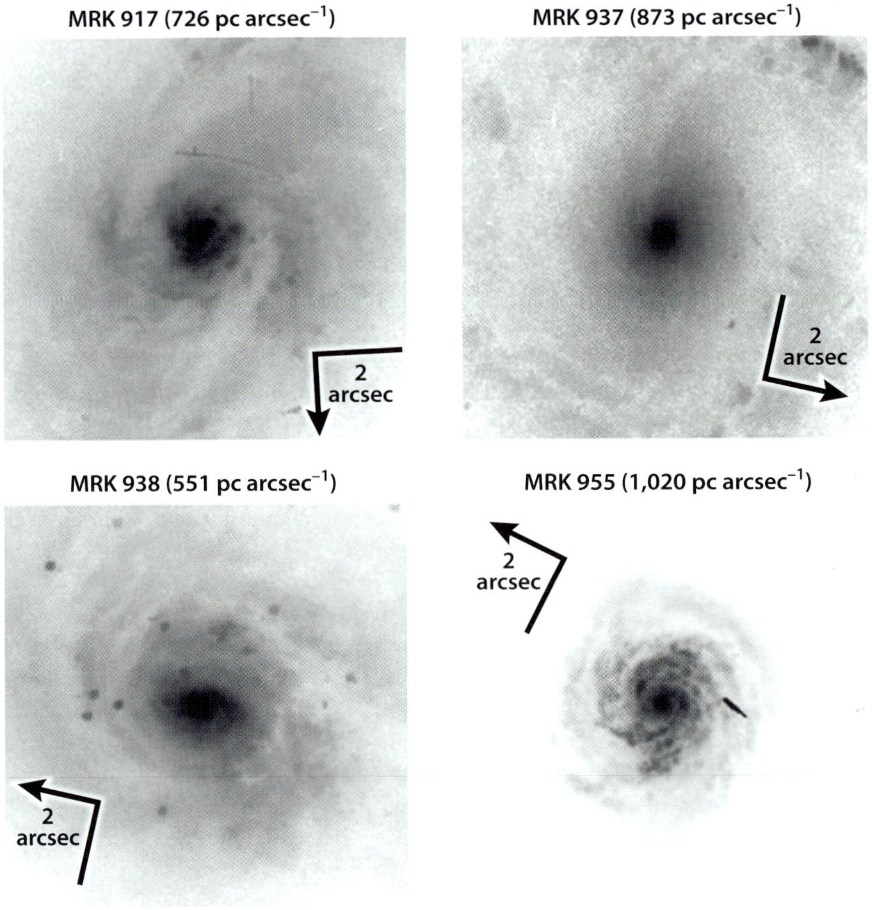

Figure 20

Hubble Space Telescope optical images of the central regions of four typical Type 2 Seyfert galaxies from Malkan et al. (1998). The arrowhead points north and the bar points east, and the angular scale is indicated. Note that the linear scale (parsecs per arcsecond) quoted in each case assumed $H_0 = 50$ km s^{-1} Mpc^{-1}. In all these examples the central few-kiloparsec-scale region shows the spiral arms and dust lanes typical of pseudobulges.

In fact, hot gas has been widely argued to be the fueling source of the jet-mode radio-AGNs (e.g., Hardcastle et al. 2007). These radio-AGNs are found predominantly in massive galaxies, often at the centers of groups and clusters, with associated hot gaseous X-ray-emitting haloes. These hot gas halos offer not only a potential source of fueling gas for the radio-AGN, but also a confining medium for the radio source to expand against and a depository for the bulk kinetic energy emitted in the radio jet: This allows for the possibility of the radio-AGN feedback cycle that is broadly supported by both observations and galaxy formation models [as we discuss in Section 6; see also the reviews by Cattaneo et al. (2009) and Fabian (2012)].

The rate at which hot gas accretes on a BH was first calculated by Bondi (1952) for a spherically symmetrical geometry with negligible angular momentum or magnetic fields. The Bondi accretion rate from a gas cloud of uniform density (ρ) and pressure is $\dot{M}_{\rm B} = 4\pi\lambda(GM_{\rm BH})^2\rho/c_s^3$, where λ is an order-unity scaling factor and c_s is the sound speed. In reality, the gas density around a BH is far from uniform, but the Bondi formula remains approximately correct if ρ and c_s are calculated at the Bondi radius (the radius within which the gravitational potential energy of the BH dominates over the thermal energy of the surrounding gas). Allen et al. (2006) showed that for a sample of 9 nearby systems, a strong correlation exists between the calculated Bondi accretion rate and the jet mechanical energy estimated from the cavities inflated in the surrounding gas. Both the slope and normalization of the relation were found to be close to unity, indicating that simple Bondi accretion of hot gas could be powering these objects.

More recently, however, Russell et al. (2013b) have found a much weaker correlation between Bondi accretion rates and jet powers. Furthermore, for the most luminous radio sources, Bondi accretion rate estimates fall more than two orders of magnitude short of providing the required jet powers (e.g., Cao & Rawlings 2004, McNamara et al. 2011). To produce such powerful sources with direct Bondi accretion would require the BH masses to be far higher than currently estimated (requiring BH masses significantly above $10^{10}\,M_{\odot}$). Such large BH masses would, however, help to explain how the nuclei of these powerful objects can be radiatively inefficient (i.e., below $\sim$1% Eddington; see Hlavacek-Larrondo & Fabian 2011).

A more plausible alternative to direct Bondi accretion of hot gas is that the gas cools out of the hot phase prior to accretion. The most powerful jet-mode radio sources are invariably located in dense galaxy groups or clusters, frequently with strong cooling flows. Hydrodynamic simulations (most recently by Gaspari et al. 2013) that include gas cooling, turbulence, and feedback from AGN heating reveal that when the hot gas is cooling, accretion on to the BH occurs predominantly in a cold and chaotic manner. Cold clouds and filaments form out of the cooling gas in the simulations, decouple from the hot gas, and enter a quasi free-fall regime. Multiple cloud-cloud collisions reduce the angular momentum of the cold gas and lead to stochastic accretion at rates that can be two orders of magnitude higher than the Bondi rate. This is sufficient to produce the observed jet powers of even the most luminous jet-mode sources. Hillel & Soker (2013) suggest that accretion of cooled gas may also dominate over direct Bondi accretion in less extreme systems as well: They argue that even the stellar winds of stars in orbit near the central SMBH would violate the zero-pressure assumption of the Bondi formulism and that the cooling shocked gas from such winds would probably accrete in cold clumps.

Observationally, cool clouds and filaments are almost universally observed in galaxies, groups, and clusters with cooling times significantly below the Hubble time, but not in systems with higher gas entropy (e.g., Hu et al. 1985, Cavagnolo et al. 2008). As extensively reviewed by Fabian (2012), the cooled material is predominantly cold molecular gas, which can extend tens of kiloparsecs around the BCG. This is in good agreement with numerical simulations. The observed masses of cooled molecular gas are comfortably high enough to provide sufficient accretion energy to fuel the inflated cavities (e.g., McNamara et al. 2011); in many objects they are two or more orders of magnitude higher, indicating that the bulk of the cooled molecular gas does not feed the BH but rather fuels star-formation activity, which can also be seen in these systems (e.g., Rafferty et al. 2006, O'Dea et al. 2008).

McNamara et al. (2011) show that the mass of cooled molecular gas does not correlate well with the radio jet power. This implies that either the efficiency with which the cooled gas gets funneled down to the BH must vary greatly (which is at least plausible, given that only a small fraction of the gas reaches the BH) or that another factor as well as the accretion rate is important in determining the jet power. An obvious hidden factor is BH spin. Blandford & Znajek (1977) showed that when material accretes onto rapidly rotating BHs threaded by high-power magnetic fields, the spin energy of the BH can additionally be extracted. Even if the Blandford-Znajek mechanism does not operate, the jet power can still be a strong function of the BH spin, because frame-dragging from the BH's rotation can contribute to the twisting of the magnetic field lines, amplifying any outflow generated by the Blandford-Payne model (Blandford & Payne 1982; see also Punsly & Coroniti 1990, Meier 1999).

BH spin has been widely cited as a possible explanation for the wide range of radio-loudness observed in the QSO population (the so-called spin paradigm by Wilson & Colbert 1995). According to current models, rapidly spinning BHs are required to produce sufficient jet power to account for the most radio luminous sources (e.g., Meier 2001, Nemmen et al. 2007). Using these models, it is possible to reproduce both the QSO radio loudness distribution and the SDSS-derived local radio luminosity functions of both jet-mode and radiative-mode radio sources (Martínez-Sansigre & Rawlings 2011). This requires a bimodal distribution of BH spins, with BHs either close to maximally spinning or with very low spin parameters. Nevertheless, despite the success of these models, the role of BH spin in powering radio jets remains unproven. Even at the scale of X-ray binaries, there is ongoing debate in the literature as to whether the power of the radio jet produced depends upon the spin of the BH (e.g., Russell et al. 2013a, Steiner et al. 2013).

In conclusion, evidence is strong that a majority of low Eddington fraction AGNs are fueled directly or indirectly from hot gas. The accretion flows are radiatively inefficient, and a significant fraction of the energetic output occurs through jet outflows, the strength of which may be enhanced by rapidly spinning BHs. For many galaxies, the fueling hot gas arises from recycled material from stellar mass loss. In more massive systems (the most massive ellipticals and galaxies at the centers of groups and clusters), extended hot gas halos provide a more abundant gas source, and large-scale cooling flows can greatly enhance the accretion rate. In such systems the extended hot gas halo also provides a working surface for a jet and a medium to confine the expanding radio lobes, minimizing adiabatic losses and leading to luminous radio emission. The hot halo also acts as a repository for the radio jet energy, offering the ideal conditions for a radio-AGN feedback cycle (see Section 6).

6. FEEDBACK PROCESSES

As recently reviewed by Cattaneo et al. (2009) and in this journal by Fabian (2012), feedback from AGNs is currently invoked in both semianalytic models and numerical simulations to successfully reproduce the observed properties of massive galaxies (e.g., Di Matteo et al. 2005, Springel et al. 2005, Bower et al. 2006, Croton et al. 2006, Hopkins et al. 2006, Ciotti et al. 2010). Although AGN feedback is generally assumed to be negative (inhibiting star formation and/or BH growth), it could also be positive in nature (e.g., Ishibashi & Fabian 2012, Zinn et al. 2013, Silk 2013). Feedback from AGNs is generally invoked in two flavors, which relate to the two fundamental modes of AGN activity. The first is a mode of powerful AGN-driven winds that are postulated to occur in galaxies with actively growing BHs and are argued to be responsible for the termination of star formation and the migration of the galaxy from the blue star-forming main sequence to the red sequence. It may also be responsible for setting up the correlation between BH mass and bulge properties (see the review by Kormendy & Ho 2013). We discuss this in Section 6.1. Complementary to this is the requirement to keep the galaxies as red and dead once they have arrived at the red sequence by preventing further gas cooling. Recurrent low-luminosity radio-AGN activity is often posited to be responsible for this kinetic mode of feedback, which is often referred to as maintenance-mode or radio-mode feedback. We examine this in Section 6.2.

6.1. Feedback and Radiative-Mode AGNs

We have reviewed above the evidence that the growth of SMBHs is related to the formation of massive stars in the central (few-kiloparsec-scale) region of the host galaxy. Thus, we first summarize the evidence for feedback associated with short-lived massive stars and then the evidence for feedback from the BH itself.

6.1.1. Feedback from massive stars. Could the feedback provided by massive stars formed in the central region of the host galaxy affect the fueling of the BH or the evolution of the host galaxy? To set the scale, note that for a normal Kroupa/Chabrier stellar initial mass function, massive stars return about 1.4×10^{49} ergs in kinetic energy per solar mass of star formation (based on Starburst 99; Leitherer et al. 1999). This amount of kinetic energy per unit mass is equivalent to a characteristic velocity of about 1,200 km s^{-1} (far larger than the galaxy escape velocity). Provided that this energy can be efficiently coupled to the surrounding gas, it could indeed result in significant feedback.

There is certainly compelling evidence that this can occur. The most dramatic effect is the production of galactic winds that can be traced in emission and absorption in both the local (e.g., Heckman et al. 1990) and high-redshift (e.g., Steidel et al. 2010, Shapley 2011) Universe. These outflows are driven by the energy and/or momentum supplied by massive stars through their stellar winds, their supernova ejecta, and their radiation (e.g., Chevalier & Clegg 1985, Murray et al. 2011). Although there are detailed multiwaveband studies of relatively small samples of local galactic winds (see the review by Veilleux et al. 2005), the largest census of galactic winds in the local Universe was undertaken by Chen et al. (2010) using SDSS galaxy spectra to measure the interstellar absorption feature produced by neutral Na atoms that trace the outflow. Their work provided statistically robust evidence that local galactic winds are wide-angle bipolar flows along the galaxy minor axis and are associated with galaxies having high SFRs per unit area (cf. **Figure 21**).

Investigations of starburst-driven outflows in the local Universe have only been able to directly detect the flows out to radii of order 10 kpc (e.g., Grimes et al. 2005). However, recent work using absorption-line spectra of background QSOs shows that the wind affects the properties of the surrounding gaseous halo of a starburst galaxy all the way out to the galaxy's virial radius (Borthakur et al. 2013). This is consistent with similar observations of Lyman-break galaxies at $z \sim 3$ (Steidel et al. 2010). Although global feedback from massive stars clearly exists, the evidence that it limits the growth of BHs is only indirect so far. We have described above the result from Wild et al. (2010) that showed that in a large SDSS-derived sample of starburst and poststarburst galaxies, the growth of the SMBHs was delayed by about 200 Myr relative to the onset of the starburst. Wild et al. showed that the turn-on in BH fueling in these galaxies occurred at the time at which the primary source of mass loss from stars in the galaxy's central region switched from supernovae (with ejecta velocities far above the local escape velocity) to intermediate-mass AGB stars (with ejecta velocities far below the local escape velocity). On this basis they speculated that supernova feedback was preventing or inhibiting the accretion of gas by the BH (**Figure 18**). This idea had been previously suggested both theoretically (Norman & Scoville 1988) and on the basis of detailed studies of small samples of local AGNs (Davies et al. 2007).

It is also possible that feedback from massive stars is responsible for the saturation effect found by Kauffmann & Heckman (2009): Once the specific sSFR in the bulge exceeds a value of $\sim 10^{-10}$ years^{-1}, the distribution of the Eddington ratio for the BH assumes a universal log-normal form that does not change at still higher SFRs. Although higher SFRs are associated with a larger gas supply, they are also associated with higher levels of feedback. Kauffmann & Heckman (2009) presented an alternative interpretation: The saturation effect could be understood if the BH regulated its own growth at a fixed average rate of a few percent of the Eddington limit when its fuel supply was plentiful. In a subsequent paper, Hopkins & Hernquist (2009) compared their observational results with their models for self-regulated BH growth, in which feedback produces

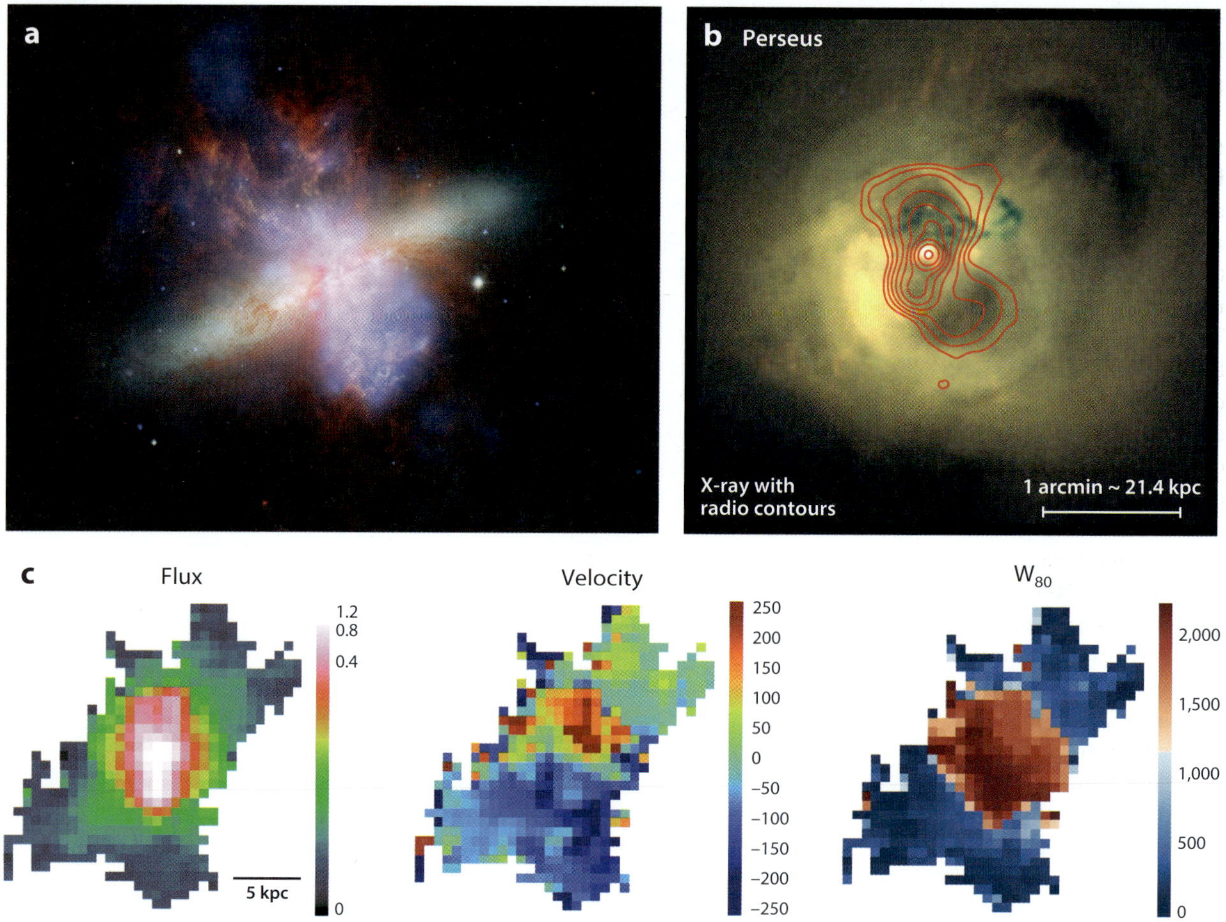

Figure 21

Illustrative examples of feedback effects at work. (*a*) The "Great Observatories" composite image of the starburst-driven galactic wind in M82: Visible light is shown in yellow-green, IR emission in red, Hα emission in orange, and X-ray emission in blue (credit: NASA/JPL-Caltech/STScI/CXC/UofA/). (*b*) The jet-mode feedback in the Perseus cluster (data from Fabian et al. 2006). The color scaling shows the X-ray emission, colored such that hard X-rays are in blue and soft X-rays in red. The contours indicate the radio emission, which can be seen to have evacuated cavities in the X-ray gas. (*c*) AGN-driven feedback from a high-luminosity radiative-mode AGN (reproduced from Liu et al. 2013b with permission). They show, respectively, the surface brightness (in units of 10^{-14} erg s^{-1} cm^{-2} arcsec^{-2}), relative velocity (in kilometers per second), and velocity width (W_{80} is the velocity width in kilometers per second containing 80% of the emitted line flux) of the [OIII] emission line in the Type 2 QSO SDSS J0319-0019.

a self-regulating phase after the AGN reaches some peak luminosity/BH mass and begins to expel gas and shut down accretion, and found good agreement at the bright end. On that note, we now describe the direct observational evidence for black hole–driven feedback in radio-quiet local AGNs.

6.1.2. Feedback from the black holes in radiative-mode AGNs. As reviewed by Fabian (2012) there is ample direct evidence that radio-quiet AGNs drive outflows. This is primarily based on the frequent presence of blue-shifted absorption lines in the optical, UV, and soft X-ray spectra of Type 1 AGNs (see Crenshaw et al. 2003 and references therein). These trace highly ionized gas with outflow velocities ranging from several hundred to several thousand kilometers per second in low-luminosity AGNs (Type 1 Seyfert nuclei) to tens of thousands of kilometers per second in powerful QSOs (the broad absorption line systems).

The uncertainty in the feedback effect of these flows arises owing to the difficulty in measuring their basic physical properties. Simple arguments imply that the mass (energy) outflow rate should be proportional to the product of the outflow's total column density times its size times its solid angle times its velocity (cubed). The strong absorption lines measured in the AGN-driven outflows are almost certainly saturated, making it difficult to determine even the ionic column density (let alone a total gas column density). Moreover, the sizes of these outflows have been largely unconstrained. The lack of the detection of such high-velocity outflows in the spectra of typical Type 2 Seyferts (e.g., Krug et al. 2010) is evidence either that the flows in Seyferts do not extend beyond the central few tens of parsecs or that their column densities drop below detectable values on these larger radial scales. It is therefore not surprising that the spectroscopic data in the SDSS on a vast number of local Type 2 AGNs has not uncovered any direct evidence that they are driving large-scale outflows that cannot be attributed to the feedback from central star formation described in Section 6.1.1 above.

Another line of evidence for outflows in typical radio-quiet AGNs comes from mapping the physical and dynamical state of the ionized gas seen in the NLR. This has been reviewed by Veilleux et al. (2005) in this journal, and we refer the reader there for details. In a number of local Seyfert galaxies the ionized gas on a scale of about a kiloparsec from the AGN is clearly being impacted by the collimated outflows (radio jets), which are commonly present even in radio-quiet Seyferts (e.g., Wilson & Heckman 1985; Rosario et al. 2010a,b). It is difficult to accurately estimate the total outflow rates in these cases, but the kinetic energy flux appears to be a small fraction of the AGN bolometric luminosity in typical local Seyfert galaxies (see Veilleux et al. 2005 and references therein).

Although direct evidence for black hole–driven feedback in typical local AGNs is lacking, recently there have been several exciting discoveries that imply that the most luminous AGNs can indeed drive highly energetic outflows that could have a significant impact on both the fueling of the BH and the evolution of the host galaxy. These are AGNs with bolometric luminosities well above the knee in the local AGN luminosity function ($L_* \sim 10^{45}$ erg s^{-1} $\sim 10^{11.5}$ $L_\odot$). As such they are rare, and only rather small samples have been studied so far.

The first such discovery is the prevalence of outflows of molecular gas in ultraluminous infrared galaxies (ULIRGs) with AGN bolometric luminosities above about 3×10^{45} erg s^{-1}. This evidence is of two types. First, the PACS instrument on the *Herschel Space Observatory* has been used to observe the far-IR emission and absorption features due to the OH molecule. Initial results were presented by Fischer et al. (2010) and Sturm et al. (2011). Most recently, Veilleux et al. (2013) and Spoon et al. (2013) summarized observations of an expanded sample of ULIRGs and QSOs at $z < 0.3$. For each object, the fractions of the bolometric luminosity due to a starburst and AGN were determined following the analysis by Veilleux et al. (2009). The OH feature is detected in absorption in most objects, with the majority of these showing evidence for molecular outflows in the form of blue-shifted OH absorption lines. The equivalent widths of the OH absorption features (tracing the column density) do not depend upon the AGN luminosity or the ratio of the AGN/starburst luminosities. The outflow velocities do not depend upon the starburst luminosity, but they do increase with both increasing AGN luminosity and AGN/starburst luminosity ratio. Characterizing the outflow speed by the midpoint in the line profiles yields outflow speeds of order 100 km s^{-1} in the starburst-dominated systems and about 400 km s^{-1} in the AGN-dominated cases, but the most rapidly outflowing gas can reach speeds of order 10^3 km s^{-1}. As noted above, deriving an outflow rate requires determination of a size for the flow. Sturm et al. (2011) have used a radiative transfer code (González-Alfonso et al. 2014) applied to the observed far-IR spectrum to estimate the size scales and, hence, the outflow rates in five ULIRGs. For an assumed OH/H_2 ratio, the implied outflow rates range from 10^2 to 10^3 $M_\odot$ year^{-1} (comparable with, or larger than, the SFR).

Complementing these probes are measurements of molecular gas in emission in the millimeterwave regime. The molecular outflow in the prototypical AGN-dominated ULIRG Mrk 231 has been mapped in CO (Feruglio et al. 2010, Cicone et al. 2012) and in HCO+, HCN, and HNC (Aalto et al. 2012) using the Plateau de Bure millimeter wave interferometer. The flow is spatially resolved with outflow speeds of up to 800 km s^{-1} detectable over a kiloparsec-scale region and an estimated outflow rate of about 700 $M_\odot$ year^{-1}, in broad agreement with the estimates based on the OH absorption line seen with *Herschel*. Complementary information on this outflow is provided by the spatially-resolved kinematics of neutral atomic gas traced by the NaI absorption line (Rupke & Veilleux 2011). Similar results using CO emission as a tracer have now been found for a small sample of other AGN-dominated ULIRGs (Cicone et al. 2014). They find that the kinetic energy in the molecular outflow is a few percent of the AGN bolometric luminosity and that the momentum flux is about 20 times larger than $L_{\rm bol}/c$. The primary uncertainty affecting these estimates is the conversion factor used to convert the CO luminosity into an H_2 mass.

These outflow rates are impressive, but the flows can only be traced out to modest radii (kiloparsec scale). Recently, the effects of AGN outflows on the global properties of the host galaxy have been discovered in samples of Type 2 QSOs—obscured AGNs with bolometric luminosities in excess of 10^{46} erg s^{-1} (Zakamska et al. 2004). Greene et al. (2011) reported on the analysis of long-slit optical spectroscopy probing the kinetics of the ionized gas surrounding 15 Type 2 QSOs at $z \sim 0.1$ to 0.45. They found broad emission lines of [OIII] 5007 (full width at 20% of the maximum line intensity of $\sim$500 to 1,500 km s^{-1}) that extended across the entire extent of the galaxy (radii of order 10 kpc). More recently Liu et al. (2013a,b) used the Gemini integral field unit to fully map out the 2-D distributions of the [OIII] and Hβ emission lines around a sample of 11 radio-quiet Type 2 QSOs at $z \sim 0.3$ to 0.65 with typical bolometric luminosities of 10^{47} erg s^{-1}. They obtain similar results to Greene et al. (2011), but their complete spatial coverage enables them to better characterize the structure and kinematics (e.g., **Figure 21**). They find that a simple model of a spherically symmetric outflow at a velocity of 500 to 1,000 km s^{-1} reproduces the data. The estimated outflow rates are uncertain because the density (and hence the mass) of the emitting material is not directly measured. It appears that the entire ISM of the galaxy could be expelled over a timescale of only a few tens of millions of years, implying a kinetic energy flow of order 2% of the bolometric luminosity (similar to the estimates in the molecular outflows described above).

In summary, there is now some persuasive evidence that the most powerful radiative-mode AGNs in the contemporary Universe can drive outflows capable of severely impacting their host galaxies. Evidence for global outflows in typical low-redshift radiative-mode AGNs is lacking.

6.2. Feedback from Radio Sources

As discussed in Section 5.3, radio-AGNs can be fueled from their surrounding hot gaseous haloes, either directly through the Bondi process or via gas that has cooled out of the hot phase in a cooling flow. The bulk of the energy generated from the accretion is channeled into the powerful radio jets, which drive their way outward. The radio jets are confined by the hot halo gas and produce expanding radio lobes filled with relativistic plasma, inflating bubbles or cavities in the hot gas. The bulk of the jet energy is deposited locally, through the mechanical work required to inflate these cavities. Because the repository of the AGN energy is the same gas as that which fuels the AGN, this offers the necessary conditions for a self-regulating AGN feedback cycle.

The most extreme examples of radio-AGN feedback occur in the massive galaxies at the centers of groups and clusters (see **Figure 21**). In these systems the radiative cooling time of the surrounding gas can be substantially shorter than the Hubble time. In the absence of a heating source, a cooling flow would be expected to develop, whereby the temperature in the central regions of the cluster drops and gas flows inward at rates of hundreds or even thousands of solar masses per year (see the review by Fabian 1994). In reality, however, observations show that the gas temperatures in the cluster cores are at most a factor of 3 below the value at large radii, and the amount of cooling gas is only about 10% of the cooling flow prediction (e.g., Peterson et al. 2001, 2003; David et al. 2001; Tamura et al. 2001).

A heating source must be balancing the radiative cooling losses to prevent the gas from cooling further, and this reduces the SFRs in the central galaxy by an order of magnitude (though not to zero—these galaxies can still have SFRs of tens of solar masses per year; e.g., O'Dea et al. 2008). The heating source is almost certainly the jet outflows from the central BH: The prevalence of radio-AGN activity is high (>70%) in central cluster galaxies (Burns 1990), and results from the SDSS show that it is enhanced relative to other galaxies of the same mass (Best et al. 2007). Bubbles and cavities in the X-ray gas are also seen in a high percentage of clusters, reaching nearly 100% in those clusters with cooling times below 3 Gyr (Dunn & Fabian 2006, Fabian 2012). As discussed in Section 2.3.2, the mechanical energy flow rates of the radio jets are calculable either from the observed synchrotron emission or from the pV work required to inflate the radio lobes. Analysis of these cooling clusters indicates that the energy input from the radio jets is in rough balance with the energy losses from radiative cooling within the cooling radius (see the review by McNamara & Nulsen 2007).

For feedback processes to be efficient, it is not only the total available energy that is important but also how and where that energy is deposited. Radio jets are highly anisotropic structures, but gas cooling and inflow need to be switched off in all directions. It is the expanding radio bubbles, rather than direct jet interactions, that must provide the bulk of the energy transfer to the ambient medium. This process is observed in the form of weak shocks or sound waves surrounding some radio bubbles; the most striking example is the Perseus cluster, in which several concentric ripples in X-ray pressure are seen (Fabian et al. 2003, 2006). These ripples have been associated with repeated episodes of bubble-blowing activity from the central AGN. The energy in the sound waves or weak shocks is gradually dissipated over length scales of ~100 kpc (depending upon the gas viscosity). This gentle feedback process is able to transfer the radio jet energy into its environment in a largely isotropic manner without disturbing the intracluster medium to such an extent that the observed temperature and metal abundance gradients would be destroyed by mixing.

The radio-AGN activity appears to be switched on nearly all of the time in cooling flow clusters and must be able to adapt quickly to changes in its environment in order to maintain the close heating-cooling balance. This quick adaption is not possible for standard Bondi accretion, because the infall time of the hot gas from kiloparsec scales down to the BH is more than a billion years (Soker 2006). However, cold clump accretion occurs on a much faster timescale, especially in a viscous accretion flow, and therefore allows the BH to react quickly to changes in the state of its environment (Pizzolato & Soker 2010, Gaspari et al. 2013). This allows for efficient self-regulation of the feedback process.

Although BCGs offer the most striking examples of radio-AGN feedback, which can be individually studied in detail, these massive cooling-flow clusters are relatively rare. The data from large sky surveys have been invaluable in extending analysis to less extreme systems and providing a direct statistical comparison of radio-AGN feedback in systems of different scales. Best et al. (2006) investigated the role of radio-AGN feedback in elliptical galaxies using data from the SDSS. On the assumption that all massive galaxies would go through recurrent short-lived radio-AGN outbursts, they considered the observed prevalence of powerful ($P_{1.4\,\mathrm{GHz}} > 10^{23}$ W Hz^{-1}) radio-loud AGNs as a function of BH mass and interpreted this as a measure of the duty cycle of powerful radio-AGN activity for galaxies of that BH mass. They combined this with the BH mass–dependent radio luminosity function to estimate, probabilistically, the fraction of its time that a BH of given mass would spend producing a radio source of a given radio luminosity. Using the radio luminosity–to–jet kinetic power conversion (Section 2.3.2) then allowed a calculation of the average rate of jet kinetic energy production of the BH. Note that most of the heating is produced by radio sources of $P_{1.4\,\mathrm{GHz}} \sim 10^{24}-10^{25}$ W Hz^{-1}. The heating contribution from periods of very low radio luminosity, where in this simple picture the radio AGN is considered to be off, can be neglected. Best et al. derived the heating rate for galaxies of different BH mass and compared these results with the X-ray luminosities (i.e., cooling rates) of the hot haloes around the galaxies. They found that for elliptical galaxies of all masses, the radio-AGN heating rate provided a good balance to the gas cooling rate. AGN heating is therefore able to counter-balance cooling and thus explain the old, red, and dead nature of the ellipticals.

The Best et al. (2006) analysis actually underestimated the jet-energy production rate, as they adopted only pV instead of $4pV$ for the cavity energies. Also, for gas cooling rates, they scaled from the total X-ray luminosity rather than considering only the luminosity within the cooling radius (which is the only region where cooling needs to be compensated to avoid catastrophic cooling collapse). Therefore, the radio-AGNs provide, in a time-averaged sense, significantly more energy than is required to suppress cooling on galaxy scales. This was shown more clearly by Best et al. (2007), who compared the time-averaged AGN heating rate against the rate of radiative cooling within the cooling radius for systems of different masses. They suggested that the average heating-to-cooling ratio decreased by two orders of magnitude when going from massive ellipticals/small groups to the most massive clusters. Updated observational determination of the BH mass–dependent radio luminosity function and the radio luminosity–to–jet power conversion have decreased this scale dependence, but it remains well in excess of an order of magnitude. Thus, if the radio-AGN heating rate is at the level that balances cooling in the most massive clusters (or at least the cool-core subset of these), then in smaller systems an order of magnitude more jet mechanical energy is produced than required.

Similar indications have been found by Nulsen et al. (2007; see also Fabian 2012, his figure 5), who found that the pV energy stored in cavities seen in lower-mass systems can be an order of magnitude above that required to balance cooling losses. This result is easily understood, because in individual galaxies or small groups a large fraction of the jet kinetic energy may get deposited on much larger scales than the cooling radius: The jets themselves can extend beyond the cooling flow region, and even if they don't then the bubbles they produce will still rise buoyantly to beyond the cooling radius, whereas the weak shocks produced by the expanding cavities dissipate their energy on similarly large (hundred-kiloparsec) scales (e.g., Fabian et al. 2005).

In lower mass systems the duty cycle of radio source activity is also low; the AGN is turned off for most of the time. The heating-cooling balance is a pseudoequilibrium process whereby in the brief periods in which the radio source does turn on, the instantaneous energetic output is far above that required to balance cooling. This leads to overheating of the gas, seen for example as a temperature difference observed between galaxy groups and low-mass clusters with and without radio-AGNs (Croston et al. 2005, Magliocchetti & Brüggen 2007; see also Shen et al. 2008). After the radio-AGN outburst, the system experiences a period during which accretion rates are suppressed by the turbulent motions and heating that was induced, leading to effectively an off state for the BH. This behavior is seen in hydrodynamic simulations (e.g., Gaspari et al. 2013) and is also suggested both by the episodic history of star formation found by Chen et al. (2013b) in massive galaxies ($M_* > 10^{11}\,M_\odot$) and by the apparent anticorrelation they find between a recent burst of star formation and the current presence of a radio source. This off period lasts longer in less massive systems because the lower binding energy and gas sound speed lead to a longer recovery time before gas cooling and accretion recommence. The BH therefore acts as a cosmic thermostat, adjusting quickly to its surroundings and switching on powerful radio-AGN activity whenever the gas temperature drops sufficiently to allow significant cooling to occur. Rare intermittent activity (duty cycle 0.001–0.01) suffices to control gas cooling in typical massive ellipticals (of stellar mass $\sim 10^{11}\,M_\odot$), with the system oscillating around an equilibrium state. In contrast, near-constant activity (duty cycle ~ 1) is required to control gas cooling in the most massive cooling flow clusters.

7. THE SITUATION AT HIGH REDSHIFT

The focus of this review has been the contemporary Universe as probed through large surveys (most notably, the SDSS). It is then natural to ask whether the results we have reviewed apply only to the present day or whether a qualitatively similar picture also applies at higher redshift.

7.1. The Representative AGN Population

To begin, let us use the cosmic evolution of the bolometric AGN luminosity function to frame the issues. Although it is customary to refer to the redshift range of 2 to 3 as the QSO era, this is actually rather misleading. It is true that the comoving number density of the most powerful AGNs ($L_{\rm bol} > 10^{14}\,L_\odot$) peaked during this epoch. These are associated with the formation of the most massive SMBHs ($>10^{9.5}\,M_\odot$), and as such comprise only a minor fraction of the BH mass in the local fossil record (Shankar et al. 2009). A much better way to think about the growth of the BH population is to use the data plotted in **Figure 1** to calculate the history of the cumulative growth of the relic mass in SMBHs over time. This leads to the striking result that the growth of BHs is actually a protracted process: 25%, 50%, and 75% of the present-day relic mass was formed by cosmic ages of about 3.3 Gyr ($z \sim 2$), 5 Gyr ($z \sim 1.3$), and 7 Gyr ($z \sim 0.75$), respectively. The history of the mass build-up in galaxies via star formation is very similar (e.g., Shankar et al. 2009). So in a big picture sense, the physics that connected (connects) the growth of galaxies and BHs must have been in place over a significant fraction of the history of the Universe, and the bulk of this coevolution took place at moderate redshifts.

To focus our discussion, we consider the population of AGNs and their host galaxies over the redshift range of 0.5 to 2.5 (during which time about 75% of the current BH mass was created by accretion). The shape of the bolometric luminosity function at $z \sim 2$ (Hopkins et al. 2007b, Shankar et al. 2009) implies that most of the BH growth then occurred for objects near L_*. More quantitatively, half the BH growth occurred for AGNs with bolometric luminosities between $L_{\rm bol} \sim 10^{12.5}$ and $10^{13.3}\,L_\odot$ (or hard X-ray luminosities of 2×10^{44} to 8×10^{44} erg s^{-1}). At $z \sim 0.5$, the same calculation implies that half the BH growth occurred in AGNs with bolometric (hard X-ray) luminosities between $10^{11.2}$ and $10^{12.3}\,L_\odot$ (3×10^{43} to 2×10^{44} erg s^{-1}). We use these fiducial values for redshift and luminosity to define what we henceforth call representative AGNs, by which we mean the objects most responsible for the creation through accretion of the mass locked up in BHs in the present fossil record.

7.2. The Host Galaxies of Radiative-Mode AGNs

Most of the information about AGN host galaxies at high z comes from either deep surveys of fields that probe comoving volumes too small to sample the representative AGN population or observations of individual AGNs of exceptionally high luminosity. It is only over the past few years that we have learned quite a bit about the properties of the host galaxies of the representative AGNs. For the most part these surveys have selected the AGNs and quantified their properties based on (rest-frame) hard X-ray data. One potentially important caveat is that it is possible that the host galaxies of X-ray-selected AGNs systematically differ from the host galaxies of AGNs selected in the IR or optical (Hickox et al. 2009, Koss et al. 2011, Juneau et al. 2013).

The most detailed investigation of the dependence of AGN properties on the mass of the host galaxy was undertaken by Aird et al. (2012) for a population of X-ray-detected objects over the range $z = 0.2$ to 1.0. They find that the distribution of AGN X-ray luminosity normalized by the stellar mass of the host galaxy (M_*) is essentially independent of M_* over mass bins with centers from $M_* = 10^{9.75}$ to $10^{11.75}$ $M_\odot$. Within the uncertainties, this is largely consistent with the [OIII]-based SDSS results at $z \sim 0.1$ we have plotted in **Figure 16**, except that our plot shows a strong drop in normalized luminosity at high galaxy mass. As we noted, the normalized sum of L_{rad} plus L_{mech} is roughly independent of M_* out to the highest masses. In fact, Chen et al. (2013b) found strong evolution in the emission-line AGNs associated with very massive galaxies ($M_* > 10^{11.4}$ $M_\odot$) between $z \sim 0.1$ and 0.6. They found that the equivalent width of the [OIII] emission line (a rough proxy for L_{rad}/M_*) increased by an order of magnitude and that the emission lines changed from LINER-like (jet mode) at low z to Seyfert-like (radiative mode) at higher z. This is at least qualitatively consistent with the difference between the results by Aird et al. (2012) and what is seen in the contemporary Universe.

Aird et al. (2012) also conclude that their sample of AGNs shows a universal distribution in Eddington ratio that is independent of galaxy or BH mass. This would be inconsistent with the picture of the downsizing of the population of BHs because the implied timescales for BH mass–doubling through accretion would be independent of BH or galaxy mass. However, their result is based on their assumption of a uniform ratio of BH–to–total galaxy stellar mass ($M_{BH}/M_* = 0.002$) as a function of mass. As we showed explicitly in **Figure 9*b***, in the SDSS sample at $z \sim 0.1$ the ratio of BH-to-stellar mass increases by a factor of about 1.5 dex (from 0.0001 to 0.003) over the range in stellar mass bins studied by Aird et al. (centered from $\log M_* = 9.75$ to 11.75). If the relationship between BH and galaxy mass is similar at $z \sim 0.1$ and 0.6, the Aird et al. results would then be entirely consistent with the BH down-sizing picture. This is supported by the conclusions of Kelly & Shen (2013) based on their analysis of the SDSS QSO population.

Mainieri et al. (2011) studied the host galaxies of a sample of 142 X-ray-selected Type 2 AGNs from the *XMM-Newton* COSMOS field. The typical redshifts are in the range of 0.8 to 2, and X-ray luminosities are 10^{44} to 10^{45} erg s^{-1}. They find few detected AGNs in host galaxies with stellar masses less than 10^{10} $M_\odot$, and they find typical stellar masses of $\sim 10^{10.5}$ to $10^{11.3}$ $M_\odot$ (similar to local Seyfert galaxies). A stacked spectrum in the rest-frame optical shows values of the 4,000-Å break and high-order Balmer absorption lines corresponding to a very young star-forming population. The masses (and the SFRs derived from *Herschel* far-IR photometry) imply that the sSFRs are similar to those of normal galaxies on the star-forming main sequence at the same redshifts (see **Figure 22**). Most host galaxies appear to be early-type galaxies (with a significant bulge component), and only a minority show either a prominent disk or a disturbed morphology indicative of a major merger.

Schawinski et al. (2012) investigated the host-galaxy properties of a population of IR-selected heavily obscured AGNs at $z \sim 2$ with typical bolometric luminosities of $\sim 10^{12.5}$ $L_\odot$. Their HST imaging reveals that most have a significant disk component and only a small fraction appear to be major mergers. At lower redshifts ($z \sim 0.7$), Cisternas et al. (2011) analyzed HST images of the hosts of AGNs with X-ray luminosities of 10^{43} to 10^{44} erg s^{-1}. They found that at least 85% of the host galaxies have normal undisturbed morphologies (similar to their control sample of non-AGN galaxies).

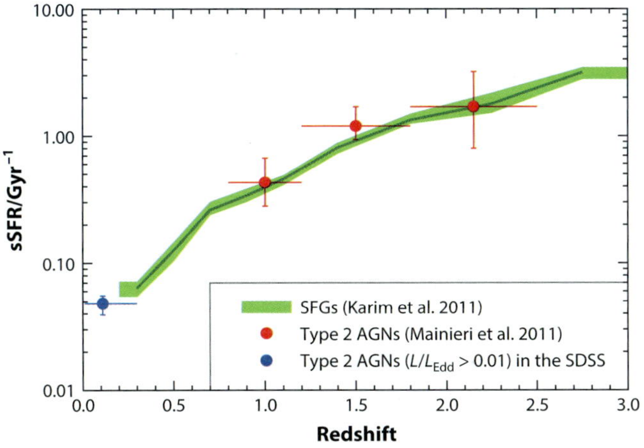

Figure 22

The specific star-formation rate (sSFR = SFR/M_*) of radiative-mode AGN hosts compared with normal star-forming galaxies (SFGs) as a function of redshift. The green-shaded region shows the evolution of the sSFR of galaxies along the star-forming main sequence derived by Karim et al. (2011). The red points show the sSFR of the host galaxies of X-ray-selected Type 2 AGNs from the *XMM-Newton* COSMOS field derived by Mainieri et al. (2011). The blue point indicates the average sSFR of emission-line-selected Type 2 AGNs with $L/L_{\text{Edd}} > 0.01$ (i.e., broadly radiative mode) from the Sloan Digital Sky Survey (SDSS). At all redshifts, the sSFRs of radiative-mode AGN hosts are consistent with those of galaxies on the star-forming main sequence.

Santini et al. (2012) investigated samples of X-ray-selected AGNs from the GOODS N and S field and the COSMOS field over the range $z \sim 0.5$ to 2.5. Only the COSMOS field has sufficient volume to probe the population of representative AGNs. Based on *Herschel* far-IR photometry, the SFRs and sSFRs of the AGN hosts in this field are enhanced by a factor of about three relative to the full population of normal galaxies at the same redshift. However, the SFRs in the AGN hosts are consistent with those of the star-forming population of normal galaxies (i.e., the AGNs avoid the quiescent galaxy population). At $z \sim 0.7$, Bournaud et al. (2012) found a high incidence rate of Seyfert-like AGNs in a sample of star-forming galaxies with clumpy disks. They argue that the fueling of AGNs at high z is primarily through inflows triggered by violent disk instabilities rather than major mergers (Elmegreen et al. 2008b, Bournaud et al. 2011).

Rosario et al. (2013) have examined the relationship between SFR (from the far-IR luminosity) and AGN hard X-ray luminosity over the range from $z \sim 0$ to 2.5. They find that at $z < 0.8$ there is a correlation between the AGN luminosity and SFR for AGNs with bolometric luminosities above $\sim 10^{10.5}$ $L_\odot$. At higher redshifts the SFR no longer correlates with AGN luminosity, but rises with increasing redshift (from $\sim$10 $M_\odot$ per year at $z \sim 1$ to $\sim$50 $M_\odot$ per year at $z \sim 2$). Again, these SFRs are consistent with the properties of typical massive star-forming galaxies at these epochs. Harrison et al. (2012a) reach similar conclusions. These results regarding the complex connection between AGNs and star formation may be related to findings of Kauffmann & Heckman (2009) at low z. As summarized above, they found a threshold value for the sSFR for effective BH growth. Once this threshold is exceeded, BH growth is switched on, but the rate of growth does not increase if the SFR rises further.

To summarize, the hosts of the representative population of radiative-mode AGNs out to $z \sim 2$ are mostly structurally normal galaxies with SFRs corresponding to those of the typical star-forming population of galaxies at that epoch. These results are consistent with most other investigations of the less powerful AGN population over the same range in redshift (e.g., Pierce et al. 2007; Gabor et al. 2009; Hickox et al. 2009; Silverman et al. 2009, 2011; Schawinski et al. 2011; Kocevski et al. 2012; Mullaney et al. 2012a,b; Aird et al. 2012, 2013; Bohm et al. 2013; Chen et al. 2013a; Juneau et al. 2013; Rosario et al. 2013; Villforth et al. 2014). These results are also consistent with the more detailed picture we have presented for radiative-mode AGNs in the contemporary Universe.

7.3. Evolution of the Radio-Loud AGN Population

The cosmic evolution of the radio luminosity function has been studied for many years, since early work on understanding radio source counts indicated that high luminosity sources undergo substantially stronger cosmic evolution than low luminosity sources (Longair 1966). Seminal work by Dunlop & Peacock (1990) clearly established that the space density of both flat-spectrum (i.e., beamed) and steep-spectrum powerful radio sources is 2–3 orders of magnitude higher by $z \sim 2$ than in the contemporary Universe. Beyond this redshift the space density peaks or declines slowly. In contrast, the low luminosity radio sources show only a factor of <2 increase in space density out to $z \sim 0.5$ (e.g., Sadler et al. 2007, Donoso et al. 2009). Rigby et al. (2011) characterized this differential evolution, showing that the redshift at which the radio source space density peaks is strongly luminosity dependent with low luminosity sources showing a peak space density at $z < 1$.

As discussed earlier, the radio-AGN population comprises both jet-mode (radiatively inefficient) AGNs and radiative-mode (radio-loud QSO/Seyfert-like) AGNs. Best & Heckman (2012) looked at the cosmic evolution of these two populations separately within the narrow redshift range probed by the SDSS and concluded that they showed different cosmic evolution. This has been confirmed by P.N. Best, L.M. Ker, C. Simpson, E.E. Rigby & J. Sabater (in preparation), who show that out to $z \sim 0.7$ the radiative-mode population shows a factor $\sim$8 increase in space density at all radio luminosities. The jet-mode sources, however, show only a small increase in space density out to $z = 0.5$, followed by a decline thereafter. These results are easily understood in terms of our earlier discussions on the fueling of these AGNs. The radiative-mode radio-AGNs evolve in broadly the same manner as the upper-mass end of the radiative-mode radio-quiet population, increasing with redshift as the availability of an abundant supply of cold gas increases. If the jet-mode AGNs are largely associated with cooling hot-gas haloes, then their evolution should be closer to that of massive galaxies and, thus, slowly declining toward higher redshifts.

These results indicate that the luminosity-dependent cosmic evolution of the radio luminosity function as a whole is largely driven by the different cosmic evolution of these two AGN populations. As a result of this differential evolution, the dominant radio-AGN population also changes with redshift. Locally, except at the highest radio luminosities, the jet-mode population dominates the radio-AGN number counts. By $z > 1$, however, the QSO/Seyfert-like sources begin to dominate at all (currently observable) luminosities. It is for this reason that the properties of radio-loud and radio-quiet AGNs appear much more comparable in the early Universe than in the contemporary Universe.

Early work to extend the study of the host galaxies of the jet-mode AGNs to higher redshifts paints a picture consistent with that in the contemporary Universe. Tasse et al. (2008) and Donoso et al. (2009) both investigated the relationship between galaxy stellar mass and radio-AGN prevalence at $z \sim 0.5$. If consideration is restricted to only the jet-mode AGNs, then a steep power-law dependence similar to that for the contemporary Universe (Section 4.3) is found. Simpson et al. (2013) recently extended that analysis to $z \sim 1$, confirming the same results. Tasse et al. (2008) also showed that the jet-mode population was typically found in galaxy overdensities (groups or clusters). These results indicate that the feedback balance between radiative cooling and radio-AGN heating, seen in the contemporary Universe, was already in place in some environments when the Universe was half of its current age.

7.4. AGN Feedback at High *z*

Feedback from AGNs in the form of ionizing radiation clearly plays a fundamental role in ionizing and heating the intergalactic medium over much of the history of the Universe. Here, we briefly summarize the evidence regarding the existence and physical properties of bulk outflows of mass and kinetic energy driven by AGNs. This evidence is based on detailed investigations of small samples of AGNs.

In a series of papers, Arav and collaborators (Arav et al. 2008, Korista et al. 2008, Moe et al. 2009, Bautista et al. 2010, Dunn et al. 2010, Aoki et al. 2011, Edmonds et al. 2011, Arav et al. 2013, Borguet et al. 2013) have been able to determine the main physical properties of outflows in high-z QSOs as traced by UV absorption lines. The QSOs have bolometric luminosities of $\sim 10^{13.3}$ to $10^{14.7}$ $L_\odot$ and $z \sim 0.6$ to 3 (they are considerably more luminous than the representative AGN population). Of the eight outflows they have analyzed, four are extremely energetic with kinetic energy fluxes of 10^{45} to 10^{46} erg s^{-1}, representing $\sim$1 to 5% of L_{bol}. The typical outflows speeds are about 3,000 to 8,000 km s^{-1}, and the inferred size scales of the outflows range from several hundred parsecs to several kiloparsecs. Similar work by this group on less luminous AGNs at lower redshift has so far found only very weak outflows with kinetic energy fluxes $\ll L_{bol}$ (Edmonds et al. 2011, Borguet et al. 2012).

Alexander et al. (2010) and Harrison et al. (2012b) have analyzed spatially resolved maps of the kinematics of the ionized gas surrounding nine radio-quiet AGNs at $z \sim 1.4$ to 3.4. The objects were selected from a parent sample of submillimeter galaxies on the basis of unusually strong and broad [OIII] 5007 emission lines. Evidence for galactic-scale outflows was found in the five AGNs with bolometric luminosities of about 10^{13} $L_\odot$ (roughly L_* at these redshifts), but not in the less luminous cases. The strong rest-frame far-IR emission in these objects implies that they have much higher SFRs than typical AGNs at these redshifts ($\sim 10^3$ $M_\odot$ year^{-1} compared with $\sim$30 to 100 $M_\odot$ year^{-1}, respectively). Thus, it is not entirely clear whether the outflows are driven by the AGN or the starburst in these objects.

Using similar techniques, galaxy-scale outflows of ionized gas have been mapped around a small sample of extremely radio-loud QSO-like AGNs at $z \sim 2$ (Nesvadba et al. 2006, 2008). These outflows appear to be driven by the mechanical energy carried by the powerful radio jets in these systems. At the highest redshifts, Maiolino et al. (2012) have mapped an outflow from an exceptionally luminous ($L_{bol} \sim 10^{14}$ $L_\odot$) QSO at $z \sim 6.4$ using the [CII] far-IR emission line. They estimate an outflow rate of several thousand solar masses per year at a velocity of order 10^3 km s^{-1}. We conclude that direct evidence for AGN feedback at high z is limited so far to objects of exceptionally high luminosity or other unusual characteristics. Little is known about whether the representative population of high-z AGNs produces powerful global outflows. Again, this is all consistent with the results derived in the contemporary Universe.

8. SUMMARY AND IMPLICATIONS

Let us begin this final section by briefly summarizing what we regard as the most robust conclusions about the demographics of the populations of AGNs and their host galaxies in the contemporary Universe. We then consider the degree to which these results may be extended to higher redshift and conclude with some speculation about the implications these conclusions may have for our understanding of the formation and evolution of galaxies.

8.1. Some Robust Conclusions

In the contemporary Universe there are two distinct families of AGNs. We have designated them radiative-mode and jet-mode AGNs. The radiative-mode population of AGNs is characterized by the production of significant amounts of radiation with typical bolometric luminosities of 1 to 100% of the Eddington limit. Historically these have been called Seyfert galaxies or QSOs depending upon their luminosity. The AGN properties can be explained within the theoretical context of a radiatively efficient geometrically thin accretion disk. In the contemporary Universe these AGNs characteristically have BH masses of 10^6 to $10^8\,M_\odot$ and reside in host galaxies with typical stellar masses 10^{10} to a few $\times 10^{11}\,M_\odot$ and high stellar surface mass densities ($\mu_* > 10^{8.5}\,M_\odot\,kpc^{-2}$). Although these densities are characteristic of early-type disk galaxies, the AGN hosts (unlike most such galaxies) have significant amounts of ongoing star formation in the central few kiloparsecs. When examined at high spatial resolution these inner regions can usually be classified as pseudobulges. Major mergers or strong tidal interactions are not the dominant fueling mechanism for the central star formation and the AGN. Internal secular processes for the radial transport of gas are evidently more important. This summary pertains to the typical AGNs that are the sites of the majority of the growth of SMBHs today. Things may differ for the rare objects with the highest luminosities.

The radiative-mode AGNs are associated with populations of both lower-mass BHs and lower-mass (pseudo)bulges that have volume-averaged mass-doubling times (via accretion and star formation, respectively) of roughly the Hubble time. The volume-averaged ratio of star formation–to–BH growth in these central regions is of order a thousand (similar to the ratio of stellar mass and BH mass in current classical bulges and elliptical galaxies). These are evidently living systems that have not yet reached their evolutionary end point. Although there is indirect evidence that feedback is limiting the growth of the BH, there is little direct evidence of AGN-generated feedback for typical objects. In contrast, there is recent evidence that this may occur in the rare low-z AGNs having the highest luminosity ($L_{bol} \sim 10^{12-14}\,L_\odot$). The association of AGNs with central star formation implies that feedback from massive stars should be present, but its effect on the fueling of the AGN has not been well-quantified.

The second, jet-mode, AGN class is characterized by the production of collimated energetic outflows (jets), but it has little detected emitted radiation (characteristic luminosity below about 1% of the Eddington limit). Historically these have been called (low-excitation) radio galaxies. The more luminous LINERs are also likely to be members of this population. The AGN properties can be explained within the theoretical context of a radiatively inefficient accretion flow. The probability of a galaxy or BH producing a jet of a given radio luminosity is a steep function of mass: Typical radio galaxies are structurally normal massive ellipticals ($M_* \sim 10^{11}$ to $10^{12}\,M_\odot$) with BH masses greater than $10^8\,M_\odot$. There is no evidence for substantial recent star formation in typical radio galaxies (although there can be for radio-AGNs in the BCGs of extreme cooling flow clusters). There is mostly indirect evidence that the AGN is fueled by the accretion of hot gas seen in X-ray observations. Feedback produced as the jets interact with this hot gas can be directly observed. Although it is energetically possible for this feedback to prevent the gas from cooling rapidly, the details remain uncertain. These AGNs are associated with the populations of both higher mass BHs and spheroids that have mass-doubling times that are orders of magnitude longer than the Hubble time. These are effectively dead systems in which the bulk of the star formation and BH growth occurred at high redshift.

In examining the situation at higher redshift, we have emphasized the importance of focusing on the representative population of AGNs that is responsible for the creation of the majority of the mass in relic SMBHs today. These are AGNs in the redshift range $z \sim 0.5$ to 2.5 and having luminosities near the knee in the luminosity function. These representative AGNs appear to be hosted by moderately massive galaxies ($M_* > 10^{10}\,M_\odot$) with SFRs that are typical of the star-forming galaxies at that epoch (they are located on the star-forming main sequence). They are typically found in structurally normal galaxies rather than in highly disturbed systems (major mergers). The population of radio-loud AGNs remains prevalent among massive galaxies. Evidence for feedback in the form of powerful AGN-driven outflows is limited so far to objects of either exceptional luminosity or other unusual properties (very high radio luminosity or SFR).

All of this sounds similar to the situation in the contemporary Universe. The main differences are quantitative rather than qualitative: (*a*) The characteristic luminosities of the actively growing BHs are larger than the present day (by a factor of about 3 at $z \sim 0.5$ and $\sim$30 at $z \sim 2$), and the characteristic masses of these BHs are also correspondingly larger, by factors that are model-dependent (e.g., Shankar et al. 2013); and (*b*) the sSFRs of host galaxies on the main sequence are higher than the current value by factors of $\sim$3 at $z = 0.5$ and $\sim$20 at $z = 2$ (e.g., Elbaz et al. 2011). Thus, to zeroth order, the situation at higher z seems like a scaled-up version of the contemporary Universe.

There is also (what we regard as) persuasive indirect evidence that the physical processes that link the formation and evolution of galaxies and BHs at high z are still at play today. As we noted in the introduction, the ratio of the rate of build-up of galaxies through star formation to the rate of the build-up of BHs via accretion has been broadly independent of cosmic time over at least the last $\sim$10 Gyrs with a value of $\sim$1,500 (Shankar et al. 2009, 2013; Hopkins & Beacom 2006). This is hard to understand unless the underlying physics has remained basically the same over this time (see Heckman et al. 2004 and Mullaney et al. 2012b for further discussion of this argument). Indeed, as **Figure 23** shows, this ratio applies not just to the total rates but over all values for the mass of the SMBH.

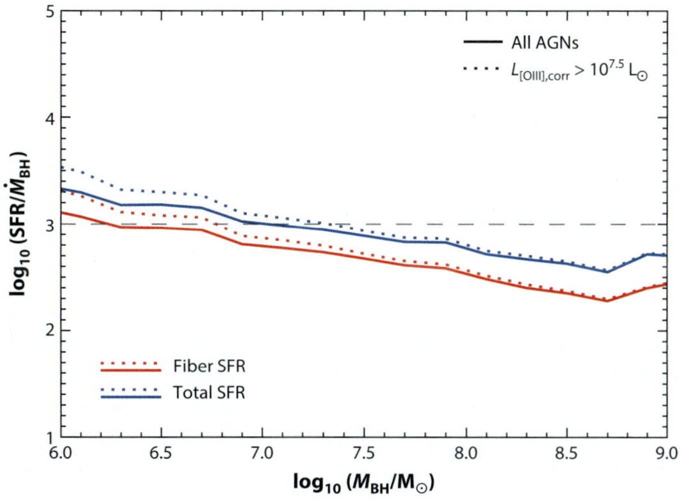

Figure 23

Ratio of the total star-formation rate (SFR) per unit volume in galaxies to the total accretion rate per unit volume onto black holes as traced by Type 2 AGNs plotted as a function of the black hole mass (see Heckman et al. 2004). The sample studied consists of all galaxies from the Sloan Digital Sky Survey main galaxy sample with stellar surface mass density above the characteristic value $\mu_* = 10^{8.5}\,M_\odot\,\text{kpc}^{-2}$. Red lines show the result if the SFR is calculated within the fiber aperture for each galaxy and the blue lines show the result using estimates of the total SFR. Dotted lines show results restricted to the AGNs with (reddening-corrected and star-formation-corrected) $L_{[OIII]} > 10^{7.5}\,L_\odot$. The horizontal dashed line shows the fiducial value of stellar to black hole mass in bulges and elliptical galaxies.

8.2. Implications and Some Speculation

Our current understanding of the formation and evolution of galaxies is that the build-up of the stellar content of galaxies via star formation is driven primarily by the accretion of gas from the cosmic web. At least since $z = 2$ or 3, the bulk of the star formation occurs in galaxies lying along a strongly evolving star-forming main sequence (a tight relationship between SFR and stellar mass). Although major mergers of gas-rich systems do occur and lead to strong bursts of star formation, they are responsible for only a minority (of order 10%) of the total cosmic star formation. This picture is predicted by numerical simulations and is consistent with observations of the evolving population of star-forming galaxies (e.g., Dekel et al. 2009, Bouche et al. 2010, Elbaz et al. 2011, Wuyts et al. 2011b, Whitaker et al. 2012, Lilly et al. 2013).

Empirically, the accretion-driven mass growth of the population of SMBHs is related to the growth (via star formation) of the inner region of the host galaxy. Evidently, both rates track the rate at which gas is accreted by the galaxy (with relative efficiencies that differ by about three orders of magnitude). Mergers are not the dominant mechanism by which gas flows to these inner regions and, hence, are not the dominant mechanism for growing the population of SMBHs. The gas flows responsible for this may instead be driven by rather slow secular processes in the contemporary Universe (e.g., Kormendy & Kennicutt 2004, Athanassoula 2008) or more violent and rapid instabilities in clumpy gas-rich disks at high z (e.g., Genzel et al. 2014, Bournaud et al. 2010, Elmegreen et al. 2008a, Dekel et al. 2009). This picture would imply that the representative population of actively growing SMBHs (AGNs) would live at the centers of disk-like structures (rather than ellipticals or classical bulges).

This simple picture leaves out one critical fact: We know from observations that the effect of this coevolution over cosmic time must ultimately place the relic (dead) BHs in dynamically hot structures (ellipticals and classical bulges) that contain little cold gas and, correspondingly, little star formation. In the classical picture for the coevolution of galaxies and BHs this is explained economically by positing that the SMBH grew in the aftermath of a major merger that created an elliptical/classical bulge and that feedback from the rapidly growing BH blew away the remaining gas (e.g., Hopkins et al. 2006, 2008, Di Matteo et al. 2008). How can this be explained in the more sedate scenario we are advocating? Let us try a two-part answer.

First, though low-mass bulges can be built through secular processes and disk instabilities (e.g., Athanassoula 2008, Bournaud et al. 2011, Hopkins et al. 2012), major mergers are the only way to build the population of massive classical bulges and elliptical galaxies. In this case, the sedate scenario would imply that these major mergers only modestly augment the masses of gas accreted by the BH and of gas turned into stars. The $M_{\rm BH}$ versus σ relationship would have to be essentially prebuilt before a major merger takes place (consistent with the recent simulations by Anglés-Alcázar et al. 2013). A major merger mostly rearranges the stars in the pre-existing disks (and their black-hole-containing pseudobulges) into a classical bulge or elliptical galaxy and a more massive (merged) BH. Subsequent dry merging between classical bulges or ellipticals can further populate the upper end of the $M_{\rm BH}$ versus σ relation (e.g., Kormendy & Ho 2013).

The second part of the answer requires an explanation for the dry-up of the reservoir of cold gas and the subsequent quenching of star formation. In current models, this shutdown in star formation in a typical massive galaxy is connected to the transition from an accretion mode in which gas reaches the galaxy in a cold form (gas temperature much less than the virial temperature) to a mode where the accreted gas is heated to roughly the virial temperature upon arrival and then slowly cools (via radiation) and is gradually accreted. This transition is predicted to occur when the galaxy halo mass grows via mergers and exceeds some critical value. Dekel et al. (2009) argue that this critical mass corresponds to the observed transition between the high-mass mostly quiescent galaxy population and the lower-mass galaxy population that defines the star-forming main sequence (see also Bouche et al. 2010, Lilly et al. 2013, Mutch et al. 2013). Although current models still need to incorporate AGN feedback to quench star formation, it is possible that this is a result of the difficulty in incorporating the relevant physical processes with the necessary fidelity in current state-of-the-art numerical simulations. In this case, late feedback from AGNs might only be required to be strong enough to suppress the late cooling flows of hot gas and keep the quiescent galaxy suitably red and dead. This may be accomplished by the low-luminosity radio-AGNs, through the self-regulating feedback loop we have described in Section 6.2 above (the so-called maintenance mode of AGN feedback).

As we have emphasized, this sedate model is designed to match the detailed picture we have of the contemporary Universe and the emerging picture we have of the representative population of AGNs and their host galaxies at higher redshift. What about the more extreme cases? More specifically, one might ask about the physical mechanism(s) responsible for imposing a characteristic AGN luminosity at a given epoch $[L_*(z)]$ above which the number density of AGNs steeply falls. In part this must reflect the knee in the BHMF. However, there is also evidence that the most powerful AGNs (at both high z and low z) are different from their less luminous kin. There are indications that the fraction of AGN host galaxies that are major mergers increases at the highest AGN luminosities (e.g., Liu et al. 2009, Mainieri et al. 2011, Treister et al. 2012). This is at least qualitatively consistent with the models presented by Hopkins & Hernquist (2009), Draper & Ballantyne (2012), Shankar et al. (2012), and Hopkins et al. (2013), which partition the amount of BH growth associated with mergers and secular processes. We have also seen that the most luminous AGNs at low and high redshift can drive powerful outflows that could affect the fueling of the BH and even the bulk properties of the gas in the host galaxy and its halo. It is tempting to speculate that the knee of the AGN luminosity function marks the transition from a sedate model of fueling by accretion and subsequent internal processes with weak/no AGN feedback to a dramatic model of fueling by mergers with strong AGN feedback.

In this review, we have stressed the importance of large surveys of the contemporary Universe in revealing the way in which galaxies and their SMBHs coevolve(d). We fully expect that the next generation of large ground- and space-based surveys will extend this type of robust and sophisticated understanding to the era when the bulk of the mass in galaxies and BHs was assembled.

DISCLOSURE STATEMENT

The authors are not aware of any affiliations, memberships, funding, or financial holdings that might be perceived as affecting the objectivity of this review.

ACKNOWLEDGMENTS

We thank our collaborators on the papers discussed in this review. We also thank the dedicated teams that made possible the large surveys that have provided so much illumination. Additionally, we thank our editor, Sandy Faber, whose insightful comments and suggestions made this a better review. P.N.B. is grateful for support from the UK Science and Technology Facilities Council.

LITERATURE CITED

Aalto S, Garcia-Burillo S, Muller S, et al. 2012. *Astron. Astrophys.* 537:A44
Aird J, Coil A, Moustakas J, et al. 2012. *Ap. J.* 749:90
Aird J, Coil A, Moustakas J, et al. 2013. *Ap. J.* 775:41

Alexander EM, Swinbank AM, Smail I, McDermid R, Nesvadba NPH. 2010. *MNRAS* 402:2211–20
Allen SW, Dunn RJH, Fabian AC, Taylor GB, Reynolds CS. 2006. *MNRAS* 372:21–30
Alonso MS, Lambas DG, Tissera P, Coldwell G. 2007. *MNRAS* 375:1017–24
Anderson SF, Voges W, Margon B, et al. 2003. *Astron. J.* 126:2209–29
Anglés-Alcázar D, Özel F, Davé R. 2013. *Ap. J.* 770:5
Antonucci R. 1993. *Annu. Rev. Astron. Astrophys.* 31:473–521
Antonucci R. 2012. *Astron. Astrophys. Trans.* 27:557–602
Aoki K, Oyabu S, Dunn JP, et al. 2011. *Publ. Astron. Soc. Jpn.* 63:457–67
Arav N, Borguet B, Chamberlain C, Edmonds D, Danforth C. 2013. *Ap. J.* Submitted. arXiv:1305.2181
Arav N, Moe M, Constantini E, et al. 2008. *Ap. J.* 681:954–64
Athanassoula E. 2005. *MNRAS* 358:1477–88
Athanassoula E. 2008. *MNRAS* 390:L69–72
Baade W, Minkowski R. 1954. *Ap. J.* 119:206–14
Bailey ME. 1980. *MNRAS* 191:195–206
Baldry IK, Glazebrook K, Brinkmann J, et al. 2004. *Ap. J.* 600:681–94
Baldwin JA, Phillips MM, Terlevich R. 1981. *Publ. Astron. Soc. Pac.* 93:5–19
Balick B, Heckman TM. 1982. *Annu. Rev. Astron. Astrophys.* 20:431–68
Balogh ML, Morris SL, Yee HKC, Carlberg R, Ellingson E. 1999. *MNRAS* 307:463–79
Barthel PD, Arnaud KA. 1996. *MNRAS* 283:L45
Baumgartner WH, Tueller J, Markwardt CB, et al. 2013. *Ap. J. Suppl.* 207:19
Bautista MA, Dunn JP, Arav N, et al. 2010. *Ap. J.* 713:25–31
Becker RH, White RL, Hefland DJ. 1995. *Ap. J.* 450:559–77
Bell AR. 1978. *MNRAS* 182:443–55
Bell EF, de Jong RS. 2001. *Ap. J.* 550:212–29
Bell EF, McIntosh DH, Katz N, Weinberg MD. 2003. *Ap. J. Lett.* 585:L117–20
Bentz MC, Denney KD, Grier CJ, et al. 2013. *Ap. J.* 767:149
Best PN. 2009. *Astron. Nachr.* 330:184–89
Best PN. 2004. *MNRAS* 351:70–82
Best PN, Heckman TM. 2012. *MNRAS* 421:1569–82
Best PN, Kaiser CR, Heckman TM, Kauffmann G. 2006. *MNRAS* 368:L67–71
Best PN, Kauffmann G, Heckman TM, Ivezić Ž. 2005a. *MNRAS* 362:9–24
Best PN, Kauffmann G, Heckman TM, et al. 2005b. *MNRAS* 362:25–40
Best PN, von der Linden A, Kauffmann G, Heckman TM, Kaiser CR. 2007. *MNRAS* 379:894–908
Bîrzan L, McNamara BR, Nulsen PEJ, Carilli CL, Wise MW. 2008. *Ap. J.* 686:859–90
Bîrzan L, Rafferty DA, McNamara BR, Wise MW, Nulsen PEJ. 2004. *Ap. J.* 607:800–9
Blandford RD, Payne DG. 1982. *MNRAS* 199:883–903
Blandford RD, Znajek RL. 1977. *MNRAS* 179:433–56
Blanton MR, Hogg DW, Bahcall NA, et al. 2003. *Ap. J.* 594:186–207
Blundell KM, Rawlings S. 2000. *Astron. J.* 119:1111–22
Bock DC-J, Large MI, Sadler EM. 1999. *Astron. J.* 117:1578–93
Bohm A, Wisotzki L, Bell EF, et al. 2013. *Astron. Astrophys.* 549:A46
Böhringer H, Voges W, Fabian AC, Edge AC, Neumann DM. 1993. *MNRAS* 264:L25–28
Bondi H. 1952. *MNRAS* 112:195–204
Borguet B, Arav N, Edmonds D, Chamberlain C, Benn C. 2013. *Ap. J.* 762:49
Borguet B, Edmonds D, Arav N, Dunn J, Kriss GA. 2012. *Ap. J.* 751:107
Borthakur S, Heckman T, Strickland D, Wild V, Schiminovich D. 2013. *Ap. J.* 768:18
Bouche N, Dekel A, Genzel R, et al. 2010. *Ap. J.* 718:1001–18
Bournaud F, Dekel A, Teyssier R, et al. 2011. *Ap. J. Lett.* 741:L33
Bournaud F, Elmegreen BG, Teyssier R, Block DL, Puerari I. 2010. *MNRAS* 408:1088–99
Bournaud F, Juneau S, Le Floc'h E, et al. 2012. *Ap. J.* 757:81
Bower RG, Benson AJ, Malbon R, et al. 2006. *MNRAS* 370:645–55
Brandt WN, Hasinger G. 2005. *Annu. Rev. Astron. Astrophys.* 43:827–59
Brinchmann J, Charlot S, White S, et al. 2004. *MNRAS* 351:1151–79

Brown TM, Ferguson HC, Stanford SA, Deharveng JM. 1998. *Ap. J.* 504:113–38
Brown TM, Smith E, Ferguson H, et al. 2008. *Ap. J.* 682:319–35
Bruzual AG. 1983. *Ap. J.* 273:105–27
Bruzual G, Charlot S. 2003. *MNRAS* 344:1000–28
Buchanan C, Gallimore JF, O'Dea CP, et al. 2006. *Astron. J.* 132:401–19
Burbidge EM, Burbidge GR, Prendergast KH. 1959. *Ap. J.* 130:26–45
Burns JO. 1990. *Astron. J.* 99:14–30
Cao X. 2010. *Ap. J.* 725:388–93
Cao X, Rawlings S. 2004. *MNRAS* 349:1419–27
Capetti A, Baldi RD. 2011. *Astron. Astrophys.* 529:A126
Cattaneo A, Best PN. 2009. *MNRAS* 395:518–23
Cattaneo A, Faber SM, Binney J, et al. 2009. *Nature* 460:213–19
Cavagnolo KW, Donahue M, Voit GM, Sun M. 2008. *Ap. J. Lett.* 683:L107–10
Cavagnolo KW, McNamara BR, Nulsen PEJ, et al. 2010. *Ap. J.* 720:1066–72
Chen CT, Hickox RC, Alberts S, et al. 2013a. *Ap. J.* 773:3
Chen Y-M, Kauffmann G, Heckman TM, et al. 2013b. *MNRAS* 429:2643–54
Chen Y-M, Tremonti CA, Heckman TM, et al. 2010. *Astron. J.* 140:445–61
Chevalier RA, Clegg AW. 1985. *Nature* 317:44–45
Churazov E, Brüggen M, Kaiser CR, Böhringer H, Forman W. 2001. *Ap. J.* 554:261–73
Churazov E, Sazonov S, Sunyaev R, et al. 2005. *MNRAS* 363:L91–95
Cicone C, Feruglio C, Maiolino R, et al. 2012. *Astron. Astrophys.* 543:A99
Cicone C, Maiolino R, Sturm E, et al. 2014. *Astron. Astrophys.* 562:A21
Cid Fernandes R, González Delgado RM, Schmitt H, et al. 2004a. *Ap. J.* 605:105–26
Cid Fernandes R, Gu Q, Melnick J, et al. 2004b. *MNRAS* 355:273–96
Cid Fernandes R, Heckman T, Schmitt H, et al. 2001. *Ap. J.* 558:81–108
Cid Fernandes R, Mateus A, Sodre L, Stasinska G, Gomes JM. 2005. *MNRAS* 358:363–78
Cid Fernandes R, Stasińska G, Mateus A, Vale Asari N. 2011. *MNRAS* 413:1687–99
Ciotti L, Ostriker JP. 1997. *Ap. J. Lett.* 487:L105–8
Ciotti L, Ostriker JP, Proga D. 2010. *Ap. J.* 717:707–23
Cisternas M, Gadotti DA, Knapen JH, et al. 2013. *Ap. J.* 776:50
Cisternas M, Jahnke K, Inskip KJ, et al. 2011. *Ap. J.* 726:57
Colless M, Dalton G, Maddox S, et al. 2001. *MNRAS* 328:1039–63
Condon JJ. 1989. *Ap. J.* 338:13–23
Condon JJ. 1992. *Annu. Rev. Astron. Astrophys.* 30:575–611
Condon JJ, Cotton WD, Greisen EW, et al. 1998. *Astron. J.* 115:1693–716
Conselice CJ, Bershady MA, Jangren A. 2000. *Ap. J.* 529:886–910
Cooray A, Sheth R. 2002. *Phys. Rep.* 372:1–129
Crenshaw DN, Kraemer SB, George IM. 2003. *Annu. Rev. Astron. Astrophys.* 41:117–67
Croft S, de Vries W, Becker RH. 2007. *Ap. J. Lett.* 667:L13–16
Croom SM, Boyle BJ, Shanks T, et al. 2005. *MNRAS* 356:415–38
Croom SM, Richards GT, Shanks T, et al. 2009. *MNRAS* 399:1755–72
Croston JH, Hardcastle MJ, Birkinshaw M. 2005. *MNRAS* 357:279–94
Croton DJ, Springel V, White S, et al. 2006. *MNRAS* 365:11–28
Daly RA, Sprinkle TB, O'Dea CP, Kharb P, Baum SA. 2012. *MNRAS* 423:2498–2502
David LP, Durisen RH, Cohn HN. 1987. *Ap. J.* 316:505–16
David LP, Nulsen PEJ, McNamara BR, et al. 2001. *Ap. J.* 557:546–59
Davies RI, Müller Sánchez F, Genzel R, et al. 2007. *Ap. J.* 671:1388–412
de Vries WH, Becker RH, White RL. 2006. *Astron. J.* 131:666–79
Dekel A, Sari R, Ceverino D. 2009. *Ap. J.* 703:785–801
Diamond-Stanic AM, Rieke GH. 2012. *Ap. J.* 746:168
Diamond-Stanic AM, Rieke GH, Rigby JR. 2009. *Ap. J.* 699:623–31
Di Matteo T, Colberg J, Springel V, Hernquist L, Sijacki D. 2008. *Ap. J.* 636:33–53
Di Matteo T, Springel V, Hernquist L. 2005. *Nature* 433:604–7

Dong R, Greene JE, Ho LC. 2012. *Ap. J.* 761:73
Donley JL, Rieke GH, Perez-Gonzalez PG, Barro G. 2008. *Ap. J.* 687:111–32
Donoso E, Best PN, Kauffmann G. 2009. *MNRAS* 392:617–29
Donoso E, Li C, Kauffmann G, Best PN, Heckman TM. 2010. *MNRAS* 407:1078–89
Donoso E, Yan L, Tsai C, et al. 2012. *Ap. J.* 748:80
Draper AR, Ballantyne DR. 2012. *Ap. J.* 751:72
Dressler A, Gunn JE. 1983. *Ap. J.* 270:7–19
Dunlop JS, Peacock JA. 1990. *MNRAS* 247:19–42
Dunn JP, Bautista M, Arav N, et al. 2010. *Ap. J.* 709:611–31
Dunn RJH, Fabian AC. 2006. *MNRAS* 373:959–71
Dunn RJH, Fabian AC, Taylor GB. 2005. *MNRAS* 364:1343–53
Edmonds D, Borguet B, Arav N, et al. 2011. *Ap. J.* 739:7
Elbaz D, Dickinson M, Hwang HS, et al. 2011. *Astron. Astrophys.* 533:A119
Ellison SL, Patton DR, Mendel JT, Scudder JM. 2011. *MNRAS* 418:2043–53
Elmegreen BG, Bournaud F, Elmegreen DM. 2008a. *Ap. J.* 688:67–77
Elmegreen BG, Bournaud F, Elmegreen DM. 2008b. *Ap. J.* 684:829–34
Fabello S, Kauffmann G, Catinella B, et al. 2011. *MNRAS* 416:1739–44
Fabian AC. 1994. *Annu. Rev. Astron. Astrophys.* 32:277–318
Fabian AC. 2012. *Annu. Rev. Astron. Astrophys.* 50:455–89
Fabian AC, Reynolds CS, Taylor GB, Dunn RJH. 2005. *MNRAS* 363:891–96
Fabian AC, Sanders JS, Allen SW, Crawford CS, Iwasawa K, et al. 2003. *MNRAS* 344:L43–47
Fabian AC, Sanders JS, Taylor GB, Allen SW, Crawford CS, et al. 2006. *MNRAS* 366:417–28
Falcke H, Körding E, Markoff S. 2004. *Astron. Astrophys.* 414:895–903
Fanaroff BL, Riley JM. 1974. *MNRAS* 167:P31
Fardal MA, Katz N, Weinberg DH, Davé R. 2007. *MNRAS* 379:985–1002
Fender RP, Belloni TM, Gallo E. 2004. *MNRAS* 355:1105–18
Ferrarese L, Ford H. 2005. *Space Sci. Rev.* 116:523–624
Ferrarese L, Merritt D. 2000. *Ap. J. Lett.* 539:L9–12
Feruglio C, Maiolino R, Piconcelli E, et al. 2010. *Astron. Astrophys.* 518:L155
Filippenko A, Ho L, Sargent W. 1993. *Ap. J. Lett.* 410:L75–78
Fiore F, Puccetti S, Grazian A, et al. 2012. *Astron. Astrophys.* 537:A16
Fischer J, Sturm E, González-Alfonso E, et al. 2010. *Astron. Astrophys.* 518:L41
Fisher DB, Drory N. 2011. *Ap. J. Lett.* 733:L47
Forman W, Jones C, Churazov E, et al. 2007. *Ap. J.* 665:1057–66
Gabor JM, Impey CD, Jahnke K. 2009. *Ap. J.* 691:705–22
Gadotti DA. 2009. *MNRAS* 393:1531–52
Gallo E, Fender RP, Pooley GG. 2003. *MNRAS* 344:60–72
Gandhi P, Horst H, Smette A, Honig S, Comastri A. 2009. *Astron. Astrophys.* 502:457–72
Gaspari M, Ruszkowski M, Oh SP. 2013. *MNRAS* 432:3401–22
Gebhardt K, Bender R, Bower G, et al. 2000. *Ap. J. Lett.* 539:L13–16
Gelbord JM. 2003. *Probing circumnuclear material in Seyfert galaxies with X-ray spectroscopy*. PhD Thesis, Johns Hopkins Univ.
Gendre MA, Best PN, Wall JV. 2010. *MNRAS* 404:1719–32
Gendre MA, Best PN, Wall JV, Ker LM. 2013. *MNRAS* 430:3086–101
Genzel R, Förster Schreiber NM, Lang P, et al. 2014. *Ap. J.* 785:75
Giacconi R, Gursky H, Paolini F. 1962. *Phys. Rev. Lett.* 9:439–43
Gisler GR. 1978. *MNRAS* 183:633–43
González-Alfonso E, Fischer J, Graciá-Carpio J, et al. 2014. *Astron. Astrophys.* 561:A27
González Delgado RM, Cid Fernandes R, Pérez E, et al. 2004. *Ap. J.* 606:127–43
González Delgado RM, Pérez E, Cid Fernandes R, Schmitt H. 2008. *Astron. J.* 135:747–65
Goto T. 2006. *MNRAS* 369:1765–72
Goulding AD, Alexander DM. 2009. *MNRAS* 398:1165–93
Graham AW, Driver SP, Allen PD, Liske J. 2007. *MNRAS* 387:198–210

Graham AW, Scott N. 2013. *Ap. J.* 764:151
Greene JE, Ho LC. 2007a. *Ap. J.* 670:92–104
Greene JE, Ho LC. 2007b. *Ap. J.* 667:131–48
Greene JE, Zakamska NL, Liu X, et al. 2009. *Ap. J.* 702:441–59
Greene JE, Zakamska N, Ho LC, Barth AJ. 2011. *Ap. J.* 732:9
Grier CJ, Martini P, Watson LC, et al. 2013. *Ap. J.* 773:90
Grimes JP, Heckman T, Stickland D, Ptak A. 2005. *Ap. J.* 628:187–204
Gürkan G, Hardcastle MJ, Jarvis MJ. 2014. *MNRAS* 438:1149–61
Haines CP, Pereira MJ, Sanderson AJR, et al. 2012. *Ap. J.* 754:97
Hainline KN, Hickox R, Greene JE, Myers AD, Zakamska NL. 2013. *Ap. J.* 774:145
Hao L, Strauss MA, Fan X, et al. 2005a. *Astron. J.* 129:1795–808
Hao L, Strauss MA, Tremonti CA, et al. 2005b. *Astron. J.* 129:1783–94
Hardcastle MJ, Evans DA, Croston JH. 2007. *MNRAS* 376:1849–56
Haring N, Rix HW. 2004. *Ap. J. Lett.* 604:L89–92
Harrison CM, Alexander DM, Mullaney JR, et al. 2012a. *Ap. J. Lett.* 760:L15
Harrison CM, Alexander DM, Swinbank AM, et al. 2012b. *MNRAS* 426:1073–96
Hasinger G, Miyaji T, Schmidt M. 2005. *Astron. Astrophys.* 441:417–34
Heckman TM. 1980. *Astron. Astrophys.* 87:152–64
Heckman TM, Armus, L, Miley GK. 1990. *Ap. J. Suppl.* 74:833–68
Heckman TM, Balick B, Sullivan WT III. 1978. *Ap. J.* 224:745–60
Heckman TM, Blitz L, Wilson AS, Armus L, Miley GK. 1989. *Ap. J.* 342:735–58
Heckman TM, Kauffmann G, Brinchmann J, et al. 2004. *Ap. J.* 613:109–18
Heckman TM, Ptak A, Hornschemeier A, Kauffmann G. 2005. *Ap. J.* 634:161–68
Heywood I, Blundell KM, Rawlings S. 2007. *MNRAS* 381:1093–102
Hickox RC, Jones C, Forman W, et al. 2009. *Ap. J.* 696:891–919
Hicks EKS, Davies RI, Maciejewski W, et al. 2013. *Ap. J.* 768:107
Hillel S, Soker N. 2013. *MNRAS* 430:1970–75
Hine RG, Longair MS. 1979. *MNRAS* 188:111–30
Hlavacek-Larrondo J, Fabian AC. 2011. *MNRAS* 413:313–21
Ho LC. 2005. *Ap. Space Sci.* 300:219–25
Ho LC. 2008. *Annu. Rev. Astron. Astrophys.* 46:475–539
Ho LC, Darling J, Greene JE. 2008. *Ap. J.* 681:128–40
Hopkins AM, Beacom JF. 2006. *Ap. J.* 651:142–54
Hopkins PF, Hernquist L. 2009. *Ap. J.* 694:599–609
Hopkins PF, Hernquist L, Cox TJ, et al. 2006. *Ap. J. Suppl.* 163:1–49
Hopkins PF, Hernquist L, Cox TJ, Keres D. 2008. *Ap. J. Suppl.* 175:356–89
Hopkins PF, Hernquist L, Cox TJ, Robertson B, Krause E. 2007a. *Ap. J.* 669:45–66
Hopkins PF, Keres D, Murray N, Quataert E, Hernquist L. 2012. *MNRAS* 427:968–78
Hopkins PF, Kocevski DD, Bundy K. 2013. *Ap. J.* In press. arXiv:1309.6321
Hopkins PF, Quataert E. 2011. *MNRAS* 415:1027–50
Hopkins PF, Richards GT, Hernquist L. 2007b. *Ap. J.* 654:731–53
Hu EM, Cowie LL, Wang Z. 1985. *Ap. J. Suppl.* 59:447–98
Hubble EP. 1925. *Popul. Astron.* 33:252–55
Hwang HS, Park C, Elbaz D, Choi Y-Y. 2012. *Astron. Astrophys.* 538:A15
Ibar E, Cirasuolo M, Ivison RJ, et al. 2008. *MNRAS* 386:953–62
Ishibashi W, Fabian AC. 2012. *MNRAS* 427:2998–3005
Ivezić Ž, Menou K, Knapp GR, et al. 2002. *Astron. J.* 124:2364–400
Janssen RMJ, Röttgering HJA, Best PN, Brinchmann J. 2012. *Astron. Astrophys.* 541:A62
Jiang YF, Greene JE, Ho LC, Xiao T, Barth AJ. 2011. *Ap. J.* 742:68
Juneau S, Dickinson M, Bournaud F, et al. 2013. *Ap. J.* 764:176
Karim A, Schinnerer E, Martínez-Sansigre A, et al. 2011. *Ap. J.* 730:61
Kaspi S, Smith PS, Netzer H, et al. 2000. *Ap. J.* 533:631–49
Kauffmann G, Haehnelt M. 2000. *MNRAS* 311:576–88

Kauffmann G, Heckman TM. 2009. *MNRAS* 397:135–47
Kauffmann G, Heckman TM, Best PN. 2008. *MNRAS* 384:953–71
Kauffmann G, Heckman TM, Budavari T, et al. 2007. *Ap. J. Suppl.* 173:357–76
Kauffmann G, Heckman TM, Tremonti C, et al. 2003a. *MNRAS* 346:1055–77
Kauffmann G, Heckman TM, White SDM, et al. 2003b. *MNRAS* 341:54–69
Kauffmann G, White SDM, Heckman TM, et al. 2004. *MNRAS* 353:713
Kelly BC, Merloni A. 2012. *Adv. Astron.* 2012:7–34
Kelly BC, Shen Y. 2013. *Ap. J.* 764:45
Kerr R. 1963. *Phys. Rev. Lett.* 11:237–38
Kewley LJ, Dopita MA, Sutherland RS, Heisler CA, Trevena J. 2001. *Ap. J.* 556:121–40
Kewley LJ, Groves B, Kauffmann G, Heckman T. 2006. *MNRAS* 372:961–76
Kocevski DD, Faber SM, Mozena M, et al. 2012. *Ap. J.* 744:148
Kollmeier JA, Onken CA, Kochanek CS, et al. 2006. *Ap. J.* 648:128–39
Korista KT, Bautista MA, Arav N, et al. 2008. *Ap. J.* 688:108–15
Kormendy J, Ho LC. 2013. *Annu. Rev. Astron. Astrophys.* 51:511–653
Kormendy J, Kennicutt RC. 2004. *Annu. Rev. Astron. Astrophys.* 42:603–83
Koss M, Mushotzky R, Treister E, et al. 2012. *Ap. J. Lett.* 746:L22
Koss M, Mushotzky R, Veilluex S, et al. 2011. *Ap. J.* 739:57
Koss M, Mushotzky R, Veilleux S, Winter L. 2010. *Ap. J.* 716:L125–30
Krolik J. 1999. *Active Galactic Nuclei*. Princeton, NJ: Princeton Univ. Press
Krug HB, Rupke DSN, Veilleux S. 2010. *Ap. J.* 708:1145–61
Lackner CN, Gunn JE. 2012. *MNRAS* 421:2277–302
Lacy M, Storrie-Lombardi LJ, Sajina A, et al. 2004. *Ap. J. Suppl.* 154:166–69
Lagos CDP, Lacey CG, Baugh CM, Bower R, Benson A. 2011. *MNRAS* 416:1566–84
Laing RA, Jenkins CR, Wall JV, Unger SW. 1994. In *Spectrophotometry of a Complete Sample of 3CR Radio Sources: Implications for Unified Models*, ed. GV Bicknell, MA Dopita, PJ Quinn. *ASP Conf. Ser.* 54:201–8. San Francisco: ASP
LaMassa S, Heckman TM, Ptak A, Urry CM. 2013. *Ap. J. Lett.* 765:L33
LaMassa S, Heckman TM, Ptak A, et al. 2010. *Ap. J.* 720:786–810
LaMassa S, Heckman TM, Ptak A, et al. 2012. *Ap. J.* 758:1
Larson RB. 2010. *Nat. Phys.* 6:96–98
Larson RB, Tinsley BM. 1978. *Ap. J.* 219:46–59
Lauer T, Faber SM, Gebhardt K, et al. 2005. *Astron. J.* 129:2138–85
Lee G-H, Woo J-H, Lee MG, et al. 2012. *Ap. J.* 750:141
Leitherer C, Schaerer D, Goldader J, et al. 1999. *Ap. J. Suppl.* 123:3–40
Li C, Kauffmann G, Heckman TM, White SDM, Jing YP. 2008a. *MNRAS* 385:1903–14
Li C, Kauffmann G, Heckman TM, White SDM, Jing YP. 2008b. *MNRAS* 385:1915–22
Li C, Kauffmann G, Wang L, et al. 2006. *MNRAS* 373:457–68
Lilly SJ, Carollo CM, Pipino A, Renzini A, Peng Y. 2013. *Ap. J.* 772:119
Lin Y-T, Shen Y, Strauss MA, Richards GT, Lunnan R. 2010. *Ap. J.* 723:1119–38
Lintott C, Schawinski K, Bamford S, et al. 2011. *MNRAS* 410:166–78
Liu G, Zakamska NL, Greene JE, Nesvadba NPH, Liu X. 2013a. *MNRAS* 430:2327–45
Liu G, Zakamska NL, Greene JE, Nesvadba NPH, Liu X. 2013b. *MNRAS* 436:2576–97
Liu X, Shen Y, Strauss M. 2012. *Ap. J.* 745:94
Liu X, Zakamska NL, Greene JE, et al. 2009. *Ap. J.* 702:1098–117
Longair MS. 1966. *MNRAS* 133:421–36
Longair MS, Seldner M. 1979. *MNRAS* 189:433–53
Lupton RH, Ivezić Z, Gunn JE, et al. 2002. In *Survey and Other Telescope Technologies and Discoveries*, ed. JA Tyson, S Wolff. *Proc. SPIE Conf. Ser.* 4836:350–56. Bellingham, WA: SPIE
Maccarone TJ. 2003. *Astron. Astrophys.* 409:697–706
Machalski J, Condon JJ. 1999. *Ap. J. Suppl.* 123:41–78
Machalski J, Godlowski W. 2000. *Astron. Astrophys.* 360:463–71
Madau P, Dickinson M. 2014. *Annu. Rev. Astron. Astrophys.* 52:415–86

Magliocchetti M, Brüggen M. 2007. *MNRAS* 379:260–74
Magliocchetti M, Maddox SJ, Lahav O, Wall JV. 1998. *MNRAS* 300:257–68
Mahadevan R. 1997. *Ap. J.* 477:585–601
Mainieri V, Bongiorno, Merloni A, et al. 2011. *Astron. Astrophys.* 535:A80
Maiolino R, Gallerani S, Neri R, et al. 2012. *MNRAS* 425:L66–70
Maiolino R, Ruiz M, Rieke GH, Papadopoulos P. 1997. *Ap. J.* 485:552–69
Malkan M, Gorjian V, Tam R. 1998. *Ap. J. Suppl.* 117:25–88
Mandelbaum R, Li C, Kauffmann G, White SDM. 2009. *MNRAS* 393:377–92
Mannering EJA, Worrall DM, Birkinshaw M. 2011. *MNRAS* 416:2869–81
Marconi A, Hunt LK. 2003. *Ap. J. Lett.* 589:L21–24
Marconi A, Risaliti G, Gilli R, et al. 2004. *MNRAS* 351:169–85
Martin DC, Fanson J, Schiminovich D, et al. 2005. *Ap. J. Lett.* 619:L1–6
Martínez-Sansigre A, Rawlings S. 2011. *MNRAS* 414:1937–64
Martini P, Dicken D, Storchi-Bergmann T. 2013. *Ap. J.* 766:121
Martini P, Kelson DD, Kim E, Mulchaey JS, Athey AA. 2006. *Ap. J.* 644:116–32
Mauch T, Sadler EM. 2007. *MNRAS* 375:931–50
McConnell NJ, Ma CP. 2013. *Ap. J.* 764:184
McLure RJ, Dunlop JS. 2002. *MNRAS* 331:795–804
McMahon RG, White RL, Helfand DJ, Becker RH. 2002. *Ap. J. Suppl.* 143:1–23
McNamara BR, Nulsen PEJ. 2007. *Annu. Rev. Astron. Astrophys.* 45:117–75
McNamara BR, Rohanizadegan M, Nulsen PEJ. 2011. *Ap. J.* 727:39
Meier DL. 1999. *Ap. J.* 522:753–66
Meier DL. 2001. *Ap. J. Lett.* 548:L9–12
Merloni A, Heinz S. 2007. *MNRAS* 381:589–601
Merloni A, Heinz S. 2008. *MNRAS* 388:1011–30
Merloni A, Heinz S, Di Matteo T. 2003. *MNRAS* 345:1057–76
Miley G. 1980. *Annu. Rev. Astron. Astrophys.* 18:165–218
Miley GK, Neugebauer G, Soifer BT. 1985. *Ap. J. Lett.* 293:L11–14
Miller CJ, Nichol RC, Gómez PL, Hopkins AM, Bernardi M. 2003. *Ap. J.* 597:142–56
Miller JM, Raymond J, Reynolds CS, et al. 2008. *Ap. J.* 680:1359–77
Mirabel IF, Wilson AS. 1984. *Ap. J.* 277:92–105
Moe M, Arav N, Bautista MA, Korista KT. 2009. *Ap. J.* 706:525–34
Mullaney JR, Daddi E, Béthermin M, et al. 2012a. *Ap. J. Lett.* 753:L30
Mullaney JR, Pannella M, Daddi E, et al. 2012b. *MNRAS* 419:95–115
Murphy KD, Yaqoob T. 2009. *Ap. J.* 701:635–41
Murray N, Menard B, Thompson TA. 2011. *Ap. J.* 735:66
Mutch SJ, Croton DJ, Poole GB. 2013. *MNRAS* 435:2445–59
Nagar NM, Falcke H, Wilson AS. 2005. *Ap. J.* 435:521–43
Narayan R. 2002. In *Lighthouses of the Universe*, ed. M Gilfanov, R Sunyaev, E Churazov, pp. 405–29. Berlin/Heidelberg: Springer-Verlag
Narayan R. 2005. *Astrophys. Space Sci.* 300:177–88
Narayan R, Yi I. 1994. *Ap. J. Lett.* 428:L13–16
Narayan R, Yi I. 1995. *Ap. J.* 444:710–35
Neilsen J, Lee JC. 2009. *Nature* 458:481–84
Nemmen RS, Bower RG, Babul A, Storchi-Bergmann T. 2007. *MNRAS* 377:1652–62
Nesvadba NPH, Lehnert MD, De Breuck C, Gilbert AM, van Breugel W. 2008. *Astron. Astrophys.* 491:407–24
Nesvadba NPH, Lehnert MD, Eisenhauer F, et al. 2006. *Ap. J.* 650:661–68
Netzer H. 2009. *MNRAS* 399:1907–20
Netzer H. 2013. *The Physics and Evolution of Active Galactic Nuclei*. Cambridge, UK: Cambridge Univ. Press
Netzer H, Mainieri V, Rosati P, Trakhtenbrot B. 2006. *Astron. Astrophys.* 453:525–33
Norman CA, Scoville NZ. 1988. *Ap. J.* 332:124–34
Nulsen PEJ, Jones C, Forman WR, et al. 2007. In *ESO Astrophys. Symp.*, ed. H Böhringer, GW Pratt, A Finoguenov, P Schuecker, pp. 210–15. Berlin/Heidelberg: Springer-Verlag

Nusser A, Silk J, Babul A. 2006. *MNRAS* 373:739–46
O'Dea CP, Baum SA, Privon G, et al. 2008. *Ap. J.* 681:1035–45
O'Dea CP, Daly RA, Kharb P, Freeman KA, Baum SA. 2009. *Astron. Astrophys.* 494:471–88
Osterbrock DE, Ferland G. 2005. *Astrophysics of Gaseous Nebulae and Active Galactic Nuclei.* Mill Valley, CA: Univ. Sci. Books
Padovani P, Miller N, Kellermann KI, et al. 2011. *Ap. J.* 740:20
Panter B, Heavens AF, Jimenez R. 2003. *MNRAS* 343:1145–54
Panter B, Jimenez R, Heavens AF, Charlot S. 2007. *MNRAS* 378:1550–64
Pasquali A, Kauffmann G, Heckman TM. 2005. *MNRAS* 361:1121–30
Peterson BM. 1997. *An Introduction to Active Galactic Nuclei.* Cambridge, UK: Cambridge Univ. Press
Peterson BM. 2013. *Space Sci. Rev.* In press. doi: 10.1007/s11214-013-9987-4
Peterson JR, Bleeker JAM, Ferrigno C, et al. 2001. *Astron. Astrophys.* 365:L104–9
Peterson JR, Kahn SM, Paerels FBS, et al. 2003. *Ap. J.* 590:207–24
Pettini M, Pagel BEJ. 2004. *MNRAS* 348:L59–63
Pierce CM, Lotz JM, Laird ES, et al. 2007. *Ap. J. Lett.* 660:L19–22
Pizzolato F, Soker N. 2010. *MNRAS* 408:961–74
Ponti G, Fender RP, Begelman MC, et al. 2012. *MNRAS* 422:L11–15
Porciani C, Magliocchetti M, Norberg P. 2004. *MNRAS* 355:1010–30
Punsly B, Coroniti FV. 1990. *Ap. J.* 350:518–35
Qiao E, Liu BF. 2009. *Publ. Astron. Soc. Jpn.* 61:403–10
Quataert E. 2001. In *Probing the Physics of Active Galactic Nuclei by Multiwavelength Monitoring*, ed. BM Peterson, RW Pogge, RS Polidan. *ASP Conf. Ser.* 224:71–85. San Francisco: ASP
Rafferty DA, McNamara BR, Nulsen PEJ, Wise MW. 2006. *Ap. J.* 652:216–31
Rawlings S, Saunders R. 1991. *Nature* 349:138–40
Reichard TA, Heckman TM, Rudnick G, et al. 2009. *Ap. J.* 691:1005–20
Remillard RA, McClintock JE. 2006. *Annu. Rev. Astron. Astrophys.* 44:49–92
Rengelink RB, Tang Y, de Bruyn AG, et al. 1997. *Astron. Astrophys. Suppl.* 124:259–80
Reviglio P, Hefland DJ. 2006. *Ap. J.* 650:717–26
Richards G, Fan X, Newberg H, et al. 2002. *Astron. J.* 123:2945–75
Richards G, Lacy M, Storrie-Lombardi LJ, et al. 2006. *Ap. J. Suppl.* 166:470–97
Rigby EE, Best PN, Brookes MH, et al. 2011. *MNRAS* 416:1900–15
Rigby JR, Diamond-Stanic AM, Aniano G. 2009. *Ap. J.* 700:1878–83
Rosario DJ, Santini P, Lutz D, et al. 2012. *Astron. Astrophys.* 545:A45
Rosario DJ, Shields GS, Taylor GB, Salviander S, Smith KL. 2010a. *Ap. J.* 716:131–43
Rosario DJ, Trakhtenbrot B, Lutz D, et al. 2013. *Astron. Astrophys.* 560:A72
Rosario DJ, Whittle M, Nelson CH, Wilson AS. 2010b. *Ap. J. Lett.* 711:L94–98
Rosenfield P, Johnson LC, Girardi L, et al. 2012. *Ap. J.* 755:131
Ross NP, Shen Y, Strauss MA, et al. 2009. *Ap. J.* 697:1634–55
Roy AL, Norris RP, Kesteven MJ, Troup ER, Reynolds JE. 1998. *MNRAS* 301:1019–30
Rupke M, Veilleux S. 2011. *Ap. J. Lett.* 729:L27
Russell DM, Gallo E, Fender RP. 2013a. *MNRAS* 431:405–14
Russell HR, McNamara BR, Edge AC, et al. 2013b. *MNRAS* 430:530–53
Sabater J, Best PN, Argudo-Fernández M. 2013. *MNRAS* 430:638–51
Sadler EM, Cannon RD, Mauch T, et al. 2007. *MNRAS* 381:211–27
Sadler EM, Ekers RD, Mahony E, Mauch T, Murphy T. 2014. *MNRAS* 438:790–824
Sadler EM, Jackson CA, Canoon RD, et al. 2002. *MNRAS* 329:227–45
Sadler EM, Jenkins CR, Kotanyi CG. 1989. *MNRAS* 240:591–635
Sadler EM, McIntyre VJ, Jackson CA, Cannon RD. 1999. *Publ. Astron. Soc. Aust.* 16:247–56
Saintonge A, Kauffmann G, Cramer C, et al. 2011. *MNRAS* 415:32–60
Saintonge A, Tacconi LJ, Fabello S, et al. 2012. *Ap. J.* 758:73
Salim S, Charlot S, Rich RM, et al. 2005. *Ap. J. Lett.* 619:L39–42
Sandage A. 1965. *Ap. J.* 141:1560–80
Sanders DB, Soifer BT, Elias JH, Neugebauer G, Matthews K. 1988a. *Ap. J. Lett.* 328:L35–39

Sanders DB, Soifer BT, Elias JH, et al. 1988b. *Ap. J.* 325:74–91
Santini P, Rosario DJ, Shao L, et al. 2012. *Astron. Astrophys.* 540:A109
Sargent MT, Schinnerer E, Murphy E, et al. 2010. *Ap. J. Suppl.* 186:341–77
Sarzi M, Shields JC, Schawinski K, et al. 2010. *MNRAS* 402:2187–210
Satyapal S, Böker T, McAlpine W, et al. 2009. *Ap. J.* 704:439–52
Satyapal S, Vega D, Dudik RP, Abel NP, Heckman T. 2008. *Ap. J.* 677:926–42
Schawinski K, Simmons BD, Urry CM, Treister E, Glikman E. 2012. *MNRAS* 425:L61–65
Schawinski K, Thomas D, Sarzi M, et al. 2007. *MNRAS* 382:1415–31
Schawinski K, Treister E, Urry CM, et al. 2011. *Ap. J. Lett.* 727:L31
Schawinski K, Urry CM, Virani S, et al. 2010. *Ap. J.* 711:284–302
Scheuer P. 1974. *MNRAS* 166:513–28
Schiminovich D, Wyder T, Martin DC, et al. 2007. *Ap. J. Suppl.* 173:315–41
Schmidt M. 1963. *Nature* 197:1040
Seljak U. 2000. *MNRAS* 318:203–13
Sellwood JA. 2014. *Rev. Mod. Phys.* 86:1–46
Seyfert CK. 1943. *Ap. J.* 97:28–40
Shabala SS, Godfrey LEH. 2013. *Ap. J.* 769:129
Shankar F, Marulli F, Mathur S, Bernardi M, Bournaud F. 2012. *Astron. Astrophys.* 540:A23
Shankar F, Weinberg D, Miralda-Escudé J. 2009. *Ap. J.* 690:20–41
Shankar F, Weinberg D, Miralda-Escudé J. 2013. *MNRAS* 428:421–46
Shao L, Kauffmann G, Cheng L, Wang J, Heckman TM. 2013. *MNRAS* 436:3451–63
Shapley A. 2011. *Annu. Rev. Astron. Astrophys.* 49:525–80
Shen S, Kauffmann G, von der Linden A, White SDM, Best PN. 2008. *MNRAS* 389:1074–86
Shen S, White SDM, Mo HJ, et al. 2006. *MNRAS* 369:1639–53
Shen Y, Strauss MA, Ross NP, et al. 2009. *Ap. J.* 697:1656–73
Sheth RK, Mo HJ, Tormen G. 2001. *MNRAS* 323:1–12
Shimasaku K, Fukugita M, Doi M, et al. 2001. *Astron. J.* 122:1238–50
Shlosman I, Begelman MC, Frank J. 1990. *Nature* 345:679–86
Silk J. 2013. *Ap. J.* 772:112
Silverman JD, Kampczyk P, Jahnke K, et al. 2011. *Ap. J.* 743:2
Silverman JD, Lamareille F, Maier C, et al. 2009. *Ap. J.* 696:396–410
Simmons BD, Lintott C, Schawinski K, et al. 2013. *MNRAS* 429:2199–211
Simoes Lopes R, Ramiro D, Storchi-Bergmann T, de Fatima Saraiva M, Martini P. 2007. *Ap. J.* 655:718–34
Simpson C, Rawlings S, Ivison RJ, et al. 2012. *MNRAS* 421:3060–83
Simpson C, Westoby P, Arumugam V, et al. 2013. *MNRAS* 433:2647–56
Soker N. 2006. *New Astron.* 12:38–46
Soltan A. 1982. *MNRAS* 200:115–22
Spinoglio L, Malkan MA. 1989. *Ap. J.* 342:83–99
Spoon HWW, Farrah D, Lebouteiller V, et al. 2013. *Ap. J.* 775:127
Springel V, Di Matteo T, Hernquist L. 2005. *MNRAS* 361:776–94
Stasinska G, Cid Fernandes R, Mateus A, Sodre L, Asari NV. 2006. *MNRAS* 371:972–82
Stasinska G, Vale Asari N, Cid Fernandes R, et al. 2008. *MNRAS* 391:L29–33
Steidel CC, Erb DK, Shapley AE, et al. 2010. *Ap. J.* 717:289–322
Steiner JF, McClintock JE, Narayan R. 2013. *Ap. J.* 762:104
Stern D, Assef RJ, Benford DJ, et al. 2012. *Ap. J.* 753:30
Stern D, Eisenhardt P, Gorjian V, et al. 2005. *Ap. J.* 631:163–68
Stern J, Laor A. 2012a. *MNRAS* 426:2703–18
Stern J, Laor A. 2012b. *MNRAS* 423:600–31
Strauss MA, Weinberg DH, Lupton RH, et al. 2002. *Astron. J.* 124:1810–24
Sturm E, González-Alfonson E, Veilleux S, et al. 2011. *Ap. J. Lett.* 733:L16
Tamura T, Bleeker JAM, Peterson JR, et al. 2001. *Astron. Astrophys.* 365:L87–92
Tasse C, Best PN, Röttgering HJA, Le Borgne D. 2008. *Astron. Astrophys.* 490:893–904
Terlevich R, Melnick J. 1985. *MNRAS* 213:841–56

Terlevich R, Tenorio-Tagle G, Franco J, Melnick J. 1992. *MNRAS* 255:713–28
Toffolatti L, Franceschini A, Danese L, de Zotti G. 1987. *Astron. Astrophys.* 184:7–15
Treister E, Schawinski K, Urry CM, Simmons BD. 2012. *Ap. J. Lett.* 758:L39
Tremaine S, Gebhardt K, Bender R, et al. 2002. *Ap. J.* 574:740–53
Trump JR, Impey CD, Kelly BC, et al. 2009. *Ap. J.* 700:49–55
Tundo E, Bernardi M, Hyde JB, Sheth RK, Pizella A. 2007. *Ap. J.* 663:53–60
Ueda Y, Hiroi K, Isobe N, et al. 2011. *Publ. Astron. Soc. Jpn.* 63:937–45
Urry CM, Padovani P. 1995. *Publ. Astron. Soc. Pac.* 107:803–45
Veilleux S, Cecil G, Bland-Hawthorn J. 2005. *Annu. Rev. Astron. Astrophys.* 43:769–826
Veilleux S, Meléndez M, Sturm E, et al. 2013. *Ap. J.* 776:27
Veilleux S, Osterbrock DE. 1987. *Ap. J. Suppl.* 63:295–310
Veilleux S, Rupke DSN, Kim D-C, et al. 2009. *Ap. J. Suppl.* 182:628–66
Villforth C, Hamann F, Rosario DJ, et al. 2014. *MNRAS* 439:3342–56
Vitale M, Zuther J, García-Marín M, et al. 2012. *Astron. Astrophys.* 546:A17
Voges W, Aschenbach B, Boller T, et al. 1999. *Astron. Astrophys.* 349:389–405
von der Linden A, Best PN, Kauffmann G, White SDM. 2007. *MNRAS* 379:867–93
Whitaker K, van Dokkum PG, Brammer G, Franx M. 2012. *Ap. J. Lett.* 754:L29
Wild V, Heckman T, Charlot C. 2010. *MNRAS* 405:933–47
Wild V, Kauffmann G, Heckman T, et al. 2007. *MNRAS* 381:543–72
Willott CJ, Rawlings S, Blundell KM, Lacy M. 1999. *MNRAS* 309:1017–33
Wilson AS, Colbert EJM. 1995. *Ap. J.* 438:62–71
Wilson AS, Heckman TM. 1985. In *Astrophysics of Active Galaxies and Quasi-Stellar Objects*, ed. JS Miller, pp. 39–110
Winter L, Lewis KT, Koss M, et al. 2010. *Ap. J.* 710:503–39
Woo J-H, Schulze A, Park D, et al. 2013. *Ap. J.* 772:49
Worthey G, Ottaviani DL. 1997. *Ap. J. Suppl.* 111:377–86
Wright EL, Eisenhardt PPM, Mainser AK, et al. 2010. *Astron. J.* 140:1868–81
Wuyts S, Förster Schreiber NM, Lutz D, et al. 2011a. *Ap. J.* 738:106
Wuyts S, Förster Schreiber NM, van der Wel A, et al. 2011b. *Ap. J.* 742:96
Xu C, Livio M, Baum S. 1999. *Astron. J.* 118:1169–76
Yan R, Blanton MR. 2012. *Ap. J.* 747:61
Yan R, Newman JA, Faber SM, et al. 2006. *Ap. J.* 648:281–98
Yee HKC, Green RF. 1984. *Ap. J.* 280:79–90
Yip CW, Connolly AJ, Szalay AS, et al. 2004. *Astron. J.* 128:585–609
Yu Q, Tremaine S. 2002. *MNRAS* 335:965–76
Yuan F, Narayan R. 2004. *Ap. J.* 612:724–28
Yuan F, Narayan R. 2014. *Annu. Rev. Astron. Astrophys.* 52:529–88
Zakamska N, Strauss MA, Heckman T, Ivezić Z, Krolik J. 2004. *Astron. J.* 128:1002–16
Zinn PC, Middelberg E, Norris RP, Dettmar RJ. 2013. *Ap. J.* 774:66

Numerical Relativity and Astrophysics

Luis Lehner[1] and Frans Pretorius[2]

[1]Perimeter Institute for Theoretical Physics, Waterloo, Ontario N2L 2Y5, Canada; email: llehner@perimeterinstitute.ca

[2]Department of Physics, Princeton University, Princeton, New Jersey 08544; email: fpretori@princeton.edu

Annu. Rev. Astron. Astrophys. 2014. 52:661–94

First published online as a Review in Advance on June 16, 2014

The *Annual Review of Astronomy and Astrophysics* is online at astro.annualreviews.org

This article's doi:
10.1146/annurev-astro-081913-040031

Keywords

black holes, neutron stars, gravitational waves, gamma-ray bursts

Abstract

Throughout the Universe many powerful events are driven by strong gravitational effects that require general relativity to fully describe them. These include compact binary mergers, black hole accretion, and stellar collapse, where velocities can approach the speed of light and extreme gravitational fields ($\Phi_{\rm Newt}/c^2 \simeq 1$) mediate the interactions. Many of these processes trigger emission across a broad range of the electromagnetic spectrum. Compact binaries further source strong gravitational wave emission that could directly be detected in the near future. This feat will open up a gravitational wave window into our Universe and revolutionize our understanding of it. Describing these phenomena requires general relativity, and—where dynamical effects strongly modify gravitational fields—the full Einstein equations coupled to matter sources. Numerical relativity is a field within general relativity concerned with studying such scenarios that cannot be accurately modeled via perturbative or analytical calculations. In this review, we examine results obtained within this discipline, with a focus on its impact in astrophysics.

1. INTRODUCTION

Strong gravitational interactions govern many of the most fascinating astrophysical phenomena and lie behind some of the most spectacular predictions of general relativity, such as black holes and neutron stars. These objects produce extreme gravitational fields and are believed to be responsible for the most energetic events in our Universe. Indeed, models for gamma-ray bursts, quasars, AGNs, pulsars, and a class of ultra-high-energy cosmic rays all peg these still poorly understood compact objects as putative central engines. Observations across the electromagnetic spectra, soon to be combined with gravitational signals produced by merging binaries, should provide important insights into their natures. Of course, such understanding can only be gained by contrasting theoretical models that include all the relevant physics with the full front of observations.

It is important to distinguish two subclasses of strongly gravitating systems. The first is where the self-gravitation of any matter/gas/plasma interacting with a compact object or binary is sufficiently weak such that the gravitational back reaction can be ignored or treated perturbatively. Such systems can be analyzed by studying the dynamics of matter on a given fixed background geometry. Examples include accreting black holes and tidal disruption of main sequence stars by a supermassive black hole. Widely separated compact binary systems also belong to this subclass, and suitable post-Newtonian (PN) expansions can be adopted to account for the slowly varying gravitational field and its effects.

By contrast, the second subclass is where the interaction is strong and can fundamentally affect the gravitational field of the system. In this case, a fully relativistic, self-gravitating study must be performed. To this end the Einstein equations, coupled to any relevant matter fields, must be employed. This task is complex due to the involved nature of Einstein's equations (a nonlinear, strongly coupled system of equations) in which analytical solutions are only known in highly specialized scenarios. Consequently, numerical simulations are required, and the discipline that concentrates on the development and application of numerical solutions of Einstein's equations is known as numerical relativity (NR).

This discipline has, over several decades, steadily progressed to the current epoch in which studies of relevance to astrophysics can now be performed that address questions of fundamental theoretical interest and that make contact with observations. [NR is also being used to address problems in cosmological contexts. Applications that require NR, including bubble collisions (Johnson, Peiris & Lehner 2012; Wainwright et al. 2014), the evolution near the bounce in cyclic models (Garfinkle et al. 2008, Xue et al. 2013), and certain aspects of cosmic string dynamics (Laguna & Garfinkle 1989), are still at either a speculative level or being explored. In contrast, the paradigm applicable to most of present day observational cosmology can effectively be addressed with exact Friedman-Lamaitre-Robison-Walker solutions and perturbations about them, and do not require NR.] Of particular interest, spurred by a hope of imminent gravitational wave observation, are systems capable of producing strong gravitational emission. Detectors include ground-based interferometers, such as LIGO (Laser Interferometer Gravitational-Wave Observatory), VIRGO, and KAGRA (Kamioka Gravitational wave detector) (Abbott et al. 2009, Accadia et al. 2011, Somiya 2012) targeting the $\simeq$10Hz–1 kHz frequency band; a pulsar timing network (see, e.g., Int. Pulsar Timing Array 2013) sensitive to the 300 pHz–100 nHz window; and possible future spaced-based missions [such as NGO (*New Gravitational-Wave Observatory*) or eLISA (*Evolved Laser Interferometer Space Antenna*); see, e.g., Amaro-Seoane et al. 2012] sensitive to $\simeq$10 μHz–0.1 Hz. Compact binary systems, involving black holes or neutron stars, are the most natural sources and have thus been the focus of most recent efforts (see, e.g., Andersson et al. 2013 for a recent overview). In this article, we review the key messages obtained by NR relevant to astrophysics. The discipline is still in the midst of rapid development over an increasing breadth of applications, promising even more exciting future discoveries of astrophysical import.

2. BRIEF REVIEW OF TECHNIQUES, METHODS, AND INFORMATION OBTAINABLE FROM GRAVITATIONAL WAVES

Understanding gravity in highly dynamical/strongly gravitating regimes requires solving Einstein's equations. This provides the metric tensor, g_{ab}, which encodes gravitational effects in geometrical terms. The covariant character of the equations encode the equivalence principle, hence there is no preferred frame of reference to write the particular form of the metric for a given physical geometry. This further implies that the field equations determining g_{ab} do not lend themselves to a well-defined initial value problem unless the spacetime is foliated into a series of surfaces that provide a notion of "time." One can then cast Einstein's equations in a form that provides a recipe to evolve the intrinsic metric of each slice with time in what has been called geometrodynamics. There are several options to carry out this program (see, e.g., the discussion by Lehner 2001), though the most common one is to define these surfaces to be spacelike. This is also most closely related to Newtonian mechanics and, hence, provides useful intuition in astrophysical scenarios; furthermore, with some additional assumptions about the coordinates, the familiar Newtonian potential can easily be extracted from the metric for weakly gravitating systems. Current efforts most commonly employ one of two particular reformulations of Einstein's equations: the generalized harmonic or the BSSN (Barmgarte-Shapiro-Shibata-Nakamura) formulations (for a recent review, see Sarbach & Tiglio 2012). These equations are hyperbolic with characteristics given by the speed of light (regardless of the state of the system, as opposed to the familiar case of hydrodynamics in which perturbations propagate with speeds tied to the state of the fluid). When coupling in matter sources, the equations of relativistic hydrodynamics (or magnetohydrodynamics) on a dynamical, curved geometry must also be considered. The relevant equations can be expressed in a way fully consistent with standard approaches to integrate the Einstein equations (for a review on this topic, see Font 2008).

With the equations defined, they can be discretized for numerical integration. For the systems considered here, a crucial observation is that simulations must be carried out in full generality. This means that time and spatial variations are equally important, and a disparate range of scales need to be resolved (ranging from at least the size of each compact object, through the scale in which gravitational waves are produced, and to the asymptotic region where they are measured). The associated computational cost is quite high, and typical simulations run on hundreds to thousands of processors for hours to weeks, even when efficient resolution of the relevant spatio-temporal scales can be achieved using (for example) adaptive mesh refinement. It is beyond the scope of this review to describe the techniques employed in detail, though we briefly mention them and point to some relevant literature for further details. [See also a few textbooks on the subject written in recent years by Alcubierre (2008); Bona, Bona-Casas & Palenzuela-Luque (2009); Baumgarte & Shapiro (2010)].

1. *Spatial discretization.* As far as the gravitational field itself is concerned, solutions are generally smooth (except at singularities) provided smooth initial data are defined because the equations of motions are linearly degenerate (i.e., do not induce shocks from smooth initial data). High-order finite difference approximations (e.g., Gustafsson, Kreiss & Oliger 1995) or spectral decompositions (e.g., Boyd 1989, Grandclement & Novak 2009) allow for a high degree of accuracy. When matter and therefore the hydrodynamic equations are involved, finite volume methods and high-resolution shock capturing schemes can be used to determine the future evolution of the fluid variables (e.g., LeVeque 1992).
2. *Time integration.* The method of lines can be straightforwardly implemented once spatial derivatives are computed.

3. *Constraint enforcement.* For systems of interest, several constraints are typically involved. Those coming from Einstein's equations themselves are a nonlinear coupled set of partial differential equations. In general scenarios, these constraints are difficult to enforce directly; instead, a strategy of constraint damping is adopted, whereby the equations of motion are modified in a suitable manner via the addition of constraints. The resulting system is thus not different from the initial one when the constraints are satisfied, otherwise the numerical evolution should damp these violations as time progresses. This desirable behavior can be rigorously shown to hold in perturbations off flat spacetime (Brodbeck et al. 1999, Gundlach et al. 2005) and also "experimentally" verified in simulations involving black holes and neutron stars (e.g., Pretorius 2005, 2006; Anderson et al. 2008a; Chawla et al. 2010). This technique—whereby the equations are suitably modified to control constraints—has also been extended to other relevant systems of equations. For instance, when considering magnetohydrodynamics or electrodynamics, to control the no-monopole constraint (Neilsen, Hirschmann & Millward 2006; Palenzuela, Lehner & Yoshida 2010).
4. *Mesh structure, resolution, and adaptivity.* As mentioned, several different physical scales need to be resolved. For efficient implementation, techniques like adaptive mesh refinement and multiple patches are in common use (e.g., Berger & Oliger 1984; Schnetter, Hawley & Hawke 2004; Lehner, Reula & Tiglio 2005; Lehner, Liebling & Reula 2006; Duez et al. 2008; East, Pretorius & Stephens 2012b).
5. *Parallelization.* The equations involved are of hyperbolic type, and they lend themselves naturally to a relatively straightforward parallelization. Several computational infrastructures have been developed for NR purposes, e.g., BAM, Cactus (**http://www.cactuscode.org**) and the Einstein Toolkit (Löffler et al. 2011), HAD (**http://had.liu.edu**), Whisky, and SACRA (Baiotti, Shibata & Yamamoto 2010).

3. BRIEF DESCRIPTION OF THE DYNAMICS OF A BINARY SYSTEM

Here we review salient properties of the early phase of binary evolution in general relativity to set the stage for subsequent discussion of the nonlinear regime uncovered by numerical simulations. For further details, the interested reader can consult Hughes (2009).

An isolated compact binary evolves owing to the emission of gravitational waves, and consequently a bound system will eventually merge. The end state of compact binary mergers (i.e., binary black holes, black hole–neutron star binaries, and all except the least massive binary neutron stars) is a single Kerr black hole (provided cosmic censorship holds, and there are no indications yet that it fails for mergers in four-dimensional, asymptotically flat spacetime). At large separations, in which the local velocity of each object in the binary is small (relative to the speed of light c), a PN expansion (e.g., Blanchet 2002), where objects are taken as point-particles without internal dynamics, suffices to accurately describe the system. As the orbit shrinks, the faithfulness of such an expansion decreases as velocities become $O(c)$. If either compact object is a neutron star tidal effects may be important; these can be modeled within the PN framework, though again the accuracy of the expansion degrades approaching tidal disruption, which can occur near merger for stellar mass binaries. During this late stage of inspiral, a full numerical solution must be employed to obtain an accurate description of the dynamics of the geometry and matter. Once a single black hole forms, very shortly afterward (on the order of a few light-crossing times of the Schwarzschild radius) the spacetime can accurately be modeled by black hole perturbation theory, and to a good approximation the matter can be evolved on a stationary black hole background. In the standard jargon of the field, the three different stages just described are often referred to as the PN inspiral, nonlinear, and ringdown stages.

The nonlinear phase can further be subdivided into a late inspiral, plunge, and early post-merger epoch (see, e.g., Buonanno, Cook & Pretorius 2007; Buonanno, Kidder & Lehner 2008). In the first subphase the binary is still in an orbit; though velocities are high, the orbital frequency quickly sweeps upward, and neutron star tidal dynamics can become relevant (if the companion is a neutron star or black hole with mass $\lesssim 20\,M_{\odot}$). The second subphase refers to a rapid increase in the magnitude of the inward radial velocity leading to merger. The plunge is related to the phenomenon of the innermost stable circular orbit (ISCO) of a black hole and is, thus, most apparent in a high-mass-ratio compact binary. The last subphase begins when either a black hole or a hypermassive neutron star forms and lasts while either object is too "distorted" for a straightforward perturbative approach to be applicable. As mentioned above, a black hole settles into a stationary state very rapidly, and hence from a computational perspective there is little to gain in switching to a perturbative treatment to measure the ringdown waves. By contrast, in certain ranges of parameter space a hypermassive neutron star can last for several seconds before collapsing to a black hole, which for the full coupled Einstein-matter equations would be too expensive to evolve at present [a rough estimate of the cost is $O(1{,}000)$ CPU hours per millisecond at "modest" resolution].

3.1. Properties of Gravitational Wave Emission

During the early inspiral stage, in which velocities are much smaller than c, to leading order the emission of gravitational waves is proportional to the acceleration of the reduced (trace free) quadrupole moment tensor $Q_{ij}(t)$ of the system (this is textbook material, though for a couple of recent review articles see Flanagan & Hughes 2005, Buonanno 2007):

$$h_{ij}^{\mathrm{TT}}(t,\vec{x}) = \frac{2G}{rc^4}\frac{\partial^2 Q_{kl}(t-r)}{\partial t^2}\left[\perp_i^k \perp_j^l - \frac{1}{2}\perp^{kl}\perp_{ij}\right] \tag{1}$$

In the above, h_{ij}^{TT} is the perturbation of the spatial components of the Minkowski metric η_{ij} in the transverse traceless (TT) gauge, written in a Cartesian coordinate system $(t,\vec{x}) = (t, x_i)$. In this gauge there are no space-time or time-time perturbations of $\eta_{\mu\nu}$; i.e., $h_{tt}^{\mathrm{TT}} = h_{tj}^{\mathrm{TT}} = 0$. The center of mass of the source is at the origin, and the above expression assumes the perturbation is measured at a distance $r = |\vec{x}|$ much greater than the characteristic size of the source, here $\sim r_{\mathrm{p}}$, the periapse of the orbit. (We use the periapse here rather than, say, the semimajor axis, because for highly eccentric systems described later the dominant gravitational wave emission only occurs around periapse passage. Thus r_p more conveniently characterizes the relevant scale of gravitational wave emission for all eccentricities.) The projection tensor $\perp_{ij} = \delta_{ij} - \hat{n}_i\hat{n}_j$, with $\hat{n}_i = x_i/r$, i.e., $\hat{n}_i$, is the unit spatial vector from the source to the observer at location x_i. The above expression is (to leading order) valid in an expanding Universe if the distance r is replaced by the luminosity distance D_l and time is dilated by a factor $1+z$, where z is the redshift between the source and the observer.

The projection in Equation 1 encodes the property from general relativity that there are only two linearly independent propagating degrees of freedom, called the cross and plus polarizations. Thus the tensor h_{ij}^{TT} only has two independent nonzero components, which are called h_+ and $h_{\times}$. To illustrate, ignoring back reaction, a binary on a circular Keplerian orbit with orbital frequency $\omega = \sqrt{2GM/r_p^3}$ produces a radiation pattern

$$h_+(t,r,\theta,\phi) = \frac{4G}{rc^4}\mathcal{M}^{5/3}(2\omega)^{2/3}\cos(2\omega t+\phi)\left[\frac{1+\cos^2\theta}{2}\right], \tag{2}$$

$$h_{\times}(t,r,\theta,\phi) = \frac{4G}{rc^4}\mathcal{M}^{5/3}(2\omega)^{2/3}\sin(2\omega t+\phi)\cos(\theta); \tag{3}$$

here the so-called chirp mass $\mathcal{M} = \eta^{3/5}M$, the symmetric mass ratio $\eta = m_1 m_2/M^2$, θ is the angle between the observer's line of sight and the axis normal to the plane of the binary, and ϕ is the (arbitrary) initial azimuthal phase.

The above expressions highlight several properties about gravitational emission from compact objects relevant for detection. First, gravitational wave detectors are directly sensitive to the amplitude of the metric perturbation and not the energy it carries. The former decays as $1/r$, whereas the latter decays as $1/r^2$ (and being proportional to the square of the third time derivative of Q_{ij}); hence an n-fold improvement in the sensitivity of detectors results in an n^3 increase in the observable volume of the Universe. The "advanced" upgrades to the first generation of ground-based interferometric detectors (that will be completed near the end of the decade) are expected to achieve an order-of-magnitude increase in sensitivity over initial LIGO readings, increasing the range over which binary neutron stars could be observed to hundreds of megaparsecs and binary black holes to billions of parsecs (Abadie et al. 2010). Note, however, that these distances assume matched filtering is used to search for signals that would otherwise be buried in detector noise. For this to maximize both detection prospects and parameter extraction requires template waveforms that are phase-accurate to within a fraction of a cycle over the most sensitive band of the detectors [which for Advanced LIGO (adLIGO) ranges from ~10 Hz to ~1 kHz]. Over the past two decades this has been the primary goal of the source modeling community; it is being achieved using high-order perturbative methods for the early inspiral, numerical solution for late inspiral and early merger, and perturbations off a single black hole afterward.

Second, the emission is clearly not isotropic. Only plus-polarized waves are radiated along the equator, and the amplitude is half of that radiated along the pole orthogonal to the orbit. Thus the distance to which a source can be observed strongly depends on its relative orientation to the detector. Importantly, however, this radiation is not strongly beamed, and so even nonideal orientations of the source to the detector can yield detectable signals.

Third, though these expressions only hint at a couple, there are several degeneracies in the signal that could limit accurate extraction of all relevant parameters from a detection. Under radiation reaction the orbit shrinks, and a binary will sweep across a range of frequencies ω, terminating at merger where $\omega_m \approx c^3/GM$. If ω_m is not in band (such as with a binary neutron star merger), at leading order there is essentially complete degeneracy between the chirp mass and the distance to the source. If an electromagnetic counterpart could be observed and a redshift determined, the degeneracy would break. Higher-order effects, in particular if the black holes spin or the masses are unequal, excite higher gravitational wave multipoles that can further lift degeneracies. This demonstrates the need to understand the full details of the gravitational wave emission and, if matter is involved, possible electromagnetic counterparts. And as is discussed more throughout this review, such multimessenger observations could bring us a wealth of information beyond just measuring binary parameters.

3.2. Priors on Binary Parameters

Merger simulations are computationally expensive, taking of order 10^4–10^5 CPU hours for a simulation of the last $O(10)$ orbits of a quasi-circular inspiral of a binary black hole system. This may not sound too extreme, though remember this is just a single point in an eight-dimensional parameter space—mass ratio, six components of the two spin vectors, and eccentricity. The cost goes up with nonvacuum binaries for several reasons. First, in addition to gravity the relevant matter equations (relativistic hydrodynamics at least) need to be solved for. Second, the effective parameter space grows larger. This is in part to characterize unknown physics such as the equation of state (EOS) of matter at nuclear densities and in part because of new initial conditions, for example, a neutron star's magnetic field configuration. Third, computational fluid dynamics algorithms are typically lower order (to be able to deal with shocks and surfaces) than the high-order finite difference or pseudospectral methods used to solve the Einstein equations, hence higher resolution is required for similar accuracy to a comparable vacuum merger.

The preceding discussion highlights that compact object merger simulations are too demanding to perform a naive, uniform sampling of parameter space to guide the construction of gravitational wave template banks. A promising approach to achieve a more optimal sampling uses the reduced basis method (Field et al. 2011), though regardless of the method one can ask, What priors can be placed on the range of parameters from either theoretical or observational considerations?

A typical neutron star likely has a mass within the range of ≈ 1 to $2.5\ \mathrm{M}_\odot$ and a radius (which for a given mass is determined by the EOS) in the range of ≈ 8 to 15 km, and they are thought to have low spins (see, e.g., Lattimer & Prakash 2010). For black holes, an obvious theoretical restriction on the spin magnitude is that $|a| \leq 1$. Observations of candidate black holes, assuming general relativity is correct and black holes satisfy the bound, are beginning to provide estimates of spins ranging across all possible magnitudes $|a| \in [0, 1]$ (McClintock et al. 2011; McClintock, Narayan & Steiner 2013). The framework of classical general relativity does not provide predictable power when naked singularities arise, and without any theoretical/observational guidance perhaps the best one can do with gravitational waves is to seek for inconsistencies from the predictions of general relativity using something akin to the parameterized post-Einsteinian approach (Yunes & Pretorius 2009). [For tests of gravity with electromagnetic signals see, e.g., Broderick et al. (2014).] Theoretical models suggest the relative orientation of spins are not uniform, due either to properties of the progenitor binary for stellar mass systems or to interactions with surrounding matter or spin-orbit resonant effects during inspiral (Bogdanovic, Reynolds & Miller 2007; Gerosa et al. 2013). Nevertheless, neither theory nor observation provides a sufficiently compelling case to dismiss the full range of spins allowed by general relativity. For stellar mass black holes, masses are expected to range from a few to possibly hundreds of solar masses; supermassive black holes lie at least within the range 10^6–$10^{10}\ \mathrm{M}_\odot$, and evidence is mounting for intermediate-mass black holes between this range (see, e.g., Greene & Ho 2004; Farrell et al. 2009; Godet et al. 2009; Shankar 2009; Davis et al. 2011; Kamizasa, Terashima & Awaki 2012; Casares & Jonker 2014). Consequently, these ranges are sufficiently broad that the mass ratio q is essentially unconstrained, in particular for the closer-to-comparable mass binaries that would require full numerical solution.

One parameter has been argued as being constrainable, especially for stellar mass binaries: the orbital eccentricity. The reason for this is that the back reaction of gravitational wave emission on the orbit tends to reduce eccentricity. To leading order under radiation reaction, the following is a decent approximation to the relationship between periapse and eccentricity [see Peters (1964) and Peters & Mathews (1963) for the derivation and full expression]:

$$r_\mathrm{p} \approx r_{\mathrm{p}0} \frac{1 + e_0}{1 + e} \left(\frac{e}{e_0} \right)^{12/19}, \tag{4}$$

where $r_{\mathrm{p}0}$, e_0 are the initial periapse and eccentricity, respectively. For the moderate initial eccentricities expected when the progenitor of the black hole binary is a stellar binary, $e \sim (r_\mathrm{p}/r_{\mathrm{p}0})^{19/12}$. Such a binary enters the adLIGO band at $r_p \sim 10^2$ km, whereas expected values for $r_{\mathrm{p}0}$ are several orders of magnitude larger (see, e.g., Kalogera et al. 2007), hence e will be completely negligible here. This has focused the majority of work on mergers on the quasi-circular $e = 0$ case. However, there are other mechanisms to form binaries, and some could lead to systems that have high eccentricity while emitting in the LIGO band. These mechanisms include dynamical capture from gravitational wave emission during a close two-body encounter in a dense cluster (O'Leary, Kocsis & Loeb 2009; Lee, Ramirez-Ruiz & Van de Ven 2010), a merger induced during a binary–single star interaction in a similar environment (Samsing, MacLeod & Ramirez-Ruiz 2014), and Kozai-resonant enhancement of eccentricity in a hierarchical triple system (Wen 2003; Kushnir et al. 2013; Seto 2013; Antognini et al. 2014; Antonini, Murray & Mikkola 2014). Event rates are highly uncertain for both classes of binaries [see Abadie et al. (2010) for a review of quasi-circular inspiral systems, and East et al. (2013), Kocsis & Levin (2012), Lee, Ramirez-Ruiz & Van de Ven (2010), O'Leary, Kocsis & Loeb (2009), and Tsang (2013) for discussions of dynamical capture systems], and though quasi-circular inspirals are likely dominant, eccentric mergers may not be completely irrelevant as often assumed in the gravitational wave community. The formation mechanism for supermassive black hole binaries is different (being driven by mergers of the host galaxies of individual black holes), though similarly there are arguments that in some cases non-negligible eccentricity might remain until merger (Roedig & Sesana 2012).

The difficulty with eccentricity is that it is not "merely" an additional parameter but changes the qualitative properties of a merger in a manner that challenges both source modeling and data analysis strategies. With regard to modeling, the orbital period increases significantly with e for a given r_p, making numerical simulations of multiorbit mergers very expensive. Perturbative methods have not yet been developed to high order for large eccentricity orbits (though see Bini & Damour 2012). Taken together it may be unreasonable to expect templates accurate enough for data analysis using matched filtering any time soon, and different (though suboptimal) strategies may need to be developed, for example, power stacking (East et al. 2013). This implies that for practical purposes there are two "classes" of binaries—quasi-circular inspirals and large eccentricity, small initial pericenter mergers.

Having discussed broad considerations relevant to the three classes of binaries (black hole–black hole, black hole–neutron star, neutron star–neutron star), we now discuss salient features of each class uncovered through numerical simulations.

4. BINARY BLACK HOLES

Due to the "no-hair" property of event horizons in four-dimensional Einstein gravity, isolated single black holes in our Universe are expected to be described almost exactly by the Kerr family of solutions. This is a two-parameter family, labeled by the total gravitational mass M and angular momentum J. (An isolated black hole can also have a conserved charge, though in astrophysical settings black holes should be uncharged to excellent approximation. "Exotic" matter fields could also support additional "hair," though we do not consider such fields here.) The latter is more conveniently described by a dimensionless spin parameter, $a = J/M^2$. As mentioned, an event horizon is only present if $|a| \leq 1$, otherwise the solution exhibits a naked singularity. If such a situation could arise (violating the so-called cosmic censorship conjecture), classical general relativity would describe neither the exterior solution nor the dynamics of the object in our Universe. This would offer a prime opportunity to study quantum gravity, though unfortunately to date no theoretical studies of plausible astrophysical processes involving dynamical, strong-field gravity, including gravitational collapse and compact object mergers, have found a naked singularity arising dynamically. [Though see Jacobson & Sotiriou (2009) for an intriguing suggestion that near extremal black holes could be "over spun" and Shapiro & Teukolsky (1991), who suggest that collapse of matter with negligible self-pressure and in a highly prolate configuration could lead to naked singularities. It is also well known that naked singularities can arise in spherical collapse of ideal fluids (see, e.g., Joshi 2013) or critical collapse in a larger class of matter models (see, e.g., Gundlach 2003). However, these examples are either nongeneric (whether by imposed symmetries or fine tuning of initial data) or arise in matter that is of arguable relevance to collapse in astrophysical settings (Wald 1997).]

Thus, technical details aside, the study of vacuum binary black hole mergers in general relativity is a well-defined problem characterized by a relatively small set of parameters: the mass ratio q of the binary, the two initial spin vectors $\vec{s}_1, \vec{s}_2$ of each black hole, the initial eccentricity e_0, and the size of the orbit (parameterized, for example, by the initial pericenter distance r_{p0}). There is no intrinsic scale in vacuum Einstein gravity, hence there is a trivial map from any solution with a given set of these parameters to a desired total mass M of the binary. In the remainder of this section we present results from the numerical solution of the Einstein field equations for vacuum mergers, discuss some astrophysical consequences, and briefly comment on issues related to testing general relativity from gravitational wave observations of vacuum mergers. For other review articles discussing similar topics see Centrella et al. (2010), Hannam (2013), and Pretorius (2009).

4.1. Results and Applications of Merger Simulations

A couple of important qualitative questions about the merger process have largely been answered. The first relates to cosmic censorship: A broad swath of parameter space has been explored (see, for example, Hinder et al. 2013), and no naked singularities have been found. Furthermore, to the level of scrutiny the solutions have been subjected, the late time behavior is consistent with a spacetime approaching a Kerr solution via quasi-normal mode decay. (In theory the quasi-normal ringdown should transition to a power-law decay at very late times. This has been verified for perturbed single black holes. Binary simulations have not yet been carried that far beyond merger, though the motivation for doing so is minimal as the amplitude of these power-law tails is too small to be observable.) The second relates to the existence of new "phases" of the merger outside the purview of the perturbative treatments governing the inspiral and ringdown. One line of reasoning argues that owing to the nonlinearity of the field equations and the fact that the late stages of a merger occur in the most dynamical, strong-field regime of the theory, these phases would be natural places to expect novel physics. The opposing argument, which turned out to better describe the simulation results, is that the merger is effectively a highly dissipative process that occurs deep within the gravitational potential well of the combined objects, and very little of the spacetime dynamics that occurs there will leave an imprint on the waves radiated outward. Or stated another way, perturbative methods have been extended to quite high order in v/c in both the conservative and dissipative dynamics of a binary, and black hole perturbation theory begins with an exact strong-field solution; these together capture the "essential" nonlinearities of the problem. As a consequence of this rather smooth behavior a convenient approach to constructing templates is the effective one-body (EOB) method (Buonanno & Damour 1999), where resummed PN inspiral waveforms are smoothly attached to quasi-normal ringdown modes via a transition function calibrated by numerical simulations (for some recent papers, see Ajith et al. 2011; Damour, Nagar & Bernuzzi 2013; Pan et al. 2014). Most of the work in this regard has been conducted on nonprecessing orbits (i.e., any net spin angular momentum is aligned with the orbital angular momentum) or lower-spin black holes, and it remains to be seen how well this technique may work for highly spinning black holes in precessing orbits. See **Figure 1** for two examples of gravitational wave emission from merger simulations and, for one of them, a match to EOB calculations.

The science gleaned from numerical simulations of vacuum binary mergers has therefore mostly been in details of the process. Important numbers of relevance to astrophysics include the total energy and angular momentum radiated during merger (and consequently the final mass and spin of the remnant black hole) and the recoil or "kick" velocity of the final black hole to balance net linear momentum radiated. We cannot possibly list these numbers for all cases simulated to date (but we do give some citations to relevant literature for further information). To give a sense of

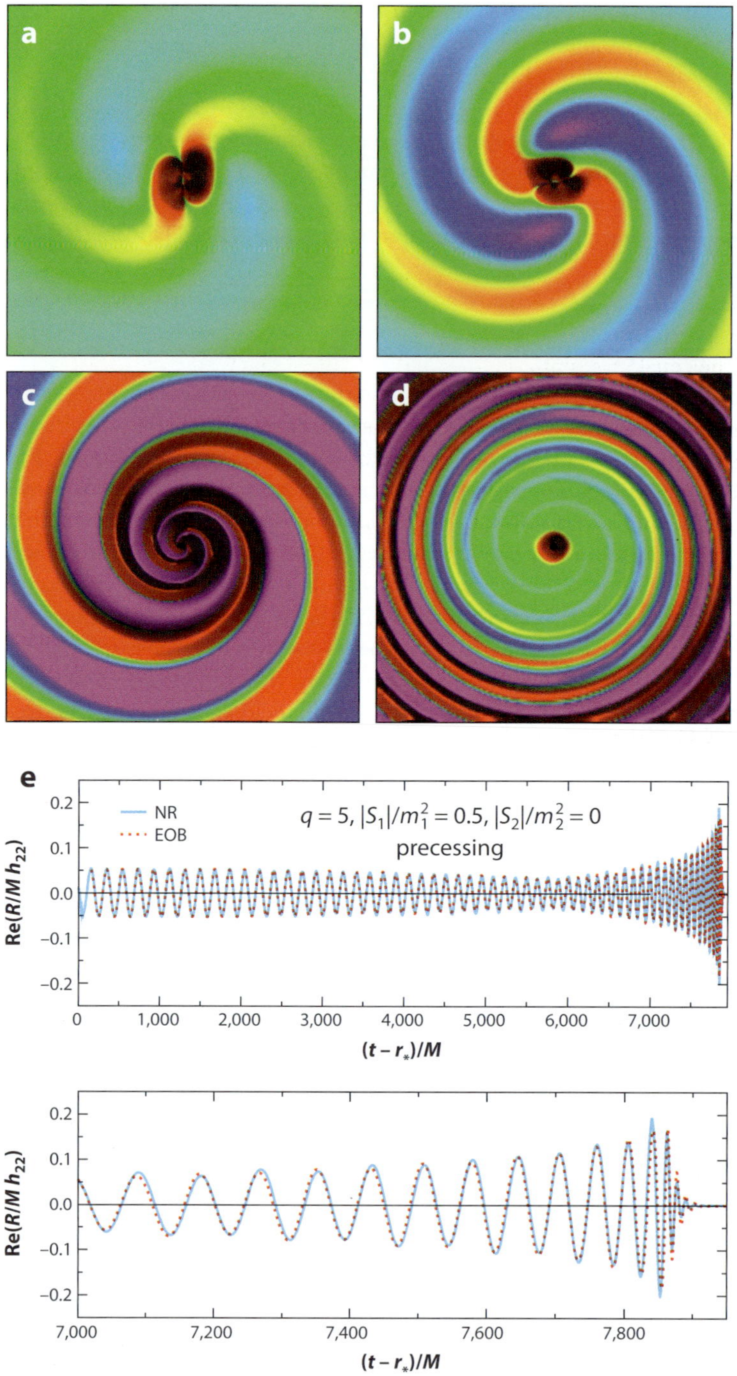

a
b
c
d
e
NR
EOB
q = 5, |S1|/m1^2 = 0.5, |S2|/m2^2 = 0
precessing
Re(R/M h22)
(t − r*)/M

the physics and order of magnitude of the numbers, we highlight a few key results and present some of the simpler fitting formulas.

4.1.1. Energy radiated. One can think of the energy radiated during a merger as coming from two sources: the gravitational binding energy liberated during inspiral and energy in the geometry of the merger remnant formed during the collision that is emitted as the horizon settles down to its stationary Kerr state (on timescales comparable with the final orbital period). In the extreme-mass-ratio limit the former dominates, and the total radiated energy equals the magnitude of the binding energy at the ISCO; for a quasi-circular inspiral this ranges from ~3.8–42% of the rest mass of the small black hole, depending on the spin of the large black hole; the lower (upper) limit is a retrograde (prograde) equatorial orbit about an extremal Kerr black hole (the Schwarzschild case gives 5.7%). As the mass ratio decreases (i.e., the masses become comparable), the emitted energy increases, and the amount coming from the ringdown grows to a comparable fraction approaching the equal mass limit. Here numerical simulations are required to compute the exact numbers, and it is more useful to quote the value as a percentage of the total gravitational energy M (the gravitational mass of the system as measured by an asymptotic observer). A useful formula interpolating between the analytic extreme-mass-ratio limit (Equation 5, *top line*) and empirical fits to numerical data (Equation 5, *bottom line*) was derived by Barausse, Morozova & Rezzolla (2012; see Lousto et al. 2010 and Tichy & Marronetti 2008 for a couple of alternative formulas):

$$\begin{aligned}\frac{E_{\rm rad}}{M} \approx\ & \eta(1-4\eta)[1-\tilde{E}_{\rm ISCO}(\tilde{a})] \\ & + 16\eta^2[p_0 + 4p_1\tilde{a}(\tilde{a}+1)]. \end{aligned} \tag{5}$$

Here $\tilde{a} = \vec{L}\cdot(\vec{S}_1+\vec{S}_2)/M^2$ is the projection of the sum of the black hole spin vectors onto the orbital angular momentum prior to merger, $p_0 \approx 0.048$, $p_1 \approx 0.017$; and $\tilde{E}_{\rm ISCO}(\tilde{a})$ is the energy of the effective ISCO of the system. This formula fits existing numerical simulation results to within better than a percent in most cases (see Barausse, Morozova & Rezzolla 2012 for comparisons and more details).

4.1.2. Final spin. There are numerous formulas characterizing the final spin of the merger remnant that have been constructed via fits to NR results [e.g., Barausse & Rezzolla (2009) and Tichy & Marronetti (2008); see also Lousto et al. (2010) for PN-inspired functions and Boyle, Kesden & Nissanke (2008) for a prescription based on a so-called spin expansion that uses symmetry arguments to economize the formulas). Here we give a simple first-principles-derived expression from

Figure 1

(*a*) A depiction of the gravitational waves emitted during the merger of two equal-mass (approximately) nonspinning black holes. Reprinted from Buonanno, Cook & Pretorius (2007) with permission. Shown is a color map of the real component of the Newman-Penrose scalar Ψ_4 multiplied by r along a slice through the orbital plane, which far from the black holes is proportional to the second time derivative of the plus polarization (*green* is 0, toward *violet* is positive, and toward *red* is negative). (*a,b*) The dominantly quadrupolar inspiral waves ~150 M and ~75 M before merger, respectively. (*c*) Near the peak of the wave emission at merger, and (*d*) the ringdown waves propagating out ~75 M after merger. The size of each box is around $(100\ \mathrm{M})^2$. (*e*) Gravitational wave emission measured from a numerical relativity simulation of a binary black hole merger (NR) overlaid with an NR-calibrated effective one-body (EOB) calculation. Reprinted from Taracchini et al. (2014) with permission. The binary has a mass ratio $q = 5$, the more massive black hole has a dimensionless spin of $a = 0.5$ with direction of the spin axis initially in the plane of the orbit, and the second less-massive black hole is nonrotating. That the amplitude of the wave is not monotonically increasing during inspiral is a manifestation of the modulation induced by precession of the orbital plane.

Buonanno, Kidder & Lehner (2008) that captures the basic physics and agrees reasonably well with numerical results. The following formula is valid for spins aligned with the orbital angular momentum (see Kesden, Lockhart & Phinney 2010 and others cited above for generalizations to precessing binaries):

$$a_f M \approx L_{\mathrm{orb}}(r_{\mathrm{ISCO}}, a_f) + m_1 a_1 + m_2 a_2. \tag{6}$$

Here $a_f M$ is the spin angular momentum of the remnant (with M approximated by $m_1 + m_2$), $m_1 a_1$ and $m_2 a_2$ are the spin angular momenta of the initial black holes, and $L_{\mathrm{orb}}(r_{\mathrm{ISCO}}, a_f)$ is the orbital angular momentum of a reduced-mass particle equivalent of the system evaluated at the ISCO of a Kerr black hole using the parameters of the remnant. The interpretation of this is straightforward: The system radiates angular momentum until the plunge to merger, after which the majority of the remaining spin plus orbital angular momentum is subsumed by the final black hole. Some angular momentum is radiated during ringdown, but this is taken into account in the above formula through the use of the effective ISCO of the remnant black hole. For interest, a quasi-normal mode with frequency ω_m and azimuthal multiple number m has the following relationship between the energy and angular momentum it carries: $J_{\mathrm{rad}} \approx (m/\omega_m)E_{\mathrm{rad}}$. $m = 2$ is the dominant mode, and for example a Schwarzschild black hole has $\omega_2 \approx 0.38/M$ (Berti, Cardoso & Starinets 2009). Though it is not possible to clearly differentiate the quasi-normal part of the wave from the emission that precedes it, a rough estimate is that of order 1–2% of the net energy is emitted in the ringdown for comparable mass mergers.

Equation 6 predicts the final spin to within a few percent in many cases. For example, it gives $a_f/M \approx 0.663$ for the merging of equal mass, nonspinning black holes; comparing with NR simulations, an initial estimate was $a_f/M \approx 0.70$ (Pretorius 2005); the latest high-accuracy simulations refine it to $a_f/M \approx 0.6865$ (Scheel et al. 2009).

4.1.3. Recoil. A recoil in the remnant, namely a velocity postmerger relative to the initial binary center of mass, can arise when there is asymmetric beaming of radiation during the merger. Asymmetry comes from unequal masses and black hole spins. The formulas describing the recoil can be rather involved (see, for example, Lousto & Zlochower 2011), so here we briefly mention some of the salient features and numbers. Nonspinning binaries with mass ratio q different from unity give rise to a recoil in the plane of the binary, reaching a maximum of $\sim$175 km s^{-1} (Baker et al. 2006, Berti et al. 2007, González et al. 2007, Herrmann et al. 2007) for $q \approx 1/3$. Spin introduces additional asymmetry in the radiation by causing the orbital plane to precess and "bob," which can induce a recoil both in and orthogonal to the plane of the binary. The magnitude of the out-of-plane recoil is sinusoidally modulated by the effectively random initial phase of the binary. Spin can also allow the onset of a plunge to occur at higher frequency and, hence, give higher gravitational wave luminosity, which further amplifies the recoil. The bobbing motion (see Pretorius 2009 for an intuitive description) is associated with the largest recoils, which remarkably can reach several thousand kilometers per second for appropriately aligned high-magnitude spins (Campanelli et al. 2007, González et al. 2007, Lousto & Zlochower 2011); see **Figure 2*a*** for examples.

These largest velocities are well in excess of the escape velocities of even the most massive galaxies. That observational evidence suggests most galaxies harbor central supermassive black holes, together with hierarchical structure formation models of the growth of these galaxies, implies that mergers with very large recoils are rare. If mergers themselves are common, and black holes can have sizable spins as implied by current observations (Reynolds 2013), then the typical recoil must be significantly less than the maximum theoretically possible. One possible explanation for this would be if most mergers take place in gas-rich environments, because then torques induced by circumbinary material will tend to align the spins of the black holes with the

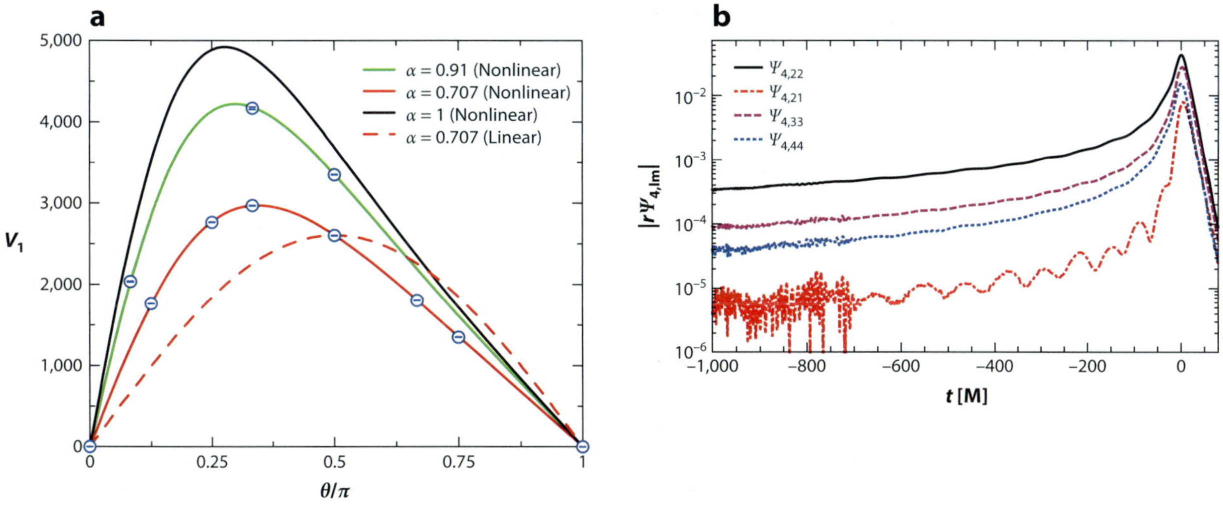

Figure 2

(*a*) Recoil velocities from equal-mass, spinning binary black hole merger simulations (*circles*) together with analytical fitting functions. Each black hole has the same spin magnitude α, equal but opposite components of the spin vector within the orbital plane; θ is the initial angle between each spin vector and the orbital angular momentum. The dashed line corresponds to a fitting formula that depends linearly on the spins, whereas solid lines add nonlinear spin contributions. Reprinted from Lousto & Zlochower 2011 with permission. (*b*) Strength of different modes in the gravitational waves from a binary black hole merger with mass ratio $M_1/M_2 = 3$, and spins $a_1 = 0.75, a_2 = 0$. This system exhibits a marked precession that complicates the multipolar mode structure seen in a fixed observer's frame. However, a transformation to a coprecessing "quadrupole aligned" frame, as shown in this figure, illustrates that the main radiation channel is still the $l = 2, m = 2$ mode. Reprinted from Schmidt et al. (2011).

overall orbital angular momentum, a configuration that has significantly lower maximum recoil (see, e.g., Bogdanovic, Reynolds & Miller 2007; Dotti et al. 2010). Another possible explanation that operates even in vacuum comes from PN calculations that include spin-orbit coupling, which shows a tendency for the black hole spins to align (antialign) with each other if the spin of the more massive black hole is initially at least partially aligned (antialigned) with the orbital angular momentum (Kesden, Sperhake & Berti 2010).

4.1.4. Tests of general relativity. A further opportunity offered by gravitational wave observations of merging binaries is to test dynamical, strong-field gravity. With obvious caveats associated with our present lack of understanding of dark matter and dark energy, general relativity has so far been shown to provide a consistent description of gravity in all observations and experiments that are constrained by its predictions (see, e.g., Will 2006). Lacking here are tests in the most nonlinear regime of the theory, in particular where black holes can form. Certainly there is no doubt about the existence of massive, dark, ultracompact objects, and observations of (for example) X-ray emission from stellar mass candidates or properties of AGNs are consistent with these phenomena being powered by Kerr black holes. However, that horizon scales cannot be quite resolved at present [this may change within a few years through VLBA observations of our Galaxy's central black hole, SGA* (Broderick, Loeb & Narayan 2009; Broderick et al. 2011) as well as nearby M87 (Doeleman et al. 2012)], together with complexities of the matter physics responsible for the emission, prevents precise determination of local properties of the spacetime. Binary black hole mergers, in particular stellar mass systems that are expected to occur largely in vacuum, offer a unique opportunity to study pure, strong-field gravity. General relativity's ability to predict the entire waveform, which is uniquely determined by a small set of numbers (M, q, e, $\vec{s}_1$, $\vec{s}_2$, and detector-source orientation parameters) and can consist of hundreds or even thousands of cycles in the LIGO band, can in principle allow for stringent self-consistency tests on high signal-to-noise–ratio (SNR) events.

However, there are several issues that complicate general relativity's promise in testing in the near future. First, given the lack of events from the initial LIGO observing runs, it is unlikely that adLIGO will observe a very loud event. Hence viable tests may require statistical analysis of a number of low-SNR events, and little work has yet been done to suggest how this might be carried out [see Agathos et al. (2014) for a recent proposal; related work on constraining the nuclear EOS using multiple mergers events involving neutron stars is also beginning to be investigated—see the discussion in Section 5]. Second, detection and parameter estimation relies on matched filtering with templates. If the only templates used are those constructed using general relativity, then all information about possible deviations will be projected out. (Note also that it is unlikely that general relativity templates will completely "miss" all events even if there are strong-field deviations. This is because binary pulsar observations confirm the leading order radiative dynamics of general relativity.) If the event has a high SNR, there should be a detectable residual excess power, but again for the typical SNRs expected for adLIGO this is unlikely. [This is not the case for possible space-based detectors, like eLISA and its exquisite SNR, which allows for detecting supermassive binary black holes mergers with masses in the range $10^4 M_\odot < M < 10^7 M_\odot$ out to redshifts of $z \simeq 20$ with a SNR ≥ 10. For a recent review see Seoane et al. (2013).] Compounding the problem, despite the large number of proposed alternatives or modifications to general relativity (see, for example, Will 1993, 2006), almost none have yet been presented that (*a*) are consistent with general relativity in the regimes where it is well tested, (*b*) predict observable deviations in the dynamical strong field relevant to vacuum mergers, and (*c*) possess a classically well-posed initial value problem to be amenable to numerical solution in the strong field. The notable exceptions are a subset of scalar tensor theories, though these require a time-varying cosmological scalar field for binary black hole systems (Horbatsch & Burgess 2012) or one or more neutron stars in the merger (see Section 5). Thus there is little guidance on what reasonable strong-field deviations one might expect. Proposed solutions to (at least partially) circumvent these problems include the parameterized post-Einsteinian and related frameworks (Yunes & Pretorius 2009, Agathos et al. 2014) and modified PN waveforms (Arun et al. 2006), as well as exploiting properties of the uniqueness of Kerr and its quasi-normal mode structure (Collins & Hughes 2004; Berti, Cardoso & Will 2006).

4.1.5. Eccentric binaries. As mentioned above, the majority of the work on binary black hole mergers from the relativity community has focused on quasi-circular inspiral, except for a handful of recent studies (Pretorius & Khurana 2007; Healy, Levin & Shoemaker 2009; Gold & Bruegmann 2013). One of the interesting results is that so-called zoom-whirl orbital dynamics is possible for comparable-mass binaries. In the test particle limit, zoom-whirl orbits are perturbations of the class of unstable circular geodesics that exist within the ISCO. They exhibit extreme sensitivity to initial conditions, in which sufficiently fine-tuned data can give an arbitrary number of near-circular "whirls" at periapse for a fixed eccentricity. Away from the test particle limit, gravitational wave emission adds dissipation to the system; however, what the simulations show is that even in the comparable mass limit the dissipation is not strong enough to eradicate zoom-whirl dynamics, but merely limits how long it can persist. Perhaps the most interesting consequences of high-eccentricity mergers could arise when neutron stars are involved; this is discussed in Section 5.

For the vacuum problem, aside from providing information on binary formation channels, high-eccentricity events could in principle offer the most stringent tests of strong-field gravity.

The reason is due to the nature of these orbits compared with quasi-circular inspiral: Significantly higher velocities are reached at periapses passage, a larger fraction of the total power is radiated in this high v/c regime, and the long time between periapse bursts imply that small deviations in emission could result in large dephasing of the waveform from one burst to the next. However, to date no studies have addressed in any quantitative manner how well general relativity can be constrained using eccentric mergers.

4.2. Further Physics

With the vacuum merger problem essentially under control, the field is now more closely examining the impact a merging black hole binary can have on its astrophysical environment. The most pertinent scenario is a supermassive binary merger, and questions relate to how the rapidly changing gravitational field, ensuing gravitational waves, and possible recoil could perturb surrounding gas, plasma, electromagnetic fields, stars, etc. Here we briefly discuss some of the more interesting and potentially observable phenomena revealed by recent studies.

4.2.1. Prompt counterparts to supermassive binary black hole mergers within circumbinary disks. First studies of the interaction of binary black holes with surrounding electromagnetic fields and plasma were presented by Palenzuela et al. (2010) and Palenzuela, Lehner & Leibling (2010). Though not modeled there, the expected source of these fields and plasma would be a circumbinary disk. More recent work has begun to self-consistently model the disk as well (Farris et al. 2012, Gold et al. 2014). Recall that in the case of a single black hole, the Blandford-Znajek (BZ) mechanism (Blandford & Znajek 1977) indicates the plasma (coming from an accretion disk) is able to tap rotational energy from the black hole and power an energetic Poynting flux. Tantalizing observational evidence linking the strength of radio signals and black hole spin has been presented by McClintock, Narayan & Steiner (2013). In the context of binary black holes, simulations demonstrated that the spacetime helps stir electromagnetic field lines and that, akin to the BZ mechanism, the plasma is able to tap translational and rotational energy from the system to produce dual jets (Palenzuela et al. 2010; Palenzuela, Lehner & Leibling 2010; see also **Figure 3**). These jets would act as spacetime tracers, and their behavior can be modeled reasonably well by an extension of the BZ formula. That is, prior to merger, the luminosity from the system obeys $L \simeq B^2 \sum_{i=1,2} [\Omega_H(i)^2 + \kappa v(i)^2]$, where $\Omega_H(i)$, $v(i)$, and κ are the angular rotational velocities of the horizons, the black hole velocities, and a relative strength parameter, respectively. [The dependence on Ω_H acquires higher-order corrections close to maximally spinning cases (Tchekhovskoy, Narayan & McKinney 2010).] (The value of $\kappa \simeq 100$ and indicates that black holes must be moving at $\gtrsim 0.1c$ for a nontrivial contribution unless, of course, they are nonspinning.) Notice that unlike the original BZ effect, even if the black holes are nonspinning there could be a sizable luminosity due to the contribution from the translational kinetic energies of the black holes (which can reach ≈ 0.2–$0.3c$ near merger for quasi-circular inspiral). After merger, a single jet arises with luminosity $L \simeq B^2(\Omega_{H_{\rm final}} + \kappa v^2_{\rm recoil})$ (though the second term is subleading unless the final black hole has negligible spin, as $v_{\rm recoil}$ is at most $\approx 0.015c$). This behavior implies interesting possibilities for detection of gravitational and electromagnetic waves associated with a merger embedded in a circumbinary disk (see, e.g., Mösta et al. 2012, O'Shaughnessy et al. 2011).

4.2.2. Postmerger consequences of supermassive binary black hole mergers. The merger event can have several interesting consequences due to the large amounts of energy radiated and (for appropriate spins and mass ratios) the recoil of the final black hole; we briefly mention a few here. For recent reviews of this and other astrophysical consequences, see Komossa

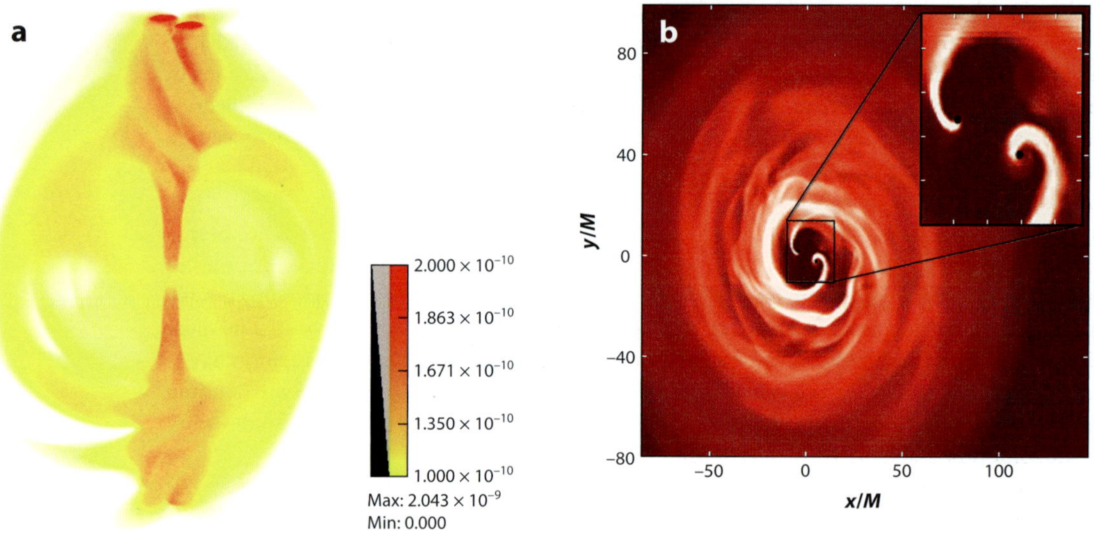

Figure 3

(*a*) Poynting flux produced by the interaction of an orbiting black hole binary with a surrounding magnetosphere. The "braided" jet structure is induced by the orbital motion of the black holes. Reprinted from Palenzuela, Lehner & Leibling (2010) with permission. (*b*) Rest-mass density induced by a supermassive black hole binary interacting with a magnetized disk prior to when the binary "decouples" from the disk, namely when the gravitational wave back reaction timescale becomes smaller than the viscous timescale. Reprinted from Farris et al. (2012) with permission.

(2012) and Schnittman (2013). With respect to timescales in the disk these effects occur essentially instantaneously. This near-impulsive perturbation of the gravitational potential in the outer parts of the accretion disk could lead to the formation of strong shocks, producing observable electromagnetic emission on timescales of a month to a year afterward (Lippai, Frei & Haiman 2008); subsequent inward migration of the disk could induce possible observable emissions (Milosavljević & Phinney 2005). The most favored orientations for recoils can produce velocities large enough to significantly displace the remnant from the galactic core or even eject the black hole from the host galaxy altogether (though as discussed above this is likely quite rare). If the system has a circumbinary accretion disk, the recoil would carry the inner part of the disk with it, and this could be observable in Doppler-displaced emission lines relative to the galactic rest frame (Komossa, Zhou & Lu 2008). Earlier studies have suggested that, prior to merging, the accretion rate, and hence the luminosity of the nucleus, would be low as the relatively slow migration of the inner edge of the accretion disk decouples from the rapidly shrinking orbit of the binary. Postmerger then, AGN-like emission could be reignited once the inner edge of the disk reaches the new ISCO of the remnant black hole. This should be displaced from the galactic center if a large recoil occurred and could be observable in nearby galaxies (see, for example, Loeb 2007). However, more recent simulations of circumbinary disks using ideal magnetohydrodynamics for the matter shows that complete decoupling does not occur, and relatively high accretion rates can be maintained all the way to merger (Bode et al. 2012, Farris et al. 2012, Noble et al. 2012) and afterward (e.g., Shapiro 2010). The binary orbit in this case causes a modulation in the luminosity of the system, which may be observable. A last effect we mention is that a displaced central black hole should also have its loss-cone refilled, increasing the frequency of close encounters with stars and their subsequent tidal disruption by the black hole, with rates as high as 0.1 year^{-1}; the disruption could produce observable electromagnetic emission (Stone & Loeb 2011).

4.2.3. Binary black hole mergers and galaxy formation. Galaxy formation models have also been exploited to understand the outcome of binary black hole mergers. There is strong stellar, gas-dynamical (Kormendy & Richstone 1995, Magorrian et al. 1998, Kormendy & Ho 2013), and electromagnetic (Gebhardt et al. 2000, Peterson et al. 2004) evidence for the existence of massive black holes at the centers of galaxies. These central black holes play a fundamental role in our current paradigm of galaxy formation and evolution; for example, they are required to explain quasar and AGN emission (Soltan 1982), as well as cosmic downsizing (Bower et al. 2006; Croton et al. 2006; Scannapieco, Silk & Bouwens 2005). In the ΛCDM model galaxies merge into increasingly larger ones as cosmic time proceeds, and consequently their massive black holes are expected to merge, initially via processes such as dynamical friction, with gravitational wave emission only taking over in the very late stages. Results from NR simulations have been utilized to follow the evolution of these black holes through coalescence. More specifically, a number of works studied the mass and spin evolution of supermassive black holes through cosmic time (Volonteri, Sikora & Lasota 2007; Berti & Volonteri 2008; Fanidakis et al. 2011; Volonteri et al. 2013), in some cases also accounting for the recoil velocity of the merger remnant (Barausse 2012, Lousto et al. 2012). It has also been suggested that if a space-based detector such as eLISA becomes available, measurements of the mass ratios of black hole binaries and the precession effects predicted by PN/NR calculations would help to discriminate between competing models of galaxy formation (Sesana et al. 2011, Barausse 2012).

5. NONVACUUM BINARIES

As in the binary black hole case, nonvacuum binaries present a well-defined problem; however they need a larger set of parameters to characterize. First, the matter physics introduces a scale, so that unlike the vacuum case the total mass of the system cannot be factored out. Thus the set of parameters needed to describe the orbit are now the masses m_1 and m_2 of the compact objects, their two initial spin vectors $\vec{s}_1, \vec{s}_2$ (which, however, are expected to be small for neutron stars), and the initial eccentricity e_0 and size of the orbit (again which we parameterized by the initial pericenter distance r_{p0}). Second, for neutron star matter one must specify the EOS (which for a given mass star determines its radius), and each star's magnetization (strength and dipole direction).

The presence of (magnetized) matter in the system, that is strongly affected by the rapidly varying geometry during coalescence, can naturally induce electromagnetic and neutrino emission in concert with the gravitational waves. A prime example is short gamma-ray bursts (sGRBs), and the evidence is mounting that nonvacuum binary mergers provide the central engine for these spectacular astrophysical phenomena (e.g., Janka et al. 1999; Lee & Ramirez-Ruiz 2007; Metzger & Berger 2012; Berger, Fong & Chornock 2013; Piran, Nakar & Rosswog 2013; Tanvir et al. 2013; Berger 2014). Thus, in addition to obtaining predictions for the gravitational wave signatures from these events (see **Figure 4** for some examples), research using simulations is also focused on gaining a theoretical understanding of their connections to sGRBs and related phenomena.

Widely separated nonvacuum binaries display the same behavior as binary black holes. Here internal details play a negligible role, as their effects first appear in a PN expansion at order $(v/c)^{10}$. Closer-to-merger tidal forces introduce subtle deviations at first, growing to quite large deformations at the point of contact in a binary neutron star system; for a black hole–neutron star system, such forces can even lead to the disruption of the star prior to merger. For binary neutron stars, if the total mass in the remnant is more than the maximum mass allowed by the EOS, a black hole will eventually form. The intermediate state is called a hypermassive neutron star (HMNS) and is temporarily supported by rotation and thermal pressure. An interesting question then, as illustrated in **Figure 4**, is, How long does the HMNS last? Once a black hole forms,

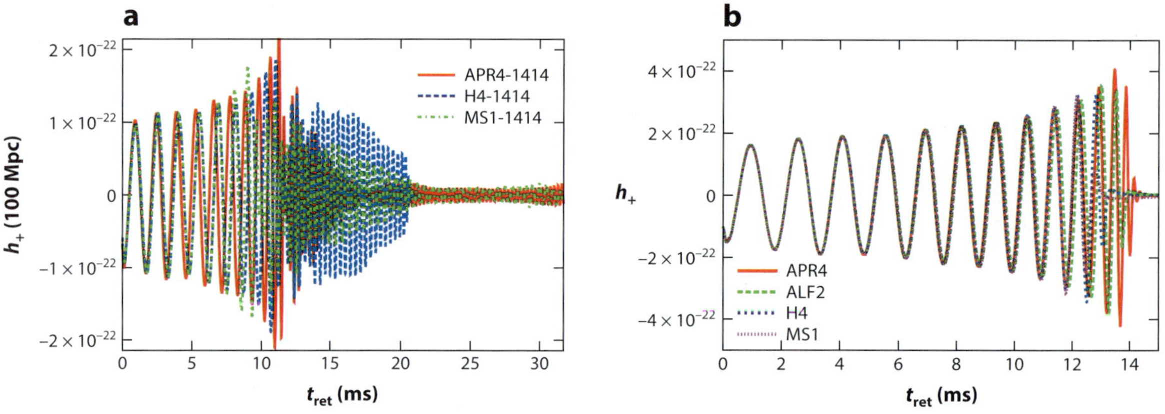

Figure 4

Examples of the "plus" polarization component of gravitational waves from binary neutron star mergers, measured 100 Mpc from the source along the direction of the orbital angular momentum. The different curves correspond to different choices of the equations of state (EOSs) of the neutron star matter, labeled APR4, ALF2, H4, and MS1. For a 1.4-$M_\odot$ neutron star, the APR4, ALF2, H4, and MS1 EOSs give radii of 11.1, 12.4, 13.6, and 14.4 km, respectively. (*a*) Mergers of an equal-mass binary neutron star system (with $m_1 = m_2 = 1.4\,M_\odot$). A hypermassive neutron star is formed at merger, but how long it survives before collapsing to a black hole strongly depends on the EOS. The H4 case collapses to a black hole $\approx$10 ms after merger; the APR4 and MS1 cases have not yet collapsed $\simeq$35 ms after merger, when the simulations were stopped (the MS1 EOS allow a maximum total mass of $2.8\,M_\odot$, so this remnant may be stable). The striking difference in gravitational wave signatures is self-evident. Reprinted from Hotokezaka et al. (2013) with permission. (*b*) Emission from black hole–neutron star mergers, with $m_{BH} = 4.05\,M_\odot$, $m_{NS} = 1.35\,M_\odot$. Variation with EOSs is primarily due to coalescence taking place earlier for larger radii neutron stars (from Kyutoku, Ioka & Shibata 2013).

and following a black hole–neutron star merger, an accretion disk can form. If there is sufficient mass in the disk, this could be the beginning of a jet that would eventually produce a sGRB. A host of other electromagnetic emission is likely as a consequence of these nonvacuum mergers, as is neutrino emission. In the following sections, we discuss these in more detail, highlighting the information gained from NR simulations. For other review articles in this subject area, see Duez (2010) and Pfeiffer (2012).

5.1. Binary Neutron Star Mergers

Fully general relativistic studies of binary neutron stars have been an active area of research for over a decade. (For a small sample of recent results in this area see, e.g., Anderson et al. 2008b; Stephens, Shapiro & Liu 2008; Rezzolla et al. 2010; Sekiguchi et al. 2011; Kiuchi et al. 2012; Hotokezaka et al. 2013; Kaplan et al. 2014; Palenzuela et al. 2013a; Read et al. 2013). The initial focus of the research was directed toward understanding broad characteristics of the gravitational wave emission, and consequently rather simple treatments of the matter were employed (typically an ideal fluid with polytropic EOS). These efforts gave a rather robust understanding of the qualitative dynamics of the system and prepared a solid foundation to increase the realism of the matter modeling in the simulations. In recent years the addition of new physical ingredients have included more realistic EOSs, magnetic fields and plasmas, and some simplistic treatments of neutrino and radiation physics. In this section, we review the more interesting developments relevant to astrophysics.

As in the case of binary black holes, several important qualitative questions about the merger process have been elucidated. The first relates to behavior postmerger and, for a sufficiently massive remnant, the onset of collapse to a black hole. A large swath of parameters centered around the observationally favored initial neutron star masses of $\approx 1.4\,M_{\odot}$, and consistent with the highest-mass neutron stars observed to date ($\approx 2.\,M_{\odot}$; see Antoniadis et al. 2013 and Demorest et al. 2010 and related discussion by Lattimer & Prakash 2010), show an HMNS forms. From a fundamental gravity point of view, an interesting observation is that this intermediate state can have a highly dense central region and an effective angular momentum higher than the Kerr bound. Yet, obeying the cosmic censorship conjecture, the object does not evolve to a nakedly singular solution but is able to efficiently transport angular momentum outward to eventually allow a black hole to form. The black hole settles down to an approximate Kerr solution surrounded by some amount of material in a disk. A crucially important question then is, What is the timescale for collapse? This timescale depends sensitively on the initial conditions and several physical properties: the individual neutron star masses and eccentricity of the binary (which influences the initial distribution of mass among the HMNS and the bound and ejected material), the EOS, neutrino and photon cooling, and thermal pressure, as well as diverse mechanisms for angular momentum transport. The reason this is such an important question is that the timescale is in principle observable, either directly via the gravitational wave emission (as illustrated in **Figure 4**, though note that the frequency of the postmerger waves are sufficiently high that the adLIGO detectors will not be sensitive to them except for a highly unlikely nearby event) or indirectly through details of the counterpart electromagnetic/neutrino emission.

Beyond these broad qualitative issues, theoretical studies have been aimed to analyze in detail the coalescence process and characteristics of the gravitational wave emission. As mentioned, early stages of the dynamics are well captured by PN treatments. Approaching merger tidal effects do start to influence the evolution of the orbit, which would be reflected in the gravitational wave emission and could be detected via delicate data analysis (Hinderer et al. 2010; Damour, Nagar & Villain 2012). Another interesting premerger consequence of tidal forces during a quasi-circular inspiral is they can induce resonant oscillations in the interface modes (i-modes) between the neutron star crust and core that grow large enough to shatter the star's crust, leading to a potentially observable precursor electromagnetic outburst (Tsang et al. 2012). For highly eccentric close encounters, the tidal force is impulsive in nature. This can similarly shatter the crust (Tsang 2013) and will excite f-mode oscillations in the star (Stephens, East & Pretorius 2011; Gold et al. 2012). The f-modes do emit gravitational waves, though at frequencies that are too high and at amplitudes too weak for likely direct detection with adLIGO.

For low eccentricity encounters the stars merge at an orbital frequency that can be estimated by the point at which the stars come into contact, i.e., $\Omega_c \simeq [(m_1+m_2)(R_1+R_2)^3]^{1/2}$. At this stage, the stars are traveling at a considerable fraction ($\simeq$10–20%) of the speed of light, resulting in a violent collision. In the contact region, shock heating is responsible for a considerable amount of mass thrown outward (some of which becomes unbound) in a rather spheroidal shape. Also, strong shearing in this region induces Kelvin-Helmholtz instabilities, and strong differential rotation develops in the newly formed HMNS. The temperature of the HMNS can reach values of $\simeq$30–50 Mev, and magnetic fields can grow by several orders of magnitude [via winding, tapping kinetic energy, and possibly magnetorotational instabilities (MRIs), though for this latter process resolutions currently used are still far from that required to adequately resolve it]. Tidal tails form during the earlier stages of the merger and distribute material in the vicinity of the equatorial plane. As mentioned, because the total mass of the binary likely exceeds the maximum mass that a stable, nonrotating, and cold star might achieve, the subsequent behavior of the HMNS divides into two possible cases: prompt or delayed collapse.

In the prompt collapse case, thermal support and differential rotation are unable to overcome the gravitational attraction, and a black hole forms essentially in a free fall timescale. This takes

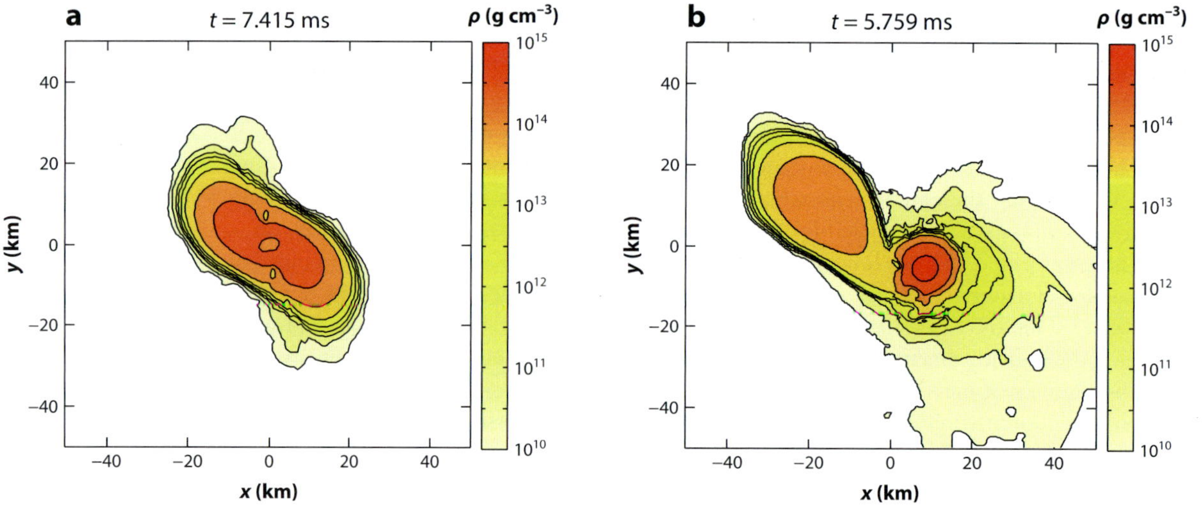

Figure 5

Equatorial density profiles $\simeq$3 ms after merger from (*a*) an equal and (*b*) unequal mass binary neutron star system. Panel *a* corresponds to a system with $m_1 = m_2 = 1.643\,M_\odot$ baryonic mass. Panel *b* corresponds to $m_1 = 1.304\,M_\odot$ and $m_2 = 1.805\,M_\odot$ baryonic masses. Reprinted from Rezzolla et al. (2010) with permission.

place in binaries with relatively large total mass $M_{\mathrm{tot}} \gtrsim 2.6\text{–}2.8\,M_\odot$, though the exact value depends intimately on the EOS.

If the collapse does not occur promptly, the postmerger dynamics differ depending on whether the merger involved equal masses or not. If it is the former case, the newly formed object resembles a dumbbell composed of two cores (the remnants of the individual stars), which gradually turns into an ellipsoidal object as a result of angular momentum transport—primarily via hydrodynamics effects—and angular momentum loss via gravitational waves. For example, **Figure 5** illustrates the equatorial density of the remnant following an equal mass binary merger and another that had $m_1/m_2 = 0.7$. The gravitational waves from a postmerger system have a characteristic frequency in the range $2 \lesssim f \lesssim 4$ kHz which is proportional (and relatively close) to the Keplerian angular velocity $(M_{\mathrm{HMNS}}/R^3_{\mathrm{HMNS}})^{1/2}$ (where M_{HMNS}, R_{HMNS} are the mass and radii of the HMNS, respectively). If the stars have different masses, the stronger tidal forces induced by the more massive star deforms the companion, stripping the outer layers and forming an envelope about the newly formed HMNS. This HMNS now displays two asymmetric cores and behaves as if the more massive core has a satellite that deforms dynamically as time progresses. Regardless of the mass ratio, a significant amount of material is estimated to lie beyond the ISCO of the black hole that will eventually form, resulting in an accretion disk with mass on the order of $0.01\text{–}0.3\,M_\odot$. Typically the more massive disks correlate with longer times to black hole formation, a behavior intuitively expected as there is more time for angular momentum to be transferred outward to the envelope.

Simulations have also shed light on the processes, and timescales, for such angular momentum transfer. The most important one is hydrodynamical, which begins to operate efficiently after the merger due to the strong torques induced by asymmetries in the HMNS. Other significant mechanisms for this transfer are tied to electromagnetic effects: winding and MRIs can do so by linking the central to outer regions of the HMNS and introducing an effective viscosity in the system. The angular momentum transport timescale due to winding is of the order of

$\tau_{\rm wind} \simeq R_{\rm HMNS}/v_{\rm A}$, with the Alfvén velocity $v_{\rm A} \simeq B/\sqrt{\rho}$. A few general relativistic simulations have pointed out that the strength of B can increase [in agreement with analogous results obtained in pseudo-Newtonian (Price & Rosswog 2006) or shearing box studies (Obergaulinger, Aloy & Müller 2010)] from typical premerger values of $10^{10-12} G$ to 10^{15-16} G via compression, winding, and transfer of hydrodynamical kinetic energy to electromagnetic energy via Kelvin-Helmholtz instabilities (Anderson et al. 2008b; Giacomazzo, Rezzolla & Baiotti 2011), which imply timescales $\tau_{\rm wind} \simeq 10$–100 ms. We stress, however, that present computational resources are still not adequate to give a thorough analysis of this process. For transport driven by the MRIs, simulations are even more challenging, so this is still a largely unexplored process within general relativistic simulations of binary neutron star mergers. Nevertheless, estimates indicate $\tau_{\rm MRI} \sim 100$ ms for putative magnetic field strengths of $B \simeq 10^{15}$ G. Therefore, either transport mechanism can operate on timescales $\gtrsim$10–100 ms and aid in expediting the collapse. Cooling via neutrino and radiation transport reduces thermal-pressure support, so it also helps to shorten the time to collapse. However, the timescale for cooling to operate in a significant manner is on the order of seconds. Currently, simulations incorporating both electrodynamics and cooling are actively being pursued and refined.

Beyond the intricate details of the merger and postmerger behavior, there is strong interest in exploring binary neutron star mergers as progenitors of sGRBs and other electromagnetically observable signals. There is already tantalizing observational evidence for the connection between nonvacuum compact binaries mergers and sGRBs (see Berger 2014 for a recent review), strengthened by compelling theoretical models that suggest a merger yielding a rapidly accreting black hole could serve as the central energy source through hydrodynamical/plasma or neutrino processes (Eichler et al. 1989; Narayan, Paczynski & Piran 1992). Other models for the origin of at least a class of sGRBs include magnetars produced by binary neutron star mergers, binary white dwarf mergers or accretion-induced collapse of a white dwarf (Levan et al. 2006; Metzger, Quataert & Thompson 2008), and the collapse of an accreting neutron star to a black hole (MacFadyen, Ramirez-Ruiz & Zhang 2005). Simulations of these systems are providing valuable information to test these models. For instance, once collapse occurs, an initial hyperaccreting stage is observed, followed by a longer fall back accretion phase with the characteristic $t^{-5/3}$ power-law dependence expected from analytic calculations (Rees 1988). Beyond the burst itself, electromagnetic emission arising from the interaction of ejected material with the ambient medium, or through radioactive decay of r-process elements formed in this material shortly after merger, have been proposed (e.g., Metzger & Berger 2012; Piran, Nakar & Rosswog 2013). A candidate for this latter "kilonova" event has recently been observed (Berger, Fong & Chornock 2013; Tanvir et al. 2013). The timescale for this class of emission can be as long as days or weeks following merger, hence it is not amenable to ab initio simulations. However, results from simulations are consistent with properties assumed in these models to give observable signals; in particular, ejected material of order $\lesssim 10^{-4} M_{\rm eject}/{\rm M}_\odot \lesssim 10^{-2}$ traveling with velocities $\simeq$0.1–0.3c has been seen in noneccentric scenarios (with somewhat larger amounts/higher velocities possible in eccentric mergers). Finally, a magnetar with magnetic field strength likely in excess of 10^{15} G indeed forms during the merger of magnetized neutron stars, though its lifetime is typically $\lesssim$100 ms except for the stiffest of EOSs and low-mass binaries (Anderson et al. 2008b; Giacomazzo, Rezzolla & Baiotti 2011). We conclude this section with a few miscellaneous topics related to binary neutron star mergers.

5.1.1. Magnetosphere interactions. Neutron stars have among the strongest magnetic fields in the Universe. As in the case of pulsars, they are surrounded by a magnetosphere that arises naturally as argued by Goldreich & Julian (1969). It is thus natural to expect that a binary interaction can trigger behavior related to that observed in pulsars (Lipunov & Panchenko 1996), though

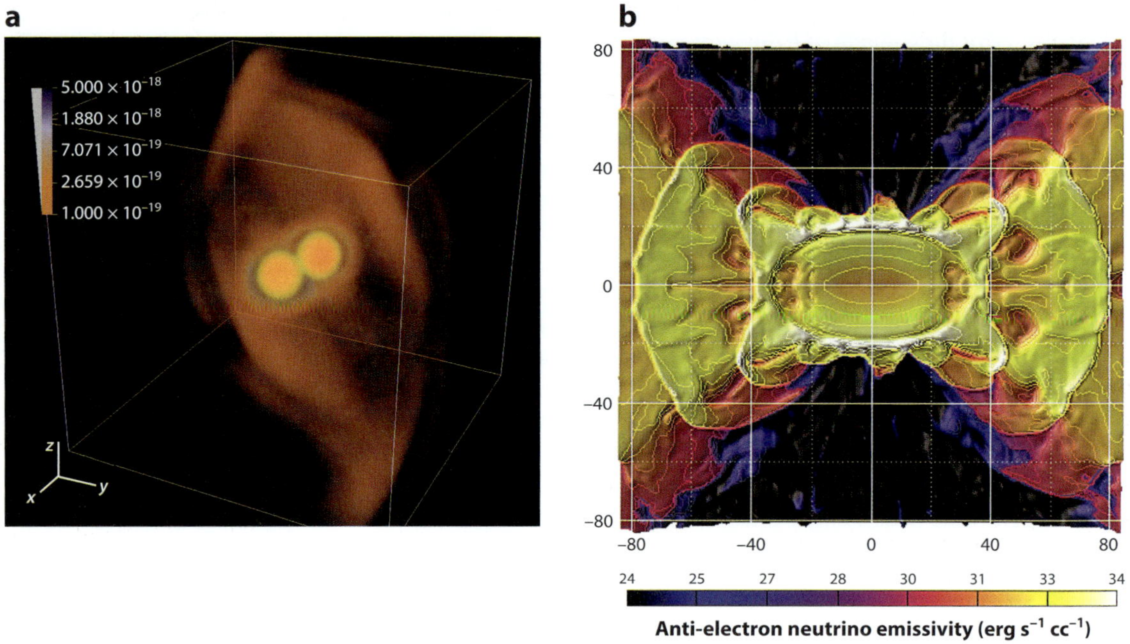

Figure 6

(*a*) Poynting flux produced by the magnetospheric interaction of orbiting, magnetized (with $B = 10^{12}$ G), equal-mass ($m_1 = m_2 = 1.4\,M_\odot$) neutron stars $\simeq$1.5 ms before merger. Reprinted from Palenzuela et al. (2013b) with permission. (*b*) Anti-electron neutrino luminosity in the x–z plane, 15 ms after an equal-mass ($m_1 = m_2 = 1.5\,M_\odot$) binary neutron star merger. Reprinted from Sekiguchi et al. (2011) with permission.

in this case with a tight connection to the orbital dynamics. This has recently been studied by Palenzuela et al. (2013a,b), showing that close to the merger event a strong Poynting flux is emitted ($L \simeq 10^{40-43} B_{11}^2$ erg s^{-1}); see **Figure 6*a*** for an illustration. As anticipated, many features common to those of pulsars are seen: the existence of gaps in the estimated charge density, shear layers, the development of a current sheet, and a striped structure in the toroidal magnetic field. In the binary case, however, these features bear tight imprints of the binary's behavior. For instance, as the orbit tightens, a ramp-up in Poynting luminosity ensues, and the current sheet structure displays a spiral pattern tied to the orbital evolution of the system. This could provide an important electromagnetic counterpart to the gravitational waves. In addition, the HMNS—which is likely highly magnetized as a result of the collision—can also trigger a strong Poynting flux as it collapses to a black hole (Lehner et al. 2012). The luminosity can be as large as $L \simeq 10^{49}(B/10^{15}G)^2$ erg s^{-1}, but shuts off abruptly as the black hole forms.

5.1.2. Neutrino emissions. Incipient works are beginning to incorporate estimates of neutrino effects in the system. This is because, as mentioned, the typical lifetime of the HMNS would likely be limited to $\lesssim 100$ ms, whereas the timescale for neutrino cooling is on the order of seconds, as a first approximation a full (costly) radiation-transport scheme need not be employed. Instead, a simplified strategy known as a "leakage scheme" (Ruffert et al. 1997) has become the starting point. The leakage scheme ignores transport from the diffusion of neutrinos as well as neutrino momentum transfer. What it does model is the possible equilibration of neutrinos; it adopts an opaque, hot stellar matter model to describe local neutrino sources and sinks and accounts for charged-current β processes, electron-positron pair-annihilation, and plasmon decays. At low optical depth the scheme uses reaction-rate calculations to estimate the local production and emission of neutrinos. In contrast, at high optical depths it assumes neutrinos are at their equilibrium abundances and that neutrino/energy losses occur at the diffusion timescale. In between, a suitable interpolation is adopted. Early efforts employing this scheme indicate a binary neutron star merger can produce strong neutrino luminosity of order $\simeq L_\nu \simeq 10^{54}$ erg s^{-1} (Sekiguchi et al. 2011). **Figure 6*b*** illustrates the antielectron neutrino luminosity shortly after an equal-mass ($m_1 = m_2 = 1.5\,M_\odot$) merger.

5.1.3. Eccentric binaries. Binaries that emit observable gravitational waves while the orbit has high eccentricity show significant qualitative and quantitative differences in properties of the merger compared with equivalent-mass quasi-circular inspirals. Because there is more angular momentum in the binary when the two stars collide, typically more mass is stripped off, some fraction of which is ejected and the rest forms an accretion disk (East & Pretorius 2012, Gold et al. 2012). This has consequences for the magnitude of ejecta-powered counterparts, abundance of heavy elements produced through r-processes, and the range of initial neutron star masses that can lead to sufficiently massive disks to power an sGRB. The larger rotational energy also implies longer lifetimes for HMNS remnants. As mentioned above, close encounters prior to merger could induce sufficient strain in each neutron star to shatter its crust, leading to precursor electromagnetic emission (Tsang 2013). Furthermore, f-modes will be excited in each star. This changes the energetics of the orbit and indirectly affects the subsequent gravitational wave emission. The f-modes will also emit gravitational waves directly, though because of their relatively low amplitudes and high frequencies (around 1.5 kHz), they will not be observable with adLIGO-era detectors. Regarding the dominant emission from the orbital motion, as with eccentric binary black hole mergers, the challenge for detection is to construct waveform models accurate enough to use in template-based searches or devise alternative strategies (issues of rates aside). Also, even though the integrated energy released is order-of-magnitude comparable to a quasi-circular inspiral, more of it is radiated at higher frequencies in close periapse, high-eccentricity mergers. For a binary neutron star this occurs outside adLIGO's most sensitive range, making such a system unlikely to be detected beyond $\approx$50 Mpc even with matched filtering (East et al. 2013).

5.1.4. Alternative gravity theories. Binary neutron stars are also good candidates to test alternative theories of gravity, in particular those that predict deviations depending upon the coupling of matter to geometry. Scalar-tensor theories posit the existence of a scalar field that, together with geometry, mediates gravitational phenomena. A subclass of these theories allows a phenomenon known as *scalarization*, whereby a sufficiently compact star spontaneously develops a scalar charge that modifies its gravitational interaction with other stars and allows for dipole radiation from the system (Damour & Esposito-Farese 1992). Though observations of binary pulsar systems tightly constrain these theories (Antoniadis et al. 2013), recent numerical work has shown that within the allowable region of parameter space strong departures from general relativity can occur late in the inspiral (Barausse et al. 2013). These differences are triggered close to the merger epoch (yet while the gravitational wave frequencies are still well within the reach of near-future detectors) and significantly modify the dynamics, causing an earlier onset of the plunge (Barausse et al. 2013, Palenzuela et al. 2014, Shibata et al. 2014).

5.2. Black Hole–Neutron Star Mergers

The remaining binary that is a target for Earth-based gravitational wave detectors is composed of a black hole and a neutron star. Here again, the regime in which the objects are widely separated is well described by a PN approximation, and the binary's dynamics proceeds as with the other cases discussed above. However, depending on the relation between two key radii—the tidal radius (R_T) and the radius of the ISCO (R_{ISCO})—markedly different behavior is expected near merger. These radii, to leading order, depend on the black hole mass and spin (for R_{ISCO}) as well as the binary-mass ratio, the star's mass, and EOS (for R_T). Approximately, the tidal radius $R_T \propto R_{NS}(3M_{BH}/M_{NS})^{1/3}$; R_{ISCO} is $6M_{BH}$ for a nonspinning black hole, decreasing (increasing) to M_{BH} ($9M_{BH}$) for a prograde (retrograde) orbit about a maximally spinning Kerr black hole. The importance of these two radii stems from the intuition that a plunge precedes tidal disruption if $R_{ISCO} > R_T$, and the opposite otherwise. This distinction is crucial, as in the former case there would be little difference in the gravitational wave signal compared with a binary black hole merger having the same masses (Foucart et al. 2013). By contrast, if disruption occurs, at its onset gravitational wave emission is sharply suppressed, not only allowing differentiation from the binary black hole case but also presenting clues about the star's EOS as this influences the frequency at which the disruption takes place (for a given neutron star mass). It is easy to convince oneself that the disruption possibility favors high spins/comparable masses while the plunging behavior favors low spins/high mass ratios. Note also that there are fewer channels for electromagnetic emission if disruption does not occur; in particular sGRBs and kilonova require it.

These observations about the nature of black hole–neutron star mergers are clearly born out in simulations. Early studies began with polytropic EOSs and nonspinning black holes, which have steadily progressed to incorporate more realistic EOSs and have covered a fair range of mass ratios and black hole spins (Foucart 2012; Kyutoku, Ioka & Shibata 2013, 2014). New physics is also being modeled, as we discuss above with binary neutron stars (because, of course, the same code infrastructure is used for both). Nevertheless, the same caveats concerning neutron star–neutron star binaries apply to black hole–neutron star systems in that simulations have not yet covered the full range of possible parameters, nor are sufficient computational resources available at present to adequately model all the relevant scales and microphysics.

Regarding the systems in which $R_{ISCO} < R_T$ and disruption occurs, for quasi-circular mergers numerical simulations have found as much as $0.3\,M_\odot$ of material outside the ISCO following merger (with the largest amounts coming from the low mass ratio/high prograde spin cases). These results have informed fitting formula predicting the amount of disk mass (Foucart 2012), which in turn can be used to estimate the spin of the final black hole (Pannarale 2014). Usually a larger fraction of stripped material is bound and subsequently accretes onto the black hole, though as much as $\approx 0.05\,M_\odot$ can be ejected from the system, moving with speeds $\approx 0.2c$. Typical maximum temperatures following disruption reach $\simeq 80$ Mev. The tail regions are substantially cooler, at about $\simeq 10$–100 Kev, though shocks can reheat this material to $\simeq 1$–3 Mev. Interestingly, if the black hole spin and orbital angular momentum direction are misaligned, strong differences arise. For inclinations $\gtrsim 30°$ a very low-mass disk seems capable of forming, with most of the material outside R_{ISCO} and following highly eccentric trajectories having large semimajor axes (Foucart 2012; Kyutoku, Ioka & Shibata 2013, 2014). Based on these trajectories it is estimated this material will return to the black hole to accrete at a rate governed by the familiar law $\dot{M} \propto t^{-5/3}$ (Lee & Ramirez-Ruiz 2007, Chawla et al. 2010, Deaton et al. 2013). Interestingly, the behavior of this material has characteristics consistent with kilonova models. However, the ejecta distribution is mainly around the orbital plane as opposed to the rather spheroidal one arising in binary neutron star mergers. In many cases, the amount of material capable of forming a disk is consistent with estimates for triggering sGRBs.

At the other end of the spectrum with low spins and/or high mass ratios in which $R_T < R_{ISCO}$, the star plunges into the black hole with little or no material left behind. For low-spin black holes this outcome happens for $m_{BH}/m_{NS} \gtrsim 6$; higher spins (or eccentricity, discussed below) can push this to somewhat larger mass ratios. For instance, for $m_{BH} = 10\,M_{\odot}$, $m_{NS} = 1.4\,M_{\odot}$, significant disruption only takes place for $a_{BH}/m_{BH} \gtrsim 0.9$ (Foucart et al. 2012). Without disruption, electromagnetic and neutrino counterparts such as sGRBs and kilonova are not expected to occur, though as we discuss below there may still be electromagnetic emission if the neutron star has a strong enough magnetic field. Naturally, as in the binary neutron star case, a plethora of phenomena can be triggered by the system's dynamics, and diverse works are proceeding to examine these scenarios, several of which we describe here.

5.2.1. Magnetized stars. A few studies have explored the behavior of the system when the neutron star is magnetized. Although electromagnetic effects are too subleading to alter the orbit and gravitational wave emission, the binary's dynamics can affect properties of the electromagnetic field after merger. In particular, the resulting field topology in the newly formed accretion disk is relevant to assessing whether a jet can be launched from the system. Notice that in the absence of spin-orbit-induced precession near the onset of disruption, the initial poloidal field gets twisted to a mainly toroidal configuration, implying further processes, such as dynamo/MRIs (Balbus & Hawley 1991), would need to take over to reinstate a poloidal configuration for an efficient jet mechanism to operate (McKinney & Blandford 2009). The orbit will precess if the black hole is spinning and the spin axis is misaligned with the orbital angular momentum; then the resulting magnetic field topology after disruption has both poloidal and toroidal components. This might aid in giving rise to favorable configurations for jet launching (see the discussion by Foucart 2012). Full simulations accurately resolving all of these effects for both precessing and nonprecessing configurations have yet to be performed, owing to their heavy computational requirements.

5.2.2. Magnetosphere interactions. As mentioned when discussing black holes interacting with a magnetosphere, the latter is able to tap kinetic energy—rotational or translational—from the black hole if there is a relative motion between them. Such a scenario naturally arises during the inspiral of a magnetized neutron star with a black hole (the binary will not be tidally locked, and so there will be relative motion of the black hole through the magnetic field lines sourced by the neutron star). Basic estimates using a simple "unipolar induction model" indicate the possibility of a strong Poynting luminosity produced by the system (Hansen & Lyutikov 2001, McWilliams & Levin 2011). First simulations in this direction have recently been completed, obtaining consistent values with $L \simeq 10^{41} B_{12}^2$ (Paschalidis, Etienne & Shapiro 2013). Though lacking a detailed account of how this Poynting flux could be converted to observable photons, this offers the possibility of an electromagnetic counterpart preceding the merger.

5.2.3. Neutrino emissions. As in the case of binary neutron stars, simulations are just beginning to incorporate neutrino effects, again using the leakage scheme. **Figure 7** shows an example of the neutrino luminosity from one such ongoing investigation (a follow-up study to Deaton et al. 2013). They found that the merger of a 1.4-$M_{\odot}$ star with a black hole having mass $7\,M_{\odot}$ and spin parameter $a/M \equiv J/M^2 = 0.9$ (so significant disruption takes place) yields a peak neutrino luminosity on the order of $\simeq 10^{54}$ erg s^{-1} shortly after disruption, decreasing by an order of magnitude after 50 ms.

5.2.4. Eccentric binaries. In eccentric black hole–neutron star encounters, similar quantitative and qualitative differences arise compared with quasi-circular inspiral as discussed above for the other systems (the possibility of zoom-whirl orbital dynamics, neutron star crust cracking, and/or excitation of f-modes during close encounters; typically larger amounts of ejecta and accretion disk mass; etc.). In addition, because the effective ISCO for eccentric orbits of particles orbiting

Figure 7

Luminosity from all neutrino species at $\simeq$12.5 ms after the merger of a black hole ($M_{BH} = 7\,M_{\odot}$) with a neutron star ($M_{NS} = 1.4\,M_{\odot}$, described by the "LS220" equation of state). The emission region coincides roughly with the disk; namely densities $\rho > 3 \times 10^9$ g cm^{-3} are approximately within the white region, $\rho > 2 \times 10^{10}$ g c^{-3} the red/orange regions, and the maximum density in the disk is $\rho \simeq 6 \times 10^{11}$ g cm^{-3}. Figure from F. Foucart for the SXS Collaboration.

a nonextremal black hole is closer to the event horizon (e.g., for a Schwarzschild black hole the geodesic ISCO moves in from $r = 6M$ to $r = 4M$ going from $e = 0$ to 1), the limit for the onset of observable tidal disruption moves to slightly higher mass ratios. (Note by "observable" tidal disruption we mean disruption that can influence the gravitational waveform and the matter ejected or left to accrete. In theory the location of the event horizon sets the ultimate location for this, though for black hole–neutron star interactions the effective ISCO appears to be a better proxy. The reason is that once a matter parcel crosses the ISCO, barring the rise of strong nongravitational forces, it will reach the event horizon on the order of the time it takes light to cross the black hole. Though to date simulations have not included all the relevant matter microphysics, it is unlikely that effects triggered by tidal disruption, e.g., shock heating, shearing of magnetic fields, etc., could grow large enough on such a short timescale to prevent immediate accretion of the matter.) Furthermore, in systems where tidal disruption begins outside the ISCO there is the possibility of multiple partial disruptions and accretion episodes prior to the final disruption/merger (Stephens, East & Pretorius 2011, East, Pretorius & Stephens 2012a).

6. GRAVITATIONAL COLLAPSE TO A NEUTRON STAR OR BLACK HOLE

Considerable efforts have been undertaken to study gravitational collapse to a neutron star or a black hole, in particular within the context of core-collapse supernovae. Here, stars with masses in the range $10\,M_{\odot} \lesssim M \lesssim 100\,M_{\odot}$ at zero-age main sequence form cores that can exceed the Chandrasekhar mass and become gravitationally unstable. This leads to collapse that compresses the inner core to nuclear densities, at which point the full consequences of general relativity must be accounted for. Depending upon the mass of the core, it can "bounce" or collapse to a black hole. **Figure 8** displays representative snapshots of the behavior of a collapsing 75-$M_{\odot}$ star at different times. The collapse forms a protoneutron star that later collapses to a black hole. In the case of a bounce, an outward propagating shock wave is launched that collides with still infalling material and stalls. Observations of core-collapse supernovae imply some mechanism is capable of reviving the shock, which is then able to plow through the stellar envelope and blow up the star. This process is extremely energetic, releasing energies on the order of 10^{53} erg, the majority of which is emitted in neutrinos (for a recent review, see Ott 2009). For several decades now, the primary motivation driving theoretical and numerical studies has been to understand what process (or combination of processes) mediates such revival and how. Several suspects have been identified: heating by neutrinos, (multidimensional) hydrodynamical instabilities, magnetic fields, and nuclear burning (see, e.g., Burrows et al. 2007, Janka et al. 2007). With the very disparate time- and space-scales involved, a multitude of physically relevant effects to consider, and the intrinsic cost to accurately model them (e.g., radiation transport is a seven-dimensional problem), progress has been slow. Moreover, electromagnetic observations do not provide much guidance to constrain possible mechanisms as they cannot peer deep into the central engine. In contrast, observations of gravitational waves and neutrinos have the potential to do so, provided the exploding star is sufficiently nearby. Thus, in addition to exploring mechanisms capable of reviving the stalled shock, simulations have also concentrated on predicting specific gravitational wave and neutrino signatures. Modeling gravity using full general relativity has only been undertaken recently (Ott et al. 2013), though prior to this some of the more relevant relativistic affects were incorporated (e.g., Dimmelmeier, Font & Müller 2002; Obergaulinger et al. 2006; Müller, Janka & Dimmelmeier 2010; Wongwathanarat, Janka & Mueller 2013). Although the full resolution of the problem is still likely years ahead, interesting insights into fundamental questions and observational prospects have been garnered. For example, simulations have shown that in rotating core-collapse scenarios, gravitational waves can be produced and their characteristics are strongly dependent on properties of the collapse: the precollapse central angular velocity, the development of nonaxisymmetric rotational instabilities, postbounce convective overturn, the standing accretion shock instability, protoneutron star pulsations, etc. If a black hole forms, gravitational wave emission is mainly determined by the quasi-normal modes of the newly formed black hole. The typical frequencies of gravitational radiation can lie in the range $\simeq$100–1,500 Hz and so are potential sources for advanced Earth-based gravitational wave detectors (though the amplitudes are sufficiently small that it would need to be a Galactic event). As mentioned, the characteristics of these waveforms depend on the details of the collapse and, hence, could allow us to distinguish the mechanism inducing the explosion. Neutrino signals have also been calculated, revealing possible correlations between oscillations of gravitational waves and variations in neutrino luminosities. However, current estimates suggest neutrino detections would be difficult for events taking place at kiloparsec distances (Ott et al. 2013).

7. FURTHER FRONTIERS

Beyond comparable-mass ratios and radii binaries, NR simulations are starting to explore binaries involving higher mass ratios or less dense stars: black hole–white dwarf, neutron star–white dwarf, intermediate mass black holes and main sequence stars, black hole binaries involving intermediate and stellar masses (Haas et al. 2012), etc. Here, more rapid progress is hampered by the computational cost, as it is considerably higher to simulate the larger range of spatial and temporal scales over which the interesting dynamics takes place. Several approaches have been suggested to address, at least in part, this difficulty. These include the use of implicit-explicit methods to tackle large-mass-ratio binary black holes (Lau, Lovelace & Pfeiffer 2011), a suitable rescaling of physical parameters to model neutron star–white dwarf binaries (Paschalidis et al. 2011), a "background

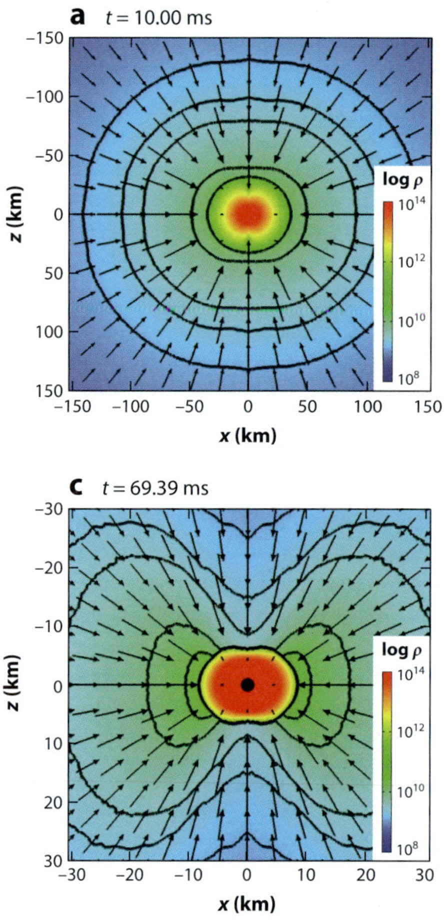

Figure 8

Density color maps of the meridional plane of a collapsing 75-$M_\odot$ star, superposed with velocity vectors at various times after bounce (and note the different scale of panel *a* from the rest). The collapse first forms (*a*,*b*) a protoneutron star that later collapses to (*c*,*d*) a black hole. Reprinted from Ott et al. (2011) with permission.

subtraction" technique to study extreme-mass-ratio systems in which the solution of the dominant gravitational body is known (East & Pretorius 2013), and a reformulation of the problem in terms of a PN approximation incorporating both black hole and matter effects to allow straightforward modification of existing "Newtonian-based" astrophysical codes (Barausse & Lehner 2013). These are illustrative examples indicating how the field is progressing beyond traditional boundaries. To date, however, as far as astrophysical applications are concerned, the predominant focus of NR has been the compact binary problem [and more recently including the study of core-collapse supernovae (Ott et al. 2013)]. Complementary efforts have been directed toward understanding fundamental questions about strongly gravitating settings, some of which have clear astrophysical implications. One example is the question of whether gravitational collapse always leads to a black hole, which is described by the Kerr solution, or to a naked singularity. Although the cases studied so far have indeed shown the black hole result to be the case, especially in astrophysically relevant contexts, counter examples have been constructed. In $d = 4$ spacetime dimensions these include collapsing matter configurations finely tuned to the threshold of black hole formation [in so-called Type II critical collapse (Choptuik 1993); see, e.g., Gundlach (2003) for a review and Joshi (2013) for spherical collapse of ideal fluids]. Due to the fine tuning required to reach Planck-scale curvatures visible outside an event horizon, it is unlikely critical collapse occurs naturally in the Universe (though see Niemeyer & Jedamzik 1998 for arguments suggesting it would be relevant if certain primordial black hole formation scenarios occurred). By contrast, this is not the case in higher dimensions in which simulations of a class of black holes (black strings) have shown that violations of cosmic censorship can arise generically (Lehner & Pretorius 2010). This not only highlights that Einstein gravity still holds secrets that could be revealed by theoretical studies but also that surprises of astrophysical significance might be in store if our Universe were in fact higher dimensional.

8. FINAL COMMENTS

In this review, we have described the status of NR applied in astrophysical contexts. We have focused our presentation on events in which strong gravitational interactions require the full Einstein field equations to unravel all details of the phenomena. Due to page limitations, we have had to choose a representative subset of all relevant activities; nevertheless we hope it is clear that the field has "come of age." Yet there is still much to learn, and continued efforts will refine NR's predictions and application in astrophysics.

DISCLOSURE STATEMENT

The authors are not aware of any affiliations, memberships, funding, or financial holdings that might be perceived as affecting the objectivity of this review.

ACKNOWLEDGMENTS

We want to thank our collaborators on work mentioned here: M. Anderson, E. Barausse, W. East, E. Hirschman, S. Liebling, S. McWilliams, P. Motl, D. Neilsen, C. Palenzuela, F. Ramazanoglou, B. Stephens, and N. Yunes. We also acknowledge helpful discussions with K. Belczynski, E. Berger, D. Brown, A. Buonanno, C. Fryer, T. Janka, W. Lee, B. Metzger, R. Narayan, E. Poisson, E. Quataert, E. Ramirez-Ruiz, S. Rosswog, A. Spitkovsky, and J. Stone. For figures, we thank A. Buonanno, F. Foucart, B. Giacommazzo, R. Gold, M. Hannam, K. Hotokezaka, K. Kyutoku, C. Lousto, C. Ott, H. Pfeiffer, S. Shapiro, and M. Shibata. We also thank the participants of the "Chirps, Mergers and Explosions: The Final Moments of Coalescing Compact Binaries" workshop at the Kavli Institute for Theoretical Physics for stimulating discussions. This work was supported in part by an NSERC Discovery Grant and CIFAR (LL), NSF grants PHY-1065710 and PHY1305682, and the Simons Foundation (FP). Research at Perimeter Institute is supported by the Government of Canada through Industry Canada and by the Province of Ontario through the Ministry of Research and Innovation.

LITERATURE CITED

Abadie J, Abbott BP, Abbott R, Abernathy M, Accadia T, et al. 2010. *Class. Quant. Gravity* 27:173001

Abbott BP, Abbott R, Adhikari R, Ajith P, Allen B, et al. 2009. *Rep. Prog. Phys.* 72:076901

Accadia T, Acernese F, Antonucci F, Astone P, Ballardin G, et al. 2011. *Class. Quant. Gravity* 28:114002

Agathos M, Del Pozzo W, Li TGF, Broeck CVD, Veitch J, et al. 2014. *Phys. Rev. D* 89:082001

Ajith P, Hannam M, Husa S, Chen Y, Bruegmann B, et al. 2011. *Phys. Rev. Lett.* 106:241101
Alcubierre M. 2008. *Introduction to 3+1 Numerical Relativity*. Oxford, UK: Oxford Univ. Press
Amaro-Seoane P, Aoudia S, Babak S, Binetruy P, Berti E, et al. 2012. *GW Notes* 6:4-110
Anderson M, Hirschmann EW, Lehner L, Liebling SL, Motl PM, et al. 2008a. *Phys. Rev. D* 77:024006
Anderson M, Hirschmann EW, Lehner L, Liebling SL, Motl PM, et al. 2008b. *Phys. Rev. Lett.* 100:191101
Andersson N, Baker J, Belczynski K, Bernuzzi S, Berti E, et al. 2013. *Class. Quant. Gravity* 30:193002
Antognini JM, Shappee BJ, Thompson TA, Amaro-Seoane P. 2014. *MNRAS* 439(1):1079–91
Antoniadis J, Freire PCC, Wex N, Tauris TM, Lynch RS, et al. 2013. *Science* 340:448
Antonini F, Murray N, Mikkola S. 2014. *MNRAS* 439(1):1079–91
Arun KG, Iyer BR, Qusailah MSS, Sathyaprakash BS. 2006. *Class. Quant. Gravity* 23:L37–43
Baiotti L, Shibata M, Yamamoto T. 2010. *Phys. Rev. D* 82:064015
Baker JG, Centrella J, Choi D-I, Koppitz M, van Meter JR, Miller MC. 2006. *Ap. J. Lett.* 653:L93–96
Balbus SA, Hawley JF. 1991. *Ap. J.* 376:214–33
Barausse E. 2012. *MNRAS* 423:2533–57
Barausse E, Lehner L. 2013. *Phys. Rev. D* 88:024029
Barausse E, Morozova V, Rezzolla L. 2012. *Ap. J.* 758:63
Barausse E, Palenzuela C, Ponce M, Lehner L. 2013. *Phys. Rev. D* 87:081506
Barausse E, Rezzolla L. 2009. *Ap. J. Lett.* 704:L40–44
Baumgarte TW, Shapiro SL. 2010. *Numerical Relativity: Solving Einstein's Equations on the Computer.* Cambridge, UK: Cambridge Univ. Press
Berger E. 2014. *Annu. Rev. Astron. Astrophys.* 52:43–105
Berger E, Fong W, Chornock R. 2013. *Ap. J. Lett.* 774:L23
Berger MJ, Oliger J. 1984. *J. Comp. Phys.* 53:484
Berti E, Cardoso V, Gonzáles JA, Sperhake U, Hannam M, et al. 2007. *Phys. Rev. D* 76:064034
Berti E, Cardoso V, Starinets AO. 2009. *Class. Quant. Gravity* 26:163001
Berti E, Cardoso V, Will CM. 2006. *Phys. Rev. D* 73:064030
Berti E, Volonteri M. 2008. *Ap. J.* 684:822–28
Bini D, Damour T. 2012. *Phys. Rev. D* 86:124012
Blanchet L. 2002. *Living Rev. Relativ.* 5:3. **http://www.livingreviews.org/lrr-2002-3**
Blandford RD, Znajek RL. 1977. *MNRAS* 179:433–45
Bode T, Bogdanovic T, Haas R, Healy J, Laguna P, Shoemaker DM. 2012. *Ap. J.* 744:45
Bogdanovic T, Reynolds CS, Miller MC. 2007. *Ap. J. Lett.* 661:L147–50
Bona C, Bona-Casas C, Palenzuela-Luque C. 2009. In *Elements of Numerical Relativity and Relativistic Hydrodynamics*, ed. C Bona, C Palenzuela-Luque, C Bona-Casas. *Lect. Notes Phys.*, Vol. 783. Berlin: Springer-Verlag
Bower RG, Benson AJ, Malbon R, Helly JC, Frenk CS, et al. 2006. *MNRAS* 370:645–55
Boyd JP. 1989. *Chebyshev and Fourier Spectral Methods.* New York: Springer-Verlag
Boyle L, Kesden M, Nissanke S. 2008. *Phys. Rev. Lett.* 100:151101
Brodbeck O, Frittelli S, Hübner P, Reula OA. 1999. *J. Math. Phys.* 40:909–23
Broderick AE, Fish VL, Doeleman SS, Loeb A. 2011. *Ap. J.* 738:38
Broderick AE, Johannsen T, Loeb A, Psaltis D. 2014. *Ap. J.* 784:7–21
Broderick AE, Loeb A, Narayan R. 2009. *Ap. J.* 701:1357–66
Buonanno A. 2007. *Proc. Les Houches Summer Sch., Part. Phys. Cosmol.: Fabr. Spacetime, Les Houches, France*, Jul. 31–Aug. 25, 2006
Buonanno A, Cook GB, Pretorius F. 2007. *Phys. Rev. D* 75:124018
Buonanno A, Damour T. 1999. *Phys. Rev. D* 59:084006
Buonanno A, Kidder LE, Lehner L. 2008. *Phys. Rev. D* 77:026004
Burrows A, Dessart L, Ott CD, Livne E. 2007. *Phys. Rep.* 442:23–37
Campanelli M, Lousto CO, Zlochower Y, Merritt D. 2007. *Ap. J. Lett.* 659:L5–8
Casares J, Jonker P. 2014. *Space Sci. Rev.* In press. doi: 10.1007/s11214-013-0030-6
Centrella J, Baker JG, Kelly BJ, van Meter JR. 2010. *Annu. Rev. Nucl. Part. Sci.* 60:75–100
Chawla S, Anderson M, Besselman M, Lehner L, Liebling SL, et al. 2010. *Phys. Rev. Lett.* 105:111101
Choptuik MW. 1993. *Phys. Rev. Lett.* 70:9–12

Collins NA, Hughes SA. 2004. *Phys. Rev. D* 69:124022
Croton DJ, Springel V, White SDM, De Lucia G, Frenk CS, et al. 2006. *MNRAS* 365:11–28
Damour T, Esposito-Farese G. 1992. *Class. Quant. Gravity* 9:2093–176
Damour T, Nagar A, Bernuzzi S. 2013. *Phys. Rev. D* 87:084035
Damour T, Nagar A, Villain L. 2012. *Phys. Rev. D* 85:123007
Davis SW, Narayan R, Zhu Y, Barret D, Farrell SA, et al. 2011. *Ap. J.* 734:111
Deaton MB, Duez MD, Foucart F, O'Connor E, Ott CD, et al. 2013. *Ap. J.* 776:47
Demorest PB, Pennucci T, Ransom SM, Roberts MSE, Hessels JWT. 2010. *Nature* 467:1081–83
Dimmelmeier H, Font JA, Müller E. 2002. *Astron. Astrophys.* 388:917–35
Doeleman SS, Fish VL, Schenck DE, Beaudoin C, Blundell R, et al. 2012. *Science* 338:355
Dotti M, Volonteri M, Perego A, Colpi M, Ruszkowski M, Haardt F. 2010. *MNRAS* 402:682–90
Duez MD. 2010. *Class. Quant. Gravity* 27:114002
Duez MD, Foucart F, Kidder LE, Pfeiffer HP, Scheel MA, Teukolsky SA. 2008. *Phys. Rev.* D 78:104015
East WE, McWilliams ST, Levin J, Pretorius F. 2013. *Phys. Rev. D* 87:043004
East WE, Pretorius F. 2012. *Ap. J. Lett.* 760:L4
East WE, Pretorius F. 2013. *Phys. Rev. D* 87:101502
East WE, Pretorius F, Stephens BC. 2012a. *Phys. Rev. D* 85:124009
East WE, Pretorius F, Stephens BC. 2012b. *Phys. Rev. D* 85:124010
Eichler D, Livio M, Piran T, Schramm DN. 1989. *Nature* 340:126–28
Fanidakis N, Baugh CM, Benson AJ, Bower RG, Cole S, et al. 2011. *MNRAS* 410:53–74
Farrell S, Webb N, Barret D, Godet O, Rodrigues J. 2009. *Nature* 460:73–75
Farris BD, Gold R, Paschalidis V, Etienne ZB, Shapiro SL. 2012. *Phys. Rev. Lett.* 109:221102
Field SE, Galley CR, Herrmann F, Oschner E, Tiglio M. 2011. *Phys. Rev. Lett.* 106:221102
Flanagan EE, Hughes SA. 2005. *New J. Phys.* 7:204
Font JA. 2008. *Living Rev. Relativ.* 11:7
Foucart F. 2012. *Phys. Rev. D* 86:124007
Foucart F, Buchman L, Duez MD, Grudich M, Kidder LE, et al. 2013. *Phys. Rev. D* 88:064017
Foucart F, Duez MD, Kidder LE, Scheel MA, Szilagyi B, et al. 2012. *Phys. Rev. D* 85:044015
Garfinkle D, Lim WC, Pretorius F, Steinhardt PJ. 2008. *Phys. Rev. D* 78:083537
Gebhardt K, Kormendy J, Ho LC, Bender R, Bower G, et al. 2000. *Ap. J. Lett.* 543:L5–8
Gerosa D, Kesden M, Berti E, O'Shaughnessy R, Sperhake U. 2013. *Phys. Rev. D* 87:104028
Giacomazzo B, Rezzolla L, Baiotti L. 2011. *Phys. Rev. D* 83:044014
Godet O, Barret D, Webb N, Farrell S, Gehrels N. 2009. *Ap. J. Lett.* 705:L109–12
Gold R, Bernuzzi S, Thoerfelder M, Brügmann B, Pretorius F. 2012. *Phys. Rev. D* 86:121501
Gold R, Brügmann B. 2013. *Phys. Rev. D* 88:064051
Gold R, Paschalidis V, Etienne ZB, Shapiro SL, Pfeiffer HP. 2014. *Phys. Rev. D* 89(6):064060
Goldreich P, Julian WH. 1969. *Ap. J.* 157:869–80
González JA, Sperhake U, Brügmann B, Hannam MD, Husa S. 2007. *Phys. Rev. Lett.* 98:091101
Grandclement P, Novak J. 2009. *Living Rev. Relativ.* 12:1
Greene JE, Ho LC. 2004. *Ap. J.* 610:722–36
Gundlach C. 2003. *Phys. Rep.* 376:339–405
Gundlach C, Martin-Garcia JM, Calabrese G, Hinder I. 2005. *Class. Quant. Gravity* 22:3767–74
Gustafsson B, Kreiss HO, Oliger J. 1995. *Time Dependent Problems and Difference Methods*. New York: Wiley
Haas R, Shcherbakov RV, Bode T, Laguna P. 2012. *Ap. J.* 749:117
Hannam M. 2013. *Gen. Relativ. Gravit.* Accepted. arXiv:1312.3641
Hansen BM, Lyutikov M. 2001. *MNRAS* 322:695
Healy J, Levin J, Shoemaker D. 2009. *Phys. Rev. Lett.* 103:131101
Herrmann F, Hinder I, Shoemaker D, Laguna P. 2007. *Class. Quant. Gravity* 24:33
Hinder I, Buonanno A, Boyle M, Etienne ZB, Healy J, et al. 2013. *Class. Quant. Gravity* 31:025012
Hinderer T, Lackey BD, Lang RN, Read JS. 2010. *Phys. Rev. D* 81:123016
Horbatsch M, Burgess C. 2012. *J. Cosmol. Astropart. Phys.* 1205:010
Hotokezaka K, Kiuchi K, Kyutoku K, Muranushi T, Sekiguchi Y-I, et al. 2013. *Phys. Rev. D* 88:044026
Hughes S. 2009. *Annu. Rev. Astron. Astrophys.* 47:107–57

Int. Pulsar Timing Array. 2013. *Class. Quant. Gravity* 30:224010
Jacobson T, Sotiriou TP. 2009. *Phys. Rev. Lett.* 103:141101
Janka HT, Eberl T, Ruffert M, Fryer CL. 1999. *Ap. J. Lett.* 527:L39–42
Janka HT, Langanke J, Marek A, Martínez-Pinedo G, Müller B. 2007. *Phys. Rep.* 442:38–74
Johnson MC, Peiris HV, Lehner L. 2012. *Phys. Rev. D* 85:083516
Joshi PS. 2013. In *Spring Handbook of Spacetime*, ed. A Ashtekar, V Petkov. In press. New York: Springer arXiv:1311.0449
Kalogera V, Belczynski K, Kim C, O'Shaughnessy RW, Willems B. 2007. *Phys. Rep.* 442:75–108
Kamizasa N, Terashima Y, Awaki H. 2012. *Ap. J.* 751(1):39
Kaplan JD, Ott CD, O'Connor EP, Kiuchi K, Roberts L, Duez M. 2014. *Ap. J.* 790:19
Kesden M, Lockhart G, Phinney ES. 2010. *Phys. Rev. D* 82:124045
Kesden M, Sperhake U, Berti E. 2010. *Ap. J.* 715:1006–11
Kiuchi K, Sekiguchi Y, Kyutoku K, Shibata M. 2012. *Class. Quant. Gravity* 29:124003
Kocsis B, Levin J. 2012. *Phys. Rev. D* 85:123005
Komossa S. 2012. *Adv. Astron.* 2012:364973
Komossa S, Zhou H, Lu H. 2008. *Ap. J. Lett.* 678:L81–84
Kormendy J, Ho LC. 2013. *Annu. Rev. Astron. Astrophys.* 51:511–653
Kormendy J, Richstone D. 1995. *Annu. Rev. Astron. Astrophys.* 33:581
Kushnir D, Katz B, Dong S, Livne E, Fernández R. 2013. *Ap. J. Lett.* 778:L37
Kyutoku K, Ioka K, Shibata M. 2013. *Phys. Rev. D* 88:041503
Kyutoku K, Ioka K, Shibata M. 2014. *MNRAS* 437:L6
Laguna P, Garfinkle D. 1989. *Phys. Rev. D* 40:1011–16
Lattimer JM, Prakash M. 2010. In *From Nuclei to Stars*, ed. S Lee, p. 275. Singapore: World Sci.
Lau SR, Lovelace G, Pfeiffer HP. 2011. *Phys. Rev. D* 84:084023
Lee WH, Ramirez-Ruiz E. 2007. *New J. Phys.* 9:17
Lee WH, Ramirez-Ruiz E, Van de Ven G. 2010. *Ap. J.* 720:953
Lehner L. 2001. *Class. Quant. Gravity* 18:R25–86
Lehner L, Liebling SL, Reula OA. 2006. *Class. Quant. Gravity* 23:S421–46
Lehner L, Palenzuela C, Liebling SL, Thompson C, Hanna C. 2012. *Phys. Rev. D* 86:104035
Lehner L, Pretorius F. 2010. *Phys. Rev. Lett.* 105:101102
Lehner L, Reula O, Tiglio M. 2005. *Class. Quant. Gravity* 22:5283–322
Levan AJ, Wynn GA, Chapman R, Davies MB, King AR, et al. 2006. *MNRAS* 368:L1–5
LeVeque RJ. 1992. *Numerical Methods for Conservation Laws*. Basel: Birkhauser Verlag
Lippai Z, Frei Z, Haiman Z. 2008. *Ap. J. Lett.* 676:L5–8
Lipunov VM, Panchenko IE. 1996. *Astron. Astrophys.* 312:937–40
Loeb A. 2007. *Phys. Rev. Lett.* 99:041103
Löffler F, Faber J, Bentivegna E, Bode T, Diener P, et al. 2011. *Class. Quant. Gravity* 29:115001
Lousto CO, Campanelli M, Zlochower Y, Nakano H. 2010. *Class. Quant. Gravity* 27:114006
Lousto CO, Zlochower Y. 2011. *Phys. Rev. Lett.* 107:231102
Lousto CO, Zlochower Y, Dotti M, Volonteri M. 2012. *Phys. Rev. D* 85:084015
MacFadyen AI, Ramirez-Ruiz E, Zhang W. 2005. astro-ph/0510192
Magorrian J, Tremaine S, Richstone D, Bender R, Bower G, et al. 1998. *Astron. J.* 115:2285–305
McClintock JE, Narayan R, Davis SW, Gou L, Kulkarni A, et al. 2011. *Class. Quant. Gravity* 28:114009
McClintock JE, Narayan R, Steiner JF. 2013. *Space Sci. Rev.* In press. doi:10.1007/s11214-013-0003-9
McKinney JC, Blandford RD. 2009. *MNRAS* 394:L126–30
McWilliams ST, Levin JJ. 2011. *Ap. J.* 742:90
Metzger BD, Berger E. 2012. *Ap. J.* 746:48
Metzger BD, Quataert E, Thompson TA. 2008. *MNRAS* 385:1455–60
Milosavljević M, Phinney ES. 2005. *Ap. J. Lett.* 622:L93–96
Mösta P, Alic D, Rezzolla L, Zanotti O, Palenzuela C. 2012. *Ap. J. Lett.* 749:L32
Müller B, Janka HT, Dimmelmeier H. 2010. *Ap. J. Suppl. Ser.* 189:104–33
Narayan R, Paczynski B, Piran T. 1992. *Ap. J. Lett.* 395:L83
Neilsen D, Hirschmann EW, Millward RS. 2006. *Class. Quant. Gravity* 23:S505

Niemeyer JC, Jedamzik K. 1998. *Phys. Rev. Lett.* 80:5481–84
Noble SC, Mundim BC, Nakano H, Krolik JH, Campanelli M, et al. 2012. *Ap. J.* 755:51
Obergaulinger M, Aloy MA, Dimmelmeier H, Müller E. 2006. *Astron. Astrophys.* 457:209–22
Obergaulinger M, Aloy MA, Müller E. 2010. *Astron. Astrophys.* 515:A30
O'Leary RM, Kocsis B, Loeb A. 2009. *MNRAS* 395:2127–46
O'Shaughnessy R, Kaplan D, Sesana A, Kamble A. 2011. *Ap. J.* 743:136
Ott CD. 2009. *Class. Quant. Gravity* 26:063001
Ott CD, Abdikamalov E, Moesta P, Haas R, Drasco S, et al. 2013. *Ap. J.* 768:115
Ott CD, Reisswig C, Schnetter E, O'Connor E, Sperhake U, et al. 2011. *Phys. Rev. Lett.* 106:161103
Palenzuela C, Barausse E, Ponce M, Lehner L. 2014. *Phys. Rev. D* 89:044024
Palenzuela C, Garrett T, Lehner L, Liebling SL. 2010. *Phys. Rev. D* 82:044045
Palenzuela C, Lehner L, Liebling SL. 2010. *Science* 329:927–30
Palenzuela C, Lehner L, Liebling SL, Ponce M, Anderson M, et al. 2013a. *Phys. Rev. D* 88:043011
Palenzuela C, Lehner L, Ponce M, Liebling SL, Anderson M, et al. 2013b. *Phys. Rev. Lett.* 111:061105
Palenzuela C, Lehner L, Yoshida S. 2010. *Phys. Rev. D* 81:084007
Pan Y, Buonanno A, Taracchini A, Kidder LE, Mroué AH, et al. 2014. *Phys. Rev. D* 89:084006
Pannarale F. 2014. *Phys. Rev. D* 89:044045
Paschalidis V, Etienne ZB, Shapiro SL. 2013. *Phys. Rev. D* 88:021504
Paschalidis V, Liu YT, Etienne Z, Shapiro SL. 2011. *Phys. Rev. D* 84:104032
Peters PC. 1964. *Phys. Rev.* 136:B1224–32
Peters PC, Mathews J. 1963. *Phys. Rev.* 131:435–40
Peterson BM, Ferrarese L, Gilbert KM, Kaspi S, Malkan MA, et al. 2004. *Ap. J.* 613:682–99
Pfeiffer HP. 2012. *Class. Quant. Gravity* 29:124004
Piran T, Nakar E, Rosswog S. 2013. *MNRAS* 430:2121–36
Pretorius F. 2005. *Phys. Rev. Lett.* 95:121101
Pretorius F. 2006. *Class. Quant. Gravity* 23:S529–52
Pretorius F. 2009. In *Physics of Relativistic Objects in Compact Binaries: From Birth to Coalescence*, ed. M Colpi, P Casella, V Gorini, U Moschella, A Possenti, pp. 305–69. Heidelberg: Springer-Verlag
Pretorius F, Khurana D. 2007. *Class. Quant. Gravity* 24:S83–108
Price DJ, Rosswog S. 2006. *Science* 312:719–22
Read JS, Baiotti L, Creighton JDE, Friedman JL, Giacomazzo B, et al. 2013. *Phys. Rev. D* 88:044042
Rees MJ. 1988. *Nature* 333:523–28
Reynolds CS. 2013. *Class. Quant. Gravity* 30:244004
Rezzolla L, Baiotti L, Giacomazzo B, Link D, Font JA.2010. *Class. Quant. Gravity* 27:114105
Roedig C, Sesana A. 2012. *J. Phys. Conf. Ser.* 363:012035
Ruffert M, Janka HT, Takahashi K, Schäfer G. 1997. *Astron. Astrophys.* 319:122–53
Samsing J, MacLeod M, Ramirez-Ruiz E. 2014. *Ap. J.* 784:71
Sarbach O, Tiglio M. 2012. *Living Rev. Relativ.* 15:9
Scannapieco E, Silk J, Bouwens R. 2005. *Ap. J. Lett.* 635:L13–16
Scheel MA, Boyle M, Chu T, Kidder LE, Matthews KD, Pfeiffer HP. 2009. *Phys. Rev. D* 79:024003
Schmidt P, Hannam M, Husa S, Ajith P. 2011. *Phys. Rev. D* 84:024046
Schnetter E, Hawley SH, Hawke I. 2004. *Class. Quant. Gravity* 21:1465–88
Schnittman JD. 2013. *Class. Quant. Gravity* 30:244007
Sekiguchi Y, Kiuchi K, Kyutoku K, Shibata M. 2011. *Phys. Rev. Lett.* 107:051102
Seoane PA, Aoudia S, Audley H, Auger G, Babak S, et al. 2013. *The gravitational universe.* White Pap., The eLISA Space Gravitational Wave Observatory, Hanover, Ger. arXiv:1305.5720
Sesana A, Gair J, Berti E, Volonteri M. 2011. *Phys. Rev. D* 83:044036
Seto N. 2013. *Phys. Rev. Lett.* 111:061106
Shankar F. 2009. *New Astron. Rev.* 53:57–77
Shapiro SL. 2010. *Phys. Rev. D* 81:024019
Shapiro SL, Teukolsky SA. 1991. *Phys. Rev. Lett.* 66:994–97
Shibata M, Taniguchi K, Okawa H, Buonanno A. 2014. *Phys. Rev. D* 89(8):084005
Soltan A. 1982. *MNRAS* 200:115–22

Somiya K. 2012. *Class. Quant. Gravity* 29:124007
Stephens BC, East WE, Pretorius F. 2011. *Ap. J. Lett.* 737:L5
Stephens BC, Shapiro SL, Liu YT. 2008. *Phys. Rev. D* 77:044001
Stone N, Loeb A. 2011. *MNRAS* 412:75–80
Tanvir NR, Levan AJ, Fruchter AS, Hjorth J, Hounsell RA, et al. 2013. *Nature* 500:547–49
Taracchini A, Buonanno A, Pan Y, Hinderer T, Boyle M, et al. 2014. *Phys. Rev. D* 89:061502
Tchekhovskoy A, Narayan R, McKinney JC. 2010. *Ap. J.* 711:50–63
Tichy W, Marronetti P. 2008. *Phys. Rev. D* 78:081501
Tsang D. 2013. *Ap. J.* 777:103
Tsang D, Read JS, Hinderer T, Piro AL, Bondarescu R. 2012. *Phys. Rev. Lett.* 108:011102
Volonteri M, Sikora M, Lasota JP. 2007. *Ap. J.* 667:704–13
Volonteri M, Sikora M, Lasota JP, Merloni A. 2013. *Ap. J.* 775:94
Wainwright CL, Johnson MC, Peiris HV, Aguirre A, Lehner L, Liebling SL. 2014. *J. Cosmol. Astropart. Phys.* 1403:030
Wald RM. 1997. *Phys. Rev. D* 56:6467–74
Wen L. 2003. *Ap. J.* 598:419–30
Will CM. 1993. *Theory and Experiment in Gravitational Physics.* Cambridge, UK: Cambridge Univ. Press
Will CM. 2006. *Living Rev. Relativ.* 9:3. **http://www.livingreviews.org/lrr-2006-3**
Wongwathanarat A, Janka HT, Mueller E. 2013. *Astron. Astrophys.* 552:A126
Xue B, Garfinkle D, Pretorius F, Steinhardt PJ. 2013. *Phys. Rev. D* 88:083509
Yunes N, Pretorius F. 2009. *Phys. Rev. D* 80:122003

Cumulative Indexes

Contributing Authors, Volumes 41–52

Article Titles, Volumes 41–52

Prefatory Chapters

Sun

Interstellar Medium

Galaxy and Galaxies

Active Galactic Nuclei

Cosmology

Instrumentation and Techniques, and the History of the Development of New Areas of Research